PRINCIPLES
OF SYMMETRY,
DYNAMICS, AND
SPECTROSCOPY

PRINCIPLES OF SYMMETRY, DYNAMICS, AND SPECTROSCOPY

WILLIAM G. HARTER
University of Arkansas
Fayetteville, Arkansas

A Wiley-Interscience Publication
JOHN WILEY & SONS, INC.
New York / Chichester / Brisbane / Toronto / Singapore

This text is printed on acid-free paper.

Copyright © 1993 by John Wiley & Sons, Inc.

All rights reserved. Published simultaneously in Canada.

Reproduction or translation of any part of this work
beyond that permitted by Section 107 or 108 of the
1976 United States Copyright Act without the permission
of the copyright owner is unlawful. Requests for
permission or further information should be addressed to
the Permissions Department, John Wiley & Sons, Inc.

Library of Congress Cataloging in Publication Data:

Harter, William G., 1943–
    Principles of symmetry, dynamics, and spectroscopy/William G.
  Harter.
      p. cm.
    "A Wiley-Interscience publication."
    Includes index.
    ISBN 0-471-05020-2
    1. Symmetry (Physics)   2. Molecular dynamics.   3. Representations
of groups.   4. Spectrum analysis.   I. Title.
QC174.H37 1993
539.7′25—dc20   92-11123
                                                    CIP

Printed in the United States of America

10 9 8 7 6 5 4 3 2 1

To Margot, Alex, Thomas, and Daniel

PREFACE

This is a text for students of physics and physcial chemistry on the principles and applications of symmetry analysis and spectroscopy to atomic and molecular systems. In the general physics and chemistry community, these subjects have come to be known as "group theory." However, this is a misnomer and there is relatively little group theory in this book. Group theory has become a well-established area of pure mathematics since sometime after the tragic death of its founder, Evariste Galois (1811–1832), in a duel. Mathematical treatments of group theory deal with abstract structures and classifications of groups that go well beyond what is found in the current text or most other books on applied group theory. Most of the latter follow the approach and subject contained in the pioneering 1928 work, *Group Theory and Quantum Mechanics*, by Hermann Weyl (Dover 1931) and the well-known texts, *Group Theory* by Eugene Wigner (Academic Press 1959), or *Group Theory and Its Application to Physical Problems* by Morton Hammermesh, (Addison-Wesley 1962).

Perhaps a better name for the mathematical part of our subject is applied group *representation* theory. This lies in the general mathematical areas of group *algebra* and *Fourier analysis*. Calling our subject by its proper name is important. The mathematical techniques of algebra and Fourier analysis are powerful ones. They lead to a more direct approach to the subject, beginning with Chapters 1 and 2. A simpler and more physical treatment of symmetry analysis, which respects its roots in algebra and wave mechanics, is perhaps one of the main things that distinguishes this text.

Ever since Schrödinger used the pejorative *gruppenpest*, the subject of group theory in physics has from time to time received a bad name and deservedly so. It is particularly obnoxious when it is used for taxonomy or mindless application of esoteric (but often meaningless) names that do not help in the understanding of physics or chemistry. It is *not* the intent of this text to encourage this sort of obscurantia.

Rather it is the physical description of the symmetry algebra and spectroscopy that is emphasized from the beginning and throughout this text. We attempt to present the mathematical analysis as a natural consequence of physical reality instead of the other way around. This approach clarifies the relation between mathematics and physics and motivates its own development. It is hoped this leads to increased understanding of physics as well as more computational power and results in not only better physics but also better mathematics.

The development begins by describing the smallest and simplest structures or symmetries which are later linked through *subgroup chains* to the larger ones. It is important to understand the smaller building blocks. Early in Chapter 2, considerable time is devoted to the double pendulum and the two-state system. This involves the elementary bilateral or C_2 symmetry which, incidentally, is the only symmetry our human bodies approximate. Two-fold symmetry provides a fundamental paradigm that recurs as the basis of U(2) symmetry in Chapter 5 and in analysis of spin resonance, optical polarization, and laser driven transitions in Chapters 7 and 8. It also introduces the key physical idea of symmetry; anything with symmetry must have at least two identical units and if these units are harmonic oscillators they must be precisely in *resonance* with each other.

Resonance is the basis of electromagnetic spectroscopy and probably the single most important phenomena in all of physics. Without it we would be deaf, dumb, and blind and physics as we know it would not exist. Most forces in nature are so weak they can be effective only if amplified by some resonance phenomenon. In typical atomic transitions, this amplification or quality factor is about 50 million; in molecules it can be more than 10^{10} or 10^{11}. Resonance between high quality oscillators requires that they match with a correspondingly high degree of precision or symmetry.

Given the importance of the relation between resonance and symmetry it seems worthwhile to describe its fundamental properties. This begins in Chapter 2 where the effects of symmetry breaking are introduced using both classical and quantum examples of resonance. The relation between classical and quantum resonance is continued with a discussion of the Lorentz theory for classical resonance of a *single* oscillator in Chapter 6. This provides a background for the modern SU(2) theory of two-level quantum resonance in Chapters 7 and 8.

Resonance between three, four, or more oscillators is described in later parts of Chapters 2 and in Chapters 3 and 4. Chapter 2 introduces the concept of symmetry-defined wave modes and shows how symmetry determines whether waves move or stand and where they stand. The concept of symmetry breaking and quenching of moving waves is introduced first for classical mechanical waves and then for quantum mechnical waves. All concepts are introduced by examples which range in complexity from simple pendulums, Bohr orbitals and 1-D lattice modes to more complex problems involving atomic orbitals in solids and rotations and vibrations of polyatomic molecules.

A unique feature of this test is its discussion and comparison of the two types of symmetry breaking. The standard type of external or *applied* symmetry breaking is exemplified by Zeeman or Stark splitting in which externally applied perturbation reduces the symmetry of a Hamiltonian and splits energy level degeneracy. The other type, which is called *internal* or *spontaneous* symmetry breaking, does the opposite; it unsplits or clusters energy levels into manifolds of greater degeneracy or near-degeneracy. These correspond to *induced* representations and the idea of the resonance effects in *local* and *global* symmetries which is introduced in Chapter 4.

Perhaps the simplest pedogogically unequivocal example of spontaneous symmetry breaking occurs in cubic or octahedral symmetry and is treated in Chapter 4. However, it is just as likely for symmetries as simple as for C_3 or C_2. The treatment of internal symmetry breaking arose from the author's work on laser molecular spectroscopy, however the general topic is having a much grander application in high energy physics.

Connected with the two kinds of symmetry breaking are the two kinds of symmetry groups. Chapter 3 and beyond show that there are two sides to the symmetry analysis of a physical body such as a molecule, atom, or nucleus. Roughly speaking there is an outside or *external* symmetry consisting of *laboratory* based operations, and an inside or *internal* symmetry consisting of *body* based operations. A detailed realization of this concept for spherical symmetry is treated in Chapter 5 in connection with rotor quantum states. However, the idea of what one calls the "inside" and "outside" or the observed and observer is a very general one that lies at the foundation of quantum theory and symmetry analysis.

A key mathematical object that embodies these ideas is the quantum symmetry *projection operator* P^α_{mn}. The *P*-operators are used in three ways in this book. First, they have well-known applications for computing quantum states. Second, (this is not so well known) they can be made to derive themselves and their related mathematical quantities. Third, (this is, seldom, if ever, used) the projection operators serve as conceptual objects which clearly delineate internal and external symmetry properties. The in and out of *P*-operators is first shown in Chapter 3. However, the most well known example is the quantum rotor projector P^j_{mn} introduced in Chapter 5 for which *m* and *n* are lab and body projections of angular momentum *j*.

Derivation of the algebra of matrix representations and projection operators begins in Chapter 1. Four axioms or postulates for the matrix approach to quantum mechanics are reviewed at the beginning of the first chapter. This approach to quantum theory is similar to that given in Volume III of the famous *Feynman Lectures* (Addison-Wesley 1965). These four physical axioms are related to the four mathematical axioms of group theory at the beginnning of the second chapter so the role of group theory in formulating quantum mechanics problems is seen immediately. However, it is group algebra and spectral decomposition by projection operators that helps to *solve* these problems and these techniques are the main topics of the first four chapters.

The mathematical development of representation theory in this text is backwards compared to that of most other applied symmetry treatments. According to the usual order one introduces Schur's lemmas, character or representation orthogonality, and finally the application of symmetry projection operators. Instead, this text develops projection operator and spectral decomposition first. Then the representation and character orthogonality follow more naturally and Schur's lemmas are trivial consequences. There is evidence in the literature that this is more like the way the subject was created even if it is not the way it is generally presented. (The standard approach is given in Appendix G.) This provides students with alternatives to an approach that just fights the last wars and may be a better way to prepare them for future engagements.

The current text has only brief discussions of the Pauli principle which is probably the single most mysterious principle or axiom in all of modern physics. Also relativistic spin and orbit theory is absent as are most of the author's research efforts on unitary and permutation groups involving multi-electronic atoms and molecules and molecular spin–vibration–rotation interactions. Also, there is only a brief introduction to Racah recoupling coefficients.

A number of MacIntosh computer animation and simulation programs have been written by the author to accompany this text. Transformation matrix algebra and Dirac eigenvector properties are displayed in a program called *MatrIt*. The properties of a driven and damped harmonic oscillator is simulated in a program called *OscillIt*. Two coupled oscillators, C_2 symmetry properties and the analogous two-state quantum systems are simulated in the program called *Color U(2)*. The latter contains a detailed development of Stokes–Poincaré polarization mechanics and Rabi spin resonance. Waves in multiple coupled oscillators and C_n symmetry analysis is the topic of the program *WaveIt*. Dispersion relations and elementary band theory are described there. Quantum wave mechanics and scattering in one-dimensional barriers, wells, and lattices are the subjects of the program called *BandIt*. Classical coulomb orbits for one- and two-center atomic and molcular systems are stimulated in a program call *CoulIt*. Hamilton's classical action ($S = \int p \, dq$) is displayed using color in *CoulIt* and *Color U(2)*. Finally, a program called *SpinIt* displays body and laboratory views (in stereo 3D) of rigid and semi-rigid rotors and coupled spins and rotors. The related chapters for these programs are listed below.

MatrIt	Chapter 1
OscillIt	Chapters 1, 2, and 6
Color U(2)	Chapters 2, 5, 7, and 8
WaveIt	Chapters 2 and 3
BandIt	Chapters 2 and 3
CoulIt	Chapters 5, 7, and 8
SpinIt	Chapters 5, 7, and 8

Information about where to obtain these programs can be obtained from the author.

WILLIAM G. HARTER

Fayetteville, Arkansas

ACKNOWLEDGMENTS

For the past five past years, the manuscript for this text has been written for use in my graduate physics courses in quantum theory, symmetry analysis, atomic, molecular, and optical physics at the University of Arkansas. Some of the subject manner was developed in courses and special lectures I gave before at the Georgia Institute of Technology and at the National Bureau of Standards in Boulder, Colorado. The Boulder lectures were made possible by Ken Evenson and the Time and Frequency Division which arranged a video taped mini course in 1977.

Some of the material for the text was produced for classes I gave while in the Quantum Electronics Department at the University of Campinas in Brazil. Subsequently, a set of lecture notes was assembled there by Chris Patterson and James Moore, who arranged to have them printed in 1978.

My first opportunity to attempt a physically motivated development of symmetry principles came in 1964 while still an undergraduate at Hiram College. This grew into a series of informal summer sources on group and algebraic methods at the NASA Lewis Research Center in Cleveland, Ohio. With the help and encouragement of Professors Donald Dooley and Dale Dreisbach at Hiram and NASA physicists David Lockwood, John Heighway, and Howard Volkin, a series of seminars were run in the summers of 1964 to 1970. These seminars included a number of topics in nuclear, atomic, molecular, and solid-state theory and provided a wonderful opportunity to develop new ideas that is seldom available today. (Subsequently, NASA needed to exercise fiscal constraint and put money into much more important (and expensive) directions such as increased management.)

The part of this text that deals with angular momentum was developed at the University of Southern California. I am grateful for many helpful comments from Chris Patterson, John Caird, Martin Gunderson, and others who were graduate students at the time. Also, at Georgia Tech, helpful general suggestions were given by Ron Fox, John Wood, Peter Shulz, and Turgay Uzer. I have received a great deal of encouragement and helpful feedback during the manuscript development from the larger physics and chemistry community. Among those who have helped me are Bill Klemperer, Jim Moore, Ken Evenson, Rick Heller, David Tannor, Nelson DeLeon, Rod McDowell, Allen Pine, Peter Milonni, David Rowe, and Julio Gea-Banacloche. I would also like to thank Rod McDowell for the use of unpublished diagrams of vibrational ladders which he produced with Jay Ackerhalt, and Allen Pine for use of examples of his high-resolution laser spectra. My special thanks for long and rewarding collaborations go to my former doctoral advisees Chris Patterson and David Weeks.

The early drafts of the first three chapters were typed by my wife, Margot, who has also drawn a number of figures and helped in so many ways during the arduous final stages. Chapters 4–6 were typed by Audrey Ralston at Georgia Tech. The final chapters and all revisions were typed by Sandra Johnsen at the University of Arkansas. Many of the illustrations in Chapters 4–6 were drawn by the Los Alamos National Laboratory drafting group. The Euler angle machine photos were done by Vincent Malette and Joanie Geiser.

A large part of the research, computer, and financial support is from National Science Foundation grants from the Division of Theoretical Physics and from the Division of Theoretical and Computational Chemistry. The current support is from NSF grant CHE-92-21371. In 1988, support was also received from the Arkansas Science and Technology Authority.

W. G. H

CONTENTS

1 A Review of Matrix Algebra and Quantum Mechanics   1

  1.1 Matrices Used in Quantum Mechanics / 1
      A. The Transformation Matrix / 2
      B. Row and Column Matrices: State Vectors / 9
      C. Hamiltonian Matrices and Operators / 12

  1.2 Matrix Diagonalization / 15
      A. Obtaining Eigenvalues: The Secular Equation / 17
      B. Obtaining Eigenvectors: The Hamilton-Cayley Equation / 18

  1.3 Some Other Matrices Used in Quantum Mechanics / 28
      A. Scattering Matrix / 29
      B. Density Matrix / 30

  1.4 Some Matrices Used in Classical Mechanics / 30
      A. Force, Mass, and Acceleration Matrices / 31
      B. Potential Energy and Hamiltonian Functions / 35

  1.5 Wave Functional Analysis and Continuous Vector Spaces / 37
      A. Functional Scalar Products / 38
      B. Orthonormality and Completeness / 38
      C. Differential Operators and Conjugates / 40
      D. Resolvants / 41

Appendix A. Elementary Vector Notations and Theory / 43

Appendix B. Linear Equations, Matrices, Determinants, and Inverses / 47

Appendix C. Proof that Hermitian and Unitary Matrices and Diagonalizable / 52

Additional Reading / 54

Problems / 54

2 BASIC THEORY AND APPLICATIONS OF SYMMETRY REPRESENTATIONS (ABELIAN SYMMETRY GROUPS)   59

2.1 Symmetry Groups / 60
2.2 Representing Symmetry and Symmetry Groups / 62
2.3 Solving a Problem with Symmetry Analysis (C_2) / 69
2.4 Irreducible Representations / 79
2.5 Partially Solving a Problem with Symmetry Analysis (C_2) / 82
2.6 An Example with Slightly Higher Symmetry (C_3) / 84
2.7 More Examples with C_n Symmetry: One-Dimensional Lattices / 91
    A. Symmetry Breaking / 97
    B. Galloping Waves / 100
    C. Comparison with Fourier Analysis / 104
2.8 Other Types of Abelian Symmetry / 104
2.9 Theory of Commuting Idempotents / 109
2.10 General Theory of Abelian Groups / 110
2.11 Some Abelian Point Symmetries / 111
2.12 Symmetry Analysis for Quantum Mechanics / 115
    A. Bohr Levels and Bloch Waves: C_{12} Clocktane Orbitals / 115
    B. C_2 Symmetry and the Two-Level System / 129
    C. C_2 Symmetry Analysis and Scattering Theory / 132
    D. Crossing Matrices and One-Dimensional Tunneling / 137

Additional Reading / 141

Problems / 143

3 BASIC THEORY AND APPLICATIONS OF SYMMETRY REPRESENTATIONS (NON-ABELIAN SYMMETRY GROUPS)   151

3.1 Simplest Examples of Non-Abelian Symmetry / 151
    A. Trigonal Symmetries C_{3v} and D_3 / 151
    B. Reflections and Hamilton's Turns / 154

C. Laboratory and Body Reference Frames / 157
  D. Tetragonal Symmetries C_{4v} and D_4 / 158
3.2 First Stage of Non-Abelian Symmetry Analysis / 159
  A. Class Operators / 159
  B. Idempotent Analysis of Class Algebras / 162
  C. Does the Class Algebra Reduction Work in General? / 164
3.3 Second Stage of Non-Abelian Symmetry Analysis / 164
3.4 Theory and Application of Elementary Operators and Irreps / 170
  A. Group Space and the Regular Representation / 170
  B. Deriving and Using P_{ij} and $\mathcal{D}_{ij}$ / 175
3.5 Character Formulas / 184
  A. Derivation of Irrep Characters / 186
  B. Applications of Characters / 187
3.6 D_n and C_{nv} Symmetry and Bloch Waves / 190
  A. Tetragonal Symmetry / 190
  B. Hexagonal Symmetry / 196
  C. Higher D_n Symmetries: D_{nh} and D_{nd} / 202
3.7 Ammonia (NH$_3$) Vibrational Modes / 205
Appendix D. Mathematical Properties of $P_i^\alpha g P_j^\alpha$ / 217
            D.1. Linear Dependence of $P_i g P_i, P_i g' P_i, \ldots$ / 217
            D.2. Linear Dependence of $P_i g P_j, P_i g' P_j, \ldots$ / 218
            D.3. Existence Proof of $P_j^\alpha g P_j^\alpha$ / 219
Additional Reading / 221
Problems / 222

4 THEORY AND APPLICATIONS OF HIGHER FINITE SYMMETRY AND INDUCED REPRESENTATIONS 227

4.1 Octahedral Symmetries and Their Characters / 227
  A. Octahedral Group (O) / 228
  B. Full Octahedral Group (O_h) / 234
  C. Full Tetrahedral Symmetry T_d / 236
  D. Partial Tetrahedral Symmetries T and T_h / 236
4.2 Irreducible Representations of Octahedral Symmetry / 237
  A. Subgroup Chains and Idempotent Splitting / 237
  B. More Subgroup Correlations / 231
  C. Conjugate and Normal Subgroups / 254

xviii CONTENTS

    4.3 Introduction to Symmetry Breaking and Induced Representations / 255
- A. Octahedral Models and Induced Representations / 256
- B. Model Solving and Induced Representation Reduction / 259
- C. The Frobenius Reciprocity Theorem and Factored P-Operators / 264
- D. Spontaneous or Internal Symmetry Breaking / 269
- E. External Symmetry Breaking / 271

    4.4 Vibrations of Octahedral Hexafluoride Molecules / 286
- A. Projection Analysis / 286
- B. Solving Equations of Motion / 292
- C. Classical Canonical Coordinates / 301
- D. Elementary Quantum Theory of Vibrations / 302

Additional Reading / 308
Problems / 308

5 REPRESENTATIONS OF CONTINUOUS ROTATION GROUPS AND APPLICATIONS   315

    5.1 Basic Theory of Orthogonal Groups O_2 and O_3 / 316

    5.2 Representations of R_2, O_2 and Related Symmetry Groups / 321

    5.3 Parameters of R_3 / 524
- A. Euler Angles $(\alpha\beta\gamma)$ / 324
- B. Darboux or ω-Axis Angles $[\phi\theta\omega]$ / 330
- C. Hamilton's Rules and the Rotation Slide Rule / 333

    5.4 Irreducible Representations of R_3 and O_3 / 338
- A. Generators and Infinitesimal Rotations / 338
- B. Physical Interpretation of Generators / 342
- C. Irreps of Angular-Momentum and Generator Operators / 345
- D. Irreducible Representations of Rotation Operators / 348

    5.5 Some Applications of R_3 Representations / 353
- A. $\mathscr{D}^{1/2}$-Spinor Representations and Hamilton's Turns / 353
- B. Spin-j Polarization Experiments / 357
- C. Symmetry Analysis of Quantum Rotors / 362
- D. Spherical Harmonics and Rotational Wave Functions / 373

E. Explicit Relations for Rotation Operators and Generators / 377

5.6 Rotational Level Splitting in Finite Symmetry / 382
    A. Cubic Symmetry Correlations $O_3 \supset O_h$ / 383
    B. Cubic Eigenstates and Wave Functions / 386
    C. Multipole Functions and Polynomials / 392
    D. Multipole Expansions / 394
    E. Level Splitting for Molecular Rotors / 396
    F. $R_3 \supset D_6$ Correlations and Level Splitting / 400

5.7 Half-Integer j-Level Splitting in Finite Symmetry / 404
    A. Ray Representations of D_6 / 404
    B. Ray Representations of Other D_n Groups / 411
    C. Ray Representations of Octahedral Symmetry / 413

5.8 Some Higher Continuous Symmetries: R_4 and U_3 / 415
    A. The Coulomb Symmetry / 415
    B. Harmonic Oscillator Symmetry / 426

Appendix E. Derivation of Angular-Momentum Representations / 434

Additional Reading / 437

Problems / 438

6 THEORY AND APPLICATIONS OF SYMMETRY REPRESENTATION PRODUCTS (FINITE GROUPS)   443

6.1 Two-Particle States and Products of Representations / 444
    A. Noninteracting Particles / 446
    B. Interacting Particles / 450
    C. Subgroup Chain Labeling for Coupling Coefficients / 454

6.2 General Concepts and Matrix Relations for Coupling Coefficients / 464
    A. Products Involving Invariants or Scalars / 464
    B. Symmetry Relations / 465
    C. Product Analysis with Repeated Irreps / 466
    D. Orthonormality and Completeness / 466

6.3 Vectors and Tensors in 3-Space / 467
    A. Symmetry-Defined Unit Vectors / 467
    B. Symmetry-Defined Unit Tensors / 469
    C. Symmetry-Defined Coordinates and Polynomials / 470
    D. Symmetry-Defined Bulk Behavior in Solids / 471

E. Symmetry Theory of Tensor Relations (Elastic Constants) / 476
F. Symmetry-Defined Electric and Magnetic Fields / 489

6.4 Theory of Quantum Operators and Irreducible Tensors / 492
   A. Symmetry-Defined Operators / 493
   B. The Wigner-Eckart Theorem and Transition Selection Rules / 497

6.5 Classical Approach to Optical Resonance and Selection Rules / 503
   A. Introduction to the Effect of Light on Matter / 503
   B. Introduction to the Effect of Matter on Light / 512
   C. Two Types of Spectroscopy for Vibrational Analysis / 523

6.6 Multiply Excited Quantum Vibration States / 532

6.7 Introduction to Theory of Symmetry Stability / 538
   A. Symmetry Analysis of Vibronic Hamiltonians / 538
   B. The Jahn-Teller Theorem / 541
   C. Dynamic Jahn-Teller and Renner Effects / 543

Additional Reading / 550
Problems / 551

7 THEORY AND APPLICATION OF SYMMETRY REPRESENTATION PRODUCTS (CONTINUOUS ROTATION GROUPS)   553

7.1 Introduction to R_3 and U_2 Coupling Coefficients / 553
   A. Two-Particle Spin States: Hydrogen Hyperfine Structure / 554
   B. Two-Electron Atomic Configurations / 561
   C. Spin-Orbital Coupling / 564
   D. Geometrical Interpretation of Angular-Momentum Coupling / 567

7.2 Mathematical Relations for Coupling and Wigner $3j$ Coefficients / 570
   A. Scalars, Vectors, and Tensors / 570
   B. General R_3 Scalar Coupling / 573
   C. Fundamental Coupling Definitions and Symmetry Relations: The Wigner $3j$ Coefficient / 574
   D. Clebsch-Gordon and Wigner Coefficient Formulas / 578

7.3 Rotational Tensor Operators and the Wigner-Eckart Theorem / 584
  A. Construction of R_3 Tensor Operators / 585
  B. The Wigner-Eckart Theorem for R_3 / 597
  C. Evaluation of Crystal Field Splitting / 598
  D. Evaluation of Reduced Matrix Elements / 603
7.4 Rotational Level Splitting for High J: Semiclassical Angular Momentum Mechanics / 608
  A. Rigid Rotors ($D_\infty \supset D_2$ Symmetry) / 609
  B. Semirigid Spherical Tops Octahedral (O) Symmetry / 618
7.5 Rotating Spinor Systems and Two-Dimensional Oscillator Analogies / 628
  A. Euler-Angle Definition of Spinor States / 630
  B. Axis-Angle Definition of Spinor Operators / 631
  C. Rotational Angle Parameterization for a Classical two-dimensional Harmonic Oscillator / 634
  D. Polarization Ellipsometry Coordinates / 636
  E. Generalized Lissajous Trajectories and Related Dynamics / 641
  F. Rotational Energy (RE) Surface Description of Anharmonic Vibrations / 649
7.6 Molecular Electronic Structure / 654
  A. Electronic Models for Diatomic Molecules / 654
  B. How to "Point" Electronic Orbitals / 674
Additional Reading / 682

8 SYMMETRY ANALYSIS FOR SEMICLASSICAL AND QUANTUM MECHANICS: DYNAMICS WITH HIGH QUANTA   686

8.1 Contact Transformations, Actions, and Semiclassical Wave Fronts / 687
  A. Contact Transformations
  B. Action Functions / 690
  C. Generator of Classical Trajectories and Wave Fronts / 695
  D. Semiclassical Approximation for Schrödinger Equation / 700
  E. Huygen's Principle and Semiclassical Mechanics / 701
8.2 Coherent Harmonic Oscillator States / 703

8.3 Coherent Wave Generation of Eigenfunctions / 711
    A. Wave-Packet Propagation and Spectral Quantization / 717

8.4 Semiclassical Radiation Theory for Spectroscopy / 722
    A. Lagrangians and Hamiltonians for Electromagnetic Interactions / 722
    B. Semiclassical Radiation Perturbation Theory / 730

8.5 The Two-Level System Approximation / 738
    A. Two-State Schrödinger Equations / 739
    B. Spin and Crank Vector Visualizations of Two-State Hamiltonian / 740
    C. Rotating-Wave Solutions / 743
    D. Block-Siegert Corrections / 748
    E. Damping and the Bloch Equations / 750
    F. Dressed Eigenstates / 753

8.6 Quantum Electromagnetic Fields and Transitions / 761
    A. Electromagnetic Fields and Operators / 762
    B. Electromagnetic Quantum States and Transitions / 767

8.7 Spectra of Atom in Laser Cavity / 777
    A. Jaynes-Cummings Hamiltonian / 777
    B. Jaynes-Cummings Eigensolutions / 781
    C. Transitions in the Jaynes-Cummings Model / 783

Additional Reading / 786

APPENDIX F   FORMULAS AND TABLES OF GROUP REPRESENTATIONS AND RELATED QUANTITIES   **790**

APPENDIX G   SCHUR'S LEMMA AND IRREDUCIBLE REPRESENTATIONS AND ORTHOGONALITY   **827**

INDEX   **837**

PRINCIPLES OF SYMMETRY, DYNAMICS, AND SPECTROSCOPY

CHAPTER 1

A REVIEW OF MATRIX ALGEBRA AND QUANTUM MECHANICS

The study of symmetry and its application to spectroscopy involves the mathematics of operators, vectors, and matrices, and this chapter contains a review of matrix theory and some of its applications. The review includes a few concepts and procedures which may not be well known to physicists, but which are needed in the development of this book. Also, we use the review to establish much of our symbolism, conventions, and definitions. Finally, there are some previews of the contents of the rest of the book.

1.1 MATRICES USED IN QUANTUM MECHANICS

The matrix description of quantum mechanics was first developed by Heisenberg and has since become widely used. This description differs from some others mostly by the organization of its bookkeeping, wherein numbers of physical interest are stored in arrays called MATRICES. Such an array is shown in Eq. (1.1.1):

$$\mathcal{M} = \begin{pmatrix} \mathcal{M}_{11} & \mathcal{M}_{12} & \mathcal{M}_{13} & \cdots \\ \mathcal{M}_{21} & \mathcal{M}_{22} & \mathcal{M}_{23} & \cdots \\ \vdots & \vdots & \vdots & \end{pmatrix}. \quad (1.1.1)$$

We will use script notation $\mathcal{M}, \mathcal{N}, \ldots$ for a whole matrix, and let the same letter with subscripts $\mathcal{M}_{ij}$ designate the (complex) number which is the COMPONENT of the matrix, located where the ith row meets the jth column.

2   A REVIEW OF MATRIX ALGEBRA AND QUANTUM MECHANICS

Many, but not all, of the matrices which we shall use will be finite-square or $(n \times n)$ matrices, that is, matrices with n rows and n columns, where $n = 1, 2, 3, \ldots$ is finite. [A (1×1) matrix is just a number.] This is because many physical situations can be described by some relation between a set of n quantum states and some other set of n different states. Consider now some examples of such matrices.

A. Transformation Matrix

Quite often in quantum mechanics we can describe a particle or a system of particles by stating whether they are in one or another of some set of n quantum states $1, 2, 3, \ldots, n$. The most famous example of this might be the states (1): spin "up" and (2): spin "down" for an electron.

Now someone else can come along and describe the same system with a different set of states $1', 2', 3', \ldots$, and n', which are just as complete in their description as the first set. For example, $1'$: spin "north" and $2'$: spin "south" are acceptable states for the electron, too, as we will see shortly.

However, you must know the relation between any two equivalent descriptions. This relation is given entirely by a TRANSFORMATION MATRIX $\mathscr{T}(b' \leftarrow b)$, the meaning of which we shall review:

$$\mathscr{T}(b' \leftarrow b) = \begin{pmatrix} \langle 1'|1 \rangle & \langle 1'|2 \rangle & \cdots & \langle 1'|n \rangle \\ \langle 2'|1 \rangle & \langle 2'|2 \rangle & \cdots & \langle 2'|n \rangle \\ \vdots & \vdots & & \\ \langle n'|1 \rangle & \langle n'|2 \rangle & \cdots & \langle n'|n \rangle \end{pmatrix}. \quad (1.1.2)$$

For the time being we shall exhibit the $\mathscr{T}$ matrix for going between "up-down" (1-2) and "north-south" ($1'$-$2'$) directions shown in Figure 1.1.1 for electronic spin. (The figure shows up-down tilted to avoid "geo-chauvinism" or undue favoring of a particular inhabited latitude.) The $\mathscr{T}$ matrix that follows and generalizations of it are derived in Chapter 5, and in fact most of symmetry theory is devoted to finding some sort of $\mathscr{T}$ matrix:

$$\mathscr{T}(b' \leftarrow b) = \begin{pmatrix} \langle 1'|1 \rangle & \langle 1'|2 \rangle \\ \langle 2'|1 \rangle & \langle 2'|2 \rangle \end{pmatrix} = \begin{pmatrix} \cos\dfrac{\theta}{2} & -i\sin\dfrac{\theta}{2} \\ -i\sin\dfrac{\theta}{2} & \cos\dfrac{\theta}{2} \end{pmatrix}. \quad (1.1.3)$$

It is important to understand the meaning of a $\mathscr{T}$ matrix. It is conventional to say that component $\langle i'|j \rangle$ gives the "PROBABILITY AMPLITUDE for a system in state j to be found in state $i' \ldots$". Now one has a perfect right to ask what that means, but many students find it difficult to get a satisfying answer. However, it is not so difficult to learn that $\langle i'|j \rangle$ *squared*,

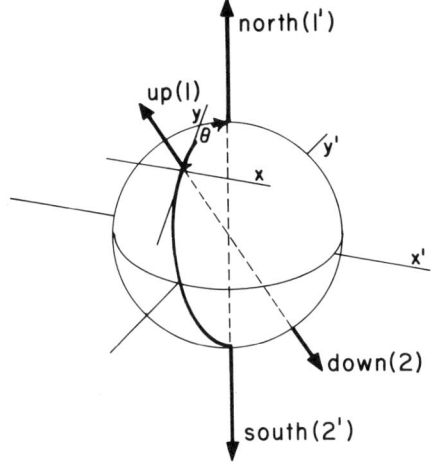

Figure 1.1.1 Alternative axes for spin state definition. The up-down axis (z) is rotated by angle θ from the north-south axis (z').

i.e., $|\langle i'|j\rangle|^2$, is the fraction of, or probability for, systems in state j to end up in state i' when forced to make the choice. The difficulty is that it is apparently impossible to predict the outcome of an individual event. To know this with certainty we must really wait for the individual particle or system to make its choice. The prediction $|\langle i'|j\rangle|^2$ is only the probability for the individual outcome [(i') from (j)] or the average percentage for choices in a large number of individual experiments.

A device that forces a particle to make a choice is called an *analyzer*. An example of a spin-state analyzer is the magnetic Stern-Gerlach device which Feynman has described in his *Lectures on Physics* (Vol. III, p. 5-1). The analyzer accepts a beam of spinning particles and displaces each one up or down according to the projection of the particles' spin on the vertical axis of the analyzer. For electrons only two possible projections are possible. One possible projection is along the axis which is up [state (1)] for an up-down analyzer, or north [state (1')] for a north-south pointing analyzer, and so on. The other possibility is opposite to the axis, i.e., down [state (2)], or south [state (2')], etc. Given an incoming beam of particles in state (j), one finds *on the average* that a fraction $|\langle i'|j\rangle|^2 = |\mathcal{T}_{i'j}|^2$ of them will come out of the (i') exit of the (1')-(2') or north-south analyzer.

For example, the fraction of spin-up (1) electrons that would actually choose to point south (2') in a north-south analyzer (see Figure 1.1.2) is $|\langle 2'|1\rangle|^2 = \sin^2(\theta/2)$ given $\langle 2'|1\rangle$ in Eq. (1.1.3). The remaining fraction $1 - \sin^2(\theta/2) = \cos^2(\theta/2)$ end up choosing north (1').

Understanding the amplitude $\langle i'|j\rangle$, apart from its square, is more difficult. Let us state some axioms which most people like to have the amplitudes obey, and discuss some thought experiments which may help us to make them more plausible. The first axiom has already been introduced.

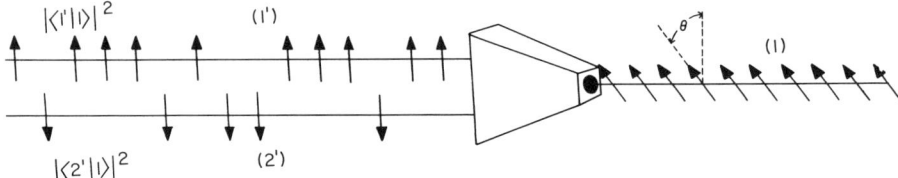

Figure 1.1.2 Spin analyzer experiment. An incident beam of particles in state (up) comes in on the right-hand side. The north-south analyzer forces each particle to choose between state 1′ (north) or state 2′ (south). On the average a fraction $|\langle 2'|1\rangle|^2 = \sin^2(\theta/2)$ choose the latter state 2′ and emerge from the bottom of the analyzer. The remaining fraction $|\langle 1'|1\rangle|^2 = \cos^2(\theta/2)$ come out at the top in the state 1′.

Axiom 1 $\langle i'|j\rangle^* \langle i'|j\rangle = |\langle i'|j\rangle|^2$ is the probability for occurrence in state i' of a system originally in state j which is forced to choose between $1', 2', \ldots, i', \ldots$.

One should note that in classical wave and polarization theory the absolute square $(|A|^2)$ of an amplitude denotes *intensity*. In quantum theory it gives probability, which is analogous.

The second axiom deals with complex conjugation ($*$) and order reversal.

Axiom 2 $\langle i'|j\rangle^* = \langle j|i'\rangle$.

In Axiom 2 one supposes that a reverse process obtained by playing time backwards is accomplished mathematically by complex conjugation. For example, a plane-wave amplitude $\langle x, t|k, \omega\rangle = e^{i(kx-\omega t)}$ reversed is $\langle x, t|k, \omega\rangle^* = e^{i([-k]x - \omega[-t])}$. This axiom is a direct carryover from classical wave theory.

The next axiom tells how states within a given set are related.

Axiom 3 The amplitudes connecting a set of states $\{1, 2, \ldots, n\}$ with itself are 1 or 0; that is, $\langle i|j\rangle = \delta_{ij} \equiv \begin{Bmatrix} 1 & i = j \\ 0 & i \neq j \end{Bmatrix}$.

Axiom 3 is closely related to the following experimental result. If some electrons that originally chose the state of spin north (1′) are once again given the chance in another analyzer to choose north or south (Figure 1.1.3), then 100% will choose north. This means $|\langle 1'|1'\rangle|^2 = 1$ and $|\langle 2'|1'\rangle|^2 = 0$. One then assumes that the phases are such that $\langle 1'|1'\rangle = 1$. The same applies to the (2′) state.

Axiom 3 means that the transformation matrix of states $\{1, 2, \ldots, n\}$ to themselves, or any other set $\{1', 2', \ldots, n\}$ to itself must be the IDENTITY or

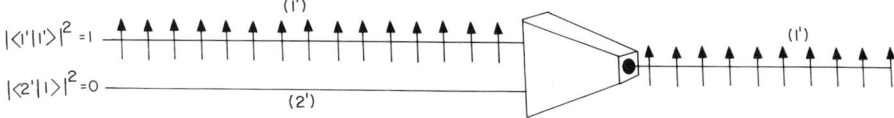
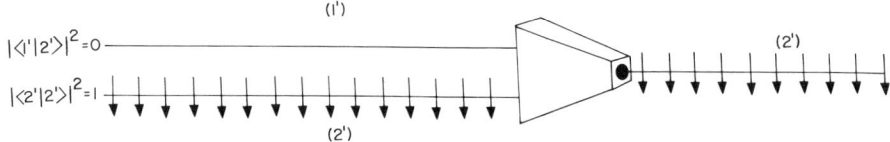

Figure 1.1.3 Analyzer experiment associated with Axiom 3. If the incident beam contains only particles in state 1' (north) or separate (south) then they will be unchanged by a north-south analyzer.

UNIT MATRIX, 1:

$$\begin{pmatrix} \langle 1|1\rangle & \langle 1|2\rangle & \cdots \\ \langle 2|1\rangle & \langle 2|2\rangle & \cdots \\ \vdots & \vdots & \end{pmatrix} = \begin{pmatrix} \langle 1'|1'\rangle & \langle 1'|2'\rangle & \cdots \\ \langle 2'|1'\rangle & \langle 2'|2'\rangle & \cdots \\ \vdots & \vdots & \end{pmatrix} = \begin{pmatrix} 1 & 0 & \cdots \\ 0 & 1 & \cdots \\ \vdots & \vdots & \end{pmatrix} = 1. \tag{1.1.4}$$

The conditions $\langle i|j\rangle = \delta_{ij}$ are often called the ORTHONORMALITY conditions for these states, as we shall explain later.

The fourth and final axiom contains the crucial aspects of quantum wave mechanics.

Axiom 4 Given three sets of states $\{1, 2, \ldots, n\}$, $\{1', 2', \ldots, n'\}$, and $\{1'', 2'', \ldots, n''\}$ all relations of the following form must hold between their amplitudes:

$$\langle k''|i'\rangle = \sum_{j=1}^{n} \langle k''|j\rangle\langle j|i'\rangle.$$

Axiom 4 may be hard to swallow, but at least it is easy to see that it is consistent with Axioms 1, 2, and 3. A sum of probabilities should be unity if all states are accounted for, and according to 1 and 2 this sum is the following:

$$1 = \sum_{i'=1}^{n} |\langle i'|j\rangle|^2 = \sum_{i'=1}^{n} \langle j|i'\rangle\langle i'|j\rangle, \tag{1.1.5a}$$

and by Axiom 4 this equals

$$1 = \langle j|j \rangle, \quad (1.1.5b)$$

which checks with Axiom 3. However, one probably needs more than this to trust Axiom 4.

Two analyzer experiments meant to test Axiom 4 are depicted in Figures 1.1.4(a) and 1.1.4(b). The idea is that a beam of electrons all in a definite state (1') [Figure 1.1.4(a)] will not be affected by a separation into some other states, say (1) and (2), if this is followed by a *coherent* recombination as indicated in the center of Figure 1.1.4(b). [A time-reversed analyzer indicated by a box whose direction is reversed in the figure is used to reform the particles into a single beam.] The two experiments shown give the same *average* number of electrons in the final states $(1''), (2'')$ on the left, if the recombination is coherent. If the recombination is incoherent, then the average number of electrons in final state (k'') would approach the sum of squares $(\sum |\langle k''|j \rangle \langle j|1' \rangle|^2)$ instead of the square of the sum $(|\sum \langle k''|j \rangle \langle j|1' \rangle|^2)$. This would happen if some device in the (1-2) analyzer determined whether each electron went through in the (1) or in the (2) state. This is because the device would add a random phase to amplitude $\langle k''|j \rangle$, i.e., a different phase for each particle that went through. Then only the positive-definite terms in the square of the sum

$$\left|\sum \langle k''|j \rangle \langle j|1' \rangle\right|^2 = \sum (\text{positive definite}) + \sum (\text{phase sensitive})$$

would survive after averaging over many particles, i.e., just the sum of squares or the sum of probabilities for each j:

$$\sum (\text{positive definite}) = \sum_j |\langle k''|j \rangle \langle j|1' \rangle|^2.$$

Meanwhile, the phase-sensitive "interference" terms

$$\sum (\text{phase sensitive}) = \sum\sum_{j \neq k} (\langle k''|j \rangle \langle j|1' \rangle)^* (\langle k''|k \rangle \langle k|1' \rangle)$$

would average to zero.

Axiom 4 can also be visualized in terms of basic wave mechanics. It is essentially a restatement of the principle of Huygens (circa 1660). Imagine a lightwave in state i' (Figure 1.1.5), going through a film to be (possibly) in some state k''. If each of the points j on the film absorb the light, but then rebroadcast it starting with the right amplitude and phase $\langle j|i' \rangle$ for that point, then Axiom 4 says state k'' will be achieved just as well as it would have if the film had not been there at all. This principle will be discussed in Chapter 8.

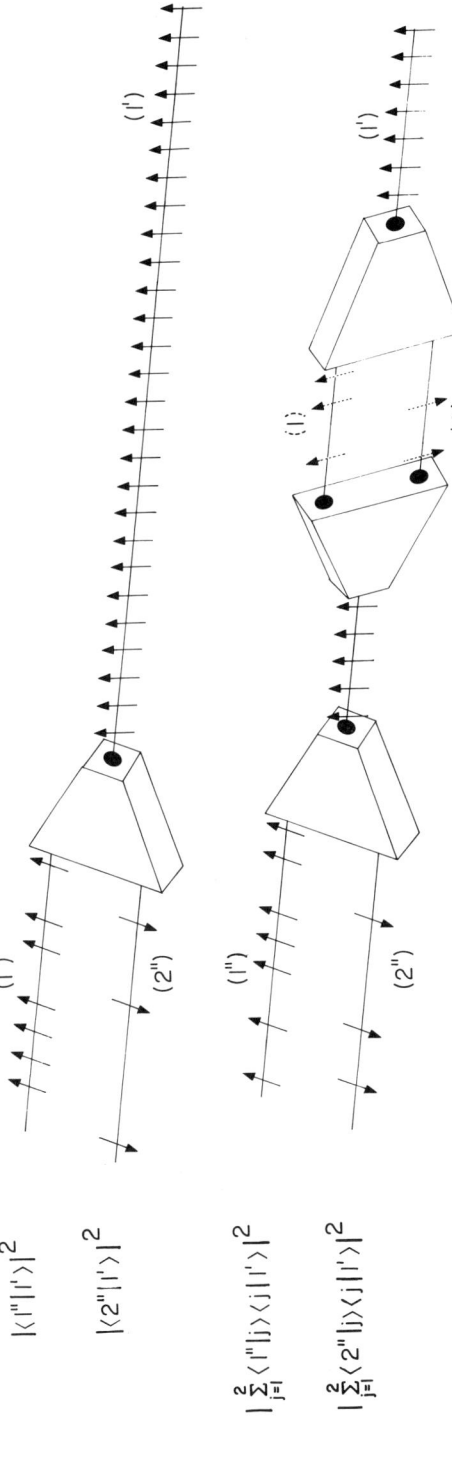

$|\langle 1''|1'\rangle|^2$

$|\langle 2''|1'\rangle|^2$

$|\sum_{j=1}^{2}\langle 1''|j\rangle\langle j|1'\rangle|^2$

$|\sum_{j=1}^{2}\langle 2''|j\rangle\langle j|1'\rangle|^2$

Figure 1.1.4 Analyzer experiments associated with Axiom 4. (a) Undisturbed beam in state 1' (north) is analyzed by a 1''-2'' analyzer. (b) The second experiment is the same as the first except the 1' beam is first analyzed by a 1-2 analyzer followed by a reversed (∗) analyzer which recombines the split beam. Axiom 4 says that such an arrangement will not effect the beam.

8 A REVIEW OF MATRIX ALGEBRA AND QUANTUM MECHANICS

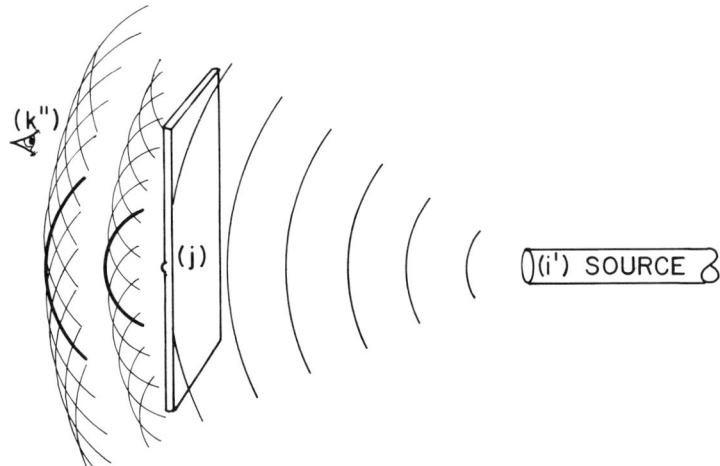

Figure 1.1.5 Relating Huygen's principle to Axiom 4.

Axiom 4 is the same as mathematical relations called the COMPLETENESS CONDITIONS for states $\{1, 2, \ldots, n\}$, and we shall soon discuss these from several points of view. Such relations are central to the development of symmetry analysis.

Furthermore, Axiom 4 is the real reason for ever studying matrix mathematics in the first place. Let us rewrite Axiom 4 using the script notation introduced in Eq. (1.1.2). Axiom 4 then corresponds to the standard definition of INNER MATRIX MULTIPLICATION (we discuss outer multiplication later):

$$\mathscr{T}_{k''i'}(b'' \leftarrow b') = \sum_{j=1}^{n} \mathscr{T}_{k''j}(b'' \leftarrow b)\mathscr{T}_{ji'}(b \leftarrow b'), \quad (1.1.6a)$$

$$\mathscr{T}(b'' \leftarrow b') = \mathscr{T}(b'' \leftarrow b)\mathscr{T}(b \leftarrow b'). \quad (1.1.6b)$$

Just writing two script letters next to each other, as in Eq. (1.1.6b), implies that the special sum in Eq. (1.1.6a) is to be performed.

Continuing the review of matrix mathematics, we define the TRANSPOSE CONJUGATE† of matrix $\mathscr{M}$ by

$$\mathscr{M}^{\dagger} = \begin{pmatrix} \mathscr{M}_{11} & \mathscr{M}_{12} & \cdots & \mathscr{M}_{1n} \\ \mathscr{M}_{21} & \mathscr{M}_{22} & \cdots & \mathscr{M}_{2n} \\ \vdots & \vdots & & \vdots \\ \mathscr{M}_{n1} & \mathscr{M}_{n2} & \cdots & \mathscr{M}_{nn} \end{pmatrix}^{\dagger} = \begin{pmatrix} \mathscr{M}_{11}^{*} & \mathscr{M}_{21}^{*} & \cdots & \mathscr{M}_{n1}^{*} \\ \mathscr{M}_{12}^{*} & \mathscr{M}_{22}^{*} & \cdots & \mathscr{M}_{n2}^{*} \\ \vdots & \vdots & & \vdots \\ \mathscr{M}_{12}^{*} & \mathscr{M}_{2n}^{*} & \cdots & \mathscr{M}_{nn}^{*} \end{pmatrix}. \quad (1.1.7)$$

Observe that the transformation matrix $\mathscr{T}$ satisfies Eq. (1.1.8), which is written in several different forms, starting with Axiom 4:

$$\sum_{i'=1}^{n} \langle k|i'\rangle\langle i'|j\rangle = \delta_{kj}, \qquad \sum_{i=1}^{n} \langle j'|i\rangle\langle i|k'\rangle = \delta_{k'j'} \qquad (1.1.8a)$$

$$\sum_{i'=1}^{n} \mathscr{T}^*_{i'k}(b' \leftarrow b)\mathscr{T}_{i'j}(b' \leftarrow b) = \delta_{kj}, \qquad \sum_{i=1}^{n} \mathscr{T}_{j'i}(b' \leftarrow b)\mathscr{T}^*_{k'i}(b' \leftarrow b) = \delta_{k'j'},$$

$$(1.1.8b)$$

$$\mathscr{T}^\dagger(b' \leftarrow b)\mathscr{T}(b' \leftarrow b) = 1 = \mathscr{T}(b' \leftarrow b)\mathscr{T}^\dagger(b' \leftarrow b). \qquad (1.1.8c)$$

Any matrix satisfying an equation of the last form, Eq. (1.1.8c), is called a UNITARY MATRIX. In fact, when the product of any two finite matrices $\mathscr{A}$ and $\mathscr{B}$ is equal to 1, then each is said to be the INVERSE of the other, written $\mathscr{A} = \mathscr{B}^{-1}$ or $\mathscr{B} = \mathscr{A}^{-1}$ (Appendix B reviews the calculation of inverses.) The unitary matrix has the very convenient property that its inverse is just its transpose conjugate ($\mathscr{T}^{-1} = \mathscr{T}^\dagger$).

For any real matrix $\mathscr{R}$ ($\mathscr{R}$ has no complex components) the transpose conjugate $\mathscr{R}^\dagger$ simply equals the TRANSPOSE $\mathscr{R}^T$ defined by

$$\mathscr{R}^T = \begin{pmatrix} \mathscr{R}_{11} & \mathscr{R}_{12} & \cdots & \mathscr{R}_{1n} \\ \mathscr{R}_{21} & \mathscr{R}_{22} & \cdots & \mathscr{R}_{2n} \\ \mathscr{R}_{n1} & \mathscr{R}_{n2} & \cdots & \mathscr{R}_{nn} \end{pmatrix}^T \equiv \begin{pmatrix} \mathscr{R}_{11} & \mathscr{R}_{21} & \cdots & \mathscr{R}_{n1} \\ \mathscr{R}_{12} & \mathscr{R}_{22} & \cdots & \mathscr{R}_{n2} \\ \mathscr{R}_{1n} & \mathscr{R}_{2n} & \cdots & \mathscr{R}_{nn} \end{pmatrix}. \qquad (1.1.9)$$

Any real transformation matrix will satisfy Eq. (1.1.10). A matrix satisfying this equation is said to be an ORTHOGONAL matrix:

$$\mathscr{T}^T = \mathscr{T}^{-1}, \qquad (1.1.10a)$$

$$\sum_i \mathscr{T}_{ij}\mathscr{T}_{ik} = \delta_{jk} = \sum_i \mathscr{T}_{ji}\mathscr{T}_{ki}. \qquad (1.1.10b)$$

B. Row ($n \times 1$) and Column ($1 \times n$) Matrices: State Vectors

One of Dirac's famous ideas was to associate the jth column of $\mathscr{T}(b' \leftarrow b)$ with a unit KET vector $|j\rangle$ and the jth row of $\mathscr{T}^\dagger(b' \leftarrow b)$ with a unit BRA vector $\langle j|$ as in Eq. (1.1.11), which follows. (A review of elementary vectors properties is in Appendix A.) The arrow ($\leftrightarrow$) means "is associated with."

$$\dot{j} = \begin{pmatrix} \langle 1'|j\rangle \\ \langle 2'|j\rangle \\ \vdots \\ \langle n'|j\rangle \end{pmatrix} \leftrightarrow |j\rangle, \qquad \dot{j}^\dagger = \overbrace{\langle j|1'\rangle\langle j|2'\rangle \cdots \langle j|n'\rangle} \leftrightarrow \langle j|. \qquad (1.1.11)$$

These columns and rows of numbers represent the state j in terms of the states $1', 2', \ldots$ and n'. In the first equation (1.1.11) the numbers $j_{1'} \equiv \langle 1'|j\rangle$, $j_{2'} \equiv \langle 2'|j\rangle, \ldots, j_{n'}$, are the components of vector $|j\rangle$ in the BASIS $|1'\rangle, |2'\rangle, \ldots,$ and $|n'\rangle$, and constitute a $n \times 1$ column matrix j. Similarly $(j^\dagger)_{m'} \equiv \langle j|m'\rangle$ are components of a $1 \times n$ row matrix $j^\dagger$ that represents $\langle j|$ in the basis $\langle 1'|, \langle 2'|,$ and $\langle n'|$.

We shall see as we go along why it is necessary to define two types of vectors $\langle |$ and $| \rangle$ that do about the same thing. However, the main point to note is that each vector $|j\rangle$ is a linear combination of any complete set of vectors $\{|1'\rangle, |2'\rangle, \ldots, |n'\rangle\}$ [Eq. (1.1.12a)] or $\{|1''\rangle, |2''\rangle, \ldots, |n''\rangle\}$, or even of the set $\{|1\rangle, |2\rangle, \ldots, |n\rangle\}$ to which it belongs, although the latter is trivially simple. These equations are Dirac's abstraction of Axiom 4 wherein the "bra" $\langle k|$ is removed from the bracket $\langle k|j\rangle$ to give KET vector $|j\rangle$:

$$|j\rangle = \sum_{i'=1}^{n} |i'\rangle\langle i'|j\rangle \equiv \sum_{i'=1}^{n} j_{i'}|i'\rangle, \tag{1.1.12a}$$

$$|j\rangle = \sum_{i''=1}^{n} |i''\rangle\langle i''|j\rangle \equiv \sum_{i''=1}^{n} j_{i''}|i''\rangle, \tag{1.1.12b}$$

$$|j\rangle = \sum_{i=1}^{n} |i\rangle\langle i|j\rangle = \sum_{i=1}^{n} \delta_{ij}|i\rangle = |j\rangle, \tag{1.1.12c}$$

For example, in the vector equation that follows we see that the state vector of spin up ($|1\rangle$) is, for small θ, a lot of spin north ($|1'\rangle$) plus a little of spin south ($|2'\rangle$):

$$|1\rangle = |1'\rangle\langle 1'|1\rangle + |2'\rangle\langle 2'|1\rangle$$

$$= \cos\frac{\theta}{2}|1'\rangle - i\sin\frac{\theta}{2}|2'\rangle \leftrightarrow \begin{pmatrix} \cos\dfrac{\theta}{2} \\ -i\sin\dfrac{\theta}{2} \end{pmatrix}. \tag{1.1.13}$$

But, in its own basis it is just spin up:

$$|1\rangle = |1\rangle\langle 1|1\rangle + |2\rangle\langle 2|1\rangle$$

$$= 1|1\rangle + 0|2\rangle \leftrightarrow \begin{pmatrix} 1 \\ 0 \end{pmatrix}.$$

This "mixing" of states to make other states is the embodiment of the quantum SUPERPOSITION PRINCIPLE.

The coefficients $\langle 1'|j\rangle, \langle 2'|j\rangle, \ldots$ in the column matrix j, Eq. (1.1.11), are said to be a REPRESENTATION of $|j\rangle$ in the $|1'\rangle, |2'\rangle, \ldots$ basis. Strictly speaking, it is not correct to replace the arrow $j \leftrightarrow |j\rangle$ by an equality sign, although you will see it done in many works. When you do set $j = |j\rangle$, there is no way to tell if the stack of numbers in j is supposed to be $\langle 1'|j\rangle, \langle 2'|, |j\rangle, \ldots$ or $\langle 1''|j\rangle, \langle 2''|j\rangle, \ldots$, or something else entirely. The notation $|j\rangle$ is reserved for the unique ABSTRACT VECTOR and the physical state it denotes.

One defines the SCALAR PRODUCT of a bra $\langle j|$ with a ket $|k''\rangle$ by Eq. (1.1.14a) so that it equals the bracket $\langle j|k''\rangle$. This is represented in any basis as an inner matrix product between a $(1 \times n)$ row matrix and $(n \times 1)$ column matrix, as in Eq. (1.1.14b):

$$\langle j|k''\rangle = \sum_{i'=1}^{n} \langle j|i'\rangle\langle i'|k''\rangle, \qquad (1.1.14a)$$

$$j \cdot k'' = \overbrace{\langle j|1'\rangle\langle j|2'\rangle \cdots \langle j|n'\rangle} \begin{pmatrix} \langle 1'|k''\rangle \\ \langle 2'|k''\rangle \\ \vdots \\ \langle n'|k''\rangle \end{pmatrix}. \qquad (1.1.14b)$$

(The dot is a common notation for the scalar or inner product. See Appendix A.)

The following definition for the transpose conjugate of vectors corresponds to the one for square matrices given previously:

$$|j\rangle^\dagger = \langle j| \leftrightarrow \begin{pmatrix} \langle 1'|j\rangle \\ \langle 2'|j\rangle \\ \vdots \\ \langle n'|j\rangle \end{pmatrix}^\dagger = \overbrace{\langle 1'|j\rangle^* \langle 2'|j\rangle^* \cdots \langle n'|j\rangle^*}$$

$$= \overbrace{\langle j|1'\rangle\langle j|2'\rangle \cdots \langle j|n'\rangle}. \qquad (1.1.15)$$

Either the bra vector $\langle \Psi|$ or the ket vector $|\Psi\rangle$ defines a physical state Ψ. The scalar product requires one of each. $\langle \Psi|\Phi\rangle$ gives the amplitude for a system in state Φ to choose state Ψ. We shall always obey the axioms by requiring that $\langle \Phi|\Psi\rangle = \langle \Psi|\Phi\rangle^*$ and $\langle \Psi|\Psi\rangle = 1 = \langle \Phi|\Phi\rangle$.

The connection between transformation matrix mechanics and wave mechanics is made when each state $|i\rangle$ or $|\Psi\rangle$ is associated with a wave function

$$\langle x|i\rangle \equiv \psi_i(x) \quad \text{or} \quad \langle x|\Psi\rangle \equiv \Psi(x). \qquad (1.1.16)$$

(See Section 1.5.)

C. Hamiltonian Matrices and Operators

Probably the best-known example of a matrix in quantum mechanics is the HAMILTONIAN matrix $\mathcal{H}_{ij}$. Among other things, this matrix defines the time behavior of a state $|\Psi(t)\rangle$ through the time-dependent Schrödinger equation in Eq. (1.1.17): (Planck's angular constant is $\hbar = h/2\pi = 1.05 \times 10^{-34}$ Joule second.)

$$i\hbar\frac{\partial}{\partial t}\langle i|\Psi(t)\rangle = \sum_{j=1}^{n}\mathcal{H}_{ij}\langle j|\Psi(t)\rangle, \tag{1.1.17a}$$

$$i\hbar\frac{\partial}{\partial t}\langle i|\Psi(t)\rangle = \sum_{j=1}^{n}\langle i|\mathbf{H}|j\rangle\langle j|\Psi(t)\rangle. \tag{1.1.17b}$$

The Dirac notation for matrix components $\mathcal{H}_{ij}$ is used in the second equation. The boldface symbol denotes an ABSTRACT OPERATOR **H**. In fact, the abstract Schrödinger equation is obtained [Eq. (1.1.17c)], by removing all reference to basis $\{|1\rangle, |2\rangle, \ldots, |n\rangle\}$ in Eq. (1.1.17c):

$$i\hbar\frac{\partial}{\partial t}|\Psi(t)\rangle = \mathbf{H}|\Psi(t)\rangle, \tag{1.1.17c}$$

$$i\hbar\frac{\partial}{\partial t}\begin{pmatrix}\langle 1|\Psi(t)\rangle \\ \langle 2|\Psi(t)\rangle \\ \vdots \\ \langle n|\Psi(t)\rangle\end{pmatrix} = \begin{pmatrix}\mathcal{H}_{11} & \mathcal{H}_{12} & \cdots & \mathcal{H}_{1n} \\ \mathcal{H}_{21} & \mathcal{H}_{22} & \cdots & \mathcal{H}_{2n} \\ \vdots & & & \\ \mathcal{H}_{n1} & \mathcal{H}_{n2} & \cdots & \mathcal{H}_{nn}\end{pmatrix}\begin{pmatrix}\langle 1|\Psi(t)\rangle \\ \langle 2|\Psi(t)\rangle \\ \vdots \\ \langle n|\Psi(t)\rangle\end{pmatrix}$$

$$= \begin{pmatrix}\sum_{j}\mathcal{H}_{1j}\langle j|\Psi(t)\rangle \\ \sum_{j}\mathcal{H}_{2j}\langle j|\Psi(t)\rangle \\ \vdots \\ \sum_{j}\mathcal{H}_{nj}\langle j|\Psi(t)\rangle\end{pmatrix}. \tag{1.1.17d}$$

Finally, Eq. (1.1.17d) gives a matrix representation of the entire equation. It shows the inner product of the $(n \times n)$ matrix representing $\mathcal{H}$ with matrix or vector which represents $|\Psi(t)\rangle$, and the resulting new vector on the extreme right of the equation.

MATRICES USED IN QUANTUM MECHANICS 13

(a) Elementary Operators Roughly speaking, operators change old vectors into new and different ones, and matrices represent this process from some "viewpoint" or basis. We can see this clearly when we apply the abstract Axiom 4 completeness relations to the abstract operator:

$$\mathbf{H} = \sum_i \sum_j |i\rangle\langle i|\mathbf{H}|j\rangle\langle j| = \sum_i \sum_j \mathscr{H}_{ij}|i\rangle\langle j|. \quad (1.1.18)$$

Equation (1.1.18) shows clearly how $\mathscr{H}$, or any quantum operator for that matter, is a linear combination of ELEMENTARY OPERATORS $\mathbf{e}_{ij} = |i\rangle\langle j|$. (Sometimes these are called UNIT TENSOR OPERATORS.)

Each elementary operator can perform the elementary operation of Eq. (1.1.9a), converting $|j\rangle$ into $|i\rangle$, or Eq. (1.1.19b), converting bra $\langle i|$ into $\langle j|$, or Eq. (1.1.19c), zeroing all other vectors:

$$\mathbf{e}_{ij}|j\rangle = |i\rangle\langle j|j\rangle = |i\rangle, \quad (1.1.19a)$$

$$\langle i|\mathbf{e}_{ij} = \langle i|i\rangle\langle j| = \langle j|, \quad (1.1.19b)$$

$$\mathbf{e}_{ij}|k\rangle = 0 = \langle k|\mathbf{e}_{ij}, \quad k \neq j. \quad (1.1.19c)$$

The $\mathbf{e}_{ij}$, and particularly the $\mathbf{e}_{ii}$, are sometimes called PROJECTION OPERATORS because the effect of $\mathbf{e}_{ii}$ on any vector $|\Psi\rangle$ is to yield its projection or component along the ith basis vector $|i\rangle$, while discarding the rest of $|\Psi\rangle$:

$$\mathbf{e}_{ii}|\Psi\rangle = |i\rangle\langle i|\Psi\rangle = \psi_i|i\rangle. \quad (1.1.20)$$

Note that the representation of $\mathbf{e}_{ij}$ using "its own" basis $\{|1\rangle, |2\rangle, \ldots\}$ is simply a matrix with zeros everywhere except for a 1 at the i-j position.

$$\mathbf{e}_{11} \leftrightarrow e_{11} = \begin{pmatrix} 1 & 0 & \cdots & 0 \\ 0 & 0 & \cdots & 0 \\ \vdots & & & \vdots \\ 0 & 0 & \cdots & 0 \end{pmatrix}, \quad \mathbf{e}_{12} \leftrightarrow e_{12} = \begin{pmatrix} 0 & 1 & \cdots & 0 \\ 0 & 0 & \cdots & 0 \\ \vdots & & & \vdots \\ 0 & 0 & \cdots & 0 \end{pmatrix}, \ldots$$

(1.1.21)

The mathematical name given for the "point-to-point" combination of Dirac's vectors ($|i\rangle\langle j|$) is the OUTER PRODUCT (or sometimes TENSOR PRODUCT) of the two vectors. (Recall that "back-to-back" $\langle i|j\rangle$ is the inner product or scalar product, which is just a number). Consider a more complicated example of an operator: $|\Psi\rangle\langle\Phi|$. In the $\{|1\rangle|2\rangle \cdots |n\rangle\}$ basis we have the following representation of this object (the symbol $\otimes$ is a notation for outer product of matrices):

$$|\Psi\rangle\langle\Phi| \leftrightarrow \begin{pmatrix} \langle 1|\Psi\rangle \\ \langle 2|\Psi\rangle \\ \vdots \\ \langle n|\Psi\rangle \end{pmatrix} \otimes \overbrace{\langle\Phi|1\rangle\langle\Phi|2\rangle \cdots \langle\Phi|n\rangle}$$

$$= \begin{pmatrix} (\langle 1|\Psi\rangle\langle\Phi|1\rangle) & (\langle 1|\Psi\rangle\langle\Phi|2\rangle) & \cdots & (\langle 1|\Psi\rangle\langle\Phi|n\rangle) \\ (\langle 2|\Psi\rangle\langle\Phi|1\rangle) & (\langle 2|\Psi\rangle\langle\Phi|2\rangle) & \cdots & (\langle 2|\Psi\rangle\langle\Phi|n\rangle) \\ \vdots & \vdots & & \\ (\langle n|\Psi\rangle\langle\Phi|1\rangle) & (\langle n|\Psi\rangle\langle\Phi|2\rangle) & \cdots & (\langle n|\Psi\rangle\langle\Phi|n\rangle) \end{pmatrix}.$$

(1.1.22)

Later (Chapter 6) we shall discuss products $|i\rangle|j\rangle$ and $\langle i|\langle j|$. However, by now we have seen enough to appreciate the great efficiency for notation of Dirac's bra-kets.

Nevertheless, one should be certain to keep in mind the difference between abstract quantities like $|i\rangle$, $|\Psi\rangle\langle\Phi|$, and **H** on one hand, and the representations of them on the other. The latter contain numbers which any physics must ultimately have, and in fact you will never really get a "look" at an abstract object unless you write some representation of it. However, abstract quantities are unique and correspond to some unique physical reality (supposedly), while the numbers in the representations depend upon your choice for a basis, that is, upon your viewpoint.

(b) Change of Basis It is quite common to want to change your representations from one basis $\{|1\rangle, |2\rangle, \ldots, |n\rangle\}$ to another "better" basis $\{|1'\rangle, |2'\rangle, \ldots, |n'\rangle\}$. In fact, most of this book will be concerned with the question of what is the best available basis for a given problem and how to find it.

The change can be made immediately if you know the transformation matrix $\mathscr{T}(b' \leftarrow b)$. For example, let us derive now a formula for each new component $\langle k'|\mathbf{H}|l'\rangle = \mathscr{H}_{k'l'}$ in terms of the old components $\langle i|\mathbf{H}|j\rangle = \mathscr{H}_{ij}$ and $\mathscr{T}(b' \leftarrow b)$ components. We get this by simply applying twice the Axiom 4 completeness relations for the old basis:

$$\mathscr{H}_{k'l'} \equiv \langle k'|\mathbf{H}|l'\rangle = \sum_{i=1}^{n}\sum_{j=1}^{n} \langle k'|i\rangle\langle i|\mathbf{H}|j\rangle\langle j|l'\rangle, \quad (1.1.23a)$$

$$\mathscr{H}_{k'l'} = \sum_{i=1}^{n}\sum_{j=1}^{n} \mathscr{T}_{k'i}(b' \leftarrow b)\mathscr{H}_{ij}\mathscr{T}_{jl'}^{\dagger}(b' \leftarrow b), \quad (1.1.23b)$$

$$\mathscr{H}' = \mathscr{T}(b' \leftarrow b)\mathscr{H}\mathscr{T}^{\dagger}(b' \leftarrow b). \quad (1.1.23c)$$

In Eq. (1.1.23b) we have used the definition of conjugation ($\mathcal{T}_{jl'}^{\dagger} = \mathcal{T}_{l'j}^{*}$) and Axiom 2 ($\mathcal{T}_{l'j}^{*} = \langle l'|j\rangle^{*} = \langle j|l'\rangle$).

Note that the transformation matrix $\mathcal{T}$ and its conjugate $\mathcal{T}^{\dagger}$ act left and right in Eq. (1.1.14c) in order to change the old representation $\mathcal{H}$ into the new one. Note also that the kth row of the left matrix ($\mathcal{T}$) contains the new bra vector components $\langle k'| \to \langle \overline{k'|1\rangle, \langle k'|2}\rangle, \ldots$ referred to the old representation. Similarly, the lth column of the right matrix ($\mathcal{T}^{\dagger}$) contains the components of the new ket vector $|l'\rangle$.

1.2 MATRIX DIAGONALIZATION

What is the best basis to express a Hamiltonian operator? Well, the Schrödinger equation, Eq. (1.1.17), would be a lot easier to solve if we could find a basis $\{|e_1\rangle, |e_2\rangle, \ldots\}$ in which all the coupling terms $\mathcal{H}_{ij}$ vanished except for the diagonal ones, i.e., a basis that satisfied Eq. (1.2.1):

$$\langle e_i|\mathbf{H}|e_j\rangle = \delta_{ij}\varepsilon_j = \begin{cases} 0, & \text{for } i \neq j \\ \varepsilon_j, & \text{for } i = j. \end{cases} \quad (1.2.1)$$

Then Eq. (1.1.17) would reduce to n uncoupled and easily solved equations:

$$i\hbar\frac{\partial}{\partial t}\langle e_1|\Psi(t)\rangle = \varepsilon_1\langle e_1|\Psi(t)\rangle; \quad \langle e_1|\Psi(t)\rangle = e^{-i\varepsilon_1 t/\hbar}\langle e_1|\Psi(0)\rangle,$$
$$i\hbar\frac{\partial}{\partial t}\langle e_2|\Psi(t)\rangle = \varepsilon_2\langle e_2|\Psi(t)\rangle; \quad \langle e_2|\Psi(t)\rangle = e^{-i\varepsilon_2 t/\hbar}\langle e_2|\Psi(0)\rangle$$
$$\vdots \qquad\qquad\qquad\qquad \vdots$$

$$(1.2.2)$$

Now it is always possible in principle to find such a basis for any Hamiltonian, though many times it is not so easy. In fact we prove in Appendix C that it is possible to find this diagonal basis for any UNITARY matrix $\mathcal{U}$ (unitary means $\mathcal{U}^{\dagger} = \mathcal{U}^{-1}$) or for any SELF-CONJUGATE or HERMITIAN matrix $\mathcal{H}$ (Hermitian means $\mathcal{H}^{\dagger} = \mathcal{H}$). All proper Hamiltonians and all so-called "observable" operators fall into the latter category.

The state vectors $|e_i\rangle$ satisfying Eq. (1.2.1) are called EIGENVECTORS and the numbers ε_i are called EIGENVALUES. (The German heritage apparently enters into the names: "eigen" means "own.") The eigenvalues are the magnitudes of the energy quanta for the physical system. The eigenvectors describe the stationary ground and excited states.

Finding $|e_1\rangle, |e_2\rangle, \ldots$ and $\varepsilon_1, \varepsilon_2, \ldots$ given an arbitrary matrix is called the EIGENVALUE or DIAGONALIZATION PROBLEM. The abstract versions of Eqs. (1.2.1) shown in Eqs. (1.2.3a) and (1.2.3b) are called the

16 A REVIEW OF MATRIX ALGEBRA AND QUANTUM MECHANICS

EIGENVALUE EQUATIONS. These are expanded in the old basis $\{|1\rangle, |2\rangle, \ldots\}$ in Eqs. (1.2.3c) and (1.2.3d) and are represented in matrix form by Eqs. (1.2.3e) and (1.2.3f). (Presumably we start off with an "old" basis and solve the equations for the "new" eigenbasis $\{|e_1\rangle, |e_2\rangle, \ldots\}$.)

$$\mathbf{H}|e_j\rangle = \varepsilon_j |e_j\rangle, \qquad (1.2.3a)$$

$$\langle e_j | \mathbf{H} = \varepsilon_j \langle e_j |, \qquad (1.2.3b)$$

$$\sum_m \langle l | \mathbf{H} | m \rangle \langle m | e_j \rangle = \varepsilon_j \langle l | e_j \rangle, \qquad (1.2.3c)$$

$$\sum_l \langle e_j | l \rangle \langle l | \mathbf{H} | m \rangle = \varepsilon_j \langle e_j | m \rangle, \qquad (1.2.3d)$$

$$\begin{pmatrix} \mathscr{H}_{11} & \mathscr{H}_{12} & \cdots & \mathscr{H}_{1n} \\ \mathscr{H}_{21} & \mathscr{H}_{22} & \cdots & \mathscr{H}_{2n} \\ \vdots & & & \vdots \\ \mathscr{H}_{n1} & \mathscr{H}_{n2} & & \mathscr{H}_{nn} \end{pmatrix} \begin{pmatrix} \langle 1|e_j\rangle \\ \langle 2|e_j\rangle \\ \vdots \\ \langle n|e_j\rangle \end{pmatrix} = \varepsilon_j \begin{pmatrix} \langle 1|e_j\rangle \\ \langle 2|e_j\rangle \\ \vdots \\ \langle n|e_j\rangle \end{pmatrix}, \qquad (1.2.3e)$$

$$\overbrace{\langle e_j|1\rangle \langle e_j|2\rangle \cdots \langle e_j|n\rangle} \begin{pmatrix} \mathscr{H}_{11} & \mathscr{H}_{12} & \cdots & \mathscr{H}_{1n} \\ \mathscr{H}_{21} & \mathscr{H}_{22} & \cdots & \mathscr{H}_{2n} \\ \vdots & & & \\ \mathscr{H}_{n1} & \mathscr{H}_{n2} & \cdots & \mathscr{H}_{nn} \end{pmatrix}$$

$$= \varepsilon_j \overbrace{\langle e_j|1\rangle \langle e_j|2\rangle \cdots \langle e_j|n\rangle}. \qquad (1.2.3f)$$

The equations contain several unknowns. For the right-handed eigenequations ($\mathscr{H}|e_j\rangle = \varepsilon_j|e_j\rangle$) one needs to find the ket components $\langle i|e_j\rangle$, i.e., all n^2 components of the $\mathscr{T}$ matrix, as well as n eigenvalues ε_j. The solutions to the left-handed bra equation ($\langle e_j|\mathscr{H} = \varepsilon_j\langle e_j|$) are derived easily from the kets, as we will see.

If an eigenvector's components $\langle 1|e_j\rangle, \langle 2|e_j\rangle, \ldots$ were known, we could find the corresponding eigenvalue ε_j immediately from Eq. (1.2.3c) or Eq. (1.2.3e). This, incidentally, is what symmetry theory can do, as explained in the next chapters. In the right situations it finds eigenvectors and transformations $\mathscr{T}(i \leftarrow e)$. However, for now let us suppose that they are still unknowns.

If the eigenvalues are known, there are several ways to get the eigenvectors or $\mathscr{T}(i \leftarrow e)$, and a description of these occupies most of the remainder of this review. But first let us review the equation that determines eigenvalues.

A. Obtaining Eigenvalues: The Secular Equation

Equation (1.2.3c) has been rewritten in Eq. (1.2.4a) by putting everything on the left side. The same has been done for the matrix form, Eq. (1.2.4b).

$$\sum_{m}(\langle l|\mathbf{H}|m\rangle - \delta_{ml}\varepsilon_j)\langle m|e_j\rangle = 0, \quad (1.2.4a)$$

$$\begin{pmatrix} \mathcal{H}_{11} - \varepsilon_j & \mathcal{H}_{12} & \cdots & \mathcal{H}_{1n} \\ \mathcal{H}_{21} & \mathcal{H}_{22} - \varepsilon_j & \cdots & \mathcal{H}_{2n} \\ \vdots & & & \vdots \\ \mathcal{H}_{n1} & \mathcal{H}_{n2} & \cdots & \mathcal{H}_{nn} - \varepsilon_j \end{pmatrix} \begin{pmatrix} \langle 1|e_j\rangle \\ \langle 2|e_j\rangle \\ \vdots \\ \langle n|e_j\rangle \end{pmatrix} = \begin{pmatrix} 0 \\ 0 \\ \vdots \\ 0 \end{pmatrix}. \quad (1.2.4b)$$

We shall take an example of an eigenvalue problem involving the matrix $H = \begin{pmatrix} 4 & 1 \\ 3 & 2 \end{pmatrix}$ and solve it along with the general case. For each equation like Eq. (1.2.4b) we will give a corresponding example indicated by Eq. $(. \ .)_x$, like

$$\begin{pmatrix} 4 - \varepsilon_j & 1 \\ 3 & 2 - \varepsilon_j \end{pmatrix} \begin{pmatrix} \langle 1|e_j\rangle \\ \langle 2|e_j\rangle \end{pmatrix} = \begin{pmatrix} 0 \\ 0 \end{pmatrix}. \quad (1.2.4b)_x$$

Note that the example could never be a Hamiltonian matrix since it is not Hermitian, but it serves as a good example for mathematical purposes.

Now Eqs. (1.2.4) can have a nonzero solution only if the determinant of the matrix vanishes. (See Appendix B.) It is the resulting equation, Eq. (1.2.5), that will determine the eigenvalues, and it is usually called the SECULAR EQUATION. Note that it is the same for bra and ket eigenequations.

$$\det \begin{vmatrix} \mathcal{H}_{11} - \varepsilon & \mathcal{H}_{12} & \cdots & \mathcal{H}_{1n} \\ \mathcal{H}_{21} & \mathcal{H}_{22} - \varepsilon & \cdots & \mathcal{H}_{2n} \\ \vdots & & & \\ \mathcal{H}_{n1} & \mathcal{H}_{n2} & & \mathcal{H}_{nn} - \varepsilon \end{vmatrix}$$

$$\equiv S(\varepsilon) \equiv \varepsilon^n + a_1\varepsilon^{n-1} + \cdots + a_{n-1}\varepsilon + a_n = 0. \quad (1.2.5)$$

$$\det \begin{vmatrix} 4 - \varepsilon & 1 \\ 3 & 2 - \varepsilon \end{vmatrix} \equiv S(\varepsilon) = \varepsilon^2 - 6\varepsilon + 5 = 0. \quad (1.2.5)_x$$

It can be a tedious job to solve an nth-degree secular equation. It is easy enough for small matrices like our example, for which the roots are ($\varepsilon_1 = 1$) and ($\varepsilon_2 = 5$). But one should know that no formula exists for roots of quintic, sextic, or higher-degree equations. All nth-degree polynomial equations have exactly n roots somewhere in the complex plane, but no formula exists to give them all for $n \geq 5$ (Abels's theorem). In general finding the eigenvalues ε_j can be the hardest part of the problem, if the eigenvectors are not known.

B. Obtaining Eigenvectors: The Hamilton-Cayley Equation

One very helpful fact is that any matrix satisfies its own secular equation. This is the content of the following theorem. The resulting equation, Eq. (1.2.6), is called the HAMILTON-CAYLEY EQUATION (HCEq).

Hamilton-Cayley Theorem:: If in the secular equation

$$S(\varepsilon) = \det|\mathcal{H} - \varepsilon\mathbf{1}| = \varepsilon^n + a_1\varepsilon^{n-1} + \cdots + a_{n-1}\varepsilon + a_n = 0$$

one replaces the terms $a_m\varepsilon^{n-m}$ by the matrices $a_m\mathcal{H}^{n-m}$, the constant term a_n by the matrix $a_n\mathbf{1}$, and the 0 by the zero matrix $\mathbf{0}$, then the resulting matrix equation, Eq. (1.2.6), is valid:

$$S(\mathcal{H}) = \mathcal{H}^n + a_1\mathcal{H}^{n-1} + \cdots + a_{n-1}\mathcal{H} + a_n\mathbf{1} = \mathbf{0}, \qquad (1.2.6)$$

$$S\left[\begin{pmatrix} 4 & 1 \\ 3 & 2 \end{pmatrix}\right] = \begin{pmatrix} 4 & 1 \\ 3 & 2 \end{pmatrix}^2 - 6\begin{pmatrix} 4 & 1 \\ 3 & 2 \end{pmatrix} + 5\begin{pmatrix} 1 & 0 \\ 0 & 1 \end{pmatrix}$$

$$= \begin{pmatrix} 19 & 6 \\ 18 & 7 \end{pmatrix} - \begin{pmatrix} 24 & 6 \\ 18 & 12 \end{pmatrix} + \begin{pmatrix} 5 & 0 \\ 0 & 5 \end{pmatrix} = \begin{pmatrix} 0 & 0 \\ 0 & 0 \end{pmatrix}. \quad (1.2.6)_x$$

We can prove this theorem by using the fact that the product of a matrix $\mathcal{M}$ with its ADJUNCT matrix $\mathcal{M}^{\text{adj}}$ equals $\det\mathcal{M}$ times a unit matrix. [As explained in Appendix B, $\mathcal{M}^{\text{adj}}_{ij}$ is $(-1)^{i+j}$ multiplied by the determinant of $\mathcal{M}$ taken with its jth row and ith column missing.] This gives the following:

$$\sum_j \mathcal{M}^{\text{adj}}_{ij}\mathcal{M}_{jk} = \sum_j \mathcal{M}_{ij}\mathcal{M}^{\text{adj}}_{jk} = \delta_{ik}(\det\mathcal{M}).$$

Setting $\mathcal{M} = \mathcal{H} - \varepsilon\mathbf{1}$ we have

$$\mathcal{M}^{\text{adj}}(\mathcal{H} - \varepsilon\mathbf{1}) = \det(\mathcal{H} - \varepsilon\mathbf{1})\mathbf{1} = (\mathcal{H} - \varepsilon\mathbf{1})\mathcal{M}^{\text{adj}},$$

$$\mathcal{M}^{\text{adj}}(\mathcal{H} - \varepsilon\mathbf{1}) = S(\varepsilon)\mathbf{1} = (\mathcal{H} - \varepsilon\mathbf{1})\mathcal{M}^{\text{adj}}. \qquad (1.2.7)$$

Now replacement of the number ε by the matrix $\mathcal{H}$ is well defined, and doing so proves the theorem: $S(\mathcal{H}) = \mathbf{0}$.

MATRIX DIAGONALIZATION   19

(a) Eigenvector Projectors If the roots ε_j are known, we may use the factored form of the Eq. (1.2.6) to derive the eigenvectors:

$$(\mathscr{H} - \varepsilon_1 \mathbf{1})(\mathscr{H} - \varepsilon_2 \mathbf{1}) \cdots (\mathscr{H} - \varepsilon_n \mathbf{1}) = \mathbf{0}, \qquad (1.2.8)$$

$$\left[\begin{pmatrix} 4 & 1 \\ 3 & 2 \end{pmatrix} - 1 \begin{pmatrix} 1 & 0 \\ 0 & 1 \end{pmatrix}\right] \cdot \left[\begin{pmatrix} 4 & 1 \\ 3 & 2 \end{pmatrix} - 5 \begin{pmatrix} 1 & 0 \\ 0 & 1 \end{pmatrix}\right] = \begin{pmatrix} 0 & 0 \\ 0 & 0 \end{pmatrix}. \qquad (1.2.8)_x$$

The procedure depends upon whether or not any of the n roots ε_j are equal. We study first the cases in which all ε_j are distinct. This includes our example $(\varepsilon_1 = 1, \varepsilon_2 = 5)$.

Case i: All n Roots ε_i Are Distinct

With all roots having different values one may select n RELATIVELY PRIME factors ρ_j ("relatively prime" means "no common factor shared by all") from the factored HCEq, Eq. (1.2.8), by deleting first one $(\mathscr{H} - \varepsilon_j \mathbf{1})$ factor, then the next, and so on:

$$\rho_1 = \mathbf{1} \cdot (\mathscr{H} - \varepsilon_2 \mathbf{1}) \cdot \qquad (\cdots) \cdot (\mathscr{H} - \varepsilon_n \mathbf{1}),$$

$$\rho_2 = (\mathscr{H} - \varepsilon_1 \mathbf{1}) \cdot \mathbf{1} \cdot \qquad (\cdots) \cdot (\mathscr{H} - \varepsilon_n \mathbf{1}),$$

$$\vdots$$

$$\rho_n = (\mathscr{H} - \varepsilon_1 \mathbf{1}) \cdot (\mathscr{H} - \varepsilon_2 \mathbf{1}) \cdot (\cdots) \cdot \mathbf{1}. \qquad (1.2.9)$$

$$\rho_1 = \mathbf{1} \cdot \left[\begin{pmatrix} 4 & 1 \\ 3 & 2 \end{pmatrix} - 5 \begin{pmatrix} 1 & 0 \\ 0 & 1 \end{pmatrix}\right] = \begin{pmatrix} -1 & 1 \\ 3 & -3 \end{pmatrix},$$

$$\rho_2 = \left[\begin{pmatrix} 4 & 1 \\ 3 & 2 \end{pmatrix} - 1 \begin{pmatrix} 1 & 0 \\ 0 & 1 \end{pmatrix}\right] \cdot \mathbf{1} = \begin{pmatrix} 3 & 1 \\ 3 & 1 \end{pmatrix}. \qquad (1.2.9)_x$$

Now we see that these ρ_j contain the eigenvectors, if we rewrite the HCEq, Eq. (1.2.8), in terms of ρ_j as is done in the following. Equations (1.2.10b) and (1.2.10c) are essentially the eigenvalue equations (1.2.3e) and (1.2.3f) with matrix ρ_j in place of the eigenvector:

$$(\mathscr{H} - \varepsilon_j \mathbf{1}) \rho_j = \mathbf{0}, \qquad (1.2.10a)$$

$$\mathscr{H} \rho_j = \varepsilon_j \rho_j, \qquad (1.2.10b)$$

$$\rho_j \mathscr{H} = \varepsilon_j \rho_j. \qquad (1.2.10c)$$

According to the last equations, each column of ρ_j must satisfy the eigenvalue equation for a ket vector $|e_j\rangle$, and each row must satisfy this equation

20 A REVIEW OF MATRIX ALGEBRA AND QUANTUM MECHANICS

for a bra vector $\langle e_j|$. In Eq. (1.2.10)$_x$ we write only the first columns and rows of ρ_j:

$$\begin{pmatrix} 4 & 1 \\ 3 & 2 \end{pmatrix} \begin{pmatrix} -1 \\ 3 \end{pmatrix} = 1 \begin{pmatrix} -1 \\ 3 \end{pmatrix},$$

$$\begin{pmatrix} 4 & 1 \\ 3 & 2 \end{pmatrix} \begin{pmatrix} 3 \\ 3 \end{pmatrix} = 5 \begin{pmatrix} 3 \\ 3 \end{pmatrix},$$

$$\begin{pmatrix} -1 & 1 \end{pmatrix} \begin{pmatrix} 4 & 1 \\ 3 & 2 \end{pmatrix} = 1 \begin{pmatrix} -1 & 1 \end{pmatrix},$$

$$\begin{pmatrix} 3 & 1 \end{pmatrix} \begin{pmatrix} 4 & 1 \\ 3 & 2 \end{pmatrix} = 5 \begin{pmatrix} 3 & 1 \end{pmatrix}. \qquad (1.2.10b)_x$$

Now if a vector satisfies an eigenvalue equation, so will two times that vector, or any "length" of that vector including zero. It is conventional to normalize all our eigenvectors to give $\langle e_j | e_j \rangle = 1$.

This normalization can be done if we divide each ρ_j by the right factor to make it into an IDEMPOTENT matrix $\mathscr{P}_j$. (Something is called idempotent if it will give itself back when multiplied by itself. **1** is an example: $\mathbf{1} \cdot \mathbf{1} = \mathbf{1}$.) The following equation follows from the definition (1.2.9) of ρ_i, i.e., all $(\mathscr{H} - \varepsilon_l \mathbf{1})$ factors except the ith:

$$\rho_i \rho_j = \prod_{l \ne i}(\mathscr{H} - \varepsilon_l \mathbf{1}) \rho_j = \prod_{l \ne j}(\varepsilon_j - \varepsilon_l) \rho_j, \qquad \text{for } i = j$$

$$= \mathbf{0}, \qquad \text{for } i \ne j. \quad (1.2.11)$$

This implies that the $\mathscr{P}_j$ defined by Eq. (1.2.12) are idempotent $[\mathscr{P}_j^2 = \mathscr{P}_j]$. Note also that different ρ_j or $\mathscr{P}_j$ will be orthogonal ($\mathscr{P}_i \mathscr{P}_j = 0$ for $i \ne j$):

$$\mathscr{P}_j \equiv \frac{\prod_{l \ne j}(\mathscr{H} - \varepsilon_l \mathbf{1})}{\prod_{l \ne j}(\varepsilon_j - \varepsilon_l)}, \qquad (1.2.12)$$

$$\mathscr{P}_1 = \frac{\begin{pmatrix} -1 & 1 \\ 3 & -3 \end{pmatrix}}{1 - 5} = \begin{pmatrix} \frac{1}{4} & -\frac{1}{4} \\ -\frac{3}{4} & \frac{3}{4} \end{pmatrix}, \qquad \mathscr{P}_2 = \frac{\begin{pmatrix} 3 & 1 \\ 3 & 1 \end{pmatrix}}{5 - 1} = \begin{pmatrix} \frac{3}{4} & \frac{1}{4} \\ \frac{3}{4} & \frac{1}{4} \end{pmatrix}.$$

$$(1.2.12)_x$$

Now, because $\mathscr{P}_j \mathscr{P}_j = \mathscr{P}_j$, the desired normalization factor N_j is found on the diagonal of $\mathscr{P}_j$ at the intersection of the row and column that were picked to be the eigenbra $\langle e_j |$ and eigenket $|e_j\rangle$, respectively. The vectors given in Eq.

MATRIX DIAGONALIZATION

(1.2.13) satisfy the normalization condition $\langle e_i | e_j \rangle = \delta_{ij}$.

$$\langle e_j | \leftrightarrow \frac{\overbrace{\text{chosen row}}}{\sqrt{N_j}}, \qquad |e_j\rangle \leftrightarrow \frac{1}{\sqrt{N_j}}\begin{pmatrix}\text{chosen}\\ \text{column}\end{pmatrix}, \qquad (1.2.13)$$

$$\langle e_1 | \leftrightarrow \frac{\overbrace{\tfrac{1}{4} \quad -\tfrac{1}{4}}}{\sqrt{\tfrac{1}{4}}}, \qquad |e_1\rangle \leftrightarrow \frac{1}{\sqrt{\tfrac{1}{4}}}\begin{pmatrix}\tfrac{1}{4}\\ -\tfrac{3}{4}\end{pmatrix}$$

$$\langle e_2 | \leftrightarrow \frac{\overbrace{\tfrac{3}{4} \quad \tfrac{1}{4}}}{\sqrt{\tfrac{3}{4}}}, \qquad |e_2\rangle \leftrightarrow \frac{1}{\sqrt{\tfrac{3}{4}}}\begin{pmatrix}\tfrac{3}{4}\\ \tfrac{3}{4}\end{pmatrix}. \qquad (1.2.13)_x$$

According to the convention just given, the normalization is divided equally between the left (bra) and right (ket) eigenvectors. This allows the left and right eigenvectors of a Hermitian (or unitary) operator to be related by transpose conjugation ($\langle e_i | = |e_i\rangle^\dagger$) as proved in Problem 1.2.7. However, this relation does not hold between left and right eigenvectors of a non-Hermitian matrix such as the (2 × 2) matrix, which is our example. In fact, its bra and ket vectors point in quite different "directions" as seen in Figure 1.2.1. (Compare $|e_j\rangle$ with $\langle e_j|$.) Nevertheless, $\langle e_1|$ is perpendicular to $|e_2\rangle$ and $\langle e_2|$ is perpendicular to $|e_1\rangle$, as guaranteed by Eq. (1.2.11). The figure gives a picture of a DUAL SPACE consisting of left or bra eigenvectors $\langle\;|$ on one hand, and the "reciprocal space" of right or ket vectors $|\;\rangle$ on the other.

Finally, notice that the $\mathscr{P}_j$ matrices are just representations of the PROJECTION OPERATORS $|e_j\rangle\langle e_j|$ discussed around Eq. (1.1.20):

$$\mathbf{e}_{e_1 e_1} = |e_1\rangle\langle e_1| \leftrightarrow \begin{pmatrix}\tfrac{1}{2}\\ -\tfrac{3}{2}\end{pmatrix} \otimes \overbrace{\begin{pmatrix}\tfrac{1}{2} & -\tfrac{1}{2}\end{pmatrix}} = \begin{pmatrix}\tfrac{1}{4} & -\tfrac{1}{4}\\ -\tfrac{3}{4} & \tfrac{3}{4}\end{pmatrix} = \mathscr{P}_1,$$

$$\mathbf{e}_{e_2 e_2} = |e_2\rangle\langle e_2| \leftrightarrow \begin{pmatrix}\tfrac{\sqrt{3}}{2}\\ \tfrac{\sqrt{3}}{2}\end{pmatrix} \otimes \overbrace{\begin{pmatrix}\tfrac{\sqrt{3}}{2} & \tfrac{1}{2\sqrt{3}}\end{pmatrix}} = \begin{pmatrix}\tfrac{3}{4} & \tfrac{1}{4}\\ \tfrac{3}{4} & \tfrac{1}{4}\end{pmatrix} = \mathscr{P}_2. \qquad (1.2.14)$$

Case ii: Some Roots ε_j are Equal

If a matrix has some repeated roots in its secular equation, then it is possible that the matrix itself will satisfy a polynomial equation of lower degree than the HCEq. The lowest degree equation satisfied by any matrix or operator is called the MINIMAL EQUATION (MEq).

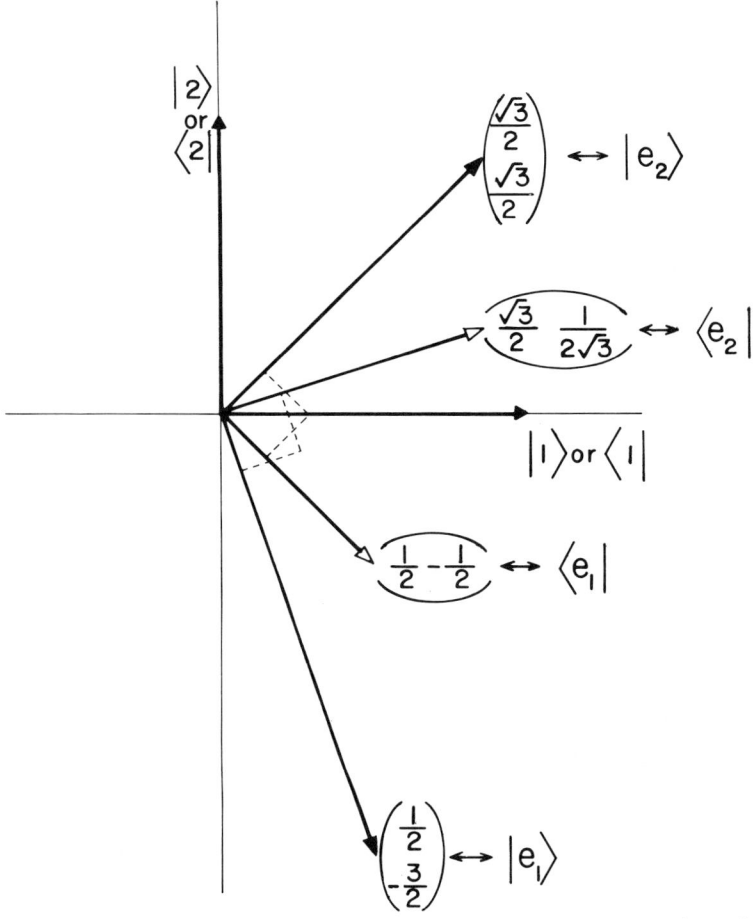

Figure 1.2.1 Left ($\langle e_j |$) and right ($| e_j \rangle$) eigenvectors of the matrix $\begin{pmatrix} 4 & 1 \\ 3 & 2 \end{pmatrix}$.

If the MEq has repeated roots, too, then the matrix is not diagonalizable. It can still be brought to an almost diagonal Jordan form, but we shall not need to discuss this. Most matrices which will be discussed in this text are Hermitian or unitary, and they are therefore diagonalizable. (See Appendix C.) The roots of their MEq are distinct even if there are any number of repeated roots in the HCEq.

For all such cases the theory from Eq. (1.2.8) to Eq. (1.2.12) applies except we deal with the MEq instead of the HCEq. The projection matrix defined in the following obeys the same rules ($\mathscr{P}_i \mathscr{P}_j = \delta_{ij} \mathscr{P}_i$):

$$\mathscr{P}_{\varepsilon_j} = \frac{\prod\limits_{\varepsilon_l \neq \varepsilon_j} (\mathscr{H} - \varepsilon_l \mathbf{1})}{\prod\limits_{\varepsilon_l \neq \varepsilon_j} (\varepsilon_j - \varepsilon_l)}. \qquad (1.2.15)$$

For example, the matrix

$$\mathcal{H} = \begin{pmatrix} 0 & 0 & 0 & 1 \\ 0 & 0 & 1 & 0 \\ 0 & 1 & 0 & 0 \\ 1 & 0 & 0 & 0 \end{pmatrix}$$

has a fourth-degree HCEq $[(H - 1)^2(H + 1)^2 = 0]$ but only a second-degree MEq $[(H - 1)(H + 1) = 0]$. Hence, only two idempotents need to be constructed. These are the following:

$$\mathcal{P}_1 = \frac{\mathcal{H} - (-1)\mathbf{1}}{1 - (-1)} = \begin{pmatrix} \frac{1}{2} & 0 & 0 & \frac{1}{2} \\ 0 & \frac{1}{2} & \frac{1}{2} & 0 \\ 0 & \frac{1}{2} & \frac{1}{2} & 0 \\ \frac{1}{2} & 0 & 0 & \frac{1}{2} \end{pmatrix},$$

$$\mathcal{P}_{-1} = \frac{\mathcal{H} - \mathbf{1}}{(-1 - 1)} = \begin{pmatrix} \frac{1}{2} & 0 & 0 & -\frac{1}{2} \\ 0 & \frac{1}{2} & -\frac{1}{2} & 0 \\ 0 & -\frac{1}{2} & \frac{1}{2} & 0 \\ -\frac{1}{2} & 0 & 0 & \frac{1}{2} \end{pmatrix}. \quad (1.2.16)$$

However, these idempotents contain more than one independent eigenvector apiece. $\mathcal{P}_1$ has two: its first row: $\langle e_1 | \to \overline{1/\sqrt{2} \quad 0 \quad 0 \quad 1/\sqrt{2}}$ and its second row: $\langle e_1' | \to \overline{0 \quad 1/\sqrt{2} \quad 1/\sqrt{2} \quad 0}$. $\mathcal{P}_{-1}$ has two also. In general, a $\mathcal{P}_{\varepsilon_j}$ will have a number of independent eigenvectors equal to the number of times ε_j is repeated in the HCEq.

Any linear combination of rows (or of columns) of a $\mathcal{P}_{\varepsilon_j}$ matrix will be an eigenbra (eigenket) vector, so these are not uniquely defined when ε_j is repeated. Furthermore, the vectors $|e_j^\alpha\rangle$ from row α may not be orthogonal to $|e_j^\beta\rangle$ from column β likewise for columns. However, the same trick works again: $\langle e_j^\alpha | e_j^\beta \rangle$ is the component $(\mathcal{P}_{\varepsilon_j})_{\alpha\beta}$ of the idempotent projector. So Gram-Schmidt orthogonalization and normalization can be done directly.

(b) Completeness and Orthogonality of Projectors The orthonormality relation, Eq. (1.2.17), and its consequences were demonstrated in the preceding section:

$$\mathcal{P}_{\varepsilon_i} \mathcal{P}_{\varepsilon_j} = \delta_{\varepsilon_i \varepsilon_j} \mathcal{P}_{\varepsilon_i}. \quad (1.2.17)$$

What needs to be shown now is the completeness relation, Eq. (1.2.18), for these same projection operators:

$$\mathcal{P}_{\varepsilon_i} + \mathcal{P}_{\varepsilon_j} + \cdots + \mathcal{P}_{\varepsilon_n} = 1 \quad \text{(sum over disctinct } \varepsilon_j\text{)}, \quad (1.2.18)$$

$$\begin{pmatrix} \frac{1}{4} & -\frac{1}{4} \\ -\frac{3}{4} & \frac{1}{4} \end{pmatrix} + \begin{pmatrix} \frac{3}{4} & \frac{1}{4} \\ \frac{1}{4} & \frac{1}{4} \end{pmatrix} = \begin{pmatrix} 1 & 0 \\ 0 & 1 \end{pmatrix}. \quad (1.2.18)_x$$

Proving Eq. (1.2.18) abstractly is easy for most mathematicians, if not for most physicists. Most mathematics students have encountered the same idea repeatedly in number theory, then in algebra, and so forth. In number theory or algebra one proves that integers or polynomials $\{p_1 p_2 \cdots p_n\}$ are relatively prime if and only if there exits some integers or polynomials $\{a_1 a_2 \cdots a_n\}$ that $\Sigma_i a_i p_i = 1$. The same proof works for the relatively prime matrix polynomials $\mathscr{P}_i$ defined in Eq. (1.2.9), or for the normalized $\mathcal{P}_j$ matrices:

$$\sum \alpha_j \mathcal{P}_j = 1.$$

Now operation with $\mathcal{P}_j$ gives $\alpha_j \mathcal{P}_j = \mathcal{P}_j$ and Eq. (1.2.18) is proved.

A less abstract proof follows from elementary algebra. Note that for any two numbers ε_1 and ε_2 we have that

$$\frac{m - \varepsilon_2}{\varepsilon_1 - \varepsilon_2} + \frac{m - \varepsilon_1}{\varepsilon_2 - \varepsilon_1} = \frac{\varepsilon_1 - \varepsilon_2}{\varepsilon_1 - \varepsilon_2} = 1.$$

For any three numbers ε_1, ε_2, and ε_3 we have that

$$\frac{(m - \varepsilon_2)(m - \varepsilon_1)}{(\varepsilon_1 - \varepsilon_2)(\varepsilon_1 - \varepsilon_3)} + \frac{(m - \varepsilon_3)(m - \varepsilon_3)}{(\varepsilon_2 - \varepsilon_1)(\varepsilon_2 - \varepsilon_3)}$$
$$+ \frac{(m - \varepsilon_1)(m - \varepsilon_2)}{(\varepsilon_3 - \varepsilon_1)(\varepsilon_3 - \varepsilon_2)} = \frac{(\varepsilon_1 - \varepsilon_2)(\varepsilon_1 - \varepsilon_3)(\varepsilon_2 - \varepsilon_3)}{(\varepsilon_1 - \varepsilon_2)(\varepsilon_1 - \varepsilon_3)(\varepsilon_2 - \varepsilon_3)} = 1,$$

and so forth. In other words, one may substitute arbitrary values for the ε_j in Eqs. (1.2.12) and (1.2.15) for the $\mathcal{P}_j$ and still have them satisfy the completeness relation in Eq. (1.2.18)! (On the other hand, we have seen that the orthonormality relations hold only if the ε_j are the eigenvalues of matrix $\mathcal{H}$.)

It is interesting to note a similarity between the form of the completeness relation for projection operators of matrices and the Lagrangian interpolation formula which is used in numerical analysis. Suppose a function $f(x)$ of a real variable has n values $f(x_1), f(x_2), \ldots, f(x_n)$, which are known at n distinct points $x_1, x_2, \ldots, x_n$. Then the following $(n-1)$th-degree polynomial $L(x)$ is constructed to exactly equal $f(x)$ at each point x_j.

$$L(x) = \sum_{j=1}^{n} f(x_j) \frac{\prod_{k \neq j}^{n}(x - x_k)}{\prod_{k \neq j}^{n}(x_j - x_k)}.$$

The coefficient of $f(x_j)$ in the preceding expression for $L(x)$ should remind you of the formula for P_j given by Eq. (1.2.12) if x is replaced by an n-dimensional matrix and x_j are its eigenvalues.

Clearly, $L(x) = f(x)$ at each point $x = x_j$. How well $L(x)$ represents $f(x)$ for $x \neq x_j$ depends upon how much $f(x)$ deviates from an $(n - 1)$th degree polynomial in between these points. Simple functions such as $f(x) = 1$ or $f(x) = x$ have the following polynomial representations for $n > 1$:

$$1 = \sum_{j=1}^{n} \frac{\prod_{k \neq j}^{n}(x - x_k)}{\prod_{k \neq j}^{n}(x_j - x_k)}, \qquad x = \sum_{j=1}^{n} x_j \frac{\prod_{k \neq j}^{n}(x - x_k)}{\prod_{k \neq j}^{n}(x_j - x_k)}.$$

These have precisely the same form as the completeness relation (1.2.18) and the spectral decomposition relation which is discussed in the following. [See Eq. (1.2.21).]

Finally, we observe that, while Eqs. (1.2.17) and (1.2.18) appear very dissimilar, they are more similar if you represent them with Dirac's notation. The completeness relation for basis $|e_j\rangle$ is represented by the following in the basis $\{|1\rangle, |2\rangle, \ldots\}$:

$$\mathbf{1} = \sum_j |e_j\rangle\langle e_j|, \qquad (1.2.19a)$$

$$\langle l|\mathbf{1}|m\rangle = \delta_{lm} = \sum_j \langle l|e_j\rangle\langle e_j|m\rangle. \qquad (1.2.19b)$$

The orthonormality condition for basis $|e_j\rangle$ is similarly represented:

$$\langle e_i|e_j\rangle = \delta_{ij} = \sum_l \langle e_i|l\rangle\langle l|e_j\rangle. \qquad (1.2.20)$$

The similarity between Eqs. (1.2.20) and (1.2.19) is now more evident.

(c) Spectral Decomposition of Matrices Any diagonalizable matrix $\mathscr{H}$ will have a set of idempotent projection matrices $\mathscr{P}_k$ satisfying the orthonormality [Eq. (1.2.17)] and completeness [Eq. (1.2.18)] relations. Now operating on Eq. (1.2.18) with the matrix $\mathscr{H}$ gives Eq. (1.2.21), which is called the

SPECTRAL DECOMPOSITION of $\mathcal{H}$ [eigenvalue equations (1.2.10) are used]:

$$\mathcal{H} = \varepsilon_i \mathcal{P}_{\varepsilon_i} + \varepsilon_j \mathcal{P}_{\varepsilon_j} + \cdots + \varepsilon_n \mathcal{P}_{\varepsilon_n} \quad \text{(sum over distinct } \varepsilon_j\text{)}, \quad (1.2.21)$$

$$\begin{pmatrix} 4 & 1 \\ 3 & 2 \end{pmatrix} = \begin{pmatrix} \frac{1}{4} & -\frac{1}{4} \\ -\frac{3}{4} & \frac{3}{4} \end{pmatrix} + 5 \begin{pmatrix} \frac{3}{4} & \frac{1}{4} \\ \frac{3}{4} & \frac{1}{4} \end{pmatrix}. \quad (1.2.21)_x$$

This decomposition provides a very elegant way to manipulate a matrix. For example, if you wanted the 50th power of the matrix $\begin{pmatrix} 4 & 1 \\ 3 & 2 \end{pmatrix}$ there is no need to begin multiplying it by itself 49 times. Instead, the spectral decomposition gives the answer quickly:

$$\begin{pmatrix} 4 & 1 \\ 3 & 2 \end{pmatrix}^{50} = (\varepsilon_1 \mathcal{P}_1 + \varepsilon_2 \mathcal{P}_2)^{50}$$

$$= \varepsilon_1^{50} \mathcal{P}_1^{50} + 50 \varepsilon_1^{49} \mathcal{P}_1^{49} \varepsilon_2 \mathcal{P}_2 + \cdots + \varepsilon_2^{50} \mathcal{P}_2^{50}$$

$$= \varepsilon_1^{50} \mathcal{P}_1 + \varepsilon_2^{50} \mathcal{P}_2$$

$$= 1 \begin{pmatrix} \frac{1}{4} & -\frac{1}{4} \\ -\frac{3}{4} & \frac{3}{4} \end{pmatrix} + 5^{50} \begin{pmatrix} \frac{3}{4} & \frac{1}{4} \\ \frac{3}{4} & \frac{1}{4} \end{pmatrix}$$

$$= \frac{1}{4} \begin{pmatrix} 1 + 3 \cdot 5^{50} & 5^{50} - 1 \\ 3 \cdot 5^{50} - 3 & 3 + 5^{50} \end{pmatrix}. \quad (1.2.22)$$

The property $\mathcal{P}_i \mathcal{P}_j = \delta_{ij} \mathcal{P}_j$ reduces a complicated expression to a very simple one. It is this type of analysis that is central to the development of symmetry analysis and its applications, which are given in the next chapters.

(d) Simultaneous Diagonalization of Commuting Matrices When any two Hermitian matrices $\mathcal{M}$ and $\mathcal{N}$ commute with each other $[\mathcal{MN} = \mathcal{NM}]$ then there must exist a complete set of base vectors that are eigenvectors of both matrices. This is constructed easily using the spectral decompositions and completeness relations for each matrix. These are shown in Eqs. (1.2.23), where it is assumed that the number of distinct eigenvalues of $\mathcal{M}$ is m, while for $\mathcal{N}$ the number is n:

$$\mathbf{1} = \mathcal{P}^1(\mathcal{M}) + \mathcal{P}^2(\mathcal{M}) + \cdots + \mathcal{P}^m(\mathcal{M}), \quad (1.2.23\text{a})$$

$$\mathcal{M} = \mu_1 \mathcal{P}^1(\mathcal{M}) + \mu_2 \mathcal{P}^2(\mathcal{M}) + \cdots + \mu_m \mathcal{P}^m(\mathcal{M}), \quad (1.2.23\text{b})$$

$$\mathbf{1} = \mathcal{P}^1(\mathcal{N}) + \mathcal{P}^2(\mathcal{N}) + \cdots + \mathcal{P}^n(\mathcal{N}), \quad (1.2.23\text{c})$$

$$\mathcal{N} = \nu_1 \mathcal{P}^1(\mathcal{N}) + \nu_2 \mathcal{P}^2(\mathcal{N}) + \cdots + \nu_n \mathcal{P}^n(\mathcal{N}). \quad (1.2.23\text{d})$$

Now if the matrices commute, so will their respective idempotents, since these are just polynomials of the matrices. Because of this, we get immediately a complete set of idempotents and hence the eigenvectors for the two of them by simply multiplying the separate completeness relations:

$$\mathbf{1} = \mathbf{1} \cdot \mathbf{1} = [\mathscr{P}^1(\mathscr{M}) + \mathscr{P}^2(\mathscr{M}) + \cdots + \mathscr{P}^m(\mathscr{M})]$$
$$\times [\mathscr{P}^1(\mathscr{N}) + \mathscr{P}^2(\mathscr{N}) + \cdots + \mathscr{P}^n(\mathscr{N})], \quad (1.2.24a)$$

$$\mathbf{1} = \mathscr{P}^1(\mathscr{M})\mathscr{P}^1(\mathscr{N}) + \mathscr{P}^1(\mathscr{M})\mathscr{P}^2(\mathscr{N}) + \cdots + \mathscr{P}^2(\mathscr{M})\mathscr{P}^1(\mathscr{N}) + \cdots, \quad (1.2.24b)$$

$$\mathbf{1} = \mathscr{P}^{1,1} + \mathscr{P}^{1,2} \cdots + \cdots + \mathscr{P}^{2,1} + \cdots + \cdots. \quad (1.2.24c)$$

Each of the resulting terms must satisfy eigenvalue equations for $\mathscr{M}$ and $\mathscr{N}$:

$$\mathscr{M}\mathscr{P}^{\alpha,\beta} = \mathscr{M}\mathscr{P}^\alpha(\mathscr{M})\mathscr{P}^\beta(\mathscr{N}) = \mu_\alpha \mathscr{P}^{\alpha,\beta},$$
$$\mathscr{N}\mathscr{P}^{\alpha,\beta} = \mathscr{N}\mathscr{P}^\alpha(\mathscr{M})\mathscr{P}^\beta(\mathscr{N}) = \mathscr{N}\mathscr{P}^\beta(\mathscr{N})\mathscr{P}^\alpha(\mathscr{M}) = \nu_\beta \mathscr{P}^{\alpha,\beta}.$$

Each term $\mathscr{P}^{\alpha,\beta}$ that is not zero is clearly one of a set of complete and orthogonal idempotents. However, many of the terms are usually zero in practice.

In fact, if $\mathscr{M}$ is an $m \times m$ matrix with m distinct eigenvalues, then there will be exactly m nonzero terms on the right of (1.2.24b) or (1.2.24c), and each of these will be identical to one of the $\mathscr{P}^\alpha(\mathscr{M})$. Having two or more nonzero terms $\mathscr{P}^{\alpha,\beta}, \mathscr{P}^{\alpha,\beta'}, \ldots$ amounts to having two orthogonal eigenvectors with eigenvalue μ_α where we said we only had one.

In other words, the reduction of $\mathscr{M}$ will, in the case of distinct μ_α, serve immediately to reduce all the other matrices $\mathscr{N}$ which commute with $\mathscr{M}$. For example, the fact that matrix $\mathscr{N} = \begin{pmatrix} 5 & -1 \\ -3 & 7 \end{pmatrix}$ commutes with the matrix $\mathscr{M} = \begin{pmatrix} 4 & 1 \\ 3 & 2 \end{pmatrix}$, which we have reduced, implies that $\mathscr{N}$ is reduced by the same idempotents. Solving its secular equation is unnecessary:

$$\mathscr{M} = \begin{pmatrix} 4 & 1 \\ 3 & 2 \end{pmatrix} = 1 \cdot \begin{pmatrix} \frac{1}{4} & -\frac{1}{4} \\ -\frac{3}{4} & \frac{3}{4} \end{pmatrix} + 5 \cdot \begin{pmatrix} \frac{3}{4} & \frac{1}{4} \\ \frac{3}{4} & \frac{1}{4} \end{pmatrix},$$

$$\mathscr{N} = \begin{pmatrix} 5 & -1 \\ -3 & 7 \end{pmatrix} = 8 \cdot \begin{pmatrix} \frac{1}{4} & -\frac{1}{4} \\ -\frac{3}{4} & \frac{3}{4} \end{pmatrix} + 4 \cdot \begin{pmatrix} \frac{3}{4} & \frac{1}{4} \\ \frac{3}{4} & \frac{1}{4} \end{pmatrix}.$$

This idea is extremely important in the development of symmetry analysis, as we shall see early in the next chapter.

(e) Obtaining Eigenvectors from Adjuncts To find the eigenvectors $|e_j\rangle$ or $\langle e_j|$ using the projection matrices one needed all eigenvalues

28   A REVIEW OF MATRIX ALGEBRA AND QUANTUM MECHANICS

$\varepsilon_1 \varepsilon_2 \cdots \varepsilon_n$ except ε_j. A competitive method exists which gives the vectors corresponding to a certain eigenvalue ε_j when only that ε_j is known.

We rewrite Eq. (1.2.7) in the following and set ε equal to the eigenvalue ε_j:

$$\mathscr{M}^{\mathrm{adj}}(\mathscr{H} - \varepsilon_j \mathbf{1}) = S(\varepsilon_j)\mathbf{1} = (\mathscr{H} - \varepsilon_j \mathbf{1})\mathscr{M}^{\mathrm{adj}}.$$

Then the secular equation must hold $[S(\varepsilon_j) = 0]$ and we are left with Eqs. (1.2.25):

$$\mathscr{M}^{\mathrm{adj}}(\mathscr{H} - \varepsilon_j \mathbf{1}) = \mathbf{0} = (\mathscr{H} - \varepsilon_j \mathbf{1})\mathscr{M}^{\mathrm{adj}}, \qquad (1.2.25\mathrm{a})$$

$$\mathscr{H}\mathscr{M}^{\mathrm{adj}} = \varepsilon_j \mathscr{M}^{\mathrm{adj}}, \qquad (1.2.25\mathrm{b})$$

$$\mathscr{M}^{\mathrm{adj}}\mathscr{H} = \varepsilon_j \mathscr{M}^{\mathrm{adj}}. \qquad (1.2.25\mathrm{c})$$

We see that the rows and columns of the adjunct matrix $\mathscr{M}^{\mathrm{adj}}$ derived from $(\mathscr{H} - \varepsilon_j \mathbf{1}) = \mathscr{M}$ must satisfy the eigenvalue equations of bras and kets, respectively, according to Eqs. (1.2.25b) and (1.2.25c).

$$\begin{pmatrix} 4 & 1 \\ 3 & 2 \end{pmatrix}\begin{pmatrix} 1 & -1 \\ -3 & 3 \end{pmatrix} = 1\begin{pmatrix} 1 & -1 \\ -3 & 3 \end{pmatrix},$$

$$\begin{pmatrix} 4 & 1 \\ 3 & 2 \end{pmatrix}\begin{pmatrix} -3 & -1 \\ -3 & -1 \end{pmatrix} = 5\begin{pmatrix} -3 & -1 \\ -3 & -1 \end{pmatrix}, \qquad (1.2.25\mathrm{b})_x$$

$$\begin{pmatrix} 1 & -1 \\ -3 & 3 \end{pmatrix}\begin{pmatrix} 4 & 1 \\ 3 & 2 \end{pmatrix} = 1\begin{pmatrix} 1 & -1 \\ -3 & 3 \end{pmatrix}$$

$$\begin{pmatrix} -3 & -1 \\ -3 & -1 \end{pmatrix}\begin{pmatrix} 4 & 1 \\ 3 & 2 \end{pmatrix} = 5\begin{pmatrix} -3 & -1 \\ -3 & -1 \end{pmatrix}. \qquad (1.2.25\mathrm{c})_x$$

Comparison of the resulting vectors (derived by inspecting the examples in Eq. (1.2.15b)$_x$ and (1.2.15c)$_x$ with the previously derived ones [Eq. (1.2.13)$_x$ or Figure 1.2.1] shows that they differ only by an overall factor. (Normalization is still needed.)

1.3   SOME OTHER MATRICES USED IN QUANTUM MECHANICS

Different types of quantum mechanical problems can be made more convenient to solve by using various matrix expressions. The mathematics is the same but the physical meaning can be quite different. We shall come to deal with the following examples of matrices and operators in later chapters.

A. Scattering Matrix

Imagine that free electrons can approach or leave some central "scattering" region in which they are not entirely free, and suppose that they can use n different paths or channels to go in and out. Their wave functions $\psi_l(r)$ in each channel l where they are free can be written as a linear combination of an outgoing part and an ingoing part with coefficients A_l^o and A_l^I, respectively:

$$\Psi_l(r) = A_l^o e^{ikr} + A_l^I e^{-ikr}.$$

Now the boundary conditions associated with the junctions of each channel with the scattering region will, in general, give a set of linear relations between these coefficients. The most well-known relation involves the S-MATRIX in Eq. (1.3.1) and expresses outgoing coefficients A_l^o in terms of ingoing ones A_l^I.

$$\begin{pmatrix} A_1^o \\ A_2^o \\ \vdots \\ A_n^o \end{pmatrix} = \begin{pmatrix} \mathscr{S}_{11} & \mathscr{S}_{12} & \cdots & \mathscr{S}_{1n} \\ \mathscr{S}_{21} & \mathscr{S}_{22} & \cdots & \mathscr{S}_{2n} \\ \vdots & \vdots & & \vdots \\ \mathscr{S}_{n1} & \mathscr{S}_{n2} & & \mathscr{S}_{nn} \end{pmatrix} \begin{pmatrix} A_1^I \\ A_2^I \\ \vdots \\ A_n^I \end{pmatrix}. \quad (1.3.1)$$

Generally we require conservation of probability, Eq. (1.3.2), in such problems; that is, we make sure that no particles go in and get lost in the scattering region:

$$\sum_{l=1}^{n} A_l^{I*} A_l^I = \sum_{j=1}^{n} A_j^{o*} A_j^o. \quad (1.3.2)$$

Expanding Eq. (1.3.2) using Eq. (1.3.1) we obtain

$$\sum_{l=1}^{n} A_l^{I*} A_l^I = \sum_{j=1}^{n} \sum_{k=1}^{n} \sum_{l=1}^{n} \mathscr{S}_{jk}^* A_K^{I*} \mathscr{S}_{jl} A_l^I = \sum_{k=1}^{n} \sum_{l=1}^{n} \sum_{j=1}^{n} \mathscr{S}_{jk}^* \mathscr{S}_{jl} A_k^{I*} A_l^I. \quad (1.3.3)$$

This must be true for all choices of A_i^I. This yields Eq. (1.3.4), i.e., it shows $\mathscr{S}$ must be UNITARY:

$$\sum_{j=1}^{n} \mathscr{S}_{jk}^* \mathscr{S}_{jl} = \delta_{kl}, \quad (1.3.4a)$$

$$\sum_{j=1}^{n} \mathscr{S}_{kj}^\dagger \mathscr{S}_{jl} = \delta_{kl}, \quad (1.3.4b)$$

$$\mathscr{S}^\dagger \mathscr{S} = 1. \quad (1.3.4c)$$

30 A REVIEW OF MATRIX ALGEBRA AND QUANTUM MECHANICS

As in the case of the Hamiltonian, there is considerable advantage to using the eigenvectors of $\mathscr{S}$. This is especially true when symmetry analysis can provide them, as we will discuss in the following chapter.

B. Density Matrix

So far we have been describing states of systems by vectors $|\Psi\rangle$ and $\langle\Psi|$, or, more explicitly, by amplitude components $\langle i|\Psi\rangle = \langle\Psi|i\rangle^*$ in some basis:

$$|\Psi\rangle = \sum_i |i\rangle\langle i|\Psi\rangle, \tag{1.3.5a}$$

$$\langle\Psi| = \sum_j \langle\Psi|j\rangle\langle j|. \tag{1.3.5b}$$

However, it will sometimes be very convenient to describe the state of a system instead by an operator or explicitly by a matrix. In such a formalism, the state in Eq. (1.3.5) would be denoted by a single DENSITY OPERATOR $|\Psi\rangle\langle\Psi|$ or DENSITY MATRIX ρ, Eq. (1.3.6), that represents the density operator.

$$|\Psi\rangle\langle\Psi| \leftrightarrow \rho = \begin{pmatrix} \langle 1|\Psi\rangle\langle\Psi|1\rangle & \langle 1|\Psi\rangle\langle\Psi|2\rangle & \cdots & \langle 1|\Psi\rangle\langle\Psi|n\rangle \\ \langle 2|\Psi\rangle\langle\Psi|1\rangle & \langle 2|\Psi\rangle\langle\Psi|2\rangle & \cdots & \langle 2|\Psi\rangle\langle\Psi|n\rangle \\ \vdots & & & \\ \langle n|\Psi\rangle\langle\Psi|1\rangle & \langle n|\Psi\rangle\langle\Psi|2\rangle & \cdots & \langle n|\Psi\rangle\langle\Psi|n\rangle \end{pmatrix}. \tag{1.3.6}$$

We note immediately that ρ is HERMITIAN $\rho_{ij} = \rho_{ji}^*$.

The diagonal terms $\langle i|\Psi\rangle\langle\Psi|i\rangle$ give the quantum mechanical probability that state will be found in base state i. As shown in later chapters this formalism is very useful when one wishes to consider the probabilities associated with thermal averages along with the unavoidable probabilities of quantum mechanics.

1.4 SOME MATRICES USED IN CLASSICAL MECHANICS

The concepts associated with classical matrix applications are generally more down-to-earth than those of the preceding quantum mechanical applications. Nevertheless, Dirac's bra-ket notation is useful in practically any application of matrices. We review now some ways that matrices and Dirac notation will enter our forthcoming discussions of symmetry analysis for classical problems.

A. Force, Mass, and Acceleration Matrices

A classical system of jiggling springs and masses can be described by some number of coordinates $\chi_1, \chi_2, \cdots, \chi_n(t)$ and the same number of velocities $\dot{\chi}_1, \dot{\chi}_2, \cdots, \dot{\chi}_n$. Each mass may need one or more coordinates depending on the nature of the system and its constraints.

Furthermore, we shall suppose that for each coordinate χ_j, there is a corresponding component of applied force F_j on the mass m located at χ_j, and this F_j obeys Newton's equation

$$F_j = m\alpha_j = m\ddot{\chi}_j. \quad (1.4.1)$$

F_j is the component along χ_j of the vector sum of all force applied to m by attached springs.

For example, suppose the two-mass system in Figure 1.4.1 is described by two coordinates (χ_1, χ_2). These coordinates give the positions of mass m_1 and m_2 with respect to their equilibrium positions indicated by dotted lines. Suppose also that the tension of each spring is proportional to the difference between its actual length and its length at equilibrium. (This is Hooke's law.) The constants of proportionality k_j are written over each spring in Figure 1.4.1.

Then the forces F_1 and F_2 are given by the following matrix equation, which relates them to the coordinates χ_1 and χ_2 (note sign!):

$$\begin{pmatrix} F_1 \\ F_2 \end{pmatrix} = -\begin{pmatrix} k_1 + k_{12} & -k_{12} \\ -k_{12} & k_2 + k_{12} \end{pmatrix} \begin{pmatrix} \chi_1 \\ \chi_2 \end{pmatrix} \equiv -\begin{pmatrix} \mathscr{F}_{11} & \mathscr{F}_{12} \\ \mathscr{F}_{21} & \mathscr{F}_{22} \end{pmatrix} \begin{pmatrix} \chi_1 \\ \chi_2 \end{pmatrix}. \quad (1.4.2)$$

The matrix in such an equation is called a FORCE MATRIX. This matrix is always Hermitian for a "conservative" system, as we shall prove below.

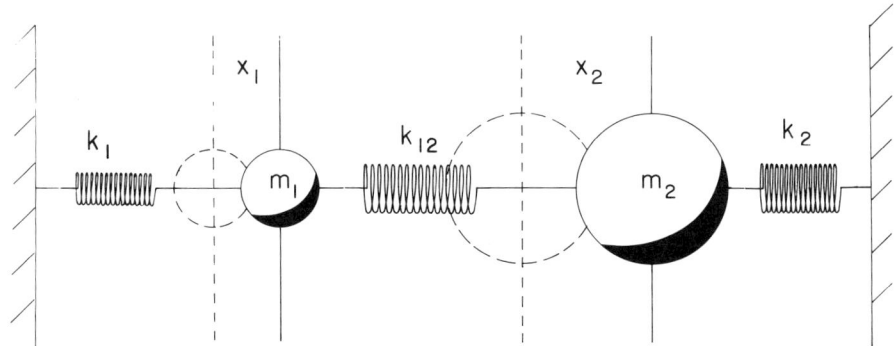

Figure 1.4.1 Example of coupled oscillator system.

32   A REVIEW OF MATRIX ALGEBRA AND QUANTUM MECHANICS

Now the forces on each mass are related to acceleration by Newton's equation, Eq. (1.4.1). This equation can be stated in matrix form also, if one defines a mass or INERTIA MATRIX m as in Eq. (1.4.3). This matrix is Hermitian, too, since it is diagonal and real:

$$\begin{pmatrix} F_1 \\ F_2 \end{pmatrix} = \begin{pmatrix} m_1 & 0 \\ 0 & m_2 \end{pmatrix} \begin{pmatrix} \ddot{x}_1 \\ \ddot{x}_2 \end{pmatrix} \equiv \begin{pmatrix} m_{11} & m_{12} \\ m_{21} & m_{22} \end{pmatrix} \begin{pmatrix} \ddot{x}_1 \\ \ddot{x}_2 \end{pmatrix}. \qquad (1.4.3)$$

Sometimes it will be convenient to define a third type of matrix relation involving an ACCELERATION MATRIX $a = -m^{-1}\mathscr{F}$ such as is written in Eq. (1.4.4):

$$\begin{pmatrix} \ddot{x}_1 \\ \ddot{x}_2 \end{pmatrix} = -\begin{pmatrix} m_{11} & m_{12} \\ m_{21} & m_{22} \end{pmatrix}^{-1} \begin{pmatrix} \mathscr{F}_{11} & \mathscr{F}_{12} \\ \mathscr{F}_{21} & \mathscr{F}_{22} \end{pmatrix} \begin{pmatrix} x_1 \\ x_2 \end{pmatrix}$$

$$= -\begin{pmatrix} \dfrac{k_1 + k_{12}}{m_1} & \dfrac{-k_{12}}{m_1} \\ \dfrac{-k_{12}}{m_2} & \dfrac{k_2 + k_{12}}{m_2} \end{pmatrix} \begin{pmatrix} x_1 \\ x_2 \end{pmatrix}$$

$$\equiv -\begin{pmatrix} a_{11} & a_{12} \\ a_{21} & a_{22} \end{pmatrix} \begin{pmatrix} x_1 \\ x_2 \end{pmatrix}. \qquad (1.4.4)$$

It is the eigenvalues and eigenvectors of this matrix that give the ELEMENTARY RESONANT FREQUENCIES and RESONANT MODES of a vibrating system. We can see this very clearly if we reformulate the problem in Dirac notation.

Let us denote the coordinates by $x_1 \equiv \langle 1|x\rangle$ and $x_2 \equiv \langle 2|x\rangle$ where the "state" $|i\rangle$ stands for that state in which mass i is displaced to $x_1 = 1$ unit and the others are at equilibrium $x_j = 0$ ($j \neq i$). State $|x\rangle$ stands for that state in which both masses are off equilibrium by $\langle 1|x\rangle = x_1$ and $\langle 2|x\rangle = x_2$, respectively. Normalization ($\langle i|i\rangle = 1$) establishes the unit distance.

The mathematical procedures are practically the same as they are for the quantum mechanical Schrodinger equation. [Recall Eqs. (1.1.17)–(1.2.3).] The matrices $\mathscr{F}$, m, and a can all be thought of as representations of the abstract operators **F**, **m**, and **a** in the basis $|1\rangle, |2\rangle$. For example, Eq. (1.4.4)

SOME MATRICES USED IN CLASSICAL MECHANICS

is rewritten as follows:

$$\begin{pmatrix} \langle 1|\ddot{x}\rangle \\ \langle 2|\ddot{x}\rangle \end{pmatrix} = -\begin{pmatrix} \langle 1|\mathbf{a}|1\rangle & \langle 1|\mathbf{a}|2\rangle \\ \langle 2|\mathbf{a}|2\rangle & \langle 2|\mathbf{a}|2\rangle \end{pmatrix}\begin{pmatrix} \langle 1|x\rangle \\ \langle 2|x\rangle \end{pmatrix}. \qquad (1.4.5)$$

Then an eigenbasis $|a_1\rangle, |a_2\rangle$ is found such that the matrix a in this new system is diagonal, with eigenvalues a_1 and a_2:

$$\langle a_i|\mathbf{a}|a_j\rangle = a_i\delta_{ij}, \qquad (1.4.6)$$

$$\langle a_i|a_j\rangle = \delta_{ij}, \qquad (1.4.7a)$$

$$\sum_i |a_i\rangle\langle a_i| = 1. \qquad (1.4.7b)$$

[These vectors can be made according to the techniques of Section 1.2 so they satisfy, along with their left-handed companions $\langle a_1|, \langle a_2|$, Eqs. (1.4.6) and (1.4.7).]

In this basis Eq. (1.4.5) will be very easy to solve since it is uncoupled now:

$$\langle a_j|\ddot{x}\rangle = -a_j\langle a_j|x\rangle.$$

The new coordinates $\langle a_j|x\rangle$, which are often called NORMAL COORDINATES, each oscillate with their particular elementary resonance frequency $\omega_j = \sqrt{a_j}$. There are quite a number of examples of these coordinates in the next few chapters.

For more complicated problems it is sometimes convenient to avoid dealing with the a matrix until the very end of a calculation. One of the problems with a is that, unlike $\mathcal{F}$ and m, it can be non-Hermitian. [See Eq. (1.4.4).] When $a \neq a^\dagger$ then left ($|\ \rangle$) and right ($\langle\ |$) eigenvectors are not related by ($\dagger$) conjugation (i.e., $\langle a_j| \neq |a_j\rangle^\dagger$) as in Figure 1.2.1.

However, there are other reasons why dealing with an equation like Eq. (1.4.8b) can be easier than working with $\mathbf{a}$ in Eq. (1.4.8a):

$$\mathbf{a}|e_j\rangle = \mathbf{m}^{-1}\mathbf{F}|e_j\rangle = a_j|e_j\rangle, \qquad (1.4.8a)$$

$$\mathbf{F}|e_j\rangle = a_j\mathbf{m}|e_j\rangle, \qquad (1.4.8b)$$

$$(\mathbf{F} - a_j\mathbf{m})|e_j\rangle = 0. \qquad (1.4.8c)$$

This happens when, for various reasons, we want to use a nonorthogonal basis $|A_1\rangle, |A_2\rangle, \ldots$ in which $\langle A_i|A_j\rangle \neq \delta_{ij}$, but $\langle A_i| = |A_i\rangle^\dagger$. In this case we would be seeking an eigenvector $|e_j\rangle$ of the form

$$|e_j\rangle = \varepsilon_1|A_1\rangle + \varepsilon_2|A_2\rangle + \cdots, \qquad (1.4.9a)$$

which satisfies a GENERALIZED EIGENVALUE equation

$$\mathbf{F}(\varepsilon_1|A_i\rangle + \varepsilon_2|A_2\rangle + \cdots) = a_j\mathbf{m}(\varepsilon_1|A_1\rangle + \varepsilon_2|A_2\rangle + \cdots). \quad (1.4.9b)$$

Forming the scalar product of this in turn with $\langle A_1| = |A_1\rangle^\dagger$, $\langle A_2| = |A_2\rangle^\dagger$ $\cdots$ results in the following generalized matrix eigenvalue equation in which the force and mass matrices are Hermitian:

$$\begin{pmatrix} \langle A_1|\mathbf{F}|A_1\rangle & \langle A_1|\mathbf{F}|A_2\rangle & \cdots \\ \langle A_2|\mathbf{F}|A_1\rangle & \langle A_2|\mathbf{F}|A_2\rangle & \cdots \end{pmatrix} \begin{pmatrix} \varepsilon_1 \\ \varepsilon_2 \\ \vdots \end{pmatrix}$$

$$= a_j \begin{pmatrix} \langle A_1|\mathbf{m}|A_1\rangle & \langle A_1|\mathbf{m}|A_2\rangle & \cdots \\ \langle A_2|\mathbf{m}|A_1\rangle & \langle A_2|\mathbf{m}|A_2\rangle & \cdots \end{pmatrix} \begin{pmatrix} \varepsilon_1 \\ \varepsilon_2 \\ \vdots \end{pmatrix}. \quad (1.4.10)$$

Now it is easy to see that the left and right eigenvectors $\langle e_j|$ and $|e_j\rangle$ satisfying Eqs. (1.4.11a) and (1.4.11b) must be related by Eq. (1.4.11c):

$$(\mathbf{F} - a_j\mathbf{m})|e_j\rangle = 0, \quad (1.4.11a)$$

$$\langle e_j|(\mathbf{F} - a_j\mathbf{m}) = 0, \quad (1.4.11b)$$

$$\langle e_j| = |e_j\rangle^\dagger. \quad (1.4.11c)$$

The components $(\varepsilon_1\varepsilon_2 \cdots)$ or $(\varepsilon_1^*\varepsilon_2^* \cdots)$ of $|e_j\rangle$ or $\langle e_j|$ are, respectively, proportional to the columns or rows of the adjunct matrix $\langle \mathbf{F} - a_m\mathbf{m}\rangle^{\mathrm{ADJ}}$. [Recall Section 1.2.B(e).]

Rewriting Eq. (1.4.11b) shows the relation between the vector $\langle e_j| = |e_j\rangle^\dagger$ and $\langle a_j|$ in Eq. (1.4.6):

$$\langle e_j|\mathbf{F} = \langle e_j|\mathbf{ma} = a_j\langle e_j|\mathbf{m}. \quad (1.4.12)$$

Equation (1.4.12) indicates that $\langle a_j|$ is proportional to $\langle e_j|\mathbf{m}$, while a similar comparison with Eq. (1.4.11a) shows that $|e_j\rangle$ is proportional to $|a_j\rangle$. We shall choose our normalization of $|e_j\rangle$ such that

$$\langle e_j|\mathbf{m}|e_k\rangle = \delta_{jk}, \quad (1.4.13a)$$

whence

$$\langle e_j|\mathbf{F}|e_k\rangle = a_j\langle e_j|\mathbf{m}|e_k\rangle = a_j\delta_{jk}. \quad (1.4.13b)$$

This **m** normalization gives in turn a peculiar form for the completeness relation:

$$1 = \sum_i |e_i\rangle\langle e_i| \mathbf{m} = \mathbf{1}^\dagger = \sum_i \mathbf{m}|e_i\rangle\langle e_i|. \quad (1.4.14)$$

B. Potential Energy and Hamiltonian Functions

Returning to Eq. (1.4.2), we can see that the work done while changing coordinate χ_1 to $\chi_1 + d\chi_1$ and $\chi_2 + d\chi_2$ is given by Eq. (1.4.15):

$$dW = \overparen{d\chi_1 \; d\chi_2}\begin{pmatrix} F_1 \\ F_2 \end{pmatrix} = \overparen{d\chi_1 \; d\chi_2}\begin{pmatrix} \mathscr{F}_{11} & \mathscr{F}_{12} \\ \mathscr{F}_{21} & \mathscr{F}_{22} \end{pmatrix}\begin{pmatrix} \chi_1 \\ \chi_2 \end{pmatrix}. \quad (1.4.15)$$

If, as we supposed, the components $\mathscr{F}_{ij}$ are constants, then we have by integration the following expression for potential energy:

$$W \equiv V(\chi_1 \chi_2) = \tfrac{1}{2}\overparen{\chi_1 \chi_2}\begin{pmatrix} \mathscr{F}_{11} & \mathscr{F}_{12} \\ \mathscr{F}_{21} & \mathscr{F}_{22} \end{pmatrix}\begin{pmatrix} \chi_1 \\ \chi_2 \end{pmatrix}$$

$$= \tfrac{1}{2}\sum_i \sum_j \langle x|i\rangle\langle i|\mathbf{F}|j\rangle\langle j|x\rangle = \tfrac{1}{2}\langle x|\mathbf{F}|x\rangle. \quad (1.4.16)$$

Note that a Taylor expansion of a general potential function $V(\chi_1 \chi_2)$ gives a much more complicated result. (We ignore the zeroth- and first-order terms which can be taken to be zero if $\chi_i = 0$ is the equilibrium point.)

$$V(\chi_1 \chi_2) = \frac{1}{2!} \sum_{ij} \chi_i \chi_j \frac{\partial^2 V}{\partial \chi_i \partial \chi_j}\bigg|_{\chi=0}$$

$$+ \frac{1}{3!} \sum_{ijk} \chi_i \chi_j \chi_k \frac{\partial^3 V}{\partial \chi_i \partial \chi_j \partial \chi_k}\bigg|_{\chi=0} + \cdots. \quad (1.4.17)$$

However, if only the quadratic terms are nonzero, then we prove that F is Hermitian for potential-driven or conservative systems:

$$\mathscr{F}_{ij} = \frac{\partial^2 V}{\partial \chi_i \partial \chi_j}\bigg|_{\chi=0} = \mathscr{F}_{ji}. \quad (1.4.18)$$

A function $H(q, p)$, called the HAMILTONIAN FUNCTION, can be constructed in terms of canonical momenta $p(j)$ and coordinates $q(j)$ as defined in the following. Equations (1.4.13) and (1.4.14) are used.

$$H(q, p) = T + V = \tfrac{1}{2}\langle \dot{x}|\mathbf{m}|\dot{x}\rangle + \tfrac{1}{2}\langle x|\mathbf{F}|x\rangle$$

$$= \tfrac{1}{2} \sum_j \big[\langle \dot{x}|\mathbf{m}|e_j\rangle\langle e_j|\mathbf{m}|\dot{x}\rangle + \langle x|\mathbf{m}|e_j\rangle\langle e_j|\mathbf{F}|x\rangle\big].$$

This may be written as

$$H(q,p) = \tfrac{1}{2}\sum_j \left[p(j)^2 + a_j|q(j)|^2\right], \quad (1.4.19a)$$

where

$$q(j) = \langle e_j|\mathbf{m}|x\rangle = \langle a_j|x\rangle, \quad (1.4.19b)$$

and

$$p(j) = \langle e_j|\mathbf{m}|\dot{x}\rangle = \dot{q}(j). \quad (1.4.19c)$$

It is easy to see that the latter obey Hamilton's equations:

$$\frac{\partial H}{\partial q(j)} = a_j q(j) = -\dot{p}(j), \quad \frac{\partial H}{\partial p(j)} = p(j) = \dot{q}(j). \quad (1.4.20)$$

This is a necessary prerequisite for any quantum theory in which each momentum $p(j)$ is replaced by an operator $(\hbar/i)\partial/\partial q(j)$ to make the Hamiltonian operator into the Schrödinger equation.

However, for classical vibration problems it is often more convenient to deal directly with Newton's equations. The F matrix is usually easy to obtain by inspection. For example, for the system in Figure 1.4.2 we would obtain the component $\langle i|\mathbf{F}|j\rangle$ by simply computing the product of the projections of each spring on coordinate axis $\chi_i = \langle i|x\rangle$ and $\chi_j = \langle j|x\rangle$ for all springs they share, and summing these products:

$$-F_1 = \langle 1|\mathbf{F}|1\rangle\langle 1|x\rangle + \langle 1|\mathbf{F}|2\rangle\langle 2|x\rangle$$
$$+ \langle 1|\mathbf{F}|3\rangle\langle 3|x\rangle + \langle 1|\mathbf{F}|4\rangle\langle 4|x\rangle + \cdots,$$
$$F_1 = (k_1 \cos^2\phi_1 + k_1' \cos^2\phi_1' + k_{12}\cos^2\theta_1)\chi_1$$
$$+ (k_1 \cos\phi_2 \cos\phi_1 + k_2' \cos\phi_1 + k_{12}\cos\theta_2\cos\theta_1)\chi_2$$
$$+ (k_{12}\cos\theta_3\cos\theta_1)\chi_3 + (k_{12}\cos\theta_4\cos\theta_1)\chi_4 + \cdots.$$

One should always keep in mind that classical equations such as Eqs. (1.4.2) and (1.4.4) are approximate descriptions which assume small χ_j. For large χ_j one will need to include anharmonic $\chi_j^3, \chi_j\chi_k^2, \chi_j^4, \ldots$, etc., to model the behavior of a highly stretched spring and changing geometry. This is especially true for more complicated problems in two or three dimensions, as is represented by Figure 1.4.2, or for models of molecular systems.

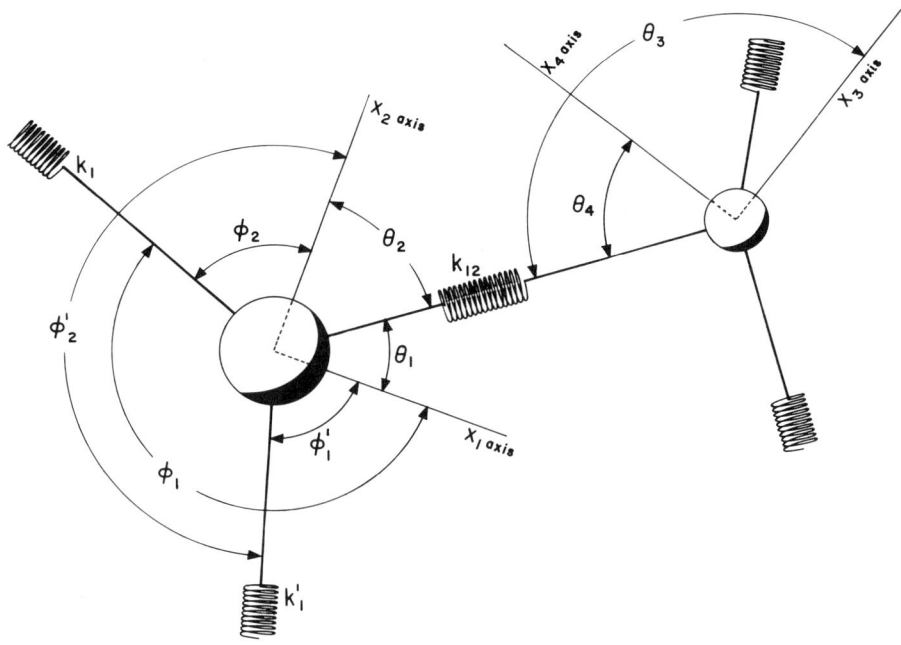

Figure 1.4.2 Example of more complicated spring-mass system.

1.5 WAVE FUNCTIONAL ANALYSIS AND CONTINUOUS VECTOR SPACES

Dirac notation was designed to be used in vector spaces that had a continuous infinity of dimensions. The symbol $\langle x|\Psi\rangle$ is Dirac's notation for a complex wavefunction $\Psi(x)$ of a continuous coordinate x. The complex conjugate $\Psi^*(x)$ is the reversed bra-ket $\langle \Psi|x\rangle$:

$$\langle x|\Psi\rangle = \Psi(x) \qquad \langle \Psi|x\rangle = \Psi^*(x). \tag{1.5.1}$$

Several modifications of the formalism are needed to accommodate this continuously infinite ket vector space $\{\cdots |x\rangle \cdots\}$ or its conjugate bra vector space $\{\cdots \langle x| \cdots\}$. These are known in mathematics as *Banach spaces* and the description of wavefunctions by these vectors falls under the mathematical heading of *functional analysis*. This is a large and complicated subject. Nevertheless, the basic ideas of completeness and orthogonality remain intact. A comparison between discrete and continuous vector spaces is given now.

A. Functional Scalar Products

Scalar products between continuously infinite bra and ket vectors involve what is known as Dirac's delta function $\delta(x, y)$. In place of the usual Kronecker delta δ_{ab} relation,

$$\langle a|b\rangle = \delta_{ab} = \begin{cases} 1 & \text{if } a = b \\ 0 & \text{if } a \neq b, \end{cases} \quad (1.5.2a)$$

one uses the Dirac delta relation,

$$\langle x|y\rangle = \delta(x, y) = \begin{cases} \infty & \text{if } x = y \\ 0 & \text{if } x \neq y. \end{cases} \quad (1.5.2b)$$

Replacing unity (1) by infinity (∞) in the scalar product is necessary because the sum over discrete dimensions is replaced by an integral. It is the sum or integral that needs to equal unity. It is easy to see that

$$\sum_b \langle a|b\rangle = \sum_b \delta_{ab} = 1,$$

but the corresponding Dirac sum is not so obvious:

$$\int dy \langle x|y\rangle = \int dy\, \delta(x, y) = 1.$$

The delta function is zero everywhere except at $x = y$ where it goes to infinity in such a way as to have unit area. Both kinds of delta functions are designed to extract a particular component from a sum. The following discrete sum

$$\langle a|\Psi\rangle = \sum_b \langle a|b\rangle\langle b|\Psi\rangle = \sum_b \delta_{ab}\Psi_b = \Psi_a \quad (1.5.3a)$$

is analogous to the following integral:

$$\langle x|\Psi\rangle = \int dy \langle x|y\rangle\langle y|\Psi\rangle = \int dy\, \delta(x, y)\Psi(y) = \Psi(x). \quad (1.5.3b)$$

B. Orthonormality and Completeness

The completeness relation (1.2.19a) for a discrete space $\{|1\rangle, |2\rangle \cdots\}$ has the form

$$1 = \sum_b |b\rangle\langle b|. \quad (1.5.4a)$$

It is replaced by the following integral for a continuous space $\{ \cdots |y\rangle \cdots \}$:

$$\mathbf{1} = \int dy |y\rangle\langle y|. \qquad (1.5.4b)$$

For many applications to quantum mechanics there is a discrete set of eigenstates $\{|\Psi_1\rangle, |\Psi_2\rangle \cdots \}$ but their wavefunctions range over a continuous set of position states $\{ \cdots |x\rangle \cdots \}$. The wavefunctions $\Psi_a(x)$ can be thought of as components of a transformation matrix $\langle x|\Psi_a\rangle$ between these two bases. Orthogonality and completeness of the transformation is analogous to the discrete space relations (1.2.19b) and (1.2.20), respectively,

$$\langle x|y\rangle = \delta(x, y) = \sum_a \langle x|\Psi_a\rangle\langle\Psi_a|y\rangle,$$

$$\langle\Psi_a|\Psi_b\rangle = \delta_{ab} = \int dz \langle\Psi_a|z\rangle\langle z|\Psi_b\rangle. \qquad (1.5.5a)$$

Here is the same equation in wavefunction notation:

$$\langle x|y\rangle = \delta(x, y) = \sum_a \Psi_a^*(x)\Psi_a(y),$$

$$\langle\Psi_a|\Psi_b\rangle = \delta_{ab} = \int dz\, \Psi_a^*(z)\Psi_b(z). \qquad (1.5.5b)$$

Other applications to quantum mechanics involve a continuous set of eigenstates $\{ \cdots |\Psi_k\rangle \cdots \}$ defined by wavefunctions over a continuous set of position states $\{ \cdots |x\rangle \cdots \}$. For example, plane wave functions $\Psi_k(x) = e^{ikx}/\sqrt{2\pi}$ can be thought of as components of a Fourier transformation matrix $\langle x|\Psi_k\rangle$ between position and momentum space. The orthogonality and completeness of this transformation are similar relations:

$$\langle x|y\rangle = \delta(x, y) = \int dk \langle x|\Psi_k\rangle\langle\Psi_k|y\rangle,$$

$$\langle\Psi_k|\Psi_{k'}\rangle = \delta(k, k') = \int dz \langle\Psi_k|z\rangle\langle z|\Psi_{k'}\rangle. \qquad (1.5.6a)$$

Here is the same equation for plane waves:

$$\langle x|y\rangle = \delta(x, y) = \frac{1}{2\pi}\int dk\, e^{-ikx}e^{iky},$$

$$\langle\Psi_k|\Psi_{k'}\rangle = \delta(k, k') = \frac{1}{2\pi}\int dz\, e^{-ikz}e^{ik'z}. \qquad (1.5.6b)$$

Orthonormality for one basis is the same type of equation as completeness for the other.

C. Differential Operators and Conjugates

Analogies between discrete and continuous vector spaces can be extended to include linear differential operators. For example, the following differential operator M acting on a function $f(x)$,

$$Mf(x) = \mu(x)\frac{d^2f}{dx^2} + \nu(x)\frac{df}{dx} + \lambda(x)f \tag{1.5.7}$$

can be expressed as a Dirac matrix element

$$\langle x|\mathbf{M}|f\rangle = \int dy \langle x|\mathbf{M}|y\rangle\langle y|f\rangle, \tag{1.5.8}$$

where $\langle x|\mathbf{M}|y\rangle$ is expressed in terms of Dirac delta function derivatives.

$$\langle x|\mathbf{M}|y\rangle = \mu(x)\frac{d^2\delta(x,y)}{dy^2} + \nu(x)\frac{d\delta(x,y)}{dy} + \lambda(x)\delta(x,y). \tag{1.5.9}$$

The Dirac delta derivative extracts the value of a derivative of a function, that is, the value of the function minus its value at an infinitesimally nearby point. Integration by parts gives

$$\int_{-\infty}^{\infty} dx\, f(x)\frac{d\delta(x,y)}{dx} = f(x)\delta(x,y)\Big|_{-\infty}^{\infty} - \int_{-\infty}^{\infty} dx\, \frac{df(x)}{dx}\delta(x,y) = -f'(y),$$

$$\int_{-\infty}^{\infty} dy\, f(y)\frac{d\delta(x,y)}{dy} = f'(x). \tag{1.5.10}$$

The transpose conjugate operator $\mathbf{M}^\dagger$ has matrix element $\langle x|\mathbf{M}^\dagger|y\rangle = \langle y|\mathbf{M}|x\rangle^*$:

$$\langle x|\mathbf{M}^\dagger|y\rangle = \langle y|\mathbf{M}|x\rangle^*$$
$$= \mu^*(y)\frac{d^2\delta(y,x)}{dy^2} + \nu^*(y)\frac{d\delta(y,x)}{dy} + \lambda^*(y)\delta(y,x). \tag{1.5.11}$$

The result of applying the operator $\mathbf{M}^\dagger$ using (1.5.8) and (1.5.11) is the following,

$$\mathbf{M}^\dagger f(x) = \frac{d^2}{dx^2}(\mu^*(x)f(x)) - \frac{d}{dx}(\nu^*(x)f(x)) + \lambda^*(x)f(x). \tag{1.5.12}$$

This is called the *adjoint* differential operator. If the operator is self-adjoint ($\mathbf{M} = \mathbf{M}^\dagger$) as are all Hamiltonian operators then the operation above is equal to the original one. This places restrictions on the form of the functions $\mu(x)$, $\nu(x)$, and $\lambda(x)$.

D. Resolvants

Given an operator $\mathbf{M}$ there is a useful function of that operator which is called the *resolvant* $R_\lambda(\mathbf{M}) = -Q_\lambda(\mathbf{M})$. It uses a parameter λ,

$$Q_\lambda(\mathbf{M}) = \frac{1}{\lambda \mathbf{1} - \mathbf{M}} = -R_\lambda(\mathbf{M}). \tag{1.5.13}$$

If $\lambda = 0$, the first resolvant R_0 is the inverse $\mathbf{M}^{-1}$ of $\mathbf{M}$ and Q_0 is the negative inverse $-\mathbf{M}^{-1}$,

$$R_0(\mathbf{M}) = \mathbf{M}^{-1}.$$

The resolvant $Q_\lambda(\mathbf{M})$ expands into a series of $\mathbf{M}/\lambda$ powers,

$$Q_\lambda(\mathbf{M}) = \frac{1}{\lambda(1 - \mathbf{M}/\lambda)} = \frac{1}{\lambda}\left(1 + \frac{\mathbf{M}}{\lambda} + \frac{\mathbf{M}^2}{\lambda^2} + \cdots\right).$$

Applying Cauchy's integral theorem to the complex variable λ yields contour integral expressions for each term,

$$\frac{1}{2\pi i}\oint_L d\lambda \frac{1}{\lambda \mathbf{1} - \mathbf{M}} = \mathbf{1}, \quad \frac{1}{2\pi i}\oint_L d\lambda \frac{\lambda}{\lambda \mathbf{1} - \mathbf{M}} = \mathbf{M},$$

$$\frac{1}{2\pi i}\oint_L d\lambda \frac{\lambda^2}{\lambda \mathbf{1} - \mathbf{M}} = \mathbf{M}^2, \ldots. \tag{1.5.14}$$

This gives a symbolic expression for any function $f(\mathbf{M})$ of the operator $\mathbf{M}$ in terms of an integral of the function $f(\lambda)$ times the resolvant:

$$f(\mathbf{M}) = \frac{1}{2\pi i}\oint_L d\lambda \frac{f(\lambda)}{\lambda \mathbf{1} - \mathbf{M}} = \frac{1}{2\pi i}\oint_L d\lambda\, f(\lambda) Q_\lambda(\mathbf{M}). \tag{1.5.15}$$

To use this we need to know where the poles of the resolvant $Q_\lambda(\mathbf{M})$ are. It is a rational function of $\lambda\mathbf{1} - \mathbf{M}$ divided by the secular polynomial $S(\lambda) = \det|\lambda\mathbf{1} - \mathbf{M}|$,

$$Q_\lambda(\mathbf{M}) = (\lambda\mathbf{1} - \mathbf{M})^{-1} = \frac{(\lambda\mathbf{1} - \mathbf{M})^{\text{ADJ}}}{\det|\lambda\mathbf{1} - \mathbf{M}|} = \frac{a(\lambda, \mathbf{M})}{S(\lambda)}. \tag{1.5.16}$$

This implies that the resolvant's poles are located at its eigenvalues

$\{\lambda_1, \lambda_2, \lambda_3, \ldots\}$, that is, the roots of $S(\lambda) = 0$. The adjunct function can have no singularities of its own. This means the contour loop L in the contour integral equations above can be deformed into a set of small circles $\{l_1, l_2, \ldots\}$ surrounding each distinct eigenvalue in the complex plane. Each integral then breaks into a sum over distinct eigenvalues.

The first integral expression for the unit operator becomes an operator completeness relation like (1.2.18) which we discussed before:

$$\mathbf{1} = \frac{1}{2\pi i} \oint_L d\lambda \, \frac{1}{\lambda \mathbf{1} - \mathbf{M}} = \sum_{\lambda_a} \mathbf{P}_{\lambda_a}. \tag{1.5.17}$$

Here the projection operators have the form

$$\mathbf{P}_{\lambda_a} = \frac{1}{2\pi i} \oint_{l_a} d\lambda \, \frac{1}{\lambda \mathbf{1} - \mathbf{M}} = \frac{1}{2\pi i} \oint_{l_a} d\lambda \, Q_\lambda(\mathbf{M}). \tag{1.5.18}$$

If all the eigenvalues are distinct then each eigenvalue is a simple pole. In that case it can be shown that the **P**-operators are orthogonal idempotents as in Eq. (1.2.17). Then the **M** operator and its powers have spectral decompositions like Eq. (1.2.21):

$$\mathbf{M} = \frac{1}{2\pi i} \int_L d\lambda \, \frac{\lambda}{\lambda \mathbf{1} - \mathbf{M}} = \sum_{\lambda_a} \lambda_a \mathbf{P}_{\lambda_a}. \tag{1.5.19}$$

However, for degenerate eigenvalue ($\lambda_a = \lambda_b$) it may be necessary to include extra terms in a spectral decomposition such as in the following:

$$\mathbf{M} = \sum_{\lambda_a} \lambda_a \frac{1}{2\pi i} \oint_{l_a} d\lambda \, \frac{\lambda}{\lambda \mathbf{1} - \mathbf{M}}$$

$$= \sum_{\lambda_a} \left[\lambda_a \frac{1}{2\pi i} \oint_{l_a} d\lambda \, \frac{1}{\lambda \mathbf{1} - \mathbf{M}} + \frac{1}{2\pi i} \oint_{l_a} d\lambda \, \frac{\lambda - \lambda_a}{\lambda \mathbf{1} - \mathbf{M}} \right]$$

$$= \sum_{\lambda_a} (\mathbf{P}_{\lambda_a} + \mathbf{N}_{\lambda_a}), \quad \text{where: } \mathbf{N}_{\lambda_a} = \frac{1}{2\pi i} \oint_{l_a} d\lambda \, \frac{\lambda - \lambda_a}{\lambda \mathbf{1} - \mathbf{M}}. \tag{1.5.20}$$

The extra term $\mathbf{N}_\lambda$ is zero if the resolvant still has a simple pole at that eigenvalue. Otherwise it is a nilpotent operator, that is, some power of it is zero. If the eigenvalue is a second order pole of the resolvant, then $(\mathbf{N}_\lambda)^2$ is zero:

$$(\mathbf{N}_{\lambda_a})^2 = \frac{1}{2\pi i} \oint_{l_a} d\lambda \, \frac{(\lambda - \lambda_a)^2}{\lambda \mathbf{1} - \mathbf{M}} = 0. \tag{1.5.21}$$

Resolvants provide a powerful tool for spectral analysis of operators.

APPENDIX A. ELEMENTARY VECTOR NOTATIONS AND THEORY

In Chapter 1 we introduced Dirac's vector notation $|\ \rangle$ and $\langle\ |$. Here we relate this notation to the older notation of vector analysis.

The dirac vector in Eq. (A.1a) can be written in the older notation as Eq. (A.1b):

$$|v\rangle = |x\rangle\langle x|v\rangle + |y\rangle\langle y|v\rangle, \quad (A.1a)$$

$$\mathbf{v} = \hat{x}v_x + \hat{y}v_y, \quad (A.1b)$$

$$\begin{pmatrix} v_x \\ v_y \end{pmatrix} = \begin{pmatrix} \langle x|v\rangle \\ \langle y|v\rangle \end{pmatrix}. \quad (A.1c)$$

It can be represented by a column vector in Eq. (A.1c) or as is shown also in Eq. (1.1.11). Figure A.1 gives a geometrical representation of the vectors **v** and **w** in a two-dimensional space where their properties can be seen clearly.

In Figure A.2 we show the meaning of multiplying a vector by a scalar by comparing the vectors **v**, 2**v**, and $-\mathbf{v}$, where $\mathbf{v} = -3\hat{x} + 4\hat{y}$.

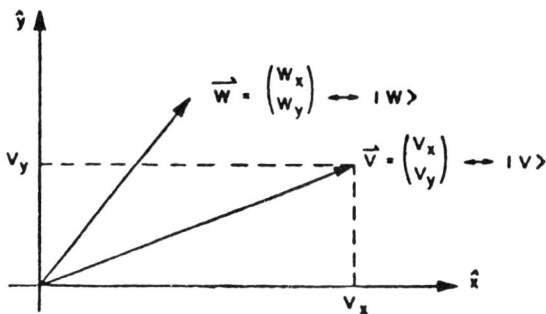

Figure A.1

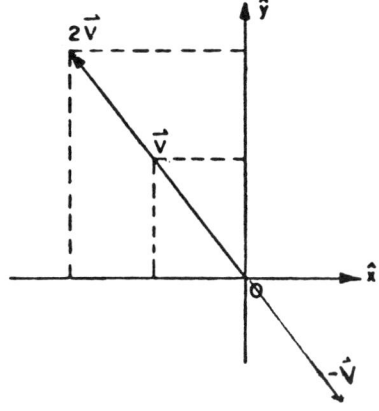

Figure A.2

44   A REVIEW OF MATRIX ALGEBRA AND QUANTUM MECHANICS

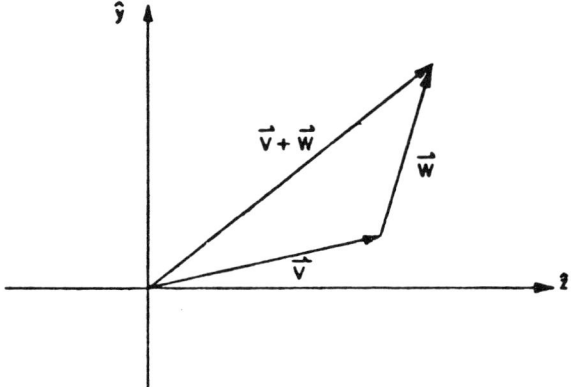

Figure A.3

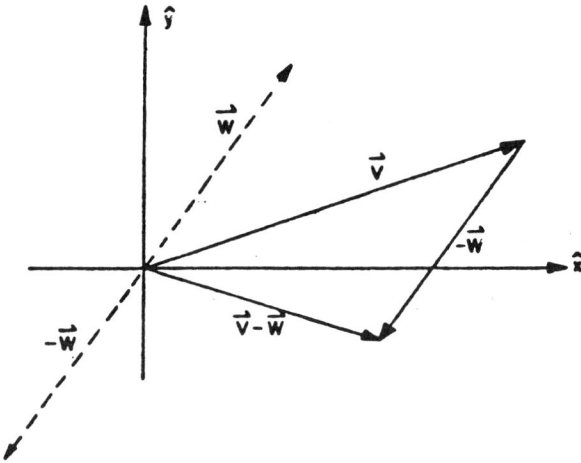

Figure A.4

The geometrical interpretation of the addition of two vectors, Eq. (A.2), is shown in Figure A.3, and Figure A.4 shows the formation of $\mathbf{v} - \mathbf{w}$:

$$\mathbf{v} + \mathbf{w} = \begin{pmatrix} v_x \\ v_y \end{pmatrix} + \begin{pmatrix} w_x \\ w_y \end{pmatrix} = \begin{pmatrix} v_x + w_x \\ v_y + w_y \end{pmatrix}. \tag{A.2}$$

Any two vectors in our two-dimensional space that are not proportional to each other could, by linear combination, form all other vectors in the space. This is an important concept that can be generalized to n-dimensional space.

APPENDIX A 45

The length or magnitude $|\mathbf{v}|$ of a vector $\mathbf{v}$ is given by the Pythagorian theorem [Eq. (A.3)]:

$$|\mathbf{v}| = \left(v_x^2 + v_y^2\right)^{1/2}. \tag{A.3}$$

Equation (A.3) is the square root of the scalar product defined by Eq. (1.1.13) and Eq. (A.4):

$$|\mathbf{v}| = (\mathbf{v} \cdot \mathbf{v})^{1/2} = \widetilde{v_x v_y} \begin{pmatrix} v_x \\ v_y \end{pmatrix} = \left(v_x^2 + v_y^2\right)^{1/2}. \tag{A.4}$$

The geometrical significance of the scalar product of two arbitrary vectors $\mathbf{v}$ and $\mathbf{w}$ in Figure A.5 is given now.

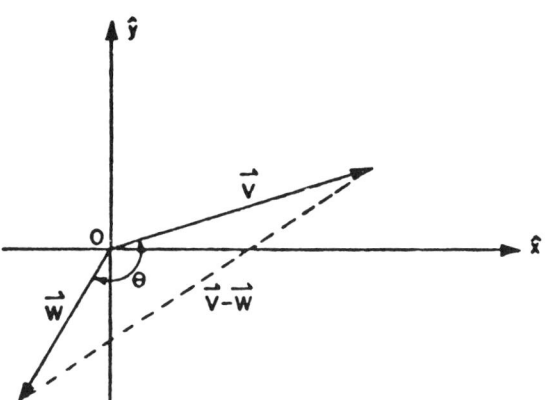

Figure A.5

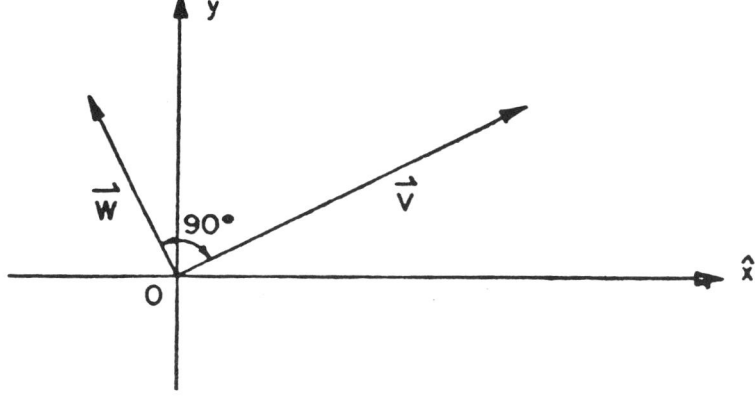

Figure A.6

Applying the law of cosines to the triangle described by the two vectors we arrive at Eq. (A.5a):

$$|\mathbf{v} - \mathbf{w}|^2 = |\mathbf{v}|^2 + |\mathbf{w}|^2 - 2|\mathbf{v}||\mathbf{w}|\cos\theta,$$
$$(\mathbf{v} - \mathbf{w}) \cdot (\mathbf{v} - \mathbf{w}) = \mathbf{v} \cdot \mathbf{v} + \mathbf{w} \cdot \mathbf{w} - 2|\mathbf{v}||\mathbf{w}|\cos\theta. \quad (A.5a)$$

Expanding the left-hand side of Eq. (A.5a), we have

$$(\mathbf{v} - \mathbf{w}) \cdot (\mathbf{v} - \mathbf{w}) = \mathbf{v} \cdot \mathbf{v} + \mathbf{w} \cdot \mathbf{w} - 2\mathbf{v} \cdot \mathbf{w}, \quad (A.5b)$$

and comparing Eq. (A.5a) with Eq. (A.5b), we derive the desired geometrical interpretation of $\mathbf{v} \cdot \mathbf{w}$, which is given in Eq. (A.6):

$$\mathbf{v} \cdot \mathbf{w} = |\mathbf{v}||\mathbf{w}|\cos\theta. \quad (A.6)$$

To require that the vectors v and w be normalized simply means that they must be unit vectors as in Eq. (A.7):

$$\mathbf{w} \cdot \mathbf{w} = 1 = \mathbf{v} \cdot \mathbf{v}. \quad (A.7)$$

The vectors $\mathbf{x}_1 \mathbf{x}_2 \cdots \mathbf{x}_n$ are said to by LINEARLY INDEPENDENT if the relation

$$\sum_{i=1}^{} a_i \mathbf{x}_i = 0 \quad (A.8)$$

necessarily implies that all $a_i = 0$.

If in Eq. (A.8) there exist at least two nonvanishing a_i, then it is said that the vectors $\mathbf{x}_1 \mathbf{x}_2 \cdots \mathbf{x}_n$ are linearly dependent.

The maximum number of linearly independent vectors in a space is called the dimension of the space.

For completeness, we present the mathematical definitions of a linear vector space.

Given that $|i\rangle$, $|j\rangle$, and $|k\rangle$ are vectors in the vector space and α and β are complex numbers, then the following is true:

(1) Given any two vectors in the space then their sum is in the space. This is called closure.
(2) The commutative law of addition holds; i.e.,

$$|i\rangle + |j\rangle = |j\rangle + |i\rangle. \quad (A.9a)$$

(3) The associative law of addition holds; i.e.,

$$(|i\rangle + |j\rangle) + |k\rangle = |i\rangle + (|j\rangle + |k\rangle). \quad (A.9b)$$

(4) There exists a null element 0 such that

$$0 + |i\rangle = |i\rangle + 0 = |i\rangle. \tag{A.9c}$$

(5) There exists an element $|i'\rangle$ for every $|i\rangle$ such that

$$|i\rangle + |i'\rangle = 0. \tag{A.9d}$$

(6) The associative law of multiplication holds; i.e.,

$$\alpha(\beta|i\rangle) = \alpha\beta|i\rangle. \tag{A.9e}$$

(7) The distributive law with respect to the addition of complex number holds; i.e.,

$$(\alpha + \beta)|i\rangle = \alpha|i\rangle + \beta|i\rangle. \tag{A.9f}$$

(8) The distributive law with respect to the additions of vectors holds; i.e.,

$$\alpha(|i\rangle + |j\rangle) = \alpha|i\rangle + \alpha|j\rangle. \tag{A.9g}$$

The mathematical definitions of the inner or scalar product follow:

(1) $\quad\quad\quad\quad \langle i|(|j\rangle + |k\rangle) = \langle i|j\rangle + \langle i|k\rangle.$ (A.10a)

(2) $\quad\quad\quad\quad \langle i|(\alpha|j\rangle = \alpha\langle i|j\rangle.$ (A.10b)

(3) $\quad\quad\quad\quad \langle i|j\rangle = \langle j|i\rangle^{*}.$ (A.10c)

(4) $\quad\quad\quad\quad \langle i|i\rangle \geq 0.$ (A.10d)

APPENDIX B. LINEAR EQUATIONS, MATRICES, DETERMINANTS, AND INVERSES

The simplest nontrivial example of a matrix equation is the linear equation, Eq. (B.1), involving two unknowns x_1 and x_2 (we assume the quantities $\mathscr{M}_{ij}$ and y_n are known constants). This is written in three ways to display the matrix notation:

$$\begin{aligned} \mathscr{M}_{11} x_1 + \mathscr{M}_{12} x_2 &= y_1, \\ \mathscr{M}_{21} x_1 + \mathscr{M}_{22} x_2 &= y_2, \end{aligned} \tag{B.1a}$$

$$\begin{pmatrix} \mathscr{M}_{11} & \mathscr{M}_{12} \\ \mathscr{M}_{21} & \mathscr{M}_{22} \end{pmatrix} \begin{pmatrix} x_1 \\ x_2 \end{pmatrix} = \begin{pmatrix} y_1 \\ y_2 \end{pmatrix}, \tag{B.1b}$$

$$\mathscr{M} x = y. \tag{B.1c}$$

48   A REVIEW OF MATRIX ALGEBRA AND QUANTUM MECHANICS

This equation is solved if and when the INVERSE $\mathcal{M}^{-1}$ of matrix $\mathcal{M}$ is found such that Eq. (B.2) holds, where 1 is the IDENTITY matrix:

$$\mathcal{M}^{-1}\mathcal{M} = \begin{pmatrix} \mathcal{M}_{11} & \mathcal{M}_{12} \\ \mathcal{M}_{21} & \mathcal{M}_{22} \end{pmatrix}^{-1} \begin{pmatrix} \mathcal{M}_{11} & \mathcal{M}_{12} \\ \mathcal{M}_{21} & \mathcal{M}_{22} \end{pmatrix} = \begin{pmatrix} 1 & 0 \\ 0 & 1 \end{pmatrix} \equiv 1. \quad (B.2)$$

This will give a solution because the product $(1 \cdot x)$ of the identity matrix with any vector x is equal to x, and so Eqs. (B.1) and (B.2) together give Eq. (B.3):

$$\mathcal{M}^{-1}\mathcal{M}x = 1x = x = \mathcal{M}^{-1}y. \quad (B.3)$$

Simple algebra gives, for the inverse of $\mathcal{M}$,

$$\begin{pmatrix} \mathcal{M}_{11} & \mathcal{M}_{12} \\ \mathcal{M}_{21} & \mathcal{M}_{22} \end{pmatrix}^{-1} = \begin{pmatrix} \dfrac{\mathcal{M}_{22}}{\mathcal{M}_{11}\mathcal{M}_{22} - \mathcal{M}_{12}\mathcal{M}_{21}} & \dfrac{-\mathcal{M}_{12}}{\mathcal{M}_{11}\mathcal{M}_{22} - \mathcal{M}_{12}\mathcal{M}_{21}} \\ \dfrac{-\mathcal{M}_{21}}{\mathcal{M}_{11}\mathcal{M}_{22} - \mathcal{M}_{22}\mathcal{M}_{21}} & \dfrac{\mathcal{M}_{11}}{\mathcal{M}_{11}\mathcal{M}_{22} - \mathcal{M}_{12}\mathcal{M}_{21}} \end{pmatrix}.$$

We derive now a general formula for the inverse of an $n \times n$ matrix.

Such a formula follows from the properties of the DETERMINANT (det $\mathcal{M}$) of a matrix $\mathcal{M}$. The determinant is defined by Eq. (B.4), where, for convenience, we use a different notation for matrix components:

$$\begin{aligned}
\det \mathcal{M} &= \det \begin{vmatrix} a_1 & a_2 & a_3 & a_4 & \cdots \\ b_1 & b_2 & b_3 & b_4 & \cdots \\ c_1 & c_2 & c_3 & c_4 & \cdots \\ d_1 & d_2 & d_3 & d_4 & \cdots \\ \vdots & \vdots & \vdots & \vdots & \end{vmatrix} \\
&= \sum_{\substack{\text{permutations } P \\ \text{of } 1\,2\,3\,\cdots\,n}} (-1)^{\text{parity}} P(a_1 b_2 c_3 d_4 \cdots) \\
&= (a_1 b_2 c_3 d_4 \cdots) - (a_1 b_3 c_2 d_4 \cdots) + (a_1 \cdots \\
&\quad - (a_2 b_1 c_3 d_4 \cdots) + (a_2 b_3 c_1 d_4 \cdots) + (a_2 \cdots \\
&\quad + (a_3 b_1 c_2 d_4 \cdots) - (a_3 b_2 c_1 d_4 \cdots) + (a_3 \cdots \\
&\vdots
\end{aligned} \quad (B.4)$$

A permutation operation P in Eq. (B.4) is said to have parity $= -1$ if it is accomplished by an odd number of "hops" of one number over another

$\cdots ijkl \cdots \to \cdots ik\overleftarrow{}jl \cdots$ For example, in permuting $4123 \cdots$ to $1234 \cdots$ we use three $\overrightarrow{}$ hops: $4123 \to 1\overleftarrow{4}23 \to 1\overleftarrow{2}43 \to 12\overleftarrow{3}4$. On the other hand a permutation is said to have parity $\overrightarrow{} = +1$ if it is $\overrightarrow{}$ done in an even number of hops.

From Eq. (B.4) we can understand the expansion of a determinant in terms of smaller subdeterminants called MINORS, as shown in the following. We abbreviate the sum over permutation by $\Sigma(-)$.

$$\det \mathcal{M} = \det \begin{vmatrix} a_1 & a_2 & a_3 & a_4 & \cdots \\ b_1 & b_2 & b_3 & b_4 & \cdots \\ c_1 & c_2 & c_3 & c_4 & \cdots \\ d_1 & d_2 & d_3 & d_4 & \cdots \\ \vdots & \vdots & \vdots & \vdots & \end{vmatrix}$$

$$= a_1 \sum_{2,3,4,\ldots,n}(-)(b_2 c_3 d_4 \cdots) - a_2 \sum_{1,3,4,\ldots,n}(-)(b_1 c_3 d_4 \cdots)$$

$$+ a_3 \sum_{1,2,4,\ldots,n}(-)(b_1 c_2 d_4 \cdots) \cdots$$

$$= a_1 \det \begin{vmatrix} b_2 & b_3 & b_4 & \cdots \\ c_2 & c_3 & c_4 & \cdots \\ d_2 & d_3 & d_4 & \cdots \end{vmatrix} - a_2 \det \begin{vmatrix} b_1 & b_3 & b_4 & \cdots \\ c_1 & c_3 & c_4 & \cdots \\ d_1 & d_3 & d_4 & \cdots \end{vmatrix}$$

$$+ a_3 \det \begin{vmatrix} b_1 & b_2 & b_4 & \cdots \\ c_1 & c_2 & c_4 & \cdots \\ d_1 & d_2 & d_4 & \cdots \end{vmatrix}$$

$$= a_1 \mu_{11} - a_2 \mu_{12} + a_3 \mu_{13} - \cdots . \tag{B.5}$$

A minor μ_{ij} of a matrix $\mathcal{M}$ is the determinant obtained from $\mathcal{M}$ after erasing its ith row and its jth column. In Eq. (B.5) we used the minors of the first row of $\mathcal{M}$. However, any row or column of $\mathcal{M}$ could be similarly used to give the general equation (B.6):

$$\det \mathcal{M} = \sum_i (-1)^{i+j} \mathcal{M}_{ij} \mu_{ij}, \quad j = 1, 2, \ldots, \text{ or } n, \tag{B.6a}$$

$$\det \mathcal{M} = \sum_j (-1)^{i+j} \mathcal{M}_{ij} \mu_{ij}, \quad i = 1, 2, \ldots, \text{ or } n. \tag{B.6b}$$

Now suppose we define a matrix called the ADJUNCT matrix $\mathcal{M}^{\text{adj}}$ of $\mathcal{M}$ by Eq. (B.7) (note switching of i and j in μ_{ji}!):

$$\mathcal{M}_{ij}^{\text{adj}} = (-1)^{i+j} \mu_{ji}. \tag{B.7}$$

$\mathscr{M}^{\text{adj}}$ together with the original matrix $\mathscr{M}$ obey Eq. (B.8), as can be seen by studying an explicit representation of it in Eq. (B.8c).

$$\sum_j \mathscr{M}_{ij} \mathscr{M}^{\text{adj}}_{jk} = (\det \mathscr{M}) \delta_{ik}, \qquad (B.8a)$$

$$\mathscr{M} \mathscr{M}^{\text{adj}} = (\det \mathscr{M}) 1, \qquad (B.8b)$$

$$\begin{pmatrix} a_1 & a_2 & a_3 & a_4 & \cdots \\ b_1 & b_2 & b_3 & b_4 & \cdots \\ c_1 & c_2 & c_3 & c_4 & \cdots \\ d_1 & d_2 & d_3 & d_4 & \cdots \end{pmatrix}$$

$$\times \begin{pmatrix} \Sigma(-)b_2 c_3 d_4 \cdots & -\Sigma(-)a_2 c_3 d_4 \cdots & \Sigma(-)a_2 b_3 d_4 \cdots \\ -\Sigma(-)b_1 c_3 d_4 \cdots & \Sigma(-)a_1 c_3 d_4 \cdots & -\Sigma(-)a_1 b_3 d_4 \cdots \\ \Sigma(-)b_1 c_2 d_4 \cdots & -\Sigma(-)a_1 c_2 d_4 \cdots & \Sigma(-)a_1 b_2 d_4 \cdots \\ -\Sigma(-)b_1 c_2 d_3 \cdots & \Sigma(-)a_1 c_2 d_3 \cdots & -\Sigma(-)a_1 b_2 d_3 \cdots \end{pmatrix}$$

$$= \begin{pmatrix} \det\begin{vmatrix} a_1 & a_2 & a_3 & a_4 & \cdots \\ b_1 & b_2 & b_3 & b_4 & \cdots \\ c_1 & c_2 & c_3 & c_4 & \cdots \\ d_1 & d_2 & d_3 & d_4 & \cdots \\ \vdots & \vdots & \vdots & \vdots \end{vmatrix} & \det\begin{vmatrix} a_1 & a_2 & a_3 & a_4 & \cdots \\ a_1 & a_2 & a_3 & a_4 & \cdots \\ c_1 & c_2 & c_3 & c_4 & \cdots \\ d_1 & d_2 & d_3 & d_4 & \cdots \\ \vdots & \vdots & \vdots & \vdots \end{vmatrix} & \det\begin{vmatrix} a_1 & a_2 & a_3 & a_4 \\ b_1 & b_2 & b_3 & b_4 \\ a_1 & a_2 & a_3 & a_4 \\ d_1 & d_2 & d_3 & d_4 \\ \vdots & \vdots & \vdots & \vdots \end{vmatrix} & \cdots \\[2em] \det\begin{vmatrix} b_1 & b_2 & b_3 & b_4 & \cdots \\ c_1 & c_2 & c_3 & c_4 & \cdots \\ c_1 & c_2 & c_3 & c_4 & \cdots \\ d_1 & d_2 & d_3 & d_4 & \cdots \\ \vdots & \vdots & \vdots & \vdots \end{vmatrix} & \det\begin{vmatrix} a_1 & a_2 & a_3 & a_4 & \cdots \\ b_1 & b_2 & b_3 & b_4 & \cdots \\ c_1 & c_2 & c_3 & c_4 & \cdots \\ d_1 & d_2 & d_3 & d_4 & \cdots \\ \vdots & \vdots & \vdots & \vdots \end{vmatrix} & \det\begin{vmatrix} a_1 & a_2 & a_3 & a_4 \\ b_1 & b_2 & b_3 & b_4 \\ b_1 & b_2 & b_3 & b_4 \\ d_1 & d_2 & d_3 & d_4 \\ \vdots & \vdots & \vdots & \vdots \end{vmatrix} & \cdots \\[2em] \det\begin{vmatrix} c_1 & c_2 & c_4 & c_4 & \cdots \\ b_1 & b_2 & b_3 & b_4 & \cdots \\ c_1 & c_2 & c_3 & c_4 & \cdots \\ d_1 & d_2 & d_3 & d_4 & \cdots \\ \vdots & \vdots & \vdots & \vdots \end{vmatrix} & \det\begin{vmatrix} a_1 & a_2 & a_3 & a_4 & \cdots \\ c_1 & c_2 & c_3 & c_4 & \cdots \\ c_1 & c_2 & c_3 & c_4 & \cdots \\ d_1 & d_2 & d_3 & d_4 & \cdots \\ \vdots & \vdots & \vdots & \vdots \end{vmatrix} & \det\begin{vmatrix} a_1 & a_2 & a_3 & a_4 \\ b_1 & b_2 & b_3 & b_4 \\ c_1 & c_2 & c_3 & c_4 \\ d_1 & d_2 & d_3 & d_4 \\ \vdots & \vdots & \vdots & \vdots \end{vmatrix} & \cdots \end{pmatrix}$$

$$= \begin{pmatrix} \det \mathscr{M} & 0 & 0 & \cdots \\ 0 & \det \mathscr{M} & 0 & \cdots \\ 0 & 0 & \det \mathscr{M} & \cdots \\ \vdots & \vdots & \vdots & \end{pmatrix}. \qquad (B.8c)$$

In the last line we use the fact that any determinant with two identical rows must vanish. Now the general formula for the inverse $\mathscr{M}^{-1}$ follows if we can divide Eq. (B.8) by det $\mathscr{M}$:

$$\mathscr{M}^{-1} = \mathscr{M}^{\mathrm{adj}}/(\det \mathscr{M}). \tag{B.9}$$

This is the case when det $\mathscr{M} \neq 0$. Then from Eq. (B.3) we may write the general solution to the linear equation $\mathscr{M}x = y$, which is called KRAMER'S RULE:

$$x = \mathscr{M}^{-1}y = \frac{\mathscr{M}^{\mathrm{adj}}}{\det \mathscr{M}}y = \begin{pmatrix} \det \begin{vmatrix} y_1 & a_2 & a_3 & a_4 & \cdots \\ y_2 & b_2 & b_3 & b_4 & \cdots \\ y_3 & c_2 & c_3 & c_4 & \cdots \\ y_4 & d_2 & d_3 & d_4 & \cdots \end{vmatrix} \\ \\ \det \begin{vmatrix} a_1 & y_1 & a_3 & b_4 & \cdots \\ b_1 & y_2 & b_3 & b_4 & \cdots \\ c_1 & y_3 & c_3 & c_4 & \cdots \\ d_1 & y_4 & d_3 & d_4 & \cdots \end{vmatrix} \\ \\ \det \begin{vmatrix} a_1 & a_2 & y_1 & a_4 & \cdots \\ b_1 & b_2 & y_2 & b_4 & \cdots \\ c_1 & c_2 & y_3 & c_4 & \cdots \\ d_1 & d_2 & y_4 & d_4 & \cdots \end{vmatrix} \end{pmatrix} \cdot (\det \mathscr{M})^{-1}.$$

(B.10)

A linear equation with $y = 0$ is called a HOMOGENEOUS equation. Equation (B.11) is an example:

$$\mathscr{M} \cdot x = 0 \tag{B.11}$$

Now if det $\mathscr{M} \neq 0$ then according to Kramer's rule the only solutions are zero vectors $x = 0$. However, if det $\mathscr{M} = 0$ there will exist nonzero solutions as shown in Chapter 1, Section 1.2.B. A matrix $\mathscr{M}$ is said to be SINGULAR if det $\mathscr{M}$ is zero and NONSINGULAR if det $\mathscr{M}$ is nonzero.

In Appendix A, a set of vectors $|a\rangle, |b\rangle, |c\rangle, \ldots$ were defined to be linearly dependent if and only if a relation of the form of Eq. (B.12) could exist for values of the coefficients $\alpha, \beta, \gamma, \ldots$ not all zero:

$$\alpha|a\rangle + \beta|b\rangle + \gamma|c\rangle + \cdots = 0. \tag{B.12}$$

By taking the scalar product of this relation with each vector in turn, one

52 A REVIEW OF MATRIX ALGEBRA AND QUANTUM MECHANICS

derives a simple matrix equation which can be used to test for linear dependence.

$$\alpha\langle a|a\rangle + \beta\langle a|b\rangle + \gamma\langle a|c\rangle + \cdots = 0,$$
$$\alpha\langle b|a\rangle + \beta\langle b|b\rangle + \gamma\langle b|c\rangle + \cdots = 0,$$
$$\alpha\langle c|a\rangle + \beta\langle c|b\rangle + \gamma\langle c|c\rangle + \cdots = 0,$$

$$\begin{pmatrix} \langle a|a\rangle & \langle a|b\rangle & \langle a|c\rangle & \cdots \\ \langle b|a\rangle & \langle b|b\rangle & \langle b|c\rangle & \cdots \\ \langle c|a\rangle & \langle c|b\rangle & \langle c|c\rangle & \cdots \\ \vdots & \vdots & \vdots & \end{pmatrix} \begin{pmatrix} \alpha \\ \beta \\ \gamma \\ \vdots \end{pmatrix} = \begin{pmatrix} 0 \\ 0 \\ 0 \\ \vdots \end{pmatrix}.$$

Clearly nonzero $\alpha, \beta, \gamma, \ldots$ exist only if the determinant of the matrix $\langle i|j\rangle$ is zero.

APPENDIX C. PROOF THAT HERMITIAN AND UNITARY MATRICES ARE DIAGONALIZABLE

If a matrix $\mathscr{M}$ satisfied a minimal equation (MEq) with no repeated roots r_j, then we showed in Section 1.2.B(a) how a set of relatively prime polynomials $\mathscr{P}_{r_j}$ of this matrix could be made to obey the completeness and orthonormality relations of Eqs. (1.2.17) and (1.2.18). This amounted to giving a complete set of eigenvectors for the matrix, and guaranteed that $\mathscr{M}$ was DIAGONALIZABLE, or spectrally decomposable.

Now suppose instead that a repeated root r_j did appear in the MEq, as shown in Eq. (C.1):

$$(\mathscr{M} - r_1 1) \cdots (\mathscr{M} - r_j 1)(\mathscr{M} - r_j 1) \cdots (\mathscr{M} - r_n 1) = 0. \quad (C.1)$$

This implies that matrix $\mathscr{N}$ in Eq. (C.2), made from the minimal polynomial by lifting just one redundant factor $(\mathscr{M} - r_j 1)$, must be a nonzero matrix whose square is zero (such a matrix is called a NILPOTENT):

$$\mathscr{N} = (\mathscr{M} - r_1 1) \cdots (1)(\mathscr{M} - r_j 1) \cdots (\mathscr{M} - r_n 1) \neq 0, \quad (C.2a)$$
$$\mathscr{N}\mathscr{N} = 0. \quad (C.2b)$$

Now, this is impossible if $\mathscr{M} \equiv \mathscr{H} \equiv \mathscr{H}^\dagger$ is Hermitian, since then (we use here the fact that all roots of $\mathscr{H} = \mathscr{H}^\dagger$ must be real... see Problem 1.2.7) $\mathscr{N}$ must also be Hermitian; Eq. (C.2b) contradicts Eq. (C.2a) if $\mathscr{N}_{ij} = \mathscr{N}_{ji}^*$. As seen in Eq. (C.3), we must have

$$(\mathscr{N}\mathscr{N})_{ii} = 0 = \sum_j \mathscr{N}_{ij}\mathscr{N}_{ji} = \sum_j \mathscr{N}_{ij}\mathscr{N}_{ij}^* = \sum_j |\mathscr{N}_{ij}|^2 \quad (C.3)$$

Hence a Hermitian matrix must be diagonalizable, i.e., decomposable, as in Eq. (C.4a) to a sum of idempotents $\mathscr{P}_{r_i}$ which satisfy completeness relations [Eq. (C.4b)]. By construction [recall Eq. (1.2.12)] the $\mathscr{P}_{r_i}$ are Hermitian [Eq. (C.4c)]:

$$\mathscr{H} = \sum_{r_i} r_i \mathscr{P}_{r_i} \qquad (r_i^* = r_i), \tag{C.4a}$$

$$1 = \sum_{r_i} \mathscr{P}_{r_i}, \tag{C.4b}$$

$$\mathscr{P}_{r_i}^\dagger = \mathscr{P}_{r_i} = \prod_{r_l \neq r_i} (\mathscr{H} - r_l 1) \Big/ \prod_{r_l \neq r_i} (r_i - r_l). \tag{C.4c}$$

The same applies to any ANTI-HERMITIAN matrix $\mathscr{M} = \mathscr{A} = -\mathscr{A}^\dagger$, since $i\mathscr{A}$ is Hermitian, as given in Eqs. (C.5). Note that $\mathscr{A}$ eigenvalues are purely imaginary:

$$\mathscr{A} = \sum_{a_j} a_j \mathscr{P}_{a_j} \qquad (a_j^* = -a_j), \tag{C.5a}$$

$$1 = \sum_{a_j} \mathscr{P}_{a_j}, \tag{C.5b}$$

$$\mathscr{P}_{a_j}^\dagger = \mathscr{P}_{a_j}. \tag{C.5c}$$

Finally, we can see that the same can be done for a unitary matrix $\mathscr{U}^\dagger \mathscr{U} = 1 = \mathscr{U} \mathscr{U}^\dagger$. First observe that the Hermitian and anti-Hermitian matrices $\mathscr{H}$ and $\mathscr{A}$, made from $\mathscr{U}$ in Eq. (C.6), will commute ($\mathscr{H}\mathscr{A} = \mathscr{A}\mathscr{H}$):

$$\mathscr{H} = \mathscr{U} + \mathscr{U}^\dagger, \qquad \mathscr{A} = \mathscr{U} - \mathscr{U}^\dagger. \tag{C.6}$$

Therefore, the product of the completeness relations of Eqs. (C.5b) and (C.4b) must yield a new set of idempotents $\mathscr{P}_{r_i, a_j} = \mathscr{P}_{r_i} \mathscr{P}_{a_j}$, which can simultaneously decompose $\mathscr{H}$, $\mathscr{A}$, and any combinations thereof, including $\mathscr{U} = (\mathscr{H} + \mathscr{A})/2$.

$$1 \cdot 1 = \left(\sum_{r_i} \mathscr{P}_{r_i} \right) \left(\sum_{a_j} \mathscr{P}_{a_j} \right) = \sum_{r_i a_j} \mathscr{P}_{(r_i a_j)}, \tag{C.7a}$$

$$\mathscr{U} = \sum_{r_i a_j} u_{ij} \mathscr{P}_{(r_i a_j)}, \qquad u_{ij} = (r_i + a_j)/2 = \frac{1}{u_{ij}^*}, \tag{C.7b}$$

$$\mathscr{P}_{(r_i a_j)}^\dagger = \mathscr{P}_{(r_i a_j)}. \tag{C.7c}$$

(Many of the products $\mathscr{P}_{r_i} \mathscr{P}_{a_j}$ may be zero, but whatever is left must be complete.) The eigenvalues of $\mathscr{U}$ are of the form $u = e^{i\phi}$ if $\mathscr{U}$ is unitary.

ADDITIONAL READING

One of the best introductions to the fundamental quantum theory is found in the third and final volume of the *Feynman Lectures*. One should read all three volumes, but the final volume is quite self-contained and may be worth the price of the whole set.

R. P. Feynman, R. B. Leighton, and M. Sands, *The Feynman Lectures on Physics* (Addison-Wesley, Reading, MA, 1967), Volumes I–III).

Feynman's unusual approach to quantum theory introduces the Dirac bra-ket transamplitudes $\langle b|a \rangle$ and their matrix mechanics in a physical way. Two- or three-state systems are used as examples in the primary development. Then infinite-dimensional amplitudes or wave functions $\langle x|\Psi \rangle = \Psi(x)$ and differential operators are treated subsequently.

Another text which begins by discussing amplitude mechanics is by Gordan Baym.

G. Baym, *Lectures on Quantum Mechanics* (Benjamin Advanced Book Program, Reading, MA, 1973).

Interpretation of quantum states is debated in *Physics Today* (April 1993, page 13).

A discussion of spectral decomposition of matrices and the algebra of projection operators should be found in any good book on mathematical methods for physics or chemistry. However, most books written for physicists do not cover this subject very well, if at all. One notable exception which has recently been published is a text by Hasani.

Sadri Hasani, *Foundations of Mathematical Physics* (Allyn and Bacon, Boston, 1991).

This text also describes the spectral decomposition and resolvants of differential operators.

It is easier to find treatments of matrix spectral decomposition in the mathematical literature. An excellent readable treatment is found on p. 155 of the following older but well known text. It features the spectral theorem as the main result of the book.

Paul R. Halmos, *Finite Dimensional Vector Spaces* (Van Nostrand, Princeton, 1958).

PROBLEMS

Section 1.1

1.1.1 (Beam stoppers) Consider an incoming beam of spin-$\frac{1}{2}$ particles approaching the apparatus shown in the following diagram. A "stopper" can block the intermediate (2)" beam if inserted as shown in the diagram.

Use the following notation for the states having different analyzer angle θ.

$$|1\rangle = |\text{up}\rangle \quad |1'\rangle = |N\rangle \quad |1''\rangle = |\text{right}\rangle$$
$$|2\rangle = |\text{dn}\rangle \quad |2'\rangle = |S\rangle \quad |2''\rangle = |\text{left}\rangle$$
$$\theta = 0, \quad 0 < \theta < \pi/2, \quad \theta = \pi/2.$$

Compute the percentage of particles that get to counter (1)...   without stopper   with stopper

for an incoming beam polarized in state: $|1\rangle$

... in state: $|2\rangle$

... in state: $|1'\rangle$

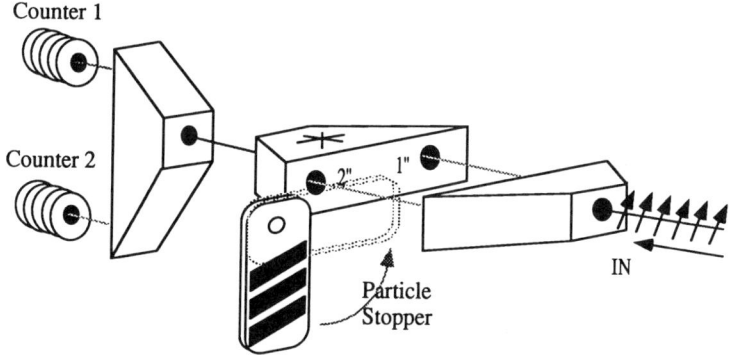

1.1.2 (Pointing spins the "right" way)
  (a) How many of the spins that were originally up ($|1\rangle$) going into this apparatus* will end up "right"? Give answer as a function of n and see how it behaves for $n = 1, 2, 3, \ldots, \infty$.
  (b) Can you design an apparatus* that duplicates the effect of the elementary operator e_{22}? How about e_{12}?

---

*"Apparatus" means some arrangement of Stern-Gerlach analyzers and stoppers.

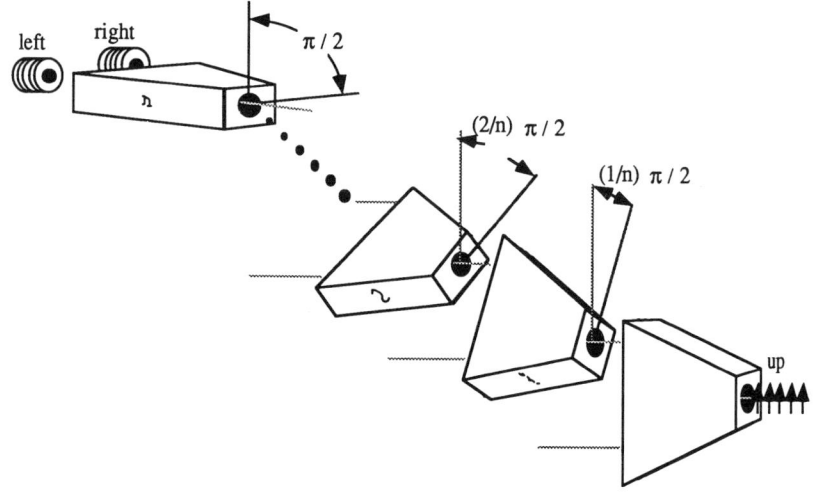

1.1.3 (a) Prove that $(AB)^\dagger = B^\dagger A^\dagger$ using components A_{ij} and B_{ij} of $n \times n$ matrices A and B.

(b) Prove that $(A|x\rangle)^\dagger = \langle x|A^\dagger$.

(c) Expand to find $(ABCD)^\dagger = ?$, $(\langle x|A|y\rangle)^\dagger = ?$, $(|x\rangle\langle y|)^\dagger ?$, $(\mathbf{e}_{12})^\dagger ?$

Section 1.2

1.2.1 (Back to your roots)

For the matrix $M = \begin{pmatrix} 2 & -1 \\ -6 & 1 \end{pmatrix}$ find

(a) Eigenvalues of M.

(b) Spectral decomposition of M.

(c) The bra and ket eigenvectors of M.

(d) All the square roots of M: $\sqrt{M} = \sqrt{\begin{pmatrix} 2 & -1 \\ -6 & 1 \end{pmatrix}}$

1.2.2 Do the same as in Problem 1.2.1 for the matrix $M = \begin{pmatrix} 11 & 7 & -9 \\ 7 & 11 & -9 \\ -9 & -9 & 27 \end{pmatrix}$.

1.2.3 (Secular behavior)

The polynomial form for the secular equation of a general $n \times n$ matrix M is

$$S(\varepsilon) = \det|M - \varepsilon \mathbf{1}|$$
$$= \varepsilon^n + a_1 \varepsilon^{n-1} + a_2 \varepsilon^{n-2} + \cdots + a_{n-1}\varepsilon + a_0 = 0.$$

(a) Derive general formulas for the coefficients $a_1, a_2, a_3, \ldots, a_{n-1}, a_n$ in terms of the eigenvalues $\{\varepsilon_1, \varepsilon_2, \varepsilon_3, \ldots, \varepsilon_n\}$ of matrix M.

(b) Derive general formulas for the coefficients $a_1, a_2, a_3, \ldots, a_{n-1}, a_n$ in terms of the components $\{M_{ij}\}$ of matrix M. (Hint: Use the ε-tensor definition

$$\det|M| = \sum_{i_1 i_2 \cdots i_m} \varepsilon_{i_1 i_2 \cdots i_m} M_{1, i_1} M_{2, i_2} \cdots M_{n, i_n}$$

for determinants and expand the determinantal secular expression.)

(c) Do the coefficients a_j change if there is a change of basis?

[*Note: The results of (a) and (b) are particularly important and will be helpful for later theory and problems throughout the book.*]

1.2.4 (Multiple degeneracy) The matrix $\rho = \rho \begin{pmatrix} 1 & 1 & 1 & 1 \\ 1 & 1 & 1 & 1 \\ 1 & 1 & 1 & 1 \\ 1 & 1 & 1 & 1 \end{pmatrix}$

represents a simple example of what is called a *pairing* operator in nuclear and superconductivity theory. It has number of repeated or degenerate eigenvalues.

(a) Find the secular equation, the Hamiltonian Cayley equation, and the minimal equation of ρ. (*Hint: Problem 1.2.3 is useful.*) Find the eigenvalues.

(b) Compute the projection operators for each distinct eigenvalue and write the spectral decomposition for this matrix.

(c) Find all the square roots $\sqrt{\rho}$ of matrix ρ.

1.2.5 (Knowing spectral decomposition backwards and forwards)

(a) Use what we have discussed about spectral decomposition and $\otimes$ products to do an eigenvalue problem backwards. Find the matrix $M = M^\dagger$ which has the following eigensolutions:

$$M\begin{pmatrix} 1 \\ -1 \\ 0 \end{pmatrix} = \begin{pmatrix} 1 \\ -1 \\ 0 \end{pmatrix}, \quad M\begin{pmatrix} 1 \\ 1 \\ -2 \end{pmatrix} = \begin{pmatrix} -1 \\ -1 \\ 2 \end{pmatrix}, \quad M\begin{pmatrix} 1 \\ 1 \\ 1 \end{pmatrix} = \begin{pmatrix} 0 \\ 0 \\ 0 \end{pmatrix}.$$

(b) Find M^{100}.

1.2.6 (Commuting observables)

(a) Use the techniques in Chapter 1 (Section 1.2Bd) to find a single set of projection operators which spectrally decompose both matrices:

$$M = \begin{pmatrix} 2 & 1 & 0 \\ 1 & 2 & 0 \\ 0 & 0 & 3 \end{pmatrix} \quad \text{and} \quad N = \begin{pmatrix} 3 & -1 & -1 \\ -1 & 3 & -1 \\ -1 & -1 & 3 \end{pmatrix}.$$

(b) Use the result of (a) to find a single transformation T which diagonalizes M and N.

1.2.7 What are the conjugation relations (if any) for eigenvalues (E_j and E_j^*), projection operators ($\mathbf{P}_{E_j}$ and $\mathbf{P}_{E_j}^\dagger$), and eigenvectors ($|E_j\rangle$ and $\langle E_j|$) in the cases that operators are (a) Hermitian: $\mathbf{H} = \mathbf{H}^\dagger$, (b) anti-Hermitian: $\mathbf{A}^\dagger = -\mathbf{A}$, or (c) unitary: $\mathbf{U}^\dagger = \mathbf{U}^{-1}$. Check your conclusions by spectrally decomposing and diagonalizing the following examples:

(a) $\langle H \rangle = \begin{pmatrix} 1/\sqrt{2} & -i/\sqrt{2} \\ i/\sqrt{2} & -1/\sqrt{2} \end{pmatrix}.$ (b) $\langle A \rangle = \begin{pmatrix} 0 & i/\sqrt{2} \\ i/\sqrt{2} & 0 \end{pmatrix}.$

(c) $\langle U \rangle = \begin{pmatrix} 1/\sqrt{2} & 1/\sqrt{2} \\ i/\sqrt{2} & -i/\sqrt{2} \end{pmatrix}.$

1.2.8 Does $M^\dagger M = 1$ imply that $MM^\dagger = 1$, as well?

(a) Prove or disprove in the case that M is a finite $n \times n$ matrix.

(b) What if M is an *infinite* matrix? (Examples of infinite matrices are the representations of creation and destruction operators $\mathbf{a}$ and $\mathbf{a}^\dagger$ in the quantum harmonic oscillator eigenstates $\{|0\rangle, |1\rangle = \mathbf{a}^\dagger|0\rangle, |2\rangle = \mathbf{a}^\dagger|1\rangle, \ldots\}$.)

CHAPTER 2

BASIC THEORY AND APPLICATIONS OF SYMMETRY REPRESENTATIONS (ABELIAN SYMMETRY GROUPS)

In the preceding review of matrices the ideas of projection operators and spectral decompositions were introduced. In this chapter we shall see how frequency spectra of physical systems are analyzed in terms of mathematical spectral decompositions. Mathematical concepts will be introduced in this and following chapters by analyzing the simplest physical models which exhibit them. In this way the mathematical and physical ideas can be closely related. It is hoped that this particular pedagogical approach to the theory of spectra will be easy to understand.

Symmetry is a key mathematical and physical concept in the classical and quantum theory of spectra. Symmetry analysis and group theory were first applied by Eugene Wigner and Herman Weyl shortly after the invention of quantum mechanics. Since then applications of symmetry analysis have been made to virtually all types of spectroscopy. Spectra, ranging in energy from radio frequency ($\sim 10^3$ Hz) to x ray ($\sim 10^{18}$ Hz), have given information about atoms, molecules, and solids. Higher-frequency γ radiation ($> 10^{20}$ Hz) has been used to study nuclear spectra. A most widely publicized application of symmetry principles concerns very high energy "elementary particle" spectra where researchers are thinking about energies in excess of 10^{12} eV or 10^{26} Hz. (1 eV is equivalent to 2.42×10^{14} Hz.)

Meanwhile the application of laser devices has reopened atomic and molecular spectroscopy. Instead of obtaining higher and higher frequency, laser spectroscopists are obtaining ever-increasing frequency *resolution*. This means finer spectral details are seen and more detailed models of atomic and molecular processes are needed. This has stimulated the development of new

60 BASIC THEORY AND APPLICATIONS OF SYMMETRY REPRESENTATIONS

symmetry analysis techniques, some of which are discussed in later chapters of this book.

However, the fundamental ideas of symmetry analysis are simple and basic to all present theories. The beginnings of most symmetry mathematics involves several mutually commuting operators. We have seen in the preceding chapter that the eigenvalue spectrum of one of several commuting operators may help to solve the others. This is the main mathematical idea which will be developed in this chapter.

2.1 SYMMETRY GROUPS

First we shall explain how a physicist can say "symmetry" precisely. Consider a simple fan blade such as you might see on the ceilings of bar and hotel rooms in the tropical areas. This is shown in Figure 2.1.1. Everybody would probably agree that this blade has some symmetry, but the question is: How much?

To answer this we ask, "In how many positions could the fan blade be put so that it would still look the same in a drawing like Figure 2.1.1?" We list these below and draw them in Figure 2.1.2. (In the latter figure some markings "left" and "right" have been added. They spoil the symmetry but allow you to distinguish the different positions.)

1 : THE ORIGINAL POSITION Don't touch the fan blade.
R_z: THE HALF-TURN POSITION Rotate it by 180° around its axle or the z axis.
R_y: THE OVERTURNED POSITION Overturn it 180° around the y axis.
R_x: THE FLIPPED POSITION Flip it 180° around the x axis.

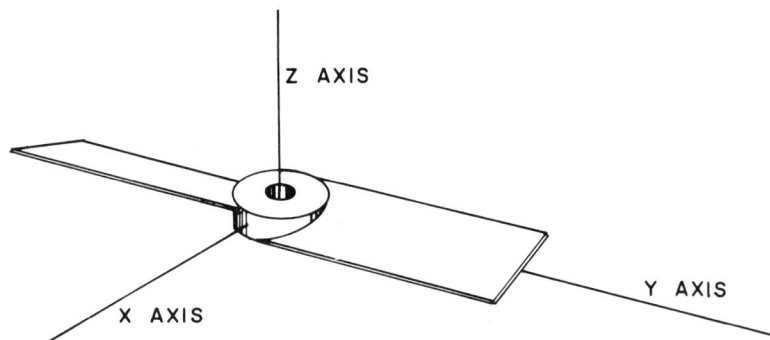

Figure 2.1.1 Fan blade. This is an example of an object which has an Abelian symmetry D_2.

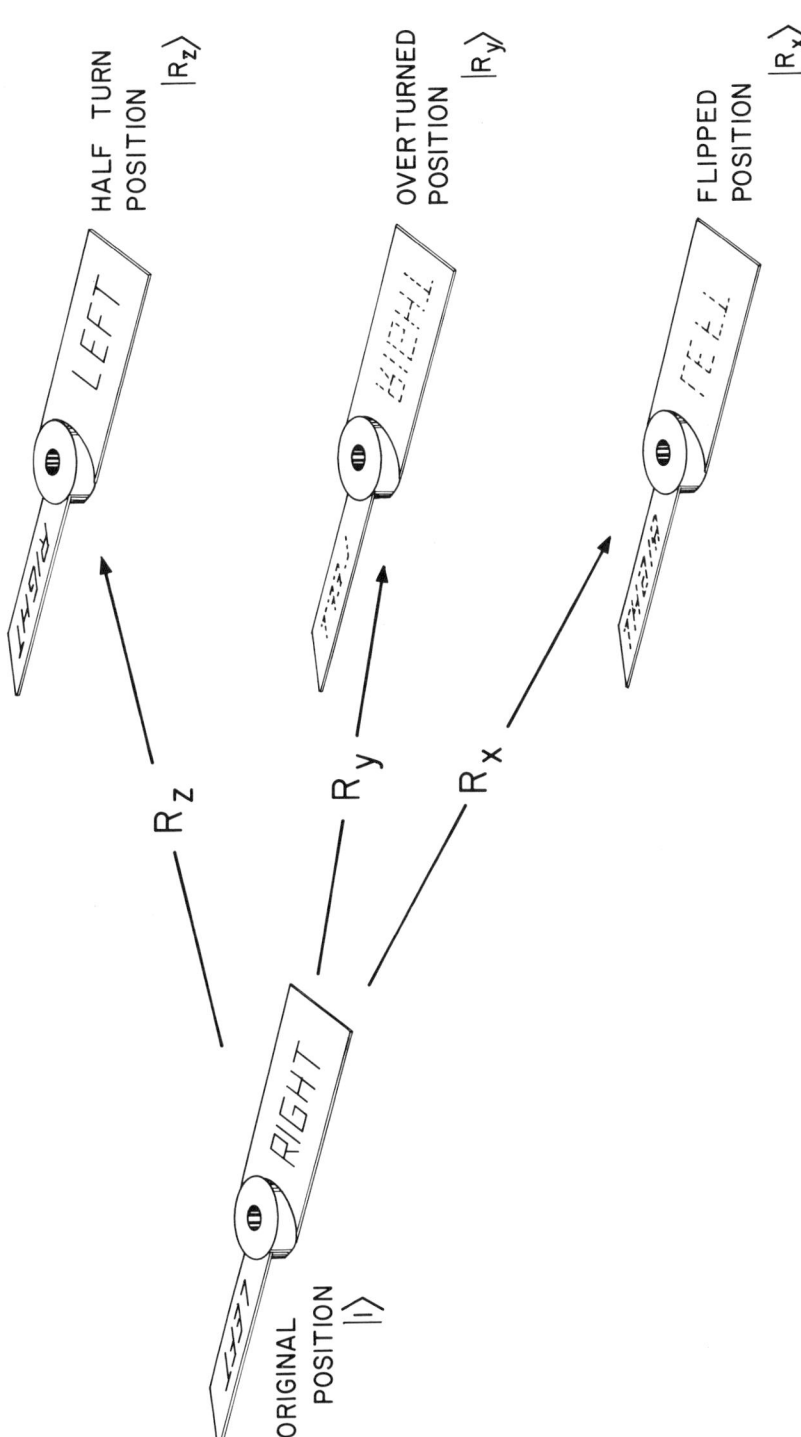

Figure 2.1.2 Symmetry operations. The 180° rotations R_x, R_y, and R_z around the x, y, and z axes, respectively, all leave the fan blade in similar positions.

The first two possibilities are obvious enough; however, the third (R_y) and the fourth (R_x) might be a little surprising at first. Note that a fan blade like this one works just as well when it is installed "backwards."

Now we see there are basically two ways to keep track of this symmetry. According to one we can tally up the allowed positions or position states $|1\rangle$, $|R_x\rangle$, $|R_y\rangle$, and $|R_z\rangle$ that look the same. According to the other, we tally up the operations (turn, flip, etc.) or OPERATORS $\{1, R_x, R_y, R_z\}$ that change one allowed position state to another. These operations have a number of mathematical properties which we shall study shortly. Most important is the idea of combination or GROUP MULTIPLICATION.

For example, if we do an R_y (overturn) followed by a R_z (half turn), all with respect to fixed spatial axes x, y, and z, then what do we get? Examination of Figure 2.1.2 shows that the same position state shows up which would have been obtained by just doing R_x (flip) by itself. Let us express these observations by the following operator and position state equation:

$$R_z R_y |1\rangle = R_z |R_y\rangle = |R_x\rangle = R_x |1\rangle. \tag{2.1.1}$$

Now Eq. (2.1.1) is true no matter whether you start with $|1\rangle$ or the other states, and so we may write it abstractly as Eq. (2.1.2):

$$R_z R_y = R_x. \tag{2.1.2}$$

In this way we make a multiplication table or GROUP TABLE such as is shown below. Here all possible products are accounted for:

	1	R_x	R_y	R_z
1	1	R_x	R_y	R_z
R_x	R_x	1	R_z	R_y
R_y	R_y	R_z	1	R_x
R_z	R_z	R_y	R_x	1

(2.1.3)

This is ultimately how symmetry is coded, through abstract mathematical properties of symmetry operators. Now we shall see how this type of mathematics enters a physical problem.

2.2 REPRESENTING SYMMETRY AND SYMMETRY GROUPS

We start by analyzing in detail one of the simplest examples of a physical system having one of the simplest symmetries. Consider the two identical and

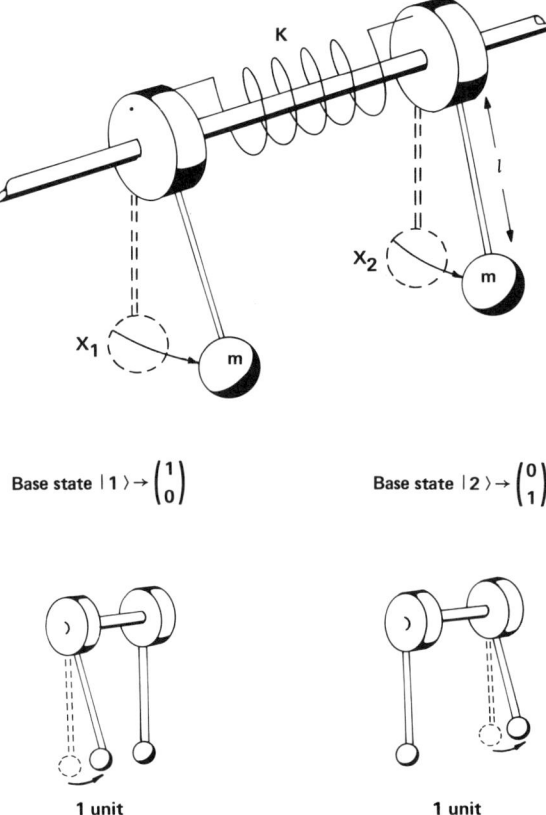

Figure 2.2.1 Coupled pendulums. This is an example of a mechanical system having one of the simplest Abelian symmetries $C_2 = \{1, R\}$. The base states are related by the symmetry operation R of reversal or reflection according to $|2\rangle = R|1\rangle$.

coupled torsion pendulums in Figure 2.2.1. Their motion will be described by the classical matrix equations if the coordinates ($\chi_1 = \langle 1|x\rangle$, $\chi_2 = \langle 2|x\rangle$) are not too great, as was explained in Chapter 1. The unit position base states $|1\rangle$ and $|2\rangle$ are indicated on the right-hand side of Figure 2.2.1. Newton's equation of motion is given abstractly by

$$|\ddot{x}(t)\rangle = -\mathbf{a}|x(t)\rangle. \qquad (2.2.1a)$$

It is represented in the $\{|1\rangle, |2\rangle\}$ basis by the following matrix equation of motion:

$$\begin{pmatrix} \langle 1|\ddot{x}(t)\rangle \\ \langle 2|\ddot{x}(t)\rangle \end{pmatrix} = -\begin{pmatrix} \langle 1|\mathbf{a}|1\rangle & \langle 1|\mathbf{a}|2\rangle \\ \langle 2|\mathbf{a}|1\rangle & \langle 2|\mathbf{a}|2\rangle \end{pmatrix}\begin{pmatrix} \langle 1|x(t)\rangle \\ \langle 2|x(t)\rangle \end{pmatrix} \qquad (2.2.1b)$$

64 BASIC THEORY AND APPLICATIONS OF SYMMETRY REPRESENTATIONS

or

$$\begin{pmatrix} \ddot{x}_1 \\ \ddot{x}_2 \end{pmatrix} = -\begin{pmatrix} a+b & -a \\ -a & a+b \end{pmatrix}\begin{pmatrix} x_1 \\ x_2 \end{pmatrix}. \quad (2.2.1c)$$

where $a \equiv 2k/ml^2$, and $b \equiv g/l$.

The constants are gravity (g), coupling spring constant (k), pendulum length (l), and mass (m). The classical position state vector $|x\rangle$ is represented in the $\{|1\rangle, |2\rangle\}$ basis by the ket

$$|x\rangle = |1\rangle\langle 1|x\rangle + |2\rangle\langle 2|x\rangle$$
$$= |1\rangle x_1 + |2\rangle x_2,$$

or by the two-component column vector in Eq. (2.2.1). This classical application of Dirac notation for vectors and operators ($\langle i|a|j\rangle$) was introduced in Section 1.4.A.

The symmetry of this device is fairly obvious. The pendulums are identical, and if someone switches them in the middle of the night no one should be able to tell the difference. The problem is to formulate this fact in a mathematical description.

This may be done by defining a symmetry operator R that reflects or switches the pendulum states according to the following equations (see also Figure 2.2.1):

$$R|1\rangle = |2\rangle,$$
$$R|2\rangle = |1\rangle. \quad (2.2.2a)$$

This allows R to be represented in the $\{|1\rangle, |2\rangle\}$ basis by the following matrix:

$$\begin{pmatrix} \langle 1|R|1\rangle & \langle 1|R|2\rangle \\ \langle 2|R|1\rangle & \langle 2|R|2\rangle \end{pmatrix} = \begin{pmatrix} 0 & 1 \\ 1 & 0 \end{pmatrix}. \quad (2.2.2b)$$

Then for every state $|x\rangle$ of the system, there is a reflected state $R|x\rangle$ which is represented as follows:

$$R|x\rangle \leftrightarrow \begin{pmatrix} \langle 1|R|1\rangle & \langle 1|R|2\rangle \\ \langle 2|R|1\rangle & \langle 2|R|2\rangle \end{pmatrix}\begin{pmatrix} \langle 1|x\rangle \\ \langle 2|x\rangle \end{pmatrix} = \begin{pmatrix} 0 & 1 \\ 1 & 0 \end{pmatrix}\begin{pmatrix} x_1 \\ x_2 \end{pmatrix} = \begin{pmatrix} x_2 \\ x_1 \end{pmatrix}. \quad (2.2.3)$$

Now the mathematical statement of the physical symmetry is: THE EQUATION FOR A REFLECTED STATE $R|x\rangle$ IS THE SAME AS IT WAS FOR THE ORIGINAL STATE $|x\rangle$. This is written as follows:

$$R|\ddot{x}\rangle = -\mathbf{a} \cdot R|x\rangle, \quad (2.2.4a)$$

$$\begin{pmatrix} \ddot{x}_2 \\ \ddot{x}_1 \end{pmatrix} = -\begin{pmatrix} \langle 1|\mathbf{a}|1\rangle & \langle 1|\mathbf{a}|2\rangle \\ \langle 2|\mathbf{a}|1\rangle & \langle 2|\mathbf{a}|2\rangle \end{pmatrix}\begin{pmatrix} x_2 \\ x_1 \end{pmatrix}. \quad (2.2.4b)$$

Note that Eq. (2.2.4b) does not agree with Eq. (2.2.1b) for arbitrary choice of $\langle i|a|j\rangle$, but that it does agree for the constants we have chosen. In fact we can now deduce the constraints that operator $\langle a \rangle$ must satisfy when reflection symmetry is present.

First, Eq. (2.2.5) below follows directly from the equation of motion $|\ddot{x}\rangle = -\mathbf{a}|x\rangle$ no matter whether R symmetry is present or not:

$$R|\ddot{x}\rangle = -R \cdot \mathbf{a}|x\rangle, \qquad (2.2.5a)$$

$$\begin{pmatrix} 0 & 1 \\ 1 & 0 \end{pmatrix}\begin{pmatrix} \ddot{x}_1 \\ \ddot{x}_2 \end{pmatrix} = \begin{pmatrix} \ddot{x}_2 \\ \ddot{x}_1 \end{pmatrix} = -\begin{pmatrix} 0 & 1 \\ 1 & 0 \end{pmatrix}\begin{pmatrix} \langle 1|a|1\rangle & \langle 1|a|2\rangle \\ \langle 2|a|1\rangle & \langle 2|a|2\rangle \end{pmatrix}\begin{pmatrix} x_1 \\ x_2 \end{pmatrix}$$

$$= -\begin{pmatrix} \langle 2|a|1\rangle & \langle 2|a|2\rangle \\ \langle 1|a|1\rangle & \langle 1|a|2\rangle \end{pmatrix}\begin{pmatrix} x_1 \\ x_2 \end{pmatrix}. \qquad (2.2.5b)$$

But the presence of R symmetry gives Eq. (2.2.4), which together with Eq. (2.2.5) gives the following:

$$\mathbf{a} \cdot R = R \cdot \mathbf{a}, \qquad (2.2.6a)$$

$$\begin{pmatrix} \langle 1|a|1\rangle & \langle 1|a|2\rangle \\ \langle 2|a|1\rangle & \langle 2|a|2\rangle \end{pmatrix}\begin{pmatrix} 0 & 1 \\ 1 & 0 \end{pmatrix} = \begin{pmatrix} 0 & 1 \\ 1 & 0 \end{pmatrix}\begin{pmatrix} \langle 1|a|1\rangle & \langle 1|a|2\rangle \\ \langle 2|a|1\rangle & \langle 2|a|2\rangle \end{pmatrix}, \qquad (2.2.6b)$$

$$\begin{pmatrix} \langle 1|a|2\rangle & \langle 1|a|1\rangle \\ \langle 2|a|2\rangle & \langle 2|a|1\rangle \end{pmatrix} = \begin{pmatrix} \langle 2|a|1\rangle & \langle 2|a|2\rangle \\ \langle 1|a|1\rangle & \langle 1|a|2\rangle \end{pmatrix}. \qquad (2.2.6c)$$

The last equation shows that R symmetry of a requires that $\langle 1|a|1\rangle = \langle 2|a|2\rangle$ and $\langle 1|a|2\rangle = \langle 2|a|1\rangle$. However, the first equation (2.2.6a) is a general abstract definition of a physical symmetry.

Definition 1 A symmetry operator commutes with any operator that is part of an equation of motion for a physical system having that symmetry.

In quantum mechanics, all symmetry operators commute with the Hamiltonian according to this definition.

A second definition of symmetry operators involves the state vectors or basis of a physical system. We shall require that the inner product $\langle x|y\rangle$ of

66   BASIC THEORY AND APPLICATIONS OF SYMMETRY REPRESENTATIONS

any two vectors $|x\rangle$ and $|y\rangle$ shall be equal to that of the transformed vectors $R|x\rangle$ and $R|y\rangle$ as given in Eq. (2.2.7):

$$\langle x|y\rangle = \langle x|R^\dagger R|y\rangle. \tag{2.2.7}$$

We demand this for quantum and classical descriptions, alike. As explained in Chapter 1 [see, for example, Eqs. (1.1.8)], these last requirements imply that a linear symmetry operator R and its representations $\mathscr{R}_{ij} = \langle i|r|j\rangle$ will be unitary.

Definition 2 Symmetry operators and their representations are unitary.

$$RR^\dagger = 1 = R^\dagger R, \tag{2.2.8a}$$

$$(\mathscr{R}_{ij})^\dagger = \mathscr{R}_{ji}^* = (\mathscr{R}_{ij})^{-1}. \tag{2.2.8b}$$

In fact, the matrices $\langle i|R|j\rangle \equiv \langle i|j'\rangle$ will be seen to have the properties of the transformation matrix defined in Chapter 1.

Combining Eq. (2.2.8a) with Eq. (2.2.6b) gives the most commonly written expression for symmetry:

$$\mathbf{a} = R \cdot \mathbf{a} \cdot R^\dagger. \tag{2.2.9}$$

In other words, the operator **a** which has R symmetry is invariant or unchanged when transformed by R. [Recall the form of matrix operator transformations in Eq. (1.1.23c).] The set of all symmetry operators R satisfying Eq. (2.2.9) for a given **a** is called a GROUP by mathematicians or the SYMMETRY GROUP of **a** by physicists. The mathematical axioms for a group, which are shown on the left of Table 2.2.1, were introduced abstractly by a mathematician named Galois long before matrix applications like quantum mechanics came along. One may see by examining the right side of Table 2.2.1 that the mathematical axioms are closely related to the physical axioms 1–4 for transformation matrices given in Chapter 1.

A lot of mathematical work has involved the determination of all possible abstract groups with a given order or number of elements. It is not so easy to come up with a multiplication table which satisfies all four group definitions. For example, Eq. (2.2.10) is a multiplication table of a group of order 6, but Eq. (2.2.11) is a multiplication table of something that is not a group (possibly, this set should be called a "heap" instead), since $(ab)c = a \neq$

TABLE 2.2.1 Demonstrating that the Set of All Symmetry Transformation Operators Is a Group

Group Definitions	Testing the Symmetry Operators
1. If Q and R are in a group then so is their product $P = QR$ (called CLOSURE rule).	1. If $R\mathbf{a} = \mathbf{a}R$ and $Q\mathbf{a} = \mathbf{a}Q$ then $RQ\mathbf{a} = \mathbf{a}RQ$, so RQ is a symmetry operator, too. (Recall Axiom 4.)
2. If Q, R, and S are in a group, then $Q(RS) = (QR)S$ (called ASSOCIATIVITY rule).	2. The operators we discuss satisfy associativity since they are defined by matrices.
3. There exists an IDENTITY element 1 such that $R1 = R = 1R$ for all R in the group.	3. The unit operator commutes with any operator, including $\mathbf{a}$, so it is a symmetry operator. (Recall Axiom 3.)
4. For each R in the group there is an INVERSE R^{-1} such that $RR^{-1} = R^{-1}R = 1$.	4. If $R\mathbf{a} = \mathbf{a}R$, then $R^\dagger = R^{-1}$, which exists by Axioms 2–4, is a symmetry operation, too, because $R^\dagger R \mathbf{a} R^\dagger = R^\dagger \mathbf{a} R R^\dagger$ or $\mathbf{a} R^\dagger = R^\dagger \mathbf{a}$.

$a(bc) = c$ breaks rule 2. *Associativity* is a very restrictive property which one tends to take for granted.

$$\text{Group} \quad \begin{array}{c|cccccc} & 1 & A & B & C & D & F \\ \hline 1 & 1 & A & B & C & D & F \\ A & A & B & C & D & F & 1 \\ B & B & C & D & F & 1 & A \\ C & C & D & F & 1 & A & B \\ D & D & F & 1 & A & B & C \\ F & F & 1 & A & B & C & D \end{array} \qquad (2.2.10)$$

$$\text{Not a group} \quad \begin{array}{c|cccccc} & 1 & a & b & c & d & f \\ \hline 1 & 1 & a & b & c & d & f \\ a & a & 1 & d & b & f & c \\ b & b & d & 1 & f & c & a \\ c & c & b & f & 1 & a & d \\ d & d & f & c & a & 1 & b \\ f & f & c & a & d & b & 1 \end{array} \qquad (2.2.11)$$

However, the determination of all possible groups will not concern us at first, since the elementary symmetry groups such as $C_2 = \{1, R\}$ for the

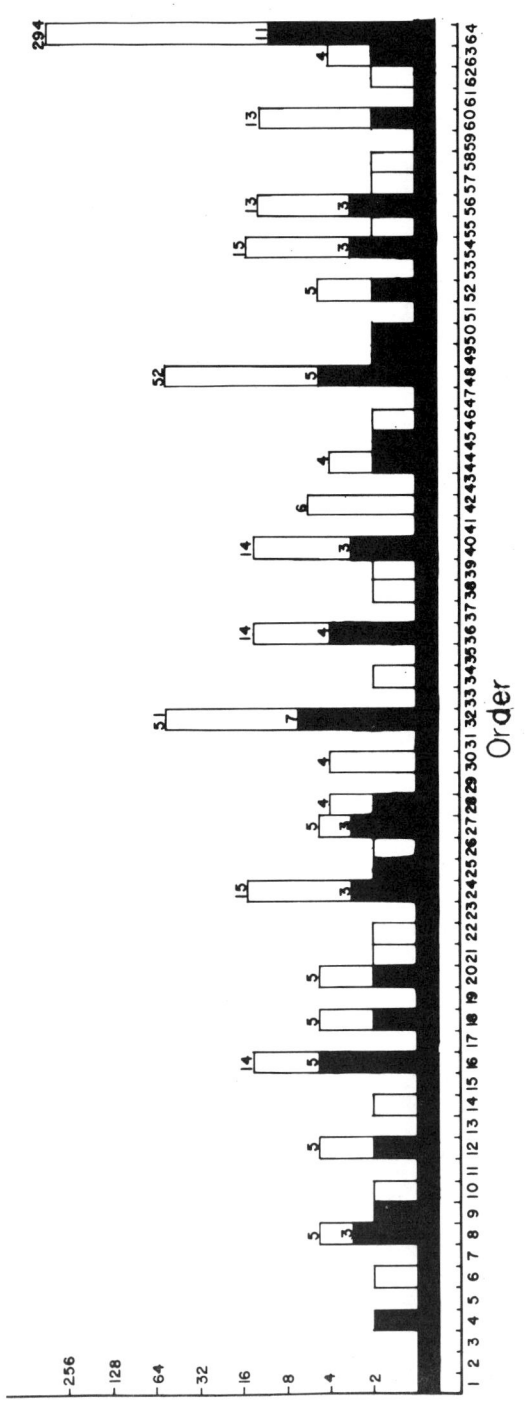

Figure 2.2.2 Plot of number of groups versus their order ($°G$) for $°G \leq 64$. The number of Abelian groups is indicated by the dark lines.

pendulums will be given by multiplication tables [viz., Eq. (2.2.12)] or by some other definition. (C_2 means "cyclic group of order 2." Actually the symmetry group we are using for the pendulums is called C_h or C_v, as will be explained in Section 2.11.)

$$
\begin{array}{c|cc}
 & 1 & R \\
\hline
1 & 1 & R \\
R & R & 1
\end{array}
\quad \text{Pendulum symmetry group } C_2. \quad (2.2.12)
$$

Nevertheless, it is instructive to see the plot in Figure 2.2.2 of the number of different groups that exist for order less than 65. Physicists have only applied a fraction of these finite groups so far. Nobody knows yet what all the rest can mean.

The following introduction to the uses of group theory begins with the application of some ABELIAN GROUPS including C_2. Abelian simply means commutative; that is, $ab = ba$ for all a, b in the Abelian group. (The name itself is in memory of the mathematician Abel.) Following this, the next three chapters contain theory and applications of the more general non-Abelian finite groups. This is followed by theory and applications of continuous symmetry groups of infinite order. However, the basic structure which we are about to derive in this chapter for some simple examples is basic to all symmetry theory.

2.3 SOLVING A PROBLEM WITH SYMMETRY ANALYSIS (C_2)

In order to use symmetry to simplify any physical problem, it is necessary to express the symmetry information, i.e., the group of symmetry operators, in a more digestible form. Let us now introduce the mechanics of this form using our simple example of two pendulums, Figure 2.2.1, which has the symmetry group $C_2 = \{1, R\}$.

First we observe that the operator R and any representation thereof must satisfy an equation of the form $R^2 = 1$, which can be rewritten as

$$R^2 - 1 = 0. \qquad (2.3.1)$$

This has two roots: $r_+ = 1$ and $r_- = -1$.

Now, whenever you see an operator or matrix satisfying an nth-degree polynomial equation, you should remember what to do. Following Chapter 1, you use the roots $r_1 r_2 \cdots r_n$ to construct a set of idempotent projection operators using the following equation, which comes from Eq. (1.2.15):

$$P^{r_j} = \prod_{r_l \ne r_j} (R - r_l \mathbf{1}) \bigg/ \prod_{r_l \ne r_j} (r_j - r_l). \qquad (2.3.2)$$

70   BASIC THEORY AND APPLICATIONS OF SYMMETRY REPRESENTATIONS

For this example we have $r_+ = 1$ and $r_- = -1$, whence Eq. (2.3.2) gives the following:

$$P^+ = (1+R)/2, \qquad P^- = (1-R)/2. \qquad (2.3.2)_x$$

Such operators must be orthogonal idempotents for the same reasons that their matrix counterparts were orthogonal in Chapter 1 [recall Eq. (1.2.17)]:

$$P^{r_j} P^{r_k} = \delta_{r_j r_k} P^{r_j}. \qquad (2.3.3)$$

This is easily verified for the example being treated:

$$P^+ P^+ = P^+, \qquad P^+ P^- = 0 = P^- P^+, \qquad P^- P^- = P^-. \qquad (2.3.3)_x$$

Now the representation $\langle i|P^r|j\rangle$ of P^r in the basis $\{|1\rangle, |2\rangle\}$ will help us reduce the equation of motion. From Eq. (2.2.2) the following representations result:

$$\begin{pmatrix} \langle 1|P^+|1\rangle & \langle 1|P^+|2\rangle \\ \langle 2|P^+|1\rangle & \langle 2|P^+|2\rangle \end{pmatrix} = \begin{pmatrix} \tfrac{1}{2} & \tfrac{1}{2} \\ \tfrac{1}{2} & \tfrac{1}{2} \end{pmatrix},$$

$$\begin{pmatrix} \langle 1|P^-|1\rangle & \langle 1|P^-|2\rangle \\ \langle 2|P^-|1\rangle & \langle 2|P^-|2\rangle \end{pmatrix} = \begin{pmatrix} \tfrac{1}{2} & -\tfrac{1}{2} \\ -\tfrac{1}{2} & \tfrac{1}{2} \end{pmatrix}.$$

By taking the first column of each of these matrices one obtains representations of the eigenkets $|e_+\rangle = P^+|1\rangle\sqrt{2}$, and $|e_-\rangle = P^-|1\rangle\sqrt{2}$:

$$\begin{pmatrix} \langle 1|e_+\rangle \\ \langle 2|e_+\rangle \end{pmatrix} = \begin{pmatrix} 1/\sqrt{2} \\ 1/\sqrt{2} \end{pmatrix}, \qquad \begin{pmatrix} \langle 1|e_-\rangle \\ \langle 2|e_-\rangle \end{pmatrix} = \begin{pmatrix} 1/\sqrt{2} \\ -1/\sqrt{2} \end{pmatrix}. \qquad (2.3.4)$$

Here the normalization coefficient is the inverse square root of the diagonal component $[(\langle 1|P^\pm|1\rangle)^{-1/2} = \sqrt{2}]$ in each case. The vectors $|e_\pm\rangle$ are orthonormal eigenvectors of the symmetry operator:

$$R|e_+\rangle = |e_+\rangle, \qquad R|e_-\rangle = -|e_-\rangle.$$

They are also the eigenvectors of the acceleration matrix $\langle \mathbf{a} \rangle$ in Eq. (2.2.1):

$$\begin{pmatrix} a+b & -a \\ -a & a+b \end{pmatrix} \begin{pmatrix} 1/\sqrt{2} \\ 1/\sqrt{2} \end{pmatrix} = b \begin{pmatrix} 1/\sqrt{2} \\ 1/\sqrt{2} \end{pmatrix},$$

$$\begin{pmatrix} a+b & -a \\ -a & a+b \end{pmatrix} \begin{pmatrix} 1/\sqrt{2} \\ -1/\sqrt{2} \end{pmatrix} = (2a+b) \begin{pmatrix} 1/\sqrt{2} \\ -1/\sqrt{2} \end{pmatrix}.$$

This may be written abstractly as

$$\mathbf{a}|e_+\rangle = \alpha_+|e_+\rangle, \qquad \mathbf{a}|e_-\rangle = \alpha_-|e_-\rangle,$$

where the eigenvalues are

$$\alpha_+ = b, \qquad \alpha_- = 2a + b,$$
$$= g/l, \qquad = 2k/ml^2 + g/l.$$

The eigenvalues may change if the parameters k, m, or l vary, but the eigenvectors are fixed by the symmetry.

This shows one of the main ideas of symmetry analysis. Easily reducible symmetry operators will help reduce more complicated operators which commute with the symmetry operators. One starts with base states

$$|1\rangle = \mathbf{1}|1\rangle, \qquad |2\rangle = R|1\rangle$$

defined by the action of C_2 group operators $\mathbf{1}$ and R on the first state $|1\rangle$. This basis is convenient for deriving the equation of motion ($|\ddot{x}\rangle = -\mathbf{a}|x\rangle$). Then the idempotents $[P^+ = (\mathbf{1} + R)/2]$ and $[P^- = (\mathbf{1} - R)/2]$ are applied to the first state to give new base states

$$|e_+\rangle = P^+|1\rangle\sqrt{2} = (\mathbf{1} + R)|1\rangle/\sqrt{2}, \qquad |e_-\rangle = P^-|1\rangle\sqrt{2} = (\mathbf{1} - R)|1\rangle/\sqrt{2}$$
$$= (|1\rangle + |2\rangle)/\sqrt{2} \qquad\qquad\qquad = (|1\rangle - |2\rangle)/\sqrt{2}.$$

This basis is convenient for *solving* the equation of motion because off-diagonal components vanish:

$$\langle e_+|\mathbf{a}|e_-\rangle = \langle 1|P^+\mathbf{a}P^-|1\rangle$$
$$= \langle 1|\mathbf{a}P^+P^-|1\rangle = 0. \qquad (2.3.5)$$

Here the first definition of symmetry is used ($R\mathbf{a} = \mathbf{a}R$ implies $P^+\mathbf{a} = \mathbf{a}P^+$) along with orthogonality ($P^+P^- = 0$). Hence $|e_+\rangle$ and $|e_-\rangle$ must be eigenvectors of $\mathbf{a}$ as indeed they are according to the representation in Eqs. (2.3.5). This is the main idea behind the applications of group theory.

However, one should note that this whole theory beginning with Eq. (2.3.1) is really outside of the area known by mathematicians as group theory. As soon as linear combinations of group operators, viz., ($\frac{1}{2}\mathbf{1} - \frac{1}{2}R$), $17R$, or 0 are considered one obtains a group ALGEBRA or RING. Elements of a group algebra satisfy the rules for a vector space (see Appendix A) whose dimension is the order of the group.

To complete the problem one writes the equations of motion in the $\{|e_+\rangle, |e_-\rangle\}$ basis. The $\langle e_+|$ component of the motion is determined by

$$\langle e_+|\ddot{x}\rangle = -\langle e_+|\mathbf{a}|x\rangle$$
$$= -\langle e_+|\mathbf{a}|e_+\rangle\langle e_+|x\rangle - \langle e_+|\mathbf{a}|e_-\rangle\langle e_-|x\rangle,$$
$$\langle e_+|\ddot{x}\rangle = -\alpha_+\langle e_+|x\rangle. \tag{2.3.6}$$

This is an equation for an amplitude oscillating with angular frequency $\omega_+ = \sqrt{\alpha_+}$. The solution is

$$\langle e_+|x(t)\rangle = A_+\cos(\omega_+ t + B_+), \tag{2.3.7a}$$

where the constants A_+ and B_+ depend on initial conditions. Similarly, the $\langle e_-|$ amplitude oscillates according to

$$\langle e_-|x(t)\rangle = A_-\cos(\omega_- t + B_-), \tag{2.3.7b}$$

with generally higher frequency $\omega_- = \sqrt{\alpha_-}$. The general solution is a combination of these obtained by using completeness:

$$|x(t)\rangle = |e_+\rangle\langle e_+|x(t)\rangle \qquad\qquad + |e_-\rangle\langle e_-|x(t)\rangle,$$
$$\begin{pmatrix}x_1(t)\\x_2(t)\end{pmatrix} = \begin{pmatrix}1/\sqrt{2}\\1/\sqrt{2}\end{pmatrix}A_+\cos(\omega_+ t + B_+) + \begin{pmatrix}1/\sqrt{2}\\-1/\sqrt{2}\end{pmatrix}A_-\cos(\omega_- t + B_-).$$
$$\tag{2.3.8}$$

The two terms $|e_+\rangle$ and $|e_-\rangle$ denote the familiar ELEMENTARY RESONANCES or NORMAL MODES of the system pictured in Figure 2.3.1. The figure also shows elementary examples of SPECTRA with two "lines." Two peaks or lines appear in plots of the response of the system to a harmonic driving force of frequency ω for three different values of the coupling constant k. A particular resonance $|e_j\rangle$ will be excited whenever ω^2 comes close to its eigenvalue α_j. (See Problem 2.3.1.)

These double pendulums have been sold in novelty shops from time to time. They are capable of performing a "beat trading" motion that can be quite slow for low values of k. By setting initial conditions $\dot{x}_1(0) = 0 = \dot{x}_2(0) = x_2(0)$ and $x_1(0) = 1$ (i.e., by selecting equal amounts of modes $|e_+\rangle$ and $|e_-\rangle$; $A^+ = 1/\sqrt{2} = A^-$ and $B^+ = 0 = B^-$), we get the alternating suppression or "beats" of activity for one pendulum and then the other, as depicted in Figure 2.3.2(a). The pendulums trade beats at a frequency equal to the difference between the eigenfrequencies [$\omega(\text{beat}) = \sqrt{\alpha_-} - \sqrt{\alpha_+}$].

A perfect beating is only possible if C_2 symmetry is present. The last point we would like to make here is that plain physical appearance does not always indicate the physical symmetry. For example, the Wilberforce pendulum in

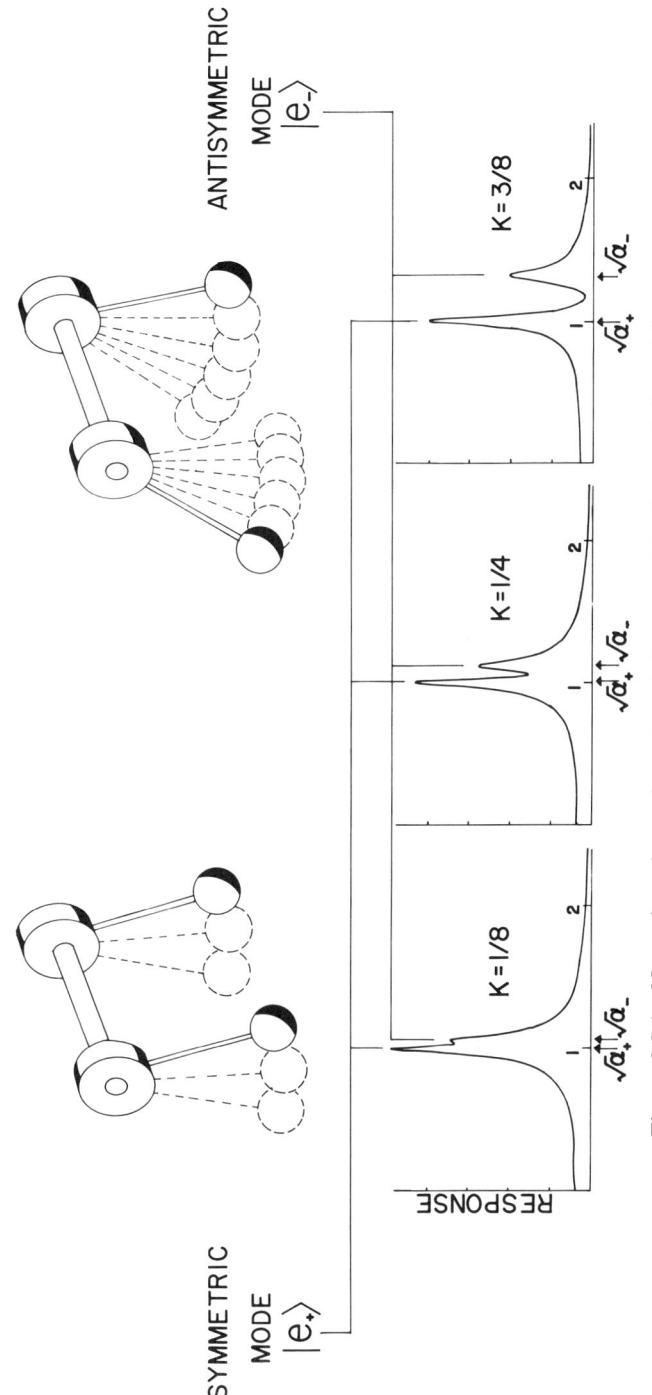

Figure 2.3.1 Normal modes and spectra of C_2 symmetrically coupled pendulums.

74 BASIC THEORY AND APPLICATIONS OF SYMMETRY REPRESENTATIONS

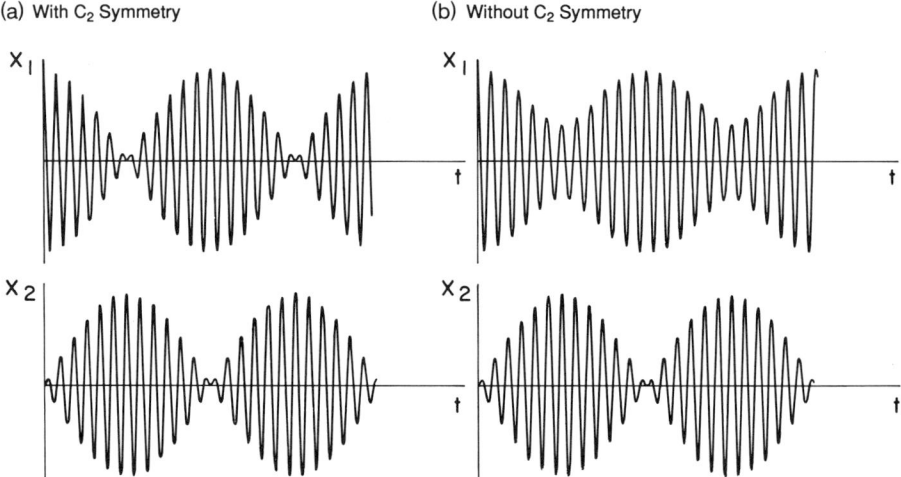

Figure 2.3.2 Beating results when two eigenmodes are excited simultaneously. Perfect transfer of motion between coordinates x_1 and x_2 occurs only if C_2 symmetry is present. Transfer occurs with a beat frequency equal to the difference of the eigenfrequencies.

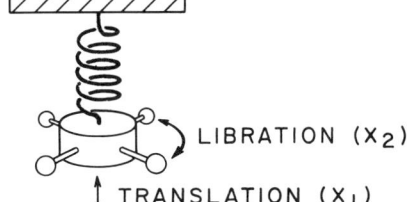

Figure 2.3.3 Wilberforce pendulum. A system can have a C_2 physical symmetry even if its geometrical form is asymmetrical.

Figure 2.3.3 certainly does not "look" C_2 symmetric, but it will execute its perfect trade of beats only if its parameters are adjusted so that it has C_2 physical symmetry.

Having used symmetry projection to help solve coupled oscillator problems, one should compare other methods for studying such problems. Throughout this book we will try to show the advantages of using a variety of approaches to help one gain a solid physical picture of the systems being studied. Also, since the coupled oscillator is such an important paradigm for many physical systems, we should pause here to consider some alternative views of it.

(a) The Configuration Space View: Lissajous Figures One may recast the problem of two equivalent one-dimensional oscillators into the identical problem of a single two-dimensional oscillator. If one plots one oscillator

coordinate versus the other (i.e., x_1 versus x_2) the resulting two-dimensional space is called configuration space. However, x_1 and x_2 could just as well be coordinates x and y, respectively, of a single oscillating mass in ordinary space.

As explained in Section 1.4.B forces are obtained from the derivatives of the potential function of the form

$$V(x_1, x_2) = \tfrac{1}{2}(\mathscr{F}_{11}x_1^2 + 2\mathscr{F}_{12}x_1x_2 + \mathscr{F}_{22}x_2^2). \tag{2.3.9}$$

The curves $V(x_1, x_2)$ = constant are tipped ellipses, and some examples are drawn next to the potential surface in Figure 2.3.4. They are the equipotential or level lines for the two-dimensional valley. One can imagine that they are the topography lines for the "Bare Valley" ski resort which has the most gentle beginner slope running NE to SW along the long axes of the ellipses, and the steepest path running NW to SE along the short axes. If the potential has C_2 symmetry then these axes are exactly at $+45°$ and $-45°$, respectively, to the x_1 axis.

Using the potential map one can see that a particle starting out on one or the other of the elliptical axes will oscillate back and forth on that axis forever. This motion is indicated in the right-hand and left-hand parts of Figure 2.3.5. The major and minor axes correspond to the symmetric low-frequency $(+)$ mode and the antisymmetric high-frequency $(-)$ mode, respectively, in Figure 2.3.1. If a particle starts out in between these eigenaxes, say on the x_1 axis as shown in the central figure, then its trajectory will be a curve which is called a Lissajous figure. Some examples of Lissajous curves are given in Figure 2.3.6. Let us consider first the case in Figure 2.3.6a for which the equipotential lines would be nearly circular, and the coupling constant (a) in Eq. (2.2.1c) or force $\mathscr{F}_{12}$ in (2.3.9) is very small. This will

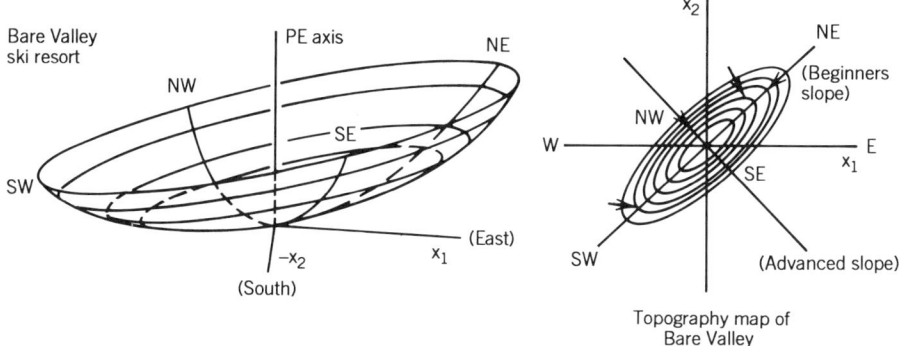

Figure 2.3.4 Topography plot of two-dimensional oscillator potential with C_2 symmetry.

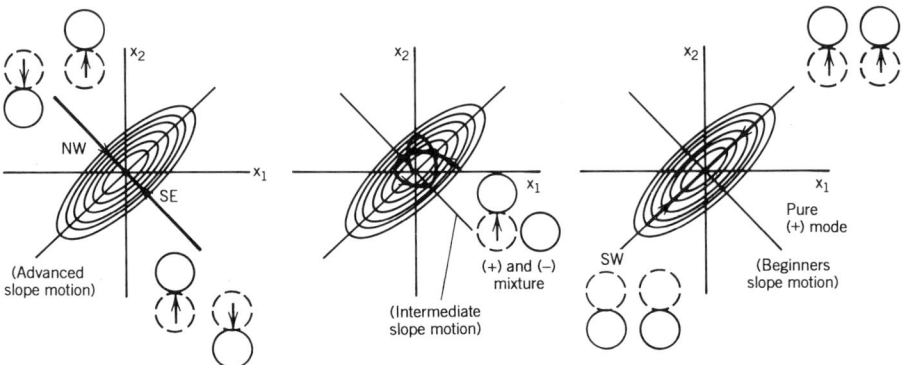

Figure 2.3.5 Motions corresponding to pure and mixed models.

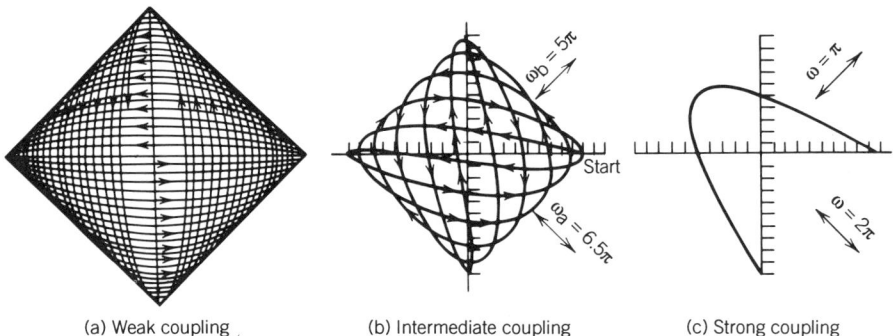

Figure 2.3.6 Lissajous plots of mixed mode motions.

provide an example of the very important phenomenon of *resonance*, whereby the effects of small forces can be greatly amplified.

A particle which starts at rest on the x_1 axis is deflected slightly upward toward the positive x_2 axis by the gradient as it falls from the right to the left. It subsequently follows a counterclockwise path which resembles an ellipse whose x_2 amplitude or minor axis increases slightly each period. Meanwhile, the major axis or x_1 amplitude decreases until the two are about the same and the trajectory encloses a maximum area which is nearly circular. From the equation of motion it follows that the x_2 oscillation is experiencing a coupling force which equals ax_1. One may consider the quantities x_2 and x_1 as distance and applied force, respectively, for the x_2 oscillator, and vice versa for x_1. Hence, the area enclosed by the Lissajous curve is a direct measure of the work done on the x_2 coordinate by the x_1 motion, where the work is

$$\int F_2 (\text{by } 1) \, dx_2 = a \int x_1 \, dx_2. \tag{2.3.10}$$

The rate of energy transfer is greatest when the Lissajous curve is most nearly a circle and the phase of the driven oscillator is nearly 90° behind the driver. This is a resonance condition, and is discussed at length in Section 6.5.

Eventually x_1 becomes exhausted while x_2 reaches its maximum amplitude. At this point x_1 falls behind x_2 in phase by nearly 90°, so that x_2 becomes the driver and x_1 the driven. The Lissajous curve now is orbiting around in a clockwise or negative sense until x_2 becomes exhausted and the curve passes near its starting point on the x_1 axis. This marks the end of one beat period in Figure 2.3.2(a).

Beat periods become shorter for stronger coupling, and the difference between eigenfrequencies becomes greater. An example of larger coupling is shown in Figure 2.3.6(b), and a more extreme example is shown in Figure 2.3.6(c) (for which $\omega_+ = 2\pi$ and $\omega_- = \pi$) and there a complete beating takes place in exactly two seconds. In each of the latter two examples the ratio of the two eigenfrequencies is exactly a ratio of two integers. When the ratio of eigenfrequencies is a rational number, i.e., a ratio of relatively prime integers,

$$\omega_+/\omega_- = n_+/n_-, \tag{2.3.11}$$

then the Lissajous trajectory will be closed and must repeat perfectly after a period of time

$$t_{LJ} = 2\pi n_+/\omega_+ = 2\pi n_-/\omega_-$$
$$= n_+ t_+ \qquad = n_- t_-. \tag{2.3.12}$$

Since the beat period is given by

$$t_{BEAT} = 1/((1/t_+) - (1/t_-))$$
$$= t_+ t_-/(t_- - t_+),$$

it then follows that the Lissajous period is

$$t_{LJ} = (n_+ - n_-) t_{BEAT}. \tag{2.3.13}$$

If the frequency ratio is irrational then t_{LJ} is infinite. For the example in Figure 2.3.6(b) the ratio is $(n_+/n_- = \frac{13}{10})$, and so a complete Lissajous cycle takes three beats. The arrows in the figure represent direction of the path for the first $1(\frac{1}{2})$ beats. You should trace the motion from the extreme right-hand point on the x_1 axis to the top of the x_2 axis. After this the trajectory simply retraces its way back to the starting point. For the "most rational" example in Figure 2.3.6 the two kinds of periods are equal. In nonlinear or anharmonic vibrational problems the idea of proximity to rationality is an important one. (See Problem 2.3.3.)

(b) The Phase Space View: Phasors One may think of a harmonic oscillator as a clock and let its sweep second hand be a complex vector which represents the oscillator phase. If the complex number

$$\mathscr{E} = Ce^{-i\omega t} = C\cos(\omega t) - iC\sin(\omega t)$$

is used to represent an oscillator, then the real and imaginary parts correspond to the position and the frequency-scaled velocity (Im $\mathscr{E}$ = $-C\sin(\omega t) = v/\omega$) of the oscillator. A vector whose abscissa and ordinate are the real and imaginary part, respectively, of $\mathscr{E}$ is called a PHASOR. As time advances the phasor rotates clockwise like a second hand, and traces a circular trajectory in a rescaled phase space of the oscillator.

To use the phasor picture for the two coupled pendulums we let each coordinate x_1 and x_2 be represented by a separate phasor clock. This is one way to represent the four-dimensional phase space of the two coupled oscillators. However, one first needs to find ways to set the clocks so that they run like clocks and maintain constant frequency and amplitude. The normal modes $|e_+\rangle$ and $|e_-\rangle$ each correspond to such a setting. If the clocks are set with equal phase and amplitude this corresponds to the $|e_+\rangle$ mode in which the clocks run synchronously at frequency $\omega_+/(2\pi)$. For the $|e_-\rangle$ mode the clocks are set with opposite ($\pm 180°$) phase and they both run at frequency $\omega_-/(2\pi)$.

An arbitrary clock setting corresponds to a combination of the $|e_+\rangle$ and $|e_-\rangle$ modes. The first frame on the left-hand side of Figure 2.3.7 shows a sum of equal amounts of $|e_+\rangle$ and $|e_-\rangle$ settings at $t = 0$. As time advances the $(+)$ and $(-)$ components each turn synchronously at their respective rates, which are taken to be 0.5 and 1.0 Hz, respectively. The phasor vector sums of the $(+)$ and $(-)$ clocks are shown at $\frac{1}{4}$-second intervals at the bottom of each frame. One can see that the x_1 phasor is roughly 90° ahead of the x_2 phasor until the former vanishes at $t = 1$ sec, and this corresponds to resonant transfer of energy from x_1 to x_2. By $t = 2$ sec the x_1 coordinate will have recovered all the energy it had at the beginning of the beat period as shown in the Lissajous plot of this example, which is Figure 2.3.6(c).

The example just treated is one of very strong coupling. The energy transfer is accomplished in one or two oscillations of the coordinates. The process is more like a series of jarring collisions than a gentle but persistent persuasion of resonance. Nevertheless, in the harmonic limit for which the equations of motion are linear the description of strong coupling is the same as that of weak coupling for which the beats take a long time.

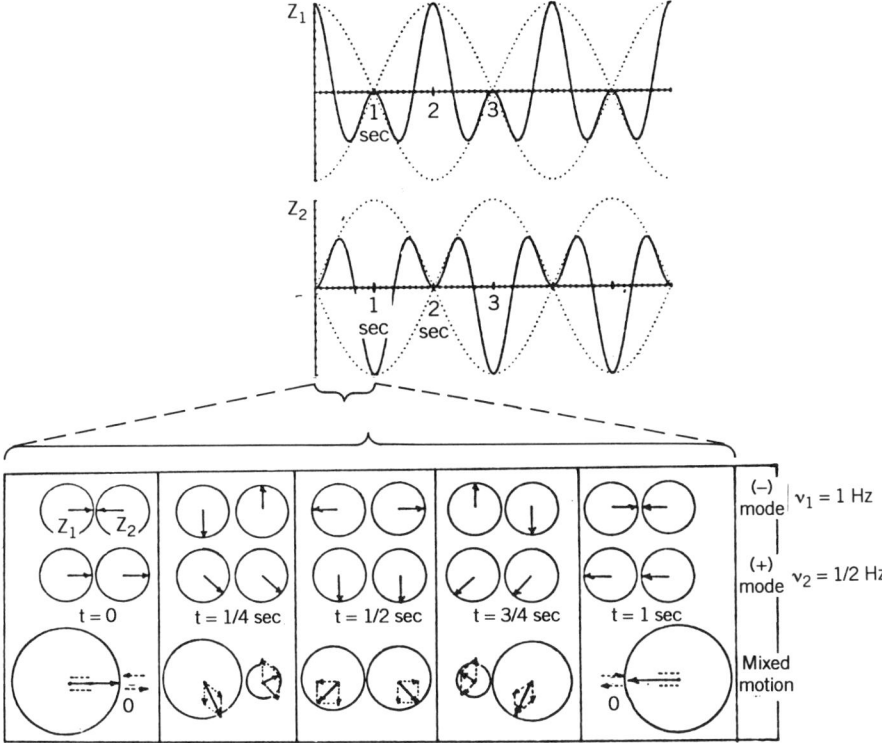

Figure 2.3.7 Phasor diagram of strongly coupled oscillation.

We shall use phasors to describe resonance and wave phenomena throughout this book. They are convenient for displaying the motions of three or more oscillators while the configuration space view can be rather limiting as the number of dimensions goes up. When discussing any pair of directly coupled phasors one should always remember that the one which is behind in phase by some amount between 0° and 180° is receiving energy on the average from the one which is ahead.

2.4 IRREDUCIBLE REPRESENTATIONS

The orthonormality of P^+ and P^- has just been used to solve an equation of motion. Now we study some of the mathematics associated with the completeness relation [Eq. (2.4.1)] and spectral decomposition [Eq. (2.4.2)], which will later help us to do more complicated problems:

$$1 = P^+ + P^-, \tag{2.4.1}$$
$$R = P^+ - P^-. \tag{2.4.2}$$

TABLE 2.4.1

	$g =$	1	R
$1 = D^+(1)P^+ + D^-(1)P^-$	$D^+(g) =$	1	1
$R = D^+(R)P^+ + D^-(R)P^-$	$D^-(g) =$	1	−1

These equations are analogous to the matrix equations (1.2.18) and (1.2.21), respectively, which were derived in the preceding review chapter. The coefficients or eigenvalues in these equations are called IRREDUCIBLE REPRESENTATIONS or CHARACTERS of the Abelian group $C_2 = \{1, R\}$. Table 2.4.1 is called a CHARACTER TABLE. Equations (2.4.1) and (2.4.2) have been rewritten just to show the notation.

In Chapter 3 we shall distinguish between irreducible representations and characters. In general the latter are the traces of the former, but for 1×1 matrices the trace of the matrix is the same as the matrix itself. Since all the elements of an Abelian group are mutually commuting unitary operators it will always be possible to simultaneously reduce the entire group to combinations of projectors multiplied by eigenvalues using the procedure in Section 1.2.B(d). (see Appendix C). So each element g of Abelian group $G = \{1, g, g', \ldots\}$ can therefore be spectrally decomposed into a sum of products of eigenvalues $D^\alpha(g)$ and idempotents P^α as follows:

$$\begin{aligned} 1 &= D^\alpha(1)P^\alpha + D^{\alpha'}(1)P^{\alpha'} + \cdots = P^\alpha + P^{\alpha'} + \cdots, \\ g &= D^\alpha(g)P^\alpha + D^{\alpha'}(g)P^{\alpha'} + \cdots, \\ g' &= D^\alpha(g')P^\alpha + D^{\alpha'}(g')P^{\alpha'} + \cdots. \end{aligned} \quad (2.4.3)$$

As shown in Section 1.2.B(b) the P^α are orthonormal ($P^\alpha P^{\alpha'} = \delta^{\alpha\alpha'}P^\alpha$) and complete ($1 = P^\alpha + P^{\alpha'} + \cdots$). The coefficients of such decompositions must obey the following rule: The product $D^\alpha(g)D^\alpha(g')$ of an irreducible representation α of group elements must equal the same irreducible representation $D^\alpha(gg')$ of the product gg'. We see that this follows from the properties of the idempotents as follows:

$$gg' = \left(\sum_\alpha D^\alpha(g)P^\alpha\right)\left(\sum_{\alpha'} D^{\alpha'}(g')P^{\alpha'}\right) = \sum_\alpha D^\alpha(g)D^\alpha(g')P^\alpha,$$

by using the definition

$$gg' = \sum_\alpha D^\alpha(gg')P^\alpha,$$

whence

$$D^\alpha(g)D^\alpha(g') = D^\alpha(gg'). \quad (2.4.4)$$

IRREDUCIBLE REPRESENTATIONS

In other words, the sets of irreducible representations will "imitate" an Abelian group with just (1×1) matrices, i.e., the numbers in a character table.

But it is more important, probably, to observe that the irreducible representations of any group are a complete set of "building blocks" for any representation of that group. Every representation has to be "made" from them and only them. Suppose somebody brings to you a representation of group C_2 without telling you where it came from. For example, let us be given a representation of C_2 by matrices $\mathscr{E}(1)$ and $\mathscr{E}(R)$ defined by

$$\left\{ \mathscr{E}(1) = \begin{pmatrix} 1 & 0 & 0 \\ 0 & 1 & 0 \\ 0 & 0 & 1 \end{pmatrix} \mathscr{E}(R) = \begin{pmatrix} 0 & 0 & 1 \\ 0 & 1 & 0 \\ 1 & 0 & 0 \end{pmatrix} \right\}. \tag{2.4.5}$$

Now we check to make sure that it is a representation of our group, i.e., that $\mathscr{E}(ab) = \mathscr{E}(a)\mathscr{E}(b)$ for all a and b in the group. (Here it is enough to check that $\mathscr{E}(R^2) = \mathscr{E}(R)\mathscr{E}(R) = (1)$.) This guarantees that the completeness and orthonormality relations for the representations $\mathscr{E}(P^\alpha)$ of the idempotents must hold as well. The $\mathscr{E}(P^\alpha)$ matrices follow from Eqs. $(2.3.2)_x$ and $(2.4.5)$:

$$\mathscr{E}(P^+) = \begin{pmatrix} \frac{1}{2} & 0 & \frac{1}{2} \\ 0 & 1 & 0 \\ \frac{1}{2} & 0 & \frac{1}{2} \end{pmatrix} \quad \mathscr{E}(P^-) = \begin{pmatrix} \frac{1}{2} & 0 & -\frac{1}{2} \\ 0 & 0 & 0 \\ -\frac{1}{2} & 0 & \frac{1}{2} \end{pmatrix} \tag{2.4.6}$$

$$\mathscr{T} = \begin{pmatrix} \frac{1}{\sqrt{2}} & 0 & \frac{1}{\sqrt{2}} \\ 0 & 1 & 0 \\ \frac{1}{\sqrt{2}} & 0 & -\frac{1}{\sqrt{2}} \end{pmatrix}. \tag{2.4.7}$$

This implies that the transformation matrix made from orthonormal columns of $\mathscr{E}(P^\alpha)$ as just shown will diagonalize the representation as seen in Eq. (2.4.8):

$$\mathscr{T}^\dagger \mathscr{E}(1) \mathscr{T} = \begin{pmatrix} 1 & 0 & 0 \\ 0 & 1 & 0 \\ 0 & 0 & 1 \end{pmatrix} = \begin{pmatrix} D^+(1) & 0 & 0 \\ 0 & D^+(1) & 0 \\ 0 & 0 & D^-(1) \end{pmatrix},$$

$$\mathscr{T}^\dagger \mathscr{E}(R) \mathscr{T} = \begin{pmatrix} 1 & 0 & 0 \\ 0 & 1 & 0 \\ 0 & 0 & -1 \end{pmatrix} = \begin{pmatrix} D^+(R) & 0 & 0 \\ 0 & D^+(R) & 0 \\ 0 & 0 & D^-(R) \end{pmatrix}. \tag{2.4.8}$$

This is because each nonzero column of a representation of P^α must be an eigenvector of the representation of all symmetry operators R in an Abelian group, each with eigenvalue $D^\alpha(R)$:

$$R \cdot P^\alpha = D^\alpha(R) P^\alpha \leftrightarrow \mathcal{Q}(R)\mathcal{Q}(P^\alpha) = D^\alpha(R)\mathcal{Q}(P^\alpha). \quad (2.4.9)$$

The completeness relation guarantees that one has accounted for all possibilities. So every representation $\mathcal{Q}(g)$ of any Abelian group operator g must be reducible to a string of (1×1) irreducible representations $D^\alpha(g)$ on the diagonal. The notation for the reduction given in Eq. (2.4.8) is given in the following using the DIRECT SUM sign $\oplus$:

$$\mathcal{T}^\dagger \mathcal{Q}(g) \mathcal{T} = D^+(g) \oplus D^+(g) \oplus D^-(g). \quad (2.4.10)$$

2.5 PARTIALLY SOLVING A PROBLEM WITH SYMMETRY ANALYSIS (C_2)

It is probably a good idea now to see an "imperfect" application of symmetry analysis in order to see some of the limitations of this theory from the start. The pendulum system drawn in Figure 2.5.1 has the same symmetry $C_2 = \{1, R\}$ which we have been discussing. Operator R is defined in terms of the base kets by $R|1\rangle = |3\rangle$, $R|2\rangle = |2\rangle$, and $R|3\rangle = |1\rangle$. That is, we may

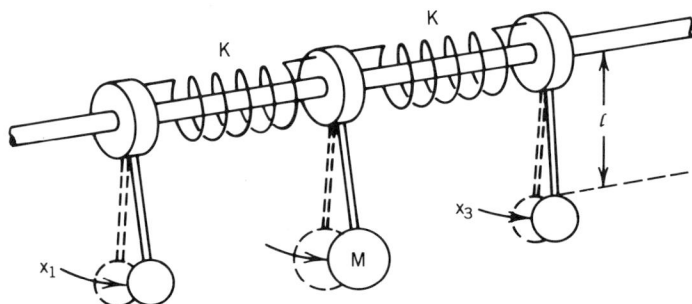

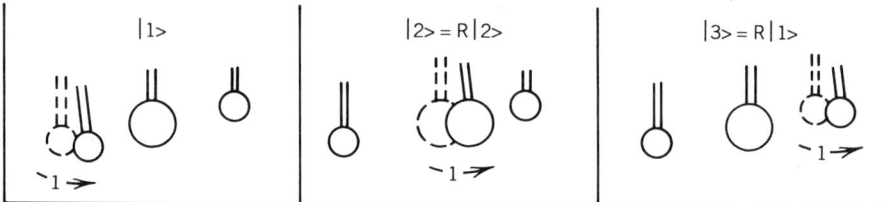

Figure 2.5.1 Coupled pendulums. This is a more complicated example of C_2 symmetry. Only two of the three bases are connected by the C_2 symmetry operator.

switch the two outside pendulums, but the middle one has to be left alone because it is different in mass and location.

The equation of motion is written out in three different forms:

$$|\ddot{x}(t)\rangle = -\mathbf{a}|x(t)\rangle, \tag{2.5.1a}$$

$$\begin{pmatrix} \langle 1|\ddot{x}(t)\rangle \\ \langle 2|\ddot{x}(t)\rangle \\ \langle 3|\ddot{x}(t)\rangle \end{pmatrix} = -\begin{pmatrix} \langle 1|a|1\rangle & \langle 1|a|2\rangle & \langle 1|a|3\rangle \\ \langle 2|a|1\rangle & \langle 2|a|2\rangle & \langle 2|a|3\rangle \\ \langle 3|a|1\rangle & \langle 3|a|2\rangle & \langle 3|a|3\rangle \end{pmatrix} \begin{pmatrix} \langle 1|x(t)\rangle \\ \langle 2|x(t)\rangle \\ \langle 3|x(t)\rangle \end{pmatrix}, \tag{2.5.1b}$$

$$\begin{pmatrix} \ddot{x}_1 \\ \ddot{x}_2 \\ \ddot{x}_3 \end{pmatrix} = -\begin{pmatrix} a+b & -a & 0 \\ -A & 2A+b & -A \\ 0 & -a & a+b \end{pmatrix} \begin{pmatrix} x_1 \\ x_2 \\ x_3 \end{pmatrix}. \tag{2.5.1c}$$

The constants in the acceleration matrix are $a = k/ml^2$, $b = g/l$, and $A = k/Ml^2$.

The representation of the C_2 symmetry operator R in the $\{|1\rangle, |2\rangle, |3\rangle\}$ basis is (see Figure 2.5.1)

$$\begin{pmatrix} \langle 1|R|1\rangle & \langle 1|R|2\rangle & \langle 1|R|3\rangle \\ \langle 2|R|1\rangle & \langle 2|R|2\rangle & \langle 2|R|3\rangle \\ \langle 3|R|1\rangle & \langle 3|R|2\rangle & \langle 3|R|3\rangle \end{pmatrix} = \begin{pmatrix} 0 & 0 & 1 \\ 0 & 1 & 0 \\ 1 & 0 & 0 \end{pmatrix},$$

which is precisely the $\mathscr{E}$ representation treated in the preceding section. [Compare Eq. (2.4.5) with the preceding one.] There we found that a change of basis from $\{|1\rangle, |2\rangle, |3\rangle\}$ to $\{|e_+\rangle, |e'_+\rangle, |e_-\rangle,\}$, represented as

$$|e_+\rangle \to \begin{pmatrix} \langle 1|e_+\rangle \\ \langle 2|e_+\rangle \\ \langle 3|e_+\rangle \end{pmatrix} = \begin{pmatrix} 1/\sqrt{2} \\ 0 \\ 1/\sqrt{2} \end{pmatrix}, \quad |e'_+\rangle \to \begin{pmatrix} \langle 1|e'_+\rangle \\ \langle 2|e'_+\rangle \\ \langle 3|e'_+\rangle \end{pmatrix} = \begin{pmatrix} 0 \\ 1 \\ 0 \end{pmatrix},$$

$$|e_-\rangle \to \begin{pmatrix} \langle 1|e_-\rangle \\ \langle 2|e_-\rangle \\ \langle 3|e_-\rangle \end{pmatrix} \begin{pmatrix} 1/\sqrt{2} \\ 0 \\ -1/\sqrt{2} \end{pmatrix}, \tag{2.5.2}$$

caused the $\mathscr{E}$ representation to assume a reduced or diagonalized form. [Recall Eq. (2.4.8).] Applying the same change of basis to the acceleration matrix in Eq. (2.5.1) we see that a partial, but not total, reduction of it occurs:

$$\mathscr{T}^\dagger a \mathscr{T} = \begin{pmatrix} a+b & -\sqrt{2}\,a & 0 \\ -\sqrt{2}\,A & A+b & 0 \\ 0 & 0 & a+b \end{pmatrix}. \tag{2.5.3}$$

84 BASIC THEORY AND APPLICATIONS OF SYMMETRY REPRESENTATIONS

Symmetry analysis guarantees that components like $\langle e_+|\mathbf{a}|e_-\rangle$, or $\langle e_-|\mathbf{a}|e_+\rangle$, etc., are zero, but it cannot guarantee that a coupling like $\langle e'_+|\mathbf{a}|e_+\rangle$ will go away. In fact, it does not—it equals $-\sqrt{2}\,a$ in Eq. (2.5.3).

So for this problem the $(-)$ eigenvector is fixed, but the $(+)$ eigenvectors still remain to be solved, and the results will depend upon the values of the constants g, k, 1, m, and M. Symmetry analysis with the group C_2 can do no more than separate the $(+)$ type modes from the $(-)$ types.

In a general symmetry analysis you will determine the REPETITION FREQUENCY f^α or the number of times each irreducible representation D^α appears in the reduction of the physical group representation. This tells you how much work is left after the symmetry analysis is completed: Each $f^\alpha \times f^\alpha$ matrix may need to be reduced. The final reduction is completed using standard techniques reviewed in Chapter 1, or may be accomplished numerically on a computer. (Or, perhaps, you may find a higher symmetry!)

2.6 AN EXAMPLE WITH SLIGHTLY HIGHER SYMMETRY (C_3)

If the three pendulums are coupled in a more symmetric way, as in Figure 2.6.1, it will be possible once again to deduce the complete solutions to the equation of motion

$$\begin{pmatrix}\ddot{x}_1\\ \ddot{x}_2\\ \ddot{x}_3\end{pmatrix} = -\begin{pmatrix}2a+b & -a & -a\\ -a & 2a+b & -a\\ -a & -a & 2a+b\end{pmatrix}\begin{pmatrix}x_1\\ x_2\\ x_3\end{pmatrix} \qquad (2.6.1)$$

using symmetry analysis techniques. The constants in the equation are $b = g/l$ and $a = s/ml^2$, where s is the coupling spring constant and g, l, and m are gravity, length, and mass constants, respectively. The basic coordinates x_1, x_2, and x_3 are defined by Figure 2.6.1. Symmetry operators include the cyclic exchange or 120° rotation operator labeled r which transforms base states as follows:

$$r|1\rangle = |2\rangle, \qquad r|2\rangle = |3\rangle, \qquad r|3\rangle = |1\rangle,$$

the double exchange operator r^2 which does the transformations,

$$r^2|1\rangle = |3\rangle, \qquad r^2|2\rangle = |1\rangle, \qquad r^2|3\rangle = |2\rangle,$$

and the identity **1**, which changes nothing. ($\mathbf{1}|i\rangle = |i\rangle$). There are some other symmetry operators such as "reflections" σ_{12}, σ_{23}, and σ_{31}, where, for example,

$$\sigma_{12}|1\rangle = |2\rangle, \qquad \sigma_{12}|2\rangle = |1\rangle, \qquad \sigma_{12}|3\rangle = |3\rangle.$$

However, let us put off discussing these until Chapter 3.

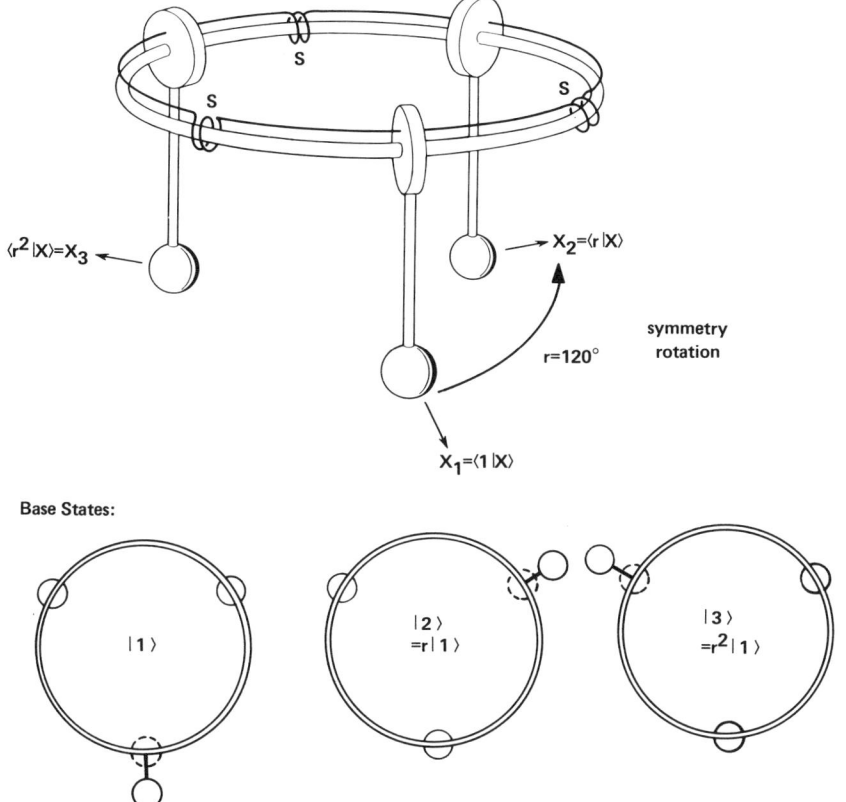

Figure 2.6.1 C_3 symmetric coupled pendulums.

The three operators $\{1, r, r^2\}$ form a group called C_3. The multiplication table is

$$\begin{array}{c|ccc} & 1 & r & r^2 \\ \hline 1 & 1 & r & r^2 \\ r & r & r^2 & 1 \\ r^2 & r^2 & 1 & r \end{array}. \qquad (2.6.2)$$

Now one can use the minimal equations ($r^3 = 1$, or $r^3 - 1 = 0$) to produce the idempotents associated with the group elements. The minimal equation for r is factored into the three roots of unity:

$$0 = (r^3 - 1) = (r - \varepsilon_1 \mathbf{1})(r - \varepsilon_2 \mathbf{1})(r - \varepsilon_3 \mathbf{1}),$$

where

$$\varepsilon_1 \equiv 1, \quad \varepsilon_2 \equiv e^{2\pi i/3}, \quad \varepsilon_3 \equiv e^{-2\pi i/3}.$$

Using Eq. (2.3.2) we obtain three idempotents P^α:

$$P^1 = \frac{(r - \varepsilon_2 \mathbf{1})(r - \varepsilon_3 \mathbf{1})}{(\varepsilon_1 - \varepsilon_2)(\varepsilon_1 - \varepsilon_3)} = \frac{1}{3}(\mathbf{1} + r + r^2),$$

$$P^2 = \frac{(r - \varepsilon_1 \mathbf{1})(r - \varepsilon_3 \mathbf{1})}{(\varepsilon_2 - \varepsilon_1)(\varepsilon_2 - \varepsilon_3)} = \frac{1}{3}(\mathbf{1} + \varepsilon_3 r + \varepsilon_2 r^2),$$

$$P^3 = \frac{(r - \varepsilon_1 \mathbf{1})(r - \varepsilon_2 \mathbf{1})}{(\varepsilon_3 - \varepsilon_1)(\varepsilon_3 - \varepsilon_2)} = \frac{1}{3}(\mathbf{1} + \varepsilon_2 r + \varepsilon_3 r^2). \quad (2.6.3)$$

The inverses of Eq. (2.6.3) are the completeness and spectral decomposition relations in the following. These follow from the theory given in Sections 1.2.B(b) and (c).

$$\mathbf{1} = P^1 + P^2 + P^3 \equiv D^1(1)P^1 + D^2(1)P^2 + D^3(1)P^3,$$

$$r = P^1 + \varepsilon_2 P^2 + \varepsilon_3 P^3 \equiv D^1(r)P^1 + D^2(r)P^2 + D^3(r)P^3,$$

$$r^2 = P^1 + \varepsilon_3 P^2 + \varepsilon_2 P^3 \equiv D^1(r^2)P^1 + D^2(r^2)P^2 + D^3(r^2)P^3. \quad (2.6.4)$$

The eigenvalues $D^\alpha(g)$ in the decompositions above are the irreducible representations (we abbreviate this "irrep," henceforward) of C_3. There are three kinds of irreps as tabulated in the following using Eq. (2.6.4). Two other standard notations for the irreps of C_3 are shown on the left-hand side of the character table (see a "phasor" version of this table at the top of Figure 2.6.2):

	$g =$	1	r	r^2
$D^{0_3}(g) = D^A(g) = D^1(g) =$		1	1	1
$D^{1_3}(g) = D^\varepsilon(g) = D^2(g) =$		1	$e^{2\pi i/3}$	$e^{-2\pi i/3}$
$D^{2_3}(g) = D^{\varepsilon^*}(g) = D^3(g) =$		1	$e^{-2\pi i/3}$	$e^{2\pi i/3}$

$$(2.6.5)$$

The C_3 example shows many of the properties of general cyclic C_n groups $\{1, r, r^2, \ldots, r^{n-1}\}$. In general there will be n distinct roots to the minimal equation $r^n = 1$ of C_n. These n roots will be labeled as in (2.6.5) by

$$\varepsilon_{k+1} = e^{i2\pi k/n} \equiv D^{k_n}(r) \qquad (k = 0, 1, 2, \ldots, n-1). \qquad (2.6.6a)$$

It is easy to show that each root yields an idempotent projection operator of the form (2.6.3) or in general

$$P^{k_n} = \frac{1}{n}\sum_g D^{k_n^*}(g)g = \frac{1}{n}\sum_{m=0}^{n-1} e^{-i2\pi km/n} r^m, \qquad (2.6.6b)$$

where the index k_n will be read as k-modulo-n in Section 2.7. One quickly verifies that (2.6.3) and (2.6.6b) satisfy the eigenequation $rP^{k_n} = D^{k_n}(r)P^{k_n}$. The spectral decomposition (2.6.3) has the general form

$$g = \sum_{k=0}^{n-1} D^{k_n}(g) P^{k_n}, \qquad (2.6.6c)$$

where the sum is over the (n) irreps $D^{k_n}(g)$ of C_n.

The application of the C_3 idempotents proceeds in the same manner as was done before in Sections 2.3 and 2.5. We first construct a representation $\{\mathscr{R}(1), \mathscr{R}(r), \mathscr{R}(r^2)\}$ in the following of the symmetry operators $\{1, r, r^2\}$ from the definition of basic coordinates shown in Figure 2.6.1:

$$\mathscr{R}(1)\begin{pmatrix} x_1 \\ x_2 \\ x_3 \end{pmatrix} = \begin{pmatrix} 1 & 0 & 0 \\ 0 & 1 & 0 \\ 0 & 0 & 1 \end{pmatrix}\begin{pmatrix} x_1 \\ x_2 \\ x_3 \end{pmatrix} = \begin{pmatrix} x_1 \\ x_2 \\ x_3 \end{pmatrix},$$

$$\mathscr{R}(r)\begin{pmatrix} x_1 \\ x_2 \\ x_3 \end{pmatrix} = \begin{pmatrix} 0 & 0 & 1 \\ 1 & 0 & 0 \\ 0 & 1 & 0 \end{pmatrix}\begin{pmatrix} x_1 \\ x_2 \\ x_3 \end{pmatrix} = \begin{pmatrix} x_3 \\ x_1 \\ x_2 \end{pmatrix},$$

$$\mathscr{R}(r^2)\begin{pmatrix} x_1 \\ x_2 \\ x_3 \end{pmatrix} = \begin{pmatrix} 0 & 1 & 0 \\ 0 & 0 & 1 \\ 1 & 0 & 0 \end{pmatrix}\begin{pmatrix} x_1 \\ x_2 \\ x_3 \end{pmatrix} = \begin{pmatrix} x_2 \\ x_3 \\ x_1 \end{pmatrix}. \qquad (2.6.7)$$

Then we construct and use the representations $\mathscr{R}(P^\alpha)$ of the idempotents:

$$\mathscr{R}(P^1) = \tfrac{1}{3}\begin{pmatrix} 1 & 1 & 1 \\ 1 & 1 & 1 \\ 1 & 1 & 1 \end{pmatrix}, \qquad \mathscr{R}(P^2) = \tfrac{1}{3}\begin{pmatrix} 1 & \varepsilon_2 & \varepsilon_3 \\ \varepsilon_3 & 1 & \varepsilon_2 \\ \varepsilon_2 & \varepsilon_3 & 1 \end{pmatrix},$$

$$\mathscr{R}(P^3) = \tfrac{1}{3}\begin{pmatrix} 1 & \varepsilon_3 & \varepsilon_2 \\ \varepsilon_2 & 1 & \varepsilon_3 \\ \varepsilon_3 & \varepsilon_2 & 1 \end{pmatrix}. \qquad (2.6.8)$$

From select columns and rows of (P^α), namely, the first of each in this case, we construct a matrix transformation $\mathcal{T}^\dagger$ that does the following: (a) $\mathcal{T}^\dagger$ reduces all $\mathcal{R}(g)$ to a direct sum of irreps. For example, $\mathcal{T}^\dagger \mathcal{R}(r) \mathcal{T}$ is given by

$$\begin{pmatrix} 1/\sqrt{3} & 1/\sqrt{3} & 1/\sqrt{3} \\ 1/\sqrt{3} & e^{+2\pi i/3}/\sqrt{3} & e^{-2\pi i/3}/\sqrt{3} \\ 1/\sqrt{3} & e^{-2\pi i/3}/\sqrt{3} & e^{+2\pi i/3}/\sqrt{3} \end{pmatrix} \begin{pmatrix} 0 & 0 & 1 \\ 1 & 0 & 0 \\ 0 & 1 & 0 \end{pmatrix} \begin{matrix} |e_1\rangle & |e_2\rangle & |e_3\rangle \\ \begin{pmatrix} 1/\sqrt{3} & 1/\sqrt{3} & 1/\sqrt{3} \\ 1/\sqrt{3} & e^{-2\pi i/3}/\sqrt{3} & e^{+2\pi i/3}/\sqrt{3} \\ 1/\sqrt{3} & e^{+2\pi i/3}/\sqrt{3} & e^{-2\pi i/3}/\sqrt{3} \end{pmatrix} & \begin{matrix} |1\rangle \\ |2\rangle \\ |3\rangle \end{matrix} \end{matrix}$$

$$= \begin{pmatrix} 1 & 0 & 0 \\ 0 & e^{2\pi i/3} & 0 \\ 0 & 0 & e^{-2\pi i/3} \end{pmatrix} = \begin{pmatrix} D^1(r) & 0 & 0 \\ 0 & D^2(r) & 0 \\ 0 & 0 & D^3(r) \end{pmatrix}. \tag{2.6.9}$$

That is, $\mathcal{R}(g)$ is reduced or diagonalized to a direct sum ($\oplus$)

$$\mathcal{T}^\dagger \mathcal{R}(g) \mathcal{T} = D^1(g) \oplus D^2(g) \oplus D^3(g)$$

of three different irreps. [Compare the foregoing with Eqs. (2.4.8) and (2.4.10).] The $\mathcal{T}^\dagger$ matrix gives the change of basis between "old" $\{|1\rangle, |2\rangle, |3\rangle\}$ and "new" $\{|e_1\rangle, |e_2\rangle, |e_3\rangle\}$ bases. Note that the representation of the latter in terms of the former are given by the columns of $\mathcal{T}$. They are indicated in the upper right-hand side of Eq. (2.6.9).

(b) $\mathcal{T}^\dagger$ diagonalizes the acceleration matrix appearing in Eq. (2.6.1):

$$\mathcal{T}^\dagger \begin{pmatrix} 2a+b & -a & -a \\ -a & 2a+b & -a \\ -a & -a & 2a+b \end{pmatrix} \mathcal{T} = \begin{pmatrix} b & 0 & 0 \\ 0 & 3a+b & 0 \\ 0 & 0 & 3a+b \end{pmatrix}.$$

In other words, the kets $|e_1\rangle$, $|e_2\rangle$, and $|e_3\rangle$ represented by the columns of the $\mathcal{T}$ matrix are eigenvectors of the acceleration operator as well.

This is as it should be if the acceleration operator (**a**) commutes with C_3 symmetry operators. The three projected kets,

$$|e_1\rangle = P^1|1\rangle\sqrt{3}, \qquad |e_2\rangle = P^2|1\rangle\sqrt{3}, \qquad |e_3\rangle = P^3|1\rangle\sqrt{3},$$

belong to three different irreps as do their companion bras,

$$\langle e_1| = \langle 1|P^1\sqrt{3}, \qquad \langle e_2| = \langle 1|P^2\sqrt{3}, \qquad \langle e_3| = \langle 1|P^3\sqrt{3}.$$

[Here self conjugacy $(P^{\alpha\dagger} = P^\alpha)$ of idempotents obtained from a unitary operator is used (see Appendix C).] Since the idempotents are orthogonal the matrix $\langle e_\alpha | \mathbf{a} | e_\beta \rangle$ must be diagonal:

$$\langle e_\alpha | \mathbf{a} | e_\beta \rangle = \langle 1 | P^\alpha \mathbf{a} P^\beta | 1 \rangle$$

$$= \langle 1 | \mathbf{a} P^\alpha P^\beta | 1 \rangle = 0, \quad \text{if } \alpha \neq \beta.$$

No new concepts outside of those given in Sections 2.3–2.5 are involved here, except that you might be surprised that two of the eigenvalues (for modes 2 and 3) are identical. This degeneracy is a direct consequence of that reflection symmetry, which we ignored. We will discuss this at length in Chapter 3, but for now let us examine the form of the eigenvector solutions.

One may take the liberty of combining the complex $|e_2\rangle$ and $|e_3\rangle$ eigenvectors into their real and imaginary parts since they are degenerate:

$$|e_{\text{Re}}\rangle = \frac{|e_2\rangle + |e_3\rangle}{\sqrt{2}} \rightarrow \frac{1}{\sqrt{6}} \begin{pmatrix} 1 & + & 1 \\ e^{-2\pi i/3} & + & e^{2\pi i/3} \\ e^{2\pi i/3} & + & e^{-2\pi i/3} \end{pmatrix}$$

$$= \frac{1}{\sqrt{6}} \begin{pmatrix} 2 \\ 2\cos(2\pi/3) \\ 2\cos(2\pi/3) \end{pmatrix} = \frac{1}{\sqrt{6}} \begin{pmatrix} 2 \\ -1 \\ -1 \end{pmatrix},$$

$$|e_{\text{Im}}\rangle = \frac{|e_2\rangle - |e_3\rangle}{i\sqrt{2}} \rightarrow \frac{-i}{\sqrt{6}} \begin{pmatrix} 1 & - & 1 \\ e^{-2\pi i/3} & - & e^{2\pi i/3} \\ e^{2\pi i/3} & - & e^{-2\pi i/3} \end{pmatrix}$$

$$= \frac{1}{\sqrt{6}} \begin{pmatrix} 0 \\ -2\sin(2\pi/3) \\ 2\sin(2\pi/3) \end{pmatrix} = \frac{1}{\sqrt{2}} \begin{pmatrix} 0 \\ -1 \\ 1 \end{pmatrix}.$$

Because of the frequency degeneracy between modes $|e_2\rangle$ and $|e_3\rangle$ the resulting real combinations are eigenvectors, too, both having eigenfrequency $\sqrt{3a+b}$. Real eigenvectors are easier to picture, as shown in Figure 2.6.2. For eigenvectors with real components, the only relative phases possible are + (in phase) and − (out of phase). Therefore real vectors correspond to so-called STANDING WAVES or modes. Complex vectors allow for arbitrary phases, so the disturbance can appear to be marching from mass to mass. Therefore, complex vectors correspond to MOVING WAVES. One way to visualize moving waves is to draw phasor clocks for each oscillator (see Figure 2.6.2) and image successive time steps. To algebraically determine time behavior of mode $|e_2\rangle$ one needs the quantity $\text{Re}[e^{-i\omega_2 t}|e_2\rangle]$, where

90 BASIC THEORY AND APPLICATIONS OF SYMMETRY REPRESENTATIONS

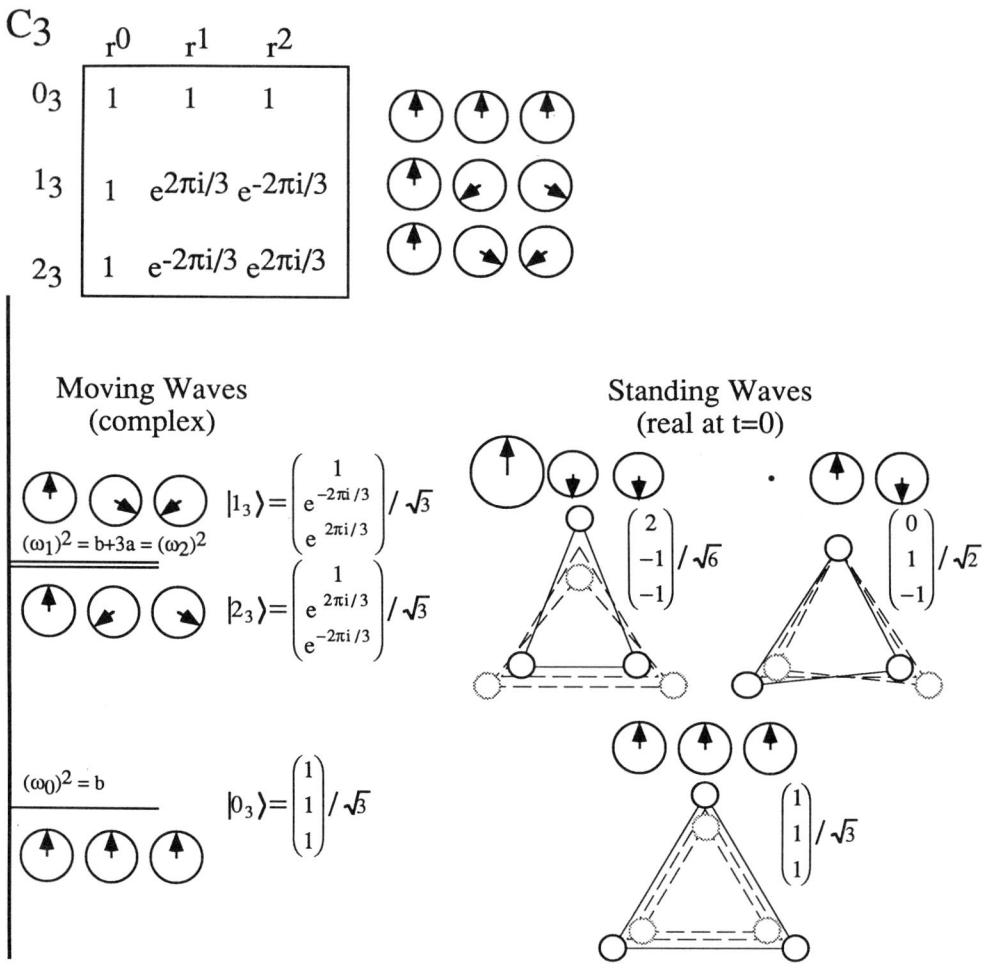

Figure 2.6.2 Normal modes or eigenwaves and spectrum of the C_3 symmetric coupled pendulums.

$\omega_2 = \omega_3 = \sqrt{3a + b}$ is the eigenfrequency. This is represented by the following:

$$\frac{1}{\sqrt{3}}\begin{pmatrix} \cos \omega_2 t \\ \cos(-2\pi/3 - \omega_2 t) \\ \cos(2\pi/3 - \omega_2 t) \end{pmatrix} = \frac{\cos \omega_2 t}{2\sqrt{3}}\begin{pmatrix} 2 \\ -1 \\ -1 \end{pmatrix} + \frac{\sin \omega_2 t}{2}\begin{pmatrix} 0 \\ -1 \\ 1 \end{pmatrix}.$$

This may be written as

$$\text{Re}\left[e^{-i\omega_2 t}|e_2\rangle\right] = (\cos \omega_2 t |e_{\text{Re}}\rangle + \sin \omega_2 t |e_{\text{Im}}\rangle)/\sqrt{2}.$$

This represents a clockwise rotation of the vibrational wave with angular eigenfrequency $\omega_2 = \sqrt{3a+b}$. Similarly the motion of the $|e_3\rangle$ mode is given by

$$\text{Re}\left[e^{-i\omega_3 t}|e_3\rangle\right] = (\cos\omega_3 t|e_{\text{Re}}\rangle - \sin\omega_3 t|e_{\text{Im}}\rangle)/\sqrt{2},$$

which is a *counter*clockwise moving wave of the same frequency.

The low-frequency ($\omega_1 = \sqrt{b}$) modes $|e_1\rangle$ has its time behavior represented by $\text{Re}[e^{-i\omega_1 t}|e_1\rangle]$ or

$$\text{Re}\begin{pmatrix} e^{-i\omega_1 t} \\ e^{-i\omega_1 t} \\ e^{-i\omega_1 t} \end{pmatrix} \Big/ \sqrt{3} = \frac{\cos\omega_1 t}{\sqrt{3}}\begin{pmatrix} 1 \\ 1 \\ 1 \end{pmatrix}.$$

This is a standing wave like $|e_{\text{Re}}\rangle$ and $|e_{\text{Im}}\rangle$. In standing waves no energy is being transferred from one pendulum to the other, and there is no net circulation of energy or momentum around the ring. However, the *combination* of two standing waves will cause very obvious interpendulum motion. For example, the combination of $|e_1\rangle$ and $\sqrt{2}|e_{\text{Re}}\rangle$ moves according to $\text{Re}[e^{-i\omega_1 t}|e_1\rangle + e^{-i\omega_2 t}\sqrt{2}|e_{\text{Re}}\rangle]$, which is represented by

$$\text{Re}\begin{pmatrix} e^{-i\omega_1 t} + 2e^{-i\omega_2 t} \\ e^{-i\omega_1 t} - e^{-i\omega_2 t} \\ e^{-i\omega_1 t} - e^{-i\omega_2 t} \end{pmatrix} \Big/ \sqrt{3} = \begin{pmatrix} \cos\omega_1 t + 2\cos\omega_2 t \\ \cos\omega_1 t - \cos\omega_2 t \\ \cos\omega_1 t - \cos\omega_2 t \end{pmatrix} \Big/ \sqrt{3}.$$

Here the time behavior is characterized by a motion or "beating" which goes back and forth with a frequency equal to the *difference* between ω_1 and ω_2. It will be analogous to the beat trading shown in Figure 2.3.2.

2.7 MORE EXAMPLES WITH C_n SYMMETRY: ONE-DIMENSIONAL LATTICES

The graph in Figure 2.2.2 shows that there is always at least one Abelian group of any order n. This "fundamental" symmetry group is the cyclic group $C_n \equiv \{1, r, r^2, r^3, \ldots, r^{n-1}\}$. Let us now use symmetry analysis to treat examples of C_n symmetry for arbitrary n.

92 BASIC THEORY AND APPLICATIONS OF SYMMETRY REPRESENTATIONS

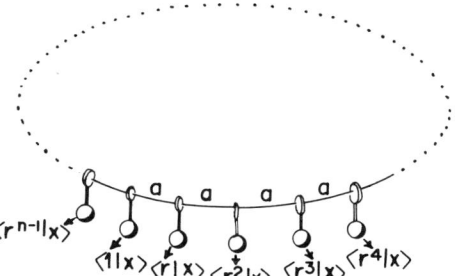

Figure 2.7.1 C_n symmetric coupled pendulums.

The first example shown in Figure 2.7.1 is a ring of n identical pendulums coupled end to end. Suppose its equation of motion is

$$-\begin{pmatrix} \langle 1|\ddot{x} \rangle \\ \langle r|\ddot{x} \rangle \\ \langle r^2|\ddot{x} \rangle \\ \vdots \\ \langle r^{n-1}|\ddot{x} \rangle \end{pmatrix}$$

$$= \begin{pmatrix} \langle 1|\mathbf{a}|1 \rangle & \langle 1|\mathbf{a}|r \rangle & \cdots & \langle 1|\mathbf{a}|r^{n-1} \rangle \\ \langle r|\mathbf{a}|1 \rangle & \langle r|\mathbf{a}|r \rangle & \cdots & \cdot \\ \langle r^2|\mathbf{a}|1 \rangle & \langle r^2|\mathbf{a}|r^2 \rangle & \cdots & \cdot \\ \vdots & & & \\ \langle r^{n-1}|\mathbf{a}|1 \rangle & \cdot & \cdot & \langle r^{n-1}|\mathbf{a}|r^{n-1} \rangle \end{pmatrix} \begin{pmatrix} \langle 1|x \rangle \\ \langle r|x \rangle \\ \langle r^2|x \rangle \\ \vdots \\ \langle r^{n-1}|x \rangle \end{pmatrix}$$

$$= \begin{pmatrix} 2a+b & -a & 0 & \cdots & 0 & -a \\ -a & 2a+b & -a & & & 0 \\ 0 & -a & 2a+b & & & 0 \\ \vdots & & & & & \\ -a & 0 & 0 & & & 2a+b \end{pmatrix} \begin{pmatrix} \langle 1|x \rangle \\ \langle r|x \rangle \\ \langle r^2|x \rangle \\ \vdots \\ \langle r^{n-1}|x \rangle \end{pmatrix},$$

(2.7.1)

where the constants a and b in the acceleration matrix are defined as they were for the preceding C_3 example. Also the base states $\{|1\rangle, |r\rangle, |r^2\rangle, \ldots\}$ and coordinates $\{\langle 1|x\rangle, \langle r|x\rangle, \langle r^2|x\rangle, \ldots\}$ are labeled by C_n operators ($|r\rangle = r|1\rangle$, etc.), as indicated in the figure.

All symmetry operators are products of the rotation r by $(2\pi/n)$ radians. This operator satisfies a minimal equation ($r^n = 1$) or ($r^n - 1 = 0$). The eigenvalues or roots of this equation are the n roots of unity (e^{ik_n}), where ($k_n = 2\pi k/n$). These complex numbers are therefore the irreps of C_n:

$$D^{k_n}(r) = e^{ik_n} \quad (k_n \equiv 2\pi k/n, \quad k = 0, 1, 2, \ldots, n-1). \quad (2.7.2)$$

MORE EXAMPLES WITH C_n SYMMETRY: ONE-DIMENSIONAL LATTICES 93

One may visualize these numbers by drawing them as vectors in the complex plane. This is done for $n = 1$–6 in Figure 2.7.2. Note that the D^{k_n} and $D^{-k_n} = (D^{k_n})^*$ are usually complex-conjugate pairs. Only $k = 0$ and $k = n/2$, for even n, give real D^{k_n}. An irrep or character table is drawn using phasors to represent each $D^{k_n}(r^p)$ irrep of $C_1, C_2, \ldots,$ and C_n in Figure 2.7.2.

For each irrep D^{k_n} there is an eigenvector $|k_n\rangle$ of the acceleration operator a which we may derive by using projection operator P^{k_n} [recall (2.6.6b)]:

$$|k_n\rangle \equiv P^{k_n}|1\rangle\sqrt{n}$$

$$= \sum_{g=1,r,\ldots} D^{k^*_n}(g)|g\rangle/\sqrt{n} \to \begin{pmatrix} 1 \\ \exp(-ik_n) \\ \exp 2(-ik_n) \\ \vdots \\ \exp(n-1)(-ik_n) \end{pmatrix} \Bigg/ \sqrt{n}. \quad (2.7.3)$$

By substituting this representation of $|k_n\rangle$ into the equation (2.7.1) of motion, one obtains the eigenvalue of operator $\langle \mathbf{a} \rangle$:

$$\mathbf{a}|k_n\rangle = [2a(1 - \cos k_n) + b]|k_n\rangle = \omega^2(k_n)|k_n\rangle. \quad (2.7.4)$$

This is the square $[\omega^2(k_n)]$ of the resonant frequency. Note that the eigenvalue for $|k_n\rangle$ is the same for its complex conjugate partner $|-k_n\rangle$. Therefore different but equally valid eigenvectors can be made by combining each complex pair to form real cosine and sine standing-wave states $|c_n^k\rangle$ and $|s_n^k\rangle$:

$$|c_n^k\rangle \equiv \frac{|k_n\rangle + |k_{-n}\rangle}{\sqrt{2}} \to \begin{pmatrix} 1 \\ \cos k_n \\ \cos 2k_n \\ \vdots \\ \cos(n-1)k_n \end{pmatrix} \sqrt{2/n}, \quad (2.7.5a)$$

$$|s_n^k\rangle \equiv \frac{-|k_n\rangle + |k_{-n}\rangle}{i\sqrt{2}} \to \begin{pmatrix} 0 \\ \sin k_n \\ \sin 2k_n \\ \vdots \\ \sin(n-1)k_n \end{pmatrix} \sqrt{2/n}. \quad (2.7.5b)$$

The real cosine or sine states may be easier to picture. We see in Figure 2.7.3 that the $|c_n^k\rangle(|s_n^k\rangle)$ state is obtained if a cosine wave (sine wave) with exactly k crests is drawn to fit in the interval occupied by n connecting springs. The sine or cosine wave amplitudes oscillate with frequency $\omega(k_n)$, but the crests and nodes are fixed in standing-wave solutions. To envision the

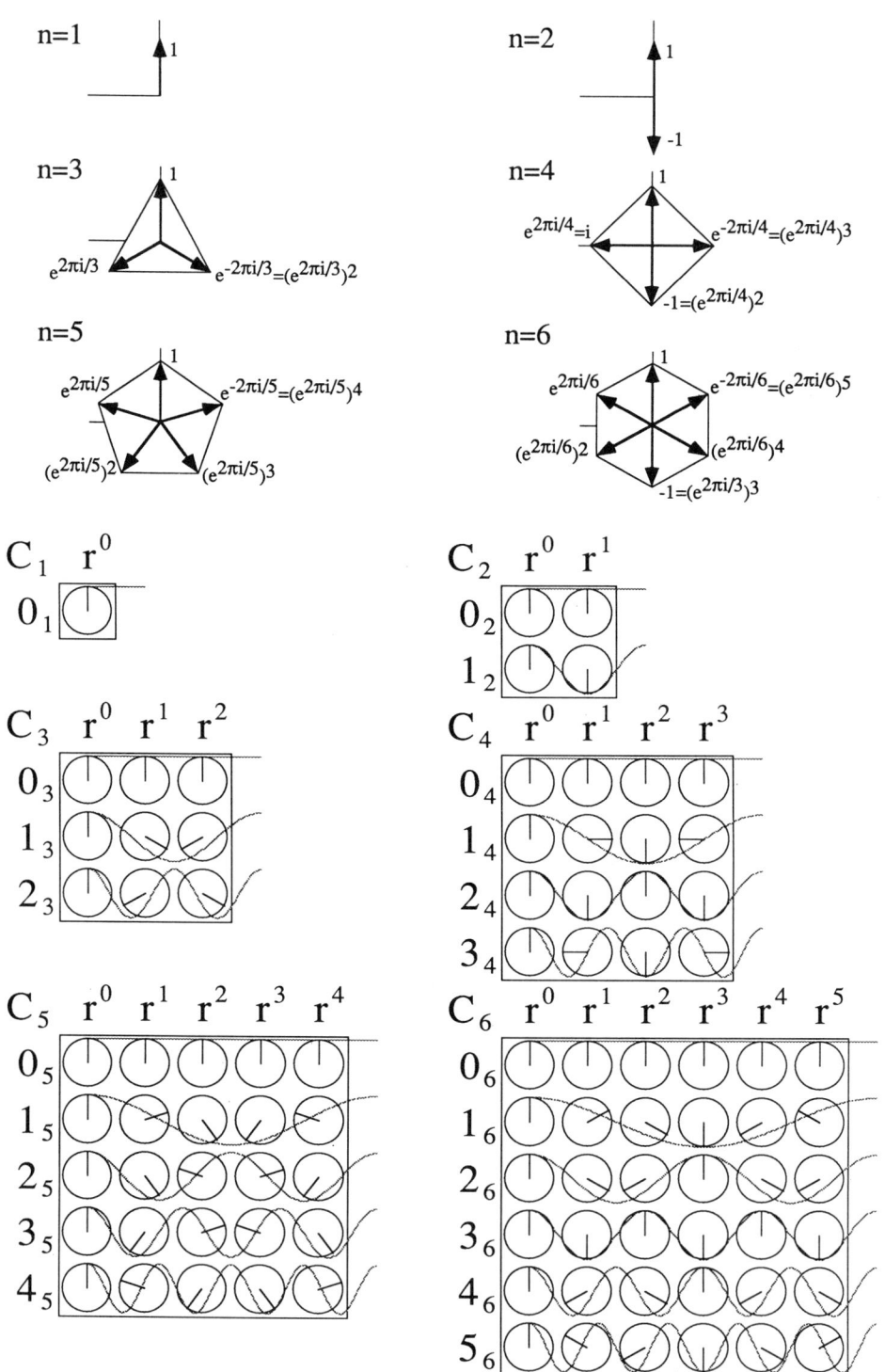

Figure 2.7.2 Representing C_n irreps by complex roots of unity ($n = 2, 3, \ldots, 6$).

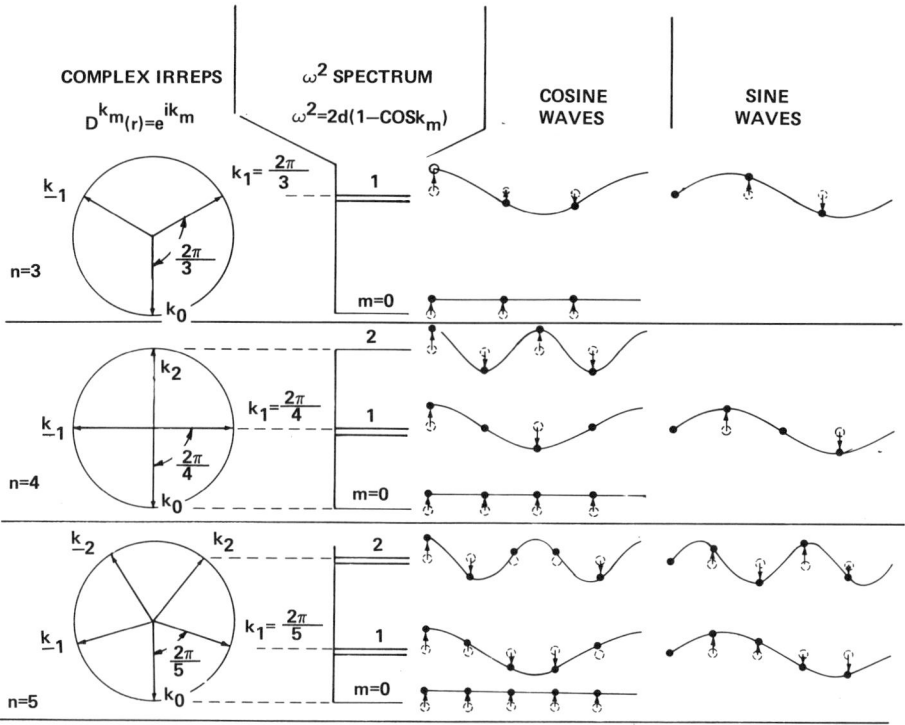

Figure 2.7.3 Standing eigenwaves and frequency spectra for C_3, C_4, C_5 symmetric coupled pendulums.

moving-wave state $|k_n\rangle$ imagine the amplitude of the cosine wave is fixed while it moves rigidly around the ring with velocity $\omega(k_n)/k_n$. (Compare Figure 2.7.3 for $n = 3$ with the C_3 solution in the preceding section.)

There is also an easy way to picture the $\omega^2(k_n)$ eigenvalue spectrum. The term $(\cos k_n)$ in Eq. (2.7.3) is the projection of kth vertex of a regular n-polygon as shown in Figure 2.7.3. (The constant term b is set equal to zero in the figure.)

As n becomes larger, the allowed points in the function $\omega^2(k_n)$ become closer together, as seen in the $n = 24$ example in Figure 2.7.4. As $n \to \infty$, the continuous band spectrum of the one-dimensional lattice is approached. Setting $b = 0$, one obtains the standard formula for angular frequency ω in terms of wave number k_n. This formula is called a lattice DISPERSION RELATION:

$$\omega(k_n) = \sqrt{2a(1 - \cos k_n)} = 2\sqrt{a}\,\sin\left(\frac{k_n}{2}\right)$$

$$= \sqrt{a}\,k_n - \frac{\sqrt{a}\,(k_n)^3}{24} + \cdots. \quad (2.7.6)$$

For small k_n this formula gives the phase velocity $c \equiv \omega/k_n = \sqrt{a}$ for long

96 BASIC THEORY AND APPLICATIONS OF SYMMETRY REPRESENTATIONS

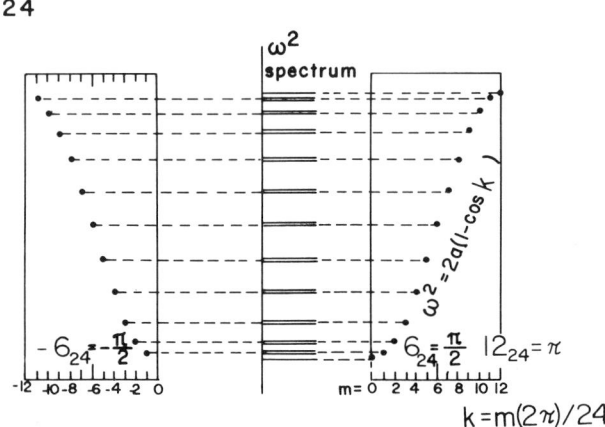

Figure 2.7.4 ω^2 spectrum of C_{24} symmetric coupled pendulums.

acoustical waves. (Here the unit of distance is the interval occupied by one connecting spring.)

Now consider waves with large k_n that lie outside the range $|k| \leq n/2$. For example, $|-3_5\rangle$ or $|7_5\rangle$ are wave states which lie outside this range for $n = 5$. However, from Figure 2.7.5 we see that $|-3_5\rangle$ and $|7_5\rangle$ waves are indistinguishable from the $|2_5\rangle$ wave as far as the five oscillating pendulums are concerned. They are also indistinguishable as far as C_n irreps are concerned. The n integers k in the range $-n/2 < k < n/2 \,[-(n-1)/2 < k < (n-1)/2]$ for n even (n odd) correspond in a one-to-one fashion with the n irreps of C_n. (The corresponding range of k_n is called the FIRST BRILLOUIN ZONE in the theory of lattice waves.) Any of the $(k \pm n)_n, (k \pm 2n)_n, \ldots$ outside this range will just duplicate a C_n irrep for a k_n which is inside:

$$D^{(k \pm Nn)_n} = e^{2i\pi(k \pm Nn)/n} = e^{2i\pi(k/n \pm N)} = e^{2i\pi k/n} = e^{ik_n} = D^{k_n}. \quad (2.7.7)$$

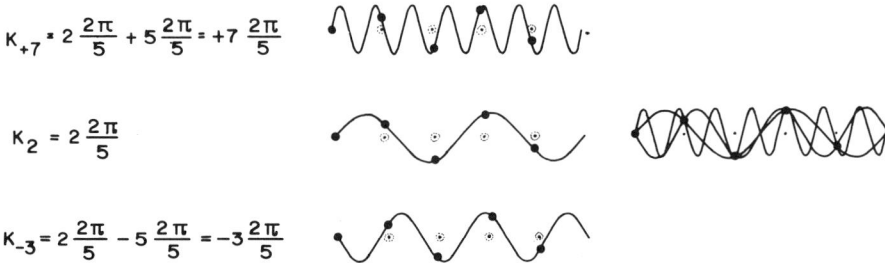

Figure 2.7.5 Higher wave-number solutions for C_5 symmetric coupled pendulums. The waves shown have all give the same motion of the pendulums.

MORE EXAMPLES WITH C_n SYMMETRY: ONE-DIMENSIONAL LATTICES

A C_n system is like an n-nary computer register: It can only count to $n - 1$ and then start over with zero. The register takes a number $m = k \pm Nn$ and reads out only m-modulo-$n = k$. A wave state $|m_n\rangle$ will have m wavelengths in the 2π interval of the full circle, but the n-pendulum system cannot distinguish this state from the state $|k_n\rangle$ with only k wavelengths where $k = m$-modulo-$n = k$-mod-n.

A. Symmetry Breaking

A physical system could tell the difference between a k_n wave and a $(k \pm n)_n$ wave if it has some kind of "detector" such as another mass between each lattice point. As an example consider the system in Figure 2.7.6. This is a copy of the one in Figure 2.7.1, except every even spring is weakened ($\underline{a} < a$) and every odd spring is made stiffer ($\bar{a} > a$). Also let there be $n = 24$ pendulums. The acceleration matrix $\langle \mathbf{a} \rangle$ now takes the following form:

$$\begin{pmatrix} \langle 1|\mathbf{a}|1\rangle & \langle 1|\mathbf{a}|r\rangle & \langle 1|\mathbf{a}|r^2\rangle & \cdots & \langle 1|\mathbf{a}|r^{23}\rangle \\ \langle r|\mathbf{a}|1\rangle & \langle r|\mathbf{a}|r\rangle & \langle r|\mathbf{a}|r^2\rangle & \cdots & \langle r|\mathbf{a}|r^{23}\rangle \\ \vdots & \vdots & \vdots & & \vdots \end{pmatrix}$$

$$= \begin{pmatrix} \bar{a} + \underline{a} + b & -\underline{a} & 0 & \cdots & -\bar{a} \\ -\underline{a} & \bar{a} + \underline{a} + b & -\bar{a} & \cdots & 0 \\ \vdots & \vdots & \vdots & & \vdots \end{pmatrix}. \quad (2.7.8)$$

Changing the springs "breaks" the symmetry from C_{24} down to C_{12}. Now only the operations $\{1, r^2, r^4, \ldots\} = C_{12}$ will commute with a if $\bar{a} \neq \underline{a}$. The operators $\{r, r^3, \ldots\}$ are no longer symmetry operations. As explained in the preceding section (recall Figure 2.7.1) the irreps of C_{12} are labeled by 12 wave numbers $6_{12}, 5_{12}, 4_{12}, \ldots, -4_{12}, -5_{12}$, which lie in the first Brillouin zone. However, now some waves outside this zone are physically very different. For example, $|-7_{12}\rangle$ is very different from $|5_{12}\rangle$ even though they both

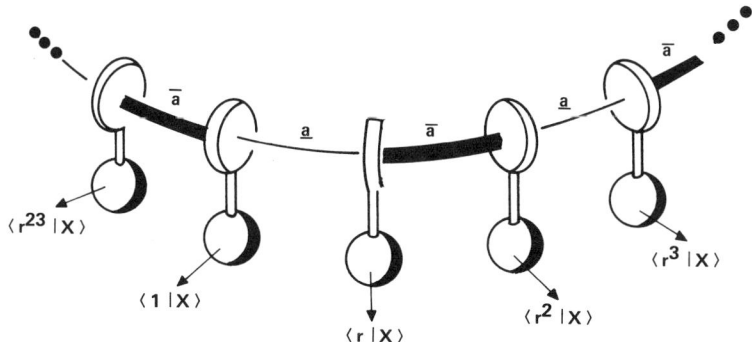

Figure 2.7.6 C_{12} symmetry breaking of C_{24} symmetric coupled pendulums.

belong to irrep $D^{5_{12}}$. The same goes for $|-8_{12}\rangle$ and $|4_{12}\rangle$, which belong to $D^{4_{12}}$, and so on. Most irreps of C_{12} appear twice in the representation of the basis. So once again (recall Section 2.5) we encounter a situation where symmetry theory does not give the complete solution immediately. We still must diagonalize a 2×2 matrix for the pairs of states which share a common irrep $D^{k_{12}}$. The resulting complete solution indicates a splitting of the C_{24} band into two separated bands as shown in Figure 2.7.7. We derive this now.

We first obtain the two independent states produced by the projector P^{k_n} of C_{12} acting first on $|1\rangle$ then on $|r\rangle$.

$$|k_n\rangle = P^{k_n}|1\rangle\sqrt{12} = (|1\rangle + e^{-ik_n}|r^2\rangle + e^{-2ik_n}|r^4\rangle + \cdots)/\sqrt{12},$$
$$|k'_n\rangle = P^{k_n}|r\rangle\sqrt{12} = (|r\rangle + e^{-ik_n}|r^3\rangle + e^{-2ik_n}|r^5\rangle + \cdots)/\sqrt{12}.$$

Then the representation of **a** in the $\{|k_n\rangle, |k'_n\rangle\}$ basis is found, using Hermiticity $(P^{k_n^\dagger} = P^{k_n})$, symmetry $(P^{k_n}\mathbf{a} = \mathbf{a}P^{k_n})$, idempotency $(P^{k_n}P^{k_n} = P^{k_n})$ of P^{k_n}, and the acceleration matrix in (2.7.8):

$$\langle k_n|\mathbf{a}|k_n\rangle = \langle 1|P^{k_n}\mathbf{a}P^{k_n}|1\rangle \cdot 12 = \langle 1|\mathbf{a}P^{k_n}|1\rangle \cdot 12$$
$$= \langle 1|\mathbf{a}|1\rangle + e^{-ik_n}\langle 1|\mathbf{a}|r^2\rangle + \cdots$$
$$= \underline{a} + \bar{a} + 0 + \cdots .$$

The other components are found in the same way:

$$\langle k'_n|\mathbf{a}|k_n\rangle = \langle r|\mathbf{a}|1\rangle + e^{-ik_n}\langle r|\mathbf{a}|r^2\rangle + \cdots$$
$$= -\underline{a} - e^{-ik_n}\bar{a},$$
$$\langle k_n|\mathbf{a}|k'_n\rangle = -\underline{a} - e^{ik_n}\bar{a},$$
$$\langle k'_n|\mathbf{a}|k'_n\rangle = \underline{a} + \bar{a}.$$

The resulting (2×2) matrix for a given k_n is

$$\langle \mathbf{a}\rangle^{k_n} = \begin{pmatrix} \bar{a} + \underline{a} & -(\bar{a}e^{ik_n} + \underline{a}) \\ -(\bar{a}e^{-ik_n} + \underline{a}) & \bar{a} + \underline{a} \end{pmatrix} \begin{matrix} |k_n\rangle \\ |k'_n\rangle \end{matrix}. \quad (2.7.9)$$

From this the eigenvalues are easily found by a secular equation solution:

$$\omega^2 = \bar{a} + \underline{a} \pm (\bar{a}^2 + 2\bar{a}\underline{a}\cos k_n + \underline{a}^2)^{1/2}. \quad (2.7.10)$$

These eigenvalues are plotted as black dots in Figure 2.7.7(a) for $\bar{a} = 3a/2$ and $\underline{a} = a/2$ and are connected with dotted lines to the circles which represent the C_{24} symmetry case where $\bar{a} = a = \underline{a}$. When comparing the $n = 24$ case (Figure 2.7.2) with the $n = 12$ case, it must be remembered that we define $k_n = k(2\pi/n)$.

The most striking effect of the breaking of symmetry down to C_{12} is the splitting of $\{|6\rangle, |-6\rangle\}$ degeneracy in Figure 2.7.7(b). With C_{24} symmetry

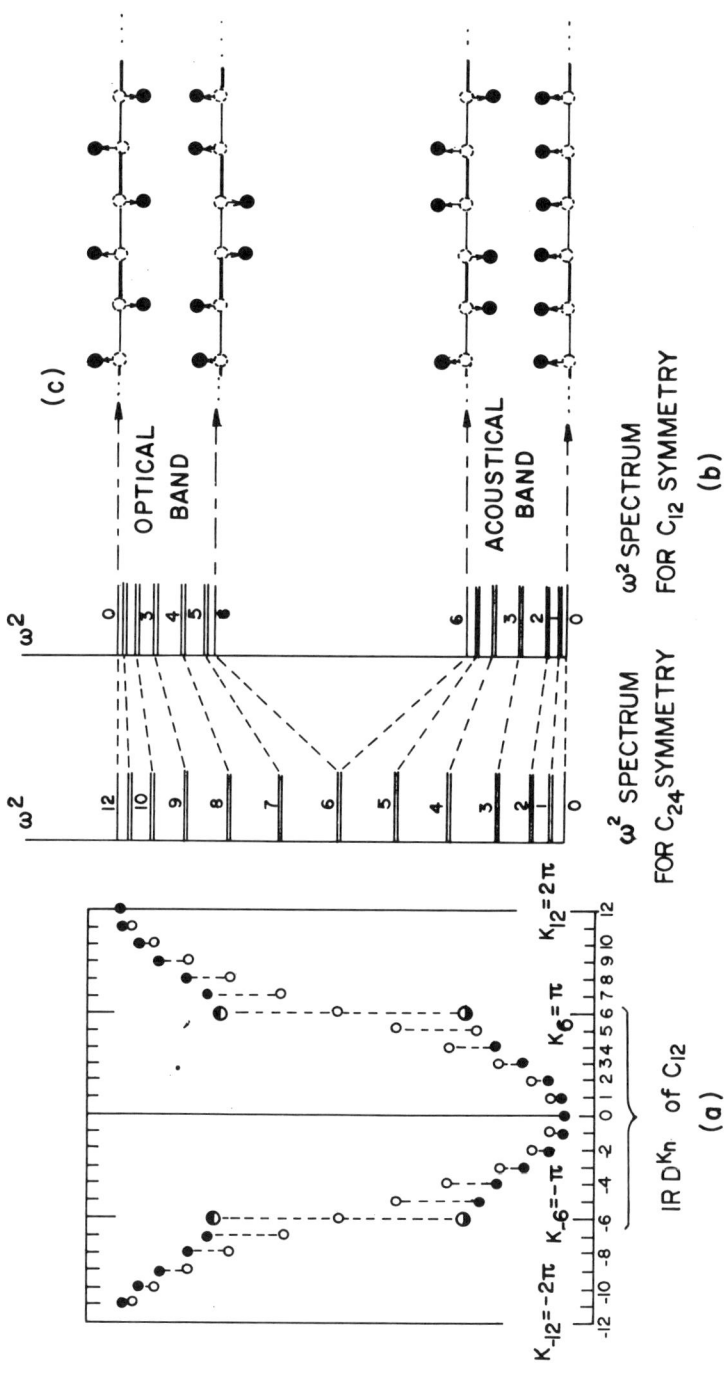

Figure 2.7.7 Band splitting due to C_{24}–C_{12} symmetry breaking.

still in effect these two wave states were degenerate. However, they belong to the same irrep $D^{6_{12}}$ of C_{12}. An acceleration matrix having only C_{12} may mix them. Indeed, equal combinations of them are needed to diagonalize the matrix

$$\langle \mathbf{a} \rangle^{6_{12}} = \begin{pmatrix} \bar{a} + \underline{a} & \bar{a} - \underline{a} \\ \bar{a} - \underline{a} & \bar{a} + \underline{a} \end{pmatrix} \begin{matrix} |6_{12}\rangle \\ |-6_{12}\rangle \end{matrix}. \qquad (2.7.11\text{a})$$

The correct combinations are the sine and cosine standing-wave states:

$$|c_{12}^6\rangle = (|6_{12}\rangle + |-6_{12}\rangle)/\sqrt{2} \qquad (\text{eigenvalue} = 2\bar{a}),$$
$$|s_{12}^6\rangle = -i(|6_{12}\rangle - |-6_{12}\rangle)/\sqrt{2} \qquad (\text{eigenvalue} = 2\underline{a}). \qquad (2.7.11\text{b})$$

So if $\bar{a} > \underline{a}$ then ($k = 6$) eigenstates *must* be standing waves. This is because the standing wave which tends to twist each stiff ($\bar{a}$) spring [see Figure (2.7.7(c)] necessarily has a higher frequency than the one which tends to twist each soft ($\underline{a}$) spring. So the frequency of these modes depends on the position of the wave nodes. Note that the other states which were moving waves under C_{24} symmetry remain so under C_{12} (see Problem 2.7.3). Their frequencies are shifted in Figure 2.7.7(b) but the degeneracies are not split. The frequency of these modes does not depend on the positions of wave nodes.

B. Galloping Waves

Elementary accounts of wave mechanics are often devoted almost exclusively to either pure moving waves or else pure standing waves. In fact, there exists a doubly infinite continuum of different types of monochromatic waves which lie between these two extreme types. The general monochromatic (single-frequency) wave function has the form

$$\psi(x, t) = (I e^{ikx} + R e^{-ikx}) e^{-i\omega t}, \qquad (2.7.12)$$

where the (generally complex) amplitudes for the incident (from the left) and the reflected (from the right) are I and R, respectively. If one of the amplitudes is zero, then x represents a purely moving wave, and if their magnitudes are equal ($|I| = |R|$) then it is a pure standing wave.

The waves which result for arbitrary I and R will be called GALLOPING waves here because of a peculiar motion which they exhibit. A quantity known as the STANDING-WAVE RATIO given by the definition

$$\text{SWR} = \Delta = (I - R)/(I + R). \qquad (2.7.13)$$

serves as a measure of properties of the general galloping wave.

MORE EXAMPLES WITH C_n SYMMETRY: ONE-DIMENSIONAL LATTICES **101**

Two examples of the galloping-wave motion are shown in Figures 2.7.8(a) and 2.7.8(b) for SWR = $\frac{1}{3}$ and SWR = $\frac{1}{19}$. The upper portion of each figure contains superimposed plots or snapshots of the real part of the wave function at 10 equally spaced instants during one period. The lower portion of each figure shows over 20 instants for the same wave plotted in a space-time (x, ct) frame. The latter should be viewed as three-dimensional perspective plots with the third dimension representing wave amplitude emerging obliquely from the page.

You should notice in the upper portion of each figure that the waves all are constrained to slide through a fixed envelope given by the magnitude of

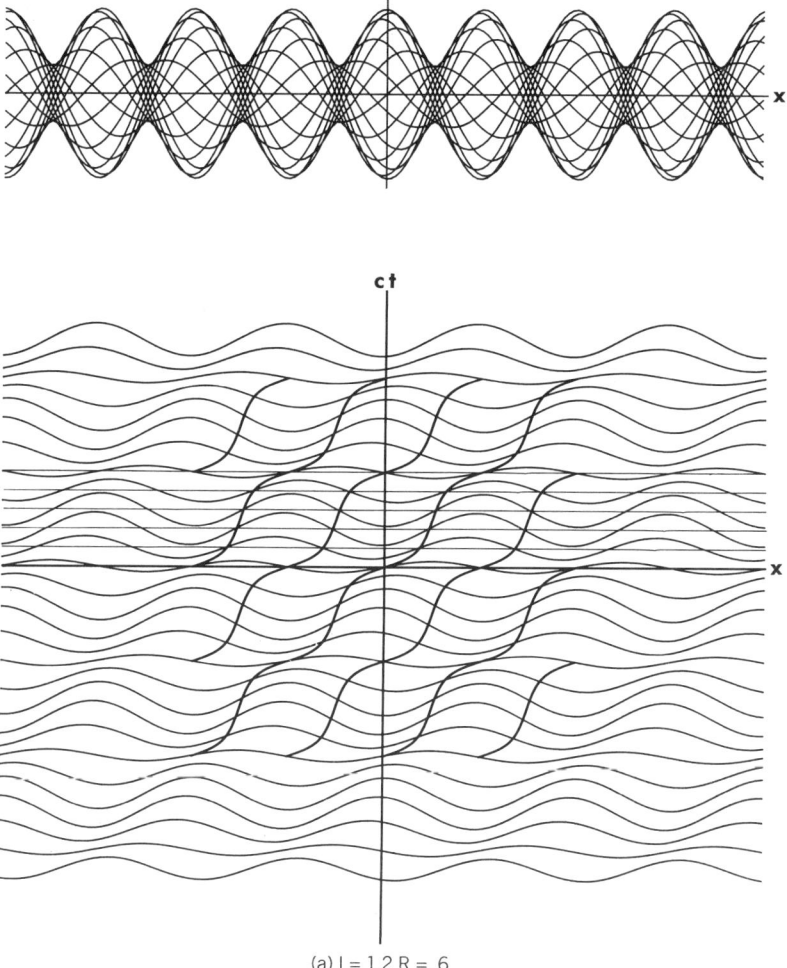

(a) I = 1.2 R = .6

Figure 2.7.8 Space-time trajectories of the zero points of a galloping wave for two values of the standing wave ratio: (a) SWR = $\Delta = \frac{1}{3}$, (b) SWR = $\Delta = \frac{1}{19}$. Wave forms for equal space-time intervals are superimposed at the top of each figure.

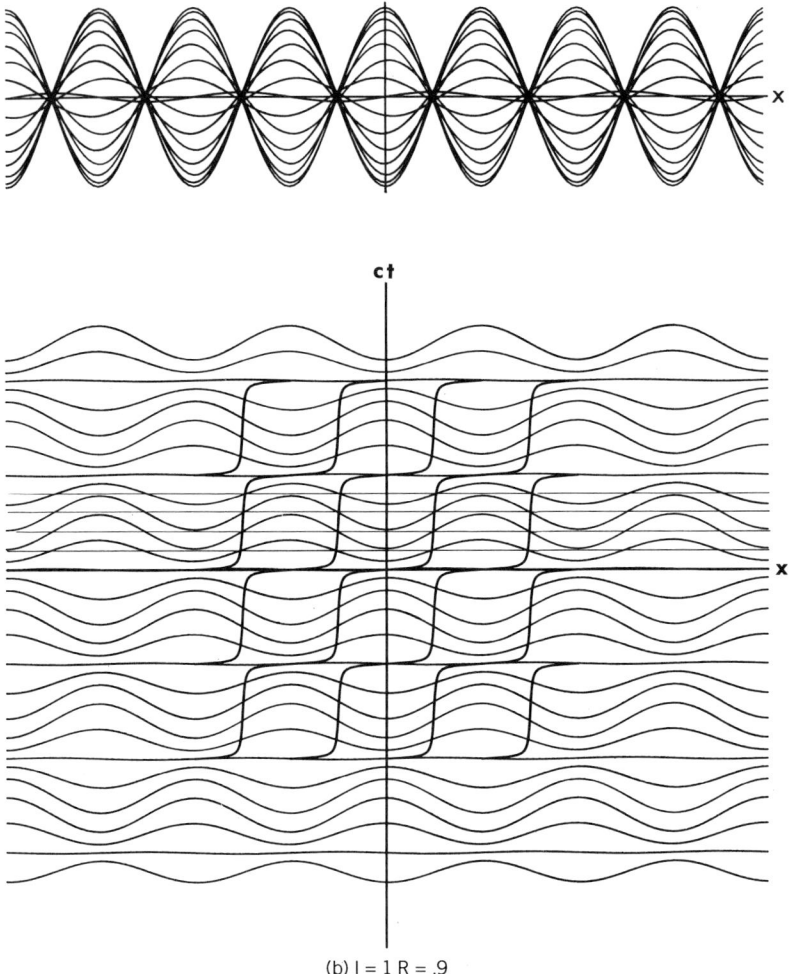

(b) I = 1 R = .9

Figure 2.7.8 (*Continued*).

the wave function (2.7.12), which is

$$|\psi(x)| = [\psi^*\psi]^{1/2} = \left[|I|^2 + |R|^2 + 2|I|\,|R|\cos(2kx)\right]^{1/2}, \quad (2.7.14)$$

where I and R are here chosen to be real. The SWR is a ratio of the maximum value of $|x(x)|$ to its minimum value.

It is interesting to see what the wave does when it shrinks from its maximum value and slides through the minima in the $|\psi(x)|$ envelope. It appears to speed up through the minima then slow down as though it were catching its breath while it has maximum amplitude. You can see oscillation

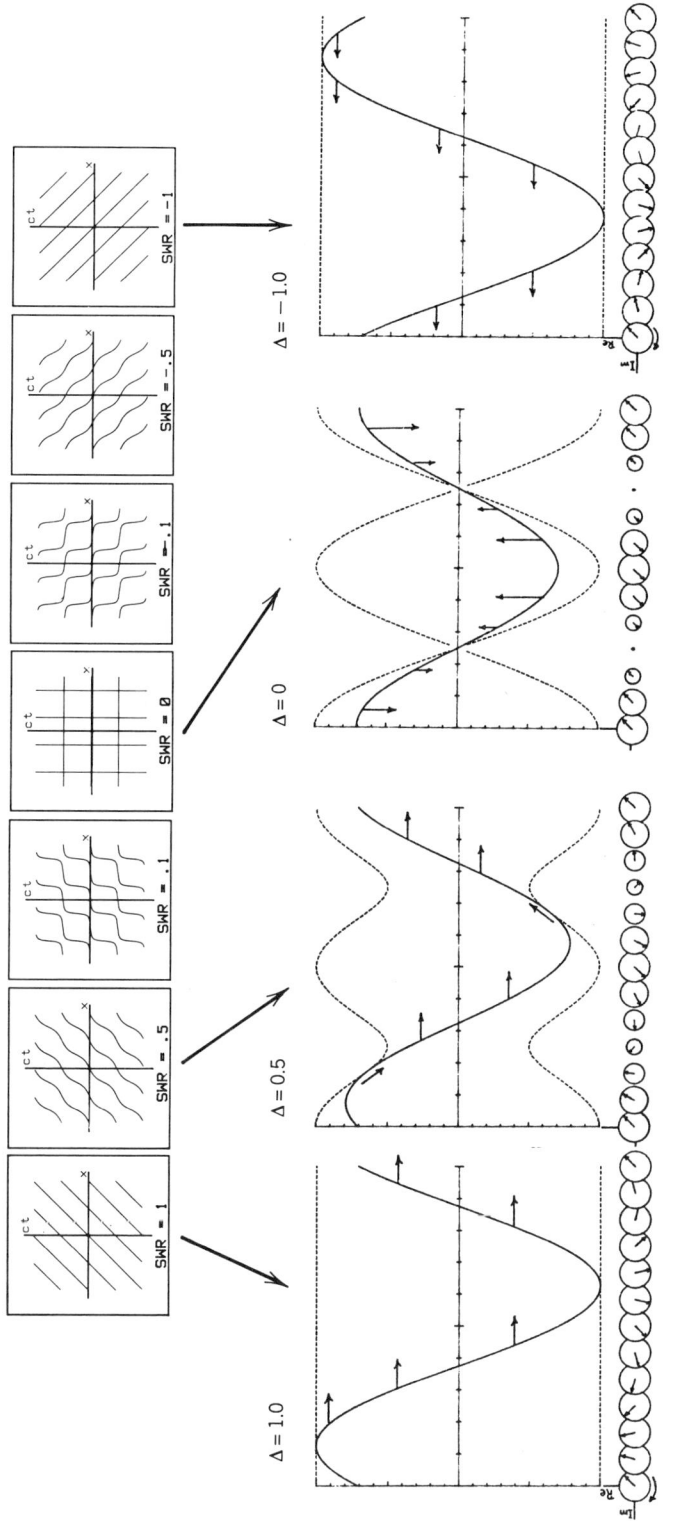

Figure 2.7.9 Varying types of monochromatic waves.

104 BASIC THEORY AND APPLICATIONS OF SYMMETRY REPRESENTATIONS

of the phase velocity defined by the space-time trajectories of the zeros in the main figures, and this provides a measure of the galloping. As the SWR deviates more from unity the galloping becomes more extreme. This is seen by comparing the zero-point motion in Figure 2.7.8(a) with those in Figure 2.7.8(b). If the latter represented a light wave then the phase velocity ranges from $c/19$ to $19c$ and back again twice in one period! You should show that, in general, the phase velocity ranges between $c(\text{SWR})$ and $c/(\text{SWR})$ (Problem 2.7.2).

A set of zero trajectory plots are drawn for a range of SWR values between 1 and -1 in Figure 2.7.9. Below four of the drawings are shown representative wave plots and phasor diagrams. If you are interested in working with the dynamics of waves in optics, quantum theory, or other areas of physics it probably would not hurt to familiarize yourself with these elementary pictures. Also, they provide some interesting relativistic effects if one considers the effects of Doppler shifting the R wave up and the I-wave component down in frequency.

C. Comparison with Fourier Analysis

The analysis of any system in terms of waves of varying frequency and wavelength is generally called FOURIER ANALYSIS. We have seen that representation theory of C_n is essentially Fourier analysis. A complete set of waves which one associates with irreps of C_n is used to describe a multitude of oscillator properties.

Later we shall be interested in waves of many different types traveling through more complicated spaces and topologies. It will be possible to describe these with irreps of more complicated symmetries than C_n. Ordinary Fourier analysis will not be as much help then, but it is still useful to think of representation theory of any symmetry as a generalized Fourier analysis.

2.8 OTHER TYPES OF ABELIAN SYMMETRY

Consider the example in Figure 2.8.1 of a mechanical system. Let the classical equation of motion be given by the following:

$$\begin{pmatrix} \langle 1|\ddot{x}\rangle \\ \langle 2|\ddot{x}\rangle \\ \langle 3|\ddot{x}\rangle \\ \langle 4|\ddot{x}\rangle \end{pmatrix} = \begin{pmatrix} A & a & b & c \\ a & A & c & b \\ b & c & A & a \\ c & b & a & A \end{pmatrix} \begin{pmatrix} \langle 1|x\rangle \\ \langle 2|x\rangle \\ \langle 3|x\rangle \\ \langle 4|x\rangle \end{pmatrix}. \qquad (2.8.1a)$$

The components in the acceleration matrix depend upon the spring constants $\{k_a, k_b, k_c\}$ and geometry according to the theory outlined in Chapter 1

OTHER TYPES OF ABELIAN SYMMETRY   105

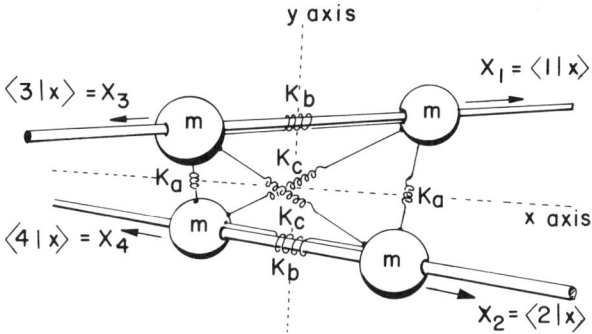

Figure 2.8.1 D_2 symmetric coupled oscillators.

(Section 1.4.B):

$$A = -(k_a \cos^2(a,b) + k_b + k_c \cos^2(b,c))/m,$$
$$a = -k_a \cos^2(a,b)/m,$$
$$b = -k_b/m,$$
$$c = -k_c \cos^2(b,c)/m. \quad (2.8.1b)$$

For small vibrations these components may be assumed constant.

The physical symmetry of this contraption turns out to be the same as that of the fan blade shown in Figure 2.1.1. This symmetry is called D_2 in a notation of crystallographers, which stands for "dihedral of two intersecting planes." The group D_2 is composed of 180° rotations around each of three orthogonal axes x, y, and z. The D_2 multiplication table is given by Eq. (2.1.3).

There can be no mistaking the group D_2 for the other group of order 4 (the chart in Figure 2.2.2 indicates two groups of order 4 exist), which is the cyclic group C_4. All the squares (R^2) of D_2 elements R are equal to the identity, while C_4 has elements corresponding to 90° rotations and for which the squares are not the identity.

Strictly speaking, the only way to be sure that D_2 is a physical symmetry of the model in Figure 2.8.1 is to test its representation in the assumed basis. Equations (2.8.2a) and (2.8.2b) define this representation, and you can check that it does indeed commute with the acceleration matrix. The abstract definitions obtained by inspecting Figure 2.8.1 are as follows:

$$\begin{aligned}
1|1\rangle &= |1\rangle, & R_x|1\rangle &= |2\rangle, & R_y|1\rangle &= |3\rangle, & R_z|1\rangle &= |4\rangle, \\
1|2\rangle &= |2\rangle, & R_x|2\rangle &= |1\rangle, & R_y|2\rangle &= |4\rangle, & R_z|2\rangle &= |3\rangle, \\
1|3\rangle &= |3\rangle, & R_x|3\rangle &= |4\rangle, & R_y|3\rangle &= |1\rangle, & R_z|3\rangle &= |2\rangle, \\
1|4\rangle &= |4\rangle, & R_x|4\rangle &= |3\rangle, & R_y|4\rangle &= |2\rangle, & R_z|4\rangle &= |1\rangle. & (2.8.2a)
\end{aligned}$$

The resulting matrix representation in the $\{|1\rangle|2\rangle|3\rangle|4\rangle\}$ basis is then given by

$$\mathcal{R}(1) = \begin{pmatrix} 1 & \cdot & \cdot & \cdot \\ \cdot & 1 & \cdot & \cdot \\ \cdot & \cdot & 1 & \cdot \\ \cdot & \cdot & \cdot & 1 \end{pmatrix}, \quad \mathcal{R}(R_x) = \begin{pmatrix} \cdot & 1 & \cdot & \cdot \\ 1 & \cdot & \cdot & \cdot \\ \cdot & \cdot & \cdot & 1 \\ \cdot & \cdot & 1 & \cdot \end{pmatrix},$$

$$\mathcal{R}(R_y) = \begin{pmatrix} \cdot & \cdot & 1 & \cdot \\ \cdot & \cdot & \cdot & 1 \\ 1 & \cdot & \cdot & \cdot \\ \cdot & 1 & \cdot & \cdot \end{pmatrix}, \quad \mathcal{R}(R_z) = \begin{pmatrix} \cdot & \cdot & \cdot & 1 \\ \cdot & \cdot & 1 & \cdot \\ \cdot & 1 & \cdot & \cdot \\ 1 & \cdot & \cdot & \cdot \end{pmatrix}. \quad (2.8.2b)$$

Now it is possible to decompose this symmetry group into sums of idempotents as was done before with the C_n groups. The minimal equations of R_x ($R_x^2 = 1$) and R_y ($R_y^2 = 1$) give two idempotents each:

$$P^{x+} = (1 + R_x)/2, \quad P^{y+} = (1 + R_y)/2,$$
$$P^{x-} = (1 - R_x)/2, \quad P^{y-} = (1 - R_y)/2. \quad (2.8.3)$$

Each pair satisfies a completeness relation by itself:

$$1 = P^{x+} + P^{x-}, \quad (2.8.4a)$$
$$1 = P^{y+} + P^{y-}. \quad (2.8.4b)$$

However, D_2 has four linearly independent operators, and so clearly neither pair is enough by itself to spectrally decompose the whole group. The trick is to use the two together by simply multiplying Eq. (2.8.4a) with Eq. (2.8.4b):

$$1 \cdot 1 = (P^{x+} + P^{x-})(P^{y+} + P^{y-})$$
$$1 = P^{x+}P^{y+} + P^{x+}P^{y-} + P^{x-}P^{y+} + P^{x-}P^{y-}. \quad (2.8.5a)$$

This gives a completeness relation involving four new idempotents:

$$P^1 \equiv P^{x+}P^{y+} = (1 + R_x + R_y + R_z)/4,$$
$$P^2 \equiv P^{x-}P^{y+} = (1 - R_x + R_y - R_z)/4,$$
$$P^3 \equiv P^{x+}P^{y-} = (1 + R_x - R_y - R_z)/4,$$
$$P^4 \equiv P^{x-}P^{y-} = (1 - R_x - R_y + R_z)/4. \quad (2.8.5b)$$

The resulting four idempotents must be orthogonal as well as complete, since

OTHER TYPES OF ABELIAN SYMMETRY   107

their commuting factors are orthogonal. For example, the product

$$P^1 P^2 = (P^{x+}P^{y+})(P^{x-}P^{y+})$$
$$= P^{x+}P^{x-}P^{y+}P^{y+}$$
$$= 0$$

is nullified by the (x) factors. Furthermore, the four idempotents are simultaneously eigenoperators of R_x, R_y, as well as $R_z = R_x R_y$. For example, we have

$$R_x P^3 = R_x P^{x+}P^{y-}, \qquad R_y P^3 = R_y P^{x+}P^{y-}, \qquad R_z P^3 = R_z P^{x+}P^{y-}$$
$$= P^3 \qquad\qquad = P^{x+}R_y P^{y-} \qquad = R_x R_y P^{x+}P^{y-}$$
$$\qquad\qquad = -P^3 \qquad\qquad = R_x P^{x+} R_y P^{y-}$$
$$\qquad\qquad\qquad\qquad\qquad = -P^3.$$

Hence, all four D_2 operators may be simultaneously spectrally decomposed:

$$1 = P^1 + P^2 + P^3 + P^4,$$
$$R_x = P^1 - P^2 + P^3 - P^4,$$
$$R_y = P^1 + P^2 - P^3 - P^4,$$
$$R_z = P^1 - P^2 - P^3 + P^4. \qquad (2.8.6)$$

The coefficients in this expansion or within the parentheses of Eq. (2.8.5b) are the irreps $D^\alpha(g)$ of the group D_2. These are summarized by the equations and table that follow. The conventional (A, B) notation for the irreps is given also:

$$P^\alpha = \tfrac{1}{4}\left(\sum_g D^{\alpha*}(g)g\right)$$

$$g = \sum_{\alpha=1}^{4} D^\alpha(g) P^\alpha$$

Conventional notation $g =$	1	R_x	R_y	R_z
$D^{A_1}(g) = D^1(g) =$	1	1	1	1
$D^{B_2}(g) = D^2(g) =$	1	-1	1	-1
$D^{B_1}(g) = D^3(g) =$	1	1	-1	-1
$D^{A_2}(g) = D^4(g) =$	1	-1	-1	1

$$(2.8.7)$$

108 BASIC THEORY AND APPLICATIONS OF SYMMETRY REPRESENTATIONS

The eigenvector solutions to the equation (2.8.1) of motion are found using the irrep idempotents as follows:

$$|e^{A_1}\rangle \equiv |e^1\rangle = P^1|1\rangle\sqrt{4} = (|1\rangle + |2\rangle + |3\rangle + |4\rangle)/2,$$

$$|e^{B_2}\rangle \equiv |e^2\rangle = P^2|1\rangle\sqrt{4} = (|1\rangle - |2\rangle + |3\rangle - |4\rangle)/2,$$

$$|e^{B_1}\rangle \equiv |e^3\rangle = P^3|1\rangle\sqrt{4} = (|1\rangle + |2\rangle - |3\rangle - |4\rangle)/2,$$

$$|e^{A_2}\rangle \equiv |e^4\rangle = P^4|1\rangle\sqrt{4} = (|1\rangle - |2\rangle - |3\rangle + |4\rangle)/2,$$

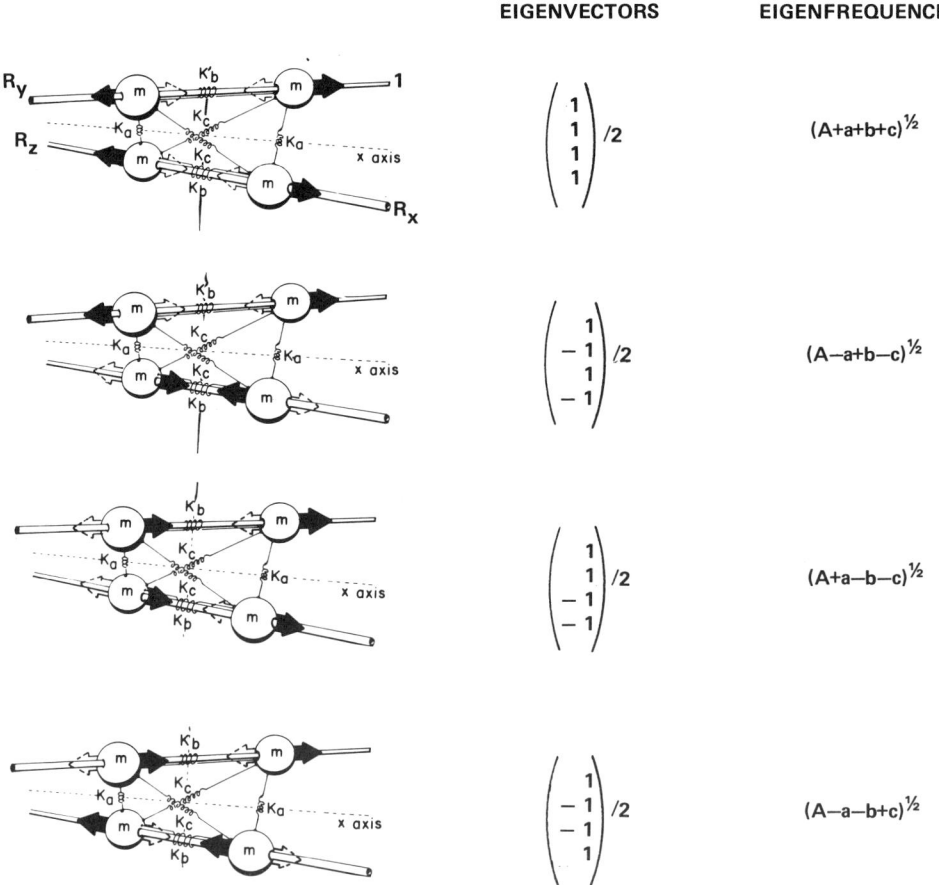

Figure 2.8.2 Normal modes, eigenvectors, and eigenfrequencies of D_2 symmetric coupled system.

where the original state definitions $|1\rangle = \mathbf{1}|1\rangle$, $|2\rangle = R_x|1\rangle$, $|3\rangle = R_y|1\rangle$, and $|4\rangle = R_z|1\rangle$ of Eq. (2.8.2a) are used. These states are pictured in Figure 2.8.2, and their eigenfrequencies from Eq. (2.8.1) are written next to each drawing.

2.9 THEORY OF COMMUTING IDEMPOTENTS

We can easily see that the trick which gave the D_2 irreps, must also work for any Abelian finite group. Suppose one element g of the Abelian group satisfies a minimal equation $g^n = \mathbf{1}$. From the theory of Section 2.6 this equation yields a set $\{p^1, p^2, \ldots, p^n\}$ of n orthogonal idempotents. Let another element h yield another set of $\{q^1, q^2, \ldots, q^m\}$. The p^j and q^k idempotents give eigenoperator expansions of g and h, respectively,

$$g = \sum_{j=1}^{n} g_j p^j, \qquad h = \sum_{k=1}^{m} h_k q^k, \tag{2.9.1}$$

while either set of idempotents satisfy the completeness relation:

$$\sum_{j=1}^{n} p^j = \mathbf{1} = \sum_{k=1}^{m} q^k. \tag{2.9.2}$$

Multiplying Eq. (2.9.2) by p^j may result in the "splitting" of p^j into a sum of operators $p^j q^k$:

$$\begin{aligned} p^j &= p^j q^1 + p^j q^2 + \cdots + p^j q^m \\ &= \cdots + P^1 + P^2 + \cdots + P^r \qquad (r \le m). \end{aligned} \tag{2.9.3}$$

The nonzero terms $\{P^1, P^2, \ldots, P^r\}$ in these sums must satisfy orthonormality and completeness relations, too. This follows from Eqs. (2.9.1) and (2.9.2), since the p^m and q^k commute with each other ($p^m q^k = q^k p^m$) since they are polynomials of the commuting elements g and h. Furthermore, the resulting set provides a spectral decomposition of g and h simultaneously [recall Section 1.2.B(d), where matrices were treated in this way]:

$$\begin{aligned} gP^a &= D^a(g)P^a; & g &= D^1(g)P^1 + D^2(g)P^2 + \cdots + D^r(g)P^r, \\ hP^a &= D^a(h)P^a; & h &= D^1(h)P^1 + D^2(h)P^2 + \cdots + D^r(h)P^r. \end{aligned} \tag{2.9.4}$$

This splitting process can be repeated, using idempotent expansions generated by other group operators $\{k, l, \ldots\}$ which are not products of powers of g and h. Finally, this process must yield exactly as many nonzero idempotents as group elements because of linear independence. These final IRREDUCIBLE idempotents cannot split anymore because if one of them did, there would be one more linearly independent operator than there are group elements. But this is impossible, since any operator in the group algebra is a linear combination of group operators. The number of group operators is $^{\circ}G$, the order of the group.

Furthermore, these irreducible or "unsplittable" idempotents are unique. Suppose two sets [Eq. (2.9.5)] have been found by using different group elements:

$$1 = P^1 + P^2 + \cdots + P^{\circ G} = P^{1'} + P^{2'} + \cdots + P^{\circ G'}. \quad (2.9.5)$$

Multiplying both sides by P^j gives

$$P^j = P^j P^{1'} + P^j P^{2'} + \cdots + P^j P^{\circ G'}, \quad (2.9.6)$$

which can have only one nonzero term, since P^j is unsplittable. Let us say that term is $P^j = P^j P^{l'}$. Then multiplying Eq. (2.9.5) by $P^{l'}$ gives Eq. (2.9.7), which proves each idempotent set is unique no matter how you get it:

$$P^{l'} = P^j P^{l'} = P^j. \quad (2.9.7)$$

2.10 GENERAL THEORY OF ABELIAN GROUPS

There is an easy way to express the group D_2 and many others that arise in physics. D_2 is nothing but $C_2 \times C_2$. Let us now define what is meant by this "multiplication" ($\times$) of groups.

Definition A group G is said to be an OUTER PRODUCT $H \times K$ of subgroups $H = \{1, h_1, h_2, \ldots\}$ and $K = \{1, k_1, k_2, \ldots\}$ if the following holds:

(1) Every element g in G is written uniquely as a product $g = h_i k_j$ of one element from H and one element from K.
(2) Each h_i in H commutes with each k_j in K $\cdots$ $h_i k_j = k_j h_i$.

In $D_2 = \{1, R_x, R_y, R_z\}$ we find subgroups $H = \{1, R_x\} = C_2$ and $K = \{1, R_y\} = C_2$. It is easy to verify that the criteria (1) and (2) apply so that $D_2 = \{1, R_x\} \times \{1, R_y\} = C_2 \times C_2$. Of course, (2) applies automatically in Abelian groups, but the definition applies even if H and K are not Abelian.

This can be very convenient if you know the irreps of factors H and K of $G = H \times K$. Then the irreps of G can be obtained immediately, as shown for

$D_2 = C_2 \times C_2$ in the following:

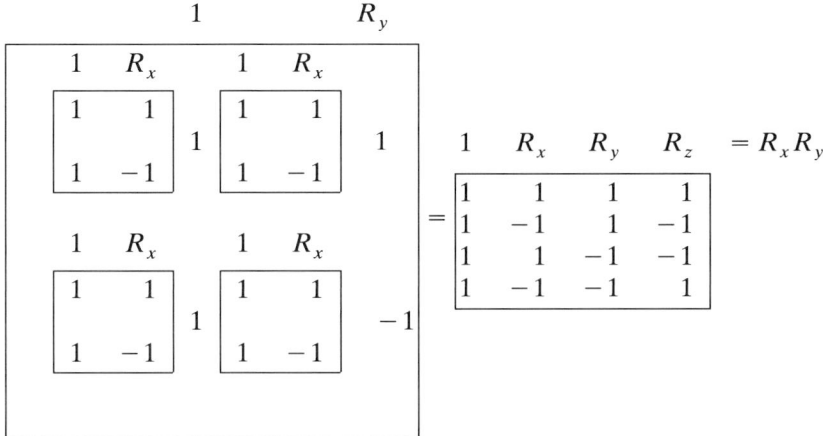

Furthermore, there is a famous theorem about Abelian finite groups which follows from the idempotent analysis in the preceding section. The theorem says: (1) Every finite Abelian group can be expressed as outer products of cyclic groups, and (2) $C_p \times C_q$ and C_{pq} are the same group if and only if integers p and q are relatively prime. This theorem allows one to name all the Abelian groups of an arbitrary finite order. For example, there is only one Abelian group of order 6, $C_3 \times C_2$, which is just the same as C_6. (We have already mentioned that $C_2 \times C_2$ is not the same as C_4.) For another example, all the groups below are different, and completely account for all Abelian groups of order $72 = 2^3 3^2$ (check Figure 2.2.2):

$$C_2 \times C_2 \times C_2 \times C_3 \times C_3 \qquad C_2 \times C_2 \times C_2 \times C_9$$
$$C_4 \times C_2 \times C_3 \times C_3 \qquad C_4 \times C_2 \times C_9$$
$$C_8 \times C_3 \times C_3 \qquad C_8 \times C_9 \sim C_{72}.$$

2.11 SOME ABELIAN POINT SYMMETRIES

A POINT SYMMETRY GROUP is a group of geometrical operations that keep at least one point fixed. These may include ROTATIONS around an axis through this point, or less familiar operations like REFLECTION or INVERSION. A CRYSTAL point symmetry group is a possible site symmetry at each lattice point in some infinite crystal lattice. Only onefold, two-fold (180°), threefold (120°), fourfold (90°), and sixfold (60°) axial symmetry can exist in a crystal. Hence the number of crystal point groups is restricted. It turns out there are just the 32 crystal point groups. These are named in Figure 2.11.1. Surprisingly, exactly half of them are Abelian. These 16

112 BASIC THEORY AND APPLICATIONS OF SYMMETRY REPRESENTATIONS

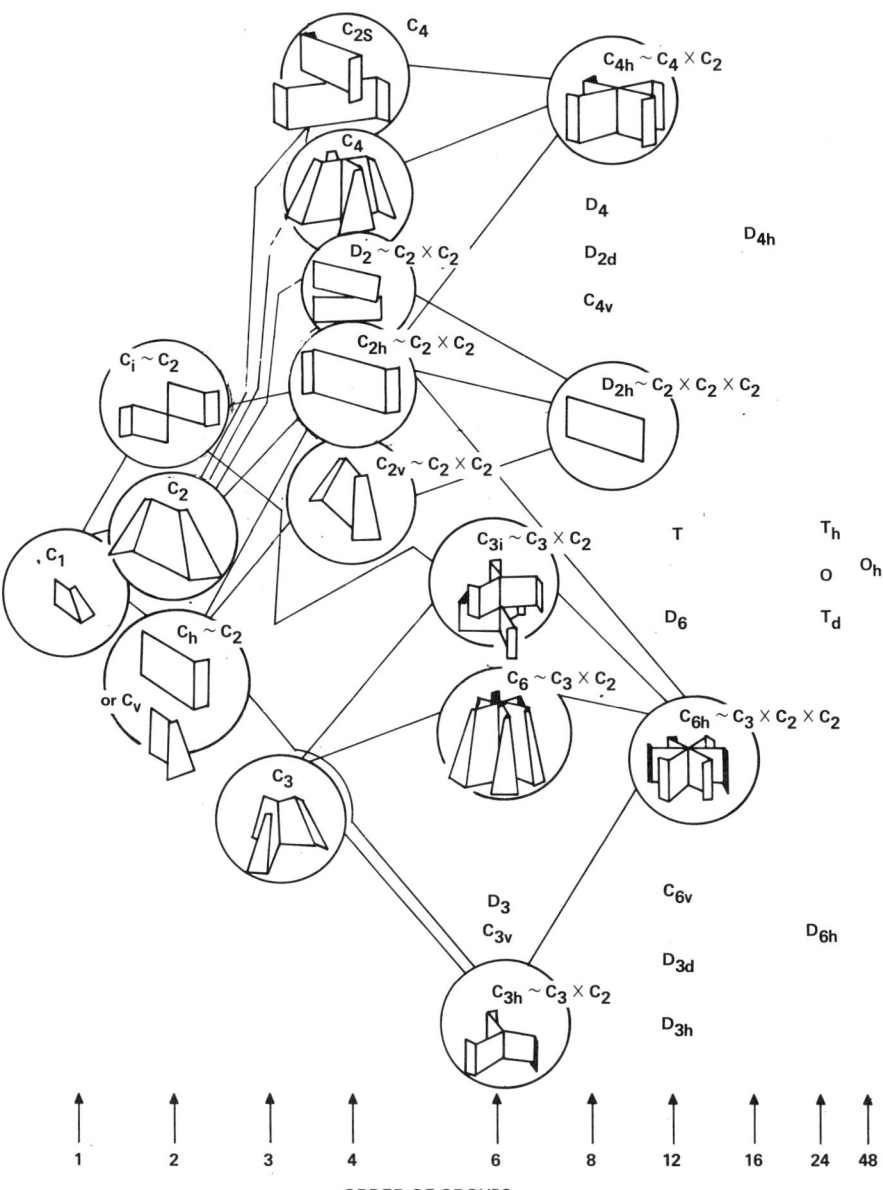

Figure 2.11.1 Abelian crystal point groups. Sixteen of the 32 crystal point groups are Abelian and are illustrated by models drawn in circles.

Abelian crystal point symmetries are indicated by drawings showing objects having the symmetry. (The 16 non-Abelian symmetries are drawn in Figure 3.1.1 at the beginning of Chapter 3.)

In Figure 2.11.1 the simplest objects having reflection symmetry are shown next to the symbols C_h or C_v. An h or v refers to a horizontal (xy) or vertical $((yz)$ or $(xz))$ mirror reflection plane which you can imagine bisecting the object. If you were to somehow map each point of the object into its mirror image point it would still look the same. The reflection operation is not as easily demonstrated as the rotation operation, but it is just as valid as a symmetry operator. For example, the horizontal reflection σ_{xy} through the (xy) plane changes a 3-vector (x, y, z) to $(x, y, -z)$. Similarly, vertical reflections σ_{xz} and σ_{yz} would change the same vector to $(x, -y, z)$ and $(-x, y, z)$, respectively. The representations of these reflections in the $\{x, y, z\}$ Cartesian vector basis are as follows:

$$\mathscr{V}(\sigma_{xy}) = \begin{pmatrix} 1 & 0 & 0 \\ 0 & 1 & 0 \\ 0 & 0 & -1 \end{pmatrix}, \quad \mathscr{V}(\sigma_{xz}) = \begin{pmatrix} 1 & 0 & 0 \\ 0 & -1 & 0 \\ 0 & 0 & 1 \end{pmatrix},$$

$$\mathscr{V}(\sigma_{yz}) = \begin{pmatrix} -1 & 0 & 0 \\ 0 & 1 & 0 \\ 0 & 0 & 1 \end{pmatrix}. \quad (2.11.1)$$

Note that the symmetries C_{6h}, C_{4h}, D_{2h}, C_{3h}, C_{2h}, and C_h all have horizontal reflection symmetry. The symmetries C_{2v} and C_v have vertical reflection symmetry. The confusion between groups C_h and C_v comes about since no rotation axis serves to uniquely define "vertical." (Similarly, one could relabel D_{2h} to be D_{2v} just as well.)

The simplest object having inversion symmetry is shown by the symbol C_i in Figure 2.11.1. If each point of the object at $(\mathbf{r})$ is mapped through the origin at the centroid of the object, to the position $(-\mathbf{r})$, it will still look the same. This is another operation that we can only "do" mathematically. The inversion operator is usually labeled by I. The representation of I is

$$\mathscr{V}(I) = \begin{pmatrix} -1 & 0 & 0 \\ 0 & -1 & 0 \\ 0 & 0 & -1 \end{pmatrix} \quad (2.11.2)$$

in the $\{x, y, z\}$ basis. Other symmetries that include inversion are the groups C_{2h}, C_{3i}, C_{4h}, D_{2h}, and C_{6h}, which are connected by lines leading away from C_i. Any symmetry group A connected by a line in Figure 2.11.1 to a larger symmetry group B is contained in B. A is said to be a SUBGROUP of B $(A \subset B)$. You may read $(\subset)$ as "is less than" or, better, "is contained in."

The three symmetry groups C_i, C_2, and C_h are exactly the same as far as their mathematical properties are concerned. Their character tables or irreps look exactly the same. However, there are many conventional notations which give different names to the same irreps depending on which symmetry is being labeled. Some examples are shown by the following:

	C_i							1	I	(inversion)
			C_2					1	R	(180° rotation)
						C_h	1	σ		(reflection)
0_2	A_g	(+)	A_1		A	A'	1	1		
	or		or							
1_2	A_u	(−)	A_2		B	A''	1	−1		

(2.11.3)

When different symmetry groups G and G' can be shown to share the same mathematical properties, they are said to be ISOMORPHIC symmetry groups. This is denoted by $G \sim G'$ in Figure 2.11.1. In the last example one has $C_i \sim C_h \sim C_2$. There are many different symmetries, but only a fraction of these are really different groups. For example, all the Abelian symmetries of order 6, namely, C_{3i}, C_6, and C_{3h}, are isomorphic to $C_3 \times C_2$, as explained in the preceding section. However, to define these symmetries more precisely one can write $C_{3i} = C_3 \times C_i$, $C_6 = C_3 \times C_2$, and $C_{3h} = C_3 \times C_h$.

Similarly, the symmetries $D_2 = C_2 \times C_2$, $C_{2h} = C_2 \times C_i$, and $C_{2v} = C_2 \times (C_v \sim C_h)$ are all isomorphic to $C_2 \times C_2$. To understand these groups a little better note that a product of inversion and a horizontal reflection is

$$\mathscr{V}(I)\mathscr{V}(\sigma_{xy}) = \begin{pmatrix} -1 & 0 & 0 \\ 0 & -1 & 0 \\ 0 & 0 & -1 \end{pmatrix} \begin{pmatrix} 1 & 0 & 0 \\ 0 & 1 & 0 \\ 0 & 0 & -1 \end{pmatrix} = \begin{pmatrix} -1 & 0 & 0 \\ 0 & -1 & 0 \\ 0 & 0 & 1 \end{pmatrix}$$
$$= \mathscr{V}(R_z(180°)), \qquad (2.11.4)$$

which is a 180° rotation around the *vertical* z axis. Therefore, the elements $\{1, R_z, I, \sigma_{xy}\}$ make up C_{2h}. On the other hand, we have $C_{2v} = \{1, R_z, \sigma_{xy}, \sigma_{yz}\}$, since

$$\sigma_{xz} \cdot \sigma_{yz} = R_z. \qquad (2.11.5)$$

Finally, note that C_{2s} is isomorphic to C_4. The former contains a 90° ROTATION-INVERSION operation in place of the simple 90° rotation operation of the latter. In other words, if the C_{2s} model is inverted before or after a 90° rotation, it will retain its original appearance. C_{2s} is often labeled C_{2i}.

In conclusion, one notes that the essential mathematical properties of all 16 Abelian crystal point groups are derived from those of three simple cyclic groups C_2, C_3, and C_4.

2.12 SYMMETRY ANALYSIS FOR QUANTUM MECHANICS

The mathematics of symmetry analysis for quantum mechanics is mostly the same as it was for the classical problems which were used to introduce it. This is especially true here, since we have made a point of using Dirac notation for classical problems. However, it is important to see how the physical interpretations of the similar mathematical solutions can be very different. We study below some of the simplest examples of quantum problems involving symmetry.

A. Bohr Levels and Bloch Waves: C_{12} Clocktane Orbitals

In this section we will consider an electron confined to orbit in a circular ring. The effects of 12 equivalent potential wells set around the ring at the 1 o'clock, 2 o'clock, ..., 11 o'clock, and 12 o'clock positions will be discussed using C_{12} symmetry analysis. Three cases will be treated, including (a) the Bohr-orbital case where no potential wells exist, (b) the nearly free electron case where very shallow or weak potential wells exist, and (c) the tight-binding or very hindered rotation case in which the potential wells are very deep. The potentials for cases (a), (b), and (c) are drawn at the top of Figures 2.12.1(a), 2.12.1(b), and 2.12.1(c), respectively.

This is a highly simplified model of the molecular orbitals of a 12-fold "clocktane" molecule. It will be the basis of quite a number of concepts developed in this book.

(a) Bohr Orbitals and Free Rotation If an electron orbits freely around a circular ring the time-dependent Schrödinger equation is

$$i\hbar \frac{\partial \psi}{\partial t} = H_0 \psi = -(\hbar^2/2\mu)\frac{\partial^2 \psi}{\partial x^2} = -(\hbar^2/2\mu r^2)\frac{\partial^2 \psi}{\partial \phi^2}, \quad (2.12.1)$$

where the independent variable may be azimuthal angle ϕ or else circumferential distance

$$x = r\phi \quad (2.12.2)$$

around a ring of radius r. The solutions to the Schrödinger eigenvalue equation

$$H_0 |m\rangle = \varepsilon_m |m\rangle \quad (2.12.3)$$

are represented by the one-dimensional plane-wave functions,

$$\langle \phi | m \rangle = \frac{e^{im\phi}}{\sqrt{2\pi}} = \frac{e^{ikx}}{\sqrt{2\pi}} \equiv \langle x | k \rangle, \quad (2.12.4)$$

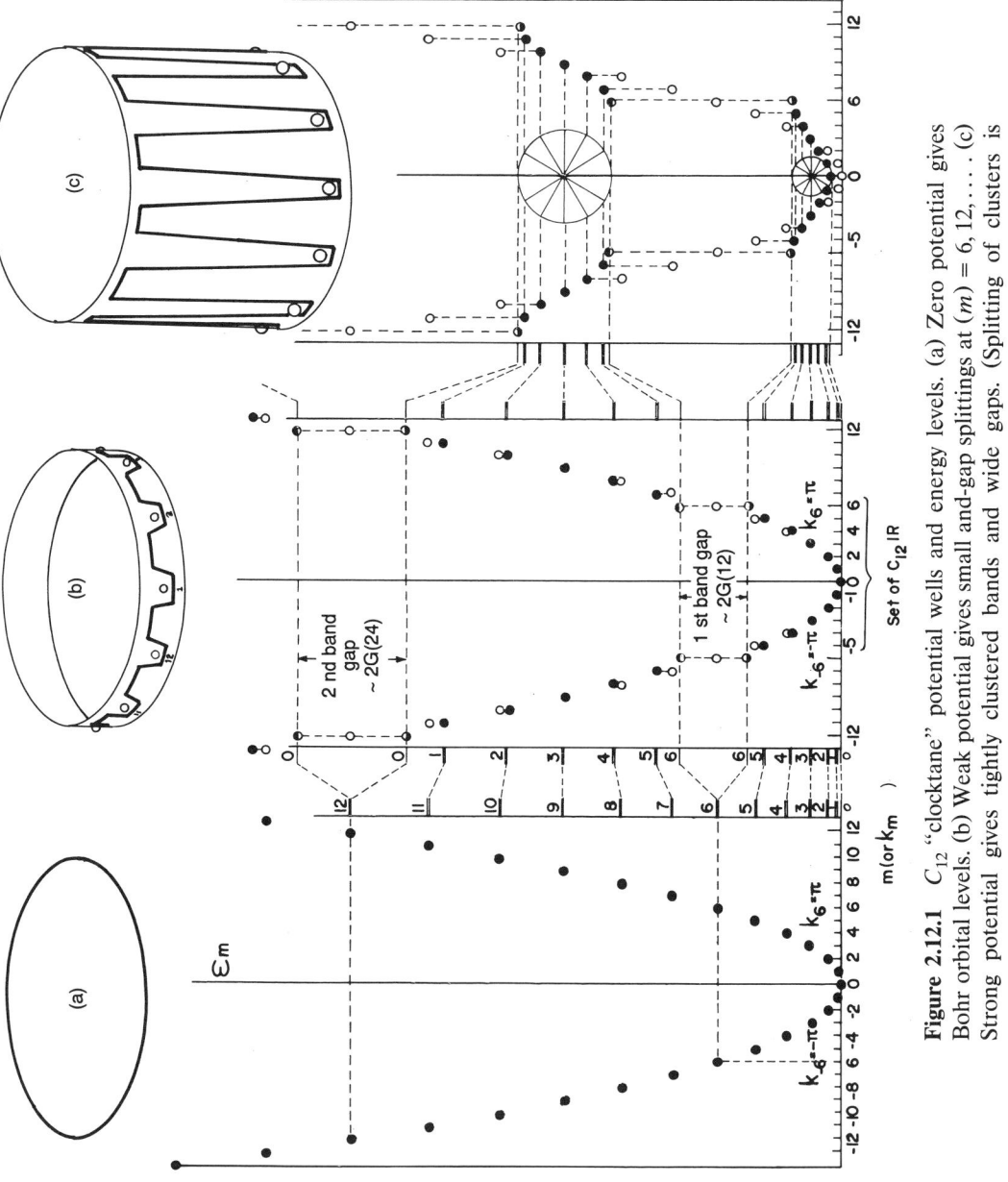

Figure 2.12.1 C_{12} "clocktane" potential wells and energy levels. (a) Zero potential gives Bohr orbital levels. (b) Weak potential gives small and-gap splittings at $(m) = 6, 12, \ldots$ (c) Strong potential gives tightly clustered bands and wide gaps. (Splitting of clusters is

116

having energy eigenvalues

$$\varepsilon_m = \frac{\hbar^2}{2\mu r^2}m^2 = \frac{\hbar^2}{2\mu}k^2, \qquad (2.12.5)$$

where $m = 0, \pm 1, \pm 2, \ldots, +\infty$ is quantized by the circular boundary conditions $\langle \phi | m \rangle = \langle \phi + 2\pi | m \rangle$. Eigenstates may be labeled using either the angular-momentum quantum number m or else the plane-wave number

$$k = m/r. \qquad (2.12.6)$$

The energy eigenvalues ε_m from Eq. (2.12.5) are plotted in Figure 2.12.1. Note that all energy levels except $m = 0$ are doubly degenerate. Moving-wave states $|m\rangle$ and $|-m\rangle$ move around the ring in opposite directions, since by Eqs. (2.12.1)–(2.12.4)

$$\langle \phi, t | \pm m \rangle = \frac{e^{i(\pm m\phi - \varepsilon_m t/\hbar)}}{\sqrt{2\pi}}; \qquad (2.12.7)$$

however, the energy ε_m is the same for $(+m)$ and $(-m)$. Therefore, one may construct sine and cosine standing-wave states

$$\langle \phi | c_m \rangle = \frac{\langle \phi | m \rangle + \langle \phi | -m \rangle}{\sqrt{2}} = (\pi)^{-1/2} \cos m\phi, \qquad (2.12.8a)$$

$$\langle \phi | s_m \rangle = \frac{\langle \phi | m \rangle - \langle \phi | -m \rangle}{i\sqrt{2}} = (\pi)^{-1/2} \sin m\phi, \qquad (2.12.8b)$$

which have the same energy ε_m but stationary nodes and crests:

$$\langle \phi, t | c_m \rangle = (\pi)^{-1/2} e^{-i\varepsilon_m t/\hbar} \cos m\phi, \qquad (2.12.9a)$$

$$\langle \phi, t | s_m \rangle = (\pi)^{-1/2} e^{-i\varepsilon_m t/\hbar} \sin m\phi. \qquad (2.12.9b)$$

The position of the nodes makes no difference in the energy if the potential is constant.

(b) Weak C_{12} Potential and Nearly Free Rotations Consider the effect of a C_{12} symmetric perturbation V_{12} added to the Hamiltonian:

$$H = H_0 + V_{12}. \qquad (2.12.10)$$

Let the perturbation consist of $(n = 12)$ identical potential wells separated by angle $(\Delta\phi = 2\pi/n = \pi/6)$ or by circumferential distance d, where

$$d = 2\pi r/n = \pi r/6, \qquad (2.12.11)$$

as shown in the upper portion of Figure 2.12.1(b). The matrix elements $\langle m'|H|m\rangle$ of this new Hamiltonian must be zero unless states $|m\rangle$ and $|m'\rangle$ belong to the same irrep D^{k_n} of $C_n = C_{12}$. Using Eq. (2.12.4) one finds the matrix element to be

$$\langle m'|H|m\rangle = \langle m'|H_0|m\rangle + \langle m'|V_{12}|m\rangle$$

$$= \delta_{m'm}\varepsilon_m + \frac{1}{2\pi}\int d\phi\, e^{-im'\phi}V_{12}e^{im\phi}$$

$$= \delta_{m'm}m^2 E + \frac{1}{2\pi}\int d\phi\, e^{-i(m'-m)\phi}V_{12}, \quad (2.12.12)$$

where $E \equiv \hbar^2/2\mu r^2$. A portion of the H-matrix is displayed in Table 2.12.1 on p. 121 for $E = 1$. Note that all off-diagonal ($m' \neq m$) components are zero unless

$$|m' - m| = n, 2n, 3n, \ldots$$
$$= 12, 24, 36, \ldots, \quad (2.12.13)$$

or

$$m' = (m) \text{ modulo } (n).$$

Consider two ways to see this. First, the only nonzero Fourier components which a C_{12} symmetric potential V_{12} could have would be a 12th ($e^{\pm 12i\phi}$), 24th ($e^{\pm 24i\phi}$), etc. Note that the second term in Eq. (2.12.12) equals the $(m' - m)$th Fourier component

$$G(m' - m) = \frac{1}{2\pi}\int d\phi\, e^{-i(m'-m)\phi}V_{12} \quad (2.12.14)$$

of V_{12}. Hence, Eq. (2.12.13) follows. For a second proof note that m and m' belong to the same irrep D^{k_n} of C_n if Eq. (2.7.7) holds. This implies Eq. (2.12.13). Note that the definition of wavevector k_n in Eq. (2.12.6) is

$$k_n = m/r = (2\pi m/nd) \quad (2.12.15)$$

when Eq. (2.12.11) is used. This agrees with the original definition in Eq. (2.7.2) if the unit of distance is the lattice interval ($d = 1$).

The introduction of potential V_{12} may change the eigenstate $|m\rangle$ of H_0 into a new eigenstate $|e(m)\rangle$ of $(H_0 + V_{12})$, which is a combination:

$$|e(m)\rangle = \psi_m|m\rangle + \psi_{m-12}|m - 12\rangle + \psi_{m-24}|m - 24\rangle + \cdots$$
$$+ \psi_{m+12}|m + 12\rangle + \psi_{m+24}|m + 24\rangle + \cdots \quad (2.12.16)$$

of all the states $|m - Nn\rangle$ labeled by the same irrep $D^{m_{12}}$ of C_{12}. Finding the perturbed eigensolutions ψ_j and the eigenvalues can be difficult in general.

SYMMETRY ANALYSIS FOR QUANTUM MECHANICS   119

However, if V_{12} is weak enough one may ignore all but two terms in Eq. (2.12.16), and consider only pairs of states with nearly equal unperturbed energies $\varepsilon_{m'} \sim \varepsilon_m = m^2 E$. For example, $|5\rangle$ and $|-7\rangle$ only differ by $(7^2 - 5^2)E = 24E$, while the next possible contender in Eq. (2.12.16) is $|17\rangle$, which differs by $E(17^2 - 5^2) = 264E$. As long as $V_{12} \ll 264E$ we can ignore $|17\rangle$. The same is true for the pair $\{|7\rangle, |-5\rangle\}$. Approximate eigensolutions made from just these pairs are found by diagonalizing (2×2) H-submatrices derived from Equation (2.12.12) or extracted from Table 2.12.1:

$$\langle H \rangle_{5,-7} = \begin{pmatrix} |5\rangle & |-7\rangle \\ 25E & G(-12) \\ G(12) & 49E \end{pmatrix}, \quad \langle H \rangle_{-5,7} = \begin{pmatrix} |-5\rangle & |7\rangle \\ 25E & G(12) \\ G(-12) & 49E \end{pmatrix}.$$

Note that the eigenvalues (ε) of these two matrices are identical:

$$\varepsilon(\pm 7) = 37E + \left(144E^2 + |G(12)|^2\right)^{1/2} \cong 49E + |G(12)|^2/24E$$

$$\varepsilon(\pm 5) = 37E - \left(144E^2 + |G(12)|^2\right)^{1/2} \cong 25E - |G(12)|^2/24E.$$

(2.12.17)

So for small $|G(12)|^2$ the perturbation V_{12} shifts the $|\pm 7\rangle$ doublet level up slightly, and the $|\pm 5\rangle$ doublet level down by the same amount. This is shown between the (a) and (b) parts of Figure 2.12.1.

The splitting of the $|\pm 6\rangle$ doublet in the same part of Figure 2.12.1 deserves special attention. Now there is only one (2×2) matrix to diagonalize, and it is found at the center of Table 2.12.1:

$$\langle H \rangle_{\pm 6} = \begin{pmatrix} |6\rangle & |-6\rangle \\ 36E & G(-12) \\ G(12) & 36E \end{pmatrix}.$$

Its eigenvectors and eigenvalues are the following:

$$|c_6\rangle = \frac{|6\rangle + |-6\rangle}{\sqrt{2}}, \quad \text{with eigenvalues } \varepsilon(c_6) = 36E + G(12),$$

$$i|s_6\rangle = \frac{|6\rangle - |-6\rangle}{\sqrt{2}}, \quad \text{with eigenvalue } \varepsilon(s_6) = 36E - G(12). \quad (2.12.18)$$

Here it is assumed that $G(12) = G(-12) \leq 0$. This corresponds to picking origin ($x = 0 = \phi$) in the center of an attractive $[V(0) < 0]$ symmetric $[V(x) = V(-x)]$ potential well. Once again (recall Figure 2.7.7) the $(m = \pm 6)$ doublet level is split by a C_{12} perturbation and standing-wave eigenstates

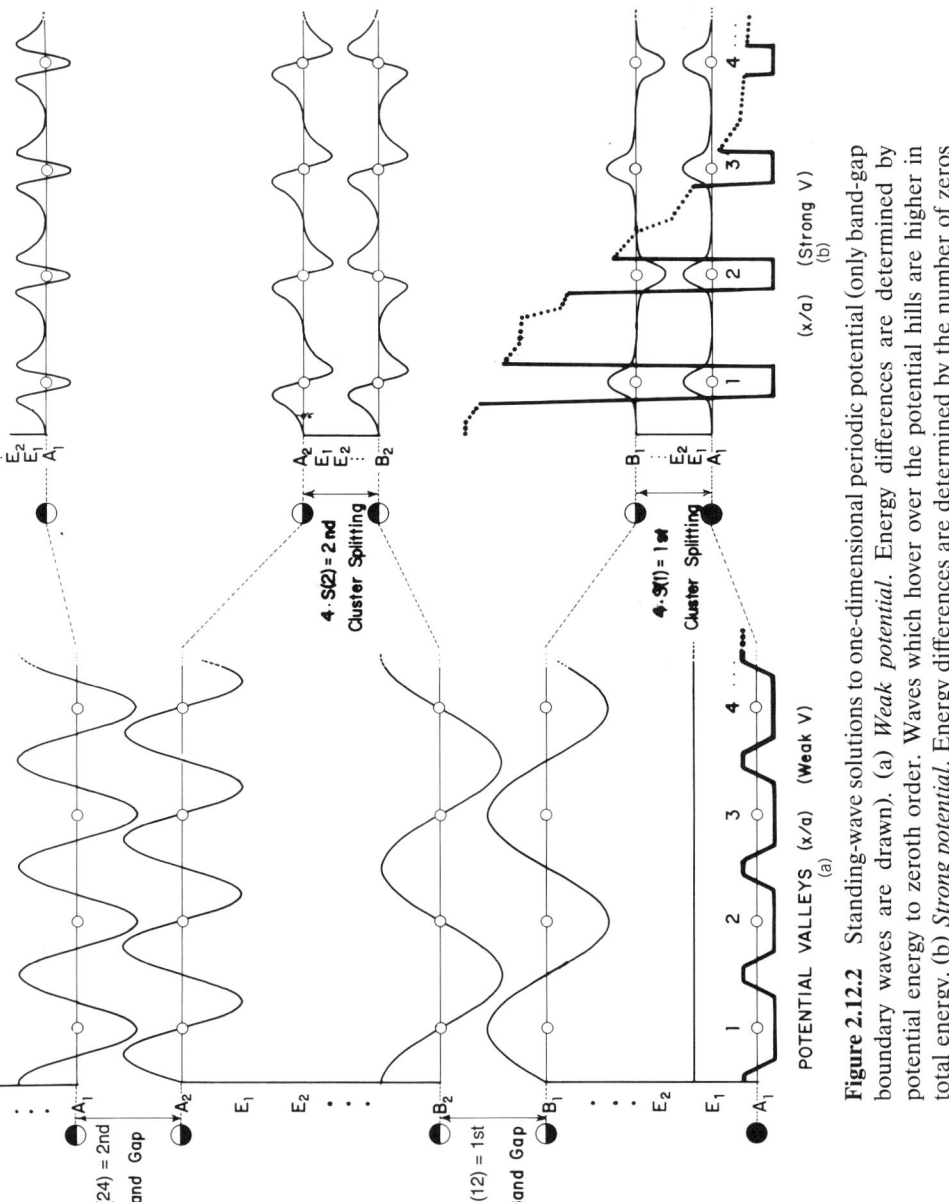

Figure 2.12.2 Standing-wave solutions to one-dimensional periodic potential (only band-gap boundary waves are drawn). (a) *Weak potential*. Energy differences are determined by potential energy to zeroth order. Waves which hover over the potential hills are higher in total energy. (b) *Strong potential*. Energy differences are determined by the number of zeros of nodes in the wave. Waves with more nodes are higher in total energy even if the nodes fall in the center of potential hills

TABLE 2.12.1 H-matrix in plane wave basis for weak C_{12} potential $V(\phi) = V_{12} \ll 1$

m =	0	1	-1	2	-2	3	-3	4	-4	5	-5	6	-6	7	-7	8	-8	9	-9	10	-10	11	-11	12	-12	13
0	0	.	.	.	.	.	.	.	.	.	.	.	.	.	.	.	.	.	.	.	.	.	.	G	G*	.
1	.	1	.	.	.	.	.	.	.	.	.	.	.	.	.	.	.	.	.	.	.	.	.	.	G*	G
-1	.	.	1	.	.	.	.	.	.	.	.	.	.	.	.	.	.	.	.	.	.	G	.	.	.	.
2	.	.	.	4	.	.	.	.	.	.	.	.	.	.	.	.	.	.	.	G*	.	.	.	.	.	.
-2	.	.	.	.	4	.	.	.	.	.	.	.	.	.	.	.	.	.	.	.	G	.	.	.	.	.
3	.	.	.	.	.	9	.	.	.	.	.	.	.	.	.	.	.	G*	.	.	.	.	.	.	.	.
-3	.	.	.	.	.	.	9	.	.	.	.	.	.	.	.	.	G	.	.	.	.	.	.	.	.	.
4	.	.	.	.	.	.	.	16	.	.	.	.	.	.	.	G*	.	.	.	.	.	.	.	.	.	.
-4	.	.	.	.	.	.	.	.	16	.	.	.	.	.	G	.	.	.	.	.	.	.	.	.	.	.
5	.	.	.	.	.	.	.	.	.	25	.	.	.	G*	.	.	.	.	.	.	.	.	.	.	.	.
-5	.	.	.	.	.	.	.	.	.	.	25	.	G	.	.	.	.	.	.	.	.	.	.	.	.	.
6	.	.	.	.	.	.	.	.	.	.	.	36	G*	.	.	.	.	.	.	.	.	.	.	.	.	.
-6	.	.	.	.	.	.	.	.	.	.	.	G	36	.	.	.	.	.	.	.	.	.	.	.	.	.
7	.	.	.	.	.	.	.	.	.	G*	.	.	.	49	.	.	.	.	.	.	.	.	.	.	.	.
-7	.	.	.	.	.	.	.	.	G	.	.	.	.	.	49	.	.	.	.	.	.	.	.	.	.	.
8	.	.	.	.	.	.	.	G*	.	.	.	.	.	.	.	64	.	.	.	.	.	.	.	.	.	.
-8	.	.	.	.	.	.	G	.	.	.	.	.	.	.	.	.	64	.	.	.	.	.	.	.	.	.
9	.	.	.	.	.	G*	.	.	.	.	.	.	.	.	.	.	.	81	.	.	.	.	.	.	.	.
-9	.	.	.	.	G	.	.	.	.	.	.	.	.	.	.	.	.	.	81	.	.	.	.	.	.	.
10	.	.	.	G*	.	.	.	.	.	.	.	.	.	.	.	.	.	.	.	100	.	.	.	.	.	.
-10	.	.	G	.	.	.	.	.	.	.	.	.	.	.	.	.	.	.	.	.	100	.	.	.	.	.
11	.	G*	.	.	.	.	.	.	.	.	.	.	.	.	.	.	.	.	.	.	.	121	.	.	.	.
-11	G	.	.	.	.	.	.	.	.	.	.	.	.	.	.	.	.	.	.	.	.	.	121	.	.	.
12	G*	.	.	.	.	.	.	.	.	.	.	.	.	.	.	.	.	.	.	.	.	.	.	144	.	H*
-12	G*	.	.	.	.	.	.	.	.	.	.	.	.	.	.	.	.	.	.	.	.	.	.	.	144	H
13	.	.	.	.	.	.	.	.	.	.	.	.	.	.	.	.	.	.	.	.	.	.	.	H*	.	169

where: $G = \frac{1}{2\pi}\int d\phi\, e^{i12\phi} V(\phi)$, $G^* = \frac{1}{2\pi}\int d\phi\, e^{-i12\phi} V(\phi)$, $H = \frac{1}{2\pi}\int d\phi\, e^{i24\phi} V(\phi)$, $H^* = \frac{1}{2\pi}\int d\phi\, e^{-i24\phi} V(\phi)$, ⋯
$= G(12)$ $= G(-12)$ $= G(24)$ $= G(-24)$

TABLE 2.12.2 H-matrix in local wave basis for strong C_{12} potential $V_{12} \gg 1$

	1	2	3	4	5	6	7	8	9	10	11	12	1'	2'	3'	4'	5'	6'	⋯
1	H	S	.	.	.	.	.	.	.	.	.	S	.	.	.	.	.	.	
2	S	H	S	.	.	.	.	.	.	.	.	.	.	.	.	.	.	.	
3	.	S	H	S	.	.	.	.	.	.	.	.	.	.	.	.	.	.	
4	.	.	S	H	S	.	.	.	.	.	.	.	.	.	.	.	.	.	
5	.	.	.	S	H	S	.	.	.	.	.	.	.	.	.	.	.	.	
6	.	.	.	.	S	H	S	.	.	.	.	.	.	.	.	.	.	.	
7	.	.	.	.	.	S	H	S	.	.	.	.	.	.	.	.	.	.	
8	.	.	.	.	.	.	S	H	S	.	.	.	.	.	.	.	.	.	
9	.	.	.	.	.	.	.	S	H	S	.	.	.	.	.	.	.	.	
10	.	.	.	.	.	.	.	.	S	H	S	.	.	.	.	.	.	.	
11	.	.	.	.	.	.	.	.	.	S	H	S	.	.	.	.	.	.	
12	S	.	.	.	.	.	.	.	.	.	S	H	.	.	.	.	.	.	
1'	.	.	.	.	.	.	.	.	.	.	.	.	H'	S	.	.	.	S	
2'	.	.	.	.	.	.	.	.	.	.	.	.	S	H'	S	.	.	.	
3'	.	.	.	.	.	.	.	.	.	.	.	.	.	S	H'	S	.	.	
4'	.	.	.	.	.	.	.	.	.	.	.	.	.	.	S	H'	S	.	
5'	.	.	.	.	.	.	.	.	.	.	.	.	.	.	.	S	H'	S	
6'	.	.	.	.	.	.	.	.	.	.	.	.	.	.	.	.	S	H'	
⋮																			

where: H = 1st band center, H' = 2nd band center,   ⋯
$S = S(1)$ = 1st band tunneling, $S = S(2)$ = 2nd band tunneling,   ⋯

become mandatory. The cosine wave stands so its crests and troughs fall in the 12 attractive wells of the potential. The cosine wave is shown by the B_1 wave in Figure 2.12.2(a). It is drawn below the (B_2) sine wave of state $|s_6\rangle$. The sine wave has higher energy, since its crests fall in the high-potential regions, while its nodes stand in the low-potential wells. The energy difference between $|c_6\rangle$ and $|s_6\rangle$ states is called the first BAND GAP. It equals twice the 12th Fourier component of V_{12}:

$$\varepsilon(c_6) - \varepsilon(s_6) = 2G(12). \qquad (2.12.19)$$

When moving-wave orbitals are forced to combine to make standing-wave eigenstates, one generally says that the orbital momentum has been QUENCHED. The same thing happens to the ($m = \pm 12$) pair of states. The cosine state $|c_{12}\rangle$ is labeled A_1 in Figure 2.12.2(a) and drawn above the sine wave state $|s_{12}\rangle$, which is labeled A_2. The splitting of this second band gap depends on the 24th Fourier coefficient of V_{12}; that is,

$$\varepsilon(c_{12}) - \varepsilon(s_{12}) = 2G(24), \qquad (2.12.20)$$

in the approximation which assumed V_{12} is weak. We have assumed $G(24)$ is positive in the figure, but it is easy to make a potential which gives the opposite sign. However, the energy level structure is the same for either sign.

The m-levels within the bands where m is not divisible by ($n/2 = 6$) may belong to moving-wave states. From these states one may construct waves which move around the ring or else make standing waves which stand anywhere on the ring. However, even moving waves with energies near a band edge or m-values approaching $n/2$, n, $3n/2$, etc., may exhibit very sluggish motion. A measure of a wave state's ability to move is derived from the wave ENERGY DISPERSION RELATION $\varepsilon(k_m)$. This is determined (approximately) by the eigenvalues of the Hamiltonian (sub)matrix

$$\langle H \rangle_{m,m-n} = \begin{pmatrix} |m\rangle & |m-n\rangle \\ m^2 E & G(n) \\ G(n) & (m-n)^2 E \end{pmatrix}$$

for small potentials. A good measure of wave motion is found by computing the wave GROUP VELOCITY

$$v_g = \frac{d\omega}{dk} = (\hbar)^{-1} \frac{d\varepsilon(k_n)}{dk_n}. \qquad (2.12.21)$$

It is important not to confuse v_g with the PHASE VELOCITY $v_p = \omega/k$; V_g is proportional to the slope of the energy eigenvalue function plotted versus k_m or m as in Figure 2.12.1. Note how the slopes of the curves which would connect the dark circles in the figure tend toward zero near the band gaps or Brillouin boundaries. As the number (n) of potential wells increases each band becomes a quasicontinuum of energy levels. Then the wave states vary continuously from moving waves to standing waves as the energy approaches a band edge.

(c) Very Hindered Rotation: Bloch Waves and Tunneling For energy eigenstates that lie deep in strong C_n symmetric potentials, many higher terms in Eq. (2.12.16) may be significant. In general the wave function for this state assumes the form of a BLOCH wave:

$$\langle \phi | e(m) \rangle = \sum_{N=-\infty}^{\infty} \psi_{m+Nn} \langle \phi | m+Nn \rangle$$

$$= e^{im\phi} u_e(\phi) \equiv e^{ik_n x} u_e(x), \qquad (2.12.22)$$

where $u(\phi)$ is a local or BLOCH FUNCTION:

$$u_e(\phi) = \sum_{N=-\infty}^{\infty} \psi_{m+Nn} e^{iNn\phi} = u_e(\phi + 2\pi/n), \qquad (2.12.23)$$

which has the C_n symmetry of the periodic potential. (This result is also called FLOQUET'S THEOREM.)

For deeply bound eigenfunctions it is useful to make another approximation for the Bloch solution. With high barriers between each of the n potential wells it makes sense to use the eigenfunctions of each individual well as a basis. Let $|1\rangle$ be an eigenstate for the first well obtained by assuming that it is the only well on the ring. (One can imagine filling in all the other wells with a constant potential equal to the maximum of the potential hills.) Then let the basis $\{|1\rangle, |2\rangle, |3\rangle, \ldots, |n\rangle\}$ of states be defined each by a C_n group operation on the first one; i.e., $|1\rangle = \mathbf{1}|1\rangle$, $|2\rangle = r|1\rangle$ (see Figure 2.12.3), $|3\rangle = r^2|1\rangle, \ldots, |n\rangle = r^{n-1}|1\rangle$. Finally, let the Schrödinger eigenvalue equation $H|\psi\rangle = \varepsilon|\psi\rangle$ be represented in the $\{|1\rangle, |2\rangle, \ldots, |n\rangle\}$ basis by

$$\begin{pmatrix} H & -S & 0 & \cdots & -S \\ -S & H & -S & \cdots & 0 \\ 0 & -S & H & \cdots & 0 \\ \vdots & \vdots & \vdots & & \vdots \\ -S & 0 & 0 & \cdots & H \end{pmatrix} \begin{pmatrix} \langle 1|\psi\rangle \\ \langle 2|\psi\rangle \\ \langle 3|\psi\rangle \\ \vdots \\ \langle n|\psi\rangle \end{pmatrix} = \varepsilon \begin{pmatrix} \langle 1|\psi\rangle \\ \langle 2|\psi\rangle \\ \langle 3|\psi\rangle \\ \vdots \\ \langle n|\psi\rangle \end{pmatrix}, \qquad (2.12.23)$$

There is one such submatrix for each band as shown in Table 2.12.2.

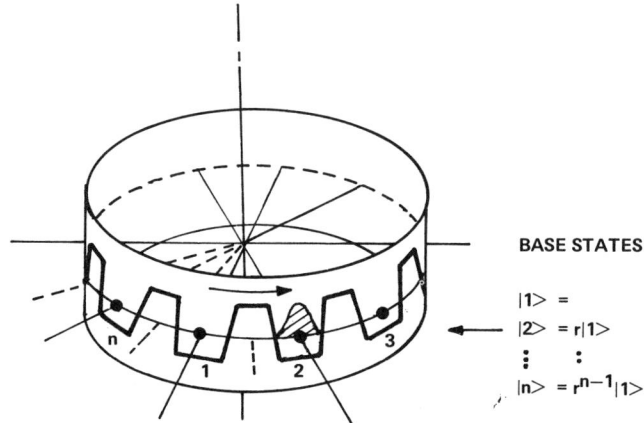

Figure 2.12.3 C_n symmetric ring potential and localized base state $|2\rangle = r|1\rangle$.

where H is the single-well energy, and $-S$ is the tunneling amplitude between nearest neighboring wells in the n-well problem. Amplitude S is proportional to the rate at which a particle originally in well (j) may "sneak" next door to well $(j + 1)$ or $(j - 1)$. Remember, $|j\rangle$ is an eigenstate only if neighboring wells are unavailable to accept the electron; i.e., if $S = 0$.

Except for a change in notation, the matrix in Eq. (2.12.23) is identical to the acceleration matrix in Eq. (2.7.1). The C_n projectors P^{k_n} applied to $|1\rangle$ give moving-wave eigenvectors:

$$|k_n\rangle = P^{k_n}|1\rangle\sqrt{n} = (|1\rangle + e^{-ik_n}|2\rangle + e^{-2ik_n}|3\rangle + \cdots)/\sqrt{n}, \quad (2.12.24)$$

which have eigenvalues

$$\varepsilon_n = H - 2S \cos k_n, \quad (2.12.25)$$

where the distance between potential wells is set equal to unity ($d = 1$). The latter equation gives the well-known cosine dispersion function $\varepsilon_n(k_n)$ for bands of tightly bound electrons. This result is indicated by the 12-fold polygonal projections which are drawn next to the energy levels in Figure 2.12.1(c). The bandwidth ($4S$) between the band edge states $|k_1\rangle$ and $|k_6\rangle$ is schematically exaggerated for the two bands in the figure. In the limit of strong potentials the bandwidths should be a tiny fraction of the band gaps. When propagation is reduced to a slow "oozing" or tunneling at rate S and this corresponds to a small group velocity and nearly zero slopes along $\varepsilon(k_n)$.

As in the treatment of pendulum waves the sine ($|s_n\rangle$) and cosine ($|c_n\rangle$) wave states are easier to picture:

$$|c_n^k\rangle = (|k_n\rangle + |-k_n\rangle)/\sqrt{2} = (|1\rangle + \cos k_n|2\rangle + \cdots)(2/n)^{1/2},$$
$$|s_n^k\rangle = -i(|k_n\rangle - |-k_n\rangle)/\sqrt{2} = (\sin k_n|2\rangle + \cdots)(2/n)^{1/2}.$$

Examples for $n = 3$ and 4 are shown in Figure 2.12.4. These are the analog of the pendulum standing waves depicted in Figure 2.7.3. Figure 2.12.4 is an attempt to display simple molecular orbitals. However, the "blob" wave functions drawn there cannot tell one much about the local potential well wave functions $\{\langle x|1\rangle, \langle x|2\rangle, \ldots, \langle x|n\rangle\}$. The waves in the figure depict the $e^{ik_n x}$ (or $\cos k_n x$) part of the Bloch wave [Eq. (2.12.22)], not the local $[u_e(x) = \langle x|1\rangle]$ part. The former varies from state to state within a given band, while the latter varies from band to band.

A more detailed picture of Bloch waves is shown in Figure 2.12.2(b). Note that the local Bloch function $[u(x) = \langle x|1\rangle]$ has the same shape in all wells for all states within a given band. It consists of zero, one, two, ... rapid oscillations within each well for the first, second, and third bands, respectively. The local functions are modulated by a more or less slowly varying envelope function ($e^{ik_n x}$ or $\cos k_n x$) to give each Bloch wave.

The node structure of Bloch waves is an important consideration. According to an elementary theorem of quantum mechanics, a one-dimensional wave with more nodes always has more total energy. (This was proved by

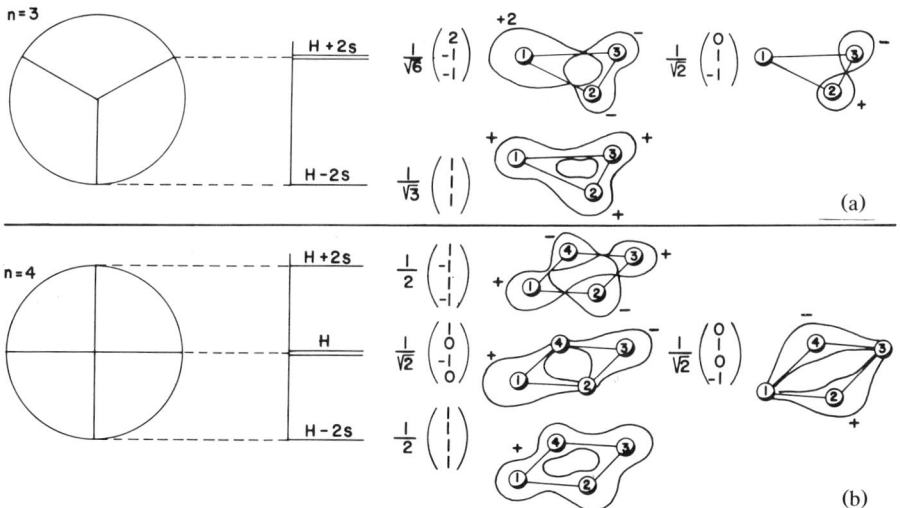

Figure 2.12.4 Sketches of molecular orbital wave functions and tunneling spectra for homocyclic potentials. (a) C_3 symmetry, (b) C_4 symmetry.

Schrödinger). Therefore the B_1 wave ($m = n/2$) in Figure 2.12.2(b), which has n nodes, must be higher in energy than all the states with lesser $m = 0, 1, 2, \ldots$, which have zero, two, four, ... nodes, respectively. The A_1 wave ($m = 0$), which has no nodes, is lowest in energy. This is paradoxical, since the A_1 wave has nonzero amplitudes in the centers of the high-potential regions where the B_1 wave is zero. (Schrödinger's theorem is *not* a trivial result.) Therefore the negative sign of $(-S)$ in Eq. (2.12.25) is right for the lowest band, since then the ($k = 0$) or A_1 wave is the lowest state.

In the second band, however, (S) changes sign as well as magnitude. The B_2 wave belongs to the ($m = n/2$) irrep of C_n, as does the B_1 wave on top of the first band. (B can stand for Brillouin or band boundary.) Starting with the B_2 wave, which has n nodes, one proceeds upward in energy with $n + 2, n + 4, \ldots, 2n - 2$, and finally $2n$ nodes while the m label *decreases*: $m = n/2 - 1, n/2 - 2, \ldots, 1$, and finally $m = 0$ for the A_2 wave on top of the second band. Notice that adding one more node inside each potential well increases the energy by roughly one whole band gap. Increasing the number of nodes between wells may increase the energy by only one bandwidth at the most.

(d) Intermediate Potentials: Exact Solutions It is instructive to study exact solutions to the Schrödinger eigenvalue equation,

$$\frac{d^2\psi}{d\phi^2} + (2\mu r^2/\hbar^2)(\varepsilon - V(\phi))\psi = 0, \qquad (2.12.26)$$

for a mass μ particle on a ring of radius r subject to a C_n symmetric potential $V(\phi) = V(\phi + 2\pi/n)$. Two cases that have been analyzed are the cosine potential,

$$V(\phi) = -V\cos(n\phi), \qquad (2.12.27)$$

and the n-square-well potential

$$V(\phi) = U(n\phi), \qquad (2.12.28)$$

where,

$$U(\theta) = \begin{cases} 0, & -\pi/2 < \theta < \pi/2, \\ U, & \pi/2 < \theta < 3\pi/2. \end{cases}$$

These two cases represent two opposite extremes. The cosine potential has only one nonzero Fourier coefficient, namely, $G(n)$, since it is, after all, just one cosine wave. The n-square well, on the other hand, has the most slowly converging Fourier series of any n-well potential of finite depth; i.e., $G[(2N + 1)n] = (-1)^N 4\pi/(2N + 1) \{N = 0, 1, 2, \ldots\}$.

For the cosine potential the Schrödinger equation (2.12.26) becomes MATHIEU'S equation

$$\frac{d^2\psi}{dt^2} + A[\varepsilon + V\cos(2t)]\psi = 0, \qquad (2.12.29)$$

where $t \equiv \phi n/2$ and

$$A = 8\mu r^2/n^2\hbar^2.$$

Mathieu's equation is treated in most standard texts on mathematical physics. The eigenvalues ε of the A and B standing-wave solutions are plotted in Figure 2.12.5 for a range of well depth V. Recall that the A levels belong to m-values for which $m = 0$ modulo n; the B levels belong to $m = n/2$ modulo n. (No B levels exist for odd n.) A and B levels are the boundaries between bands and gaps. Notice that the first, second, third, and Nth gaps originate at $\varepsilon = 1, 4, 9, \ldots$ and N^2, respectively, in the units $[2m^2\hbar^2/n^2\mu r^2]$ of the figure axis for $V = 0$. [This corresponds to our case (a)]. Notice that only the first gap is open for small V [case (b)]. This is because $G(Nn)$ is zero except for $N = 1$. Notice that the A and B levels run together and the bandwidths shrink rapidly for levels $E = \varepsilon$ that are much less than V. This corresponds to our case (c), which was treated in the preceding section. Each band between the A and B curves in Figure 2.12.5 contains $(n/2 - 1)[(n/2 - 1/2)]$ doubly degenerate moving-wave levels for n even (odd). These are not drawn, since they depend upon your choice for n.

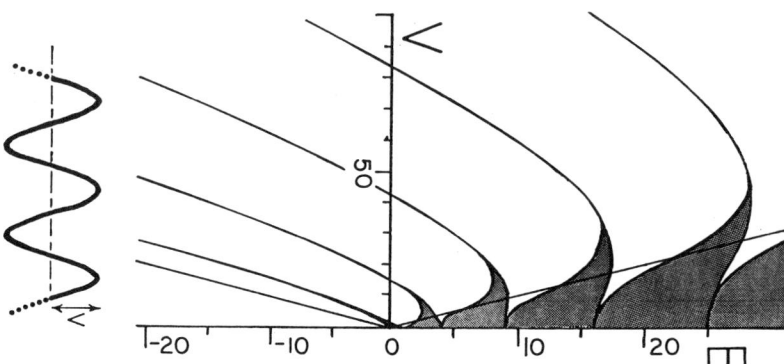

Figure 2.12.5 Energy versus V plots for band-gap edges for sinusoidal potential (Mathieu equation).

The solutions for the n-square-well potential are the well-known KRONIG-PENNEY bands. The A- and B-band boundary levels are plotted in Figure 2.12.6 using Kronig-Penney equations which are given in many quantum mechanics and solid state texts. The energy E is measured from the bottom of the wells instead of from a point halfway up, as it was in the preceding figure. In this way it is easy to see that the case (c) "tight bands" approach asymptotes at $E = 4, 16, 36, \ldots$. These correspond to the infinite-square-well energies.

Aside from this, the main thing which distinguishes the square-well solutions is the remarkable crossing of A and B levels on the case (b) or nearly free side of the spectrum where $E > U$. This comes about whenever the energy E and the potential U are adjusted so that an integral number n_w of half-waves fit into each well while another integral number n_p of half-waves

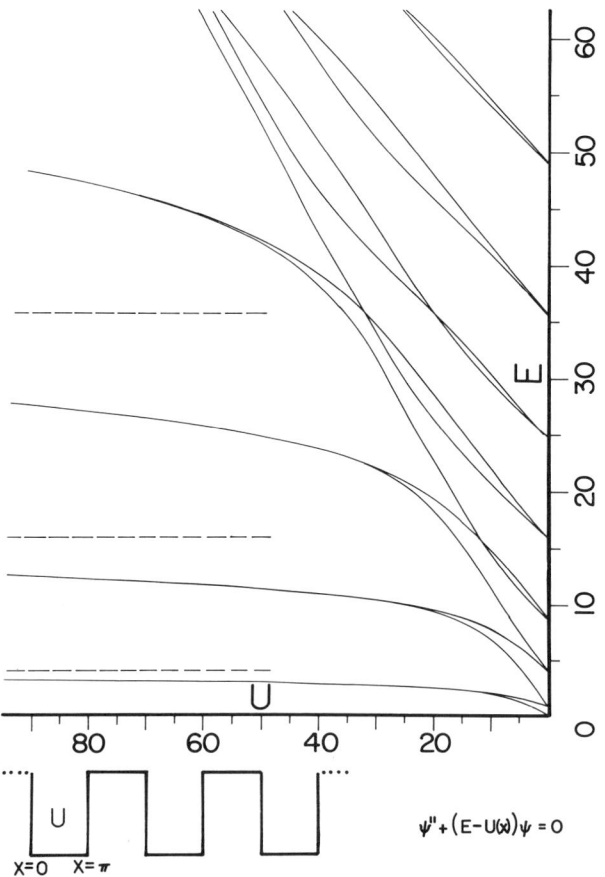

Figure 2.12.6 Energy versus U plots for band-gap edges for square-well potential (Kronig-Penney equation).

fit on the plateau between wells. For example, with $n_w = 3$ and $n_p = 2$ there is a crossing at $E = (2n_w)^2 = 36$ and $U = (2n_w)^2 - (2n_p)^2 = 20$ in Figure 2.12.6. The double degeneracy at this point corresponds to sine and cosine waves with the same energy.

(e) Comparison: When Is Parameter Fitting Useless? The approximate modeling of subsection (c) treats each band subspace as though it was all alone. It provides no information about where a given band is or how wide it is. Location and width depend on the "unknown" parameters H and S, respectively. Furthermore, the approximate theory fails to account for the mixing of states in one band with those belonging to other bands. One effect of this mixing would be the alteration of the cosine level spacing predicted by Eq. (2.12.25).

However, an approximate modelist could say, "But, wait... I have some other parameters besides H and S, namely, T, for next-nearest-neighbor tunneling, and U for next-next...." Then he could give you a formula that would "explain" the modified spacing (see Problem 2.12.3). Eventually you can pull out as many parameters as there are level spaces; however, this is no more useful than a 52-parameter theory for weekly rainfall averages of 1927.

The approximate tunneling models are quite valid when V/E is large or when the energy gap between subspaces is large. The signal for impending failure is the need for too many parameters to reproduce the correct energy level spacing.

A more detailed formulation of the S parameter in the very hindered case is given in Section 2.12.D using semiclassical WKBJ formulas.

B. C_2 Symmetry and the Two-Level System

The C_2-like groups only have two irreps corresponding to even [A or $(+)$] and odd [B or (-1)]. Hence all the energy levels $\varepsilon^{(A, B)}$ in the four lowest spectral bands of a two-well potential of depth $(0 < V < 100)$ are given by the curves in Figures 2.12.5 and 2.12.6 for cosine and square wells, respectively. In either figure the lower-lying spectrum for large V consists of pairs of nearly degenerate levels. This corresponds to case (c) [recall Section 2.12.A(c)] in which the two wells are nearly isolated from each other by large barriers of height U or V.

In this limit let each pair of states be described by a Schrödinger equation:

$$i\hbar \frac{\partial}{\partial t}|\psi\rangle = H|\psi\rangle, \quad (2.12.30a)$$

represented by

$$i\hbar \frac{\partial}{\partial t}\begin{pmatrix}\langle 1|\psi\rangle \\ \langle 2|\psi\rangle\end{pmatrix} = \begin{pmatrix} H & -S \\ -S & H \end{pmatrix}\begin{pmatrix}\langle 1|\psi\rangle \\ \langle 2|\psi\rangle\end{pmatrix} \quad (2.12.30b)$$

in the basis $|1\rangle$ and $|2\rangle$ for which the particle is localized in wells 1 and 2, respectively. The Schroïdinger eigenequation,

$$H|\psi\rangle = \varepsilon|\psi\rangle, \tag{2.12.31a}$$

$$\begin{pmatrix} H & -S \\ -S & H \end{pmatrix} \begin{pmatrix} \langle 1|\psi\rangle \\ \langle 2|\psi\rangle \end{pmatrix} = \varepsilon \begin{pmatrix} \langle 1|\psi\rangle \\ \langle 2|\psi\rangle \end{pmatrix} \tag{2.12.31b}$$

is a special case of Eq. (2.12.23). The eigensolutions may be obtained using C_2 projectors $P^+ = P^A$ and $P^- = P^B$:

$$|\psi^{(+)}\rangle = P^+|1\rangle\sqrt{2} = (|1\rangle + |2\rangle)/\sqrt{2}, \quad \text{eigenvalue } \varepsilon^{(+)} = H - S,$$
$$|\psi^{(-)}\rangle = P^-|1\rangle\sqrt{2} = (|1\rangle - |2\rangle)/\sqrt{2}, \quad \text{eigenvalue } \varepsilon^{(-)} = H + S.$$
$$\tag{2.12.32}$$

The time behavior of eigenstate $|\psi^{(+)}\rangle$ and $|\psi^{(-)}\rangle$ according to Eq. (2.12.30) is simple harmonic phase oscillation at angular eigenfrequency $(H - S)/\hbar$ and $(H + S)/\hbar$, respectively,

$$|\psi^{(+)}(t)\rangle = e^{(H-S)t/i\hbar}|\psi^{(+)}(0)\rangle,$$
$$|\psi^{(-)}(t)\rangle = e^{(H+S)t/i\hbar}|\psi^{(-)}(0)\rangle.$$

This is analogous to the behavior of the (+) and (−) resonant modes of the two-pendulum system described in Section 2.3.

One of the most well-known applications of these solutions involves the ammonia (NH_3) inversion doublet levels. The base states are imagined to be $|1\rangle = |up\rangle$ and $|2\rangle = |dn\rangle$ for which the N atom lies in a potential well above and below the H_3 plane, respectively, as shown in Figure 2.12.7. If the N atom can tunnel between $|up\rangle$ and $|dn\rangle$ there is an energy splitting between eigenstates $|(+)\rangle$ and $|(-)\rangle$. The splitting is equal to $2S$, and the tunneling frequency is the difference frequency $\omega = 2S/\hbar$. Tunneling is analogous to "beat trading" between two pendulums in Figure 2.3.2.

Consider what happens to the ammonia doublet states when an electric field is applied along the direction of (N)-atom motion. The field breaks the C_2 symmetry (to be precise the reflection symmetry of NH_3 is called C_h symmetry according to Section 2.11 and makes the up-field state $|up\rangle$ energetically less favorable than the down-field state $|dn\rangle$. The Hamiltonian matrix becomes[1]

$$\begin{pmatrix} \langle up|H|up\rangle & \langle up|H|dn\rangle \\ \langle dn|H|up\rangle & \langle dn|H|dn\rangle \end{pmatrix} = \begin{pmatrix} H - pE & -S \\ -S & H + pE \end{pmatrix}, \tag{2.12.33}$$

where E is the field strength and p is the dipole moment of the N atom. The effect of the E field on the energy eigenvalues is to make them go further

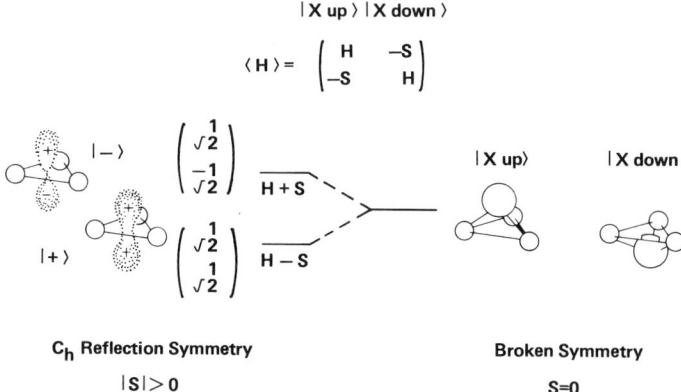

Figure 2.12.7 Two-state model for ammonia (NH$_3$) inversion. The case ($S = 0$) corresponds to internal or "spontaneous" symmetry breaking.

apart, as shown in Figure 2.12.8. The figure shows a plot of the eigenvalues ε:

$$\varepsilon = H \pm \left[S^2 + p^2 E^2\right]^{1/2}. \tag{2.12.34}$$

For small E field ($pE \ll S$) ε is given by

$$\varepsilon = H \pm \left[S + p^2 E^2/(2S)\right],$$

while for large E field ($pE \gg S$) ε is linear in E:

$$\varepsilon = H \pm pE.$$

The large E field eigenstates are $|\text{up}\rangle$ and $|\text{dn}\rangle$, the original base states. This is indicated by the drawings next to the energy trajectories in Figure 2.12.8.

The hyperbola in the figure is the simplest example of what spectroscopists call an "avoided level crossing." One can imagine that the straight-line trajectory for $|\text{up}\rangle$ is crossing that of $|\text{dn}\rangle$. If $S = 0$ that is exactly what would occur. However, with nonzero tunneling $|\text{up}\rangle$ and $|\text{dn}\rangle$ get mixed up to make $|\psi^{(+)}\rangle$ and $|\psi^{(-)}\rangle$ at $E = 0$. What starts out being the $|\text{dn}\rangle$ trajectory curves around at ($E = 0$) and goes out on the $|\text{up}\rangle$ trajectory, and vice versa. It can be shown that if you vary E slowly ($\dot{E} \ll S$) from large positive to large negative values then state $|\text{dn}\rangle$ does indeed turn into state $|\text{up}\rangle$ and vice versa. This is called ADIABATIC FOLLOWING. However, if you make the same change suddenly ($\dot{E} \gg S$) the initial state $|\text{up}\rangle$ or $|\text{dn}\rangle$ will not have time to change. The effect will be to jump the curves and cross over to the other branch of the hyperbola. In this way the field will change the energy of the system. Two-state dynamics are discussed in Sections 7.5 and 8.5.

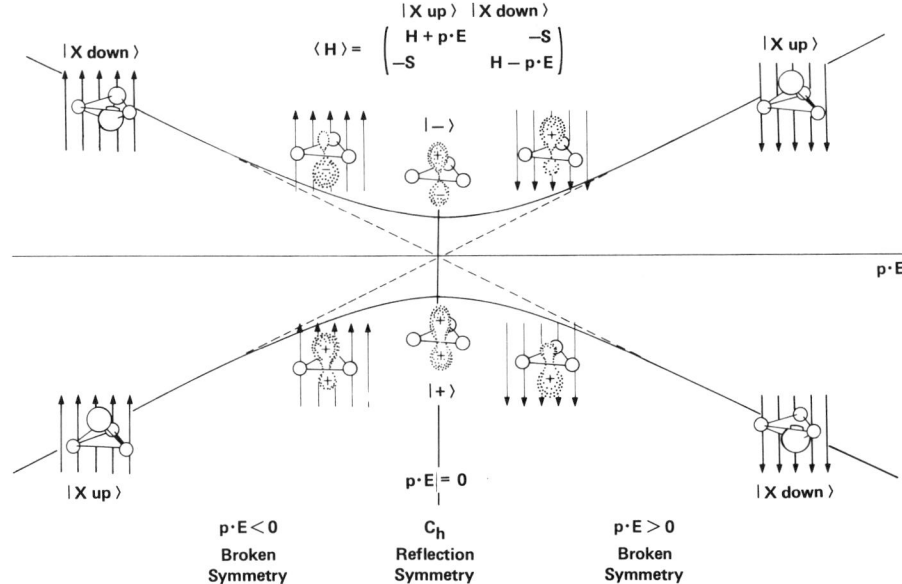

Figure 2.12.8 Effect of axial electric field on ammonia two-state eigensolutions. Cases with nonzero field ($p \cdot E \neq 0$) correspond to external or applied symmetry breaking.

C. C_2 Symmetry Analysis and Scattering Theory

Consider the scattering wave functions ψ_1 and ψ_2 for a single particle outside of a square-well potential as indicated by Figure 2.12.9:

$$\psi_1(x) = I_1 e^{ikx} + O_1 e^{-ikx}, \quad (x < -a)$$
$$\psi_2(x) = I_2 e^{-ikx} + O_2 e^{ikx} \quad (x > a). \tag{2.12.35}$$

The outside or scattering waves must match the "inside" wave

$$\psi_\omega = A e^{ilx} + B e^{-ilx} \quad (-a < x < a), \tag{2.12.36}$$

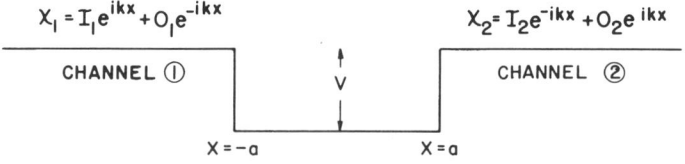

Figure 2.12.9 C_2 symmetric scattering potential.

where their respective wave vectors are given by

$$l = \sqrt{2mE/h}, \quad k = \sqrt{2m(E-V)}h. \quad (2.12.37)$$

The match is made through boundary conditions:

$$\psi_1(-a) = \psi_\omega(-a), \quad \psi_\omega(a) = \psi_2(a),$$
$$\psi_1'(-a) = \psi_\omega'(-a), \quad \psi_\omega'(a) = \psi_2'(a),$$

on either side of the well. Substituting these conditions and eliminating constants A and B yields a linear "S-matrix" relation between the ingoing amplitudes I_j in channels $j = 1$ and 2, and the corresponding outgoing amplitudes O_j as follows:

$$\begin{pmatrix} O_1 \\ O_2 \end{pmatrix} = \begin{pmatrix} (i/D)(l^2 - k^2)\sin 2la & 2lk/D \\ 2lk/D & (i/D)(l^2 - k^2)\sin 2la \end{pmatrix} \begin{pmatrix} I_1 \\ I_2 \end{pmatrix}, \quad (2.12.38a)$$

where

$$D \equiv -e^{2ika}(2lk\cos 2la - i(l^2 + k^2)\sin 2la). \quad (2.12.38b)$$

In general the presence of any penetrable potential well would yield an S-matrix relation

$$\begin{pmatrix} O_1 \\ O_2 \end{pmatrix} = \begin{pmatrix} S_{11} & S_{12} \\ S_{21} & S_{22} \end{pmatrix} \begin{pmatrix} I_1 \\ I_2 \end{pmatrix}, \quad (2.12.39)$$

where the components S_{ij} are complicated functions of energy E and the potential. As explained in Section 2.1.3A, the S matrix is unitary if the particles are not destroyed or created by the well.

Without looking at the explicit form of the S matrix, we can tell from the left-to-right (C_2) symmetry analysis [following Section 2.3, Eq. (2.3.4)] that the following vectors must be the eigenvectors of this matrix:

$$\mathscr{X}^+ = \begin{pmatrix} 1/\sqrt{2} \\ 1/\sqrt{2} \end{pmatrix}, \quad \mathscr{X}^- = \begin{pmatrix} 1/\sqrt{2} \\ -1/\sqrt{2} \end{pmatrix}. \quad (2.12.40)$$

Furthermore, since the S matrix is unitary [recall Eq. (1.3.4)] we know that the eigenvalues will be of the form $e^{i\mu_+}$ and $e^{i\mu_-}$. Just this much information by itself can simplify the visualization of the scattering process. Suppose the incoming amplitudes are proportional to the components of $\mathscr{X}^+$ as follows:

$$\begin{pmatrix} I_1 \\ I_2 \end{pmatrix} = A\mathscr{X}^+ = \begin{pmatrix} A/\sqrt{2} \\ A/\sqrt{2} \end{pmatrix}. \tag{2.12.41}$$

Then the following must be the resulting outgoing amplitudes:

$$\begin{pmatrix} O_1 \\ O_2 \end{pmatrix} = e^{i\mu_+} \begin{pmatrix} A/\sqrt{2} \\ A/\sqrt{2} \end{pmatrix}. \tag{2.12.42}$$

The resulting waves in channels 1 and 2 are then given by

$$\begin{pmatrix} \psi_1 \\ \psi_2 \end{pmatrix} = \begin{pmatrix} I_1 e^{ikx} + O_1 e^{-ikx} \\ I_2 e^{-ikx} + O_2 e^{ikx} \end{pmatrix} = \frac{A}{\sqrt{2}} \begin{pmatrix} e^{ikx} + e^{-i(kx - \mu_+)} \\ e^{-ikx} + e^{i(kx + \mu_+)} \end{pmatrix}$$

$$= \sqrt{2} A e^{i(\mu_+/2)} \begin{pmatrix} \cos\left(kx - \dfrac{\mu_+}{2}\right) \\ \cos\left(kx + \dfrac{\mu_+}{2}\right) \end{pmatrix}. \tag{2.12.43}$$

This represents the (+) or EVEN STANDING-WAVE scattering solution. Similarly, the choice of the ingoing amplitudes given by

$$\begin{pmatrix} I_1 \\ I_2 \end{pmatrix} = B\mathscr{X}^- = \begin{pmatrix} B/\sqrt{2} \\ -B/\sqrt{2} \end{pmatrix}$$

yields the (−) or ODD STANDING-WAVE scattering solution:

$$\begin{pmatrix} \psi_1 \\ \psi_2 \end{pmatrix} = \begin{pmatrix} I_1 e^{ikx} + O_1 e^{-ikx} \\ I_2 e^{-ikx} + O_2 e^{ikx} \end{pmatrix} = \sqrt{2} B e^{i(\mu_-/2)} \begin{pmatrix} \cos\left(kx - \dfrac{\mu_-}{2}\right) \\ -\cos\left(kx + \dfrac{\mu_-}{2}\right) \end{pmatrix}. \tag{2.12.44}$$

Equations (2.12.43) and (2.12.44) give the form of the waves for what are called EIGENCHANNEL SCATTERING STATES or PARTIAL WAVES of even (+) and odd (−) symmetry, respectively.

In Figure 2.12.10 these waves are sketched to show how the two types evolve with a steady increase of V. Note that the behavior of either type is related to the behavior of its EIGENPHASE SHIFT $\delta_\alpha \equiv \mu_\alpha/2$. Notice that phase shift δ_+, for example, is fairly constant, while V varies until one

SYMMETRY ANALYSIS FOR QUANTUM MECHANICS 135

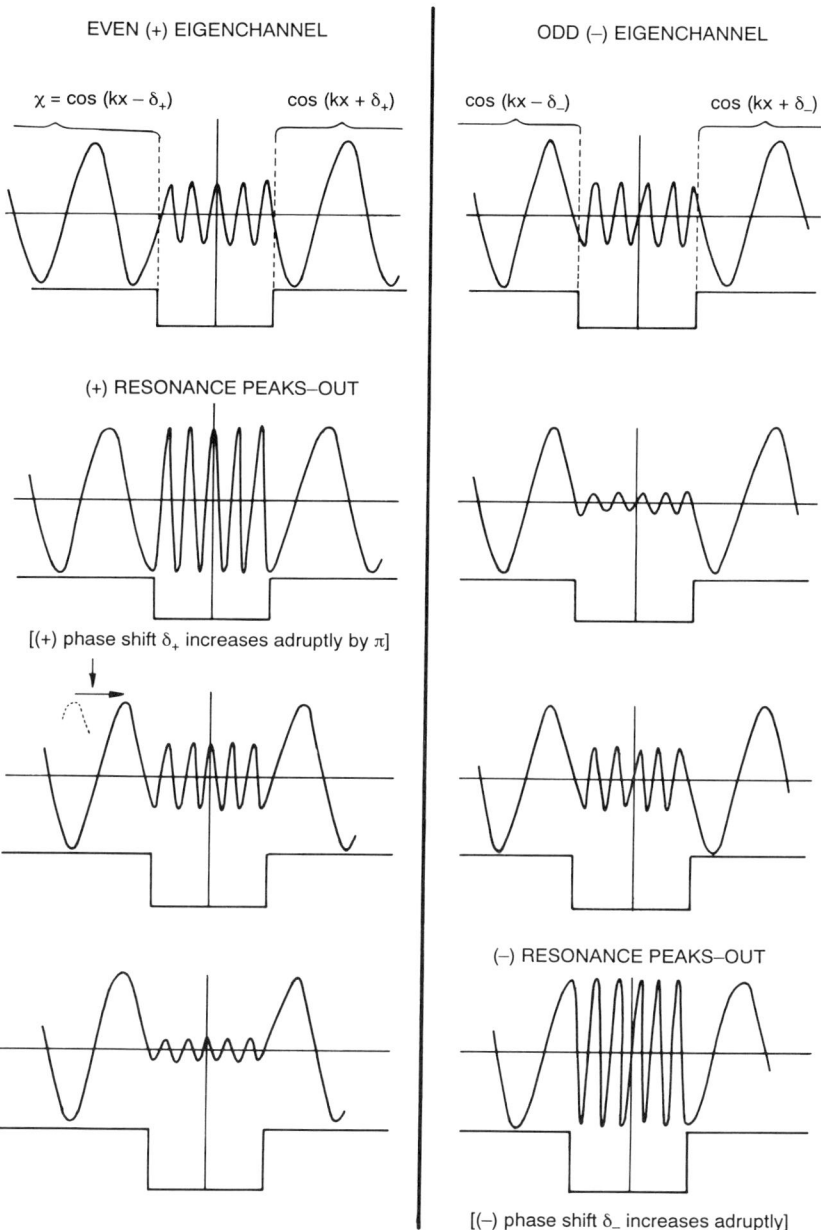

Figure 2.12.10 Scattering eigensolutions for C_2 symmetric square-well potential.

approaches a scattering RESONANCE of type (+). This occurs when an integral number of wavelengths fit in the well. In passing the (+) resonance the shift δ_+ quickly jumps by about π as another crest is more or less abruptly swallowed by the well. The same applies to (−) resonances except that they occur when a half-integral number of waves fit in the well.

Now ordinary transmission-reflection scattering can be expressed in terms of the (+) and (−) standing waves since they are a complete basis set. The most general scattering function can be represented by

$$\Psi = A^+ \mathscr{X}^+ + A^- \mathscr{X}^- \qquad (2.12.45)$$

using a combination of standing-wave solutions:

$$x^- = \mathscr{X}^+ \cos(kr + \delta_+), \qquad x^- = \mathscr{X}^- \cos(kr + \delta_-)$$

$$= \begin{pmatrix} 1/\sqrt{2} \\ 1/\sqrt{2} \end{pmatrix} \cos(kr + \delta_+), \quad = \begin{pmatrix} 1/\sqrt{2} \\ -1/\sqrt{2} \end{pmatrix} \cos(kr + \delta_-) \quad (2.12.46)$$

and complex coefficients A^+ and A^-. Note that the "radial factors" $\cos(kr + \delta_\alpha)$ are functions of r, where $r = x$ for $x > a$ and $r = -x$ for $x < -a$.

The two components of the "angular factors" or vectors ($\mathscr{X}$) refer to channel (1) (left) or (2) (right), respectively. For example, a scattering state that had only outgoing particles in channel (2) would be represented by the following wave function:

$$\Psi = A^+ \mathscr{X}^+ \cos(kr + \delta_+) + A^- \mathscr{X}^- \cos(kr + \delta_-)$$

$$= \begin{pmatrix} A^+ \cos(kr + \delta_+) + A^- \cos(kr + \delta_-) \\ A^+ \cos(kr + \delta_+) - A^- \cos(kr + \delta_-) \end{pmatrix} 1/\sqrt{2} = \begin{pmatrix} ? = \psi_1 \\ \alpha e^{ikr} \end{pmatrix}.$$

$$(2.12.47)$$

The last equation gives the boundary condition, which yields a relation between A^+ and A^-:

$$\left(\frac{A^+ e^{i\delta_+} - A^- e^{i\delta_-}}{2\sqrt{2}} \right) e^{ikr} + \left(\frac{A^+ e^{-i\delta_+} - A^- e^{-i\delta_-}}{2\sqrt{2}} \right) e^{-ikr} = \alpha e^{ikr},$$

$$A^+ = A^- e^{i(\delta_+ - \delta_-)}. \qquad (2.12.48)$$

Substituting Eq. (2.12.48) into the first component of Eq. (2.12.47) gives the following:

$$\psi_1 = 2A^-[\cos(\delta_+ - \delta_-)e^{ikr}e^{i\delta_+} + e^{-ikr}e^{-i\delta_-}]. \qquad (2.12.49)$$

This shows that the reflected wave e^{ikr} in channel (1) will vanish when $\delta_+ - \delta_- = n\pi/2$. This corresponds to the perfect transmission that occurs at a resonance and is called the RAMSAUER-TOWNSEND effect in the theory of scattering.

D. Crossing Matrices and One-Dimensional Tunneling

The boundary conditions in piecewise constant potential provide relations between the amplitudes of right-hand and left-hand moving-wave solutions. If x_{12} is at the boundary between two different regions then continuity of the wave function ψ and its derivative $\psi' = d\psi/dx$ gives

$$R_1 e^{g_1 x_{12}} + L_1 e^{-g_1 x_{12}} = R_2 e^{g_2 x_{12}} + L_2 e^{-g_2 x_{12}},$$

$$R_1 g_1 e^{g_1 x_{12}} - L_1 g_1 e^{-g_1 x_{12}} = R_2 g_2 e^{g_2 x_{12}} - L_2 g_2 e^{-g_2 x_{12}},$$

where the exponential factors corresponding to the two sides are

$$g_1 = \left[\frac{2m}{\hbar}(V_1 - E)\right]^{1/2}, \qquad g_2 = \left[\frac{2m}{\hbar}(V_2 - E)\right]^{1/2}.$$

These factors are real (for $V_j > E$) or imaginary (for $V_j < E$). These relations may be solved to give the *crossing-matrix* relation

$$\begin{pmatrix} R_1 \\ L_1 \end{pmatrix} = \begin{pmatrix} C_{11} & C_{12} \\ C_{21} & C_{22} \end{pmatrix} \begin{pmatrix} R_2 \\ L_2 \end{pmatrix}, \qquad (2.12.50a)$$

where the C-matrix components are the following:

$$C_{11} = e^{-(g_1 - g_2)x_{12}}\left(\frac{g_1 + g_2}{2g_1}\right), \qquad C_{12} = e^{-(g_1 + g_2)x_{12}}\left(\frac{g_1 - g_2}{2g_1}\right),$$

$$C_{21} = e^{(g_1 + g_2)x_{12}}\left(\frac{g_1 - g_2}{2g_1}\right), \qquad C_{22} = e^{(g_1 - g_2)x_{12}}\left(\frac{g_1 + g_2}{2g_1}\right). \qquad (2.12.50b)$$

The crossing matrix for two or more boundaries is simply the matrix product of the C matrices for each boundary in the order in which they occur. For the square-well example in Figure 2.12.9 the crossing matrix relation is

$$\begin{pmatrix} I_1 \\ O_1 \end{pmatrix} = \begin{pmatrix} C_{11} & C_{12} \\ C_{21} & C_{22} \end{pmatrix} \begin{pmatrix} O_2 \\ I_2 \end{pmatrix}$$

where:

$$C_{11} = e^{i2ka}\left[\cos 2la - i\frac{l^2 + k^2}{2kl}\sin 2la\right], \quad C_{12} = -i\frac{l^2 - k^2}{2kl}\sin 2la$$

$$C_{21} = i\frac{(l^2 - k^2)}{2kl}\sin 2la, \quad C_{22} = e^{-i2ka}\left[\cos 2la + i\frac{l^2 + k^2}{2kl}\sin 2la\right].$$

(2.12.51)

Note that the identification of amplitudes $O_2 = R_2$ and $I_2 = L_2$, which is consistent with Eq. (2.12.35).

For an arbitrary one-dimensional potential system the C matrix and S matrix may be related by solving their respective relations (2.12.50) and (2.12.39):

$$\begin{pmatrix} S_{11} & S_{12} \\ S_{21} & S_{22} \end{pmatrix} = \begin{pmatrix} C_{21}/C_{12} & \det C/C_{11} \\ 1/C_{11} & -C_{12}/C_{11} \end{pmatrix} \quad (2.12.52a)$$

$$\begin{pmatrix} C_{11} & C_{12} \\ C_{21} & C_{22} \end{pmatrix} = \begin{pmatrix} -\det S^*/S_{21}^* & S_{11}^*/S_{21}^* \\ S_{11}/S_{21} & -\det S/S_{21} \end{pmatrix}. \quad (2.12.52b)$$

Note that it is generally true that $\det C = 1$, and the S matrix is unitary.

The crossing-matrix methods can be extended to treat WKBJ approximate solutions of the form

$$\psi = (Re^{\theta} + Le^{-\theta})/N^{1/2}, \quad (2.12.53a)$$

where the exponents

$$\theta = \int k(x)\,dx \quad (2.12.53b)$$

and normalization

$$N = \frac{2\pi\hbar^2}{m}k(x) \quad (2.12.53c)$$

depend on the potential through the wave vector

$$k(x) = \left[\frac{2m}{\hbar^2}(E - V(x))\right]^{1/2}, \quad (2.12.53d)$$

which may be real or imaginary. The WKBJ approximations are most accurate when the potential varies slowly compared to wave function. The crossing matrices are used to connect the amplitudes at neighboring classical turning points where the wave function reaches an inflection point and the WKBJ solution fails.

For example, the crossing matrix which relates the amplitudes across a potential barrier from points a to b in Figure 2.12.11 is approximately given by

$$C_{\text{barrier}} = \begin{pmatrix} [1 + \theta^2]^{1/2} & i\theta \\ -i\theta & [1 + \theta^2]^{1/2} \end{pmatrix}, \quad (2.12.54a)$$

where

$$\theta = e^{\int_a^b |k(x)| dx}. \quad (2.12.54b)$$

There are discrepancies between various texts and papers concerning the form of C_{barrier}. However, these become unimportant for high barriers ($\theta \gg 1$). For crossing the valley in Figure 2.12.11 the C matrix is

$$C_{\text{valley}} = \begin{pmatrix} e^{i\alpha} & 0 \\ 0 & e^{-i\alpha} \end{pmatrix}, \quad (2.12.55a)$$

where

$$\alpha = \int_b^c k(x) \, dx. \quad (2.12.55b)$$

Combinations of these matrices can be used to describe the entire C_n symmetric potential system in Figure 2.12.11. The product of the $2n$ C-matrix factors must yield the identity matrix in order to satisfy C_n symmetric periodic boundary conditions. We have

$$\left[C_{\text{barrier}} \ C_{\text{valley}} \right]^n = 1$$

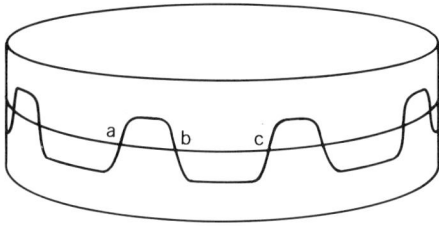

Figure 2.12.11 General C_n symmetric potential function.

or

$$\begin{pmatrix} [1+\theta^2]^{1/2}e^{i\alpha} & i\theta e^{-i\alpha} \\ -i\theta e^{-i\alpha} & [1+\theta^2]^{1/2}e^{-i\alpha} \end{pmatrix}^n = \begin{pmatrix} 1 & 0 \\ 0 & 1 \end{pmatrix}. \quad (2.12.56)$$

This requires that the eigenvalues of $C = C_{\text{barrier}}C_{\text{valley}}$ be conjugate pairs of the nth roots of the unity ($e^{\pm i2\pi m/n}$, where $m = 0, 1, 2, \ldots, n-1$). This leads to the following trace relation:

$$[1+\theta^2]^{1/2}\cos\alpha = \cos\left(\frac{2\pi m}{n}\right). \quad (2.12.57)$$

We know that if each valley existed by itself it would have a series of energy levels ($E_0, E_1, E_2, \ldots, E_j, \ldots$). For a single isolated potential valley the jth energy level would satisfy a quantization condition of the form

$$\alpha(E_j) = \int_b^c k(x)\,dx = (j+\tfrac{1}{2})\pi \quad (j = 0, 1, 2, \ldots). \quad (2.12.58)$$

This amounts to a phase change across the valley of $(j + 1/2)\pi$ corresponding $(j + 1)$ half-waves minus a fraction of a half-wave which would protrude beyond the valley turning points if the wave could continue on either side. It is remarkable that this protruding fraction is approximately one quarter of a half-wave, that is, a $\pi/4$ phase decrement on either side. This WKBJ approximation holds for a wide range of potential slopes. The square well, on the other hand, has a protruding phase shift δ that varies continuously with the barrier height. (See Problem 2.12.2.) The $\pi/4$ phase shift is found in the standard WKBJ analysis and the resulting zero-point quantum term $\tfrac{1}{2}$ is called a Maslov constant.

All this assumes the absence of tunneling. With n valleys there will be a cluster of n levels around each single valley level E_j. A Taylor expansion around each level E_j gives the phase as a function of nearby energy E:

$$\alpha(E) = \alpha(E_j) + \frac{\partial\alpha}{\partial E}(E-E_j)$$

$$= \left(j+\frac{1}{2}\right)\pi + \left(\frac{\pi}{\hbar\omega_{\text{classical}}}\right)(E-E_j). \quad (2.12.59)$$

Here we used a relation between the action S and the phase shift $\alpha = S/2\hbar$ and the classical angular frequency,

$$\frac{\partial S}{\partial E} = \frac{2\pi}{\omega_{\text{classical}}}.$$

A general proof of this relation is given in Section 8.1 [Eq. (8.1.31)]. An elementary derivation is easily made, also. Substituting the approximation (2.12.59) into the boundary condition (2.12.57) yields

$$\frac{\cos(2\pi m/n)}{[1+\theta^2]^{1/2}} = \cos\alpha = \cos\left[\left(j+\frac{1}{2}\right)\pi + \left(\frac{\pi}{\hbar\omega_{\text{classical}}}\right)(E-E_j)\right]$$

$$= \cos\left(j\pi + \frac{\pi}{2}\right)\cos\left[\frac{\pi}{\hbar\omega_{\text{classical}}}(E-E_j)\right]$$

$$- \sin\left(j\pi + \frac{\pi}{2}\right)\sin\left[\frac{\pi}{\hbar\omega_{\text{classical}}}(E-E_j)\right]$$

$$\cong -(-1)^j \frac{\pi}{\hbar\omega_{\text{classical}}}(E-E_j).$$

This leads to an energy-level cluster splitting formula for the (m)th level in the (j)th cluster:

$$E = E_j - (-1)^j \frac{\hbar\omega_{\text{classical}}}{\pi}[1+\theta^2]^{-1/2}\cos k_m \quad \left(k_m = \frac{2\pi m}{n}\right). \quad (2.12.60)$$

This corresponds to an approximate formula for the tunnelling or "sneak" factor S in Eq. (2.12.25), where $2S$ is the magnitude of the *intra*cluster splitting.

$$S = (-1)^j \frac{\hbar\omega_{\text{classical}}}{2\pi}\theta^{-1} = (-1)^j \hbar\nu_{\text{classical}} e^{-\int_a^b |(2m/\hbar^2)(V-E)|^{1/2}\,dx}.$$

Note that the S amplitude depends on two factors. The first is the classical frequency $\nu_{\text{classical}}$ which is the number of times the particle "knocks at the door" of the barrier each second. (The energy $h\nu_{\text{classical}}$ is the *inter*cluster splitting.) The other factor is the exponential of the tunneling integral. The tunneling factor decreases rapidly with the height $(V-E)$ of the barrier above the energy level. The sign of the amplitude alternates from cluster to cluster according to $(-i)^j$. This is consistent with the observation that bands in Figure 2.12.2 would have A-type waves on the lower boundary and B-type waves on the upper boundary for even j and vice versa for odd j.

ADDITIONAL READING

For a discussion of wave motion and Fourier analysis it is hard to beat two of the original texts by Leon Brillouin

Leon Brillouin, *Wave Propagation and Group Velocity* (Academic, New York, 1960); *Wave Propagation in Periodic Structures*, 2nd ed. (Dover, New York, 1953).

An excellent modern book on Fourier theory with lots of interesting mathematics, examples of physical effects, and historical notes on engineering applications is by Körner.

T. W. Körner, *Fourier Analysis* (Cambridge University Press, Cambridge, 1988).

A good introduction to oscillation and wave mechanics is available from the Berkeley Course volume by Frank S. Crawford.

F. S. Crawford, *Waves*, Berkeley Physics Course, Vol. 3, (McGraw-Hill, New York, 1968).

A wonderful application to boat wakes by the same author is in the *American Journal of Physics*.

F. S. Crawford, "Elementary derivation of the wake pattern of a boat," *Am. J. Phys.*, **52**, 782 (1984).

An article on the two-Fourier-component waves and their galloping motion is the following:

W. G. Harter, J. Evans, R. Vega, and S. Wilson, "Galloping waves and their relativistic properties," *Am. J. Phys.*, **53**, 671 (1985).

The problem of scattering and tunneling in potential wells has been treated arduously if not clearly. The supposedly definitive work on WKB approximations is by Froman and Froman.

N. Froman and P. O. Froman, *JWKB Approximation, Contributions to the Theory* (North-Holland, Amsterdam, 1965).

They find a number of problems with tunneling amplitudes found in quantum mechanics texts such as the text by Merzbacher.

E. Merzbacher, *Quantum Mechanics* (Wiley, New York, 1970) p. 126.

Despite this the Merzbacher text has one of the clearer introductions to crossing and scattering matrices.

The most widely used barrier tunneling factors come from an article by Miller and Good, *Phys. Rev.*, **91**, 174 (1953).

The semiclassical description of barrier tunneling chemical physics is described quite clearly by William H. Miller.

W. H. Miller, *J. Phys. Chem.*, **83**, 960 (1979).

This has been applied extensively. For a modern application see the following and references contained therein.

J. M. Robbins, S. C. Creagh, and R. G. Littlejohn, *Phys. Rev. A* **39**, 2838 (1989).

An earlier description of barrier tunneling and reflection is found in the following reference:

M. S. Child, *J. Mol. Spectrose*, **53**, 280 (1974).

This contains a graphical description of quantum mechanics of multiple potentials with barriers and valleys connected with a variety of topologies. See also

M. S. Child, in *Nonadiabatic Transitions, Atom-Molecule Collision Theory*, R. B. Bernstein (ed.), (Plenum, New York, 1979).

R. P. Bell, *The Tunnel Effect in Chemistry* (Chapman and Hall, London, 1980).

PROBLEMS

Section 2.3

2.3.1 Use group postulates 1 to 4 to answer or prove: If $ab = c$ for a, b, and c in a group G, can $ad = c$, too, where $d \neq b$? How many times can a group element appear in a given row or a given column of a group table?

2.3.2 If a set $\delta = \{a, b, c, \ldots, g, .\}$ of operators satisfying postulates 1 and 2 has a "left identity" 1_L (such that $1_L g = g$) for all g, and a "left inverse" g_L^{-1} for each g (such that $g_L^{-1} g = 1_L$), can you prove that the set is a group; i.e., rules 3 and 4 are satisfied?

2.3.3 Consider the oscillator Hamiltonian

$$H = \frac{A}{2}(p_1^2 + q_1^2) + \frac{D}{2}(p_2^2 + q_2^2) + B(q_1 q_2 + p_1 p_2) + C(q_1 p_2 - q_2 p_1).$$

(a) Write out Hamilton's equations of motion which give time derivatives of p_1, p_2, q_1, and q_2 in terms of A, B, C, D.

(b) For $C = 0$ determine the acceleration matrix $\langle \mathbf{a} \rangle$ in Newton's equations of motion.

(c) What (if any) constraints on A, B, and D are needed to yield a C_2 symmetric system?

(d) Give a simple formula for the normal mode angular frequencies $\omega \uparrow$ and $\omega \downarrow$ in the C_2 symmetric case.

Assume $A = 1.0$, $B = 0.3$, $C = 0$, $D = 1.0$ [units of (radian) H_z] in the following problems.

(e) Compute the time period between beat maxima (in seconds!) How many wiggles per beat?

(f) Compute the time period between perfect recurrences of all variables
(Poincaré period). How many beats per recurrence?

2.3.4 If the two-pendulum C_2 symmetric system is acted upon by a periodic driving force, suppose the force on pendulum 1 is equal to $f_1 \cos(\omega t)$ and the force on pendulum 2 is equal to $f_2 \cos(\omega t)$. Suppose also that there are frictional forces. Let the operator equation of motion be

$$|\ddot{x}\rangle + \mathbf{d}|\dot{x}\rangle + \mathbf{a}|x\rangle = |f\rangle,$$

$$\text{where } |x\rangle = \begin{pmatrix} x_1 \\ x_2 \end{pmatrix}, \quad |f\rangle = \begin{pmatrix} f_1 \cos(\omega t) \\ f_2 \cos(\omega t) \end{pmatrix},$$

$$\text{and } \mathbf{d} = \begin{pmatrix} d & e \\ e & d \end{pmatrix}, \quad \mathbf{a} = \begin{pmatrix} a & b \\ b & a \end{pmatrix}.$$

(a) Derive and plot the pendulum steady-state response amplitude as a function of the stimulus frequency ω for $a = 3.0$, $b = 10.$, $d = 0.1 = e$, $f_1 = 1.0$, and $f_2 = 0.0$.

(b) Do the same for $a = 3.0$, $b = 1.0$, $d = 0.1 = e$, $f_1 = 1.0$, and $f_2 = 1.0$.

(c) Do the same for $a = 3.0$, $b = 1.0$, $d = 0.1 = e$, $f_1 = 1.0$, and $f_2 = -1.0$.

Section 2.6

2.6.1 Let $\{|1\rangle, |2\rangle = R|1\rangle\}$ be the basis in which the representation of a general C_2 symmetric Hamiltonian is $\langle H_2 \rangle = \begin{pmatrix} a & b \\ c & d \end{pmatrix}$. Let $\{|1\rangle, |2\rangle = r|1\rangle, |3\rangle = r^2|1\rangle\}$ be the basis in which the representation of a general C_3 symmetric Hamiltonian is

$$\langle H_3 \rangle = \begin{pmatrix} A & B & C \\ D & E & F \\ G & H & I \end{pmatrix}.$$

(a) Derive equations relating a, b, c, and d.

(b) Derive equations relating $A, B, C, \ldots$ and I.

Can any of these quantities be complex if $H^\dagger = H$?

2.6.2 Could an unsymmetrical system such as is shown in the diagram ever behave something like a C_2 symmetric system? (Suppose we're allowed to assign different scale factors to different coordinates.)

Suppose that $m_1 = 10m_2$, $k_1 = 20$, and $k_{12} = 1$. What (if any) value of k_2 mimics C_2 behavior?

Suppose, instead that $m_1 = 10m_2$, $k_1 = 8$, and $k_{12} = 1$. Is the C_2 symmetry physically possible? What, if any, restrictions are there for the constants?

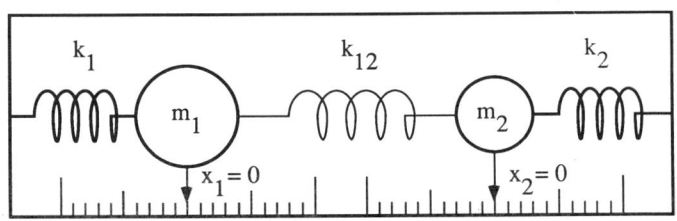

2.7.1 Consider C_n symmetric coupled pendulum systems described by Eq. (2.7.1). Give a detailed account of the eigensolutions for $n = 4, 5$, and 8, as follows.

(a) Label the coordinates with C_n group operators.

(b) Write the C_n character table and projection operators.

(c) Sketch the moving-wave modes of vibration using phasors. (Label them with C_n group representation labels.)

(d) Sketch the standing-wave modes of vibration using phasors and sine or cosine curves.

(e) Calculate the eigenfrequencies for $a = 1.0$ and $b = 0.0$.

(f) Discuss the effect of nonzero constant ($b \neq 0$) on the dispersion function. How does the phase and group velocity change for low wave number? Draw the dispersion function and indicate the first Brillouin zone.

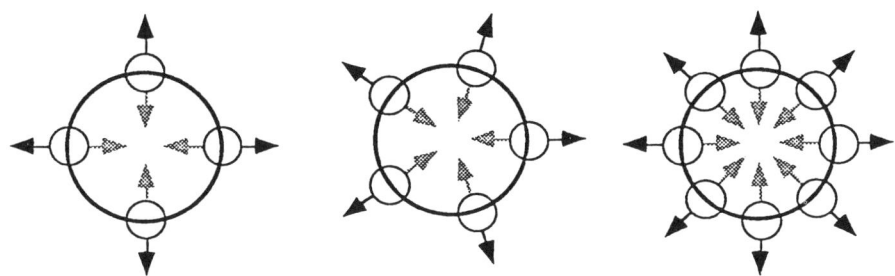

2.7.2 Derive the wave speed for a nondispersive wave given by Eq. (2.7.12) in terms of its phase velocity $c = \omega/k$, time t, the incident amplitude I, and reflected amplitude R. Give maximum and minimum speed in terms of the SWR quantity $\Delta = (I - R)/(I + R)$.

2.7.3 Consider the C_8 symmetric coupled pendulum system (use results of Problem 2.7.1) when the coupling strengths alternate between $\bar{a} > a$ and $\underline{a} < a$ as described around Eq. (2.7.8).

(a) Derive the eigenfrequencies for $\bar{a} = 1.2$ and $\underline{a} = 0.8$, and sketch the modes of vibration. Pay particular attention to the frequency doublet levels (or level) which split(s) when $\bar{a} \neq \underline{a}$.

2.7.4 Use a "higher symmetry embedding" technique to find the normal modes of the four-pendulum ($m = 4$) system shown in the top and side views of the figure. What higher C_n ($m > n$) symmetry ring would have the same motions for a subset of its pendulums with the walls in

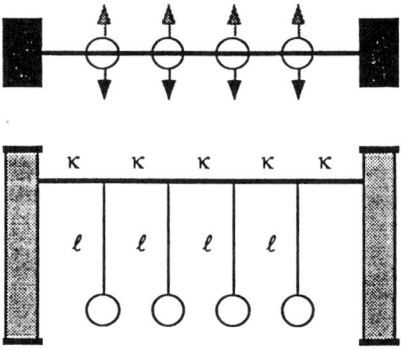

the figure corresponding to nodal pendulums? Use this system to give formulas for the mode frequencies in terms of torsional spring constants (κ) and lengths (l) indicated in the figure. Generalize this for $5, 6, \ldots, m$ pendulums between walls.

2.7.5 Consider a set of oscillators that have C_3 symmetry but only C_3 symmetry and no higher symmetry. Suppose there is some kind of amplifier by which oscillator x_1 can perturb oscillator x_2, and oscillator x_2 can perturb oscillator x_3, and oscillator x_3 can perturb oscillator x_1 more than vice versa. In other words, let us suspend Newton's third law and make the action and reaction forces unequal. This corresponds to the following equation of motion, in which the coupling constants s and t are unequal:

$$\begin{pmatrix} \ddot{x}_1 \\ \ddot{x}_2 \\ \ddot{x}_3 \end{pmatrix} = \begin{pmatrix} a & s & t \\ t & a & s \\ s & t & a \end{pmatrix} \begin{pmatrix} x_1 \\ x_2 \\ x_3 \end{pmatrix}$$

Discuss this system. Does it really have C_3 symmetry? What are its mode solutions? Which are stable?

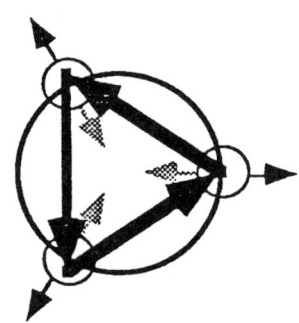

Double Trouble

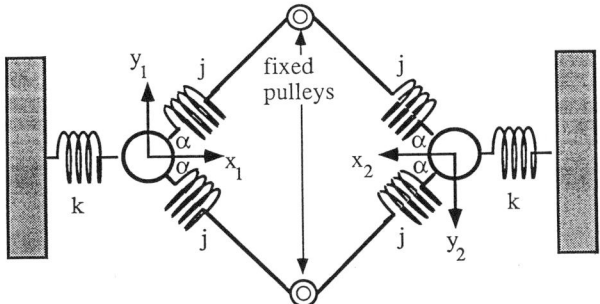

2.8.1 (a) Describe and label a symmetry group that would be appropriate and adequate to help solve the equations of motion for the four-freedom oscillator system shown. Label the base states using group operators and derive a representation R of this group. (Show that it obeys your group table.)

(b) Write the Lagrangian and acceleration matrix in terms of x_1, x_2, y_1, y_2, mass m, and spring constants j, k for small oscillations. (Treat angle α as a constant.) Write the equations of motion.

(c) Reduce the representation R and the acceleration matrix using symmetry projectors of the group. Sketch and describe the normal modes and their frequencies for each type of projection or irrep for the case $j = 1.0$, $k = 2.0$, and $\alpha = 45°$.

(d) Suppose each mass could move in *three* directions ($x_{1,2}$, $y_{1,2}$, and $z_{1,2}$) and the symmetry was bilateral in the z direction, as well. What symmetry group or groups would be appropriate? [Give the answer in terms of $(C_p \times C_q \times \cdots$, etc.) for some $p, q, \ldots$.]

(e) Give the character table of the largest group you found in answer (d) and tell which irreps would be used to label normal modes of the three-dimensional double-mass oscillator.

2.8.2 Calculate the number of nonisomorphic Abelian groups of order n and express them in terms of outer products of cyclic groups of prime order for the following values of n.

(a) $n = 8$.
(b) $n = 9$.
(c) $n = 12$.
(d) $n = 16$.
(e) $n = 24$.

Write out character tables for $n = 8$ Abelian groups.

2.8.3 The Fourier series for an infinite one-dimensional square wave $S(x)$ has the form

$$S(x) = 4/\pi \left[\cos(x) - \tfrac{1}{3}\cos(3x) + \tfrac{1}{5}\cos(5x) - \tfrac{1}{7}\cos(7x) + \cdots \right],$$

where $S(s) = +1$ if $\cos(x) > 0$ and $S(x) = -1$ otherwise. (Derive this series.) Consider a 12-mass ring of C_{12} symmetry subject to square standing-wave initial conditions as shown in the diagram. Calculate the coefficients of a *finite* series of C_{12}-defined modes that would give these initial conditions and compare them to the first few coefficients just given.

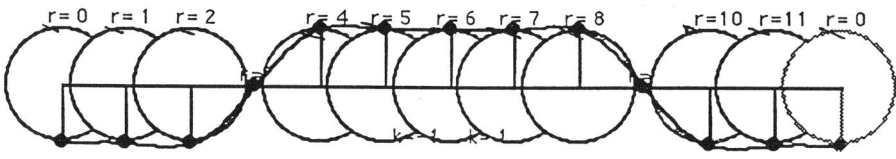

What is needed to make a square *moving* wave?

2.12.1 The Kronig-Penney bands in Figure 2.12.6 exhibit "accidental" degeneracies for three nonzero values of the potential barrier height V with energy $E < 50$. [Bohr units $= \hbar^2 \pi^2 / 2\mu(a+b)^2$, where μ is mass, a is well width, and b is barrier width. here let $a = b$.]

(a) What are these special values of V and E in Bohr units? Sketch the wave functions for each of the three accidental cases and explain the degeneracies.

(b) List the other cases (if any) for $a = b$ and $E < 75$ Bohr units. What degeneracies occur just above the top of Figure 2.12.6?

(c) Give a formula for finding these degeneracies for general a and b values.

2.12.2 Prove the straight-line-sine (k, δ)-space geometric solution in the diagram for the one-dimensional square well. The diagram uses special values of the mass ($m = \tfrac{1}{2}$), well width ($x = 2$), and Planck's constant ($\hbar = 1$). Show how this construction gives information about the eigenfunctions as well as the eigenvalues. Derive it for general width ($x = 2a$), particle mass (m), and Planck's constant. Check it against Figure 2.12.6 for $V = 25$. (Explain *why* you can check it against Figure 2.12.6.)

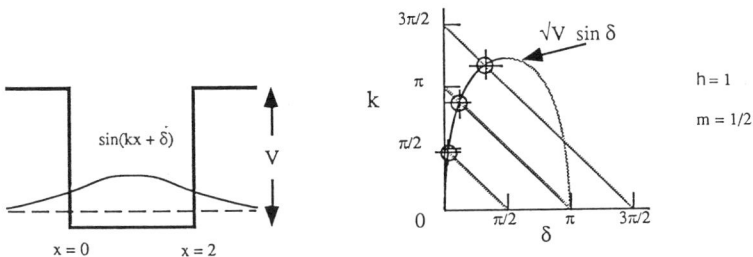

2.12.3 Draw to scale the approximate (use Figure 2.12.6 again) E spectrum ($0 < E < 80$) for the following potential wells. (Schrödinger equation $\psi'' + (E - U(x))\psi = 0$):

(a) C_4 symmetric square wells. [See diagram (a).]

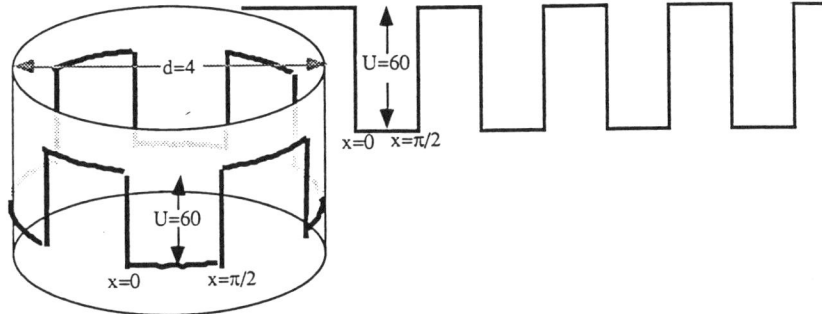

(*a*) Four equilvalent wells on a circle

(*b*) Four equivalent wells in a straight line

(b) Four equivalent square wells in a straight line. [See diagram (b)].

(c) Give a detailed magnified view and approximate accounting of the lowest two bands or "clusters" of fine structure levels and a rough sketch of the eigenfunction in each case.

2.12.4 (Discussion). The spectrum of $1, 2, 3, \ldots$ or n square potential wells on an infinite one-dimensional line is usually thought to be a *discrete* portion of energy below the top of the barriers, followed by a *continuum* for energy above the wells. However, for $n = \infty$ it is well known that the continuum breaks into bands and gaps above *and* below the top of the wells. Do these gaps suddenly appear at $n = \infty$? Discuss and find a way to reconcile this apparent discrepancy in the spectral properties.

BASIC THEORY AND APPLICATIONS OF SYMMETRY REPRESENTATIONS

and find a way to reconcile this apparent discrepancy in the spectral properties.

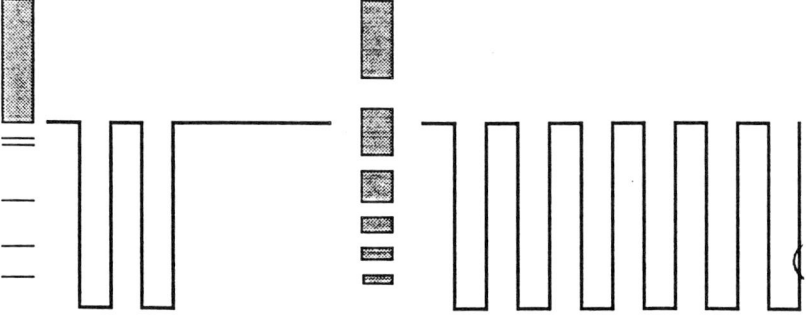

CHAPTER 3

BASIC THEORY AND APPLICATIONS OF SYMMETRY REPRESENTATIONS (NON-ABELIAN SYMMETRY GROUPS)

So far we have considered symmetry operations which commute with each other ($gh = hg$), i.e., groups which are Abelian. It has been shown how one can expand all operators of any Abelian group in combinations of a single set of orthogonal idempotent operators. Several examples have been given in which the Abelian idempotent operators help to solve physical problems.

However, many of the most interesting problems involve non-Abelian symmetry groups, that is, groups having some operators which do not commute. In Figure 3.1.1 the non-Abelian crystal point symmetries are listed and modeled. Note that half of the possible crystal symmetries and all of the larger groups are non-Abelian.

It will now be shown how to generalize the idea of the idempotent expansions to include these more complicated symmetry groups and related physical problems. The simplest examples that illustrate each point will be used in the development of this theory.

3.1 SIMPLEST EXAMPLES OF NON-ABELIAN SYMMETRY

A. Trigonal Symmetries C_{3v} and D_3

The smallest non-Abelian symmetries are the groups C_{3v} and D_3 with each having just six elements. C_{3v} is the full symmetry of the three-pendulum system which was discussed in Section 2.6. The theory there involved only the symmetry C_3, which is a subgroup of C_{3v} (the line between C_3 and C_{3v} in Figure 3.1.1 indicates that C_3 is contained in C_{3v}; i.e., $C_3 \subset C_{3v}$). The operators of C_{3v} which were neglected in Section 2.6 are the three vertical

152 BASIC THEORY AND APPLICATIONS OF SYMMETRY REPRESENTATIONS

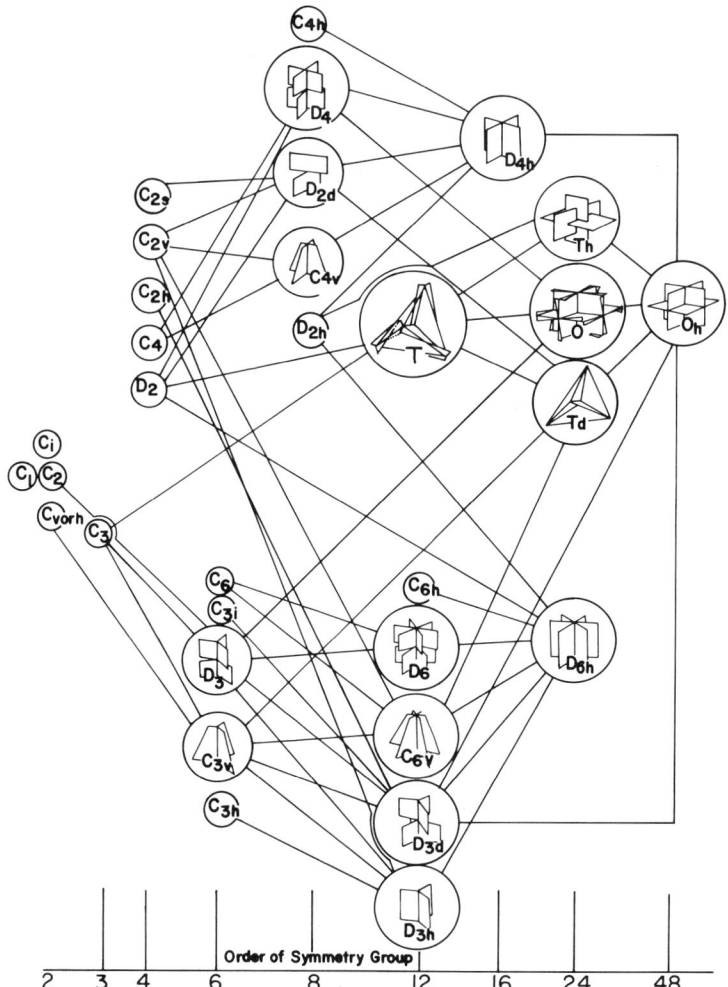

Figure 3.1.1 Crystal point symmetry groups. Models are sketched in circles for the 16 non-Abelian groups. (See also Figure 2.11.1.)

mirror-plane reflection operators labeled by σ_1, σ_2, and σ_3 in Figure 3.1.2. (See also Figure 3.1.3.) On the left-hand side of Figure 3.1.2 there are pictures of the pendulum bobs frozen in some nonequilibrium position state. Then the effect of the reflection operators on the positions of the pendulums are shown by the pictures in the center of Figure 3.1.2. The effect of σ_2 followed by σ_1 is a counterclockwise 120° rotation (r^2). These results may be expressed by the group product relations ($\sigma_1\sigma_2 = r$) and ($\sigma_2\sigma_1 = r^2$).

For non-Abelian groups, the order of the operations makes a big difference. We shall define a group product ab to mean b acts first, followed by a

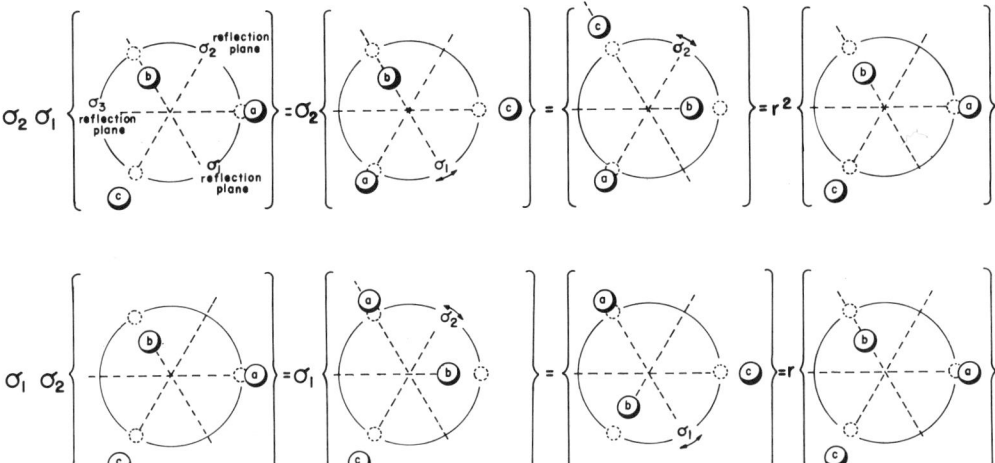

Figure 3.1.2 Effects of plane reflections on pendulum mass configuration. The effects of the products of reflection operations σ_1 and σ_2 taken in different orders are shown. Note that the reflection planes are fixed on the page, while the pendulum masses get moved.

on an operand sitting to the right of the operations. The complete multiplication table for all the C_{3v} operations is

	1	r	r^2	σ_1	σ_2	σ_3
1	1	r	r^2	σ_1	σ_2	σ_3
r	r	r^2	1	σ_3	σ_1	σ_2
r^2	r^2	1	r	σ_2	σ_3	σ_1
σ_1	σ_1	σ_2	σ_3	1	r	r^2
σ_2	σ_2	σ_3	σ_1	r^2	1	r
σ_3	σ_3	σ_1	σ_2	r	r^2	1

(3.1.1)

By replacing the reflections σ_1, σ_2, and σ_3 with 180° rotations ρ_1, ρ_2, and ρ_3 around axes normal to the σ_j planes one obtains the symmetry group D_3. Objects of D_3 and C_{3v} symmetry are compared in Figure 3.1.3. Note that each reflection operation σ_j in C_{3v} is related to its rotational counterpart ρ_j in D_3 by the equations

$$\sigma_j = I\rho_j = \rho_j I, \qquad \rho_j = I\sigma_j = \sigma_j I, \qquad (3.1.2)$$

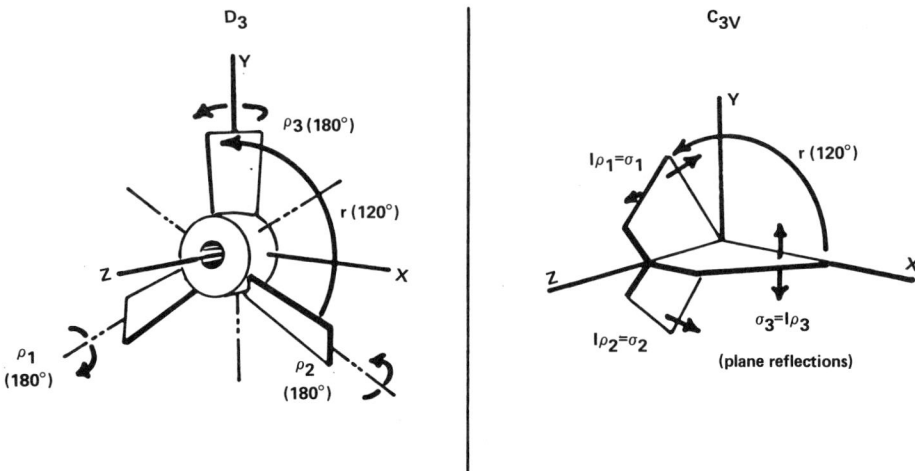

Figure 3.1.3 Pictorial comparison of D_3 and C_{3v} symmetry. A propeller having D_3 symmetry is shown next to a three-plane paddle having C_{3v} symmetry. The group operations are labeled by arrows, which indicate the effect they have. For example, ρ_3 is a 180° rotation around the y axis, while $I\rho_3 = \sigma_3$ is a reflection through the xz plane. (Here axes are fixed and the objects rotate.)

involving inversion I. [Recall Eq. (2.11.4).] Since I commutes with all point-group operations, the relation between D_3 and C_{3v} is easily established. For example, the product $(\sigma_1\sigma_2 = r)$ implies that $(I\rho_1 I\rho_2 = r)$ or $(\rho_1\rho_2 = r)$, since $I^2 = 1$. In this way one sees that D_3 has the same multiplication table, aside from a change in notation, as C_{3v}; i.e., D_3 is isomorphic to C_{3v}. Mathematicians have shown that there is only one abstract non-Abelian group of order 6.

B. Reflections and Hamilton's Turns

There are some easy ways to visualize and compute point-group products. Consider first the product of mirror reflections. Imagine two planes: an xz plane, and another ϕz plane which makes an angle ϕ with the xz plane while intersecting it along the z axis. The matrix representations of reflections σ_{xz} and $\sigma_{\phi z}$ may be obtained from geometry as shown in Figure 3.1.4. The effect of mirror reflections on the unit vectors $|x\rangle$ and $|y\rangle$ is shown. The reflection planes intersect along the z axis, which is normal to the figure. Hence, we have $\sigma_{\phi z}|z\rangle = |z\rangle$ for all angles ϕ.

The product of reflections represented in the $\{|x\rangle, |y\rangle\}$ basis is as follows:

$$\langle \sigma_{\phi z}\sigma_{xz}\rangle = \begin{pmatrix} \cos 2\phi & \sin 2\phi \\ \sin 2\phi & -\cos 2\phi \end{pmatrix}\begin{pmatrix} 1 & 0 \\ 0 & -1 \end{pmatrix}$$
$$= \begin{pmatrix} \cos 2\phi & -\sin 2\phi \\ \sin 2\phi & \cos 2\phi \end{pmatrix} = \langle R[2\phi]\rangle.$$

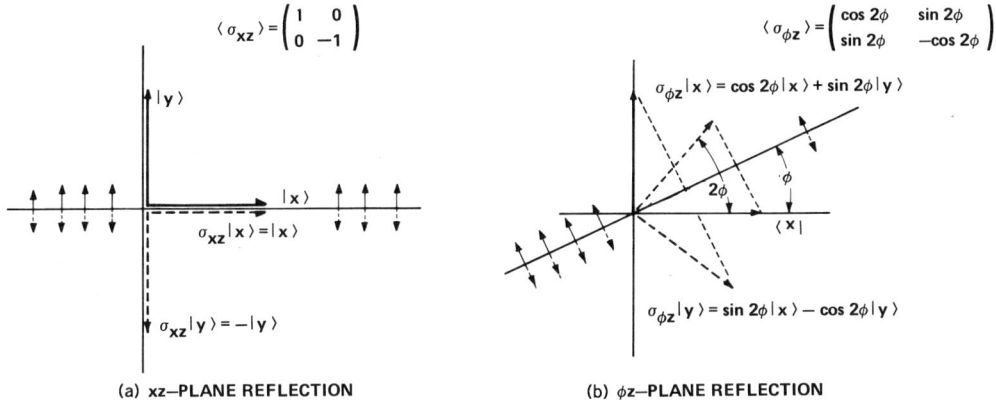

Figure 3.1.4 Representations of mirror plane reflection operators (a) $\langle\sigma_{xz}\rangle$, (b) $\langle\sigma_{\phi z}\rangle$.

The result is a representation of a counterclockwise rotation by angle 2ϕ in the xy plane or around the z axis of intersection. This proves that a rotation $R[\omega]$ by angle ω about an axis $\hat{\omega}$ is a product of two planar reflections. The first reflection may be through any plane that contains $\hat{\omega}$. The second reflection is through a plane containing $\hat{\omega}$ and making an angle $\omega/2$ with the first plane as shown in Figure 3.1.5. The $\omega/2$ arc between the two normals $\mathbf{N}_1$ and $\mathbf{N}_2$ in the figure is called a HAMILTON-TURN vector.

The Hamilton-turn vector can be positioned anywhere on its great circle or equatorial arc. It is useful for computing products of rotations around different axes. Suppose a rotation $R[\omega']$ about axis ω' follows $R[\omega]$. Then

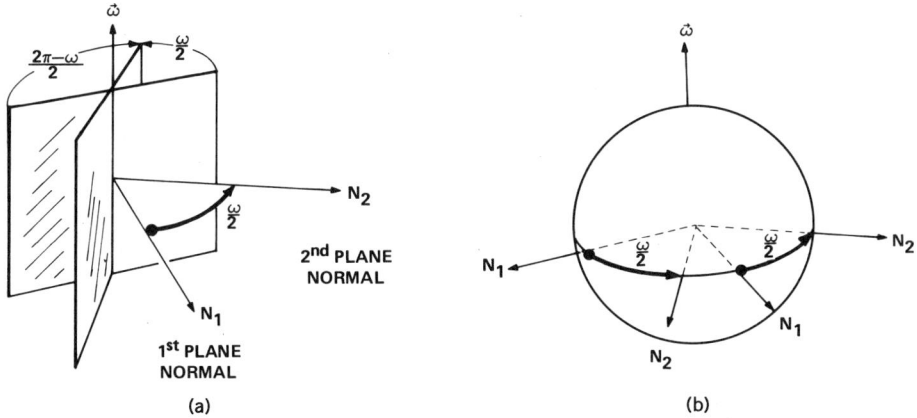

Figure 3.1.5 Hamilton's representation of rotation $R[\omega]$. (a) Two planes intersect with half the angle of the rotation. (b) Hamilton-turn great circle arc vectors on a unit sphere represent the rotation about the axis ω, which is orthogonal to their plane.

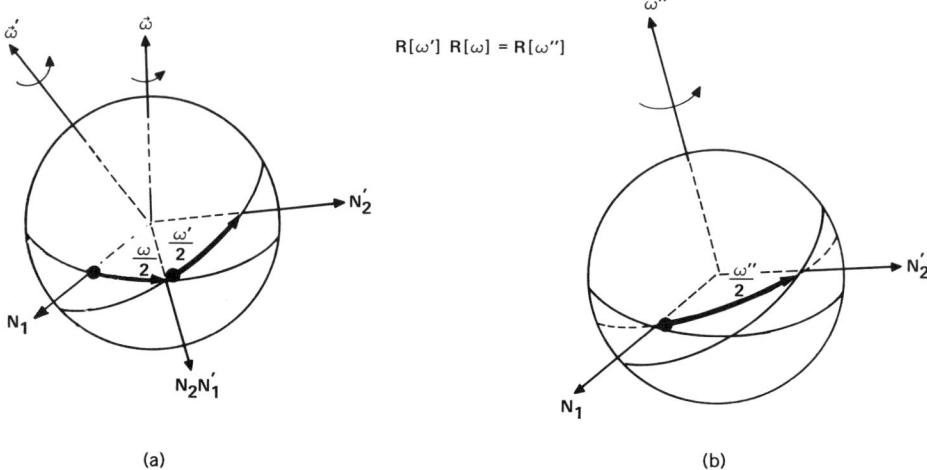

Figure 3.1.6 Hamilton's construction for rotation group product ($R[\omega']R[\omega] = R[\omega'']$). (a) The head of arc vector ($\omega/2$) for first rotation $R[\omega]$ meets tail of arc vector ($\omega'/2$) for second rotation $R[\omega']$. (b) Resultant great circle arc vector ($\omega''/2$) defines product rotation $R[\omega'']$.

one may move their Hamilton-turn vectors into a head-to-tail position at the arc intersection as shown in Figure 3.1.6(a). Then the $\mathbf{N}_2$ and $\mathbf{N}_1'$ normals and planes coincide and the corresponding reflections cancel. The great circle arc $\omega''/2$ between the first normal ($\mathbf{N}_1$) for $R[\omega]$, and the second normal ($\mathbf{N}_2'$) for $R[\omega']$ is the turn vector for the group product:

$$R[\omega''] = R[\omega']R[\omega].$$

This is shown in Figure 3.1.6(b). In other words, spherical vector addition with half-angle arcs is a key to understanding rotational point-group products. Furthermore, all point-group operations are either rotations (R) or rotations with an inversion (I) attached. Since inversion commutes with all rotations ($R \cdot I = I \cdot R$) and squares to unity ($I^2 = 1$), the entire point-group calculation reduces to Hamilton-turn addition.

In Chapter 5 the half-angle ($\omega/2$) will be connected with spin-$\frac{1}{2}$. [Recall the half-angles that occur in Eq. (1.1.2).] Electrons and half-integral spin particles have the disturbing property of turning up with a negative (-1) phase after a "full" rotation by 2π or $360°$. For these particles a rotation by (ω) is (-1) times a rotation by $-(2\pi - \omega)$. The Hamilton-turn method distinguishes the latter rotation by a clockwise supplementary arc of length ($\pi - \omega/2$), as shown in Figure 3.1.5. This aids in the understanding of "double-group" theory, which will be discussed in Chapter 5.

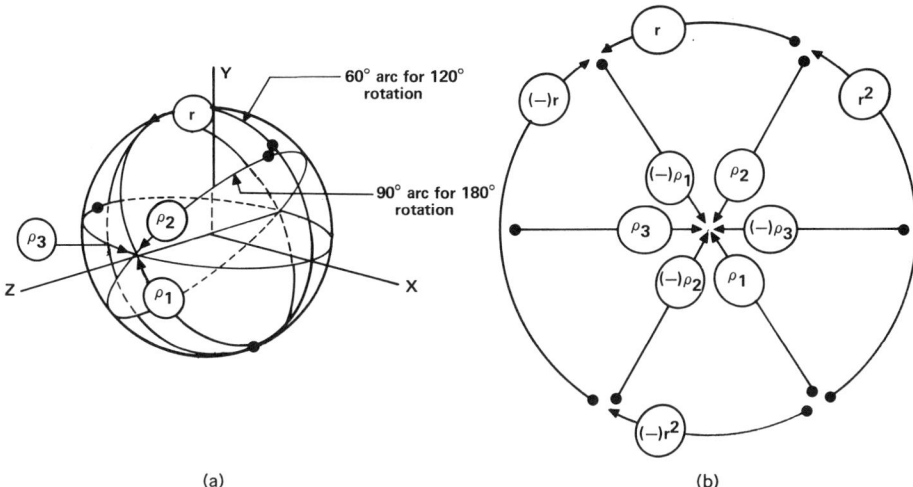

Figure 3.1.7 Geometrical definition of symmetry group D_3. (a) Hamilton arc vectors are drawn for rotations r, i_1, and i_3. (b) Group nomogram is obtained by projecting (a) onto the xy plane.

The D_3 or C_{3v} group table of Eq. (3.1.1) may be replaced by a geometrical nomogram. The Hamilton-turn arcs for D_3 rotations are drawn into place in Figure 3.1.7(a), following the conventions established by Figure 3.1.3. A view down the z axis is shown in Figure 3.1.7(b). Note that the supplementary arcs designated by a $(-)$ symbol are included there. Vector addition is accomplished by visually sliding the arc vectors into desired positions. For example, note the product

$$(-)\rho_1 r = \rho_2,$$

which is given by the uppermost sector of Figure 3.1.7(b). By ignoring the $(-)$ and changing ρ_j to σ_j one obtains the product $(\sigma_1 r = \sigma_2)$ in agreement with the C_{3v} table. [Recall Eq. (3.1.1).]

C. Laboratory and Body Reference Frames

For future work it is important to notice that it makes a difference how one defines the "reference frame" for molecular point-group operations. The rotation axes and reflection planes can be imagined to be attached to the stars, or more prosaically, to a fixed "laboratory" coordinate system x, y, and z. This was done in Figure 3.1.2. Consider now an alternative definition of operators $\underline{\sigma}_i$ made by fixing all these axes and planes to the "body" frame of the pendulums, as in Figure 3.1.8. However, note that a change of the multiplication rules will result, since each operator changes the laboratory position of all planes or axes except its own.

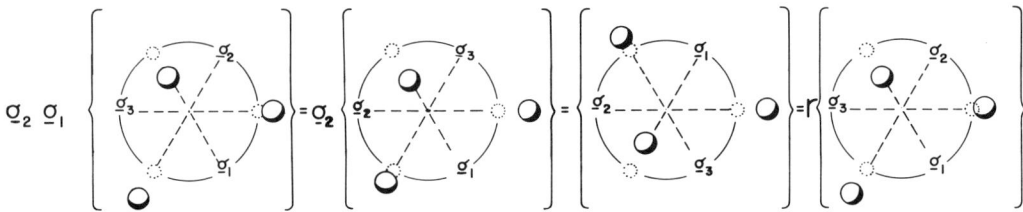

Figure 3.1.8 Effects of body-fixed plane reflections on pendulum mass configuration.

At the beginning of Chapter 5 a convenient relation is established between operators defined in laboratory and body frames. It is shown that if each body operator is identified with the *inverse* of a laboratory operator, then the original multiplication rules are recovered.

For the Abelian "fan-blade" group D_2 (recall Figure 2.1.1), or for any symmetry involving just inversions or rotations by 180°, it makes no difference whether the rotation axes are imbedded in the object or fixed in space as far as the multiplication table is concerned.

D. Tetragonal Symmetries C_{4v} and D_4

C_{4v} and D_4 are isomorphic crystal point symmetry groups of order 8. They are symmetries of a square pyramid and a square fan blade, respectively, as shown in Figure 3.1.9.

You should use this analogous symmetry to help you understand the theory of Sections 3.2–3.6. Each time some theory or application of C_{3v} symmetry is given you should work out the same for C_{4v} without looking at the answers in Section 3.6.

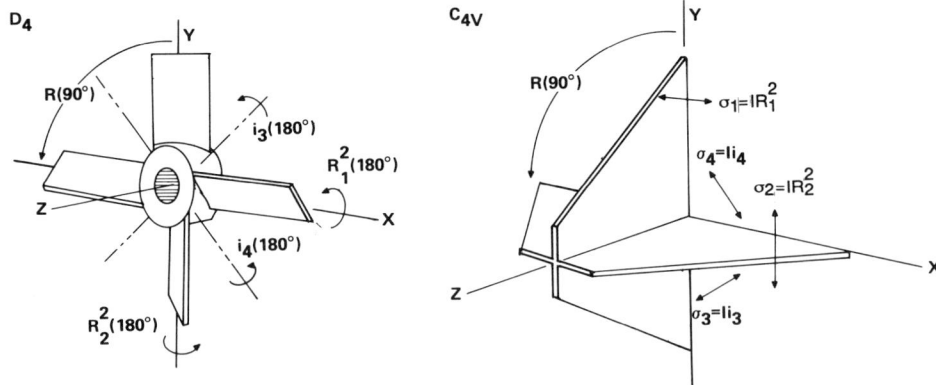

Figure 3.1.9 Pictorial comparison of D_4 and C_{4v} symmetry.

This sort of on-the-job training will make the difficult non-Abelian algebra easier to learn and remember. You should begin by constructing a Hamilton nomogram and group table for D_4 and C_{4v}. (See Problem 3.3.1.).

3.2 FIRST STAGE OF NON-ABELIAN SYMMETRY ANALYSIS

The idempotent theory developed in Chapters 1 and 2 always works when the operators commute, but it appears at first not to be applicable to non-Abelian groups. However, an interesting trick allows the theory to work for the noncommutative cases, as well. First one assembles the elements of the group C_{3v} into categories of operations which look more or less alike. Let category c_1 be the identity $\mathbf{1}$ by itself, category $c_2 = \{r, r^2\}$ be the 120° rotations, and category $c_3 = \{\sigma_1, \sigma_2, \sigma_3\}$ include all reflections. Then, if one defines the operators c_j to be the sums,

$$c_1 = \mathbf{1}, \quad c_2 = r + r^2, \quad c_3 = \sigma_1 + \sigma_2 + \sigma_3, \quad (3.2.1\text{a})$$

of operators within each category, it turns out that these c_j commute with each other $[c_j c_k = c_k c_j]$. Their multiplication structure is obtained by selectively "condensing" the group table in Eq. (3.1.1). The c_j multiplication table that follows is derived by simply counting up the elements in the rectangular sections of Eq. (3.1.1):

	c_1	c_2	c_3
c_1	c_1	c_2	c_3
c_2	c_2	$2c_1 + c_2$	$2c_3$
c_3	c_3	$2c_3$	$3c_1 + 3c_2$

(3.2.1b)

Since the c_j operators are mutually commutative, it is possible to generate a set of idempotents from them. This is done in Section 3.2.B, but first, a general theory of c_j operators is given in the following section.

A. Class Operators

In general the categories c_j are called CLASSES, the c_j are CLASS OPERATORS, and their multiplication structure represents the CLASS ALGEBRA. The formal definition of a class is the following.

Definition Elements g and g' are in the same class c_j of group G if there are any elements h in G such that $g = hg'h^{-1}$.

The observation that categories of elements such as $\{\sigma_1, \sigma_2, \sigma_3\}$ "look alike" means that, for example, σ_2 or σ_3 can be replaced by σ_1. One may state this more precisely by group equations such as $(r\sigma_3 r^{-1} = \sigma_1)$. One translation of this equation is "r transforms σ_3 into σ_1." The transformation of σ_3 by r has the fore-and-aft form which operator transformations should have. [Recall Eq. (1.1.23).] Figure 3.1.3 shows rather clearly that r transforms σ_3 into σ_1. Another way to read $(r\sigma_3 r^{-1} = \sigma_1)$ is to note that by turning around with r^{-1} then doing σ_3, and finally returning with r, one obtains the same operation as σ_1.

It is easy to see that a class operator $c_g = g + g' + \cdots$ made by summing all the elements of a class must commute with every element h of the group in question. The following equation:

$$h^{-1}c_g h = h^{-1}(g + g' + \cdots)h = c_g \qquad (3.2.2)$$

follows from the class definition and the fact that $h^{-1}gh = h^{-1}g'h$ if and only if $g = g'$. Inverting this yields

$$c_g h = h c_g. \qquad (3.2.3)$$

Roughly speaking, the classes show the symmetry of the symmetry group. Equation (3.2.3) guarantees that each c_g commutes with all symmetry operators. In addition, any product $c_g c_l$ or linear combination $\alpha c_g + \beta c_l$ will commute with everything in the group.

The Hamilton-turn diagrams are helpful for describing class structure. Suppose rotation operator R' is obtained from another rotation operator R through the transformation by a third rotation T; i.e.,

$$R' = TRT^{-1}. \qquad (3.2.4)$$

Then the Hamilton arcs of R and R' must be opposite sides of a spherical parallelogram, as indicated by shading as shown in Figure 3.2.1. Note that one diagonal of the parallelogram consists of two t arcs. The arc length of this diagonal is equal to the angle of the T rotation according to Hamilton's half-angle rule.

Given a transformation equation (3.2.4), it is possible that several or many transformation operators $\{T, U, V, \ldots\}$ transform R into R'. For each operator Q which commutes with R, one has

$$QRQ^{-1} = R. \qquad (3.2.5)$$

Substituting this into Eq. (3.2.4) gives

$$R' = TQRQ^{-1}T^{-1} = URU^{-1}, \qquad (3.2.6a)$$

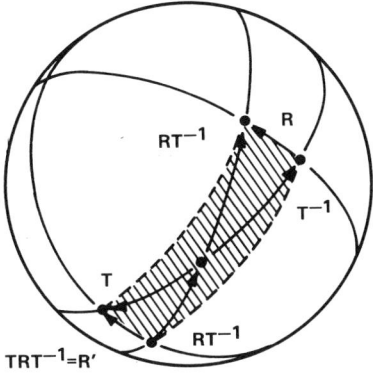

Figure 3.2.1 Showing class equivalence using Hamilton's vectors. Operation R is equivalent to $R' = TRT^{-1}$.

where

$$U = TQ,$$
$$U^{-1} = Q^{-1}T^{-1}. \tag{3.2.6b}$$

Thus each operator Q that commutes with R gives one more operator $U = TQ$ which transforms R into R'.

The set of all operators Q in a group G which commute with a given R form a subgroup $\{N_R = 1, Q, \ldots\}$ of G. This N_R is the "symmetry group" of R and is called its NORMALIZER. N_R includes R and all powers of R, and sometimes several inversion or reflection operators as well. The set of all operators $TN_R = \{T, TQ, \ldots\} = \{T, U, \ldots\}$ made by left multiplication by T on N_R is called the (left) COSET of subgroup N_R. Coset TN_R contains T and all operators which transform R into a particular operator $R' = TRT^{-1}$ in its class.

For example, the normalizer of r in C_{3v} is the subgroup $C_3 = \{1, r, r^2\}$. The subgroup C_3 and its coset $\sigma_1 C_3 = \{\sigma_1, \sigma_2, \sigma_3\}$ divide C_{3v} into two parts, and so there are only two elements $\{r, r^2\}$ in the class c_r of r. For another example, the normalizer of σ_1 in C_{3v} is the subgroup $C_v = \{1\sigma_1\}$. The subgroup $C_v = \{1, \sigma_1\}$ and cosets $rC_v = \{r, \sigma_3\}$ and $r^2 C_v = \{r^2, \sigma_2\}$ divide the group C_{3v} into three parts, and so there are three elements $\{\sigma_1, \sigma_2, \sigma_3\}$ in the class c_σ of σ_1.

It is left as an exercise to prove generally that a normalizer of operator h contains a number $(°N_h)$ of elements which evenly divides the order $(°G)$ of the group. Furthermore, the fraction

$$°c_h = (°G)/(°N_h) \tag{3.2.7}$$

is the number of elements in the class c_h of h. These statements belong to LAGRANGE'S theorems.

For the analysis of a symmetry group or algebra it is helpful to define the ALL-COMMUTING or CENTRAL operators $\{\mathbb{C}, \mathbb{C}', \ldots\}$. These are any operators

$$\mathbb{C} = \sum_g \gamma_g g \qquad (3.2.8)$$

in the group algebra which commute with all group operators $[\mathbb{C} h = h\mathbb{C}]$. By averaging over all group operators using commutativity $[\mathbb{C} = h\mathbb{C} h^{-1}]$ we obtain

$$\mathbb{C} = (1/{}^\circ G) \sum_h h\mathbb{C} h^{-1} = (1/{}^\circ G) \sum_g \gamma_g \left(\sum_h hgh^{-1} \right). \qquad (3.2.9)$$

By appealing to Lagrange's theorem and Eq. (3.2.7) one derives

$$\mathbb{C} = (1/{}^\circ G) \sum_g \gamma_g {}^\circ N_g c_g = \sum_g (\gamma_g / {}^\circ c_g) c_g. \qquad (3.2.10)$$

This proves that any all-commuting operator $\mathbb{C}$ must be a linear combination of class operators. The $\mathbb{C}$'s make up the commutative class algebra for which the c_g's are a basis. Since any product $c_g c_h$ belongs to the algebra it must be a combination,

$$c_g c_h = \sum_j \gamma_{gh}^j c_j, \qquad (3.2.11)$$

of c_g's, as well. The coefficients γ_{gh}^j are called algebraic STRUCTURE CONSTANTS.

B. Idempotent Analysis of Class Algebras

It is possible to write all the class operators c_g as a combination of a single set of idempotents P^α, as was done for Abelian groups of operators in Chapter 2. The only difference here is that the minimal equations are more complicated.

For example, using the C_{3v} class algebra table [Eq. (3.2.1b)] we find the minimal equation of c_3. This is the lowest-degree equation that involves just powers of c_3. [It may also involve $c_1 = 1$; however, c_1 may be thought of as the zeroth power $(c_3)^0$.] The degree of the following equation is not yet high enough, since an unwanted $3c_2$ term appears:

$$(c_3)^2 = 3c_1 + 3c_2.$$

FIRST STAGE OF NON-ABELIAN SYMMETRY ANALYSIS   163

Multiplying by c_3 again gives the desired minimal equation

$$(c_3)^3 = 3c_3 + 3c_2c_3 = 9c_3. \qquad (3.2.12)$$

Note that the degree of a minimal equation for class operators cannot exceed the number of classes, since that is the dimension of the algebra. The cubic minimal equation for c_3 factors into the following form:

$$(c_3 - 3\mathbf{1})(c_3 + 3\mathbf{1})(c_3 - 0\mathbf{1}) = 0. \qquad (3.2.13)$$

The three roots $c_3^{(1)} = 3$, $c_3^{(2)} = -3$, and $c_3^{(3)} = 0$ yield three idempotents $\mathbb{P}^1$, $\mathbb{P}^2$, and $\mathbb{P}^3$, respectively, when substituted into the general formula (1.2.15)

$$\mathbb{P}^\alpha = \prod_{\gamma \neq \alpha} (c_3 - c_3^{(\gamma)}\mathbf{1}) / \prod_{\gamma \neq \alpha} (c_3^{(\alpha)} - c_3^{(\gamma)}). \qquad (3.2.14)$$

The desired idempotents are given as follows.

$$\mathbb{P}^1 = [(c_3 + 3\mathbf{1})(c_3 - 0)] / [(3 + 3)(3 - 0)] = \left[(c_3)^2 + 3c_3\right]/18,$$
$$\mathbb{P}^1 = (c_1 + c_2 + c_3)/6 = (1 + r + r^2 + \sigma_1 + \sigma_2 + \sigma_3)/6,$$
$$\mathbb{P}^2 = (c_1 + c_2 - c_3)/6 = (1 + r + r^2 - \sigma_1 - \sigma_2 - \sigma_3)/6,$$
$$\mathbb{P}^3 = (2c_1 - c_2)/3 = (2\mathbf{1} - r - r^2)/3. \qquad (3.2.15)$$

The $\mathbb{P}^\alpha$ are called ALL-COMMUTING, CENTRAL, or CLASS idempotents of C_{3v}. The original class operators can be expanded in terms of all-commuting idempotents according to spectral decomposition, where

$$c_j = \sum_\alpha c_j^{(\alpha)} \mathbb{P}^\alpha,$$
$$c_j \mathbb{P}^\alpha = c_j^{(\alpha)} \mathbb{P}^\alpha. \qquad (3.2.16)$$

The eigenvalues $c_1^{(\alpha)}$ and $c_2^{(\alpha)}$ are found by multiplying $\mathbb{P}^\alpha$ and c_1 and c_2, respectively, and $c_3^{(\alpha)}$ was given following Eq. (3.2.13). The C_{3v} class spectral decomposition has the following form:

$$\mathbf{1} \equiv c_1 = \mathbb{P}^1 + \mathbb{P}^2 + \mathbb{P}^3,$$
$$c_2 = 2\mathbb{P}^1 + 2\mathbb{P}^2 - \mathbb{P}^3,$$
$$c_3 = 3\mathbb{P}^1 - 3\mathbb{P}^2. \qquad (3.2.17)$$

C. Does the Class Algebra Reduction Work in General?

We can easily prove that a decomposition such as the one for C_{3v} is possible for all finite groups. Since class operators commute, one can follow the procedure which worked for Abelian groups (recall Section 2.9), provided no minimal equation contains a repeated root. But suppose a root is repeated; that is, suppose that

$$(c - c^1 \mathbf{1})(\cdots)(c - c^r \mathbf{1})^m (\cdots) = 0 \qquad (3.2.18)$$

was the MEq of c with $m \geq 2$ repeated roots. This would imply that one could construct a nonzero operator n,

$$n = (c - c^1 \mathbf{1})(\cdots)(c - c^r \mathbf{1})^{m-1} (\cdots) \neq 0, \qquad (3.2.19)$$

which is NILPOTENT, i.e., an operator whose square is zero:

$$n^2 = 0. \qquad (3.2.20)$$

If nilpotent n acts on any combination $G = \Sigma \gamma_g g$ of group operators it yields an operator nG which is also nilpotent (note that $nG = Gn$ since n is in the class algebra):

$$nGnG = Gn^2 G = 0. \qquad (3.2.21)$$

Now setting $G = n^\dagger$, and following the same arguments which are stated in Appendix C, we conclude that $n^\dagger n$ and finally n must be zero. A Hermitian operator cannot be nilpotent without being zero. Hence, a class operator can never have repeated roots; therefore, the class spectral decomposition is always possible.

Finally, the all-commuting or class idempotents can be seen to be unique for the same reasons that applied to Abelian groups (cf. Section 2.9). Also, they may be shown to be Hermitian (see Problem 1.2.7):

$$\mathbb{P}^{\alpha \dagger} = \mathbb{P}^\alpha. \qquad (3.2.22)$$

3.3 SECOND STAGE OF NON-ABELIAN SYMMETRY ANALYSIS

It is not possible that every operator in a non-Abelian group is a combination of the same set of idempotents. If this were so, then every operator would commute. Stated another way, the representations of noncommuting operators cannot *all* be transformed to diagonal matrices at once.

However, every non-Abelian group has some Abelian subgroups. For example, C_{3v} has the subgroup $C_3 = \{\mathbf{1}, r, r^2\}$, $C_v = \{\mathbf{1}, \sigma_1\}$, $C_v' = \{\mathbf{1}, \sigma_2\}$, or $C_v'' = \{\mathbf{1}, \sigma_3\}$, which were mentioned before. (The lines leading back to

Abelian groups in Figure 2.1.1 indicate all the possibilities for each crystal symmetry.) The idea is to find the largest possible set of mutually commuting operators. Abelian subgroup operators commute with each other and with the all-commuting class operators. A reflection subgroup C_v and class operators taken together give four independent commuting operators σ_j, c_1, c_2, and c_3. By taking the cyclic subgroup C_3 one gets another set of four independent operators r, c_1, c_2, and c_3. (Note that r^2 is not linearly independent, since $r^2 = c_2 - r$.] The maximum number of independent mutually commuting operators which can be found in a group algebra is called the RANK of the algebra. The rank is the maximum number of operators that can be represented at once by diagonal matrices. It is up to the physicist to decide which Abelian subgroups he would like to represent by diagonal matrices. While one choice may be as good as another from a mathematical viewpoint, it can make a big difference in ease of computation and clarity of physical insight. Two very different choices C_3 or C_v exist within the C_{3v} symmetry. Since r in C_3 does not commute with σ_j in C_v, one cannot choose both. It will turn out that the C_3 choice yields moving-wave solutions, while a C_v choice yields standing-wave solutions.

For example, let us choose $C_v'' = \{1, \sigma_3\}$. This group has the familiar completeness relation

$$1 = P^+ + P^- = \tfrac{1}{2}(1 + \sigma_3) + \tfrac{1}{2}(1 - \sigma_3). \tag{3.3.1}$$

When this is multiplied by the completeness relation [Eq. (3.2.17)] for the all-commuting idempotents [Eq. (3.2.15)] as is done in the following, a new and different set of idempotents will result:

$$1 = (P^+ + P^-)(\mathbb{P}^1 + \mathbb{P}^2 + \mathbb{P}^3), \tag{3.3.2a}$$

$$1 = P^+\mathbb{P}^1 + P^-\mathbb{P}^1 + P^+\mathbb{P}^2 + P^-\mathbb{P}^2 + P^+\mathbb{P}^3 + P^-\mathbb{P}^3, \tag{3.3.2b}$$

$$1 = \mathbb{P}^1 + 0 + 0 + \mathbb{P}^2 + P_1^3 + P_2^3 \tag{3.3.2c}$$

In the last line we use the C_{3v} group table [Eq. (3.1.1)] to work out the products. The first and fourth products yield the original all-commuting $\mathbb{P}^1$ and $\mathbb{P}^2$, while the second and third products yield nothing. However, the last two terms yield a SPLITTING of the all-commuting $\mathbb{P}^3$ into two new and orthogonal idempotents,

$$P_1^3 = P^+\mathbb{P}^3 = \tfrac{1}{6}(1 + \sigma_3)(21 - r - r^2) = \tfrac{1}{6}(21 - r - r^2 - \sigma_1 - \sigma_2 + 2\sigma_3), \tag{3.3.3a}$$

$$P_2^3 = P^-\mathbb{P}^3 = \tfrac{1}{6}(1 - \sigma_3)(21 - r - r^2) = \tfrac{1}{6}(21 - r - r^2 + \sigma_1 + \sigma_2 - 2\sigma_3), \tag{3.3.3b}$$

where

$$\mathbb{P}^3 = P_1^3 + P_2^3. \quad (3.3.3c)$$

This procedure has yielded four idempotents $\mathbb{P}^1$, $\mathbb{P}^2$, P_1^3, and P_2^3, which may be used to expand four mutually commuting operators c_1, c_2, c_3, and σ_3. The four idempotents are called IRREDUCIBLE or unsplittable idempotents, since they cannot be split further into sums of orthogonal idempotents. In general a group algebra of rank r has exactly r irreducible idempotents.

However, C_{3v} has six linearly independent operators in all. In order to expand the whole C_{3v} group we will need two more operators of some kind. It is not possible to find two more orthogonal idempotents, since C_{3v} is non-Abelian. The question is: What form should these two additional operators take?

We begin to see the answer to this by performing the following expansion of general C_{3v} operator g using Eq. (3.3.2c) twice:

$$g = \mathbf{1}g\mathbf{1} = \left(\mathbb{P}^1 + \mathbb{P}^2 + P_1^3 + P_2^3\right)g\left(\mathbb{P}^1 + \mathbb{P}^2 + P_1^3 + P_2^3\right), \quad (3.3.4a)$$

$$g = \mathbb{P}^1 g \mathbb{P}^1 + \mathbb{P}^2 g \mathbb{P}^2 + P_1^3 g P_1^3 + P_1^3 g P_2^3 + P_2^3 g P_1^3 + P_2^3 g P_2^3 \quad (3.3.4b)$$

In writing Eq. (3.3.4b) we ignore vanishing cross-terms such as

$$\mathbb{P}^1 g \mathbb{P}^2 = g \mathbb{P}^1 \mathbb{P}^2 = 0, \qquad \mathbb{P}^1 g P_1^3 = g \mathbb{P}^1 P_1^3 = 0,$$

which involve idempotents which are all-commuting. The first two terms involving all-commuting $\mathbb{P}^1$ and $\mathbb{P}^2$ in Eq. (3.3.4b) can be rewritten as

$$\mathbb{P}^1 g \mathbb{P}^1 = g \mathbb{P}^1 = \mathscr{D}^1(g) \mathbb{P}^1, \qquad \mathbb{P}^2 g \mathbb{P}^2 = g \mathbb{P}^2 = \mathscr{D}^2(g) \mathbb{P}^2, \quad (3.3.5)$$

using commutivity ($\mathbb{P}^1 g = g \mathbb{P}^1$), idempotency, $\mathbb{P}^1 \mathbb{P}^1 = \mathbb{P}^1$, and the group eigenvector properties of $\mathbb{P}^1$ and $\mathbb{P}^2$, which follow from Eq. (3.2.15). Substituting, we find the eigenvalues

$$\mathscr{D}^1(1) = \mathscr{D}^1(r) = \mathscr{D}^1(r^2) = 1, \qquad \mathscr{D}^1(\sigma_1) = \mathscr{D}^1(\sigma_2) = \mathscr{D}^1(\sigma_3) = 1,$$
$$\mathscr{D}^2(1) = \mathscr{D}^2(r) = \mathscr{D}^2(r^2) = 1, \qquad \mathscr{D}^2(\sigma_1) = \mathscr{D}^2(\sigma_2) = \mathscr{D}^2(\sigma_3) = -1,$$
$$(3.3.6)$$

and thereby account for the first two of the C_{3v} irreps. So far this is the same as an Abelian irrep derivation.

However, the third C_{3v} irrep is different. Its properties are derived from the last four terms of the form $P_i^3 g P_j^3$ in Eq. (3.3.4b). The diagonal terms $P_i^3 g P_i^3$ turn out to be proportional to the original irreducible idempotents as follows:

$$P_1^3 g P_1^3 = \mathscr{D}_{11}^3(g) P_1^3, \qquad P_2^3 g P_2^3 = \mathscr{D}_{22}^3(g) P_2^3.$$

The new terms $P_i^3 g P_j^3$ turn out to be proportional to two IRREDUCIBLE NILPOTENT projectors $P_{ij}^3 = P_{ji}^{3\dagger}$,

$$P_1^3 g P_2^3 = \mathscr{D}_{12}^3(g) P_{12}^3, \qquad P_2^3 g P_1^3 = \mathscr{D}_{21}^3(g) P_{21}^3,$$

which satisfy

$$P_{12}^3 P_{21}^3 = P_1^3, \qquad P_{21}^3 P_{12}^3 = P_2^3,$$

as well as nilpotency,

$$P_{12}^3 P_{12}^3 = 0 = P_{21}^3 P_{21}^3.$$

The proportionality coefficients $\mathscr{D}_{ij}^\alpha(g)$ expressed in matrix form

$$\mathscr{D}^\alpha(g) = \begin{pmatrix} \mathscr{D}_{11}^\alpha(g) & \mathscr{D}_{12}^\alpha(g) \\ \mathscr{D}_{21}^\alpha(g) & \mathscr{D}_{22}^\alpha(g) \end{pmatrix}$$

comprise the third irrep $\mathscr{D}^3$ of C_{3v}:

$$\mathscr{D}^3(1) = \begin{pmatrix} 1 & 0 \\ 0 & 1 \end{pmatrix}, \quad \mathscr{D}^3(r) = \begin{pmatrix} -1/2 & -\sqrt{3}/2 \\ \sqrt{3}/2 & -1/2 \end{pmatrix},$$

$$\mathscr{D}^3(r^2) = \begin{pmatrix} -1/2 & \sqrt{3}/2 \\ -\sqrt{3}/2 & -1/2 \end{pmatrix}, \quad \mathscr{D}^3(\sigma_1) = \begin{pmatrix} -1/2 & -\sqrt{3}/2 \\ -\sqrt{3}/2 & 1/2 \end{pmatrix},$$

$$\mathscr{D}^3(\sigma_2) = \begin{pmatrix} -1/2 & \sqrt{3}/2 \\ \sqrt{3}/2 & 1/2 \end{pmatrix}, \quad \mathscr{D}^3(\sigma_3) = \begin{pmatrix} 1 & 0 \\ 0 & -1 \end{pmatrix}. \tag{3.3.7}$$

These coefficients will be derived in the following section. It is more important now to appreciate the general form of non-Abelian group reduction.

Each non-Abelian group operator g may be expressed as a combination

$$g = \sum_\alpha \sum_{ij}^{l^\alpha} \mathscr{D}_{ij}^\alpha(g) P_{ij}^\alpha \tag{3.3.8}$$

of ELEMENTARY OPERATORS P_{ij}^α weighted by coefficients $\mathscr{D}_{ij}^\alpha$. Diagonal operators

$$P_{ii}^\alpha \equiv P_i^\alpha \tag{3.3.9}$$

168 BASIC THEORY AND APPLICATIONS OF SYMMETRY REPRESENTATIONS

are just the l^α irreducible idempotents which were split from all-commuting $\mathbb{P}^\alpha$:

$$\mathbb{P}^\alpha = P^\alpha_{11} + P^\alpha_{22} + \cdots + P^\alpha_{l^\alpha l^\alpha}. \tag{3.3.10}$$

Off-diagonal operators are defined by

$$P^\alpha_{ij} = P^\alpha_i g P^\alpha_j / \mathscr{D}^\alpha_{ij}(g) = P^{\alpha^\dagger}_{ji}, \tag{3.3.11}$$

where nonzero $P^\alpha_i g P^\alpha_j$ may be found and normalized by coefficients $\mathscr{D}^\alpha_{ij}(g)$ to give

$$P^\alpha_{ij} P^\beta_{kl} = \delta^{\alpha\beta} \delta_{jk} P^\alpha_{il}. \tag{3.3.12}$$

The orthonormality relation (3.3.12) can be understood if you always remember that P^α_{ij} carries orthogonal idempotents P^α_i and P^α_j as "bodyguards" on its left and right, respectively. When P^α_{ij} runs into P^β_{kl}, then P^α_{ij}'s right-hand bodyguard P^α_j encounters the left-hand bodyguard P^β_k of P^β_{kl}. Annihilation occurs unless the guards are identical ($\alpha = \beta$ and $j = k$). Furthermore, one can show (Appendix D) that all operators $P^\alpha_i G P^\alpha_j$ with identical sets of bodyguards are indistinguishable, except for proportionality factors $\mathscr{D}^\alpha_{ij}(G)$; i.e.,

$$P^\alpha_i G P^\alpha_j = \mathscr{D}^\alpha_{ij}(G) P^\alpha_{ij},$$

where $G = \Sigma_g \gamma_g g$ is any linear combination of group operators. If it happens that G is annihilated by its bodyguards, then let $\mathscr{D}^\alpha_{ij}(G) = 0$. The coefficients $\mathscr{D}$ are chosen so that Eq. (3.3.12) holds, whence they also satisfy group matrix representation relations:

$$\sum_j^{l^\beta} \mathscr{D}^\beta_{ij}(g) \mathscr{D}^\beta_{jk}(h) = \mathscr{D}^\beta_{ik}(gh). \tag{3.3.13}$$

To prove the matrix relations first note the effect which a group operator (h) would have on the left bodyguard of P^β_{kl}. Equation (3.3.8) gives

$$h P^\beta_{kl} = \sum_\alpha \sum_{i,j}^{l^\alpha} \mathscr{D}^\alpha_{ij}(h) P^\alpha_{ij} P^\beta_{kl},$$

and Eq. (3.3.12) gives

$$h P^\beta_{kl} = \sum_i^{l^\beta} \mathscr{D}^\beta_{ik}(h) P^\beta_{il}. \tag{3.3.14}$$

This is the left-hand transformation rule, which holds for all group operators, including products such as (gh):

$$ghP_{kl}^{\beta} = \sum_{i}^{l^{\beta}} \mathscr{D}_{ik}^{\beta}(gh) P_{il}^{\beta}. \tag{3.3.15}$$

However, this product is also equal to left operation by g onto Eq. (3.3.14):

$$ghP_{kl}^{\beta} = \sum_{j}^{l^{\beta}} \mathscr{D}_{jk}^{\beta}(h) gP_{jl}^{\beta}$$

$$= \sum_{j}^{l^{\beta}} \mathscr{D}_{jk}(h) \sum_{i}^{l^{\beta}} \mathscr{D}_{ij}^{\beta}(g) P_{il}^{\beta}$$

$$= \sum_{i}^{l^{\beta}} \left[\sum_{j}^{l^{\beta}} \mathscr{D}_{ij}^{\beta}(g) \mathscr{D}_{jk}^{\beta}(h) \right] P_{il}^{\beta}.$$

Equating the coefficient in brackets to $\mathscr{D}_{ik}^{\beta}(gh)$ in Eq. (3.3.15) yields Eq. (3.3.13).

This is how a non-Abelian group is analyzed; each operator g is a combination of elementary operators, P_{ij}^{α}, whose multiplicative properties are, as the name implies, elementary. For example, each C_{3v} operator is a combination involving six terms:

$$g = \mathscr{D}^1(g)\mathbb{P}^1 + \mathscr{D}^2(g)\mathbb{P}^2 + \mathscr{D}_{11}^3(g)P_{11}^3 + \mathscr{D}_{12}^3(g)P_{12}^3 \\ + \mathscr{D}_{21}^3(g)P_{21}^3 + \mathscr{D}_{22}^3(g)P_{22}^3 \tag{3.3.16}$$

and three sets of irreps. The analysis of non-Abelian groups is a generalization of that for Abelian groups. For Abelian groups all irreps are 1×1 matrices, whereas a non-Abelian group must have some irreps $\mathscr{D}^{\alpha}$ which are $l^{\alpha} \times l^{\alpha}$ matrices with $l^{\alpha} \geq 2$. For each (α) there are $(l^{\alpha})^2$ elementary operators. Since the number of elementary operators must equal the order $(^{\circ}G)$ of the group, we have

$$^{\circ}G = \sum_{\alpha} (l^{\alpha})^2. \tag{3.3.17}$$

In the case of C_{3v} this number is $6 = 1^2 + 1^2 + 2^2$. The number of different types (α) of irreps equals the number of all-commuting idempotents ($\mathbb{P}^{\alpha} = \sum_i P_{ii}^{\alpha}$), which is the number of classes. In the following sections we shall show how the $\mathbb{P}^{\alpha}$, viz., Eq. (3.2.15), give directly the dimensions l^{α} of the irreps.

Now, while Abelian group irreps and elementary operators are all uniquely defined, those with $l_\alpha \geq 2$ for non-Abelian groups are not. For example, in C_{3v} we would have split $\mathbb{P}^3$ with idempotents from $C_3 = \{1, r, r^2\}$ instead of $C_v = \{1, \sigma_3\}$. However, the number l^α of P_i^α split from a given $\mathbb{P}^\alpha$ is fixed. Applying the three C_3 projectors $\{P^{0_3}, P^{1_3}, P^{2_3}\}$ to the all-commuting $\mathbb{P}^3$ yields only two nonzero idempotents,

$$P^{0_3}\mathbb{P}^3 = P^3(1 + r + r^2)/3 = 0,$$
$$P_1^3 = P^{1_3}\mathbb{P}^3 = \mathbb{P}^3(1 + \varepsilon^* r + \varepsilon r^2)/3 = (1 + \varepsilon^* r + \varepsilon r^2)/3,$$
$$P_2^3 = P^{2_3}\mathbb{P}^3 = \mathbb{P}^3(1 + \varepsilon r + \varepsilon^* r)/3 = (1 + \varepsilon r + \varepsilon^* r^2)/3. \quad (3.3.18)$$

Nevertheless, they differ markedly from the two operators obtained using $C_v = \{1, \sigma_3\}$ projectors in Eq. (3.3.3). They lead to a different but equivalent set of $\mathscr{D}^3$ irreps. We shall return to the significance of different choices shortly.

3.4 THEORY AND APPLICATION OF ELEMENTARY OPERATORS AND IRREPS

In order to demonstrate the theory and application of the non-Abelian symmetry C_{3v}, we shall eventually solve the equations for the mechanical system in Figure 3.4.1. However, the main objective of this section will be to derive and prove general relations which are followed by the elementary operators of a non-Abelian finite group $G = \{1, g, g', \ldots\}$. The analysis and application of multidimensional ($l^\alpha \geq 2$) irreps has a number of subtle features which are not present in Abelian symmetry analysis. However, once these are understood the application of the theory is just as easy as it was in the Abelian cases. You will soon be able to solve some very complicated problems with relatively little computation.

A. Group Space and the Regular Representation

In Figure 3.4.1 the coordinates $\langle g|x \rangle$ and unit vectors $|g\rangle$ are chosen in perfect correspondence to the symmetry group C_{3v}. Each group operation g acting on the first coordinate $\langle 1|x \rangle$ or vector $|1\rangle$ gives one of the others according to Eq. (3.4.1).

$$g|1\rangle = |g\rangle, \quad (3.4.1a)$$

$$\langle 1|g^\dagger = \langle g| = \langle 1|g^{-1}, \quad (3.4.1b)$$

$$\langle g|x\rangle = \langle 1|g^\dagger|x\rangle. \quad (3.4.1c)$$

THEORY AND APPLICATION OF ELEMENTARY OPERATORS AND IRREPS 171

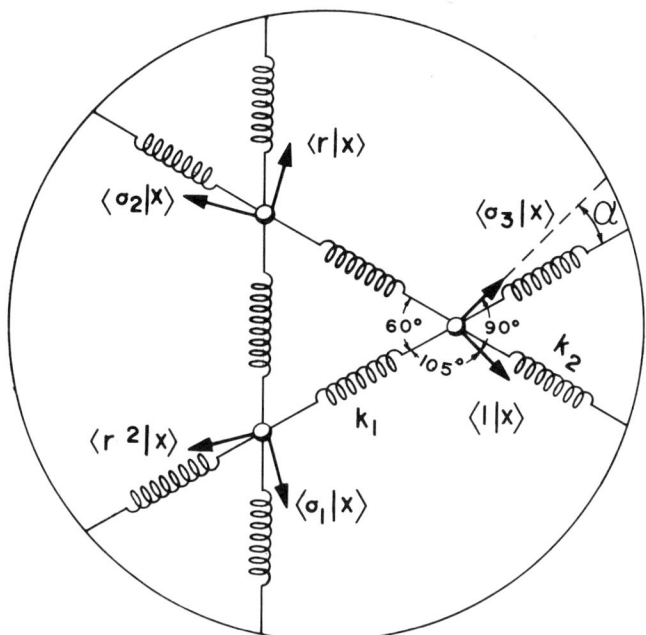

Figure 3.4.1 Spring-mass model with C_{3v} symmetry. One coordinate degree of freedom exists for each symmetry operator in C_{3v}.

These vectors are defined to be orthonormal, as expressed by the following:

$$\langle g|h\rangle = \delta_{g,h} = \delta_{g^{-1}h,1}, \tag{3.4.2a}$$

$$\langle 1|g^{\dagger}h|1\rangle = \delta_{g^{-1}h,1} = \langle 1|g^{-1}h|1\rangle. \tag{3.4.2b}$$

The group basis $\{|1\rangle, |g\rangle, \ldots\}$ spans what is called the REGULAR REPRESENTATION $\mathscr{R}(g)$. This is defined in the following with the help of the two preceding equations:

$$\mathscr{R}_{hf}(g) = \langle h|g|f\rangle = \langle 1|h^{\dagger}gf|1\rangle = \delta_{h^{-1}gf,1} = \delta_{g,hf^{-1}}. \tag{3.4.3}$$

The regular representation exists quite independently of any physical system such as the one in Figure 3.4.1. In fact, you may construct each matrix $\mathscr{R}(g)$ directly from any group table in which the top row is arranged so that each position contains the inverse $f^{\dagger} = f^{-1}$ of the occupant of the corresponding left column position, as in the following. $\mathscr{R}(g)$ is then obtained by transcribing one (1) wherever g shows up and zero, $(0) \equiv (\cdot)$, elsewhere. For example,

$\mathcal{R}(r)$ is constructed as follows:

$$
\begin{array}{c|cccccc}
 & 1 & r^2 & r & \sigma_1 & \sigma_2 & \sigma_3 \\
\hline
1 & 1 & r^2 & r & \sigma_1 & \sigma_2 & \sigma_3 \\
r & r & 1 & r^2 & \sigma_3 & \sigma_1 & \sigma_2 \\
r^2 & r^2 & r & 1 & \sigma_2 & \sigma_3 & \sigma_1 \\
\sigma_1 & \sigma_1 & \sigma_3 & \sigma_2 & 1 & r & r^2 \\
\sigma_2 & \sigma_2 & \sigma_1 & \sigma_3 & r^2 & 1 & r \\
\sigma_3 & \sigma_3 & \sigma_2 & \sigma_1 & r & r^2 & 1
\end{array}
\rightarrow \mathcal{R}(r) =
\begin{pmatrix}
 & & 1 & & & \\
1 & & & & & \\
 & 1 & & & & \\
 & & & & 1 & \\
 & & & & & 1 \\
 & & & 1 & &
\end{pmatrix}
\begin{matrix} |1\rangle & |r\rangle & |r^2\rangle & |\sigma_1\rangle & |\sigma_2\rangle & |\sigma_3\rangle \end{matrix}.
$$

(3.4.4)

Note that the trace of the regular representation of all group operators g is zero, except for the identity $g = \mathbf{1}$, which has a trace equal to the order $^\circ G$ of the group:

$$\text{Trace } \mathcal{R}(g) = \delta_{g,\mathbf{1}} {}^\circ G. \tag{3.4.5}$$

This is very important for later derivations.

Note also that group space vectors can be defined with respect to operator combinations $G = \Sigma \gamma_g g$ of group operators:

$$|G\rangle = G|1\rangle = (\gamma_1 \mathbf{1} + \gamma_g g + \cdots)|1\rangle = \gamma_1|1\rangle + \gamma_g|g\rangle \cdots, \tag{3.4.6a}$$

$$\langle G| = \langle 1|G^\dagger = \langle 1|(\gamma_1^* \mathbf{1} + \gamma_g^* g + \cdots) = \gamma_1^* \langle 1| + \gamma_g^* \langle g| + \cdots. \tag{3.4.6b}$$

For example, consider the operators $P_i^\alpha g P_j^\alpha = \mathcal{D}_{ij}^\alpha(g) P_{ij}^\alpha$. If you change (g) in $P = P_i^\alpha g P_j^\alpha$ it is unchanged according to Appendix D except for an overall factor $\mathcal{D}_{ij}^\alpha(g)$. Let us suppose $\mathcal{D}_{ij}^\alpha(g)$ are still unknown, and define $|\hat{P}\rangle$ so that the overall factor is absorbed in normalization of the vector. Let

$$|\hat{P}_{ij}^\alpha\rangle \equiv P_{ij}^\alpha|1\rangle / (N_{ij}^\alpha)^{1/2}, \tag{3.4.7a}$$

where the normalization factor N_{ij}^α is determined by

$$\langle \hat{P}_{ij}^\alpha | \hat{P}_{ij}^\alpha \rangle = 1 \tag{3.4.7b}$$

to within a phase.

For C_{3v} one can easily calculate the operators $P_i^\alpha g P_j^\alpha$ for select g, and by normalizing one finds the corresponding vectors $|\hat{P}_{ij}^\alpha\rangle$. Actually, the first four operators have been derived once before. [See Eqs. (3.3.3) and equations following (3.3.4).] The latter two equations are worked out in the following using the group table in Eq. (3.4.4):

$$\mathbb{P}^1\mathbf{1}\mathbb{P}^1 = \mathbb{P}^1 = \tfrac{1}{6}(1 + \mathbf{r} + \mathbf{r}^2 + \boldsymbol{\sigma}_1 + \boldsymbol{\sigma}_2 + \boldsymbol{\sigma}_3),$$
$$\mathbb{P}^2\mathbf{1}\mathbb{P}^2 = \mathbb{P}^2 = \tfrac{1}{6}(1 + \mathbf{r} + \mathbf{r}^2 - \boldsymbol{\sigma}_1 - \boldsymbol{\sigma}_2 - \boldsymbol{\sigma}_3),$$
$$P_{11}^3 \equiv P_1^3 \mathbf{1} P_1^3 = P_1^3 = \tfrac{1}{6}(2 \cdot 1 - \mathbf{r} - \mathbf{r}^2 - \boldsymbol{\sigma}_1 - \boldsymbol{\sigma}_2 + 2\boldsymbol{\sigma}_3),$$
$$P_{22}^3 \equiv P_2^3 \mathbf{1} P_2^3 = P_2^3 = \tfrac{1}{6}(2 \cdot 1 - \mathbf{r} - \mathbf{r}^2 + \boldsymbol{\sigma}_1 + \boldsymbol{\sigma}_2 - 2\boldsymbol{\sigma}_3),$$
$$\mathscr{D}_{12}^3(\boldsymbol{\sigma}_2) P_{12}^3 = P_1^3 \boldsymbol{\sigma}_2 P_2^3 = \tfrac{1}{4}(0 - \mathbf{r} + \mathbf{r}^2 - \boldsymbol{\sigma}_1 + \boldsymbol{\sigma}_2 + 0),$$
$$\mathscr{D}_{21}^3(\boldsymbol{\sigma}_2) P_{21}^3 = P_2^3 \boldsymbol{\sigma}_2 P_1^3 = \tfrac{1}{4}(0 + \mathbf{r} - \mathbf{r}^2 - \boldsymbol{\sigma}_1 + \boldsymbol{\sigma}_2 + 0).$$

By assembling these coefficients into vectors and then normalizing so that $\langle \hat{P} | \hat{P} \rangle = 1$, one obtains the following:

$$\langle \hat{P}^1 | \leftrightarrow \tfrac{1}{\sqrt{6}} \overline{1\ 1\ 1\ 1\ 1\ 1}, \qquad \langle \hat{P}_{22}^3 | \leftrightarrow \tfrac{1}{2\sqrt{3}} \overline{2 - 1 - 1\ 1\ 1 - 2},$$

$$\langle \hat{P}^2 | \leftrightarrow \tfrac{1}{\sqrt{6}} \overline{1\ 1\ 1 - 1 - 1 - 1}, \qquad \langle \hat{P}_{12}^3 | \leftrightarrow \tfrac{1}{2} \overline{0 - 1\ 1 - 1\ 1\ 0},$$

$$\langle \hat{P}_{11}^3 | \leftrightarrow \tfrac{1}{2\sqrt{3}} \overline{2 - 1 - 1 - 1 - 1\ 2}, \qquad \langle \hat{P}_{21}^3 | \leftrightarrow \tfrac{1}{2} \overline{0\ 1 - 1 - 1\ 1\ 0}.$$

$$(3.4.8)$$

Pictures of these vectors are drawn in Figure 3.4.2, as they represent motions of the system in Figure 3.4.1. Let us now study these pictures in order to understand the P_{ij}^α operators.

The first two motions corresponding to $|\mathbb{P}^1\rangle$ (expansion) and $|\mathbb{P}^2\rangle$ (rotation or libration) have obvious symmetry properties. The interpretation of the $|P_{ij}^3\rangle$ is more subtle, particularly in regard to the left index (i) and the right index (j). An index equal to 1 or 2 denotes symmetry or antisymmetry, respectively. [Recall Eqs. (3.3.3).] The states $|P_{1j}^3\rangle$ and $|P_{2j}^3\rangle$ are symmetric and antisymmetric, respectively, to reflection σ_3 through the x plane of Figure 3.4.2. The $|P_{ij}^3\rangle$ states are eigenvectors of σ_3. Since σ_3 runs into the left-hand bodyguard P_i^3, the left-hand index determines the overall reflection parity of the state. The eigenrelation

$$\sigma_3 |P_{ij}^3\rangle = \sigma_3 P_{ij}^3 |1\rangle = (-1)^{i-1} |P_{ij}^3\rangle$$

follows from the definition

$$\sigma_3 P_i^3 = -(-1)^i P_i^3 = P_i^3 \sigma_3$$

of the irreducible idempotents.

174 BASIC THEORY AND APPLICATIONS OF SYMMETRY REPRESENTATIONS

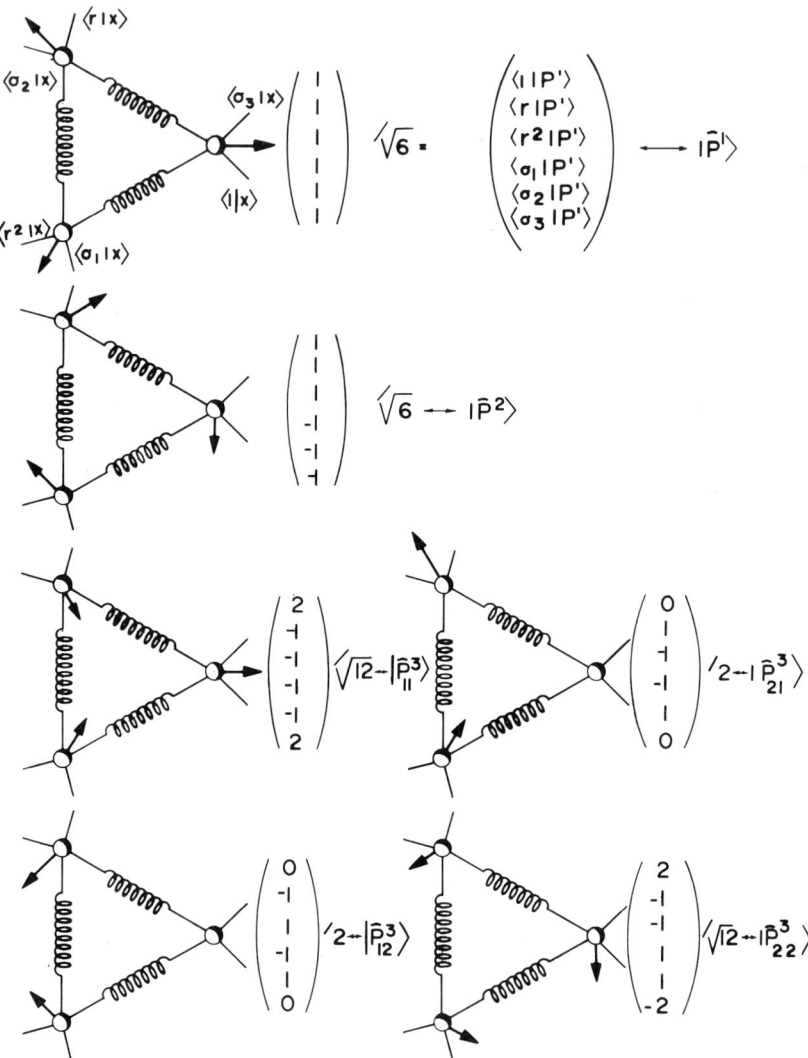

Figure 3.4.2 States of motion corresponding to C_{3v} P operators.

However, the right-hand index determines the *local* reflection symmetry properties of the motion. Note that each mass in state $|P_{i1}^3\rangle$ remains on its local reflection plane, while each mass in state $|P_{i2}^3\rangle$ moves perpendicularly to its local plane if it moves at all. State $|P_{i1}^3\rangle$ and $|P_{i2}^3\rangle$ are made of locally symmetric and antisymmetric motions, respectively. The right-hand index determines local symmetry since the right-hand bodyguard P_j^3 first encounters the local state $|1\rangle$ in the projection

$$P_{ij}^3|1\rangle = P_{ij}^3 P_j|1\rangle = P_{ij}^3|P_j\rangle,$$

where

$$P_j = (1 - (-1)^j \sigma_3)/2.$$

P_{ij}^3 acting on $|P_j\rangle$ gives a combination of states $g|P_j\rangle$ corresponding to similar motion of the other masses. But each state $g|P_j\rangle$ obeys a local symmetry equation

$$g\sigma_3 g^{-1} g|P_j\rangle = -(-1)^j g|P_j\rangle$$

equivalent to the initial equation

$$\sigma_3 |P_j\rangle = -(-1)^j |P_j\rangle.$$

Note that vectors $|\hat{P}_{ij}^\alpha\rangle$ in Figure 3.4.2 are mutually orthogonal. This is a general result. To prove it first note that the scalar product in the following is zero if $i \neq k$, or $\alpha \neq \beta$:

$$\langle \hat{P}_{ij}^\alpha | \hat{P}_{kl}^\beta \rangle = \langle 1|P_{ij}^{\alpha\dagger} P_{kl}^\beta|1\rangle / (N_{ij}^\alpha N_{kl}^\beta)^{1/2} = \langle 1|P_{ji}^\alpha P_{kl}^\alpha|1\rangle \delta^{\alpha\beta}/(N_{ij}^\alpha N_{kl}^\alpha)^{1/2}$$
$$= \langle 1|P_{jl}^\alpha|1\rangle \delta_{ik} \delta^{\alpha\beta}/(N_{ij}^\alpha N_{kl}^\alpha)^{1/2}. \quad (3.4.9)$$

Now the matrix $\mathscr{R}(P_{jl}^\alpha)$ is nilpotent for $j \neq l$, since $\mathscr{R}(P_{jl}^\alpha)\mathscr{R}(P_{jl}^\alpha) = \mathscr{R}(P_{jl}^\alpha P_{jl}^\alpha) = \mathscr{R}(0)$. Therefore, $\mathscr{R}(P_{jl}^\alpha)$ has all zero eigenvalues and zero trace. Therefore $\langle 1|P_{jl}^\alpha|1\rangle$ must be zero if $j \neq l$ according to Eq. (3.4.5), since P_{jl}^α, being traceless, does not have any component of the unit operator $\mathbf{1}$. This gives the scalar product result,

$$\langle \hat{P}_{ij}^\alpha | \hat{P}_{kl}^\alpha \rangle = \langle 1|P_{jj}^\alpha|1\rangle \delta^{\alpha\beta} \delta_{ik} \delta_{jl}/N_{ij}^\alpha = \delta^{\alpha\beta} \delta_{ik} \delta_{jl}. \quad (3.4.10a)$$

In general, normalization does not depend on the left-hand indices:

$$N_{ij}^\alpha \equiv N_j^\alpha = \langle 1|P_{jj}^\alpha|1\rangle. \quad (3.4.10b)$$

We find a simple formula for N_j^α in the next section.

B. Deriving and Using P_{ij} and $\mathscr{D}_{ij}(g)$

If we change our basis from the old regular representation vectors $\{\cdots |g\rangle \cdots\}$ to the new ones $\{\cdots |\hat{P}_{ij}^\alpha\rangle \cdots\}$, some simplifications will occur. The transformation will reduce or "almost diagonalize" the regular representation $\mathscr{R}$ matrices, and it will reduce the acceleration matrix $\langle a \rangle$ which describes the motion of the system in Figure 3.4.1.

To understand these two occurrences is to understand the heart of non-Abelian symmetry analysis, and so let us study them carefully. In addition, a general method for deriving irreps $\mathscr{D}_{ij}^\alpha$ will be shown, as well as

176   BASIC THEORY AND APPLICATIONS OF SYMMETRY REPRESENTATIONS

some justification of the previous assumptions made about irreps and the elementary P_{ij} operators.

(a) Reducing the Regular Representation   The representation of a symmetry operator g in this new basis $\{\cdots|P_{ij}^\alpha\rangle\cdots\}$ takes a simpler form, as shown by the following:

$$\langle \hat{P}_{ij}^\alpha|g|\hat{P}_{kl}^\beta\rangle = \langle 1|P_{ji}^\alpha g P_{kl}^\beta|1\rangle/\sqrt{N_j^\alpha N_l^\beta}$$

$$= \langle 1|P_{ji}^\alpha P_i^\alpha g P_k^\alpha P_{kl}^\alpha|1\rangle \delta^{\alpha\beta}/\sqrt{N_j^\alpha N_l^\alpha}$$

(using idempotency: $P_i^\alpha = P_i^\alpha P_i^\alpha$)

$$= \mathscr{D}_{ik}^\alpha(g)\langle 1|P_{ji}^\alpha P_{ik}^\alpha P_{kl}^\alpha|1\rangle \delta^{\alpha\beta}/\sqrt{N_j^\alpha N_l^\alpha} \quad [\text{using Eq. (3.3.11)}]$$

$$= \mathscr{D}_{ik}^\alpha(g)\langle 1|P_{jl}^\alpha|1\rangle \delta^{\alpha\beta}/\sqrt{N_j^\alpha N_l^\alpha} \quad [\text{using Eq. (3.3.12)}].$$

Finally, using the preceding equation (3.4.10) there results

$$\langle \hat{P}_{ij}^\alpha|g|\hat{P}_{kl}^\beta\rangle = \mathscr{D}_{ik}^\alpha(g)\delta^{\alpha\beta}\delta_{jl}. \tag{3.4.11}$$

It is instructive to see an example of this wherein the $\mathscr{D}_{ik}^\alpha$ are evaluated. Consider the C_{3v} vector $r|P_{11}^3\rangle$ below. The group multiplication table gives the following vector and its regular representation.

$$r|\hat{P}_{11}^3\rangle = r(2\mathbf{1} - r - r^2 - \sigma_1 - \sigma_2 + 2\sigma_3)|1\rangle/2\sqrt{3}$$
$$= (2r - r^2 - \mathbf{1} - \sigma_3 - \sigma_1 + 2\sigma_2)|1\rangle/2\sqrt{3}$$
$$= (-|1\rangle + 2|r\rangle - |r^2\rangle - |\sigma_1\rangle + 2|\sigma_2\rangle - |\sigma_3\rangle)/2\sqrt{3} \leftrightarrow \begin{pmatrix} -1 \\ 2 \\ -1 \\ -1 \\ 2 \\ -1 \end{pmatrix}\bigg/2\sqrt{3}$$

$$\tag{3.4.12a}$$

By computing the scalar products of this vector with $|\hat{P}_{11}^3\rangle$ and $|\hat{P}_{21}^3\rangle$ we see that it is the following combination of them:

$$\begin{pmatrix} -1 \\ 2 \\ -1 \\ -1 \\ 2 \\ -1 \end{pmatrix}\bigg/2\sqrt{3} = -\frac{1}{2}\begin{pmatrix} 2 \\ -1 \\ -1 \\ -1 \\ -1 \\ 2 \end{pmatrix}\bigg/2\sqrt{3} + \frac{\sqrt{3}}{2}\begin{pmatrix} 0 \\ 1 \\ -1 \\ -1 \\ 1 \\ 0 \end{pmatrix}\bigg/2 \quad (3.4.12b)$$

THEORY AND APPLICATION OF ELEMENTARY OPERATORS AND IRREPS   177

This is an example of the left-hand transformation rule first stated in Eq. (3.3.12):

$$r|\hat{P}_{11}^3\rangle = \left(-\frac{1}{2}\right)|\hat{P}_{11}^3\rangle + \left(\frac{\sqrt{3}}{2}\right)|\hat{P}_{21}^3\rangle = \mathcal{D}_{11}^3(r)|\hat{P}_{11}^3\rangle + \mathcal{D}_{21}^3(r)|\hat{P}_{21}^3\rangle. \quad (3.4.12c)$$

The rule for the second component is derived similarly:

$$r|\hat{P}_{21}^3\rangle = \left(\frac{-\sqrt{3}}{2}\right)|\hat{P}_{11}^3\rangle + \left(-\frac{1}{2}\right)|\hat{P}_{21}^3\rangle = \mathcal{D}_{12}(r)|\hat{P}_{11}^3\rangle + \mathcal{D}_{22}(r)|\hat{P}_{21}^3\rangle.$$
(3.4.13)

Together, the last two equations give the (3)-type irrep matrix of operator r for 120° rotation:

$$\mathcal{D}^3(r) = \begin{pmatrix} \mathcal{D}_{11}^3(r) & \mathcal{D}_{12}^3(r) \\ \mathcal{D}_{21}^3(r) & \mathcal{D}_{22}^3(r) \end{pmatrix} = \begin{pmatrix} -1/2 & -\sqrt{3}/2 \\ \sqrt{3}/2 & -1/2 \end{pmatrix}. \quad (3.4.14)$$

In general a set of vectors $\{|P_{1j}^\alpha\rangle, |P_{2j}^\alpha\rangle, \ldots\}$ are said to be PARTNERS IN AN (α) IR BASIS when their symmetry transformation is the following:

$$g|P_{ij}^\alpha\rangle = \sum_{k=1}^{l^\alpha = 2} \mathcal{D}_{ki}^\alpha(g)|P_{kj}^\alpha\rangle. \quad (3.4.15)$$

Note that the transformation (g) mixes only partners belonging to the same right-hand index j of local symmetry.

In Figure 3.4.3 there is shown a physical picture of the sort of transformation in Eq. (3.4.12c). This picture shows how an x motion that is rotated 120° by r is a combination of itself and a y motion with coefficients $(-1/2)$ and $(\sqrt{3}/2)$ respectively. At this point you may realize that we have just gone through what could very well be the world's most complicated derivation of $\cos(120°) = -\frac{1}{2}$ and $\sin(120°) = \sqrt{3}/2$. The same goes for the other $\mathcal{D}^3$ components which were given in Eq. (3.4.14) and (3.3.7).

This shows one of the main ideas of symmetry analysis. Only a few key numbers such as the $\mathcal{D}$ coefficients are needed for analysis in a virtual infinity of different problems. The necessary coefficients are usually easy to derive from simple physical considerations. The P-operator mathematics exists to expedite the theory and its applications. It also provides derivations for cases in which physical intuition fails.

Generally, one will be given the $\mathcal{D}_{ij}^\alpha$ coefficients and need to derive the P_{ij}^α operators instead of the other way around. Let us now derive a formula for P_{ij}^α. To begin this, observe the form of the regular representation of operators

178 BASIC THEORY AND APPLICATIONS OF SYMMETRY REPRESENTATIONS

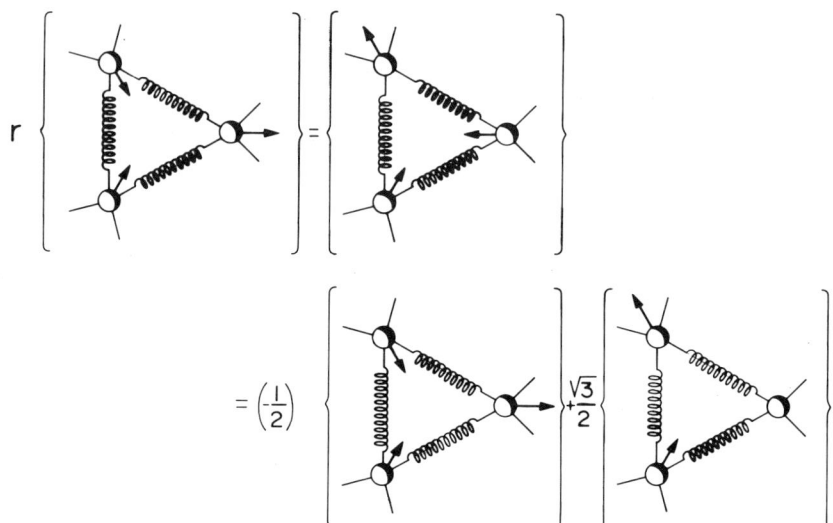

Figure 3.4.3 Pictorial representation of transformation $r|P_{11}^3\rangle = -\frac{1}{2}|P_{11}^3\rangle + (\sqrt{3}/2)P_{21}^3$.

in the basis of the $\{\cdots|P_{ij}^\alpha\rangle\cdots\}$ according to Eq. (3.4.11):

$$\mathscr{R}(g) = \begin{array}{c} \\ \langle P^1| \\ \langle P^2| \\ \langle P_{11}^3| \\ \langle P_{21}^3| \\ \langle P_{12}^3| \\ \langle P_{22}^3| \end{array} \begin{pmatrix} |P^1\rangle & |P^2\rangle & |P_{11}^3\rangle & |P_{21}^3\rangle & |P_{12}^3\rangle & |P_{22}^3\rangle \\ \mathscr{D}^1(g) & \cdot & \cdot & \cdot & \cdot & \cdot \\ \cdot & \mathscr{D}^2(g) & \cdot & \cdot & \cdot & \cdot \\ \cdot & \cdot & \mathscr{D}_{11}^3(g) & \mathscr{D}_{12}^3(g) & \cdot & \cdot \\ \cdot & \cdot & \mathscr{D}_{21}^3(g) & \mathscr{D}_{22}^3(g) & \cdot & \cdot \\ \cdot & \cdot & \cdot & \cdot & \mathscr{D}_{11}^3(g) & \mathscr{D}_{12}^3(g) \\ \cdot & \cdot & \cdot & \cdot & \mathscr{D}_{21}^3(g) & \mathscr{D}_{22}^3(g) \end{pmatrix}.$$

(3.4.16)

In particular, the elementary operators in their own basis have a very simple form:

$$\langle \hat{P}_{im}^\alpha | P_{kl}^\beta | \hat{P}_{jn}^\gamma \rangle = \delta^{\alpha\beta}\delta^{\gamma\beta}\delta_{ik}\delta_{jl}\delta_{mn}.$$

The operator P_{ij}^3 is represented by unity at the (ij) position in each $\mathscr{D}^3$ block matrix and zeros elsewhere:

$$\mathcal{R}(P^3_{11}) = \begin{bmatrix} 0 & \cdot & \cdot & \cdot & \cdot & \cdot \\ \cdot & 0 & \cdot & \cdot & \cdot & \cdot \\ \cdot & \cdot & 1 & 0 & \cdot & \cdot \\ \cdot & \cdot & 0 & 0 & \cdot & \cdot \\ \cdot & \cdot & \cdot & \cdot & 1 & 0 \\ \cdot & \cdot & \cdot & \cdot & 0 & 0 \end{bmatrix},$$

$$\mathcal{R}(P^3_{12}) = \begin{bmatrix} 0 & \cdot & \cdot & \cdot & \cdot & \cdot \\ \cdot & 0 & \cdot & \cdot & \cdot & \cdot \\ \cdot & \cdot & 0 & 1 & \cdot & \cdot \\ \cdot & \cdot & 0 & 0 & \cdot & \cdot \\ \cdot & \cdot & \cdot & \cdot & 0 & 1 \\ \cdot & \cdot & \cdot & \cdot & 0 & 0 \end{bmatrix}, \dots . \qquad (3.4.17)$$

All-commuting idempotents $\mathbb{P}^\alpha$ are represented by unit matrices in each $\mathscr{D}^\alpha$ block.

$$\mathcal{R}(\mathbb{P}^3) = \begin{bmatrix} 0 & \cdot & \cdot & \cdot & \cdot & \cdot \\ \cdot & 0 & \cdot & \cdot & \cdot & \cdot \\ \cdot & \cdot & 1 & 0 & \cdot & \cdot \\ \cdot & \cdot & 0 & 1 & \cdot & \cdot \\ \cdot & \cdot & \cdot & \cdot & 1 & 0 \\ \cdot & \cdot & \cdot & \cdot & 0 & 1 \end{bmatrix},$$

$$\mathcal{R}(\mathbb{P}^2) = \begin{bmatrix} 0 & \cdot & \cdot & \cdot & \cdot & \cdot \\ \cdot & 1 & \cdot & \cdot & \cdot & \cdot \\ \cdot & \cdot & 0 & 0 & \cdot & \cdot \\ \cdot & \cdot & 0 & 0 & \cdot & \cdot \\ \cdot & \cdot & \cdot & \cdot & 0 & 0 \\ \cdot & \cdot & \cdot & \cdot & 0 & 0 \end{bmatrix}, \dots .$$

This form should be taken as one of the defining relations between irreps and the elementary operators. It is expressed more concisely by the following general equation:

$$\mathscr{D}^\alpha_{ij}(P^\beta_{kl}) = \delta^{\alpha\beta}\delta_{ik}\delta_{jl}. \qquad (3.4.18)$$

This is needed in order to derive a formula for the elementary operators P^α_{jk} in terms of symmetry operators. Let us express an operator A, which is a sum of symmetry operators, as follows. [This uses the regular representation trace

180 BASIC THEORY AND APPLICATIONS OF SYMMETRY REPRESENTATIONS

relation in Eq. (3.4.5).]

$$A = \sum_g \{\text{TRACE } \mathscr{R}(g^\dagger A)/{}^\circ G\}g.$$

For $A = P_{jk}^\alpha$ one needs the trace of $\mathscr{R}(g^\dagger P_{jk}^\alpha)$. According to the left-transformation rule (3.4.15) one has

$$g^\dagger P_{jk}^\alpha = \sum_i^{l^\alpha} \mathscr{D}_{ij}^\alpha(g^\dagger) P_{ik}^\alpha = \sum_i \mathscr{D}_{ji}^{\alpha*}(g) P_{ik}^\alpha.$$

Substituting this in the preceding equation gives

$$A = P_{jk}^\alpha = \sum_g \sum_i^{l^\alpha} \mathscr{D}_{ji}^{\alpha*}(g)\{\text{Trace } \mathscr{R}(P_{ik}^\alpha)/{}^\circ G\}g.$$

A trace is independent of the choice of basis. The form of representations of P_{ik}^α in Eq. (3.4.17) make it clear that

$$\text{Trace } \mathscr{R}(P_{ik}^\alpha) = l^\alpha \delta_{ik}.$$

Substituting this in the equation for P_{jk}^α gives the desired result:

$$P_{jk}^\alpha = (l^\alpha/{}^\circ G) \sum_g \mathscr{D}_{jk}^{\alpha*}(g) g. \qquad (3.4.19)$$

The normalization formula promised after Eq. (3.4.10) is found by using the fact that $\mathscr{D}_{ij}^\alpha(1) = \delta_{ij}$ in Eq. (3.4.19):

$$N_{ij}^\alpha = \langle 1|P_{jj}^\alpha|1\rangle = (l^\alpha/{}^\circ G) \equiv N^\alpha. \qquad (3.4.20)$$

(b) Reducing an Equation of Motion The first of six coupled equations of motion for the system in Figure 3.4.1 is given by

$$-\langle 1|\ddot{x}\rangle = \sum_g \langle 1|\mathbf{a}|g\rangle\langle g|x\rangle, \qquad (3.4.21a)$$

where the first row of the acceleration matrix is the following:

$|g\rangle =$	$|1\rangle$	$|r\rangle$	$|r^2\rangle$	$|\sigma_1\rangle$	$|\sigma_2\rangle$	$|\sigma_3\rangle$
$\langle 1|\mathbf{a}|g\rangle =$	$\dfrac{(k_1+k_2)}{m}$	$\dfrac{k_1}{4m}$	$\dfrac{k_1}{4m}$	$\dfrac{k_1(2-\sqrt{3})}{4m}$	$\dfrac{k_1(2+\sqrt{3})}{4m}$	$\dfrac{k_1+2k_2\sin 2\alpha}{2m}$

$$(3.4.21b)$$

A representation of the acceleration operator (**a**) in the $|P^\alpha_{ij}\rangle$ basis is derived as follows

$$\langle \hat{P}^\alpha_{ij}|\mathbf{a}|\hat{P}^\beta_{kl}\rangle = \langle 1|P^\alpha_{ji}\mathbf{a}P^\beta_{kl}|1\rangle/\sqrt{N^\alpha N^\beta}$$

$$= \langle 1|\mathbf{a}P^\alpha_{ji}P^\beta_{kl}|1\rangle/\sqrt{N^\alpha N^\beta} \quad \text{(because of symmetry of } \mathbf{a} \text{ all } g \text{ commutute, } \mathbf{a}g = g\mathbf{a}\text{)}$$

$$= \langle 1|\mathbf{a}P^\alpha_{jl}|1\rangle \delta_{ik}\delta^{\alpha\beta}/N^\alpha \quad \text{[using elementary operator properties in Eq. (3.3.12)].}$$

Finally, the Eqs. (3.4.19) and (3.4.20) for P_j and N give the reduced acceleration matrix components,

$$\langle \hat{P}^\alpha_{ij}|\mathbf{a}|\hat{P}^\beta_{kl}\rangle = \left(\sum_g \mathscr{D}^{\alpha*}_{jl}(g)\langle 1|\mathbf{a}|g\rangle\right)\delta_{ik}\delta^{\alpha\beta}, \qquad (3.4.22)$$

which are shown explicitly in the following matrix:

$$m\langle \mathbf{a}\rangle = \begin{array}{c} \\ \end{array} \begin{array}{cccccc} |P^1\rangle & |P^2\rangle & |P^3_{11}\rangle & |P^3_{12}\rangle & |P^3_{21}\rangle & |P^3_{22}\rangle \end{array}$$

	$\|P^1\rangle$	$\|P^2\rangle$	$\|P^3_{11}\rangle$	$\|P^3_{12}\rangle$	$\|P^3_{21}\rangle$	$\|P^3_{22}\rangle$
$\mathbf{a}^{(1)}$	·	·	·	·	·	$\|P^1\rangle$
·	$\mathbf{a}^{(2)}$	·	·	·	·	$\|P^2\rangle$
·	·	$\mathbf{a}^{(3)}_{11}$	$\mathbf{a}^{(3)}_{12}$	·	·	$\|P^3_{11}\rangle$
·	·	$\mathbf{a}^{(3)}_{21}$	$\mathbf{a}^{(3)}_{22}$	·	·	$\|P^3_{12}\rangle$
·	·	·	·	$\mathbf{a}^{(3)}_{11}$	$\mathbf{a}^{(3)}_{12}$	$\|P^3_{21}\rangle$
·	·	·	·	$\mathbf{a}^{(3)}_{21}$	$\mathbf{a}^{(3)}_{22}$	$\|P^3_{22}\rangle$

(3.4.23a)

where:

$$\mathbf{a}^{(1)} = 3k_1 + k_2(1 + \sin 2\alpha), \qquad \mathbf{a}^{(2)} = k_2(1 - \sin 2\alpha)$$

$$\mathbf{a}^{(3)}_{11} = \frac{3k_1}{4} + k_2(1 + \sin 2\alpha), \qquad \mathbf{a}^{(3)}_{12} = \frac{3k_1}{4}$$

$$\mathbf{a}^{(3)}_{21} = \frac{3k_1}{4}, \qquad \mathbf{a}^{(3)}_{22} = \frac{3k_1}{4} + k_2(1 - \sin 2\alpha). \quad (3.4.23b)$$

Note that the entire representation comes from just the first row of the original $\langle \mathbf{a}\rangle$ matrix in Eq. (3.4.21). Note also that the new representation leaves one with only a 2 × 2 matrix (this is repeated twice) to solve from the original 6 × 6 $\langle \mathbf{a}\rangle$ matrix. The $|P^1\rangle$ and $|P^2\rangle$ are already eigenvectors.

It is instructive to compare Eq. (3.4.11) with Eq. (3.4.22) and the matrix in Eq. (3.4.16) with the matrix in Eq. (3.4.23). In Eq. (3.4.11) one observes the

182 BASIC THEORY AND APPLICATIONS OF SYMMETRY REPRESENTATIONS

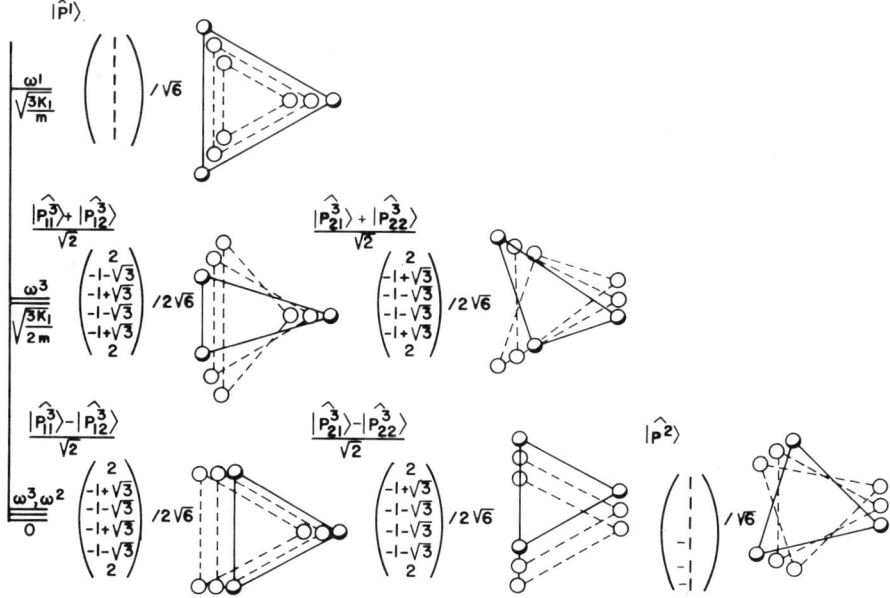

Figure 3.4.4 Standing-wave eigenvectors of C_{3v} spring-mass system.

factor δ_{jl} in the matrix element of symmetry operator g between $\langle P_{ij}^{\alpha}|$ and $|P_{kl}^{\beta}\rangle$, while in Eq. (3.4.22) the same matrix element of acceleration operator has a δ_{ik}. In the matrix Eq. (3.4.16) we find two identical block-diagonal matrices (D^3), where the first block connects states in one set $\{|P_{11}^3\rangle, |P_{21}^3\rangle\}$ of partners, while the second block connects the states in another set $\{|P_{12}^3\rangle, |P_{22}^3\rangle\}$ [recall Eq. (3.4.15)]. In the matrix Eq. (3.4.23) we also observe two identical block-diagonal matrices; only now the first block connects type-1 partners ($|P_{11}^3\rangle$ and $|P_{12}^3\rangle$), while the second block connects type-2 partners ($|P_{21}^3\rangle$ and $|P_{22}^3\rangle$).

When there is a repeated irrep $\mathscr{D}^3$ in any symmetry-analysis problem one has only to deal with the matrix connecting independent partners of type 1. This is so since the matrix for type 1 must be identical to that for type 2, ..., and type l^{α}. The degeneracy of each (α) eigenvalue will then be equal to the dimension l^{α} of irrep (α), since the eigenvalue equations are identical as well.

Figure 3.4.4 shows the eigenvectors of Eq. (3.4.23) with $k_2 = 0$. For this case there are three states with eigenvalue zero: the x, y translations and rotation. In molecular physics these are called NONGENUINE vibrations. The remaining three GENUINE vibration states are similar to those drawn in Figure 2.6.2. Indeed, the two degenerate states arising from the irreps D^{13} and D^{23} of C_3 now correspond to states arising from just one irrep D^3 of C_{3v}. The degeneracy exists because C_3 does not represent the full symmetry of the system.

(c) Moving-Wave or Circular Motions The C_{3v} idempotents and irreps used in the preceding sections were derived by choosing representations of the reflection subgroup $C_v = \{1, \sigma_3\}$ to be diagonal. The problem can be done just as well using irreps for which rotation subgroup $C_3 = \{1, r, r^2\}$ is diagonal. However, instead of repeating the entire derivation, let us simply transform the C_v defined irrep $\mathscr{D}^3(r)$ [Eq. (3.4.14)] to diagonal form. Using the methods of Section 1.2 we quickly find that

$$\mathscr{D}^{3^0}(r) \equiv \mathscr{T}^\dagger \mathscr{D}^3(r)\mathscr{T} = \begin{pmatrix} \varepsilon & 0 \\ 0 & \varepsilon^* \end{pmatrix} = \begin{pmatrix} \mathscr{D}^{1_3}(r) & 0 \\ 0 & \mathscr{D}^{2_3}(r) \end{pmatrix} \quad (3.4.24a)$$

reduces to a direct sum of C_3 irreps: $\varepsilon = e^{-2\pi i/3} \equiv \mathscr{D}^{1_3}(r)$ and $\varepsilon^* \equiv \mathscr{D}^{2_3}(r)$, using transformation

$$\mathscr{T} = \begin{pmatrix} 1/\sqrt{2} & 1/\sqrt{2} \\ -i/\sqrt{2} & i/\sqrt{2} \end{pmatrix}. \quad (3.4.24b)$$

Applying the same transformation to the $\mathscr{D}^3(\sigma_j)$ in Eq. (3.3.7) gives all nondiagonal reflection representations. $\mathscr{D}^{3^0}(\sigma_j) = \mathscr{T}^\dagger \mathscr{D}^3(\sigma_j)\mathscr{T}$, where

$$\mathscr{D}^{3^0}(\sigma_1) = \begin{pmatrix} 0 & \varepsilon^* \\ \varepsilon & 0 \end{pmatrix}, \quad \mathscr{D}^{3^0}(\sigma_2) = \begin{pmatrix} 0 & \varepsilon \\ \varepsilon^* & 0 \end{pmatrix}, \quad \mathscr{D}^{3^0}(\sigma_3) = \begin{pmatrix} 0 & 1 \\ 1 & 0 \end{pmatrix}. \quad (3.4.24c)$$

The preceding equations (3.4.24) define the CIRCULAR or MOVING-WAVE irreps $\mathscr{D}^{3^0}$ of C_{3v}. If you use them to solve the preceding eigenvibration problem the same eigenvalue must result, since $\mathscr{D}^3$ and $\mathscr{D}^{3^0}$ are equivalent. However, the columns of the transformation $\mathscr{T}$ in Eq. (3.4.24b) give quite different eigenvectors:

$$\left| \begin{matrix} (3) \\ 1_3 \end{matrix} \right\rangle = (1/\sqrt{2})\left| \begin{matrix} (3) \\ x \end{matrix} \right\rangle - (i/\sqrt{2})\left| \begin{matrix} (3) \\ y \end{matrix} \right\rangle,$$

$$\left| \begin{matrix} (3) \\ 2_3 \end{matrix} \right\rangle = (1/\sqrt{2})\left| \begin{matrix} (3) \\ x \end{matrix} \right\rangle + (i/\sqrt{2})\left| \begin{matrix} (3) \\ y \end{matrix} \right\rangle \quad (3.4.25)$$

if $|_x^{(3)}\rangle$ and $|_y^{(3)}\rangle$ are eigenvectors obtained in Figure 3.4.4. The new eigenstates correspond to a beautiful circular motion such as is depicted in Figure 3.4.5. However, this is just an $(x + iy)$ combination of x- and y-linear motion. Since x and y motions were degenerate any combination of them is allowed. This is related to the fact that degenerate $(l^\alpha \geq 2)$ irreps of non-Abelian groups are not uniquely defined.

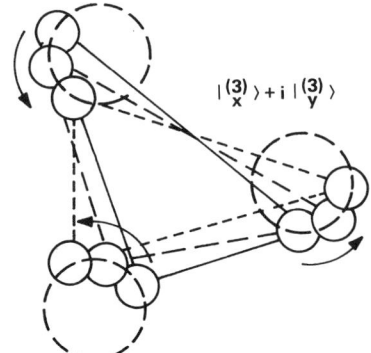

Figure 3.4.5 Moving-wave or circular form of type-3 eigenvector.

The circular $(x \pm iy)$ bases will be excited by circular polarized radiation, as we will see in Chapters 4 and 6. They also result in the application of magnetic fields, Jahn-Teller distortions, and for rotating molecules.

3.5 CHARACTER FORMULAS

For Abelian symmetry analysis the number of repetitions or FREQUENCY of a given irrep $\mathscr{D}^\alpha$ is the number of states of that type still to be separated by means other than symmetry projection. The same is true for multidimensional irreps of non-Abelian groups because different substrates or partners of any irrep will give rise to identical equations. One has only to solve one set for each irrep (α) regardless of its dimension l^α.

Let us now derive some simple formulas for frequency of an irrep of a finite group in a given representation. These are called CHARACTER formulas. Characters of multidimensional irreps are the traces of the $\mathscr{D}^\alpha$ matrices as defined in the following:

$$\chi^\alpha(g) = \text{TRACE}\, \mathscr{D}^\alpha(g) = \sum_{i=1}^{l^\alpha} \mathscr{D}_{ii}^\alpha(g). \tag{3.5.1}$$

For Abelian groups irreps and characters are the same thing, since then $l^\alpha = 1$ always. Nevertheless, every formula given in this section can be applied to either Abelian or non-Abelian groups.

One property of the characters is that they are equal for any two symmetry operators g and g' that are equivalent or from the same class. If g is in the same class with g', then $g' = h^{-1}gh$ for some other symmetry operators h.

Representing this gives

$$\chi^\alpha(g') = \sum_i^{l^\alpha} \mathscr{D}_{ii}^\alpha(g') = \sum_{i,j,k}^{l^\alpha} \mathscr{D}_{ij}^\alpha(h^{-1}) \, \mathscr{D}_{jk}^\alpha(g) \, \mathscr{D}_{ki}^\alpha(h)$$

$$= \sum_{j,k}^{l^\alpha} \left(\sum_i^{l^\alpha} \mathscr{D}_{ki}^\alpha(h) \, \mathscr{D}_{ij}^\alpha(h^{-1}) \right) \mathscr{D}_{jk}^\alpha(g)$$

$$= \sum_j^{l^\alpha} \mathscr{D}_{jj}^\alpha(g),$$

and finally

$$\chi_{g'}^\alpha \equiv \chi^\alpha(g') = \chi^\alpha(g), \qquad (3.5.2)$$

for all g in the class c_g.

Recall the completeness relations between the all-commuting idempotents $\mathbb{P}^\alpha$ and the irreducible or elementary idempotents P_{ii}^α:

$$\mathbb{P}^\alpha = \sum_i P_i^\alpha = \sum_i P_{ii}^\alpha.$$

This leads to a formula for the all-commuting idempotents $\mathbb{P}^\alpha$ in terms of characters using the formula for P_{ii}^α given in Eq. (3.4.19):

$$\mathbb{P}^\alpha = \sum_i P_{ii}^\alpha = (l^\alpha/{}^\circ G) \sum_g \sum_i \mathscr{D}_{ii}^{\alpha*}(g) g$$

$$= (l^\alpha/{}^\circ G) \sum_g \chi^{\alpha*}(g) g.$$

Since the characters are the same for equivalent operators, the preceding sum can be reduced to a sum over just one element from each class $c_g = (g + g' + \cdots)$,

$$\mathbb{P}^\alpha = (l^\alpha/{}^\circ G)\left[\chi_g^{\alpha*}(g + g' + \cdots) + \chi_h^{\alpha*}(h + h' + \cdots) + \cdots \right],$$

or

$$\mathbb{P}^\alpha = (l^\alpha/{}^\circ G) \sum_{\substack{\text{classes} \\ c_g}} \chi_g^{\alpha*} c_g. \qquad (3.5.3)$$

This sum is a key to the derivation and application of characters.

A. Derivation of Irrep Characters

The algebra of classes discussed in Section 3.2 gives the all-commuting idempotents

$$\mathbb{P}^\alpha = \sum_{\substack{\text{classes} \\ c_g}} p_g^\alpha c_g \tag{3.5.4}$$

as a sum of classes. For the example of C_{3v} we obtained

$$\mathbb{P}^1 = \tfrac{1}{6}c_1 + \tfrac{1}{6}c_2 + \tfrac{1}{6}c_3,$$
$$\mathbb{P}^2 = \tfrac{1}{6}c_1 + \tfrac{1}{6}c_2 - \tfrac{1}{6}c_3,$$
$$\mathbb{P}^3 = \tfrac{2}{3}c_1 - \tfrac{1}{3}c_2, \tag{3.5.4}_x$$

in Eq. (3.2.15). Relating the p_g^α coefficients in Eqs. (3.5.4) to characters in Eq. (3.5.3) gives

$$\chi_g^{\alpha*} = p_g^\alpha (l^\alpha / {}^\circ G)^{-1}. \tag{3.5.5}$$

To use this one must first determine the dimension l^α which is the trace or character of the irrep of the unit class:

$$\chi_1^\alpha = l^\alpha = \text{Trace } \mathscr{D}^\alpha(1). \tag{3.5.6}$$

Solving Eq. (3.5.5) gives

$$l^\alpha = ({}^\circ G p_1^{\alpha*})^{1/2}. \tag{3.5.7}$$

The first column of Eq. $(3.5.4)_x$ gives

$$l^1 = \left(6 \cdot \tfrac{1}{6}\right)^{1/2} = 1,$$
$$l^2 = \left(6 \cdot \tfrac{1}{6}\right)^{1/2} = 1,$$
$$l^3 = \left(6 \cdot \tfrac{2}{3}\right)^{1/2} = 2, \tag{3.5.7}_x$$

which is the first column of the C_{3v} character table in Eq. (3.5.8). The other characters follow from Eqs. $(3.5.4)_x$ and (3.5.5):

$$\begin{array}{r|ccc}
j = & 1 & 2\colon (r, r^2) & 3\colon (\sigma_1 \sigma_2 \sigma_3) \\
\hline
\chi_j^{A'} = \chi_j^{A_1} = \chi_j^1 = & 1 & 1 & 1 \\
\chi_j^{A''} = \chi_j^{A_2} = \chi_j^2 = & 1 & 1 & -1 \\
\chi_j^E = \chi_j^3 = & 2 & -1 & 0
\end{array}. \tag{3.5.8}$$

The standard notation $(1) \equiv A_1$, $(2) \equiv A_2$, and $(3) \equiv E$ for D_3 and C_{3v} irreps will be used from now on. The notation $(1) \equiv A'$ and $(2) \equiv A'$ is sometimes used instead for C_{3v} irreps.

B. Applications of Characters

The power of character theory is great, since it is independent of your choice of basis. Since a matrix trace is invariant to such choice it does not matter how a Hamiltonian is represented or which equivalent version of irreps you chose. Furthermore, sums over symmetry operators are replaced by sums over classes of operators, which can amount to a considerable saving of labor.

Here one of the simplest C_{3v} examples consisting of the three-pendulum system of Section 2.6 will be reexamined using character theory. The character method is not noticeably easier for such a simple problem, but it is a good pedagodgical example.

(a) Deriving Irrep Frequency Suppose you want to find out which and how many irreps would appear on the diagonal of a given representation $\mathscr{R}$ upon reduction. That is, suppose you want to know the FREQUENCY f^α of irreps $\mathscr{D}^\alpha$ in the complete reduction,

$$\mathscr{T}^\dagger \mathscr{R}(g) \mathscr{T} = f^\alpha \mathscr{D}^\alpha(g) \oplus f^\beta \mathscr{D}^\beta(g) \oplus \cdots, \quad (3.5.9)$$

of $\mathscr{R}$. This reduction holds for all combinations of g including the all-commuting idempotents $\mathbb{P}^\alpha$. Substituting $\mathbb{P}^\alpha$ for g and taking the trace gives

$$\text{Trace } \mathscr{R}(\mathbb{P}^\alpha) = f^\alpha \cdot l^\alpha, \quad (3.5.10)$$

since $\mathscr{D}^\beta(\mathbb{P}^\alpha)$ is a unit matrix if $\alpha = \beta$ and zero otherwise. The equation for f^α,

$$f^\alpha = (1/l^\alpha) \text{ Trace } \mathscr{R}(\mathbb{P}^\alpha)$$
$$= (1/{}^\circ G) \sum_{\substack{\text{classes} \\ c_g}} \chi_g^{\alpha *} \text{ Trace } \mathscr{R}(c_g),$$

$$f^\alpha = (1/{}^\circ G) \sum_{\substack{\text{classes} \\ c_g}} \chi_g^{\alpha *} {}^\circ c_g \text{ Trace } \mathscr{R}(g), \quad (3.5.11)$$

follows from Eq. (3.5.3), where ${}^\circ c_g$ is the order of class c_g and g is any element in the class.

For the example we need only the traces of 1, r, and , say, σ_3. The first trace equals the number of pendulum coordinates: Trace $\mathscr{R}(1) = 3$. The r trace vanishes: Trace $\mathscr{R}(r) = 0$, since all coordinates are moved by r. One coordinate sits on each reflection plane; hence Trace $\mathscr{R}(\sigma_3) = 1$. The A_1

frequency is given by

$$f^{A_1} = (1/{}^\circ G)\left[\chi_1^{A_1^*} {}^\circ c_1 \cdot 3 + \chi_r^{A_1^*} {}^\circ c_r \cdot 0 + \chi_\sigma^{A_1^*} {}^\circ c_\sigma \cdot 1\right]$$
$$= (1/{}^\circ G)\left[3\chi_1^{A_1^*} + 3\chi_\sigma^{A_1^*}\right] = 1, \qquad (3.5.11a)_x$$

where the character table in Eq. (3.5.8) is used. Similarly, one has the other frequencies:

$$f^{A_2} = \tfrac{1}{6}\left[3\chi_1^{A_2^*} + 3\chi_\sigma^{A_2^*}\right] = 0, \qquad (3.5.11b)_x$$
$$f^E = \tfrac{1}{6}\left[3\chi_1^{E^*} + 3\chi_\sigma^{E^*}\right] = 1. \qquad (3.5.11c)_x$$

From this one learns that the C_{3v} pendulum system has one A_1 level and one E level. The latter is degenerate since $l^E = 2$.

(b) Deriving Eigenvalues Let us use C_{3v} characters to rederive the eigenvalues of the three-pendulum acceleration matrix

$$\langle \mathbf{a} \rangle = \begin{pmatrix} \langle 1|\mathbf{a}|1\rangle & \langle 1|\mathbf{a}|r\rangle & \langle 1|\mathbf{a}|r^2\rangle \\ \cdot & \cdot & \cdot \\ \cdot & \cdot & \cdot \end{pmatrix} = \begin{pmatrix} 2a+b & -a & -a \\ \cdot & \cdot & \cdot \\ \cdot & \cdot & \cdot \end{pmatrix}, \qquad (3.5.12)$$

which was solved in Chapter 2. In this case one may assume that the basis which diagonalizes $\langle \mathbf{a}\rangle$ would bring $\langle \mathbb{P}^\alpha\rangle$ to diagonal form also. This form of $\langle \mathbb{P}^\alpha\rangle$ would have a unit submatrix at the same diagonal positions that hold α-type eigenvalues $a^\alpha = \left\langle {}_i^\alpha \middle| \mathbf{a} \middle| {}_i^\alpha \right\rangle$ of the $\langle \mathbf{a}\rangle$ matrix. Zeros would occur at all other positions of the $\langle \mathbb{P}^\alpha\rangle$ matrix. Given that frequencies f^{A_1} and f^E are unit from Eqs. (3.5.11)$_x$ one may derive the $\langle \mathbf{a}\rangle$ eigenvalues from the formula

$$a^\alpha = (1/l^\alpha)\,\text{Trace}\,\langle \mathbf{a}\mathbb{P}^\alpha\rangle$$
$$= (1/{}^\circ G)\sum_g \chi^{\alpha *}(g)\,\text{Trace}\,\langle \mathbf{a}g\rangle.$$

Once again the sum may be simplified since the trace of $\langle t\mathbf{a}gt^{-1}\rangle$ is equal to the trace of $\langle \mathbf{a}g\rangle$. Since all symmetry operators commute with $\mathbf{a}$, one has that

$$\text{Trace}\,\langle t\mathbf{a}gt^{-1}\rangle = \text{Trace}\,\langle \mathbf{a}tgt^{-1}\rangle = \text{Trace}\,\langle \mathbf{a}g'\rangle$$

does not depend on which element of class $c_g = \{g, g', \ldots\}$ is used. There-

CHARACTER FORMULAS 189

fore the a^α formula reduces to a sum

$$a^\alpha = (1/{}^\circ G) \sum_{\substack{\text{classes} \\ c_g}} \chi_g^{\alpha*} c_g \text{ Trace } \langle \mathbf{a} g \rangle \qquad (3.5.13)$$

over just one element from each class.

The trace in the formula can be evaluated in such a way that only the first row $\{\langle 1|\mathbf{a}|1\rangle, \langle 1|\mathbf{a}|r\rangle, \ldots\}$ of an $\langle \mathbf{a} \rangle$ matrix is required. For the matrix in Eq. (3.5.12) we have

$$\text{Trace } \langle \mathbf{a} g \rangle = \langle 1|\mathbf{a} g|1\rangle + \langle r|\mathbf{a} g|r\rangle + \langle r^2|\mathbf{a} g|r^2\rangle$$
$$= \langle 1|\mathbf{a} g|1\rangle + \langle 1|r^{-1}\mathbf{a} g r|1\rangle + \langle 1|r^{-2}\mathbf{a} g r^2|1\rangle$$
$$= \langle 1|\mathbf{a}|g\rangle + \langle 1|\mathbf{a}|r^{-1}g r\rangle + \langle 1|\mathbf{a}|r^{-2}g r^2\rangle. \qquad (3.5.14)$$

Substituting in turn $g = 1$, r, and σ_3, one finds

$$\text{Trace } \langle \mathbf{a} \rangle = 3\langle 1|\mathbf{a}|1\rangle = 3(2a + b),$$
$$\text{Trace } \langle \mathbf{a} r \rangle = 3\langle 1|\mathbf{a}|r\rangle = -3a,$$
$$\text{Trace } \langle \mathbf{a}\sigma_3 \rangle = \langle 1|\mathbf{a}|1\rangle + \langle 1|\mathbf{a}|r\rangle + \langle 1|\mathbf{a}|r^2\rangle = b,$$

where the symmetry definitions $\sigma_3|1\rangle = |1\rangle$, $\sigma_2|1\rangle = |r\rangle$, and $\sigma_1|1\rangle = |r^2\rangle$ of pendulum coordinates are used in the last line. Substituting the traces into Eq. (3.5.13) gives

$$a^{A_1} = \tfrac{1}{6}[1 \cdot 3(2a + b) + 1 \cdot 2 \cdot (-3a) + 3b] = b,$$
$$a^E = \tfrac{1}{6}[2 \cdot 3(2a + b) - 1 \cdot 2 \cdot (-3a) + 0] = 3a + b,$$

in agreement with the previous calculation [recall Eq. (2.6.10)].

If an irrep had been repeated with a frequency $f^\alpha \geq 2$ the character procedure may still be applied. However, one can only derive the *average* of the a^α eigenvalues $\{a^\alpha, a^{\alpha'}, a^{\alpha''}, \ldots\}$:

$$\langle \mathbf{a} \rangle_{\text{average}}^\alpha \equiv 1/f^\alpha \sum a^{\alpha'}$$
$$= (1/{}^\circ G f^\alpha) \sum_{\substack{\text{classes} \\ c_g}} \chi_g^{\alpha*} {}^\circ c_g \text{ Trace } \langle \mathbf{a} g \rangle. \qquad (3.5.15)$$

To find the individual eigenvalues and eigenvectors requires full P_{ij}^α projection operator techniques in general.

3.6 D_n AND C_{nv} SYMMETRY AND BLOCH WAVES

A discussion of Bloch waves and C_n symmetry was given in Section 2.12. Here the D_n and C_{vn} symmetry analysis of Bloch waves will be given. This should provide a clear physical picture of the meaning of various D_n irreps for arbitrary n as well as some other spectroscopic concepts.

A. Tetragonal Symmetry

The tetragonal symmetries C_{4v} or D_{4n} were introduced in Section 3.1 and Figure 3.1.9. There it was suggested that the reader perform all the derivations that were done for C_{3v} or D_3. (See Problem 3.1.1.) The results are given in the following and are followed by an interpretation of the irrep bases in terms of Bloch waves.

The D_4 Hamilton monogram is shown in Figure 3.6.1, following the conventions established before in Figure 3.1.9. This facilitates the computation of the group table, which is given in the following. Note that the C_{4v} group table is obtained by replacing transverse 180° rotations $\{R_1^2, R_2^2, i_3, i_4\}$ by vertical-plane reflections $\{\sigma_1 = IR_1^2, \sigma_2 = IR_2^2, \sigma_3 = Ii_3, \sigma_4 = Ii_4\}$.

1	R^2	R	R^3	R_1^2	R_2^2	i_3	i_4
R^2	1	R^3	R	R_2^2	R_1^2	i_4	i_3
R	R^3	R^2	1	i_3	i_4	R_2^2	R_1^2
R^3	R	1	R^2	i_4	i_3	R_1^2	R_2^2
R_1^2	R_2^2	i_4	i_3	1	R^2	R^3	R
R_2^2	R_1^2	i_3	i_4	R^2	1	R	R^3
i_3	i_4	R_1^2	R_2^2	R	R^3	1	R^2
i_4	i_3	R_2^2	R_1^2	R^3	R	R^2	1

(3.6.1a)

1	c_2	c_R		c_1		c_3
	1	c_R		c_1		c_3
		$21 + 2c_2$	$2c_3$		$2c_1$	
			$21 + 2c_2$		$2c_R$	
					$21 + 2c_2$	

(3.6.1b)

The class algebra table (3.6.1b) follows. Note that transverse 180° rotations R_1^2 and R_2^2 around x and y axes, respectively, belong in a different class than diagonal rotations i_3 and i_4.

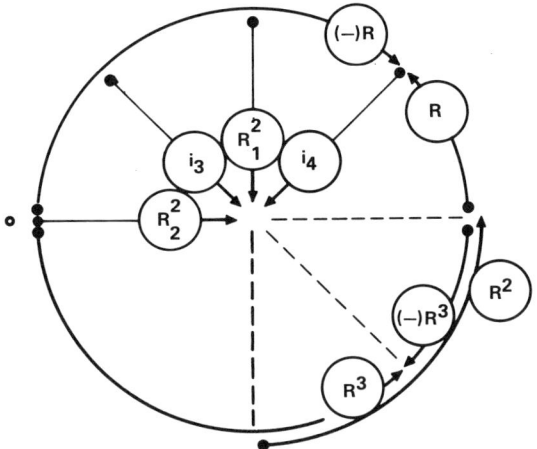

Figure 3.6.1 Hamilton turn nomogram for D_4 symmetry.

At least two minimal equations are needed to reduce the five-dimensional class algebra since all the c_j belong to one-, two-, or three-dimensional subalgebras. One finds $c_R^3 = 4c_R$ and $c_1^3 = 4c_1$, and combining the resulting idempotents gives the following five all-commuting idempotents:

$$\mathbb{P}^1 = \tfrac{1}{8}(\mathbf{1} + c_2 + c_R + c_1 + c_3),$$
$$\mathbb{P}^2 = \tfrac{1}{8}(\mathbf{1} + c_2 - c_R + c_1 - c_3),$$
$$\mathbb{P}^3 = \tfrac{1}{8}(\mathbf{1} + c_2 + c_R - c_1 - c_3),$$
$$\mathbb{P}^4 = \tfrac{1}{8}(\mathbf{1} + c_2 - c_R - c_1 + c_3),$$
$$\mathbb{P}^5 = \tfrac{1}{2}(\mathbf{1} - c_2). \tag{3.6.2}$$

Following Section (3.5.A) one converts the preceding $\mathbb{P}^\alpha$ to irrep characters:

C_{4v}				$(\sigma_1\sigma_2)$	$(\sigma_3\sigma_4)$
D_4	$j=1,$	$(R^2),$	$(R,R^3),$	$(R_1^2 R_2^2),$	$(i_3 i_4)$
$\chi^{A'} = \chi^{A_1} = \chi_j^1 =$	1	1	1	1	1
$\chi^{B'} = \chi^{B_1} = \chi_j^2 =$	1	1	-1	1	-1
$\chi^{A''} = \chi^{A_2} = \chi_j^3 =$	1	1	1	-1	-1
$\chi^{B''} = \chi^{B_2} = \chi_j^4 =$	1	1	-1	-1	1
$\chi^{E} = \chi_j^E = \chi_j^5 =$	2	-2	0	0	0

(3.6.3)

Conventional notation for D_4 irreps is given on the left-hand side of (3.6.3) along with an optional notation sometimes used for the isomorphic C_{4v} symmetry.

The A and B irreps all have dimension $l^{A_j} = 1 = l^{B_j}$, and so the idempotents $\mathbb{P}^{A_j}$ and $\mathbb{P}^{B_j}$ are irreducible as well as all-commuting. However, the E irrep has dimension $l^E = 2$, so the $\mathbb{P}^E$ idempotent must split it in two. How one splits $\mathbb{P}^E$ depends on one's choice of Abelian subgroups. One choice is the subgroup $C_4 = \{1, R, R^2, R^4\}$, which is analogous to the C_3 choice in the C_{3v} analysis. Multiplying $\mathbb{P}^E$ by a C_4 unit decomposition gives

$$\mathbb{P}^E = \mathbb{P}^E \mathbf{1} = \mathbb{P}^E(P^{0_4} + P^{1_4} + P^{2_4} + P^{3_4})$$
$$= 0 + \mathbb{P}^E P^{1_4} + 0 + \mathbb{P}^E P^{3_4}$$
$$= P^E_{1_4} + P^E_{3_4}, \qquad (3.6.4a)$$

where

$$P^E_{1_4} = (1 - R^2 + iR - iR^3)/4,$$
$$P^E_{3_4} = (1 - R^2 - iR + iR^3)/4. \qquad (3.6.4b)$$

This results in a complex set of E irreps analogous to the circular irreps of C_{3v}. Two examples are

$$\mathscr{D}^{\circ E}(R) = \begin{pmatrix} -i & 0 \\ 0 & i \end{pmatrix}, \quad \mathscr{D}^{\circ E}(R_1^2) = \begin{pmatrix} 0 & 1 \\ 1 & 0 \end{pmatrix}. \qquad (3.6.5)$$

(Note that products of two elements can generate the whole group D_4; $\mathscr{D}^{\circ E}$ is defined by representing just two GENERATORS.) Since the circular or C_4-defined irreps $\mathscr{D}^{\circ E}$ are complex they give rise to moving-wave eigenstates in physical applications.

Standing-wave irreps result if one chooses to diagonalize one of the transverse 180° rotations, say R_1^2, or the Abelian subgroup $C_2 = \{1, R_1^2\}$. The unit decomposition of C_2 splits P^E as follows:

$$\mathbb{P}^E = \mathbb{P}^E \mathbf{1} = \mathbb{P}^E(P^1 + P^2)$$
$$= P^E_1 + P^E_2, \qquad (3.6.6a)$$

where

$$P^E_1 = (1 - R^2 + R_1^2 - R_2^2)/4,$$
$$P^E_2 = (1 - R^2 - R_1^2 + R_2^2)/4. \qquad (3.6.6b)$$

The resulting E irreps are the following real matrices.

$$\mathscr{D}^E(1) = \begin{pmatrix} 1 & 0 \\ 0 & 1 \end{pmatrix}, \qquad \mathscr{D}^E(R^2) = \begin{pmatrix} -1 & 0 \\ 0 & -1 \end{pmatrix},$$

$$\mathscr{D}^E(R) = \begin{pmatrix} 0 & -1 \\ 1 & 0 \end{pmatrix}, \qquad \mathscr{D}^E(R^3) = \begin{pmatrix} 0 & 1 \\ -1 & 0 \end{pmatrix},$$

$$\mathscr{D}^E(R_1^2) = \begin{pmatrix} 1 & 0 \\ 0 & -1 \end{pmatrix}, \qquad \mathscr{D}^E(R_2^2) = \begin{pmatrix} -1 & 0 \\ 0 & 1 \end{pmatrix},$$

$$\mathscr{D}^E(i_3) = \begin{pmatrix} 0 & 1 \\ 1 & 0 \end{pmatrix}, \qquad \mathscr{D}^E(i_4) = \begin{pmatrix} 0 & -1 \\ -1 & 0 \end{pmatrix}. \qquad (3.6.7)$$

Note that the whole subgroup $D_2 = \{1 R_1^2 R_2^2 R^2\}$ is represented by diagonal matrices. Stated another way, the standing-wave E irrep is reduced with respect to $D_2 = \{1 R_1^2 R_2^2 R^2\}$

$$\mathscr{D}^E(h \text{ in } D_2) = \left[\begin{array}{c|c} D^{B_1}(h) & 0 \\ \hline 0 & D^{B_2}(h) \end{array}\right] = D^{B_1}(h) \oplus D^{B_2}(h), \qquad (3.6.8)$$

as well as $C_2 = \{1, R_1^2\}$.

$$\mathscr{D}^E(h \text{ in } C_2) = \left[\begin{array}{c|c} D^1(h) & 0 \\ \hline 0 & D^2(h) \end{array}\right] = D^1(h) \oplus D^2(h). \qquad (3.6.9)$$

(Recall the irreps of D_2 labeled in Section 2.8.) Contrast this with the moving-wave irrep $\mathscr{D}^{\circ E}$, which is reduced instead with respect to $C_4 = \{1, R, R^2, R^3\}$.

$$\mathscr{D}^{\circ E}(h \text{ in } C_4) = \left[\begin{array}{c|c} D^{1_4}(h) & 0 \\ \hline 0 & D^{3_4}(h) \end{array}\right] = D^{1_4}(h) \oplus D^{3_4}(h). \qquad (3.6.10)$$

The standing Bloch waves provide a simple picture of the D_4 irreps. The waves drawn in Figure 3.6.2 are a special case of the Bloch solutions described in Section 2.12.A and Figure 2.12.2. Note that (B) labels waves on the first Brillouin band boundaries as before, while A waves stand at the zeroth and second boundaries. Note that the subscripts 1 and 2 denote symmetry and antisymmetry, respectively, to the reversal R_1^2 around the first potential well; i.e., A_1 and B_1 are symmetric waves, while A_2 and B_2 are antisymmetric. In the character table (3.6.3) this is expressed by entries

$$\chi_{R_1^2}^{A_1} = 1 = \chi_{R_1^2}^{B_1}; \qquad \chi_{R_1^2}^{A_2} = -1 = \chi_{R_1^2}^{B_2}. \qquad (3.6.11)$$

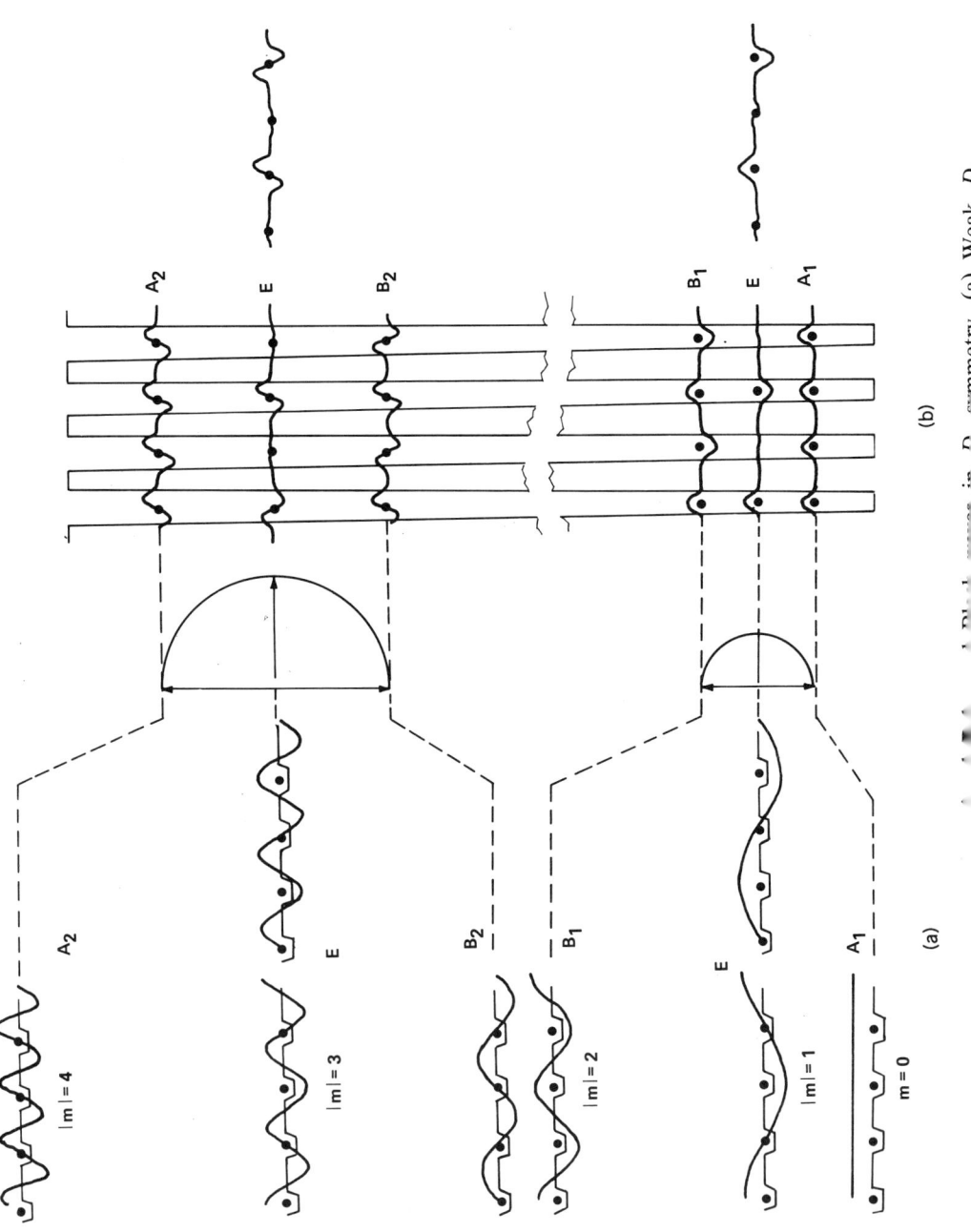

(a) (b)

One could label standing cosine and sine standing-wave states similarly as χ_1^E and χ_2^E, respectively. However, do not forget that the E states are degenerate partner bases of a single irrep E. The E waves can slide their nodes to halfway or anywhere between wells without costing them energy, while the A and B waves cannot. Note that the B waves change sign under operation R (or R^3), which moves exactly one well spacing forward or backward. In the character table (3.6.3) this corresponds to the entries

$$\chi_R^{B_1} = -1 = \chi_R^{B_2}. \tag{3.6.12}$$

The A waves are unaffected by R or R^3, i.e.,

$$\chi_R^{A_1} = 1 = \chi_R^{A_2}. \tag{3.6.13}$$

As the potential wells of a D_4 symmetric potential grow deeper the form of the waves changes. This was explained in Section 2.12.A(c). However, as long as the D_4 symmetry is maintained the symmetry properties of a given level do not change. The states labeled A_1, B_1, etc., on the left-hand side of Figure 3.6.2 transform into corresponding states of the same label on the right-hand side. This transformation involves adding many higher harmonic-wave states which have that same symmetry label. The resulting spectrum will consist of repeating and alternating $(A_1 E B_1)$, $(B_2 E A_2)$, $(A_1 E B_1)$, $(B_2 E A_2)$, ... clusters of D_4 levels in the limit of very deep wells. The lower-energy clusters become more nearly degenerate as the wells become deeper and their tunneling amplitudes S nearly vanish. Recall the discussion of Section 2.12.A(c). There D_n symmetry was implicitly assumed.

It remains to see what happens to a D_4 spectrum if symmetry is reduced to $D_4 = \{1, R, R^2, R^3\}$ or $C_2 = \{1, R_1^2\}$. A magnetic field placed along the z axis would reduce D_4 or C_{4v} symmetry to C_4, since transverse rotations R_j^2 or reflections σ_j do not leave such a field invariant. Similarly, if a field is put along the x axis, only C_2 symmetry remains.

The E irrep of C_{4v} or D_4 is not an irreducible representation of C_4. According to Eq. (3.6.10) it reduces to irreps of C_4,

$$\mathscr{D}^E \downarrow C_4 \simeq D^{1_4} \oplus D^{3_4}, \tag{3.6.14}$$

corresponding to two moving waves of wave number or momentum $m = 1$ and $m = 2 = -1$ modulo 4. The arrow ($\downarrow$) indicates SUBDUCTION or symmetry reduction, wherein a representation of a larger group is restricted to a subgroup. If the representation becomes reducible, or reduced as in Eq. (3.6.10), then the degeneracy of levels labeled by that representation may split. In this case Eq. (3.6.14) predicts that E levels will split in two as in Figure 3.6.3. This is elementary ZEEMAN SPLITTING for which the right-handed moving wave ($m = 1 \bmod 4$) has different energy than the left-handed ($m = -1 \bmod 4$) wave.

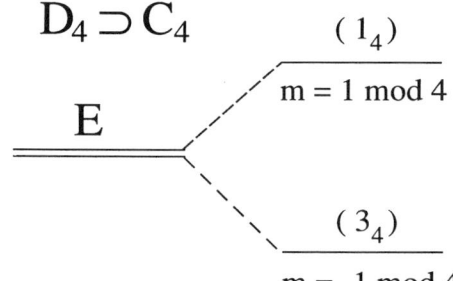

Figure 3.6.3 Zeeman splitting of E level ($E \downarrow C_4 = (1_4) \oplus (3_4)$).

This is an important concept in symmetry analysis of spectroscopy. Energy-level splitting is associated with splitting of idempotents, viz., $\mathscr{D}^E \downarrow C_4 \simeq D^{1_4} + D^{3_4}$ is associated with $P^E = P^E_{1_4} + P^E_{3_4}$. One says that the E irrep of C_{4v} or D_4 is CORRELATED with the 1_4 and 3_4 irreps of subgroup C_4. Similarly, other irreps in (3.6.3) are correlated with C_4 irreps; though, of course, one-dimensional levels cannot split. For example, comparing the first three columns of the table with C_4 irreps gives

$$\mathscr{D}^{A_1} \downarrow C_4 = \mathscr{D}^{A_2} \downarrow C_4 = D^{0_4} = D^A, \qquad (3.6.15a)$$

$$\mathscr{D}^{B_1} \downarrow C_4 = \mathscr{D}^{B_2} \downarrow C_4 = D^{2_4} = D^B. \qquad (3.6.15b)$$

This means A_1 and A_2 waves which could not combine in C_{4v} or D_4 symmetry can do so under the lower symmetry, since they belong to the same irrep $A \equiv 0_4$ of C_4. The same applies to B_1 and B_2 waves. It means that waves that had to be standing waves before may combine with complex coefficients in the presence of a magnetic field and get moving again. Indeed, one of the effects of magnetic fields it to improve the "circulation" of otherwise quenched orbitals.

B. Hexagonal Symmetry

C_{6v} and D_6 symmetry groups are particularly easy to treat, since they are outer products of groups that we have already solved. Since they are both symmetries of hexagonal objects (see Figure 3.6.4) they contain a 60° rotation (h) around the z axis. This implies the existence of a 180° rotation (h^3) which commutes with the transverse 180° rotations (ρ_j) in D_6 or reflections ($\sigma_j = I\rho_j$) in C_{6v}. Hence it is permissible to write

$$\begin{aligned} D_3 \times C_2 &= \{1, h^2, h^4, \rho_1, \rho_2, \rho_3\} \times \{1, h^3\} \\ &= \{1, h^2, h^4, \rho_1, \rho_2, \rho_3, h^3, h^5, h, \rho_1 h^3, \rho_2 h^3, \rho_3 h^3\} \\ &= D_6 \end{aligned} \qquad (3.6.16)$$

according to the definition of the outer product given in Section 2.10. Each

D_n AND C_{nv} SYMMETRY AND BLOCH WAVES   **197**

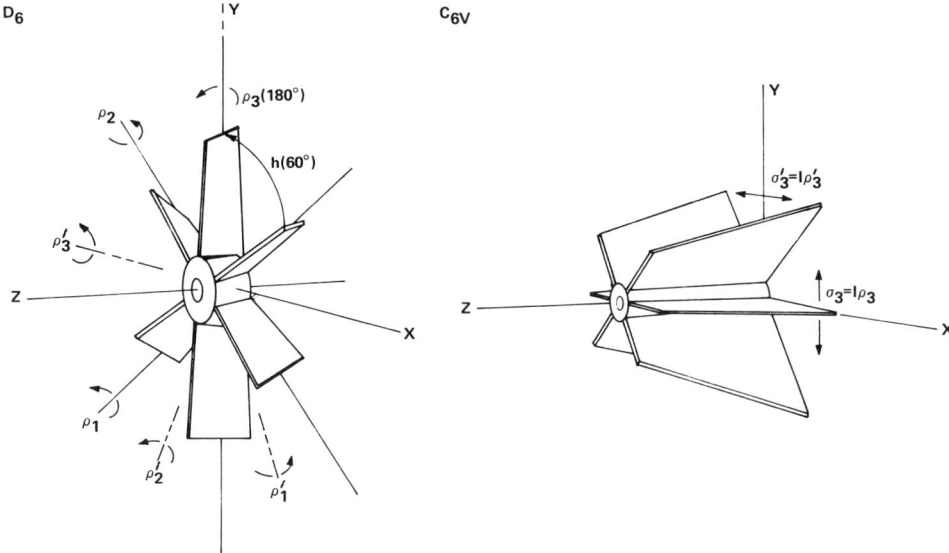

Figure 3.6.4 Pictorial comparison of D_6 and C_{6v} symmetry. Note that the 180° rotations in D_6 which are marked with primes (ρ'_j) have axes of rotation which are orthogonal to those of the corresponding unprimed rotations (ρ_j). The same applies to the C_{6v} reflections.

element in D_6 is a unique product of elements in D_3 and $C_2 = \{1, h^3\}$. The $\pm 60°$ rotations are $h = h^4 h^3$ and $h^5 = h^2 h^3$, respectively. The new 180° rotations are $\rho'_1 \equiv \rho_1 h^3$, $\rho'_2 \equiv \rho_2 h^3$, and $\rho'_3 \equiv \rho_3 h^3$, and their axes are indicated in Figure 3.6.4.

The cross-product definition allows the immediate construction of the D_6 character table from those in D_3 and C_2:

D_3	1	$(h^2 h^4)$	(ρ_i)
A_1	1	1	1
A_2	1	1	-1
E	2	-1	0

$\times$

C_2	1	h^3
A	1	1
B	1	-1

D_6	1	$(h^2 h^4)$	(ρ_i)	h^3	(hh^5)	(ρ'_i)
A_1	1	1	1	1	1	1
A_2	1	1	-1	1	1	-1
$= E_2$	2	-1	0	2	-1	0
B_1	1	1	1	-1	-1	-1
B_2	1	1	-1	-1	-1	1
E_1	2	-1	0	-2	1	0

. (3.6.17)

The process is the same as was used in Section 2.10 for Abelian groups, except now there are two-dimensional irreps involved. The bottom row of the character table is the trace of the irrep $\mathscr{D}^{E_1}$ given by the cross-product relation

$$\mathscr{D}_{ij}^{E_1} \text{ (of } D_6) = \mathscr{D}_{ij}^{E} \text{ (of } D_3) \mathscr{D}^{B} \text{ (of } C_2). \tag{3.6.18}$$

For example, using standing-wave irreps of D_3 one has

$$\mathscr{D}^{E_1}(h^4) = \mathscr{D}^{E}(h^4)\mathscr{D}^{B}(1), \qquad \mathscr{D}^{E_1}(h) = \mathscr{D}^{E}(h^4)\mathscr{D}^{B}(h^3)$$

$$= \begin{pmatrix} -\dfrac{1}{2} & \dfrac{\sqrt{3}}{2} \\ -\dfrac{\sqrt{3}}{2} & -\dfrac{1}{2} \end{pmatrix} \qquad = \begin{pmatrix} \dfrac{1}{2} & -\dfrac{\sqrt{3}}{2} \\ \dfrac{\sqrt{3}}{2} & \dfrac{1}{2} \end{pmatrix} \tag{3.6.18}_x$$

for the z rotation by $\omega_z = 240°$ and $\omega_z = 60°$, respectively. Note that either one is given by the rotation matrix formula

$$\mathscr{D}^{E_1}(\omega_z) = \begin{pmatrix} \cos\omega_z & -\sin\omega_z \\ \sin\omega_z & \cos\omega_z \end{pmatrix} \tag{3.6.19}$$

for x and y components for a vector. The other two-dimensional irrep $\mathscr{D}^{E_2}$ is given by the cross-product relation

$$\mathscr{D}_{ij}^{E_2} \text{ (of } D_6) = \mathscr{D}_{ij}^{E} \text{ (of } D_3) \mathscr{D}^{A} \text{ (of } C_2). \tag{3.6.20}$$

For the z rotations by $\omega_z = 240°$ (h^4) and $\omega_z = 60°$ (h) one has identical matrices:

$$\mathscr{D}^{E_2}(h^4) = \begin{pmatrix} -1/2 & \sqrt{3}/2 \\ -\sqrt{3}/2 & -1/2 \end{pmatrix} = \mathscr{D}^{E_2}(h). \tag{3.6.20}_x$$

The matrix formula which gives the $\mathscr{D}^{E_2}$ z rotations is

$$\mathscr{D}^{E_2} = \begin{pmatrix} \cos 2\omega_z & \sin 2\omega_z \\ -\sin 2\omega_z & \cos 2\omega_z \end{pmatrix}, \tag{3.6.21}$$

and it is appropriate for irreducible tensor rotations, as will be explained in Chapter 6.

For now it is easier to appreciate the difference between E_1 and E_2 irreps, and the A and B irreps as well, by appealing to Bloch-wave structure.

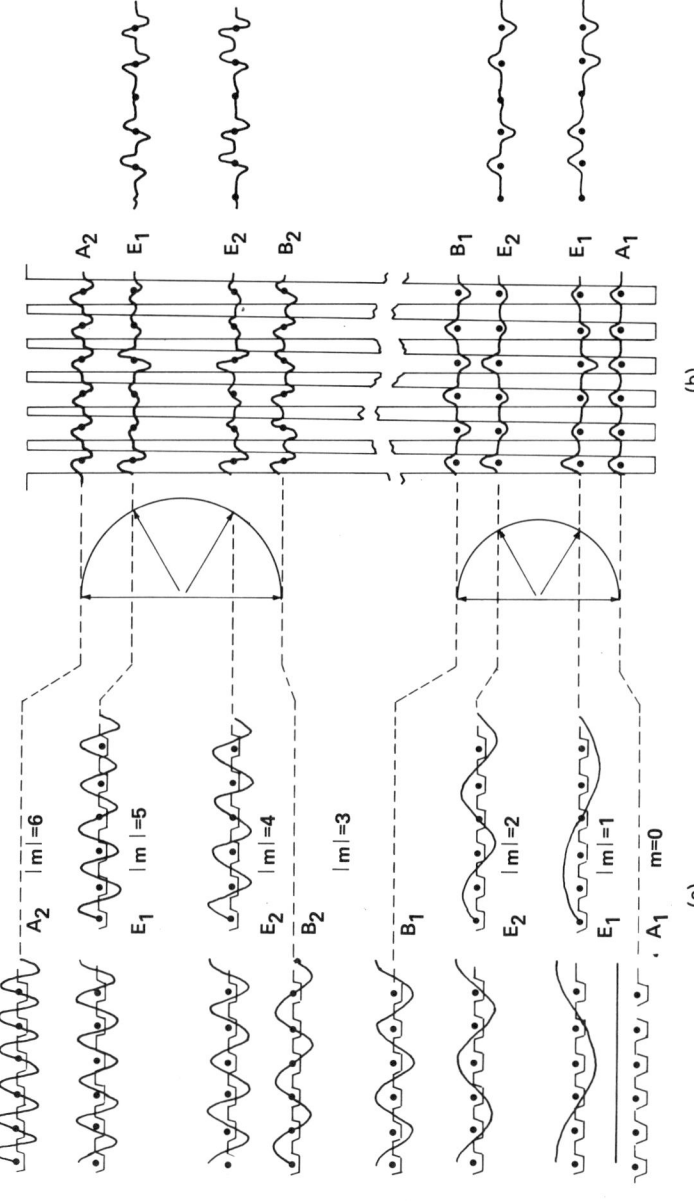

Figure 3.6.5 One-dimensional Bohr and Bloch waves in D_6 symmetry. (a) Weak D_6 potential. (b) Strong D_6 potential.

The standing-wave solutions ($\sin m\phi$) and ($\cos m\phi$) for weak D_6 symmetric potentials are drawn in Figure 3.6.5(a) for $0 \leq m \leq 6$. Note that E_1 and E_2 label the $m = 1$ and $m = 2$ waves, respectively, and these states are doubly degenerate. E_1 and E_2 also label the $m = 5$ and $m = 4$ waves, since $5 = -1$ mod 6 and $4 = -2$ mod 6, respectively.

The B_1 and B_2 waves each have $m = 3$ waves on the ring, but are not degenerate. A source of possible confusion concerns the question of which has higher energy. Figure 3.6.5(a) shows the B_1 wave with antinodes or maxima in potential valleys. Hence, it has lower energy than the B_2 wave, which sits on the potential hills. This implies that B_1 is symmetric to 180° rotations around axes centered in valleys and antisymmetric to rotations around axes centered in hills. The B_1 characters from Eq. (3.6.17) for transverse 180° rotations are

$$\chi_\rho^{B_1} = 1, \qquad \chi_{\rho'}^{B_1} = -1. \tag{3.6.22}$$

This indicates an implicit assumption about how the ρ_j rotation axes are related to the potential: they lie in *valleys*. Then the local parity in each valley determines A and B indices. (1) labels evenness or symmetry, and (2) labels oddness or antisymmetry. The A_1 and B_1 characters,

$$\chi_\rho^{A_1} = \chi_\rho^{B_1} = 1, \tag{3.6.23}$$

are positive, while the A_2 and B_2 characters,

$$\chi_\rho^{A_2} = \chi_\rho^{B_2} = -1, \tag{3.6.24}$$

are negative. Whenever possible we shall let valley operators label the symmetry. This is particularly important for large potentials, as indicated in Figure 3.6.5(b). Then the rapidly varying parts of the wave function are squeezed into the valleys, and their shapes determine the energy of a whole band. A_1 and B_1 levels are upper or lower energy bounds for a band of waves made of locally even wave functions in valleys. A_2 and B_2 levels are the same for locally odd wave functions.

Finally, remember that A and B levels refer to even or odd behavior under axial or cyclic rotation (h or h^5) from one well to the next. The A waves are identical in each well and have unit characters:

$$\chi_h^{A_1} = \chi_h^{A_2} = 1. \tag{3.6.25}$$

The B waves change sign from well to well and have negative characters:

$$\chi_h^{B_1} = \chi_h^{B_2} = -1. \qquad (3.6.26)$$

The correlation of D_6 irreps with those of subgroup $C_6 = \{1, h, h^2, h^3, h^4, h^5\}$ and with those of subgroup $C_2 = \{1, \rho_3\}$ can be deduced by characters. From the preceding equations we deduce that

$$\mathscr{D}^{A_1} \downarrow C_6 = \mathscr{D}^{A_2} \downarrow C_6 = D^A \text{ (of } C_6) = D^{0_6},$$

$$\mathscr{D}^{B_1} \downarrow C_6 = \mathscr{D}^{B_2} \downarrow C_6 = D^B \text{ (of } C_6) = D^{3_6}.$$

this is analogous to Eqs. (3.6.15) for $D_4 \downarrow C_4$ correlation. Similarly, we deduce from Eqs. (3.6.23) and (3.6.24) that

$$\mathscr{D}^{A_1} \downarrow C_2 = \mathscr{D}^{B_1} \downarrow C_2 = D^+ \text{ (of } C_2) = D^{0_2},$$

$$\mathscr{D}^{A_2} \downarrow C_2 = \mathscr{D}^{B_2} \downarrow C_2 = D^- \text{ (of } C_2) = D^{1_2}$$

is the $C_2 = \{1, \rho_3\}$ correlation for A and B irreps of D_6.

It is convenient to summarize all such correlations into CORRELATION TABLES such as the following for $D_6 \downarrow C_6$ and $D_6 \downarrow C_2$:

	D^{0_6}	D^{1_6}	D^{2_6}	D^{3_6}	D^{4_6}	D^{5_6}
$\mathscr{D}^{A_1} \downarrow C_6 =$	1	.	.	.	.	.
$\mathscr{D}^{A_2} =$	1	.	.	.	.	.
$\mathscr{D}^{B_1} =$	.	.	.	1	.	.
$\mathscr{D}^{B_2} =$	.	.	.	1	.	.
$\mathscr{D}^{E_1} =$	.	1	.	.	.	1
$\mathscr{D}^{E_2} =$	.	.	1	.	1	.

$$(3.6.27a)$$

	D^{0_2}	D^{1_2}
$\mathscr{D}^{A_1} \downarrow C_2 =$	1	.
$\mathscr{D}^{A_2} =$	.	1
$\mathscr{D}^{B_1} =$	1	.
$\mathscr{D}^{B_2} =$	.	1
$\mathscr{D}^{E_1} =$	1	1
$\mathscr{D}^{E_2} =$	1	1

$$(3.6.27b)$$

The last two rows in either table indicate Zeeman splittings such as

$$\mathcal{D}^{E_1} \downarrow C_6 = D^{16} + D^{56}.$$

Finally, note that C_n or D_n symmetry with odd n cannot have B waves. Neither can D_n have two classes ρ_j and ρ'_j of valley and hill 180° rotations, since hills are opposite valleys on the odd-n symmetric ring. You should be able to construct diagrams like Figure 3.6.5 for the irreps A_1, A_2, E_1, and E_2 of D_5, and the irreps A_1, A_2, E_1, E_2, and E_3 of D_7.

C. Higher D_n Symmetries: D_{nh} and D_{nd}

Anyone who has resolved D_2, D_3, and D_4 symmetries and their representations can quickly understand all the other dihedral or D-type symmetries. This includes the crystal point symmetries D_{2h}, D_{3h}, D_{4h}, D_{6h}, D_{2d}, and D_{3d}. (D_{2h} is Abelian.) However, the noncrystal or molecular point symmetries $D_{5h}, D_{7h}, D_{8h}, \ldots$, etc., or $D_{4d}, D_{5d}, D_{6d}, \ldots$, etc., are no more difficult.

The D_{nh} symmetries all contain a horizontal or xy-plane reflection operation:

$$\sigma_h = \sigma_{xy} = IR_z^2 \, (180°).$$

This operation commutes with all R_z rotations and all transverse 180° ρ_j operations. Hence, the group D_{nh} can be written as an outer product,

$$D_{nh} = D_n \times C_h = D_n \times \{1, \sigma_h\}. \tag{3.6.28}$$

For even n the rotation $R_z \, (180°)$ and the product $R_z \sigma_h = I$, i.e., inversion, must be in D_{nh}. Then one can write

$$D_{nh} = D_n \times C_i = D_n \times \{1, I\} \quad (n \text{ even}). \tag{3.6.29}$$

D_{nh} symmetry can be pictured as two parallel and identical regular n polygons placed one above the other. Some examples are shown in Figure 3.6.6(a). To have D_{nh} symmetry each vertex of the upper polygon must lie directly above a vertex of the lower one. If one polygon is rotated halfway by diagonal angle $\frac{1}{2}(2\pi/n)$ then the symmetry D_{nd} results. As shown in Figure 3.6.6(b) the center of each polygonal face of the upper D_{nd} polygon lies above a diagonal or vertex of the lower one.

From Figure 3.6.6 one can appreciate that two types of D_{nd} groups emerge. For n odd ($n \geq 3$) the symmetry contains inversion and can be

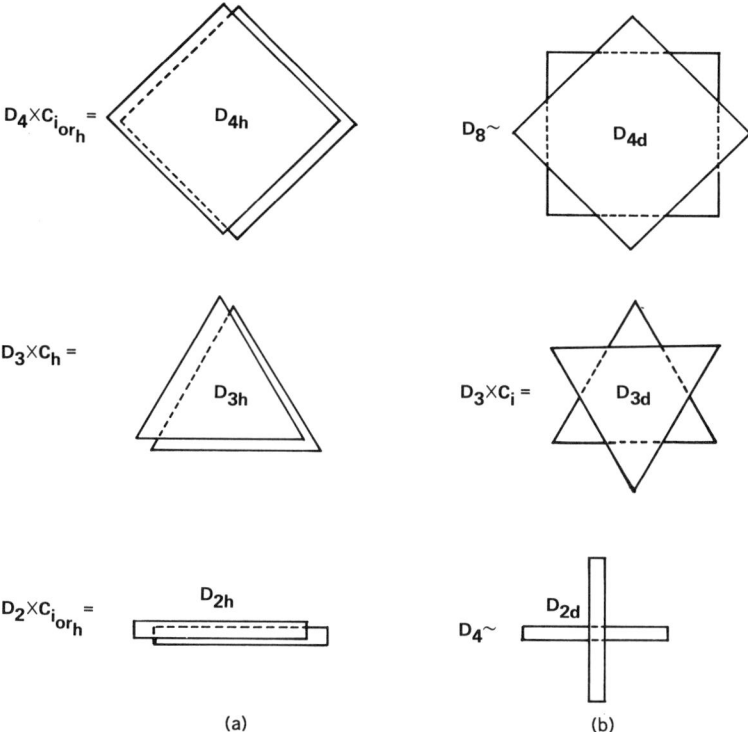

Figure 3.6.6 Comparison of horizontial and diagonal dihedral symmetries. (a) Horizontial D_{nh} symmetries. (b) Diagonal D_{nd} symmetries.

written

$$D_{nd} = D_n \times C_i = D_n \times \{1, I\} \quad (n \text{ odd}). \quad (3.6.30)$$

For n even ($n \geq 2$) D_{nd} is isomorphic ($\sim$) to a higher D_{2n} or C_{2nv} group:

$$D_{nd} \sim D_{2n} \quad (n \text{ even}). \quad (3.6.31)$$

For example, D_{2d} contains the operations

$$D_{2d} = \{1, R^2, IR, IR^3, IR_1^2, IR_2^2, i_3, i_4\}. \quad (3.6.32)$$

D_{2d} is the same as D_4 except for the extra inversion attached to the R and R_1^2 class elements. It is then easy to see that D_{2d} has the same group table as D_4 or C_{4v}. Hence, one set of irreps can be used for all three of these groups.

The presence of inversion symmetry in even-n D_{nh} and odd-n D_{nd} symmetries should be emphasized. Whenever possible the irrep labels should indicate inversion parity by subindices X_g for even and X_u for odd, where $X = A_i$, B_i, or E_m labels the rest of the symmetry. For example, the D_{2h}

irreps are obtained quickly for those of D_2 and C_i using Eq. (3.6.29):

$$D_2 = \begin{array}{c|cccc} & 1 & R_z^2 & R_y^2 & R_x^2 \\ \hline A_1 & 1 & 1 & 1 & 1 \\ B_2 & 1 & -1 & 1 & -1 \\ A_2 & 1 & 1 & -1 & -1 \\ B_2 & 1 & -1 & -1 & 1 \end{array} \times \begin{array}{c} g \\ u \end{array} \quad C_i = \begin{array}{c|cc} & 1 & I \\ \hline & 1 & 1 \\ & 1 & -1 \end{array}$$

$$D_{2h} = \begin{array}{c|cccc|cccc} & 1 & R_z^2 & R_y^2 & R_x^2 & I & IR_z^2 & IR_y^2 & IR_x^2 \\ \hline A_{1g} & 1 & 1 & 1 & 1 & 1 & 1 & 1 & 1 \\ B_{1g} & 1 & -1 & 1 & -1 & 1 & -1 & 1 & -1 \\ A_{2g} & 1 & 1 & -1 & -1 & 1 & 1 & -1 & -1 \\ B_{2g} & 1 & -1 & -1 & 1 & 1 & -1 & -1 & 1 \\ A_{1u} & 1 & 1 & 1 & 1 & -1 & -1 & -1 & -1 \\ B_{1u} & 1 & -1 & 1 & -1 & -1 & 1 & -1 & 1 \\ A_{2u} & 1 & 1 & -1 & -1 & -1 & -1 & 1 & 1 \\ B_{2u} & 1 & -1 & -1 & 1 & -1 & 1 & 1 & -1 \end{array}. \quad (3.6.33)$$

Similarly, the characters of $D_{3d} = D_3 \times C_i$ follow:

$$D_3 = \begin{array}{c|ccc} & 1 & (r, r^2) & (\rho_1 \rho_2 \rho_3) \\ \hline A_1 & 1 & 1 & 1 \\ A_2 & 1 & 1 & -1 \\ E & 2 & -1 & 0 \end{array} \times \begin{array}{c} g \\ u \end{array} \quad C_i = \begin{array}{c|cc} & 1 & I \\ \hline & 1 & 1 \\ & 1 & -1 \end{array}$$

$$D_{3d} = \begin{array}{c|ccc|ccc} & 1 & r & \rho_i & I & Ir & I\rho_i \\ \hline A_{1g} & 1 & 1 & 1 & 1 & 1 & 1 \\ A_{2g} & 1 & 1 & -1 & 1 & 1 & -1 \\ E_g & 2 & -1 & 0 & 2 & -1 & 0 \\ A_{1u} & 1 & 1 & 1 & -1 & -1 & -1 \\ A_{2u} & 1 & 1 & -1 & -1 & -1 & 1 \\ E_u & 2 & -1 & 0 & -2 & 1 & 0 \end{array}. \quad (3.6.34)$$

Now you should have no trouble producing the character table for the largest crystallographic D group

$$D_{6h} = D_6 \times C_i = D_3 \times C_2 \times C_i. \quad (3.6.35)$$

The odd-n D_{nh} symmetries have horizontal reflection symmetry but no inversion center. This may be shown in a prime (X') and double-prime (X'') notation for even and odd reflection symmetry, respectively. For example, using Eq. (3.6.28) one has D_{3h} characters:

$$
\begin{array}{c|ccc}
D_3 = & 1 & (r, r^2) & (\rho_1 \rho_2 \rho_3) \\
\hline
A_1 & 1 & 1 & 1 \\
A_2 & 1 & 1 & -1 \\
E & 2 & -1 & 0
\end{array}
\times
\begin{array}{c|cc}
C_h = & 1 & \sigma_h \\
\hline
A' & 1 & 1 \\
A'' & 1 & -1
\end{array}
$$

$$
= \begin{array}{c|cccccc}
D_{3h} = & 1 & r & \rho_i & \sigma_h & \sigma_h r & \sigma_h \rho_i \\
\hline
A'_1 & 1 & 1 & 1 & 1 & 1 & 1 \\
A'_2 & 1 & 1 & -1 & 1 & 1 & -1 \\
E' & 2 & -1 & 0 & 2 & -1 & 0 \\
A''_1 & 1 & 1 & 1 & -1 & -1 & -1 \\
A''_2 & 1 & 1 & -1 & -1 & -1 & 1 \\
E'' & 2 & -1 & 0 & -2 & 1 & 0
\end{array}. \quad (3.6.36)
$$

However, the actual irreps and characters of D_{3h} are the same as those of the isomorphic group $D_{3d} \sim D_{3h}$ in the previous equation (3.6.34). Only the notation and physical action of the operations is different.

The economy of symmetry mathematics should be evident by now. Just by learning irreps of C_2, C_3, C_4, D_3, and D_4 one becomes able to treat 27 of the 32 crystal point groups listed in Figure 3.1.1.

3.7 AMMONIA (NH$_3$) VIBRATIONAL MODES

A detailed C_{3v} projection analysis of ammonia (NH$_3$) molecular vibrational modes will be presented in this section. A reader who is studying symmetry analysis for the first time may want to skip ahead to Chapter 4. References contained at the end of the chapter should be consulted after studying this section.

This section will be devoted to a comparatively simple spring-mass model of the NH$_3$ molecule. Only two spring constants j and k will be used for the N—H and H—H bonds, respectively, as shown in Figure 3.7.1. However, all models of NH$_3$ must take account of 12 coordinates for the four masses and acceleration or energy operators which involve the coordinates in the Hamiltonian. Therefore, it is important to develop procedures which will use symmetry analysis efficiently and find the easiest way to solve *any* model Hamiltonian.

For complex problems that are not nearly perfect or regular representations of the symmetry, one can usually save a lot of algebra by dealing

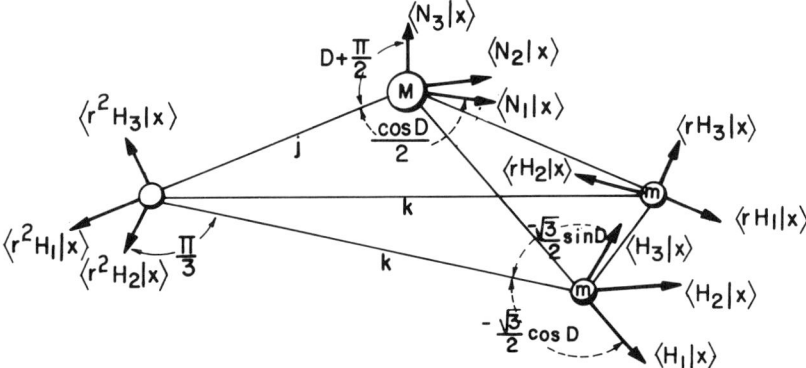

Figure 3.7.1 Spring-mass model for ammonia (NH_3) molecule. Dashed arrows indicate coordinate bond angles by giving the cosine of the angle. Base (H—H) bands have spring constants k, while the pyramidal (N—H) bands have constant j.

separately with the force operator F and the inertia operator m instead of diagonalizing directly the acceleration operator $a = m^{-1}F$. In this way it is possible to keep unpleasant algebraic factors out of the denominators. Also, one does not have to orthogonalize or normalize the basis if so doing would involve still more algebra. The procedures for dealing with F and m operators were introduced in Section 1.4.

In order to construct the reduced forms of the (F) and (m) representations, one needs the irrep bases $P_{ij}^\alpha |x_a\rangle$ obtained from C_{3v} irreps $\alpha = 1$ (or A_1), 2 (or A_2), and 3 (or E). Twelve base vectors $P_{ij}^\alpha |x_a\rangle$ are shown on the left-hand side of Figure 3.7.2 with their representative vibrational displacements. To make these vectors it is convenient to first collect coordinates or state vectors into SYMMETRY ORBITS. Orbits are sets of vectors which can be transformed directly into one another by symmetry operations. For example, the three downward radially pointing hydrogen bases $\{|H_1\rangle, |rH_1\rangle = r|H_1\rangle, |r^2H_1\rangle = r^2|H_1\rangle\}$ form an orbit. Operating on this orbit with the C_{3v} projectors P^1, P_{11}^E, and P_{21}^E gives three states which, when normalized, are the following:

$$P^1|H_1\rangle\sqrt{3} = (|H_1\rangle + |rH_1\rangle + |r^2H_1\rangle)/\sqrt{3},$$
$$P_{11}^3|H_1\rangle\sqrt{3/2} = (2|H_1\rangle - |rH_1\rangle - |r^2H_1\rangle)/\sqrt{6},$$
$$P_{21}^3|H_1\rangle\sqrt{3/2} = (|rH_1\rangle - |r^2H_1\rangle)/\sqrt{2}. \tag{3.7.1}$$

Note that operators P^2, P_{12}^E, and P_{22}^E will annihilate the state $|H_1\rangle$ since it is symmetric to local plane reflection:

$$|H_1\rangle = \sigma_3|H_1\rangle = P_1|H_1\rangle, \tag{3.7.2}$$

where

$$P_1 = (1 + \sigma_3)/2.$$

Exactly the same holds for the three upward radially pointing hydrogen bases $\{|H_3\rangle, |rH_3\rangle, |r^2H_3\rangle\}$. The states $P_{ii}^\alpha|H_{1 \text{ or } 3}\rangle$ are shown in Figure 3.7.2, along with the remaining states projected from the orbits $\{|H_2\rangle, |rH_2\rangle, |r^2H_2\rangle$, $\{|N_1\rangle, r|N_1\rangle\}$, and $\{|N_3\rangle\}$. The angular hydrogen base $|H_2\rangle$ is locally antisymmetric; i.e.,

$$|H_2\rangle = -\sigma_3|H_2\rangle = P_2|H_2\rangle, \quad (3.7.3)$$

where

$$P_2 = (1 - \sigma_3)/2. \quad (3.7.4)$$

Hence, only operators P^2, P_{12}^3, and P_{22}^3 give nonzero states:

$$P^2|H_2\rangle\sqrt{3} = (|H_2\rangle + |rH_2\rangle + |r^2H_2\rangle)/\sqrt{3},$$
$$P_{12}^3|H_2\rangle\sqrt{3/2} = (-|rH_1\rangle + |r^2H_2\rangle)/\sqrt{2},$$
$$P_{22}^3|H_2\rangle\sqrt{3/2} = (2|H_2\rangle - |rH_2\rangle - |r^2H_2\rangle)/\sqrt{6}. \quad (3.7.5)$$

The in-σ_3-plane nitrogen base $|N_1\rangle$ is ready-made as a type-3 base along with partner $|N_2\rangle$:

$$P_{11}^3|N_1\rangle = |N_1\rangle, \quad P_{21}^3|N_1\rangle = |N_2\rangle; \quad (3.7.6)$$

the same is true for the scalar (1)-type base

$$P^1|N_3\rangle = |N_3\rangle. \quad (3.7.7)$$

The states on the left-hand side of Figure 3.7.2 are not genuine vibration states since they all have either translational moment or momentum, rotational momentum, or both. Here we measure rotational momentum around the N atom so that it is easy to calculate. Seven of the states have zero rotation, and so it is easy to combine these into genuine or constrained states or zero linear momentum. From the three type-1 states it is possible to make one state of pure translational motion of the rigid molecule. We are only interested in the two remaining genuine or constrained states:

$$|c^1\rangle = MP^1|H_1\rangle\sqrt{3} + \sqrt{3}m \sin D \, P^1|N_3\rangle, \quad (3.7.8)$$
$$|c^1\rangle' = MP^1|H_3\rangle\sqrt{3} - \sqrt{3}m \cos D \, P^1|N_3\rangle, \quad (3.7.9)$$

shown on the right-hand side of Figure 3.7.2. Similarly, the zero-rotation

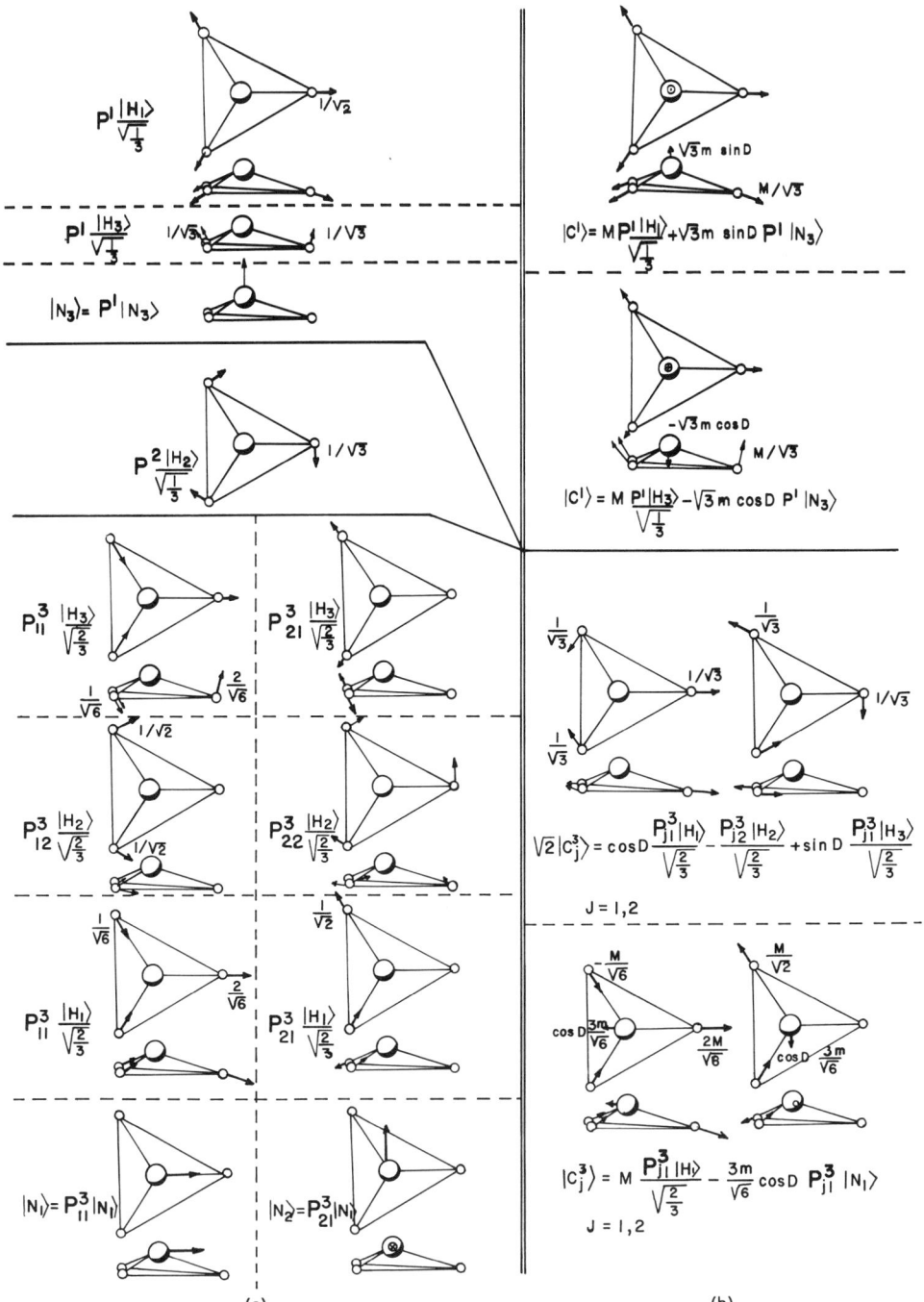

Figure 3.7.2 Symmetry defined motions of NH_3. (a). Primitive projected states obtained in Eqs. (3.7.1)–(3.7.6) are shown. (b) Genuine vibrational states are constrained to have zero translation and rotation.

type-3 states combine to give zero-translation constrained partner states $|c_i^3\rangle'$:

$$|c_1^3\rangle' = MP_{11}^3|H_1\rangle\sqrt{3/2} - 3m\cos D|N_1\rangle/\sqrt{6}, \quad (3.7.10a)$$

$$|c_2^3\rangle' = MP_{21}^3|H_1\rangle\sqrt{3/2} - 3m\cos D|N_2\rangle/\sqrt{6}. \quad (3.7.10b)$$

Lastly, one can make type-3 states of rigid rotation around x and y axes. We shall ignore this motion until Chapter 5, and consider only genuine vibrational motion which includes the preceding $|c\rangle$ states and the following ones:

$$|c_1^3\rangle = \left(\cos D\, P_{11}^3|H_1\rangle - P_{12}^3|H_2\rangle + \sin D\, P_{11}^3|H_3\rangle\right)\sqrt{3/2}, \quad (3.7.11a)$$

$$|c_2^3\rangle = \left(\cos D\, P_{21}^3|H_1\rangle - P_{22}^3|H_2\rangle + \sin D\, P_{21}^3|H_3\rangle\right)\sqrt{3/2}. \quad (3.7.11b)$$

All the $|c\rangle$ states satisfy the constraint equations

$$\mathbf{T} = m\sum_{i=1}^{3}\mathbf{c}_{H_i} + M\mathbf{c}_N = 0, \quad (3.7.12a)$$

$$\mathbf{A} = m\sum_{i=1}^{3}\mathbf{r}_{H_i}\times\mathbf{c}_{H_i} + M\mathbf{r}_N\times\mathbf{c}_N = 0, \quad (3.7.12b)$$

when $\mathbf{c}_p$ is the displacement from equilibrium of atom p. Note that the translational constraint ($\mathbf{T} = 0$) guarantees that any new choice of origin for the angular constraint would still give zero $\mathbf{A}$. If $\mathbf{A} = 0$, then

$$\mathbf{T}\times\mathbf{d} = m\sum_{i=1}^{3}\mathbf{c}_{H_i}\times\mathbf{d} + M\mathbf{c}_N\times\mathbf{d} = 0$$

implies that

$$\mathbf{A}' = m\sum_{i=1}^{3}(\mathbf{r}_{H_i} + \mathbf{d})\times\mathbf{c}_{H_i} + M(\mathbf{r}_N + \mathbf{d})\times\mathbf{c}_N = 0$$

is zero also.

The preceding constraints and C_{3v} symmetry analysis reduce the genuine vibration problem from NH_3 to the calculation and solution of two 2×2 matrices:

$$\langle Q\rangle^1 = \begin{pmatrix} \langle c^1|Q|c^1\rangle & \langle c^1|Q|c^{1'}\rangle \\ \langle c^{1'}|Q|c^1\rangle & \langle c'|Q|c^{1'}\rangle \end{pmatrix}, \quad \langle Q\rangle^3 = \begin{pmatrix} \langle c_1^3|Q|c_1^3\rangle & \langle c_1^3|Q|c_1^{3'}\rangle \\ \langle c_1^{3'}|Q|c_1^3\rangle & \langle c_1^{3'}|Q|c_1^{3'}\rangle \end{pmatrix},$$

representing $Q = F$ and $Q = m$ operators. Note that vectors $|c^\alpha\rangle$ and $|c^{\alpha'}\rangle$ are not orthogonal, and so we are committed to a separate treatment of the F and m operators. One needs only the rows of these matrices that correspond to the first vector of each orbit, namely, $\langle H_1|, \langle H_2|, \langle H_3|, \langle N_1|,$ and $\langle N_3|$, as given in Eqs. (3.7.13) below:

$$\langle F \rangle =$$

	$\|H_1\rangle$	$r\|H_1\rangle$	$r^2\|H_1\rangle$	$\|H_2\rangle$	$r\|H_2\rangle$	$r^2\|H_2\rangle$	$\|H_3\rangle$	$r\|H_3\rangle$	$r^2\|H_3\rangle$	$\|N_1\rangle$	$\|N_2\rangle$	$\|N_3\rangle$
$\langle H_1\|$	$\frac{3k}{2}\cos^2 D + j$	$\frac{3k}{4}\cos^2 D$	$\frac{3k}{4}\cos^2 D$	0	$\frac{\sqrt{3}k}{4}\cos D$	$-\frac{\sqrt{3}k}{s4}\cos D$	$\frac{3k}{2}\sin D\cos D$	$\frac{3k}{4}\sin D\cos D$	$\frac{3k}{4}\sin D\cos D$	$-j\cos D$	0	$j\sin D$
$\langle H_2\|$	0	$-\frac{\sqrt{3}k}{4}\cos D$	$\frac{\sqrt{3}k}{4}\cos D$	$\frac{k}{2}$	$-\frac{k}{4}$	$-\frac{k}{4}$	0	$-\frac{\sqrt{3}k}{4}\sin D$	$\frac{\sqrt{3}k}{4}\sin D$	0	0	0
$\langle H_3\|$			$\frac{3k}{4}\sin D\cos D$	0	$\frac{\sqrt{3}k}{4}\sin D$	$-\frac{\sqrt{3}k}{4}\sin D$	$\frac{3k}{2}\sin^2 D$	$\frac{3k}{4}\sin^2 D$	$\frac{3k}{4}\sin^2 D$	0	0	0
$\langle N_1\|$				0	0	0	0	0	0		0	0
$\langle N_3\|$										$\frac{3j}{2}\cos^2 D$	0	$3j\sin^2 D$

(3.7.13a)

$$\langle m \rangle =$$

	$\|H_1\rangle$			$\|H_2\rangle$			$\|H_3\rangle$			$\|N_1\rangle$		$\|N_3\rangle$
$\langle H_1\|$	m	·	·	·	·	·	·	·	·	·	·	·
$\langle H_2\|$	·	·	·	m	·	·	·	·	·	·	·	·
$\langle H_3\|$	·	·	·	·	·	·	m	·	·	·	·	·
$\langle N_1\|$	·	·	·	·	·	·	·	·	·	M	·	·
$\langle N_3\|$	·	·	·	·	·	·	·	·	·	·	·	M

(3.7.13b)

The calculation of the representations of operators $Q = F$ or $Q = m$ in the $|c_j^\alpha\rangle$ basis can be made easier by using the properties of the elementary operators. Let us begin with the vector $|c_1^{3'}\rangle$ from Eq. (3.7.10):

$$\langle c_1^{3'}|Q|c_1^{3'}\rangle$$
$$= M(3/2)^{1/2}\langle H_1|P_{11}^3 Q\big[M(3/2)^{1/2}P_{11}^3|H_1\rangle - 3m\cos D|N_1\rangle/\sqrt{6}\big]$$
$$- (3m/\sqrt{6})\cos D\langle N_1|Q\big[M(3/2)^{1/2}P_{11}^3|H_1\rangle - 3m\cos D|N_1\rangle/\sqrt{6}\big].$$

Using the idempotent properties ($P_{11}^3 P_{11}^3 = P_{11}^3$), commutivity ($P_{11}^3 Q = QP_{11}^3$), and Eq. (3.7.5), the following expression results:

$$\langle c_1^{3'}|Q|c_1^{3'}\rangle = M^2 \sum_{n=0}^{1} \mathscr{D}_{11}^{3*}(r^n)\langle H_1|Q|r^n H_1\rangle - 3Mm\cos D\langle H_1|Q|N_1\rangle$$
$$+ \frac{9m^2}{6}\cos^2 D\langle N_1|Q|N_1\rangle. \qquad (3.7.15)$$

Here the formula (3.4.19) for P_{11}^3 is used again. Now substituting the required F-matrix components for Eq. (3.7.13a) gives the desired component:

$$\langle c_1^{3'}|F|c_1^{3'}\rangle = M^2(j + (3k/4)\cos^2 D) + 3jM\cos^2 D + jm^2(9/4)\cos^4 D$$
$$= M^2\Big[(3k/4)\cos^2 D + j\big[(3m/2M)\cos^2 D + 1\big]^2\Big]. \qquad (3.7.16)$$

Similarly, the m-matrix component is

$$\langle c_1^{3'}|m|c_1^{3'}\rangle = Mm^2\big[1 + (3m/2M)\cos^2 D\big]. \qquad (3.7.17)$$

The off-diagonal $\langle c_1^3|Q|c_1^{3'}\rangle$ is produced in the same way:

$$\langle c_1^3|Q|c_1^{3'}\rangle = (M/\sqrt{8})\cos D \sum_{n=0}^{2} \langle H_1|Q|r^n H_1\rangle \mathscr{D}_{11}^{3*}(r^n)$$
$$- (3m/\sqrt{8})\cos^2 D\langle H_1|Q|N_2\rangle$$
$$- (M/\sqrt{8}) \sum_{n=0}^{2} \langle H_2|Q|r^n H_1\rangle \mathscr{D}_{21}^{3*}(r^n)$$
$$+ (3m/\sqrt{8})\cos D\langle H_2|Q|N_2\rangle$$
$$+ (M/\sqrt{8})\sin D \sum \langle H_3|Q|r^n H_1\rangle \mathscr{D}_{11}^{3*}(r^n)$$
$$- (3m/\sqrt{8})\sin D\cos D\langle H_3|Q|N_1\rangle.$$

The off-diagonal F component

$$\langle c_1^3 | F | c_1^{3'} \rangle = (M/\sqrt{2})\cos D[(3k/2) + j(1 + (3m/2M)\cos^2 D)] \quad (3.7.18)$$

is the coupling between the two vibration coordinates. The off-diagonal m component,

$$\langle c_1^3 | m | c_1^{3'} \rangle = (Mm/\sqrt{2})\cos D, \quad (3.7.19)$$

indicates that $|c_1^3\rangle$ and $|c_1^{3'}\rangle$ are not orthogonal. The type-3 calculation is completed with the $\langle c_1^3 | Q | c_1^3 \rangle$ components since $Q^\dagger = Q$ for $Q = F$ and $Q = m$. The resulting F matrix is

$$\langle F \rangle^3 = \begin{pmatrix} 3k/2 + (j/2)\cos^2 D & M \cos D(3k/2 + jH) \\ M \cos D(3k/2 + jH) & M^2[(3k/4)\cos^2 D + jH^2] \end{pmatrix}, \quad (3.7.20)$$

where

$$H \equiv 1 + (3m/2M)\cos^2 D,$$

while the m matrix takes the form

$$\langle m \rangle^3 = \begin{pmatrix} m & (mM/\sqrt{2})\cos D \\ (mM/\sqrt{2})\cos D & mM^2 H \end{pmatrix}. \quad (3.7.21)$$

The calculation for the type-1 or A_1 modes proceeds in the same way but without the duplicity of the two-dimensional type-3 modes. The force operator is represented in the $\{|c^1\rangle, |c^{1'}\rangle\}$ basis by the following matrix:

$$\langle F \rangle^1 = \begin{pmatrix} M^2(3k\cos^2 D + jJ) & 3JM^2(k - jm/M)\sin D \cos D \\ 3JM^2(k - jm/M)\sin D \cos D & M^2(3k + j(2m/M)^2 \cos^2 D)\sin^2 D \end{pmatrix}, \quad (3.7.22)$$

where

$$J = 1 + (3m/M)\cos^2 D.$$

The mass operator is represented in the same basis by the following matrix

$$\langle m \rangle^1 = \begin{pmatrix} mM^2K & -3(m/M)\sin D \cos D \\ -3(m/M)\sin D \cos D & J \end{pmatrix}, \quad (3.7.23)$$

where

$$K \equiv 1 + (3m/M)\sin^2 D.$$

Combining Eqs. (3.7.20) and (3.7.21) gives the acceleration matrix for type-3 modes:

$$\langle a \rangle^3 = \langle m^{-1}F \rangle^3 = \begin{pmatrix} 3k/m & (3k/m\sqrt{8})M \cos D \\ (j/mM\sqrt{8})\cos D & Hj/m \end{pmatrix}. \quad (3.7.24)$$

The solutions to the secular equation,

$$\lambda_3^2 - S_3\lambda_3 + P_3 = 0,$$

where

$$S_3 = (3k/2 + Hj)/m, \quad (3.7.25a)$$

$$P_3 = (3kj/2m^2)(H - (1/2)\cos^2 D) \quad (3.7.25b)$$

are squared eigenfrequencies of type-3 vibrations:

$$\lambda_{3\pm} = (\omega_{3\pm})^2 = \left(S_3 \pm \sqrt{S_3^2 - 4P_3}\right)/2. \quad (3.7.26)$$

Similarly, the type-1 acceleration matrix is

$$\langle a \rangle^1 = \langle m^{-1}F \rangle^1 = \begin{pmatrix} (3k/m)\cos^2 D + Kj/m & 3(k/m - j/M)\sin D \cos D \\ 3(k/m)\sin D \cos D & 3(k/m)\sin^2 D \end{pmatrix}.$$

$$(3.7.27)$$

Its eigenfrequencies are given by

$$(\omega_{1\pm})^2 = \left(S_1 \pm \sqrt{S_1^2 - 4P_1}\right)/2, \quad (3.7.28)$$

where

$$S_1 = (3k + Kj)/m \qquad (3.7.29a)$$

$$P_1 = 3kj(1/m^2 + 3/mM)\sin^2 D. \qquad (3.7.29b)$$

Now the physics begins as one decides what to do with these equations. One of the most interesting applications involves determining the angle D by "listening" to the vibrations of NH_3 and comparing this with what has been found by other methods such as "looking" with x rays. To do this first note that the last term in the secular equations are the products of the roots $P_1 = (\omega_{1+}\omega_{1-})^2$ and $P_3 = (\omega_{3+}\omega_{3-})^2$, respectively. The ratio R in the following is a function only of the angle D and masses $m = 1$ amu and $M = 14$ amu of hydrogen and nitrogen, respectively,

$$R = (\omega_{1+}\omega_{1-}/\omega_{3+}\omega_{3-})^2$$

$$R = P_1/P_3 = 2(1 + 3m/M)\sin^2 D / \left[1 + (3m/2M - 1/2)\cos^2 D\right] \qquad (3.7.30)$$

Solving for angle D gives

$$D = \sin^{-1}\left[R(1 + 3m/M)/[4(1 + 3m/M) - R(1 - 3m/M)]\right]^{1/2}. \qquad (3.7.31)$$

The frequencies $\omega_{\alpha\pm}$ needed to calculate R and D are obtained from spectra such as the infrared adsorption data shown in Figure 3.7.3. The numbers which are quoted in Herzberg's text are as follows:

$\nu_{1+} = 3337$ cm^{-1}, $\qquad \nu_{3+} = 3414$ cm^{-1},

$\omega_{1+} = 6.290 \times 10^{14}$ rad/sec, $\qquad \omega_{3+} = 6.435 \times 10^{14}$ rad/sec,

$\nu_{1-} = 908$ cm^{-1}, $\qquad \nu_{3-} = 1628$ cm^{-1},

$\omega_{1-} = 1.712 \times 10^{14}$ rad/sec, $\qquad \omega_{3-} = 3.069 \times 10^{14}$ rad/sec.

Determination of the frequencies required careful analysis of many such spectra and an understanding of the rotational structure which surrounds each "line" in Figure 3.7.3. The development of laser devices has given a more clear picture of rotational structure and led to better understanding of it, as will be explained later in the book.

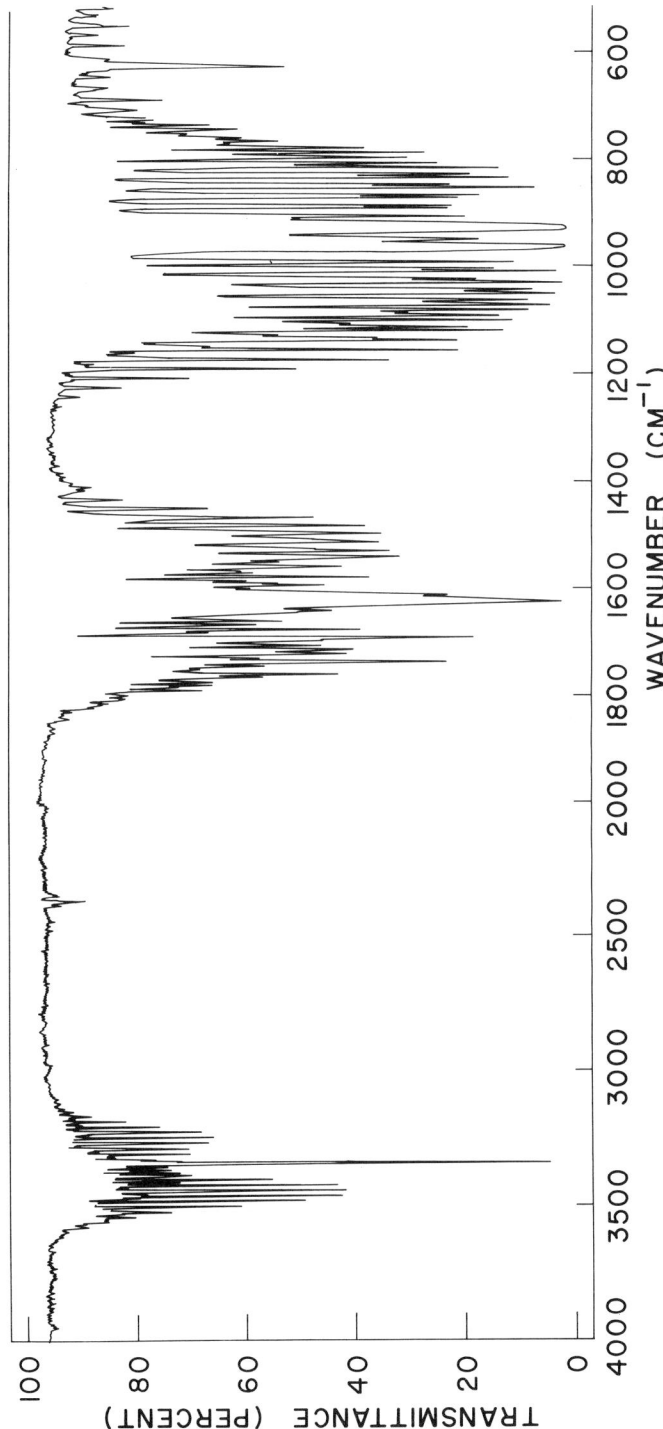

Figure 3.7.3 Infrared spectra of ammonia (NH$_3$). Tabulated values given by Herzberg are listed beside each line. It appears that the (3 +) line is lost in the rotational sidebands of the (1 +) line.

Substituting the observed frequencies into Eqs. (3.7.30) and (3.7.31), one obtains $D = 17°$, which agrees fairly well with the angle 20° found by other means. ($D = 20°$ corresponds to an H—N—H angle of 109°.)

One may try to deduce the spring constants k and j from the spectra just given. One procedure is to use the sums $S_\alpha = (\omega_{\alpha+})^2 + (\omega_{\alpha-})^2$, which are the second factors in the secular equations (3.7.25a) and (3.7.29a).

Solving these gives the following:

$$j/m = (2S_3 - S_1)/1 + 3(m/M)\cos 2D),$$

$$3k/2m = S_1(1 + (3m/2M)\cos^2 D) - S_3(1 + (3m/M)\sin^2 D)$$

Substitution of the experimental $\omega^{(\alpha)}$, $D = 20°$, and the H mass ($m = 1.67 \times 10^{-27}$ kg) gives numerical results ($j = 850$ nt/m) and ($k = -62$ nt/m), which indicates some inaccuracy in the $k - j$ spring model. The negative value for k is physically impossible.

However, this model can still be used to estimate and visualize some things if it is approached more carefully than in the foregoing. Instead let us plot in Figure 3.7.4 the eigenvalues ω_α as functions of the ratio k/j for $D = 20°$. Near $k/j = 0.3$ we find that the ratio of the two lowest functions becomes equal to the observed ratio ω_{3-}/ω_{1-}. From this point one may compare the theory to the experiment for the higher levels. Notice that the ω_{1+} prediction is 20% too high but ω_{3-} is pretty close.

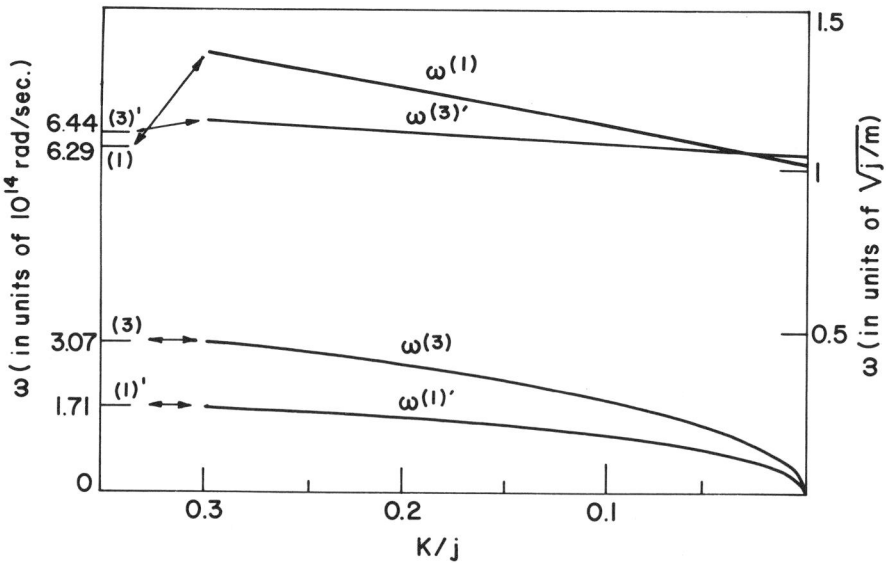

Figure 3.7.4 Plots of k-j model eigenvalues. Experimental values are shown on the left-hand side.

It is possible to make a better theory by including bond-bending constants, i.e., so-called "covalence forces," or other parameters. Indeed, we shall prove by symmetry product analysis in Chapter 6 that the number of parameters at your disposal in any problem of this sort will be exactly enough to make any eigenvalue $(\omega^{(\alpha)}, \omega^{(\alpha)'}, \ldots)$ and eigenvector combination $(a|_i^{(\alpha)}\rangle + b|_i^{(\alpha)'}\rangle + \cdots)$ that is possible. However, the use of too many parameters would give one a rather empty victory unless one goes on to use the parameters for analyzing spectra of other molecules having similar H—H bonds. This approach has been carried out successfully for many cases.

Note in Figure 3.7.4 that ω_{1+} and ω_{3+} are fairly constant and close to $(j/m)^{1/2}$. This is mainly because M is larger than m, so the high modes just amount to hydrogens vibrating more or less independently on j springs. This gives a rough estimate for j: $j = m(6.3 \times 10^{14})^2 = 660$ nt/m. Incidentally, the degeneracy at $k/j = 0.035$ does not, as far as we now, have anything to do with symmetry. Such crossings are called ACCIDENTAL DEGENERACIES.

APPENDIX D. MATHEMATICAL PROPERTIES OF $P_i^\alpha g P_j^\alpha$

This appendix is devoted to closing mathematical loopholes in order to assure that the techniques developed in Chapter 3 are applicable to any finite group $G = \{\cdots g \cdots\}$ of unitary operators. Sections D.1 and D.2 contain proofs that the operators $\{P_i^\alpha g P_j^\alpha, P_i^\alpha g' P_j^\alpha, \ldots\}$ for all elements $\{g, g', \ldots\}$ in G are simply proportional to each other if the idempotents P_i^α and P_j^α cannot be split. Section D.3 contains a proof that each $P_i^\alpha g P_j^\alpha \neq 0$ exists for some g if the $P^\alpha = P_1^\alpha + P_2^\alpha + \cdots$ are complete sets of all-commuting idempotents.

D.1 LINEAR DEPENDENCE OF $P_i^\alpha g P_i^\alpha, P_i^\alpha g' P_i^\alpha, \ldots$

Consider the set $\{p_g \equiv P_i^\alpha g P_i^\alpha, p_{g'} \equiv P_i^\alpha g' P_i^\alpha, \cdots\}$ of all group elements "guarded" by a Hermitian irreducible idempotent $P_i^\alpha = P_i^\alpha P_i^\alpha = P_i^{\alpha\dagger}$. The assumption that P_i^α is an idempotent implies that this set is a closed subalgebra, since any product $p_g p_{g'}$ must be a linear combination of some p_g. Furthermore, P_i^α acts as the identity or unit element ($p_g P_i^\alpha = p_g$) for all p_g. The assumption that P_i^α is Hermitian implies that the algebra contains a conjugate,

$$p_g^\dagger = P_i^{\alpha\dagger} g^\dagger P_i^{\alpha\dagger} = P_i^\alpha g^\dagger P_i^\alpha, \qquad (D.1)$$

for each p_g. Hence, the p_g basis can be replaced by a Hermitian basis of

218 BASIC THEORY AND APPLICATIONS OF SYMMETRY REPRESENTATIONS

operators

$$h_{g^+} \equiv (p_g + p_g^\dagger)/2, \qquad h_{g^-} \equiv (p_g - p_g^\dagger)/2i.$$
$$= h_{g^+}^\dagger \qquad\qquad\qquad = h_{g^-}^\dagger. \qquad (D.2)$$

Each h satisfies some minimal equation

$$(h - \eta_1 P_i^\alpha)(h - \eta_2 P_i^\alpha) \cdots (h - \eta_n P_i^\alpha) = 0, \qquad (D.3)$$

which contains no repeated eigenvalues ($\eta_i \neq \eta_j$ if $i \neq j$). Note that P_i^α is serving here as the unit element.

Finally, the assumption that P_i^α is irreducible implies that a minimal equation (D.3) must have just one factor

$$(h - \eta_1 P_i^\alpha) = 0, \qquad (D.4)$$

or that

$$h = \eta_1 P_i^\alpha.$$

If two or more factors were present in Eq. (D.3) then P_i^α could be split into two or more orthogonal idempotents ($P_i^\alpha = P + P' + \cdots$), contrary to the assumption. Equation (D.4) holds for all h, all p_g, and all combinations of p_g, as well. One has

$$p_g = P_i^\alpha g P_i^\alpha = \rho P_i^\alpha, \qquad (D.5)$$

where the eigenvalue ρ is given the notation

$$\rho = \mathscr{D}_{ii}^\alpha(g) \qquad (D.6)$$

in Section 2.3.

D.2 LINEAR DEPENDENCE OF $\{P_i^\alpha g P_j^\alpha, P_i^\alpha g' P_j^\alpha, \ldots\}$

Suppose a "guarded" operator $N = P_i^\alpha g P_j^\alpha$ is nonzero. Then so is its conjugate $N^\dagger = P_j^\alpha g P_i$ and their products:

$$NN^\dagger = P_i^\alpha g P_j^\alpha g^\dagger P_i^\alpha = \lambda P_i^\alpha, \qquad (D.7a)$$

$$N^\dagger N = P_j g^\dagger P_i^\alpha g P_j^\alpha = \lambda' P_j^\alpha. \qquad (D.7b)$$

The λP_i^α form of each product follows from Eq. (D.5), where λ and λ' must be real and positive definite. For any other group operators g' one has

$$P_i^\alpha g' P_j^\alpha g^\dagger P_i^\alpha = \nu P_i^\alpha, \tag{D.8}$$

according to Eq. (D.5). Combining the preceding two equations gives

$$0 = \lambda \nu P_i^\alpha - \nu \lambda P_i^\alpha$$
$$= \lambda P_i^\alpha g' P_j^\alpha g^\dagger P_i^\alpha - \nu P_i^\alpha g P_j^\alpha g^\dagger P_i^\alpha,$$
$$0 = \left(\lambda P_i^\alpha g' P_j^\alpha - \nu P_i^\alpha g P_j^\alpha\right) P_j^\alpha g^\dagger P_i^\alpha. \tag{D.9}$$

Attaching gP_j on the right and using Eq. (D.7b) gives

$$0 = \left(\lambda P_i^\alpha g' P_j^\alpha - \nu P_i^\alpha g P_j^\alpha\right) P_j^\alpha g^\dagger P_i^\alpha g P_j^\alpha,$$
$$0 = \left(\lambda P_i^\alpha g' P_j^\alpha - \nu P_i^\alpha g P_j^\alpha\right) \lambda'. \tag{D.10}$$

Assuming λ and λ' are nonzero gives

$$P_i^\alpha g' P_j^\alpha = \nu/\lambda \, P_i^\alpha g P_j^\alpha, \tag{D.11}$$

Therefore, the irreducible "guards" convert all $\{g, g', \ldots\}$ to the same operator to within a proportionality factor ν/λ. Next we prove that $P_i^\alpha g P_j^\alpha$ is nonzero for some g; i.e., $\lambda \neq 0$. Of course, if all $\{P_i^\alpha g P_j^\alpha, P_i^\alpha g' P_j^\alpha, \ldots\}$ were zero one would not have needed to prove that they are proportional.

D.3 EXISTENCE PROOF OF $P_i^\alpha g P_j^\alpha$

Suppose $P_1^\alpha g P_2^\alpha = 0$ for all group operators g. Then we have

$$P_1^\alpha G P_2^\alpha = 0, \tag{D.12}$$

where

$$G = \sum_g \gamma_g g$$

is any combination of operators. In other words we suppose that P_1^α and P_2^α cannot serve together as bodyguards for any operator g without annihilation.

Let the set $\{P_1^\alpha, P_i^\alpha, P_{i'}^\alpha, \ldots\}$ of irreducible idempotents contain those which still may serve as bodyguards with P_1^α, i.e., let

$$P_i^\alpha = \lambda P_i^\alpha h P_1^\alpha h^\dagger P_i^\alpha \tag{D.13}$$

for some h. Let the second set $\{P_2^\alpha, P_j^\alpha, P_{j'}^\alpha, \cdots\}$ contain those which may serve with P_2^α; i.e., let

$$P_j^\alpha = \mu P_j^\alpha k P_2^\alpha k^\dagger P_j^\alpha \tag{D.14}$$

for some k. Combining Eqs. (D.12)–(D.14) implies annihilation,

$$P_i^\alpha g P_j^\alpha = \lambda\mu P_i^\alpha h P_1^\alpha h^\dagger P_i^\alpha g P_j^\alpha k P_2^\alpha k^\dagger P_j^\alpha = 0, \tag{D.15}$$

for all g, due to the factor $P_1^\alpha G P_2^\alpha$. In other words, any idempotent P_i^α in the P_1^α subset annihilates any operator containing a member of the P_2^α subset, and vice versa. This implies that the (α) part of the decomposition

$$g = \mathbf{1}g\mathbf{1} = \sum_\alpha \sum_{m,n} P_m^\alpha g P_n^\alpha$$

splits into (at least) two separate parts

$$g = \sum_\alpha \left(\sum_{i,i'} P_i^\alpha g P_{i'}^\alpha + \sum_{j,j'} P_j^\alpha g P_{j'}^\alpha + \cdots \right)$$
$$= \sum_\alpha \left(P_{(1)}^\alpha g P_{(1)}^\alpha + P_{(2)}^\alpha g P_{(2)}^\alpha + \cdots \right),$$

involving idempotents

$$P_{(1)}^\alpha = P_1^\alpha + P_i^\alpha + P_{i'}^\alpha + \cdots,$$
$$P_{(2)}^\alpha = P_2^\alpha + P_j^\alpha + P_{j'}^\alpha + \cdots,$$

which are sums of the separate sets. These new idempotents satisfy commutation relations

$$P_{(a)}^\alpha g = P_{(a)}^\alpha g P_{(a)}^\alpha = g P_{(a)}^\alpha$$

with all group elements g.

However, this leads to a contradiction. The original all-commuting idempotent

$$P^\alpha = P_1^\alpha + P_2^\alpha + \cdots P_{l^\alpha}^\alpha$$
$$= P_{(1)}^\alpha + P_{(2)}^\alpha + \cdots$$

should not produce any new all-commuting $P_{(a)}^\alpha$ when splitting into irreducible idempotents. (The dimension of the class algebra is fixed.) Hence, any two irreducible idempotents P_1^α and P_2^α from the same family must give some nonzero $P_1^\alpha g P_2^\alpha$.

ADDITIONAL READING

An introduction to Hamilton's turns is given by Biedenharn and Louck.

L. C. Biedenharn and J. D. Louck, in G. C. Rota, Ed., *Angular Momentum in Quantum Mechanics*, Encyclopedia of Mathematics and Its Applications, (Addison-Wesley, Reading, MA, 1981).

The original text by Hamilton (not readily available) is a little difficult for most modern readers.

W. R. Hamilton, *Lectures on Quaternions*, (Dublin, 1853).

The slide rule base upon the turns is described in the author's *American Journal of Physics* article

W. G Harter and N. doc Santos, "Double group theory on the half shell and the two-level systems," *am. J. Phys.*, **46**, 251 (1978).

A complete description of group algebra is found in a mathematics text by Curtis and Reiner.

C. W. Curtis and I. Reiner, *Representation Theory of finite Groups and Associative Algebras*, (Wiley, New York, 1965).

The discussion and bibliography for the NH_3 vibrational spectra found in Herzberg's classic texts is still probably the most complete discussion of molecular spectroscopy.

G. Herzberg, *Molecular Spectra and Molecular Structure:* I. Spectra of Diatomic Molecules, II. Infrared and Raman Spectra of Polyatomic Molecules, III. Electronic Structure of Polyatomic Molecules (Van Nostrand-Reinhold, New York, 1950).

The XY_3 analysis start on p. 155 of Vol. II.

The standard reference for molecular normal coordinate analysis is

F. B. Wilson, V. C. Decius, and P. C. Cross; *Molecular Vibrations* (McGraw-Hill, New York, 1955).

222 BASIC THEORY AND APPLICATIONS OF SYMMETRY REPRESENTATIONS

Two more modern references are

S. Califano *Vibrational States* (Wiley, New York, 1976).

D. Papousek and M. R. Alier, *Molecular Vibrational-Rotational Spectra*, 1982 *Studies in Physical and Theoretical Chemistry*, Vol. 17 (Elsevier Amsterdam, New York, 1982).

PROBLEMS

Section 3.1

3.1.1 The subsets $(\mathcal{H}, \mathcal{H}g_1, \mathcal{H}g_2, \ldots)$ or $(\mathcal{H}, g_1\mathcal{H}, g_2\mathcal{H}, \ldots)$ of group $\mathcal{G} = \{1, g, g', \ldots\}$ are called RIGHT or LEFT COSETS, respectively, of subgroup $\mathcal{H} = \{1, h, \ldots\} \subset \mathcal{G}$. The subsets $(\mathcal{H}, \mathcal{H}', \mathcal{H}g_1\mathcal{H}', \mathcal{H}g_2\mathcal{H}', \ldots)$ are called DOUBLE COSETS of subgroups $\mathcal{H}$ and $\mathcal{H}'$.
   (a) Construct right cosets for subgroup $C_v = \{1, \sigma_3\}$ of C_{3v}.
   (b) Construct left cosets for subgroup $C_v = \{1, \sigma_3\}$ of C_{3v}.
   (c) Construct right cosets for subgroup $C_3 = \{1, r, r^2\}$ of C_{3v}.
   (d) Construct left cosets for subgroup $C_3 = \{1, r, r^2\}$ of C_{3v}.
   (e) Construct double cosets for subgroup $C_v = \{1, \sigma_1\}$ and $C' = \{1, \sigma_2\}$ of C_{3v}.
   (f) Construct double cosets for subgroup $C_v = \{1, \sigma_1\}$ and $C_3 = \{1, r, r^2\}$ of C_{3v}.

3.1.2 A subgroup $\mathcal{N}$ of $\mathcal{G}$ Is called a NORMAL or INVARIANT subgroup of group $\mathcal{G}$ if $g\mathcal{N}g^{-1} = \mathcal{N}$ for any g in $\mathcal{G}$.
   (a) If $C_v = \{1, \sigma_3\}$ a normal subgroup of C_{3v}?
   (b) Is $C_3 = \{1, r, r^2\}$ a normal subgroup of C_{3v}?
   (c) Prove that if the left and right cosets of a subgroup $\mathcal{H}$ are equal, then $\mathcal{H}$ is a normal subgroup. Check this using the results of Problem 3.1.1.

Section 3.2

3.2.1 Use Hamilton's turns or nomograms to derive a multiplication table for each of the following symmetries which includes the $\pm$ signs that would be needed to describe rotations of electrons. Do any of the resulting tables give mutually commuting operators?
   (a) D_2.
   (b) D_3.
   (c) D_4.

3.2.2 Let $^\circ\mathcal{N}(g)$ be the number of elements of a group that commute with element g. Let $^\circ C_g$ be the order of the class of g in group $\mathcal{G}$. What is the product of these two numbers?

Evaluate these numbers for the groups D_2, D_3, and D_4.

Section 3.3

3.3.1 The four flat-bladed fan of arbitrary pitch has D_4 symmetry.
  (a) Name or label all the rotations in D_4 and tell what they do.
  (b) Draw a sketch of something with C_{4v} symmetry and relate C_{4v} symmetry to D_4.
  (c) Construct a Hamilton-turn nomogram for D_4. Could it be used for C_{4v} as well?
  (d) Construct a group multiplication table for D_4. Arrange it according to *classes*.
  (e) Construct a class algebra table using part (d).
  (f) Spectrally decompose the class algebra. Find $\mathbf{P}^\alpha = \Sigma d_j^\alpha c_j$ and $c_j = \Sigma c_j^\alpha \mathbf{P}^\alpha$.
  (g) Split the all-commuting idempotents found in part (f) into irreducible idempotents of the D_4 algebra with the help of the idempotents from the subgroup $D_2 = \{1, \mathbf{R}_x(180°), \mathbf{R}_y(180°), \mathbf{R}_z(180°)\}$.
  (h) Split the all-commuting idempotents found in part (f) into irreducible idempotents of the D_4 algebra with the help of the idempotents from the subgroup $C_4 = \{1, R_z(90°), R_z(180°), R_z(270°)\}$.

3.3.2 The quaternion group $Q = \{1, \underline{1}, i, \underline{i}, j, \underline{j}, k, \underline{k}\}$ is defined by Hamilton's hypercomplex relations: $ij = k$, $ji = \underline{k}$ (and cyclically); $ii = jj = kk = \underline{ii} = \underline{jj} = \underline{kk} = 1$; $i\underline{1} = \underline{i} = \underline{1}i$ (and cyclically); $\underline{1}\,\underline{1} = 1$; and 1 is identity.
  (a) Using Hamilton's turns show that this corresponds to the spin-$\frac{1}{2}$ version of the D_2 group with $\mathbf{i} = \mathbf{R}_x$, $\underline{\mathbf{i}} = -\mathbf{R}_x$, $\mathbf{j} = \mathbf{R}_y$, ..., etc.
  (b) Redo parts (d), (e), and (f) in Problem 3.3.1 for the quaternion group.
  (c) Use the subgroup $\{1, k, \underline{1}, \underline{k}\}$ in Q to split any all-commuting idempotents in the Q algebra that might still need splitting.

Section 3.4

3.4.1 The D_4 symmetric spring-mass model consists of four mass-m particles held by diagonal, external, and side springs of constants d, e, and s, respectively, as shown in the diagram.

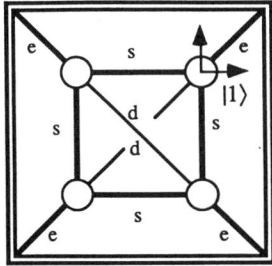

(a) Label the eight x-y-plane coordinates using group operators "named" in Problem 3.3.1.
(b) Write the first row ($\langle 1|a|1\rangle \cdots \langle 1|a|8\rangle \cdots$) of the acceleration matrix in terms of the mass and spring constants.
(c) Use the irreducible $D_2 \subset D_4$ defined idempotents of Problem 3.3.1 to obtain the irreducible operators $\mathbf{P}_i^\alpha \mathbf{g} \mathbf{P}_j^\alpha$. Find all the linearly independent vectors $\mathbf{P}_i^\alpha |1\rangle$ and $\mathbf{P}_i^\alpha \mathbf{g} \mathbf{P}_j^\alpha |1\rangle$ for the system. Draw pictures of the normalized displacements or distortions which they represent. Use these results to find a complete set of irreducible projection operators $\mathbf{P}_{ij}^\alpha$.
(d) Write a table of all the $D_2 \subset D_4$ defined irreducible representations D^α. [Use part (c) of this problem if necessary.]
(e) Derive a table of all the $C_4 \subset D_4$ defined irreducible representations D^α.
(f) Obtain the block-diagram reduced forms of the acceleration matrix $\langle a \rangle$ using symmetry projectors from parts (c) or (d).
(g) Solve any residual block matrices obtained in part (f). Sketch the vibrational normal mode distortions and corresponding frequency levels for constants $e = 0$, $s = 5$, $d = 3$, $m = 1$.
(h) Draw pictures which differ from part (g) of the normal modes for $e = 1$, $s = 5$, $d = 3$, $m = 1$.

3.4.2 Obtain a set of irreducible representations for the quaternion group, following results you obtained in Problem 3.3.2. Is Q isomorphic to D_4?

Section 3.5

3.5.1 Given an n-dimensional class algebra ($\mathbf{c}_i \mathbf{c}_j = \mathbf{c}_j \mathbf{c}_i = \sum_{k=1}^n C_{ij}^k \mathbf{c}_k$) one may construct a "regular" representation; $R(\mathbf{c}_1)R(\mathbf{c}_j) = R(\mathbf{c}_i \mathbf{c}_j)$ of the *algebra* using the structure constants C_{ij}^k. (This is *not* to be confused with *group* regular representation.)

(a) Show how to do this. Make and test such a 3 × 3 representation for C_{3v} classes.

(b) The eigenvalues $(c_j^\alpha, c_j^\beta, c_j^\gamma, \ldots)$ of each matrix $R(c_j)$ can be used to derive characters $(\chi_j^\alpha, \chi_j^\beta, \ldots)$. Give formulas for χ_j^α in terms of c_j^α (and other quantities) and test them using C_{3v}.

3.5.2 The irreducible representation-orthogonality relation is

$$\sum_{\text{all } g \in \mathscr{G}} \mathscr{D}_{ij}^\alpha(g) \mathscr{D}_{kl}^\beta(g) = {}^\circ\mathscr{G}/l^\alpha \delta^{\alpha\beta} \delta_{ik} \delta_{jl}.$$

(a) Prove this by considering $\mathscr{D}_{ij}^\alpha(P_{kl}^\beta) = ?$

(b) Discuss the form of the completeness relation that would compliment this orthogonality relation, and derive it, too.

(c) Derive an irreducible character completeness relation. $\sum_\alpha \chi_j^{\alpha*} \chi_k^\alpha = ?$

(d) Derive an irreducible character orthogonality relation. $\sum_j {}^\circ C_j \chi_j^{\alpha*} \chi_j^\beta = ?$ Here ${}^\circ C_j$ is the order of the (j)th class.

3.5.3 (Schur's lemmas)
Suppose an $l^\alpha \times l^\alpha$ matrix $\mathscr{L}$ commutes with all matrices $\{\mathscr{D}^\alpha(g) \cdots \mathscr{D}^\alpha(g') \cdots \}$, where $\mathscr{D}^\alpha$ is an irreducible representation; i.e., $\mathscr{L}\mathscr{D}^\alpha(g) = \mathscr{D}^\alpha(g)\mathscr{L}$ for all **g**.

(a) What special properties must $\mathscr{L}$ have? [Hint: Consider $\mathscr{D}^\alpha(P_{ij}^\alpha)$.]

(b) What special properties should an $l^\alpha \times l^\beta$ matrix $\mathscr{M}$ have in order to satisfy $\mathscr{M}\mathscr{D}^\beta(g) = \mathscr{D}^\alpha(g)\mathscr{M}$ for all g, where α and β label inequivalent irreducible representations. [Consider $D^\alpha(P^\beta)$.]

3.5.4 An element g is called a COMMUTATOR element of a group $\mathscr{G}$ if it can be written $\mathbf{g} = \mathbf{a}^{-1}\mathbf{b}^{-1}\mathbf{ab}$ for some **a** and **b** in $\mathscr{G}$. The COMMUTATOR SUBGROUP ($\mathscr{H}_c$) of $\mathscr{G}$ is the smallest subgroup which contains all the commutator elements of $\mathscr{G}$.

(a) If **g** is a commutator, is $\mathbf{g}^{-1}$ also one? Is **1** one?

(b) If **g** is a commutator, is every member of its class one, too?

(c) prove that class sum $\mathbf{c}_j$ contains commutator elements if and only if there is at least one other class sum $\mathbf{c}_k$ such that $\mathbf{c}_j \mathbf{c}_k = \cdots + n\mathbf{c}_k + \cdots$, where $n \neq 0$.

(d) Find the commutator subgroup of C_{3v}.

3.5.5 A representation R is said to be FAITHFUL to a group if $R(\mathbf{g}) = R(\mathbf{g}')$ for all **g** and **g'** that are group elements. A representation R is said to be FAITHFUL to a GROUP ALGEBRA if $R(\mathbf{x}) = R(\mathbf{y})$ implies $\mathbf{x} = \mathbf{y}$ for all possible linear combinations x and y of group elements.

Note: In the second case we are talking about the whole group algebra and require that $R(a) = \mathbf{0}$ if and only if $\mathbf{a} = \mathbf{0}$.

(a) What is the lowest possible dimension of a representation that is faithful to the group C_{3v}? To the group C_{4v}?

(b) What is the lowest possible dimension of a representation that is faithful to the algebra of group C_{3v}? To the algebra of group C_{4v}?

(c) What is the maximum number of mutually commuting group elements in C_{3v}? In group C_{4v}?

(d) What is the maximum number of mutually commuting operators in the group algebra of C_{3v}? In the group algebra of C_{4v}?

(e) Are either of the answers to (c) or (d) related to those of either (a) or (b)? Which and why?

3.5.6 Prove the order $°\mathcal{G}$ of a group must be evenly divisible by each of the following numbers. [Part (c) is the hardest to prove.]

(a) The order $°\mathcal{H}$ of any of its subgroups.

(b) The order $°C_g$ of any of its classes.

(c) The dimension l^α of any of its irreducible representations.

CHAPTER 4

THEORY AND APPLICATIONS OF HIGHER FINITE SYMMETRY AND INDUCED REPRESENTATIONS

All but five of the 32 crystal point symmetries (Recall Figure 3.1.1) have been analyzed in terms of cyclic groups C_2, C_3, and C_4 and dihedral groups D_2, D_3, and D_4. The remaining five symmetries T, T_h, T_d, O, and O_h are called tetrahedral, cubic, or octahedral symmetries. To analyze these one needs only to concentrate on two of them: the octahedral group O and its tetrahedral subgroup T. It will be shown that the group T_h is simply $T_h = T \times C_i$, O_h is simply $O \times C_i$, and T_d is isomorphic to O.

The derivation and application of octahedral irreps can be done using the P-operator techniques discussed in the preceding chapters. However, there are some additional and important relations which help simplify the analysis of high symmetry. In this chapter the relations between subgroups C_n and D_n of octahedral symmetry will be exploited in the reduction of octahedral representations. The theory of induced representations will be introduced. This theory has become very important in understanding high-resolution laser spectra of symmetric molecules, among other things. Examples which will be treated in this section include the elements of spectral cluster theory and the classical vibrational spectra of SF_6 and UF_6 molecules. An introduction to the quantum theory of molecular vibrations is given also.

4.1 OCTAHEDRAL SYMMETRIES AND THEIR CHARACTERS

Let us begin with a synopsis of the point symmetries that may be found in a cubic crystal. They are also fairly common symmetries for polyatomic molecules.

228 HIGHER FINITE SYMMETRY AND INDUCED REPRESENTATION

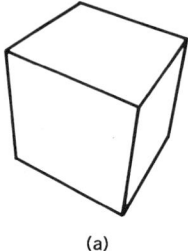
(a)

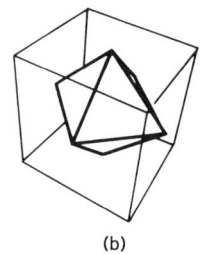
(b)

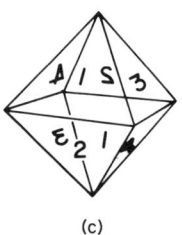
(c)

Figure 4.1.1 Objects having octahedral (O) symmetry. (a) The cube or hexahedron. (b) The octahedron. The cube is transformed into the octahedron by placing vertices of one in the center of the faces of the other. (c) (4! = 24) permutations of four integers correspond to the 24 equivalent positions in which the octahedron may be placed.

A. The Octahedral Group (O)

The rotational symmetry of a regular cube or hexahedron in Figure 4.1.1(a) or of a regular octahedron in Figures 4.1.1(b) and 4.1.1(c) is called O or octahedral symmetry. The symmetry of the octahedron can be stated by naming all its equivalent orientational positions, or by naming all the rotational operations which can change one position into another. For simple counting purposes the former is probably easier. The octahedron has eight equivalent faces to show. Since each one can be rotated to three equivalent positions there must be $8 \cdot 3 = 24$ different rotational positions in all. (Equivalently, there are six cube faces or octahedron corners with four rotations for each, or 12 edges times two rotations each, all products which give the number 24.) It is interesting to note that each rotation corresponds to one of the $24 = 4!$ permutations of the digits $(1, 2, 3, 4)$ as depicted in Figure 4.1.1(c).

The labeling of the 24 symmetry operators is a little more difficult, but easy enough if done one class at a time. Figure 4.1.2 labels the operators from the class of (a) the identity 1, (b) the 120° rotations, r_j and r_j^2, (c) the 180° rotations R_j^2 around χ_1, χ_2, χ_3 axes, (d) the 90° rotations, R_j and R_j^3, and (e) the 180° rotations i_j around edges. Each symbol next to an axis labels a counterclockwise rotation of the octahedron by a specific angle associated with its class. Axes are all meant to be fixed in the space of the laboratory rather than the body of the octahedron.

The O-group multiplication may be found easily using the Hamilton arc vectors described in Section 3.1.B. The octahedral arcs are shown for different classes in Figure 4.1.3(a). A stereogram of the superimposed arc paths is given below that in Figure 4.1.3(b). A projected view of hemisphere looking down the x_3 or z axis is shown in Figure 4.1.4. Each of the 24 rotations is labeled on one or more arc vectors. For now you should ignore

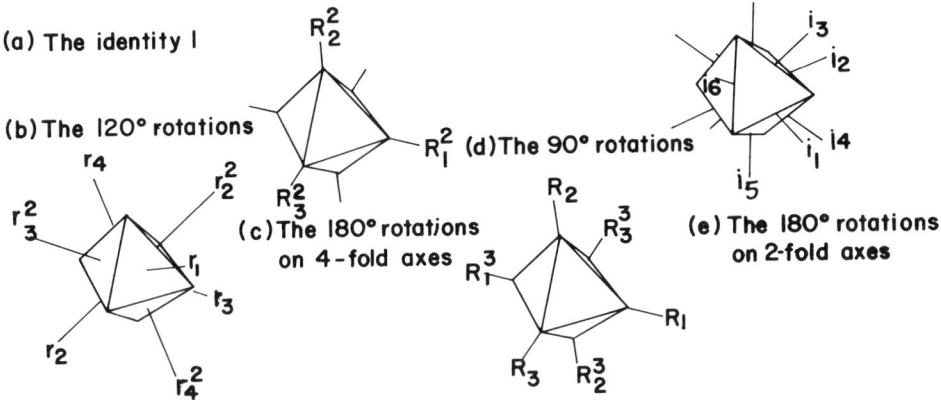

Figure 4.1.2 The five classes of octahedral operations. (a) The identity class (no rotation). (b) The threefold rotations (120°). (c) The tetragonal twofold rotations (180°). (d) The fourfold rotations (90°). (e) The diagonal twofold rotations (180°).

the minus signs next to some labels. These will be used in Chapter 5 to treat spin-$\frac{1}{2}$ rotations.

All $(24)^2 = 576$ octahedral products $R_b R_a = R_{ba}$ can be read from the diagram in Figure 4.1.4. The head-to-tail addition of the a vector to the b vector yield the ba vector, as indicated by the triangle above the diagram. Some products are already set up, as are, for example, $r_3 R_2 = R_1$ or $R_1^2 i_4 = R_3$. Other products require adjustment of one or two vectors so the head of the a vector meets the tail of the b vector. The resulting octahedral multiplication table is given in the table section.

Relatively few multiplications are needed to construct the algebra of the octahedral classes:

$$c_1 = 1,$$
$$c_r = r_1 + r_2 + r_3 + r_4 + r_1^2 + r_2^2 + r_3^2 + r_4^2,$$
$$c_{R^2} = R_1^2 + R_2^2 + R_3^2,$$
$$c_R = R_1 + R_2 + R_3 + R_1^3 + R_2^3 + R_3^3,$$
$$c_i = i_1 + i_2 + i_3 + i_4 + i_5 + i_6. \qquad (4.1.1)$$

For example, only three multiplications are needed to determine the class membership of the 18 elements in the product

$$\begin{aligned} c_{R^2} c_i &= R_1^2 i_1 + \cdots = R_2 + \cdots \\ &\quad + R_2^2 i_1 + \cdots \quad i_2 + \cdots \\ &\quad + R_3^2 i_1 + \cdots \quad R_2^3 + \cdots . \end{aligned} \qquad (4.1.2)$$

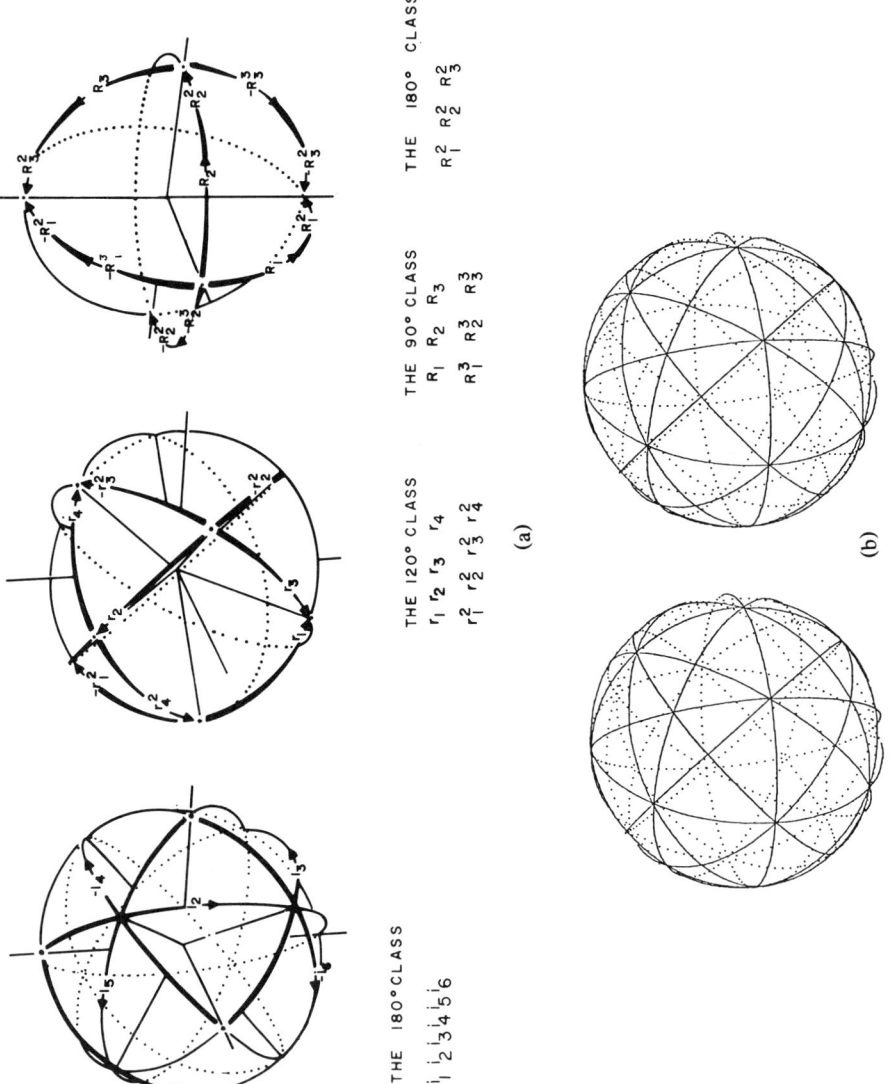

Figure 4.1.3 Hamilton arcs for octahedral rotations (a) Separate classes. (b) Stereogram of

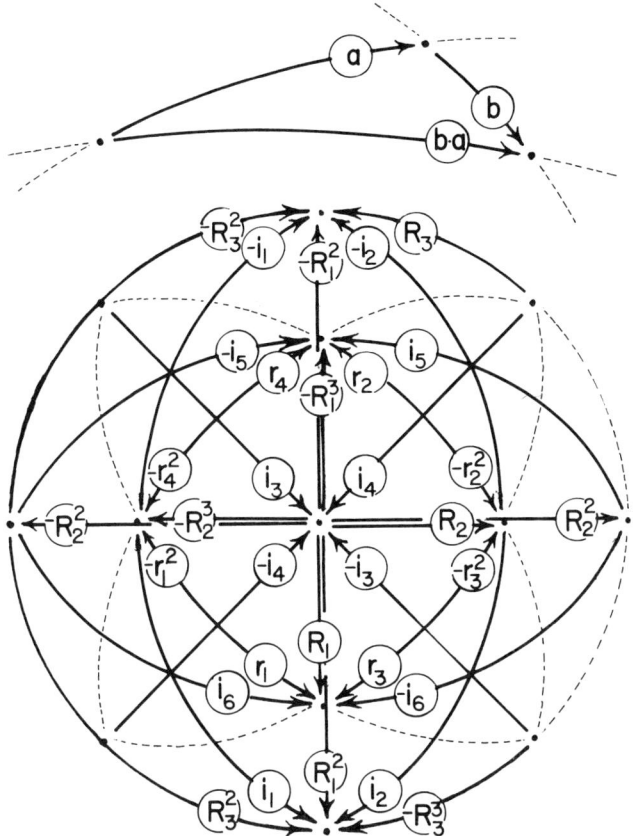

Figure 4.1.4 Octahedral group nomogram.

Since there are two members from c_R for each one from c_i in the first column of Eq. (4.1.2) one must conclude that the 18 products consist of 12 R_j's and 6 i_j's, or

$$c_{R^2}c_i = 2c_R + c_i = c_i c_{R^2}. \tag{4.1.3}$$

This simplification of any class product,

$$\begin{aligned}
c_g c_h &= g_1 h_1 + g_2 h_1 + \cdots = g_1 h_1 + tg_1 h_1 t^{-1} + \cdots \\
&\quad + g_1 h_2 + g_2 h_2 + \cdots \quad g_1 h_2 + tg_1 h_2 t^{-1} + \cdots \\
&\quad + g_1 h_3 + g_2 h_3 + \cdots \quad g_1 h_3 + tg_1 h_3 t^{-1} + \cdots \\
&\quad + \cdots \quad \cdots \quad \cdots \quad \cdots,
\end{aligned} \tag{4.1.4}$$

is always possible since the second column (or row) of the product is just a transformation,

$$tg_1 h_j t^{-1} = tg_1 t^{-1} th_j t^{-1}$$
$$= g_2 \quad h'_j, \quad (4.1.5)$$

of the first one. According to the definition of classes (see Section 3.2.A) the transformation converting g_1 to g_2 only rearranges the h_j's in class $c_h = tc_h t^{-1}$. Hence the proportion of a given class found in each column (or row) of Eq. (4.1.4) must be equal.

The complete set of octahedral class products $c_g c_h = c_h c_g$ are given by the following class algebra table:

1	c_r	c_{R^2}	c_R	c_i
c_r	$81 + 4c_r + 8c_{R^2}$	$3c_r$	$4c_R + 4c_i$	$4c_R + 4c_i$
c_R		$31 + 2c_{R^2}$	$c_R + 2c_i$	$2c_R + c_i$
c_R			$61 + 3c_r + 2c_{R^2}$	$3c_r + 4c_{R^2}$
c_i				$61 + 3c_r + 2c_{R^2}$

(4.1.6)

This class algebra can have no more than five linearly independent powers of any element. So it must be possible to construct a minimal equation for c_i by combining the six powers $\{c_i^0, c_i, c_i^2, c_i^3, c_i^4, c_i^5\}$:

$$c_i^0 = 1, \qquad c_i^3 = 16c_R + 20c_i,$$
$$c_i^1 = c_i, \qquad c_i^4 = 48c_r + 64c_{R^2} + 20c_i^2,$$
$$c_i^2 = 61 + 3c_r + 2c_{R^2}, \qquad c_i^4 = 1201 + 108c_r + 104c_{R^2}, \quad (4.1.7)$$
$$c_i^5 = 320c_R + 20c_i^4 + 256c_i$$
$$= 2400\,1 + 2160c_r + 2080c_{R^2} + 320c_R + 256c_i.$$

Combining the expressions for c_i^3 and c_i^5 gives a minimal equation

$$c_i^5 - 40c_i^3 + 144c_i = 0,$$
$$(c_i^2 - 36)(c_i^2 - 4)c_i = 0,$$
$$(c_i + 6)(c_i - 6)(c_i + 2)(c_i - 2)c_i = 0. \quad (4.1.8)$$

[Note: For typographical convenience we write $(c_i + r\mathbf{1})$ as $(c_i + r)$.] The roots $\{2, -2, 0, 6, -6\}$ correspond to cubic irreps which are labeled $(\alpha) =$

$\{T_2, T_1, E, A_1, A_2\}$, respectively. The Equations (1.2.12) or (1.2.15) from Chapter 1 for idempotents then gives the following for the root $c^{T2} = 2$:

$$\mathbb{P}^{T_2} = \frac{(c_i + 2)(c_i - 0)(c_i - 6)(c_i + 6)}{(2 + 2)(2 - 0)(2-6)(2 + 6)} = [(c_i + 2)c_i(c_i^2 - 36)]/(-256)$$
$$= [c_i^4 + 2c_i^3 - 36c_i^2 - 72c_i]/(-256)$$
$$= [31 - c_{R^2} - c_R + c_i]/8. \quad (4.1.9)$$

The complete set of all-commuting idempotents is given now:

$$\mathbb{P}^{A_1} = [1 + c_r + c_{R^2} + c_R + c_i]/24,$$
$$\mathbb{P}^{A_2} = [1 + c_r + c_{R^2} - c_R - c_i]/24,$$
$$\mathbb{P}^E = [21 - c_r + 2c_{R^2}]/12,$$
$$\mathbb{P}^{T_1} = [31 \quad -c_{R^2} + c_R - c_i]/8,$$
$$\mathbb{P}^{T_2} = [31 \quad -c_{R^2} - c_R + c_i]/8. \quad (4.1.10)$$

The coefficients in the brackets are the octahedral irrep characters χ_j^α according to the theory of Section (3.5.A). The character table is as follows:

		$g = 1$	r, r^2	R^2	R, R^3	i
$\Gamma_1 =$	$\chi_g^{A_1} =$	1	1	1	1	1
$\Gamma_2 =$	$\chi_g^{A_2} =$	1	1	1	-1	-1
$\Gamma_3 =$	$\chi_g^E =$	2	-1	2	0	0
$\Gamma_4 =$	$\chi_g^{F_1} = \chi_g^{T_1} =$	3	0	-1	1	-1
$\Gamma_5 =$	$\chi_g^{F_2} = \chi_g^{T_2} =$	3	0	-1	-1	1

$$(4.1.11)$$

This book will use the notation A, E, and T for the single ($\chi_1^A \equiv l^A = 1$), double ($\chi_1^E \equiv l^E = 2$), and triple ($\chi_1^T \equiv l^T = 3$) dimensions of the respective irreps. The subscripts 1 and 2 indicate the even and odd character of 90° rotations R or R^3:

$$\chi_R^{A_1} = 1 = \chi_R^{T_1},$$
$$\chi_R^{A_2} = -1 = \chi_R^{T_2}. \quad (4.1.12)$$

Some alternative notation which occurs in the reference literature is indicated in Eq. (4.1.11).

B. Full Octahedral Group (O_h)

The full octahedral group contains all the operations of $O = \{1, r_1, r_2, \ldots, R_1, R_2, \ldots, i_1, i_2, \ldots\}$ twice; once with inversion and once without. Since the inversion operator (I) commutes with all rotation operators, one may write O_h as an outer product:

$$O_h = O \times C_i = O \times \{1, I\}. \tag{4.1.13}$$

The 48 elements of O_h are listed in Figure 4.1.5. O_h includes subgroups T, T_h, and T_d as well as O. One way to account for the 48 operations is to consider all transformations of the orthogonal Cartesian unit vectors $\{|x\rangle \equiv$

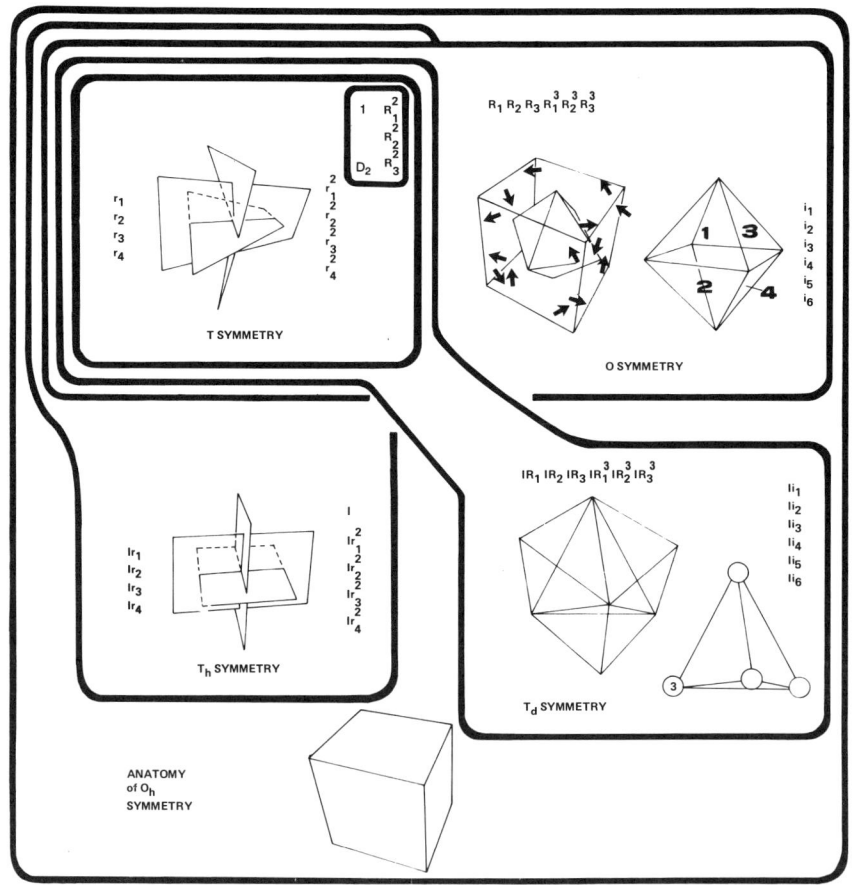

Figure 4.1.5 The full octahedral group (O_h) and four non-Abelian subgroups T, T_h, T_d, and O. The Abelian D_2 subgroup of T is indicated also.

$|x_1\rangle, |y\rangle \equiv |x_2\rangle, |z\rangle \equiv |x_3\rangle\}$ which leave them pointing along Cartesian axes. This includes all $3! = 6$ permutations. For example, the $120°$ rotation r_1 gives

$$\{r_1|x_1\rangle = |x_2\rangle, r_1|x_2\rangle = |x_3\rangle, r_1|x_3\rangle = |x_1\rangle\}. \quad (4.1.14)$$

O_h also includes $2^3 = 8$ possible sign changes. For example, inversion I gives

$$\{I|x_1\rangle = -|x_1\rangle, I|x_2\rangle = -|x_2\rangle, I|x_3\rangle = -|x_3\rangle\}. \quad (4.1.15)$$

Altogether, O_h contains $3! \, 2^3 = 48$ such operations. The geometrical significance of the operations will be discussed in the explanation of the tetrahedral subgroups and then again in Sections 4.2.A(a) and 4.2.A(b).

The full octahedral or O_h characters follow easily from those of O just as D_{4h} characters follow from those of D_4. The cross-product relation (4.1.13) is the key:

	$g=1$	$r_1 \cdots$	$R_1^2 \cdots$	$R_1 \cdots$	$i_1 \cdots$	I	$Ir_1 \cdots$	$IR_1^2 \cdots$	$IR_1 \cdots$	$Ii_1 \cdots$
$(\alpha) = A_{1g}$	1	1	1	1	1	1	1	1	1	1
A_{2g}	1	1	1	-1	-1	1	1	1	-1	-1
E_g	2	-1	2	0	0	2	-1	2	0	0
T_{1g}	3	0	-1	1	-1	3	0	-1	1	-1
T_{2g}	3	0	-1	-1	1	3	0	-1	-1	1
A_{1u}	1	1	1	1	1	-1	-1	-1	-1	-1
A_{2u}	1	1	1	-1	-1	-1	-1	-1	1	1
E_u	2	-1	2	0	0	-2	1	-2	0	0
T_{1u}	3	0	-1	1	-1	-3	0	1	-1	1
T_{2u}	3	0	-1	-1	1	-3	0	1	1	-1

$$(4.1.16)$$

The g and u subindices stand for even (positive) and odd (negative) inversion parity, respectively. The following gives the form of the simple relation between O_h and O characters.

$$\chi_R^{T_{2g}} = \chi_R^{T_2}, \quad \chi_{IR}^{T_{2g}} = \chi_R^{T_2},$$

$$\chi_R^{T_{2u}} = \chi_R^{T_2}, \quad \xi_{IR}^{T_{2u}} = -\chi_R^{T_2}. \quad (4.1.17)$$

The same relation will hold between O_h and O irreps:

$$\mathscr{D}^{\alpha_g}(R) = \mathscr{D}^{\alpha}(R), \quad \mathscr{D}^{\alpha_g}(IR) = \mathscr{D}^{\alpha}(R),$$

$$\mathscr{D}^{\alpha_u}(R) = \mathscr{D}^{\alpha}(R), \quad \mathscr{D}^{\alpha_u}(IR) = -\mathscr{D}^{\alpha}(R). \quad (4.1.18)$$

C. Full Tetrahedral Symmetry T_d

The full symmetry of a regular tetrahedron is called T_d. It contains 24 operators $\{1, r_j \ldots, r_j^2 \ldots, R_j^2 \ldots, IR_j \ldots, Ii_j \ldots\}$ listed in the T_d section of Figure 4.1.5. Note that the tetrahedron edges are face diagonals of a cube. The tetrahedron is invariant to reflections Ii_j through diagonal planes, but not the horizontal reflections IR_j^2. The tetrahedron is also invariant to 90° rotation inversions IR_j or IR_j^3 around the Cartesian axes. Hence, T_d contains D_{2d} subgroups. Also T_d contains the 120° rotations r_i or r_i^2 around (111) axes and 180° rotations R_i^2 around the Cartesian axes.

The groups O and T_d are isomorphic. For each group product $R_1 R_2 = r_1$ or $r_1 i_1 = R_1^3$ in O, there is a corresponding product $IR_1 IR_2 = r_1$ or $r_1 Ii_1 = IR_1^3$ in T_d, respectively, since inversion I is all-commuting. Hence, the character table (4.1.11) serves as well for T_d if we relabel elements R_j and i_j as IR_j and Ii_j. Furthermore, by numbering the tetrahedron vertices as shown in Figure 4.1.5, the correspondence between T_d operations and the 4! permutations of four integers is established.

D. Partial Tetrahedral Symmetries T and T_h

The purely rotational subgroup of T_d is called T, and it is the symmetry of the three twisted planes drawn in Figure 4.1.5. If the planes are flat and orthogonal the symmetry doubles to T_h shown in the same figure. T_h contains horizontal plane reflections IR_j^2 as well as 120° rotation inversions Ir_j and Ir_j^2. T_h is an outer product of T and C_i:

$$T_h = T \times C_i = T \times \{1, I\}. \qquad (4.1.19)$$

T and T_h are the only non-Abelian crystal point groups that have any classes of elements separated from their inverses. The counterclockwise 120° rotations $r_1 r_2 r_3 r_4$ are now in a separate class from their inverses $r_1^2 r_2^2 r_3^2 r_4^2$. No operation t in T or T_h exists such that $t^{-1} r_i t = r_i^2$. This is one case where some "look-alike" elements do not belong to the same class.

Group T has four classes c_1, c_r, c_{r^2}, and c_{R^2}; hence, it has four types of irreducible representations. The derivation of the all-commuting idempotents and the irrep character table (4.1.20) is left as an exercise.

	$g = 1$	$r_1 \cdots$	$r_1^2 \cdots$	$R_1^2 \cdots$	
$\chi_g^A =$	1	1	1	1	
$\chi_g^\varepsilon =$	1	$e^{2\pi i/3}$	$e^{-2\pi i/3}$	1	(4.1.20)
$\chi_g^\varepsilon =$	1	$e^{-2\pi i/3}$	$e^{2\pi i/3}$	1	
$\chi_g^T =$	3	0	0	-1	

The T_h characters follow easily using Eq. (4.1.19).

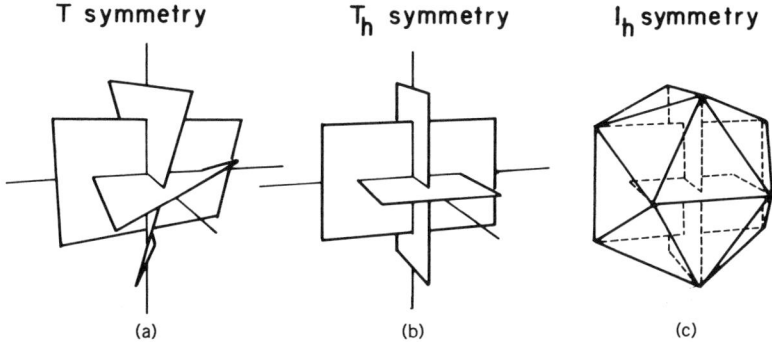

Figure 4.1.6 Examples of icosahedral (I_h) symmetry and tetrahedral (T_h) symmetry.

It should be evident from Figure 4.1.5 that T_h is not the symmetry of a tetrahedron. Instead, it is a subgroup of the highest point symmetry in three space. If the rectangles in the T_h portion of Figure 4.1.5 have a length:width "golden" ratio of

$$r = (1 + \sqrt{5})/2 = 1.618\ldots,$$

then their vertices describe an ICOSAHEDRON. This 20-sided regular solid is shown in Figure 4.1.6. Icosahedral symmetry is not a crystal point group since it has fivefold symmetry axes. The importance of this symmetry in physics has been realized only recently with the discovery of the structure of viruses, quasi-crystals, and the molecule C_{60} which has the geodesic dome structure made famous by the architect Buckminster Fuller. C_{60} is called "buckminsterfullerene" in honor of Fuller's artistry.

4.2 IRREDUCIBLE REPRESENTATIONS OF OCTAHEDRAL SYMMETRY

The construction and labeling of octahedral symmetry representations will be treated now. By considering carefully the subgroups of O and O_h it is possible to simplify some of the octahedral group algebra.

A. Subgroup Chains and Idempotent Splitting

The dimensions (l^α) of the O irreps $\mathscr{D}^\alpha$ are determined by the characters

$$l^\alpha = \chi_1^\alpha = \text{Trace } \mathscr{D}^\alpha(1)$$

238 HIGHER FINITE SYMMETRY AND INDUCED REPRESENTATION

in Eqs. (4.1.10) and (4.10.11). Each all-commuting idempotent $\mathbb{P}^\alpha$ can be split into l^α irreducible idempotents $P_j^\alpha \equiv P_{jj}^\alpha$,

$$\mathbb{P}^\alpha = P_1^\alpha + P_2^\alpha + \cdots + P_{l^\alpha}^\alpha, \tag{4.2.1}$$

according to the theory of Section 3.3. In the group O we expect the following splittings:

$$\mathbb{P}^E = P_1^E + P_2^E, \tag{4.2.1a}_x$$

$$\mathbb{P}^{T_1} = P_1^{T_1} + P_2^{T_1} + P_3^{T_1}, \tag{4.2.1b}_x$$

$$\mathbb{P}^{T_2} = P_1^{T_2} + P_2^{T_2} + P_3^{T_2}, \tag{4.2.1c}_x$$

while P^{A_1} and P^{A_2} remain unsplit since $l^{A_1} = 1 = l^{A_2}$.

One way to do this splitting is to use subgroup idempotents. Recall that we used the C_2 subgroup idempotents to split the C_{3v} (or D_3) all-commuting idempotent P^E [recall Eq. (3.3.3)]:

$$\mathbb{P}^E = \mathbb{P}^E \mathbf{1} = \mathbb{P}^E(P^+ + P^-)$$
$$= P_1^E + P_2^E. \tag{4.2.2}$$

However, later on a different splitting was effected by C_3 subgroup idempotents [recall Eq. (3.3.15)]:

$$\mathbb{P}^E = \mathbb{P}^E \mathbf{1} = \mathbb{P}^E(P_{0_3} + P_{1_3} + P_{2_3})$$
$$= P_{1_3}^E + P_{2_3}^E. \tag{4.2.3}$$

This choice led to circular or moving-wave eigenstates.

Octahedral symmetry has more subgroups and correspondingly more different types of idempotent splittings and wave solutions. Besides D_3, one may notice that D_4 and T are subgroups of O according to Figure 3.1.1. If one chooses to split O with D_4 idempotents then two subchoices remain. One is free to use $D_4 \supset C_4$ defined idempotents [recall Eq. (3.6.4)],

$$\mathbb{P}^E = \mathbb{P}^E \mathbf{1} = P_{1_4}^E + P_{3_4}^E, \tag{4.2.4}$$

or else $D_4 \supset D_2 \supset C_2$ defined idempotents [recall Eqs. (3.6.6)–(3.6.8)],

$$\mathbb{P}^E = \mathbb{P}^E \mathbf{1} = P_{B_1}^E + P_{B_2}^E. \tag{4.2.5}$$

Each choice for a chain of subgroups such as $O \supset D_4 \supset D_2$, $O \supset D_4 \supset C_4$, $O \supset D_3 \supset C_2$, or $O \supset D_3 \supset C_3$ corresponds to different idempotent splittings and different (but equivalent) irreps. Let us consider each choice in turn.

(a) Tetragonal Standing-Wave Irreps ($O_h \supset D_{4h} \supset D_{2h}$) Suppose we require O irreps which are diagonal with respect to subgroup $D_2 = \{1, R_1^2, R_2^2, R_3^2\}$, and reduced with respect to tetragonal subgroup $D_4 = \{1, R_3, R_3^2, R_3^3, R_1^2, R_2^2, i_3, i_4\}$. These irreps are most commonly used in solid-state applications. The formal derivation of them involves splitting by the combination of idempotent relations for D_4:

$$1 = P^E + P^{A_1} + P^{A_2} + P^{B_1} + P^{B_2} \quad (4.2.6)$$

and D_2:

$$1 = p^{A_1} + p^{A_2} + p^{B_1} + p^{B_2}. \quad (4.2.7)$$

Here it will be necessary to label D_2 idempotents with lower-case p; for example, let

$$p^{B_2} = (1 - R_3^2 + R_2^2 - R_1^2)/4,$$

where D_2 characters in Eq. (3.6.33) are used. This will distinguish them from D_4 idempotents with the same superscript, such as

$$P^{B_2} = (1 + R_3^2 - R_3 - R_3^3 - R_1^2 - R_2^2 + i_3 + i_4)/8.$$

[P^{B_2} follows from D_4 characters in Eq. (3.6.3).]

The splitting combination is

$$1 = (P^E + P^{A_1} + P^{A_2} + P^{B_1} + P^{B_2})(p^{A_1} + p^{A_2} + p^{B_1} + p^{B_2})$$

$$= (P^E p^{B_1} + P^E p^{B_2} + P^{A_1} p^{A_1} + P^{A_2} p^{A_2} + P^{B_1} p^{A_1} + P^{B_2} p^{A_2})$$

$$= P_1^E \quad + P_2^E \quad + P^{A_1} \quad + P^{A_2} \quad + P^{B_1} \quad + P^{B_2}. \quad (4.2.8)$$

This involves six irreducible D_4 idempotents. [Recall Eqs. (3.6.2) and (3.6.3).] Operating with this on the octahedral all-commuting idempotent $\mathbb{P}^{T_1}$ gives the desired three-term splitting,

$$\mathbb{P}^{T_1} = \mathbb{P}^{T_1} 1 = \mathbb{P}^{T_1}(P_1^E + P_2^E + P^{A_1} + P^{A_2} + P^{B_1} + P^{B_2})$$

$$= P_1^{T_1} + P_2^{T_1} + 0 + P_3^{T_1} + 0 + 0, \quad (4.2.9)$$

where some group multiplication algebra yields the irreducible idempotents:

$$P_1^{T_1} \equiv \mathbb{P}^{T_1}P_1^E = \mathbb{P}^{T_1}P^E p^{B_1} = \left(1 + R_1^2 - R_2^2 - R_3^2 + R_1 + R_1^3 - i_5 - i_6\right)/8, \tag{4.2.10a}$$

$$P_2^{T_1} \equiv \mathbb{P}^{T_1}P_2^E = \mathbb{P}^{T_1}P^E p^{B_2} = \left(1 - R_1^2 + R_2^2 - R_3^2 + R_2 + R_2^3 - i_1 - i_2\right)/8, \tag{4.2.10b}$$

$$P_3^{T_1} \equiv \mathbb{P}^{T_1}P^{A_2} = \mathbb{P}^{T_1}P^{A_2}p^{A_2} = \left(1 - R_1^2 - R_2^2 + R_3^2 + R_3 + R_3^3 - i_3 - i_4\right)/8. \tag{4.2.10c}$$

This formal idempotent production introduces the idea of SUBGROUP CHAIN LABELING of irrep idempotents and bases. Instead of blindly numbering them "one," "two," and "three," one uses the previously established irrep labels of the subgroup chain $D_4 \supset D_2$ to label the three T_1 components. Let us label the T_1 idempotents (4.2.10) as follows:

$$P_1^{T_1} \equiv P_{B_1}^{[T_1]\,E}, \qquad P_2^{T_1} \equiv P_{B_2}^{[T_1]\,E}, \qquad P_3^{T_1} \equiv P_{A_2}^{[T_1]\,A_2}, \tag{4.2.11}$$

where each link in the descending chain of labels stands for a corresponding subgroup irrep in the $O \supset D_4 \supset D_2$ chain.

$$P_m^{[\mu]} \equiv P_n^{[\mu]\,\nu} \quad \begin{matrix} [\mu] \text{ labels } & O \text{ irrep} \\ \nu \text{ labels } & D_4 \text{ irrep} \\ n \text{ labels } & D_2 \text{ irrep} \end{matrix} \tag{4.2.12}$$

This chain labeling tells exactly how the $\mathscr{D}^{T_1}$ irrep is reduced or diagonalized when restricted or subduced ($\downarrow$) to subgroups D_4 or D_2.

$$\mathscr{D}^{T_1} \downarrow D_4 = \begin{pmatrix} (\mathscr{D}^E) & & 0 \\ & & 0 \\ 0 & 0 & (D^{A_2}) \end{pmatrix}, \tag{4.2.13a}$$

$$\mathscr{D}^{T_1} \downarrow D_2 = \begin{pmatrix} (D^{B_1}) & 0 & 0 \\ 0 & (D^{B_2}) & 0 \\ 0 & 0 & (D^{A_2}) \end{pmatrix}. \tag{4.2.13b}$$

IRREDUCIBLE REPRESENTATIONS OF OCTAHEDRAL SYMMETRY

The irrep matrices $\mathscr{D}^{T_1}_{(g)}$ may be derived formally from the irreducible idempotents $P_j^{T_1}$ by constructing the guarded elements $P_j^{T_1} g P_j^{T_1}$, normalizing them, and using the elementary operator relation $P_i^{T_1} g P_j^{T_1} = \mathscr{D}^{T_1}_{ij}(g) P_{ij}^{T_1}$. This was done for C_{3v} in Sections 3.4.A and 3.4.B. Also, this particular T_1-type representation can also be derived from simple Cartesian vector properties, as we will see shortly. The $\mathscr{D}^{T_1}$ are listed below for all 24 O operations.

$$\mathscr{D}^{T_1}(1) = \begin{vmatrix} 1 & \cdot & \cdot \\ \cdot & 1 & \cdot \\ \cdot & \cdot & 1 \end{vmatrix} \quad R_1^2 = \begin{vmatrix} 1 & \cdot & \cdot \\ \cdot & -1 & \cdot \\ \cdot & \cdot & -1 \end{vmatrix} \quad r_1 = \begin{vmatrix} \cdot & \cdot & 1 \\ 1 & \cdot & \cdot \\ \cdot & 1 & \cdot \end{vmatrix} \quad r_2 = \begin{vmatrix} \cdot & \cdot & -1 \\ 1 & \cdot & \cdot \\ \cdot & -1 & \cdot \end{vmatrix} \quad r_1^2 = \begin{vmatrix} \cdot & 1 & \cdot \\ \cdot & \cdot & 1 \\ 1 & \cdot & \cdot \end{vmatrix} \quad r_2^2 = \begin{vmatrix} \cdot & 1 & \cdot \\ \cdot & \cdot & -1 \\ -1 & \cdot & \cdot \end{vmatrix}$$

$$\mathscr{D}^{T_1}(R_3^2) = \begin{vmatrix} -1 & \cdot & \cdot \\ \cdot & -1 & \cdot \\ \cdot & \cdot & 1 \end{vmatrix} \quad R_2^2 = \begin{vmatrix} -1 & \cdot & \cdot \\ \cdot & 1 & \cdot \\ \cdot & \cdot & -1 \end{vmatrix} \quad r_4 = \begin{vmatrix} \cdot & \cdot & 1 \\ -1 & \cdot & \cdot \\ \cdot & -1 & \cdot \end{vmatrix} \quad r_3 = \begin{vmatrix} \cdot & \cdot & -1 \\ -1 & \cdot & \cdot \\ \cdot & 1 & \cdot \end{vmatrix} \quad r_3^2 = \begin{vmatrix} \cdot & -1 & \cdot \\ \cdot & \cdot & 1 \\ -1 & \cdot & \cdot \end{vmatrix} \quad r_4^2 = \begin{vmatrix} \cdot & -1 & \cdot \\ \cdot & \cdot & -1 \\ 1 & \cdot & \cdot \end{vmatrix}$$

$$\mathscr{D}^{T_1}(R_3) = \begin{vmatrix} \cdot & -1 & \cdot \\ 1 & \cdot & \cdot \\ \cdot & \cdot & 1 \end{vmatrix} \quad i_4 = \begin{vmatrix} \cdot & -1 & \cdot \\ -1 & \cdot & \cdot \\ \cdot & \cdot & -1 \end{vmatrix} \quad i_1 = \begin{vmatrix} \cdot & \cdot & 1 \\ \cdot & -1 & \cdot \\ 1 & \cdot & \cdot \end{vmatrix} \quad i_2 = \begin{vmatrix} \cdot & \cdot & -1 \\ \cdot & -1 & \cdot \\ -1 & \cdot & \cdot \end{vmatrix} \quad R_1^3 = \begin{vmatrix} 1 & \cdot & \cdot \\ \cdot & \cdot & 1 \\ \cdot & -1 & \cdot \end{vmatrix} \quad R_1 = \begin{vmatrix} 1 & \cdot & \cdot \\ \cdot & \cdot & -1 \\ \cdot & 1 & \cdot \end{vmatrix}$$

$$\mathscr{D}^{T_1}(R_3^3) = \begin{vmatrix} \cdot & 1 & \cdot \\ -1 & \cdot & \cdot \\ \cdot & \cdot & 1 \end{vmatrix} \quad i_3 = \begin{vmatrix} \cdot & 1 & \cdot \\ 1 & \cdot & \cdot \\ \cdot & \cdot & -1 \end{vmatrix} \quad R_2 = \begin{vmatrix} \cdot & \cdot & 1 \\ \cdot & 1 & \cdot \\ -1 & \cdot & \cdot \end{vmatrix} \quad R_2^3 = \begin{vmatrix} \cdot & \cdot & -1 \\ \cdot & 1 & \cdot \\ 1 & \cdot & \cdot \end{vmatrix} \quad i_6 = \begin{vmatrix} -1 & \cdot & \cdot \\ \cdot & \cdot & 1 \\ \cdot & 1 & \cdot \end{vmatrix} \quad i_5 = \begin{vmatrix} -1 & \cdot & \cdot \\ \cdot & \cdot & -1 \\ \cdot & -1 & \cdot \end{vmatrix}.$$

(4.2.14)

The eight matrices in the first two columns belong to the subduced representation $\mathscr{D}^{T_1} \downarrow D_4$. Notice that they have the block-diagonal reduced form of Eq. (4.2.13a). The four matrices $\mathscr{D}^{T_1} \downarrow D_2 = \{ \cdots \mathscr{D}(R_j^2) \}$ are in the diagonal form of Eq. (4.2.13b).

The octahedral irrep $\mathscr{D}^{T_2}$ is very similar to $\mathscr{D}^{T_1}$. The character table (4.1.11) of O indicates that one can obtain $\mathscr{D}^{T_2}$ from $\mathscr{D}^{T_1}$ simply by a change of sign for the elements $\{R_1, R_2, \ldots, R_3^3, i_1, i_2, \ldots, i_6\}$ in the "second half" of O, i.e., for the coset $R_1 T$ of subgroup T:

$$\mathscr{D}^{T_2}(R_j) = -\mathscr{D}^{T_1}(R_j), \quad \mathscr{D}^{T_2}(i_j) = -\mathscr{D}^{T_1}(i_j). \quad (4.2.15)$$

This definition of $\mathscr{D}^{T_2}$ is quite convenient for many purposes. However, it changes the $\mathscr{D}^E$ representation of the elements R_3, R_3^3, i_3, and i_4 in subgroup D_4. Effectively, it introduces a phase change $\{|{}^E_1\rangle, |{}^E_2\rangle\} \to \{|{}^E_1\rangle, -|{}^E_2\rangle\}$ in the E basis. When it is necessary to avoid this we shall use the

242 HIGHER FINITE SYMMETRY AND INDUCED REPRESENTATION

following "kosher" $\mathscr{D}^{T_2}$ irrep:

$$\mathscr{D}^{T_2}(1) = \begin{vmatrix} 1 & \cdot & \cdot \\ \cdot & 1 & \cdot \\ \cdot & \cdot & 1 \end{vmatrix} \quad R_1^2 = \begin{vmatrix} 1 & \cdot & \cdot \\ \cdot & -1 & \cdot \\ \cdot & \cdot & -1 \end{vmatrix} \quad r_1 = \begin{vmatrix} \cdot & \cdot & 1 \\ \cdot & -1 & \cdot \\ -1 & \cdot & \cdot \end{vmatrix} \quad r_2 = \begin{vmatrix} \cdot & \cdot & -1 \\ \cdot & -1 & \cdot \\ \cdot & \cdot & 1 \end{vmatrix} \quad r_1^2 = \begin{vmatrix} \cdot & -1 & \cdot \\ \cdot & \cdot & -1 \\ 1 & \cdot & \cdot \end{vmatrix} \quad r_2^2 = \begin{vmatrix} \cdot & -1 & \cdot \\ \cdot & \cdot & 1 \\ -1 & \cdot & \cdot \end{vmatrix}$$

$$\mathscr{D}^{T_2}(R_3^2) = \begin{vmatrix} -1 & \cdot & \cdot \\ \cdot & -1 & \cdot \\ \cdot & \cdot & 1 \end{vmatrix} \quad R_2^2 = \begin{vmatrix} -1 & \cdot & \cdot \\ \cdot & 1 & \cdot \\ \cdot & \cdot & -1 \end{vmatrix} \quad r_4 = \begin{vmatrix} \cdot & \cdot & 1 \\ 1 & \cdot & \cdot \\ \cdot & 1 & \cdot \end{vmatrix} \quad r_3 = \begin{vmatrix} \cdot & \cdot & -1 \\ 1 & \cdot & \cdot \\ \cdot & -1 & \cdot \end{vmatrix} \quad r_3^2 = \begin{vmatrix} \cdot & 1 & \cdot \\ \cdot & \cdot & -1 \\ -1 & \cdot & \cdot \end{vmatrix} \quad r_4^2 = \begin{vmatrix} \cdot & 1 & \cdot \\ \cdot & \cdot & 1 \\ 1 & \cdot & \cdot \end{vmatrix}$$

$$\mathscr{D}^{T_2}(R_3) = \begin{vmatrix} \cdot & -1 & \cdot \\ 1 & \cdot & \cdot \\ \cdot & \cdot & -1 \end{vmatrix} \quad i_4 = \begin{vmatrix} \cdot & \cdot & -1 \\ \cdot & -1 & \cdot \\ -1 & \cdot & \cdot \end{vmatrix} \quad i_1 = \begin{vmatrix} \cdot & \cdot & -1 \\ \cdot & 1 & \cdot \\ 1 & \cdot & \cdot \end{vmatrix} \quad i_2 = \begin{vmatrix} \cdot & \cdot & 1 \\ \cdot & 1 & \cdot \\ -1 & \cdot & \cdot \end{vmatrix} \quad R_1^3 = \begin{vmatrix} -1 & \cdot & \cdot \\ \cdot & \cdot & 1 \\ \cdot & -1 & \cdot \end{vmatrix} \quad R_1 = \begin{vmatrix} -1 & \cdot & \cdot \\ \cdot & \cdot & -1 \\ \cdot & 1 & \cdot \end{vmatrix}$$

$$\mathscr{D}^{T_2}(R_3^3) = \begin{vmatrix} \cdot & 1 & \cdot \\ -1 & \cdot & \cdot \\ \cdot & \cdot & -1 \end{vmatrix} \quad i_3 = \begin{vmatrix} \cdot & 1 & \cdot \\ 1 & \cdot & \cdot \\ \cdot & \cdot & 1 \end{vmatrix} \quad R_2 = \begin{vmatrix} \cdot & \cdot & -1 \\ \cdot & -1 & \cdot \\ 1 & \cdot & \cdot \end{vmatrix} \quad R_2^3 = \begin{vmatrix} \cdot & \cdot & 1 \\ \cdot & -1 & \cdot \\ -1 & \cdot & \cdot \end{vmatrix} \quad i_6 = \begin{vmatrix} 1 & \cdot & \cdot \\ \cdot & \cdot & 1 \\ \cdot & 1 & \cdot \end{vmatrix} \quad i_5 = \begin{vmatrix} 1 & \cdot & \cdot \\ \cdot & \cdot & -1 \\ \cdot & -1 & \cdot \end{vmatrix}.$$

(4.2.16)

The $O \supset D_4 \supset D_2$ subgroup chain labeling of the T_2 bases is given by the following:

$$P_1^{T_2} \equiv P_{B_1}^{E\ [T_2]}, \quad P_2^{T_2} \equiv P_{B_2}^{E\ [T_2]}, \quad P_3^{T_2} \equiv P_{A_2}^{B_2\ [T_2]}. \quad (4.2.17)$$

The two-dimensional irrep $\mathscr{D}^E$ of O can be produced formally by the same D_4 splitting combination (4.2.8). The splitting of P^E goes as follows:

$$\mathbb{P}^E = \mathbb{P}^E 1 = \mathbb{P}^E \left(P_1^E + P_2^E + P^{A_1} + P^{A_2} + P^{B_1} + P^{B_2} \right)$$

$$= \quad 0 \quad + 0 \quad + P_1^E \quad + 0 \quad + P_2^E + 0, \quad (4.2.18a)$$

where

$$P_1^E \equiv \mathbb{P}^E P^{A_1} = \mathbb{P}^E P^{A_1} p^{A_1} = 3\left(P^{A_1} \text{ (of } D_4) - \tfrac{1}{2} r_1 P^{A_1} - \tfrac{1}{2} r_1^2 P^{A_1} \right)/4, \quad (4.2.18b)$$

$$P_2^E \equiv \mathbb{P}^E P^{B_1} = \mathbb{P}^E P^{B_1} p^{A_1} = 3\left(P^{B_1} \text{ (of } D_4) - \tfrac{1}{2} r_1 P^{B_1} - \tfrac{1}{2} r_1^2 P^{B_1} \right)/4. \quad (4.2.18c)$$

IRREDUCIBLE REPRESENTATIONS OF OCTAHEDRAL SYMMETRY

The following is a list of the $\mathscr{D}^E$ irreps of O symmetry:

$$\mathscr{D}^E(1) \quad R_1^2 = \quad r_1 = \quad r_2 = \quad r_1^2 = \quad r_2^2 =$$

$$\begin{vmatrix} 1 & 0 \\ 0 & 1 \end{vmatrix} \quad \begin{vmatrix} 1 & 0 \\ 0 & 1 \end{vmatrix} \quad \begin{vmatrix} \frac{-1}{2} & \frac{-\sqrt{3}}{2} \\ \frac{\sqrt{3}}{2} & \frac{-1}{2} \end{vmatrix} \quad \begin{vmatrix} \frac{-1}{2} & \frac{-\sqrt{3}}{2} \\ \frac{\sqrt{3}}{2} & \frac{-1}{2} \end{vmatrix} \quad \begin{vmatrix} \frac{-1}{2} & \frac{\sqrt{3}}{2} \\ \frac{-\sqrt{3}}{2} & \frac{-1}{2} \end{vmatrix} \quad \begin{vmatrix} \frac{-1}{2} & \frac{\sqrt{3}}{2} \\ \frac{-\sqrt{3}}{2} & \frac{-1}{2} \end{vmatrix}$$

$$\mathscr{D}^E(R_3^2) \quad R_2^2 = \quad r_4 = \quad r_3 = \quad r_3^2 = \quad r_4^2 =$$

$$\begin{vmatrix} 1 & 0 \\ 0 & 1 \end{vmatrix} \quad \begin{vmatrix} 1 & 0 \\ 0 & 1 \end{vmatrix} \quad \begin{vmatrix} \frac{-1}{2} & \frac{-\sqrt{3}}{2} \\ \frac{\sqrt{3}}{2} & \frac{-1}{2} \end{vmatrix} \quad \begin{vmatrix} \frac{-1}{2} & \frac{-\sqrt{3}}{2} \\ \frac{\sqrt{3}}{2} & \frac{-1}{2} \end{vmatrix} \quad \begin{vmatrix} \frac{-1}{2} & \frac{\sqrt{3}}{2} \\ \frac{-\sqrt{3}}{2} & \frac{-1}{2} \end{vmatrix} \quad \begin{vmatrix} \frac{-1}{2} & \frac{\sqrt{3}}{2} \\ \frac{-\sqrt{3}}{2} & \frac{-1}{2} \end{vmatrix}$$

$$\mathscr{D}^E(R_3) \quad i_4 = \quad i_1 = \quad i_2 = \quad R_1^3 = \quad R_1 =$$

$$\begin{vmatrix} 1 & 0 \\ 0 & -1 \end{vmatrix} \quad \begin{vmatrix} 1 & 0 \\ 0 & -1 \end{vmatrix} \quad \begin{vmatrix} \frac{-1}{2} & \frac{\sqrt{3}}{2} \\ \frac{\sqrt{3}}{2} & \frac{1}{2} \end{vmatrix} \quad \begin{vmatrix} \frac{-1}{2} & \frac{\sqrt{3}}{2} \\ \frac{\sqrt{3}}{2} & \frac{1}{2} \end{vmatrix} \quad \begin{vmatrix} \frac{-1}{2} & \frac{-\sqrt{3}}{2} \\ \frac{-\sqrt{3}}{2} & \frac{1}{2} \end{vmatrix} \quad \begin{vmatrix} \frac{-1}{2} & \frac{-\sqrt{3}}{2} \\ \frac{-\sqrt{3}}{2} & \frac{1}{2} \end{vmatrix}$$

$$\mathscr{D}^E(R_3^3) \quad i_3 = \quad R_2 = \quad R_2^3 = \quad i_6 = \quad i_5 =$$

$$\begin{vmatrix} 1 & 0 \\ 0 & -1 \end{vmatrix} \quad \begin{vmatrix} 1 & 0 \\ 0 & -1 \end{vmatrix} \quad \begin{vmatrix} \frac{-1}{2} & \frac{\sqrt{3}}{2} \\ \frac{\sqrt{3}}{2} & \frac{1}{2} \end{vmatrix} \quad \begin{vmatrix} \frac{-1}{2} & \frac{\sqrt{3}}{2} \\ \frac{\sqrt{3}}{2} & \frac{1}{2} \end{vmatrix} \quad \begin{vmatrix} \frac{-1}{2} & \frac{-\sqrt{3}}{2} \\ \frac{-\sqrt{3}}{2} & \frac{1}{2} \end{vmatrix} \quad \begin{vmatrix} \frac{-1}{2} & \frac{-\sqrt{3}}{2} \\ \frac{-\sqrt{3}}{2} & \frac{1}{2} \end{vmatrix}$$

(4.2.19)

Notice that the part which represents $D_4 = \{1, R_3^2, R_3, R_3^3, R_1^2, R_2^2, i_3, i_4\}$ is diagonal:

$$\mathscr{D}^E \downarrow D_4 = \begin{pmatrix} D^{A_1} & 0 \\ 0 & D^{B_1} \end{pmatrix},$$

while $D_2 = \{1, R_1^2, R_2^2, R_3^2\}$ is represented entirely by unit matrices:

$$\mathscr{D}^E \downarrow D_2 = \begin{pmatrix} D^{A_1} & 0 \\ 0 & D^{A_1} \end{pmatrix} = \begin{pmatrix} 1 & 0 \\ 0 & 1 \end{pmatrix}.$$

Notice also that each member of a D_2 coset, such as $r_1 D_2 = \{r_1 r_2 r_3 r_4\}$ is represented by the same matrix. In fact, the six cosets of D_2 are each represented by one E irrep (3.3.7) of D_3.

The irreps of full octahedral symmetry differ from the O irreps only by ($\pm$) factors according to Eq. (4.1.18). Consider, for example, the irrep $\mathscr{D}^{T_{1u}}$ given for Ir_1, i_4, and R_1^2 in Figure 4.2.1. The matrices for i_4 and R_1^2 are the same as in Eq. (4.2.14). However, a u representation of any element with inversion attached has its sign changed from Eq. (4.2.14).

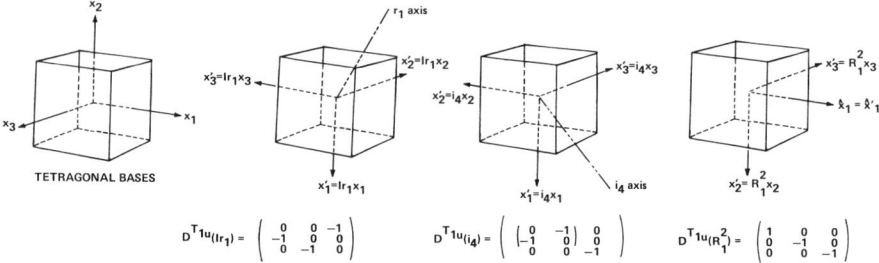

Figure 4.2.1 Tetragonal vector bases and T_{1u} irreps of Ir_1, i_4, and R_1^2 operators (octahedral operators are defined in Figure 4.1.2).

The irrep $\mathscr{D}^{T_{1u}}$ can be easily understood, since it represents the effect of O_h transformations on Cartesian unit vectors $\{\hat{x} = \hat{x}_1, \hat{y} = \hat{x}_2, \hat{z} = \hat{x}_3\}$. The effects of Ir_1, i_4, and R_1^2 are shown in Figure 4.2.1. It is easy to visualize a transformed unit vector such as

$$Ir_1|x_1\rangle = -|x_2\rangle,$$

which gives the corresponding T_{1u} irrep component

$$\mathscr{D}^{T_{1u}}_{21}(Ir_1) = \langle x_2|Ir_1|x_1\rangle = -1. \qquad (4.2.20)$$

Notice also the difference between horizontal-plane reflections such as

$$\mathscr{D}^{T_{1u}}(IR_1^2) = \begin{pmatrix} -1 & \cdot & \cdot \\ \cdot & 1 & \cdot \\ \cdot & \cdot & 1 \end{pmatrix}_{(D_{4h} \supset D_{2h})}, \qquad (4.2.21)$$

which are diagonal in this representation, and diagonal-plane reflections such as

$$\mathscr{D}^{T_{1u}}(Ii_4) = \begin{pmatrix} \cdot & 1 & \cdot \\ 1 & \cdot & \cdot \\ \cdot & \cdot & 1 \end{pmatrix}_{(D_{4h} \supset D_{2h})}, \qquad (4.2.22)$$

which are not.

The $O_h \supset D_{4h} \supset D_{2h}$ subgroup chain labeling follows from that of $O \supset D_4 \supset D_2$. The labeling of the T_{1u} vector bases goes as follows:

$$|\hat{x}_1\rangle = \begin{vmatrix} T_{1u} \\ 1 \end{vmatrix} \equiv \begin{pmatrix} [T_{1u}] \\ E_u \\ B_{1u} \end{pmatrix}, \quad |\hat{x}_2\rangle = \begin{vmatrix} T_{1u} \\ 2 \end{vmatrix} \equiv \begin{pmatrix} [T_{1u}] \\ E_u \\ B_{2u} \end{pmatrix}, \quad |\hat{x}_3\rangle = \begin{vmatrix} T_{1u} \\ 3 \end{vmatrix} \equiv \begin{pmatrix} [T_{1u}] \\ A_{2u} \\ A_{2u} \end{pmatrix}.$$
(4.2.23)

(b) Trigonal Standing-Wave Irreps ($O_h \supset D_{3d} \supset C_{2v}$) Suppose we require O irreps which are diagonal with respect to subgroup $C_2 = \{1, i_4\}$ and

IRREDUCIBLE REPRESENTATIONS OF OCTAHEDRAL SYMMETRY 245

reduced with respect to subgroup $D_3 = \{1, r_1, r_1^2, i_2, i_4, i_5\}$. The formal construction of these irreps involves splitting with the idempotent combination

$$\begin{aligned}
1 &= (P^E + P^{A_2} + P^{A_1})(P^A + P^B) \\
&= P^E P^A + P^E P^B + P^{A_2} P^B + P^{A_1} P^A \\
&= P_1^E + P_2^E + P^{A_2} + P^{A_1},
\end{aligned} \tag{4.2.24}$$

made from D_3 and C_2 operators.

For example, the following is the splitting of P^{T_1}:

$$\begin{aligned}
P^{T_1} &= \mathbb{P}^{T_1}(P_1^E + P_2^E + P^{A_2} + P^{A_1}) \\
&= \mathbb{P}^{T_1} P_1^E + \mathbb{P}^{T_1} P_2^E + \mathbb{P}^{T_1} P^{A_2} + 0 \\
&= P_1^{T_1} + P_2^{T_1} + P_3^{T_1}.
\end{aligned} \tag{4.2.25a}$$

This defines the $D_3 \supset C_2$ subgroup correlation and base labeling

$$P_1^{T_1} \equiv \mathbb{P}^{T_1} P^E P^A, \qquad P_2^{T_1} \equiv \mathbb{P}^{T_1} P^E P^B, \qquad P_3^{T_1} \equiv \mathbb{P}^{T_1} P^{A_2} P^B. \tag{4.2.25b}$$

Another way to obtain trigonal $\mathscr{D}^{T_1}$ representations involves the Cartesian unit vectors $\{\hat{v}_1, \hat{v}_2, \hat{v}_3\}$ defined by

$$\hat{v}_1 = \begin{pmatrix} \frac{1}{\sqrt{2}} \\ -\frac{1}{\sqrt{2}} \\ 0 \end{pmatrix}, \quad \hat{v}_2 = \begin{pmatrix} \frac{1}{\sqrt{6}} \\ \frac{1}{\sqrt{6}} \\ -\frac{2}{\sqrt{6}} \end{pmatrix}, \quad \hat{v}_3 = \begin{pmatrix} \frac{1}{\sqrt{3}} \\ \frac{1}{\sqrt{3}} \\ \frac{1}{\sqrt{3}} \end{pmatrix}, \tag{4.2.26}$$

which are drawn in Figure 4.2.2. These are defined so $\hat{v}_3$ points along the trigonal (111) direction or r_1 axis, and $\hat{v}_1$ lies along the i_4 axis. $\hat{v}_2$ is normal

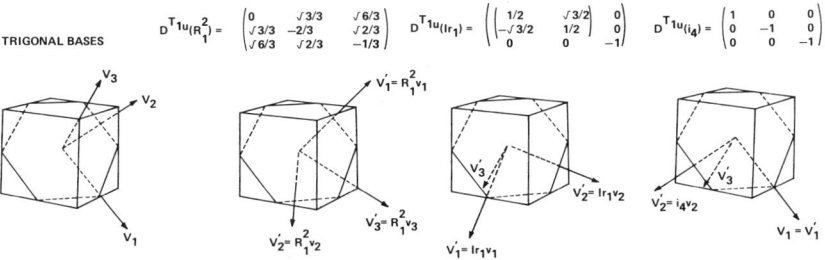

Figure 4.2.2 Trigonal vector bases and T_{1u} irreps of R_1^2, Ir_1, and i_4 operators.

246 HIGHER FINITE SYMMETRY AND INDUCED REPRESENTATION

to $\hat{v}_1$ and $\hat{v}_3$. Clearly, $\hat{v}_1$ and $\hat{v}_2$ behave like E partners under 120° r_1 rotations. Furthermore, $\hat{v}_2$ and $\hat{v}_3$ are antisymmetric ($-$) to 180° i_4 rotations, while $\hat{v}_1$ is symmetric ($+$). The full $O_h \supset D_{3d} \supset C_{2v}$ labeling of these vector bases is

$$|v_1\rangle \equiv \left|\begin{matrix}[T_{1u}]\\1\end{matrix}\right\rangle = \left|\begin{matrix}[T_{1u}]\\E_u\\A_u\end{matrix}\right\rangle, \quad |v_2\rangle \equiv \left|\begin{matrix}T_{1u}\\2\end{matrix}\right\rangle = \left|\begin{matrix}[T_{1u}]\\E_u\\B_u\end{matrix}\right\rangle, \quad |v_3\rangle \equiv \left|\begin{matrix}T_{1u}\\3\end{matrix}\right\rangle = \left|\begin{matrix}[T_{1u}]\\A_{2u}\\B_u\end{matrix}\right\rangle.$$

(4.2.27)

This is in agreement with Eq. (4.2.25b).

The vector transformations and $\mathscr{D}^{T_{1u}}$ representations of R_1^2, Ir_1, and i_4 are shown in Figure 4.2.2. The entire trigonal representation $\mathscr{D}^{T_1}$ of O is given in the following equation.

$\mathscr{D}^{T_1}(1) = \begin{vmatrix} 1 & \cdot & \cdot \\ \cdot & 1 & \cdot \\ \cdot & \cdot & 1 \end{vmatrix}$
$i_4 = [12] \begin{vmatrix} 1 & \cdot & \cdot \\ \cdot & -1 & \cdot \\ \cdot & \cdot & -1 \end{vmatrix}$
$R_1^2 = [13][24] \begin{vmatrix} \cdot & \frac{\sqrt{3}}{3} & \frac{\sqrt{6}}{3} \\ \frac{\sqrt{3}}{3} & \frac{-2}{3} & \frac{\sqrt{2}}{3} \\ \frac{\sqrt{6}}{3} & \frac{\sqrt{2}}{3} & \frac{-1}{3} \end{vmatrix}$
$R_3 = [1423] \begin{vmatrix} \cdot & \frac{-\sqrt{3}}{3} & \frac{-\sqrt{6}}{3} \\ \frac{\sqrt{3}}{3} & \frac{2}{3} & \frac{-\sqrt{2}}{3} \\ \frac{\sqrt{6}}{3} & \frac{-\sqrt{2}}{3} & \frac{1}{3} \end{vmatrix}$

$r_1 = [132] \begin{vmatrix} \frac{-1}{2} & \frac{-\sqrt{3}}{2} & \cdot \\ \frac{\sqrt{3}}{2} & \frac{-1}{2} & \cdot \\ \cdot & \cdot & 1 \end{vmatrix}$
$i_5 = [13] \begin{vmatrix} \frac{-1}{2} & \frac{-\sqrt{3}}{2} & \cdot \\ \frac{-\sqrt{3}}{2} & \frac{1}{2} & \cdot \\ \cdot & \cdot & -1 \end{vmatrix}$
$r_4 = [234] \begin{vmatrix} \frac{1}{2} & \frac{-\sqrt{3}}{6} & \frac{\sqrt{6}}{3} \\ \frac{-\sqrt{3}}{2} & \frac{-1}{6} & \frac{\sqrt{2}}{3} \\ \cdot & \frac{-\sqrt{8}}{3} & \frac{-1}{3} \end{vmatrix}$
$i_6 = [24] \begin{vmatrix} \frac{-1}{2} & \frac{\sqrt{3}}{6} & \frac{-\sqrt{6}}{3} \\ \frac{\sqrt{3}}{6} & \frac{-5}{6} & \frac{-\sqrt{2}}{3} \\ \frac{-\sqrt{6}}{3} & \frac{-\sqrt{2}}{3} & \frac{1}{3} \end{vmatrix}$

$r_1^2 = [123] \begin{vmatrix} \frac{-1}{2} & \frac{\sqrt{3}}{2} & \cdot \\ \frac{-\sqrt{3}}{2} & \frac{-1}{2} & \cdot \\ \cdot & \cdot & 1 \end{vmatrix}$
$i_2 = [23] \begin{vmatrix} \frac{-1}{2} & \frac{\sqrt{3}}{2} & \cdot \\ \frac{\sqrt{3}}{2} & \frac{1}{2} & \cdot \\ \cdot & \cdot & -1 \end{vmatrix}$
$r_2^2 = [142] \begin{vmatrix} \frac{-1}{2} & \frac{-\sqrt{3}}{6} & \frac{\sqrt{6}}{3} \\ \frac{\sqrt{3}}{3} & \frac{5}{6} & \frac{\sqrt{2}}{3} \\ \frac{-\sqrt{6}}{3} & \frac{\sqrt{2}}{3} & \frac{-1}{3} \end{vmatrix}$
$R_2^3 = [1342] \begin{vmatrix} 1 & \frac{\sqrt{3}}{6} & \frac{-\sqrt{6}}{3} \\ \frac{-\sqrt{3}}{2} & \frac{1}{6} & \frac{-\sqrt{2}}{3} \\ \cdot & \frac{\sqrt{8}}{3} & \frac{1}{3} \end{vmatrix}$

IRREDUCIBLE REPRESENTATIONS OF OCTAHEDRAL SYMMETRY   **247**

Notice that group operators which were diagonalized in the tetragonal irrep (4.2.14) become undiagonalized in (4.2.28), and vice versa. For example, the tetragonal irrep (4.2.21) of the horizontal-plane reflection becomes

$$\mathscr{D}^{T_{1u}}(IR_1^2) = \begin{pmatrix} \cdot & -\dfrac{\sqrt{3}}{3} & -\dfrac{\sqrt{6}}{3} \\ -\dfrac{\sqrt{3}}{3} & \dfrac{2}{3} & -\dfrac{\sqrt{2}}{3} \\ -\dfrac{\sqrt{6}}{3} & -\dfrac{\sqrt{2}}{3} & \dfrac{1}{3} \end{pmatrix}_{(D_{3d} \supset C_{2v})} \quad (4.2.29)$$

in the trigonal basis, but the trigonal irrep

$$\mathscr{D}^{T_{1u}}(Ii_4) = \begin{pmatrix} -1 & \cdot & \cdot \\ \cdot & 1 & \cdot \\ \cdot & \cdot & 1 \end{pmatrix}_{(D_{3d} \supset C_{2v})} \quad (4.2.30)$$

of the diagonal-plane reflection becomes diagonalized.

$R_2^2 = [14][23]$

$$\begin{vmatrix} \cdot & -\dfrac{\sqrt{3}}{3} & -\dfrac{\sqrt{6}}{3} \\ -\dfrac{\sqrt{3}}{3} & -\dfrac{2}{3} & \dfrac{\sqrt{2}}{2} \\ -\dfrac{\sqrt{6}}{3} & \dfrac{\sqrt{2}}{3} & -\dfrac{1}{3} \end{vmatrix}$$

$R_3^3 = [1324]$

$$\begin{vmatrix} \cdot & \dfrac{\sqrt{3}}{3} & \dfrac{\sqrt{6}}{3} \\ -\dfrac{\sqrt{3}}{3} & \dfrac{2}{3} & -\dfrac{\sqrt{2}}{3} \\ -\dfrac{\sqrt{6}}{3} & -\dfrac{\sqrt{2}}{3} & \dfrac{1}{3} \end{vmatrix}$$

$R_3^2 = [12][34]$

$$\begin{vmatrix} -1 & \cdot & \cdot \\ \cdot & \dfrac{1}{3} & -\dfrac{\sqrt{8}}{3} \\ \cdot & -\dfrac{\sqrt{8}}{3} & -\dfrac{1}{3} \end{vmatrix}$$

$i_3 = [34]$

$$\begin{vmatrix} -1 & \cdot & \cdot \\ \cdot & -\dfrac{1}{3} & \dfrac{\sqrt{8}}{3} \\ \cdot & \dfrac{\sqrt{8}}{3} & \dfrac{1}{3} \end{vmatrix}$$

$r_2 = [124]$

$$\begin{vmatrix} -\dfrac{1}{2} & \dfrac{\sqrt{3}}{6} & -\dfrac{\sqrt{6}}{3} \\ -\dfrac{\sqrt{3}}{3} & \dfrac{5}{6} & \dfrac{\sqrt{2}}{3} \\ \dfrac{\sqrt{6}}{3} & \dfrac{\sqrt{2}}{3} & -\dfrac{1}{3} \end{vmatrix}$$

$R_1 = [1234]$

$$\begin{vmatrix} \dfrac{1}{2} & -\dfrac{\sqrt{3}}{6} & \dfrac{\sqrt{6}}{3} \\ \dfrac{\sqrt{3}}{2} & \dfrac{1}{6} & -\dfrac{\sqrt{2}}{3} \\ \cdot & \dfrac{\sqrt{8}}{3} & \dfrac{1}{3} \end{vmatrix}$$

$r_3 = [143]$

$$\begin{vmatrix} \dfrac{1}{2} & \dfrac{\sqrt{3}}{2} & \cdot \\ \dfrac{\sqrt{3}}{6} & -\dfrac{1}{6} & -\dfrac{\sqrt{8}}{3} \\ -\dfrac{\sqrt{6}}{3} & \dfrac{\sqrt{2}}{3} & -\dfrac{1}{3} \end{vmatrix}$$

$R_1^3 = [1432]$

$$\begin{vmatrix} \dfrac{1}{2} & \dfrac{\sqrt{3}}{2} & \cdot \\ -\dfrac{\sqrt{3}}{6} & \dfrac{1}{6} & \dfrac{\sqrt{8}}{3} \\ \dfrac{\sqrt{6}}{3} & -\dfrac{\sqrt{2}}{3} & \dfrac{1}{3} \end{vmatrix}$$

$r_3^2 = [134]$

$$\begin{vmatrix} \dfrac{1}{2} & \dfrac{\sqrt{3}}{6} & -\dfrac{\sqrt{6}}{3} \\ \dfrac{\sqrt{3}}{2} & -\dfrac{1}{6} & \dfrac{\sqrt{2}}{3} \\ \cdot & -\dfrac{\sqrt{8}}{3} & -\dfrac{1}{3} \end{vmatrix}$$

$i_1 = [14]$

$$\begin{vmatrix} -\dfrac{1}{2} & -\dfrac{\sqrt{3}}{6} & \dfrac{\sqrt{6}}{3} \\ -\dfrac{\sqrt{3}}{6} & -\dfrac{5}{6} & -\dfrac{\sqrt{2}}{3} \\ \dfrac{\sqrt{6}}{3} & -\dfrac{\sqrt{2}}{3} & \dfrac{1}{3} \end{vmatrix}$$

$r_4^2 = [243]$

$$\begin{vmatrix} \dfrac{1}{2} & -\dfrac{\sqrt{3}}{2} & \cdot \\ -\dfrac{\sqrt{3}}{6} & -\dfrac{1}{6} & -\dfrac{\sqrt{8}}{3} \\ \dfrac{\sqrt{6}}{3} & \dfrac{\sqrt{2}}{3} & -\dfrac{1}{3} \end{vmatrix}$$

$R_2 = [1243]$

$$\begin{vmatrix} \dfrac{1}{2} & -\dfrac{\sqrt{3}}{2} & \cdot \\ \dfrac{\sqrt{3}}{6} & \dfrac{1}{6} & \dfrac{\sqrt{8}}{3} \\ -\dfrac{\sqrt{6}}{3} & -\dfrac{\sqrt{2}}{3} & \dfrac{1}{3} \end{vmatrix}$$

$$(4.2.28)$$

The transformation matrix which relates the T_{1u} irreps represented in tetragonal bases to trigonal irreps is

$$\mathcal{T}(4 \leftarrow 3) = \begin{pmatrix} \langle x_1|v_1\rangle & \langle x_1|v_2\rangle & \langle x_1|v_3\rangle \\ \langle x_2|v_1\rangle & \langle x_2|v_2\rangle & \langle x_2|v_3\rangle \\ \langle x_3|v_1\rangle & \langle x_3|v_2\rangle & \langle x_3|v_3\rangle \end{pmatrix} = \begin{pmatrix} \frac{1}{\sqrt{2}} & \frac{1}{\sqrt{6}} & \frac{1}{\sqrt{3}} \\ -\frac{1}{\sqrt{2}} & \frac{1}{\sqrt{6}} & \frac{1}{\sqrt{3}} \\ 0 & -\frac{2}{\sqrt{6}} & \frac{1}{\sqrt{3}} \end{pmatrix}. \quad (4.2.31)$$

The three columns of $\mathcal{T}(4 \leftarrow 3)$ are the vectors (4.2.26). The equivalence transformation between the trigonal $(D_{3d} \supset C_{2v})$ and tetragonal $(D_{4h} \supset D_{2h})$ matrices has the following form:

$$\mathcal{T}(4 \leftarrow 3)(\mathcal{D}^{T_{1u}})_{(D_{3d} \supset C_{2v})} \mathcal{T}^\dagger(4 \leftarrow 3) = (\mathcal{D}^{T_{1u}})_{(D_{4h} \supset D_{2h})},$$
$$\mathcal{T}(4 \leftarrow 3)^\dagger (\mathcal{D}^{T_{1u}})_{(D_{4h} \supset D_{2h})} \mathcal{T}(4 \leftarrow 3) = (\mathcal{D}^{T_{1u}})_{(D_{3d} \supset C_{2v})}. \quad (4.2.32)$$

The trigonal $\mathcal{D}^{T_2}$ irreps can be obtained by simply changing the signs of half the matrices according to Eq. (4.2.15). However, if one requires "kosher" $D_3 \supset C_2$ subgroup labeling, then the irreps in the table section in the back of this book should be used. The $O_h \supset D_{3d} \supset C_{2v}$ labeling of these is

$$\left|\begin{matrix} T_{2p} \\ 1 \end{matrix}\right\rangle = \left|\begin{matrix} [T_{2p}] \\ A_{1p} \\ A_p \end{matrix}\right\rangle, \quad \left|\begin{matrix} T_{2p} \\ 2 \end{matrix}\right\rangle = \left|\begin{matrix} [T_{2p}] \\ E_p \\ A_p \end{matrix}\right\rangle, \quad \left|\begin{matrix} T_{2p} \\ 3 \end{matrix}\right\rangle = \left|\begin{matrix} [T_{2p}] \\ E_p \\ B_p \end{matrix}\right\rangle, \quad (4.2.33)$$

where $p = g$ or u is the inversion parity.

The trigonal E-type irreps have the same form as the tetragonal irreps (4.2.19). The E bases can be labeled

$$\left|\begin{matrix} E_p \\ 1 \end{matrix}\right\rangle \equiv \left|\begin{matrix} [E_p] \\ A_{1p} \\ A_{1p} \end{matrix}\right\rangle_{(D_{4h}D_{2h})} = \left|\begin{matrix} [E_p] \\ E_p \\ A_p \end{matrix}\right\rangle_{(D_{3d}C_{2v})}, \quad \left|\begin{matrix} E_p \\ 2 \end{matrix}\right\rangle \equiv \left|\begin{matrix} [E_p] \\ B_{1p} \\ A_{1p} \end{matrix}\right\rangle_{(D_{4h}D_{2h})} = \left|\begin{matrix} [E_p] \\ E_p \\ B_p \end{matrix}\right\rangle_{(D_{3d}C_{2v})}, \quad (4.2.34)$$

where $p = u$ or g is inversion parity. They go equally well with $D_4 \supset D_2$ or $D_3 \supset C_2$ subgroup chains.

(c) Tetragonal Moving-Wave Irreps ($O \supset D_4 \supset C_4$) The introduction of magnetic fields and molecular rotations will require that cyclic rotation groups be represented by diagonal matrices. A simple transformation of the tetragonal irreps in Section 4.2.A(a) causes the irreps of the cyclic subgroup $C_4 = \{1, R_3, R_3^2, R_3^3\}$ to be diagonal. For example, the $\mathcal{D}^{T_1}$ irrep takes the following form:

$$(4.2.35)$$

The transformation which disagonalizes $\mathscr{D}^{T_1}(R_3)$ from Eq. (4.2.14) is

$$\mathscr{T}^\dagger \begin{pmatrix} 0 & -1 & \cdot \\ 1 & 0 & \cdot \\ \cdot & \cdot & 1 \end{pmatrix} \mathscr{T} = \begin{pmatrix} i & \cdot & \cdot \\ \cdot & -i & \cdot \\ \cdot & \cdot & 1 \end{pmatrix} = \mathscr{D}^{T_1^0}(R_3), \qquad (4.2.36)$$

$$\mathscr{T} = \begin{pmatrix} \langle x_1 | 1_4^E \rangle & \langle x_1 | 3_4^E \rangle & \cdot \\ \langle x_2 | 1_4^E \rangle & \langle x_2 | 3_4^E \rangle & \cdot \\ \cdot & \cdot & \langle x_3 | 0_4^{A_2} \rangle \end{pmatrix} = \begin{pmatrix} \dfrac{-1}{\sqrt{2}} & \dfrac{1}{\sqrt{2}} & \cdot \\ \dfrac{-i}{\sqrt{2}} & \dfrac{-i}{\sqrt{2}} & \cdot \\ \cdot & \cdot & 1 \end{pmatrix}.$$

You may recall seeing this form of transformation before [viz., Eqs. (3.4.24)], and one should certainly expect to see it again. It is the transformation between standing-wave or linear (x, y) polarization states and moving-wave or circular $(x \pm iy)$ polarization states. The following is the labeling of the circular T states by subgroup chain $O \supset D_4 \supset C_4$.

$$\left| \begin{matrix} T_1 \\ 1_4 \end{matrix} \right\rangle = \left| \begin{matrix} [T_1] \\ E \\ 1_4 \end{matrix} \right\rangle, \quad \left| \begin{matrix} T_1 \\ 3_4 \end{matrix} \right\rangle = \left| \begin{matrix} [T_1] \\ E \\ 3_4 \end{matrix} \right\rangle, \quad \left| \begin{matrix} T_1 \\ 0_4 \end{matrix} \right\rangle = \left| \begin{matrix} [T_1] \\ A_2 \\ 0_4 \end{matrix} \right\rangle. \qquad (4.2.37)$$

The corresponding labeling for the T_2 bases is similar.

$$\left| \begin{matrix} T_2 \\ 1_4 \end{matrix} \right\rangle = \left| \begin{matrix} [T_2] \\ E \\ 1_4 \end{matrix} \right\rangle, \quad \left| \begin{matrix} T_2 \\ 3_4 \end{matrix} \right\rangle = \left| \begin{matrix} [T_2] \\ E \\ 3_4 \end{matrix} \right\rangle, \quad \left| \begin{matrix} T_2 \\ 2_4 \end{matrix} \right\rangle = \left| \begin{matrix} [T_2] \\ B_2 \\ 2_4 \end{matrix} \right\rangle. \qquad (4.2.38)$$

The third T_2 component belongs to 2_4 waves angular quantum $m = 2 \bmod 4$. The first two components belong to $(m = \pm 1 \bmod 4)$ waves, as do the corresponding T_1 components.

The corresponding labeling of the E, A_2, and A_1 bases is as follows:

$$\left\{ \left| \begin{matrix} E \\ 0_4 \end{matrix} \right\rangle = \left| \begin{matrix} [E] \\ A_1 \\ 0_4 \end{matrix} \right\rangle, \quad \left| \begin{matrix} E \\ 2_4 \end{matrix} \right\rangle = \left| \begin{matrix} [E] \\ B_1 \\ 2_4 \end{matrix} \right\rangle \right\}, \quad |A_2\rangle = \left| \begin{matrix} [A_2] \\ B_1 \\ 2_4 \end{matrix} \right\rangle, \quad |A_1\rangle = \left| \begin{matrix} [A_1] \\ A_1 \\ 0_4 \end{matrix} \right\rangle.$$

$$(4.2.39) \qquad\qquad (4.2.40) \qquad\qquad (4.2.41)$$

Notice that the E- and $A_{1,2}$-type bases do not require transformation from their standing-wave forms, since these are already diagonal representations of C_4.

(d) Trigonal Moving-Wave Irreps ($O \supset D_3 \supset C_3$) The standing-to-moving-wave transformation can be used to diagonalize the representations of cyclic subgroup $C_3 = \{1, r_1, r_1^2\}$. This produces complex irreps listed Table F.2.7.

(e) Idempotent Splitting Corresponds to Level Splitting Each class idempotent $\mathbb{P}^\alpha$ is a sum of l^α irreducible idempotents $P_{ii}^\alpha \equiv P_i^\alpha$ ($i = 1 \cdots l^\alpha$). Each P_i^α can be thought of as a projection operator for a single state Ψ_i^α belonging to one of l^α degenerate energy levels ε_i^α in an (α)-type multiplet. For example, O symmetry gives rise to T_1-type triplets, T_2-type triplets, E-type doublets, and A_1- or A_2-type singlets.

If we reduce the O symmetry to one of its subgroups such as D_4, D_2, or C_4 then each of these levels may split according to the way that their corresponding class idempotents split. For example, the $O \supset D_4$ splitting relation (4.2.9) gives the following (nonzero) terms:

$$\mathbb{P}^{T_1} = \mathbb{P}^{T_1} P_1^E + \mathbb{P}^{T_1} P_2^E + \mathbb{P}^{T_1} P^{A_2} = \mathbb{P}^{T_1} P^E + \mathbb{P}^{T_1} P^{A_2}$$
$$= P_{11}^{T_1} + P_{22}^{T_1} + P_{33}^{T_1}.$$

The first line corresponds to a T_1 triplet level splitting into an E-type doublet and an A_2-type singlet under D_4 symmetry. The E-type doublet remains degenerate until we reduce the symmetry to one of the D_4 subgroups D_2, C_4, or lower. If C_4 is used the resulting pair of levels emerging from an E doublet are labeled 1_4 and 3_4 according to (4.2.4), while $D_4 \supset D_2$ symmetry breaking would split E into B_1- and B_2-type levels according to (4.2.5).

A level-splitting diagram which traces all the level correlations of the $O \supset D_4 \supset D_2$ chain is shown in Figure 4.2.3(a). The companion figure [4.2.3(b)] traces the $O \supset D_4 \supset C_4$ correlations. Without their labels the two figures appear identical! However, we have noted that the C_4 symmetry breaking gives moving-wave states and is analogous to magnetic Zeeman or rotational Coriolis splitting. The D_2 symmetry breaking, on the other hand, gives standing-wave states and is analogous to electric Stark or anisotropy splitting. It is helpful to use the level-splitting diagrams to keep track of both the mathematics and physics associated with idempotent splitting and subgroup correlations. Another way to do this uses correlation tables as shown in the following section.

B. More Subgroup Correlations

There are quite a few more possible O_h subgroup chains besides the four main types discussed in the preceding section. However, one or more of the four types of irreps can easily be adapted to most other labeling schemes.

The "road map" for any link in a subgroup chain is the CORRELATION TABLE. The correlation table was introduced in Eqs. (3.6.27) for hexagonal

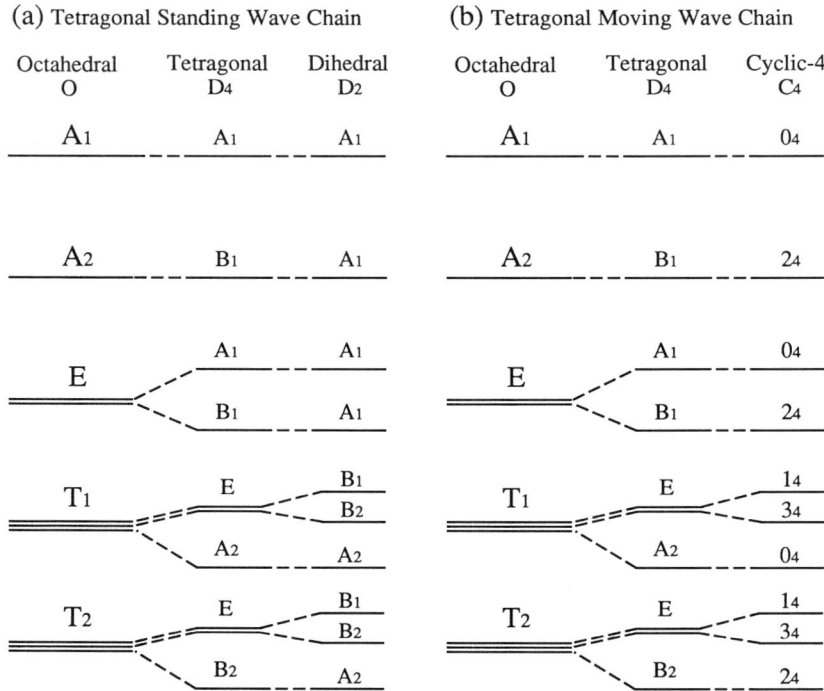

Figure 4.2.3 Level-splitting diagram for two of the tetragonal subgroup chains in octahedral symmetry.

or D_6 subgroups. A correlation table lists the frequencies f^A of irreps D^A of subgroup H in a *subduced* irrep $\mathscr{D}^\alpha \downarrow H$ of a larger group $G \supset H$. These frequencies can be obtained by comparing character tables of G and H and using the theory of Section 3.5.B(a).

For example, the $O \supset D_4$ and $O \supset C_4$ correlations used in the preceding section are summed up by the following tables:

$\downarrow D_4$	A_1	A_2	B_1	B_2	E
$\mathscr{D}^{A_1}$	1	·	·	·	·
$\mathscr{D}^{A_2}$	·	·	1	·	·
$\mathscr{D}^{E}$	1	·	1	·	·
$\mathscr{D}^{T_1}$	·	1	·	·	1
$\mathscr{D}^{T_2}$	·	·	·	1	1

$\downarrow C_4$	0_4	1_4	2_4	3_4
$\mathscr{D}^{A_1}$	1	·	·	·
$\mathscr{D}^{A_2}$	·	·	1	·
$\mathscr{D}^{E}$	1	·	1	·
$\mathscr{D}^{T_1}$	1	1	·	1
$\mathscr{D}^{T_2}$	·	1	1	1

(4.2.42a)   (4.2.42b)

IRREDUCIBLE REPRESENTATIONS OF OCTAHEDRAL SYMMETRY

Comparison of the C_4 characters

$$
\begin{array}{c|cccc}
 & g=1 & R & R^2 & R^3 \\
\hline
\chi^{0_4} = & 1 & 1 & 1 & 1 \\
\chi^{1_4} = & 1 & -i & -1 & i \\
\chi^{2_4} = & 1 & -1 & 1 & -1 \\
\chi^{3_4} = & 1 & i & -1 & -i
\end{array}, \quad (4.2.43)
$$

with the O-character table (4.1.11) yielding the $O \supset C_4$ correlations (4.2.42b). For example, the $\mathscr{D}^{T_1} \downarrow C^4$ characters are

$$
\chi_g^{T_1} = \begin{array}{c|cccc} g=1 & R & R^2 & R^3 \\ \hline 3 & 1 & -1 & 1 \end{array}
$$

from the O table. It is easy to see that

$$\chi_g^{T_1} = \chi^{0_4} + \chi^{1_4} + \chi^{3_4}$$

for each g in the C_4 table (4.2.43). This gives the T_1 row of (4.2.42b).

The trigonal $O \supset D_3$ and $O \supset C_3$ correlations are found in the same way:

$\downarrow D_3$	A_1	A_2	E
$\mathscr{D}^{A_1}$	1	·	·
$\mathscr{D}^{A_2}$	·	1	·
$\mathscr{D}^{E}$	·	·	1
$\mathscr{D}^{T_1}$	·	1	1
$\mathscr{D}^{T_2}$	1	·	1

$\downarrow C_3$	0_3	1_3	2_3
$\mathscr{D}^{A_1}$	1	·	·
$\mathscr{D}^{A_2}$	1	·	·
$\mathscr{D}^{E}$	·	1	1
$\mathscr{D}^{T_1}$	1	1	1
$\mathscr{D}^{T_2}$	1	1	1

(4.2.44a) \qquad\qquad (4.2.44b)

For a number of applications it will be convenient to have correlation tables for subgroup chains involving $O_h \supset C_{2v}$, $O_h \supset C_{3v}$, and $O_h \supset C_{4v}$. The character table of C_{2v} has the $C_2 \times C_2$ form as explained in Sections 2.10 and 2.11:

$$
\begin{array}{c|cccc}
 & g=1 & i_4 & IR_3^2 & Ii_3 \\
\hline
\chi_g^{A'} = & 1 & 1 & 1 & 1 \\
\chi_g^{B'} = & 1 & -1 & 1 & -1 \\
\chi_g^{A''} = & 1 & 1 & -1 & -1 \\
\chi_g^{B''} = & 1 & -1 & -1 & 1
\end{array}. \quad (4.2.45)
$$

The C_{3v} and C_{4v} characters were given by their isomorphic D_3 and D_4 characters in Eqs. (3.5.8) and (3.6.3), respectively. By comparing these with O_h characters one easily derives the following correlations:

$\downarrow C_{2v}$	A'	B'	A''	B''
$\mathscr{D}^{A_{1g}}$	1	·	·	·
$\mathscr{D}^{A_{2g}}$	·	1	·	·
$\mathscr{D}^{E_g}$	1	1	·	·
$\mathscr{D}^{T_{1g}}$	·	1	1	1
$\mathscr{D}^{T_{2g}}$	1	·	1	1
$\mathscr{D}^{A_{1u}}$	·	·	1	·
$\mathscr{D}^{A_{2u}}$	·	·	·	1
$\mathscr{D}^{E_u}$	·	·	1	1
$\mathscr{D}^{T_{1u}}$	1	1	·	1
$\mathscr{D}^{T_{2u}}$	1	1	1	·

(4.2.46a)

$\downarrow C_{3v}$	A'	A''	E
$\mathscr{D}^{A_{1g}}$	1	·	·
$\mathscr{D}^{A_{2g}}$	·	1	·
$\mathscr{D}^{E_g}$	·	·	1
$\mathscr{D}^{T_{1g}}$	·	1	1
$\mathscr{D}^{T_{2g}}$	1	·	1
$\mathscr{D}^{A_{1u}}$	·	1	·
$\mathscr{D}^{A_{2u}}$	1	·	·
$\mathscr{D}^{E_u}$	·	·	1
$\mathscr{D}^{T_{1u}}$	1	·	1
$\mathscr{D}^{T_{2u}}$	·	1	1

(4.2.46b)

$\downarrow C_{4v}$	A'	B'	A''	B''	E
$\mathscr{D}^{A_{1g}}$	1	·	·	·	·
$\mathscr{D}^{A_{2g}}$	·	1	·	·	·
$\mathscr{D}^{E_g}$	1	1	·	·	·
$\mathscr{D}^{T_{1g}}$	·	·	1	·	1
$\mathscr{D}^{T_{2g}}$	·	·	1	1	1
$\mathscr{D}^{A_{1u}}$	·	·	1	·	·
$\mathscr{D}^{A_{2u}}$	·	·	·	1	·
$\mathscr{D}^{E_u}$	·	·	1	1	·
$\mathscr{D}^{T_{1u}}$	1	·	·	·	1
$\mathscr{D}^{T_{2u}}$	·	1	·	·	1

(4.2.46c)

C. Conjugate and Normal Subgroups

When correlating a C_{nv} or D_n subgroup it is sometimes necessary to specify *which* subgroup. For example, consider the two subgroups $\{1, R_3^2, i_3, i_4\}$ and $\{1, R_3^2, R_1^2, R_2^2\}$. Both subgroups contain three 180° rotations around orthogonal axes. Both should be labeled D_2 and both have the same set of irreps:

$$D_2' = \{1 \quad R_3^2 \quad i_3 \quad i_4\}$$
$$D_2 = \{1 \quad R_3^2 \quad R_1^2 \quad R_2^2\}$$

	1	1	1	1
A_1	1	1	1	1
B_1	1	−1	1	−1
A_2	1	1	−1	−1
B_2	1	−1	−1	1

(4.2.47)

However, they have quite different correlations with the octahedral irreps as the following tables show:

$\downarrow \{1R_3^2 i_3 i_4\}$	A_1	B_1	A_2	B_2
$\mathscr{D}^{A_1}$	1	·	·	·
$\mathscr{D}^{A_2}$	·	·	1	·
$\mathscr{D}^{E}$	1	·	1	·
$\mathscr{D}^{T_1}$	·	1	1	1
$\mathscr{D}^{T_2}$	1	1	·	1

(4.2.48a)

$\downarrow \{1R_3^2 R_2^2 R_1^2\}$	A_1	B_1	A_2	B_2
$\mathscr{D}^{A_1}$	1	·	·	·
$\mathscr{D}^{A_2}$	1	·	·	·
$\mathscr{D}^{E}$	2	·	·	·
$\mathscr{D}^{T_1}$	·	1	1	1
$\mathscr{D}^{T_2}$	·	1	1	1

(4.2.48b)

The D'_2 subgroup containing i_3 and i_4 could be used by itself to label each O irrep uniquely. The other D_2 subgroup cannot provide E-irrep labeling, since E is correlated twice with the same irrep A_1. Indeed, the labeling chain $O \supset D_4 \supset D_2$ discussed in Section 4.2.A(b) requires the D_4 link in order to work for all O irreps.

Whenever we choose a particular subgroup H for labeling we generally ignore several CONJUGATE SUBGROUPS $H^t = tHt^{-1}, H^u = uHu^{-1}, \ldots$ obtained by transforming H by group operators. For example, $D'_2 = \{1, R_3^2, i_3, i_4\}$ has two other distinct conjugate subgroups $D_2^{R_1} = \{1, R_2^2, i_1, i_2\}$ and $D_2^{R_2} = \{1, R_1^2, i_5, i_6\}$. Most of the subgroups of octahedral symmetry have several conjugates. By focusing on one choice from several conjugates we single out a particular axis or direction in the octahedral symmetry. However, each choice must give the same correlation table. This is true since the correlations depend only on the characters which are independent of transformations $g' = tgt^{-1}$ within classes.

The other subgroup $D_2 = \{1, R_1^2, R_2^2, R_3^2\}$ is an example of a self-conjugate or NORMAL subgroup of the octahedral group. A normal subgroup $N \subset G$ is one for which $gNg^{-1} = N$ for all transformations g in G. Normal subgroups are distinguished by being "unique" in their group.

4.3 INTRODUCTION TO SYMMETRY BREAKING AND INDUCED REPRESENTATIONS

Many spectroscopic applications of symmetry analysis involve effects in which higher-symmetry systems are changed into lower-symmetry ones. These effects are called SYMMETRY-BREAKING EFFECTS, and there are two basic types of symmetry breaking. One type is EXTERNAL or "applied" symmetry breaking, wherein a system of higher symmetry is perturbed by an outside force of lower symmetry. This perturbation generally causes splitting of spectral degeneracy, as in the example of Zeeman splitting in Figure 3.6.3 or band-gap splitting in Figure 2.7.7. Another type is INTERNAL or "spontaneous" symmetry breaking, in which a system tends to "stick" in a low-symmetry state when resonance or tunneling between equivalent states becomes negligible. This sticking generally goes along with *increased* spectral degeneracy, as in the example of NH_3 with $S = 0$ in Figure 2.12.7, or band collapse represented in Figure 2.12.1(c).

Either type of symmetry breaking involves the subgroup correlations which were introduced in Section 4.2 and the preceding chapter. In this section some applications will exhibit the relationship between the two opposing types of symmetry breaking: one which reduces spectral degeneracy, and the other which increases it. This will involve a relation between *subduced* representations ($\mathscr{D}^\alpha \downarrow H$) and new type of representation called the *induced* representation, denoted by ($D^A \uparrow G$).

A. Octahedral Models and Induced Representations

Let us consider two different physical problems which are mathematically very similar. One problem is classical and involves an octahedral arrangement of six vibrating masses constrained to slide on coordinate axes $\{\pm x, \pm y, \pm z\}$, as shown in Figure 4.3.1(a). The other problem is quantum mechanical and involves a particle which spends most of its time in any of the six potential valleys centered around the $\{\pm x, \pm y, \pm z\}$ axes in Figure 4.3.1(b). The figure shows a plot of the potential versus angular direction with high-potential directions indicated by mountains and low ones by valleys. In Chapters 5–7 we will study such angular potentials in greater detail, but for now a qualitative picture of one example is all that is needed.

In either problem there will be six base states $\{|1\rangle, |2\rangle, \ldots, |6\rangle\}$. In the spring-mass problem ket $|j\rangle$ means mass j is stretched one unit outward from its equilibrium oposition, while the other masses are fixed at their respective equilibrium or resting points. In the quantum-mechanical problem ket $|j\rangle$ stands for a state for which the probability of finding the particle in the jth valley is unity. In other words, the wave function

$$\langle \theta \phi | j \rangle = \psi_j(\theta \phi) \qquad (4.3.1)$$

of the jth base state is localized around the polar angles of the jth axis or valley in Figure 4.3.1(b).

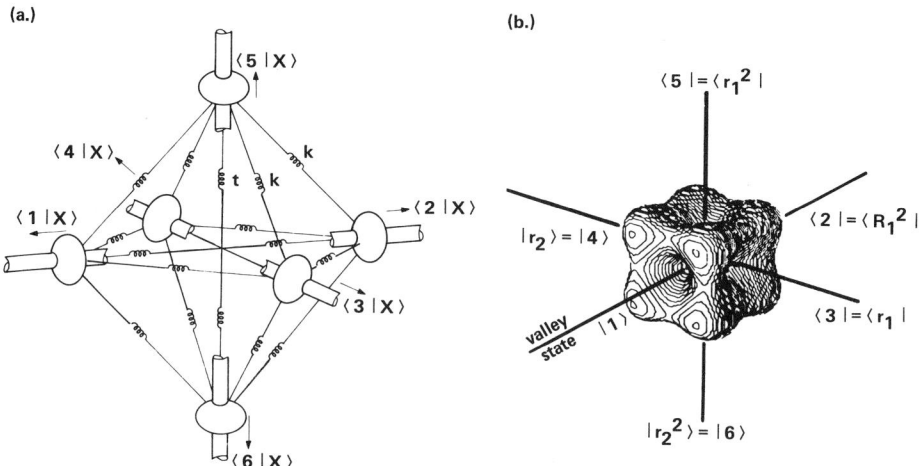

Figure 4.3.1 Examples of physical systems with octahedral symmetry. (a) Coupled oscillating beads sliding on octahedral axes are described by six classical coordinates $x_j = \langle j | x \rangle$. (b) A six-state quantum system could describe a particle capable of tunneling between six equivalent potential valleys.

INTRODUCTION TO SYMMETRY BREAKING AND INDUCED REPRESENTATIONS 257

Either problem involves eigensolutions of matrices which have virtually the same form. The spring-mass problem involves an acceleration matrix:

$$-\begin{pmatrix} \langle 1|\mathbf{a}|1\rangle & \langle 1|\mathbf{a}|2\rangle & \cdots & \langle 1|\mathbf{a}|6\rangle \\ \langle 2|\mathbf{a}|1\rangle & \langle 2|\mathbf{a}|2\rangle & \cdots & \langle 2|\mathbf{a}|6\rangle \\ \cdot & & & \\ \cdot & & & \\ \cdot & & & \\ \langle 6|\mathbf{a}|1\rangle & \langle 6|\mathbf{a}|2\rangle & \cdots & \langle y|\mathbf{a}|6\rangle \end{pmatrix} = \begin{pmatrix} h & t & s & s & s & s \\ t & h & s & s & s & s \\ s & s & h & t & s & s \\ s & s & t & h & s & s \\ s & s & s & s & h & t \\ s & s & s & s & t & h \end{pmatrix},$$

(4.3.2a)

where components

$$h = 2k + t,$$
$$s = k/2 \qquad (4.3.2\text{b})$$

are related to spring constants k and t of nearest- and next-nearest-neighbor connections, respectively, in Figure 4.3.1(a). The quantum problem involves a Hamiltonian matrix:

$$\begin{pmatrix} \langle 1|\mathbf{H}|1\rangle & \langle 1|\mathbf{H}|2\rangle & \cdots & \langle 1|\mathbf{H}|6\rangle \\ \langle 2|\mathbf{H}|1\rangle & \langle 2|\mathbf{H}|2\rangle & \cdots & \langle 2|\mathbf{H}|6\rangle \\ \cdot & & & \cdot \\ \cdot & & & \cdot \\ \cdot & & & \cdot \\ \langle 6|\mathbf{H}|1\rangle & \langle 6|\mathbf{H}|2\rangle & \cdots & \langle 6|\mathbf{H}|6\rangle \end{pmatrix} = \begin{pmatrix} H & T & S & S & S & S \\ T & H & S & S & S & S \\ S & S & H & T & S & S \\ S & S & T & H & S & S \\ S & S & S & S & H & T \\ S & S & S & S & T & H \end{pmatrix}.$$

(4.3.3)

The diagonal components are the energy-expectation values H of the localized wave functions (4.3.1). The off-diagonal components S and T are tunneling amplitudes between nearest- and next-nearest-neighboring valleys, respectively, in Figure 4.3.1(b).

Instead of labeling the bases $\{|1\rangle, |2\rangle, \ldots, |6\rangle\}$ with numbers, let us use octahedral group operators. Let each state be labeled as follows:

$$|g\rangle = g|1\rangle, \qquad (4.3.4)$$

by the operation g which converts the first state,

$$|1\rangle = \mathbf{1}|1\rangle, \qquad (4.3.5)$$

into the gth state $|g\rangle$. This group labeling scheme has been used in Chapters 2 and 3. However, now there are four times as many octahedral operations as there are base vectors. Indeed, the first state $|1\rangle$ of the spring-mass problem could be labeled by R_3, R_3^2, and R_3^3 as well as by $\mathbf{1}$. State $|1\rangle$ is unchanged by 3-axis rotations, since it is just a displacement along this axis:

$$|1\rangle = \mathbf{1}|1\rangle = R_3|1\rangle = R_3^2|1\rangle = R_3^3|1\rangle. \qquad (4.3.6)$$

In other words the first spring-mass state $|1\rangle$ is invariant to its local $C_4 = \{1, R_3, R_3^2, R_3^3\}$ axial subgroup. Let us rewrite this using the invariant C_4 idempotent,

$$P^A \equiv P^{0_4} = (\mathbf{1} + R_3 + R_3^2 + R_3^3)/4,$$

as follows:

$$|1\rangle = P^{0_4}|1\rangle = (\mathbf{1} + R_3 + R_3^2 + R_3^3)|1\rangle/4. \qquad (4.3.7)$$

The relations (4.3.6) or (4.3.7) will be called LOCAL SYMMETRY CONDITIONS of $|1\rangle$. They imply that each of the other states $|g\rangle = g|1\rangle$ could be labeled

$$|g\rangle = g|1\rangle = gR_3|1\rangle = gR_3^2|1\rangle = gR_3^3|1\rangle,$$
$$|g\rangle = |gR_3\rangle = |gR_3^2\rangle = |gR_3^3\rangle, \qquad (4.3.8)$$

i.e., by any element gR_3^h in the gth coset gC_4 of the local subgroup.

In Figure 4.3.1(a) we have chosen one element g from each coset:

$$R_1^2(\mathbf{1}, R_3, R_3^2, R_3^3) = (R_1^2, i_4, R_2^2, i_3),$$
$$r_1(\mathbf{1}, R_3, R_3^2, R_3^3) = (r_1, i_1, r_4, R_2),$$
$$r_2(\mathbf{1}, R_3, R_3^2, R_3^3) = (r_2, i_2, r_3, R_2^3),$$
$$r_1^2(\mathbf{1}, R_3, R_3^2, R_3^3) = (r_1^2, R_1^3, r_3^2, i_6),$$
$$r_2^2(\mathbf{1}, R_3, R_3^2, R_3^3) = (r_2^2, R_1, r_4^2, i_5), \qquad (4.3.9)$$

of $C_4 = \{1, R_3, R_3^2, R_3^3\}$ to label the base states

$$\{|1\rangle = |1\rangle, |2\rangle = |R_1^2\rangle, |3\rangle = |r_1\rangle, |4\rangle = |r_2\rangle, |5\rangle = |r_1^2\rangle, |6\rangle = |r_2^2\rangle\}. \qquad (4.3.10)$$

Furthermore, let us suppose that the base states of the quantum problem satisfy the 0_4 local symmetry conditions as well. Then they can be labeled just like the classical bases as indicated in Figure 4.3.1(b). Imposing these conditions implies that the wave functions (4.3.1) in the valleys have the fourfold (0_4) symmetry of their locality. Later on we shall consider wave functions having other types of local symmetry.

INTRODUCTION TO SYMMETRY BREAKING AND INDUCED REPRESENTATIONS   259

Vectors (4.3.10) which satisfy 0_4 local symmetry conditions are the basis of what is called an INDUCED REPRESENTATION $\mathscr{I} = D^{0_4} \uparrow O$ of octahedral group O induced by irrep D^{0_4} of subgroup C_4. Let us construct the induced representation of octahedral rotation i_4, for example. Operation on bases (4.3.10) gives

$$
\begin{aligned}
i_4|1\rangle &= i_4|1\rangle, & i_4|2\rangle &= i_4 R_1^2|1\rangle, & i_4|3\rangle &= i_4 r_1|1\rangle, & i_4|4\rangle &= i_4 r_2|1\rangle, & i_4|5\rangle &= i_4 r_1^2|1\rangle, & i_4|6\rangle &= i_4 r_2^2|1\rangle, \\
&= R_1^2|1\rangle, & &= R_3^3|1\rangle, & &= i_5|1\rangle, & &= i_6|1\rangle, & &= i_2|1\rangle, & &= i_1|1\rangle, \\
&= |2\rangle, & &= |1\rangle, & &= |6\rangle, & &= |5\rangle, & &= |4\rangle, & &= |3\rangle,
\end{aligned}
$$
(4.3.11)

where the octahedral multiplication nomogram, Eqs. (4.3.9) and (4.3.8), were used in turn. Expressing these results in matrix form gives

$$
\mathscr{I}^{0_4 \uparrow 0}(i_4) = \begin{pmatrix}
\langle 1|i_4|1\rangle & \langle 1|i_4|2\rangle & \cdots & \langle 1|i_4|6\rangle \\
\langle 2|i_4|1\rangle & \langle 2|i_4|2\rangle & & \cdot \\
\cdot & & & \cdot \\
\cdot & & & \cdot \\
\langle 6|i_4|1\rangle & \langle 6|i_4|2\rangle & \cdots & \langle 6|i_4|6\rangle
\end{pmatrix}
$$

$$
= \begin{pmatrix}
\cdot & 1 & \cdot & \cdot & \cdot & \cdot \\
1 & \cdot & \cdot & \cdot & \cdot & \cdot \\
\cdot & \cdot & \cdot & \cdot & \cdot & 1 \\
\cdot & \cdot & \cdot & \cdot & 1 & \cdot \\
\cdot & \cdot & \cdot & 1 & \cdot & \cdot \\
\cdot & \cdot & 1 & \cdot & \cdot & \cdot
\end{pmatrix}.
$$
(4.3.12)

Note that the transformed base states $\{g|1\rangle, g|2\rangle, \ldots\}$ such as (4.3.11), can be derived more quickly simply by inspecting Figure 4.3.1 and using Figure 4.1.2, which defines octahedral g. The $0_4 \uparrow O$ induced representations must be reduced now to solve the problems diagrammed in Figure 4.3.1. Later on we will discuss other kinds of induced representations.

B. Model Solving and Induced Representation Reduction

The Newton equations of motion for the classical spring-mass problem are second-order differential equations

$$m \frac{\partial^2}{\partial t^2} |x\rangle = \mathbf{a}|x\rangle,$$

or

$$m \frac{\partial^2 \langle j|x\rangle}{\partial t^2} = \sum_k \langle j|\mathbf{a}|k\rangle \langle k|x\rangle, \qquad (4.3.13)$$

coupled by the acceleration matrix (4.3.2). The Schrödinger equations of the quantum problem are first-order differential equations,

$$i\hbar \frac{\partial}{\partial t}|\psi\rangle = \mathbf{H}|\psi\rangle$$

or

$$i\hbar \frac{\partial \langle j|\psi\rangle}{\partial t} = \sum_k \langle j|\mathbf{H}|k\rangle\langle k|\psi\rangle, \quad (4.3.14)$$

coupled by the Hamiltonian matrix (4.3.3). However, either set of equations becomes decoupled if one first solves time-independent eigenvalue equations,

$$\mathbf{a}|e_j\rangle = \alpha_j|e_j\rangle, \quad (4.3.15)$$

for the spring-mass problem and Schrödinger's equation,

$$\mathbf{H}|e_j\rangle = E_j|e_j\rangle, \quad (4.3.16)$$

for the quantum problem. In either problem the eigenvalues determine the energy spectrum. The quantum eigenvalues,

$$E_j = \hbar\omega_j, \quad (4.3.17)$$

are proportional to the spectral frequencies of the quantum system. The square roots

$$(\alpha_j/m)^{1/2} = \omega_j, \quad (4.3.18)$$

of the classical eigenvalues give the resonant frequencies of the spring-mass system.

The eigenvectors $|e_j\rangle$ can be obtained for either octahedral problem by applying the octahedral elementary P operators to the first ket $|1\rangle$,

$$|e_j^\alpha\rangle = P_{jk}^\alpha|1\rangle/(N_k^\alpha)^{1/2}, \quad (4.3.19a)$$

where normalization factor N_k^α is chosen so that

$$\langle e_j^\alpha|e_j^\alpha\rangle = 1,$$

or

$$N_k^\alpha = \langle 1|P_{jk}^{\alpha\dagger}P_{jk}^\alpha|1\rangle = \langle 1|P_{kk}^\alpha|1\rangle. \quad (4.3.19b)$$

There are 24 octahedral P operators, but only six $|e_j^\alpha\rangle$ states are possible. One needs a way to tell beforehand whether a projected state $P_{jk}^\alpha|1\rangle$ is going to have a nonzero norm N_k^α.

The key is to use only those P_{jk}^α whose right-hand "bodyguards" P_k^α are compatible with the state $|1\rangle$. Since the state $|1\rangle$ has local C_4 symmetry condition (4.3.7) of type 0_4, only those C_4-defined P_k^α with $k = 0_4$ are

INTRODUCTION TO SYMMETRY BREAKING AND INDUCED REPRESENTATIONS 261

compatible. Among the C_4-defined subgroup labels in Eqs. (4.2.37)–(4.2.41), the $k = 0_4$ label appears under three different octahedral irrep labels, namely, $\alpha = A_1$, E, and T_1. For $\alpha = A_1$ and $k = 0_4 = j$ [Eq. (4.2.41)] there is one eigenstate,

$$\begin{aligned}|e_{0_4}^{A_1}\rangle &= P_{0_4}^{A_1}|1\rangle/(N^{A_1})^{1/2} \\ &= \tfrac{1}{24}\sum_g \mathscr{D}^{A_1*}(g)g|1\rangle/(N^{A_1})^{1/2} \\ &= (|1\rangle + |2\rangle + |3\rangle + |4\rangle + |5\rangle + |6\rangle)/(6)^{1/2}.\end{aligned} \quad (4.3.20)$$

The formula (3.4.19) for P^α and the irreps $\mathscr{D}^{A_1}(g) \equiv 1$ from the octahedral characters (4.1.11) are used. For $\alpha = E$ and $k = 0_4$ [Eq. (4.2.39)] there is an eigenstate for each value of $j = 0_4$ and 2_4. For $j = 0_4$ we have

$$\begin{aligned}|e_{0_4}^E\rangle &= P_{0_4 0_4}^E |1\rangle/(N^E)^{1/2} \\ &= \tfrac{2}{24}\sum_g \mathscr{D}_{0_4 0_4}^{E*}(g)g|1\rangle/(N^E)^{1/2} \\ &= \tfrac{2}{24}\big[(1 + R_3 + R_3^2 + R_3^3) + (R_1^2 + i_4 + R_2^2 + i_3) \\ &\quad - \tfrac{1}{2}(r_1 + i_1 + r_4 + R_2) - \tfrac{1}{2}(r_2 + i_2 + r_3 + R_2^3) \\ &\quad - \tfrac{1}{2}(r_1^2 + R_1^3 + r_3^2 + i_6) - \tfrac{1}{2}(r_2^2 + R_1 + r_4^2 + i_5)\big]|1\rangle/(N^E)^{1/2},\end{aligned} \quad (4.3.21)$$

where formula (3.4.19) is used again, this time with the $\mathscr{D}^E$ irrep (4.2.19). (Recall that tetragonal standing- and moving-wave E irreps are the same.) The g sum is collected into the six C_4 cosets which label the bases $\{|1\rangle, |2\rangle, \ldots, |6\rangle\}$. Thus we have

$$|e_{0_4}^E\rangle = (2|1\rangle + 2|2\rangle - |3\rangle - |4\rangle - |5\rangle - |6\rangle)/(2\sqrt{3}). \quad (4.3.22)$$

The second E partner with $j = 2_4$ is given:

$$\begin{aligned}|e_{2_4}^E\rangle &= P_{2_4 0_4}^E |1\rangle/(N^E)^{1/2} \\ &= \frac{2}{24}\sum_g \mathscr{D}_{2_4 0_4}^{E*}(g)g|1\rangle/(N^E)^{1/2} \\ &= \frac{2}{24}\Big[\frac{\sqrt{3}}{2}(r_1 + i_1 + r_4 + R_2) + \frac{\sqrt{3}}{2}(r_2 + i_2 + r_3 + R_2^3) \\ &\quad - \frac{\sqrt{3}}{2}(r_1^2 + R_1^3 + r_3^2 + i_6) - \frac{\sqrt{3}}{2}(r_2^2 + R_1 + r_4^2 + i_5)\Big]|1\rangle/(N^E)^{1/2},\end{aligned}$$

$$|e_{2_4}^E\rangle = (|3\rangle + |4\rangle - |5\rangle - |6\rangle)/2. \quad (4.3.23)$$

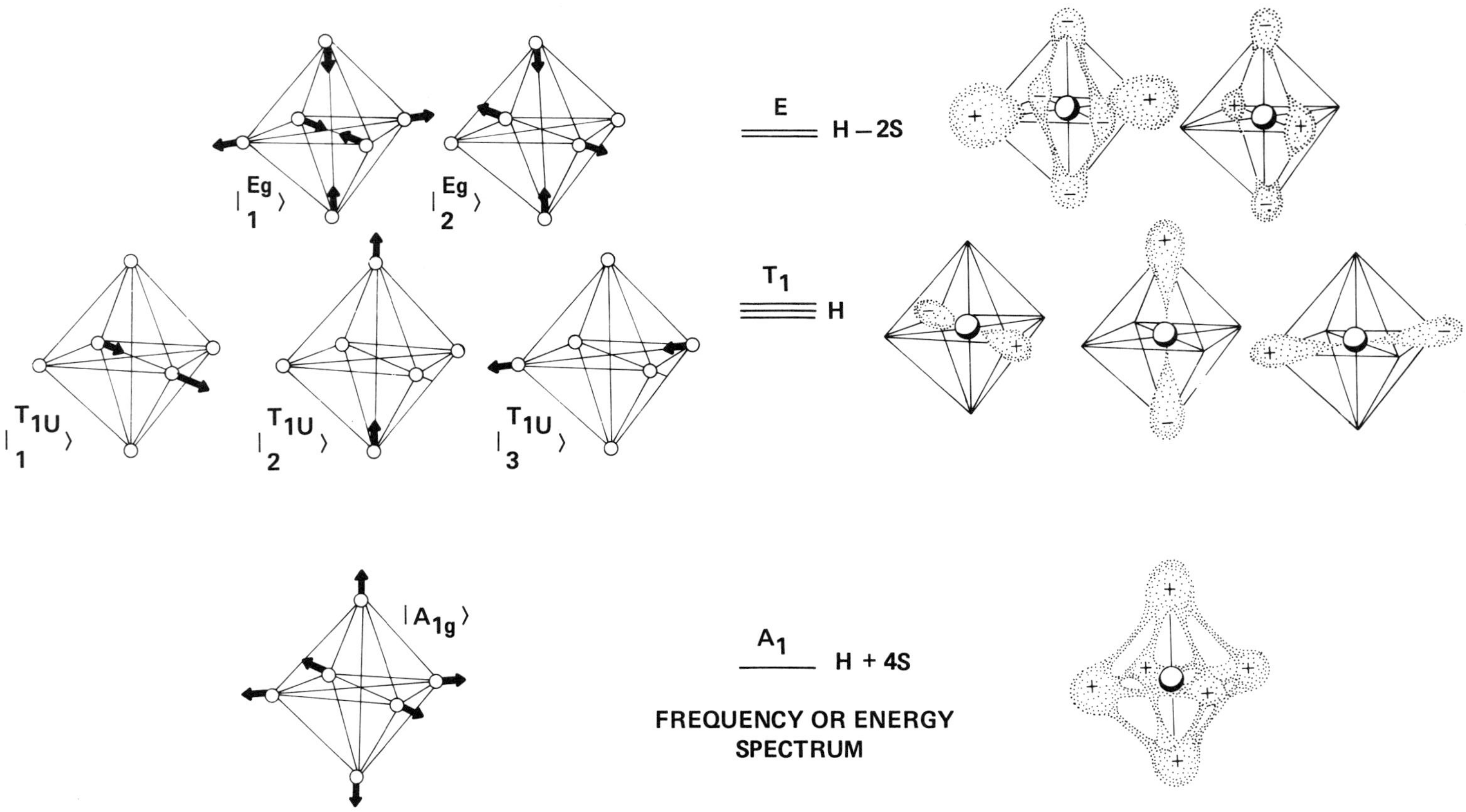

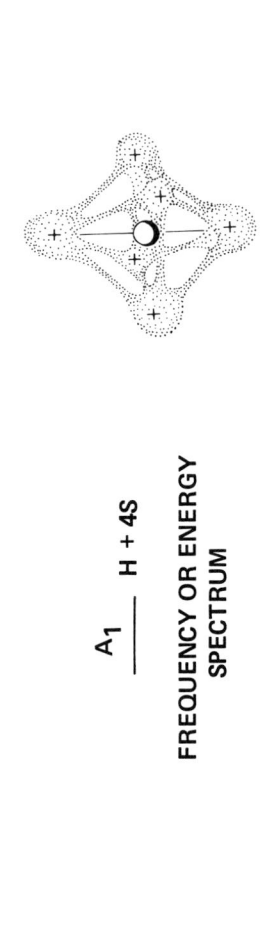

Figure 4.3.2 Eigensolutions for octahedral systems. (a) Normal vibration modes for classical system. (b) Eigenwaves for quantum system. (Note that tunneling amplitude is negative: $S < 0$.)

INTRODUCTION TO SYMMETRY BREAKING AND INDUCED REPRESENTATIONS 263

Finally, for $\alpha = T_1$ and $k = 0_4$ [Eq. (4.2.37)], there are three more eigenstates, one for each value of $j = 1_4, 3_4$, and 0_4. Putting these together with the A_1 and E eigenvectors gives all six eigensolutions for the octahedral problems. Sketches of the quantum eigenwaves are given in Figure 4.3.2(b). [The corresponding classical vibration modes are shown in Figure 4.3.2(a).] The T_1 solution for $j = 1_4$ is

$$|e_{1_4}^{T_1}\rangle = P_{1_4 0_4}^{T_1}|1\rangle/(N^{T_1})^{1/2}$$

$$= \frac{3}{24}\sum_g \mathscr{D}_{1_4 0_4}^{T_1 *}(g)g|1\rangle/(N^{T_1})^{1/2}$$

$$= \frac{3}{24}\left[-\frac{1}{\sqrt{2}}(r_1 + i_1 + r_4 + R_2) + \frac{1}{\sqrt{2}}(r_2 + i_2 + r_3 + R_2^3)\right.$$

$$\left. - \frac{i}{\sqrt{2}}(r_1^2 + R_1^3 + r_3^2 + i_6) + \frac{i}{\sqrt{2}}(r_2^2 + R_1 + r_4^2 + i_5)\right]|1\rangle/(N^{T_1})^{1/2},$$

$$|e_{1_4}^{T_1}\rangle = (-|3\rangle + |4\rangle - i|5\rangle + i|6\rangle)/2. \qquad (4.3.24)$$

The tetragonal moving-wave irreps (4.2.35) were used. Similarly, the other T_1 partners with $j = 3_4$ and 0_4 are found:

$$|e_{3_4}^{T_1}\rangle = (|3\rangle - |4\rangle - i|5\rangle + i|6\rangle)/2, \qquad (4.3.25)$$

$$|e_{0_4}^{T_1}\rangle = (|1\rangle - |2\rangle)/\sqrt{2}. \qquad (4.3.26)$$

The T_1-type eigenfunctions sketched in Figure 4.3.2 are the real standing-wave partners,

$$|e_x^{T_1}\rangle = \left(-|e_{1_4}^{T_1}\rangle + |e_{3_4}^{T_1}\rangle\right)/\sqrt{2} = (|3\rangle - |4\rangle)/\sqrt{2}, \qquad (4.3.27a)$$

$$|e_y^{T_1}\rangle = i\left(|e_{1_4}^{T_1}\rangle + |e_{3_4}^{T_1}\rangle\right)/\sqrt{2} = (|5\rangle - |6\rangle)/\sqrt{2}, \qquad (4.3.27b)$$

$$|e_z^{T_1}\rangle = |e_{0_4}^{T_1}\rangle = (|1\rangle - |2\rangle)/\sqrt{2}, \qquad (4.3.27c)$$

obtained through the transformation (4.2.36).

It is easy to verify that the vectors (4.3.20)–(4.3.27) are all eigenvectors of the Hamiltonian matrix (4.3.3). The energy spectrum,

$$E^{A_1} = H + T + 4S,$$
$$E^{T_1} = H - T,$$
$$E^E = H + T - 2S, \qquad (4.3.28)$$

is indicated by singlet (A_1), triplet (T_1), and doublet (E) levels next to the wave functions in Figure 4.3.2. For negligible transaxial tunneling ($T \sim 0$) the

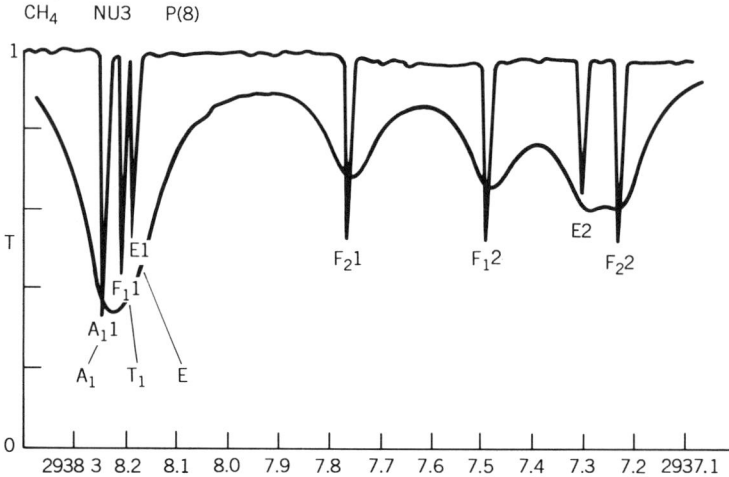

Figure 4.3.3 Evidence of an (A_1T_1E) spectral cluster in methane laser spectra. (Courtesy of Dr. Allan Pine, MIT Lincoln Laboratories, from *Journal of Optical Society of America* **66**, 97 (1976)). The ordering and approximate spacing of the $A_1 T_1$ and E lines is consistent with that of Figure 4.3.2.

$|E^{T_1} - E^{A_1}|$ difference becomes twice the $|E^E - E^{T_1}|$ difference. This two-to-one splitting is being observed repeatedly in laser spectra of tetrahedral and octahedral molecules. One of the first resolved (A_1T_1E) "clusters" is shown in part of Allen Pine's methane spectrum in Figure 4.3.3. (The line intensities actually correspond to nuclear spin and inversion degeneracy and *not* to the octahedral degeneracies.)

Notice that the E and A_1 wave functions in Figure 4.3.2 have even inversion parity, i.e., they belong to E_g and A_{1g} irreps of O_h. The T_1 wave functions have odd parity and are therefore T_{1u}-type bases for O_h. The mathematical significance of the (A_{1g}, T_{1u}, E_g) combination will be explained in the following section.

C. Frobenius Reciprocity Theorem and Factored P Operators

It is important to understand the role of subgroup correlation here. The A_1, T_1, and E octahedral states were the only ones correlated with an 0_4 substate. They are therefore the only octahedral irreps to appear in the 0_4 column of the C_4 subgroup correlation table (4.2.42b). This is the necessary and sufficient condition that A_1, T_1, and E should appear in the reduction of $0_4 \uparrow O$. We now shall prove that, in general, the ath column of a $G \supset H$ correlation table will contain the frequencies f^α ($D^a \uparrow G$) of G irreps $\mathscr{D}^\alpha$ in an induced representation $D^a \uparrow G$. By definition, the αth row contains the frequencies $f^a(\mathscr{D}^\alpha \downarrow H)$ of H irreps D^a in a subduced representation

$\mathscr{D}^\alpha \downarrow H$. The beautiful relation

$$f^\alpha(D^a \uparrow G) = f^a(\mathscr{D}^\alpha \downarrow H) \qquad (4.3.29)$$

is called the FROBENIUS RECIPROCITY THEOREM. It is easy to understand and prove this theorem using correlated P operators by generalizing the $0_4 \uparrow O$ example in the preceding section.

Suppose a base state $|1\rangle$ satisfies a local symmetry condition,

$$|1\rangle = P^a|1\rangle, \qquad (4.3.30a)$$

where

$$P^a = (1/{}^oH)\sum_h D^{a*}(h)h \qquad (4.3.30b)$$

is an irrep projector of subgroup $H = \{1, h, h', \dots\} \subset G$ for which

$$hP^a = D^a(h)P^a \qquad (4.3.31)$$

for all h in H. Combining this with condition (4.3.30a) gives

$$h|1\rangle = D^a(h)|1\rangle. \qquad (4.3.32)$$

For example, let us consider a local moving-wave state $|1\rangle$ which satisfies an $H = C_4$ condition

$$|1\rangle = P^{1_4}|1\rangle, \qquad (4.3.30a)_x$$

where

$$P^{1_4} = \tfrac{1}{4}(1 + iR_3 - R_3^2 - iR_3^3) \qquad (4.3.30b)_x$$

and

$$R_3|1\rangle = -i|1\rangle. \qquad (4.3.32)_x$$

The sample equations are generalizations of Eqs. (4.3.6) and (4.3.7).

For each coset $l_2 H = \{l_2, l_2 h, l_2 h', \dots\}, l_3 H, \dots$ of subgroup H let there be one more orthogonal states $|m\rangle = l_m|1\rangle$. The ${}^oG/{}^oH$ orthonormal states

$$\{D^a \uparrow G\} = \{|1\rangle, |2\rangle = l_2|1\rangle, |3\rangle = l_3|1\rangle, \dots, l_{oG/oH}|1\rangle\} \qquad (4.3.33)$$

are a basis for the induced representation $D^a \uparrow G$ of "supergroup" $G =$

{1, g, ...}. The induced representation is defined by

$$\mathcal{I}_{ij}^{a \uparrow G} = \langle i|g|j \rangle = \begin{cases} D^a(l_i^{-1}gl_j), & \text{if } l_i^{-1}gl_j \text{ is in } H, \\ 0, & \text{if not} \end{cases} \quad (4.3.34)$$

For example, the $1_4 \uparrow O$ induced representation of i_4 is

$$\mathcal{I}^{1_4 \uparrow O}(i_4) = \begin{pmatrix} \cdot & i & \cdot & \cdot & \cdot & \cdot \\ -i & \cdot & \cdot & \cdot & \cdot & \cdot \\ \cdot & \cdot & \cdot & \cdot & \cdot & -i \\ \cdot & \cdot & \cdot & \cdot & -i & \cdot \\ \cdot & \cdot & \cdot & i & \cdot & \cdot \\ \cdot & \cdot & i & \cdot & \cdot & \cdot \end{pmatrix}. \quad (4.3.34)_x$$

One chooses one element l_j from each coset to label the bases of an induced representation. We shall call these chosen elements the COSET LEADERS. The same choices were made in the example above as in Eqs. (4.3.10) and (4.3.11).

To reduce $\mathcal{I}^{a \uparrow G}$ one uses irreducible idempotents $P_a^\alpha \equiv P_{aa}^\alpha$ and nilpotents P_{ba}^α of G. It is convenient to use the ones which are defined by the subgroup H, i.e., for which

$$P_k^\alpha P^a = \delta^{ka} P_a^\alpha \quad (4.3.35)$$

for all irreducible idempotents P^a of H. Now, suppose irrep D^a of H is repeated $f^a(\mathcal{D}^\alpha \downarrow H)$ times in the subduced irrep $\mathcal{D}^\alpha \downarrow H$ of G.

$$\mathcal{D}^\alpha \downarrow H = \mathcal{D}^\alpha(h) = \begin{pmatrix} D^a(h) & & & \\ & \ddots & & \\ & & D^a(h) & \\ & & & \ddots \end{pmatrix}. \quad (4.3.36)$$

Each repeat corresponds to a G idempotent $\{P_a^\alpha \cdots P_{a'}^\alpha \cdots\}$ whose H sublabel is (a), that is, a base state correlated with (a). These are the only idempotents which will not annihilate P^a or state $|1\rangle = P^a|1\rangle$ according to Eq. (4.3.35). Each one gives rise to an orthogonal set of states

$$|_{ba}^\alpha\rangle = P_{ba}^\alpha|1\rangle/(N_a^\alpha)^{1/2} = (l^\alpha/{}^\circ G)\sum_g \mathcal{D}_{ba}^{\alpha*}(g)g|1\rangle/(N_a^\alpha)^{1/2}, \quad (4.3.37)$$

and another $\mathcal{D}^\alpha$ in the reduction of $\mathcal{I}^{a \uparrow G}$. If the states $P_{ba}^\alpha|1\rangle$ exist then there are f^a repeated $\mathcal{D}^\alpha$ in $D^a \uparrow G$ which proves the Frobenius theorem (4.3.29). The norm (4.3.19b) of these states works out to be the following nonzero value:

$$\begin{aligned}
N_a^\alpha &= \langle 1|P_{aa}^\alpha|1\rangle = (l^\alpha/{}^\circ G)\sum_g \mathscr{D}_{aa}^{\alpha*}(g)\langle 1|g|1\rangle \\
&= (l^\alpha/{}^\circ G)\sum_h \mathscr{D}_{aa}^{\alpha*}(h)D^a(h) \\
&= (l^\alpha/{}^\circ G)\mathscr{D}_{aa}^{\alpha*}\left(\sum_h D^a(h)h\right) \\
&= (l^\alpha/{}^\circ G)\,{}^\circ H\mathscr{D}_{aa}^{\alpha*}(P^a), \\
N_a^\alpha &= l^\alpha\,{}^\circ H/{}^\circ G.
\end{aligned} \qquad (4.3.38)$$

Equation (4.3.32) is used to get the second line, while Eq. (4.3.30b) is used for the fourth line. The fact that $\mathscr{D}_{aa}^\alpha(P^a) = 1$ follows from the assumed form (4.3.36) of $\mathscr{D}^\alpha$. For example, by combining the representations (4.2.35) of C_4 one finds

$$\mathscr{D}^{T_1}(P^{1_4}) = \mathscr{D}^{T_1}(1 + iR_3 - R_3^2 - iR_3^3)/4 = \begin{pmatrix} 1 & 0 & 0 \\ 0 & 0 & 0 \\ 0 & 0 & 0 \end{pmatrix}, \qquad (4.3.39)$$

and similarly for $\mathscr{D}^{T_2}(P^{1_4})$. From this one concludes that $\mathscr{I}^{1_4 \uparrow 0}$ can be reduced to

$$\mathscr{T}^\dagger \mathscr{I}_{(g)}^{1_4 \uparrow 0} \mathscr{T} = \begin{pmatrix} \boxed{(\mathscr{D}_{(g)}^{T_1})} & \cdot & \cdot \\ \cdot & & \cdot \\ \cdot & \cdot & \boxed{(\mathscr{D}_{(g)}^{T_2})} \end{pmatrix}. \qquad (4.3.40)$$

Note that only T_1 and T_2 appear in the (1_4) column of correlation table (4.2.42b).

The projection operator algebra which leads to this reduction can be simplified. Suppose the sum Σ_g is replaced by a sum and subsum $\Sigma_l \Sigma_h$ over coset leaders $\{l_1 \equiv 1, l_2, \ldots\}$ and subgroup elements $\{1, h, \ldots\}$, respectively.

$$P_{jk}^a = (l^\alpha/{}^\circ G) \sum_l \sum_h \mathscr{D}_{jk}^{\alpha*}(lh)\,lh. \qquad (4.3.41)$$

The projection operation can be factored if $\mathscr{D}^\alpha$ is H-defined. The fundamental definition (3.3.11) gives

$$\begin{aligned}
\mathscr{D}_{jk}^\alpha(lh)P_{jk}^\alpha &= P_j^\alpha lh P_k^\alpha \\
&= D^k(h)P_j^\alpha l P_k^\alpha \\
&= D^k(h)\mathscr{D}_{jk}^\alpha(l)P_{jk}^\alpha,
\end{aligned} \qquad (4.3.42)$$

where Eqs. (4.3.31) and (4.3.35) were used. Then the projection

$$P_{jk}^\alpha|1\rangle = (l^\alpha/°G)\sum_l \mathscr{D}_{jk}^{\alpha*}(l)l\left[\sum_h D^{k*}(h)h|1\rangle\right]$$
$$= (l^{\alpha °}H/°G)\sum_l \mathscr{D}_{jk}^{\alpha*}(l)l[P^k|1\rangle] \qquad (4.3.43)$$

is reduced to a sum just over coset leaders l if $P^a|1\rangle = |1\rangle$.

$$|_{ja}^\alpha\rangle = P_{ja}^\alpha|1\rangle/(N_a^\alpha)^{1/2} = (l^{\alpha °}H/°G)^{1/2}\sum_l \mathscr{D}_{ja}^{\alpha*}(l)|l\rangle. \qquad (4.3.44)$$

The normalization formula (4.3.38) has been included in this result.

This coset factorization was observed in the previous examples of Eqs. (4.3.21), (4.3.23), and (4.3.24). It cuts the arithmetic labor down by a factor of $°G/°H$, and allows one to tabulate that many fewer $\mathscr{D}(g)$ matrices. It also simplifies energy matrix formulas such as

$$\langle_{ja}^\alpha|H|_{kb}^\beta\rangle = \langle 1|P_{ja}^{\alpha\dagger}\mathbf{H}P_{kb}^\beta|1\rangle/(N^\alpha N^\beta)^{1/2}$$
$$= \langle 1|\mathbf{H}P_{aj}^\alpha P_{kb}^\beta|1\rangle/(N^\alpha N^\beta)^{1/2}$$
$$= \delta^{\alpha\beta}\delta_{jk}\langle 1|\mathbf{H}P_{ab}^\alpha|1\rangle/N^\alpha. \qquad (4.3.45)$$

Here G symmetry of Hamiltonian **H** was assumed so that $P_{aj}^\alpha = P_{ja}^{\alpha\dagger}$ could commute through it. The coset factorization formula (4.3.43) reduces the nonzero components to a sum

$$\langle_{ja}^\alpha|\mathbf{H}|_{jb}^\alpha\rangle = \sum_l \langle 1|\mathbf{H}|l\rangle \mathscr{D}_{ab}^{\alpha*}(l) \qquad (4.3.46)$$

just over coset leaders $\{l_1 \equiv 1, l_2, \ldots, l_{oG/oH}\}$. For example, the energy formulas (4.3.28) can be rederived just from one row of matrix (4.3.3). Given

$$\langle 1|\mathbf{H}|l\rangle = (H, T, S, S, S, S),$$
$$l = (1, R_1^2, r_1, r_2, r_1^2, r_2^2),$$

one has for the E levels (here $\mathscr{D}^{E*} = \mathscr{D}^E$ is real)

$$\langle_{j_{04}}^E|\mathbf{H}|_{j_{04}}^E\rangle = H\mathscr{D}_{0_40_4}^E(1) + T\mathscr{D}_{0_40_4}^E(R_1^2)$$
$$+ S\left(\mathscr{D}_{0_40_4}^E(r_1) + \mathscr{D}_{0_40_4}^E(r_2) + \mathscr{D}_{0_40_4}^E(r_1^2) + \mathscr{D}_{0_40_4}^E(r_2^2)\right)$$
$$= H - T + 2S, \qquad (4.3.46)_x$$

and similarly for the T_1 and A_1 levels.

The $(A_1 T_1 E)$ induced representation is labeled $(A_{1g} T_{1u} E_g)$ when O_h inversion is included. Indeed, the first column of the $O_h \supset C_{4v}$ correlation table (4.2.46c) gives just these irreps. By extending the local symmetry of state $|1\rangle$ in Section 4.3.B from 0_4 (of C_4) to A' (of C_{4v}) one obtains the induced representation

$$D^{A'} \uparrow O_h \sim \mathscr{D}^{A_{1g}} \oplus \mathscr{D}^{T_{1u}} \oplus \mathscr{D}^{E_g}. \tag{4.3.47}$$

By specifying the local C_{4v} reflection symmetry one predicts the overall O_h inversion symmetry of the induced wave states in Figure 4.3.2.

The moving-wave irreps 1_4 and 3_4 of C_4 are part of a subduced E irrep of C_{4v}. [Recall Eq. (3.6.10).] The induced representations $1_4 \uparrow O$ and $3_4 \uparrow O$ both contribute a $T_1 + T_2$ pair of O irreps. Therefore it should not be surprising that $E \uparrow O_h$ contains two such pairs distinguished by opposite parities. The E column of (4.2.46c) gives

$$D^E \uparrow O_h \sim \mathscr{D}^{T_{1g}} + \mathscr{D}^{T_{2g}} + \mathscr{D}^{T_{1u}} + \mathscr{D}^{T_{2u}}. \tag{4.3.48}$$

Indeed, the Frobenius theorem applies as well to multiply degenerate representations such as D^E. A physical example of an $E \uparrow O_h$ representation will appear in the SF_6 vibration problem in Section 4.4. (It is instructive to rework the analysis of this subsection assuming degenerate D^a. See Problem 4.3.2.)

D. Spontaneous or Internal Symmetry Breaking

When all the tunneling coefficients, S as well as T, go to zero the A_1, T_1, and E levels in Figure 4.3.2 collapse into a sixfold degeneracy. Then any combination of the base states are eigenvectors, including each one of the original base states $\{|1\rangle, |2\rangle, \ldots, |6\rangle\}$. One can then imagine that the system is simply six identical disconnected wells, each with a C_4 or C_{4v} local symmetry. The meaning of the "O_h connection" is lost, even though the Hamiltonian still really has O_h symmetry. This is a simple example of spontaneous symmetry breaking. Single C_4 symmetric states like $|1\rangle$ become "frozen in" within a system that intrinsically has a much higher O_h symmetry. With this freezing or breakdown comes higher (sixfold) degeneracy corresponding to an induced representation.

More complex examples of spontaneous symmetry breaking include ferromagnetism, order-disorder phase transitions, Jahn-Teller distortions, and many other effects both real and imagined in other areas of physics such as high-energy theory. We shall discuss some of these in detail later. For now let us review a simpler example.

By applying the Frobenius theorem to the D_6 correlations (3.6.27), it is possible to understand better the spontaneous symmetry breaking there. The columns of the $D_6 \supset C_2$ correlation table give two kinds of induced representations. The first is

$$D^{0_2} \uparrow D_6 \sim A_1 + E_1 + E_2 + B_1, \qquad (4.3.49a)$$

which corresponds to C_2 symmetric waves frozen in twofold symmetric valleys. The second is

$$D^{1_2} \uparrow D_6 \sim B_2 + E_2 + E_1 + A_2, \qquad (4.3.49b)$$

which corresponds to C_2 *anti*symmetric waves frozen in twofold valleys. These waves were sketched in Figure 3.6.5, and it is easy to see how they belong to induced representations. In the absence of communication between valleys each set becomes degenerate. Then wave functions which are stuck in a single valley may be eigensolutions.

One should note that the induced representations $D^{a_6} \uparrow D_6$ are useful for the opposite extreme in which all valleys disappear. The first and fourth columns of $D_6 \supset C_6$ correlations (3.6.27a) give

$$\begin{aligned} D^{0_6} \uparrow D_6 &\sim A_1 + A_2, \\ D^{3_6} \uparrow D_6 &\sim B_1 + B_2. \end{aligned} \qquad (4.3.50)$$

These are the band-gap degeneracies that occur when the potential is constant. The wave states belonging to k_m for which $m = \pm 1 \bmod 6$ and $m = \pm 2 \bmod 6$ are bases for representations

$$\begin{aligned} D^{1_6} \uparrow D_6 &= E_1, \\ D^{2_6} \uparrow D_6 &= E_2, \end{aligned} \qquad (4.3.51)$$

respectively. These are both induced and irreducible representations of D_6.

Finally, one should observe that the regular representation $\mathscr{R}$ of group G (recall definition of $\mathscr{R}$ vis-à-vis $G = C_{3v}$ in Section 3.4) is an induced representation D^{0_1} (of C_1)$\uparrow G$ induced by the smallest subgroup C_1. For example, the octahedral regular representation is

$$\mathscr{R} = D^{0_1} \uparrow O = \mathscr{D}^{A_1} \oplus \mathscr{D}^{A_2} \oplus 2\mathscr{D}^{E} \oplus 3\mathscr{D}^{T_1} \oplus 3\mathscr{D}^{T_2}.$$

According to the Frobenius theorem the frequency of repetition for each irrep must equal its dimension, since C_1 has only one kind of irrep.

The simplest regular or induced representation is $D^{0_1} \uparrow C_2$ which was the basis of the ammonia two-state inversion model in Section 2.12. In the absence of tunneling or external field perturbation this basis belongs to a twofold degenerate inversion doublet.

INTRODUCTION TO SYMMETRY BREAKING AND INDUCED REPRESENTATIONS   271

E. External Symmetry Breaking

Degeneracy of spectra for a physical system may correspond to irreducible representations of its symmetry group or to combinations of irreps induced by subgroup irreps. It is generally possible to split any degenerate or nearly degenerate level by introducing external forces of lower symmetry. The splitting or "level-crossing" effect of an electric field on the ammonia doublet was shown in Figure 2.12.8. Let us now consider some more complex level splitting and crossing using the octahedral states as a basis.

(a) Electric Quadrupole or D_{4h} Splitting

Suppose the classical or quantum octahedral systems in Figure 4.3.1 are subjected to a D_{4h} perturbation as indicated in Figure 4.3.4. This perturbation could be two additional identical springs attached to coordinate states $|1\rangle$ and $|2\rangle$ along the z axis of the classical oscillator. Or it could be a change in the masses of the two particles on the z axis. For the quantum system let us imagine a quadrupole electric potential $V_Q V(z^2, x^2 + y^2)$ such as one might find at the center of a homogeneously charged x-y slab. Later, in Chapters 5–7, the definition of "quadrupolarity" will be made more precise. For now we imagine any field that has D_{4h} symmetry, i.e., 90° and 180° rotations around the z axis as well as "vertical" xz- or yz-plane reflections and "horizontal" xy-plane reflections.

Let the Hamiltonian (4.3.3) be perturbed to the form

$$\langle H + V_Q \rangle = \begin{pmatrix} & |1\rangle & |2\rangle & |3\rangle & |4\rangle & |5\rangle & |6\rangle \\ H+Q & 0 & S & S & S & S \\ 0 & H+Q & S & S & S & S \\ S & S & H & 0 & S & S \\ S & S & 0 & H & S & S \\ S & S & S & S & H & 0 \\ S & S & S & S & 0 & H \end{pmatrix}, \quad (4.3.52)$$

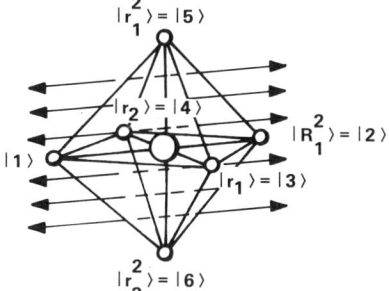

Figure 4.3.4 D_{4h} symmetric Q perturbation of octahedrally symmetric system.

where Q is the change in energy of potential wells 1 and 2 due to the perturbation. (For simplicity, we assume next-nearest-neighbor tunneling T is zero.) Let a similar change be made to the acceleration matrix (4.3.2) for the spring-mass problem. More complicated D_{4h}-symmetric perturbations are possible, but the V_Q is sufficient for exhibiting the symmetry-breaking effects.

The first task in any symmetry-breaking problem is to see which levels split. According to the $D_4 \subset O$ correlation (4.2.42a) we have the following for the A_1, T_1, and E levels in Figure 4.3.2:

$$\mathscr{D}^{A_{1g}} \downarrow D_{4h} = A_{1g}, \tag{4.3.53a}$$

$$\mathscr{D}^{T_{1g}} \downarrow D_{4h} = A_{2u} \oplus E_u, \tag{4.3.53b}$$

$$\mathscr{D}^{E_g} \downarrow D_{4h} = A_{1g} \oplus B_{1g}. \tag{4.3.53c}$$

(Here, the inversion parity subindices u and g have been added to label the $D_{4h} \supset O_h$ correlation completely.) This implies that V_Q may split the doublet E_g level into A_{1g} and B_{1g} singlets, while the T_{1u} triplet breaks into an A_{2u} singlet and a E_u doublet. Furthermore, there are now two D_{4h} singlet-A_{1g} levels, one from octahedral A_{1g} and one from E_g.

The next task is to find eigenstates and eigenvalues of the new Hamiltonian $H + V_Q$. The correlation (4.3.53) implies that four of the six octahedral states are already eigenvectors if they are D_{4h} defined. Using $O_h \supset D_{4h} \supset D_{2h}$ chain labels [Section 4.2.A(a)] for these states, it follows that the states (from Figure 4.3.2)

$$\left\langle \left\| \begin{array}{c} T_{1u} \\ [E_u] \\ B_{1u} \end{array} \right\rangle \right. = \begin{pmatrix} 0 \\ 0 \\ 1/\sqrt{2} \\ -1/\sqrt{2} \\ 0 \\ 0 \end{pmatrix}, \quad \left\langle \left\| \begin{array}{c} T_{1u} \\ [E_u] \\ B_{2u} \end{array} \right\rangle \right. = \begin{pmatrix} 0 \\ 0 \\ 0 \\ 0 \\ 1/\sqrt{2} \\ -1/\sqrt{2} \end{pmatrix},$$

$$\left\langle \left\| \begin{array}{c} T_{1u} \\ [A_{2u}] \\ A_{2u} \end{array} \right\rangle \right. = \begin{pmatrix} 1/\sqrt{2} \\ -1/\sqrt{2} \\ 0 \\ 0 \\ 0 \\ 0 \end{pmatrix}, \quad \left\langle \left\| \begin{array}{c} E_g \\ [B_{1g}] \\ A_{1g} \end{array} \right\rangle \right. = \begin{pmatrix} 0 \\ 0 \\ 1/2 \\ 1/2 \\ -1/2 \\ -1/2 \end{pmatrix} \tag{4.3.54}$$

must be eigenvectors of $H + V_Q$. This is necessary, since $H + V_Q$ has D_{4h} symmetry, and the $D_{4h} \supset D_{2h}$ labels are all distinct. (Note that the D_{4h} label is bracketed to indicate the maximal symmetry of $H + V_Q$.) On the other hand, the states

$$\left\langle \left\| \begin{matrix} E_g \\ [A_{1g}] \\ A_{1g} \end{matrix} \right\rangle \right. = \begin{pmatrix} 2 \\ 2 \\ -1 \\ -1 \\ -1 \\ -1 \end{pmatrix} \bigg/ \sqrt{12}, \quad \left\langle \left\| \begin{matrix} A_{1g} \\ [A_{1g}] \\ A_{1g} \end{matrix} \right\rangle \right. = \begin{pmatrix} 1 \\ 1 \\ 1 \\ 1 \\ 1 \\ 1 \end{pmatrix} \bigg/ \sqrt{6} \quad (4.3.55)$$

have the same D_{4h} labels, and are therefore to be coupled through a 2×2-matrix representation of $H + V_Q$.

The representation of $H + V_Q$ in this basis is easily computed by combining Eqs. (4.3.52), (4.3.54), and (4.3.55),

$$\langle H + V_Q \rangle = \begin{matrix} & \left|\begin{matrix} T_{1u} \\ [E_u] \\ B_{1u} \end{matrix}\right\rangle & \left|\begin{matrix} T_{1u} \\ [E_u] \\ B_{2u} \end{matrix}\right\rangle & \left|\begin{matrix} T_{1u} \\ [A_{2u}] \\ A_{2u} \end{matrix}\right\rangle & \left|\begin{matrix} A_{1g} \\ [A_{1g}] \\ A_{1g} \end{matrix}\right\rangle & \left|\begin{matrix} E_g \\ [A_{1g}] \\ A_{1g} \end{matrix}\right\rangle & \left|\begin{matrix} E_g \\ [B_{1g}] \\ A_{1g} \end{matrix}\right\rangle \\ \begin{bmatrix} H & \cdot & \cdot & \cdot & \cdot & \cdot \\ \cdot & H & \cdot & \cdot & \cdot & \cdot \\ \cdot & \cdot & H + Q & \cdot & \cdot & \cdot \\ \cdot & \cdot & \cdot & H + 4S + Q/3 & \sqrt{2}\,Q/3 & \cdot \\ \cdot & \cdot & \cdot & \sqrt{2}\,Q/3 & H - 2S + 2Q/3 & \cdot \\ \cdot & \cdot & \cdot & \cdot & \cdot & H - 2S \end{bmatrix} \end{matrix}$$

(4.3.56)

The eigenvalues of the new matrix are easily found. They are plotted in Figure 4.3.5 as a function of Q for a fixed negative values of S. Note the splitting or level shifting of the A_{1g}, T_{1u}, and E_g levels around $Q = 0$ in the center of the figure. Sublevels belonging to states

$$\left| \begin{matrix} T_{1u} \\ 3 \end{matrix} \right\rangle = \left| \begin{matrix} T_{1u} \\ [A_{2u}] \end{matrix} \right\rangle, \quad |A_{1g}\rangle = \left| \begin{matrix} A_{1g} \\ [A_{1g}] \end{matrix} \right\rangle, \quad \text{and} \quad \left| \begin{matrix} E_g \\ 1 \end{matrix} \right\rangle = \left| \begin{matrix} E_g \\ [A_{1g}] \end{matrix} \right\rangle$$

have shifts proportional to Q for small Q. These are called FIRST-ORDER energy shifts or splitting. The slopes of the energy-level trajectories are 1, $\frac{1}{3}$, and $\frac{2}{3}$, respectively, for these sublevels. The slopes correspond to the diagonal terms Q, $Q/3$, and $2Q/3$ in the Hamiltonian matrix (4.3.56).

The A_{1g} sublevels curve away from each other when $|Q|$ becomes comparable to $|S|$, and the off-diagonal components $\sqrt{2}\,Q/3$ take effect. Then the two different A_{1g}-substates become mixed up in order to be eigenvectors of the (2×2) submatrix in Eq. (4.3.56). It is instructive to examine the A_{1g} eigenvectors when $|Q|$ is much greater than $|S|$. Then the A_{1g} eigenvalue

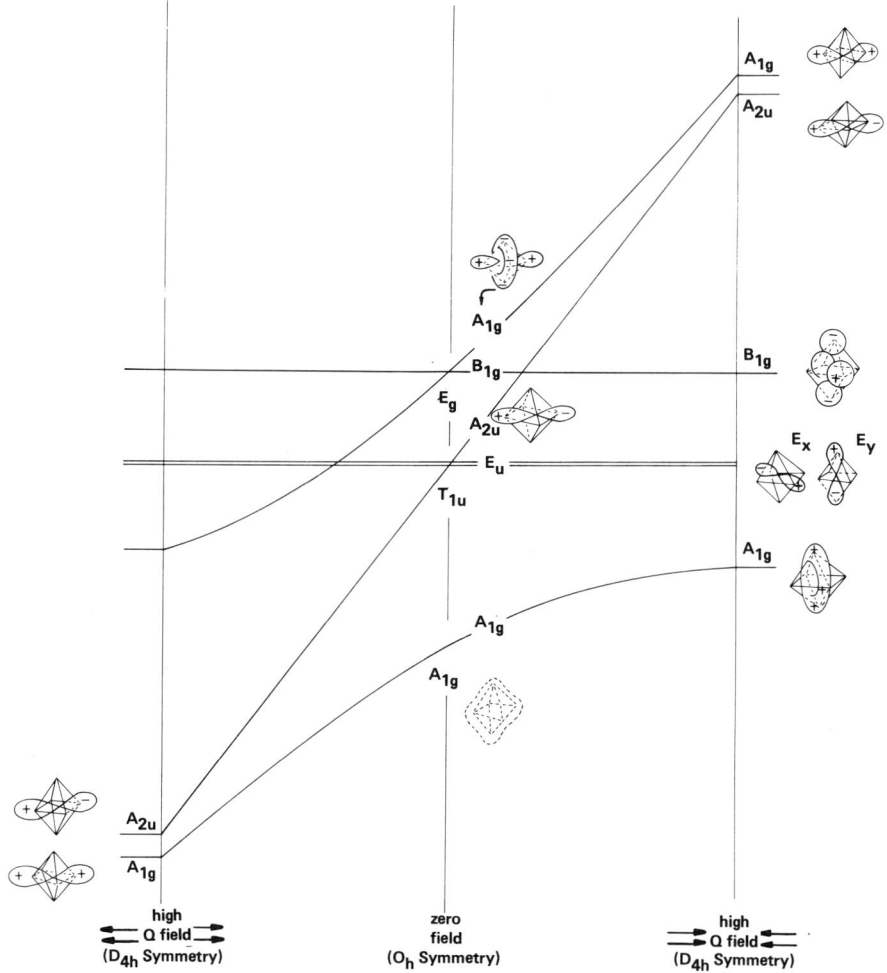

Figure 4.3.5 Energy-level correlation for D_{4h} **Q** perturbation. Eigenvalues and eigenfunctions change as the O_h symmetry in broken to D_{4h} by the **Q** perturbation.

plots asymptotically approach straight lines, and the eigenvectors are nearly independent of Q. The $[A_{1g}]$ eigenvectors of submatrix

$$m = \begin{pmatrix} H + Q/3 & \sqrt{2}\,Q/3 \\ \sqrt{2}\,Q/3 & H + 2Q/3 \end{pmatrix} \qquad (4.3.57)$$

in Eq. (4.3.56) with $S \sim 0$ are easily found. The vector belonging to eigen-

INTRODUCTION TO SYMMETRY BREAKING AND INDUCED REPRESENTATIONS

value $H + Q$ is

$$|(H+Q)[A_{1g}]\rangle = \frac{1}{\sqrt{3}}\begin{vmatrix} A_{1g} \\ [A_{1g}] \end{vmatrix} + \frac{\sqrt{2}}{\sqrt{3}}\begin{vmatrix} E_g \\ [A_{1g}] \end{vmatrix}, \quad (4.3.58a)$$

which is represented by

$$\frac{1}{\sqrt{3}}\begin{pmatrix} 1 \\ 1 \\ 1 \\ 1 \\ 1 \\ 1 \end{pmatrix} \bigg/ \sqrt{6} + \frac{\sqrt{2}}{\sqrt{3}}\begin{pmatrix} 2 \\ 2 \\ -1 \\ -1 \\ -1 \\ -1 \end{pmatrix} \bigg/ \sqrt{12} = \begin{pmatrix} 1 \\ 1 \\ 0 \\ 0 \\ 0 \\ 0 \end{pmatrix} \bigg/ \sqrt{2} \quad (4.3.58b)$$

in the original $\{|1\rangle, |2\rangle, \ldots, |6\rangle\}$ basis. $|(H+Q)[A_{1g}]\rangle$ corresponds to a positively phased wave on the two ends ($|1\rangle$ and $|2\rangle$) of the z axis as shown in the lower left-hand and upper right-hand sides of Figure 4.3.5. The salient $[A_{1g}]$ trajectories asymptotically approach the A_{2u} level as $|Q|$ grows. The orthogonal A_{1g} eigenvector whose energy remains close to H for high $|Q|$ is

$$|(H)[A_{1g}]\rangle = \frac{\sqrt{2}}{\sqrt{3}}\begin{vmatrix} A_{1g} \\ [A_{1g}] \end{vmatrix} - \frac{1}{\sqrt{3}}\begin{vmatrix} E_g \\ [A_{1g}] \end{vmatrix}, \quad (4.3.59a)$$

which is represented by

$$\frac{\sqrt{2}}{\sqrt{3}}\begin{pmatrix} 1 \\ 1 \\ 1 \\ 1 \\ 1 \\ 1 \end{pmatrix} \bigg/ \sqrt{6} - \frac{1}{\sqrt{3}}\begin{pmatrix} 2 \\ 2 \\ -1 \\ -1 \\ -1 \\ -1 \end{pmatrix} \bigg/ \sqrt{12} = \begin{pmatrix} 0 \\ 0 \\ 1 \\ 1 \\ 1 \\ 1 \end{pmatrix} \bigg/ 2. \quad (4.3.59b)$$

The wave function of this A_{1g} state is confined to the octahedral "equator" ($|3\rangle, |4\rangle, |5\rangle$, and $|6\rangle$) as shown in Figure 4.3.5. Particles residing on the equator are not effected by the Q perturbation.

It is instructive to transform the A_{1g} submatrix in Eq. (4.3.56) to the basis [(4.3.58) and (4.3.59)] of high $|Q|$ eigenstates.

$$\begin{pmatrix} \frac{1}{\sqrt{3}} & \frac{\sqrt{2}}{\sqrt{3}} \\ \frac{\sqrt{2}}{\sqrt{3}} & -\frac{1}{\sqrt{3}} \end{pmatrix} \begin{pmatrix} \frac{H+4S+Q}{3} & \frac{\sqrt{2}Q}{3} \\ \frac{\sqrt{2}Q}{3} & \frac{H-2S+2Q}{3} \end{pmatrix} \begin{pmatrix} \frac{1}{\sqrt{3}} & \frac{\sqrt{2}}{\sqrt{3}} \\ \frac{\sqrt{2}}{\sqrt{3}} & -\frac{1}{\sqrt{3}} \end{pmatrix} = \begin{pmatrix} H+Q & S\sqrt{8} \\ S\sqrt{8} & H+2S \end{pmatrix}.$$

(4.3.60)

Then the tunneling amplitude S appears in the off-diagonal position as well as in the lower diagonal. The lower diagonal $H + 2S$ gives the expectation value of the $|(H)[A_{1g}]\rangle$ state. The states $|(H)[A_{1g}]\rangle$, $|[E_g]\rangle$, and $|[B_{1g}]\rangle$ belong to a D_{4h} subcluster of eigenvalues $H + 2S$, H, and $H - 2S$, respectively. They all have equatorial wave functions in Figure 4.3.5 and resemble the D_4 molecular orbitals depicted in Figure 2.12.4. Furthermore, the $\{A_{1g}, A_{2u}\}$ pairs in Figure 4.3.5 are analogous to the inversion doublet discussed in Section 2.12. However, in the limit of high Q there is no $A_{1g} - A_{2u}$ splitting as long as next-nearest-neighbor tunneling parameter T is assumed zero.

Nevertheless, the evolution of A_{1g} wave functions and energy trajectories with Q in Figure 4.3.5 is analogous to the avoided crossing discussed in Section 2.12. Here the A_{1g} waves change continuously from polar waves on one side of the Q axis to equatorial waves on the other.

(b) Electric Dipole or C_{4v} Stark Splitting Suppose the classical or quantum octahedral systems in Figure 4.3.1 are subjected to a C_{4v} perturbation as indicated in Figure 4.3.6. This perturbation could be two different z springs attached to coordinate states $|1\rangle$ and $|2\rangle$ of the oscillator. For the quantum system let us imagine a dipolar electric potential $V_D = V(z, x^2 + y^2)$ such as one might find in a uniform E field pointing along the z axis. Such a field has C_{4v} symmetry, i.e., 90° and 180° rotations around the z axis and "vertical" x-z, y-z, and diagonal plane reflections. It no longer has the x-y plane reflection symmetry which was present in the preceding D_{4h} example of V_Q.

Let the Hamiltonian (4.3.3) be perturbed to the form

$$\langle H + V_D \rangle = \begin{pmatrix} H-D & 0 & S & S & S & S \\ 0 & H+D & S & S & S & S \\ S & S & H & 0 & S & S \\ S & S & 0 & H & S & S \\ S & S & S & S & H & 0 \\ S & S & S & S & 0 & H \end{pmatrix}, \quad (4.3.61)$$

with columns labeled $|1\rangle, |2\rangle, |3\rangle, |4\rangle, |5\rangle, |6\rangle$,

where D is the change in energy of potential wells 1 and 2 due to the electric field. Again, let us ignore the double-tunneling parameter T.

The level splitting associated with a reduction to C_{4v} symmetry depends upon the $O_h \supset C_{4v}$ correlation (4.2.46c). This gives

$$\mathscr{D}^{A_{1g}} \downarrow C_{4v} = A', \quad (4.3.62a)$$

$$\mathscr{D}^{T_{1u}} \downarrow C_{4v} = A' \oplus E, \quad (4.3.62b)$$

$$\mathscr{D}^{E_g} \downarrow C_{4v} = A' \oplus B', \quad (4.3.62c)$$

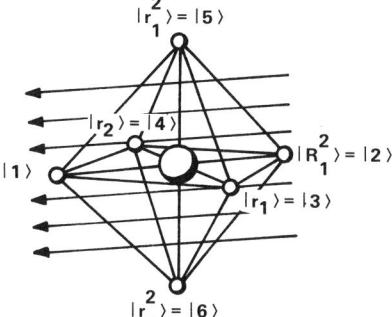

Figure 4.3.6 C_{4v} symmetric D perturbation of octahedrally symmetric system.

which is qualitatively similar to the D_{4h} splitting (4.3.53). The important difference is that now there are *three* states belonging to the same label $[A']$ of the C_{4v} symmetry of the new Hamiltonian. The states

$$\left\langle\begin{array}{c}A_{1g}\\ [A']\\ A'\end{array}\right|\right\rangle = \begin{pmatrix}1\\1\\1\\1\\1\\1\end{pmatrix}\Big/\sqrt{6}, \quad \left\langle\begin{array}{c}E_g\\ [A']\\ A'\end{array}\right|\right\rangle = \begin{pmatrix}2\\2\\-1\\-1\\-1\\-1\end{pmatrix}\Big/\sqrt{12}, \quad \left\langle\begin{array}{c}T_{1u}\\ [A']\\ A'\end{array}\right|\right\rangle = \begin{pmatrix}1/\sqrt{2}\\1/\sqrt{2}\\0\\0\\0\\0\end{pmatrix}$$

(4.3.63)

may all be mixed by a general C_{4v} symmetric perturbation. On the other hand, each of the states

$$\left\langle\begin{array}{c}E_g\\ [B']\\ A'\end{array}\right|\right\rangle = \begin{pmatrix}0\\0\\1/2\\1/2\\-1/2\\-1/2\end{pmatrix}, \quad \left\langle\begin{array}{c}T_{1u}\\ [E]\\ B'\end{array}\right|\right\rangle = \begin{pmatrix}0\\0\\1/\sqrt{2}\\-1/\sqrt{2}\\0\\0\end{pmatrix}, \quad \left\langle\begin{array}{c}T_{1u}\\ [E]\\ B''\end{array}\right|\right\rangle = \begin{pmatrix}0\\0\\0\\0\\1/\sqrt{2}\\-1/\sqrt{2}\end{pmatrix}$$

(4.3.64)

is distinguished by one of its $C_{4v} \supset C_{2v}$ labels, and therefore each one is an eigenstate of $\langle H + V_D\rangle$. Note that the same base states are used here as in the preceding D_{4h} example. Because of inversion commutation the $D_{4h} \supset D_{2h}$ base states are the same as $C_{4v} \supset C_{2v} = \{1, R_3^2, IR_2^2, IR_1^2\}$ base states. Therefore, the states [(4.3.54) and (4.3.55)] were simply relabeled by C_{4v} and C_{2v} irreps in the preceding equations. The relevant C_{2v} characters are given by the following:

$$
\begin{array}{c|cccc}
 & 1 & R_3^2 & IR_2^2 & IR_1^2 \\
\hline
A' & 1 & 1 & 1 & 1 \\
B' & 1 & -1 & 1 & -1 \\
A'' & 1 & 1 & -1 & -1 \\
B'' & 1 & -1 & -1 & 1
\end{array}
\qquad (4.3.65)
$$

The representation of $H + V_D$ in the basis [(4.3.63) and (4.3.64)] is the following:

$$
\langle H + V_D \rangle =
\begin{array}{c|cccccc}
 & \left|\begin{matrix} T_{1u} \\ [E] \\ B' \end{matrix}\right\rangle & \left|\begin{matrix} T_{1u} \\ [E] \\ B'' \end{matrix}\right\rangle & \left|\begin{matrix} T_{1u} \\ [A'] \\ A' \end{matrix}\right\rangle & \left|\begin{matrix} A_{1g} \\ [A'] \\ A' \end{matrix}\right\rangle & \left|\begin{matrix} E_g \\ [A'] \\ A' \end{matrix}\right\rangle & \left|\begin{matrix} E_g \\ [B'] \\ A' \end{matrix}\right\rangle \\
\hline
 & H & \cdot & \cdot & \cdot & \cdot & \cdot \\
 & \cdot & H & \cdot & \cdot & \cdot & \cdot \\
 & \cdot & \cdot & H & -\dfrac{D}{\sqrt{3}} & -\dfrac{\sqrt{2}\,D}{\sqrt{3}} & \cdot \\
 & \cdot & \cdot & -\dfrac{D}{\sqrt{3}} & H - 4S & 0 & \cdot \\
 & \cdot & \cdot & -\dfrac{\sqrt{2}\,D}{\sqrt{3}} & 0 & H - 2S & \cdot \\
 & \cdot & \cdot & \cdot & \cdot & \cdot & H - 2S
\end{array}
$$

(4.3.66)

The perturbation V_D has odd inversion symmetry, since the I operation reverses any "polar" vectors such as those of a homogeneous electric field:

$$I^\dagger V_D I = IV_D I = -V_D. \qquad (4.3.67)$$

Therefore, a matrix component of V_D between any two states of the same parity must vanish. This includes the component

$$\left\langle A_{1g} \middle| V_D \middle| \begin{matrix} E_g \\ 1 \end{matrix} \right\rangle = -\left\langle A_{1g} \middle| I^\dagger V_D I \middle| \begin{matrix} E_g \\ 1 \end{matrix} \right\rangle.$$

Since (g) denotes inversion symmetry,

$$\left(\langle A_{1g} | I^\dagger = \langle A_{1g} |, \; I \left| \begin{matrix} E_g \\ 1 \end{matrix} \right\rangle = \left| \begin{matrix} E_g \\ 1 \end{matrix} \right\rangle \right)$$

INTRODUCTION TO SYMMETRY BREAKING AND INDUCED REPRESENTATIONS 279

the component equals its negative and must therefore be zero:

$$\left\langle A_{1g} \left| V_D \right| {E_g \atop 1} \right\rangle = -\left\langle A_{1g} \left| V_D \right| {E_g \atop 1} \right\rangle = 0. \quad (4.3.68)$$

The same goes for the diagonal components

$$\left\langle {T_{1u} \atop 3} \left| V_D \right| {T_{1u} \atop 3} \right\rangle = \left\langle A_{1g} \left| V_D \right| A_{1g} \right\rangle = \left\langle {E_g \atop 1} \left| V_D \right| {E_g \atop 1} \right\rangle = 0,$$

which would not otherwise be prohibited by C_{4v} symmetry. This is quite the opposite of the V_Q perturbation which has even inversion symmetry and cannot connect any states of different parity.

Elimination of certain matrix components by symmetry analysis belongs to the subject of SELECTION RULES which are to be discussed in Chapters 6 and 7. For now let us see how the absence of D terms on the diagonal of Hamiltonian (4.3.66) affects the energy levels. In Figure 4.3.7 the Hamiltonian eigenvalues are plotted as a function of D. Note that for small D there is little or no splitting of T_{1u} or E_g. At $D = 0$ all energy trajectories have zero slope. The splitting and shifts are said to be SECOND ORDER when diagonal components of a perturbation vanish. The energies do not change until appreciable amounts of different states are mixed to make new eigenstates. The mixing is controlled by the off-diagonal components $-D/\sqrt{3}$ and $-2D/\sqrt{3}$ in the Hamiltonian (4.3.66).

As in the preceding V_Q example it is instructive to study the case in which the perturbation V_D dominates the Hamiltonian, i.e., the high-D limit. The following transformation diagonalizes the D-dependent part of the $[A']$ submatrix.

$$\begin{pmatrix} \frac{1}{\sqrt{2}} & \frac{1}{\sqrt{6}} & \frac{1}{\sqrt{3}} \\ 0 & \frac{2}{\sqrt{6}} & -\frac{1}{\sqrt{3}} \\ -\frac{1}{\sqrt{2}} & \frac{1}{\sqrt{6}} & \frac{1}{\sqrt{3}} \end{pmatrix} \begin{pmatrix} H & -\frac{D}{\sqrt{3}} & -\frac{\sqrt{2}D}{\sqrt{3}} \\ -\frac{D}{\sqrt{3}} & H+4S & 0 \\ -\frac{\sqrt{2}D}{\sqrt{3}} & 0 & H-2S \end{pmatrix} \begin{pmatrix} \frac{1}{\sqrt{2}} & 0 & -\frac{1}{\sqrt{2}} \\ \frac{1}{\sqrt{6}} & \frac{2}{\sqrt{6}} & \frac{1}{\sqrt{6}} \\ \frac{1}{\sqrt{3}} & -\frac{1}{\sqrt{3}} & \frac{1}{\sqrt{3}} \end{pmatrix} = \begin{pmatrix} H-D & 2S & 0 \\ 2S & H+2S & 2S \\ 0 & 2S & H+D \end{pmatrix}.$$

(4.3.69)

The high-D ground eigenstate of eigenvalue $(H - D)$ is the original local base state $|1\rangle$:

$$|1\rangle = \left(\frac{1}{\sqrt{2}}\right) \left| {T_{1u} \atop [A'] \atop A'} \right\rangle + \left(\frac{1}{\sqrt{6}}\right) \left| {A_{1g} \atop [A'] \atop A'} \right\rangle + \left(\frac{1}{\sqrt{3}}\right) \left| {E_g \atop [A'] \atop A'} \right\rangle. \quad (4.3.70)$$

280 HIGHER FINITE SYMMETRY AND INDUCED REPRESENTATIONS

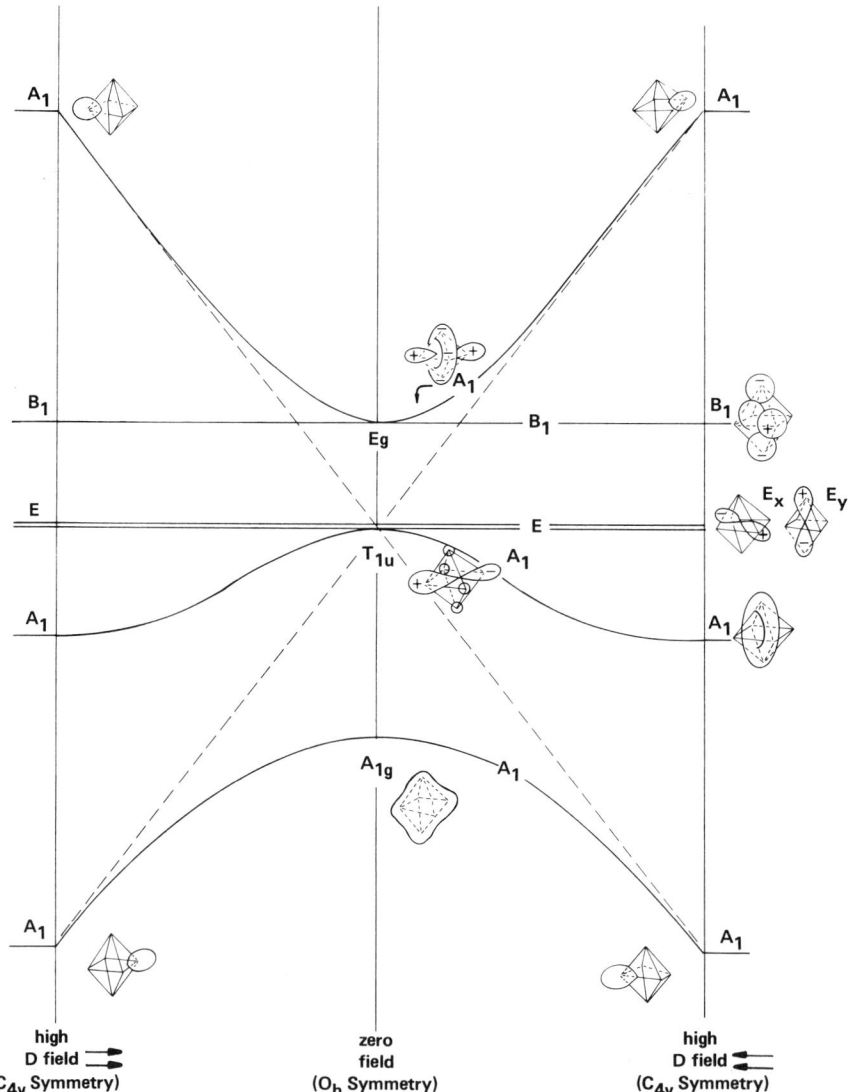

Figure 4.3.7 Energy-level correlation for D perturbation. Eigenfunction and eigenvalues change as O_h symmetry is broken to C_{4v} by the D field. No first-order splitting (for low D field) occurs.

INTRODUCTION TO SYMMETRY BREAKING AND INDUCED REPRESENTATIONS **281**

This corresponds to a wave function localized on the positive z axis as shown in the lower right-hand side of Figure 4.3.7. The uppermost eigenfunction on the same side corresponds to the eigenvector

$$|2\rangle = \left(\frac{1}{\sqrt{2}}\right)\left|\begin{matrix}T_{1u}\\ [A']\\ A'\end{matrix}\right\rangle + \left(\frac{1}{\sqrt{6}}\right)\left|\begin{matrix}A_{1g}\\ [A']\\ A'\end{matrix}\right\rangle + \left(\frac{1}{\sqrt{3}}\right)\left|\begin{matrix}E_g\\ [A']\\ A'\end{matrix}\right\rangle, \quad (4.3.71)$$

with eigenvalues $H + D$. Of course, $|1\rangle$ and $|2\rangle$ switch positions if D changes sign. The equatorial wave functions belong to the same sort of (A, E, B) subcluster seen in the preceding example.

The $[A']$ state,

$$(|3\rangle + |4\rangle + |5\rangle + |6\rangle)/2 = \left(\frac{2}{\sqrt{6}}\right)\left|\begin{matrix}A_{1g}\\ [A']\\ A'\end{matrix}\right\rangle - \left(\frac{1}{\sqrt{3}}\right)\left|\begin{matrix}E_g\\ [A']\\ A'\end{matrix}\right\rangle, \quad (4.3.72)$$

has energy expectation $H + 2S$.

(c) Magnetic Dipole or C_4 Zeeman Splitting Suppose a uniform magnetic field is placed along the fourfold z axis of the octahedral system. Then the symmetry is reduced to the Abelian cyclic group C_4. A magnetic vector **B** is a pseudovector of axial vector. B_z is reversed by vertical x-z or y-z planar reflections, and so these elements of C_{4v} are not symmetry operators for magnetic fields. The effects of a uniform magnetic field B_z are similar to those arising from a uniform rotation about the z axis, as stated in Larmor's theorem. The symmetry properties of a rotating reference frame are more obvious. It is clear that ϕz plane reflections reverse the z-rotational sense of direction and cannot be symmetry operations.

The C_4 correlation table (4.2.42b) tells what can happen to the $(A_1 T_1 E)$ levels of the O_h system.

$$\mathscr{D}^{A_1} \downarrow C_4 = 0_4,$$
$$\mathscr{D}^{T_1} \downarrow C_4 = 0_4 \oplus 1_4 \oplus 3_4, \quad (4.3.73)$$
$$\mathscr{D}^{E} \downarrow C_4 = 0_4 \oplus 2_4.$$

According to this all degeneracy is removed. This is called C_4 Zeeman splitting and is analogous to the splitting discussed in Section 3.6.A. Here a 3×3 matrix for the 0_4 states may have to be diagonalized. Transformations to moving-wave or circularly polarized tetragonal bases are helpful in problems such as these. Physical examples of C_4 splitting will be given in Section 4.4.

(d) Threefold Axial Perturbations and Basis Changing Suppose a uniform electric field is placed along the threefold (111) direction or r_1 axis of the octahedral system. Let the Hamiltonian for this field be perturbed to the following form:

$$\langle H + V_d \rangle = \begin{array}{c} \\ \end{array} \begin{array}{cccccc} |1\rangle & |2\rangle & |3\rangle & |4\rangle & |5\rangle & |6\rangle \end{array} \\ \begin{vmatrix} H-d & 0 & S & S & S & S \\ 0 & H+d & S & S & S & S \\ S & S & H-d & 0 & S & S \\ S & S & 0 & H+d & S & S \\ S & S & S & S & H-d & 0 \\ S & S & S & S & 0 & H+d \end{vmatrix}.$$

(4.3.74)

Here the down-field states $|1\rangle$, $|3\rangle$, and $|5\rangle$ have reduced energy expectation values $(H-d)$, and the up-field states $|2\rangle$, $|4\rangle$, and $|6\rangle$ have increased values $(H+d)$. The symmetry of the Hamiltonian is reduced from O_h to $C_{3v} = \{1, r_1, r_1^2, Ii_2, Ii_4, Ii_5\}$. The broken symmetry includes $\pm 120°$ rotations around the field direction and reflections through three diagonal planes which are parallel to the field direction.

The $O \supset C_{3v}$ correlations (4.2.46b) tell qualitatively how the octahedral levels will be split by a C_{3v} symmetric perturbation:

$$\mathscr{D}^{A_{1g}} \downarrow C_{3v} = A', \tag{4.3.75a}$$

$$\mathscr{D}^{T_{1u}} \downarrow C_{3v} = A' + E, \tag{4.3.75b}$$

$$\mathscr{D}^{E_g} \downarrow C_{3v} = E. \tag{4.3.75c}$$

According to this a (2×2) Hamiltonian submatrix must be solved for the pair of A' states and another one solved for the pair of E states.

In order to reduce $\langle H + V_d \rangle$ to (2×2) submatrices one must use base states labeled by C_{3v} irreps. This is where one must be careful to consistently define all bases and avoid annoying phase errors. The $(A_{1g} T_{1u} E_g)$ bases we have been using are labeled by tetragonal subgroup chain $O_h \supset D_{4h} \supset D_{2h}$. The tetragonal irreps $\mathscr{D}^{A_{1g}}$ and $\mathscr{D}^{E_g}$ are already reduced with respect to C_{3v}. From Section 4.2, and, in particular, Eq. (4.2.19) we have the following:

$$\left\{ \begin{aligned} & \mathscr{D}^{A_{1g}}(1) = 1, \quad \mathscr{D}^{A_{1g}}(r_1) = 1, \quad \mathscr{D}^{A_{1g}}(r_1^2) = 1, \quad \mathscr{D}^{A_{1g}}(Ii_2) = 1, \mathscr{D}^{A_{1g}}(Ii_4) = 1, \quad \mathscr{D}^{A_{1g}}(Ii_5) = 1 \\ & \mathscr{D}^{E_g}(1) = \begin{pmatrix} 1 & 0 \\ 0 & 1 \end{pmatrix}, \quad \mathscr{D}^{E_g}(r_1) = \begin{pmatrix} -\frac{1}{2} & -\frac{\sqrt{3}}{2} \\ \frac{\sqrt{3}}{2} & -\frac{1}{2} \end{pmatrix}, \quad \mathscr{D}^{E_g}(r_1^2) = \begin{pmatrix} -\frac{1}{2} & \frac{\sqrt{3}}{2} \\ -\frac{\sqrt{3}}{2} & -\frac{1}{2} \end{pmatrix}, \quad \mathscr{D}^{E_g}(Ii_2) = \begin{pmatrix} -\frac{1}{2} & \frac{\sqrt{3}}{2} \\ \frac{\sqrt{3}}{2} & \frac{1}{2} \end{pmatrix}, \quad \mathscr{D}^{E_g}(Ii_4) = \begin{pmatrix} 1 & 0 \\ 0 & -1 \end{pmatrix}, \quad \mathscr{D}^{E_g}(Ii_5) = \begin{pmatrix} -\frac{1}{2} & -\frac{\sqrt{3}}{2} \\ -\frac{\sqrt{3}}{2} & \frac{1}{2} \end{pmatrix}. \end{aligned} \right\}$$

(4.3.76)

INTRODUCTION TO SYMMETRY BREAKING AND INDUCED REPRESENTATIONS   283

However, the T_{1u} irreps (4.2.14) are not reduced with respect to C_{3v}:

$$\mathscr{D}^{T_{1u}}(1) = \begin{pmatrix} 1 & \cdot & \cdot \\ \cdot & 1 & \cdot \\ \cdot & \cdot & 1 \end{pmatrix}, \quad \mathscr{D}^{T_{1u}}(r_1) = \begin{pmatrix} \cdot & \cdot & 1 \\ 1 & \cdot & \cdot \\ \cdot & 1 & \cdot \end{pmatrix}, \quad \mathscr{D}^{T_{1u}}(r^2) = \begin{pmatrix} \cdot & 1 & \cdot \\ \cdot & \cdot & 1 \\ 1 & \cdot & \cdot \end{pmatrix}, \quad \mathscr{D}^{T_{1u}}(Ii_2) = \begin{pmatrix} \cdot & \cdot & 1 \\ \cdot & 1 & \cdot \\ 1 & \cdot & \cdot \end{pmatrix}, \quad \mathscr{D}^{T_{1u}}(Ii_4) = \begin{pmatrix} \cdot & 1 & \cdot \\ 1 & \cdot & \cdot \\ \cdot & \cdot & 1 \end{pmatrix}, \quad \mathscr{D}^{T_{1u}}(Ii_5) = \begin{pmatrix} 1 & \cdot & \cdot \\ \cdot & \cdot & 1 \\ \cdot & 1 & \cdot \end{pmatrix}.$$

(4.3.77)

Hence the tetragonal bases $\left\{ \left| {T_{1u} \atop 1} \right\rangle, \left| {T_{1u} \atop 2} \right\rangle, \left| {T_{1u} \atop 3} \right\rangle \right\}$ are not yet in a convenient form for a trigonal problem. One needs an A' and an E pair of the T_{1u} bases according to correlation (4.3.75b). What combination of tetragonal $\left| {T_{1u} \atop j} \right\rangle$ make a C_{3v} trigonal $\left\{ \left| {E \atop 1} \right\rangle \left| {E \atop 2} \right\rangle \right\}$ or $|A'\rangle$ bases?

By applying C_{3v} P operators to one of the T_{1u} bases the correct combinations are obtained. $|A'\rangle$ is found from

$$P^{A'} \left| {T_{1u} \atop 3} \right\rangle = (1/6)(1 + r_1 + r_1^2 + Ii_2 + Ii_4 + Ii_5) \left| {T_{1u} \atop 3} \right\rangle$$

using

$$g \left| {T_{1u} \atop 3} \right\rangle = \sum_{j=1}^{3} \mathscr{D}_{j3}^{T_{1u}}(g) \left| {T_{1u} \atop j} \right\rangle$$

and the tetragonal $\mathscr{D}^{T_{1u}}(g)$ in Eq. (4.3.77). The normalized result is

$$\left| {T_{1u} \atop A'} \right\rangle \equiv P^{A'} \left| {T_{1u} \atop 3} \right\rangle \sqrt{3} = \left(\left| {T_{1u} \atop 1} \right\rangle + \left| {T_{1u} \atop 2} \right\rangle + \left| {T_{1u} \atop 3} \right\rangle \right)/\sqrt{3}. \quad (4.3.78)$$

Similarly, the $|E'\rangle$ states are obtained by using the P^E projectors

$$P_{ij}^E = (l^E/oC_{3v}) \sum_g \mathscr{D}_{ij}^{E*}(g) g$$

and the irreps $\mathscr{D}^E(g) = \mathscr{D}^{E_g}(g)$ in Eq. (4.3.76). The normalized results are

$$\left| {T_{1u} \atop {E \atop 1}} \right\rangle \equiv P_{11}^E \left| {T_{1u} \atop 3} \right\rangle \sqrt{3/2} = \left(-\left| {T_{1u} \atop 1} \right\rangle - \left| {T_{1u} \atop 2} \right\rangle + 2\left| {T_{1u} \atop 3} \right\rangle \right)/\sqrt{6}, \quad (4.3.79a)$$

$$\left| {T_{1u} \atop {E \atop 2}} \right\rangle \equiv P_{21}^E \left| {T_{1u} \atop 3} \right\rangle \sqrt{3/2} = \left(\left| {T_{1u} \atop 1} \right\rangle - \left| {T_{1u} \atop 2} \right\rangle \right)/\sqrt{2}. \quad (4.3.79b)$$

The Hamiltonian (4.3.74) reduces to a submatrix,

$$\langle H + V_d \rangle^{A'} = \begin{pmatrix} H + 4S & -d \\ -d & H \end{pmatrix} \begin{matrix} \left| \begin{matrix} A_{1g} \\ A' \end{matrix} \right\rangle & \left| \begin{matrix} T_{1u} \\ A' \end{matrix} \right\rangle \end{matrix}, \quad (4.3.80)$$

between the A' states

$$\left| \begin{matrix} A_{1g} \\ A' \end{matrix} \right\rangle = (|1\rangle + |2\rangle + |3\rangle + |4\rangle + |5\rangle + |6\rangle)/\sqrt{6},$$

$$\left| \begin{matrix} T_{1u} \\ A' \end{matrix} \right\rangle = (|1\rangle - |2\rangle + |3\rangle - |4\rangle + |5\rangle - |6\rangle)/\sqrt{6}, \quad (4.3.81)$$

and an identical pair of submatrices,

$$\langle H + V_d \rangle^{E} = \begin{pmatrix} H - 2S & -d \\ -d & H \end{pmatrix} \begin{matrix} \left| \begin{matrix} E_g \\ E \\ 1,2 \end{matrix} \right\rangle & \left| \begin{matrix} T_{1u} \\ E \\ 1,2 \end{matrix} \right\rangle \end{matrix}, \quad (4.3.82)$$

between the E states of the first partners

$$\left| \begin{matrix} E_g \\ E \\ 1 \end{matrix} \right\rangle = (2|1\rangle + 2|2\rangle - |3\rangle - |4\rangle - |5\rangle - |6\rangle)/\sqrt{12},$$

$$\left| \begin{matrix} T_{1u} \\ E \\ 1 \end{matrix} \right\rangle = (2|1\rangle - 2|2\rangle - |3\rangle + |4\rangle - |5\rangle + |6\rangle)/\sqrt{12}, \quad (4.3.83)$$

or else between the E states of the second partners:

$$\left| \begin{matrix} E_g \\ E \\ 2 \end{matrix} \right\rangle = (|3\rangle + |4\rangle - |5\rangle - |6\rangle)/2,$$

$$\left| \begin{matrix} T_{1u} \\ E \\ 2 \end{matrix} \right\rangle = (|3\rangle - |4\rangle - |5\rangle + |6\rangle)/2 \quad (4.3.84)$$

The remaining solution and analysis is left as an exercise.

INTRODUCTION TO SYMMETRY BREAKING AND INDUCED REPRESENTATIONS 285

Note that the $D_{4h} - C_{3v}$ transformation

$$\mathcal{T} = \begin{pmatrix} -\dfrac{1}{\sqrt{6}} & \dfrac{1}{\sqrt{2}} & \dfrac{1}{\sqrt{3}} \\ -\dfrac{1}{\sqrt{6}} & -\dfrac{1}{\sqrt{2}} & \dfrac{1}{\sqrt{3}} \\ \dfrac{2}{\sqrt{6}} & 0 & \dfrac{1}{\sqrt{3}} \end{pmatrix} \begin{array}{c} \left| \begin{array}{c} E \\ 1 \end{array} \right\rangle \quad \left| \begin{array}{c} E \\ 2 \end{array} \right\rangle \quad \left| A' \right\rangle \\[2ex] \left| \begin{array}{c} T_1 \\ 1 \end{array} \right\rangle \\ \left| 2 \right\rangle \\ \left| 3 \right\rangle \end{array} \qquad (4.3.85)$$

obtained in Eqs. (4.3.78) and (4.3.79) differs slightly from the $D_{4h} - D_{3d}$ transformation (4.2.31). One must use caution when changing between D_n and C_{nv} bases.

In fact D_{3d} symmetry results if a threefold axial quadrupole field fell on the octahedral bases. The $O_h \supset D_{3d}$ correlations,

$$\mathcal{D}^{A_{1g}} \downarrow D_{3d} = A_{1g},$$
$$\mathcal{D}^{T_{1u}} \downarrow D_{3d} = A_{2u} + E_u, \qquad (4.3.86)$$
$$\mathcal{D}^{E_g} \downarrow D_{3d} = E_g,$$

follow from D_3 correlations (4.2.44). It shows that D_{3d}-defined bases would not mix at all. A D_{3d}-symmetric quadrupole Hamiltonian $\langle H + V_q \rangle$ results if one replaces all the $\pm d$'s in (4.3.74) with q's. However, such a matrix causes no splitting at all. One needs a more "severely" defined Hamiltonian of D_{3d} symmetry. For example, it is possible that the amplitudes,

$$\langle 1|H|4 \rangle = \langle 2|H|3 \rangle = \cdots = S'(q), \qquad (4.3.87)$$

for tunneling along the field might be different from the amplitudes,

$$\langle 1|H|3 \rangle = \langle 2|H|4 \rangle = \cdots = S, \qquad (4.3.88)$$

for tunneling transverse to the field. Field-dependent tunneling will effect the splitting predicted in Eq. (4.3.86) as well as changing the spectrum of other problems which have been treated previously in this section. (See Problem 4.3.6.) The variation of tunneling amplitudes with interatomic distance is important in the study of molecular bonding. Electronic energy eigenvalues of symmetric systems are generally quite sensitive to tunneling. The effects of nuclear motion on electronic energy and vice versa will be taken up in Section 6.7 and in subsequent chapters.

4.4 VIBRATIONS OF OCTAHEDRAL HEXAFLUORIDE MOLECULES

Octahedral hexafluoride molecules such as UF_6 have attracted attention because their spectral properties provide a way to separate the isotopes used in reactor fuel. Let us study the first of these properties involving its mechanical vibrations. In Figure 4.4.1 is a drawing of a mechanical model of UF_6 assuming O_h symmetry for a molecule, and a coordinate system for its motions. The larger ball represents the U atom of mass M, and the smaller balls represent F atoms of mass m. To begin with let us assume only two kinds of central force of "spring" constants: k for the F—F interaction, and j for the F—U bond. In order to analyze SF_6 it is convenient to include a bending constant b for the F—S bond, as indicated in the figure.

Obviously, a continuous infinity of choices for coordinates exists in problems like this, and one just hopes to pick a fairly convenient system. The choice made in Figure 4.4.1 is adequate for solution of the harmonic equation of motion and for demonstration of some further theoretical points concerning symmetry analysis.

A. Projection Analysis

As in previous examples it is convenient to obtain eigenvectors from vectors projected by the elementary operators

$$P_{ij}^\alpha = (l^\alpha/°G) \sum_g \mathcal{D}_{ij}^{\alpha*}(g) g \qquad (4.4.1)$$

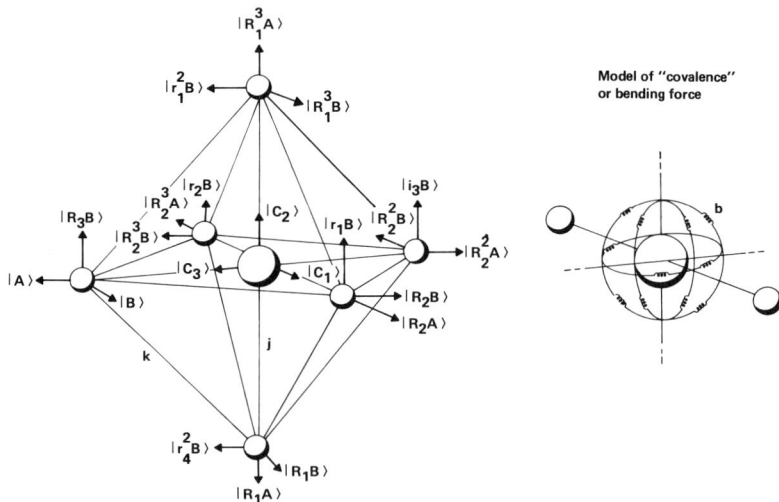

Figure 4.4.1 Octahedral hexafluoride ($UF_6, SF_6, \ldots$) molecular model Cartesian coordinates for each atom are labeled by orbit (A, B, or C) and coset leaders. ($1 = R_3, r_1, R_2, \ldots$ etc.) Spring constants are equal to (k) for (F—F) bonds and j for radial (F-central) bonds. Bending spring constant is b.

VIBRATIONS OF OCTAHEDRAL HEXAFLUORIDE MOLECULES   287

of the octahedral groups O or O_h. It is important to see how P operators can be efficiently applied. First of all, our choice of coordinates in Figure 4.4.1 shows which bases are connected by symmetry operations. Sets of base vectors like $\{|A\rangle, |R_2 A\rangle = R_2|A\rangle, \ldots\}$ or $\{|B\rangle, |R_2 B\rangle = R_2|B\rangle, \ldots\}$ or $\{|C\rangle, |R_2 C\rangle \equiv R_2|C\rangle, |R_1^3 C\rangle \equiv R_1^3|C\rangle\}$ are each called ORBITS of symmetry when any two of their bases can be connected by a symmetry operator. Elementary operators p_{ij}^α need only be applied to the first vector in each orbit. The result of operating on any others is just a linear combination of the first ones, as seen in the following:

$$P_{ij}^\alpha |gA\rangle = P_{ij}^\alpha g|A\rangle = \sum_k \mathcal{D}_{jk}^\alpha(g) P_{ik}^\alpha |A\rangle. \tag{4.4.2}$$

As explained in Sections 4.3.B and 4.3.C, only certain projectors P_{jk}^α need be applied to a given state such as $|A\rangle$. These are the ones for which the right-hand index (k) is compatible with a local C_4 symmetry condition such as

$$|A\rangle = P^A|A\rangle = (1/4)(1 + R_3 + R_3^2 + R_3^3)|A\rangle.$$

This is the same as Eq. (4.3.7) involving C_4 irrep $A \equiv 0_4$. In fact the orbit $\{|A\rangle, R_1^2|A\rangle, \ldots\}$ is just the basis of the first induced representation

$$D^{0_4} \uparrow O = A_1 \oplus T_1 \oplus E$$

introduced in Section 4.3.A. The full $C_{4v} \subset O_h$ labeling of this orbit is given by the induced representation

$$D^{A'} \uparrow O_h = A_{1g} \oplus T_{1u} \oplus E_g, \tag{4.4.3}$$

which corresponds to the (A_{1g}, T_{1u}, E_g) eigenvectors for the octahedral systems in Figure 4.3.1. The mechanical system in Figure 4.3.1 is a constrained version of the UF_6 molecule in which the $\{|B\rangle \cdots \}$ and $\{|C\rangle \cdots \}$ degrees of freedom are absent.

The $\{|B\rangle \cdots \}$ orbit contains 12 base states. The local C_2 rotational symmetry condition for $|B\rangle$ is

$$|B\rangle = P^B|B\rangle = \tfrac{1}{2}(1 - R_3^2)|B\rangle, \tag{4.4.4}$$

since 180° rotation of $|B\rangle$ around the 3 axis is just $-|B\rangle$. A more detailed local symmetry condition for $|B\rangle$ is

$$|B\rangle = P_1^E|B\rangle = \tfrac{1}{4}(1 - R_3^2 - IR_1^2 + IR_2^2)|B\rangle, \tag{4.4.5}$$

where

$$P_1^E \equiv P_{B_1}^E = \tfrac{1}{4}(1 - R_3^2 - IR_1^2 + IR_2^2) \qquad (4.4.6)$$

is a $C_{4v} \supset C_{2v}$ projection operator. The full O_h labeling of the $\{|B\rangle \cdots \}$ orbit is therefore given by the 12-dimensional induced representation

$$\mathscr{D}^E \uparrow O_h = T_{1g} \oplus T_{2g} \oplus T_{1u} \oplus T_{2u}, \qquad (4.4.7)$$

according to the E column of C_{4v} correlation (4.2.46c).

The three base states of the $\{|C_1\rangle, |C_2\rangle, |C_3\rangle\}$ orbit involve translation of the central U atom. $|C_j\rangle$ obviously satisfies the local symmetry condition of a polar vector

$$|C_j\rangle = P_{jj}^{T_{1u}}|C_j\rangle; \qquad (4.4.8)$$

i.e., they are ready-made bases of O_h irrep T_{1u} in the tetragonal basis. Altogether, the UF$_6$ coordinates belong to a 21-dimensional representation which reduces to

$$A + B + C = (A_{1g} \oplus T_{1u} \oplus E_g) \oplus (T_{1g} \oplus T_{2g} \oplus T_{1u} \oplus T_{2u}) \oplus (T_{1u})$$
$$= A_{1g} \oplus E_g \oplus T_{1g} \oplus T_{2g} \oplus T_{2u} \oplus 3T_{1u}. \qquad (4.4.9)$$

Each irrep labels a fundamental resonance. All resonances except the three T_{1u}'s are uniquely labeled.

It will be necessary to account for the 12 B-orbit vectors

$$|e_{jk}^\alpha B\rangle = P_{jk}^\alpha |B\rangle/(N_k^\alpha)^{1/2}. \qquad (4.4.10)$$

The tetragonal local symmetry condition (4.4.5) tells which k to use. If you forget how the k are labeled it is easy to represent the local symmetry projector (4.4.6) and see which components are nonzero. Using tetragonally defined irreps (4.2.14) the following representations result:

$$\mathscr{D}^{T_{1g}}(P_1^E) = (1/4)\left(\mathscr{D}^{T_{1g}}(1) - \mathscr{D}^{T_{1g}}(R_3^2) - \mathscr{D}^{T_{1g}}(IR_1^2) + \mathscr{D}^{T_{1g}}(IR_2^2)\right)$$

$$= \left[\begin{pmatrix} 1 & 0 & 0 \\ 0 & 1 & 0 \\ 0 & 0 & 1 \end{pmatrix} - \begin{pmatrix} -1 & 0 & 0 \\ 0 & -1 & 0 \\ 0 & 0 & 1 \end{pmatrix} - \begin{pmatrix} 1 & 0 & 0 \\ 0 & -1 & 0 \\ 0 & 0 & -1 \end{pmatrix} + \begin{pmatrix} -1 & 0 & 0 \\ 0 & 1 & 0 \\ 0 & 0 & -1 \end{pmatrix}\right]\Big/4 = \begin{pmatrix} 0 & 0 & 0 \\ 0 & 1 & 0 \\ 0 & 0 & 0 \end{pmatrix}$$

$$(4.4.11a)$$

$$\mathscr{D}^{T_{1u}}(P_1^E) = \left[\begin{pmatrix} 1 & 0 & 0 \\ 0 & 1 & 0 \\ 0 & 0 & 1 \end{pmatrix} - \begin{pmatrix} -1 & 0 & 0 \\ 0 & -1 & 0 \\ 0 & 0 & 1 \end{pmatrix} - \begin{pmatrix} -1 & 0 & 0 \\ 0 & 1 & 0 \\ 0 & 0 & 1 \end{pmatrix} + \begin{pmatrix} 1 & 0 & 0 \\ 0 & -1 & 0 \\ 0 & 0 & 1 \end{pmatrix}\right]\Big/4 = \begin{pmatrix} 1 & 0 & 0 \\ 0 & 0 & 0 \\ 0 & 0 & 0 \end{pmatrix}$$

$$(4.4.11b)$$

VIBRATIONS OF OCTAHEDRAL HEXAFLUORIDE MOLECULES

The first and second diagonal components, respectively, of the T_{1u} and T_{1g} matrices are nonzero. Hence, T_{1u} and T_{1g} vectors require that $k = 1$ and 2, respectively,

$$|e_{j1}^{T_{1u}}B\rangle = P_{j1}^{T_{1u}}|B\rangle/(N_1^{T_{1u}})^{1/2}, \qquad (4.4.12a)$$

$$|e_{j2}^{T_{1g}}B\rangle = P_{j2}^{T_{1g}}N_2^{T_{1g}}|B\rangle/(N_2^{T_{1g}})^{1/2}. \qquad (4.4.12b)$$

As explained in Section 4.3.C the calculation of $P_{jk}^\alpha|B\rangle$ is simple when P_{jk}^α matches the local symmetry as per Eq. (4.3.35). In writing out each projected state one needs only to sum over group elements l_j which label coordinates $\langle l_j B \rangle$ in Figure 4.4.1. This result was expressed by Eq. (4.3.44). Using the T_1 irreps (4.2.14) we derive the following:

$$|e_{11}^{T_{1u}}B\rangle = \tfrac{1}{16}\big(\mathscr{D}_{11}^{T_1}(1)|B\rangle$$
$$+ \cdots + \mathscr{D}_{11}^{T_1}(R_1)|R_1 B\rangle + \cdots + \mathscr{D}_{11}^{T_1}(R_1^3)|R_1^3 B\rangle$$
$$+ \cdots + \mathscr{D}_{11}^{T_1}(R_2^2)|R_2^2 B\rangle\big)/(N)^{1/2}$$
$$= \big(|B\rangle + |R_1 B\rangle + |R_1^3 B\rangle - |R_2^2 B\rangle\big)/2, \qquad (4.4.13a)$$

$$|e_{21}^{T_{1u}}B\rangle = \big(|R_3 B\rangle + |r_1 B\rangle + |r_2 B\rangle + |i_3 B\rangle\big)/2, \qquad (4.4.13b)$$

$$|e_{31}^{T_{1u}}B\rangle = \big(-|R_2 B\rangle + |R_2^3 B\rangle + |r_1^2 B\rangle + |r_4^2 B\rangle\big)/2. \qquad (4.4.13c)$$

Figure 4.4.2 shows the first partner state $|e_{11}^{T_{1u}}B\rangle$ to be a translation relative to the 1 axis of the octahedral equator. Shown also is the state

$$|e_{12}^{T_{1g}}B\rangle = \tfrac{1}{16}\big(\cdots + \mathscr{D}_{12}^{T_1}(R_3)|R_3 B\rangle + \cdots + \mathscr{D}_{12}^{T_1}(i_3)|i_3 B\rangle$$
$$+ \cdots + \mathscr{D}_{12}^{T_1}(r_1^2)|r_1^2 B\rangle + \cdots + \mathscr{D}_{12}^{T_1}(r_4^2)|r_4^2 B\rangle\big)/(N)$$
$$= \big(-|R_3 B\rangle + |i_3 B\rangle + |r_1^2 B\rangle - |r_4^2 B\rangle\big)/2, \qquad (4.4.14)$$

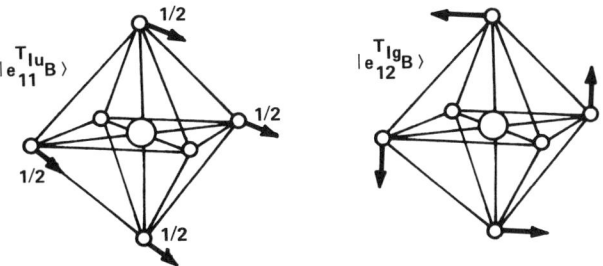

Figure 4.4.2 T_1 motions from the B orbit.

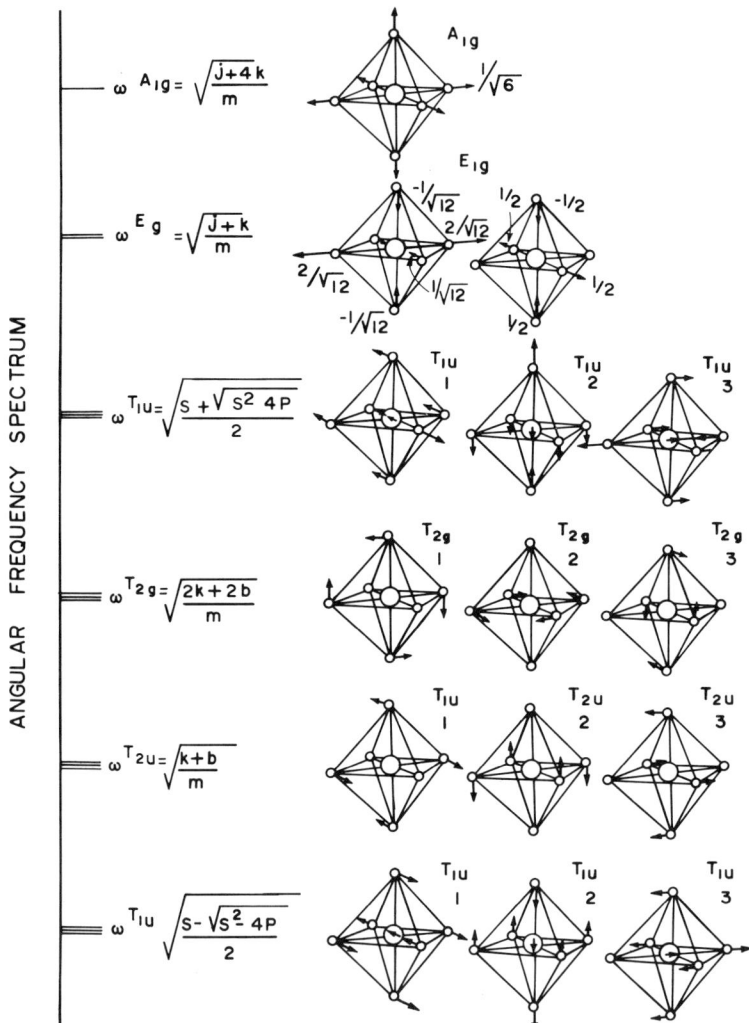

Figure 4.4.3 Hexafluoride vibrational modes and spectrum. T_{1u} modes are not drawn precisely, since their form depends upon the choice of constants and rotational perturbations. (See Figure 4.4.7.)

which corresponds to rotation around the 1 axis. This is a nongenuine vibration and so it will not be considered further until rotations are studied in subsequent chapters.

The derivation of the T_2 states

$$|e_{j1}^{T_{2u}}B\rangle = P_{j1}^{T_{2u}}|B\rangle/(N_1^{T_{2u}})^{1/2}, \qquad (4.4.15a)$$

$$|e_{j2}^{T_{2g}}B\rangle = P_{j2}^{T_{2g}}|B\rangle/(N_2^{T_{2g}})^{1/2}, \qquad (4.4.15b)$$

follows the same lines. The results are shown by the first and second triplets in Figure 4.4.3. Shown also are the A-orbit states

$$|e^{A_{1g}}A\rangle = P^{A_{1g}}|A\rangle/(N^{A_{1g}})^{1/2}, \qquad (4.4.16a)$$

$$|e_{j1}^{E_g}A\rangle = P_{j1}^{E_g}|A\rangle/(N^{E_g})^{1/2}, \qquad (4.4.16b)$$

which were derived before in Eqs. (4.3.20)–(4.3.23).

Thus we have accounted for all modes of UF_6 except those labeled by T_{1u}. Besides the T_{1u} states (4.4.13) there are axial F-atom translation states

$$|e_{j3}^{T_{1u}}A\rangle = P_{j3}^{T_{1u}}|A\rangle/(N_3^{T_{1u}})^{1/2}, \qquad (4.4.17)$$

which were derived before in Eqs. (4.3.27), and the central U-atom translation states (4.4.8). The first partner of each is shown in Figure 4.4.4a. Each of these modes causes the molecular center of mass to translate. If the arrows in Figure 4.4.4(a) stand for velocity, then the linear momentum of $|e^{T_{1u}}A\rangle$, $|e^{T_{1u}}B\rangle$, and $|e^{T_{1u}}C\rangle$ is $m\sqrt{2}$, $2m$, and M, respectively. Clearly, the combination modes

$$|c_j^{T_{1u}}\rangle = \sqrt{2}|e_j^{T_{1u}}A\rangle - |e_j^{T_{1u}}B\rangle$$

$$= 2P_{j3}^{T_{1u}}|A\rangle - 2P_{j1}^{T_{1u}}|B\rangle \qquad (4.4.18)$$

and

$$|c_j^{T_{1u'}}\rangle = M\sqrt{2}|e_j^{T_{1u}}A\rangle + 2M|e_j^{T_{1u}}B\rangle - 6m|c_j\rangle \qquad (4.4.19)$$

have zero translational momentum and do not move the molecular center of mass. They are drawn in Figure 4.4.4(b). The genuine vibrations will be

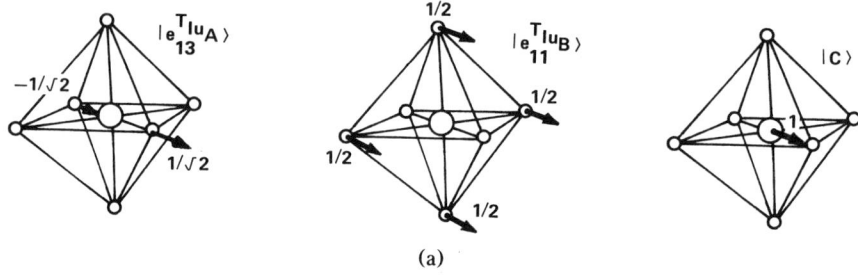

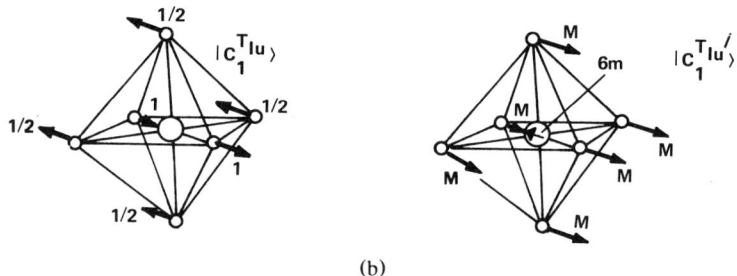

Figure 4.4.4 $\binom{T_{1u}}{1}$ motions. (a) Primitive motions. (b) Constrained motions with zero translation.

combinations of these two constrained states only. Rigid molecular translation corresponds to a third T_{1u} state,

$$|t_j^{T_{1u}}\rangle = \sqrt{2}|e_j^{T_{1u}}A\rangle + 2|e_j^{T_{1u}}B\rangle + |c_j\rangle, \qquad (4.4.20)$$

and this type of motion will not be considered further here.

B. Solving Equations of Motion

The O_h symmetry projection provides states which simplify the equation of motion

$$\mathbf{m}|\ddot{x}\rangle = -\mathbf{F}|x\rangle. \qquad (4.4.21)$$

The symmetry analysis also tells us exactly how much of the mass matrix $\langle \mathbf{m} \rangle$ and force matrix $\langle \mathbf{F} \rangle$ needs to be written. Of the $(21)^2 = 441$ possible components of $\langle \mathbf{F} \rangle$ only the 39 entries in Eq. (4.4.22) will be needed. The same is true of the $\mathbf{m}$ matrix, although its form for this problem is quite simple anyway. In general a row of the matrix is needed for each symmetry orbit. If the matrices are Hermitian as are $\langle \mathbf{F} \rangle$ and $\langle \mathbf{m} \rangle$ only the upper-diagonal parts of each row need be written,

$$\langle F \rangle =$$

	$\lvert A\rangle$	$\lvert R_1^2 A\rangle$	$\lvert R_1 A\rangle$	$\lvert R_2 A\rangle$	$\lvert R_1^3 A\rangle$	$\lvert R_2^3 A\rangle$	$\lvert B\rangle$	$\lvert r_1 B\rangle$	$\lvert r_2 B\rangle$	$\lvert r_1^2 B\rangle$	$\lvert r_4^2 B\rangle$	$\lvert R_2^2 B\rangle$	$\lvert R_1 B\rangle$	$\lvert R_2 B\rangle$	$\lvert R_3 B\rangle$	$\lvert R_1^3 B\rangle$	$\lvert R_2^3 B\rangle$	$\lvert i_3 B\rangle$	$\lvert C_1\rangle$	$\lvert C_2\rangle$	$\lvert C_3\rangle$
$\langle A\rvert$	$2k+j$	0	$k/2$	$k/2$	$k/2$	$k/2$													0	0	$-j$
$\langle B\rvert$							$k+b$	0	0	$-k/2$	$-k/2$	0	0	$k/2$	0	0	$-k/2$	0	$-b/2$	0	0
$\langle C_1\rvert$														$-(k+b)/2$	0	0	$-(k+b)/2$	0	$2(j+b)$	0	0

$$\langle m\rangle = \begin{pmatrix} m & & & \\ & m & & \\ & & \ddots & \\ & & & M \end{pmatrix} \qquad (4.4.22)$$

The $\langle \mathbf{F} \rangle$ components are derived from the spring-coordinate geometry of Figure 4.4.1 according to the small vibration approximations of Section 1.4.B. One should expect a spring-mass model to be only an approximate model for the vibrations of molecules such as UF_6 or SF_6. If more information is gained about the molecular binding then more realistic $\langle \mathbf{F} \rangle$ components can be given.

The matrix formula (4.3.46) gives the components of $\langle \mathbf{F} \rangle$ and $\langle \mathbf{m} \rangle$ matrices in terms of the "primitive" representation (4.4.22). For the bases with distinct irrep labels the resulting components are the desired eigenvalues. From the eigenvalues of $\langle \mathbf{F} \rangle$ and $\langle \mathbf{m} \rangle$ one derives the squared eigenfrequencies $\omega^2 = \langle \mathbf{m}^{-1}\mathbf{F} \rangle$. As an example consider the T_{2u} and T_{2g} eigenfrequency calculations:

$$\left(\omega^{T_{2u}}\right)^2 = (1/m) \sum_l \langle B|\mathbf{F}|lB \rangle \mathscr{D}_{11}^{T_{2u}}(l)$$

$$= \left[(k+b)\mathscr{D}_{11}^{T_2}(1) - (k/2 + b/2)\left(\mathscr{D}_{11}^{T_2}(R_2) + \mathscr{D}_{11}^{T_2}(R_2^3)\right)\right]/m$$

$$= (k+b)/m, \quad (4.4.23)$$

$$\left(\omega^{T_{2g}}\right)^2 = (1/m) \sum_l \langle B|\mathbf{F}|lB \rangle \mathscr{D}_{22}^{T_{2g}}(l)$$

$$= \left[(k+b)\mathscr{D}_{11}^{T_2}(1) - (k/2 + b/2)\left(\mathscr{D}_{22}^{T_2}(R_2) + \mathscr{D}_{22}^{T_2}(R_2^3)\right)\right]/m$$

$$= 2(k+b)/m. \quad (4.4.24)$$

Similarly, the eigenfrequencies of the radially moving motions are derived as they were in Section 4.3 [recall Eq. (4.3.36)$_x$, for example]:

$$\left(\omega^{A_{1g}}\right)^2 = (1/m) \sum_l \langle A|\mathbf{F}|lA \rangle \mathscr{D}^{A_{1g}}(l) = (4k+j)/m, \quad (4.4.25)$$

$$\left(\omega^{E_g}\right)^2 = (1/m) \sum_l \langle A|\mathbf{F}|lA \rangle \mathscr{D}_{11}^{E_g}(l) = (k+j)/m. \quad (4.4.26)$$

This accounts for all genuine vibrational motions in Figure 4.4.4 except the two T_{1u} modes.

A two-by-two $\langle \mathbf{F} \rangle$ and $\langle \mathbf{m} \rangle$ matrix representations in the $|c_j^{T_{1u}}\rangle$ and $|c_j^{T_{1u'}}\rangle$ basis [(4.4.18) and (4.4.19)] need to be derived. The first component of operator $\mathbf{Q} = \mathbf{F}$ or $\mathbf{Q} = \mathbf{m}$ is computed as follows:

$$\langle c_j^{T_{1u}}|\mathbf{Q}|c_j^{T_{1u}}\rangle = \left(2\langle A|P_{3j}^{T_{1u}} - 2\langle B|P_{1j}^{T_{1u}}\right)\mathbf{Q}\left(2P_{j3}^{T_{1u}}|A\rangle - 2P_{j1}^{T_{1u}}|B\rangle\right)$$

$$= 4\langle A|P_{33}^{T_{1u}}\mathbf{Q}|A\rangle - 4\langle A|P_{31}^{T_{1u}}\mathbf{Q}|B\rangle$$

$$- 4\langle B|P_{13}^{T_{1u}}\mathbf{Q}|A\rangle + 4\langle B|P_{11}^{T_{1u}}\mathbf{Q}|B\rangle. \quad (4.4.27)$$

The projection matrices $\langle P \rangle$ can be reduced to sums over the labeling coset leader elements l for each orbit:

$$\langle c_j^{T_{1u}}|\mathbf{Q}|c_j^{T_{1u}}\rangle = 2Q_{AA} - \sqrt{2}\,Q_{AB} + Q_{BB} - \sqrt{2}\,Q_{BA}, \quad (4.4.28)$$

where

$$Q_{AA} = \sum_l \langle A|\mathbf{Q}|lA\rangle \mathscr{D}_{33}^{T_1}(l),$$

$$Q_{BA} = Q_{AB} = (1/\sqrt{2})\sum_l \langle A|\mathbf{Q}|lA\rangle \mathscr{D}_{31}^{T_1}(l),$$

$$Q_{BB} = \sum_l \langle B|\mathbf{Q}|lB\rangle \mathscr{D}_{11}^{T_1}(l).$$

Substituting the primitive $\langle \mathbf{F}\rangle$ and $\langle \mathbf{m}\rangle$ components (4.4.22) into the preceding gives the following:

$$\begin{pmatrix} F_{AA} & F_{AB} \\ F_{BA} & F_{BB} \end{pmatrix} = \begin{pmatrix} 2k+j & -\sqrt{2}\,k \\ -\sqrt{2}\,k & k+b \end{pmatrix}, \quad \begin{pmatrix} m_{AA} & m_{AB} \\ m_{BA} & m_{BB} \end{pmatrix} = \begin{pmatrix} m & 0 \\ 0 & m \end{pmatrix}.$$

$$(4.4.29)$$

Hence, the first T_{1u} components of $\langle \mathbf{F}\rangle$ and $\langle \mathbf{m}\rangle$ are

$$\langle c_j^{T_{1u}}|\mathbf{F}|c_j^{T_{1u}}\rangle = 9k + 2j + b$$

$$\langle c_j^{T_{1u}}|\mathbf{m}|c_j^{T_{1u}}\rangle = 3m. \quad (4.4.30)$$

Similarly the other components are computed. The off-diagonal component of $\mathbf{Q} = \mathbf{F}$ or $\mathbf{Q} = \mathbf{M}$ is

$$\langle c_j^{T_{1u}}|\mathbf{Q}|c_j^{T_{1u'}}\rangle = 2MQ_{AA} + 2\sqrt{2}\,MQ_{AB} - 12m\langle A|\mathbf{Q}|C_3\rangle$$
$$- \sqrt{2}\,MQ_{BA} - 2MQ_{BB} + 12m\langle B|\mathbf{Q}|C_1\rangle,$$

which yields

$$\langle c_j^{T_{1u}}|\mathbf{F}|c_j^{T_{1u'}}\rangle = 2(j-b)(M+6m),$$

$$\langle c_j^{T_{1u}}|\mathbf{m}|c_j^{T_{1u'}}\rangle = 0. \quad (4.4.31)$$

The calculation of the second diagonal component completes each matrix since $\mathbf{F}$ and $\mathbf{m}$ are Hermitian,

$$\langle c_j^{T_{1u'}}|\mathbf{F}|c_j^{T_{1u'}}\rangle = (2j + 4b)(M + 6m)^2,$$

$$\langle c_j^{T_{1u'}}|\mathbf{m}|c_j^{T_{1u'}}\rangle = 6mM(M+6m). \quad (4.4.32)$$

The acceleration matrix $\langle \mathbf{a} \rangle = \langle \mathbf{m}^{-1}\mathbf{F} \rangle$ is then found:

$$\begin{pmatrix} \langle c^{T_{1u}} | \mathbf{a} | c^{T_{1u}} \rangle & \langle c^{T_{1u}} | \mathbf{a} | c^{T_{1u'}} \rangle \\ \langle c^{T_{1u'}} | \mathbf{a} | c^{T_{1u}} \rangle & \langle c^{T_{1u'}} | \mathbf{a} | c^{T_{1u'}} \rangle \end{pmatrix} = \begin{pmatrix} \dfrac{9k + 2j + b}{3m} & \dfrac{2(j+b)(M+6m)}{3m} \\ \dfrac{j-b}{3mM} & \dfrac{(j+2b)(M+6m)}{3mM} \end{pmatrix}.$$

(4.4.33)

The secular equation for the T_{1u} acceleration matrix is

$$\lambda^2 - S\lambda + P = 0, \qquad (4.4.34a)$$

where each root or eigenvalue

$$\lambda = \left(\omega^{T_{1u}}\right)^2 \qquad (4.4.34b)$$

is the square of an eigenfrequency, while

$$S = (3k + j + b)/m + (2j + 4b)/M \qquad (4.4.34c)$$

is the sum of eigenvalues, and

$$P = (kj + 2kb + jb)(M + 6m)/m^2 M \qquad (4.4.34d)$$

is the product of eigenvalues. The T_{1u} eigenfrequency equation is then

$$\omega_{\pm}^{T_{1u}} = \left(S \pm (S^2 - 4P)^{1/2}\right)^{1/2} / \sqrt{2}. \qquad (4.4.34e)$$

The eigenfrequencies $\omega^\alpha(k/j)$ for $b = 0$ are plotted in Figure 4.4.5 using the preceding equations and Eqs. (4.4.23)–(4.4.26). The values of (k/j) which fit the A_{1g} and upper T_{1u} lines observed in UF_6, NpF_6, and PuF_6 are indicated each by a line of circles which denote experimental results. Note that the values of k/j are small; they range between 0.1 and 0.2. For small k the spectrum breaks into two "clusters." The high-frequency cluster (A_{1g}, T_{1u}, E_g) involves radial vibrations which stretch the j spring or radial bonds. This cluster belongs to the induced representation $A' \uparrow O_h$ as explained in Section 4.3. The low-frequency cluster consists of T_{2g}, T_{1u}, T_{2u}, and T_{1g} if you count rotations. This includes all the angular motions, which arise from the B orbit and which belong to the induced representation $E \uparrow O_h$. Angular or bending motions do not affect the j springs directly, but they do stretch the k and b springs. For $k = 0 = b$ the bending or B-orbit

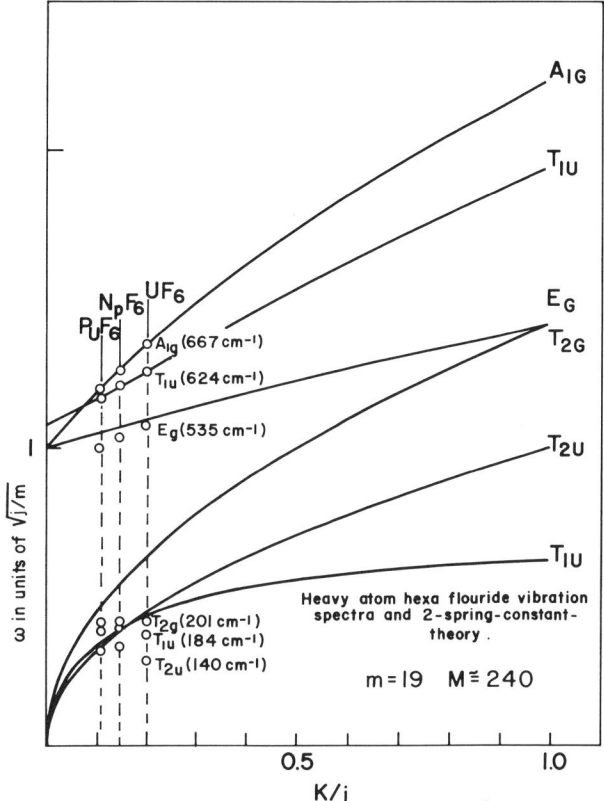

Figure 4.4.5 Heavy-atom hexafluoride vibration spectra and (k, j) spring constant theory.

cluster of levels become degenerate with zero frequency. The splitting of the A-orbit cluster (A_{1g}, T_{1u}, E_g) for $k = 0 = b$ depends on the strength of the j springs and the mass ratio m/M. In Figure 4.4.5 (m/M) is quite small, since the central atomic mass M is large compared to m. The motion of the central atom is part of the T_{1u} modes, and is involved in the coupling between them for nonzero k or b. (Notice how the T_{1u} curves "repel" each other as k varies across Figure 4.4.5.) However, even for $k = 0 = b$ there is coupling between antipodal m atoms through j springs connected by the central M mass. This coupling is analogous to the "transaxial" tunneling described by the T parameter in Eq. (4.3.28), since it leaves E_g and A_{1g} levels degenerate, but splits away the T_{1u} level.

For a lighter hexafluoride molecule such as SF_6 the distinction between radial and angular states seems to disappear, as shown in Figure 4.4.6. Now a small $(b = 0.2)$ value for the bending spring seems to be necessary to even approximately match the observed vibration frequencies. The presence of k

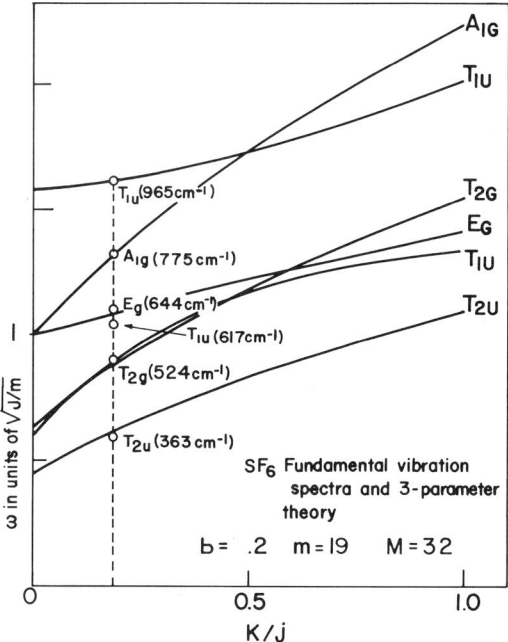

Figure 4.4.6 Sulfur hexafluoride (SF$_6$) spectra and theoretical curves. A small bonding constant ($b = 0.2$) has been included.

and b springs coupling and a light central atom makes the ideas of spectral clusters and localized vibrational modes less useful. Note, however, that the ratio $\omega^{T_{2g}} : \omega^{T_{2u}}$ is predicted to be

$$\omega^{T_{2g}}/\omega^{T_{2u}} = \sqrt{2}, \qquad (4.4.35)$$

independent of k or b according to Eqs. (4.4.23) and (4.4.24). This agrees closely with the experimental values for UF$_6$ and SF$_6$.

More knowledge of molecular potentials or a more sophisticated arrangement of "springs" is needed to precisely analyze all XY_6 vibrational spectra. This is particularly true when higher excitations or overtones are observed, and when anharmonic contributions to the potentials need to be considered. However, some information can be learned from the present spring-mass model.

For example, let us derive a prediction for the isotope shift between S^{32}F$_6$ ($M = 32$) and S^{34}F$_6$ ($M = 34$) T_{1u} modes. Setting $m = 19$, $b = 0.2j$, and $k = 0.18j$ as given in Figure 4.4.6 one needs to consider the solutions (4.4.34e) for two different mass values: $M = 32$ and $M = 34$. The ratios for the high (+) and low (−) T_{1u} frequencies are predicted to be

$$\omega_+^{T_{1u}}(M=32)/\omega_+^{T_{1u}}(34) = 1.0173, \qquad \omega_-^{T_{1u}}(M=32)/\omega_-^{T_{1u}}(34) = 1.0064, \tag{4.4.36a}$$

in close agreement with the experimental values:

$$965 \text{ cm}^{-1}/948 \text{ mm}^{-1} = 1.0179, \qquad 615 \text{ mm}^{-1}/612 \text{ mm}^{-1} = 1.0049. \tag{4.4.36b}$$

Isotope shifts for the other modes besides T_{1u} can be calculated very quickly. There is no shift in the harmonic spectra due to a change of M. A shift due to changing all the m nuclei is given by the ratio $\omega^\alpha(m')/\omega^\alpha(m) = (m'/m)^{1/2}$. A general theory of isotope shifts by Teller and Redlich is described in Herzberg's books listed at the end of Chapter 3.

The interest in the hexafluoride molecules has been centered mostly on the T_{1u} modes. These are the only modes which vibrate the central M atom and the only ones which have a useful M isotope shift. It is instructive to see more exactly how the various atoms move in the T_{1u} vibrational modes. To see this one must find the eigenvectors

$$|e_i^{T_{1u}}(\pm)\rangle = \varepsilon_\pm |c_i^{T_{1u}}\rangle + \varepsilon'_\pm |c_i^{T_{1u}'}\rangle \tag{4.4.37}$$

of the T_{1u} acceleration submatrix (4.4.33). For the $S^{32}F_6$ molecular parameter values $m = 19$, $M = 32$, $b = 0.2j$, and $k = 0.18j$, there is the following eigenequation:

$$j\begin{pmatrix} 0.0670 & 4.1 \\ 0.00044 & 0.112 \end{pmatrix}\begin{pmatrix} \varepsilon_\pm \\ \varepsilon'_\pm \end{pmatrix} = \lambda_\pm^{T_{1u}} \begin{pmatrix} \varepsilon_\pm \\ \varepsilon'_\pm \end{pmatrix}, \tag{4.4.38a}$$

where the eigenvalues

$$\lambda_+^T \lambda_+^{T_{1u}} = \left(\omega_+^{T_{1u}}\right)^2 = 0.138j, \qquad \lambda_-^{T_{1u}} = \left(\omega_-^{T_{1u}}\right)^2 = 0.0415j \tag{4.4.38b}$$

are found using Eqs. (4.4.34). The eigenvector solutions are

$$|e_i^{T_{1u}}(+)\rangle = 0.068|c_i^{T_{1u}}\rangle + 0.0012|c_i^{T_{1u}'}\rangle, \tag{4.4.39a}$$

$$|e_i^{T_{1u}}(-)\rangle = 0.114|c_i^{T_{1u}}\rangle - 0.00071|c_i^{T_{1u}'}\rangle, \tag{4.4.39b}$$

where the m = normalization condition

$$\langle(\pm)e|\mathbf{m}|e(\pm)\rangle = 1 \tag{4.4.40}$$

of Eq. (1.4.13a) is used. The first ($i = 1$) partner for the two SF_6 modes is pictured in Figure 4.4.7(a). This was obtained by combining the constrained

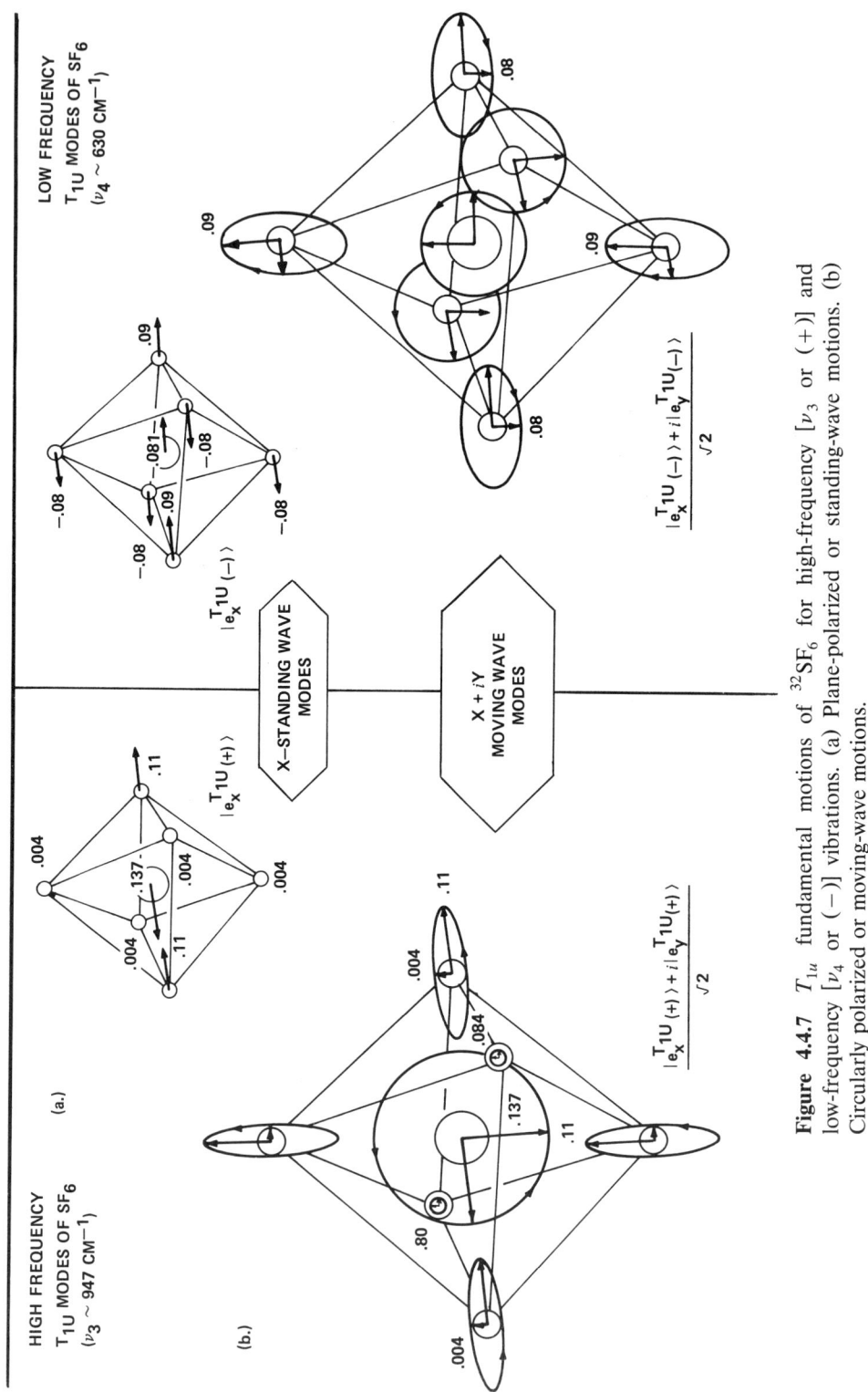

Figure 4.4.7 T_{1u} fundamental motions of $^{32}SF_6$ for high-frequency [ν_3 or (+)] and low-frequency [ν_4 or (−)] vibrations. (a) Plane-polarized or standing-wave motions. (b) Circularly polarized or moving-wave motions.

motions in Figure (4.4.3(b)) using the coefficients in Eqs. (4.4.39). Note that the displacement of the equatorial fluorine atoms for $|e_1(+)\rangle$ is

$$\langle R_1^3 B | e_1(+)\rangle = 0.068 \langle R_1^3 B | c_1 \rangle + 0.0012 \langle R_1^3 B | c_1' \rangle$$
$$= 0.068(-1/2) + 0.0012(32) = 0.004, \quad (4.4.41)$$

which is quite small compared to the values for the axial atoms. Note that the central atom opposes the axial fluorine atoms in $|e(+)\rangle$ while it goes along with them in $|e(-)\rangle$. Clearly, $|e(+)\rangle$ has higher frequency because it distorts the strong j spring more than $|e(-)\rangle$. The displacement of the equatorial atoms in $|e(-)\rangle$ is almost as large as that of the axial atoms. Bear in mind that the precise magnitudes of the T_{1u} displacements depend on the values of the molecular parameters m, M, b, j, and k. The T_{1u} motions are not entirely fixed by symmetry as are the A_{1g}, E_g, T_{2g}, and T_{2u} motions in Figure 4.4.3.

The T_{1u} partners $\{|e_1(\pm)\rangle, |e_2(\pm)\rangle, |e_3(\pm)\rangle\}$ are the same translational distortions pointing along the x, y, and z directions, respectively. These are the plane-polarized or standing-wave bases. It is important to visualize the motions of the circularly polarized or moving-wave bases,

$$\left| e_{1_4}^{T_{1u}}(\pm) \right\rangle = \left(-|e_1(\pm)\rangle - i|e_2(\pm)\rangle \right)/\sqrt{2},$$
$$\left| e_{3_4}(\pm) \right\rangle = \left(|e_1(\pm)\rangle - i|e_2(\pm)\rangle \right)/\sqrt{2}, \quad (4.4.42)$$
$$\left| e_{0_4}(\pm) \right\rangle = |e_3(\pm)\rangle,$$

shown in Figure 4.4.7(b). [This transformation of bases was discussed around Eq. (4.2.36).] Notice that each particle has an elliptical orbit. For example, the semimajor and minor orbital axes of the xy-equatorial fluorine atoms in the $|e(+)\rangle$ mode are $a = 0.11$ and $b = 0.004$. The classical angular momentum of a mass m in an elliptical oscillator orbit is

$$l = mab\omega/2.$$

The sum of these momenta for all the atoms in a molecule is the classical VIBRATIONAL ANGULAR MOMENTUM of the system. The coupling of the vibrational momentum to the angular momentum of the molecule as a whole leads to rotational or CORIOLIS splitting of a vibrational line.

C. Classical Canonical Coordinates

From now on let us demand the m-normalization conditions

$$\langle e_j^\alpha | \mathbf{m} | e_k^\beta \rangle = \delta^{\alpha\beta} \delta_{jk},$$
$$\langle e_j^\alpha | \mathbf{F} | e_k^\beta \rangle = \delta^{\alpha\beta} \delta_{jk} (\omega^\alpha)^2, \quad (4.4.43)$$

which were discussed in Section 1.4.A. The T_{1u} eigenvectors $|e_j^{T_{1u}}(\pm)\rangle$ have already been made to satisfy this. The other modes $|e_j^\alpha\rangle$ only involve motion of the mass-m fluorine atoms, and the newly normalized vectors are obtained by affixing a factor $1/\sqrt{m}$ to the old ones. It is also convenient to use CANONICAL VARIABLES (q, p) defined by

$$q_j^\alpha = \langle e_j^\alpha|\mathbf{m}|x\rangle,$$
$$p_j^\alpha = \langle e_j^\alpha|\mathbf{m}|\dot{x}\rangle. \tag{4.4.44}$$

The relation between canonical coordinates q_j^α and "ordinary" coordinates $X_{lQ} = \langle lQ|X\rangle$ is easily derived:

$$q_j^\alpha = \sum_{Q=A,B,C}\sum_l \langle e_j^\alpha|\mathbf{m}|lQ\rangle\langle lQ|x\rangle. \tag{4.4.45}$$

Here completeness of the original unit states $|lQ\rangle$ is used. The labeling operator l ranges over all states in orbit $\mathbf{Q} = A, B,$ and C.

Expressing the "ordinary" coordinates in terms of the canonical ones is probably more useful. Using the m-completeness relation (1.4.14) one derives

$$X_{lQ} = \langle lQ|x\rangle = \sum_\alpha \sum \langle lQ|e_j^\alpha\rangle\langle e_j^\alpha|\mathbf{m}|x\rangle$$
$$= \sum_\alpha \sum_j \langle lQ|e_j^\alpha\rangle q_j^\alpha. \tag{4.4.46}$$

For example, let us express $X_{R_1^3B}$ in terms of q_j^α. The coefficients $\langle lQ|e_j^\alpha\rangle$ can be read directly from Figures 4.4.4 and 4.4.7(a):

$$X_{R_1^3B} = \frac{1}{2\sqrt{m}}q_3^{T_{2g}} - \frac{1}{2\sqrt{m}}q_1^{T_{2u}} + 0.004 q_1^{T_{1u}}(+) - 0.08 q_1^{T_{1u}}(-). \tag{4.4.46}$$

The quantum theory discussed in the following section will give wave functions or statistical distributions for the q's. From this one derives an X_{lQ} wave function through Eq. (4.4.46).

D. Elementary Quantum Theory of Vibrations

One advantage of the canonical variables (4.4.44) is that they satisfy Hamilton's equations

$$\frac{\partial H}{\partial p_i^\alpha} = \dot{q}_i^\alpha = p_j^\alpha,$$
$$\frac{\partial H}{\partial q_i^\alpha} = -\dot{p}_i^\alpha = -(\omega^\alpha)^2 q_j^\alpha, \tag{4.4.47}$$

for the Hamiltonian

$$H = \sum_\alpha \sum_i \left[(p_i^\alpha)^2/2 + (\omega^\alpha q_i^\alpha)^2/2 \right]. \qquad (4.4.48)$$

The Hamiltonian describes a collection of independent oscillators each having unit mass and angular frequency ω^α. According to Dirac's quantization rules the canonical variables q_i^α and p_j^α can be replaced by Hermitian operators $q_i^\alpha = (q_i^\alpha)^\dagger$ and $p_j^\alpha = (p_j^\alpha)^\dagger$ satisfying the commutation relations

$$\left[q_i^\alpha, p_j^\beta \right] = \delta^{\alpha\beta} \delta_{ij} 1 \hbar. \qquad (4.4.49)$$

These relations and the Hamiltonian (4.4.48) are all that is needed to complete the quantum-mechanical oscillator problem. Most of the work is done once the classical normal modes are found, and the Hamiltonian is reduced to the simple form of Eq. (4.4.48).

An even simpler form can result if the following opertors are defined:

$$a_i^\alpha \equiv \left(\sqrt{\omega^\alpha}\, q_i^\alpha + \frac{i}{\sqrt{\omega^\alpha}} p_i^\alpha \right) \bigg/ \sqrt{2\hbar}, \qquad (4.4.50a)$$

$$a_i^{\alpha\dagger} \equiv \left(\sqrt{\omega^\alpha}\, q_i^\alpha - \frac{i}{\sqrt{\omega^\alpha}} p_i^\alpha \right) \bigg/ \sqrt{2\hbar}. \qquad (4.4.50b)$$

According to Eq. (4.4.49) they satisfy the following commutation relations:

$$\left[a_i^\alpha, a_j^{\beta\dagger} \right] = \delta^{\alpha\beta} \delta_{ij} 1 \hbar, \qquad (4.4.51a)$$

$$\left[a_i^\alpha, a_j^\beta \right] = 0 = \left[a_j^{\beta\dagger}, a_i^{\alpha\dagger} \right]. \qquad (4.4.51b)$$

The Hamiltonian is expressed as follows in terms of a and $a^\dagger$:

$$H = \sum_\alpha \sum_i \frac{\hbar \omega^\alpha}{2} \left(a_i^{\alpha\dagger} a_i^\alpha + a_i^\alpha a_i^{\alpha\dagger} \right) = \sum_\alpha \sum_i \hbar \omega^\alpha \left(a_i^{\alpha\dagger} a_i^\alpha + \frac{1}{2} \right). \qquad (4.4.52)$$

The following commutation relations then result:

$$\left[H, a_i^{\alpha\dagger} \right] = \hbar \omega^\alpha a_i^{\alpha\dagger}, \qquad \left[H, a_i^\alpha \right] = -\hbar \omega^\alpha a_i^\alpha. \qquad (4.4.53)$$

Operators $a^\dagger$ and a are called RAISING and LOWERING operators because of their effect on eigenstates. Suppose eigenstate $|\varepsilon\rangle$ satisfies $H|\varepsilon\rangle = \varepsilon|\varepsilon\rangle$. Then the eigenvalue of $a^\dagger|\varepsilon\rangle$ will be raised by $\hbar\omega$ as follows:

$$Ha_i^{\alpha\dagger}|\varepsilon\rangle = \left(\left[H, a_i^{\alpha\dagger}\right] + a_i^{\alpha\dagger}H\right)|\varepsilon\rangle$$
$$= (\hbar\omega^\alpha + \varepsilon)a_i^{\alpha\dagger}|\varepsilon\rangle, \qquad (4.4.54)$$

and the eigenvalue of $a|\varepsilon\rangle$ is lowered by the same amount:

$$Ha_i^\alpha|\varepsilon\rangle = (\varepsilon - \hbar\omega^\alpha)a_i^\alpha|\varepsilon\rangle. \qquad (4.4.55)$$

It is possible to produce all eigenstates $|\varepsilon\rangle$ by applying raising operators to the ground state $|0\cdots\rangle$. The ground state is defined to be that state which cannot be lowered.

$$a_i^\alpha|0\cdots\rangle \equiv 0 \quad \text{(for all } \alpha, i\text{)}. \qquad (4.4.56)$$

Then the eigenstates $|\varepsilon\rangle = |\cdots n_i^\alpha \cdots n_j^\beta \cdots\rangle$ are

$$|\cdots n_i^\alpha \cdots n_j^\beta \cdots\rangle = \cdots \left(a_i^{\alpha\dagger}\right)^{n_i^\alpha} \cdots \left(a_j^{\beta\dagger}\right)^{n_j^\beta} \cdots |0\cdots\rangle/\sqrt{N_{\cdots n \cdots n'}}, \qquad (4.4.57)$$

where the $1/\sqrt{N}$ factor normalizes the state so that

$$\langle \cdots n_i^\alpha \cdots n_j^\beta \cdots | \cdots n_i^\alpha \cdots n_j^\beta \cdots \rangle = 1. \qquad (4.4.58)$$

Derivation of the normalization and eigenvalues involves the following operator relations whose proofs are left as exercises:

$$\left[a,(a^\dagger)^n\right] = n(a^\dagger)^{n-1}, \qquad (4.4.59)$$

$$\left[a^2,(a^\dagger)^n\right] = n(n-1)(a^\dagger)^{n-2} + 2n(a^\dagger)^{n-1}a + (a^\dagger)^n a^2,$$

$$\vdots$$

$$\left[a^m,(a^\dagger)^n\right] = \sum_{r=0} (n!/(n-m+r)!)(m!/(m-r)!r!)(a^\dagger)^{n-m+r}(a)^r. \qquad (4.4.60)$$

In the preceding the irrep labels $\binom{\alpha}{i}$ are assumed to be the same and are deleted for typographical simplicity. The normalization is now derived from a special case of identity (4.4.60):

$$\langle 0 \cdots | (a)^n (a^\dagger)^n | 0 \cdots \rangle = \langle 0 \cdots | \left[(a)^n (a^\dagger)^n \right] + (a^\dagger)^n (a)^n | 0 \cdots \rangle$$
$$= n! + 0 \cdots.$$

This holds for each and every $\binom{\alpha}{i}$; hence n is given:

$$N_{\cdots n_i^\alpha \cdots n_j^\beta \cdots} = \cdots (n_i^\alpha)! \cdots (n_j^\beta)! \cdots. \qquad (4.4.61)$$

The general raising and lowering relations result, also:

$$a_i^{\alpha\dagger} | \cdots n_i^\alpha \cdots n_j^\beta \cdots \rangle = \sqrt{n_i^\alpha + 1} | \cdots n_i^\alpha + 1 \cdots n_j^\beta \cdots \rangle,$$

$$a_i^\alpha | \cdots n_i^\alpha \cdots n_j^\beta \cdots \rangle = \sqrt{n_i^\alpha} | \cdots n_i^\alpha - 1 \cdots n_j^\beta \cdots \rangle. \qquad (4.4.62)$$

Using these relations it is easy to derive the eigenvalue spectrum of the Hamiltonian (4.4.52):

$$H \cdots | n_i^\alpha \cdots n_j^\beta \cdots \rangle = \left[\sum_\gamma \sum_k (n_k^\alpha + \tfrac{1}{2}) \hbar \omega^\gamma \right] | \cdots n_i^\alpha \cdots n_j^\beta \cdots \rangle. \qquad (4.4.63)$$

The energy eigenvalues for the hexafluoride molecules are given by

$$\varepsilon = (n^{A_{1g}} + 1/2) \hbar \omega^{A_{1g}} + (n_1^{E_g} + n_2^{E_g} + 1) \hbar \omega^{E_g}$$
$$+ (n_1^{T_{1u}}(+) + n_2^{T_{1u}}(+) + n_3^{T_{1u}}(+) + 3/2) \hbar \omega^{T_{1u}}(+)$$
$$+ (n_1^{T_{2u}} + n_2^{T_{2u}} + n_3^{T_{2u}} + 3/2) \hbar \omega^{T_{2u}} + (n_1^{T_{2g}} + n_2^{T_{2g}} + n_3^{T_{2g}} + 3/2) \hbar \omega^{T_{2g}}$$
$$+ (n_1^{T_{1u}}(-) + n_2^{T_{1u}}(-) + n_3^{T_{1u}}(-) + 3/2) \hbar \omega^{T_{1u}}(-). \qquad (4.4.63)_x$$

A sketch of octahedral hexafluoride energy-levels is shown in Figure 4.4.8. See Figs. 6.6.2–4 for a more complete and accurate diagram of levels for SF_6, UF_6, and SiF_4. The levels are labeled by a conventional spectroscopic notation ν_1 for A_{1g}, ν_2 for E_g, ν_3 for T_{1u}, ν_4 for T_{2g}, and ν_5 for T_{2u}. When the $A_{1g}, E_g, \ldots$ states are doubly excited they are labeled $2\nu_1, 2\nu_2, \ldots$, etc. These are called double HARMONICS, and similarly for higher excitations.

The number of levels in a given nth harmonic depends strongly on the degeneracy of the fundamental. A second harmonic of E_g labeled $(2\nu_2)$ has three degenerate states:

$$|\ldots, n_1^{E_g}, n_2^{E_g}, \ldots (2\nu_2)\rangle = \{|\ldots, 2, 0, \ldots\rangle, |\ldots, 1, 1, \ldots\rangle, |\ldots, 0, 2, \ldots\rangle\},$$

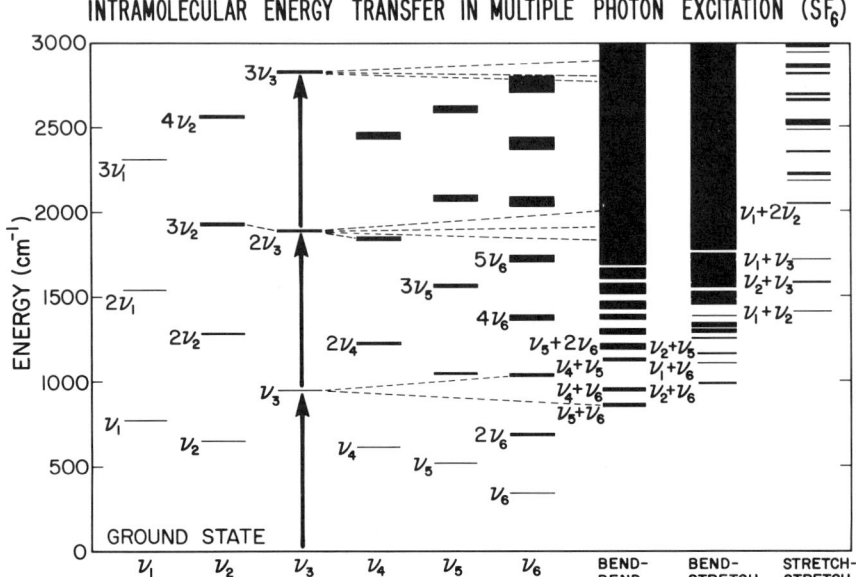

Figure 4.4.8 Sketch of SF$_6$ quantum vibration levels. The density of levels increases rapidly at higher energy. Standard spectroscopic notation is used. For example, two quanta of the $T_{1u}(+)$ or ν_3 vibration is labeled $2\nu_3$. The figure shows expected flow of energy during laser excitation of the ν_3 "ladder." (Due to Robin S. McDowell and Jay R. Ackerhalt of Los Alamos National Laboratory.)

while a second harmonic of T_{1u} ($2\nu_3$) has six degenerate states:

$$|\ldots, n_1^{T_{1u}}, n_2^{T_{1u}}, n_3^{T_{1u}}, \ldots (2\nu_3)\rangle$$
$$= \{|\ldots, 2, 0, 0, \ldots\rangle, |\ldots, 0, 2, 0, \ldots\rangle, |\ldots, 0, 0, 2, \ldots\rangle,$$
$$|\ldots, 1, 1, 0, \ldots\rangle, |\ldots, 1, 0, 1, \ldots\rangle, |\ldots, 0, 0, 1, \ldots\rangle\}.$$

In general the nth harmonic of an l^α-dimensional fundamental vibration has degeneracy $d(n)$ equal to a binomial coefficient in Pascals' triangle:

Fundamental dimension:		$l = 1$	2	3	4	5	
ground state:	$n = 0$	$d = 1$	1	1	1	1	$\cdots$
fundamental:	$n = 1$	1	2	3	4	5	$\cdots$
second harmonic:	$n = 2$	1	3	6	10	15	$\cdots$
third harmonic:	$n = 3$	1	4	10	20	35	$\cdots$
fourth harmonic:	$n = 4$	1	5	15	35	70	$\cdots$

(4.4.64)

The formula for the degeneracy is

$$d(n^\alpha) = (l^\alpha + n^\alpha - 1)!/n^\alpha!(l^\alpha - 1)!. \tag{4.4.65}$$

The general level is a combination of each vibration $\alpha, \beta, \ldots$ excited $n^\alpha, n^\rho, \ldots$ times, respectively. Its degenercy is simply the product

$$d(n^\alpha, n^\beta, \ldots) = d(n^\alpha) \, d(n^\beta) \cdots . \qquad (4.4.66)$$

The d degeneracies are split by anharmonic terms such as $q^3, q^4, q^3 q', \ldots,$ etc. as explained in Section 6.6.

Finally, let us consider briefly the wave function of the oscillator eigenstates. The ground-state conditions (4.4.56) take the form

$$\langle q \cdots | \sqrt{\omega} \, q + (i/\sqrt{\omega}) p | 0 \cdots \rangle = 0,$$

$$\sqrt{\omega} \, q + (i/\sqrt{\omega})(\hbar/i) \frac{\partial}{\partial q} \langle q \cdots | 0 \cdots \rangle = 0, \qquad (4.4.67)$$

in the coordinate representation where $p \to (\hbar/i) \partial / \partial q$. The solution to these differential equations is a product of Gaussian wave functions:

$$\langle q \cdots | 0 \cdots \rangle = \psi_{0 \ldots}(q \cdots) = \left(e^{-\omega(q)^2/2\hbar} / \right)(\cdots . \qquad (4.4.68)$$

For each coordinate $q = q^{A_{1g}}, q_1^{E_g}, q_2^{E_g}, q_1^{T_{1u}}, \ldots$ there is a Gaussian distribution. An excited state is obtained by raising a particular oscillator one unit. Starting with a Gaussian ground-state function one produces the wave function of state $|1, 0, \ldots \rangle$ as follows

$$\psi_{1,0,\ldots}(q, \ldots) = \left(\sqrt{\omega} \, q - (\hbar/\sqrt{\omega}) \frac{\partial}{\partial q} \right) \psi_{0,0,\ldots}(q, \ldots)$$

$$= \left(2\sqrt{\omega} \, q e^{-\omega(q)^2/2\hbar} / \right)(\cdots . \qquad (4.4.69)$$

Finally, the nth-excited state for one coordinate is

$$\psi_{n,0,\ldots}(q, \ldots) = \left((\omega 2^n \hbar \pi n!) H_n(q\sqrt{\omega}/\hbar) e^{-\omega(q)^2/2\hbar} \right)(\cdots, \qquad (4.4.70)$$

where $H_n(x)$ is the nth Hermite polynomial.

Oscillator wave functions $\psi(q)$ die off quickly when q exceeds the magnitude of the CLASSICAL TURNING POINT $q(\text{TP})$. This is the classical point of maximum excursion for q at which momentum $\Sigma_i (p_i^\alpha)^2$ is zero in the Hamiltonian (4.4.48):

$$H = (\omega^\alpha)^2 (q^\alpha(\text{TP}))^2 / 2;$$

setting H equal to the oscillator eigenvalue we have

$$(n^\alpha + l^\alpha/2) \hbar \omega^\alpha = (\omega^\alpha)^2 (q^\alpha(\text{TP}))^2 / 2$$

or

$$q^\alpha(\text{TP}) = [(2n^\alpha + l^\alpha) \hbar / \omega^\alpha]^{1/2}, \qquad (4.4.71)$$

where l^α is the number of partners $\{q_1^\alpha \ q_2^\alpha \ \cdots \ q_{l^\alpha}^\alpha\}$ in mode α and n^α is the oscillator quantum number.

ADDITIONAL READING

It is hard to find a simple physical treatment of induced representations and subgroup coset spaces. The references cited below are not simple to read.

G.W. Mackey, Induced Representations of Groups and Quantum Mechanics (Benjamin, New York, 1968).

A.J. Coleman, *Induced Representations and the Symmetric Group*, Queens University Papers on Mathematics (Queens University Press, Kingston, Ontario, 1965).

The first applications of induced representations to tetrahedral and octahedral fine structure are given in the papers listed below.

W.G. Harter and C.W. Patterson, *Phys. Rev. Lett.*, **38**, 224 (1977); *J. Chem. Phys.*, **66**, 4872 (1977).

W.G. Harter, C.W. Patterson, and F.J. Ja Paixao, *Rev. Mod. Phys.*, **50**, 37 (1978).

W.G. Harter and C.W. Patterson, *J. Math. Phys.*, **20**, 1453 (1979).

The excerpt of methane spectra shown in Figure 4.3.3 was given in Allen Pine's article.

A.S. Pine, *J. Opt. Soc. Am.*, **66**, 97 (1976).

A review of the spectrosocpy of octahedral and tetrahedral molecules for the laser isotope program is the following:

R.S. McDowell, C.W. Patterson and W.G. Harter, *Los Alamos Science*, **3**, 38 (1982).

A qualitative discussion of symmetry breaking is given in the following: R. Peierls, (Dirac Memorial Lecture) *Contemporary Physics*, **33**, 221 (1992)

PROBLEMS

Section 4.1

4.1.1 Generally one does not need to tabulate representations of all elements of a given group in order to define them. Some smaller number of elements $\{g_1, g_2, \ldots\}$ can be chosen whose products $\{g_1^2, g_2^2, \ldots, g_1 g_2, g_1 g_2^2, \ldots, g_1 p_{g_2} q_{g_1}, \ldots\}$ finally generate each and every element of the group. Such elements are called *generators* of the group. What are the minimal number of generators of the groups listed below? (Hint: Consider the Hamilton turns.)

(a) C_3, (b) C_{3v}, (c) D_4, (d) O, (e) O_h.

4.1.2. **(a)** How many elements of the octahedral group O commute with the 120° rotation operator r_1? Do these elements form a symmetry group? If so which and why, or why not?

(b) How many elements of the octahedral group O commute with the 90° rotation operator R_3? Do these elements form a symmetry group? If so which and why, or why not?

4.1.3 There are several D_2 symmetry subgroups in the octahedral group. Identify them and tell which if any of these are normal subgroups. Are there any normal D_3 or D_4 subgroups?

Section 4.2

4.2.1 Construct correlation tables and level-splitting diagrams for the following tetrahedral (T_d) and octahedral (O_h) subgroup chains.
- (a) $T_d \supset C_{3v} \supset C_v$.
- (b) $T_d \supset D_{2d} \supset C_{2v}$.
- (c) $T_d \supset C_{3v} \supset C_3$.
- (d) $O_h \supset D_{4h} \supset D_{2h}$.
- (e) $O_h \supset C_{4v} \supset C_4$.
- (f) $O_h \supset T_h \supset D_2$.

4.2.2 Compare the two different types of D_2 subgroup correlations within O symmetry by sketching level-splitting diagrams.

$$\begin{matrix} p_x \\ p_y \\ p_z \end{matrix} \equiv\!\!\equiv\!\!\equiv \; l=1$$

4.2.3 Describe and/or label the level splitting that would (or would not) happen to an atomic ($l = 1$)p orbital triplet placed in the center of the following "cages." (Give symmetry group and irreps for each case.)

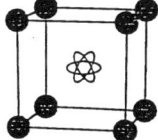
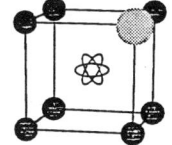
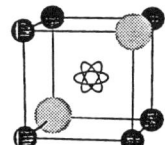

(a) Eight equal charges on cubic vertices.

(b) Same as (a) but one extra charge on cubic vertex.

(c) Same as (a) but two equal charges on cubic body diagonal.

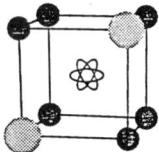
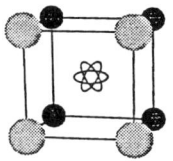
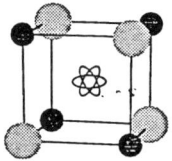

(d) Same as (a) but two equal charges on cubic face diagonal.

(e) Same as (a) but four equal charges on cubic face.

(f) Same as (a) but four extra charges on alternate vertices.

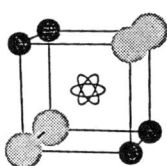
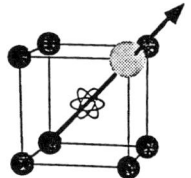

(g) Same as (a) but four equal charges on cubic body diagonals.

(h) Same as (b) but with a magnetic field along cubic diagonal.

Section 4.3

4.3.1 Let each C_3 coset $\{1C_3, g_2C_3, g_3C_3, \ldots\}$ in octahedral group O be associated with a ket vector $\{|1\rangle, |g_2\rangle, |g_3\rangle, \ldots\}$, respectively. Let a representation $\mathscr{I}(g)$ of each element g of the group O be defined by

$$g|g_k\rangle = |gg_k\rangle = |g_j h\rangle,$$

where gg_k is an element of coset $g_j C_3$, or

$$\mathscr{I}_{jk}(g) = \langle g_j|g|g_k\rangle = 1, \quad \text{if } g_j^{-1} gg_k \text{ is in subgroup } C_3,$$
$$= 0, \quad \text{otherwise.}$$

[This is the *principal C_3 induced* representation of O. It is labeled $\mathscr{I} = D^0(\text{of } C_3)\uparrow O$.]

(a) Construct matrices $\mathscr{I}(1), \mathscr{I}(r_1), \mathscr{I}(R_1^2), \mathscr{I}(R_1), \mathscr{I}(i_1)$.

(b) Use the traces of these representations to compute which and how many irreducible representations of O will appear if $\mathscr{I}$ is reduced. Compare with the results predicted using the correlation table between C_3 and O and the Frobenius reciprocity theorem.

(c) Do parts (a) and (b) for the cosets $\{1D_4, g_2D_4, g_3D_4, \ldots\}$ of the subgroup D_4 which contains the 90° z axial rotation $\mathbf{R}_3$.

4.3.2 Let each C_{3v} coset $\{1C_{3v}, g_2C_{3v}, g_3C_{3v}, \ldots\}$ in octahedral group O_h be associated with a pair of ket vectors $\{|1,1\rangle, |1,2\rangle, |g_2,1\rangle,$

$|g_2, 2\rangle, |g_3, 1\rangle, |g_3, 2\rangle, \ldots\}$, respectively. Let a representation $\mathscr{I}(g)$ of each element g of the group O_h be defined by

$$g|g_k, 1\rangle = D^E_{1,1}(g_j^{-1}gg_k)|g_j, 1\rangle + D^E_{2,1}(g_j^{-1}gg_k)|g_j, 2\rangle,$$

where gg_k is in coset $g_j C_{3v}$,

$$g|g_k, 2\rangle = D^E_{1,2}(g_j^{-1}gg_k)|g_j, 1\rangle + D^E_{2,2}(g_j^{-1}gg_k)|g_j, 2\rangle,$$

or $g_j^{-1}gg_k$ is in subgroup C_{3v},

or

$$\mathscr{I}_{j, a; k, b} = \langle g_j, a | g | g_k, b \rangle = D^E_{a,b}(g_j^{-1}gg_k), \quad \text{if } g_j^{-1}gg_k \text{ is in subgroup } C_{3v},$$

$$= 0, \quad \text{otherwise.}$$

[This is the E of C_{3v} induced representation of O_h. It is labeled $\mathscr{I} = D^E$ (of C_{3v})↑O_h.]

(a) Construct matrices $\mathscr{I}(1)$, $\mathscr{I}(r_1)$, $\mathscr{I}(IR_1^2)$, $\mathscr{I}(IR_1)$, and $\mathscr{I}(Ii_1)$.

(b) Use the traces of these representations to compute which and how many irreducible representations of O_h will appear if $\mathscr{L}$ is reduced. Compare with the results predicted using the correlation table between C_{3v} and O_h and the Frobenius reciprocity theorem.

(c) Do parts (a) and (b) for the cosets $\{1D_{4h}, g_2 D_{4h}, g_3 D_{4h}, \ldots\}$ of the subgroup D_{4h} which contains the 90° z-axial rotation R_3. (Use the E_g representation of D_{4h}.)

4.3.3 Suppose a quantum particle can tunnel between eight equilibrium positions or potential wells each being located at the vertices of a cube. Let the local energy of each well be H, and let the nearest-neighbor tunneling rates be $-S$, while the next-nearest-neighbor tunneling rates are equal to $-T$. Let the local wave function associated with each well be trigonally symmetric; that is, let: $\mathbf{r}_1|1\rangle = |1\rangle$ and similarly for each of the local states $|g_k\rangle$.

(a) Label the states using coset elements derived in Problem 4.3.1 or 4.3.2.

(b) Construct the first two rows of tunneling Hamiltonian matrix in terms of H, S, and T.

(c) Construct symmetry defined states $\mathbf{P}^\alpha_{ij}|1\rangle$ and sketch the wave functions. You may want to compare the states generated by the irreducible representations defined by different subgroup chains O_h-C_{3v}-C_v and O_h-D_{4h}-D_{2h} or other.

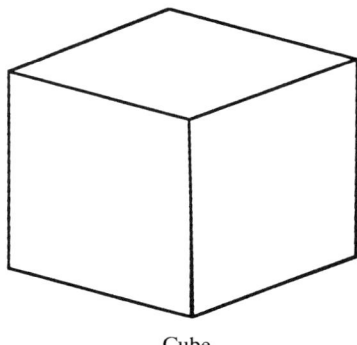
Cube

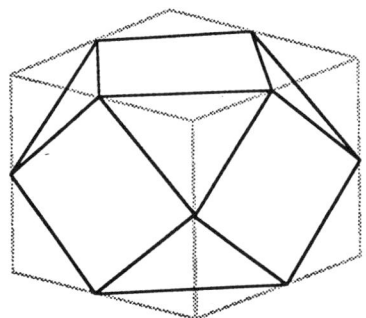
Truncated Cubic Octahedron

(d) Display eigenvalue spectrum for the case $S > 0$ and $T = 0$.

(e) Display eigenvalue spectrum for the case $S = 0$ and $T > 0$.

4.3.4 Suppose a quantum particle can tunnel between 12 equilibrium positions or potential wells each being located at the vertices of a cubic octahedron. Let the local energy of each well be H, and let the nearest-neighbor tunneling rates be $-S$, while the next-nearest-neighbor tunneling rates are equal to $-T$. Let the local wave function associated with each well be C_{2v} symmetric; that is, let $\mathbf{i}_1|1\rangle = |1\rangle$ and similarly for each of the local states $|\mathbf{g}_k\rangle$.

(a) Label the states using coset elements.

(b) Construct the first two rows of tunneling Hamiltonian matrix in terms of H, S, and T.

(c) Construct symmetry-defined states $\mathbf{P}_{ij}^\alpha|1\rangle$ and sketch the wave functions. You may want to compare the states generated by the irreducible representations defined by different subgroup chains O_h-C_{2v}-C_v and O_h-D_{4h}-D_{2h} or other.

(d) Display eigenvalue spectrum for the case $S > 0$ and $T = 0$.

(e) Display eigenvalue spectrum for the case $S = 0$ and $T > 0$.

4.3.5 The character tables for the icosahedral symmetry Y and the fivefold dihedral subgroup D_5 are listed in the following.

Y classes	0°	72°	144°	120°	180°
c_g order	1	12	12	20	15
A	1	1	1	1	1
T_1	3	$G+$	$G-$	0	-1
T_3	3	$G-$	$G+$	0	-1
G	4	-1	-1	1	0
H	5	0	0	-1	1

The numbers $G+$ and $G-$ are the "golden ratios" $G+ = (1 + \sqrt{5})/2$ and $G- = (1 - \sqrt{5})/2$.

D_5 classes	0°	72°	144°	180°
c_g order	1	2	2	5
A_1	1	1	1	1
A_2	1	1	1	-1
E_1	2	$-G-$	$-G+$	0
E_2	2	$-G+$	$-G-$	0

(a) Show how the icosahedral levels split when the symmetry is reduced to D_5 and then to C_5. Construct a correlation table between the Y group and these two subgroups.

(b) Show how the icosahedral levels split when the symmetry is reduced to D_3 and then to C_3. Construct a correlation table between the Y group and these two subgroups.

4.3.6 Compute the detailed effects of field perturbations on a charged particle tunneling through the cubic eight-well system (Recall Problem 4.3.3). Plot the energy levels as a function of the field and draw eigenfunction sketches for extreme field limits. Label all levels with appropriate (maximal) symmetry labels.

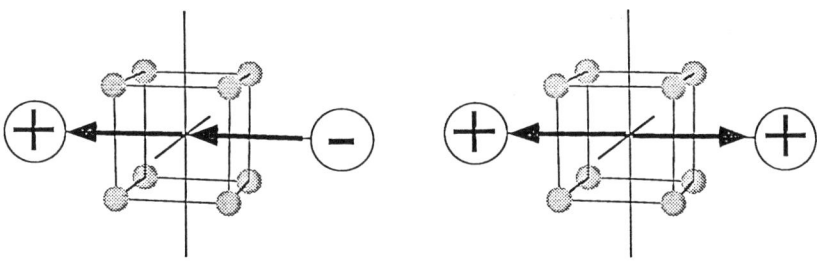

(a) Dipole field on four-fold axis.

(b) Quadrupole field on four-fold axis.

(c) Sketch (without detailed calculation) the energy level effects of the same fields on the three-fold axis. Use physical arguments and correlation results.

(d) Check the results of (c) with detailed diagonalization of the model Hamiltonians.

Section 4.4

4.4.1 Construct and solve a mechanical force model for the methane-like XY_4 tetrahedral molecule. Using the data listed below try to fit your model to SiF_4, CF_4, CD_4 or CH_4 fundamental frequencies.

SiF_4: $\nu_1 = 801$ cm^{-1}; $\nu_2 = 264$ mm^{-1}; $\nu_3, \nu_4 = 382$ cm^{-1}, 1032 cm^{-1};

CF_4: $\nu_1 = 908$ cm^{-1}; $\nu_2 = 435$ cm^{-1}; $\nu_3, \nu_4 = 632$ cm^{-1}, 1283 cm^{-1};

CD_4: $\nu_1 = 2069$ cm^{-1}; $\nu_2 = 1092$ cm^{-1}; $\nu_3, \nu_4 = 996$ cm^{-1}, 2259 cm^{-1};

CH_4: $\nu_1 = 2917$ cm^{-1}; $\nu_2 = 1534$ cm^{-1}; $\nu_3, \nu_4 = 1306$ cm^1, 3019 cm^{-1}.

4.4.2 (a) Construct and solve a mechanical force model for a cubic X_8 molecule using O_h symmetry.

(b) Label the nonzero frequency modes of the cubic molecule C_8H_8 (cubane).

CHAPTER 5

REPRESENTATIONS OF CONTINUOUS ROTATION GROUPS AND APPLICATIONS

We consider now the symmetry of a vacuum occupied by a single point or a perfect sphere. This symmetry is higher than all the point symmetries discussed earlier since it contains all their symmetry operations along with an infinite number more. A rotation about any axis through the point of origin, by any angle 1.0001°, 19°, or whatever, is now a symmetry operation. So is inversion or any reflection. The resulting symmetry group is called the three-dimensional ORTHOGONAL GROUP O_3.

The theory of the orthogonal groups O_2 and O_3 in two and three dimensions is known to physicists as ANGULAR-MOMENTUM CALCULUS. This will be the main topic of this chapter. To mathematicians O_2 and O_3 are two fundamental examples of LIE (pronounced Lee) groups, which are named after mathematician Sophus Lie. This chapter also contains a brief introduction to other important Lie groups including O_4, and the UNITARY GROUPS U_2 and U_3.

Lie groups appear to be very different from the finite or discrete symmetry groups which have been discussed so far. In the place of group tables and tables of representations for finite group algebras there will be analytic functions which give the same information for Lie groups. However, it is possible to find finite subalgebras called LIE ALGEBRAS which help one to analyze Lie groups and their representations. Then some of the algebraic methods developed in the preceding chapters can be used again.

One of the important applications of the R_3 Lie-group representations involves the theory of the free quantum rotor and the hindered quantum rotor. The theory of hindered quantum rotors is related to crystal field splitting. This theory is a generalization of the Bohr orbital level splitting

316 REPRESENTATIONS OF CONTINUOUS ROTATION GROUPS AND APPLICATIONS

treated in Section 2.12 or 3.6, and it is introduced in Section 5.6. This is a very important part of the theoretical spectroscopy which is discussed in later chapters.

5.1 BASIC THEORY OF ORTHOGONAL GROUPS O_2 AND O_3

The group O_n consists of all real linear transformations of n-dimensional vectors which leave all vector lengths and angles between all pairs of vectors unchanged. For example, consider the set of all 3×3 orthogonal transformation matrices:

$$\mathscr{O} = \begin{pmatrix} \mathscr{O}_{11} & \mathscr{O}_{12} & \mathscr{O}_{13} \\ \mathscr{O}_{21} & \mathscr{O}_{22} & \mathscr{O}_{23} \\ \mathscr{O}_{31} & \mathscr{O}_{32} & \mathscr{O}_{33} \end{pmatrix}. \quad (5.1.1)$$

Suppose that their action on column vectors

$$u = \begin{pmatrix} u_1 \\ u_2 \\ u_3 \end{pmatrix}, \quad v = \begin{pmatrix} v_1 \\ v_2 \\ v_3 \end{pmatrix}, \ldots \quad (5.1.2)$$

gives new or transformed vectors:

$$\begin{pmatrix} u'_1 \\ u'_2 \\ u'_3 \end{pmatrix} = \begin{pmatrix} \mathscr{O}_{11} & \mathscr{O}_{12} & \mathscr{O}_{13} \\ \mathscr{O}_{21} & \mathscr{O}_{22} & \mathscr{O}_{23} \\ \mathscr{O}_{31} & \mathscr{O}_{32} & \mathscr{O}_{33} \end{pmatrix} \begin{pmatrix} u_1 \\ u_2 \\ u_3 \end{pmatrix}, \quad \begin{pmatrix} v'_1 \\ v'_2 \\ v'_3 \end{pmatrix} = \begin{pmatrix} \mathscr{O}_{11} & \mathscr{O}_{12} & \mathscr{O}_{13} \\ \mathscr{O}_{21} & \mathscr{O}_{22} & \mathscr{O}_{23} \\ \mathscr{O}_{31} & \mathscr{O}_{32} & \mathscr{O}_{33} \end{pmatrix} \begin{pmatrix} v_1 \\ v_2 \\ v_3 \end{pmatrix},$$
$$(5.1.3)$$

for which all scalar products are unchanged:

$$u' \cdot u' = u \cdot u, \quad u' \cdot v' = u \cdot v, \quad v' \cdot v' = v \cdot v. \quad (5.1.4)$$

This implies the following for arbitrary u_i or v_j:

$$\sum_{j=1}^{3} u'_j v'_j = \sum_{j=1}^{3} \sum_{u=1}^{3} \sum_{l=1}^{3} (\mathscr{O}_{jk} u_k)(\mathscr{O}_{jl} v_l) = \sum_{k=1}^{3} \sum_{l=1}^{3} \left(\sum_{j=1}^{3} \mathscr{O}_{jk} \mathscr{O}_{jl} \right) u_k v_l = \sum_{n=1}^{3} u_n v_n.$$

In particular, setting $u_i = \delta_{ik}$ and $v_j = \delta_{jl}$, one finds the following constraint on the $\mathscr{O}$ matrices:

$$\sum_{j=1}^{3} \mathscr{O}_{jk} \mathscr{O}_{jl} = \delta_{kl}. \quad (5.1.5)$$

BASIC THEORY OF ORTHOGONAL GROUPS O_2 AND O_3 317

This equation implies that the columns of each $\mathscr{O}$ matrix must be three mutually orthogonal column vectors of unit length. It also implies that the transpose $\mathscr{O}^T$ of the matrix $\mathscr{O}$ must be the inverse of $\mathscr{O}$:

$$\mathscr{O}^T\mathscr{O} = 1 = \mathscr{O}\mathscr{O}^T. \tag{5.1.6}$$

Note that if the columns of any finite square matrix are components of mutually orthogonal vectors, then the same must be true of the rows:

$$\sum_{j=1}^{3} \mathscr{O}_{kj}\mathscr{O}_{lj} = \delta_{kl}. \tag{5.1.7}$$

Equations (5.1.5) and (5.1.7) are special cases of the orthogonal and completeness relations discussed in Section 1.2.B(b). [Compare with Eqs. (1.2.19) and (1.2.20).]

One may imagine that these operations $\mathscr{O}$ transform physical objects in space. Physical objects are composed of particles whose position vectors $u, v, w, \ldots$ get moved by $\mathscr{O}$, as indicated in Figure 5.1.1.

After an orthogonal transformation all the particles in the object will end up having the same interparticle distances, since all scalar products stay the same. While thinking of this one might conclude that $\mathscr{O}_3$ is the set of all rigid rotations. In fact $\mathscr{O}_3$ does contain all rotations in a subgroup labeled R_3. However, this accounts for only half of O_3. The other half consists of discontinuous or IMPROPER transformations such as INVERSION (I) through a point [see Figure 5.1.1(b)] or combinations of I with rotations. The improper transformations change left-handed objects into right-handed ones and vice versa. It is easy to snap a left-handed surgeon's glove into a right-handed one, but if you try the same transformation on most objects you will break or tear them.

In Section 2.11 we introduced another example of an improper transformation in O_3 represented by a matrix having only ± 1 on its diagonal:

$$\sigma_{xy} \rightarrow \begin{pmatrix} 1 & 0 & 0 \\ 0 & 1 & 0 \\ 0 & 0 & -1 \end{pmatrix}.$$

This transformation is inversion through a plane (the xy or 12 plane is used here) or REFLECTION. A related transformation,

$$i_z = \sigma_{xy}I \rightarrow \begin{pmatrix} -1 & 0 & 0 \\ 0 & -1 & 0 \\ 0 & 0 & 1 \end{pmatrix},$$

could be called inversion through an axis. (The z or 3 axis is used here.)

318 REPRESENTATIONS OF CONTINUOUS ROTATION GROUPS AND APPLICATIONS

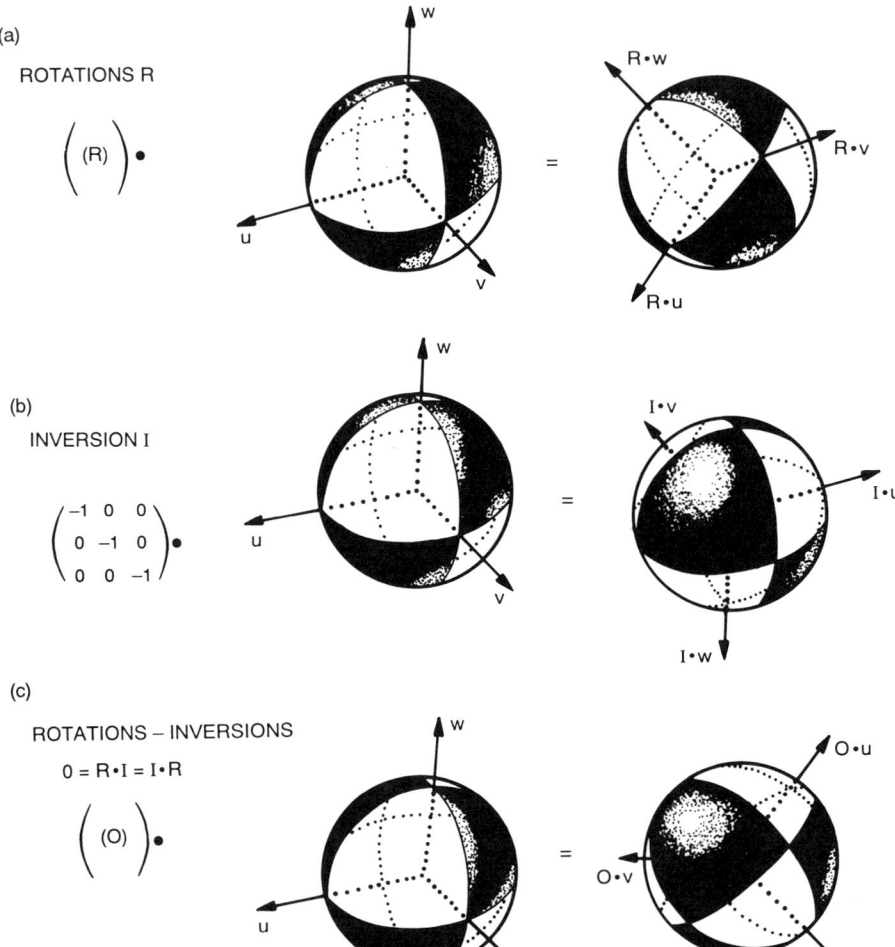

Figure 5.1.1 Action of orthogonal transformations on a unit vector triad. (a) Rotations are proper operations. (b) Inversion or (c) rotation inversions are improper operations.

However, it is actually quite proper, since it is just a 180° rotation around the z axis. To check the propriety of a general orthogonal transformation matrix, one may evaluate its determinant. The determinant gives the volume spanned by the transformed unit vectors $\hat{x}'_j = \mathscr{O}\hat{x}_j$:

$$\det \mathscr{O} = \hat{x}'_1 \cdot \hat{x}'_2 \times \hat{x}'_3 = \begin{cases} 1, & \text{if } \mathscr{O} \text{ is proper,} \\ -1, & \text{if } \mathscr{O} \text{ is improper.} \end{cases} \quad (5.1.8)$$

BASIC THEORY OF ORTHOGONAL GROUPS O_2 AND O_3 **319**

(a)

(b)

Figure 5.1.2 (a) Early Cal Tech experiment on parity and time reversal. (b) Later experiment.

You might try to determine whether proper or improper operations have been applied in Figures 5.1.2.

Note that O_3 and R_3 are not Abelian groups; that is, a general pair of elements will not necessarily commute ($\mathscr{O}\mathscr{O}' \neq \mathscr{O}'\mathscr{O}$). However, the inversion operator I is seen to commute with all elements in $\mathscr{O}_3$. This allows us to express $\mathscr{O}_3$ as an OUTER PRODUCT $R_3 \times C_i$ of the rotation group R_3 with the finite group $C_i = \{1, I\}$ as depicted in Figure 5.1.3. (The formal definition of the outer product was given in Section 2.10.) This allows one to obtain the representations of O_3 directly from products of those of R_3 and C_i.

Let us consider briefly the mathematical definition of the two-dimensional orthogonal group O_2. The four components of 2×2 orthogonal matrices $\mathscr{O}_{ij}$ obey Eqs. (5.1.5). Of these, the following three are independent:

$$\mathscr{O}_{11}\mathscr{O}_{11} + \mathscr{O}_{21}\mathscr{O}_{21} = 1, \qquad (5.1.9a)$$

$$\mathscr{O}_{12}\mathscr{O}_{12} + \mathscr{O}_{22}\mathscr{O}_{22} = 1, \qquad (5.1.9b)$$

$$\mathscr{O}_{11}\mathscr{O}_{21} + \mathscr{O}_{12}\mathscr{O}_{21} = 0. \qquad (5.1.9c)$$

320 REPRESENTATIONS OF CONTINUOUS ROTATION GROUPS AND APPLICATIONS

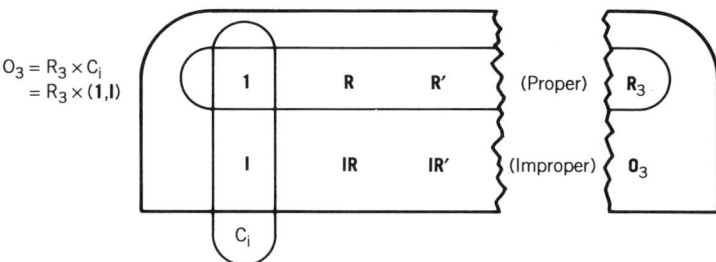

Figure 5.1.3 Diagram of orthogonal group O_3. O_3 is an outer product of rotation group R_3 and the cyclic (C_2-like) inversion group C_i.

This leaves only one undetermined variable or parameter. Equation (5.1.9a) suggests that we set $\mathcal{O}_{11} \equiv \cos\phi$ and $\mathcal{O}_{21} = \sin\phi$. Then we see that the following two types of solutions to the orthogonal equations are possible:

$$\mathcal{O} = \begin{pmatrix} \cos\phi & -\sin\phi \\ \sin\phi & \cos\phi \end{pmatrix}, \qquad (5.1.10a)$$

$$\mathcal{O}' = \begin{pmatrix} \cos\phi & \sin\phi \\ \sin\phi & -\cos\phi \end{pmatrix} = \begin{pmatrix} \cos\phi & -\sin\phi \\ \sin\phi & \cos\phi \end{pmatrix} \begin{pmatrix} 1 & 0 \\ 0 & -1 \end{pmatrix}. \quad (5.1.10b)$$

The first matrix has $\det \mathcal{O} = +1$ and corresponds to a proper rotation by angle ϕ, while the second matrix has $\det \mathcal{O}' = -1$ and corresponds to an improper inversion or reflection through the 1 axis, followed by a rotation of angle ϕ. (The second transformation is equal to a single reflection through a line with slope angle $\phi/2$.)

From this it is clear that the group R_2 of proper rotations is a subgroup of O_2, and that R_2 is Abelian. However, the reflections such as $\begin{pmatrix} 1 & 0 \\ 0 & -1 \end{pmatrix}$ do not commute with elements of R_2, and so it is not possible to express O_2 as an outer product involving R_2. Furthermore, O_2 is not Abelian. Note, also, that the improper two-dimensional reflection $\begin{pmatrix} 1 & 0 \\ 0 & -1 \end{pmatrix}$ could be accomplished by a proper three-dimensional rotation about the 1 axis.

Similar reasoning can be applied to any of the orthogonal Lie groups O_n. For example, each of the n^2 real components $\mathcal{O}_{ij}$ of O_n can assume any values that satisfy the orthogonality equations,

$$\sum_{i=1}^{3} \mathcal{O}_{ij}\mathcal{O}_{ik} = \delta_{jk}, \qquad j,k = 1,2,\ldots,n. \qquad (5.1.11)$$

Of these, exactly $n(n+1)/2$ are independent ($j \geq k = 1,2,\ldots,n$). This implies that only $n^2 - n(n+1)/2 = n(n-1)/2$ independent parameters are needed to define all elements of O_n:

$$O_n = O_1, O_2, O_3, O_4, O_5, O_6 \cdots,$$
$$\text{number of parameters} = n(n-1)/2 = 0, \ 1, \ 3, \ 6, \ 10, \ 15 \quad \cdots. \quad (5.1.12)$$

(Note that O_1 contains only two elements. No continuous parameters are needed.) In general, R_n and O_n have the same number of parameters since they differ only by the inclusion of discrete parity or inversion operations.

5.2 REPRESENTATIONS OF R_2, O_2, AND RELATED SYMMETRY GROUPS

Two-dimensional rotational symmetry R_2 is the limiting case of C_n symmetry when n approaches infinity. Let the angle of rotation $(0 \leq \phi < 2\pi)$ be the single parameter of R_2. For C_n, ϕ was restricted to integral multiples of $2\pi/n$. C_n symmetry was discussed in Sections 2.7 and 2.12.

The irreps of R_2 are obtained from those of C_n by taking the limit $n \to \infty$. They are given by

$$D^m(\phi) = e^{im\phi}, \qquad (5.2.1)$$

where all integers $m = 0, \pm 1, \pm 2, \ldots, \infty$ are allowed now. In the case of C_n, these integers had been restricted by Eq. (2.7.2) so as to give only the nth roots of unity. (Recall also Figure 2.7.2.)

Section 2.1.7 contains a discussion of the physical significance of the irreps in terms of waves. The case $m = |m|$ (positive) corresponds to an angular wave with m crests running around the circle in the counterclockwise direction, while $m = -|m|$ (negative) corresponds to the same wave running in the clockwise direction.

If the full O_2 symmetry is present, these two wave states $|+m\rangle$ and $|-m\rangle$ must have the same frequency or phase velocity. This follows since one is the (improper) reflection of the other. We may therefore combine them to make cosine ($|c_m\rangle$) and sine ($|s_m\rangle$) standing waves which are also stationary states or normal modes:

$$|c_m\rangle = (|+m\rangle + |-m\rangle)/\sqrt{2}, \qquad |s_m\rangle = (-|+m\rangle + |-m\rangle)/i\sqrt{2}. \qquad (5.2.2)$$

Then the resulting representation of the proper rotations is

$$\begin{pmatrix} \frac{1}{\sqrt{2}} & \frac{1}{\sqrt{2}} \\ -\frac{i}{\sqrt{2}} & \frac{i}{\sqrt{2}} \end{pmatrix} \begin{pmatrix} e^{im\phi} & 0 \\ 0 & e^{-im\phi} \end{pmatrix} \begin{pmatrix} \frac{1}{\sqrt{2}} & \frac{i}{\sqrt{2}} \\ \frac{1}{\sqrt{2}} & -\frac{i}{\sqrt{2}} \end{pmatrix}$$

$$= \begin{pmatrix} \cos m\phi & -\sin m\phi \\ \sin m\phi & \cos m\phi \end{pmatrix} \equiv \mathscr{D}^m(\phi). \quad (5.2.3)$$

This is the real irrep $\mathscr{D}^m$ of O_2 rotations for $m \geq 1$. These O_2 irreps are listed in the following on the left-hand side. Two of the most commonly used irrep notations are given. The equivalent complex moving-wave representations are given below the table for the standing-wave ones.

Irrep notations			Standing-Wave Representations	
			z-Axis Rotation ϕ	x-Plane Reflection $\updownarrow$
A_1	Σ^+	$m=0$	$\mathscr{D}^{\Sigma^+}(\phi) = 1$	$D^{\Sigma^+}(\updownarrow) = 1$
A_2	Σ^-	0	$\mathscr{D}^{\Sigma^-}(\phi) = 1$	$D^{\Sigma^-}(\updownarrow) = -1$
E_1	Π	± 1	$\mathscr{D}^\Pi(\phi) = \begin{pmatrix} \cos\phi & -\sin\phi \\ \sin\phi & \cos\phi \end{pmatrix}$	$\mathscr{D}^\Pi(\updownarrow) = \begin{pmatrix} 1 & 0 \\ 0 & -1 \end{pmatrix}$
E_2	Δ	± 2	$\mathscr{D}^\Delta(\phi) = \begin{pmatrix} \cos 2\phi & -\sin 2\phi \\ \sin 2\phi & \cos 2\phi \end{pmatrix}$	$\mathscr{D}^\Delta(\updownarrow) = \begin{pmatrix} 1 & 0 \\ 0 & -1 \end{pmatrix}$
E_3	Φ	± 3	$\mathscr{D}^\Phi(\phi) = \begin{pmatrix} \cos 3\phi & -\sin 3\phi \\ \sin 3\phi & \cos 3\phi \end{pmatrix}$	$\mathscr{D}^\Phi(\updownarrow) = \begin{pmatrix} 1 & 0 \\ 0 & -1 \end{pmatrix}$
$\vdots$	$\vdots$	$\vdots$	$\vdots$	$\vdots$
			Moving-Wave Representations	
			z-Axis Rotation ϕ	x-Plane Reflection $\updownarrow$
A_1	Σ^+	$m=0$	$\mathscr{D}^{\Sigma^+}(\phi) = 1$	$D^{\Sigma^+}(\updownarrow) = 1$
A_2	Σ^-	0	$\mathscr{D}^{\Sigma^-}(\phi) = -1$	$D^{\Sigma^-}(\updownarrow) = -1$
E_1	Π	± 1	$D^\Pi(\phi) = \begin{pmatrix} e^{i\phi} & 0 \\ 0 & e^{-i\phi} \end{pmatrix}$	$D^\Pi(\updownarrow) = \begin{pmatrix} 0 & 1 \\ 1 & 0 \end{pmatrix}$
E_2	Δ	± 2	$D^\Delta(\phi) = \begin{pmatrix} e^{2i\phi} & 0 \\ 0 & e^{-2i\phi} \end{pmatrix}$	$D^\Delta(\updownarrow) = \begin{pmatrix} 0 & 1 \\ 1 & 0 \end{pmatrix}$
E_3	Φ	± 3	$D^\Phi(\phi) = \begin{pmatrix} e^{3i\phi} & 0 \\ 0 & e^{-3i\phi} \end{pmatrix}$	$D^\Phi(\updownarrow) = \begin{pmatrix} 0 & 1 \\ 1 & 0 \end{pmatrix}$
			$\vdots$	$\vdots$

$$(5.2.4)$$

REPRESENTATIONS OF R_2, O_2, AND RELATED SYMMETRY GROUPS

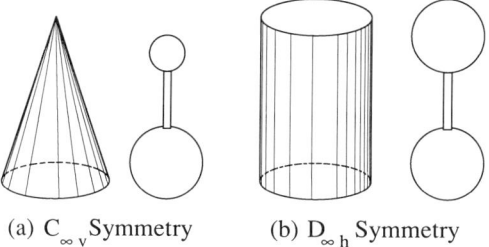

(a) $C_{\infty v}$ Symmetry (b) $D_{\infty h}$ Symmetry

Figure 5.2.1 Examples of objects with (a) $C_{\infty v}$ symmetry, and (b) $D_{\infty h}$ symmetry.

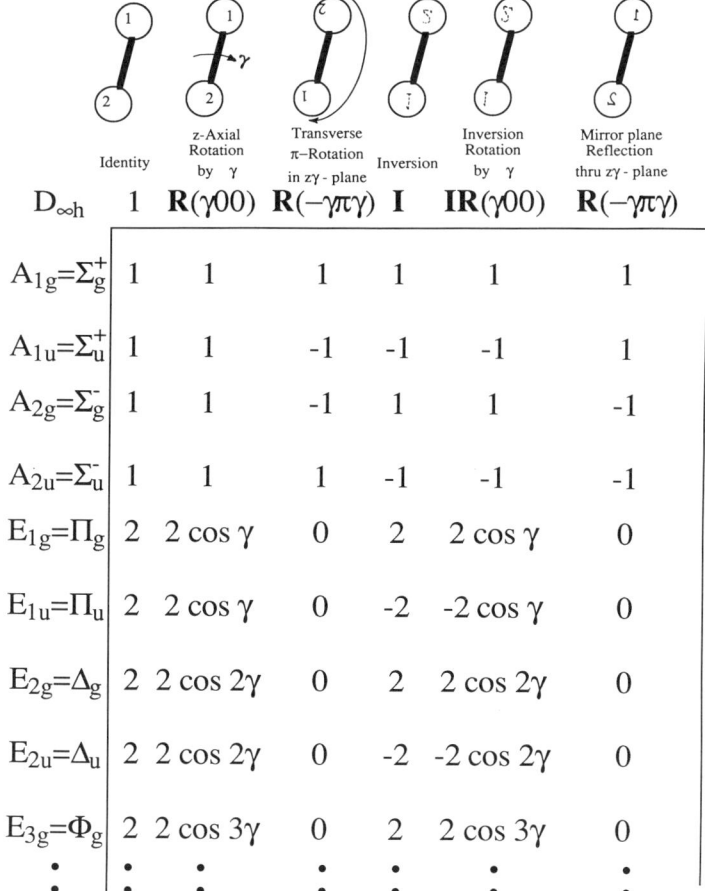

$D_{\infty h}$	Identity	z-Axial Rotation by γ	Transverse π-Rotation in zy-plane	Inversion	Inversion Rotation by γ	Mirror plane Reflection thru zy-plane
	1	$R(\gamma 00)$	$R(-\gamma \pi \gamma)$	I	$IR(\gamma 00)$	$R(-\gamma \pi \gamma)$
$A_{1g}=\Sigma_g^+$	1	1	1	1	1	1
$A_{1u}=\Sigma_u^+$	1	1	-1	-1	-1	1
$A_{2g}=\Sigma_g^-$	1	1	-1	1	1	-1
$A_{2u}=\Sigma_u^-$	1	1	1	-1	-1	-1
$E_{1g}=\Pi_g$	2	$2\cos\gamma$	0	2	$2\cos\gamma$	0
$E_{1u}=\Pi_u$	2	$2\cos\gamma$	0	-2	$-2\cos\gamma$	0
$E_{2g}=\Delta_g$	2	$2\cos 2\gamma$	0	2	$2\cos 2\gamma$	0
$E_{2u}=\Delta_u$	2	$2\cos 2\gamma$	0	-2	$-2\cos 2\gamma$	0
$E_{3g}=\Phi_g$	2	$2\cos 3\gamma$	0	2	$2\cos 3\gamma$	0
$\vdots$	$\vdots$	$\vdots$	$\vdots$	$\vdots$	$\vdots$	$\vdots$

Figure 5.2.2 Classes and characters of $D_{\infty h}$ molecular symmetry. Operations are indicated by their effect on a diatomic (X_2) molecule. The notation [$R(\alpha 00)$, etc.] for Euler angles will be explained in Section 5.3.

The spatial symmetry isomorphic to O_2 is known as $C_{\infty v}$. It is the limiting case of C_{nv} as $n \to \infty$, and is the symmetry of a right circular cone or of a diatomic or linear molecule composed of different nuclei. [See Figure 5.2.1(a).] Another related spatial symmetry is

$$D_{\infty h} = C_{\infty v} \times C_i \sim O_2 \times C_i. \tag{5.2.5}$$

This includes all the $O_2 \sim C_{\infty v}$ operations as well as all of them in combination with the 3-space inversion I. $D_{\infty h}$ is the spatial symmetry of a right circular cylinder or a homonuclear diatomic or linear molecule. [See Figure 5.2.1(b)]. The irreps of $D_{\infty h}$ follow directly from the products of $C_{\infty v}$ and C_i irreps. The standard subscript notation is used to tell whether a given irrep base is even (g) or odd (u) under the inversion operation I. The $D_{\infty h}$ irrep characters are given for the archetypical operations which are shown above the table in Figure 5.2.2. The parametric notation $R(\alpha, \beta, \gamma)$ is explained in the following section.

5.3 PARAMETERS OF R_3

Equation (5.1.12) shows that R_3 has three parameters. There are many ways to choose these parameters. The following is a discussion of two different choices which indicates the advantages of each.

A. Euler Angles ($\alpha\beta\gamma$)

Probably the most common choices for parameters are the EULER ANGLES α, β, and γ. The goniometer in Figure 5.3.1 defines the rotational position state $|\alpha\beta\gamma\rangle$ of a ball and its $\{\bar{x}, \bar{y}, \bar{z}\}$ axes by angles α, β, and γ on three dials. The α dial and β dial define the azimuth and polar angles, respectively, of the $\bar{z}$ axis. The α and β Euler angles can serve as polar coordinates $\phi = \alpha$ and $\theta = \beta$ for a radius vector $\hat{r} = \bar{z}$ in the $\bar{z}$ direction. However, the similarity between Euler angles and spatial coordinates ends there. The third (γ) Euler angle gives the twist of the $\{\bar{x}\bar{y}\bar{z}\}$ system about the $\bar{z}$ axis.

It is important to note that generally each rotational position state $|\alpha\beta\gamma\rangle$ of the ball in Figure 5.3.1 has two possible Euler angle settings; i.e.,

$$|\alpha\beta\gamma\rangle = |\alpha \pm 180°, -\beta, \gamma \pm 180°\rangle \tag{5.3.1}$$

The settings $\{\alpha = 50°, \beta = 60°, \gamma = 70°\}$ and $\{\alpha = -130°, \beta = -60°, \gamma = -110°\}$ are shown in Figures 5.3.1(b) and 5.3.1(c), respectively. They put the $\{\bar{x}, \bar{y}, \bar{z}\}$ axes in the same position relative to the lab $\{xyz\}$ system. One may choose to ignore this double-valuedness by restricting β by the relation

$$0° \leq \beta < 180° \tag{5.3.2}$$

PARAMETERS OF R_3 **325**

Euler Angular Position Coordinates

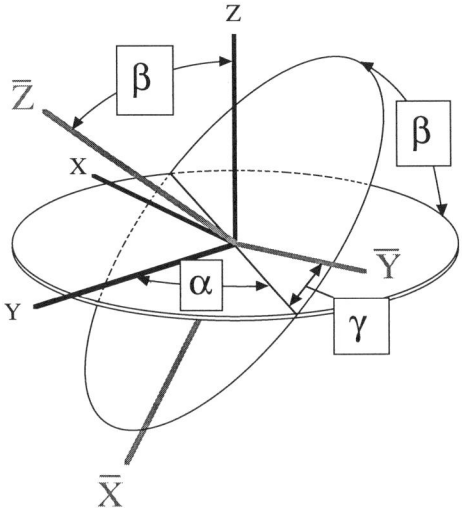

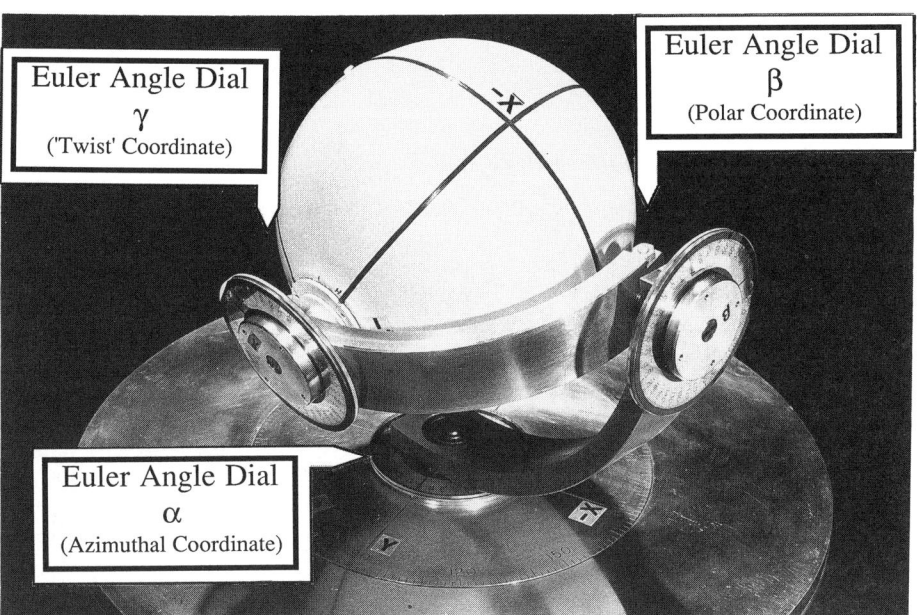

Figure 5.3.1 Definition of Euler angles as rotation position coordinates (*Photos by Vincent Malette and Joanie Geiser*.)

(b) Position State $|\alpha\beta\gamma\rangle = |\,50°\ 60°\ 70°\rangle$

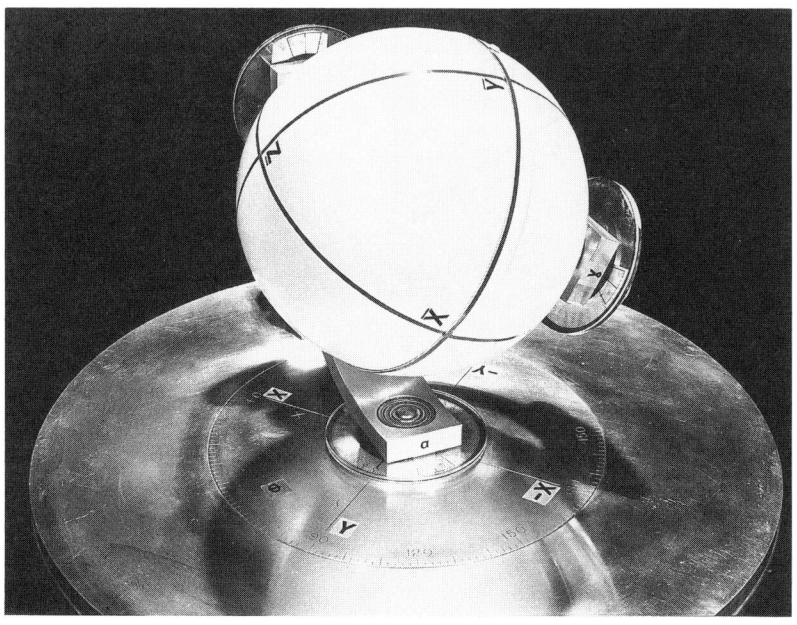

(c) Position State $|\alpha\beta\gamma\rangle = |\,-130°\ -60°\ -110°\rangle$

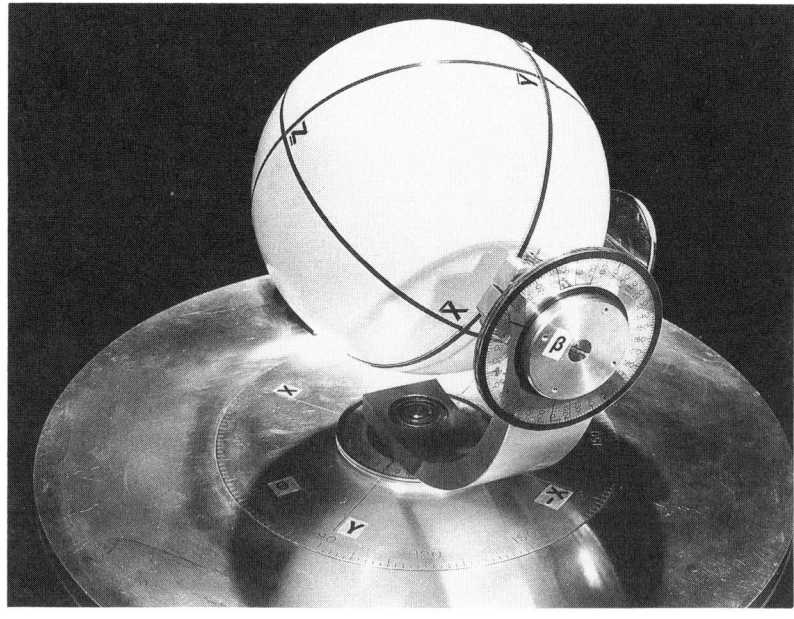

Figure 5.3.1 *(Continued).*

(d) Original State $|\alpha\beta\gamma\rangle = |\,0°\ 0°\ 0°\rangle \equiv |\,1\,\rangle$

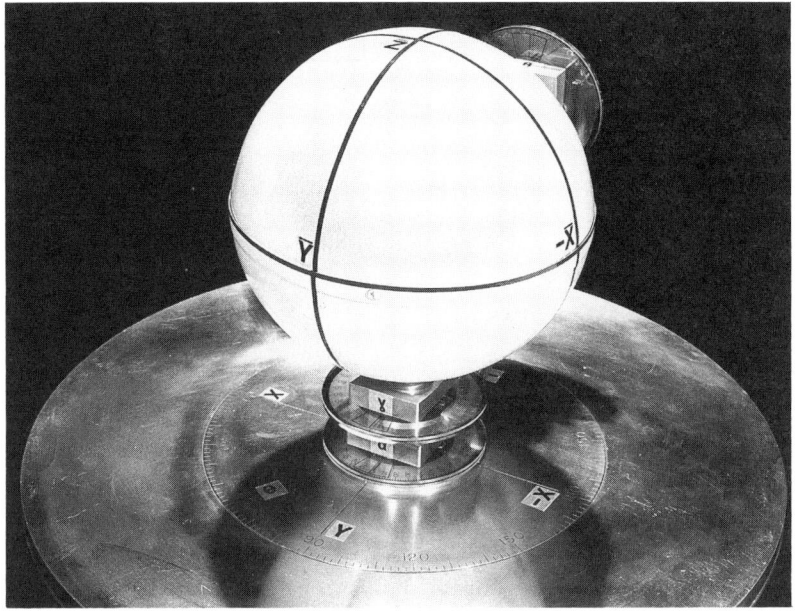

Figure 5.3.1 *(Continued).*

to positive angles. However, we shall see that the double-valuedness is an unavoidable part of any rotational coordinate system. It will turn out that an electron spin polarization state changes phase when the angles $\{\alpha, \beta, \gamma\}$ change from the values in Figure 5.3.1(b) to those in Figure 5.3.1(c). So in some sense the two positions really are different even if they look the same to the classically trained eye.

As in Chapters 2–4, it is necessary to define how a rotational position state $|\alpha\beta\gamma\rangle$ is obtained by *group operations* $R(\alpha\beta\gamma)$ acting upon an initial unrotated state $|1\rangle = |000\rangle$; i.e.,

$$|\alpha\beta\gamma\rangle = R(\alpha\beta\gamma)|1\rangle.$$

To do this, imagine that the $[\bar{x}, \bar{y}, \bar{z}]$ axes and ball are suspended at the center of a transparent shell as indicated in Figure 5.3.2. Suppose this shell supports two sliding cranks, which are fitted with suction cups, and capable of being moved parallel to the laboratory y and z axes, respectively.

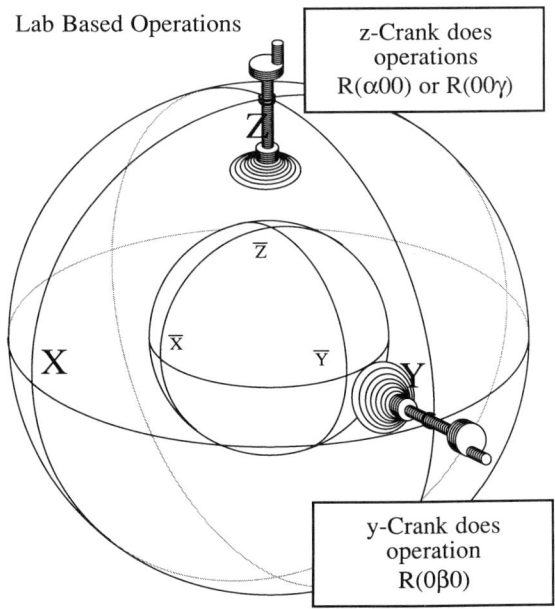

Figure 5.3.2 Laboratory definition of rotation operators $R(\alpha 00)$ and $R(0\beta 0)$. All rotations of the body can be achieved by successively working two cranks attached to a shell fixed in the laboratory. The z crank performs rotations $R(\alpha 00)$ or $R(00\gamma)$. The y crank performs rotations $R(0\beta 0)$.

One way to obtain the rotational position defined by α, β, and γ is through the following operational steps

First step: [This is labeled $R(00\gamma)$]: Attach $\bar{z}$ crank, turn it counterclockwise by angle γ, and then detach it.

Second step: [This is labeled $R(0\beta 0)$]: Attach $\bar{y}$ crank, turn it counterclockwise by angle β, and then detach it.

Third step: [This is labeled $R(\alpha 00)$]: Attach $\bar{z}$ crank, turn it counterclockwise by angle α, and then detach it.

It is conventional to write this sequence of rotations in the following way:

$$R(\alpha\beta\gamma) = R(\alpha 00) R(0\beta 0) R(00\gamma). \quad (5.3.3)$$

Then there is the following relation:

$$|\alpha\beta\gamma\rangle = R(\alpha\beta\gamma)|000\rangle = R(\alpha 00) R(0\beta 0) R(00\gamma)|000\rangle \quad (5.3.4)$$

between states and operators. As usual, the rightmost operator in a group product acts first. This sequence of rotations is shown in Figure 5.5.11 on

Body Based Operations

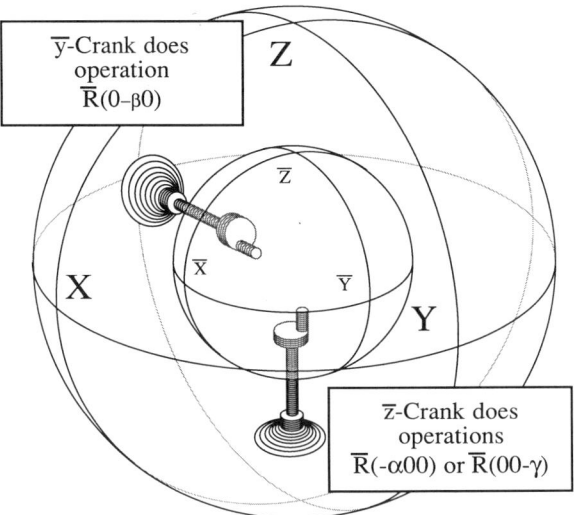

Figure 5.3.3 Body definition of rotation operators $\bar{R}(\alpha 00)$ and $\bar{R}(0\beta 0)$. All rotations of the outer (lab) shell can be achieved by working successively two cranks attached to the body. The $\bar{z}$ crank performs rotations $\bar{R}(\alpha 00)$ or $\bar{R}(00\gamma)$. The $\bar{y}$ crank performs rotations $\bar{R}(0\beta 0)$.

page 379. Note that the α and γ angles are redundant when $\beta = 0$. In the initial orientation state $|1\rangle = |000\rangle$ of Figure 5.3.1(d) the α and γ dials are parallel and coaxial. Hence, one may write

$$R(\alpha 00) = R(00\alpha) = R(\alpha + \gamma, 0, \alpha - \gamma)$$

for arbitrary α and γ.

Consider another way to define the relative orientation of the $\overline{[xyz]}$ and $[xyz]$ axes. Suppose someone on the rotating object can attach cranks to the inside of the spherical shell as indicated in Figure 5.3.3. Using the internal cranks according to the following steps also yields the $|\alpha\beta\gamma\rangle$ orientation state such as was shown first in Figure 5.3.1(a).

First step: [Labeled $\bar{R}(-\alpha 00)$]: Attach $\bar{z}$ crank, turn it *clockwise* by angle α, and then detach it.
Second step: [Labeled $\bar{R}(0 - \beta 0)$]: Attach $\bar{y}$ crank, turn it clockwise by angle β, and then detach it.
Third step: [Labeled $\bar{R}(00 - \gamma)$]: Attach $\bar{z}$ crank, turn it clockwise by angle γ, and then detach it.

Let us denote this sequence of operations as follows:

$$(\bar{R}(\alpha\beta\gamma))^{-1} = \bar{R}(00-\gamma)\bar{R}(0-\beta 0)\bar{R}(-\alpha 00)$$
$$= (\bar{R}(00\gamma))^{-1}(\bar{R}(0\beta 0))^{-1}(\bar{R}(\alpha 00))^{-1}$$
$$= (\bar{R}(\alpha 00)\bar{R}(0\beta 0)\bar{R}(00\gamma))^{-1},$$

or, more simply, as

$$\bar{R}(\alpha\beta\gamma) = \bar{R}(\alpha 00)\bar{R}(0\beta 0)\bar{R}(00\gamma). \tag{5.3.5}$$

This defines another set of rotation operators $\bar{R}(\alpha\beta\gamma)$ with the same group structure as the $R(\alpha\beta\gamma)$ in Eq. (5.3.3). However, the $\bar{R}$ and R are *not* the same operators by any means. In fact, any pair $\bar{R}(\alpha\beta\gamma)$ and $R(\alpha'\beta'\gamma')$ will always commute; i.e.,

$$\bar{R}(\alpha\beta\gamma)R(\alpha'\beta'\gamma') = R(\alpha'\beta'\gamma')\bar{R}(\alpha\beta\gamma). \tag{5.3.6}$$

(Remember, pairs of different R do not commute, neither do different $\bar{R}$ unless the angles α, β, and γ equal 180° or 0°.)

One may imagine that the operators $R(\alpha\beta\gamma)$ move the laboratory system. If you like to be dramatic you could say the $\bar{R}(\alpha\beta\gamma)$ move the entire surrounding universe. However, let us assume that the only thing that counts here is the relative state of orientation of the object relative to the lab axes. Then we must accept the relation

$$|\alpha\beta\gamma\rangle = R(\alpha\beta\gamma)|000\rangle = \bar{R}^{-1}(\alpha\beta\gamma)|000\rangle. \tag{5.3.7}$$

This says that the effect of moving the object one way must be indistinguishable from that of moving the rest of the universe the opposite or inverse way.

The application of lab- and body-defined rotation operators to the theory of quantum rotors is introduced in Section 5.5.C. The explicit matrix representation and operator construction of the rotations and their generators is discussed in Section 5.5.E. Here we are giving their abstract definitions and properties.

B. Darboux or ω-Axis Angles [φθω]

No matter how many rotations are applied, there will always be a line of points in a rotated object that end up exactly where they were initially. In other words, any rotation can be done by just one crank axis provided it can be properly positioned. The geometer Darboux developed the idea of a movable axial or rotation vector $\boldsymbol{\omega}$. Therefore it is appropriate to name the angles involved in the related parametrization accordingly.

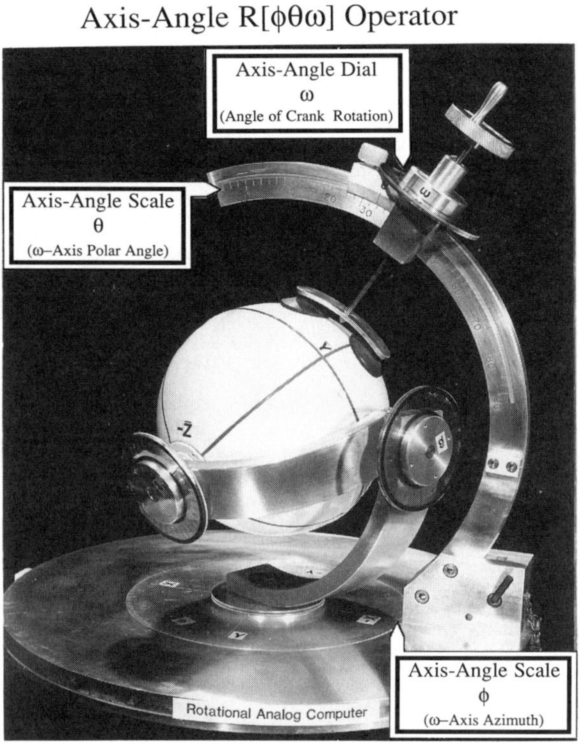

Axis-Angle R[ϕθω] Operator

Figure 5.3.4 Defining axis or Darboux angles $R[\phi\theta\omega]$. Rotation operation $R[\phi\theta\omega]$ is defined by the angle ω of rotation and the direction of the axis of rotation given by angles ϕ and θ of azimuth and polar declination, respectively. (*Photo by Vincent Malette and Joanie Geiser.*)

Darboux angles $[\phi\theta\omega]$ are defined operationally using the device pictured in Figure 5.3.4. First one sets the rotation axis or ω crank according to polar angle θ and azimuth angle ϕ. Then one attaches the ω crank to the object, turns it counterclockwise by rotation angle ω, and detaches the crank. The setting $[\phi = 80°, \theta = 33.69°]$ of the axis angles followed by a rotation by $\omega = 128.68° \cong 129°$ corresponds to the rotation operator $R[\phi\theta\omega] = R[80°, 34°, 129°]$. (Note that we shall use square brackets $[\omega]$ to distinguish Darboux angles from Euler angles.) Application of this operator to the initial state $|1\rangle = |000\rangle$ happens to yield the position state

$$|\alpha = 50°, \beta = 60°, \gamma = 70°\rangle = R[80°, 34°, 129°]|0,0,0\rangle, \quad (5.3.8)$$

of Figure 5.3.1(a). This is shown in the sequence of Figures 5.3.5(a)–5.3.5(c). We shall derive later the algebraic relations which exist between the Darboux

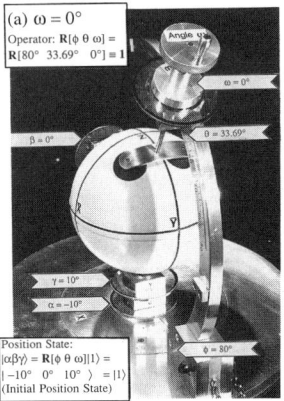

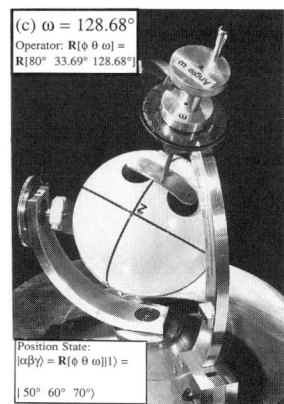
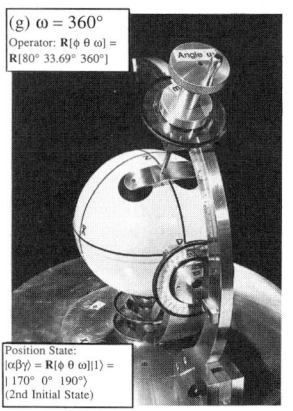
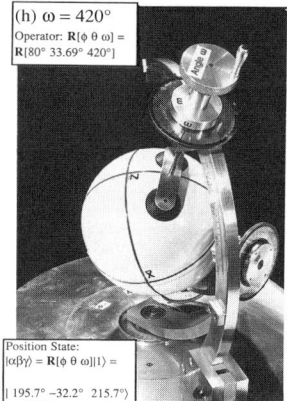
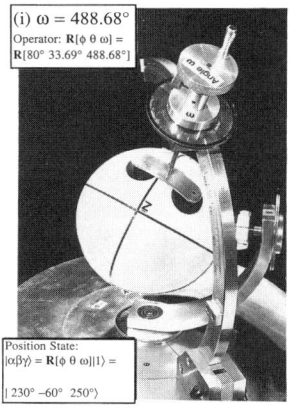

Figure 5.3.5 Sequence of rotations $R[\phi\theta\omega] = R[80°\ 33.69°\ \omega]$. Sequence repeats only after $\omega = 4\pi$ or $720°$. Postion state $|\alpha\beta\gamma\rangle = |50°\ 50°\ 70°\rangle$ is obtained after rotating by angle $\omega = 128.68°$ as shown in part (c). Continued rotation through another $360°$ [as shown in parts (d) through (i)] yields another position state $|-130°\ -60°\ -110°\rangle$ with different Euler Angles but the same orientation of coordinates. Note that the difference $(\gamma - \alpha)$ is a constant $(20°)$ throughout. (*Photos by Vincent Malette and Joanie Geiser.*)

or axis angles $[\phi\theta\omega]$ and the Euler angles $(\alpha\beta\gamma)$. In the meantime note that the sequence in Figures 5.3.5(d)–5.3.5(i) shows a continuation of the ω rotation by another $360°$. This yields the alternative Euler-angle settings of state

$$|\alpha = -130°, \beta = -60°, \gamma = -110°\rangle = R[80°, 34°, 489°]|000\rangle, \quad (5.3.9)$$

which was first shown in Figure 5.3.1(c). The goniometer and crank demon-

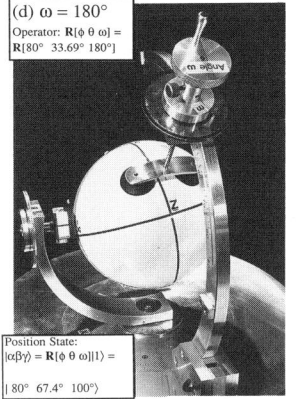

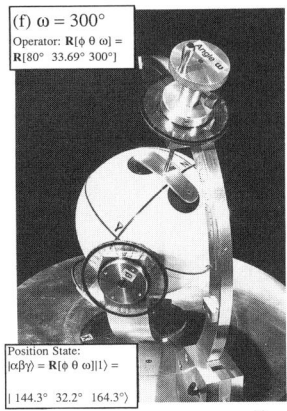
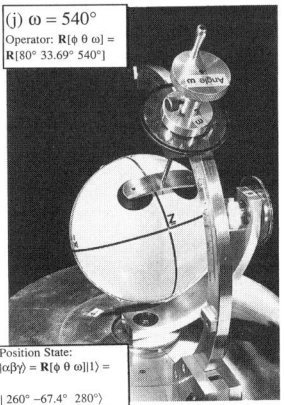
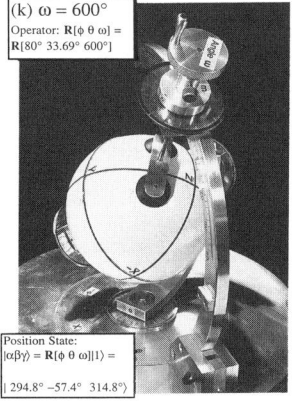
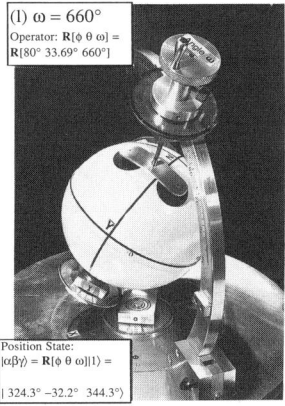

Figure 5.3.5 *(Continued).*

strate the double-valued nature of three-dimensional rotations. It also shows the complimentary relation between Euler angles and Darboux angles. Euler angles are convenient for labeling rotational *states*, while Darboux angles are more convenient for labeling rotational *operators*. Indeed, all the operators in Chapters 2–4 were defined by direction of axis and angle of rotation.

Therefore it is important to derive the mathematical relations between the two types of parameters. First we shall study a geometrical relation based upon Hamilton's rules in Section 5.3.C. Then in Section 5.5 an algebraic relation based upon representations of spin-$\frac{1}{2}$ will be given.

C. Hamilton's Rules and the Rotational Slide Rule

Hamilton's rules, as derived in the Section 3.1.B are used to find the product of two rotations $R[\omega]$ and $R[\omega']$ according to the following steps. First one

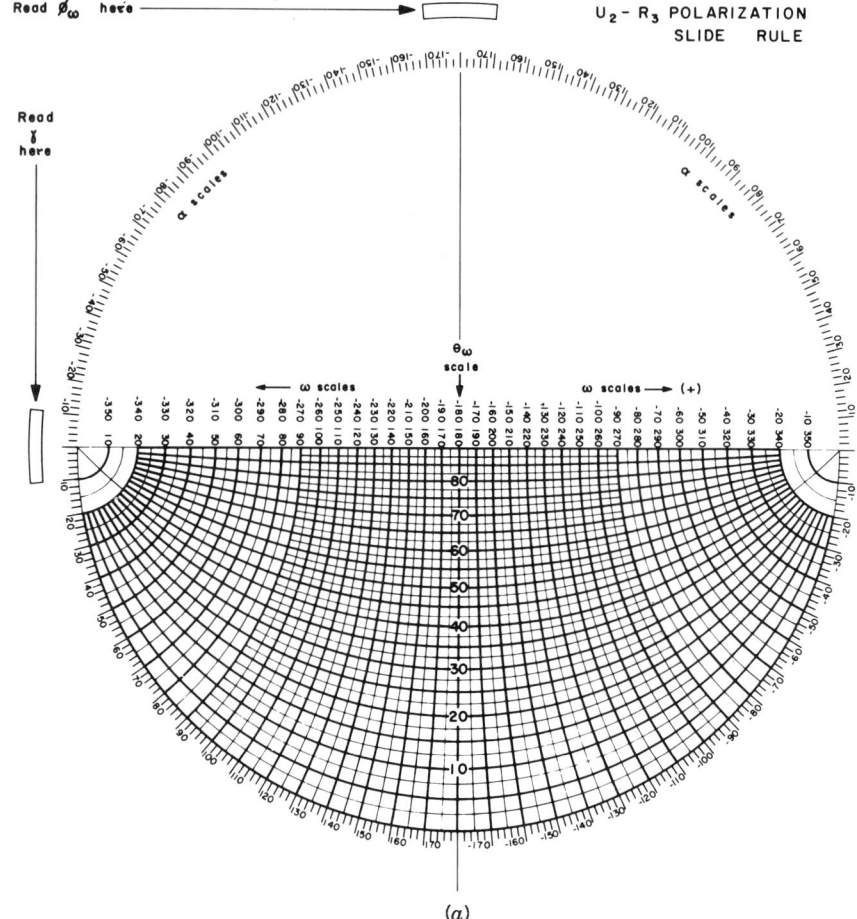

Figure 5.3.6 Rotational slide rule. (a) Lower scale, (b) upper scale.

constructs the great circle arcs perpendicular to ω and ω', respectively, on the unit sphere. Then one draws circle vectors along each arc of length $\omega/2$ and $\omega'/2$, respectively, and pointed in the directions of rotation so that the head of the arrow of the first rotation touches the tail of the arrow of the second rotation. Finally, one finds the great circle arc between the tail of the first arrow and the head of the second. The resulting vector-sum arrow defines the desired product rotation.

Figure 5.3.6 shows the slide rule that permits one to carry out the spherical vector addition accurately. The upper scale should be printed on a transparent plastic and fastened so its center rotates over the center of the lower scale. The scales are designed to facilitate group products and Euler $(\alpha\beta\gamma)$ to Darboux $[\phi\theta\omega]$ conversion.

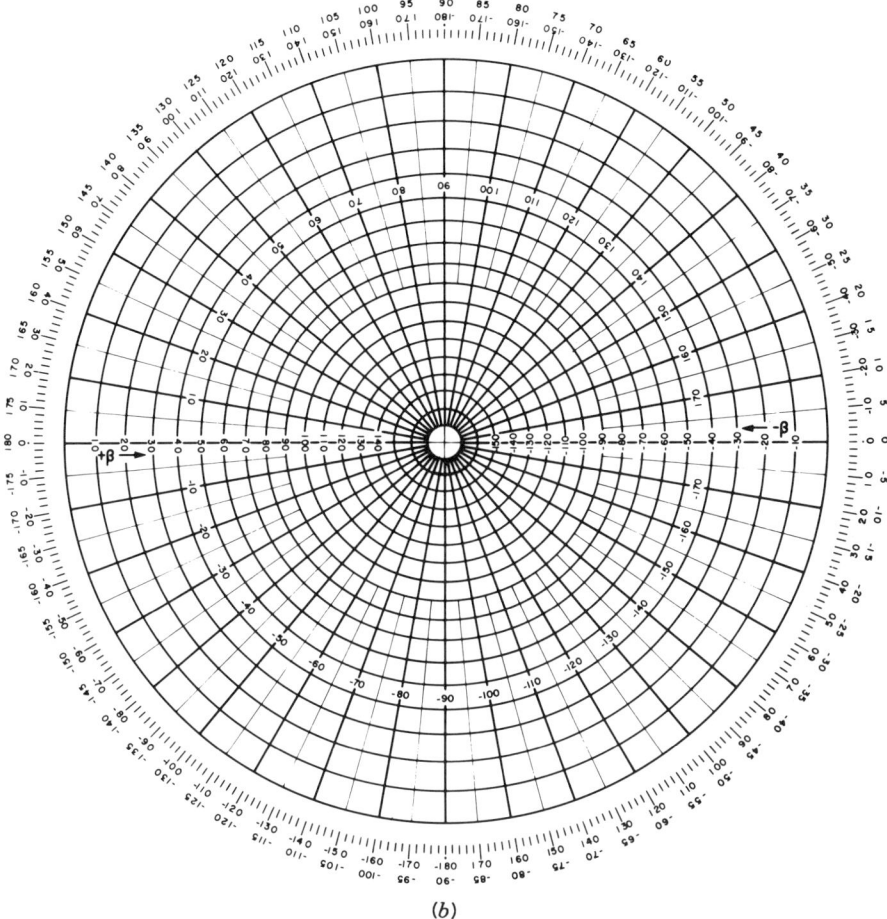

Figure 5.3.6 *(Continued)*.

To compute a product of two rotations $R[\phi\theta\omega]$ and $R[\phi'\theta'\omega']$ one first draws the arcs of the respective rotations onto the upper scale. An arc is drawn by setting the desired ϕ in the "ϕ window" (see top of lower scale) and tracing the desired (θ, ω) arc onto the upper scale using the θ and ω scales of the lower scale. It is necessary to first find the intersection of the $[\phi\theta]$ arc with the $[\phi'\theta']$ arc. Then one counts back ω degrees along the $[\phi\theta]$ arc to mark the tail of the first vector, and counts forward ω' degrees along the $[\phi'\theta']$ arc to mark the head of the second vector. The ω scale is used for each counting. Finally the slide rule is turned until the head and tail points lie along a θ line. (Interpolation may be necessary.) Then the desired answer ϕ'' in the product $R[\phi''\theta''\omega''] = R[\phi'\theta'\omega']R[\phi\theta\omega]$ is read in the ϕ window, while θ'' and ω'' are shown by their respective scales.

The upper slide-rule scale is a stereographic projection of the northern hemisphere of a globe, while the lower scale is the same projection of half of the western hemisphere. The structure of rotations as described by Hamilton makes it possible to do all products on just half of a sphere. A rotation for which ω is less than 180° corresponds to an arc of less than 90°. Any rotation with ω between 180° and 360° can be replaced by a rotation that goes the other way by angle $-(360° - \omega)$ and has an arc of $-(180° - \omega/2)$, which again is less then 90°. Whenever this conversion is made while operating on electron wave functions, it is necessary to multiply the result by (-1), as we will prove later:

$$R[\omega \cdot \cdot] = \begin{cases} R[\omega - 2\pi \cdot \cdot], & \text{for integral spin,} \\ -R[\omega - 2\pi \cdot \cdot], & \text{for half-integral spin.} \end{cases} \quad (5.3.10)$$

So anytime there appears an arc vector that extends over the edge of the

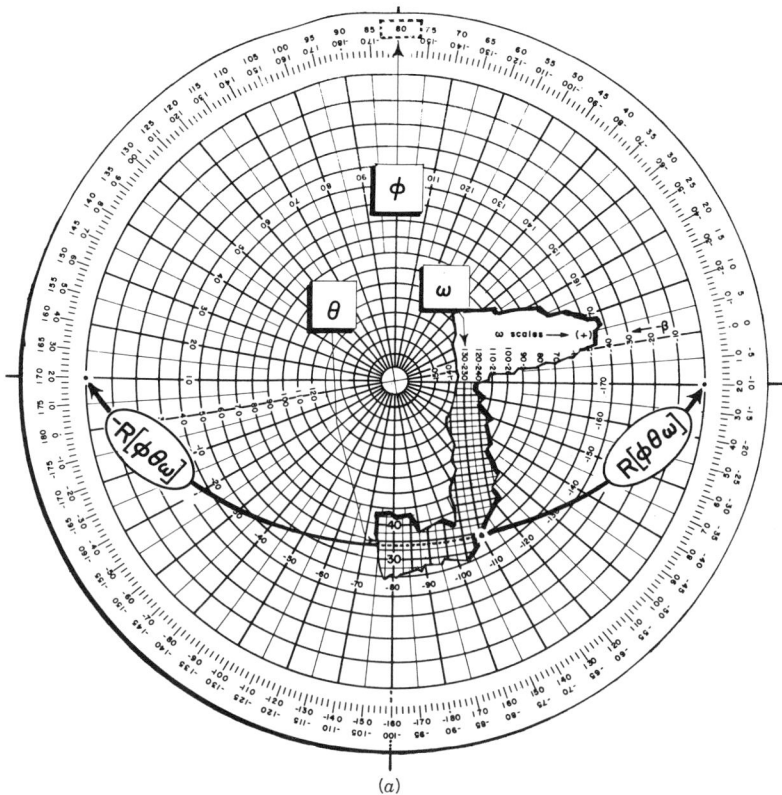

(a)

Figure 5.3.7 Setting the rotational slide rule. (a) Darboux or axis angles. (b) Euler angles.

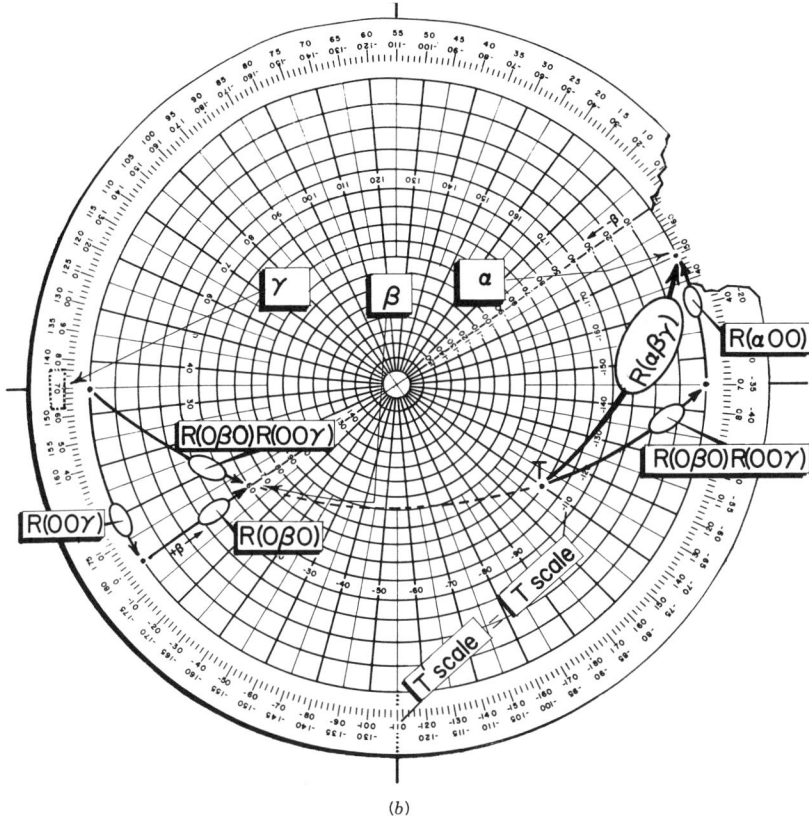

Figure 5.3.7 *(Continued).*

slide rule, we simply replace it by one of length $(360° - \omega)/2$ going the other way.

Figure (5.3.7) shows how the slide rule may be used to convert back and forth between Euler angles $(\alpha\beta\gamma)$ and axis angles $[\phi\theta\omega]$ in the equation

$$R(\alpha\beta\gamma) = R[\phi\theta\omega] \equiv R[\omega].$$

Figure 5.3.7(b) shows a given rotation $R[\omega]$ reduced by two vector sums into the product $R(\alpha 00)R(0\beta 0)R(00\gamma) = R(\alpha\beta\gamma)$ given by Eq. (5.3.3). The scales of the slide rule have been designed so that these products can be done without drawing arrows. To obtain the position shown in Figure 5.3.7(b) one must move the upper scale so that the angle between the meridian passing through the tail of $R[\omega]$ and the $+\beta$ scale is bisected by the center (θ) line of the lower scale. The "tail scales" or T scales indicated in Figure 5.3.7(b) make this easy. One simply reads the angle of $R[\omega]$'s tail using the T scale.

Then one finds this number on the T' scale, and sets it over the center line as shown in Figure 5.3.7(b). This converts an $R[\phi\theta\omega]$ operator to the equal $R(\alpha\beta\gamma)$ operator. The inverse is done by reversing the procedure. Different choices for three inputs from the six angles α, β, γ, ϕ, θ, and ω can be used to calculate the three remaining unknown angles.

Algebraic relations between the Euler and Darboux angles will be derived in Section 5.5.A. These are used in Chapter 7 to help analyze optical polarization and two-dimensional oscillator mechanics.

5.4 IRREDUCIBLE REPRESENTATIONS OF R_3 AND O_3

This section contains derivations of the all-important $\mathscr{D}^j$ matrices or irreducible representations of R_3. Starting with simple rotation vector algebra we will introduce matrix operators which generate rotation matrices. The connection with the quantum theory of angular momentum will be made. (The elementary angular-momentum relations are reviewed in Appendix E.) Finally, a formula for the irrep $\mathscr{D}^j$ is given.

A. Generators and Infinitesimal Rotations

Consider a rotation $R \cdot \mathbf{u}$ of a vector $\mathbf{u}$ by a very small (preferably infinitesimal) angle $\delta\omega$ around some axis defined by vector $\delta\boldsymbol{\omega} = \delta\omega\hat{\omega}$. This is shown by Figure 5.4.1. In vector notation one may write this transformation as

$$R[\phi\theta\delta\omega] \cdot \mathbf{u} = \mathbf{u} + \delta\boldsymbol{\omega} \times \mathbf{u}. \quad (5.4.1)$$

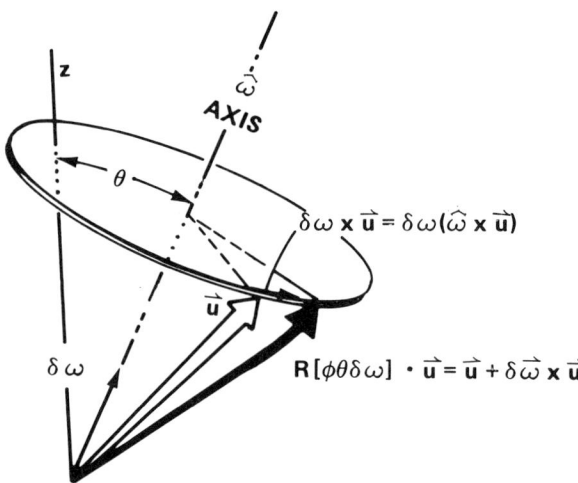

Figure 5.4.1 Small or infinitesimal rotation $R[\phi\theta\omega]$. The rotation moves vector $\mathbf{u}$ into vector $R[\phi\theta\omega]\mathbf{u}$.

This is represented in Cartesian coordinates by

$$\mathcal{R}[\phi\theta\delta\omega]\begin{pmatrix} u_x \\ u_y \\ u_z \end{pmatrix} = \begin{pmatrix} u_x \\ u_y \\ u_z \end{pmatrix} + \begin{pmatrix} \delta\omega_y u_z - \delta\omega_z u_y \\ \delta\omega_z u_x - \delta\omega_x u_z \\ \delta\omega_x u_y - \delta\omega_y u_x \end{pmatrix}, \quad (5.4.2)$$

where the components of $\delta\omega$ are

$$\delta\omega_x = \delta\omega \cos\phi \sin\theta, \quad \delta\omega_y = \delta\omega \sin\phi \sin\theta, \quad \text{and} \quad \delta\omega_z = \delta\omega \cos\theta. \quad (5.4.3)$$

We may expand Eq. (5.4.2) into matrix form:

$$\mathcal{R}[\phi\theta\delta\omega]\begin{pmatrix} u_x \\ u_y \\ u_z \end{pmatrix} = \left[\begin{pmatrix} 1 & \cdot & \cdot \\ \cdot & 1 & \cdot \\ \cdot & \cdot & 1 \end{pmatrix} + \delta\omega_x \begin{pmatrix} \cdot & \cdot & \cdot \\ \cdot & \cdot & -1 \\ \cdot & 1 & \cdot \end{pmatrix} \right.$$

$$\left. + \delta\omega_y \begin{pmatrix} \cdot & \cdot & 1 \\ \cdot & \cdot & \cdot \\ -1 & \cdot & \cdot \end{pmatrix} + \delta\omega_z \begin{pmatrix} \cdot & -1 & \cdot \\ 1 & \cdot & \cdot \\ \cdot & \cdot & \cdot \end{pmatrix} \right]\begin{pmatrix} u_x \\ u_y \\ u_z \end{pmatrix}. \quad (5.4.4)$$

The preceding equation is independent of the u_j. Hence a matrix equation holds:

$$\mathcal{R}[\phi\theta\delta\omega] = \mathbf{1} + \delta\omega_x \mathcal{G}_x + \delta\omega_y \mathcal{G}_y + \delta\omega_z \mathcal{G}_z. \quad (5.4.5a)$$

Here the matrices

$$\mathcal{G}_x \equiv \begin{pmatrix} \cdot & \cdot & \cdot \\ \cdot & \cdot & -1 \\ \cdot & 1 & \cdot \end{pmatrix}, \quad \mathcal{G}_y \equiv \begin{pmatrix} \cdot & \cdot & 1 \\ \cdot & \cdot & \cdot \\ -1 & \cdot & \cdot \end{pmatrix}, \quad \text{and} \quad \mathcal{G}_z \equiv \begin{pmatrix} \cdot & -1 & \cdot \\ 1 & \cdot & \cdot \\ \cdot & \cdot & \cdot \end{pmatrix}$$

(5.4.5b)

are the representations of the R_3 GENERATORS. The idea of this name is that you may generate any rotation in R_3 from them.

For example, in order to perform a z-axis rotation by a finite angle ω_z you may take the following limit as $\delta\omega = \omega_z/n \to 0$:

$$\mathcal{R}[00\omega_z] = \lim_{n\to\infty} (\mathcal{R}[00\omega_z/n])^n = \lim_{n\to\infty} \left(1 + \frac{\omega_z}{n}\mathcal{G}_z\right)^n = e^{\omega_z \mathcal{G}_z}. \quad (5.4.6)$$

It is then an easy matter to evaluate $e^{\omega_z \mathcal{G}_z}$ using the spectral decomposition methods derived in Section 1.2. First the eigenvalues ε_l of G_z are found by solving the secular equation:

$$\det|\mathscr{G}_z - \varepsilon 1| = \det \begin{vmatrix} -\varepsilon & -1 & \cdot \\ 1 & -\varepsilon & \cdot \\ \cdot & \cdot & -\varepsilon \end{vmatrix} = \varepsilon^3 + \varepsilon = 0.$$

Substitution of the eigenvalues $\{\varepsilon_1 = i, \varepsilon_2 = -i, \varepsilon_3 = 0\}$ into the spectral decomposition give $\mathscr{R}[00\omega_z]$.

$$\mathscr{R}[00\omega_z] = e^{\omega_z \mathscr{G}_z} = \sum_{l=1}^{3} e^{\omega_z \varepsilon_l} \prod_{k \neq l} (\mathscr{G}_z - \varepsilon_k 1) \bigg/ \prod_{k \neq l} (\varepsilon_l - \varepsilon_k),$$

$$\mathscr{R}[00_z] = e^{i\omega_z} \frac{\begin{pmatrix} 0 & -1 & 0 \\ 1 & 0 & 0 \\ 0 & 0 & 0 \end{pmatrix} \begin{pmatrix} i & -1 & 0 \\ 1 & i & 0 \\ 0 & 0 & i \end{pmatrix}}{(i-0)(i+i)}$$

$$+ e^{-i\omega_z} \frac{\begin{pmatrix} -i & -1 & 0 \\ 1 & -i & 0 \\ 0 & 0 & i \end{pmatrix} \begin{pmatrix} 0 & -1 & 0 \\ 1 & 0 & 0 \\ 0 & 0 & 0 \end{pmatrix}}{(-i-i)(-i-0)}$$

$$+ e^{0} \frac{\begin{pmatrix} -i & -1 & 0 \\ 1 & -i & 0 \\ 0 & 0 & -i \end{pmatrix} \begin{pmatrix} i & -1 & 0 \\ 1 & i & 0 \\ 0 & 0 & i \end{pmatrix}}{(0-i)(0+i)}$$

$$= e^{i\omega_z} \begin{pmatrix} 1/2 & i/2 & 0 \\ -i/2 & 1/2 & 0 \\ 0 & 0 & 0 \end{pmatrix} + e^{-i\omega_z} \begin{pmatrix} 1/2 & -i/2 & 0 \\ i/2 & 1/2 & 0 \\ 0 & 0 & 0 \end{pmatrix} + \begin{pmatrix} 0 & 0 & 0 \\ 0 & 0 & 0 \\ 0 & 0 & 1 \end{pmatrix}$$

$$= \begin{pmatrix} \cos \omega_z & -\sin \omega_z & 0 \\ \sin \omega_z & \cos \omega_z & 0 \\ 0 & 0 & 1 \end{pmatrix} \quad (5.4.7)$$

This is the standard z-rotation matrix represented in a Cartesian $\{xyz\}$ basis. The spectral decomposition used to derive it will be used again in Section 5.4.D to find more general types of R_3 representations. The idea is to use a representation of the generators to make the corresponding representation of any rotation in R_3.

There are some important properties of the generators which are useful in the construction of their representations. Note first that even rotations by very small angles do not commute. Consider the following rotation:

$$R = \underbrace{e^{-\varepsilon \mathscr{G}_y}}_{\begin{pmatrix}\text{rotation by } (-\varepsilon)\\ \text{around } y \text{ axis}\end{pmatrix}} \cdot \underbrace{e^{-\varepsilon \mathscr{G}_x}}_{\begin{pmatrix}\text{rotation by } (-\varepsilon)\\ \text{around } x \text{ axis}\end{pmatrix}} \cdot \underbrace{e^{-\varepsilon \mathscr{G}_y}}_{\begin{pmatrix}\text{rotation by } (\varepsilon)\\ \text{around } y \text{ axis}\end{pmatrix}} \cdot \underbrace{e^{\varepsilon \mathscr{G}_x}}_{\begin{pmatrix}\text{rotation by } (\varepsilon)\\ \text{around } x \text{ axis}\end{pmatrix}} \cdot$$

(5.4.8)

Let us expand this by assuming $\varepsilon \sim \frac{1}{3}$ radian and keeping only the lowest-order terms.

$$R = \left(1 - \varepsilon \mathscr{G}_y + \frac{\varepsilon^2}{2}\mathscr{G}_y^2 \cdots\right)\left(1 - \varepsilon \mathscr{G}_x + \frac{\varepsilon^2}{2}\mathscr{G}_x^2 \cdots\right)\left(1 + \varepsilon \mathscr{G}_y + \frac{\varepsilon^2}{2}\mathscr{G}_y^2 \cdots\right)$$

$$\times \left(1 + \varepsilon \mathscr{G}_x + \frac{\varepsilon^2}{2}\mathscr{G}_x^2 \cdots\right)$$

$$= 1 + \varepsilon \mathscr{G}_x + \varepsilon^2 \mathscr{G}_y \mathscr{G}_x - \varepsilon^2 \mathscr{G}_x^2 - \varepsilon^2 \mathscr{G}_y \mathscr{G}_x + \frac{\varepsilon^2}{2}\mathscr{G}_x^2$$

$$+ \varepsilon \mathscr{G}_y - \varepsilon^2 \mathscr{G}_x \mathscr{G}_y - \varepsilon^2 \mathscr{G}_y^2 + \frac{\varepsilon^2}{2}\mathscr{G}_y^2$$

$$- \varepsilon \mathscr{G}_x + \varepsilon^2 \mathscr{G}_y \mathscr{G}_x + \frac{\varepsilon^2}{2}\mathscr{G}_x^2$$

$$- \varepsilon \mathscr{G}_y + \frac{\varepsilon^2}{2}\mathscr{G}_y^2 + \cdots.$$

Cancellation of ε terms leaves only ε^2-and-higher order terms:

$$\mathscr{R} = 1 - \varepsilon^2(\mathscr{G}_x \mathscr{G}_y - \mathscr{G}_y \mathscr{G}_x) + \cdots = 1 - \varepsilon^2[\mathscr{G}_x, \mathscr{G}_y] + \cdots. \quad (5.4.9)$$

From the generator matrices given by Eq. (5.4.5b) we evaluate the commutator

$$[\mathscr{G}_x, \mathscr{G}_y] = \mathscr{G}_z. \quad (5.4.10)$$

This shows that R given by Eq. (5.4.8) is a small clockwise (negative) rotation by angle $\varepsilon^2 \sim \frac{1}{9}$ around the z axis. (This neglects an error of order ε^3.) The rotation (5.4.8) is shown graphically in Figure 5.4.2.

The other commutation relations are cyclic permutations of the first one, and can be interpreted equivalently:

$$[\mathscr{G}_y, \mathscr{G}_z] = \mathscr{G}_x, \quad [\mathscr{G}_z, \mathscr{G}_x] = \mathscr{G}_y. \quad (5.4.11)$$

These relations define what is called the LIE ALGEBRA of R_3. They are a

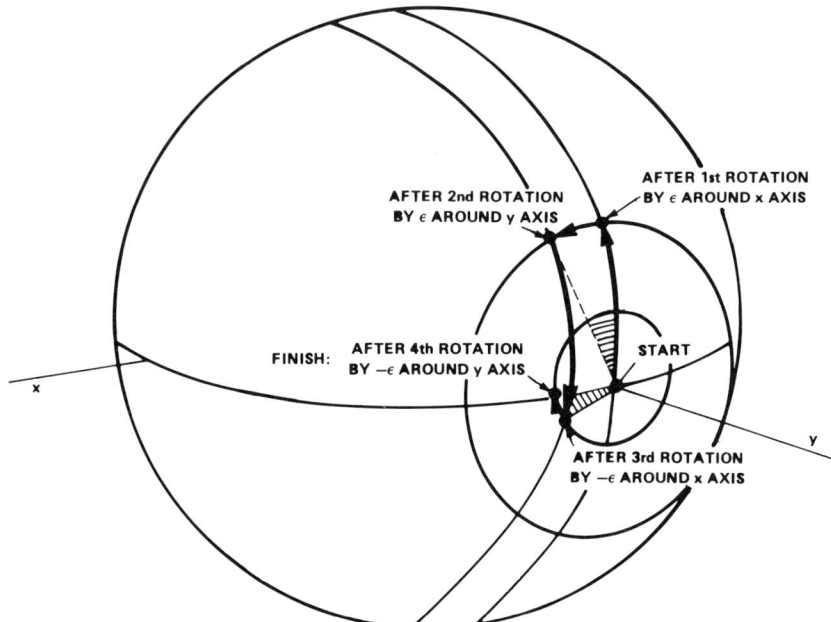

Figure 5.4.2 Effect of commutator operator $R = e^{-\varepsilon G_y}e^{-\varepsilon G_x}e^{\varepsilon G_y}e^{\varepsilon G_x}$ for $\varepsilon \approx 1/3$. Path of a vector which starts along the y axis is traced. (Note: The paths indicated by arrows are not Hamilton arcs.)

finite code for the structure of the infinite group R_3. The relations will be used to generate the irreducible representations, coupling coefficients, and other things connected with R_3.

B. Physical Interpretation of Generators

For rotations around the z axis Eq. (5.4.5) gives

$$\mathcal{R}(\delta\alpha 00) = 1 + \delta\alpha \mathcal{G}_z \qquad (5.4.12)$$

for a rotation by infinitesimal angle $\delta\alpha$. We use Euler parameters from now on unless otherwise stated. Rewriting this equation in terms of corresponding abstract quantum operators R, $\mathbf{1}$, and G_z gives

$$R(\delta\alpha 00) = \mathbf{1} + \delta\alpha G_z. \qquad (5.4.13)$$

Let us consider the effect of this rotation on an abstract position state $|xyz\rangle$ of a particle located at point (xyz). We want the particle to be rotated to the

IRREDUCIBLE REPRESENTATIONS OF R_3 AND O_3

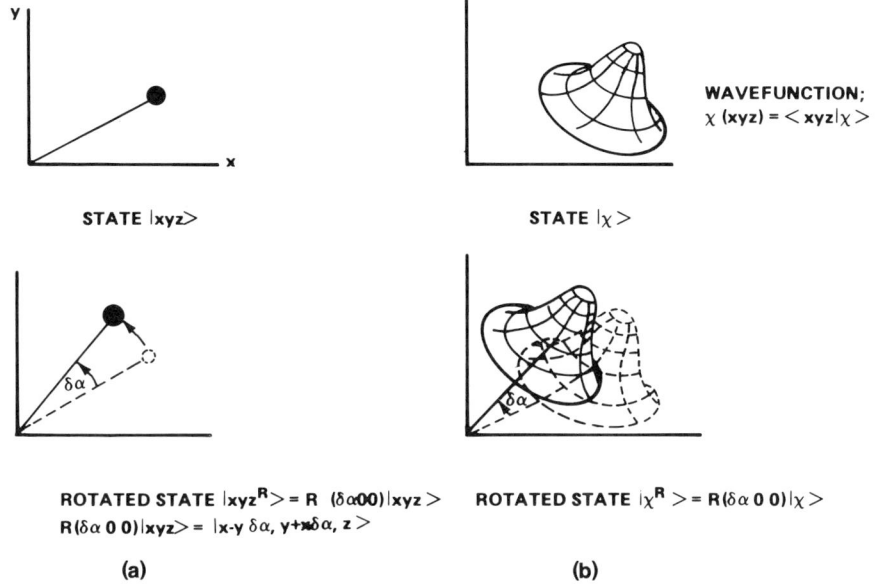

Figure 5.4.3 Effect of infinitesimal rotation operator $R(\alpha 00)$ on quantum states. (a) Localized state has Dirac-delta wave function centered at one point. (b) General state has arbitrary wave function $\chi(xyz) = \langle xyz|\chi\rangle$.

new point $(x - y\delta\alpha, y + x\delta\alpha, z)$ as shown in Figure 5.4.3(a):

$$R(\delta\alpha 00)|xyz\rangle = |x - y\delta\alpha, y + x\delta\alpha, z\rangle. \tag{5.4.14}$$

We also want the rotation operator to be unitary so that $R^\dagger(\delta\alpha 00) = R(-\delta\alpha 00)$, or

$$(R(\delta\alpha 00)|xyz\rangle)^\dagger = (|x - y\delta\alpha, y + x\delta\alpha, z\rangle)^\dagger$$
$$\langle xyz|R(-\delta\alpha 00) = \langle x - y\delta\alpha, y + x\delta\alpha, z|. \tag{5.4.15}$$

This in turn implies that the wave function $\psi(xyz) = \langle xyz|\psi\rangle$ of a general quantum state $|\psi\rangle$ transforms into the following wave function:

$$\psi'(xyz) \equiv \langle xyz|R(\delta\alpha 00)|\psi\rangle$$
$$= \langle x + y\delta\alpha, y - x\delta\alpha, z|\psi\rangle$$
$$= \psi(x + y\delta\alpha, y - x\delta\alpha, z). \tag{5.4.16}$$

The rotated point and the rotated wave function are sketched in Figure 5.4.3.

One may expand the rotated wave function into a Taylor series:

$$\psi'(xyz) = \psi(xyz) + y\frac{\partial \psi}{\partial x}\delta\alpha - x\frac{\partial \psi}{\partial y}\delta\alpha \cdots$$

$$= \left[1 - \delta\alpha\left(x\frac{\partial}{\partial y} - y\frac{\partial}{\partial x}\right)\right]\psi(xyz) + \cdots . \quad (5.4.17)$$

Now from elementary quantum mechanics the coordinate representation of the quantum operators for x and y components of momentum are given by $p_x \to (\hbar/i)\partial/\partial x$ and $p_y \to (\hbar/i)\partial/\partial y$; i.e.,

$$\langle xyz|p_x|\psi\rangle = (\hbar/i)\frac{\partial}{\partial x}\langle xyz|\psi\rangle, \quad \langle xyz|p_y|\psi\rangle = (\hbar/i)\frac{\partial}{\partial y}\langle xyz|\psi\rangle.$$

This in turn gives the following representation of the z component of angular momentum $\mathbf{J} = \mathbf{r} \times \mathbf{p}$:

$$\langle xyz|J_z|\psi\rangle = \langle xyz|xp_y - yp_x|\psi\rangle = \hbar/i\left(x\frac{\partial}{\partial y} - y\frac{\partial}{\partial x}\right)\psi(xyz). \quad (5.4.18)$$

Comparison of Eqs. (5.4.13), (5.4.17), and (5.4.18) shows that the R_3 generators must be the angular-momentum operators divided by the constant $(i\hbar)$:

$$\langle xyz|R(\delta\alpha 00)|\psi\rangle = \langle xyz|1 + \delta\alpha G_z|\psi\rangle = \langle xyz|1 + \frac{\delta\alpha J_z}{i\hbar}|\psi\rangle.$$

This holds for all components:

$$G_x = J_x/i\hbar, \quad G_y = J_y/i\hbar, \quad G_z = J_z/i\hbar. \quad (5.4.19)$$

The following commutation relations:

$$[J_x, J_y] = i\hbar J_z, \quad [J_y, J_z] = i\hbar J_x, \quad [J_z, J_x] = i\hbar J_y \quad (5.4.20)$$

follow from Eqs. (5.4.10) and (5.4.11). They also follow directly from the commutation relations of $\mathbf{x}$ and $\mathbf{p}$. ($[x_a, p_b] = i\hbar\delta_{ab}$.)

If rotations are symmetry operators for a Hamiltonian H then the generators must commute with H, too. This implies that each angular-momentum operator may be diagonalized simultaneously with H. A very important consequence of rotational symmetry is the CONSERVATION OF ANGULAR MOMENTUM. The relation between conservation and symmetry can be understood physically. If an object has no "lumps" which ruin its spherical symmetry, then there can be no "bumps" which alter the angular momentum.

C. Irreps of Angular-Momentum and Generator Operators

The following abstract relations hold between angular-momentum operators and their eigenstates $\left|{j \atop m}\right\rangle$ for $j = 0, \frac{1}{2}, 1, \ldots$ and $|m| \le j$.

$$J_z\left|{j \atop m}\right\rangle = m\hbar\left|{j \atop m}\right\rangle \tag{5.4.21}$$

$$J^2\left|{j \atop m}\right\rangle \equiv (J_x^2 + J_y^2 + J_z^2)\left|{j \atop m}\right\rangle = j(j+1)\hbar^2\left|{j \atop m}\right\rangle, \tag{5.4.22}$$

$$J_+\left|{j \atop m}\right\rangle \equiv (J_x + iJ_y)\left|{j \atop m}\right\rangle = \sqrt{(j-m)(j+m+1)}\,\hbar\left|{j \atop m+1}\right\rangle, \tag{5.4.23a}$$

$$J_-\left|{j \atop m}\right\rangle \equiv (J_x - iJ_y)\left|{j \atop m}\right\rangle = \sqrt{(j+m)(j-m+1)}\,\hbar\left|{j \atop m-1}\right\rangle. \tag{5.4.23b}$$

These are derived in Appendix E and later in this section.

The quantum numbers m and j define the z component and total angular momentum, respectively, of the eigenstates. For each allowed value of j there are $2j + 1$ partner states

$$\left\{\left|{j \atop j}\right\rangle, \left|{j \atop j-1}\right\rangle, \left|{j \atop j-2}\right\rangle, \ldots, \left|{j \atop -j}\right\rangle\right\},$$

which are a basis of an irreducible representation $\mathscr{D}^j$ of the Lie algebra of generators. By rewriting Eqs. (5.4.23) we obtain representations of the angular-momentum operators:

$$J_x = (J_+ + J_-)/2, \qquad J_y = (J_+ - J_-)/2i. \tag{5.4.24}$$

The angular-momentum representations are given here:

$$\mathscr{D}^j_{m'm}(J_x) = \left\langle{j \atop m'}\right|J_x\left|{j \atop m}\right\rangle = \frac{\delta_{m'm+1}}{2}\hbar\sqrt{(j-m)(j+m+1)}$$
$$+ \frac{\delta_{m'm-1}}{2}\hbar\sqrt{(j+m)(j-m+1)}, \tag{5.4.25a}$$

$$\mathscr{D}^j_{m'm}(J_y) = \left\langle{j \atop m'}\right|J_y\left|{j \atop m}\right\rangle = \frac{\delta_{m'm+1}}{2i}\hbar\sqrt{(j-m)(j+m+1)}$$
$$- \frac{\delta_{m'm-1}}{2i}\hbar\sqrt{(j+m)(j-m+1)}, \tag{5.4.25b}$$

$$\mathscr{D}^j_{m'm}(J_z) = \left\langle{j \atop m'}\right|J_z\left|{j \atop m}\right\rangle = \delta_{m'm}m\hbar. \tag{5.4.25c}$$

The following are some numerical examples of the irreps (in units of $\hbar$).

$$\left|\begin{matrix}j\\m\end{matrix}\right\rangle = \left|\begin{matrix}0\\0\end{matrix}\right\rangle \quad \left|\begin{matrix}1/2\\1/2\end{matrix}\right\rangle\left\langle\begin{matrix}1/2\\1/2\end{matrix}\right| \quad \left|\begin{matrix}1\\1\end{matrix}\right\rangle\left|\begin{matrix}1\\0\end{matrix}\right\rangle\left\langle\begin{matrix}1\\-1\end{matrix}\right| \quad \left|\begin{matrix}3/2\\3/2\end{matrix}\right\rangle\left|\begin{matrix}3/2\\1/2\end{matrix}\right\rangle\left\|\begin{matrix}3/2\\-1/2\end{matrix}\right\rangle\left\|\begin{matrix}3/2\\-3/2\end{matrix}\right\rangle \quad \left|\begin{matrix}2\\2\end{matrix}\right\rangle\left\|\begin{matrix}2\\1\end{matrix}\right\rangle\left|\begin{matrix}2\\0\end{matrix}\right\rangle\left\langle\begin{matrix}2\\-1\end{matrix}\right|\left|\begin{matrix}2\\-2\end{matrix}\right\rangle$$

$$\mathscr{D}^{j}(J_x) = (0) \quad \begin{pmatrix} \cdot & \frac{1}{2} \\ \frac{1}{2} & \cdot \end{pmatrix} \quad \begin{pmatrix} \cdot & \frac{\sqrt{2}}{2} & \cdot \\ \frac{\sqrt{2}}{2} & \cdot & \frac{\sqrt{2}}{2} \\ \cdot & \frac{\sqrt{2}}{2} & \cdot \end{pmatrix} \quad \begin{pmatrix} \cdot & \frac{\sqrt{3}}{2} & \cdot & \cdot \\ \frac{\sqrt{3}}{2} & \cdot & 1 & \cdot \\ \cdot & 1 & \cdot & \frac{\sqrt{3}}{2} \\ \cdot & \cdot & \frac{\sqrt{3}}{2} & \cdot \end{pmatrix} \quad \begin{pmatrix} \cdot & 1 & \cdot & \cdot & \cdot \\ 1 & \cdot & \frac{\sqrt{3}}{2} & \cdot & \cdot \\ \cdot & \frac{\sqrt{3}}{2} & \cdot & \frac{\sqrt{3}}{2} & \cdot \\ \cdot & \cdot & \frac{\sqrt{3}}{2} & \cdot & 1 \\ \cdot & \cdot & \cdot & 1 & \cdot \end{pmatrix} \cdots, \quad (5.4.25a)_x$$

$$\mathscr{D}^{j}(J_y) = (0) \quad \begin{pmatrix} \cdot & \frac{-i}{2} \\ \frac{i}{2} & \cdot \end{pmatrix} \quad \begin{pmatrix} \cdot & -\frac{\sqrt{2}\,i}{2} & \cdot \\ \frac{\sqrt{2}\,i}{2} & \cdot & -\frac{\sqrt{2}\,i}{2} \\ \cdot & \frac{\sqrt{2}\,i}{2} & \cdot \end{pmatrix} \quad \begin{pmatrix} \cdot & -\frac{\sqrt{3}\,i}{2} & \cdot & \cdot \\ \frac{\sqrt{3}\,i}{2} & \cdot & -i & \cdot \\ \cdot & i & \cdot & -\frac{\sqrt{3}\,i}{2} \\ \cdot & \cdot & \frac{\sqrt{3}\,i}{2} & \cdot \end{pmatrix} \quad \begin{pmatrix} \cdot & -i & \cdot & \cdot & \cdot \\ i & \cdot & -\frac{\sqrt{3}\,i}{2} & \cdot & \cdot \\ \cdot & \frac{\sqrt{3}\,i}{2} & \cdot & -\frac{\sqrt{3}\,i}{2} & \cdot \\ \cdot & \cdot & \frac{\sqrt{3}\,i}{2} & \cdot & -i \\ \cdot & \cdot & \cdot & i & \cdot \end{pmatrix} \cdots, \quad (5.4.25b)_x$$

$$\mathscr{D}^{j}(J_z) = (0) \quad \begin{pmatrix} \frac{1}{2} & \cdot \\ \cdot & \frac{-1}{2} \end{pmatrix} \quad \begin{pmatrix} 1 & \cdot & \cdot \\ \cdot & 0 & \cdot \\ \cdot & \cdot & -1 \end{pmatrix} \quad \begin{pmatrix} \frac{3}{2} & \cdot & \cdot & \cdot \\ \cdot & \frac{1}{2} & \cdot & \cdot \\ \cdot & \cdot & \frac{-1}{2} & \cdot \\ \cdot & \cdot & \cdot & \frac{-3}{2} \end{pmatrix} \quad \begin{pmatrix} 2 & \cdot & \cdot & \cdot & \cdot \\ \cdot & 1 & \cdot & \cdot & \cdot \\ \cdot & \cdot & 0 & \cdot & \cdot \\ \cdot & \cdot & \cdot & -1 & \cdot \\ \cdot & \cdot & \cdot & \cdot & -2 \end{pmatrix} \cdots \quad (5.4.25c)_x$$

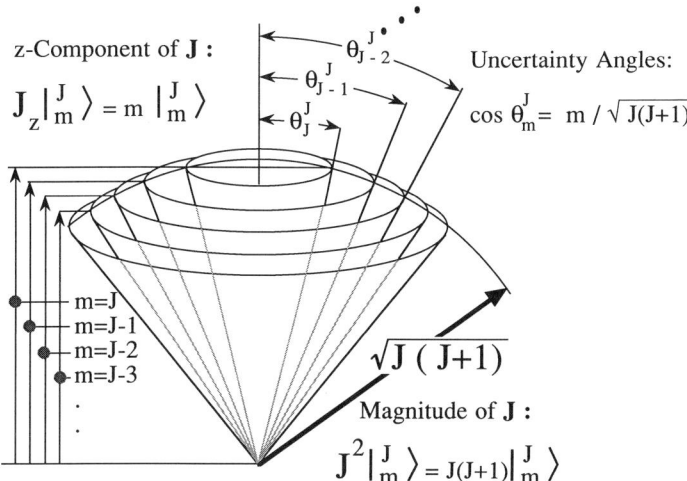

Figure 5.4.4 Angular-momentum vector cones for quantum states $\left|{}^{J}_{J}\right\rangle$, $\left|{}^{J}_{J-1}\right\rangle$, $\left|{}^{J}_{J-2}\right\rangle$,.... Note that uncertainty angle θ_m increases $\langle J_z \rangle = m$ decreases. States with the lowest $|m|$ have the highest uncertainty in the J_x and J_y components.

In order to gain a physical interpretation of Eqs. (5.4.21)–(5.4.25) one may imagine that the classical picture of an angular-momentum vector is replaced by a cone as shown in Figure 5.4.4. One represents the angular momentum of each state $\left|{}^{j}_{m}\right\rangle$ by a cone centered on the z axis with altitude $m\hbar$ according to Eq. (5.4.21) and a slant height of $\sqrt{j(j+1)}\,\hbar$ according to Eq. (5.4.22).

Since neither J_x nor J_y commute with J_z or with each other, one cannot find a state which is an eigenvector of more than one of these operators. Having chosen J_z to be diagonal one expects some uncertainty in the values of J_y and J_x. This uncertainty is indicated roughly by the cone base in Figure 5.4.4. A more quantitative description of the uncertainty in the transverse J_x and J_y components is contained in Section 5.5.B.

It was necessary to choose one J generator to be diagonal in order to complete the irrep derivation. The choice is analogous to that which we encountered in the derivation of C_{3v} irreps. There a particular generator of an Abelian subgroup (viz., σ_3 of C_v, or else r of C_3) was chosen to be diagonal. Such choices are needed to provide a subgroup labeling chain as explained in Chapter 4. For the rotation group R_3 the most commonly chosen labeling chain is simply $R_3 \supset R_2$, where R_2 is generated by J_z. Then the quantum numbers j and m in the bases $\left|{}^{j}_{m}\right\rangle$ serve as irrep labels for R_3 and R_2, respectively. Together, they label each base vector of the rotation group representations.

348 REPRESENTATIONS OF CONTINUOUS ROTATION GROUPS AND APPLICATIONS

D. Irreducible Representations of Rotation Operators

Let us represent rotated states $R(\alpha\beta\gamma)\left|{j \atop m}\right\rangle$ by cone drawings such as the one of $R(0\beta 0)\left|{j \atop m}\right\rangle$ in Figure 5.4.4. A rotated state can be written as a combination of the original $2j + 1$ unrotated partner states:

$$R(\alpha\beta\gamma)\left|{j \atop m}\right\rangle = \sum_{m'} \left|{j \atop m'}\right\rangle\left\langle {j \atop m'}\right|R(\alpha\beta\gamma)\left|{j \atop m}\right\rangle. \quad (5.4.26)$$

The coefficients

$$\mathscr{D}^j_{m'm}(R(\alpha\beta\gamma)) \equiv \left\langle {j \atop m'}\right|R(\alpha\beta\gamma)\left|{j \atop m}\right\rangle \equiv \mathscr{D}^j_{m'm}(\alpha\beta\gamma) \quad (5.4.27)$$

are the irreducible representations (irreps) which will be derived now. The derivation uses the irreps $\mathscr{D}^j(J)$ of the generators and Eq. (5.3.3), (5.4.6), and (5.4.19). The latter are summarized as follows:

$$R(\alpha\beta\gamma) = R(\alpha 00)R(0\beta 0)R(00\gamma) = e^{\alpha J_z/i\hbar}e^{\beta J_y/i\hbar}e^{\gamma J_z/i\hbar}. \quad (5.4.28)$$

The ($j = 0$) irrep is trivial since the generators are represented by zero on the left-hand side of Eq. (5.4.28):

$$\mathscr{D}^0(\alpha\beta\gamma) = 1. \quad (5.4.29)$$

This is called the SCALAR or INVARIANT irrep of R_3.

The ($j = \frac{1}{2}$) irrep involves two diagonal matrices for the α and γ angles and one nondiagonal matrix for the β rotation around the y axis. The latter can be evaluated by spectral decomposition as was done in Eq. (5.4.7):

$$\mathscr{D}^{1/2}(\alpha\beta\gamma) = e^{\frac{\alpha}{i}\begin{pmatrix}1/2 & 0 \\ 0 & -1/2\end{pmatrix}} e^{\frac{\beta}{i}\begin{pmatrix}0 & -i/2 \\ i/2 & 0\end{pmatrix}} e^{\frac{\gamma}{i}\begin{pmatrix}1/2 & 0 \\ 0 & -1/2\end{pmatrix}}$$

$$= \begin{pmatrix}e^{-i\alpha/2} & 0 \\ 0 & e^{i\alpha/2}\end{pmatrix}\begin{pmatrix}\cos\beta/2 & -\sin\beta/2 \\ \sin\beta/2 & \cos\beta/2\end{pmatrix}\begin{pmatrix}e^{-i\gamma/2} & 0 \\ 0 & e^{i\gamma/2}\end{pmatrix}$$

$$= \begin{pmatrix}e^{-i(\alpha+\gamma)/2}\cos\beta/2 & -e^{-i(\alpha-\gamma)/2}\sin\beta/2 \\ e^{i(\alpha-\gamma)/2}\sin\beta/2 & e^{i(\alpha+\gamma)/2}\cos\beta/2\end{pmatrix}. \quad (5.4.30)$$

This is called the SPINOR or FUNDAMENTAL irrep of R_3. Strictly speaking, however, $\mathscr{D}^{1/2}$ is not a representation of R_3. Note that the product of 180° rotations is represented by

$$\mathscr{D}^{1/2}(\pi 00)\mathscr{D}^{1/2}(\pi 00) = \begin{pmatrix}e^{-i\pi/2} & 0 \\ 0 & e^{i\pi/2}\end{pmatrix}\begin{pmatrix}e^{-i\pi/2} & 0 \\ 0 & e^{i\pi/2}\end{pmatrix} = \begin{pmatrix}-1 & 0 \\ 0 & -1\end{pmatrix}. \quad (5.4.31)$$

In other words, if you walk completely around, or rotate by 360° any spin state, its phase will come out negative! For half-integral j one has the following:

$$\mathscr{D}^j(R)\mathscr{D}^j(R') = \omega_{RR'}\mathscr{D}^j(RR'), \qquad (5.4.32)$$

where

$$\omega_{RR'} = \pm 1.$$

Such representations of are called PROJECTIVE or RAY REPRESENTATIONS of R_3. More details of these will be discussed in Section 5.7.

The $(j = 1)$ irrep may be calculated in the same manner:

$$\mathscr{D}^1(\alpha\beta\gamma) = \begin{pmatrix} e^{-i\alpha} & \cdot & \cdot \\ \cdot & 1 & \cdot \\ \cdot & \cdot & e^{i\alpha} \end{pmatrix} \begin{pmatrix} \frac{1+\cos\beta}{2} & \frac{-\sin\beta}{\sqrt{2}} & \frac{1-\cos\beta}{2} \\ \frac{\sin\beta}{\sqrt{2}} & \cos\beta & \frac{-\sin\beta}{\sqrt{2}} \\ \frac{1-\cos\beta}{2} & \frac{\sin\beta}{\sqrt{2}} & \frac{1+\cos\beta}{2} \end{pmatrix} \begin{pmatrix} e^{-i\gamma} & \cdot & \cdot \\ \cdot & 1 & \cdot \\ \cdot & \cdot & e^{i\gamma} \end{pmatrix}. \qquad (5.4.33)$$

This is called the VECTOR or DIPOLE representation. It is equivalent to the $\mathscr{R}$ representation of Eq. (5.4.7). If $R[00\omega_z]$ is diagonalized then $\mathscr{D}^1(\omega_z 00)$ results. The bases for $\mathscr{R}$ are the vector components $\{x, y, z\}$, but the bases for $\mathscr{D}^1$ are the circular vector components $\{-(x+iy), z, (x-iy)\}$ by inspecting the columns (or rows) of the projection matrices in Eq. (5.4.7).

The $(j = \frac{3}{2})$ and $(j = 2)$ irreps for the nondiagonal $\mathscr{R}(0\beta 0)$ rotation are

$\mathscr{D}^{3/2}(0\beta 0)$

$$= \begin{pmatrix} \cos^3\frac{\beta}{2} & -\sqrt{3}\cos^2\frac{\beta}{2}\sin\frac{\beta}{2} & \sqrt{3}\cos\frac{\beta}{2}\sin^2\frac{\beta}{2} & -\sin^3\frac{\beta}{2} \\ \sqrt{3}\cos^2\frac{\beta}{2}\sin\frac{\beta}{2} & \cos^3\frac{\beta}{2} - 2\cos\frac{\beta}{2}\sin^2\frac{\beta}{2} & \sin^3\frac{\beta}{2} - 2\cos^2\frac{\beta}{2}\sin\frac{\beta}{2} & \sqrt{3}\cos\frac{\beta}{2}\sin^2\frac{\beta}{2} \\ \sqrt{3}\cos\frac{\beta}{2}\sin^2\frac{\beta}{2} & -\sin^3\frac{\beta}{2} + 2\cos^2\frac{\beta}{2}\sin\frac{\beta}{2} & \cos^3\frac{\beta}{2} - 2\cos\frac{\beta}{2}\sin^2\frac{\beta}{2} & -\sqrt{3}\cos^2\frac{\beta}{2}\sin\frac{\beta}{2} \\ \sin^3\frac{\beta}{2} & \sqrt{3}\cos\frac{\beta}{2}\sin^2\frac{\beta}{2} & \sqrt{3}\cos^2\frac{\beta}{2}\sin\frac{\beta}{2} & \cos^3\frac{\beta}{2} \end{pmatrix}$$

$$(5.4.34)$$

$\mathscr{D}^2(0\beta 0)$

$$\begin{vmatrix} \left(\frac{1+\cos\beta}{2}\right)^2 & \left(\frac{1+\cos\beta}{2}\right)\sin\beta & \sqrt{\frac{3}{8}}\sin^2\beta & \left(\frac{1+\cos\beta}{2}\right)\sin\beta & \left(\frac{1-\cos\beta}{2}\right)^2 \\ \left(\frac{1+\cos\beta}{2}\right)\sin\beta & \left(\frac{1+\cos\beta}{2}\right)(2\cos\beta - 1) & -\sqrt{\frac{3}{2}}\sin\beta\cos\beta & \left(\frac{1-\cos\beta}{2}\right)(2\cos\beta + 1) & -\left(\frac{1-\cos\beta}{2}\right)\sin\beta \\ \sqrt{\frac{3}{8}}\sin^2\beta & \sqrt{\frac{3}{2}}\sin\beta\cos\beta & \frac{3\cos^2\beta - 1}{2} & \sqrt{\frac{3}{2}}\sin\beta\cos\beta & \sqrt{\frac{3}{8}}\sin^2\beta \\ \left(\frac{1+\cos\beta}{2}\right)\sin\beta & \left(\frac{1-\cos\beta}{2}\right)(2\cos\beta + 1) & \sqrt{\frac{3}{2}}\sin\beta\cos\beta & \left(\frac{1+\cos\beta}{2}\right)(2\cos\beta - 1) & -\left(\frac{1+\cos\beta}{2}\right)\sin\beta \\ \left(\frac{1-\cos\beta}{2}\right)^2 & \left(\frac{1-\cos\beta}{2}\right)\sin\beta & \sqrt{\frac{3}{8}}\sin^2\beta & \left(\frac{1+\cos\beta}{2}\right)\sin\beta & \left(\frac{1+\cos\beta}{2}\right)^2 \end{vmatrix}$$

$$(5.4.35)$$

Irrep $\mathscr{D}^2$ is called the TENSOR or QUADRUPOLE representations. These names will be explained in later sections.

For the $j = \frac{3}{2}$, 2, or higher it is convenient to derive a general formula for the irreps. An elegant construction due to Schwinger gives the general formula. Schwinger's construction identifies the angular-momentum states with creation operators of a two-dimensional oscillator. The spin-$\frac{1}{2}$ states are defined by

$$a_\uparrow^\dagger |00\rangle = \left| \begin{array}{c} j = 1/2 \\ m = 1/2 \end{array} \right\rangle, \quad a_\downarrow^\dagger |00\rangle = \left| \begin{array}{c} 1/2 \\ -1/2 \end{array} \right\rangle, \quad (5.4.36)$$

where the a operators satisfy the oscillator commutation relations

$$\left[a_i, a_j^\dagger \right] = \delta_{ij} 1,$$

which were given in Eqs. (4.4.51). The angular-momentum operators defined by

$$J_- \equiv a_\downarrow^\dagger a_\uparrow, \quad J_z \equiv \left(a_\uparrow^\dagger a_\uparrow - a_\downarrow^\dagger a_\downarrow \right)/2, \quad J_+ = a_\uparrow^\dagger a_\downarrow \quad (5.4.37)$$

then satisfy their commutation relations as well:

$$[J_z, J_\pm] = \pm J_\pm.$$

(Here we are using units for which $\hbar = 1$.) The general angular-momentum state is defined by

$$\left| \begin{array}{c} j \\ m \end{array} \right\rangle = \left(a_\uparrow^\dagger \right)^{n_\uparrow} \left(a_\downarrow^\dagger \right)^{n_\downarrow} |00\rangle / (n_\uparrow! n_\downarrow!)^{1/2}$$
$$= |n_\uparrow n_\downarrow\rangle, \quad (5.4.38a)$$

where

$$\begin{array}{ll} n_\uparrow = j + m, & j = (n_\uparrow + n_\downarrow)/2, \\ n_\downarrow = j - m, & m = (n_\uparrow - n_\downarrow)/2. \end{array} \quad (5.4.38b)$$

According to Eq. (4.4.62) the angular-momentum lowering operator has the following effect on $\left| \begin{array}{c} j \\ m \end{array} \right\rangle$:

$$J_- \left| \begin{array}{c} j \\ m \end{array} \right\rangle = a_\downarrow^\dagger a_\uparrow |n_\uparrow n_\downarrow\rangle = \left[(n_\uparrow)(n_\downarrow + 1) \right]^{1/2} |n_\uparrow - 1, n_\downarrow + 1\rangle. \quad (5.4.39)$$

This is an elegant rederivation of the familiar result in Eq. (5.4.23b).

$$J_- \left| \begin{array}{c} j \\ m \end{array} \right\rangle = \left[(j + m)(j - m + 1) \right]^{1/2} \left| \begin{array}{c} j \\ m - 1 \end{array} \right\rangle$$

IRREDUCIBLE REPRESENTATIONS OF R_3 AND O_3 351

Similarly, a formula for D^j can be derived from $\mathscr{D}^{1/2}$. Let us rewrite Eq. (5.4.30) as

$$R(0\beta 0)\left|\begin{array}{c}1/2\\1/2\end{array}\right\rangle = \cos\beta/2\left|\begin{array}{c}1/2\\1/2\end{array}\right\rangle + \sin\beta/2\left|\begin{array}{c}1/2\\-1/2\end{array}\right\rangle,$$

$$R(0\beta 0)\left|\begin{array}{c}1/2\\-1/2\end{array}\right\rangle = -\sin\beta/2\left|\begin{array}{c}1/2\\1/2\end{array}\right\rangle + \cos\beta/2\left|\begin{array}{c}1/2\\-1/2\end{array}\right\rangle, \quad (5.4.40)$$

then as an operator equation.

$$R(0\beta 0)a^\dagger_\uparrow \equiv a^{\dagger'}_\uparrow = \cos\beta/2\, a^\dagger_\uparrow + \sin\beta/2\, a^\dagger_\downarrow,$$
$$R(0\beta 0)a^\dagger_\downarrow \equiv a^{\dagger'}_\downarrow = -\sin\beta/2\, a^\dagger_\uparrow + \cos\beta/2\, a^\dagger_\downarrow. \quad (5.4.41)$$

This will yield the transformation properties of the general state:

$$R(0\beta 0)\left|\begin{array}{c}j\\n\end{array}\right\rangle = \frac{(a^{\dagger'}_\uparrow)^{j+n}}{\sqrt{(j+n)!}}\frac{(a^{\dagger'}_\downarrow)^{j-n}}{\sqrt{(j-n)!}}|00\rangle$$

$$= \frac{1}{\sqrt{(j+n)!(j-n)!}}(\cos\beta/2\, a^\dagger_\uparrow + \sin\beta/2\, a^\dagger_\downarrow)^{j+n}$$

$$\times(-\sin\beta/2\, a^\dagger_\uparrow + \cos\beta/2\, a^\dagger_\downarrow)^{j-n}|00\rangle. \quad (5.4.42)$$

Expanding the binomials yields a polynomial in creation operators:

$$R(0\beta 0)\left|\begin{array}{c}j\\n\end{array}\right\rangle$$

$$= \left[\frac{1}{(j+n)!(j-n)!}\right]^{1/2}$$

$$\times\left[\sum_l \frac{(j+n)!}{l!(j+n-l)!}\left(\cos\frac{\beta}{2}a^\dagger_\uparrow\right)^l\left(\sin\frac{\beta}{2}a^\dagger_\downarrow\right)^{j+n-l}\right]$$

$$\times\left[\sum_k \frac{(j-n)!}{k!(j-n-k)!}\left(-\sin\frac{\beta}{2}a^\dagger_\uparrow\right)^k\left(\cos\frac{\beta}{2}a^\dagger_\downarrow\right)^{j-n-k}\right]|00\rangle$$

$$= \left[\frac{1}{(j+n)!(j-n)!}\right]^{1/2}$$

$$\times\left[\sum_l\sum_k(-1)^k\frac{\left(\cos\frac{\beta}{2}\right)^{j-n-k+l}\left(\sin\frac{\beta}{2}\right)^{j+n-l+k}(a^\dagger_\uparrow)^{l+k}(a^\dagger_\downarrow)^{2j-l-k}}{l!(j+n-l)!k!(j-n-k)!}\right]|00\rangle.$$

$$(5.4.43)$$

By making the substitutions

$$j + m \equiv l + k, \qquad j - m \equiv 2j - l - k, \qquad l = j + m - k,$$

one recovers the recognizable form of Eq. (5.4.24).

$$R(0\beta 0)\left|\begin{array}{c}j\\n\end{array}\right\rangle = \left[\frac{1}{(j+n)!(j-n)!}\right]^{1/2}$$

$$\times \left[\sum_m \sum_k (-1)^k \frac{\left(\cos\frac{\beta}{2}\right)^{2j+m-n-2k}\left(\sin\frac{\beta}{2}\right)^{n-m+2k}(a_\uparrow^\dagger)^{j+m}(a_\downarrow^\dagger)^{j-m}}{(j+m-k)!(n-m+k)!k!(j-n-k)!}\right]|00\rangle$$

$$= \sum_m \left[\sum_k (-1)^k \frac{\sqrt{(j+n)!(j-n)!(j+m)!(j-m)!}\left(\cos\frac{\beta}{2}\right)^{2j+m-n-2k}\left(\sin\frac{\beta}{2}\right)^{n-m+2k}}{(j+m-k)!(n-m+k)!k!(j-n-k)!}\right]$$

$$\times \frac{(a_\uparrow^\dagger)^{j+m}(a_\downarrow^\dagger)^{j-m}}{\sqrt{(j+m)!}\sqrt{(j-m)!}}|00\rangle. \qquad (5.4.44)$$

Identification of the factor in brackets in the foregoing with the transformation matrix $\mathscr{D}^j$ in the following completes its derivation:

$$R(0\beta 0)\left|\begin{array}{c}j\\n\end{array}\right\rangle = \sum_m \mathscr{D}^j_{mn}(0\beta 0)\left|\begin{array}{c}j\\m\end{array}\right\rangle.$$

Addition of the factors $e^{-i\alpha m}$ and $e^{-i\gamma n}$ gives the complete formula for all R_3 representations in the $\left|\begin{array}{c}j\\m\end{array}\right\rangle$ bases:

$$\mathscr{D}^j_{mn}(\alpha\beta\gamma) = \sum_{k=0}(-1)^k \frac{\sqrt{(j+m)!(j-m)!(j+n)!(j-n)!}}{(j+m-k)!k!(j-n-k)!(n-m+k)!}$$

$$\times e^{-i(m\alpha+n\gamma)}\left(\cos\frac{\beta}{2}\right)^{2j+m-n-2k}\left(\sin\frac{\beta}{2}\right)^{n-m+2k}. \qquad (5.4.45)$$

Since the orthogonal group O_3 is the outer product $R_3 \times C_i$ it is a simple matter to obtain its irreps from those of its factor groups R_3 and C_i. Two classes of irreps arise; an even-parity class $\mathscr{D}^{j+}$ and an odd-parity class $\mathscr{D}^{j-}$.

They are defined as follows:

$$\mathscr{D}^{j+}(R) = \mathscr{D}^{j}(R), \qquad \mathscr{D}^{j+}(IR) = \mathscr{D}^{j}(R),$$
$$\mathscr{D}^{j-}(R) = \mathscr{D}^{j}(R), \qquad \mathscr{D}^{j-}(IR) = -\mathscr{D}^{j}(R). \qquad (5.4.46)$$

All group irreps defined so far are unitary matrices by construction. The $\mathscr{D}^{j}(\alpha\beta\gamma)$ matrices are unitary since the operators $R(\alpha\beta\gamma)$ are unitary:

$$R^{\dagger}(\alpha\beta\gamma) = \left(e^{\alpha J_z/i\hbar} e^{\beta J_y/i\hbar} e^{\gamma J_z/i\hbar}\right)^{\dagger} = e^{-\alpha J_z/i\hbar} e^{-\beta J_y/i\hbar} e^{-\gamma J_z/i\hbar}$$
$$= R(-\alpha - \beta - \gamma) = R^{-1}(\alpha\beta\gamma).$$

In the preceding we use the fact that the J operators are self-conjugate or Hermitian ($J_i = J_i^{\dagger}$) operators. Finally, we have

$$\mathscr{D}^{j*}_{mn}(\alpha\beta\gamma) = \mathscr{D}^{j}_{nm}(-\alpha - \beta - \gamma).$$

5.5 SOME APPLICATIONS OF R_3 REPRESENTATIONS

Much of the remainder of this book will be devoted to applications of the $\mathscr{D}^{j}$-matrix representations of the rotation group R_3. However, there are some important elementary applications which are useful for familiarizing oneself with the properties of the $\mathscr{D}^{j}$ matrices. These will be discussed now.

A. $\mathscr{D}^{1/2}$-Spinor Representations and Hamilton's Turns

Hamilton's rules for spherical vector addition of rotations were introduced in Section 3.1.B and again in Section 5.3.C. It is instructive to see how they are derived from spinor representations. Indeed, the spinor representations are closely related to Hamilton's quaternion or "hypercomplex" numbers.

In order to rederive Hamilton's rules using spinors we shall need the spinor representation in terms of Darboux or axis angles $[\phi\theta\omega]$. We already have irreps of z rotations $R(\alpha 00)$ and y rotations $R(0\beta 0)$ in Eq. (5.4.30) for the Euler parametrization. Therefore, all that is needed is to express the ω rotation $R[\phi\theta\omega]$ around axis $[\phi\theta]$ in terms of $R(\phi 00)$, $R(0\theta 0)$, and $R(\omega 00)$. In other words, we need to find how the ω rotation can be done using just y and z cranks in Figure 5.3.2.

The $R[\phi\theta\omega]$ rotation may be done using only the y and z cranks by performing the following steps. First, one moves the ω contact point so it lies under the z crank. This is accomplished by zeroing the azimuth ϕ with z crank $R(-\phi 00)$, and then zeroing the polar angle θ with y crank $R(0 - \theta 0)$. Then the ω rotation can be done by z crank $R(0\omega 0)$. Finally, one returns the ω-contact point to its original position using the reverse and inverse operator sequence $R(\phi 00)R(0\theta 0)$. The desired product of the five operations is

$$R[\phi\theta\omega] = R(\phi 00) R(0\theta 0) R(\omega 00) R(0 - \theta 0) R(-\phi 00).$$

The spinor representation of this product is

$$\mathscr{D}^{1/2}[\phi\theta\omega] =$$

$$\times \begin{pmatrix} e^{-i\phi/2} & 0 \\ 0 & e^{i\phi/2} \end{pmatrix} \begin{pmatrix} \cos\frac{\theta}{2} & -\sin\frac{\theta}{2} \\ \sin\frac{\theta}{2} & \cos\frac{\theta}{2} \end{pmatrix} \begin{pmatrix} e^{-i\omega/2} & \\ & e^{i\omega/2} \end{pmatrix} \begin{pmatrix} \cos\frac{\theta}{2} & \sin\frac{\theta}{2} \\ -\sin\frac{\theta}{2} & \cos\frac{\theta}{2} \end{pmatrix} \begin{pmatrix} e^{i\phi/2} & \\ & e^{-i\phi/2} \end{pmatrix}$$

$$= \begin{pmatrix} \cos\frac{\omega}{2} - i\cos\theta\sin\frac{\omega}{2} & -\sin\phi\sin\theta\sin\frac{\omega}{2} - i\cos\phi\sin\theta\sin\frac{\omega}{2} \\ \sin\phi\sin\theta\sin\frac{\omega}{2} - i\cos\phi\sin\theta\sin\frac{\omega}{2} & \cos\frac{\omega}{2} + i\cos\theta\sin\frac{\omega}{2} \end{pmatrix}.$$

(5.5.1)

By equating the components of $\mathscr{D}^{1/2}[\phi\theta\omega]$ to those of $\mathscr{D}^{1/2}(\alpha\beta\gamma)$ in Eq. (5.4.30) one obtains relations between Euler and Darboux angles:

$$\cos[(\alpha+\gamma)/2]\cos\beta/2 = \cos\omega/2, \tag{5.5.2a}$$
$$\sin[(\alpha+\gamma)/2]\cos\beta/2 = \sin\omega/2\cos\theta, \tag{5.5.2b}$$
$$\cos[(\gamma-\alpha)/2]\sin\beta/2 = \sin\omega/2\sin\theta\sin\phi, \tag{5.5.2c}$$
$$\sin[(\gamma-\alpha)/2]\sin\beta/2 = \sin\omega/2\sin\theta\cos\phi. \tag{5.5.2d}$$

Hamilton, and more recently, Cayley and Klein showed how to expand $\mathscr{D}^{1/2}[\phi\theta\omega]$ into QUATERNIONS q_α or PAULI SPINORS σ_α defined by

$$q_x \equiv i\sigma_x \equiv i\begin{pmatrix} 0 & 1 \\ 1 & 0 \end{pmatrix}, \quad q_y \equiv i\sigma_y \equiv i\begin{pmatrix} 0 & -i \\ i & 0 \end{pmatrix}, \quad q_z \equiv i\sigma_z \equiv i\begin{pmatrix} 1 & 0 \\ 0 & -1 \end{pmatrix},$$

$$= 2i\mathscr{D}^{1/2}(J_x), \qquad = 2i\mathscr{D}^{1/2}(J_y), \qquad = 2i\mathscr{D}^{1/2}(J_z).$$

(5.5.3)

[Note that quaternion q_α is proportional to the irrep $\mathscr{D}^{1/2}(J_\alpha)$ in Eqs. (5.5.12).] The desired expansion is

$$\mathscr{D}^{1/2}[\phi\theta\omega] = \cos\frac{\omega}{2}\begin{pmatrix} 1 & 0 \\ 0 & 1 \end{pmatrix} - i\sin\frac{\omega}{2}\left[\cos\phi\sin\theta\begin{pmatrix} 0 & 1 \\ 1 & 0 \end{pmatrix} \right.$$
$$\left. + \sin\phi\sin\theta\begin{pmatrix} 0 & -i \\ i & 0 \end{pmatrix} + \cos\theta\begin{pmatrix} 1 & 0 \\ 0 & -1 \end{pmatrix}\right]. \quad (5.5.4)$$

In operator notation this is

$$R[\phi\theta\omega] = \cos\frac{\omega}{2}\mathbf{1} - i\sin\frac{\omega}{2}[\hat{\omega}_x\sigma_x + \hat{\omega}_y\sigma_y + \hat{\omega}_z\sigma_z] \tag{5.5.5a}$$

$$= \cos\frac{\omega}{2}\mathbf{1} - i\sin\frac{\omega}{2}[\boldsymbol{\omega}\cdot\boldsymbol{\sigma}], \tag{5.5.5b}$$

where

$$\omega_x = \cos\phi \sin\theta, \qquad \omega_y = \sin\phi \sin\theta, \qquad \omega_z = \cos\theta \qquad (5.5.5c)$$

are the Cartesian components of the unit vector along the ω-crank axis. This expansion allows one to write a closed-form expression for group products such as

$$\begin{aligned}R[\phi'\theta'\omega']R[\phi\theta\omega] &= \left(\cos\frac{\omega'}{2}\mathbf{1} - i\sin\frac{\omega'}{2}\hat{\boldsymbol{\omega}}'\cdot\boldsymbol{\sigma}\right)\left(\cos\frac{\omega}{2}\mathbf{1} - i\sin\frac{\omega}{2}\hat{\boldsymbol{\omega}}\cdot\boldsymbol{\sigma}\right)\\ &= \cos\frac{\omega'}{2}\cos\frac{\omega}{2}\mathbf{1} - i\left[\cos\frac{\omega'}{2}\sin\frac{\omega}{2}\hat{\boldsymbol{\omega}} + \cos\frac{\omega}{2}\sin\frac{\omega'}{2}\hat{\boldsymbol{\omega}}'\right]\cdot\boldsymbol{\sigma}\\ &\quad - \sin\frac{\omega'}{2}\sin\frac{\omega}{2}(\hat{\boldsymbol{\omega}}'\cdot\boldsymbol{\sigma})(\hat{\boldsymbol{\omega}}\cdot\boldsymbol{\sigma}).\end{aligned} \qquad (5.5.6)$$

The third term in the foregoing can be reduced by the Pauli identity

$$(\mathbf{A}\cdot\boldsymbol{\sigma})(\mathbf{B}\cdot\boldsymbol{\sigma}) \equiv \mathbf{A}\cdot\mathbf{B} + i(\mathbf{A}\times\mathbf{B})\cdot\boldsymbol{\sigma}. \qquad (5.5.7)$$

The identity is easily proved by writing out each Pauli spinor component σ_x, σ_y, and σ_z as defined in Eq. (5.5.3). The result is the desired product

$$R[\phi'\theta'\omega']R[\phi\theta\omega] = R[\phi''\theta''\omega''], \qquad (5.5.8)$$

where

$$R[\phi''\theta''\omega''] \equiv \cos\frac{\omega''}{2}\mathbf{1} - i\sin\frac{\omega''}{2}\hat{\boldsymbol{\omega}}''\cdot\boldsymbol{\sigma} \qquad (5.5.9)$$

and

$$\cos\frac{\omega''}{2} = \cos\frac{\omega'}{2}\cos\frac{\omega}{2} - \sin\frac{\omega'}{2}\sin\frac{\omega}{2}\hat{\boldsymbol{\omega}}'\cdot\hat{\boldsymbol{\omega}}, \qquad (5.5.10a)$$

$$\hat{\boldsymbol{\omega}}''\sin\frac{\omega''}{2} = \cos\frac{\omega'}{2}\sin\frac{\omega}{2}\hat{\boldsymbol{\omega}} + \cos\frac{\omega}{2}\sin\frac{\omega'}{2}\hat{\boldsymbol{\omega}}' + \sin\frac{\omega'}{2}\sin\frac{\omega}{2}\hat{\boldsymbol{\omega}}'\times\hat{\boldsymbol{\omega}}. \qquad (5.5.10b)$$

This can be regarded as the principal structure equation for the two-dimensional UNIMODULAR UNITARY group SU_2 as well as R_3. The group of all matrices of the form given in Eq. (5.4.16) or (5.5.1) is the same as that of all unitary ($U^\dagger = U^{-1}$) and unimodular (det $U = 1$) two-by-two matrices. The structure equations are the ones which the slide rule described in Section 5.3.C is designed to solve using Hamilton's construction.

To show that Hamilton's construction is the same as the structure equations (5.5.10) one needs to analyze the spherical trigonometry. Consider a

356   REPRESENTATIONS OF CONTINUOUS ROTATION GROUPS AND APPLICATIONS

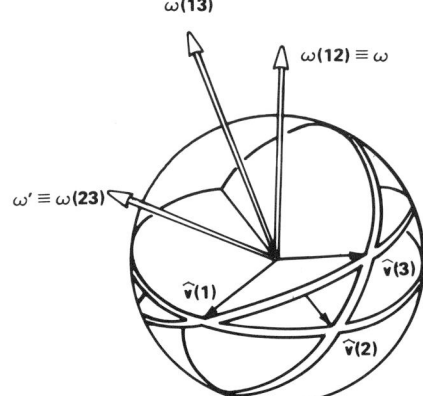

Figure 5.5.1 Hamilton arcs associated with rotation axes $\omega \equiv \omega(12)$, $\omega' \equiv \omega(23)$, and $\omega'' \equiv \omega(13)$.

spherical triangle defined by three unit normal vectors $\hat{v}(1)$, $\hat{v}(2)$, and $\hat{v}(3)$, as shown in Figure 5.5.1. Let rotation axes $\omega \equiv \omega(12)$ and $\omega' \equiv \omega(23)$ have their directions defined by

$$\hat{v}(1) \times \hat{v}(2) = \sin(12)\hat{\omega}, \quad \hat{v}(2) \times \hat{v}(3) = \sin(23)\hat{\omega}',$$

where (ij) is the arc length or angle between $\hat{v}(i)$ and $\hat{v}(j)$. By definition we can write $\hat{v}(2)$ as

$$\hat{v}(2) = \cos(12)\hat{v}(1) + \sin(12)\hat{\omega} \times \hat{v}(1), \tag{5.5.11}$$

and similarly for $\hat{v}(3)$:

$$\hat{v}(3) = \cos(23)\hat{v}(2) + \sin(23)\hat{\omega}' \times \hat{v}(2). \tag{5.5.12}$$

Substituting the expression for $\hat{v}(2)$ gives

$$\hat{v}(3) = \cos(23)\cos(12)\hat{v}(1) + [\cos(23)\sin(12)\hat{\omega} + \sin(23)\cos(12)\hat{\omega}'] \times \hat{v}(1)$$
$$+ \sin(23)\sin(12)\hat{\omega}' \times (\hat{\omega} \times \hat{v}(1)). \tag{5.5.13}$$

The third term above can be reduced by using cross-product identities:

$$\hat{\omega}' \times (\hat{\omega} \times \hat{v}(1)) \equiv (\hat{\omega}' \cdot \hat{v}(1))\hat{\omega} - (\hat{\omega}' \cdot \hat{\omega})\hat{v}(1),$$
$$(\hat{\omega}' \times \hat{\omega}) \times \hat{v}(1) \equiv (\hat{\omega}' \cdot \hat{v}(1))\hat{\omega} - (\hat{\omega} \cdot \hat{v}(1))\hat{\omega}'$$
$$= (\hat{\omega}' \cdot \hat{v}(1))\hat{\omega}.$$

Subtracting the identities gives

$$\hat{\omega}' \times (\hat{\omega} \times \hat{v}(1)) - (\hat{\omega}' \times \hat{\omega}) \times \hat{v}(1) = -(\hat{\omega}' \cdot \hat{\omega})\hat{v}(1).$$

The resulting expression for $\hat{\omega}' \times (\hat{\omega} \times \hat{v}(1))$ can be substituted into the one for $\hat{v}(3)$. There results an equation

$$\hat{v}(3) = \cos(13)\hat{v}(1) + \sin(13)\hat{\omega}'' \times \hat{v}(1), \qquad (5.5.14)$$

where

$$\cos(13) = \cos(23)\cos(12) - \sin(23)\sin(12)\hat{\omega}' \cdot \hat{\omega}, \qquad (5.5.15a)$$

and

$$\hat{\omega}'' \sin(13) = \cos(23)\sin(12)\hat{\omega} + \cos(12)\sin(23)\hat{\omega}' + \sin(23)\sin(12)\hat{\omega}' \times \hat{\omega}. \qquad (5.5.15b)$$

The form of the vector equations matches the principal structure equations (5.5.10) provided one makes the half-angle identifications

$$(12) = \omega/2, \qquad (23) = \omega'/2, \qquad (13) = \omega''/2. \qquad (5.5.16)$$

This completes the vector-based proof of Hamilton's construction.

B. Spin-j Polarization Experiments

The matrix $\mathscr{D}^{1/2}(\alpha = \pi/2, \beta = \theta, \gamma = 0)$ is the example of transformation matrix $\mathscr{T}$ which was introduced in Eq. (1.1.3). In general, $\mathscr{D}^j_{m'm}(\alpha\beta\gamma)$ gives the amplitude for a particle or system in a rotated state $R(\alpha\beta\gamma)|^j_m\rangle$, to choose the state $\left|^j_{m'}\right\rangle$ if it is somehow forced to make a choice between all the $|^j_m\rangle$. In Chapter 1 thought experiments involving $(j = \frac{1}{2})$ states were sketched. The amplitudes $\mathscr{D}_{m'm} = \langle m'|m\rangle$ between states

$$\left\{ \left|\begin{matrix}1/2\\1/2\end{matrix}\right\rangle = |1\rangle = |\text{``up''}\rangle, \quad \left|\begin{matrix}1/2\\-1/2\end{matrix}\right\rangle = |2\rangle = |\text{``down''}\rangle \right\}$$

and rotated states $\{R(\pi/2, \theta, 0)|1\rangle = |1'\rangle, R(\pi/2, \theta, 0)|2\rangle = |2'\rangle\}$ were introduced.

Figure 5.5.2 depicts a similar thought experiment involving $(j = 2)$ states or particles. It begins with states $R(\alpha\beta\gamma)|^2_1\rangle$ represented by tipped cones emerging from a β-tipped analyzer drawn on the left. We suppose that these are all forced by an untipped analyzer to choose one of the untipped states $|^2_2\rangle, |^2_1\rangle, |^2_0\rangle, |^2_{-1}\rangle$, or $|^2_{-2}\rangle$. The amplitude for choosing channel m' or state $|^2_{m'}\rangle$ is then $\mathscr{D}^2_{m'1}(0\beta0)$. The $\mathscr{D}$ functions from Eq. (5.4.35) are plotted with a broken line at the right of Figure. 5.5.2 for each m'. The intensity or probability $|\mathscr{D}^2_{m'1}|^2$ is the solid curve. Note that 100% of the particles would choose the $|^2_{-1}\rangle$ channel when $\beta = \pi$. This is right, since $|^2_1\rangle$ turned upside down must be the same as $|^2_{-1}\rangle$ to within a phase factor.

It is interesting to try to get some feeling for the behavior $\mathscr{D}$ functions. Consider a $(j = 6)$ basis. The cone pictures of all the base states are drawn in Figure 5.5.3(a).

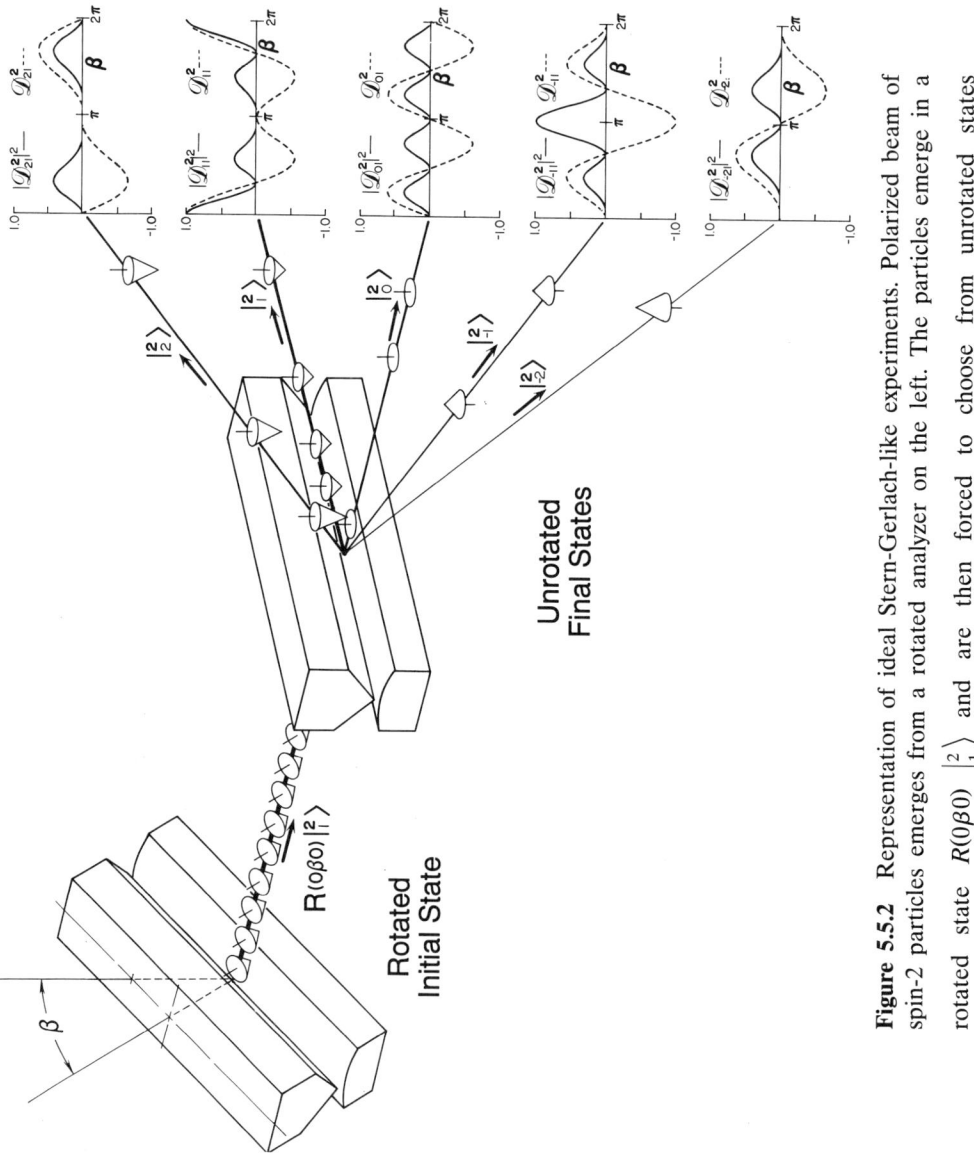

Figure 5.5.2 Representation of ideal Stern-Gerlach-like experiments. Polarized beam of spin-2 particles emerges from a rotated analyzer on the left. The particles emerge in a rotated state $R(0\beta0)\ \left|{}^2_1\right\rangle$ and are then forced to choose from unrotated states $\{\left|{}^2_2\right\rangle,\left|{}^2_1\right\rangle,\left|{}^2_0\right\rangle,\left|{}^2_{-1}\right\rangle,\left|{}^2_{-2}\right\rangle\}$ by the analyzer on the right. The amplitudes (dashed lines) and

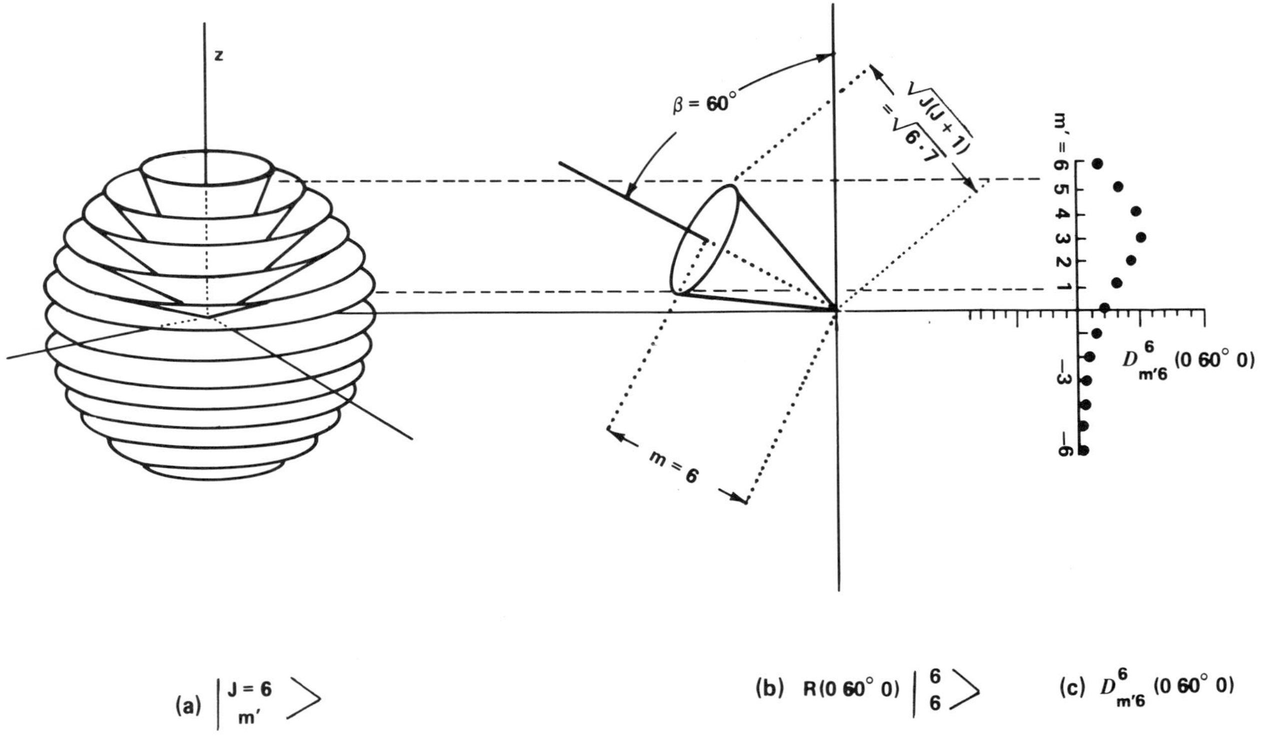

(a) $\left|\begin{array}{c}J=6\\m'\end{array}\right\rangle$ (b) $R(0\ 60°\ 0)\left|\begin{array}{c}6\\6\end{array}\right\rangle$ (c) $D^{6}_{m'6}(0\ 60°\ 0)$

Figure 5.5.3 Angular-momentum vector-cone geometry and rotation amplitudes for $J=6$. (a) Vector cones for all the states $\left\{\left|\begin{array}{c}6\\6\end{array}\right\rangle,\left|\begin{array}{c}6\\5\end{array}\right\rangle,\left|\begin{array}{c}6\\4\end{array}\right\rangle,\ldots,\left|\begin{array}{c}6\\-6\end{array}\right\rangle\right\}$ belonging to $J=6$. (b) Vector cone for rotated state $R(0\ 60°\ 0)\left|\begin{array}{c}6\\6\end{array}\right\rangle$. Dashed lines indicate projection of uncertainty of rotated cone. (c) Greatest amplitudes $D^{6}_{m'6}(0\ 60°\ 0)$ occur for the m' values which lie inside the projection of uncertainty.

359

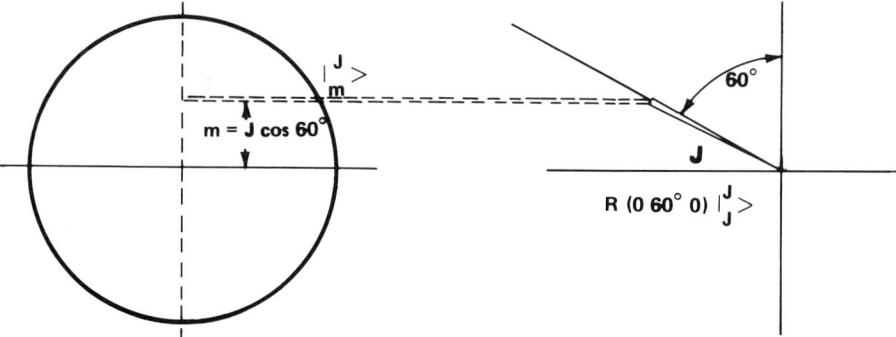

Figure 5.5.4 Vector geometry in classical limit for large $J > 10^4$. Uncertainty of $\left|{}^J_J\right\rangle$ state is negligible in the classical limit.

Consider the rotated state $R(0\ 60°\ 0)|^6_6\rangle$ shown in Figure 5.5.3(b). Let us compare the amplitudes for it to choose the unrotated states belonging to cones that lie inside the projection of its cone, namely, $|^6_5\rangle$ to $|^6_1\rangle$, to those that lie outside. For classical angular momentum the z component of a $60°$ rotated angular-momentum vector of length j is $m' = j\cos 60°$ as shown in Figure 5.5.4, simple classical vector rules apply for angular-momentum states for which the $|^j_j\rangle$-cone apex angle

$$\theta_{\text{cone}} = 2\cos^{-1}\left[j/\sqrt{j(j+1)}\right] \sim \cos^{-1}[j/(j+1/2)] \quad (5.5.17)$$

is small enough to consider the cone as a vector. Even the high value of $j = 10^4$ gives $\theta_{\text{cone}} = 1.15°$. For $j = 10^6$ we have $\theta_{\text{cone}} = 0.115°$. This should be regarded as an approximate lower bound to classical angular-momentum theory.

For lesser j one must settle for a series of quantum amplitudes given by $\mathscr{D}$ functions in Eq. (5.4.45). The exact values of $\mathscr{D}^6_{m'6}(0\ 60°\ 0)$ are plotted in Figure 5.5.3(c). Note that the amplitudes seem to form a smooth lump around the inside components. However, the amplitudes are nonzero outside the projection of the $R(0\ 60°\ 0)|^6_6\rangle$ cone.

Figure 5.5.5 shows a whole series of plots of $\mathscr{D}^{20}_{m'm}(0\beta 0)$ as a function of m' for various fixed constants β and m. In the first column each plot has $m' = 20 = j$. This gives a Gaussian-like lump similar to Figure 5.5.3(c) for $\beta \neq 0$. (For $\beta = 0$ the lump collapses into a single spike since $\left\langle {}^{\ j}_{m'}\bigg|{}^{\ j}_m\right\rangle = \delta_{m'm}$.)

For the other columns we set $m' = (j-1)$, $(j-2)$, and $(j-3)$. One observes a lump with one, two, and three nodes, respectively. For $\beta > 30°$ these lumps resemble oscillator wave functions. Their "classical turning points" or points of inflection occur at the projections of the cone edges.

This curious discrete wave behavior occurs for other angular-moment quantities such as coupling and recoupling coefficients. Ponzano and Regge [1968] and Shulten and Gordon [1975] have given a theory for this behavior in

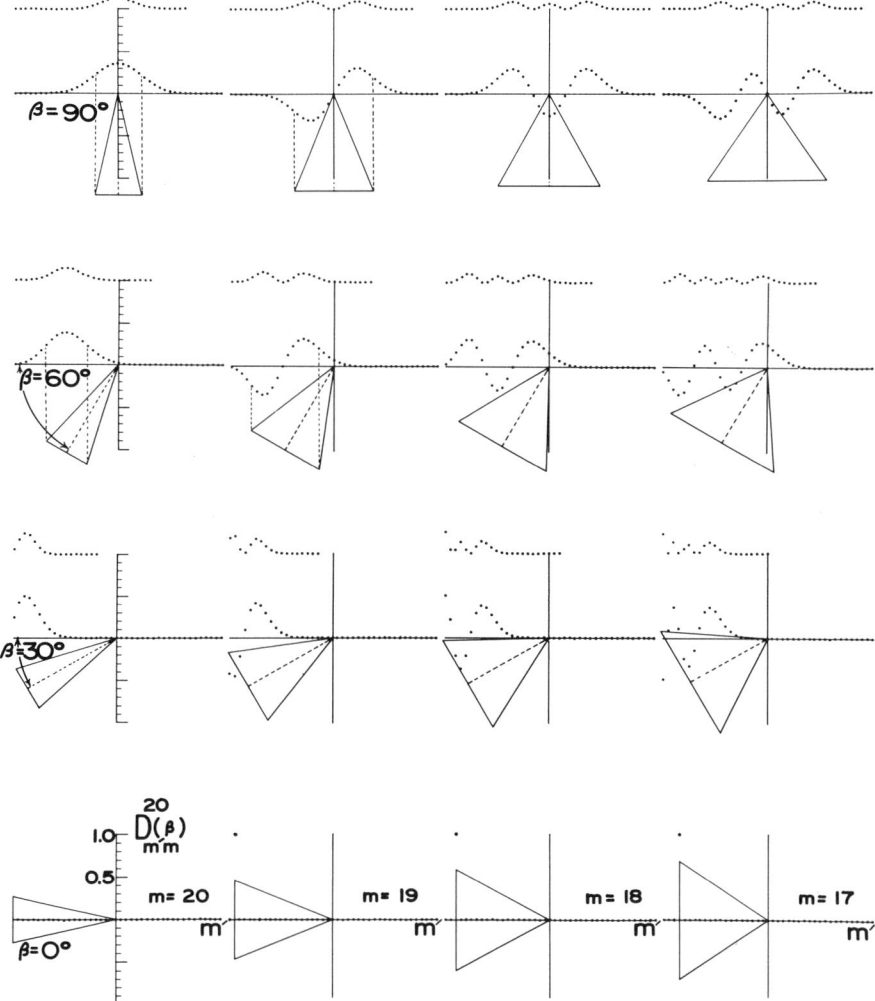

Figure 5.5.5 Rotation amplitudes, probabilities, and cone geometry for $J = 20$. In each subdiagram the amplitudes $D^{20}_{m'm}(0\beta 0)$ are plotted as a function of m' for fixed $\beta = 0°, 30°, 60°,$ and $90°$ and fixed $m = 20, 19, 18,$ and 17. The squared amplitudes or probabilities are plotted above each subdiagram.

coupling coefficients. They have derived one-dimensional potential wells which give wavelike solutions which approximate the discrete waves closely.

Finally it is interesting to note that the first moment or the J_z expectation value,

$$\begin{aligned} M' &= \langle {}^j_m | R^\dagger(0\beta 0) J_z R(0\beta 0) |^j_m \rangle \\ &= \sum_{m'=-j}^{j} m' \left| \mathscr{D}^j_{m'm}(0\beta 0) \right|^2, \end{aligned} \qquad (5.5.18)$$

of the probability distribution $|\mathscr{D}^j|^2$ gives the classical value $M' = m\cos\beta$. (This will be proved in Chapter 7.) The squares $|\mathscr{D}^j|^2$ of the $\mathscr{D}$ amplitudes are plotted above each $\mathscr{D}$ graph in Figure 5.5.5.

C. Symmetry Analysis of Quantum Rotors

Chapters 1–3 contained a development the basic ideas of symmetry analysis. The mathematics is summarized by the following. A group $\mathscr{G} = \{\cdots R \cdot \cdot R' \cdot \cdot\}$ of operators can have each element written

$$R = \sum_j \sum_m \sum_n \mathscr{D}^j_{mn}(R) P^j_{mn} \qquad (5.5.19)$$

in terms of irrep components $\mathscr{D}^j_{mn}$ and a complete set of elementary projection operators

$$P^j_{mn} = \frac{l^j}{{}^\circ\mathscr{G}} \sum_R \mathscr{D}^{j*}_{mn}(R) R, \qquad (5.5.20)$$

where l^j is the dimension of irrep $\mathscr{D}^j$, and the sum is over all group operators. ${}^\circ\mathscr{G}$ is the number of operators in $\mathscr{G}$. Elementary projectors satisfy orthonormality relations

$$P^j_{mn} P^{j'}_{m'n'} = \delta^{jj'} \delta_{nm'} P^j_{mn'}, \qquad (5.5.21)$$

and a completeness relation,

$$1 = \sum_j \sum_m^{l^j} P^j_{mm}. \qquad (5.5.22)$$

For the groups R_3, O_3, and SU_2 the formulas are the same except that the group sum $(1/{}^\circ\mathscr{G})\sum_R$ is replaced by an integral over Euler angles:

$$(1/{}^\circ\mathscr{G})\sum = 1 \to (1/8\pi^2) \int_0^{2\pi} d\alpha \int_{0 \text{ or } -\pi}^{\pi} \sin\beta\, d\beta \int_0^{2\pi} d\gamma = 1 \equiv \int d(\alpha\beta\gamma).$$

The integral over α and β (or γ and β) is analogous to a spherical polar area integral over ϕ and θ, respectively. As explained in Section 5.3.A, Euler angles α and β play the roles of azimuthal and polar angles in the laboratory coordinate frame while γ and β play the same roles in the body frame. For R_3 the range of β is 0 to π, for SU_2 it is $-\pi$ to π. A rigorous derivation of the Euler integral involves the theory of invariant group measure which is beyond the scope of this text. The integral formula for the R_3 ($0 \le \beta < \pi$) or

SU_2 ($-\pi \le \beta\pi$) projection operator is

$$P^j_{mn} = (2j + 1) \int d(\alpha\beta\gamma) \, \mathscr{D}^{j*}_{mn}(\alpha\beta\gamma) R(\alpha\beta\gamma), \qquad (5.5.23)$$

where $l^j = (2j + 1)$ is the dimension of the irreps $\mathscr{D}^j$. (See Appendix G.)

The application of P operators in the quantum theory of molecular rotation is analogous to the applications discussed in Chapters 1–4. One begins with a starting position state $|1\rangle = |000\rangle$ in which a rigid molecular body has its body coordinates $\{\bar{x}\bar{y}\bar{z}\}$ lined up with the lab coordinates $\{xyz\}$. Then the P operators are applied to this state to give symmetry-defined states

$$|{}^j_{mn}\rangle = P^j_{mn}|000\rangle/(2j + 1)^{1/2}$$

$$= \int d(\alpha\beta\gamma) \, \mathscr{D}^{j*}_{mn}(\alpha\beta\gamma)(2j + 1)^{1/2} R(\alpha\beta\gamma)|000\rangle. \qquad (5.5.24)$$

The states $|{}^j_{mn}\rangle$ have definite R_3 transformation properties as well as some simple physical properties which will be discussed shortly. Each $|{}^j_{mn}\rangle$ state is a linear combination of rotational position states

$$|\alpha\beta\gamma\rangle = R(\alpha\beta\gamma)|000\rangle$$

multiplied by amplitudes

$$r^j_{mn}(\alpha\beta\gamma) = \mathscr{D}^{j*}_{mn}(\alpha\beta\gamma)\sqrt{2l + 1}. \qquad (5.5.25)$$

The amplitude $r^j_{mn}(\alpha\beta\gamma)$ is called the QUANTUM ROTOR WAVE FUNCTION. Its absolute square $|r^j_{mn}(\alpha\beta\gamma)|^2$ gives the probability that a molecular rotor in state $|{}^j_{mn}\rangle$ would end up in rotational position state $|\alpha\beta\gamma\rangle$ if it was somehow forced to choose from all Euler angles. We shall sometimes write the r amplitude as

$$r^j_{mn}(\alpha\beta\gamma) = \langle\alpha\beta\gamma|{}^j_{mn}\rangle = \langle 000|R^\dagger(\alpha\beta\gamma)|{}^j_{mn}\rangle. \qquad (5.5.26)$$

The orthogonality of the r and $\mathscr{D}$ functions follows from that of the P operators. Rewriting Eq. (3.4.18) gives

$$\mathscr{D}^{j'}_{m'n'}(P^j_{mn}) = \delta^{jj'}\delta_{mm'}\delta_{nn'}. \qquad (5.5.27)$$

Inserting Eq. (5.5.23) gives $\mathscr{D}$ orthogonality relations

$$(2j + 1)\int d(\alpha\beta\gamma)\mathscr{D}^{j*}_{mn}(\alpha\beta\gamma)\mathscr{D}^{j'}_{m'n'}(\alpha\beta\gamma) = \delta^{jj'}\delta_{mm'}\delta_{nn'}, \qquad (5.5.28)$$

as well as r orthogonality. ($\mathscr{D}$ orthogonality is discussed in Appendix G.)

$$\int d(\alpha\beta\gamma) r^{j*}_{mn}(\alpha\beta\gamma) r^{j'}_{m'n'}(\alpha\beta\gamma) = \delta^{jj'}\delta_{mm'}\delta_{nn'}. \quad (5.5.29)$$

Let us consider now the transformation properties of the $|^j_{mn}\rangle$ states under the laboratory-based rotation operators $R(\alpha\beta\gamma)$ or body-based rotation operators $\bar{R}(\alpha\beta\gamma)$. This will include discussion of the corresponding generators $\{J_x J_y J_z\}$ and $\{\bar{J}_{\bar{x}} \bar{J}_{\bar{y}} \bar{J}_{\bar{z}}\}$, respectively. The transformation rules will follow more or less directly from the elementary left- and right-hand multiplication formulas:

$$R(\alpha\beta\gamma) P^j_{mn} = \sum_{m'} \mathscr{D}^j_{m'm}(\alpha\beta\gamma) P^j_{m'n}, \quad (5.5.30a)$$

$$P^j_{mn} R(\alpha\beta\gamma) = \sum_{n'} \mathscr{D}^j_{nn'}(\alpha\beta\gamma) P^j_{mn'}. \quad (5.5.30b)$$

The left- and right-handed formulas are verified by replacing R using Eq. (5.5.19) and then using orthogonality relations (5.5.21). Recall a similar derivation of Eq. (3.3.14).

The left-handed formula gives the standard laboratory transformation of state $|^j_{mn}\rangle$ in Eq. (5.5.24):

$$R(\alpha\beta\gamma)|^j_{mn}\rangle = R(\alpha\beta\gamma) P^j_{mn}|000\rangle/\sqrt{2j+1}$$
$$= \sum_{m'} \mathscr{D}^j_{m'm}(\alpha\beta\gamma) P^j_{m'n}|000\rangle/\sqrt{2j+1},$$
$$R(\alpha\beta\gamma)|^j_{mn}\rangle = \sum_{m'} \mathscr{D}^j_{m'm}(\alpha\beta\gamma)|^j_{m'n}\rangle. \quad (5.5.31)$$

Note that only the left m subindices of the state are involved. Using the lab-based rotation-generator relations

$$R(\alpha\beta\gamma) = R(\alpha 00) R(0\beta 0) R(00\gamma)$$
$$= e^{\alpha J_z/i\hbar} e^{\beta J_y/i\hbar} e^{\gamma J_z/i\hbar} \quad (5.5.32)$$

one can rederive the effect of lab J operators from infinitesimal rotations such as

$$R(\delta\alpha 00) = 1 - i\delta\alpha J_z/\hbar, \quad (5.5.33a)$$
$$R(0\delta\beta 0) = 1 - i\delta\beta J_y/\hbar. \quad (5.5.33b)$$

For example, one obtains

$$(1 - i\delta\alpha J_z)|^j_{mn}\rangle = \sum_{m'} \mathscr{D}^j_{m'm}(1 - i\delta\alpha J_z)|^j_{m'n}\rangle$$

$$= \sum_{m'} [\mathscr{D}^j_{m'm}(1) - i\delta\alpha \mathscr{D}^j_{m'm}(J_z)]|^j_{m'n}\rangle. \quad (5.5.34)$$

Using the representations $\mathscr{D}^j_{m'm}(1) = \delta_{m'm}$ and $\mathscr{D}^j_{m'm}(J_z) = \delta_{m'm}m\hbar$ from Eq. (5.5.12c) one obtains

$$J_z \left|^j_{mn}\right\rangle = \hbar m \left|^j_{mn}\right\rangle. \quad (5.5.35a)$$

The lab operators J_x and J_y are treated similarly:

$$J_x \left|^j_{mn}\right\rangle = \frac{\hbar}{2}(j(j+1) - m(m+1))^{1/2} \left|^j_{m+1\,n}\right\rangle$$

$$+ \frac{\hbar}{2}(j(j+1) - m(m-1))^{1/2} \left|^j_{m-1\,n}\right\rangle, \quad (5.5.35b)$$

$$J_y \left|^j_{mn}\right\rangle = \frac{-i\hbar}{2}(j(j+1) - m(m+1))^{1/2} \left|^j_{m-1\,n}\right\rangle$$

$$+ \frac{i\hbar}{2}(j(j+1) - m(m-1))^{1/2} \left|^j_{m-1\,n}\right\rangle. \quad (5.5.35c)$$

The transformation derivations for the body operators $\bar{R}(\alpha\beta\gamma)$ and their generators are only slightly more complicated:

$$\bar{R}(\alpha\beta\gamma)|^j_{mn}\rangle = \bar{R}(\alpha\beta\gamma) P^j_{mn}|000\rangle/\sqrt{2j+1} \quad (5.5.36)$$

$$= P^j_{mn} \bar{R}(\alpha\beta\gamma)|000\rangle/\sqrt{2j+1} \quad (5.5.37)$$

$$= P^j_{mn} R^\dagger(\alpha\beta\gamma)|000\rangle/\sqrt{2j+1}. \quad (5.5.38)$$

Here the commutation (5.3.6) between a body operator $\bar{R}$ and the lab-based P operator is used to give Eq. (5.5.37). Then we use initial state relation (5.3.7) between $\bar{R}$ and inverse $R^{-1} = R^\dagger$ of lab rotations. Finally, the right multiplication formula (5.5.30b) gives the body-operator transformation rules:

$$\bar{R}(\alpha\beta\gamma)|^j_{mn}\rangle = \sum_{n'} \mathscr{D}^{j\dagger}_{nn'}(\alpha\beta\gamma) P^j_{mn'}|000\rangle/\sqrt{2j+1}, \quad (5.5.39)$$

$$\bar{R}(\alpha\beta\gamma)|^j_{mn}\rangle = \sum_{n'} \mathscr{D}^{j\dagger}_{nn'}(\alpha\beta\gamma)|^j_{mn'}\rangle. \quad (5.5.40)$$

Note that the right-hand n indices are involved in the body-based transformation. The n index is analogous to the local symmetry index discussed in Section 3.4.A. Indeed, one can think of the general quantum rotor states as bases of the regular representation of R_3.

The body-based angular-momentum operators $\bar{J}_{\bar{z}}$ and $\bar{J}_{\bar{y}}$ are related to $\bar{R}$ through relations such as

$$\bar{R}(\alpha\beta\gamma) = \bar{R}(\alpha 0 0)\bar{R}(0\beta 0)\bar{R}(0 0\gamma)$$
$$= e^{\alpha \bar{J}_{\bar{z}}/\hbar} e^{\beta \bar{J}_{\bar{y}}/i\hbar} e^{\gamma \bar{J}_{\bar{z}}/i\hbar}. \qquad (5.5.41)$$

In particular, an infinitesimal $\bar{z}$ rotation is

$$\bar{R}(0 0\delta\gamma) = 1 - i\delta\gamma \bar{J}_{\bar{z}}/\hbar.$$

Substituting this into Eq. (5.3.35) and (5.5.40) gives

$$\bar{R}(0 0\delta\gamma)|^j_{mn}\rangle = \sum_{n'} \mathscr{D}^{j\dagger}_{nn'}(R(0 0\delta\gamma))|^j_{mn'}\rangle,$$
$$(1 - i\delta\gamma J_z/\hbar)|^j_{mn}\rangle = \sum_{n'} \mathscr{D}^{j\dagger}_{nn'}(1 - i\delta\gamma J_z/\hbar)|^j_{mn'}\rangle.$$

Using the definitions of unitarity ($\mathscr{D}^{j\dagger}_{nn'} = \mathscr{D}^{j*}_{n'n}$) and identity $\mathscr{D}^j_{nn'}(1) = \delta_{nn'}$, we obtain

$$\bar{J}_{\bar{z}}\left|\begin{matrix}j\\mn\end{matrix}\right\rangle = -\sum_{n'}\left|\begin{matrix}j\\mn'\end{matrix}\right\rangle \mathscr{D}^j_{n'n}(J_z^*). \qquad (5.5.42)$$

The minus sign arises from the conjugation (*):

$$\mathscr{D}^{j*}(i\delta\gamma J_z) = -i\mathscr{D}^j(\delta\gamma J_z^*).$$

The irrep formula (5.4.25c) gives the result

$$\bar{J}_{\bar{z}}\left|\begin{matrix}j\\mn\end{matrix}\right\rangle = -n\hbar\left|\begin{matrix}j\\mn\end{matrix}\right\rangle. \qquad (5.5.43)$$

The minus sign for the body-based angular-momentum components is regarded as a nuisance. It can be traced right back to the lab-body relation (5.3.7). Obviously, any body residents would report that the laboratory was whirling around with a rotational sense opposite to the sense of rotation of the body measured in the lab. Jahn [1938] chose to eliminate the minus sign by introducing REVERSED angular momentum operators

$$J_{\bar{x}} = -\bar{J}_{\bar{x}}, \qquad J_{\bar{y}} = -\bar{J}_{\bar{y}}, \qquad J_{\bar{z}} = -\bar{J}_{\bar{z}}. \qquad (5.5.44)$$

SOME APPLICATIONS OF R_3 REPRESENTATIONS

This definition leads to the following body-based equations:

$$J_{\bar{z}}\left|\begin{matrix}j\\mn\end{matrix}\right\rangle = n\hbar\left|\begin{matrix}j\\mn\end{matrix}\right\rangle, \qquad (5.5.45a)$$

$$J_{\bar{y}}\left|\begin{matrix}j\\mn\end{matrix}\right\rangle = \frac{i\hbar}{2}(j(j+1)-n(n+1))^{1/2}\left|\begin{matrix}j\\mn+1\end{matrix}\right\rangle$$

$$-\frac{i\hbar}{2}(j(j+1)-n(n-1))^{1/2}\left|\begin{matrix}j\\mn-1\end{matrix}\right\rangle, \qquad (5.5.45b)$$

$$J_{\bar{x}}\left|\begin{matrix}j\\mn\end{matrix}\right\rangle = \frac{\hbar}{2}(j(j+1)-n(n+1))^{1/2}\left|\begin{matrix}j\\mn+1\end{matrix}\right\rangle$$

$$+\frac{\hbar}{2}(j(j+1)-n(n-1))^{1/2}\left|\begin{matrix}j\\mn-1\end{matrix}\right\rangle. \qquad (5.5.45c)$$

These results complement the lab-based equations (5.5.35).

A physical interpretation of the $J_{\bar{z}}$ equation is shown partly by Figure 5.5.6. One may regard the $\langle J_{\bar{z}}\rangle = n\hbar$ as the $\bar{z}$ component of lab-based angular momentum on the body $\bar{z}$ axis. Note that the same $\left|\begin{matrix}j\\mn\end{matrix}\right\rangle$ state will also have a definite z component $\langle J_z\rangle = m\hbar$ of angular momentum on the

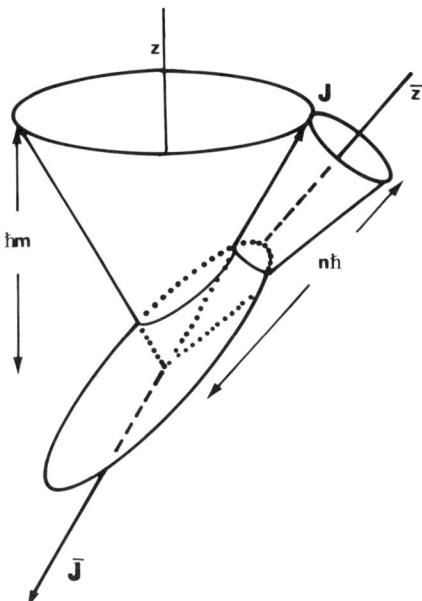

Figure 5.5.6 Laboratory and body components of rotational momentum.

laboratory z axis. Since J_z and $J_{\bar{z}}$ commute they may share the eigenstate $\left|{}^{j}_{mn}\right\rangle$.

The price one pays for reversing the momenta comes in the form of negative commutation relations:

$$[J_{\bar{x}}, J_{\bar{y}}] = -i\hbar J_{\bar{z}} \tag{5.5.46}$$

between body-component operators. The raising and lowering operators are interchanged, also:

$$(J_{\bar{x}} + iJ_{\bar{y}})\left|{}^{j}_{mn}\right\rangle = \hbar(j(j+1) - n(n-1))^{1/2}\left|{}^{j}_{m\,n-1}\right\rangle, \tag{5.5.47a}$$

$$(J_{\bar{x}} - iJ_{\bar{y}})\left|{}^{j}_{mn}\right\rangle = \hbar(j(j+1) - n(n+1))^{1/2}\left|{}^{j}_{m\,n+1}\right\rangle. \tag{5.5.47b}$$

This follows from the body-based equations (5.5.45b) and (5.5.45c) for the $\left|{}^{j}_{mn}\right\rangle$ states.

Using the same equations it is easy to show that the states $\left|{}^{j}_{mn}\right\rangle$ are eigenstates of the spherical quantum rotor Hamiltonian,

$$\begin{aligned}H_{\text{sphere}} &= (J)^2/2I = B(J)^2 \\ &= B(J_{\bar{x}}^2 + J_{\bar{y}}^2 + J_{\bar{z}}^2),\end{aligned} \tag{5.5.48}$$

where I is the rotational inertia, and

$$B = 1/2I$$

is called the rotational constant. The energy-level spectrum of the spherical rotor is given by the eigenvalues

$$E^j = \left\langle {}^{j}_{mn}\left|H_{\text{sphere}}\right|{}^{j}_{mn}\right\rangle = j(j+1)\hbar^2 B. \tag{5.5.49}$$

The spectrum consists of j levels with a spacing of

$$E_j - E_{j-1} = 2\hbar^2 Bj, \tag{5.5.50}$$

as shown on the left-hand side of Figure 5.5.7.

SOME APPLICATIONS OF R_3 REPRESENTATIONS

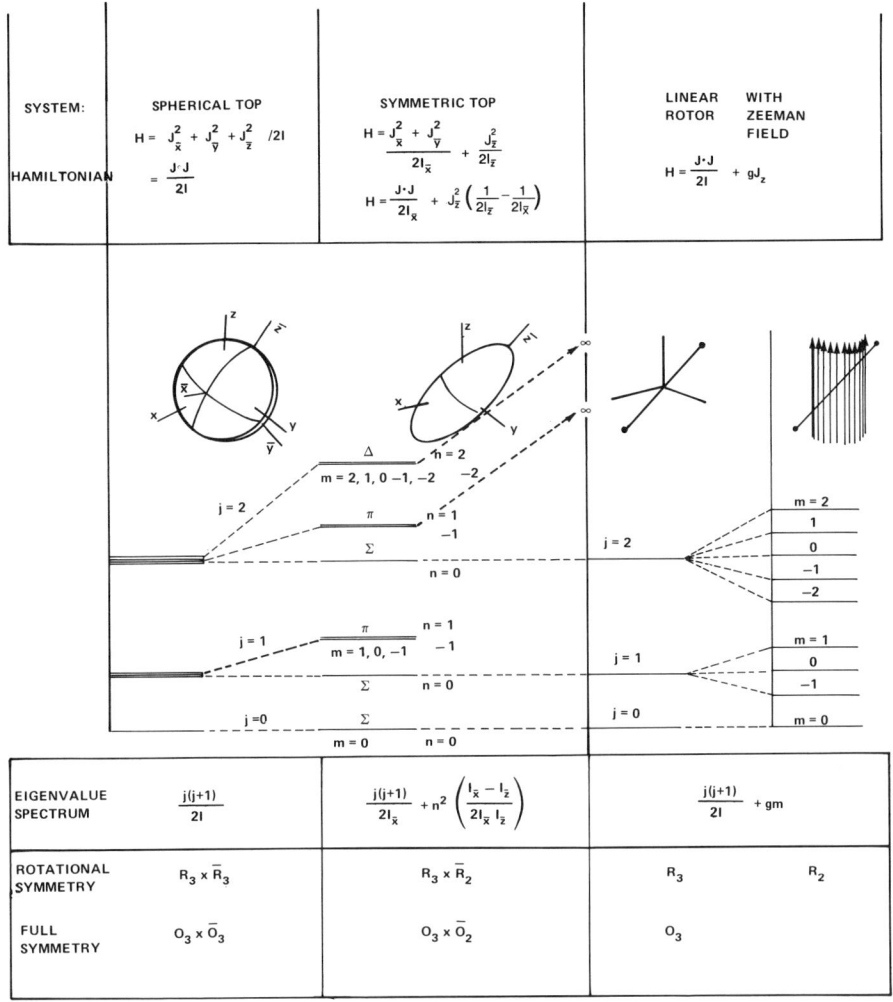

Figure 5.5.7 Hamiltonians, energy levels, and symmetry groups of various quantum rotor systems.

Each j level of the spherical rotor for $j > 0$ is degenerate with a degeneracy of $(2j + 1)^2$. In elementary quantum theory one learns that atomic levels of angular momentum j have degeneracy $(2j + 1)$ corresponding to $(2j + 1)$ states $\left|{}^{\,j}_{m}\right\rangle$ for each magnetic quantum number $m = j, j - 1, \ldots, -j + 1$, and $-j$. For the quantum rotor the z component m ranges over the same values. However, for each m value there are $(2j + 1)$ more n values $n = j, j - 1, \ldots, -j$ of the body $\bar{z}$ component. This gives a total degeneracy of

1, 9, 25, 49, ... for $j = 0, 1, 2, 3, \ldots$, respectively, in the spherical quantum rotor level spectrum.

The states $\left|{}^{\ j}_{mn}\right\rangle$ are also eigenvectors of the cylindrical or symmetric quantum rotor Hamiltonian:

$$\begin{aligned}H_{\text{sym.}} &= \left(J_{\bar{x}}^2 + J_{\bar{y}}^2\right)/2I_{\bar{x},\bar{y}} + J_{\bar{z}}^2/2I_{\bar{z}}\\ &= \left(J_{\bar{x}}^2 + J_{\bar{y}}^2 + J_{\bar{z}}^2\right)/2I_{\bar{x},\bar{y}} + J_{\bar{z}}^2\left[(1/2I_{\bar{z}}) - \left(1/2I_{\bar{x},\bar{y}}\right)\right]\\ &= B(J)^2 + J_{\bar{z}}^2[C - B].\end{aligned} \quad (5.5.51a)$$

Here the constant

$$I_{\bar{x},\bar{y}} = I_{\bar{x}} = I_{\bar{y}} = 1/2B \quad (5.5.51b)$$

is the transverse moment of inertia for a cylindrical object, and $I_{\bar{z}} = 1/2C$ is the axial inertia. If the rotor is a PROLATE symmetric or football-shaped object, then $I_{\bar{z}} < I_{\bar{x},\bar{y}}$. We shall consider this case. In the opposite case ($I_{\bar{z}} > I_{\bar{x}\bar{y}}$) one has an OBLATE symmetric or discus-shaped rotor.

The prolate symmetric rotor spectrum depends on total momentum quantum number j, the body component n, and the molecular constants B and C:

$$E_n^j = \left\langle{}^{\ j}_{mn}\left|H_{\text{sym.}}\right|{}^{\ j}_{mn}\right\rangle = j(j+1)\hbar^2 B + n^2\hbar^2(C - B). \quad (5.5.52)$$

For the prolate rotor the constant $(C - B)$ is positive and so the high-$|n|$ levels have higher energy, as shown in the left-hand central column of Figure 5.5.7. The levels for $|n| = 0, 1, 2, 3, 4$ are labeled $\Sigma, \Pi, \Delta, \Phi, \Gamma, \ldots$ according to a tradition that grew out of atomic spectroscopy. In atomic spectroscopy, transitions to $j = 0, 1, 2, 3, 4$, are labeled $s, p, d, f, g, \ldots$, which stand for "sharp," "principal," "diffuse," "fine," "goodness-knows-what," etc. (Thereafter the labeling is alphabetical except that the letter j is deleted so as not to confuse it with the quantum number j.)

Note that the letters $\Sigma, \Pi, \Delta, \ldots$ appear as an alternative labeling for irreps of $C_{\infty v}$ symmetry in Eq. (5.2.4) and $D_{\infty h}$ symmetry in Figure 5.2.1. A prolate rotor has $D_{\infty h} \supset C_{\infty v}$ body symmetry and so it is appropriate to label its energy levels with the $C_{\infty v}$ irreps. In later discussions we will consider the inversion and reflection symmetry labels as well. For now we note that a rotor which is invariant to reflections through any plane containing its $\bar{z}$ axis must have degenerate $+n$ and $-n$ levels for $n \geq 1$.

An asymmetric rotor has a Hamiltonian of the form

$$\begin{aligned}H_{\text{asym.}} &= J_{\bar{x}}^2/2I_{\bar{x}} + J_{\bar{y}}^2/2I_{\bar{y}} + J_{\bar{z}}^2/2I_{\bar{z}}\\ &= AJ_{\bar{x}}^2 + BJ_{\bar{y}}^2 + CJ_{\bar{z}}^2.\end{aligned} \quad (5.5.53)$$

SOME APPLICATIONS OF R_3 REPRESENTATIONS 371

For this the body symmetry is reduced to D_2 or D_{2h}, i.e., 180° rotation reflections around the $\bar{x}$, $\bar{y}$, and $\bar{z}$ body axes. This symmetry is Abelian and has only one-dimensional irreps. Therefore no degeneracy between states belonging to different body labeling can be expected. The representations of $H_{\text{asym.}}$ in the $\left|^{j}_{mn}\right\rangle$ basis is found using Eq. (5.5.45):

$$H_{\text{asym.}}\left|^{j}_{mn}\right\rangle$$
$$= \hbar^2\left[\left(\frac{A+B}{2}\right)(j(j+1) - n^2) + Cn^2\right]\left|^{j}_{mn}\right\rangle$$
$$+ \hbar^2\left(\frac{A-B}{4}\right)[(j+n+1)(j+n+2)(j-n)(j-n-1)]^{1/2}\left|^{j}_{m\,n+2}\right\rangle$$
$$+ \hbar^2\left(\frac{A-B}{4}\right)[(j+n)(j+n-1)(j-n+1)(j-n+2)]^{1/2}\left|^{j}_{m\,n-2}\right\rangle.$$
(5.5.54)

Clearly, eigenvector combinations of $\left|^{j}_{mn}\right\rangle$ for fixed m and various $n, n \pm 2, n \pm 4, \ldots$ are needed to diagonalize $\langle H_{\text{asym.}} \rangle$. This problem will be discussed in later chapters. Note, however, that the components of $\langle H_{\text{asym.}} \rangle$ reproduce those of $\langle H_{\text{sym.}} \rangle$ for $A = B$ and $\langle H_{\text{sph.}} \rangle$ for $A = B = C$.

So far we have seen what happens when the body symmetry is reduced from spherical R_3 symmetry to $R_2 = C_\infty$ symmetry of the cylindrical rotor (if we count reflections, then $H_{\text{sym.}}$ has $O_2 \sim C_{\infty v}$ symmetry), and finally to a finite D_2 symmetry. The n degeneracy becomes more and more split until finally there is none. However, we have not yet reduced the lab symmetry and until we do, the $(2j + 1)$-fold m-degeneracy will remain for each (j, n) level. Body-based operators $J_{\bar{x}}$, $J_{\bar{y}}$, and $J_{\bar{z}}$ have no effect on the m labels. A simple example of the effect of a lab-based operator is the Hamiltonian

$$H_Z = B(J)^2 + gJ_z. \quad (5.5.55)$$

The perturbation gJ_z is a simplified model for the ZEEMAN effect on a rotor of an external uniform magnetic field along the z axis of the laboratory. The constant g is called the gyromagnetic ratio for the rotor. The level spectrum of the Zeeman Hamiltonian is given by

$$E^j_m = \left\langle^{j}_{mn}\left|H_z\right|^{j}_{mn}\right\rangle = Bj(j+1)\hbar^2 + gm\hbar. \quad (5.5.56)$$

Each j level gets split into $2j + 1$ parts as shown in Figure 5.5.7 on the right-hand side. The magnetic field reduces the lab symmetry from spherical R_3 symmetry of a vacuum to $R_2 = C_\infty$ symmetry of the field axis. The magnetic quantum numbers m are R_2 irrep labels for the split levels.

A more symmetric example of a lab-based perturbation is given by

$$H_S + B(J)^2 + \mu J_z^2. \quad (5.5.57)$$

Here μJ_z^2 is a very simplified model for a STARK effect on a rotor of an external uniform electric field along the z axis. An electric field has a higher $C_{\infty v} \sim O_2$ symmetry since reflections through any plane parallel to the E field leaves it unchanged. Hence a degeneracy between $\pm m$ levels will exist in the level spectrum:

$$E^j_{|m|} = \left\langle \begin{matrix} j \\ mn \end{matrix} \middle| H_S \middle| \begin{matrix} j \\ mn \end{matrix} \right\rangle = Bj(j+1)\hbar^2 + \mu m^2 \hbar^2. \quad (5.5.58)$$

Hamiltonian H_S is the laboratory analogy for the body Hamiltonian $H_{\text{sym.}}$ of the symmetric rotor. H_S reduces the lab symmetry from R_3 to R_2, while the other reduces the body symmetry in the same way. You may realize now that there are two sides to any quantum rotor story: an inside and an outside. Two commuting sets of symmetry operators will exist for each rotor Hamiltonian H. The first set will be a subgroup $M^{\text{LAB}} \subset O_3^{\text{LAB}}$ of lab operators $\{ \cdots R(\alpha\beta\gamma), \ldots, IR(\alpha\beta\gamma), \ldots \}$, which commute with H. The nature of subgroup M^{LAB} will depend on the laboratory situation and the symmetry or shape of the external fields being applied. The second set will be a subgroup $\overline{N}^{\text{BODY}} \subset \overline{O}_3^{\text{BODY}}$ of body operators $\{ \ldots \overline{R}(\alpha\beta\gamma), \ldots, I\overline{R}(\alpha\beta\gamma), \ldots \}$, which commute with H. The nature of $\overline{N}^{\text{BODY}}$ depends on molecular structure and the internal symmetry or shape of the rotor. The whole rotor symmetry may be written

$$G^{\text{ROTOR}} = M^{\text{LAB}} \times \overline{N}^{\text{BODY}}, \quad (5.5.59)$$

since the two subgroups M and $\overline{N}$ are independent and mutually commuting. For example, a spherical rotor in a field-free laboratory environment is said to have $O_3 \times \overline{O}_3$ symmetry, while symmetric and asymmetric rotors have $O_3 \times \overline{O}_2$ and $O_3 \times \overline{D}_2$ symmetry, respectively, in the same environment. In the presence of a laboratory electric field the spherical rotor has $O_2 \times \overline{O}_3$ symmetry, while the symmetric and asymmetric rotors have $O_2 \times \overline{O}_2$ and $O_2 \times \overline{D}_2$ symmetry, respectively. In later chapters we will consider tetrahedral rotors in free space of symmetry $O_3 \times \overline{T}_d$ as well as rotors in various external fields with symmetry $O_2 \times \overline{T}_d$, $R_2 \times \overline{T}_d$, $D_2 \times \overline{T}_d$, and so forth. Any of some thousand combinations ($M \times \overline{N}$) of point groups are possible rotor symmetries which are subgroups of $O_3 \times \overline{O}_3$. (With just the crystal point symmetries there are $32^2 = 1024$ different combinations.) The problem of correlating R_3 and O_3 irreps with point symmetry subgroups is treated in Section 5.6.

D. Spherical Harmonics and Rotational Wave Functions

Some quantum rotors have a very simple internal structure. An orbiting electron is one example. A rotating diatomic molecule is another. For each of these examples a single vector $\bar{z}$ determines the orientation of the body. The radius vector $\mathbf{r}$ for the electron or the internuclear position vector $\mathbf{r}_1 - \mathbf{r}_2$ for a linear molecular rotor define the rotational position of these bodies. The position vectors have polar coordinates of azimuth (ϕ) and polar angle (θ) which can be identified with the first two Euler angles, respectively.

$$\alpha = \phi, \quad \beta = \theta. \tag{5.5.60}$$

The third Euler angle (γ) is superfluous since a rotation $\bar{R}(00\gamma)$ of a vector around its own direction should have no effect. Here we ignore any internal structure of the nuclei or the electron whose positions are defined by a $\bar{z}$ vector. The particles are imagined to be points.

These arguments lead to a LOCAL SYMMETRY condition for pointlike rotors:

$$R(00\gamma)|000\rangle = |000\rangle \tag{5.5.61}$$

This condition is analogous to similar conditions discussed in Chapter 4. It leads to the conclusion that

$$\left| \begin{matrix} j \\ mn \end{matrix} \right\rangle = P^j_{mn}|000\rangle/(2j+1)^{1/2} \tag{5.5.62}$$

will vanish for linear rotors unless $n = 0$. By substituting the condition into the projection we derive

$$\left| \begin{matrix} j \\ mn \end{matrix} \right\rangle = P^j_{mn} R(00\gamma)|000\rangle/(2j+1)^{1/2}$$

$$= e^{-in\gamma} \left| \begin{matrix} j \\ mn \end{matrix} \right\rangle, \tag{5.5.63}$$

which is true for all γ only if $n = 0$.

Note also that a linear rotor or orbiting electron has vanishing rotational inertia $I_{\bar{z}}$ around the radius $\bar{z}$ vector. The energy values (5.5.52) go to infinity for such a rotor for all nonzero n. On the right-hand side of Figure 5.5.7 only the $n = 0$ levels are drawn.

With $n = 0$ the $\mathscr{D}$ functions simplify considerably. Rewriting the central ($n = 0$) column of the $\mathscr{D}^{1*}$ and $\mathscr{D}^{2*}$ matrices using (5.4.33) and (5.4.35) yields the following:

$$\mathscr{D}_{m,0}^{1*}(\alpha\beta\gamma) = \begin{pmatrix} \cdot & -\dfrac{e^{+i\alpha}\sin\beta}{\sqrt{2}} & \cdot \\ \cdot & \cos\beta & \cdot \\ \cdot & \dfrac{e^{-i\alpha}\sin\beta}{\sqrt{2}} & \cdot \end{pmatrix}, \quad \begin{matrix} (m=1) \\ (m=0) \\ (m=-1) \end{matrix} \quad (5.5.64a)$$

$$\mathscr{D}_{m,0}^{2*}(\alpha\beta\gamma) = \begin{pmatrix} \cdot & \cdot & \sqrt{\dfrac{3}{8}}\,e^{+2i\alpha}\sin^2\beta & \cdot & \cdot \\ \cdot & \cdot & -\sqrt{\dfrac{3}{2}}\,e^{+i\alpha}\sin\beta\cos\beta & \cdot & \cdot \\ \cdot & \cdot & \dfrac{3\cos^2\beta - 1}{2} & \cdot & \cdot \\ \cdot & \cdot & \sqrt{\dfrac{3}{2}}\,e^{-i\alpha}\sin\beta\cos\beta & \cdot & \cdot \\ \cdot & \cdot & \sqrt{\dfrac{3}{8}}\,e^{-2i\alpha}\sin^2\beta & \cdot & \cdot \end{pmatrix}. \quad \begin{matrix} (m=2) \\ (m=1) \\ (m=0) \\ (m=-1) \\ (m=-2) \end{matrix}$$

(5.5.64b)

If you are familiar with the SPHERICAL HARMONICS Y_m^j from elementary quantum mechanics, then you will recognize them here. In fact they are special cases of the rotor wave functions for $n = 0$:

$$r_{m0}^j(\alpha\beta\gamma) = \mathscr{D}_{m0}^{j*}(\alpha\beta\gamma)(2j+1)^{1/2} = \sqrt{4\pi}\,Y_m^j(\beta\alpha). \quad (5.5.65)$$

Note that body quantum number n and the third Euler angle are not present in a linear rotor wave function.

It is important to summarize several of the roles which the $\mathscr{D}$ matrix functions play. For the physicist they are wave functions as stated in Eqs. (5.5.25) and (5.5.26):

$$r_{mn}^j(\alpha\beta\gamma) = \mathscr{D}_{mn}^{j*}(\alpha\beta\gamma)(2j+1)^{1/2} = \langle 000|R^{\dagger}(\alpha\beta\gamma)\Big|{}_{mn}^{j}\rangle$$
$$= \langle \alpha\beta\gamma|{}_{mn}^{j}\rangle, \quad (5.5.66)$$

but they are mainly rotation representations, i.e., the representation of a product $R^\dagger(\alpha\beta\gamma)R(\phi\theta\omega)$ is the product of the representations as stated here:

$$\mathscr{D}^j_{mn}(R^\dagger(\alpha\beta\gamma)R(\phi\theta\omega)) = \sum_{m'}\mathscr{D}^j_{mm'}(R^\dagger(\alpha\beta\gamma))\mathscr{D}^j_{m'n}(R(\phi\theta\omega)). \quad (5.5.67)$$

Taking the complex conjugate yields

$$\mathscr{D}^{j*}_{mn}(R^\dagger(\alpha\beta\gamma)R(\phi\theta\omega)) = \sum_{m'}\mathscr{D}^j_{m'm}(R(\alpha\beta\gamma))\mathscr{D}^{j*}_{m'n}(R(\phi\theta\omega)).$$

Using the notation of Eq. (5.5.66) one derives

$$\langle 000|R^\dagger(\phi\theta\omega)R(\alpha\beta\gamma)\left|{j \atop mn}\right\rangle = \sum_{m'}\mathscr{D}^j_{m'm}(R(\alpha\beta\gamma))\langle 000|R^\dagger(\phi\theta\omega)\left|{j \atop m'n}\right\rangle, \quad (5.5.68)$$

which simplifies to the following form:

$$\langle \phi\theta\omega|R(\alpha\beta\gamma)\left|{j \atop mn}\right\rangle = \sum_{m'}\mathscr{D}^j_{m'm}(\alpha\beta\gamma)\left\langle \phi\theta\omega\left|{j \atop m'n}\right.\right\rangle. \quad (5.5.69)$$

This agrees with the left transformation formula (5.5.31) if the bra $\langle\phi\theta\omega|$ is dropped.

A special case of this transformation formula for $n = 0$ involves the Y functions:

$$Y^{j^R}_m(\theta\phi) = \sum_{m'}\mathscr{D}^j_{m'm}(\alpha\beta\gamma)Y^j_{m'}(\theta\phi). \quad (5.5.70a)$$

Here the function

$$Y^{j^R}_m(\theta\phi) = \langle \phi\theta \cdot |R(\alpha\beta\gamma)\left|{j \atop m0}\right\rangle \quad (5.5.70b)$$

is a spherical harmonic wave function of a rotated state $R(\alpha\beta\gamma)\left|{j \atop m}\right\rangle$. The rotated harmonic is schematically represented by lobes of a wave function centered on the $\{\bar{x}\bar{y}\bar{z}\}$ axes in Figure 5.5.8. The wave function $Y^{j^R}_m$ looks the same in the rotated $\{\bar{x}\bar{y}\bar{z}\}$ axes as Y^j_m looks in the lab axes $\{xyz\}$. Y^j_m is

$$Y_m^{j'}(\theta\phi) = \Sigma D_{m'm}^{j}(\alpha\beta\gamma) Y_{m'}^{j}(\theta\phi)$$

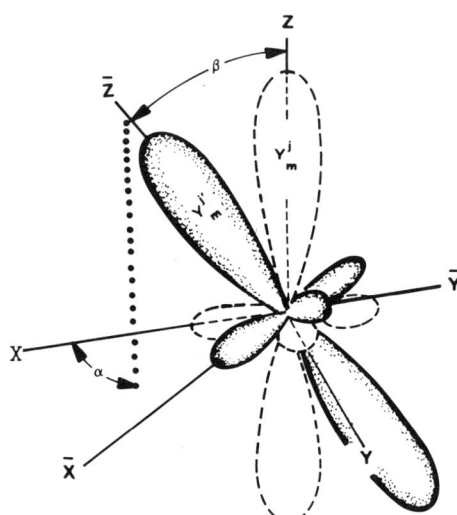

Figure 5.5.8 Transformation between rotated harmonic $Y_m^{j'}$ and unrotated wave functions. Lobes of the wave are represented by level surfaces: $Y^{j'}(\theta\phi)$ = constant.

sketched by dotted lines in the figure. The real or imaginary parts of $Y_m^j(\theta\phi)$ will have $2m$ lobes and $2m$ nodes in the interval $\{0 \leq \phi < 2\pi\}$. This can be derived from its z-rotation properties or by inspecting the standard formula:

$$Y_m^j(\theta\phi) = \left[\frac{(2j+1)(j-m)!}{4\pi(j+m)!}\right]^{1/2} (-1)^m e^{im\phi} P_j^m(\cos\theta). \quad (5.5.71)$$

The P_l^m are the well-known associated Legendre functions. The fact that $(2j+1)$ different functions $\{Y_j^j, Y_{j-1}^j, \ldots, Y_{-j}^j\}$ should combine in Eq. (5.5.70a) to make a rotated version is a wonderful thing indeed. The role of $\mathscr{D}$ functions as transformation coefficients for harmonics is important, and it is probably their best-known role.

If one sets $m = 0$ the resulting harmonic has no azimuthal dependence. Y_0^j wave functions have cylindrical or azimuthal symmetry as schematized in Figure 5.5.9. Y_0^j is a function only of its local polar angle θ,

$$Y_0^j(\theta \cdot) = \left(\frac{2j+1}{4\pi}\right)^{1/2} P_j^0(\cos\theta), \quad (5.5.72)$$

where $P_j^0(\cos\theta)$ is a standard Legendre function. The rotated harmonic Y^{jR} in Figure 5.5.9 is a function only of the polar angle Θ with respect to the

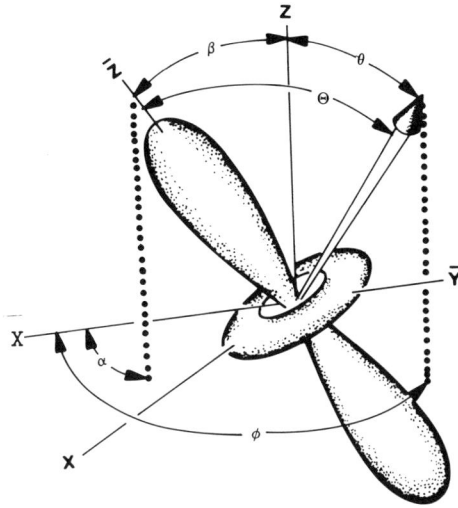

Figure 5.5.9 Azimuthally symmetric rotated harmonic $Y_0^{j'}$. Angles shown are used in the addition theorem.

tipped $\bar{z}$ axis. Substituting this into the Y transformation (5.5.70) yields

$$Y_0^{j R}(\theta\phi) = \sum_m \mathscr{D}_{m0}^j(\alpha\beta \cdot) Y_m^j(\theta\phi), \tag{5.5.73}$$

$$\left(\frac{2j+1}{4\pi}\right)^{1/2} P_j^0(\cos\Theta) = \sum_m \left(\frac{4\pi}{2j+1}\right)^{1/2} Y_m^{j*}(\beta\alpha) Y_m^j(\theta\phi), \tag{5.5.74}$$

where Eq. (5.5.65) was used, also. This yields the very important ADDITION THEOREM

$$P_j(\cos\Theta) = \left(\frac{4\pi}{2j+1}\right) \sum_{m=-j}^{j} Y_m^{j*}(\beta\alpha) Y_m^j(\theta\phi) \tag{5.5.75}$$

for spherical harmonics. However, a better name for this might have been the *multiplication* theorem, since it follows directly from the R_3 group multiplication relation (5.5.67).

E. Explicit Relations for Rotation Operators and Generators

(a) Laboratory- and Body-Defined Polar Coordinates The Euler-angle device introduced in Figure 5.3.1 is redrawn in Figure 5.5.10 to include labeling which will aid in the discussion of explicit representations of coordinate rotations. The device consists of four rigid parts or frames, and each

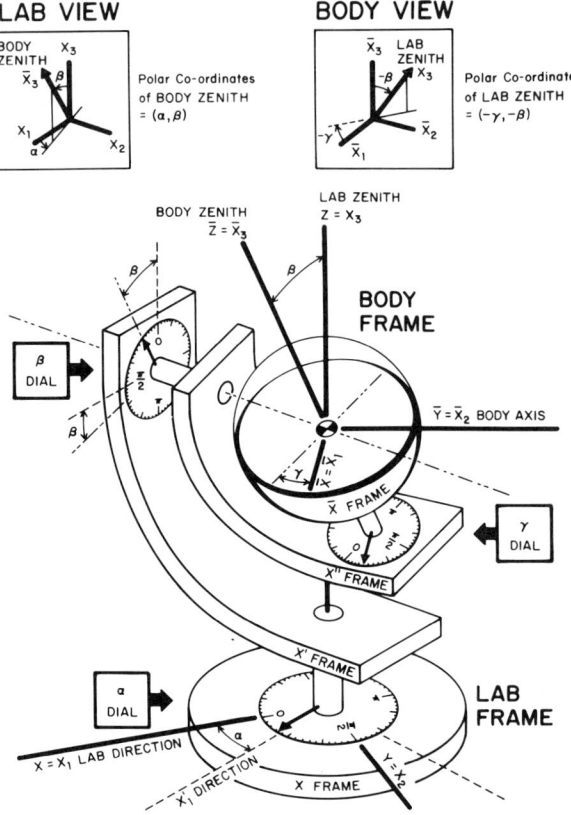

Figure 5.5.10 Mechanical definition of Euler coordinates for rotational mechanics. Laboratory and body views of the body and lab zeniths are shown in respective insets. (See also Figures 5.3.1 and 5.3.4.) [Reprinted from W. G. Harter and C. W. Patterson, *J. Chem. Phys.*, **80**, 4260 (1984).]

frame defines a coordinate axis and is supporting or supported by other frames. There is a lab $(x_1 x_2 x_3)$ frame supporting an $(x'_1 x'_2 x'_3)$ frame which in turn supports an $(x''_1 x''_2 x''_3)$ frame which finally supports the body $(\bar{x}_1 \bar{x}_2 \bar{x}_3)$ frame. Each of the frames is connected to its neighbor (or neighbors) in the sequence by pivots or bearings. Each bearing has an indicator and dial which displays one of the three Euler angles (α, β, γ). By defining Euler angles in this way one sees clearly that they are *holonomic* coordinates; that is, they depend only upon the relative orientation of the three frames and not upon the path or order of operations which yielded a given orientation.

The first two Euler angles (α, β) serve as polar coordinates for the body zenith or $\bar{x}_3$ axis in the lab frame as shown in the upper left-hand inset. The azimuth and polar angles of the lab zenith or x_3 axis in the body frame are

$(-\gamma)$ and $(-\beta)$, respectively, as indicated in the upper right-hand in set of Figure 5.5.10.

(b) Coordinate Rotation and Transformation Matrix As explained before Eq. (5.3.3), the setting of an Euler-defined position is accomplished by an ordered sequence of z, y, and z rotations. This sequence is indicated explicitly by Figure 5.5.11. The matrix representation of this sequence is given by the following:

$$\langle R(\alpha\beta\gamma)\rangle = \begin{pmatrix} \cos\alpha & -\sin\alpha & 0 \\ \sin\alpha & \cos\alpha & 0 \\ 0 & 0 & 1 \end{pmatrix} \begin{pmatrix} \cos\beta & 0 & \sin\beta \\ 0 & 1 & 0 \\ -\sin\beta & 0 & \cos\beta \end{pmatrix} \begin{pmatrix} \cos\gamma & -\sin\gamma & 0 \\ \sin\gamma & \cos\gamma & 0 \\ 0 & 0 & 1 \end{pmatrix},$$

$$= \begin{matrix} x_1 \\ x_2 \\ x_3 \end{matrix} \begin{pmatrix} \cos\alpha\cos\beta\cos\gamma - \sin\alpha\sin\gamma & -\cos\alpha\cos\beta\sin\gamma - \sin\alpha\cos\gamma & \cos\alpha\sin\beta \\ \sin\alpha\cos\beta\cos\gamma + \cos\alpha\sin\gamma & -\sin\alpha\cos\beta\sin\gamma + \cos\alpha\cos\gamma & \sin\alpha\sin\beta \\ -\sin\beta\cos\gamma & \sin\beta\sin\gamma & \cos\beta \end{pmatrix}.$$

(5.5.76)

Note that the $|\bar{x}_3\rangle$ ket or third column of (5.5.76) is the polar coordinate representation of the body zenith, as it should be. Similarly, the $\langle x_3|$ bra or third row of (5.5.76) is the representation of the lab zenith in the body frame where one could measure an azimuthal angle of $(-\gamma)$ and a polar angle $(-\beta)$ for the x_3 vector.

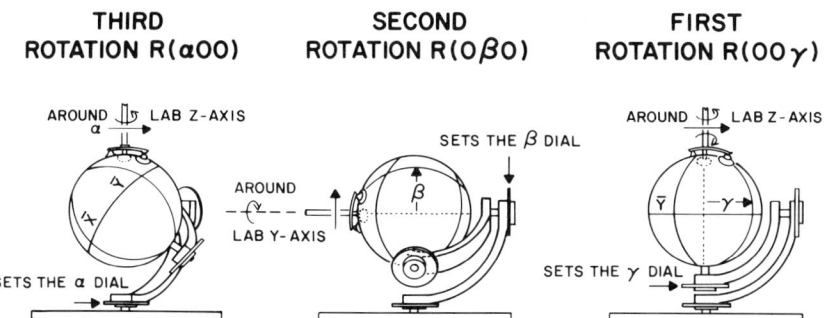

Figure 5.5.11 Operational definition of Euler coordinates. Ordered rotational sequence $R(\alpha 00)R(0\beta 0)R(00\gamma) = R(\alpha\beta\gamma)$ of lab-based operations orients the body into the $(\alpha\beta\gamma)$ position relative to the lab. Only rotations about the lab y and z axes are used. [Reprinted from W. G. Harter and C. W. Patterson *J. Chem. Phys.* **80** 4260 (1984).]

(c) Angular Velocity and Momentum in Laboratory and Body Frames

The angular velocity of the body frame relative to the laboratory may be expressed simply in terms of the angular velocities $\dot{\alpha}$, $\dot{\beta}$, and $\dot{\gamma}$ of the three dials in Figure 5.3.10. The rotation of each successive frame relative to its supporting frame depends simply upon the angular velocity of the bearing which connects them. For example, the time derivative $(d\mathbf{r}/dt)|_x$ of a vector in the lab (x) frame can be written in terms of the velocity $(d\mathbf{r}/dt)|_{x'}$ of the same vector observed in the (x') frame as follows:

$$\left.\frac{d\mathbf{r}}{dt}\right|_x = \left.\frac{d\mathbf{r}}{dt}\right|_{x'} + \boldsymbol{\omega}_\alpha \times \mathbf{r}, \tag{5.5.77a}$$

where the angular-velocity vector of the α dial is

$$\boldsymbol{\omega}_\alpha = \dot{\alpha}\mathbf{x}_3 = \dot{\alpha}\left(-\sin\beta\cos\gamma\bar{\mathbf{x}}_1 + \sin\beta\sin\gamma\bar{\mathbf{x}}_2 + \cos\beta\bar{\mathbf{x}}_3\right). \tag{5.5.77b}$$

This vector lies along the unit lab zenith $(\mathbf{x}_3)$ and is expressed in terms of the body axes using the third row of (5.5.76).

Similarly, the time derivative $(d\mathbf{r}/dt)|_{x'}$ in the (x') frame can be expressed in terms of the velocity $(d\mathbf{r}/dt)|_{x''}$ which would be observed in the (x'') frame by

$$\left.\frac{d\mathbf{r}}{dt}\right|_{x'} = \left.\frac{d\mathbf{r}}{dt}\right|_{x''} + \boldsymbol{\omega}_\beta \times \mathbf{r}, \tag{5.5.78a}$$

since these two frames are connected by the β axis which lies along the vector

$$\boldsymbol{\omega}_\beta = \dot{\beta}(-\sin\alpha\mathbf{x}_1 + \cos\alpha\mathbf{x}_2) = \dot{\beta}(\sin\gamma\bar{\mathbf{x}}_1 + \cos\gamma\bar{\mathbf{x}}_2). \tag{5.5.78b}$$

This is seen by inspecting Figure 5.5.10. Finally, the desired expression for the body frame observed velocity $(d\mathbf{r}/dt)|_{\bar{x}}$ is given in terms of $(d\mathbf{r}/dt)|_{x''}$ by

$$\left.\frac{d\mathbf{r}}{dt}\right|_{x''} = \left.\frac{d\mathbf{r}}{dt}\right|_{\bar{x}} + \boldsymbol{\omega}_\gamma \times \mathbf{r}, \tag{5.5.79a}$$

where the γ dial angular velocity vector is along the unit body zenith $\bar{\mathbf{x}}_3$:

$$\boldsymbol{\omega}_\gamma = \dot{\gamma}(\cos\alpha\sin\beta\mathbf{x}_1 + \sin\alpha\sin\beta\mathbf{x}_2 + \cos\beta\mathbf{x}_3) = \dot{\gamma}\bar{\mathbf{x}}_3. \tag{5.5.79b}$$

A combination of the preceding three relations yields lab-body velocity relations of the form

$$\left.\frac{d\mathbf{r}}{dt}\right|_x = \left.\frac{d\mathbf{r}}{dt}\right|_{\bar{x}} + \boldsymbol{\omega} \times \mathbf{r}, \tag{5.5.80a}$$

where the total angular velocity

$$\omega = \omega_\alpha + \omega_\beta + \omega_\gamma \qquad (5.5.80b)$$

is a vector sum of those for the three dials. Taking the lab expressions for (5.5.77)–(5.5.79b) we derive the following matrix equation and its inverse:

$$\begin{pmatrix} \omega_x \\ \omega_y \\ \omega_z \end{pmatrix} = \begin{pmatrix} 0 & -\sin\alpha & \cos\alpha \sin\beta \\ 0 & \cos\alpha & \sin\alpha \sin\beta \\ 1 & 0 & \cos\beta \end{pmatrix} \begin{pmatrix} \dot\alpha \\ \dot\beta \\ \dot\gamma \end{pmatrix}, \qquad (5.5.81a)$$

$$\begin{pmatrix} \dot\alpha \\ \dot\beta \\ \dot\gamma \end{pmatrix} = \begin{pmatrix} -\cos\alpha \cot\beta & -\sin\alpha \cot\beta & 1 \\ -\sin\alpha & \cos\alpha & 0 \\ \cos\alpha/\sin\beta & \sin\alpha/\sin\beta & 0 \end{pmatrix} \begin{pmatrix} \omega_x \\ \omega_y \\ \omega_z \end{pmatrix}. \qquad (5.5.81b)$$

If the body expressions are used instead then one derives the following:

$$\begin{pmatrix} \omega_{\bar x} \\ \omega_{\bar y} \\ \omega_{\bar z} \end{pmatrix} = \begin{pmatrix} -\sin\beta \cos\gamma & \sin\gamma & 0 \\ \sin\beta \sin\gamma & \cos\gamma & 0 \\ \cos\beta & 0 & 1 \end{pmatrix} \begin{pmatrix} \dot\alpha \\ \dot\beta \\ \dot\gamma \end{pmatrix}, \qquad (5.5.82a)$$

$$\begin{pmatrix} \dot\alpha \\ \dot\beta \\ \dot\gamma \end{pmatrix} = \begin{pmatrix} -\cos\gamma/\sin\beta & \sin\gamma/\sin\beta & 0 \\ \sin\gamma & \cos\gamma & 0 \\ \cot\beta \cos\gamma & -\cot\beta \sin\gamma & 1 \end{pmatrix} \begin{pmatrix} \omega_{\bar x} \\ \omega_{\bar y} \\ \omega_{\bar z} \end{pmatrix}. \qquad (5.5.82b)$$

The coefficients in these matrix relations are useful for relating the lab-based angular momenta $\{J_x = \partial L/\partial \omega_x,\ J_y = \partial L/\partial \omega_y,\ J_z = \partial L/\partial \omega_z\}$ to the body-based momenta $\{J_{\bar x} = \partial L/\partial \omega_{\bar x},\ J_{\bar y} = \partial L/\partial \omega_{\bar y},\ J_{\bar z} = \partial L/\partial \omega_{\bar z}\}$, where L is a given classical Lagrangian. The connection is made through the quantities $\{J_\alpha = \partial L/\partial \dot\alpha,\ J_\beta = \partial L/\partial \dot\beta,\ J_\gamma = \partial L/\partial \dot\gamma\}$, which are the Euler canonical momenta, and one uses the chain rule

$$J_m = \frac{\partial L}{\partial \omega_m} = \frac{\partial L}{\partial \dot\alpha}\frac{\partial \dot\alpha}{\partial \omega_m} + \frac{\partial L}{\partial \dot\beta}\frac{\partial \dot\beta}{\partial \omega_m} + \frac{\partial L}{\partial \dot\gamma}\frac{\partial \dot\gamma}{\partial \omega_m} = \frac{\partial \dot\alpha}{\partial \omega_m} J_\alpha + \frac{\partial \dot\beta}{\partial \omega_m} J_\beta + \frac{\partial \dot\gamma}{\partial \omega_m} J_\gamma.$$

Using (5.5.81)–(5.5.82b) then gives lab and body-based momentum in terms of Euler angles:

$$\begin{pmatrix} J_x \\ J_y \\ J_z \end{pmatrix} = \begin{pmatrix} -\cos\alpha \cot\beta & -\sin\alpha & \dfrac{\cos\alpha}{\sin\beta} \\ -\sin\alpha \cot\beta & \cos\alpha & \dfrac{\sin\alpha}{\sin\beta} \\ 1 & 0 & 0 \end{pmatrix} \begin{pmatrix} J_\alpha \\ J_\beta \\ J_\gamma \end{pmatrix}, \quad (5.5.83a)$$

$$\begin{pmatrix} J_{\bar{x}} \\ J_{\bar{y}} \\ J_{\bar{z}} \end{pmatrix} = \begin{pmatrix} -\dfrac{\cos\gamma}{\sin\beta} & \sin\gamma & \cot\beta \cos\gamma \\ \dfrac{\sin\gamma}{\sin\beta} & \cos\gamma & -\cot\beta \sin\gamma \\ 0 & 0 & 1 \end{pmatrix} \begin{pmatrix} J_\alpha \\ J_\beta \\ J_\gamma \end{pmatrix}. \quad (5.5.83b)$$

These relations become quantum operators if one represents the Euler operators by

$$J_\alpha = (h/i)\frac{\partial}{\partial \alpha}, \quad J_\beta = (h/i)\frac{\partial}{\partial \beta}, \quad J_\gamma = (h/i)\frac{\partial}{\partial \gamma}. \quad (5.5.84)$$

These are consistent with commutation relations

$$[J_x, J_y] = ihJ_z \text{ (and cyclically)}, \quad [J_\alpha, J_\beta] \equiv 0, \text{ etc.,}$$
$$[J_{\bar{x}}, J_{\bar{y}}] = -ihJ_{\bar{z}} \text{ (and cyclically)}, \quad [J_{\bar{x}}, J_y] \equiv 0, \text{ etc.,} \quad (5.5.85)$$

which we deduced using coordinate-free arguments. One should note that the coordinates conjugate to J_x, J_y, or J_z are nonholonomic; the $d\omega_m$ are not exact differentials. Their values depend upon the rotational path history followed by the rotor to a given orientation. Euler angles depend on only the orientation.

5.6 ROTATIONAL LEVEL SPLITTING IN FINITE SYMMETRY

Consider a rotating diatomic molecule or an orbiting electron in one of the angular-momentum states $|^l_m\rangle$ belonging to a $2l + 1$-degenerate energy level ε_l. The transformation properties of these states are defined by irreps $\mathscr{D}^l$ of symmetry group R_3 or by $\mathscr{D}^{l\pm}$ of group O_3 according to the usual transformation equations:

$$R|^l_m\rangle = \sum_{m'} \mathscr{D}^l_{m'm}(R)|^l_{m'}\rangle.$$

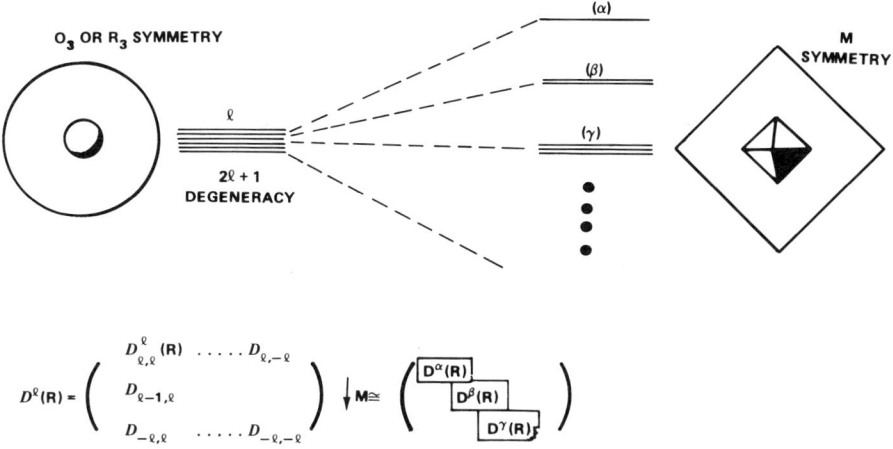

Figure 5.6.1 Molecular or crystal field splitting. l-orbital level splitting due to symmetry reduction from $O_3 \supset R_3$ to lower symmetry M corresponds to the reduction of the subduced representation $D^l \downarrow M$.

Suppose now this system is put into a lower-symmetry environment such as a crystal lattice or molecular site, where the symmetry is described by some finite group M. As a result, the energy levels will then belong to irreps $\mathscr{D}^\alpha, \mathscr{D}^\beta, \ldots$ of the new symmetry M. The corresponding degeneracies $l^\alpha, l^\beta, \ldots$ will generally be less than the free-orbital degeneracies $(2l + 1)$. Some of the orbital levels must split, as shown in Figure 5.6.1. This effect is often called CRYSTAL FIELD SPLITTING, and was first analyzed extensively by Bethe [1929].

We now see ways to determine which irreps $\mathscr{D}^\alpha$ of a symmetry M are correlated with a given irrep of O_3. We also discuss types of potential fields that cause splitting and show ways to obtain approximate eigenstates of them under certain conditions. These results will be used later to derive various model formulas for rotational spectra.

Of all the finite symmetries discussed in Chapters 2–4, we will treat the higher ones in each chain first, namely, O_h and D_{6h}. The properties of the others then follow relatively easily from the subgroup chain correlations.

A. Cubic Symmetry Correlations $O_3 \supset O_h$

The O_h character table was given in Eq. (4.1.16). To determine the orbital splittings, we will use these characters to find the frequencies f^α of O_h irrep $\mathscr{D}^\alpha$ in a given irrep $\mathscr{D}^l$ of O_3. The formula for f^α,

$$f^\alpha = (1/{}^\circ G) \sum_{\substack{\text{classes} \\ c_g}} \chi_g^{\alpha *} \, {}^\circ c_g \, \text{Trace } \mathscr{D}^l(g) \qquad (5.6.1)$$

REPRESENTATIONS OF CONTINUOUS ROTATION GROUPS AND APPLICATIONS

was given in Eq. (3.5.11). We shall apply this formula to the octahedral subgroup O of $O_h = O \times C_i$ and R_3 of $O_3 = R_3 \times C_i$. It is easy to determine the inversion parity (u or g) separately. For O we know the characters x_g^α [see upper left quadrant of Eq. (4.1.16)], the order of the classes $^\circ c_g$ (from Figure 4.1.2, we have $^\circ c_1 = 1$, $^\circ c_{120°} = 8$, $^\circ c_{180°} = 3$, $^\circ c_{90°} = 6$, and $^\circ c_{180^d} = 6$) and the order of the group ($^\circ G = 24$). We still need to find the trace of $\mathscr{D}^l$ for each rotation class.

To find the trace of an R_3 rotation by angle ω one picks the most convenient representation. The diagonal matrix

$$\mathscr{D}^l(\omega 00) = \begin{pmatrix} e^{-il\omega} & & & \\ & e^{-i(l-1)\omega} & & \\ & & \ddots & \\ & & & e^{il\omega} \end{pmatrix} \quad (5.6.2)$$

represents rotation $R(\omega 00)$ around the z axis. Any rotation $R[\omega]$ by the same angle ω around any axis $\hat{r}(\theta\phi)$ must be equivalent to $R(\omega 00)$.

$$R[\omega] = R(\phi\theta 0) R(\omega 00) R^{-1}(\theta\phi 0). \quad (5.6.3)$$

Therefore the traces of the representations must be equal.

The desired trace of Eq. (5.6.2) is evaluated by summing a geometrical series:

$$\text{Trace } \mathscr{D}^l(\omega 00) = e^{-il\omega} + e^{-i(l-1)\omega} + \cdots + e^{i(l-1)\omega} + e^{il\omega},$$

$$e^{-i\omega} \text{Trace } \mathscr{D}^l(\omega 00) = e^{-i(l+1)\omega} + e^{-il\omega} + e^{-i(l-1)\omega} + \cdots + e^{i(l-1)\omega}.$$

Subtracting the preceding two equations gives

$$(1 - e^{-i\omega}) \text{Trace } \mathscr{D}^l(\omega 00) = e^{il\omega} - e^{-i(l+1)\omega}$$

or

$$\text{Trace } \mathscr{D}^l(\omega 00) = \frac{e^{-i\omega/2}(e^{i(l+1/2)\omega} - e^{-i(l+1/2)\omega})}{e^{-i\omega/2}(e^{i\omega/2} - e^{-i\omega/2})} = \frac{\sin(l + 1/2)\omega}{\sin \omega/2}. \quad (5.6.4)$$

From this formula one derives the trace table (5.6.5a). Substitution of each row in turn into Eq. (5.6.1) gives the correlation table (5.6.5b) of frequencies. Several examples of predicted crystal field splitting are shown in Figure 5.6.2.

	Trace $\mathscr{D}^l(\omega 00)$					Single Electron Orbital Spectroscopic Labeling		Frequency of O Irreps					
l	$\omega=0°$	$\omega=120°$	$\omega=180°$	$\omega=90°$	$\omega=180°$		l	f^{A_1}	f^{A_2}	f^E	f^{T_1}	f^{T_2}	
0	1	1	1	1	1	s_g	0	1	.	.	.	.	A_{1g}
1	3	0	-1	1	-1	p_u	1	.	.	.	1	.	T_{1u}
2	5	-1	1	-1	1	d_g	2	.	.	1	.	1	$E_g + T_{2g}$
3	7	1	-1	-1	-1	f_u	3	.	1	.	1	1	$A_{2u} + T_{1u} + T_{2u}$
4	9	0	1	1	1	g_g	4	1	.	1	1	1	$A_{1g} + E_g + T_{1g} + T_{2g}$
5	11	-1	-1	-1	-1	h_u	5	.	.	1	2	1	
6	13	1	1	-1	1	i_g	6	1	1	1	1	2	
7	15	0	-1	-1	-1	k_u	7	.	1	1	2	2	
8	17	-1	1	1	1	l_g	8	1	.	2	2	2	
9	19	1	-1	1	-1	m_u	9	1	1	1	3	2	
10	21	0	1	-1	1	n_g	10	1	1	2	2	3	
11	23	-1	-1	-1	-1	o_u	11	.	1	2	3	3	
12	25	1	1	1	1	q_g	12	2	1	2	3	3	
						(5.6.5a)							
13	27	0	-1	-1	-1	r_u	13	1	1	2	4	3	
14	29	-1	1	-1	1	t_g	14	1	1	3	3	4	
15	31	1	-1	1	-1	u_u	15	1	2	2	4	4	
16	33	0	1	1	1		16	2	1	3	4	4	
17	35	-1	-1	1	-1		17	1	1	3	5	4	
18	37	1	1	-1	1		18	2	2	3	4	5	
19	39	0	-1	1	-1		19	1	2	3	5	5	
20	41	-1	1	-1	1		20	2	1	4	5	5	(5.6.5b)

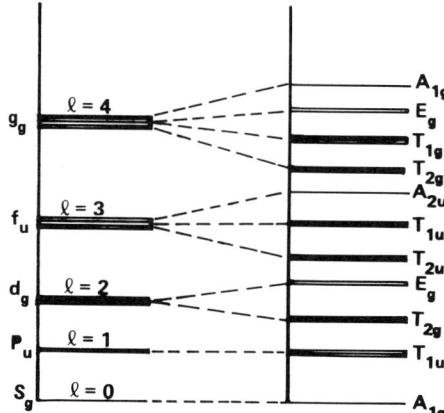

Figure 5.6.2 Octahedral splitting of s, p, d, f, and g orbitals.

The standard atomic notation $s, p, d, f, g, \ldots$ for orbital momentum $l = 0, 1, 2, 3, 4, \ldots$, respectively, is used. The inversion parities [g = even = (+), u = odd = (−)] are assigned as they apply to one-electron (hydrogenlike) wave functions $Y_m^l(\theta\phi)$, namely, odd (even) l means odd (even) parity. However, the inversion parities for multielectronic orbitals or rotating molecules do not follow this rule, in general, as we will see later. In general, a given l may have either parity depending on the physical situation. Whichever it is, the O_h-irreps will have the same parity as their l "parent" in any $O_3 \supset O_h$ correlation.

Figure 5.6.2 is a sketch of the form, though not necessarily the order, of the O_h splittings of the lower l levels, as dictated by (5.6.5b). The convenience of the symmetry analysis which gives this so quickly should be evident. We shall see in Section 5.6.F how the ordering and spacing of the split levels can also be found easily under certain conditions.

Note that the table of (5.6.5a) for $l = 0$ repeats itself after $l = 12$. This is as far as one ever needs to calculate it. $\mathscr{D}^{12}$ contains the regular representation which has zero trace for $\omega = 0$ plus one scalar ($l = 0$ or A_{1g}) irrep whose trace is all 1. [See the first row in the table of (5.6.5a).] Therefore, the general result is

$$\mathscr{D}^l \downarrow 0 \sim [\text{integer of } l/12](\mathscr{D}^{A_1} \oplus \mathscr{D}^{A_2} \oplus 2\mathscr{D}^E \oplus 3\mathscr{D}^{T_1} \oplus 3\mathscr{D}^{T_2}) \oplus \mathscr{D}^{[l \bmod 12]}.$$
(5.6.6)

B. Cubic Eigenstates and Wave Functions

The cubic symmetry correlations (5.6.5b) indicate that a fivefold degenerate $l = 2$ level should split into a doubly degenerate E level and a triply degenerate T_2 level under the influence of O-symmetry forces. Let us consider this example in some detail in order to understand its physical implications as well as some computational methods.

First one needs to find what combination of the five d-orbital eigenstates $\left\{\left|\begin{smallmatrix}2\\2\end{smallmatrix}\right\rangle, \left|\begin{smallmatrix}2\\1\end{smallmatrix}\right\rangle, \left|\begin{smallmatrix}2\\0\end{smallmatrix}\right\rangle, \left|\begin{smallmatrix}2\\-1\end{smallmatrix}\right\rangle, \left|\begin{smallmatrix}2\\-2\end{smallmatrix}\right\rangle\right\}$ go together to make octahedral base states $\left\{\left|\begin{smallmatrix}E\\1\end{smallmatrix}\right\rangle, \left|\begin{smallmatrix}E\\2\end{smallmatrix}\right\rangle, \left|\begin{smallmatrix}T_2\\1\end{smallmatrix}\right\rangle, \left|\begin{smallmatrix}T_3\\3\end{smallmatrix}\right\rangle, \left|\begin{smallmatrix}T_2\\3\end{smallmatrix}\right\rangle\right\}$. This is accomplished by applying appropriate octahedral projection operators P_{ij}^α to the orbital states:

$$\left|\begin{matrix}E\\e\end{matrix}\right\rangle = P_{ef}^E \left|\begin{matrix}2\\m\end{matrix}\right\rangle \Big/ (N^E)^{1/2}, \qquad \left|\begin{matrix}T_2\\t\end{matrix}\right\rangle = P_{tu}^{T_2} \left|\begin{matrix}2\\m\end{matrix}\right\rangle \Big/ (N^T)^{1/2}.$$

The projection factorization techniques discussed in Section 4.3.C help to simplify this projection.

Suppose the $|_m^l\rangle$ states are defined with respect to the z axis which is an octahedral fourfold symmetry axis. Then the first state $|\begin{smallmatrix}2\\2\end{smallmatrix}\rangle$ satisfies a local C_4 symmetry condition

$$\left|\begin{matrix}2\\2\end{matrix}\right\rangle = P^{2_4}\left|\begin{matrix}2\\2\end{matrix}\right\rangle = \tfrac{1}{4}(1 - R_3 + R_3^2 - R_3^3)\left|\begin{matrix}2\\2\end{matrix}\right\rangle. \qquad (5.6.7)$$

This calls for the tetragonal $(O \supset D_4 \supset C_4)$ projectors P_{ij}^α of O, and only those for which $j = 2_4$. If you forget which components correlate with (2_4) you may construct the octahedral irreps of the local symmetry projector. We have

$$\mathscr{D}^E(P^{2_4}) = \begin{pmatrix} 0 & 0 \\ 0 & 1 \end{pmatrix}, \qquad (5.6.8a)$$

$$\mathscr{D}^{T_2}(P^{2_4}) = \begin{pmatrix} 0 & 0 & 0 \\ 0 & 0 & 0 \\ 0 & 0 & 1 \end{pmatrix}, \qquad (5.6.8b)$$

from Eq. (5.6.7), (4.2.19), and (4.2.14). (Note: We shall use the T_2 irreps which follow directly from the T_1 by sign change of R and i matrices.) The position of the unit in the matrices (5.6.8) signals that the second and third components of E and T_2, respectively, are the right ones. The desired eigenstates for E will be given by

$$P_{e2}^E\left|\begin{matrix}2\\2\end{matrix}\right\rangle = \tfrac{1}{3}\sum_{R_l}\mathscr{D}_{e2}^E(R_l)R_l\left|\begin{matrix}2\\2\end{matrix}\right\rangle, \qquad (5.6.9a)$$

and for T_2 by

$$P_{t3}^{T_2}\left|\begin{matrix}2\\2\end{matrix}\right\rangle = \tfrac{1}{2}\sum_{R_l}\mathscr{D}_{t3}^{T_2}(R_l)R_l\left|\begin{matrix}2\\2\end{matrix}\right\rangle, \qquad (5.6.9b)$$

where $\mathcal{D}_{ij}^{\alpha}(R_l)$ are octahedral irreps of coset leaders R_l and

$$R_l \left| \begin{matrix} 2 \\ 2 \end{matrix} \right\rangle = \sum_m \mathcal{D}_{m2}^2(R_l(\alpha\beta\gamma)) \left| \begin{matrix} 2 \\ m \end{matrix} \right\rangle \quad (5.6.10)$$

depends on the R_3 $\mathcal{D}$ functions and the choice of R_l. Fortunately, there are only six cosets $R_l C_4$ of local subgroup $C_4 = (1, R_3, R_3^2, R_3^3)$. Hence, the sums over leaders R_l contain only six terms. Let us choose coset leaders $1 = R(000)$, $R_2^2 = R(0, \pi, 0)$, $R_1 = R(-\pi/2\pi/2\pi/2)$, $R_1^3 = R(-\pi/2 - \pi/2\pi/2)$, $R_2 = R(0, \pi/2, 0)$, and $R_2^3 = R(0, -\pi/2, 0)$. Each one of these leaders transforms a point on the local z axis onto one of the octahedral axes as explained in Section 4.3. The Euler angles $(\alpha\beta\gamma)$ are obtained by inspection of the octahedral operations in Figure 4.1.2 or by using axis-to-Euler conversion formulas (5.5.2). The representation components $\mathcal{D}_{m2}^2(\alpha\beta\gamma)$ for each of the leaders is written in the following using the $\mathcal{D}$ formula (5.4.21). Only the first column of the $\mathcal{D}$ matrix is required here. We use a single-parenthesis notation $\mathcal{D}(R = ($ to denote this.

$$\mathcal{D}_{0,2}^2(\alpha\beta\gamma) = \begin{pmatrix} e^{-2i(\alpha+\gamma)} \left(\dfrac{1+\cos\beta}{2} \right)^2 \\ e^{-i(\alpha+2\gamma)} \left(\dfrac{1+\cos\beta}{2} \right) \sin\beta \\ e^{-2i\gamma} \sqrt{\dfrac{3}{8}} \sin^2\beta \\ e^{i(\alpha-2\gamma)} \left(\dfrac{1-\cos\beta}{2} \right) \sin\beta \\ e^{2i(\alpha-\gamma)} \left(\dfrac{1-\cos\beta}{2} \right)^2 \end{pmatrix},$$

$$\mathcal{D}^2(1 = \begin{pmatrix} 1 \\ 0 \\ 0 \\ 0 \\ 0 \end{pmatrix}, \quad \mathcal{D}^2(R_2^2 = \begin{pmatrix} 0 \\ 0 \\ 0 \\ 0 \\ 1 \end{pmatrix}, \quad \mathcal{D}^2(R_1 = \begin{pmatrix} 1/4 \\ -i/2 \\ -\sqrt{3/8} \\ i/2 \\ 1/4 \end{pmatrix},$$

$$\mathcal{D}(R_1^3 = \begin{pmatrix} 1/4 \\ i/2 \\ -\sqrt{3/8} \\ -i/2 \\ 1/4 \end{pmatrix}, \quad \mathcal{D}(R_2 = \begin{pmatrix} 1/4 \\ 1/2 \\ \sqrt{3/8} \\ 1/2 \\ 1/4 \end{pmatrix}, \quad \mathcal{D}(R_2^3 = \begin{pmatrix} 1/4 \\ -1/2 \\ \sqrt{3/8} \\ -1/2 \\ 1/4 \end{pmatrix}.$$

$$(5.6.11)$$

ROTATIONAL LEVEL SPLITTING IN FINITE SYMMETRY 389

Substituting these into Eqs. (5.6.8)–(5.6.10) and using the T_2 octahedral irreps we find T_2 octahedral eigenstates:

$$\sum_R \mathscr{D}_{13}^{T_2}(R)\mathscr{D}^2(R = 0 + 0 + 0 + 0 - \begin{pmatrix} 1/4 \\ 1/2 \\ \sqrt{3/8} \\ 1/2 \\ 1/4 \end{pmatrix} + \begin{pmatrix} 1/4 \\ -1/2 \\ \sqrt{3/8} \\ -1/2 \\ 1/4 \end{pmatrix} = \begin{pmatrix} 0 \\ -1 \\ 0 \\ -1 \\ 0 \end{pmatrix},$$

$$\sum_R \mathscr{D}_{23}^{T_2}(R)\mathscr{D}^2(R = 0 + 0 + \begin{pmatrix} 1/4 \\ -i/2 \\ -\sqrt{3/8} \\ i/2 \\ 1/4 \end{pmatrix} - \begin{pmatrix} 1/4 \\ i/2 \\ -\sqrt{3/8} \\ -i/2 \\ 1/4 \end{pmatrix} + 0 + 0 = \begin{pmatrix} 0 \\ -i \\ 0 \\ i \\ 0 \end{pmatrix},$$

$$\sum_R \mathscr{D}_{33}^{T_2}(R)\mathscr{D}^2(R = \begin{pmatrix} 1 \\ 0 \\ 0 \\ 0 \\ 0 \end{pmatrix} - \begin{pmatrix} 0 \\ 0 \\ 0 \\ 0 \\ 1 \end{pmatrix} + 0 + 0 + 0 + 0 = \begin{pmatrix} 1 \\ 0 \\ 0 \\ 0 \\ 1 \end{pmatrix}.$$

Finally we normalize T_2 column vectors and write them in terms of $\left|{}_m^2\right\rangle$ kets:

$$\left|{}_1^{T_2}\right\rangle = \left(-\left|{}_1^2\right\rangle - \left|{}_{-1}^2\right\rangle\right)/\sqrt{2},$$

$$\left|{}_2^{T_2}\right\rangle = \left(-i\left|{}_1^2\right\rangle + i\left|{}_{-1}^2\right\rangle\right)/\sqrt{2},$$

$$\left|{}_3^{T_2}\right\rangle = \left(\left|{}_2^2\right\rangle - \left|{}_{-2}^2\right\rangle\right)/\sqrt{2}. \qquad (5.6.12)$$

By applying the octahedral E irreps in the same way one obtains

$$\left|{}_1^E\right\rangle = \left|{}_0^2\right\rangle,$$

$$\left|{}_2^E\right\rangle = \left(\left|{}_2^2\right\rangle + \left|{}_{-2}^2\right\rangle\right)/\sqrt{2}. \qquad (5.6.13)$$

It is instructive to examine the wave functions of the states (5.6.12) and (5.6.13) to see why their energy levels split and see the shape of the octahedral harmonics. The spherical harmonics $Y_m^l(\theta\phi)$ are functions of polar angles or dimensionless Cartesian coordinates:

$$x/r = \cos\phi \sin\theta, \qquad y/r = \sin\phi \sin\theta, \qquad z/r = \cos\theta.$$

For example, the Y_m^2 functions are as follows:

$$n_2\left\langle\theta\phi\bigg|\begin{array}{c}2\\2\end{array}\right\rangle = n_2 Y_2^2(\theta\phi) = (3/8)^{1/2} e^{2i\phi} \sin^2\theta = (3/8)^{1/2}(x+iy)^2/r^2,$$

$$n_2\left\langle\theta\phi\bigg|\begin{array}{c}2\\1\end{array}\right\rangle = n_2 Y_1^2 = -(3/2)^{1/2} e^{i\phi} \sin\theta \cos\theta = -(3/2)^{1/2}(x+iy)z/r^2,$$

$$n_2\left\langle\theta\phi\bigg|\begin{array}{c}2\\0\end{array}\right\rangle = n_2 Y_0^2 = (3\cos^2\theta - 1)/2 = (2z^2 - x^2 - y^2)/2r^2,$$

$$n_2\left\langle\theta\phi\bigg|\begin{array}{c}2\\-1\end{array}\right\rangle = n_2 Y_{-1}^2 = (3/2)^{1/2} e^{-i\phi} \sin\theta \cos\theta = (3/2)^{1/2}(x-iy)z/r^2,$$

$$n_2\left\langle\theta\phi\bigg|\begin{array}{c}2\\-2\end{array}\right\rangle = n_2 Y_{-2}^2 = (3/8)^{1/2} e^{-2i\phi} \sin^2\theta = (3/8)^{1/2}(x-iy)^2/r^2,$$

(5.6.14)

where a normalization factor $n_2 = (4\pi/5)^{1/2}$ is isolated. From Eqs. (5.6.12) and (5.6.13) the octahedral functions follows:

$$n_2\left\langle\theta\phi\bigg|\begin{array}{c}T_2\\1\end{array}\right\rangle = -n_2(Y_1^2 + Y_{-1}^2)/\sqrt{2} = (i\sqrt{3}/r^2)yz, \qquad (5.6.15a)$$

$$n_2\left\langle\theta\phi\bigg|\begin{array}{c}T_2\\2\end{array}\right\rangle = n_2(-iY_1^2 + iY_{-1}^2)/\sqrt{2} = (i\sqrt{3}/r^2)xz, \qquad (5.6.15b)$$

$$n_2\left\langle\theta\phi\bigg|\begin{array}{c}T_2\\3\end{array}\right\rangle = n_2(Y_2^2 - Y_{-2}^2)/\sqrt{2} = (i\sqrt{3}/r^2)xy, \qquad (5.6.15c)$$

$$n_2\left\langle\theta\phi\bigg|\begin{array}{c}E\\1\end{array}\right\rangle = n_2 Y_0^2 = (2z^2 - x^2 - y^2)/2r^2, \qquad (5.6.15d)$$

$$n_2\left\langle\theta\phi\bigg|\begin{array}{c}E\\2\end{array}\right\rangle = n_2(Y_2^2 + Y_{-2}^2)/\sqrt{2} = \sqrt{3}(x^2 - y^2)/2r^2. \qquad (5.6.15e)$$

Two of the octahedral wave functions are sketched in Figure 5.6.3. The second E wave function ($\psi_2^E \sim x^2 - y^2$) is drawn adjacent to the higher-energy E-doublet level, while the third T_2 wave function ($\psi_3^{T_2} \sim xy$) is sketched below. The energy levels are the eigenvalues of the lowest-degree octahedrally symmetric Hamiltonian or potential

$$V^{(4)} = D(x^4 + y^4 + z^4 - \tfrac{3}{5}r^4). \qquad (5.6.16)$$

The eigenvalues of $V^{(4)}$ and other operators will be computed in Chapter 7. However, it is easy to understand qualitatively why the states $\left|\begin{array}{c}E\\e\end{array}\right\rangle$ have higher-energy eigenvalues than those of states $\left|\begin{array}{c}T_2\\t\end{array}\right\rangle$. If one plots the equipo-

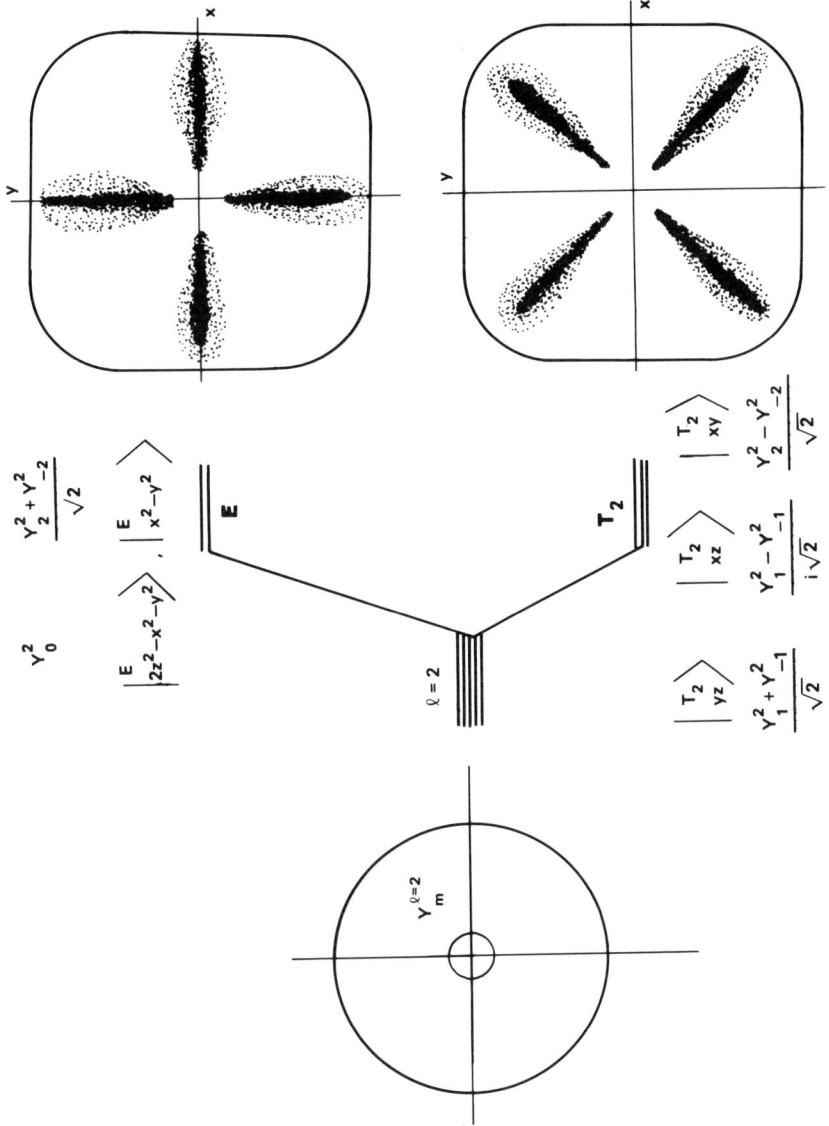

Figure 5.6.3 Detailed sketch of octahedral splitting of a d orbital. The wave functions $\left\langle\genfrac{}{}{0pt}{}{|E\rangle}{2}\right.$ and $\left\langle\genfrac{}{}{0pt}{}{|T_2\rangle}{3}\right.$ are sketched inside the equipotential contour $x^4 + y^4 =$ constant ($z = 0$).

tential surface $x^4 + y^4 + z^4 = 1$, then the xy contour assumes the supercircle shape of a rounded square as shown in Figure 5.6.3 (SUPERELLIPSE is the name given by the designer Piet Hein for any contour $x^n/a^n + y^n/b^n = 1$, where $n > 2$.) The T_2 waves avoid the higher potential regions by having their wave crest's lobes in the corners of the supercircle. The E waves have nodes near the T_2 crests and end up with their lobes pointing into the higher potential regions on the sides. We are studying here the generalization of the one-dimensional Bohr and Bloch waves and energy-band splitting. More discussion and examples will be given in Sections 5.6.D and 5.6.E.

C. Multipole Functions and Polynomials

Before continuing the discussion of waves and level splitting a short description of wave functions and polynomial classification will be given. Let us define the ELEMENTARY MULTIPOLE FUNCTIONS X_m by the following:

$$X_m^l \equiv \sqrt{\frac{4\pi}{2l+1}} \, (r)^l Y_m^l. \qquad (5.6.17)$$

The factor $(r)^l$ converts Y_m^l into a polynomial in x, y, and z of degree l. The X_m^l are listed as explicit polynomials of x, y, and z for $l = 1$–4 in the O_3 table of Appendix F. In this table one observes $(2l + 1) = 1, 3, 5, 7,$ and 9 new multipole functions appearing for each l. However, the corresponding number of independent lth-order monomials $x^a y^b z^c$ $(a + b + c = l)$ is generally greater: $(l + 1)(l + 2)/2 = 1, 3, 6, 10,$ and 15. Note on the right of the table that each lth-order monomial is a combination of X^l and products $X^{l-2}(r^2)$, $X^{l-4}(r^4)$ of lower-order multipole functions with $r^2 = x^2 + y^2 + z^2$.

Suppose you want to find how many independent combinations of lth-order monomials $x^a y^b z^c$ or multipole functions X_m^l exist which transform according to a given irrep $\mathcal{D}^\alpha$ of a given symmetry such as O_h. This problem can be approached in a number of different ways.

One direct approach uses coupling coefficients to make G symmetry-defined polynomials. This will be done in Section 6.3.C and the results are listed in the tables of Appendix F for order four or less.

On the other hand, one may use $O_3 \supset G$ correlation tables such as Eq. (5.6.5b) to see how many independent functions of each type come up for each l. Let us compare these two accountings.

For example, we may make one new scalar (A_{1g}) function of fourth-order [see the $l = 4$ row and A_1 column of Eq. (5.6.5b)] and one zeroth-order polynomial [see the $l = 0$ row and A_1 column of Eq. (5.6.5b)] or a constant multiplied by r^4. This means that there are just two linearly independent A_g^1

polynomials of fourth order. Indeed, the O_h table lists

$$IV_{(1)}^{A_{1g}} = x^4 + y^4 + z^4 \quad \text{and} \quad IV_{(2)}^{A_{1g}} = x^2y^2 + x^2z^2 + y^2z^2$$
$$= \left(r^4 - IV_{(1)}^{A_{1g}}\right)/2; \quad (5.6.18)$$

we have already encountered the first of these in Eq. (5.6.16).

Similarly, we may make two independent $\binom{E_g}{1}$ polynomials or two independent $\binom{E_g}{2}$ polynomials of fourth order. One set is new [see $l = 4$ row and E column of Eq. (5.6.5b)], and the other is a product of r^2 and the second-order polynomial [see $l = 2$ row and E column of Eq. (5.6.5b), which we derived in Eqs. (5.6.15d) and (5.6.15e)]. The E polynomials of fourth order in the O_h table of Appendix F are combinations of these two sets.

The relation between T_{1u} vector components (x, y, z) and $(l = 1)$ dipole functions

$$I_1^1 = -(x + iy)/\sqrt{2}, \quad I_0^1 = z, \quad I_{-1}^1 = (x - iy)/\sqrt{2} \quad (5.6.19)$$

is particularly important. We have seen this sort of relation several times before when relating standing-wave and moving-wave states. The $(l = 1)$-orbital level does not split in the presence of octahedral symmetry. Note that there is just one set of fourth-order polynomials of the T_1 or T_{1g} type. [One should not make the mistake of counting functions which are products of $r = (x^2 + y^2 + z^2)^{1/2}$ or r^3 with third-order $(l = 3)$ or first-order $(l = 1)T_{1u}$ functions, since these are not polynomials.]

Some O_h irreps have to wait even longer before they appear in polynomial form. According to the table of (5.6.5b) the first odd-l A_1 (i.e., A_{1u}) polynomial is of *ninth* $(l = 9)$ order.

The even-l A_1 (i.e., A_{1g}) polynomials are more plentiful. Totally invariant A_{1g} polynomials will be very important for tensor operator theory, which is given in Chapters 6 and 7. There we will learn that the number $n^{A_{1g}}(2k)$ of scalars or invariants found between $l = 0$ and $l = 2k$ is related to the kth correlation frequencies $f^\alpha(\mathcal{D}^k \downarrow M)$ as follows:

$$n^{A_{1g}}(2k) = \sum_{\substack{M \text{ irreps} \\ \alpha}} f^\alpha(f^\alpha + 1)/2. \quad (5.6.20)$$

For example, for angular momentum $k = 6$ we find the following number by summing over the seventh $(l = 6)$ row of the table of (5.6.5b):

$$n^{A_{1g}}(2k) = \sum_{\alpha = A_1, A_2, E, T_1, T_2} f^\alpha(\mathcal{D}^6 \downarrow O_h)(f^\alpha(\mathcal{D}^6 \downarrow O_h) + 1)/2$$
$$= 1 + 1 + 1 + 3 + 1 = 7 \quad (5.6.21)$$

This agrees with the sum of frequencies in the A_1 column of (5.6.5b) from $(l = 0)$ to $(l = 12)$. (Do not count the A_{1u} at $l = 9$ since that is a pseudoinvariant.)

Finally, note that the number of scalar cubic polynomials of order $2l$ is simply the number of partitions of l into 1, 2, or 3 terms. For example, 6 can be written seven ways: 6, 5 + 1, 4 + 2, 4 + 1 + 1, 3 + 3, 3 + 2 + 1, and

2 + 2 + 2 corresponding to seven scalar polynomials

$$x^{12} + y^{12} + z^{12}, x^{10}y^2 + x^{10}z^2 + x^{10}z^2, x^8y^4 + x^8z^4 + y^8z^4,$$
$$x^8y^2z^2 + x^2y^8z^2 + x^2y^2z^8, x^6y^6 + \cdots, \text{etc.} \quad (5.6.22)$$

This last counting method is a little too special to be of direct use for other symmetries besides O_h.

D. Multipole Expansions

In Section 5.6.C the ($l = 2$) eigenvectors of an octahedrally anisotropic potential $V^{(4)}$ were derived. A possible source of such a potential is drawn in Figure 5.6.4. There it is imagined that six point charges are brought in on the vertices of a crystal octahedron surrounding an orbiting electron. The electron is repelled by the charges. It finds the highest potential energy for a given radius r along the $\pm x$ (± 100), $\pm y$ (0 ± 10), and $\pm z$ (00 ± 1) axes where charges are sitting. It finds the lowest energy along the eight axes $(111), (-111), \ldots, (-1 - 1 - 1)$ in between.

To derive this anisotropic potential one needs to compute the following sum over the six charges:

$$\begin{aligned} V(r\theta\phi) &= \sum_{j=1}^{6} q_j / (|\mathbf{r}_j - \mathbf{r}|) \\ &= \sum_{j=1}^{6} q_j / \left[r_j^2 + r^2 - 2rr_j \cos \Theta_j \right]^{1/2} \\ &= \sum_{j=1}^{6} q_j / r_j \left[1 + (r/r_j)^2 - 2(r/r_j) \cos \Theta_j \right]^{1/2}, \quad (5.6.23) \end{aligned}$$

where Θ_j is the angle between the electronic radius r and radius r_j of charge q_j. (See Figure 5.6.4.) This can be rewritten using the Legendre generating function

$$1/(1 + h^2 - 2hz)^{1/2} = \sum_{l=0}^{\infty} h^l P_l(z) \quad (5.6.24)$$

and the addition theorem (5.5.75). The result is

$$\begin{aligned} V(r\theta\phi) &= \sum_j \sum_l q_j \left(r^l / r_j^{l+1} \right) P_l(\cos \Theta_j) \\ &= \sum_l \sum_{m=-l}^{l} y_m^l Y_m^l(\theta\phi) = \sum_l \sum_{m=-l}^{l} x_m^l X_m^l(\mathbf{r}_j), \quad (5.6.25a) \end{aligned}$$

where

$$x_m^l = (2l + 1/4\pi)^{1/2} y_m^l / r^l = \sum_j \left(q_j / r_j^{2l+1} \right) X_m^{l*}(\mathbf{r}_j) \quad (5.6.25b)$$

are constant coefficients which depend upon the charge locations. This is called the MULTIPOLE EXPANSION of a change distribution. Consider some examples for the charge octahedron with $r_j = R$ and $q_j = Q$. For $l = 0$

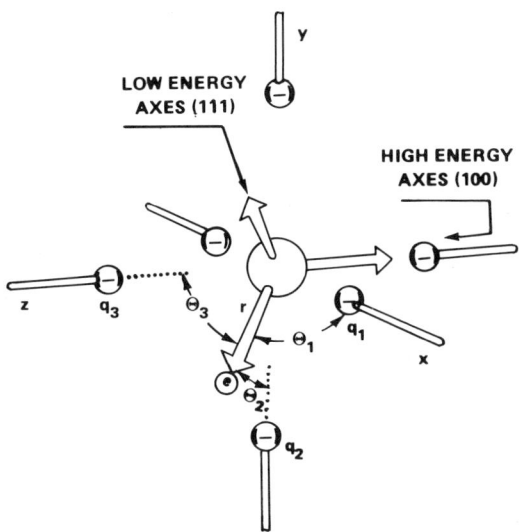

Figure 5.6.4 Octahedral arrangement of charge centers. Potential energy depends on the polar angles Θ_j between the electron and fixed charges on the octahedral axes $(\pm 1\ 0\ 0)$, $(0\ \pm 1\ 0)$, and $(0\ 0\ \pm 1)$. The least energy is had when the position vector $\mathbf{r}$ is farthest away from these axes.

we have

$$x_0^0 = (4\pi)^{1/2} 6Q/R. \qquad (5.6.26)$$

For $l = 1$, 2, and 3 the sum (5.6.25b) contributes nothing. Indeed, the $O_h \subset O_3$ correlation (5.6.5b) has told us that the next octahedral invariant will be at $l = 4$. Using the X_m^4 functions in Appendix F we find only the following coefficients are nonzero:

$$x_0^4 = (Q/8)(3R^4 + 3R^4 + 3R^4 + 3R^4 + 8R^4 + 8R^4) = (7Q/2)R^4,$$
$$x_4^4 = x_{-4}^4 = Q(35/128)^{1/2}(R^4 + R^4 + R^4 + R^4 + 0 + 0) = (\sqrt{70}\,Q/4)R^4. \qquad (5.6.27)$$

Hence the octahedral multipole expansion has the form

$$V(r\theta\phi) = (6Q/R) + (35Q/4R^5)\left[(2/\sqrt{70})(X_4^4 + X_{-4}^4) + \tfrac{2}{5}X_0^4\right] + \cdots$$
$$= (6Q/R) + (35Q/4R^5)\left[x^4 + y^4 + z^4 - \tfrac{3}{5}r^4\right] + \cdots \qquad (5.6.28)$$

if we consider only terms up to the fourth order. Sixth- and higher-order terms will be discussed in Chapter 7. The first $(l = 0)$ term in the foregoing gives the potential at the center of the octahedron. The second term contains

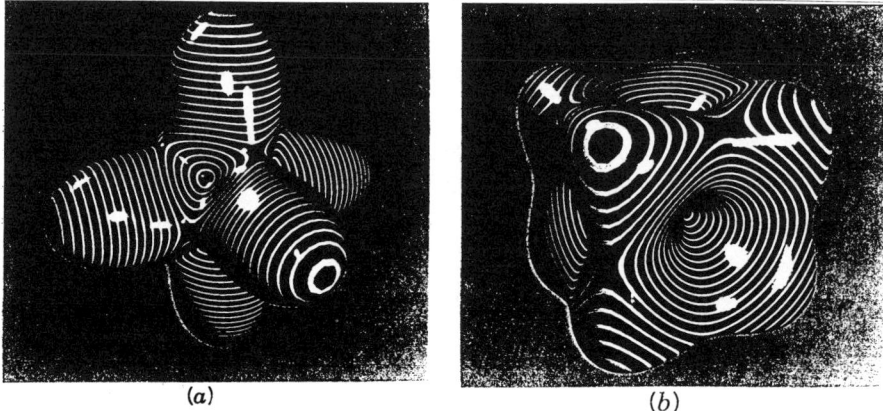

Figure 5.6.5 Octahedrally anisotropic potential surfaces of fourth-degree. (a) This surface corresponds to repulsive charges placed on the octahedral axes, while (b) corresponds to attractive charges.

the potential $V^{(4)}$ from Eq. (5.6.16). It describes the octahedral potential anisotropy just outside the center. As long as x, y, and z remain small compared to R the higher-order multipole expansion terms can be neglected.

In order to visualize the fourth-order anisotropy $V^{(4)}$ it is useful to plot it as a function of (θ, ϕ) for fixed r. Expressing $V^{(4)}$ in terms of spherical harmonics gives

$$V^{(4)} = (4\pi)^{1/2} r^4 \left[(2/\sqrt{70})(Y_4^4 + Y_{-4}^4) + \tfrac{2}{5} Y_0^4 \right]/3$$
$$= r^4 \left[35 \cos^4 \theta - 30 \cos^2 \theta + 3 + 5 \sin^4 \theta \cos 4\phi \right]/20. \quad (5.6.29)$$

The function $(a + bV^{(4)})$ is plotted radially in Figure 5.6.5 as a function of θ and ϕ for $r = 1$, $a = 1$, and $b = \pm 5$. The positive ($b = 5$) value in Figure (5.6.5a) corresponds to repulsive charges while the negative ($b = -5$) value in Figure (5.6.5b) corresponds to the attractive charges. The scale factors (a, b) are chosen to clearly exhibit the anisotropic mountain and valley features while keeping the function $(a + bV^{(4)})$ positive. To visualize the potential imagine you are walking on a planet of the shape $(a + bV^{(4)})$. Then mountains and valleys are high and low potential directions, respectively. This type of spherical plot will be used for studying other multipole functions in later chapters.

E. Level Splitting for Molecular Rotors

Molecular rotational levels may split in a way that is analogous to crystal field splitting. Consider, for example, an octahedral SF_6 molecule which is sketched

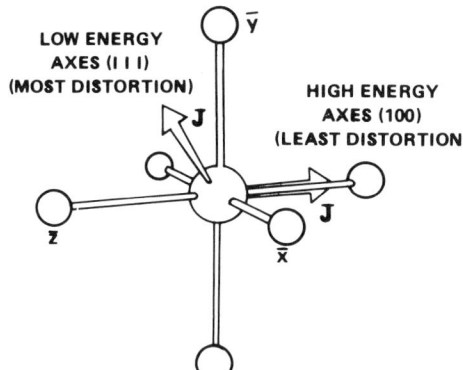

Figure 5.6.6 Octahedral rotor. Centrifugal distortion energy depends on the direction of the angular-momentum vector J relative to the octahedral axes $(\pm 1\ 0\ 0)$, $(0\ \pm 1\ 0)$, and $(0\ 0\ \pm 1)$. Greatest distortion and the least energy is had when the J vector is farthest away from these axes.

in Figure 5.6.6. If this molecule were rigid the Hamiltonian would be that of a spherical rotor as described in Eq. (5.5.48). However, the SF_6 molecule has finite spring constants and will be distorted by centrifugal forces when rotating. In Chapter 4 we showed that the radial bonds were several times stronger than the bending bonds. Therefore, we expect less centrifugal distortion when the angular-momentum vector (**J**) points along the $\bar{x}$, $\bar{y}$, or $\bar{z}$ axes than in between. If **J** is along a (111) direction as shown in Figure 5.6.6 the centrifugal distortion should be maximum. Then the centrifugal forces would exert the greatest leverage on all the bending bonds at once. Expansion causes the inertia of the molecule to increase and the energy to be lower. Hence, for a given magnitude of **J** the threefold (111) axes correspond to energy minima. If **J** points along the $\bar{x}$, $\bar{y}$, or $\bar{z}$ axes there will only be a small expansion allowed by four strong radial bonds. The fourfold symmetry axes correspond to energy maxima.

The lowest-order perturbation operator which describes anisotropic SF_6 centrifugal distortion is

$$V^{(4)}_{\text{cent}} = t\left(J_{\bar{x}}^4 + J_{\bar{y}}^4 + J_{\bar{z}}^4 - \tfrac{3}{5}J^4\right). \qquad (5.6.30)$$

This is the body-based angular-momentum analog of the crystal potential (5.6.16). If you let the fourfold axes in figure 5.6.5b be $J_{\bar{x}}$, $J_{\bar{y}}$, and $J_{\bar{z}}$ axes, then the figure is an angular plot of energy versus direction of **J** for fixed magnitude $|J| = (J_{\bar{x}}^2 + J_{\bar{y}}^2 + J_{\bar{z}}^2)^{1/2}$. For a molecule rotating freely in a laboratory vacuum the classical angular momentum is constant. However, the direction of angular momentum in the body frame may precess because of the octahedral anisotropy. An SF_6 Hamiltonian of the form $H_{\text{sphere}} + V^{(4)}_{\text{cent}}$ has $O_3 \times \bar{O}_h$ symmetry; the lab has full orthogonal O_3 symmetry, while the body symmetry is reduced to octahedral O_h symmetry of the internal structure.

The eigenstates of the octahedral centrifugal distortion Hamiltonian will have wave functions of the form

$$r_{mb}^{NB} = \sum_{n=-N}^{N} \left(\begin{array}{c|c} N & B \\ n & b \end{array} \right) r_{mn}^{N}$$

$$= \sum_{n} \left(\begin{array}{c|c} N & B \\ n & b \end{array} \right) \mathscr{D}_{mn}^{N*}(\alpha\beta\gamma)(2N+1)^{1/2}, \qquad (5.6.31)$$

where $\begin{pmatrix} B \\ b \end{pmatrix}$ stands for octahedral symmetry irrep labels $\begin{pmatrix} A_1 \\ \cdot \end{pmatrix}$, $\begin{pmatrix} A_2 \\ \cdot \end{pmatrix}$, $\begin{pmatrix} E \\ 1 \end{pmatrix}$, $\begin{pmatrix} E \\ 2 \end{pmatrix}$, $\begin{pmatrix} T_1 \\ 1 \end{pmatrix}, \ldots, \begin{pmatrix} T_2 \\ 3 \end{pmatrix}$, and N is our new notation for total orbital momentum of the nuclei in the rotating molecule. Note that only the body momentum component n is being summed. The lab component m is still a good quantum number as long as no external crystal fields exist in the lab.

In Section 5.6.C we introduced methods which gave octahedral crystal field harmonics,

$$\left\langle \theta\phi \Big| \begin{array}{c} B \\ b \end{array} \right\rangle = \sum_{m} \left(\begin{array}{c|c} N & B \\ m & b \end{array} \right) Y_m^N(\theta\phi), \qquad (5.6.32)$$

in terms of spherical hormonics. [Recall Eq. (5.6.15).] The same procedures can be used to derive the octahedral rotor functions r_b^B in terms of the body-defined angular-momentum functions r_n^N. Body-defined symmetry transformations obey the rules

$$\bar{R}(\alpha\beta\gamma) \left| \begin{array}{c} N \\ mn \end{array} \right\rangle = \sum_{n'} \mathscr{D}_{n'n}^{N*}(\alpha\beta\gamma) \left| \begin{array}{c} N \\ mn' \end{array} \right\rangle \qquad (5.6.33)$$

according to Eq. (5.5.40). The only difference in form between body- and lab-defined transformation matrices is the complex conjugation ($*$) symbol. Also, that lab transformations (5.5.31) sum over the m's while the body transformations sum over n's. Hence, the coefficient derived for crystal field problems are related to the rotor coefficients through complex conjugation:

$$\left(\begin{array}{c|c} N & B \\ n & b \end{array} \right) = \left(\begin{array}{c|c} N & B \\ n & b \end{array} \right)^*. \qquad (5.6.34)$$

The symbol B^* will be used to label the irrep $\mathscr{D}^{B^*} = (\mathscr{D}^B)^*$ of the finite subgroup. In case we are using real octahedral irreps we have $B^* = B$. For example, the E rotor eigenfunctions,

$$r_{m1}^{2E} = r_{m0}^{2},$$

$$r_{m2}^{2E} = \left(r_{m2}^{2} + r_{m-2}^{2} \right)/\sqrt{2}, \qquad (5.6.35)$$

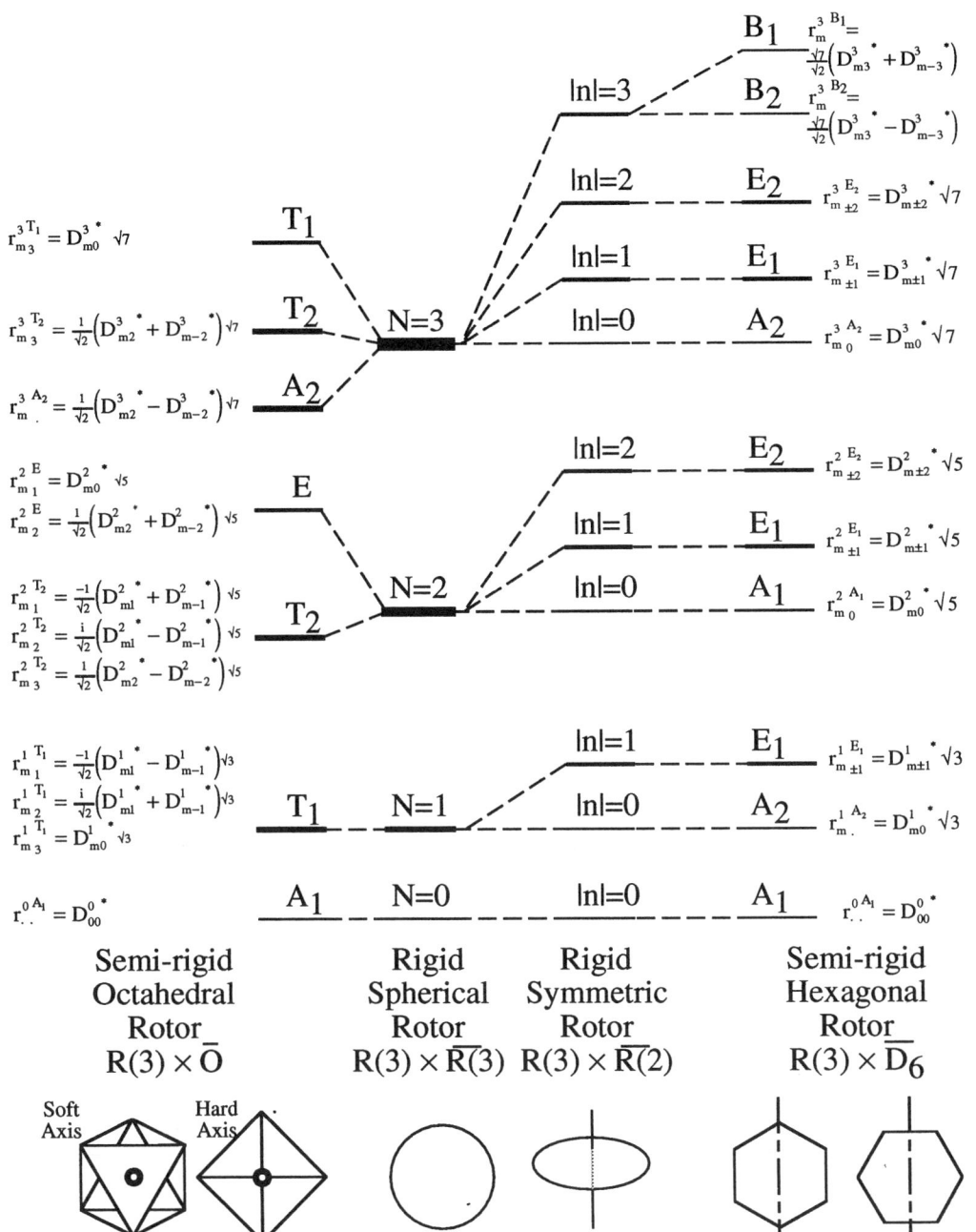

Figure 5.6.7 (a) Correlation of lowest levels of rigid spherical and symmetric rotors with semirigid rotors. (b) Centrifugal distortion of semirigid rotors. Amount of distortion depends on direction of rotation axis. Some directions are softer than others. This directional anisotropy must have the symmetry of the rotor. Hence the levels and wave functions must be defined by representations of that symmetry as shown in the figure.

follow directly from Eqs. (5.6.15d) and (5.6.15e), while the T_2 eigenfunctions

$$r_{m1}^{2\,T_2} = -(r_{m1}^2 + r_{m-1}^2)/\sqrt{2},$$

$$r_{m2}^{2\,T_2} = i(r_{m1}^2 - r_{m-1}^2)/\sqrt{2},$$

$$r_{m3}^{2\,T_2} = (r_{m2}^2 - r_{m-2}^2)/\sqrt{2}. \tag{5.6.36}$$

follow from Eqs. (5.6.15a)–(5.6.15c) after conjugation.

Figure 5.6.7 shows the ($N = 2$) splitting of E and T_2 on the left-hand side. The splitting is analogous to the crystal field splitting in Figure 5.6.3. The E levels favor the high-energy axes while the T_2 levels favor the low-energy axes. The latter correspond to greater distortion, as indicated at the bottom of Figure 5.6.7. Note that the energy splittings and the distortion are greatly exaggerated in the figure. Generally, the splittings are a small fraction of the spacing between N levels. Neither the splittings nor the distortions would be visible in drawing of this scale.

The $N = 3$ splittings are sketched on the upper left-hand side of Figure 5.6.7. The correlations (5.6.5b) predict an A_2, T_2, and T_1 level, though not necessarily in that order. The derivation of $r_{mb}^{3\,B}$ wave functions is left as an exercise. Note that all the levels shown in Figure 5.6.7 have an additional $(2N + 1)$-fold lab degeneracy which is not indicated. For example, each ($N = 2$) sublevel is fivefold degenerate with $m = 2, 1, 0, -1, -2$. To see these levels one needs to apply a lab-based perturbation such as a crystal or Zeeman field.

F. $R_3 \supset D_6$ Correlations and Level Splitting

The splittings of N levels appropriate for a hexagonal ring molecule are shown on the right-hand side of Figure 5.6.7. First the body-based symmetry is reduced from $O_3 \supset O_2$ corresponding to the rigid symmetric rotor. Then $|n|$ is a good quantum number and the levels split according to n^2 from Eq. (5.5.52). [Note that a benzenelike molecule has $(C - B) < 0$ so the levels would actually decrease with n^2.] Finally, the reduction of body symmetry from O_2 to D_6 is indicated on the extreme right-hand side of Figure 5.6.7. This could be caused by a hexagonally anisotropic centrifugal distortion potential.

The $O_3 \supset D_6$ correlations may be constructed in the same way as they were for $O_3 \supset O_h$. The D_6 characters derived and labeled in Eq. (3.6.17).

They lead to the following correlations:

	f^{A_1}	f^{A_2}	f^{E_2}	f^{B_1}	f^{B_2}	f^{E_1}
$l = 0$	1	·	·	·	·	·
1	·	·	·	·	·	1
2	1	·	1	·	·	1
3	·	1	1	1	1	1
4	1	·	2	1	1	1
5	·	1	2	1	1	2
6	2	1	2	1	1	2
7	1	2	2	1	1	3
8	2	1	3	1	1	3
9	1	2	3	2	2	3
10	2	1	4	2	2	3

(5.6.37)

It is helpful to try to understand these splittings physically. One way to do this involves the simpler Bohr orbitals and Bloch waves in Figure 3.6.5. There it is easier to see how $m = 0$, ± 1, and ± 2 waves become A, E_1, and E_2 levels, respectively, when a D_6 symmetric perturbation is present. The same sort of labeling is done for the $n = 0$, ± 1, and ± 2 levels on the right-hand side of Figure 5.6.7. Of course, the $|n|$ values can only be as high as the total momentum number N for the splitting of each N level. The ($N = 3$) level in Figure 5.6.7 has an $|n| = 3$ sublevel which splits into B_1 and B_2 levels under D_6 symmetric perturbations. These levels correspond to the first Brillouin-zone standing waves in Figure 3.6.5(a), i.e., sine and cosine waves. The crests of the cosine wave and the nodes of the sine wave stand in the potential valleys. This makes the latter have higher potential energy than the former. Since either wave has the same kinetic energy their levels split. The B_1 and B_2 rotor wave functions for $N = 3$ are

$$r_m^{3\,B_1} = (r_{m3}^3 + r_{m-3}^3)/\sqrt{2},$$
$$r_m^{3\,B_2} = (r_{m3}^3 - r_{m-3}^3)/\sqrt{2}. \quad (5.6.38)$$

In order to determine which waves are B_1 or B_2 one must specify the 180° p axis according to Eqs. (3.6.23 and 3.6.24). Whenever possible we shall let one of the p axes be the $\bar{y}$ axis so the R_3 matrix $\mathscr{D}^N$ (0 180° 0) is real. Substituting ($\alpha = 0$, $\beta = 180°$, $\gamma = 0$) into Eq. (5.4.45) yields

$$\mathscr{D}_{mn}^j(0\ 180°\ 0)$$
$$= \sum_k (-1)^k \frac{[(j+n)!(n-n)!(j+m)!(j-m)!]^{1/2}(0)^{2j+m-n-2k}}{(j+m-k)!(n-m+k)!k!(j-n-k)!}.$$

(5.6.39)

The only nonzero terms are those for which the exponent is zero; i.e., $k = j + (m - n)/2$. To keep both the denominator factors $(j + m - k) = (m + n)/2$ and $(j - n - k) = -(m + n)/2$ from being negative it is necessary to have

$$m + n = 0.$$

This yields

$$\mathscr{D}^N_{mn}(0\ 180°\ 0) = \delta_{n,\,-m}(-1)^{N+m}. \tag{5.6.40}$$

Note that for odd-N and $n = 0$ this component is negative. Therefore the $n = 0$ component of the $N = 3$ level belongs to A_2:

$$r_m^{3\,A_2} = r_{m0}^3. \tag{5.6.41}$$

The rings in Figure 5.6.8 serve as mnemonics for the $R_3 \downarrow D_6$ level splitting. The D_6 levels contained in a particular $N = J$ level lie between the arrows which are labeled with that momentum number. For example, we read $A_2 E_1 E_2 B_2 B_1$ between the $(J = 3)$ arrow outside and the $(J = 3)$ arrow on the inside of the odd-J circle. For $J = 6n + p$ ($p < 6$ and $n = 1, 2, 3, \ldots$) one must read around the circle n times before stopping. For example, $J = 8$ gives $A_1 E_1 E_2 B_1 B_2 E_2 E_1 A_2 A_1 E_1 E_2$.

For low N in most molecules the ordering given by the wheel will be correct. Furthermore, the levels $(A_1 A_2)$ or $(B_1 B_2)$ enclosed by inside teeth should be very nearly degenerate. However, this may not be the case for high N. If the centrifugal potential valleys become deep enough they may trap the angular-momentum vector on the twofold symmetry axes. Then the levels will

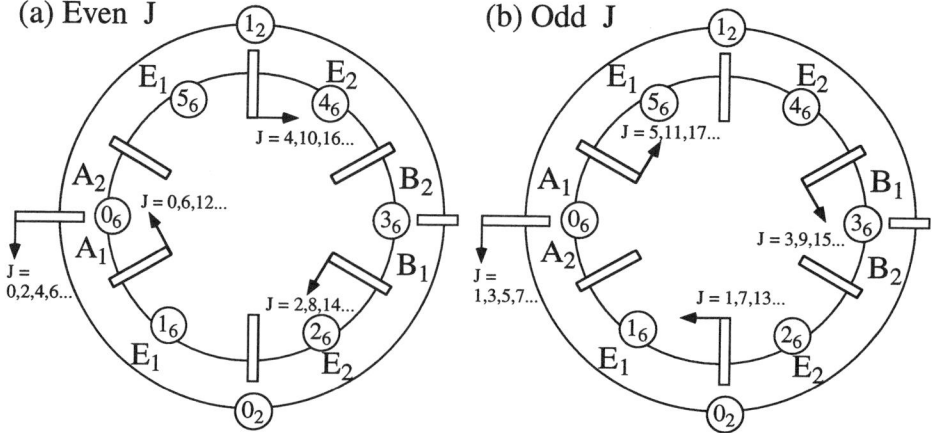

Figure 5.6.8 Mnemonic wheels for hexagonal-D_6 orbital splitting of J levels for (a) even J and (b) odd J.

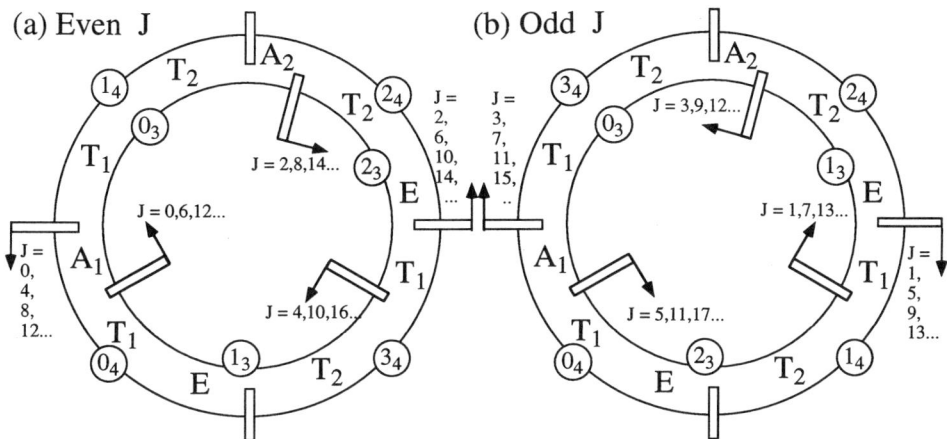

Figure 5.6.9 Mnemonic wheels for octahedral-O orbital. Splitting of J levels for (a) even J and (b) odd J.

fall into clusters $(A_1 E_1 E_2 B_1)$ or $(A_2 E_1 E_2 B_2)$ as they did in Figure 3.6.5(b). The clusters are indicated by the outside teeth on the mnemonic wheels.

Mnemonic wheels can be constructed easily for any D_n symmetry by appealing to the wave mechanics described in Sections 3.6.A and 3.6.B. Each wheel must contain the sequences $\{A_1 E_1 E_2 \cdots E_{n/2-1} B_1\}$ and $\{B_2 E_{n/2-1} \cdots E_2 E_1 A_2\}$ for even n, or $\{A_1 E_1 \cdots E_{(n-1)/2}\}$ and $\{E_{(n-1)/2} \cdots E_1 A_2\}$ for odd n.

One of the surprising results of modern spectroscopy is the existence of similar mnemonic wheels for octahedral symmetry. In fact the correlation table of (5.6.5) can be replaced by the wheels in Figure 5.6.9. The octahedral wheels are read in the same way as the D_n wheels. For example, the T_2 and E states lie between the ($J = 2$) arrows in Figure 5.6.9(a), and the A_2, T_1, and T_2 states lie between the ($J = 3$) arrows in Figure 5.6.9(b). The examples agree with the $\mathscr{D}^2 \downarrow O$ and $\mathscr{D}^3 \downarrow O$ correlations. Furthermore, the wheels give *correct ordering* for the energy levels of the $V^{(4)}$ Hamiltonians (5.6.29) or (5.6.30). This is very convenient to have for large angular-momentum J. (The ordering was not discovered until large-J computer calculations were done by Los Alamos researchers in 1976.) For large J the teeth on the wheels indicate which levels are clustered, as explained in Chapter 7. The a_3 and a_4 teeth indicate clusters belonging to induced representations $a_3 \uparrow O$ and $a_4 \uparrow O$. This is also analogous to the D_6 clusters which belong to $a_2 \uparrow D_6$ or $a_6 \uparrow D_6$ induced representations.

For $J = 2$ only part of a cluster appears. If there has been *six* states satisfying (2_4) local symmetry conditions (5.6.7) then the induced representation

$$(2_4) \uparrow O \sim A_2 \oplus T_2 \oplus E$$

404 REPRESENTATIONS OF CONTINUOUS ROTATION GROUPS AND APPLICATIONS

would result according to correlation (4.2.42b). However, A_2 is left out of the five ($J = 2$) states computed in Eqs. (5.6.15). The physical interpretation of clusters and induced representations will be taken up again in Chapter 7.

5.7 HALF-INTEGER j-LEVEL SPLITTING IN FINITE SYMMETRY

We have mentioned the peculiar double-valued transformation behavior of half-integral $j = \frac{1}{2}, \frac{3}{2}, \frac{5}{2}, \ldots$ angular-momentum states $|^j_m\rangle$. Any system that is composed of an odd number of electrons whose spins play some physical role will have to be described by this type of state. If such a system is put into a finite symmetry environment a peculiar splitting of the levels results. We develop the theory of this here, beginning with hexagonal D_6 symmetry as an example.

A. Ray Representations of D_6

Let us consider two methods for determining the explicit multivalued symmetry properties of half-integral spin representations. In the first method one simply calculates the $\mathscr{D}^{1/2}(R)$ representation using either Euler angles [recall Eq. (5.4.16)] or Darboux angles [recall Eq. (5.5.1)] and then forms all products of finite symmetry rotations. The second method uses Hamilton turns to obtain products geometrically. We shall compare the two methods while discussing the finite symmetry D_6.

Consider, for example, the two 180° rotations ρ_2 and ρ'_3 around transverse axes shown in Figure 3.6.4. The Darboux angles of ρ'_3 are ($\phi = \pi/2$, $\theta = \pi$, $\omega = \pi$) while ρ_2 has angles ($\phi = \pi/3$, $\theta = \pi$, $\omega = \pi$). Substituting these into Eq. (5.5.1) gives

$$\mathscr{D}^{1/2}(\rho'_3) = \begin{pmatrix} 0 & -1 \\ 1 & 0 \end{pmatrix} \quad \text{and} \quad \mathscr{D}^{1/2}(\rho_2) = \begin{pmatrix} 0 & -ie^{-i\pi/3} \\ -ie^{i\pi/3} & 0 \end{pmatrix}. \tag{5.7.1}$$

The products $\rho'_3 \rho_2$ and $\rho_2 \rho'_3$ are as follows:

$$\mathscr{D}^{1/2}(\rho'_3)\mathscr{D}^{1/2}(\rho_2) = \begin{pmatrix} ie^{i\pi/3} & 0 \\ 0 & -ie^{-i\pi/3} \end{pmatrix} = -\begin{pmatrix} e^{-i\pi/6} & 0 \\ 0 & e^{i\pi/6} \end{pmatrix}$$

$$\mathscr{D}^{1/2}(\rho_2)\mathscr{D}^{1/2}(\rho'_3) = \begin{pmatrix} -ie^{-i\pi/3} & 0 \\ 0 & ie^{i\pi/3} \end{pmatrix} = \begin{pmatrix} e^{-i5\pi/6} & 0 \\ 0 & e^{i5\pi/6} \end{pmatrix} \tag{5.7.2}$$

These results may be expressed in terms of z rotations h or h^5 for which the

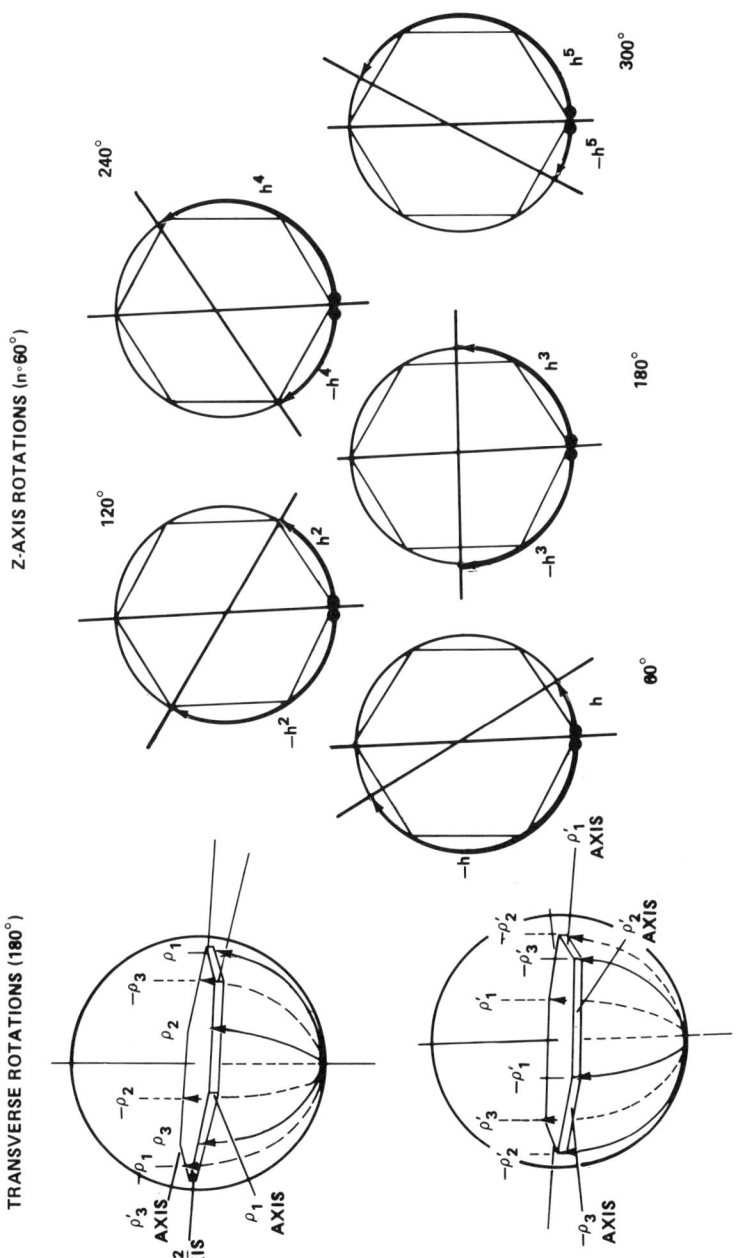

Figure 5.7.1 Hamilton-turn vectors for D_6 spin algebra.

406 REPRESENTATIONS OF CONTINUOUS ROTATION GROUPS AND APPLICATIONS

Darboux angles are ($\theta = 0 = \phi$) and $\omega = \pi/3$ or $5\pi/3$, respectively:

$$\mathscr{D}^{1/2}(\rho'_3)\mathscr{D}^{1/2}(\rho_2) = -\mathscr{D}^{1/2}(h), \qquad \mathscr{D}^{1/2}(\rho_2)\mathscr{D}^{1/2}(\rho'_3) = \mathscr{D}^{1/2}(h^5). \tag{5.7.3}$$

Notice the minus sign in the first product. The sign changes which appear in spin-$\frac{1}{2}$ representations make a big difference in the form of the eigenstates and group transformation properties of half-integral spin states.

Now it is possible to derive, and to a certain extent, visualize all such products using Hamilton-turn vector addition. One first assigns Hamilton-turn vectors to all the D_6 rotations as in Figure 5.7.1 using the rules given in Sections 3.1.B and 5.5.A. Each rotation by angle ω is replaced by sequential reflections through planes intersecting by angle $\omega/2$. A turn vector is a great circle arc drawn from the first plane to the second so that it is orthogonal to the planes. Let us call the vector g positive $(+g)$ or negative $(-g)$ if it goes counterclockwise or clockwise, respectively, when viewed from the chosen axis of rotation. For example, the 30° counterclockwise arc $(+h)$ in Figure 5.7.1 represents the 60° counterclockwise rotation operator h of D_6. The 150° clockwise arc $(-h)$ represents a 300° clockwise rotation. Ordinarily, this would be the same operation as h, but its effect on half-integral spin states gives the opposite phase. Hence, it is labeled $(-h)$.

The products $\rho'_3\rho_2$ and $\rho_2\rho'_3$ are performed "vectorially" in Figure 5.7.2. The operator that acts first (i.e., ρ_2 in $\rho'_3\rho_2$) has its head positioned so that it

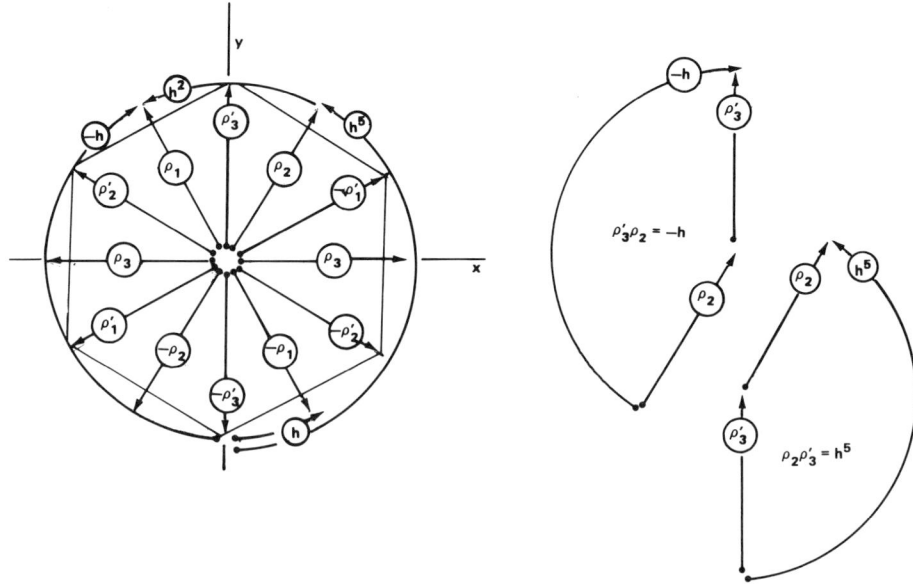

Figure 5.7.2 D_6 nomogram and example of products $(\rho'_3\rho_2 = -h)$ and $(\rho_2\rho'_3) = h^5$.

may meet the tail of the vector for the second factor. (Turn vectors can be moved anywhere on their great circles.) Then the vector sum is identified for each product.

Using either the vector sum or the matrix methods, we derive a rather different set of rules for D_6 operators which is given in the following multiplication table:

	1	h^2	h^4	ρ_1	ρ_2	ρ_3	h^3	h	h^5	ρ'_1	ρ'_2	ρ'_3
h^4		-1	$-h^2$	$-\rho_2$	$-\rho_3$	ρ_1	$-h$	h^5	$-h^3$	$-\rho'_2$	$-\rho'_3$	ρ'_1
h^2		h^4	-1	$-\rho_3$	ρ_1	ρ_2	h^5	h^3	$-h$	$-\rho'_3$	ρ'_1	ρ'_2
ρ_1		ρ_2	ρ_3	-1	$-h^2$	$-h^4$	$-\rho'_1$	ρ'_3	$-\rho'_2$	h^3	h^5	$-h$
ρ_2		ρ_3	$-\rho_1$	h^4	-1	$-h^2$	$-\rho'_2$	$-\rho'_1$	$-\rho'_3$	h	h^3	h^5
ρ_3		$-\rho_1$	$-\rho_2$	h^2	h^4	-1	$-\rho'_3$	$-\rho'_2$	ρ'_1	$-h^5$	h	h^3
h^3		h^5	$-h$	ρ'_1	ρ'_2	ρ'_3	-1	h^4	$-h^2$	$-\rho_1$	$-\rho_2$	$-\rho_3$
h^5		$-h$	$-h^3$	$-\rho'_3$	ρ'_1	ρ'_2	$-h^2$	-1	$-h^4$	ρ_3	$-\rho_1$	$-\rho_2$
h		h^3	h^5	ρ'_2	ρ'_3	$-\rho'_1$	h^4	h^2	-1	$-\rho_2$	$-\rho_3$	ρ_1
ρ'_1		ρ'_2	ρ'_3	$-h^3$	$-h^5$	h	ρ_1	$-\rho_3$	ρ_2	-1	$-h^2$	$-h^4$
ρ'_2		ρ'_3	$-\rho'_1$	$-h$	$-h^3$	$-h^5$	ρ_2	ρ_1	ρ_3	h^4	-1	$-h^2$
ρ'_3		$-\rho'_1$	$-\rho'_2$	h^5	$-h$	$-h^3$	ρ_3	ρ_2	$-\rho_1$	h^2	h^4	-1

(5.7.4)

The table shown is not that of a group, since negative $(-g)$ operators are not elements of a group. The ordinary group table of D_6 is obtained if we drop all the minus signs.

Two procedures exist for analyzing this new mathematical structure with the minus signs. The first procedure defines operators $-1, -r, -r^2, \ldots, -\rho'_3$ to be elements of a group $\{1, r, r^2, \ldots, \rho'_3, -1, -r, -r^2, \ldots, -\rho'_3\}$ called the DOUBLE GROUP or COVERING GROUP which is twice as large as D_6. This involves a multiplication table with four times the number of entries shown in (5.7.4). Some of the irreps of this group can be related to the half-integral rotation matrices.

The second procedure, which we shall follow, treats a table such as that of (5.7.4) directly, in an *algebra*. The advantage of this is that, instead of doubling or quadrupling our arithmetic, we find that it is reduced to about one-half that of the original group. This makes half-integral spin analysis half as much work instead of double trouble!

There are other physical problems that require representations which obey equations of the form

$$\mathscr{D}(g)\mathscr{D}(h) = \omega_{g,h}\mathscr{D}(gh). \qquad (5.7.5)$$

These are called RAY REPRESENTATIONS or PROJECTIVE REPRESENTATIONS of the group $\mathscr{G} = \{\cdots g \cdots h \cdots\}$. The theory of these is well worth the time it saves in a multitude of calculations.

In the present D_6 example the problem is to find irreducible representations of the algebra defined by the table of (5.7.4). This will be equivalent to finding irreducible *ray* representations (irreps) of D_6. It is necessary to reduce the algebra to a combination of irreducible P operators. This reduction occurs in two stages as it did for ordinary group algebras in Sections 3.2 and 3.3, respectively.

The first stage involves all-commuting operators,

$$C = \sum_g \gamma_g g, \tag{5.7.6}$$

which satisfy

$$Cg = gC$$

or

$$g^{-1}Cg = C, \tag{5.7.7}$$

for all g. Note here that the inverse g^{-1} is defined for the new algebra so that $g^{-1}g = 1 = gg^{-1}$, and may not be the same as the group inverse. For example, in the D_6 group the inverses of h and ρ_1 are $h^{-1} = -h^5$ and $\rho_1^{-1} = -\rho_1 = \rho_1^3$, respectively, according to the table of (5.7.4). All commuting operators in spinor algebras are linear combinations of operators of the form

$$c_h = ({}^\circ N_h / {}^\circ G) \sum_g ghg^{-1}, \tag{5.7.8}$$

where ${}^\circ N_h$ is the number of operators which commute with h and ${}^\circ G$ is the order of group G. The c_h operators are analogous to the class sums introduced in Section 3.2. However, they are quite different for the spin algebra than for the group. For example, for the D_6 spin algebra only the three c_h operators are linearly independent:

$$c_1 = 1, \quad c_2 = h^2 - h^4, \quad c_3 = h - h^5. \tag{5.7.9}$$

The others are either zero ($c_\rho = c_{\rho'} = c_{h^3} = 0$) or proportional to one of these three operators.

The algebra of the all-commuting operators $\{c_1, c_h, c_{h^2}\}$ is determined by the following multiplication rules:

	$c_1 = 1$	$c_2 = h^2 - h^4$	$c_3 = h - h^5$
c_1	c_1	c_2	c_3
c_2	c_2	$2c_1 - c_2$	c_3
c_3	c_3	c_3	$2c_1 + c_2$

(5.7.10)

These rules follow easily from those of the table of (5.7.4). The c_3 minimal equation

$$(c_3)^3 - 3c_3 = 0$$

has three roots $(\sqrt{3}, -\sqrt{3}, 0)$, which yield in turn three all-commuting idempotents.

$$P^1 = \frac{(c_3 + \sqrt{3}\,1)(c_3 - 0)}{(\sqrt{3} + \sqrt{3})(\sqrt{3} - 0)}, \qquad P^2 = \frac{(c_3 - \sqrt{3}\,1)(c_3 - 0)}{(-\sqrt{3} - \sqrt{3})(-\sqrt{3} - 0)},$$

$$= \frac{1}{3}1 + \frac{1}{6}c_2 + \frac{\sqrt{3}}{6}c_3, \qquad = \frac{1}{3}1 + \frac{1}{6}c_2 - \frac{\sqrt{3}}{6}c_3,$$

$$P^3 = \frac{(c_3 - \sqrt{3}\,1)(c_3 + 3\,1)}{(0 - \sqrt{3})(0 + \sqrt{3})}$$

$$= \frac{1}{3}1 - \frac{1}{3}c_2. \tag{5.7.11}$$

These correspond to three irreducible representations whose characters follow from the coefficients of (5.7.11) according to Eqs. (3.5.5) and (3.5.7). The latter equation gives the same dimensions:

$$l^\alpha = (12/3)^{1/2} = 2 \tag{5.7.12}$$

for all three representations. The character table is given here:

$g =$	1	h^2	h
$\chi_g^{E_1} = \chi_g^1 =$	2	1	$\sqrt{3}$
$\chi_g^{E_2} = \chi_g^2 =$	2	1	$-\sqrt{3}$
$\chi_g^{E_3} = \chi_g^3 =$	2	-2	0

(5.7.13)

A conventional notation $\{E_1 E_2 E_3\}$ for the ray representation will be used. Note that only three classes survived to be part of the class algebra and appear in the character table. The characters of the other classes are identically zero.

One could proceed from this point to an algebraic derivation of the twelve P_{ij}^α operators. Indeed, there are four such operators for each α. [Recall that the sum of squares $(l^\alpha)^2$ of dimensions must equal $^\circ G$. Here we have $2^2 + 2^2 + 2^2 = 12$.] However, it is easier to use the $\mathcal{D}^j(\alpha\beta\gamma)$ matrices for most applications.

Let us determine which D_6 ray representations are correlated with a given SU_2 or R_3 irrep $\mathscr{D}^j \downarrow D_6$. The same frequency formula (3.5.11) and trace formula (5.6.4) may be used. The traces and frequencies are given below.

	$\omega = 0$	$\omega = 120°$	$\omega = 60°$			f^{E_1}	f^{E_2}	f^{E_3}
$j = \frac{1}{2}$	2	1	$\sqrt{3}$		$j = \frac{1}{2}$	.	.	.
$\frac{3}{2}$	4	−1	$\sqrt{3}$		$\frac{3}{2}$	1	.	1
$\frac{5}{2}$	6	0	0		$\frac{5}{2}$	1	1	1
$\frac{7}{2}$	8	1	$\sqrt{3}$	,	$\frac{7}{2}$	2	1	1
$\frac{9}{2}$	10	−1	$\sqrt{3}$		$\frac{9}{2}$	2	1	2
$\frac{11}{2}$	12	0	0		$\frac{11}{2}$	2	2	2
⋮	⋮	⋮	⋮		⋮	⋮	⋮	⋮

(5.7.14)

Note that the frequency or correlation table repeats after $j = \frac{5}{2}$:

$$\mathscr{D}^j \downarrow D_6 \sim [\mathscr{D}^{E_1} + \mathscr{D}^{E_3} + \mathscr{D}^{E_2}] + \mathscr{D}^{j-3}. \qquad (5.7.15)$$

Hence, the D_6 algebra for half-integral spin has exactly half as many different types of irreps as the D_6 group. That is, there are three kinds $\{E_1 E_2 E_3\}$ of ray representations, but six kinds $\{A_1 A_2 B_1 B_2 E_1 E_2\}$ of ordinary representations for the group D_6. Most other crystal point groups also have fewer ray-irreps than irreps, as will be seen in the examples that follow. However, in all cases the sum of the squares $(l^\alpha)^2$ of dimensions equals the order of the group:

$$12 = (l^{E_1})^2 + (l^{E_2})^2 + (l^{E_3})^2 = 2^2 + 2^2 + 2^2. \qquad (5.7.16)$$

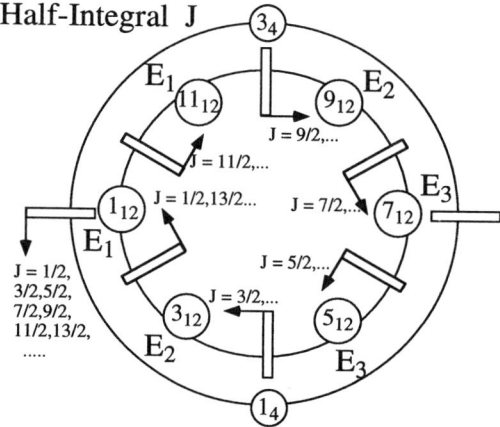

Figure 5.7.3 Mnemonic wheel for hexagonal half-integral spin splitting.

HALF-INTEGER j-LEVEL SPLITTING IN FINITE SYMMETRY   411

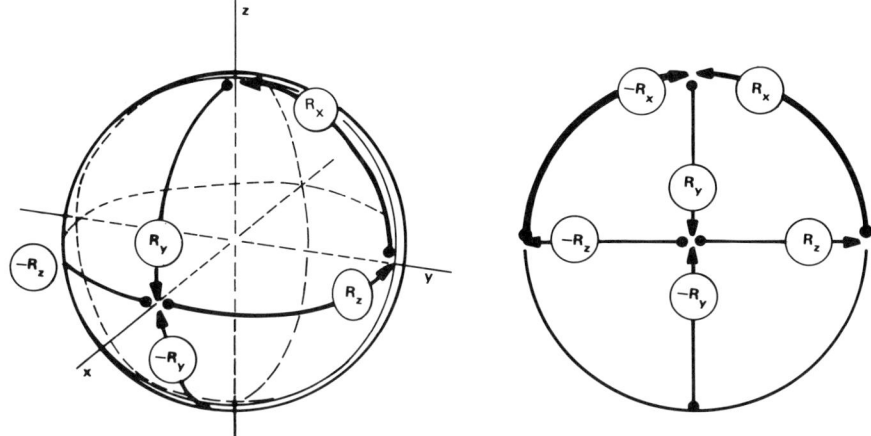

Figure 5.7.4 Hamilton arcs and vector nomogram for D_2 spin algebra.

The mnemonic wheels for the D_6 half-integral j correlations are shown in Figure 5.7.3. Similar wheels can be made for all D_n irreps.

B. Ray Representations of Other D_h Groups

The group $D_2 = C_2 \times C_2$ was the first symmetry group introduced in Chapter 2. It is an Abelian group since its three orthogonal 180° rotations commute with one another. However, the corresponding ray algebra generated by $\mathscr{D}^{1/2}$ or by Hamilton vectors is noncommutative. The D_2 Hamilton vector nomogram is shown in Figure 5.7.4. This gives the following multiplication table:

$$\begin{array}{|cccc|} \hline 1 & R_x & R_y & R_z \\ R_x & -1 & R_z & -R_y \\ R_y & -R_z & -1 & R_x \\ R_z & R_y & -R_x & -1 \\ \hline \end{array}. \qquad (5.7.17)$$

From this it is easy to show that the identity operator *alone* is all-commuting. This implies that only one irrep of D_2 exists. The character table assumes a very simple form!

$$g = 1,$$
$$\chi_g^E = \boxed{2}. \qquad (5.7.18)$$

D_2 gives the world's simplest nontrivial ray algebra. It has a single irreducible representation $\mathscr{D}^E$ given by the spinor representation (5.5.1) or (5.4.30):

$$\mathscr{D}^E(R_x) = \begin{pmatrix} 0 & -i \\ -i & 0 \end{pmatrix}, \quad \mathscr{D}^E(R_y) = \begin{pmatrix} 0 & -1 \\ 1 & 0 \end{pmatrix}, \quad \mathscr{D}^E(R_z) = \begin{pmatrix} -i & 0 \\ 0 & i \end{pmatrix}.$$

$$(5.7.19)$$

Incidentally, these are just the QUATERNION q_j matrices defined by Eq. (5.5.3). In fact, the D_2 double group is called the quaternion group $Q = \{\mathbf{1}, q_x, q_y, q_z, -\mathbf{1}, -q_x, -q_y, -q_z\}$, where

$$-R_z = q_x q_y = q_z = -q_y q_x \text{ (and cyclically)},$$

and

$$q_j^2 = -\mathbf{1}. \tag{5.7.20}$$

This group Q has a character table which resembles the one for D_4. D_4 and Q account for the two eighth-order non-Abelian groups which are counted in Figure 2.2.2. (See Problems 3.3.2 and 3.4.2.)

The derivations of spin-$\frac{1}{2}$ characters and irreps for other axial groups C_n and D_n are left as exercises. However, it is interesting to note that the C_n characters can be visualized geometrically using the nth roots of negative unity (-1). It is instructive to compare the complex vector roots of positive unity in Figure 2.7.2 with the corresponding negative ones in Figure 5.7.5. There is an essential difference between odd-n groups $C_3, C_5, \ldots$, which have a (-1) representation, and even-n groups $C_2, C_4, C_6, \ldots$, which do not. This difference carrys over to the D_n groups. D_n groups with even n have only $n/2$ different two-dimensional ray representations $\{E_{1/2}, E_{3/2}, \ldots, E_{(n-1)/2}\}$. The D_n groups with odd n have $(n-1)/2$ different two-dimensional ray representations and a conjugate pair of one-dimensional ones. The number and dimension of ray representations of D_n for odd n is the same as that of ordinary representations.

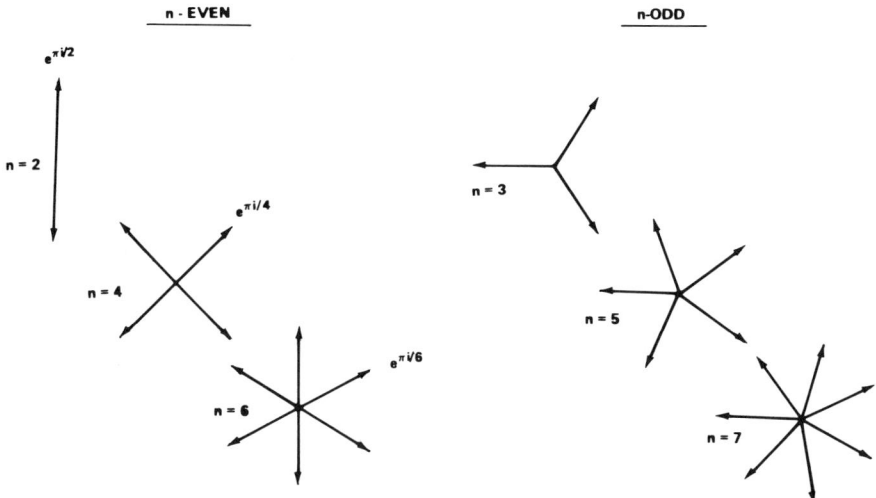

Figure 5.7.5 Representing C_n spin representations by complex nth roots of negative unity ($n = 2, 3, \ldots, 7$).

C. Ray Representations of Octahedral Symmetry

The structure of octahedral ray representations is quite different from the ordinary representations discussed in Sections 4.1 and 4.2. However, the O-group nomogram (4.1.4) gives the necessary ray algebra. Once again the 180° rotations get left out of the revised class algebra, and the only all-commuting operators are the following:

$$c_1 = 1,$$
$$c_r = r_1 + r_2 + r_3 + r_4 - r_1^2 - r_2^2 - r_3^2 - r_4^2,$$
$$c_R = R_1 + R_2 + R_3 - R_1^3 - R_2^3 - R_3^3. \qquad (5.7.21)$$

When computing the class algebra one has only to compute products which give one particular element, say $\mathbf{1}$, r_1, and R_1 from each class. The other class elements copy this behavior. The necessary products are computed in the following table using Figure 4.1.4. (Also, see Appendix F, Table F.2.1.)

	r_1	r_2	r_3	r_4	$-r_1^2$	$-r_2^2$	$-r_3^2$	$-r_4^2$	R_1	R_2	R_3	$-R_1^3$	$-R_2^3$	$-R_3^3$
r_1					1								R_1	
r_2						1								
r_3							1		R_1					
r_4								1						
$-r_1^2$	1						$-r_1$							
$-r_2^2$		1							r_1			R_1		
$-r_3^2$			1							r_1				
$-r_4^2$				1		r_1								R_1
R_1										r_1		1		
R_2											r_1		1	
R_3									r_1					1
$-R_1^3$									1					
$-R_2^3$										1				
$-R_3^3$											1			

$$(5.7.22)$$

The desired algebra multiplication table follows:

$$
\begin{array}{c|ccc}
 & c_1 = 1 & c_r & c_r \\
\hline
c_r = 1 & 1 & c_r & c_R \\
c_r & c_r & 81 + 2c_r & 4c_R \\
c_R & c_R & 4c_R & 61 + 3c_r
\end{array}. \qquad (5.7.23)
$$

414 REPRESENTATIONS OF CONTINUOUS ROTATION GROUPS AND APPLICATIONS

From this one derives the minimal equation of c_R:
$$c_R^3 - 18c_R = 0.$$
Each of the three roots $(3\sqrt{2}, -3\sqrt{2}, 0)$ are identified with one of the resulting idempotents:
$$P^{E_1} = (21 + c_r + \sqrt{2}\,c_R)/12,$$
$$P^{E_2} = (21 + c_r - \sqrt{2}\,c_R)/12,$$
$$P^G = (41 - c_r)/6. \qquad (5.7.24)$$

The usual relations (3.5.5)–(3.5.7) give the desired characters.

	$g=1$	r_1	R_1
$\chi_g^{E_1} =$	2	1	$\sqrt{2}$
$\chi_g^{E_2} =$	2	1	$-\sqrt{2}$
$\chi_g^{G} =$	4	-1	0

$(5.7.25)$

The splitting of half-integral angular-momentum j levels or the $\mathscr{D}^j \downarrow O$ correlations may be calculated easily from the characters (5.7.25). The R_3 trace formula (5.6.4) gives the table of traces (5.7.26a). From this one derives the frequencies $f^\alpha(\mathscr{D}^j \downarrow O)$ in the table of (5.7.26b).

	1 ($\omega = 0$)	r_1 ($\omega = 120°$)	R_1 ($\omega = 90°$)
$j = \tfrac{1}{2}$	2	1	$\sqrt{2}$
$\tfrac{3}{2}$	4	-1	0
$\tfrac{5}{2}$	6	0	$-\sqrt{2}$
$\tfrac{7}{2}$	8	1	0
$\tfrac{9}{2}$	10	-1	$\sqrt{2}$
$\tfrac{11}{2}$	12	0	0
$\vdots$	$\vdots$	$\vdots$	$\vdots$

$(5.7.26a)$

	f^{E_1}	f^{E_2}	f^G
$j = \tfrac{1}{2}$	1	·	·
$\tfrac{3}{2}$	·	·	1
$\tfrac{5}{2}$	·	1	·
$\tfrac{7}{2}$	1	1	1
$\tfrac{9}{2}$	1	·	2
$\tfrac{11}{2}$	1	1	2
$\vdots$	$\vdots$	$\vdots$	$\vdots$

$(5.7.26b)$

SOME HIGHER SYMMETRIES: R_4 AND U_3

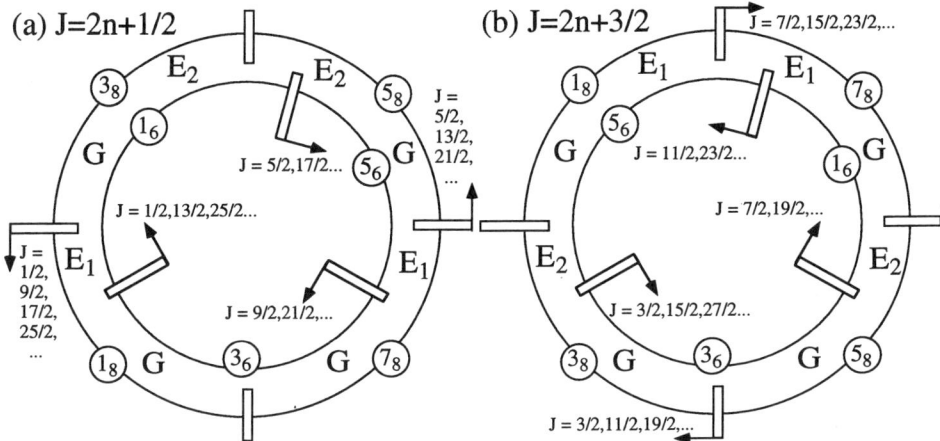

Figure 5.7.6 Mnemonic wheels for octahedral spin algebra correlation with half-integral J.

Notice that (5.7.26b) shows that $\mathscr{D}^{1/2} = \mathscr{D}^{E_1}$ and $\mathscr{D}^{3/2} = \mathscr{D}^{G}$ are both irreducible octahedral ray representations.

It is interesting to note that the preceding correlation table can be replaced by the mnemonic wheels in Figure 5.7.6. Again it happens that the ordering and clustering of half-integral eigenvalues of $V^{(4)}$ Hamiltonians are predicted by the wheels. This is very convenient to have for molecular ion problem which involve high angular momentum j.

5.8 SOME HIGHER CONTINUOUS SYMMETRIES: R_4 AND U_3

We give now a brief discussion of two Hamiltonians which have the Coulomb and oscillator potentials, respectively. The Hamiltonians possess a symmetry higher than R_3 or O_3. Their theory is very interesting but, is being developed sufficiently to have considerable practical value. We discuss it with the hope that future work will be fruitful. Also, we use the opportunity to introduce techniques for analyzing larger Lie groups, which are now being applied in modern theory. The Coulomb and oscillator symmetries were known by Pauli and others. More recently progress has been made toward applying these symmetries to atomic, molecular, and nuclear structure.

A. The Coulomb Symmetry

A particle of mass m in a Coulomb potential field k/r is described by the following Hamiltonian:

$$H = p^2/2m + k/r = \begin{cases} p^2/2m - |k|/r \text{ (attractive case)}, \\ p^2/2m + |k|/r \text{ (repulsive case)}. \end{cases} \quad (5.8.1)$$

Such a Hamiltonian obviously has spherical ($R_3 \subset O_3$) symmetry. It conserves and commutes with angular-momentum operators:

$$L_1 = x_2 p_3 - x_3 p_2, \quad L_2 = x_3 p_1 - x_1 p_3, \quad L_3 = x_1 p_2 - x_2 p_1. \quad (5.8.2)$$

However, there is an additional hidden symmetry of this Hamiltonian.

Consider a vector ε called the ECCENTRICITY or LENZ-RUNGE VECTOR:

$$\varepsilon = \mathbf{r}/r - \mathbf{L} \times \mathbf{p}/km. \quad (5.8.3)$$

(The ε vector was actually discovered by Hamilton before Lenz or Runge wrote about it.) The ε vector points along the symmetry axis of the classical elliptic or hyperbolic orbit, and its length is the eccentricity of the orbit. The relation between two kinds of Coulomb orbits and their vectors ε and $\mathbf{L}$ is sketched in Figure 5.8.1.

It is easy to show that conservation of ε and $\mathbf{L}$ is consistent with the classical orbit equation

$$r = -L^2/km(1 - \varepsilon \cos \theta). \quad (5.8.4)$$

Consider the dot product of ε and $\mathbf{r}$:

$$\varepsilon \cdot \mathbf{r} = \varepsilon r \cos \theta = r - \mathbf{L} \times \mathbf{p} \cdot \mathbf{r}/km = r + L^2/km.$$

This is the same as the orbit equation. Clearly, the eccentricity vector ε is another quantity besides $\mathbf{L} = \mathbf{r} \times \mathbf{p}$ which is conserved in a Coulomb field. Note also that its classical magnitude,

$$\begin{aligned}\varepsilon^2 = \varepsilon \cdot \varepsilon &= 1 - 2\mathbf{r} \cdot \mathbf{L} \times \mathbf{p}/kmr + (\mathbf{L} \times \mathbf{p}) \cdot (\mathbf{L} \times \mathbf{p})/k^2 m^2 \\ &= 1 + 2L^2/kmr + L^2 p^2/k^2 m^2 \\ &= 1 + 2(p^2/2m + k/r)L^2/k^2 m = 1 + (2H/k^2 m)L^2, \quad (5.8.5)\end{aligned}$$

is expressed in terms of conserved total energy H and orbital momentum $\mathbf{L}$.

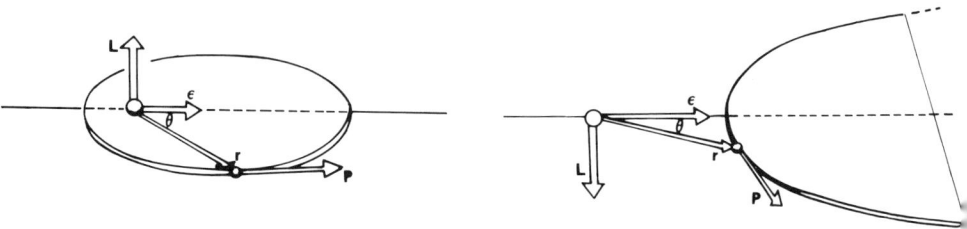

Figure 5.8.1 Location of classical eccentricity vector. The orientation of the eccentricity vector ε in an elliptical ($k < 0$) and hyperbolic repulsive ($k > 0$) Coulomb orbit is shown.

SOME HIGHER SYMMETRIES: R_4 AND U_3

The corresponding quantum eccentricity operators are defined by

$$\varepsilon_1 = x_1/r - (L_2 p_3 - L_3 p_2 - p_2 L_3 + p_3 L_2)/2km, \quad (5.8.6a)$$
$$\varepsilon_1 = x_1/r - ([L_2 p_3 - L_3 p_2] - ip_1)/2km, \quad (5.8.6b)$$
$$\varepsilon_2 = x_2/r - ([L_3 p_1 - L_1 p_3] - ip_2)/2km, \quad (5.8.6c)$$
$$\varepsilon_3 = x_3/r - ([L_1 p_2 - L_2 p_1] - ip_3)/2km. \quad (5.8.6d)$$

The eccentricity operators and the angular-momentum operators (5.8.2) will be seen to generate a symmetry higher than O_3. In constructing the ε operators it is necessary to symmetrize with respect to operators that do not commute; i.e., let $\mathbf{L} \times \mathbf{p}$ equal $(\mathbf{L} \times \mathbf{p} - \mathbf{p} \times \mathbf{L})/2$ in Eqs. (5.8.6). Then one uses commutation relations such as

$$[L_1, p_2] = [x_2 p_3, p_2] - [x_3 p_2, p_2]$$
$$= [x_2, p_2] p_3 = ip_3$$

to simplify the results wherever possible. In Chapter 7 we shall prove that commutators of L_i with any vector v_j must have the same form,

$$[L_1, v_2] = iv_3, \quad (5.8.7)$$

as the angular-momentum relations

$$[L_1, L_2] = iL_3.$$

The derivations of $\varepsilon - L$ commutators require the following commutation relations:

$$[x_i/r, p_j]\psi(x) = (x_i/r)\frac{1}{i}\frac{\partial \psi}{\partial x_j} - \frac{1}{i}\frac{\partial}{\partial x_j}(x_i \psi(x)/r) = i(\delta_{ij} + x_i x_j)\psi(x)/r$$

or

$$[x_i/r, p_j] = i(\delta_{ij} + x_i x_j)/r, \quad (5.8.8)$$

along with the derivative properties.

$$[AB, C] = A[B, C] + [A, C]B,$$
$$[A, BC] = B[A, C] + [A, B]C. \quad (5.8.9)$$

After considerable algebra the following commutation relation results:

$$[\varepsilon_1, \varepsilon_2] = -2L_3[p^2/2m + k/r]/k^2 m$$
$$= -i(2H/k^2 m)L_3. \quad (5.8.10)$$

In the case of bound states for which $\langle H \rangle = -|E|$ one may normalize ε to give

$$K_j \equiv \sqrt{-\frac{k^2 m}{2H}} \, \varepsilon_j = \sqrt{\frac{k^2 m}{2E}} \, \varepsilon_j, \tag{5.8.11}$$

whence the following commutation relations result.

$$[L_1 L_2] = iL_3, \quad [L_1, K_2] = iK_3, \quad [K_1, K_2] = iL_3$$
$$\cdots \qquad \cdots \qquad \cdots. \tag{5.8.12}$$

We now see that these relations are the ones for R_4 or O_4 symmetry.

(a) Introduction to the Lie Algebra of R_4 We showed in Section 5.4 that infinitesimal rotations are of the form

$$R(\cdot \cdot \varepsilon) = 1 + \varepsilon G = 1 + (\varepsilon/i)L,$$

where operators

$$L = iG \tag{5.8.13}$$

are the generators of rotations. The rotation operators are orthogonal:

$$R^T(\varepsilon) = R^{-1}(\varepsilon) = R(-\varepsilon).$$

This implies that

$$1 + \varepsilon G^T = 1 - \varepsilon G,$$

and that the generators are antisymmetric,

$$G^T = -G, \quad L^T = -L. \tag{5.8.14}$$

Note that since G is real, L is Hermitian:

$$L^\dagger = L^{*T} = (iG)^{*T} = iG = L.$$

Let us generalize the treatment of R_3 to give one for R_4. According to Eq. (5.1.12), R_4 has six parameters, or three more than R_3. Therefore, we expect R_4 to have six generators. Indeed, we may construct the same representations

SOME HIGHER SYMMETRIES: R_4 AND U_3 419

of the R_3 generators (5.4.5b) using 4×4 matrices: [Recall Eq. (5.4.5b).]

$$L_{32} \equiv L_x = iG_{32} \rightarrow i\mathscr{G}_{32} = \begin{pmatrix} \cdot & \cdot & \cdot & \cdot \\ \cdot & \cdot & -i & \cdot \\ \cdot & i & \cdot & \cdot \\ \cdot & \cdot & \cdot & \cdot \end{pmatrix},$$

$$L_{13} \equiv L_y = iG_{13} \rightarrow i\mathscr{G}_{13} = \begin{pmatrix} \cdot & \cdot & i & \cdot \\ \cdot & \cdot & \cdot & \cdot \\ -i & \cdot & \cdot & \cdot \\ \cdot & \cdot & \cdot & \cdot \end{pmatrix},$$

$$L_{21} \equiv L_z = iG_{21} \rightarrow i\mathscr{G}_{21} = \begin{pmatrix} \cdot & -i & \cdot & \cdot \\ i & \cdot & \cdot & \cdot \\ \cdot & \cdot & \cdot & \cdot \\ \cdot & \cdot & \cdot & \cdot \end{pmatrix}. \quad (5.8.15\text{a})$$

These use just the first three dimensions. Then let us make three more R_4 generators using the fourth dimension and components $(j, 4)$ or $(4, j)$:

$$K_{41} \equiv K_x = iG_{41} \rightarrow i\mathscr{G}_{41} = \begin{pmatrix} \cdot & \cdot & \cdot & -i \\ \cdot & \cdot & \cdot & \cdot \\ \cdot & \cdot & \cdot & \cdot \\ i & \cdot & \cdot & \cdot \end{pmatrix},$$

$$K_{42} \equiv K_y = iG_{42} \rightarrow i\mathscr{G}_{42} = \begin{pmatrix} \cdot & \cdot & \cdot & \cdot \\ \cdot & \cdot & \cdot & -i \\ \cdot & \cdot & \cdot & \cdot \\ \cdot & i & \cdot & \cdot \end{pmatrix},$$

$$K_{43} \equiv K_z = iG_{43} \rightarrow i\mathscr{G}_{43} = \begin{pmatrix} \cdot & \cdot & \cdot & \cdot \\ \cdot & \cdot & \cdot & \cdot \\ \cdot & \cdot & \cdot & -i \\ \cdot & \cdot & i & \cdot \end{pmatrix}. \quad (5.8.15\text{b})$$

Between these we find commutation relations

$$[L_x, L_y] = iL_z, \quad [L_x, K_y] = iK_z, \quad [K_x, K_y] = iL_z.$$
$$\cdots \qquad \cdots \qquad \cdots, \quad (5.8.16)$$

which match those of the operators L_i and K_j of the preceding section. These relations define the Lie algebra of R_4.

We now derive the irreducible representations of the R_4 Lie algebra. This can be done by first constructing raising and lowering operators that are eigenvectors of L_z, in much the same way that R_3 is solved in Appendix E. Let us imagine that each operator $L_x, K_x, L_y, \ldots$ corresponds to a vector $|L_x\rangle, |K_x\rangle, |L_y\rangle, \ldots$ and make a REGULAR REPRESENTATION analogous to the one for groups.

The regular representation of a Lie-algebra operator L_z is constructed by converting commutation relations,

$$[L_z, L_x] = iL_y, \qquad [L_z, K_x] = iK_y, \qquad [L_z, L_y] = -iL_x,$$

$$[L_z, K_y] = -iK_x, \qquad [L_z, L_z] = 0, \qquad [L_z, K_z] = 0,$$

into vector equations (this is permitted since commutation is a linear operation):

$$L_z|L_x\rangle = i|L_y\rangle, \qquad L_z|K_x\rangle = i|K_y\rangle, \qquad L_z|L_y\rangle = -i|L_x\rangle,$$

$$L_z|K_y\rangle = -i|K_x\rangle, \qquad L_z|L_z\rangle = 0, \qquad L_z|K_z\rangle = 0. \qquad (5.8.17a)$$

These equations define the regular representation matrix,

$$\mathscr{R}(L_z) = \begin{pmatrix} \cdot & \cdot & -i & \cdot & \cdot & \cdot \\ \cdot & \cdot & \cdot & -i & \cdot & \cdot \\ i & \cdot & \cdot & \cdot & \cdot & \cdot \\ \cdot & i & \cdot & \cdot & \cdot & \cdot \\ \cdot & \cdot & \cdot & \cdot & \cdot & \cdot \\ \cdot & \cdot & \cdot & \cdot & \cdot & \cdot \end{pmatrix}, \qquad (5.8.17b)$$

in the $\{|L_x\rangle, |K_x\rangle, \ldots\}$ basis.

The last commutation $[L_z, K_z] = 0$ indicates that a similar representation of K_z can be simultaneously diagonalized with the one for L_z. $\mathscr{R}(K_z)$ is given here:

$$\mathscr{R}(K_z) = \begin{pmatrix} \cdot & \cdot & \cdot & -i & \cdot & \cdot \\ \cdot & \cdot & -i & \cdot & \cdot & \cdot \\ \cdot & i & \cdot & \cdot & \cdot & \cdot \\ i & \cdot & \cdot & \cdot & \cdot & \cdot \\ \cdot & \cdot & \cdot & \cdot & \cdot & \cdot \\ \cdot & \cdot & \cdot & \cdot & \cdot & \cdot \end{pmatrix}. \qquad (5.8.18)$$

The simultaneous eigenvectors of $\mathscr{R}(L_z)$ and $\mathscr{R}(K_z)$ are found using the theory of Section 1.2.B(d). First we obtain the individual idempotents for

$\mathcal{R}(J_z)$ and $\mathcal{R}(K_z)$:

$$\mathcal{P}_1^J = \frac{\mathcal{R}(J_z) - (-1)\mathbf{1}}{1 - (-1)} = \tfrac{1}{2}\begin{pmatrix} 1 & \cdot & -i & \cdot & \cdot & \cdot \\ \cdot & 1 & \cdot & -i & \cdot & \cdot \\ i & \cdot & 1 & \cdot & \cdot & \cdot \\ \cdot & i & \cdot & 1 & \cdot & \cdot \\ \cdot & \cdot & \cdot & \cdot & \cdot & \cdot \\ \cdot & \cdot & \cdot & \cdot & \cdot & \cdot \end{pmatrix},$$

$$\mathcal{P}_1^K = \tfrac{1}{2}\begin{pmatrix} 1 & \cdot & \cdot & -i & \cdot & \cdot \\ \cdot & 1 & -i & \cdot & \cdot & \cdot \\ \cdot & i & 1 & \cdot & \cdot & \cdot \\ i & \cdot & \cdot & 1 & \cdot & \cdot \\ \cdot & \cdot & \cdot & \cdot & \cdot & \cdot \\ \cdot & \cdot & \cdot & \cdot & \cdot & \cdot \end{pmatrix},$$

$$\mathcal{P}_{-1}^J = \tfrac{1}{2}\begin{pmatrix} 1 & \cdot & i & \cdot & \cdot & \cdot \\ \cdot & 1 & \cdot & i & \cdot & \cdot \\ -i & \cdot & 1 & \cdot & \cdot & \cdot \\ \cdot & -i & \cdot & 1 & \cdot & \cdot \\ \cdot & \cdot & \cdot & \cdot & \cdot & \cdot \\ \cdot & \cdot & \cdot & \cdot & \cdot & \cdot \end{pmatrix},$$

$$\mathcal{P}_{-1}^K = \tfrac{1}{2}\begin{pmatrix} 1 & \cdot & \cdot & i & \cdot & \cdot \\ \cdot & 1 & i & \cdot & \cdot & \cdot \\ \cdot & -i & 1 & \cdot & \cdot & \cdot \\ -i & \cdot & \cdot & 1 & \cdot & \cdot \\ \cdot & \cdot & \cdot & \cdot & \cdot & \cdot \\ \cdot & \cdot & \cdot & \cdot & \cdot & \cdot \end{pmatrix}. \quad (5.8.19)$$

Then the products of these give the desired results. The first columns of the products $\mathcal{P}_m^J \mathcal{P}_n^K$ are

$$\mathcal{P}_1^J \mathcal{P}_1^K = \tfrac{1}{4}\begin{pmatrix} 1 \\ 1 \\ i \\ i \\ \cdot \end{pmatrix}, \quad \mathcal{P}_1^J \mathcal{P}_{-1}^K = \tfrac{1}{4}\begin{pmatrix} 1 \\ -1 \\ i \\ -i \\ \cdot \end{pmatrix},$$

$$\mathcal{P}_{-1}^J \mathcal{P}_1^K = \tfrac{1}{4}\begin{pmatrix} 1 \\ -1 \\ -i \\ i \\ \cdot \end{pmatrix}, \quad \mathcal{P}_{-1}^J \mathcal{P}_{-1}^K = \tfrac{1}{4}\begin{pmatrix} 1 \\ 1 \\ -i \\ -i \\ \cdot \end{pmatrix}.$$

They show that the following operators:

$$M_+ \equiv L_x + iL_y + K_x + iK_y, \qquad N_+ \equiv L_x + iL_y - K_x - K_y,$$
$$= L_+ + K_+, \qquad\qquad\qquad = L_+ - K_+,$$
$$N_- \equiv L_x - iL_y - K_x + iK_y, \qquad M_- \equiv L_x - iL_y + K_x - iK_y,$$
$$= L_- - K_-, \qquad\qquad\qquad = L_- + K_- \qquad (5.8.20\text{a})$$

obey the eigencommutation relations:

$$[L_z, M_+] = M_+, \qquad [L_z, N_+] = N_+, \qquad [L_z, N_-] = -N_-,$$
$$[L_z, M_-] = -M_-, \qquad [K_z, M_+] = M_+, \qquad [K_z, N_+] = -N_+,$$
$$[K_z, N_-] = N_-, \qquad [K_z, M_-] = -M_-. \qquad (5.8.20\text{b})$$

The $M_\pm$ and $N_\pm$ operators are the R_4 raising and lowering operators.

Now a very nice separation is possible if one redefines generators by

$$M_a \equiv (L_\alpha + K_\alpha)/2 \quad \text{and} \quad N_\alpha \equiv (L_\alpha - K_\alpha)/2. \qquad (5.8.21)$$

There results two *independent* sets of R_3 commutation relations

$$[M_z, M_\pm] = \pm M_\pm, \qquad [N_z, N_\pm] = \pm N_\pm, \qquad (5.8.22\text{a})$$

and

$$[M_x, M_y] = iM_z, \qquad [N_x, N_y] = iN_z, \qquad (5.8.22\text{b})$$

where the two sets of operators are mutually commuting:

$$[M_\alpha, N_\beta] = 0. \qquad (5.8.23)$$

The two commuting R_3 groups generated by M and N operators are factors of a cross-product $R_4 = R_3 \times R_3$ according to the definition in Section 2.10. This tells us that our knowledge of R_3 can be applied directly to R_4 since $R_4 = R_3 \times R_3$.

Now Cartan has developed an elegant geometrical method for treating the structure and representations of Lie algebras. We introduce some of this now as it applies to R_4. Lie-algebra operators can generally be combined to form raising and lowering operators $\{\cdots E_{+\alpha} \cdots E_{-\alpha} \cdots\}$ like the $L_\pm$ and $K_\pm$ in R_4. The E_α are eigenvectors in the sense of

$$[H_j, E_\alpha] = r_j(\alpha) E_\alpha, \qquad (5.8.24)$$

for a number of mutually commuting operators $\{H_1 H_2 \cdots H_r\}$. (The H_j are $H_1 = L_z$ and $H_2 = K_z$ in R_4.) The number of mutually commuting H_i operators is called the RANK of the algebra. Then the eigenvalues or roots

SOME HIGHER SYMMETRIES: R_4 AND U_3

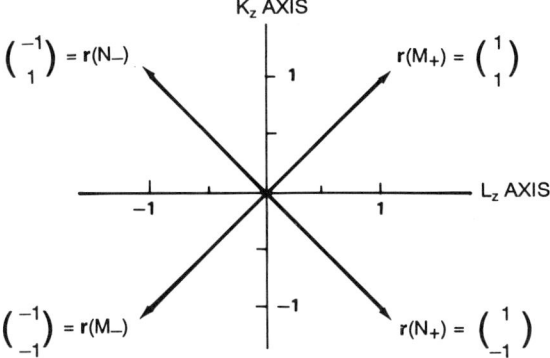

Figure 5.8.2 Root diagram of the R_4 Lie algebra.

$r_j(\alpha)$ can be taken as components of a vector called the ROOT VECTOR:

$$\mathbf{r}(\alpha) = \begin{pmatrix} r_1(\alpha) \\ r_2(\alpha) \\ \vdots \\ r_r(\alpha) \end{pmatrix} \quad (5.8.25)$$

of E_α. For example, we would have the set of four root vectors of R_4 drawn in Figure 5.8.2 according to Eqs. (5.8.20). Cartan has shown how to make root vectors into a code that defines all commutation relations and finally all irreducible representations.

For example, note that the commutation between an E_α and an E_β must give an operator whose root vector is a *sum* of $\mathbf{r}(\alpha)$ and $\mathbf{r}(\beta)$, or else zero:

$$\begin{aligned}
[H_j, [E_\alpha, E_\beta]] &= [H_j, E_\alpha E_\beta] - [H_j, E_\beta E_\alpha] \\
&= [H_j, E_\alpha] E_\beta + E_\alpha [H_j, E_\beta] - [H_j, E_\beta] E_\alpha - E_\beta [H_j, E_\alpha] \\
&= (r_j(\alpha) + r_j(\beta)) [E_\alpha, E_\beta].
\end{aligned} \quad (5.8.26)$$

This implies that a commutator of a raising operator with root $\mathbf{r}(\alpha)$ and a corresponding lowering operator with root $-\mathbf{r}(\alpha)$ must give operators with zero roots, namely H_j's. In fact, it can be shown that

$$[E_{+\alpha}, E_{-\alpha}] = N \sum_j r_j(\alpha) H_j. \quad (5.8.27)$$

Here N is a normalization constant which is usually set equal to unity. For example, from Figure 5.8.2 we have

$$[M_+, M_-] = r_{L_z}(M_+) L_z + r_{K_z}(M_+) K_z = 2 M_z. \quad (5.8.27)_x$$

(b) Irreducible Representations of R_4 If R_4 is an outer product $R_3 \times R_3$ of rotation groups, then the representations of R_4 are the outer products

$$\mathscr{D}^{j_m j_n} \equiv \mathscr{D}^{j_m} \otimes \mathscr{D}^{j_n}$$

of representations of R_3. The basis states must obey the standard angular-momentum relations twice: once for M_z and $M_\pm$ and once again for N_z and $N_\pm$:

$$M_z \begin{vmatrix} j_m & j_n \\ m & n \end{vmatrix} \rangle = m \begin{vmatrix} j_m & j_n \\ m & n \end{vmatrix} \rangle, \quad N_z \begin{vmatrix} j_m & j_n \\ m & n \end{vmatrix} \rangle = n \begin{vmatrix} j_m & j_n \\ m & n \end{vmatrix} \rangle,$$

$$M_\pm \begin{vmatrix} j_m & j_n \\ m & n \end{vmatrix} \rangle = \sqrt{(j_m \mp m)(j_n \pm m + 1)} \begin{vmatrix} j_m & j_n \\ m \pm 1 & n \end{vmatrix} \rangle,$$

$$N_\pm \begin{vmatrix} j_m & j_n \\ m & n \end{vmatrix} \rangle = \sqrt{(j_N \mp n)(j_N \pm n + 1)} \begin{vmatrix} j_m & j_n \\ m & n \pm 1 \end{vmatrix} \rangle. \quad (5.8.28)$$

For a general Lie algebra there will be similar Cartan relations of the form

$$H_j | \mathbf{m} \rangle = m_j | \mathbf{m} \rangle. \quad (5.8.29)$$

Here the eigenvalues m_j are said to be components of a WEIGHT VECTOR

$$\mathbf{m} = \begin{pmatrix} m_1 \\ m_2 \\ \vdots \\ m_r \end{pmatrix}$$

for a given state. The weight vectors for states belonging to irreps $\mathscr{D}^{j_m j_n}$ of R_4 are indicated by circles in the Figure 5.8.3.

A raising (lowering) operator E_α ($E_{-\alpha}$) applied to any state with weight m gives another state of weight $\mathbf{m} + \mathbf{r}(\alpha)$ ($\mathbf{m} - \mathbf{r}(\alpha)$), or else zero:

$$H_j E_\alpha | \mathbf{m} \rangle = (E_\alpha H_j + [H_j, E_\alpha]) | \mathbf{m} \rangle,$$

$$(m_j E_\alpha + r_j(\alpha) E_\alpha) | \mathbf{m} \rangle = (m_j + r_j(\alpha)) E_\alpha | \mathbf{m} \rangle. \quad (5.8.30)$$

The beautiful theory of Cartan gives the properties of the weight vectors and the irreps of a Lie algebra in terms of its root vectors. We will study more about this later.

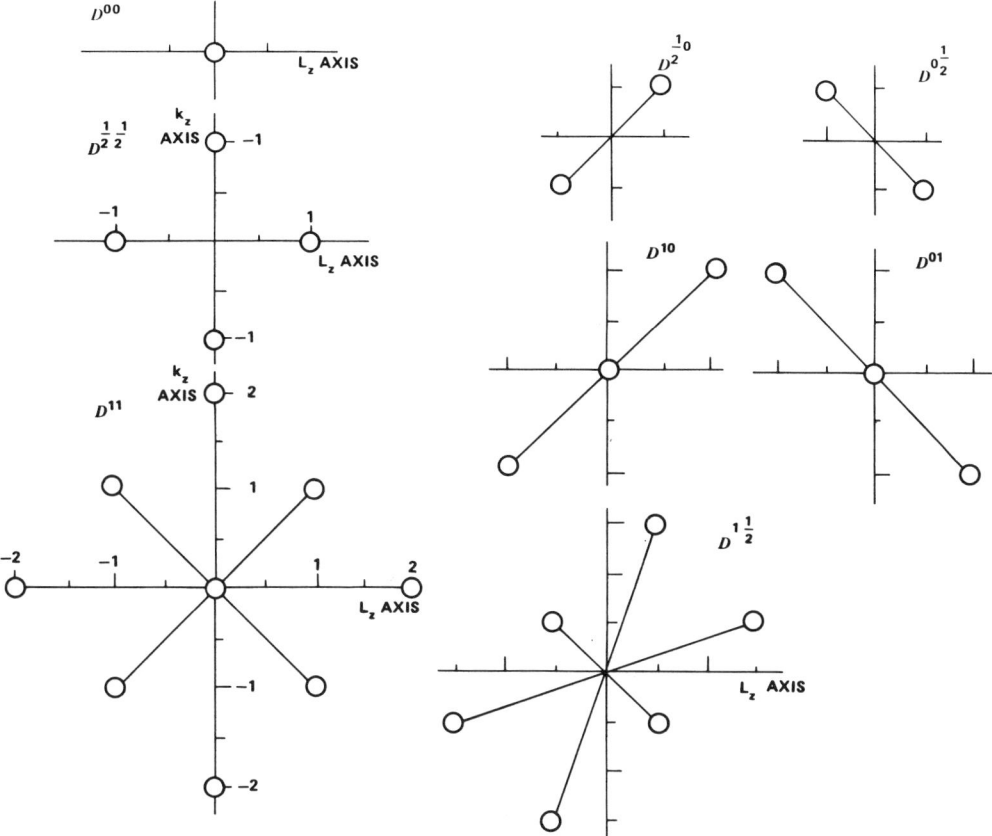

Figure 5.8.3 Examples of weight diagrams for irreducible representations for the R_4 Lie algebra. Eight examples of R_4 irreps are chosen in order to show the form of their weight diagrams.

(c) Coulomb Eigenstates The single-electron Coulomb energy levels belong only to certain irreps $\mathscr{D}^{j_m j_n}$ of R_4, namely, those for which $j_m = j_n = 0, \frac{1}{2}, 1, \ldots$ (see the left-hand side of Figure 5.8.3. This is because the vectors ε or **K** are orthogonal to L:

$$\mathbf{K} \cdot \mathbf{L} = 0,$$

which implies that

$$\mathbf{M}^2 - \mathbf{N}^2 = (\mathbf{M} + \mathbf{N}) \cdot (\mathbf{M} - \mathbf{N}) = 4\mathbf{K} \cdot \mathbf{L} = 0. \quad (5.8.31)$$

This relation holds for the operators as well, since M and N commute. Hence we have $j_n(j_n + 1) = j_m(j_m + 1)$ or $j_m = j_n$.

The first energy level is singly degenerate corresponding to the scalar irrep $\mathscr{D}^{00}$. This would be the hydrogen 1s level. The second energy level is fourfold degenerate, corresponding to the irrep $\mathscr{D}^{(1/2)(1/2)}$. From the weight diagram in Figure 5.8.3 we see that one state has orbital momentum $L_z = +1$; two states have $L_z = 0$, while a fourth state has $L_z = -1$. Together these will make the $2p$ ($l = 1$) and $2s$ ($l = 0$) levels of hydrogen. Similarly, the ninefold degenerate $\mathscr{D}^{11}$ corresponds to the degenerate $3s$, $3p$, and $3d$ ($l = 2$) states of hydrogen.

The Coulomb energy spectrum can be given in terms of the $\langle M^2 \rangle = \langle N^2 \rangle$ eigenvalues. The classical equation (5.8.5) gives

$$-K^2 = \frac{k^2 m}{2H}, \qquad \varepsilon^2 = \frac{k^2 m}{2H} + L^2,$$

or

$$H = -(k^2 m/2)/(K^2 + L^2) = (k^2 m/2)/(2M^2 + 2N^2). \quad (5.8.32)$$

However, this fails to take account of the operator noncommutivity. After considerable algebra one obtains the correct form:

$$H = -(k^2 m/2)/(2M^2 + 2N^2 + 1), \quad (5.8.33)$$

which gives the Rydberg formula if we substitute $\langle M^2 \rangle = \langle N^2 \rangle = j_M(j_M + 1)$:

$$E^{j_m} = -(k^2 m/2)/(2j_m + 1)^2. \quad (5.8.34)$$

The form of the spectrum and wave functions is sketched some pages ahead in Figure 5.8.6.

B. The Harmonic Oscillator Symmetry

A particle of mass $m = 1$ in a three-dimensional isotropic harmonic oscillator potential is described by the following Hamiltonian:

$$H = p^2/2 + \omega^2 r^2/2 = a_x^\dagger a_x + a_y^\dagger a_y + a_z^\dagger a_z + \tfrac{3}{2} 1. \quad (5.8.35a)$$

Here the creation operators

$$a_j^\dagger = \left(\sqrt{\omega}\, x_j + i p_j/\sqrt{\omega}\right)/2 \quad (5.8.35b)$$

are the standard ones introduced in Section 4.4.D.

This Hamiltonian is obviously invariant to R_3 rotations. However, the expression of it in terms of a's shows that it is also invariant to a more

SOME HIGHER SYMMETRIES: R_4 AND U_3 427

general group of transformations. The full symmetry is the unitary group of transformations in three dimensions or U_3. We have that

$$\sum_j a_j^\dagger a_j \to \sum_j \left(\sum_k U_{jk}^* a_k^\dagger\right)\left(\sum_l U_{jl} a_l\right) = \sum_k \sum_l \left(\sum_j U_{jk}^* U_{jl}\right) a_k^\dagger a_l$$

$$= \sum_k a_k^\dagger a_k$$

for all unitary transformations for which

$$\sum_j U_{jk}^* U_{jl} = \sum_j U_{kj}^\dagger U_{jl} = \delta_{kl},$$

or in abstract notation

$$U^\dagger = U^{-1}.$$

We discuss this group U_3 now.

(a) Introduction to the Lie Algebra of U_3   The requirement that an infinitesimal operator

$$U = 1 + \varepsilon G = 1 + \frac{\varepsilon}{i} L$$

be a unitary operator gives the following relations:

$$U^{-1} = U^\dagger,$$

$$1 - \varepsilon G = 1 - \frac{\varepsilon}{i} L = 1 - \frac{\varepsilon}{i} L^\dagger = 1 + \varepsilon G^\dagger$$

$$G^\dagger = -G, \qquad L^\dagger = L. \qquad (5.8.36)$$

In U_3 one admits complex operators U and generators G. However, G must be anti-Hermitian, or what is the same thing, $L = iG$ must be Hermitian. Translating this into matrix language the generators of U_3 include the three antisymmetric Hermitian generators of R_3:

$$L_{32} = iG_{32} \to i\mathscr{G}_{32}, \qquad L_{13} = iG_{13} \to i\mathscr{G}_{13}, \qquad L_{21} = iG_{21} \to i\mathscr{G}_{21}$$

$$= \begin{pmatrix} \cdot & \cdot & \cdot \\ \cdot & \cdot & -i \\ \cdot & i & \cdot \end{pmatrix}, \qquad = \begin{pmatrix} \cdot & \cdot & i \\ \cdot & \cdot & \cdot \\ -i & \cdot & \cdot \end{pmatrix}, \qquad = \begin{pmatrix} \cdot & -i & \cdot \\ i & \cdot & \cdot \\ \cdot & \cdot & \cdot \end{pmatrix}.$$

$$(5.8.37)$$

However, there are six more *symmetric* Hermitian matrices

$$K_{32} \to \mathscr{K}_{32}, \qquad K_{13} \to \mathscr{K}_{13}, \qquad K_{21} \to \mathscr{K}_{11}$$

$$= \begin{pmatrix} \cdot & \cdot & \cdot \\ \cdot & \cdot & 1 \\ \cdot & 1 & \cdot \end{pmatrix}, \qquad = \begin{pmatrix} \cdot & \cdot & 1 \\ \cdot & \cdot & \cdot \\ 1 & \cdot & \cdot \end{pmatrix}, \qquad = \begin{pmatrix} \cdot & 1 & \cdot \\ 1 & \cdot & \cdot \\ \cdot & \cdot & \cdot \end{pmatrix}$$

$$E_{11} \to \mathscr{E}_{11}, \qquad E_{22} \to \mathscr{E}_{22}, \qquad E_{33} \to \mathscr{E}_{33}$$

$$= \begin{pmatrix} 1 & \cdot & \cdot \\ \cdot & \cdot & \cdot \\ \cdot & \cdot & \cdot \end{pmatrix}, \qquad = \begin{pmatrix} \cdot & \cdot & \cdot \\ \cdot & 1 & \cdot \\ \cdot & \cdot & \cdot \end{pmatrix}, \qquad = \begin{pmatrix} \cdot & \cdot & \cdot \\ \cdot & \cdot & \cdot \\ \cdot & \cdot & 1 \end{pmatrix}.$$

(5.8.38)

It is convenient to define ELEMENTARY matrices or operators as follows:

$$E_{12} \to \mathscr{E}_{12} = \begin{pmatrix} \cdot & 1 & \cdot \\ \cdot & \cdot & \cdot \\ \cdot & \cdot & \cdot \end{pmatrix} = \frac{K_{21} - G_{21}}{2},$$

$$E_{13} \to \mathscr{E}_{13} = \begin{pmatrix} \cdot & \cdot & 1 \\ \cdot & \cdot & \cdot \\ \cdot & \cdot & \cdot \end{pmatrix} = \frac{K_{13} + G_{13}}{2},$$

$$E_{23} \to \mathscr{E}_{23} = \begin{pmatrix} \cdot & \cdot & \cdot \\ \cdot & \cdot & 1 \\ \cdot & \cdot & \cdot \end{pmatrix} = \frac{K_{32} - G_{32}}{2},$$

$$E_{21} \to \mathscr{E}_{21} = \begin{pmatrix} \cdot & \cdot & \cdot \\ 1 & \cdot & \cdot \\ \cdot & \cdot & \cdot \end{pmatrix} = \frac{K_{21} + G_{21}}{2},$$

$$E_{31} \to \mathscr{E}_{31} = \begin{pmatrix} \cdot & \cdot & \cdot \\ \cdot & \cdot & \cdot \\ 1 & \cdot & \cdot \end{pmatrix} = \frac{K_{13} - G_{13}}{2},$$

$$E_{32} \to \mathscr{E}_{32} = \begin{pmatrix} \cdot & \cdot & \cdot \\ \cdot & \cdot & \cdot \\ \cdot & 1 & \cdot \end{pmatrix} = \frac{K_{32} + G_{32}}{2}. \qquad (5.8.39)$$

The elementary operators satisfy the following commutation relations:

$$[E_{ij}, E_{kl}] = \delta_{jk} E_{il} - \delta_{li} E_{kj}. \qquad (5.8.40)$$

The elementary operators are already in a form that conforms to the Cartan root vector formalism introduced in the previous section. However, there are now three operators E_{11}, E_{22}, and E_{33} which commute with each other. Hence, there are three components for each root vector, as in the following example:

$$[E_{11}, E_{12}] = E_{12},$$
$$[E_{22}, E_{12}] = -E_{12}, \quad R(12) = \begin{pmatrix} 1 \\ -1 \\ 0 \end{pmatrix}.$$
$$[E_{33}, E_{12}] = 0,$$

This root vector is drawn in Figure 5.8.4(a), along with five others corresponding to the generators in Eq. (5.8.39). Note that the roots form a hexagon as indicated in Figure 5.8.4.

It is convenient to define combinations of the E_{11}, E_{22}, and E_{33} that correspond to the symmetry axes H_1, H_2, and H_3 of the hexagon:

$$H_1 \equiv (E_{11} - E_{33})/\sqrt{2N},$$
$$H_2 \equiv (E_{11} - 2E_{22} + E_{33})/\sqrt{6N},$$
$$H_3 \equiv (E_{11} + E_{22} + E_{33})/\sqrt{3N}. \quad (5.8.41)$$

By discarding the unit generator H_3 we obtain a Lie algebra of a group called SU_3 or A_2. By choosing the constant $N = 6$ and defining

$$E_1 \equiv E_{12}/\sqrt{6}, \quad E_2 \equiv E_{23}/\sqrt{6}, \quad E_3 \equiv E_{13}/\sqrt{6},$$
$$E_{-1} \equiv E_{21}/\sqrt{6}, \quad E_{-2} \equiv E_{32}/\sqrt{6}, \quad E_{-3} \equiv E_{31}/\sqrt{6}, \quad (5.8.42)$$

we obtain Cartan's standard form for the SU_3 algebra with the commutation relations given in terms of the roots in Figure 5.8.4(c):

$$[H_i, E_\alpha] = r_i(a) E_\alpha, \quad [E_\alpha, E_{-\alpha}] = \sum_j r_j(\alpha) H_j,$$
$$[E_\alpha, E_\beta] = \sqrt{j(k+1)/2} \, |r(\alpha)| E_{\alpha+\beta}.$$

The integers j and k are given by

j = number of times that $\mathbf{r}(\alpha)$ can be raised by $\mathbf{r}(\beta)$,

and $k - j = \dfrac{2r(\alpha) \cdot r(\beta)}{r(\beta) \cdot r(\beta)}$,

k = number of times that $\mathbf{r}(\alpha)$ can be lowered by $-r(\beta)$. (5.8.43)

Cartan has shown that these relations are valid for an important class of Lie algebras.

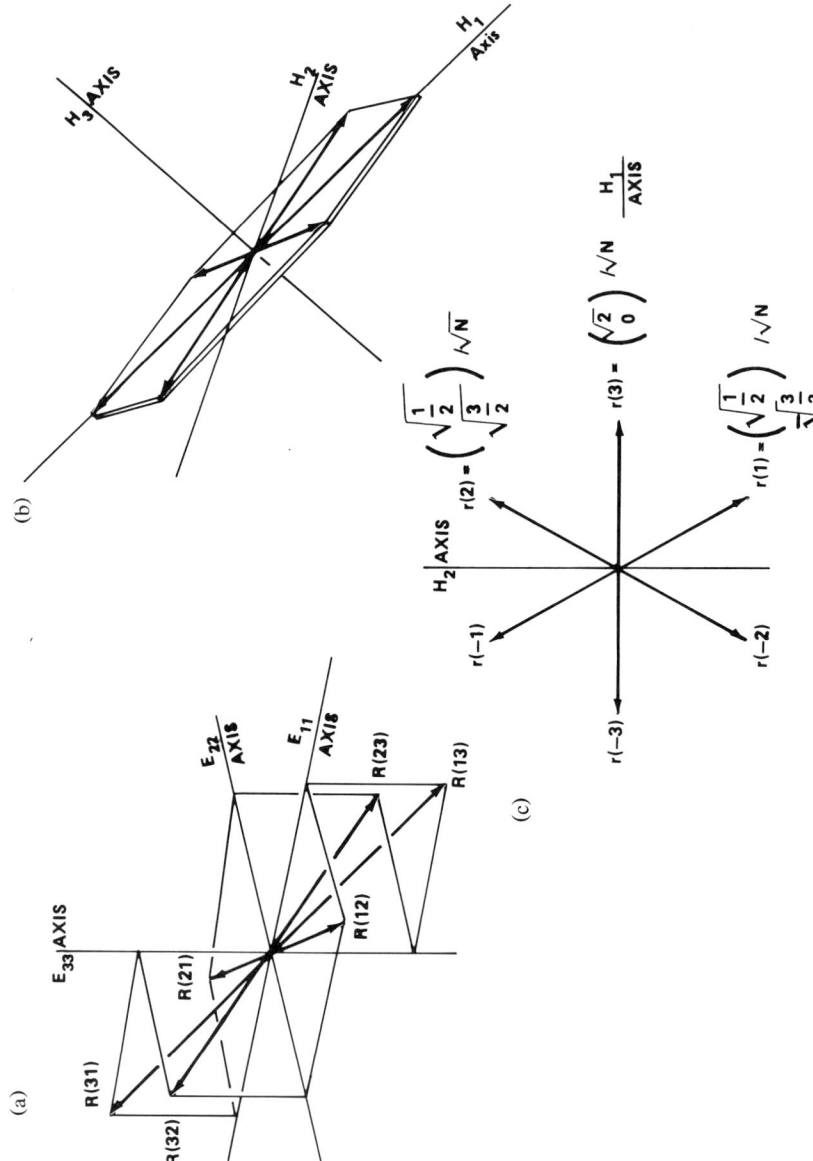

Figure 5.8.4 Root diagrams for Lie algebras of U_3 and SU_3. (a) Elementary form of U_3 root diagram is based on the commutation structure of elementary E_{ij} operators. (b) The U_3 roots $R(ij)$ lie in a hexagonal plane. (c) The normalized Cartan forms of the SU_3 roots form a hexagon.

(b) Introduction to Irreducible Representations of U_3 and SU_3 As shown in the discussion of R_4, the base states of Lie algebra irreps have weight vectors which may be raised or lowered by root vectors. In U_3 we call the vectors $r(1)$, $r(2)$, and $r(3)$ raising or POSITIVE root vectors, while $r(-1)$, $r(-2)$, and $r(-3)$ are called lowering or NEGATIVE root vectors.

Now some examples of weight vectors for SU_3 irreps are shown in Figure 5.8.5. For each set of vectors there is always one called the HIGHEST WEIGHT $M^{n_1 n_2}$ which cannot be raised without going outside the set for that irrep. The Cartan theory gives the following relations between $M^{n_1 n_2}$ and the positive root vectors $r(1)$ and $r(2)$,

$$n_1 = 2 M^{n_1 n_2} \cdot r(1) / r(1) \cdot r(1)$$
$$= \text{number of times } M \text{ can be lowered by } r(-1),$$
$$n_2 = 2 M^{n_1 n_2} \cdot r(2) / r(2) \cdot r(2)$$
$$= \text{number of times } M \text{ can be lowered by } r(-2). \quad (5.8.44)$$

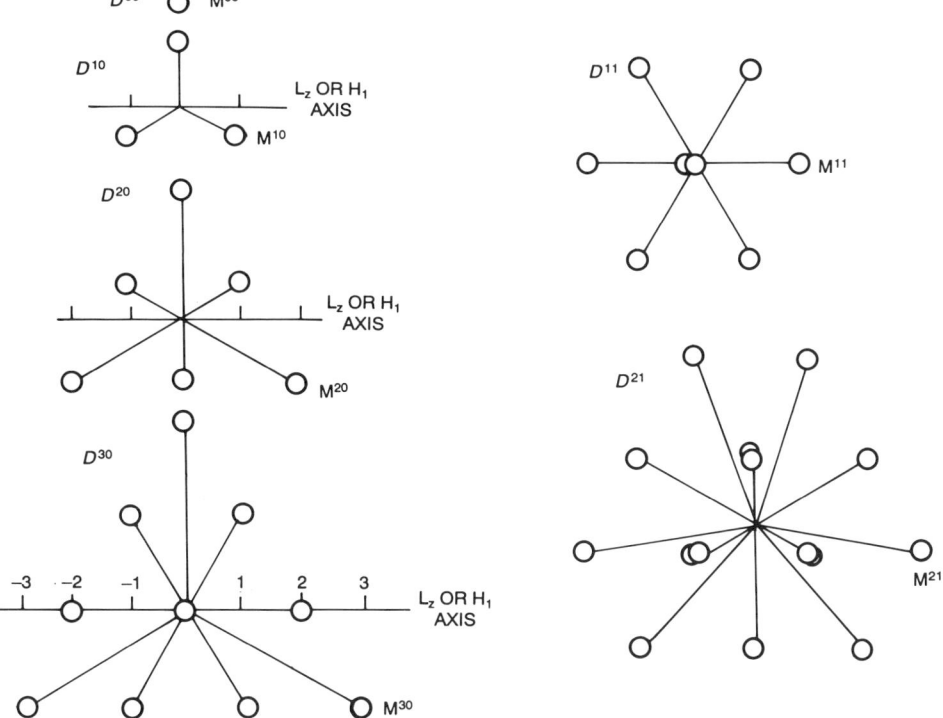

Figure 5.8.5 Examples of weight diagrams for irreducible representations of SU_3.

The vector relations determine the weight vectors of each SU_3 irrep, i.e., the eigenvalues of $\mathscr{D}^{n_1 n_2}(H_j)$. Further theory is needed to find $\mathscr{D}^{n_1 n_2}(E)$.

(c) Harmonic Oscillator Spectrum It happens that the SU_3 irreps $\mathscr{D}^{n_1 n_2}$ with $n_2 = 0$ correspond to single-particle energy levels of the three-dimensional harmonic oscillator. These correspond to weight diagrams on the left-hand side of Figure 5.8.5. The irreps with $n_2 \neq 0$ describe multiparticle states.

The first oscillator level is nondegenerate, corresponding to the irrep $\mathscr{D}^{00}$. The second level is triply degenerate corresponding to irrep $\mathscr{D}^{10}$. The states in the second level are one quantum each of excitation of motion in the x, y, or z directions. We may also describe them as angular-momentum $p(=1)$ states $\left|\begin{smallmatrix}1\\1\end{smallmatrix}\right\rangle$, $\left|\begin{smallmatrix}1\\0\end{smallmatrix}\right\rangle$ and $\left|\begin{smallmatrix}1\\-1\end{smallmatrix}\right\rangle$. In fact, we may interpret the H_1 axis as the orbital momentum (L_z) axis since operators $2\sqrt{3}\,H_1$, $\sqrt{6}\,E_1$, and $\sqrt{6}\,E_{-1}$ behave like ordinary angular-momentum operators L_z, L_+, and L_-, respectively.

The third oscillator level belongs to the $\mathscr{D}^{20}$ irrep and has a sixfold degeneracy: five $d(l=2)$ states and one $s(l=0)$ orbital, or, equivalently, the six combinations of double excitations xx, yy, zz, xy, xz, and yz. The fourth level belongs to $\mathscr{D}^{30}$ with a $f(l=3)$ and a $p(l=1)$ orbital, and so forth.

(d) Comparing Coulomb and Harmonic Oscillator Spectra It is interesting to compare the oscillator and Coulomb problems. In some ways these two high-symmetry examples are extreme opposites. The Coulomb problem treats a charged particle orbiting outside a single attractive point charge. The oscillator problem treats the same particle orbiting inside a large uniform spherical cloud of attractive charge. The Coulomb problem is a good starting point for atomic and molecular structure theory. The oscillator spectrum has been shown to offer a similar, though less easily understood, starting point for the internal structure of nuclei.

In atomic shell theory we imagine filling each Coulomb level in Figure 5.8.6(a) with one pair of electrons. The *magic numbers* like 2 or 10 are the numbers of electrons which give a closed-shell atom, i.e., $(1s)^2$ = He, $(1s)^2(2s+2p)^8$ = Ne, which have greater chemical stability. Similarly, in nuclear shell theory the magic numbers 2, 8, or 20 give the numbers of protons or neutrons which give closed-shell nuclei, i.e., $_2\text{He}_2^4$, $_8\text{O}_8^{16}$, $_{20}\text{Ca}_{20}^{40}$ which have exceptionally great binding energy. The theory of nuclear shells is based upon the oscillator spectrum in Figure 5.8.6(b).

These single-particle theories work well, up to a point. For example, in potassium [next atom after argon $((1s)^2(2s)^2(2p)^6(3s)^2(3p)^6))$] it appears that the next electron does not settle in the $3d$ shell, but rather in a $4s$ level. One way to explain this is to say that the presence of the other electrons makes a deeply diving $4s$ level have lower energy than the $3d$.

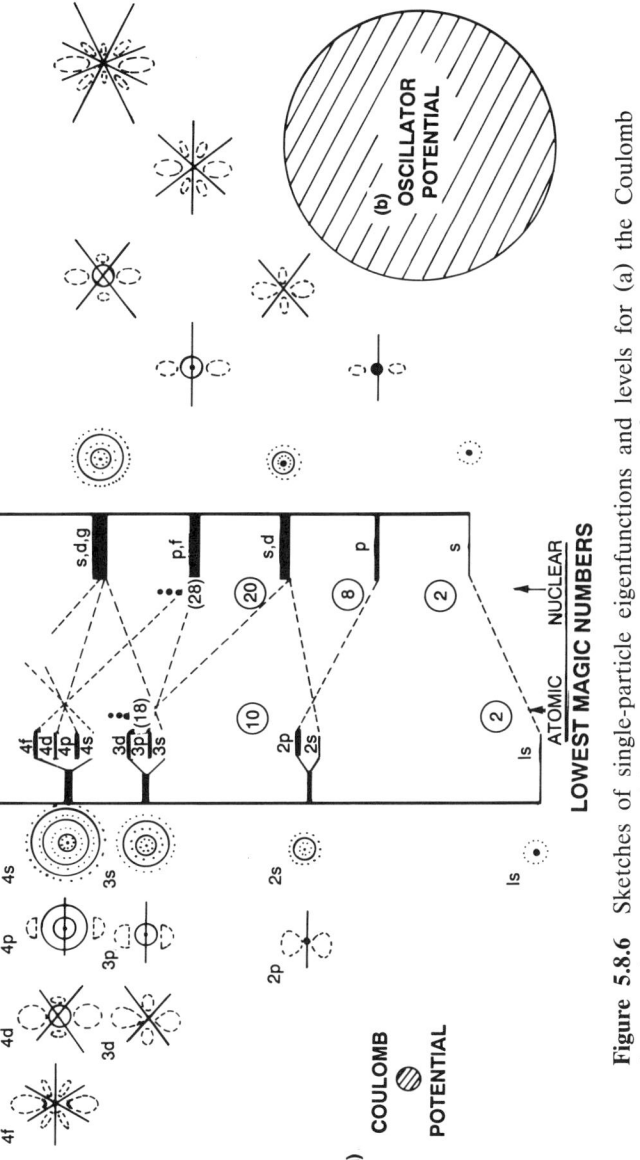

Figure 5.8.6 Sketches of single-particle eigenfunctions and levels for (a) the Coulomb ($V = k/r$) potential, and (b) the harmonic oscillator ($V = kr^2$) potential.

434 REPRESENTATIONS OF CONTINUOUS ROTATION GROUPS AND APPLICATIONS

The explanation of higher nuclear shell structure requires a more detailed theory, too, including spin-orbit and internuclear interactions. The next nuclear magic numbers turn out to be 28, 50, 82, and 126.

APPENDIX E. DERIVATION OF ANGULAR-MOMENTUM REPRESENTATIONS

In the following the matrices representing the generators or angular-momentum operators will be derived using their commutation relations. (E.1):

$$[J_1, J_2] = i\hbar J_3, \quad [J_2, J_3] = i\hbar J_1, \quad [J_3, J_1] = i\hbar J_2. \quad \text{(E.1)}$$

We will choose to find the bases in which J_3 and J_z are diagonal, and therefore it will be convenient to rewrite (E.1) in the form (E.2),

$$[J_3, J_+] = \hbar J_+, \quad [J_3, J_-] = -\hbar J_- \quad [J_+, J_-] = 2\hbar J_3, \quad \text{(E.2)}$$

where $J_+ = J_1 + iJ_2$, and $J_- = J_1 - iJ_2$ play there the role of eigenvectors under the transformation $[J_3, \ ]$. In deriving (E.2) the relations below were used, and will be needed again shortly (let us also take $\hbar = 1$):

$$J_+ J_- = J_1^2 + J_2^2 - i[J_1, J_2] = J_1^2 + J_2^2 + \hbar J_3 = J_1^2 + J_2^2 + J_3, \quad \text{(E.3)}$$

$$J_- J_+ = J_1^2 + J_2^2 - \hbar J_3 = J_1^2 + J_2^2 - J_3. \quad \text{(E.4)}$$

The operator $J^2 = J_1^2 + J_2^2 + J_3^2$ can be expressed in terms of $J_\pm$ and J_3 as in (E.5) and (E.6):

$$J^2 = J_+ J_- + J_3^2 - J_3 \quad \text{(E.5)}$$

$$= J_- J_+ + J_3^2 + J_3. \quad \text{(E.6)}$$

The preceding makes it easy to prove that J_3 commutes with J^2, as in the following, where Eq. (E.2) was used:

$$[J_3, J^2] = [J_3, J_+ J_-] = [J_3, J_+]J_- + J_+[J_3, J_-] = 0. \quad \text{(E.7)}$$

The J_3 and J^2 are commuting Hermitian operators, and so orthonormal vectors $\left|{\alpha \atop m}\right\rangle$ exist which are eigenvectors of both J^2 and J_3. It is necessary to find what values α and m can be:

$$J^2\left|{\alpha \atop m}\right\rangle = \alpha\left|{\alpha \atop m}\right\rangle, \quad J_3\left|{\alpha \atop m}\right\rangle = m\left|{\alpha \atop m}\right\rangle. \quad \text{(E.8)}$$

First, note that the operators $J_\pm$ make eigenvectors with higher or lower eigenvalues of J_3:

$$J_3 J_+ \left|\begin{matrix}\alpha\\m\end{matrix}\right\rangle = ([J_3, J_+] + J_+ J_3)\left|\begin{matrix}\alpha\\m\end{matrix}\right\rangle = (J_+ + mJ_+)\left|\begin{matrix}\alpha\\m\end{matrix}\right\rangle$$

$$= (m+1)J_+\left|\begin{matrix}\alpha\\m\end{matrix}\right\rangle, \quad (E.9)$$

$$J_3 J_- \left|\begin{matrix}\alpha\\m\end{matrix}\right\rangle = (m-1)J_-\left|\begin{matrix}\alpha\\m\end{matrix}\right\rangle. \quad (E.10)$$

These operators $J_\pm$ are called RAISING or LOWERING operators because of this effect. We shall assure that the vector $J_\pm\left|\begin{matrix}\alpha\\m\end{matrix}\right\rangle$ is proportional to $\left|\begin{matrix}\alpha\\m\pm 1\end{matrix}\right\rangle$.

Let us now inquire about just how far one may "raise" or "lower." We observe from (E.5) and (E.6) that α must be equal or greater than m or $-m$:

$$\alpha - m = \left\langle\begin{matrix}\alpha\\m\end{matrix}\right|J^2 - J_3\left|\begin{matrix}\alpha\\m\end{matrix}\right\rangle = \left\langle\begin{matrix}\alpha\\m\end{matrix}\right|J_- J_+ + J_3^2\left|\begin{matrix}\alpha\\m\end{matrix}\right\rangle = \left\langle\begin{matrix}\alpha\\m\end{matrix}\right|J_+^\dagger J_+ + J_3^\dagger J_3\left|\begin{matrix}\alpha\\m\end{matrix}\right\rangle \geq 0, \quad (E.11)$$

$$\alpha + m = \left\langle\begin{matrix}\alpha\\m\end{matrix}\right|J^2 + J_3\left|\begin{matrix}\alpha\\m\end{matrix}\right\rangle = \left\langle\begin{matrix}\alpha\\m\end{matrix}\right|J_+ J_- + J_3^2\left|\begin{matrix}\alpha\\m\end{matrix}\right\rangle = \left\langle\begin{matrix}\alpha\\m\end{matrix}\right|J_-^\dagger J_- + J_3^\dagger J_3\left|\begin{matrix}\alpha\\m\end{matrix}\right\rangle \geq 0. \quad (E.12)$$

In the last step of (E.11) and (E.12) we use the fact that diagonal elements of any operator $A^\dagger A$ is positive or zero, as seen in the following:

$$\langle a|A^\dagger A|a\rangle = \sum_b \langle a|A^\dagger|b\rangle\langle b|A|a\rangle = \sum_b |\langle a|A|b\rangle|^2 \geq 0.$$

So for a given α, there is a limit to the raising process. Suppose $m = j$ is the highest value of m for a given α. Then (E.13) must hold:

$$J_+\left|\begin{matrix}\alpha\\j\end{matrix}\right\rangle = 0. \quad (E.13)$$

Applying (E.6) and the foregoing we discover the value of α:

$$J^2\left|\begin{matrix}\alpha\\j\end{matrix}\right\rangle = (J_- J_+ + J_3^2 + J_3)\left|\begin{matrix}\alpha\\j\end{matrix}\right\rangle = j(j+1)\left|\begin{matrix}\alpha\\j\end{matrix}\right\rangle. \quad (E.14)$$

This eigenvalue $\alpha = j(j+1)$ applies to all lowered states $J_-\left|\begin{matrix}\alpha\\j\end{matrix}\right\rangle$, i.e., all states $\left|\begin{matrix}\alpha\\m\end{matrix}\right\rangle$ since J_- commutes with J^2. But, this lowering is limited, too, say to $m = j'$ as in Eq. (E.15):

$$J_- \left| \begin{matrix} \alpha \\ j' \end{matrix} \right\rangle = 0. \tag{E.15}$$

Applying (E.5) exactly as in the foregoing we get the following:

$$J^2 \left| \begin{matrix} \alpha \\ j' \end{matrix} \right\rangle = j'(j' - 1) \left| \begin{matrix} \alpha \\ j' \end{matrix} \right\rangle,$$

from which we can only conclude $j(j + 1) = j'(j' - 1)$. The two solutions $j' = j + 1$ and $j' = -j$ are indicated, but only the latter is acceptable since we said $m = j$ was the largest.

So, for a given $\alpha = j(j + 1)$, the m range in steps of 1, between $m = +j$ and $m = -j$, inclusive. We see that this is possible only if $j = 0, \frac{1}{2}, 1, \frac{3}{2}, \ldots$.

Now the representation $\mathscr{D}^j$ of $J_\pm$ in (E.16) can be produced for any j.

$$J_+ \left| \begin{matrix} j \\ m \end{matrix} \right\rangle = \mathscr{D}^j_{m+1, m}(J_+) \left| \begin{matrix} j \\ m+1 \end{matrix} \right\rangle, \quad J_- \left| \begin{matrix} j \\ m \end{matrix} \right\rangle = \mathscr{D}^j_{m-1, m}(J_-) \left| \begin{matrix} j \\ m-1 \end{matrix} \right\rangle \tag{E.16}$$

By taking the scalar product the following relation results:

$$\left| \mathscr{D}^j_{m+1, m}(J_+) \right|^2 = \left\langle \begin{matrix} j \\ m \end{matrix} \middle| J_+^\dagger J_+ \middle| \begin{matrix} j \\ m \end{matrix} \right\rangle = \left\langle \begin{matrix} j \\ m \end{matrix} \middle| J_- J_+ \middle| \begin{matrix} j \\ m \end{matrix} \right\rangle,$$

$$\left| \mathscr{D}^j_{m-1, m}(J_-) \right|^2 = \left\langle \begin{matrix} j \\ m \end{matrix} \middle| J_-^\dagger J_- \middle| \begin{matrix} j \\ m \end{matrix} \right\rangle = \left\langle \begin{matrix} j \\ m \end{matrix} \middle| J_+ J_- \middle| \begin{matrix} j \\ m \end{matrix} \right\rangle.$$

Applying (E.5) and (E.6), respectively, gives the desired result to within a phase.

$$\left| \mathscr{D}^j_{m+1, m}(J_+) \right|^2 = \left\langle \begin{matrix} j \\ m \end{matrix} \middle| J^2 - J_3^2 - J_3 \middle| \begin{matrix} j \\ m \end{matrix} \right\rangle$$
$$= (j(j+1) - m^2 - m) = (j - m)(j + m + 1),$$

$$\left| \mathscr{D}^j_{m-1, m}(J_-) \right|^2 = \left\langle \begin{matrix} j \\ m \end{matrix} \middle| J^2 - J_3^2 + J_3 \middle| \begin{matrix} j \\ m \end{matrix} \right\rangle$$
$$= (j(j+1) - m^2 + m) = (j + m)(j - m + 1).$$

It is conventional to choose positive real phases:

$$\mathscr{D}^j_{m+1, m}(J_+) = \sqrt{j(j+1) - m(m+1)},$$
$$\mathscr{D}^j_{m-1, m}(J_-) = \sqrt{j(j+1) - m(m-1)}. \tag{E.17}$$

ADDITIONAL READING

The most comprehensive treatment of quantum theory of angular-momentum and rotation-group theory are two volumes of the *Encyclopedia of Mathematics and Its Applications*.

L. C. Biedenharn and J. D. Louck, *Angular Momentum in Quantum Physics: Theory and Application*, Vol. 8, Encyclopedia of Mathematics, edited by G. C. Rota (Addison-Wesley, Reading, MA, 1981).

L. C. Biedenharn and J. D. Louck, *The Racah Wigner Algebra in Quantum Theory*, Vol. 9, Encyclopedia of Mathematics, edited by G. C. Rota (Addison-Wesley, Reading, MA, 1981).

These two volumes are valuable for their history and references as well as for their content. A shorter and more "user friendly" reference is the recent monograph by Zare.

R. N. Zare, *Angular Momentum: Understanding Spatial Aspects in Chemistry and Physics* (Wiley Interscience, New York, 1988).

This book is aimed at applications in atomic and molecular physics, particularly diatomic and nearly symmetric top molecules.

A slightly older and more advanced book along the same lines is by Judd. It also includes a treatment of $O(4)$.

Brian R. Judd, *Angular Momentum Theory for Diatomic Molecules* (Academic, New York, 1975).

Judd is also the author of another classic text on angular-momentum calculus.

B. R. Judd, *Operator Techniques in Atomic Spectroscopy* (McGraw-Hill, New York, 1963).

Several of the well-known older texts which treat angular-momentum theory are listed here.

E. Feinberg and G. E. Pake, *Notes on the Quantum Theory of Angular Momentum* (Addison-Wesley, Reading, MA, 1953).

M. E. Rose, *Elementary Theory of Angular Momentum* (Wiley, New York, 1957).

E. P. Wigner, *Group Theory*, (Academic, New York, 1959).

A. R. Edmonds, *Angular Momentum in Quantum Mechanics*, (Princeton University Press, Princeton, NJ, 1960).

D. M. Brink and G. R. Satchler, *Angular Momentum* (Oxford University Press, London, 1968).

A collection of famous papers on angular momentum are found in the following volumes. The first volume contains Schwinger's original boson algebraic treatment of angular momentum.

L. C. Biedenharn and H. Van Dam, Eds., *Quantum Theory of Angular Momentum* (Academic, New York, 1965).

E. M. Loebl, Ed., *Group Theory and Its Applications* (Academic, New York, 1968), Vols. I and II.

The quantum rotor and its angular-momentum theory probably originated with Casimir.

H. B. G. Casimir, *Rotation of a Rigid Body in Quantum Mechanics* (J. B. Walter's Vitgevers-Maatschappij, N. V. Gronigen den Haag Batavia, 1931).

References for the theory of quaternions and the $U(2)$ slide-rule wave given after Chapter 3. See also (if you can find it).

C. J. Joly, *Manual of Quaternions* (Macmillan, New York, 1965).

The application of $O(4)$ symmetry to the Coulomb problem is discussed in the following.

M. J. Englefield, *Group Theory and the Coulomb Problem* (Wiley Interscience, New York, 1972).

See also an article by Carl Wulfman in Volume II of Loebl (just listed). This led to a renaissance in the application of $O(4)$ and related methods to the spectra of doubly excited atoms.

C. E. Wulfman, *Chem. Phys. Lett.*, **23**, 370 (1973).

O. Sinanoglu and D. R. Herrick, *J. Chem. Phys.*, **62**, 886 (1975).

D. R. Herrick and M. E. Kellman, *Phys. Rev. A*, **21**, 418 (1980).

M. E. Kellman and D. R. Herrick, *Phys. Rev. A*, **22**, 1536 (1980).

D. R. Herrick, M. E. Kellman, and R. D. Poliah, *Phys. Rev. A*, **22** 1517 (1980).

P. Rehmus and R. S. Berry, *Chem. Phys.*, **38**, 257 (1978).

P. Rehmus, M. E. Kellman, and R. S. Berry, *Chem. Phys.*, **31**, 239 (1978).

D. R. Herrick, *Adv. Chem. Phys.*, **52**, 1 (1982).

PROBLEMS

Section 5.1

5.1.1 Suppose an n-by-n real matrix R is constructed with ($n \geq 2$) so that its columns consist of n mutually orthonormal real n-dimensional vectors $\Psi_1, \Psi_2 \cdots \Psi_n \cdots$. Suppose we construct two new vectors ϕ_1 and ϕ_2 by extracting first and second *rows*, respectively, of R.

(a) What (if anything) can we say about the values of scalar products $\phi_1 \cdot \phi_1$? Or $\phi_1 \cdot \phi_2$? Given R could you *quickly* give the values of scalar products $\phi_1 \cdot \Psi_1$? Or $\phi_1 \cdot \Psi_2$?

(b) Answer similarly worded questions involving *complex* matrix J and scalar products $\langle \phi_1 | \phi_1 \rangle$, $\langle \phi_1 | \phi_2 \rangle$, $\langle \phi_1 | \Psi_1 \rangle$, and $\langle \phi_1 | \Psi_2 \rangle$ of complex vectors.

(c) What type of matrices are R and J using the terminology of Chapter 1?

5.1.2 (a) What is the trace of a proper orthogonal 3-by-3 transformation matrix O_{ij} that performs a rotation of angle ϕ around the z axis? What is the trace if the same rotation is around the x axis? ...Or around the (111) axis?

(b) What is the trace of the corresponding *improper* operation?

(c) Suppose instead of using a set of Cartesian unit vectors $\{|1\rangle = e_1, |2\rangle = e_2, |3\rangle = e_3\}$ to construct matrix $O_{ij} = \langle i|O|j\rangle$ we use nonorthogonal vectors $|a\rangle = |1\rangle + |2\rangle$, $|b\rangle = |1\rangle + |3\rangle$, $|c\rangle = |2\rangle + |3\rangle$. By how much will the trace of matrix $O_{ab} = \langle a|O|b\rangle$ differ from that of O_{ij}?

5.1.3 A crystal symmetry group operation of translation t or rotation **R** may be defined using a basis of three primitive lattice vectors $\mathbf{l}_1$, $\mathbf{l}_2$ and $\mathbf{l}_3$ or three reciprocal vectors $\mathbf{r}_1 = \mathbf{l}_2 \times \mathbf{l}_3/v$, $\mathbf{r}_2 = \mathbf{l}_3 \times \mathbf{l}_1/v$, and $\mathbf{r}_3 = \mathbf{l}_1 \times \mathbf{l}_2/v$, where $\mathbf{v} = \mathbf{l}_1 \times \mathbf{l}_2 \cdot \mathbf{l}_3$. The idea is that a symmetry operation must transform each lattice vector into an *integral* linear combination of lattice vectors, i.e., another vector in the lattice. Using characters, derive which rotational angles ω are possible for pure rotations $\mathbf{R}[\omega]$ and for improper $\mathbf{R} \cdot \mathbf{I}$ rotations around any lattice point in a two- or three-dimensional lattice. Which of the following point groups correspond to possible crystal point symmetries? D_{2h}? D_{3h}? D_{4h}? D_{5h}? D_{6h}? D_{7h}? D_{2d}? D_{3d}? D_{4d}?

5.2.1 Construct a partial correlation table involving irreps listed in Fig. 5.2.2 for the reduction of symmetry from $D_{\infty h}$ to subgroups

(a) D_{2h}.
(b) D_{3h}.
(c) D_{4h}.
(d) D_{5h}.
(e) D_{6h}.

Sketch each correlation by drawing a level-splitting diagram.

Section 5.3

5.3.1 Use the Cartesian $\{|e_1\rangle, |e_2\rangle, |e_3\rangle\}$ xyz basis to make the 3-by-3 matrix $R_{ij}(\alpha\beta\gamma) = \langle e_i|R(\alpha\beta\gamma)|e_j\rangle$ representing the Euler-angle rotation operator $R(\alpha\beta\gamma)$ defined by (5.3.4).

5.3.2 Suppose you have constructed a rotation matrix $R_{ij}(\alpha 00)$ for rotation α about the z axis and another matrix $R_{ij}(0\beta 0)$ for rotation β about the y axis.

(a) How would you combine these two operations to make one which rotated by angle α around an axis z' that lies in the xz plane at angle β from z. [Hint: z' is obtained from z by rotation

$R(0\beta 0)$. Use class conjugation transformation: $r' = trt^{-1}$.] Test your construction by making a rotation by angle α around the x axis.

(b) How would you combine these two operations to make a Darboux defined rotation $R_{ij}[\phi\theta\omega]$ by angle ω around an axis at azimuth ϕ and polar angle θ.

(c) Carry out the operations found in part (b) to construct the 3-by-3 matrix $R_{ij}[\phi\theta\omega]$.

5.3.3 (a) Determine the nine direction cosines $\hat{\mathbf{x}} \cdot \hat{\bar{\mathbf{x}}} = \cos(x\bar{x})$, $\hat{\mathbf{x}} \cdot \hat{\bar{\mathbf{y}}} = \cos(x\bar{y}), \ldots,$ $\hat{\mathbf{z}} \cdot \hat{\bar{\mathbf{z}}}$ in terms of Euler angles $(\alpha\beta\gamma)$ between Cartesian systems $(\mathbf{x}, \mathbf{y}, \mathbf{z})$ and $(\bar{\mathbf{x}} = \mathbf{R}(\alpha\beta\gamma) \cdot \mathbf{x}, \bar{\mathbf{y}} = \mathbf{R} \cdot \mathbf{y}, \bar{\mathbf{z}} = \mathbf{R} \cdot \mathbf{z})$.

(b) Use these results to determine the Euler angles of the transformation

$$R(\alpha\beta\gamma) = \begin{pmatrix} \langle x|\bar{x}\rangle & \langle x|\bar{y}\rangle & \langle x|\bar{z}\rangle \\ \langle y|\bar{x}\rangle & \langle y|\bar{y}\rangle & \langle y|\bar{z}\rangle \\ \langle z|\bar{x}\rangle & \langle z|\bar{y}\rangle & \langle z|\bar{z}\rangle \end{pmatrix}$$

$$= \begin{pmatrix} 1/\sqrt{2} & 1/\sqrt{6} & 1/\sqrt{3} \\ -1/\sqrt{2} & 1/\sqrt{6} & 1/\sqrt{3} \\ 0 & -2/\sqrt{6} & 1/\sqrt{3} \end{pmatrix} = R[\phi\theta\omega]$$

(c) Find axis (Darboux) angles as well. (It may help to look at Problem 5.3.2 first.)

Section 5.4

5.4.1 Verify (5.4.30).

5.4.2 Complete the calculation of $D^1(\alpha\beta\gamma)$ and compare the resulting matrix with the result of Problem 5.3.1. Are their traces equal? Are they equivalent? What change of basis would relate them?

5.4.3 Complete the calculation of $D^2(\alpha\beta\gamma)$.

5.4.5 Suppose the spinor representation of an $U(2)$ group operation is $D^{1/2}(u) = \begin{pmatrix} u_{11} & u_{12} \\ u_{21} & u_{22} \end{pmatrix}$. Derive the formula for the angular-momentum j representation $D^j_{mn}(u)$ in terms of $u_{11}, u_{12}, u_{21}, u_{22}, j, m,$ and n.

5.4.6 Derive phase symmetry relations for D functions.

Write: $D^j_{-m-n}(0\beta 0)$ in terms of $D^j_{mn}(0\beta 0)$.

$D^j_{nm}(0\beta 0)$ in terms of $D^j_{mn}(0\beta 0)$.

$D^j_{mn}(0 - \beta 0)$ in terms of $D^j_{mn}(0\beta 0)$.

$D^j_{mn}(0\pi - \beta 0)$ in terms of $D^j_{-mn}(0\beta 0)$, and/or D^j_{m-n}.

$D^{j*}_{mn}(\alpha\beta\gamma)$ in terms of $D^j_{-m-n}(\alpha\beta\gamma)$.

Section 5.5

5.5.1 Verify the expression (5.5.1) for $D^{1/2}[(\phi\theta\omega)]$ and compare with the expression (5.4.30) for $D^{1/2}(\alpha\beta\gamma)$. Verify Euler to Darboux angle relations (5.5.2).

5.5.2 Compute the Darboux parameterized matrix $D^1[\phi\theta\omega]$. Compare to the real vector representation found in Problem 5.3.2(c).

5.5.3 (a) Write a closed-form conversion equations for Euler angles $(\alpha\beta\gamma)$ in terms of axis (Darboux) angles $[\phi\theta\omega]$ and vice versa.

(b) For the following rotations compute Euler angles and axis angles. (These octahedral operations are shown in Figure 4.1.2.)

		Euler angles			Axis angles		
		α	β	γ	ϕ	θ	ω
180° x axis	R_{1^2}	___	___	___	___	___	___
120° (111) axis:	r_1	___	___	___	___	___	___
180° diagonal:	i_1	___	___	___	___	___	___
90° x-axis:	R_1	___	___	___	___	___	___

5.5.4 Determine the eigenvectors and eigenvalues of the quantum rotor Hamiltonian (5.5.52) for all levels having total angular momentum $j = 1$ and $j = 2$ for rigid molecules having the following principal inertia moments. (Let $\hbar = 1$.) Sketch spectra of levels.
(a) $I_1 = I_2 = 1; I_3 = 2$.
(b) $I_1 = I_2 = 10; I_3 = 1$.
(c) $I_1 = 1; I_2 = 2; I_3 = 3$.

5.5.5 Express the symmetric top Hamiltonian $BJ_{\bar{x}}^2 + BJ_{\bar{y}}^2 + CJ_{\bar{z}}^2$ in terms of Euler angles and momenta $(J_\alpha J_\beta J_\gamma)$. Simplify the expression for the following cases and give classical equations of motion for each:
(a) J along lab z axis.
(b) J along body z axis.

Section 5.6

5.6.1 Construct a correlation table up to $l = 12$ for the symmetry breaking from $O(3)$ to D_6.

5.6.2 Construct a correlation table between $O(3)$ irreps $(0^+ 0^- 1^+ 1^- 2^+ \cdots)$ and (a) tetrahedral (T_d) irreps; and (b) icosahedral (I_h) irreps.

5.6.3 Construct the lowest-order spherical harmonic combination (besides $l = 0$) which has (a) tetrahedral (T_d) symmetry; and (b) icosahedral (I_h) symmetry.

5.6.4 An $l = 3$ (f) orbital is subject to an octahedral $V_4 = [x^4 + y^4 + z^4 - (2/5)r^4]$ potential.
  (a) Calculate the O_h or cubic eigenstates of this potential using octahedral projection operators.
  (b) Sketch a representative eigenfunction for each energy level and estimate which levels are highest and which are lowest.

CHAPTER 6

THEORY AND APPLICATIONS OF SYMMETRY REPRESENTATION PRODUCTS (FINITE GROUPS)

So far we have studied the theory in which representations of symmetry can be analyzed into sums ($\oplus$) of irreducible representations (irreps) and have seen how this helps to simplify certain physical problems. This chapter is devoted to a special kind of *product* ($\otimes$) of representations and the study of its physical applications. Let us begin by listing some of the applications of the product analysis which we will treat later in this chapter and elsewhere in the book.

(i) We have seen how the Cartesian coordinates $\{xyz\}$ or unit vectors $\{\hat{x}\hat{y}\hat{z}\}$ are bases for one or more irreps of a symmetry group. For example, in Section 4.2.A the vectors $\{\hat{x}\hat{y}\hat{z}\}$ were shown to be a basis of $\mathscr{D}^{T_{1u}}$ of O_h. Furthermore, the vectors $\{\hat{x}\hat{y}\}$ are a basis of $\mathscr{D}^E$ of $\mathscr{D}_4$, while $\hat{z}$ is a basis of $\mathscr{D}^{A_2}$. In Section 6.3 we shall see how product analysis makes similar assignments for polynomials $\cdots x^2 y \cdots$ of coordinates and for tensors $\cdots \hat{x}\hat{x}\hat{y} \cdots$ of arbitrary rank.

(ii) We have seen how base states $\left|{\alpha \atop 1}\right\rangle, \left|{\alpha \atop 2}\right\rangle, \ldots$ that are partners in a basis for a given irrep $\mathscr{D}^\alpha$ of a symmetry group are made by projection. Section 6.4 shows how product analysis makes all possible tensor operators out of products $\left|{\alpha \atop i}\right\rangle\left\langle{\beta \atop j}\right|$ so that they are also partners of various irreps. Certain properties of the matrix elements of these operators called selection rules are derived and the Wigner-Eckart theorem is discussed.

(iii) We have seen how the first quantum excitations of vibrations in molecules are described by a single excitation of a mode belonging to a

443

definite irrep. Section 6.6 shows how overtone or combination-tone states with two or more excitations are described by product analysis.

(iv) We have seen how the orbital states of a single electron in some symmetry environment will belong to an irrep of the symmetry. Now we shall see how two or more electrons orbiting in the same environment can be described using product analysis. This particular application of product analysis will be used to introduce its mathematical details in Sections 6.1 and 6.2.

6.1 TWO-PARTICLE STATES AND PRODUCTS OF REPRESENTATIONS

In Section 4.3.A we discussed the model for the orbital states of a particle tunneling between potential wells fixed at the vertices of a regular octahedron. In this model there were six states $|1\rangle, |2\rangle, \ldots,$ and $|6\rangle$ corresponding to orbitals more or less localized in wells $1, 2, \ldots,$ and 6, respectively, as shown in Figure 4.3.1(b). Then Figure 4.3.2(b) shows which combinations of these states would be eigenstates according to O_h symmetry.

The same model can be made to accommodate two orbiting particles. Each basis of a two-particle system has the form $|i_A, j_B\rangle \equiv |i\rangle_A |j\rangle_B$. There are $6^2 = 36$ two-particle bases:

$$\{|1,1\rangle, |1,2\rangle, \ldots, |1,6\rangle, |2,1\rangle, |2,2\rangle, \ldots, |6,6\rangle\}. \quad (6.1.1)$$

The wave functions for these base states are drawn schematically in Figure 6.1.1. The A and B particle waves are indicated by large and small circles, respectively, in the figure. Each vector $|i_A, j_B\rangle \equiv |i\rangle_A |j\rangle_B$ denotes a state in which the first particle (say particle A) is in state i and the second one (say particle B) is in state j.

One defines a scalar product between state vector $|\phi, \psi\rangle \equiv |\phi\rangle |\psi\rangle$ and vector $\langle i, j| = |i, j\rangle^\dagger = (|i\rangle|j\rangle)^\dagger = \langle j|\langle i|$ as follows:

$$\langle i, j|\phi, \psi\rangle = \langle j|\langle i|\phi\rangle|\psi\rangle = \langle i|\phi\rangle\langle j|\psi\rangle. \quad (6.1.2)$$

This quantity gives the amplitude for the system in state $|\phi, \psi\rangle$ to have particle A in state i and particle B in state j. The probability or intensity $|\langle i, j|\phi, \psi\rangle|^2$ of this coincidence is the product of the probabilities or intensities $|\langle i|\phi\rangle|^2$ and $|\langle j|\psi\rangle|^2$ for the separate events. This is consistent with the axioms for quantum amplitudes given in Chapter 1. Two-particle amplitudes of the form in Eq. (6.1.2) will obey the basic axioms 1–4 in Chapter 1 and the orthonormality and completeness relations will hold:

$$\langle i, j|k, l\rangle = \delta_{ik}\delta_{jl}, \quad (6.1.3a)$$

$$\sum_i \sum_j |i, j\rangle\langle i, j| = 1. \quad (6.1.3b)$$

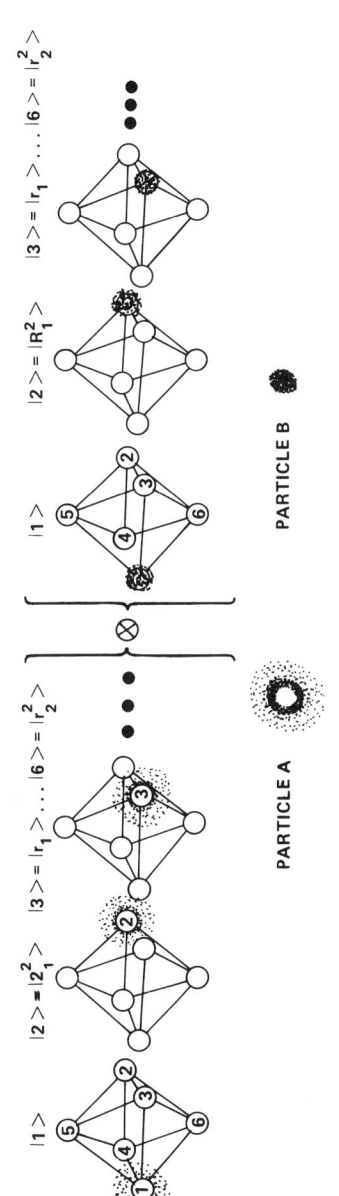
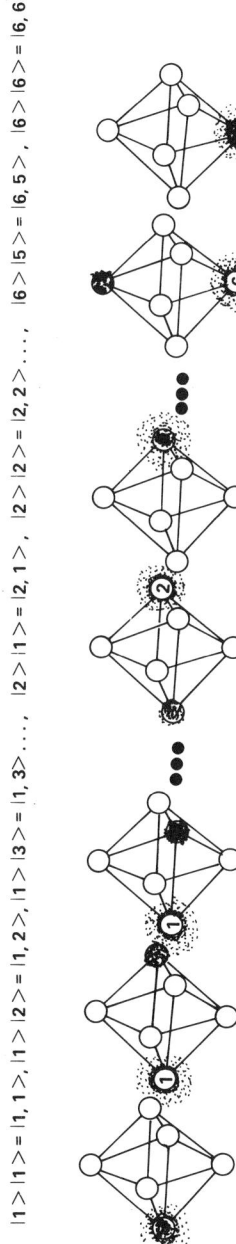

$|1\rangle|1\rangle = |1,1\rangle, |1\rangle|2\rangle = |1,2\rangle, |1\rangle|3\rangle = |1,3\rangle\ldots, \quad |2\rangle|1\rangle = |2,1\rangle, \quad |2\rangle|2\rangle = |2,2\rangle\ldots, \quad |6\rangle|5\rangle = |6,5\rangle, \quad |6\rangle|6\rangle = |6,6\rangle$

Figure 6.1.1 Sketch of two-particle wave function in octahedral environment. Thirty-six unrestricted two-particle states are possible if particles A and B are distinguishable.

However, if the two particles were identical electrons with the same spin, the states of the form $|j, j\rangle$ would be ruled out by the famous principle of Pauli. We shall discuss Pauli's principle and its consequences later. Meanwhile, we may use the basis in Eq. (6.1.1) without modification if we suppose the two particles are electrons in different spin states or they are different types of particles altogether.

A. Noninteracting Particles

If we suppose that the two particles can orbit around the octahedron without ever feeling each other's presence, then the description of the resulting symmetry analysis is simple. The symmetry operators are all combinations (a, b) of the operators a and b for particles A and B, respectively, as defined by the following:

$$(a, b)|k, l\rangle \equiv (a, b)|k\rangle|l\rangle = a|k\rangle b|l\rangle. \qquad (6.1.4a)$$

This defines a representation of the two-particle operators:

$$\langle i, j|(a, b)|k, l\rangle = \langle i|a|k\rangle\langle j|b|l\rangle$$
$$= \alpha_{ik}\beta_{jl}. \qquad (6.1.4b)$$

The total symmetry group G of the two-particle system is just the outer product of the individual symmetry groups which in this case are both O_h groups:

$$G = O_h^A \times O_h^B = \{\cdots(a, 1)\cdots(a', 1)\cdots\} \times \{\cdots(1, b')\cdots(1, b')\cdots\}$$
$$= \{\cdots(a, b)\cdots(a, b')\cdots(a', b)\cdots(a', b')\cdots\}. \qquad (6.1.5)$$

This is because the two types of operators work independently through each other. Any $(a, 1)$ commutes with any $(1, b)$ so that

$$(a, b)(a', b') = (aa', bb'). \qquad (6.1.6)$$

Therefore, the definition of direct product (Section 2.10) is correctly used in Eq. (6.1.5).

Representations of $G = O^A \times O^B$ such as are indicated in Eq. (6.1.4) or in Eq. (6.1.7) will be called DIRECT PRODUCTS ($\otimes$) of the separate factor group representations. We shall see that all irreps $\mathscr{D}^{\alpha\beta}(a, b) = \mathscr{D}^{\alpha}(a) \otimes$

$\mathscr{D}^\beta(b)$ of G can be made in this way:

$$(a,b)\left|\begin{matrix}\alpha & \beta \\ k & l\end{matrix}\right\rangle = a\left|\begin{matrix}\alpha \\ k\end{matrix}\right\rangle b\left|\begin{matrix}\beta \\ l\end{matrix}\right\rangle$$

$$= \left(\sum_{i=1}^{l^\alpha}\mathscr{D}_{ik}^\alpha(a)\left|\begin{matrix}\alpha \\ i\end{matrix}\right\rangle\right)\left(\sum_{j=1}^{l^\beta}\mathscr{D}_{jl}^\beta(b)\left|\begin{matrix}\beta \\ j\end{matrix}\right\rangle\right)$$

$$= \sum_{i=1}^{l^\alpha}\sum_{j=1}^{l^\beta}\mathscr{D}_{ik}^\alpha(a)\mathscr{D}_{jl}^\beta(b)\left|\begin{matrix}\alpha & \beta \\ i & j\end{matrix}\right\rangle, \quad (6.1.7\text{a})$$

$$\left\langle\begin{matrix}\alpha & \beta \\ i & j\end{matrix}\right|(a,b)\left|\begin{matrix}\alpha & \beta \\ k & l\end{matrix}\right\rangle = \mathscr{D}_{ik}^\alpha(a)\mathscr{D}_{jl}^\beta(b) \equiv (\mathscr{D}^\alpha(a) \otimes \mathscr{D}^\beta(b))_{ij:kl},$$
$$(6.1.7\text{b})$$

$$(\mathscr{D}^\alpha(a) \otimes \mathscr{D}^\beta(b))_{ij:kl} \equiv \mathscr{D}_{ij:kl}^{\alpha\beta}(a,b). \quad (6.1.7\text{c})$$

This direct product ($\otimes$) is sometimes called the TENSOR or KRONECKER product of matrices.

One of the difficulties of using the direct product is getting used to the double index notation. Actual construction of this product is quite easy, as demonstrated by the following example:

$$\mathscr{D}^{T_{1g}E_u}(r_2, i_5) = \mathscr{D}^{T_{1g}}(r_2) \otimes \mathscr{D}^{E_u}(i_5) = \begin{pmatrix} 0 & 0 & -1 \\ 1 & 0 & 0 \\ 0 & -1 & 0 \end{pmatrix} \otimes \begin{pmatrix} -\frac{1}{2} & -\frac{\sqrt{3}}{2} \\ -\frac{\sqrt{3}}{2} & \frac{1}{2} \end{pmatrix}.$$

The second factor $\mathscr{D}^{E_u}$ appears in blocks of the outer product multiplied by corresponding components of the first factor $\mathscr{D}^{T_{1g}}$.

$$= \begin{pmatrix} 0\begin{pmatrix} -\frac{1}{2} & -\frac{\sqrt{3}}{2} \\ -\frac{\sqrt{3}}{2} & \frac{1}{2} \end{pmatrix} & 0\begin{pmatrix} -\frac{1}{2} & -\frac{\sqrt{3}}{2} \\ -\frac{\sqrt{3}}{2} & \frac{1}{2} \end{pmatrix} & -1\begin{pmatrix} -\frac{1}{2} & -\frac{\sqrt{3}}{2} \\ -\frac{\sqrt{3}}{2} & \frac{1}{2} \end{pmatrix} \\ 1\begin{pmatrix} -\frac{1}{2} & -\frac{\sqrt{3}}{2} \\ -\frac{\sqrt{3}}{2} & \frac{1}{2} \end{pmatrix} & 0\begin{pmatrix} -\frac{1}{2} & -\frac{\sqrt{3}}{2} \\ -\frac{\sqrt{3}}{2} & \frac{1}{2} \end{pmatrix} & 0\begin{pmatrix} -\frac{1}{2} & -\frac{\sqrt{3}}{2} \\ -\frac{\sqrt{3}}{2} & \frac{1}{2} \end{pmatrix} \\ 0\begin{pmatrix} -\frac{1}{2} & -\frac{\sqrt{3}}{2} \\ -\frac{\sqrt{3}}{2} & \frac{1}{2} \end{pmatrix} & -1\begin{pmatrix} -\frac{1}{2} & -\frac{\sqrt{3}}{2} \\ -\frac{\sqrt{3}}{2} & \frac{1}{2} \end{pmatrix} & 0\begin{pmatrix} -\frac{1}{2} & -\frac{\sqrt{3}}{2} \\ -\frac{\sqrt{3}}{2} & \frac{1}{2} \end{pmatrix} \end{pmatrix}$$

$$= \begin{pmatrix} \cdot & \cdot & \cdot & \cdot & \frac{1}{2} & \frac{\sqrt{3}}{2} \\ \cdot & \cdot & \cdot & \cdot & \frac{\sqrt{3}}{2} & -\frac{1}{2} \\ -\frac{1}{2} & -\frac{\sqrt{3}}{2} & \cdot & \cdot & \cdot & \cdot \\ -\frac{\sqrt{3}}{2} & \frac{1}{2} & \cdot & \cdot & \cdot & \cdot \\ \cdot & \cdot & \frac{1}{2} & \frac{\sqrt{3}}{2} & \cdot & \cdot \\ \cdot & \cdot & \frac{\sqrt{3}}{2} & -\frac{1}{2} & \cdot & \cdot \end{pmatrix} \quad (6.1.7)_x$$

The double indices for the matrix components of this example are deployed lexicographically, i.e., $\{11, 12, \ldots, 21, 22, \ldots\}$, as shown in the following:

$$\mathcal{D}^{TE} = \mathcal{D}^T \otimes \mathcal{D}^E = \begin{pmatrix} \mathcal{D}^{TE}_{11:11} & \mathcal{D}_{11:12} & \mathcal{D}_{11:21} & \mathcal{D}_{11:22} & \mathcal{D}_{11:31} & \mathcal{D}_{11:32} \\ \mathcal{D}_{12:11} & \mathcal{D}_{12:12} & \mathcal{D}_{12:21} & \mathcal{D}_{12:22} & \mathcal{D}_{12:31} & \mathcal{D}_{12:32} \\ \mathcal{D}_{21:11} & \mathcal{D}_{21:12} & \mathcal{D}_{21:21} & \mathcal{D}_{21:22} & \mathcal{D}_{21:31} & \mathcal{D}_{21:32} \\ \mathcal{D}_{22:11} & \mathcal{D}_{22:12} & \mathcal{D}_{22:21} & \mathcal{D}_{22:22} & \mathcal{D}_{22:31} & \mathcal{D}_{22:32} \\ \mathcal{D}_{31:11} & \mathcal{D}_{31:12} & \mathcal{D}_{31:21} & \mathcal{D}_{31:22} & \mathcal{D}_{31:31} & \mathcal{D}_{31:32} \\ \mathcal{D}_{32:11} & \mathcal{D}_{32:12} & \mathcal{D}_{32:21} & \mathcal{D}_{32:22} & \mathcal{D}_{32:31} & \mathcal{D}_{32:32} \end{pmatrix}.$$

(6.1.8)

The characters of direct products of irreps are just the products of the characters, as seen in the following:

$$\chi^{\alpha\beta}(a,b) \equiv \text{TRACE } \mathcal{D}^{\alpha\beta}(a,b) \equiv \sum_{m=1}^{l^{\alpha}} \sum_{n=1}^{l^{\beta}} \mathcal{D}^{\alpha\beta}_{mn:mn}(a,b)$$

$$= \left(\sum_{m=1}^{l^{\alpha}} \mathcal{D}^{\alpha}_{mm}(a) \right) \left(\sum_{n=1}^{l^{\beta}} \mathcal{D}^{\beta}_{nn}(b) \right)$$

$$= \chi^{\alpha}(a) \chi^{\beta}(b). \quad (6.1.9)$$

It turns out that virtually all properties of a direct product group $G = O^A \times O^B$ are derived by product analysis of the corresponding factor groups. This is true for the operators in Eq. (6.1.4), irreps in Eq. (6.1.7), and characters in Eq. (6.1.9).

The Hamiltonian operator for noninteracting particles is just the sum of their two separate single-particle operators. Let us write this operator so it indicates that each term operates exclusively on only one particle:

$$H = (H_A, 1) + (1, H_B). \quad (6.1.10)$$

The irrep basis states of the form $\left| \begin{array}{cc} \alpha & \beta \\ i & j \end{array} \right\rangle$ are eigenstates of the noninteracting Hamiltonian if $\left| \begin{array}{c} \alpha \\ i \end{array} \right\rangle$ and $\left| \begin{array}{c} \beta \\ j \end{array} \right\rangle$ were eigenstates of H_A and H_B, respectively:

$$H \left| \begin{array}{cc} \alpha & \beta \\ i & j \end{array} \right\rangle = H_A \left| \begin{array}{c} \alpha \\ i \end{array} \right\rangle \left| \begin{array}{c} \beta \\ j \end{array} \right\rangle + \left| \begin{array}{c} \alpha \\ i \end{array} \right\rangle H_B \left| \begin{array}{c} \beta \\ j \end{array} \right\rangle = (\varepsilon^{\alpha} + \varepsilon^{\beta}) \left| \begin{array}{cc} \alpha & \beta \\ i & j \end{array} \right\rangle. \quad (6.1.11)$$

The energies of two noninteracting particles are simply added to give the energy of the two-particle system.

One may combine the levels and states in Figure 4.3.2(b) as shown on the left-hand side of Figure 6.1.2(a). This gives the level diagram of the noninter-

TWO-PARTICLE STATES AND PRODUCTS OF REPRESENTATIONS   449

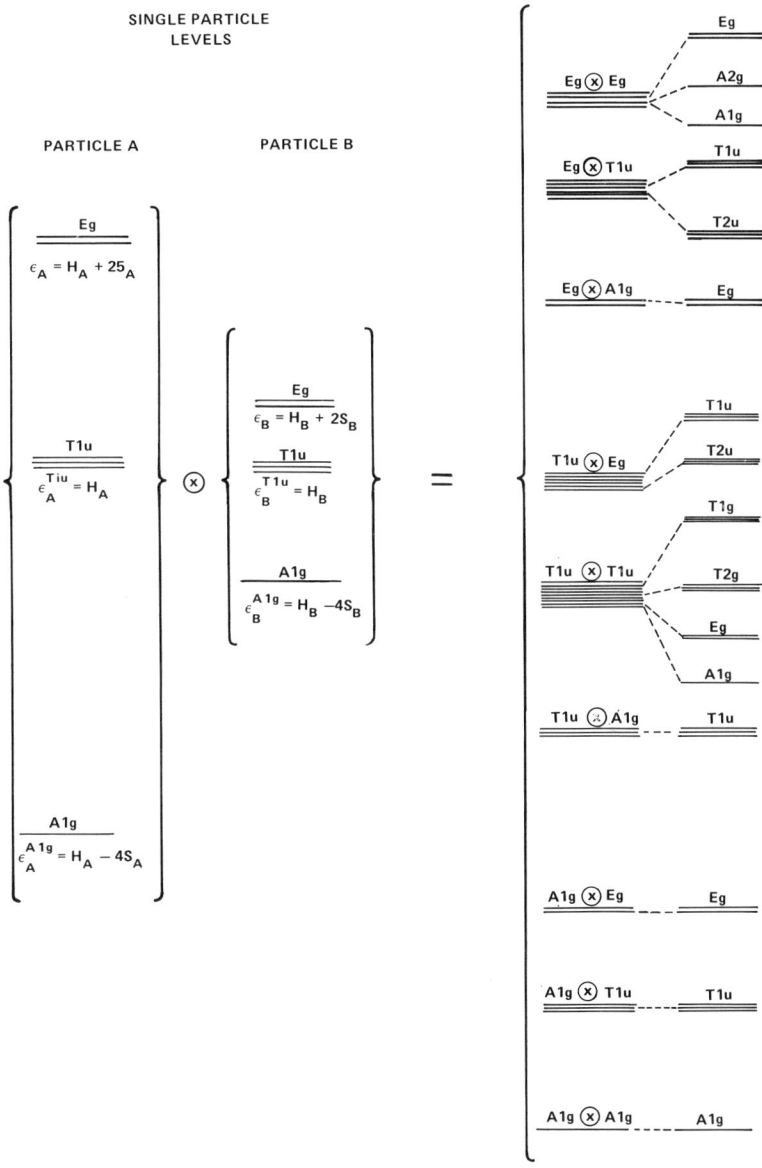

Figure 6.1.2 Eigenlevels of octahedral two-particle system. (a) Two-particle levels are labeled by irreps $(D^\alpha \otimes D^\beta)$ of cross-product symmetry $G^A \times G^B = O_h \times O_h$. (b) Interaction causes levels to split into levels labeled by irreps of $G^{\mathrm{RIGID}} = O_h^{\mathrm{RIGID}}$.

acting two-particle system. The levels have degeneracies $l^\alpha l^\beta$ equal to the dimension irreps of $G = O_h^A \times O_h^B$. Note also that state $\left|\begin{smallmatrix}\alpha & \beta \\ i & j\end{smallmatrix}\right\rangle$ would have the same energy as state $\left|\begin{smallmatrix}\beta & \alpha \\ j & i\end{smallmatrix}\right\rangle$ if particles A and B were identical electrons. This additional symmetry will be studied when we introduce the Pauli principle. In the next section we explain the level splitting shown on the extreme right-hand side of Figure 6.1.2.

B. Interacting Particles

Suppose we could gradually turn on a repulsive interaction between the two particles orbiting in the octahedral potential such as the Coulomb potential that would exist between two electrons. Mathematically, this can be described by some perturbation operator V added to the Hamiltonian in Eq. (6.1.10).

A matrix element of the form $\langle 1, 1|V|1, 1\rangle$, for example, would give the expectation value of the repulsion energy for the state $|1, 1\rangle$ in which two particles are sharing the first potential well. This energy would be more than for states like $|1, 3\rangle$ or $|1, 2\rangle$, in which the particles are not so close to each other; i.e., $\langle 1, 1|V|1, 1\rangle \geq \langle 1, 3|V|1, 3\rangle \geq \langle 1, 2|V|1, 2\rangle$. From this it follows that most of the operators in $G = O^A \times O^B = \cdots (a, b) \cdots$ will not commute with V or any Hamiltonian that contains V. Consider the operator $g = (1, R_1)$, for example. If this commutes with V, then the following contradictory relation follows:

$$\langle 1, 1|V|1, 1\rangle = \langle 1, 1|g^\dagger V g|1, 1\rangle = \langle 1, 3|V|1, 3\rangle. \qquad (6.1.12)$$

However, all the octahedral sites are still equivalent, and V should have the symmetry O_h. The symmetry operators that still commute with V are just those which move the two particles rigidly together without changing the distance between them, i.e., operators of the form (a, a). Call the resulting group O_h^{RIGID}:

$$O_h^{\text{RIGID}} = \{(1, 1) \cdots (a, a) \cdots (b, b) \cdots\}. \qquad (6.1.13)$$

Each of the irreducible product representations of the now-defunct symmetry $G = O_h^A \times O_h^B$ will be, in general, only reducible representations of the subgroup O_h^{RIGID}. The techniques of Chapter 3 can be used to reduce them, and this will give the eigenvectors of the new Hamiltonian in the limit of small V.

First, the character analysis will tell which irreps $\mathscr{D}^\gamma$ of O_h^{RIGID} are contained in $\mathscr{D}^{\alpha\beta}$, that is, the frequency $f^\gamma(\mathscr{D}^{\alpha\beta})$. Equations (3.5.11) and

(6.1.9) yield the following frequency formula:

$$f^\gamma(\mathscr{D}^{\alpha\beta}) = \frac{1}{48} \sum_{\substack{\text{classes} \\ g}} {}^\circ c_g \chi_g^{\gamma*} \text{ TRACE } \mathscr{D}^{\alpha\beta}(g,g),$$

$$= \frac{1}{48} \sum_{\substack{\text{classes} \\ g}} {}^\circ c_g \chi_g^{\gamma*} \chi_g^\alpha \chi_g^\beta. \qquad (6.1.14)$$

Substitution of the O_h characters (4.1.16) for $\alpha = \beta = T_{1u}$ yields the frequencies $f^{A_{1g}} = f^{E_g} = f^{T_{1g}} = f^{T_{2g}} = 1$. All other f^γ are zero. This means that some transformation matrix $\mathscr{C}$ exists which effects the following reduction:

$$\mathscr{C}^\dagger \mathscr{D}^{T_{1u} T_{1u}}(g,g) \mathscr{C} = \mathscr{D}^{A_{1g}}(g) \oplus \mathscr{D}^{E_g}(g) \oplus \mathscr{D}^{T_{1g}}(g) \oplus \mathscr{D}^{T_{2g}}(g). \quad (6.1.15)$$

It is easy to check that the sum of the four characters add up to the product character $\chi^{T_{1u}} \chi^{T_{1u}}$:

	$h=1$	r	R^2	R	i	I	Ir	IR^2	IR	Ii
$\chi_h^{A_{1g}} = 1$	1	1	1	1	1	1	1	1	1	
$\chi_h^{E_g} = 2$	-1	2	0	0	2	-1	2	0	0	
$\chi_h^{T_{1g}} = 3$	0	-1	1	-1	3	0	-1	1	-1	
$\chi_h^{T_{2g}} = 3$	0	-1	-1	1	3	0	-1	-1	1	
$\chi_h^{T_{1u}} \chi_h^{T_{1u}} = 9$	0	1	1	1	9	0	1	1	1	

For many cases one may find the correct combinations by trial and inspection of characters in this way. This may be quicker than using the formula (6.1.14).

The appearance of the four irreps in the $T_{1u} \times T_{1u}$ product (6.1.15) means that the (T_{1u}, T_{1u}) level in Figure 4.1.2 will split into four levels when the interaction V is present. The four levels labeled A_{1g}, E_g, T_{1g}, and T_{2g} are shown splitting away from (T_{1u}, T_{1u}) in Figure 6.1.2(b). The form of the eigenstates for these four levels is determined by columns of the $\mathscr{C}$ matrix. Each splitting in Figure 6.1.2(b) corresponds to a different $\mathscr{C}$ matrix.

Let us see how to find the transformation matrix $\mathscr{C}$ that reduces a general product $\mathscr{D}^\alpha \otimes \mathscr{D}^\beta$ and, particularly, the one in Eq. (6.1.15). Following the theory of Chapter 3, one applies the elementary operators P_{mn}^γ of $H = O_h^{\text{RIGID}}$ to the base states of $O_h^A \times O_h^B$. One obtains the following eigenstates:

$$\left| \begin{matrix} \gamma \\ m \end{matrix} (\alpha \otimes \beta) \right\rangle \equiv P_{mn}^\gamma \left| \begin{matrix} \alpha & \beta \\ k & l \end{matrix} \right\rangle \Big/ \sqrt{N_{nkl}^{\gamma\alpha\beta}}$$

$$= (l^\gamma/{}^\circ H) \sum_a \mathscr{D}_{mn}^{\gamma*}(a)(a,a) \left| \begin{matrix} \alpha & \beta \\ k & l \end{matrix} \right\rangle \Big/ \sqrt{N_{nkl}^{\gamma\alpha\beta}}, \quad (6.1.16a)$$

452   THEORY AND APPLICATIONS OF SYMMETRY REPRESENTATION PRODUCTS

where the effect of operator (a, a) is given by Eq. (6.1.7a) as

$$\left|{}^{\gamma}_{m}(\alpha \otimes \beta)\right\rangle = (l^\gamma/{}^\circ H) \sum_a \sum_{i=1}^{l^\alpha} \sum_{j=1}^{l^\beta} \mathscr{D}^{\gamma*}_{mn}(a) \mathscr{D}^{\alpha}_{ik}(a) \mathscr{D}^{\beta}_{jl}(a) \left|{}^{\alpha\ \beta}_{i\ j}\right\rangle \Big/ \sqrt{N^{\gamma\alpha\beta}_{nkl}}, \qquad (6.1.16b)$$

and where the normalization constant is the following:

$$N^{\gamma\alpha\beta}_{nkl} = \left\langle {}^{\alpha\ \beta}_{k\ l}\Big| P_{mn} \Big|{}^{\gamma\ \beta}_{k\ l}\right\rangle = (l^\gamma/{}^\circ H) \sum_a \mathscr{D}^{\gamma*}_{nn}(a) \mathscr{D}^{\alpha}_{kk}(a) \mathscr{D}^{\beta}_{ll}(a). \qquad (6.1.16c)$$

The indices n, k, and l are chosen to be the first combination which gives nonzero $N^{\gamma\alpha\beta}_{nkl}$. For example, let $\alpha = \beta = T_{1u}$, and $\gamma = T_{1g}$. Substituting tetragonally defined irreps (4.2.14) into Eq. (6.1.16) gives $N_{111} = N_{113} = N_{121} = N_{122} = 0$, and $N_{123} = \frac{1}{2}$. [Note that Eq. (6.1.16) gives the same results whether one sums over group O or $O_h = O \times C_i$.] Now with n, k, and l fixed, one tries different j, i, and m in Eq. (4.1.16b). For this example the following results are obtained:

$$\left|{}^{T_{1g}}_{1}(T_{1u} \otimes T_{1u})\right\rangle = \frac{1}{\sqrt{2}} \left|{}^{T_{1u}\ T_{1u}}_{2\ \ 3}\right\rangle - \frac{1}{\sqrt{2}} \left|{}^{T_{1u}\ T_{1u}}_{3\ \ 2}\right\rangle,$$

$$\left|{}^{T_{1g}}_{2}(T_{1u} \otimes T_{1u})\right\rangle = \frac{1}{\sqrt{2}} \left|{}^{T_{1u}\ T_{1u}}_{3\ \ 1}\right\rangle - \frac{1}{\sqrt{2}} \left|{}^{T_{1u}\ T_{1u}}_{1\ \ 3}\right\rangle,$$

$$\left|{}^{T_{1g}}_{3}(T_{1u} \otimes T_{1u})\right\rangle = \frac{1}{\sqrt{2}} \left|{}^{T_{1u}\ T_{1u}}_{1\ \ 2}\right\rangle - \frac{1}{\sqrt{2}} \left|{}^{T_{1u}\ T_{1u}}_{2\ \ 1}\right\rangle. \qquad (6.1.17)$$

The coefficients in the foregoing example are called COUPLING COEFFICIENTS or CLEBSCH-GORDON COEFFICIENTS $\mathscr{C}^{\alpha\beta\gamma}_{ijm}$. They are defined in terms of the product states $\left|{}^{\alpha\ \beta}_{i\ j}\right\rangle$:

$$\left|{}^{\gamma}_{m}(\alpha \otimes \beta)\right\rangle = \sum_{i=1}^{l^\alpha} \sum_{j=1}^{l^\beta} \mathscr{C}^{\alpha\beta\gamma}_{ijm} \left|{}^{\alpha\ \beta}_{i\ j}\right\rangle. \qquad (6.1.18a)$$

They are transformation amplitudes connecting $\left|{}^{\alpha\ \beta}_{i\ j}\right\rangle$ to states $|{}^{\gamma}_{m}(\alpha \times \beta)\rangle$

belonging to irrep $\mathscr{D}^\gamma$:

$$\mathscr{C}_{ijm}^{\alpha\beta\gamma} = \left\langle \begin{matrix} \alpha & \beta \\ i & j \end{matrix} \middle| \begin{matrix} \gamma \\ m \end{matrix}(\alpha \otimes \beta) \right\rangle = \frac{(l^\gamma/{}^\circ G)\Sigma_g \mathscr{D}_{mn}^{\gamma*}(g)\mathscr{D}_{ik}^{\alpha}(g)\mathscr{D}_{jl}^{\beta}(g)}{\sqrt{(l^\gamma/{}^\circ G)\Sigma_g \mathscr{D}_{nn}^{\gamma*}(g)\mathscr{D}_{kk}^{\alpha}(g)\mathscr{D}_{ll}^{\beta}(g)}}.$$

(6.1.18b)

The $\mathscr{C}_{ijm}^{\alpha\beta\gamma}$ are the components of the transformation matrix $\mathscr{C}$ in Eq. (6.1.15) which reduces the product representation $\mathscr{D}^\alpha \times \mathscr{D}^\beta$. The C transformation which reduces the example with $\alpha = T_{1u} = \beta$ is shown in Eq. (6.1.19). Equation (6.1.17) gives three of the columns of $\mathscr{C}$ and the other six are derived similarly. The $\mathscr{C}$ matrix is the usual format for a table of coupling coefficients as shown in Eq. (6.1.19c):

$$\sum_{i=1}^{l^\alpha}\sum_{j=1}^{l^\beta}\sum_{k=1}^{l^\alpha}\sum_{l=1}^{l^\beta} \left(\mathscr{C}_{ijm}^{\alpha\beta\gamma'}\right)^* \mathscr{D}_{ij:kl}^{\alpha\beta}(R_1, R_1) \mathscr{C}_{kln}^{\alpha\beta\gamma} = \mathscr{D}_{mn}^{\gamma}(R_1)\delta^{\gamma'\gamma},\quad (6.1.19a)$$

$$\mathscr{C}^\dagger \begin{vmatrix} 1 & \cdot & \cdot & \cdot & \cdot & \cdot & \cdot & \cdot & \cdot \\ \cdot & \cdot & -1 & \cdot & \cdot & \cdot & \cdot & \cdot & \cdot \\ \cdot & 1 & \cdot & \cdot & \cdot & \cdot & \cdot & \cdot & \cdot \\ \cdot & \cdot & \cdot & \cdot & \cdot & -1 & \cdot & \cdot & \cdot \\ \cdot & \cdot & \cdot & \cdot & \cdot & \cdot & \cdot & 1 \\ \cdot & \cdot & \cdot & \cdot & \cdot & \cdot & -1 & \cdot \\ \cdot & \cdot & \cdot & 1 & \cdot & \cdot & \cdot & \cdot & \cdot \\ \cdot & \cdot & \cdot & \cdot & -1 & \cdot & \cdot & \cdot & \cdot \\ \cdot & \cdot & \cdot & \cdot & 1 & \cdot & \cdot & \cdot & \cdot \end{vmatrix} \mathscr{C} =$$

$$\boxed{1} = \mathscr{D}^{A_{1g}}(R_1)$$

$$\boxed{\begin{matrix} -\dfrac{1}{2} & -\dfrac{\sqrt{3}}{2} \\ -\dfrac{\sqrt{3}}{2} & \dfrac{1}{2} \end{matrix}} = \mathscr{D}^{E_g}(R_1)$$

$$\boxed{\begin{matrix} 1 & \cdot & \cdot \\ \cdot & \cdot & -1 \\ \cdot & 1 & \cdot \end{matrix}} = \mathscr{D}^{T_{1g}}(R_1)$$

$$\mathscr{D}^{T_{2g}}(R_1) = \boxed{\begin{matrix} -1 & \cdot & \cdot \\ \cdot & \cdot & 1 \\ \cdot & -1 & \cdot \end{matrix}}$$

(6.1.19b)

where:

$\mathscr{C} =$

T_{1u}	⊗	T_{1u}	A_{1g}	E_g 1	E_g 2	T_{1g} 1	T_{1g} 2	T_{1g} 3	T_{2g} 1	T_{2g} 2	T_{2g} 3
T_{1u} 1		T_{1u} 1	$\dfrac{1}{\sqrt{3}}$	$\dfrac{1}{\sqrt{6}}$	$-\dfrac{1}{\sqrt{2}}$	·	·	·	·	·	·
1		2	·	·	·	·	·	$\dfrac{1}{\sqrt{2}}$	·	·	$\dfrac{1}{\sqrt{2}}$
1		3	·	·	·	·	$-\dfrac{1}{\sqrt{2}}$	·	$\dfrac{1}{\sqrt{2}}$	·	·
2		1	·	·	·	·	·	$-\dfrac{1}{\sqrt{2}}$	·	·	$\dfrac{1}{\sqrt{2}}$
2		2	$\dfrac{1}{\sqrt{3}}$	$\dfrac{1}{\sqrt{6}}$	$\dfrac{1}{\sqrt{2}}$	·	·	·	·	·	·
2		3	·	·	·	$\dfrac{1}{\sqrt{2}}$	·	·	$\dfrac{1}{\sqrt{2}}$	·	·
3		1	·	·	·	·	$\dfrac{1}{\sqrt{2}}$	·	·	$\dfrac{1}{\sqrt{2}}$	·
3		2	·	·	·	$-\dfrac{1}{\sqrt{2}}$	·	·	$\dfrac{1}{\sqrt{2}}$	·	·
3		3	$\dfrac{1}{\sqrt{3}}$	$-\dfrac{2}{\sqrt{6}}$	·	·	·	·	·	·	·

(6.1.19c)

Note that the irreps (4.2.14), (4.2.15), and (4.2.19) have been used. The coupling coefficients for the "kosher T_2" irreps (4.2.16) have slightly different phases, as we will see in the following section. Note also that the transformation equations (6.1.17) or (6.1.19) only define the coupling coefficients $\mathscr{C}_{ijm}^{\alpha\beta\gamma}$ to within an overall phase that is a function of α, β, and γ. Any l^γ-column section of a C matrix, such as in Eq. (6.1.19), can be multiplied overall by $e^{i\phi}$ without changing the effect of the transformation $\mathscr{C}$.

C. Subgroup Chain Labeling for Coupling Coefficients

The subgroup chain labeling described in Section 4.2.A is also useful for deriving coupling coefficients. The straightforward projection method used in the preceding section becomes quite laborious in some applications. Subgroup structure can be used to divide and conquer many symmetry problems involving large groups or complex representations. The numerical values of coupling coefficients involving multidimensional representations depend on which bases are chosen. We now derive coefficients for two different choices belonging to subgroup chains (a) $O \supset D_4 \supset D_2$ and (b) $O \supset D_3 \supset C_2$.

(a) Tetragonal ($O \supset D_4 \supset D_2$) Coupling Coefficients Subgroup chain calculations begin with the lowest link. Subgroup $D_2 = \{1, R_1^2, R_2^2, R_3^2\}$ is the

TWO-PARTICLE STATES AND PRODUCTS OF REPRESENTATIONS 455

lowest link in the tetragonal chain described in Section 4.2.A(a). The lowest link in all the chains is an Abelian group. The coupling coefficients for one-dimensional Abelian irreps are just 1's and 0's, and they are very easy to find. For example, the D_2 characters (4.1.43) yield the following $\otimes$-multiplication table:

$$\begin{array}{r|cccc} B = & A_1 & A_2 & B_1 & B_2 \\ \hline D^{A_1} \otimes D^B = & D^{A_1} & D^{A_2} & D^{B_1} & D^{B_2} \\ D^{A_2} \otimes D^B = & D^{A_2} & D^{A_1} & D^{B_2} & D^{B_1} \\ D^{B_1} \otimes D^B = & D^{B_1} & D^{B_2} & D^{A_1} & D^{A_2} \\ D^{B_2} \otimes D^B = & D^{B_2} & D^{B_1} & D^{A_2} & D^{A_1} \end{array}. \quad (6.1.20)$$

The combinations allowed by the table give unit coupling coefficients (viz., $D^{A_2} \otimes D^{B_1} = D^{B_2}$ implies that $\mathscr{C}^{A_2 B_1 B_2} = 1$), while all others are zero (viz., $0 = \mathscr{C}^{A_2 B_1 A_1} = \mathscr{C}^{A_2 B_1 A_2} = \mathscr{C}^{A_2 B_1 B_1}$).

The zero combinations can be eliminated immediately from coupling coefficients of the next higher subgroup link D_4. The D_4 coupling coefficients $\mathscr{C}^{\alpha\beta\gamma}_{A_2 B_1 A_1}$, $\mathscr{C}^{\alpha\beta\gamma}_{A_2 B_1 A_2}$, and $\mathscr{C}^{\alpha\beta\gamma}_{A_2 B_1 B_1}$ must vanish along with any combination of D_2 labels not allowed by the table of (6.1.20). For example, in the outer product of E bases $\left\{ \left|{E \atop 1}\right\rangle \equiv \left|{E \atop B_1}\right\rangle, \left|{E \atop 2}\right\rangle \equiv \left|{E \atop B_2}\right\rangle \right\}$ the D_4 coupling matrix $\mathscr{C}$ must perform the reduction

$$\mathscr{C}^\dagger \mathscr{D}^E \otimes \mathscr{D}^E \mathscr{C} = \mathscr{D}^{A_1} \oplus \mathscr{D}^{B_1} \oplus \mathscr{D}^{A_2} \oplus \mathscr{D}^{B_2} \quad (6.1.21a)$$

according to D_4 characters (3.6.3) and the theory of the preceding section. The D_2 irreps correlated with D_4 irreps A_1, B_1, A_2, and B_2 are A_1, A_1, A_2, and A_2, respectively. Therefore the $E \otimes E \, \mathscr{C}$ matrix must have the following form:

$\left\|{\gamma \atop k}(E \times E)\right\rangle =$	$\left\|{A_1 \atop A_1}\right\rangle$	$\left\|{B_1 \atop A_1}\right\rangle$	$\left\|{A_2 \atop A_2}\right\rangle$	$\left\|{B_2 \atop A_2}\right\rangle$	$\leftarrow D_4$ label $\leftarrow D_2$ label
$\left\|{E \atop B_1}\right\rangle\left\|{E \atop B_1}\right\rangle$	a	c	0	0	
$\left\|{E \atop B_1}\right\rangle\left\|{E \atop B_2}\right\rangle$	0	0	e	g	(6.1.21b)
$\left\|{E \atop B_2}\right\rangle\left\|{E \atop B_1}\right\rangle$	0	0	f	h	
$\left\|{E \atop B_2}\right\rangle\left\|{E \atop B_2}\right\rangle$	b	d	0	0	

Impossible D_2 products are eliminated by writing zeros and the allowed ones are indicated by unknowns $a, b, c, \ldots, h$. To solve for the unknowns we write

456 THEORY AND APPLICATIONS OF SYMMETRY REPRESENTATION PRODUCTS

the matrix equation (6.1.21) in the form

$$(\mathscr{D}^E(R) \otimes \mathscr{D}^E(R))\mathscr{C} = \mathscr{C}(\mathscr{D}^{A_1}(R) \oplus \mathscr{D}^{B_1}(R) \oplus \mathscr{D}^{A_2}(R) \oplus \mathscr{D}^{B_2}(R)).$$
(6.1.21c)

Using the D_4 irreps (3.6.7) for one operator R outside of the D_2 subgroup gives the following matrix equation:

$$(\mathscr{D}^E(R) \otimes \mathscr{D}^E(R))\,\mathscr{C}$$

$$\begin{pmatrix} 0 & 0 & 0 & 1 \\ 0 & 0 & -1 & 0 \\ 0 & -1 & 0 & 0 \\ 1 & 0 & 0 & 0 \end{pmatrix} \begin{pmatrix} a & c & 0 & 0 \\ 0 & 0 & e & g \\ 0 & 0 & f & h \\ b & d & 0 & 0 \end{pmatrix}$$

$$\mathscr{C}(\mathscr{D}^{A_1}(R) \oplus \mathscr{D}^{B_1}(r) \oplus \mathscr{D}^{A_2}(R) \oplus \mathscr{D}^{B_2}(R))$$

$$= \begin{pmatrix} a & c & 0 & 0 \\ 0 & 0 & e & g \\ 0 & 0 & f & h \\ b & d & 0 & 0 \end{pmatrix} \begin{pmatrix} 1 & 0 & 0 & 0 \\ 0 & -1 & 0 & 0 \\ 0 & 0 & 1 & 0 \\ 0 & 0 & 0 & -1 \end{pmatrix},$$

$$\begin{pmatrix} b & d & 0 & 0 \\ 0 & 0 & -f & -h \\ 0 & 0 & -e & -g \\ a & c & 0 & 0 \end{pmatrix} = \begin{pmatrix} a & -c & 0 & 0 \\ 0 & 0 & e & -g \\ 0 & 0 & f & -h \\ b & -d & 0 & 0 \end{pmatrix}. \qquad (6.1.22)$$

This determines the unknowns to within a normalization factor:

$$b = a, \qquad c = -d, \qquad f = -e, \qquad h = g.$$

By normalizing each column the desired coupling coefficient table is obtained:

$E \otimes E$	A_1 A_1	B_1 A_1	A_2 A_2	B_2 A_2
$B_1\ B_1$	$\frac{1}{\sqrt{2}}$	$\frac{1}{\sqrt{2}}$	0	0
$B_1\ B_2$	0	0	$\frac{1}{\sqrt{2}}$	$\frac{1}{\sqrt{2}}$
$B_2\ B_1$	0	0	$-\frac{1}{\sqrt{2}}$	$\frac{1}{\sqrt{2}}$
$B_2\ B_2$	$\frac{1}{\sqrt{2}}$	$-\frac{1}{\sqrt{2}}$	0	0

. (6.1.23)

TWO-PARTICLE STATES AND PRODUCTS OF REPRESENTATIONS   457

Consider one more example involving a different D_4 product and $\mathscr{C}$ matrix:

$$\mathscr{C}^{\dagger}\mathscr{D}^{E} \otimes \mathscr{D}^{A_2}\mathscr{C} = \mathscr{D}^{E}. \qquad (6.1.24)$$

This $\mathscr{C}$ matrix must have the form

$$\gamma_k(E \otimes A_2) = \left|{E \atop B_1}\right\rangle \left|{E \atop B_2}\right\rangle,$$

$$\mathscr{C} = \begin{array}{c} \left|{E \atop B_1}\right\rangle\left|{A_2 \atop A_2}\right\rangle \\ \left|{E \atop B_2}\right\rangle\left|{A_2 \atop A_2}\right\rangle \end{array} \begin{bmatrix} 0 & b \\ a & 0 \end{bmatrix}, \qquad (6.1.25)$$

according to (6.1.20). Solving Eq. (6.1.24) and (6.1.25) gives

$$\begin{pmatrix} 0 & -1 \\ 1 & 0 \end{pmatrix}\begin{pmatrix} 0 & b \\ a & 0 \end{pmatrix} = \begin{pmatrix} 0 & b \\ a & 0 \end{pmatrix}\begin{pmatrix} 0 & -1 \\ 1 & 0 \end{pmatrix},$$

$$\begin{pmatrix} -a & 0 \\ 0 & b \end{pmatrix} = \begin{pmatrix} b & 0 \\ 0 & -a \end{pmatrix}$$

for the D_4 operator R. The resulting coupling coefficient table is shown in the following, along with three others:

$E \otimes A_2$	${E \atop B_1}$ ${E \atop B_2}$		$E \otimes B_1$	${E \atop B_1}$ ${E \atop B_2}$		$E \otimes B_2$	${E \atop B_1}$ ${E \atop B_2}$
B_1 A_2	0	-1	B_1 A_1	1	0	B_1 A_2	0 1
B_2 A_2	1	0	B_2 A_1	0	-1	B_2 A_2	1 0

$$(6.1.26)$$

Now we are ready to analyze the highest link (O) in the $O \supset D_4 \supset D_2$ chain. Let us rederive the $T_1 \otimes T_1$ table of (6.1.19c) using the Kosher irreps labeled by this chain. The proper chain labels of O states are now repeated, following Eqs. (4.2.11), (4.2.17), and (4.2.18):

$$\left\{ \left| \begin{matrix} T_1 \\ E \\ B_1 \end{matrix} \right\rangle, \left| \begin{matrix} T_1 \\ E \\ B_2 \end{matrix} \right\rangle, \left| \begin{matrix} T_1 \\ A_2 \\ A_2 \end{matrix} \right\rangle \right\} \left\{ \left| \begin{matrix} T_2 \\ E \\ B_1 \end{matrix} \right\rangle, \left| \begin{matrix} T_2 \\ E \\ B_2 \end{matrix} \right\rangle, \left| \begin{matrix} T_2 \\ B_2 \\ A_2 \end{matrix} \right\rangle \right\} \left\{ \left| \begin{matrix} E \\ A_1 \\ A_1 \end{matrix} \right\rangle, \left| \begin{matrix} E \\ B_1 \\ A_1 \end{matrix} \right\rangle \right\} \left\{ \left| \begin{matrix} A_2 \\ A_2 \\ A_2 \end{matrix} \right\rangle \right\} \left\{ \left| \begin{matrix} A_1 \\ A_1 \\ A_1 \end{matrix} \right\rangle \right\}.$$

(6.1.27a)

By combining this with the D_4 couplings [(6.1.23) and (6.1.26)], one may deduce the structure of the $T_1 \otimes T_1$ coupling matrix as follows:

$T_1 \otimes T_1$	$\left\|\begin{matrix}A_1\\A_1\\A_1\end{matrix}\right\rangle$	$\left\|\begin{matrix}E\\A_1\\A_1\end{matrix}\right\rangle$	$\left\|\begin{matrix}E\\B_1\\A_1\end{matrix}\right\rangle$	$\left\|\begin{matrix}T_1\\E\\B_1\end{matrix}\right\rangle$	$\left\|\begin{matrix}T_1\\E\\B_2\end{matrix}\right\rangle$	$\left\|\begin{matrix}T_1\\A_2\\A_2\end{matrix}\right\rangle$	$\left\|\begin{matrix}T_2\\E\\B_1\end{matrix}\right\rangle$	$\left\|\begin{matrix}T_2\\E\\B_2\end{matrix}\right\rangle$	$\left\|\begin{matrix}T_2\\A_2\\A_2\end{matrix}\right\rangle$
$\left\|\begin{matrix}E\\B_1\end{matrix}\right\rangle\left\|\begin{matrix}E\\B_1\end{matrix}\right\rangle$	$\dfrac{a}{\sqrt{2}}$	$\dfrac{c}{\sqrt{2}}$	$\dfrac{e}{\sqrt{2}}$	0	0	0	0	0	0
$\left\|\begin{matrix}E\\B_1\end{matrix}\right\rangle\left\|\begin{matrix}E\\B_2\end{matrix}\right\rangle$	0	0	0	0	0	$\dfrac{h}{\sqrt{2}}$	0	0	$\dfrac{k}{\sqrt{2}}$
$\left\|\begin{matrix}E\\B_1\end{matrix}\right\rangle\left\|\begin{matrix}A_2\\A_2\end{matrix}\right\rangle$	0	0	0	0	$-f$	0	0	$-i$	0
$\left\|\begin{matrix}E\\B_2\end{matrix}\right\rangle\left\|\begin{matrix}E\\B_1\end{matrix}\right\rangle$	0	0	0	0	0	$-\dfrac{h}{\sqrt{2}}$	0	0	$\dfrac{k}{\sqrt{2}}$
$\left\|\begin{matrix}E\\B_2\end{matrix}\right\rangle\left\|\begin{matrix}E\\B_2\end{matrix}\right\rangle$	$\dfrac{a}{\sqrt{2}}$	$\dfrac{c}{\sqrt{2}}$	$-\dfrac{e}{\sqrt{2}}$	0	0	0	0	0	0
$\left\|\begin{matrix}E\\B_2\end{matrix}\right\rangle\left\|\begin{matrix}A_2\\A_2\end{matrix}\right\rangle$	0	0	0	f	0	0	i	0	0
$\left\|\begin{matrix}A_2\\A_2\end{matrix}\right\rangle\left\|\begin{matrix}E\\B_1\end{matrix}\right\rangle$	0	0	0	0	$-g$	0	0	$-j$	0
$\left\|\begin{matrix}A_2\\A_2\end{matrix}\right\rangle\left\|\begin{matrix}E\\B_2\end{matrix}\right\rangle$	0	0	0	g	0	0	j	0	0
$\left\|\begin{matrix}A_2\\A_2\end{matrix}\right\rangle\left\|\begin{matrix}A_2\\A_2\end{matrix}\right\rangle$	b	d	0	0	0	0	0	0	0

(6.1.27b)

The first, second, fourth, and fifth rows of the $T_1 \otimes T_1$ table are copied from the $E \otimes E$ table of (6.1.23) for subgroup D_4. The third and sixth rows, as well as the seventh and eighth rows, follow from the $E \otimes A_2$ ($= A_2 \otimes E$) table of (6.1.26). Finally, the ninth row follows since $A_2 \otimes A_2 = A_1$ according to Eq. (6.1.20).

The unknown coefficients $a, b, \ldots, k$ are derived in the same way as before. To save space we shall write the $\mathscr{C}$ matrix only once in the following between the $(\mathscr{D}^{T_1}(R_1) \otimes \mathscr{D}^{T_1}(R_1))$ matrix (6.1.19b) on top and the $(\mathscr{D}^{A_1}(R_1)$ $\oplus \mathscr{D}^{E}(R_1) \oplus \mathscr{D}^{T_1}(R_1) \oplus \mathscr{D}^{T_1}(R_1) \oplus \mathscr{D}^{T_2}(R_1))$ matrix below. [Note that $\mathscr{D}^{T_2}(R_1)$ differs from the nonkosher T_2 irrep given on the lower right-hand side of Eq. (6.1.19b).]

TWO-PARTICLE STATES AND PRODUCTS OF REPRESENTATIONS 459

$$
\begin{pmatrix}
1 & & & & & & & & \\
& & -1 & & & & & & \\
& 1 & & & & & & & \\
& & & & & -1 & & & \\
& & & & & & & 1 & \\
& & & & -1 & & & & \\
& & & 1 & & & & & \\
& & & & -1 & & & & \\
& & & & 1 & & & &
\end{pmatrix}
\begin{pmatrix}
a/\sqrt{2} & c/\sqrt{2} & e/\sqrt{2} & & & & & & \\
& & & & & & h/\sqrt{2} & & k/\sqrt{2} \\
& & & & -f & & & -i & \\
& & & & & & -h/\sqrt{2} & & k/\sqrt{2} \\
a/\sqrt{2} & c/\sqrt{2} & -e/\sqrt{2} & & & & & & \\
& & & f & & & i & & \\
& & & & -g & & & -j & \\
& & & g & & & j & & \\
b & d & 0 & & & & & &
\end{pmatrix}
:
$$

$$
:
\begin{pmatrix}
1 & & & & \\
& -1/2 & -\sqrt{3}/2 & & \\
& -\sqrt{3}/2 & 1/2 & & \\
& & & 1 & \\
& & & & -1 \\
& & & 1 & \\
& & & & -1 \\
& & & & & -1 \\
& & & & 1
\end{pmatrix}
\quad (6.1.28)
$$

Multiplying first and second factors and then the second and third matrices gives the following matrix equation:

$$
\begin{pmatrix}
a/\sqrt{2} & c/\sqrt{2} & e/\sqrt{2} & & & & & & \\
& & & & f & 0 & & i & \\
& & & & h/\sqrt{2} - & & & k/\sqrt{2} & \\
& & & g & & & j & & \\
b & d & 0 & & & & & & \\
& & & -g & & & -j & & \\
& & & & -h/\sqrt{2} & & & k/\sqrt{2} & \\
& & & -f & & & -i & & \\
a/\sqrt{2} & c/\sqrt{2} & -e/\sqrt{2} & & & & & &
\end{pmatrix}
=
\begin{pmatrix}
a/\sqrt{2} & C & E & & & & & & \\
& & & & h/\sqrt{2} & & k/\sqrt{2} & & \\
& & & & & f & & & i \\
& & & & -h/\sqrt{2} & & k/\sqrt{2} & & \\
a/\sqrt{2} & D & F & & & & & & \\
& & & f & & & -i & & \\
& & & & & g & & & j \\
& & & g & & & -j & & \\
b & C & -E & & & & & &
\end{pmatrix},
$$

$$(6.1.29)$$

where $C = -c/\sqrt{8} - e\sqrt{3}/\sqrt{8}$, $D = -c/\sqrt{8} + e\sqrt{3}/\sqrt{8}$, $E = -c\sqrt{3}/\sqrt{8} + e/\sqrt{8}$, $F = -c\sqrt{3}/\sqrt{8} - e/\sqrt{8}$. The following equations for the unknown constants result:

$$b = \frac{a}{\sqrt{2}} \quad \text{(for } A_1 \text{ column)},$$

$$e = -c\sqrt{3}, \quad d = \frac{c}{\sqrt{2}} \quad \text{(for } E \text{ columns)},$$

$$g = -f, \quad h = f\sqrt{2} \quad \text{(for } T_1 \text{ columns)},$$

$$k = i\sqrt{2}, \quad j = i \quad \text{(for } T_2 \text{ columns)}.$$

A single undetermined constant is left for each irrep. This is determined up to a phase by requiring normalized column vectors. Then the coupling coefficient table is written:

$T_1 \otimes T_1$		A_1 A_1 A_1	E A_1 A_1	B_1 A_1	T_1 E B_1	B_2	A_2	T_2 E B_1	B_2	A_2
B_1	B_1	$\frac{1}{\sqrt{3}}$	$\frac{1}{\sqrt{6}}$	$-\frac{1}{\sqrt{2}}$						
	B_2					$\frac{1}{\sqrt{2}}$				$\frac{1}{\sqrt{2}}$
	A_2						$-\frac{1}{\sqrt{2}}$			$-\frac{1}{\sqrt{2}}$
B_2	B_1							$-\frac{1}{\sqrt{2}}$		$\frac{1}{\sqrt{2}}$
	B_2	$\frac{1}{\sqrt{3}}$	$\frac{1}{\sqrt{6}}$	$\frac{1}{\sqrt{2}}$						
	A_2					$\frac{1}{\sqrt{2}}$			$\frac{1}{\sqrt{2}}$	
A_2	B_1					$\frac{1}{\sqrt{2}}$				$-\frac{1}{\sqrt{2}}$
	B_2				$-\frac{1}{\sqrt{2}}$			$\frac{1}{\sqrt{2}}$		
	A_2	$\frac{1}{\sqrt{3}}$	$-\frac{2}{\sqrt{6}}$	0						

(6.1.30)

The table agrees with the previously calculated $(T_1 \otimes T_1)$ coefficients (6.1.19c) in all but the second column from the right-hand side. The difference in sign there is due to the difference in sign $\left(\left|{T_2 \atop 2}\right\rangle = -\left|{T_2 \atop B_2}\right\rangle\right)$ between nonkosher and kosher bases, respectively.

In Appendix F all the O-group coupling coefficients are tabulated. The O_h coefficients follow immediately from these. One only has to remember that even-even and odd-odd products are even ($g \otimes g = u \otimes u = g$), while odd-even products are odd ($g \otimes u = u \otimes g = u$).

(b) Trigonal ($O \supset D_3 \supset C_2$) Coupling Coefficients Products of the even [(+), A, A_1, A', etc.)] and odd [(−), B, A_2, A'', etc.] irreps of C_2-like groups follow the usual odd-even rules:

$$\begin{array}{c} B = A \quad B \\ A \otimes B = \boxed{A \quad B} \\ B \otimes B = \boxed{B \quad A} \end{array} . \qquad (6.1.31)$$

The irrep bases of $D_3 = \{1, r_1, r_1^2, i_2, i_4, i_5\}$ are labeled

$$\left|{E \atop A}\right\rangle, \left|{E \atop B}\right\rangle, \left|{A_2 \atop B}\right\rangle, \left|{A_1 \atop A}\right\rangle,$$

where the lower indices are irreps of subgroup $C_2 = \{1, i_4\}$. Following the procedures given in the preceding section one deduces the form of the $E \otimes E$ coupling coefficients:

$$
\begin{array}{c|cccc}
 & A_1 & A_2 & E & E \\
E \otimes E & A & B & A & B \\
\hline
A \quad A & a & 0 & e & 0 \\
A \quad B & 0 & c & 0 & g \\
B \quad A & 0 & d & 0 & h \\
B \quad B & b & 0 & f & 0 \\
\end{array} = \mathscr{C}. \qquad (6.1.32)
$$

The irrep matrix

$$\mathscr{D}^E(r_1) = \mathscr{D}^3(r) = \begin{pmatrix} -\dfrac{1}{2} & -\dfrac{\sqrt{3}}{2} \\ \dfrac{\sqrt{3}}{2} & -\dfrac{1}{2} \end{pmatrix}$$

is taken from Eq. (3.4.14) and used to solve for the unknown coupling coefficient

$$\mathscr{D}^E(r_1) \otimes \mathscr{D}^E(r_1)\mathscr{C} = \mathscr{C}(\mathscr{D}^{A_1}(r_1) \oplus \mathscr{D}^{A_2}(r_1) \oplus \mathscr{D}^E(r_1)). \quad (6.1.33)$$

This equation is represented as follows

$$\begin{pmatrix} \frac{1}{4} & \frac{\sqrt{3}}{4} & \frac{\sqrt{3}}{4} & \frac{3}{4} \\ -\frac{\sqrt{3}}{4} & \frac{1}{4} & -\frac{3}{4} & \frac{\sqrt{3}}{4} \\ -\frac{\sqrt{3}}{4} & -\frac{3}{4} & \frac{1}{4} & \frac{\sqrt{3}}{4} \\ \frac{3}{4} & -\frac{\sqrt{3}}{4} & -\frac{\sqrt{3}}{4} & \frac{1}{4} \end{pmatrix} \begin{pmatrix} a & 0 & e & 0 \\ 0 & c & 0 & g \\ 0 & d & 0 & h \\ b & 0 & f & 0 \end{pmatrix}$$

$$= \mathscr{C} \begin{pmatrix} 1 & 0 & 0 & 0 \\ 0 & 1 & 0 & 0 \\ 0 & 0 & -\frac{1}{2} & -\frac{\sqrt{3}}{2} \\ 0 & 0 & \frac{\sqrt{3}}{2} & -\frac{1}{2} \end{pmatrix}. \quad (6.1.34)$$

Solving this equation and normalizing gives the coupling table:

$E \otimes E$		A_1 A	A_2 B	E A	E B
A	A	$\frac{1}{\sqrt{2}}$	0	$\frac{1}{\sqrt{2}}$	0
A	B	0	$\frac{1}{\sqrt{2}}$	0	$-\frac{1}{\sqrt{2}}$
B	A	0	$-\frac{1}{\sqrt{2}}$	0	$-\frac{1}{\sqrt{2}}$
B	B	$\frac{1}{\sqrt{2}}$	0	$-\frac{1}{\sqrt{2}}$	0

(6.1.35a)

The $E \otimes A_2$ table follows similarly:

$E \otimes A_2$		E A	E B
A	B	0	-1
B	B	1	0

(6.1.35b)

The D_3 tables have very similar form to the D_4 tables. However, the octahedral coupling coefficients for the $O \supset D_3 \supset C_2$ chain are very different in form from those of the $O \supset D_4 \supset D_2$ chain. For example, the $T_2 \otimes T_1$ table is given here:

$T_2 \otimes T_1$	A_1	E 1	E 2	T_2 1	T_2 2	T_2 3	T_1 1	T_1 2	T_1 3
1 1	$\frac{1}{\sqrt{3}}$	$\frac{1}{\sqrt{6}}$	.	$\frac{1}{\sqrt{6}}$	$\frac{1}{\sqrt{3}}$	.	.	.	.
1 2	.	.	$-\frac{1}{\sqrt{6}}$	.	.	$-\frac{1}{\sqrt{3}}$	.	.	$\frac{1}{\sqrt{2}}$
1 3	.	.	$-\frac{1}{\sqrt{3}}$	.	.	$\frac{1}{\sqrt{6}}$	.	$-\frac{1}{\sqrt{2}}$	.
2 1	.	.	$-\frac{1}{\sqrt{6}}$	.	.	$-\frac{1}{\sqrt{3}}$	.	.	$-\frac{1}{\sqrt{2}}$
2 2	$\frac{1}{\sqrt{3}}$	$-\frac{1}{\sqrt{6}}$	.	$\frac{1}{\sqrt{6}}$	$-\frac{1}{\sqrt{3}}$	.	.	$\frac{1}{\sqrt{2}}$	.
2 3	.	$\frac{1}{\sqrt{3}}$	.	.	.	$-\frac{1}{\sqrt{6}}$	.	$\frac{1}{\sqrt{2}}$	.
3 1	.	.	$-\frac{1}{\sqrt{3}}$	.	.	$\frac{1}{\sqrt{6}}$	.	$\frac{1}{\sqrt{2}}$	.
3 2	.	$\frac{1}{\sqrt{3}}$	.	.	.	$-\frac{1}{\sqrt{6}}$	$-\frac{1}{\sqrt{2}}$	.	.
3 3	$-\frac{1}{\sqrt{3}}$	.	.	.	$\frac{2}{\sqrt{6}}$	.	.	.	.

(6.1.36)

In this table the bases are labeled as follows according to $O \supset D_3 \supset C_2$:

$$\begin{pmatrix} A_1 \\ 1 \end{pmatrix} = \begin{pmatrix} A_1 \\ A_1 \\ A \end{pmatrix};$$

$$\begin{pmatrix} E \\ 1 \end{pmatrix} = \begin{pmatrix} E \\ E \\ A \end{pmatrix}, \quad \begin{pmatrix} E \\ 2 \end{pmatrix} = \begin{pmatrix} E \\ E \\ B \end{pmatrix};$$

$$\begin{pmatrix} T_2 \\ 1 \end{pmatrix} = \begin{pmatrix} T_2 \\ A \\ A \end{pmatrix}, \quad \begin{pmatrix} T_2 \\ 2 \end{pmatrix} = \begin{pmatrix} T_2 \\ E \\ A \end{pmatrix}, \quad \begin{pmatrix} T_2 \\ 3 \end{pmatrix}, \quad \begin{pmatrix} T_2 \\ E \\ B \end{pmatrix};$$

$$\begin{pmatrix} T_1 \\ 1 \end{pmatrix} = \begin{pmatrix} T_1 \\ E \\ A \end{pmatrix}, \quad \begin{pmatrix} T_1 \\ 2 \end{pmatrix} = \begin{pmatrix} T_1 \\ E \\ B \end{pmatrix}, \quad \begin{pmatrix} T_1 \\ 3 \end{pmatrix} = \begin{pmatrix} T_1 \\ A_2 \\ B \end{pmatrix}.$$

The other trigonal coupling tables are given in Appendix F.

6.2 GENERAL CONCEPTS AND MATRIX RELATIONS FOR COUPLING COEFFICIENTS

There are some important algebraic relations between the $\mathscr{C}_{ijk}^{\alpha\beta\gamma}$ coefficients and irreps $\mathscr{D}^{\alpha}$. These are based on important orthogonality and completeness properties. We must review these before considering detailed applications of coupling theory.

A. Products Involving Invariants or Scalars

One irreducible representation called the SCALAR or INVARIANT irrep $\mathscr{D}^0$ exists for any group $G = \{g, g', \ldots\}$. It is defined to be unity for all elements:

$$\mathscr{D}^{A_1}(g) = \mathscr{D}^0(g) = 1 \quad \text{(for all } g\text{)}. \tag{6.2.1}$$

It is labeled variously A_1, A_{1g}, A', Σ, or (0) depending on what group is being treated. The coupling coefficients for the product of the scalar with any other irrep are obviously given by

$$\mathscr{C}_{\cdot ij}^{0\alpha\beta} = \delta^{\alpha\beta}\delta_{ij} = \mathscr{C}_{i\cdot j}^{\alpha 0 \beta}. \tag{6.2.2}$$

This relation is so simple that we did not bother to write it in table form before.

A somewhat more complicated problem involves representations whose products produce scalars. In order to find which products will yield the scalar, one may use Eq. (6.1.18b) with $\mathscr{D}^{\gamma}$ set equal to 1:

$$\mathscr{C}_{ij\cdot}^{\alpha\beta 0} = \frac{(1/{}^\circ G)\Sigma_g 1 \cdot \mathscr{D}_{ik}^{\alpha}(g)\mathscr{D}_{jl}^{\beta}(g)}{(1/{}^\circ G)\Sigma_g 1 \cdot \mathscr{D}_{kk}^{\alpha}(g)\mathscr{D}_{ll}^{\beta}(g)}. \tag{6.2.3}$$

By combining Eqs. (3.4.18) and (3.4.19) (see also Appendix G) one derives the irrep orthogonality relation:

$$(l^{\alpha}/{}^\circ G) \sum_g \mathscr{D}_{ik}^{\alpha}(g)\mathscr{D}_{jl}^{\beta *}(g) = \delta^{\alpha\beta}\delta_{ij}\delta_{kl}. \tag{6.2.4}$$

Comparison with the preceding Eq. (6.2.3) shows that the only nonzero coupling coefficients in Eq. (6.2.3) occur when the product of an irrep $\mathscr{D}^{\alpha}$ is formed with its complex conjugate $\mathscr{D}^{\alpha *}$ or with itself if it is real. The irreps of finite groups fall into three categories with respect to complex conjugation. These are listed in the following:

Type 1 Irrep $\mathscr{D}^{\alpha}$ is real or else can be transformed to be real. $\mathscr{D}^{\alpha}(g)^* = \mathscr{D}^{\alpha}(g)$.

Type 2 Irreps $\mathscr{D}^{\alpha} = \mathscr{D}^{\beta *}$ and $\mathscr{D}^{\beta} = \mathscr{D}^{\alpha *}$ are a complex conjugate pair of inequivalent irreps.

Type 3 Irrep $\mathscr{D}^{\alpha}$ is always complex but equivalent to its conjugate $\mathscr{D}^{\alpha *}$.

Irreps that can be transformed to either standing-wave or moving-wave forms belong to type (1). The O, O_h, T_d, and D_n irreps are all of this type. For type (1) irreps ($\mathscr{D}^{\alpha*}(g) = \mathscr{D}^{\alpha}(g)$), we have from Eqs. (6.2.3) and (6.2.4) the following formula for making scalars:

$$\mathscr{D}_{ij}^{\alpha\beta 0} = \sqrt{1/l^\alpha}\, \frac{(l^\alpha/{}^\circ G)\sum_g \mathscr{D}_{ik}^\alpha(g)\mathscr{D}_{jl}^\beta(g)}{(l^\alpha/{}^\circ G)\sum_g \mathscr{D}_{kk}^\alpha(g)\mathscr{D}_{ll}^\beta(g)} = \sqrt{1/l^\alpha}\, \delta^{\alpha\beta}\delta_{ij}. \quad (6.2.5)$$

This checks with the results in Eq. (6.1.19c), where the scalar of O symmetry is conventionally labeled $\mathscr{D}^0 \equiv \mathscr{D}^{A_{1g}}$.

Complex one-dimensional moving-wave irreps of groups C_n, C_{nh}, and T_h groups belong to type (2). For type (2) one has the following formula, where $(\mathscr{D}^\alpha)^* = \mathscr{D}^{(\alpha^*)}$:

$$\mathscr{C}_{ij}^{\alpha\beta 0} = \sqrt{1/l^\alpha}\, \delta^{\alpha^*\beta}\delta_{ij}. \quad (6.2.6)$$

This form holds for type (3) irreps, too. The spinor and ray representations $\mathscr{D}^{n/2}$, $\mathscr{D}^G$, etc., all belong to type (2) or (3). The ones with real characters $\chi^{\alpha^*} = \chi^\alpha$ belong to type (3), and a transformation $\mathscr{J}$ exists such that $\mathscr{D}^{\alpha^*} = \mathscr{J}^\dagger \mathscr{D} \mathscr{J}$.

B. Symmetry Relations

Notice in Eq. (6.1.1) that for coupling coefficients $\mathscr{C}_{ijk}^{\alpha\alpha\gamma}$ involving the product of irrep T_{1u} with itself, either

$$\mathscr{C}_{ijk}^{\alpha\alpha\gamma} = \mathscr{C}_{jik}^{\alpha\alpha\gamma}, \quad (6.2.7a)$$

or

$$\mathscr{C}_{ijk}^{\alpha\alpha\gamma} = -\mathscr{C}_{jik}^{\alpha\alpha\gamma}, \quad (6.2.7b)$$

for all i, j, k. This follows from a more detailed treatment of permutation symmetry. Generally one says that $\mathscr{D}^\gamma$ belongs to the SYMMETRIZED SQUARE of $\mathscr{D}^\alpha$ if Eq. (6.2.7a) holds or to the ANTISYMMETRIZED SQUARE of $\mathscr{D}^\alpha$ if Eq. (6.2.7b) holds.

Other symmetry relations involving unequal irreps such as the following can be established to within a phase:

$$\mathscr{C}_{ijk}^{\alpha\beta\gamma} = \mathscr{C}_{jik}^{\beta\alpha\gamma}, \quad (6.2.8)$$

$$\mathscr{C}_{ijk}^{\alpha\beta\gamma} = \sqrt{l^\gamma/l^\alpha}\, \mathscr{C}_{jki}^{\beta\gamma\alpha}. \quad (6.2.9)$$

(Note that for the foregoing, $\mathscr{C}_{jik}^{\alpha\beta\gamma}$ would in general be meaningless.) However, these relations depend upon your choice and convention for overall phases, as we will discuss in detail in Chapter 7.

C. Product Analysis with Repeated Irreps

For most of the point groups and many other symmetry groups, we will find that reduction of any direct product will not yield any irreps $\mathscr{D}^\gamma$ more than once. These are called SIMPLY REDUCIBLE groups. There are groups that are not so simple. Two crystal point groups T and T_h are examples. For these the three-dimensional $\mathscr{D}^{T_i}$ ($i = u$ or g) appear twice in the product $\mathscr{D}^{T_i} \times \mathscr{D}^{T_j}$. For example, the T characters give the following product:

$$\mathscr{D}^{T_g} \times \mathscr{D}^{T_g} \cong \mathscr{D}^{A_g} + \mathscr{D}^\varepsilon + \mathscr{D}^{\varepsilon^*} + 2\mathscr{D}^{T_g}. \tag{6.2.10}$$

Whenever $\mathscr{D}^\gamma$ appears n times in the reduction of a product $\mathscr{D}^\alpha \times \mathscr{D}^\beta$, we will have n sets of coupling coefficients $\mathscr{C}_{ijm}^{\alpha\beta\gamma_1}, \mathscr{C}_{ijm}^{\alpha\beta\gamma_2}, \ldots, \mathscr{C}_{ijm}^{\alpha\beta\gamma_n}$. Each set can be made to give orthonormal sets of (γ) bases:

$$\left| \begin{matrix} (\gamma)_\omega \\ m \end{matrix} (\alpha\beta) \right\rangle = \sum_{i=1}^{l^\alpha} \sum_{j=1}^{l^\beta} \mathscr{C}_{ijm}^{\alpha\beta(\gamma)_\omega} \left| \begin{matrix} \alpha & \beta \\ i & j \end{matrix} \right\rangle. \tag{6.2.11}$$

These coefficients will not be uniquely defined until something is done to distinguish the n repeated states. For the two T_g in (6.2.10) one may simply reuse the $\mathscr{C}_{ijm}^{T_1 T_1 T_1}$ and $\mathscr{C}_{ijm}^{T_1 T_1 T_2}$ coefficients of the octahedral supergroup. In this case the repeating T's are distinguished by orthogonal irrep labels T_1 and T_2. However, a convenient "supergroup" may not always be available. Indeed, a general treatment of multiplicities in products constitutes an unsolved problem at present. Until this is solved one must sort and orthogonalize repeated product bases arbitrarily.

D. Orthonormality and Completeness

Any set of coupling coefficients are components of a transformation matrix. They are generally expected to be orthonormal and complete with respect to the product basis $\left\{ \left| \begin{matrix} \alpha & \beta \\ 1 & 1 \end{matrix} \right\rangle, \left| \begin{matrix} \alpha & \beta \\ 1 & 2 \end{matrix} \right\rangle, \ldots, \left| \begin{matrix} \alpha & \beta \\ l^\alpha & l^\beta \end{matrix} \right\rangle \right\}$ that is involved. Different product bases are expected to be orthogonal, as stated by the following:

$$\left\langle \begin{matrix} (\gamma)_\omega \\ m \end{matrix} \middle| \begin{matrix} (\gamma')_{\omega'} \\ m' \end{matrix} \right\rangle = \delta_{mm'} \delta^{(\gamma)_\omega (\gamma')_{\omega'}} = \sum_{i=1}^{l^\alpha} \sum_{j=1}^{l^\beta} \left\langle \begin{matrix} (\gamma)_\omega \\ m \end{matrix} \middle| \begin{matrix} \alpha\beta \\ ij \end{matrix} \right\rangle \left\langle \begin{matrix} \alpha\beta \\ ij \end{matrix} \middle| \begin{matrix} (\gamma')_{\omega'} \\ m' \end{matrix} \right\rangle$$

$$= \sum_{i=1}^{l^\alpha} \sum_{j=1}^{l^\beta} \left(\mathscr{C}_{ijm}^{\alpha\beta(\gamma)_\omega} \right)^* \left(\mathscr{C}_{ijm'}^{\alpha\beta(\gamma')_{\omega'}} \right). \tag{6.2.12a}$$

The product bases are also expected to be complete, as stated by the following:

$$\left\langle \begin{array}{c}\alpha\beta\\ij\end{array}\Big|\begin{array}{c}\alpha\beta\\i'j'\end{array}\right\rangle = \delta_{ii'}\delta_{jj'} = \sum_\gamma \sum_\omega \sum_{m=1}^{l\gamma} \left\langle \begin{array}{c}\alpha\beta\\ij\end{array}\Big|\begin{array}{c}(\gamma)_\omega\\m\end{array}\right\rangle \left\langle \begin{array}{c}(\gamma)_\omega\\m\end{array}\Big|\begin{array}{c}\alpha\beta\\i'j'\end{array}\right\rangle$$

$$= \sum_\gamma \sum_\omega \sum_{m=1}^{l\gamma} \left(\mathscr{C}_{ijm}^{\alpha\beta(\gamma)_\omega}\right)\left(\mathscr{C}_{i'j'm}^{\alpha\beta(\gamma)_\omega}\right)^*. \quad (6.2.12b)$$

The product reduction equations such as Eq. (6.1.19a) may be written in the following general form:

$$\sum_{i=1}^{l^\alpha}\sum_{j=1}^{l^\beta}\sum_{k=1}^{l^\alpha}\sum_{l=1}^{l^\beta}\left(\mathscr{C}_{ijm}^{\alpha\beta\gamma_\omega}\right)^* \mathscr{D}_{ik}^\alpha(g)\mathscr{D}_{jl}^\beta(g)\mathscr{C}_{kln}^{\alpha\beta(\gamma')_{\omega'}} = \delta^{(\gamma')_{\omega'}(\gamma)_\omega}\mathscr{D}_{mn}^\gamma(g). \quad (6.2.13)$$

By applying Eq. (6.2.12b) twice, one obtains the inverse relation:

$$\mathscr{D}_{ik}^\alpha(g)\mathscr{D}_{jl}^\beta(g) = \sum_\gamma \sum_\omega \sum_{m=1}^{l\gamma}\sum_{n=1}^{l\gamma}\mathscr{C}_{ijm}^{\alpha\beta(\gamma)_\omega}\left(\mathscr{C}_{kln}^{\alpha\beta(\gamma)_\omega}\right)^*\mathscr{D}_{mn}^\gamma(g). \quad (6.2.14)$$

Finally, the orthogonality relation Eq. (6.2.4) yields an equation that is quite analogous to Eq. (6.1.18), which was used previously to derive $\mathscr{C}_{ijm}^{\alpha\beta\gamma}$:

$$(l^\alpha/{}^\circ G)\sum_g \mathscr{D}_{ik}^\alpha(g)\mathscr{D}_{jl}^\beta(g)\mathscr{D}_{mn}^{\gamma*}(g) = \sum_\omega \mathscr{C}_{ijm}^{\alpha\beta(\gamma)_\omega}\left(\mathscr{C}_{kln}^{\alpha\beta(\gamma)_\omega}\right)^*. \quad (6.2.15)$$

This is sometimes called the FACTORIZATION LEMMA. For most crystal point groups, we may drop the sum (Σ_ω) over repeats. Also, the conjugate (*) is not needed for real coefficients.

6.3 VECTORS AND TENSORS IN 3-SPACE

The knowledge of symmetry can systematically simplify the treatment of vector and tensor quantities. We discuss how some procedures that work for any spatial symmetry with examples in the symmetries of O_h, C_{3v}, and D_{4h}. Stress-strain tensor relations in solids will be treated in detail.

A. Symmetry-Defined Unit Vectors

As a first step one must define three unit vectors of a Cartesian coordinate system and obtain a 3 × 3 representation $\mathscr{V}$ of the symmetry operators in this basis which is called the VECTOR representation. It is convenient to define unit vectors along symmetry axes whenever possible. For example, one choice for O_h or D_{4h} symmetry has the vectors $\hat{x}_j$ lying on the tetragonal axes and was shown in Figure 4.2.1(a). Figure 4.2.2(a) showed another choice of basis centered on a trigonal axis. The latter vectors $\hat{v}_j$ are also a convenient coordinate system in D_{3d} or C_{3v} symmetry.

To find the vector representation of a symmetry operator g, we imagine this operation moves the physical object with the unit vectors attached, as in Figures 4.2.1(b) and 4.2.2(b). Then, since g is a symmetry operator, the object must look as though it had not been moved at all. However, the unit vectors will be moved to new places. The vector representation $\mathscr{V}$ is the transformation matrix as defined by

$$\mathscr{V}_{ij}(g) = x_i \cdot x'_j = x_i \cdot g \cdot x_j,$$

$$\hat{x}'_j \equiv g \cdot \hat{x}_j = \sum_i \mathscr{V}_{ij}(g)\hat{x}_i. \quad (6.3.1)$$

The second step is to establish relations between the vector representation and certain irreps of the symmetry. In some cases a vector representation may be an irrep as in the case of O_h in Figure 4.2.1, where $\mathscr{V} = \mathscr{D}^{T_{1u}}$. In other cases the vector representation may be equal to a direct sum of irreps as in the case of D_3 in Figure 4.2.2, where $\mathscr{V} = \mathscr{D}^{E_u} \oplus \mathscr{D}^{A_{2u}}$. In general, you may have to transform the vector representation into an irrep or sum of irreps, as would be the case if we used the tetragonal vectors in Figure 4.2.1 to represent the D_{3d} symmetry operations in Figure 4.2.2 or vice versa. Recall the transformation (4.2.31) between the tetragonal and trigonal bases.

In any case, the final result should be three unit vectors $\hat{x}_j^\alpha$ with labels $\binom{\alpha}{j}$ of the symmetry irrep. They should have the corresponding transformation character,

$$\hat{x}_j^{\prime\alpha} = \sum_i \mathscr{D}_{ij}^\alpha(g)\hat{x}_i^\alpha. \quad (6.3.2)$$

For O_h x_j vectors and D_{3d} v_j vectors we have the following:

$$\hat{x}_1^{T_{1u}} \equiv \hat{x}_1, \qquad \hat{x}_1^3 \equiv \hat{x}_1^{E_u} \equiv \hat{v}_1,$$

$$\hat{x}_2^{T_{1u}} \equiv \hat{x}_2, \qquad \hat{x}_2^3 \equiv \hat{x}_2^{E_u} \equiv \hat{v}_2,$$

$$\hat{x}_3^{T_{1u}} \equiv \hat{x}_3, \qquad \hat{x}^1 \equiv \hat{x}^{A_{2u}} \equiv \hat{v}_3.$$

For some applications it is helpful to use the full subgroup chains $O_h \supset D_{4h} \supset D_{2h}$ or $O_h \supset D_{3d} \supset C_{2v}$ to label the vectors $\hat{x}_j$ or $\hat{v}_j$, respectively, as given in Eqs. (4.2.23) or (4.2.27):

$$\hat{x}_1 \equiv \hat{x} \begin{matrix} T_{1u}:O_h \\ E_u:D_{4h}, \\ B_{1u}:D_{2h} \end{matrix} \qquad \hat{x}_2 \equiv \hat{x} \begin{matrix} T_{1u} \\ E_u \\ B_{2u} \end{matrix}, \qquad \hat{x}_3 \equiv \hat{x} \begin{matrix} T_{1u} \\ A_{2u}, \\ A_{2u} \end{matrix} \qquad (6.3.3a)$$

$$\hat{v}_1 \equiv \hat{x} \begin{matrix} T_{1u}:O_h \\ E_u:D_{3d}, \\ A_u:C_{2v} \end{matrix} \qquad \hat{v}_2 \equiv \hat{x} \begin{matrix} T_{1u} \\ E_u \\ B_u \end{matrix}, \qquad \hat{v}_3 \equiv \hat{x} \begin{matrix} T_{1u} \\ A_{2u}. \\ B_u \end{matrix} \qquad (6.3.3b)$$

B. Symmetry-Defined Unit Tensors

Using coupling coefficients it is easy to make unit vectors $\hat{x}_i^\alpha$ into complete sets of unit tensors as follows:

$$\hat{T}_m^\gamma = \sum_i \sum_j \mathscr{C}_{ijm}^{\alpha\beta\gamma} \hat{x}_i^\alpha \hat{x}_j^\beta. \qquad (6.3.4)$$

According to the theory of coupling coefficients, these tensors must transform according to the irreps which label them. First, the x_j^α transformation (6.3.2) gives

$$\hat{T}_m^{\prime\gamma} = \sum_i \sum_j \mathscr{C}_{ijm}^{\alpha\beta\gamma} \hat{x}_i^{\prime\alpha} \hat{x}_j^{\prime\beta}$$

$$= \sum_i \sum_j \sum_{i'} \sum_{j'} \mathscr{C}_{ijm}^{\alpha\beta\gamma} \mathscr{D}_{i'i}^\alpha \mathscr{D}_{j'j}^\beta \hat{x}_{i'}^\alpha \hat{x}_{j'}^\beta.$$

Then the coupling relations (6.2.12)–(6.2.14) give

$$\hat{T}_m^{\prime\gamma} = \sum_{m'} \mathscr{D}_{m'm}^\gamma \sum_{i'} \sum_{j'} \mathscr{C}_{i'j'm'}^{\alpha\beta\gamma} \hat{x}_{i'}^\alpha \hat{x}_{j'}^\beta,$$

which is the desired transformation relation

$$\hat{T}_m^{\prime\gamma} = \sum_{m'} \mathscr{D}_{m'm}^\gamma \hat{T}_{m'}^\gamma \qquad (6.3.5)$$

for an irreducible or SYMMETRY-DEFINED tensor set $\{T_1^\gamma, T_2^\gamma, \ldots\}$.

For example, the $(\alpha \otimes \beta) = (T_{1u} \otimes T_{1u})$ tetragonal O_h coupling coefficients (6.1.19c) yield the "nonkosher" symmetry-defined unit tensors here:

$$\hat{\hat{T}}^{A_{1g}} = (\hat{x}_1\hat{x}_1 + \hat{x}_2\hat{x}_2 + \hat{x}_3\hat{x}_3)/\sqrt{3}, \quad \hat{\hat{T}}_1^{T_{2g}} = (\hat{x}_2\hat{x}_3 + \hat{x}_3\hat{x}_2)/\sqrt{2},$$

$$\hat{\hat{T}}_1^{E_g} = (\hat{x}_1\hat{x}_1 + \hat{x}_2\hat{x}_2 - 2\hat{x}_3\hat{x}_3)/\sqrt{6}, \quad \hat{\hat{T}}_2^{T_{2g}} = (\hat{x}_3\hat{x}_1 + \hat{x}_1\hat{x}_3)/\sqrt{2},$$

$$\hat{\hat{T}}_2^{E_g} = (-\hat{x}_1\hat{x}_1 + \hat{x}_2\hat{x}_2)/\sqrt{2}, \quad \hat{\hat{T}}_3^{T_{2g}} = (\hat{x}_1\hat{x}_2 + \hat{x}_2\hat{x}_1)/\sqrt{2},$$

$$\hat{\hat{T}}_1^{T_{1g}} = (\hat{x}_2\hat{x}_3 - \hat{x}_3\hat{x}_2)/\sqrt{2}, \quad \hat{\hat{T}}_2^{T_{1g}} = (\hat{x}_3\hat{x}_1 - \hat{x}_1\hat{x}_3)/\sqrt{2},$$

$$\hat{\hat{T}}_3^{T_{1g}} = (\hat{x}_1\hat{x}_2 - \hat{x}_2\hat{x}_1)/\sqrt{2}. \tag{6.3.6}$$

C. Symmetry-Defined Coordinates and Polynomials

In any of the foregoing tensors the vectors $\hat{x}_j$ may be replaced by coordinates $x = x_1$, $y = x_2$, and $z = x_3$ to give symmetry-defined polynomials provided that the transformation character of the coordinates is defined correctly. One way to do this is to let the x_j be coordinates of a field point $\mathbf{r}$ which is unaffected by symmetry operations:

$$\mathbf{r} = x_1\hat{x}_1 + x_2\hat{x}_2 + x_3\hat{x}_3 = x_1'\hat{x}_1' + x_2'\hat{x}_2' + x_3'\hat{x}_3'. \tag{6.3.7}$$

Substituting the vector transformation (6.3.1) in the foregoing gives the following. In the last step the orthogonality of the $\mathscr{V}$ matrices is used:

$$x_i' = \sum_{j=1}^{3} \mathscr{V}_{ij}(g^{-1}) x_j = \sum_{j=1}^{3} \mathscr{V}_{ji}(g) x_j. \tag{6.3.8}$$

Then the coordinates transform just like the base vectors. From Eq. (6.3.6) we get the following quadratic polynomials which transform according to definite irreps of O_h (note that the antisymmetric product T_{1g} is missing, since it is zero):

$$p^{A_{1g}} = (x^2 + y^2 + z^2)/\sqrt{3}, \quad p_1^{T_{2g}} = \sqrt{2}\,yz,$$

$$p_1^{E_g} = (x^2 + y^2 - 2z^2)/\sqrt{6}, \quad p_2^{T_{2g}} = \sqrt{2}\,xz,$$

$$p_2^{E_g} = (-x^2 + y^2)/\sqrt{2} \quad p_3^{T_{2g}} = \sqrt{2}\,xy. \tag{6.3.9}$$

Continued application of coupling coefficients then makes a complete set of cubic, quartic, or higher-degree polynomials for each irrep. These are tabulated in Appendix F.

Polynomials made in this way are sometimes called POINT SYMMETRY HARMONICS since any eigenfunction in the symmetry environment must be a combination of only those harmonics belonging to a particular irrep (γ).

D. Symmetry-Defined Bulk Behavior in Solids

(a) Stress Tensor The word "tensor" comes from its application to the study of internal tension or stress in solid bodies. Soon we will be dealing with more abstract tensors in quantum mechanics, so it is instructive to review the original application of these ideas.

Suppose some solid is weighted down with various forces and weights, and these forces are felt at each point or atom within the solid. For any plane P containing an atom, define a vector $\mathbf{F}(P)$ to be the force felt by the plane per (infinitesimal) unit of area due to all the material on one side [Figure 6.3.1(a)]. $\mathbf{F}(P)$ is the vector average of all the forces transmitted by the various springs or fibers that penetrate or touch a unit area of P. Now imagine a single set of parallel fibers or springs all under uniform tension $\mathbf{t}$ or compression $-\mathbf{t}$ [Figure 6.3.1(b)], per unit transverse area. The force on the unit area just due to these fibers is proportional to the projection $\mathbf{t} \cdot \hat{\mathbf{n}}$ of $\mathbf{t}$ on the unit normal $\hat{\mathbf{n}}$ on the chosen side:

$$\mathbf{F}_t(P) = \mathbf{t}\cos(\mathbf{t}, \hat{n}) = \mathbf{t}(\mathbf{t} \cdot \hat{\mathbf{n}})/t = (\mathbf{tt}) \cdot \hat{\mathbf{n}}/t.$$

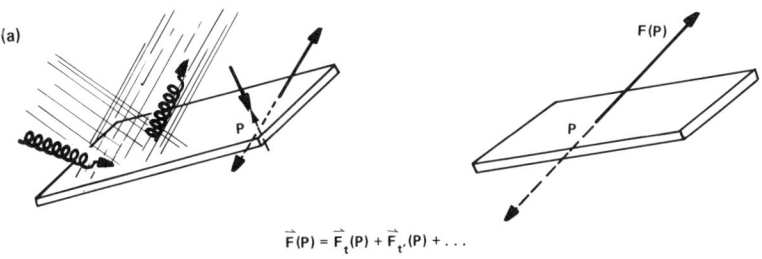

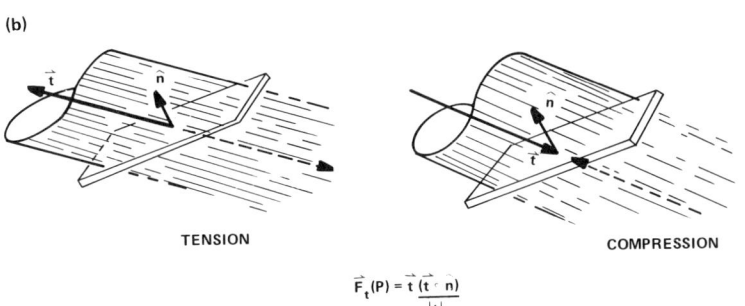

Figure 6.3.1 Representing stress in a solid. (a) The effect of many fibers is represented by the sum $\mathbf{F}(\rho)$ of all the forces pulling on one side of an infinitesimal area. (b) The force due to one set of fibers is expressed by the scalar product of a tensor dyad **tt** and the unit surface normal.

This equation simply counts the number of fibers which actually penetrate a unit area with normal $\hat{\mathbf{n}}$, and multiplies this by $t = |\mathbf{t}|$. The quantity $(\mathbf{tt})/t$ is an elementary STRESS TENSOR. The total force $\mathbf{F}(P)$ is given by an n-projected sum of these tensors for all the types of fibers present in the solid:

$$\mathbf{F}(P) = (\mathbf{tt}/t + \mathbf{t't'}/t' + \cdots) \cdot \hat{\mathbf{n}} \equiv \overset{\leftrightarrow}{T} \cdot \hat{\mathbf{n}}. \qquad (6.3.10)$$

The $\overset{\leftrightarrow}{T}$ in Eq. (6.3.10) is the stress tensor which describes the tension per area present in a small region of the solid. We see that $\overset{\leftrightarrow}{T}$ can be written as a combination of elementary unit tensors $\hat{x}_i \hat{x}_j$ or as a combination of any set of symmetry-defined tensors such as the O_h set T_m^γ in Eq. (6.3.6):

$$\overset{\leftrightarrow}{T} = \mathcal{T}_{11}\hat{x}_1\hat{x}_1 + \mathcal{T}_{12}\hat{x}_1\hat{x}_2 + \mathcal{T}_{13}\hat{x}_1\hat{x}_3,$$

$$+ \mathcal{T}_{21}\hat{x}_2\hat{x}_1 + \mathcal{T}_{22}\hat{x}_2\hat{x}_2 + \mathcal{T}_{23}\hat{x}_2\hat{x}_3$$

$$+ \mathcal{T}_{31}\hat{x}_3\hat{x}_1 + \mathcal{T}_{32}\hat{x}_3\hat{x}_2 + \mathcal{T}_{33}\hat{x}_3\hat{x}_3$$

$$\overset{\leftrightarrow}{T} = \mathcal{T}^{A_{1g}}\hat{T}^{A_{1g}} + \mathcal{T}_1^{E_g}\hat{T}_1^{E_g} + \mathcal{T}_2^{E_g}\hat{T}_2^{E_g}$$

$$+ \mathcal{T}_1^{T_{2g}}\hat{T}_1^{T_{2g}} + \mathcal{T}_2^{T_{2g}}\hat{T}_2^{T_{2g}} + \mathcal{T}_3^{T_{2g}}\hat{T}_3^{T_{2g}}$$

$$+ \mathcal{T}_1^{T_{1g}}\hat{T}_1^{T_{1g}} + \mathcal{T}_2^{T_{1g}}\hat{T}_2^{T_{1g}} + \mathcal{T}_3^{T_{1g}}\hat{T}_3^{T_{1g}}. \qquad (6.3.11)$$

In order to obtain a physical feeling for these components $\mathcal{T}_{1j}$ or $\mathcal{T}_m^\gamma$, we shall discuss their physical representations, which are shown in Figure 6.3.2.

First consider the Cartesian components $\mathcal{T}_{ij}$. Components $\mathcal{T}_{1j}$, $\mathcal{T}_{2j}$, and $\mathcal{T}_{3j}$ are the components of the force $\mathbf{F}(j)$ in Figure 6.3.2(a):

$$\mathbf{F}(j) = \overset{\leftrightarrow}{T} \cdot \hat{x}_j = \mathcal{T}_{1j}\hat{x}_1 + \mathcal{T}_{2j}\hat{x}_2 + \mathcal{T}_{3j}\hat{x}_3. \qquad (6.3.12)$$

We see that $\mathbf{F}(j)$ is the force per unit area felt by the jth face of the cube due to the outside world on the positive side of the $\hat{x}_j$ axis. It is assumed that the force $F(j)$ is the same for all parallel planes in the neighborhood, so an equal and opposite force $-\mathbf{F}(j)$ can be imagined tugging or pressing on the opposite cube face as indicated by the dotted arrows in the figure.

Now the symmetry-defined components $\mathcal{T}_m^\gamma$ tell how much of a particular type of unit stress $\hat{T}_m^\gamma$ is present. The common names of these symmetry-defined stresses are written in Figure 6.3.2(b). We shall see shortly what is the advantage of these coordinates over the $\mathcal{T}_{ij}$, but for now we can see that one set is easily found in terms of the other using the $(T_{1u} \otimes T_{1u})$ coupling

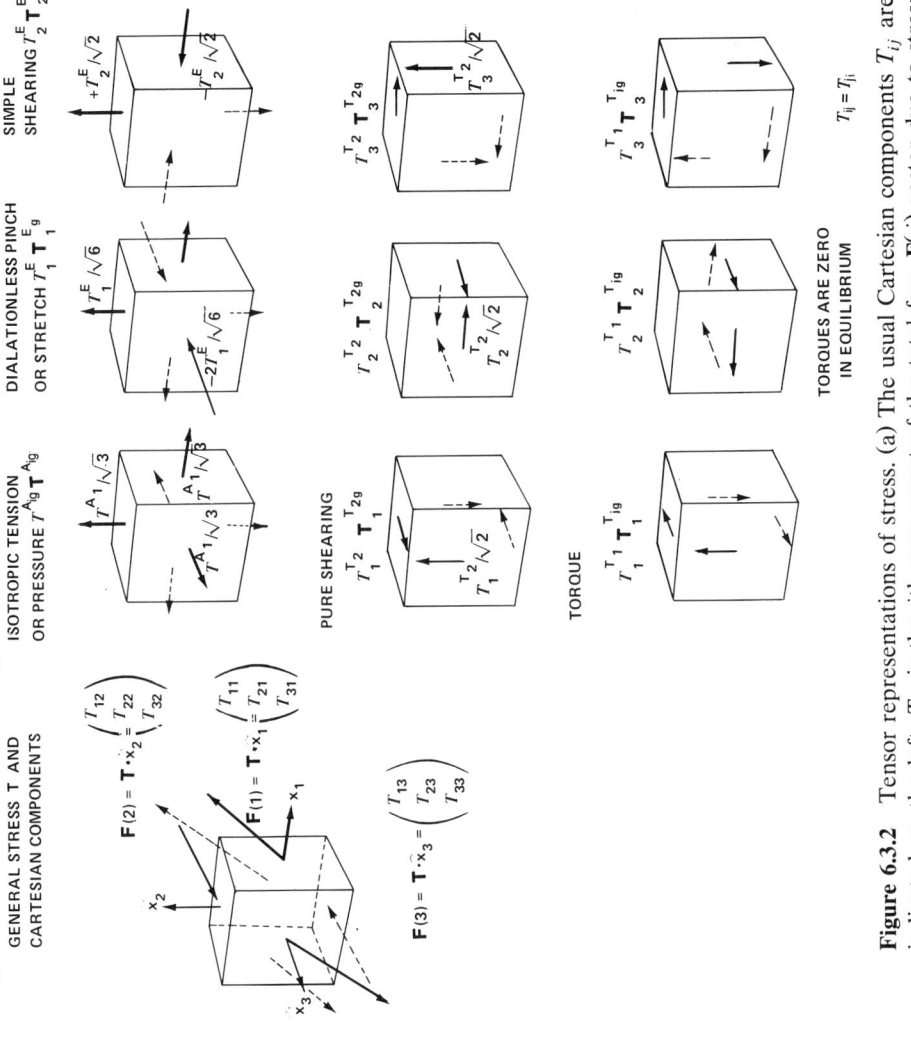

Figure 6.3.2 Tensor representations of stress. (a) The usual Cartesian components T_{ij} are indicated on the left. T_{ij} is the ith component of the total force $\mathbf{F}(j)$ vector due to stress outside the surface with normal $\hat{x}_j$. (b) The symmetry-defined stresses are shown. The magnitude of each vector is given in terms of the relevant component T_a^α.

coefficients:

$$\mathcal{T}_m^\gamma = \sum_i \sum_j \mathscr{C}_{ijm}^{T_{1u}T_{1u}\gamma} \mathcal{T}_{ij}, \qquad (6.3.13a)$$

$$\mathcal{T}_{ij} = \sum_\gamma \sum_m \mathscr{C}_{ijm}^{T_{1u}T_{1u}\gamma} \mathcal{T}_m^\gamma. \qquad (6.3.13b)$$

Note that the three torque components, i.e., the three components $\mathcal{T}_1^{T_1}$, $\mathcal{T}_2^{T_1}$, and $\mathcal{T}_3^{T_1}$ will have to be zero for any static or equilibrium stress. If the torques are zero, then the remaining symmetry-defined stress tensors or any combination of them are symmetric,

$$\mathcal{T}_{ij} = \mathcal{T}_{ji},$$

leaving just six independent stress tensor components.

(b) Strain Tensor A similar type of mathematics can be used to define the deformation or STRAIN inside a body. Suppose each point $\mathbf{r}$ inside is moved to some new point $\mathbf{r} + \mathbf{s}(\mathbf{r})$. Strain is a measure of what is happening to neighboring points $\mathbf{r} + d\mathbf{r}$ around a given $\mathbf{r}$. It tells how much the material in the neighborhood is getting crushed or stretched.

Figure 6.3.3 shows that the vector $d\mathbf{r}$ between neighboring points is transformed into $\nabla \mathbf{s} \cdot d\mathbf{r}$ in the limit of small $d\mathbf{r}$. The derivative $\nabla \mathbf{s} \equiv \overleftrightarrow{S}$ is called the STRAIN TENSOR:

$$\overleftrightarrow{S} = \nabla \mathbf{s} = \frac{\partial \mathbf{s}}{\partial x_1} \hat{x}_1 + \frac{\partial \mathbf{s}}{\partial x_2} \hat{x}_2 + \frac{\partial \mathbf{s}}{\partial x_3} \hat{x}_3,$$

$$\overleftrightarrow{S} = \frac{\partial s_1}{\partial x_1} \hat{x}_1 \hat{x}_1 + \frac{\partial s_1}{\partial x_2} \hat{x}_1 \hat{x}_2 + \frac{\partial s_1}{\partial x_3} \hat{x}_1 \hat{x}_3$$
$$+ \frac{\partial s_2}{\partial x_1} \hat{x}_2 \hat{x}_1 + \frac{\partial s_2}{\partial x_2} \hat{x}_2 \hat{x}_2 + \frac{\partial s_2}{\partial x_3} \hat{x}_2 \hat{x}_3$$
$$+ \frac{\partial s_3}{\partial x_1} \hat{x}_3 \hat{x}_1 + \frac{\partial s_3}{\partial x_2} \hat{x}_3 \hat{x}_2 + \frac{\partial s_3}{\partial x_3} \hat{x}_3 \hat{x}_3. \qquad (6.3.14a)$$

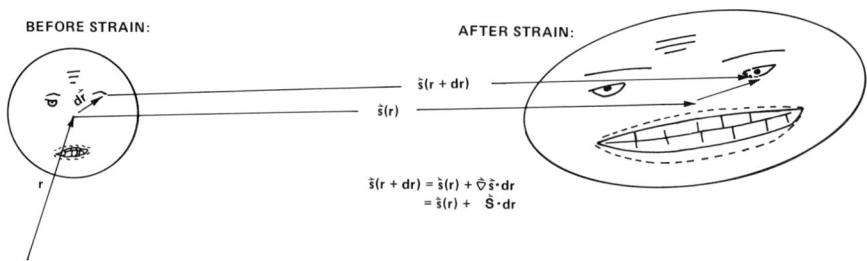

Figure 6.3.3 Defining the strain tensor $\overleftrightarrow{S} = \nabla \mathbf{s}$. The strain causes each point $\mathbf{r}$ to be moved to a new point $\mathbf{r} + \mathbf{s}(r)$.

VECTORS AND TENSORS IN 3-SPACE   475

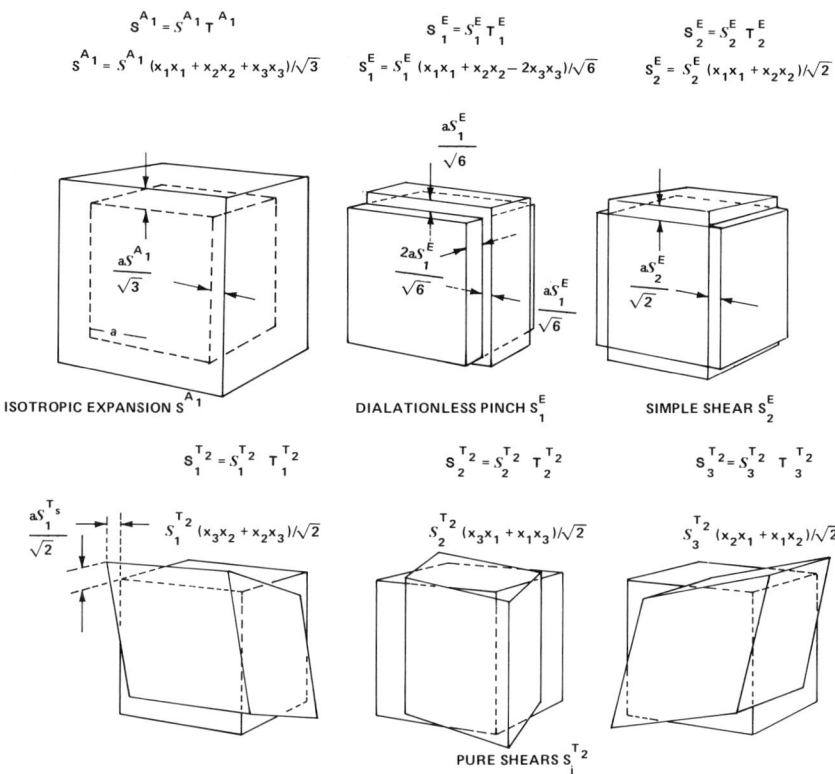

Figure 6.3.4 Symmetry-defined strains and their common names. Rotation motion $S_j^{T_1}$ is not considered to be a strain.

If $s(r) = s$ is constant, then no strain exists ($S = 0$), and we have just a uniform translation. If $s(r)$ is linear in the coordinates x_j of r, then $\overleftrightarrow{S}$ is a constant tensor, and we say that a HOMOGENEOUS STRAIN exists.

A homogeneous strain may be defined by the components $\mathscr{S}_{ij} = \partial s_i/\partial x_j$ or by any complete set of symmetry-defined components $\mathscr{S}_m^\gamma$ in exactly the same way that we defined stresses:

$$\vec{S} = \sum_{i=1}^{3}\sum_{j=1}^{3} \mathscr{S}_{ij}\hat{x}_i\hat{x}_j = \sum_\gamma \sum_m \mathscr{S}_m^\gamma \hat{T}_m^\gamma. \qquad (6.3.14b)$$

The mathematical form of $\hat{T}_m^\gamma$ is the same as before. The physical meaning of each term is different, as seen in Figure 6.3.4.

It is conventional to ignore purely rotational displacement tensors $\overleftrightarrow{S}_1^{T_1} = \mathscr{S}_1^{T_1}(-\hat{x}_3\hat{x}_2 + \hat{x}_2\hat{x}_3)$. These are generally not considered to be strains. Thus, there are only six independent strain components.

E. Symmetry Theory of Tensor Relations (Elastic Constants)

We study now generalizations of Hooke's law ($F = -kx$) for elastic solids in which six stresses are linearly related to six strains. The linear relations may involve either the standard Cartesian tensor components $\mathscr{T}_{hi}$ and $\mathscr{S}_{ik}$ and elastic constants $k_{hi,jk}$ as follows:

$$\mathscr{T}_{hi} = \sum_{j=1}^{3}\sum_{k=1}^{3} k_{hi,jk}\mathscr{S}_{jk}, \qquad (6.3.15a)$$

or else symmetry-defined components and constants as follows:

$$\mathscr{T}_m^\gamma = \sum_\delta \sum_n k_{mn}^{\gamma\delta}\mathscr{S}_n^\delta. \qquad (6.3.15b)$$

At first sight it might appear that the $6^2 = 36$ constants would be needed in either equation. However, this number can be reduced by using symmetry-defined components. For example, in the presence of O_h symmetry, we shall see that only three of the $k_{mn}^{\gamma\delta}$ are needed, namely, $k^{A_1} \equiv k^{A_1A_1}$, $k^E \equiv k_{11}^{EE} = k_{22}^{EE}$, and $k^{T_2} \equiv k_{11}^{T_2T_2} = k_{22}^{T_2T_2} = k_{33}^{T_2T_2}$. All others are zero!

Before treating examples, let us prove a general theorem about tensor relations and symmetry. Let us start with a general tensor relation, Eq. (6.3.16a), between two tensors $\mathscr{T}_{hi\cdots}$ and $\mathscr{S}_{kl\cdots}$, where $\mathscr{T}$ and $\mathscr{S}$ have m and n indices, respectively. Symmetry analysis isolates the independent constants from the set of 3^{m+n} constants $k_{hi\cdots kl\cdots}$ in the tensor relation:

$$\mathscr{T}_{hij\cdots} = \sum_k \sum_l \cdots \sum k_{hi\cdots kl\cdots}\mathscr{S}_{kl\cdots}. \qquad (6.3.16a)$$

The idea is to first rewrite this relation using the symmetry-defined components as follows:

$$\mathscr{T}(\gamma)_m^\omega = \sum_\phi \sum_\delta \sum_n k(\gamma)(\delta)_{m\;n}^{\omega\;\phi}\mathscr{S}(\delta)_n^\phi. \qquad (6.3.16b)$$

One could do this whether the symmetry was present or not. However, the new constants $k(\gamma)(\delta)_{m\;n}^{\omega\;\phi}$ assume a very simple form when the symmetry is present, as proved in the following. First, let us first review the meaning of the group labeling. Each tensor is labeled by a subgroup chain $\cdots F \supset G \supset H \cdots$ as follows:

$$\begin{aligned} \phi: &\quad \text{Nonsymmetry supergroup } F \text{ label or index,} \\ \mathscr{S}_m \equiv \mathscr{S}(\gamma): &\quad \text{Symmetry group } G \text{ label,} \\ h: &\quad \text{Symmetry subgroup } H \text{ label.} \end{aligned} \quad (6.3.17)$$

For example, the strain tensors in Figure 6.3.4 may be labeled by the tetragonal $O_h \supset D_{4h} \supset D_{2h}$ chain which was introduced in Chapter 4. The six components may be labeled as follows according to this scheme for $G = D_{4h}$:

$$\mathscr{S}^{A_{1g}} = \mathscr{S}\genfrac{}{}{0pt}{}{A_{1g}}{A_{1g}}(A_{1g}), \quad \mathscr{S}_1^{E_g} = \mathscr{S}\genfrac{}{}{0pt}{}{E_g}{A_{1g}}(A_{1g}), \quad \mathscr{S}_2^{E_g} = \mathscr{S}\genfrac{}{}{0pt}{}{E_g}{A_{1g}}(B_{1g}),$$

$$\mathscr{S}_1^{T_{2g}} = \mathscr{S}\genfrac{}{}{0pt}{}{T_{2g}}{B_{1g}}(E_b), \quad \mathscr{S}_2^{T_{2g}} = -\mathscr{S}\genfrac{}{}{0pt}{}{T_{2g}}{B_{2g}}(E_g), \quad \mathscr{S}_3^{T_{2g}} = \mathscr{S}\genfrac{}{}{0pt}{}{T_{2g}}{A_{2g}}(A_{2g}). \quad (6.3.18)$$

This scheme was introduced in Eq. (4.2.17). [Remember the slight but terrible difference between "kosher" subgroup labeled components and the "standard" T_{2g} components wherein $\mathscr{S}_2^{T_{2g}} = -\mathscr{S}\genfrac{}{}{0pt}{}{T_{2g}}{B_{2g}}(E_g)$.] In the proof of the tensor theorem which follows the (middle) symmetry-group G label is enclosed with parentheses. The upper nonsymmetry supergroup F label is superfluous as long as there are no two independent $\mathscr{S}^{(\gamma)}$ components with the same (γ). Note two $(\gamma) = (A_{1g})$ components exist for D_{4h}, and they are distinguished by the labels $\phi = A_{1g}$ and E_g, respectively, of the next higher link $(F = O_h)$ of the subgroup chain.

If the symmetry group is octahedral $G = O_h$ it may be convenient to include the full rotational group $F = O_3$ in the subgroup chain $(O_3 \supset O_h \supset D_{4h} \cdots)$ and label the strain tensors as follows:

$$\mathscr{S}^{A_{1g}} \equiv \mathscr{S}\genfrac{}{}{0pt}{}{0^+}{A_{1g}}(A_{1g}), \quad \mathscr{S}_1^{E_g} \equiv \mathscr{S}\genfrac{}{}{0pt}{}{2^+}{A_{1g}}(E_g), \quad \mathscr{S}_2^{E_g} \equiv \mathscr{S}\genfrac{}{}{0pt}{}{2^+}{B_{1g}}(E_g),$$

$$\mathscr{S}_1^{T_{2g}} \equiv \mathscr{S}\genfrac{}{}{0pt}{}{2^+}{E_g(1)}(T_{2g}), \quad \mathscr{S}_2^{T_{2g}} \equiv -\mathscr{S}\genfrac{}{}{0pt}{}{2^+}{E_g(2)}(T_{2g}), \quad \mathscr{S}_3^{T_{2g}} \equiv \mathscr{S}\genfrac{}{}{0pt}{}{2^+}{A_{2g}}(T_{2g}). \quad (6.3.19)$$

Here the connection between $J = 0^+$ and 2^+ representations of O_3 and second-order polynomials

$$\left\{(x^2 + y^2 + z^2) \sim (0^+); \left(2z^2 - x^2 - y^2, \sqrt{3}\,(x^2 - y^2), xy, xz, yz\right) \sim (2^+)\right\}$$

is used. Recall Eqs. (5.6.15) and (6.3.9).

Tensor Theorem If physical properties of a solid having G symmetry are described by the G-symmetry-defined tensor relation in Eq. (6.3.16b), then there will be at most one independent nonzero constant for each distinct pair of equal symmetry group labels $(\gamma) = (\delta)$ which appear on opposite sides of the relation. These constants obey the following rules:

$$k\begin{pmatrix}\omega \\ (\gamma) \\ m\end{pmatrix}\begin{pmatrix}\phi \\ (\delta) \\ n\end{pmatrix} = \begin{cases} 0, & \text{if } \delta \neq \gamma \text{ or } m \neq n, \\ k(\gamma)^{\omega\phi}, & \text{for each } m = n \text{ if } \delta = \gamma, \end{cases}$$

and the relation assumes the following form:

$$\mathcal{T}\begin{pmatrix}\omega \\ (\gamma) \\ m\end{pmatrix} = \sum_{\phi} k(\gamma)^{\omega\phi}\,\mathcal{S}\begin{pmatrix}\phi \\ (\gamma) \\ m\end{pmatrix} \quad (6.3.20)$$

for each m.

Proof In the presence of symmetry, we may redefine our axes in the solid as indicated in Figure 4.2.2 using any symmetry operator g, without changing the values of the constants $k\begin{pmatrix}(\gamma)\\m\end{pmatrix}\begin{pmatrix}(\delta)\\n\end{pmatrix}$ in the relation. Since the axes all end up in equivalent positions with respect to the solid, the constants cannot be different. However, the unit vectors and tensors will be transformed according to their respective irreps as follows:

$$\sum_{m'} \mathcal{D}^\gamma_{m'm}(g)\,\mathcal{T}\begin{pmatrix}\omega\\(\gamma)\\m'\end{pmatrix}^\omega = \sum_{\delta}\sum_{\phi}\sum_n k\begin{pmatrix}\omega\\(\gamma)\\m\end{pmatrix}\begin{pmatrix}\phi\\(\delta)\\n\end{pmatrix}\sum_{n} \mathcal{D}^\delta_{n'n}(g)\,\mathcal{S}\begin{pmatrix}\phi\\(\delta)\\n\end{pmatrix}.$$

Since this is true for all g in the group, we may substitute for g the elementary operators P^α_{ji} using $\mathcal{D}^\gamma_{m'm}(P^\alpha_{ji}) = \delta^{\gamma\alpha}\delta_{m'j}\delta_{mi}$, i.e., Eq. (3.4.18).

VECTORS AND TENSORS IN 3-SPACE   479

The resulting relations prove the theorem since each relation

$$\mathcal{T}(\alpha)^{\omega}_{j} = \sum_{\phi} k(\alpha)^{\omega\phi}_{i\ i} \mathcal{S}(\alpha)^{\phi}_{j}$$

is independent of the choice of subgroup indices i or j.

Equation (6.3.20) is very powerful, and it dictates strong requirements for the k coefficients. For example, if D_{4h} symmetry is present in a crystalline solid then only the following k coefficients survive in the stress-strain tetradic relation:

$\langle k(D_{4h}) \rangle$

		A_{1g} $\mathcal{S}(A_{1g})$ A_{1g}	E_g $\mathcal{S}(A_{1g})$ A_{1g}	E_g $\mathcal{S}(B_{1g})$ A_{1g}	T_{2g} $\mathcal{S}(E_g)$ B_{1g}	T_{2g} $\mathcal{S}(E_g)$ B_{2g}	T_{2g} $\mathcal{S}(A_{2g})$ A_{2g}
	A_{1g} $\mathcal{T}(A_{1g})$ A_{1g}	$k^{(A_{1g})AA}$	$k^{(A_{1g})AE}$	.	.	.	.
	E_g $\mathcal{T}(A_{1g})$ A_{1g}	$k^{(A_{1g})EA}$	$k^{(A_{1g})EE}$	.	.	.	.
=	E_g $\mathcal{T}(B_{1g})$ A_{1g}	.	.	$k^{(B_{1g})}$	.	.	.
	T_{2g} $\mathcal{T}(E_g)$ B_{1g}	.	.	.	$k^{(E_g)}$	.	.
	T_{2g} $\mathcal{T}(E_g)$ B_{2g}	.	.	.	.	$k^{(E_g)}$	.
	T_{2g} $\mathcal{T}(A_{2g})$ A_{2g}	.	.	.	.	.	$k^{(A_{2g})}$

(6.3.21)

In O_h symmetry the constants are more restricted so that only three independent constants are left out of the original 36:

$\langle k(O_h) \rangle =$

	0^+ $\mathscr{S}(A_{1g})$ A_{1g}	2^+ $\mathscr{S}(E_g)$ A_{1g}	2^+ $\mathscr{S}(E_g)$ B_{1g}	2^+ $\mathscr{S}(T_{2g})$ $E_g(1)$	2^+ $\mathscr{S}(T_{2g})$ $E_g(2)$	2^+ $\mathscr{S}(T_{2g})$ A_{2g}
0^+ $\mathscr{T}(A_{1g})$ A_{1g}	$k^{(A_{1g})}$	.	.	.	.	.
2^+ $\mathscr{T}(E_g)$ A_{1g}	.	$k^{(E_g)}$	.	.	.	.
2^+ $\mathscr{T}(E_g)$ B_{1g}	.	.	$k^{(E_g)}$	.	.	.
2^+ $\mathscr{T}(T_{2g})$ $E_g(1)$	.	.	.	$k^{(T_{2g})}$	.	.
2^+ $\mathscr{T}(T_{2g})$ $E_g(2)$	.	.	.	.	$k^{(T_{2g})}$	.
2^+ $\mathscr{T}(T_{2g})$ A_{2g}	.	.	.	.	.	$k^{(T_{2g})}$

(6.3.22)

Note that if full rotational (O_2) symmetry is present, then the constants are further restricted so that only two remain: $k^{0^+} = k^{(A_{1g})}$ and $k^{2^+} = k^{(T_{2g})}$. This is the case for amorphous or sintered solids which are isotropic on the average.

For lower-symmetry materials one expects a larger number of off-diagonal k coefficients even if symmetry-defined tensors are used. Then it is helpful to reduce the number of these coefficients by using energy considerations. If the stresses and strains are CONSERVATIVE then the energy E may be written as a function of some set of symmetry-defined strains $\{\mathscr{S}_1 \mathscr{S}_2 \cdots \}$ only. Then one has

$$dE = \sum_j \frac{\partial E}{\partial \mathscr{S}_j} d\mathscr{S}_j = -\sum_j \mathscr{T}_j d\mathscr{S}_j, \qquad (6.3.23a)$$

where one *defines* the stresses by

$$\mathscr{T}_j = -\frac{\partial E}{\partial \mathscr{S}_j}. \tag{6.3.23b}$$

This definition is supposed to be valid even for large strains for which $\mathscr{T}_j$ is not a linear function of $\mathscr{S}_i$'s. In any case small changes in the stress should be related to small changes in the strains through the following linear relation:

$$d\mathscr{T}_j = \frac{\partial \mathscr{T}_j}{\partial \mathscr{S}_i} d\mathscr{S}_i = \sum_i -\frac{\partial^2 E}{\partial \mathscr{S}_i \partial \mathscr{S}_j} d\mathscr{S}_i. \tag{6.3.24}$$

The coefficients of this relation are generalized k coefficients

$$k_{ij} = -\frac{\partial^2 E}{\partial \mathscr{S}_i \partial \mathscr{S}_j} = k_{ji}, \tag{6.3.25}$$

which will be constants only for small strains $\mathscr{S}_j = d\mathscr{S}_j \sim 0$. Otherwise they are complicated functions of $\mathscr{S}_j$'s. However, even then they satisfy the RECIPROCITY RELATIONS $k_{ij} = k_{ji}$ if the stresses are conservative. This reduces the number of off-diagonal k coefficients by one-half. Only six D_{4h} coefficients in Eq. (6.3.21) are independent if reciprocity holds, since then $k^{(A_{1g})AE} = k^{(A_{1g})EA}$.

The orthonormality and completeness relations of any symmetry definition make it easy to relate back and forth between Cartesian components and constants $k_{hi,jk}$ and the symmetry-defined quantities $k_{mn}^{(\gamma)_\omega(\delta)_\phi}$. This is true even without the presence of symmetry; however, the simplification of the constants makes this transformation very convenient.

Consider, for example, some of the Cartesian components of the O_h-symmetric elasticity relation:

$$k_{hi,jk} = \sum_\gamma \sum_m \mathscr{C}_{him}^{T_{1u}T_{1u}\gamma} \mathscr{C}_{jkm}^{T_{1u}T_{1u}\gamma} k^\gamma. \tag{6.3.26}$$

Now only the symmetric coupling coefficients are used: $(\mathscr{C}_{jkm}^{T_1T_1\gamma} = \mathscr{C}_{kjm}^{T_1T_1\gamma})$. We do not deal with the antisymmetric $\mathscr{C}_{jkm}^{T_1T_1T_1} = -\mathscr{C}_{kjm}^{T_1T_1T_1}$ coefficients since $k^\gamma = k^{T_1}$ is not considered. This gives

$$k_{hi,jk} = k_{ih,jk} = k_{hi,kj} = k_{ih,kj},$$

and leaves only 36 independent $k_{hi,jk}$. From either (6.3.26) or the reciprocity relation (6.3.25) one has

$$k_{hi,jk} = k_{jk,hi},$$

leaving 21 independent components. By substituting the values of the coupling coefficients from Eq. (6.1.19c) one obtains relations between Cartesian k coefficients and the three independent octahedrally defined coefficients k^{A_1}, k^E, and k^{T_2}:

$$k_{11,11} = k_{22,22} = k_{33,33} = k^{A_1}/3 + 2k^E/3,$$
$$k_{11,22} = k_{11,33} = k_{22,33} = k^{A_1}/3 - k^E/3,$$
$$k_{12,12} = k_{13,13} = k_{23,23} = k^{T_2}/2 + k^{T_1}/2 = k^{T_2}/2. \qquad (6.3.27)$$

In the last line we assume the rotational constant k^{T_1} is zero. The coupling coefficients immediately give us all the relations between the Cartesian components that are a result of symmetry. The inverse transformations are sometimes useful, too:

$$k^\gamma = \sum_h \sum_i \sum_j \sum_k \mathscr{C}_{him}^{T_{1u}T_{1u}\gamma} \mathscr{C}_{jkm}^{T_{1u}T_{1u}\gamma} k_{hi,jk}. \qquad (6.3.28)$$

For octahedral symmetry one has the following:

$$k^{A_1} = k_{11,11} + 2k_{11,12},$$
$$k^E = k_{11,11} - k_{11,22},$$
$$k^{T_2} = 2k_{12,12}. \qquad (6.3.28)_x$$

In order to simplify the relations in Eq. (6.3.28), one may use the relations derived in Eq. (6.3.27) to sort out the Cartesian components that were equal to others or else zero.

We now review the relation between the O_h-symmetry-defined elastic constants k^γ and some of the elastic moduli commonly used in physics and engineering.

(a) Bulk Modulus The bulk modulus B is defined in the following, where $\Delta V/V$ is relative change in volume due to an addition of a uniform pressure having force ΔF per area A:

$$\Delta F/A = B\,\Delta V/V. \qquad (6.3.29)$$

We make the correspondence between this equation of the symmetry-defined relation $\mathscr{T}^{A_1} = k^{A_1}\mathscr{S}^{A_1}$. According to Figure 6.3.2(b) the pressure is given by the following:

$$\Delta F/A = -\mathscr{T}^{A_1}/\sqrt{3}.$$

According to Figure 6.3.4 the volume strain is given by the following, where it is assumed that the strain coordinate $\mathscr{S}^{A_1}$ is small:

$$\Delta V/V = \frac{\left[(2a)^3(1 - \mathscr{S}^{A_1}/\sqrt{3})^3 - (2a)^3\right]}{[2a]^3} \sim -\sqrt{3}\mathscr{S}^{A_1}.$$

Combining the last two equations with $\mathscr{T}^{A_1} = k^{A_1}\mathscr{S}^{A_1}$ shows that k^{A_1} is three times the bulk modulus.

$$B = k^{A_1}/3 = (k_{11,11} + 2k_{11,22})/3. \tag{6.3.30}$$

(b) Shear Modulus Using the quantities labeled in Figure 6.3.5, we define the shear modulus μ as follows:

$$\Delta F/A = \mu d/h,$$
$$\Delta F/(2a)^2 = \mu d/2a. \tag{6.3.31}$$

We recognize this to belong to one of the symmetry-defined relations $\mathscr{T}^{T_2} = k^{T_2}\mathscr{S}^{T_2}$. Using Figures 6.3.2(b) and 6.3.4 for (T_2) stress and strain definitions, respectively, we obtain the following relations:

$$\mathscr{T}^{T_2}/\sqrt{2} = \Delta F/(2a)^2, \qquad \mathscr{S}^{T_2}a/\sqrt{2} = d/4.$$

Substituting these in Eq. (6.3.31), we find that k^{T_2} is twice the shear modulus:

$$\mu = k^{T_2}/2. \tag{6.3.32}$$

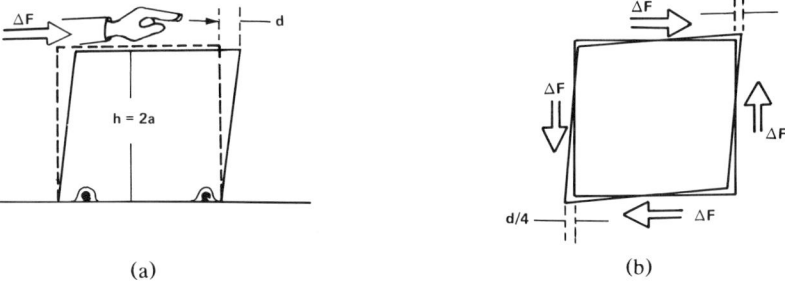

Figure 6.3.5 Defining shear of a cubical block. (a) Standard definition involves a sideways motion d of the upper surface of the block with the base fixed. (b) A more symmetrical picture of the same distortion shows that the corners each move a distance $d/4$. Here the center of gravity and the diagonal axis of the block do not move.

(c) Young's Modulus and Poisson's Ratio If a pure tension $\vec{\vec{T}} = (F/A)\hat{x}_3\hat{x}_3$ is applied to the cube, we can expect the 3-axis to lengthen. However, we may also expect a change along the transverse 1 and 2 axes since this tensor is a combination of T^{A_1} and T_1^E which affect these dimensions:

$$\vec{\vec{T}} = (F/A)\hat{x}_3\hat{x}_3 = (F/A)\left(\hat{T}^{A_1}/\sqrt{3} - 2\hat{T}_1^E/\sqrt{6}\right)$$
$$\equiv \mathcal{T}^{A_1}\hat{T}^{A_1} + \mathcal{T}_1^E\hat{T}_1^E. \tag{6.3.33}$$

We now find what the strain will be in terms of the symmetry-defined constants k^{A_1} and k^E using the following relations:

$$(F/A)/\sqrt{3} = \mathcal{T}^{A_1} = k^{A_1}\mathcal{S}^{A_1}, \quad -2(F/A)/\sqrt{6} = \mathcal{T}_1^E = k^E\mathcal{S}_1^E \tag{6.3.34}$$

This yields the strain indicated in Figure 6.3.6, where the changes Δa_L and Δa_T of the longitudinal and transverse semiaxes are given by Eq. (6.3.35):

$$\Delta a_L/a = \mathcal{S}^{A_1}/\sqrt{3} - 2\mathcal{S}_1^E/\sqrt{6} = (F/A)(1/3k^{A_1} + 4/6k^E),$$
$$\Delta a_T/a = \mathcal{S}^{A_1}/\sqrt{3} + \mathcal{S}_1^E/\sqrt{6} = (F/A)(1/3k^{A_1} - 2/6k^E). \tag{6.3.35}$$

Now, Young's modulus Y and Poisson's ratio σ are defined in the following. The definition of σ uses a negative sign since Δa_L and Δa_T always turn out to have opposite signs:

$$Y \equiv (F/A)/(\Delta a_L/a)$$
$$= 1/(1/3k^{A_1} + 2/3k^E)$$
$$= 3k^E k^{A_1}/(2k^{A_1} + k^E), \tag{6.3.36a}$$
$$\sigma \equiv -(\Delta a_T/a)(\Delta a_L/a)$$
$$= -(1/3k^{A_1} - 1/3k^E)/(1/3k^{A_1} + 2/3k^E)$$
$$= (k^{A_1} - k^E)/(2k^{A_1} + k^E). \tag{6.3.36b}$$

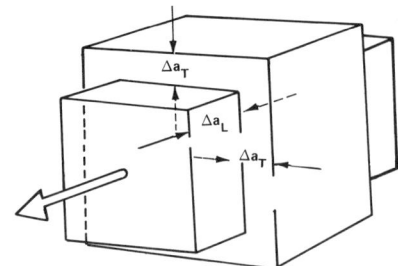

Figure 6.3.6 General strain caused by pure tension. Block elongates by $2\Delta a_L$ while the transverse dimensions shrink by $2\Delta a_T$.

Solving for the k^γ gives the following:

$$k^{A_1} = Y/(1 - 2\sigma), \qquad k^E = Y/(1 + \sigma). \qquad (6.3.37)$$

(d) Isotropic Solids In an isotropic material in which crystal order is randomized or nonexistent, we found that $k^E = k^{T_2}$. This will happen in any solid that has no preferential directions. This corresponds to the presence of the infinite rotational symmetry O_3. In order to visualize the fact that $k^E = k^{T_2}$ in this case, observe that the $\mathscr{S}_2^E$ strain in Figure 6.3.4 is the same as the $\mathscr{S}^{T_2}$ strain rotated by 45° around the 3-axis. The same applies to the corresponding stresses in Figure 6.3.2(b). If the 45° rotation is a symmetry operation in addition to all the O_h operations, then we must have $k^E = k^{T_2}$. This is the case in isotropic solids, so there are only two constants which are needed to describe their elastic properties. Combining Eqs. (6.3.30), (6.3.32), and (6.3.37) gives the following relations between them:

$$B = Y/3(1 - 2\sigma), \qquad \mu = Y/2(1 + \sigma) \qquad (6.3.38)$$

(e) More Spring Constant Theory Consider a model of solid strontium titanate $SrTiO_3$ as shown in Figure 6.3.7(a) with the interatomic nearest-neighbor forces represented by three different springs. This rather simplified rigid-ion model neglects the bending and long-range Coulomb forces as well as the ion polarizabilities—all of which are considered in more sophisticated shell models. However, we shall see that our model illustrates the group-theoretical techniques inherent in more complicated models and leads to proper order of magnitudes for the elastic constants.

Let us compute the elastic constants k^γ in terms of the interatomic spring constants j, k, and l by imposing one of the symmetry-defined strains belonging to irrep (γ) and summing the forces per unit area on the center plane in Figure 6.3.7(b) due to its nearest-neighbor atoms on one side. By

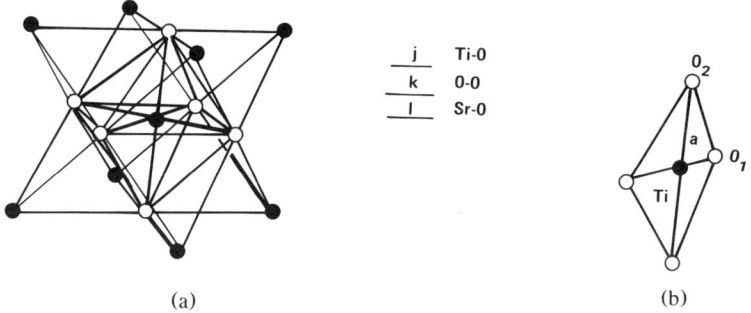

Figure 6.3.7 (a) Model for strontium titanate $SrTiO_3$. Three different springs used for interatomic bonds. (b) One-half unit cell in Sr plane has area $2a^2$. Unit cell contains one Ti atom and two O atoms [O(1) and O(2)].

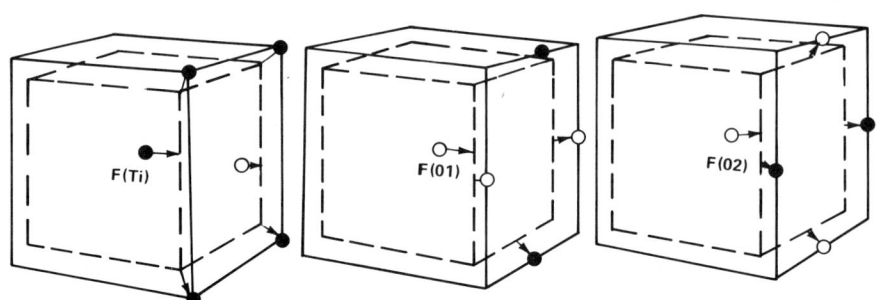

Figure 6.3.8 Atomic forces due to A_{1g} strain on (a) Ti atom, (b) O(1) atom, and (c) O(2) atom.

relating this force to the corresponding components of stress $\mathscr{T}^\gamma$, one obtains elastic constant $k^\gamma = \mathscr{S}^\gamma/\mathscr{T}^\gamma$. The central plane considered contains one Ti atom and two O atoms per unit area $4a^2$, where a is half the lattice parameter ($2a = 3.9 \times 10^{-8}$ cm) or the distance between the Ti and O atoms in this cubic lattice.

Now, an A_{1g} strain gives rise to forces **F**(Ti), **F**(01), and **F**(02) in the plane considered, as shown in Figure 6.3.8. Thus we have

$$\mathscr{T}^{A_1}/\sqrt{3} = \frac{|\mathbf{F}(Ti)| + |\mathbf{F}(01)| + |\mathbf{F}(02)|}{4a^2}, \quad (6.3.39)$$

where the various forces are computed in the following:

$$|F(Ti)| = F_1(Ti) = j\mathscr{S}^{A_1}a/\sqrt{3},$$
$$|F(01)| = F_1(01) = (2k + 2l)\mathscr{S}^{A_1}a/\sqrt{3},$$
$$|F(02)| = F_1(02) = (2k + 2l)\mathscr{S}^{A_1}a/\sqrt{3}. \quad (6.3.40)$$

[Note that $F_1(02)$ may always be derived from $F_1(01)$ by interchanging k and l.] From the foregoing two equations, we find

$$k^{A_1} = \mathscr{T}^{A_1}/\mathscr{S}^{A_1} = (j + 4k + 4l)/4a. \quad (6.3.41a)$$

Similar treatment involving the other two types of strain gives the elastic constants k^E and k^{T_2} (we may choose any component of the symmetry type):

$$k^E = (j + k + l)/4a, \quad (6.3.41b)$$
$$k^{T_2} = (2k + 2l)/4a. \quad (6.3.41c)$$

In our calculations, we could equally well have chosen a plane containing strontium atoms with the same result.

Now, it is interesting to observe that these three k^γ are not linearly independent. In particular, two of the Cartesian elastic constants, $k_{11,22}$ and $k_{12,12}$, will be equal for all k, j, and l. From Eq. (6.3.27), we find

$$k_{11,11} = (2k + j + 2l)/4a,$$

$$k_{11,22} = (k + l)/4a,$$

$$k_{12,12} = (k + l)/4a. \qquad (6.3.42)$$

If we include bending constants or "covalent" forces in our model in addition to the central or "ionic" forces, then $k_{11,22}$ no longer equals $k_{12,12}$.

Plotting the observed values of $k_{11,22}$ versus $k_{12,12}$ for various cubic solids gives us some idea of which have ionic bonds and which have covalent bonds. This is done in Figure 6.3.9. Note that NaCl and similar salts are very close to the $k_{11,22} = k_{12,12}$ line, with exception of AgCl and LiF. Most metals and crystals like diamond are considerably removed from the center line.

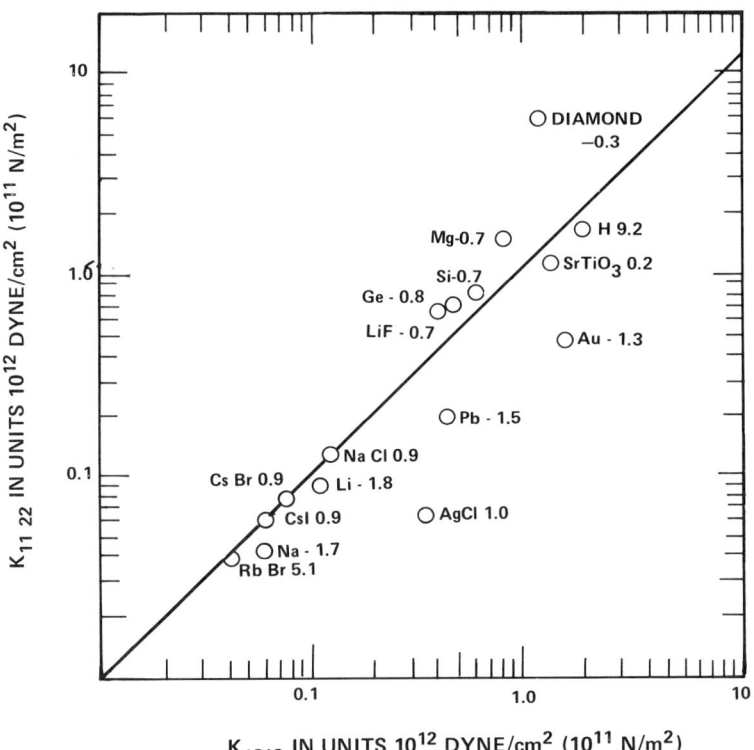

Figure 6.3.9 Plot of observed values of $k_{11,22}$ vs $k_{12,12}$ for various cubic solids.

Next to each point is written an "anisotropy" number A defined as follows:

$$A = \frac{k^E - k^{T_2}}{(k^{T_2}/2)} = \frac{k_{11,11} - k_{11,22} - 2k_{12,12}}{k_{12,12}}. \quad (6.3.43)$$

The closer this is to zero, the more closely the solid will imitate an isotropic material, as far as its bulk properties are concerned. Interestingly, diamond is the most "isotropic."

(f) Elasticity in C_{3v} Symmetry For C_{3v} one may use the trigonally defined ($O_h \supset D_{3d} \supset C_{2v}$) unit vectors shown in Figure 4.2.2: $\hat{v}_1 \equiv \hat{x}^3 \equiv \hat{x}_1^E$, $\hat{v}_2 \equiv \hat{x}^3 \equiv \hat{x}_2^E$, and $\hat{v}_3 \equiv \hat{x}^1 \equiv \hat{x}^{A_1}$. Using C_{3v} coupling coefficients (6.1.35a) we can assemble the symmetry-defined unit tensors and use them to describe stress and strain:

	$T^{(1)_1}$	$T^{(1)_2}$	$T_1^{(3)_1}$	$T_2^{(3)_1}$	$T_1^{(3)_2}$	$T_2^{(3)_2}$	T^2	$T_1^{(3)_3}$	$T_2^{(3)_3}$
v_1v_1	$\sqrt{\frac{1}{2}}$	·	$\sqrt{\frac{1}{2}}$	·	·	·	·	·	·
v_1v_2	·	·	·	$-\sqrt{\frac{1}{2}}$	·	·	$\sqrt{\frac{1}{2}}$	·	·
v_1v_3	·	·	·	·	$\sqrt{\frac{1}{2}}$	·	·	$\sqrt{\frac{1}{2}}$	·
v_2v_1	·	·	·	$-\sqrt{\frac{1}{2}}$	·	·	$-\sqrt{\frac{1}{2}}$	·	·
v_2v_2	$\sqrt{\frac{1}{2}}$	·	$-\sqrt{\frac{1}{2}}$	·	·	·	·	·	·
v_2v_3	·	·	·	·	·	$\sqrt{\frac{1}{2}}$	·	·	$\sqrt{\frac{1}{2}}$
v_3v_1	·	·	·	·	$\sqrt{\frac{1}{2}}$	·	·	$-\sqrt{\frac{1}{2}}$	·
v_3v_2	·	·	·	·	·	$\sqrt{\frac{1}{2}}$	·	·	$-\sqrt{\frac{1}{2}}$
v_3v_3	·	1	·	·	·	·	·	·	·

(columns $T_1^{(3)_3}, T_2^{(3)_3}$ are under brace "rotation")

$$(6.3.44)$$

In the foregoing table of tensors, we have separated those three tensors on the right, which correspond to rotations from the others on the left. As usual, elasticity theory makes no use of rotations.

Using the components of the tensors on the left side of the table and the tensor theorem [Eq. (6.3.20)], we find the following elasticity relations will occur in C_{3v} solids:

$$\mathcal{T}^{(1)_1} = k^{(1)_{11}}\mathcal{S}^{(1)_1} + k^{(1)_{12}}\mathcal{S}^{(1)_2}, \qquad \mathcal{T}_m^{(3)_1} = k^{(3)_{11}}\mathcal{S}_m^{(3)_1} + k^{(3)_{12}}\mathcal{S}_m^{(3)_2},$$
$$\mathcal{T}^{(1)_2} = k^{(1)_{21}}\mathcal{S}^{(1)_1} + k^{(1)_{22}}\mathcal{S}^{(1)_2}, \qquad \mathcal{T}_m^{(3)_2} = k^{(3)_{21}}\mathcal{S}_m^{(3)_1} + k^{(3)_{22}}\mathcal{S}_m^{(3)_2},$$
$$m = 1, 2. \quad (6.3.45)$$

Here the $O_h \supset D_{3d} \supset C_{3v}$ supergroup labeling has been ignored. The repeated $A_1 = (1)$ and $E = (3)$ species have simply been tagged by numbers $(\cdot)_1$ and $(\cdot)_2$.

F. Symmetry-Defined Electric and Magnetic Fields

It is commonly said that an electric field is a vector while a magnetic field is a pseudovector. It is instructive to find to which irreps of C_{3v} they belong using physical arguments. One asks what happens to an electric field **E** or a magnetic field **B** under symmetry operations of C_{3v}. To do a symmetry by "thought experiment" operation, let us imagine that everything connected with the physical object gets transformed including the source charges or currents that make **E** or **B**.

Consider C_{3v} geometry and some electric or magnetic fields $\hat{E}_3$ and $\hat{B}_3$ which point along the $\hat{v}_3$ symmetry axis. Figure 6.3.10 shows what happens to

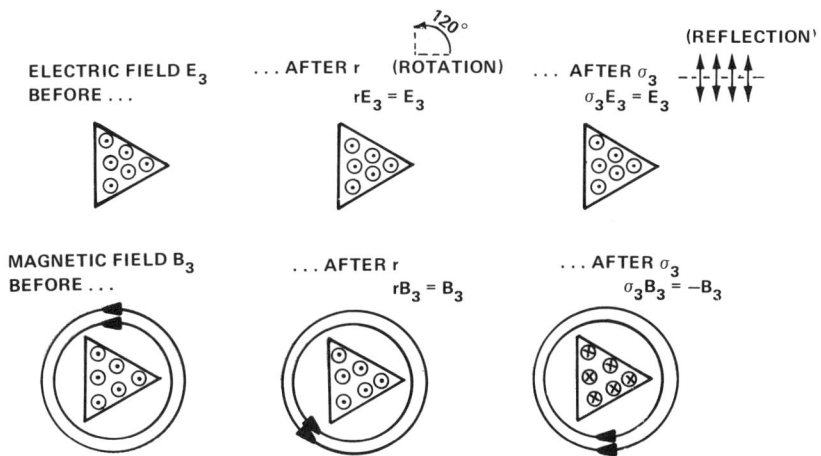

Figure 6.3.10 Effect of C_{3v} symmetry operations on electric field $\mathbf{E}_3$ and magnetic field $\mathbf{B}_3$ along symmetry axis. (a) $\mathbf{E}_3$ and $\mathbf{B}_3$ before symmetry operations. (b) After r. (c) After σ_3.

490 THEORY AND APPLICATIONS OF SYMMETRY REPRESENTATION PRODUCTS

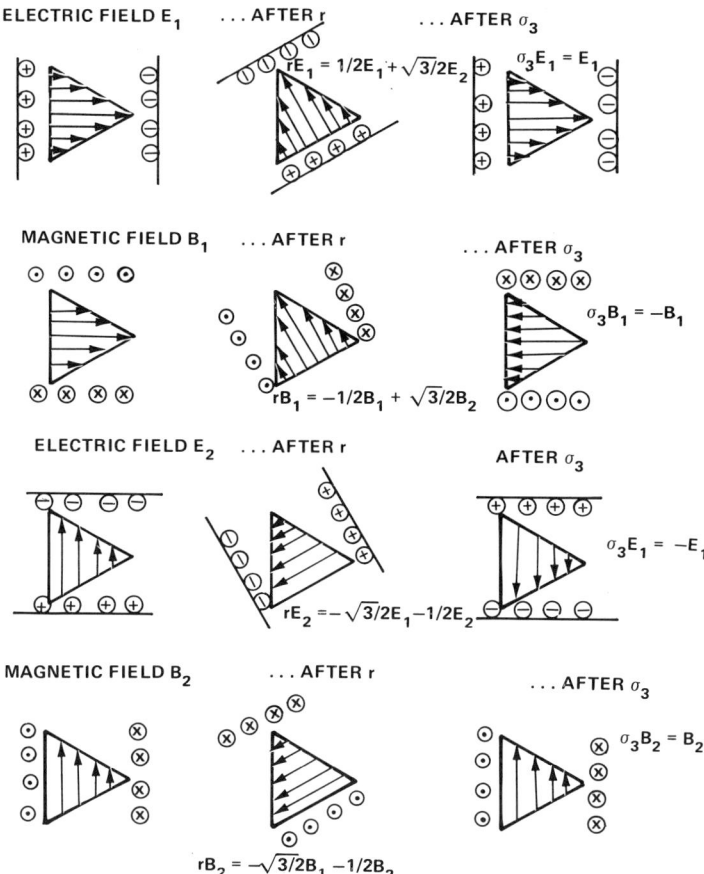

Figure 6.3.11 Effect of C_{3v} symmetry operations on electric fields $\mathbf{E}_1\mathbf{E}_2$ and magnetic fields $\mathbf{B}_1\mathbf{B}_2$ transverse to symmetry axis. (a) $\mathbf{E}_1$, $\mathbf{B}_1$, $\mathbf{E}_2$ and $\mathbf{B}_2$ before symmetry operations. (b) After r. (c) After σ_3.

$\hat{E}_3$ or $\hat{B}_3$ after operations by r (120° rotations) and σ_3 (reflection). The idea is to imagine what happens to sources or currents that give rise to $\mathbf{B}_3$ or the charges that give rise to $\mathbf{E}_3$. In Figure 6.3.10, reflection σ_3 reverses the current loop and thereby reverses $\mathbf{B}$. In Figure 6.3.11, we perform a similar analysis for fields $\mathbf{E}_1$ and $\mathbf{B}_1$ in the transverse directions.

The unit electric fields $\hat{E}_1$, $\hat{E}_2$, and $\hat{E}_3$ transform just like the unit vectors v_j in Figure 4.2.2, and so the symmetry definition of them is as follows:

$$\hat{E}_1 = \hat{E}_1^3 = \hat{E}_1^E, \qquad \hat{E}_2 = \hat{E}_2^3 = \hat{E}_2^E, \qquad \hat{E}_3 = \hat{E}^1 = \hat{E}^{A_1}. \quad (6.3.46)$$

The magnetic fields behave a little differently, as seen in the preceding figures. In Figure 6.3.10, we see that $\hat{B}_3$ transforms according to irrep $\mathscr{D}^2 = \mathscr{D}^{A_2}$ of C_{3v}. In Figure 6.3.11, we see that some combinations of $\hat{B}_1$ and $\hat{B}_2$ will serve as bases for the irrep $\mathscr{D}^3 = \mathscr{D}^E$ of C_{3v}. To find this combination one applies the elementary operators.

$$\begin{aligned} P^3_{12} \hat{B}_1 &= \tfrac{1}{3}\left[-\tfrac{3}{2} r\hat{B}_1 + \tfrac{3}{2} r^2 \hat{B}_1 - \tfrac{3}{2}\sigma_1 \hat{B}_1 + \tfrac{3}{2}\sigma_2 \hat{B}_1 \right] \\ &= \tfrac{1}{3}\left[-\tfrac{3}{2}\left(-\tfrac{1}{2}\hat{B}_1 + \tfrac{3}{2}\hat{B}_2\right) + \tfrac{3}{2}\left(-\tfrac{1}{2}\hat{B}_1 - \tfrac{3}{2}\hat{B}_2\right) \right. \\ &\quad \left. - \tfrac{3}{2}\left(\tfrac{1}{2}\hat{B}_1 + \tfrac{3}{2}\hat{B}_2\right) + \tfrac{3}{2}\left(\tfrac{1}{2}\hat{B}_1 - \tfrac{3}{2}\hat{B}_2\right) \right] \\ &= -\hat{B}_2, \\ P^3_{22} \hat{B}_1 &= \hat{B}_1. \end{aligned} \qquad (6.3.47)$$

This gives the correct symmetry definition of the three magnetic fields:

$$\hat{B}_1 = \hat{B}_2^3 = \hat{B}_2^E, \qquad \hat{B}_2 = -\hat{B}_1^3 = -\hat{B}_1^E, \qquad \hat{B}_3 = \hat{B}^2 = \hat{B}^{A_2}. \quad (6.3.48)$$

Now one may define "external" fields **E** and **B** that are totally unaffected by symmetry operations and write them in terms of the symmetry-defined fields:

$$\mathbf{E} = \sum_{\gamma, m} \mathscr{E}^\gamma_m \hat{E}^\gamma_m = \sum_{\gamma, m} \mathscr{E}^{\gamma'}_m \hat{E}^{\gamma'}_m, \qquad \mathbf{B} = \sum_{\gamma, m} \mathscr{B}^\gamma_m \hat{B}^\gamma_m = \sum_{\gamma, m} \mathscr{B}^{\gamma'}_m \hat{B}^{\gamma'}_m. \quad (6.3.49)$$

This defines components $\mathscr{E}^\gamma_m$ and $\mathscr{B}^\gamma_m$ which have the irrep transformation properties which label them.

Magnetostriction and Electrostriction

Some solids respond to external electric fields by distorting or even shattering into pieces. This effect is called ELECTROSTRICTION. Magnetic fields can cause analogous MAGNETOSTRICTION effects.

From the symmetry of the crystal, you may quickly tell what is possible by assuming a linear or tensor relation such as Eq. (6.3.50) between the field and strain components where $\varepsilon_{ij,k}$ and $\mu_{ij,k}$ are constants of electro- and magnetostriction, respectively.

$$\mathscr{S}_{ij} = \sum_k \varepsilon_{ij,k} \mathscr{E}_k, \qquad (6.3.50\text{a})$$

$$\mathscr{S}_{ij} = \sum_k \mu_{ij,k} \mathscr{B}_k. \qquad (6.3.50\text{b})$$

The situation becomes clearer if we use the symmetry-defined components and the tensor theorem. The labeling of Eq. (6.3.44) is used here:

$$\mathscr{S}^{(1)_1} = \varepsilon^{(1)_1} \mathscr{E}^1 = \varepsilon^{(1)_1} \mathscr{E}_3,$$

$$\mathscr{S}^{(1)_2} = \varepsilon^{(1)_2} \mathscr{E}^1 = \varepsilon^{(1)_2} \mathscr{E}_3,$$

$$\mathscr{S}_1^{(3)_1} = \varepsilon^{(3)_1} \mathscr{E}_1^3 = \varepsilon^{(3)_1} \mathscr{E}_1, \qquad \mathscr{S}_1^{(3)_1} = \mu^{(3)_1} \mathscr{B}_1^3 = -\mu^{(3)_1} \mathscr{B}_2,$$

$$\mathscr{S}_2^{(3)_1} = \varepsilon^{(3)_1} \mathscr{E}_2^3 = \varepsilon^{(3)_1} \mathscr{E}_2, \qquad \mathscr{S}_2^{(3)_1} = \mu^{(3)_1} \mathscr{B}_2^3 = \mu^{(3)_1} \mathscr{B}_1,$$

$$\mathscr{S}_1^{(3)_2} = \varepsilon^{(3)_2} \mathscr{E}_1^3 = \varepsilon^{(3)_2} \mathscr{E}_1, \qquad \mathscr{S}_1^{(3)_2} = \mu^{(3)_2} \mathscr{B}_1^3 = -\mu^{(3)_2} \mathscr{B}_2,$$

$$\mathscr{S}_2^{(3)_2} = \varepsilon^{(3)_2} \mathscr{E}_2^3 = \varepsilon^{(3)_2} \mathscr{E}_2, \qquad \mathscr{S}_2^{(3)_2} = \mu^{(3)_2} \mathscr{B}_2^3 = \mu^{(3)_2} \mathscr{B}_1. \qquad (6.3.51)$$

Note that each of the three electric fields $\hat{E}_1$, $\hat{E}_2$, and $\hat{E}_3$ is capable of inducing two kinds of strain described by four independent parameters in all. However, the magnetic field $\hat{B}_3$ cannot cause any strain that is linear in B_3 and only two parameters are needed to describe transverse strains. Furthermore, note that electric field $\hat{E}_2$ can cause the strain $\mathscr{S}_2^{(3)_1}$ and $\mathscr{S}_2^{(3)_2}$, while the magnetic field $\hat{B}_2$ induces the strain $-\mathscr{S}_1^{(3)_1}$ or $-\mathscr{S}_1^{(3)_2}$, and similarly for $\hat{E}_1$ and $\hat{B}_1$.

It is easy to prove that no linear striction effects can occur in cubic symmetry O or O_h. $\mathscr{E}_j$ and $\mathscr{B}_j$ can be shown to take the symmetry definitions of T_{1u} and T_{1g}, respectively, under O_h symmetry:

$$\mathscr{E}_j = \mathscr{E}_j^{T_{1u}}, \qquad \mathscr{B}_j = \mathscr{B}_j^{T_{1g}}. \qquad (6.3.52)$$

No $\mathscr{S}^{T_{1u}}$ strain exists, and there is no $\mathscr{S}^{T_{1g}}$ strain without rotations.

6.4 THEORY OF QUANTUM OPERATORS AND IRREDUCIBLE TENSORS

A common approach to physical models involves the manipulation or variation of parameters in a model operator and equation of motion. One tries to define and evaluate the most important parameters of a model in such a way that the theory gives some insight into the behavior of the system being studied. We have seen examples of parameters such as the spring constants in models of molecular vibration (cf. NH_3, UF_6, SF_6, in Chapters 3 and 4 and energy or tunneling coefficients in electronic orbital models.

Now we see how to tell exactly how many independent parameters are possible in a given model from symmetry theory and how to systematically enumerate and relate different systems of parameters. We shall do this by reviewing the general structure of operators. This will include an introduction to transition operators and selection rules and the Wigner-Eckart theorem.

A. Symmetry-Defined Operators

The mathematics of quantum mechanics reviewed in Chapter 1 revolves around sets of base vectors $\{|1\rangle, |2\rangle, \cdots |n\rangle\}$ or $\{\langle 1|, \langle 2|, \cdots \langle n|\}$ and operators $A, B, \ldots$ which transform these vectors.

As we observed in Chapter 1, all operators can be expressed as linear combinations of "elementary" or "unit" operators $e_{ij} = |i\rangle\langle j|$ made from outer products of bras and kets:

$$A = \sum_{i=1}^{n} \sum_{j=1}^{n} |i\rangle\langle i|A|j\rangle\langle j| = \sum_{i=1}^{n} \sum_{j=1}^{n} \mathscr{A}_{ij}|i\rangle\langle j|. \tag{6.4.1}$$

This is true for any basis which is complete.

In particular this is true for any symmetry-defined basis

$$\left\{\left|\begin{array}{c}\alpha\\1\end{array}\right\rangle\left|\begin{array}{c}\alpha\\2\end{array}\right\rangle \cdots \left|\begin{array}{c}\alpha\\l^{\alpha}\end{array}\right\rangle \cdots \left|\begin{array}{c}\beta\\1\end{array}\right\rangle \cdots\right\},$$

where the basis vectors are made to be irrep bases for a group of unitary symmetry operators $G = \{1, g, g', \ldots\}$ as follows:

$$g\left|\begin{array}{c}\alpha\\k\end{array}\right\rangle = \sum_{i=1}^{l^{\alpha}} \mathscr{D}_{ik}^{\alpha}(g)\left|\begin{array}{c}\alpha\\i\end{array}\right\rangle. \tag{6.4.2}$$

Now consider the transformation of a unit operator $\left|\begin{array}{c}\alpha\\k\end{array}\right\rangle\left\langle\begin{array}{c}\beta\\l\end{array}\right|$:

$$\left|\begin{array}{c}\alpha\\k\end{array}\right\rangle\left\langle\begin{array}{c}\beta\\l\end{array}\right| \to g\left|\begin{array}{c}\alpha\\k\end{array}\right\rangle\left\langle\begin{array}{c}\beta\\l\end{array}\right|g^{\dagger} = g\left|\begin{array}{c}\alpha\\k\end{array}\right\rangle\left(g\left|\begin{array}{c}\beta\\l\end{array}\right\rangle\right)^{\dagger}$$

$$= \sum_{i=1}^{l^{\alpha}} \sum_{j=1}^{l^{\beta}} \mathscr{D}_{ik}^{\alpha}(g)\left|\begin{array}{c}\alpha\\k\end{array}\right\rangle\left(\mathscr{D}_{jl}^{\beta}(g)\left|\begin{array}{c}\beta\\j\end{array}\right\rangle\right)^{\dagger}$$

$$= \sum_{i=1}^{l^{\alpha}} \sum_{j=1}^{l^{\beta}} \mathscr{D}_{ik}^{\alpha}(g)\mathscr{D}_{jl}^{\beta}(g)^{*}\left|\begin{array}{c}\alpha\\i\end{array}\right\rangle\left\langle\begin{array}{c}\beta\\j\end{array}\right|. \tag{6.4.3}$$

For real irreps $[\mathscr{D}_{jl}^{\beta}(g) = \mathscr{D}_{jl}^{\beta*}(g)]$ this last equation takes the same form as the transformation of a two-particle state. [Compare it with Eq. (6.1.7a) with $g = a = b$.] Since most multidimensional point symmetry irreps can be made real we shall assume $\mathscr{D} = \mathscr{D}^{*}$ for the following discussion.

In this case we can construct SYMMETRY-DEFINED UNIT OPERATORS V_{m}^{γ} by summing with coupling coefficients as follows:

$$V_{m}^{\gamma} \equiv \sum_{i=1}^{l^{\alpha}} \sum_{j=1}^{l^{\beta}} C_{ijm}^{\alpha\beta\gamma}\left|\begin{array}{c}\alpha\\i\end{array}\right\rangle\left\langle\begin{array}{c}\beta\\j\end{array}\right|. \tag{6.4.4}$$

According to the theory of Section 6.3.B V_m^γ will transform as follows:

$$gV_m^\gamma g^\dagger = \sum_{l=1}^{l^\gamma} \mathcal{D}_{lm}^\gamma(g) V_l^\gamma. \qquad (6.4.5)$$

These operators are a complete set from which all operators can be made if the $\left|{\alpha \atop i}\right\rangle\left\langle{\beta \atop j}\right|$ or, before that, the $|1\rangle\langle 1|, |1\rangle\langle 2|, \ldots$ are complete. There may be two or more independent operators for each irrep label (γ), so it may be necessary to distinguish between them in some way. We use (reluctantly) another subscript and write $V_m^{(\gamma)\omega}$. Any operator A must be a linear combination of the $V_m^{(\gamma)\omega}$:

$$A = \sum_\gamma \sum_\omega \sum_m \mathcal{A}_m^{(\gamma)\omega} V_m^{(\gamma)\omega}. \qquad (6.4.6)$$

Sometimes the V_m^γ are called IRREDUCIBLE TENSORIAL SETS or TENSOR OPERATORS. Their stress tensor ancestors (recall Section 6.3) are remembered in name only here.

(a) Invariant Operators Recall the Hamiltonian operators H for the octahedral orbit model in Eq. (4.3.3). H must be a linear combination of $V_m^{(\gamma)\omega}$, where (γ) are O_h IR labels. In particular, it can only be combinations of those which have $\gamma = A_{1g}$, i.e., the *invariant* irrep since H is O_h symmetric ($gHg^\dagger = H$):

$$H = \sum_\omega \mathcal{H}^{(A_{1g})\omega} V^{(A_{1g})\omega}. \qquad (6.4.7)$$

Now the question is: How many independent invariant operators $V^{(A_{1g})_1}, V^{(A_{1g})_2}, \ldots$ exist in that basis? That is exactly the number of parameters $\mathcal{H}^{(A_{1g})\omega}$ for the Hamiltonian.

According to Section 6.2.A [see Eq. (6.2.5)], just one invariant can be made from a product $\mathcal{D}^\alpha \times \mathcal{D}^\beta$ of real irreps for each $\alpha = \beta$ and none for $\alpha \neq \beta$. For the octahedral model we have $\alpha = A_{1g}, T_{1u}$, and E_g, so there are three invariants and three independent parameters.

Let us construct the first rows of these matrices using Eq. (6.2.5) and the eigenvectors from Eqs. (4.3.20), (4.3.22), (4.3.23), and (4.3.27):

$$V^{(A_{1g})_1} \equiv |A_{1g}\rangle\langle A_{1g}| \qquad \leftrightarrow \qquad \begin{array}{|cccccc|} \hline \tfrac{1}{6} & \tfrac{1}{6} & \tfrac{1}{6} & \tfrac{1}{6} & \tfrac{1}{6} & \tfrac{1}{6} \\ \hline \end{array}, \qquad (6.4.8a)$$

$$V^{(A_{1g})_2} \equiv \sum_{j=1}^{3} \left|{T_{1u} \atop j}\right\rangle\left\langle{T_{1u} \atop j}\right|/\sqrt{3} \quad \leftrightarrow \quad \begin{array}{|cccccc|} \hline \tfrac{1}{2} & \tfrac{1}{2} & 0 & 0 & 0 & 0 \\ \hline \end{array} /\sqrt{3}, \quad (6.4.8b)$$

and

$$V^{(A_{1g})_3} \equiv \sum_{j=1}^{2} \left|\begin{matrix}E_g\\j\end{matrix}\right\rangle\left\langle\begin{matrix}E_g\\j\end{matrix}\right| / \sqrt{2} \leftrightarrow \left[\begin{array}{cccccc} \frac{1}{3} & \frac{1}{3} & \frac{1}{6} & \frac{1}{6} & \frac{1}{6} & \frac{1}{6} \end{array}\right] / \sqrt{2}. \quad (6.4.8c)$$

In the discussion of this physical model two parameters were used: The local energy (H) and the nearest-neighbor tunneling amplitude $(-S)$. A third parameter exists since there are three invariants. In fact, we mentioned a next-nearest-neighbor tunneling amplitude $(-T)$. The corresponding Hamiltonian

$$H \leftrightarrow \left[\begin{array}{cccccc} H & -T & -S & -S & -S & -S \end{array}\right]. \quad (6.4.9)$$

It is instructive to write H abstractedly in terms of $V^{(A_{1g})_\omega}$. This shows the relations that exist between parameters and eigenvalues in this case:

$$H = \varepsilon^{A_{1g}} V^{(A_{1g})_1} + \varepsilon^{T_{1u}} \sqrt{3}\, V^{(A_{1g})_2} + \varepsilon^{E_g} \sqrt{2}\, V^{(A_{1g})_3},$$

$$H \leftrightarrow \left[\begin{array}{cccccc} \dfrac{\varepsilon^{A_{1g}} + 3\varepsilon^{T_{1u}} + 2\varepsilon^{E_g}}{6} & \dfrac{\varepsilon^{A_{1g}} - 3\varepsilon^{T_{1u}} + 2\varepsilon^{E_g}}{6} & \dfrac{\varepsilon^{A_{1g}} - \varepsilon^{E_g}}{6} & \dfrac{\varepsilon^{A_{1g}} - \varepsilon^{E_g}}{6} & \dfrac{\varepsilon^{A_{1g}} - \varepsilon^{E_g}}{6} & \dfrac{\varepsilon^{A_{1g}} - \varepsilon^{E_g}}{6} \end{array}\right]$$
$$(6.4.10)$$

If the ε^α were given by some experiment, then comparison of Eqs. (6.4.9) and (6.4.10) would immediately give the values of parameters H, S, and T.

This is a special example of a more general theorem.

Parameter Theorem The maximum number of independent parameters for a Hermitian matrix or physical operator is exactly enough to determine all its eigenvalues and all its eigenvectors that are not already fixed by symmetry. This number n is given by

$$n = \sum_\alpha f^\alpha (f^\alpha + 1)/2 \quad (6.4.11a)$$

for a Hermitian operator or matrix, and for an arbitrary operator by

$$n = \sum_\alpha (f^\alpha)^2, \quad (6.4.11b)$$

where f^α is the frequency of irrep (α) in the basis being considered.

Proof The number of independent invariant operators in any basis is completely determined by the product analysis of the basis. Let f^α be the number of orthogonal sets of bases

$$\left\{\left|\begin{matrix}\alpha\\1\end{matrix}\right\rangle^{(1)}, \left|\begin{matrix}\alpha\\2\end{matrix}\right\rangle^{(1)} \cdots \left|\begin{matrix}\alpha\\l^\alpha\end{matrix}\right\rangle^{(1)}\right\}, \left\{\left|\begin{matrix}\alpha\\1\end{matrix}\right\rangle^{(2)}, \left|\begin{matrix}\alpha\\2\end{matrix}\right\rangle^{(2)}, \ldots, \left|\begin{matrix}\alpha\\l^\alpha\end{matrix}\right\rangle^{(2)}\right\}, \ldots, \left\{\left|\begin{matrix}\alpha\\1\end{matrix}\right\rangle^{(f^\alpha)}, \ldots\right\}$$

that transform according to irrep (α). Using these, we may construct exactly

$(f^\alpha)^2$ invariants

$$V^{(\alpha)}{}_{\omega\phi} \equiv \sum_{j=1}^{l^\alpha} \left|{\alpha \atop j}\right\rangle^{(\omega)(\phi)} \left\langle{\alpha \atop j}\right| / \sqrt{l^\alpha}$$

and a general invariant operator I must be expressible in terms of just these, using $(f^\alpha)^2$ coefficients $\mathcal{I}^{(\alpha)}_{\omega\phi}$:

$$I \equiv \sum_\alpha \sum_\omega \sum_\phi \mathcal{I}^{(\alpha)}_{\omega\phi} V^{(\alpha)}{}_{\omega\phi}.$$

If I is Hermitian ($\mathcal{I}^{(\alpha)}_{\omega\phi} = (\mathcal{I}^{(\alpha)}_{\phi\omega})^*$) then the number of parameters is reduced to $f^\alpha(f^\alpha + 1)/2$. In this case, the $\mathcal{I}^{(\alpha)}_{\omega\phi}$ determine the eigenvectors not fixed by symmetry and vice versa.

The UF_6, SF_6, ... molecular vibration models are examples of parameter treatments. One obtains fair results using two or three parameters j, k, and b by making physical pictures involving special arrangements of springs. However, the real molecule is held together by electrons and electrostatic forces instead of springs so it is a wonder that the spring model is even close.

It is instructive to derive the maximum number of parameters for any SF_6 model that uses the same basis. From Eq. (6.4.11a) the SF_6 result is $1 + 1 + 3 + 1 + 1 = 7$. Now one could go looking for other places to stick four more springs in SF_6, but this leads to an empty victory. There are better ways to approach the problem. A generalization of the parameter analysis in Eqs. (6.4.9) and (6.4.10) will be employed later.

(b) Noninvariant Operators If a basis is n dimensional, there are $n(n+1)/2$ independent Hermitian operators that transform the basis into itself. The preceding section accounted for the invariant operators. Now, the other operators must be labeled by noninvariant irreps if they are going to be defined by symmetry. For example, the six-dimensional octahedron basis will have 21 operators, 18 of which belong to some irrep besides A_{1g}.

Electric multipole operators are examples of noninvariant tensor operators. Suppose the effect of an electromagnetic wave on the energy of a charge (q) orbiting at position $\mathbf{x}$ is given by

$$V = q\mathbf{E}(\mathbf{x}, t) \cdot \mathbf{x} = q(E_1 x_1 + E_2 x_2 + E_3 x_3) e^{i(\mathbf{k}\cdot\mathbf{x} - \omega t)}, \quad (6.4.12)$$

where x_1, x_2, and x_3 are quantum position operators for the x, y, and z coordinates. The effects of the *magnetic* field on the *momentum* of the particle will be neglected here. Expanding the exponential gives the following:

$$V = qe^{-i\omega t} \sum_j E_j x_j \left(1 + i\mathbf{k}\cdot\mathbf{x} - \tfrac{1}{2}(\mathbf{k}\cdot\mathbf{x})^2 + \cdots\right)$$

$$= qe^{-i\omega t} \left(\sum_j E_j x_j + i \sum_i \sum_j k_i E_j x_i x_j - \tfrac{1}{2} \sum_i \sum_j \sum_k k_i k_j E_k x_i x_j x_k + \cdots\right).$$

$$(6.4.13)$$

If the wavelength $\lambda = 2\pi/k$ is long compared to the excursion $\langle x_j \rangle$ of the particle, then terms involving products of k and x_j may be neglected. This leaves only the first terms of V such as

$$V_j^{T_{1u}} \equiv qe^{-i\omega t}E_j x_j, \tag{6.4.14}$$

which transforms like coordinate x_j, i.e., as $T_{1u} - (j)$. Such operators are called ELECTRIC DIPOLE operators. The dipole operator $V_j^{T_{1u}}$ expresses the effect of a j-polarized electric wave of very long wavelength. The assumption of long wavelength corresponds to the DIPOLE APPROXIMATION. Long ocean waves would make a raft bob up and down, but would have no tendency to tip or bend it in the dipole approximation. The raft could not detect the direction of slope and propagation or curvature of waves that are much longer than it is.

Terms which tend to tip the raft are like the second terms in Eq. (6.4.13) and are called QUADRUPOLE terms. The transformation behavior of $x_i x_j$ operators are those of a second-rank tensor. The O_h-defined second-rank tensor operators belong to irreps A_{1g}, E_g, T_{1g}, and T_{2g}, as explained in Section 6.3.B. The third-rank OCTAPOLE terms $x_i x_j x_k$ would belong to A_{1u}, A_{2u}, E_u, T_{1u}, and T_{2u} irreps of O_h.

Now we see how to quickly evaluate the effect of various types of irreducible tensor operators on a quantum system.

B. Wigner-Eckart Theorem and Transition Selection Rules

A powerful theorem relates tensor operator matrix elements for different components or polarizations. First we prove and explain the theorem, and then we apply it using the dipole moment in the octahedron as an example.

Wigner-Eckart Theorem If operator T_i^α belongs to a set $\{T_1^\alpha T_2^\alpha \cdots T_{j^\alpha}^\alpha\}$ which transform according to an irrep of symmetry group $G = \{1, g, g', \ldots\}$ as follows:

$$gT_i^\alpha g^\dagger = \sum_{i'} T_{i'}^\alpha \mathscr{D}_{i'i}(g)$$

then the matrix elements between irrep bases will all be of the following form:

$$\left\langle \begin{matrix}\gamma\\k\end{matrix} \middle| T_i^\alpha \middle| \begin{matrix}\beta\\j\end{matrix} \right\rangle = \sum_\omega \mathscr{C}_{ijk}^{\alpha\beta(\gamma)\omega} \langle \gamma | |T^\alpha| |\beta \rangle_\omega, \tag{6.4.15a}$$

where the $\mathscr{C}_{ijk}^{\alpha\beta(\gamma)\omega}$ are coupling coefficients and the constant $\langle \gamma | |T^\alpha| |\beta \rangle_\omega$ is independent of the components i, j, or k.

Proof First insert the operator $1 = g^\dagger g$ in the matrix element and expand the result according to the assumed transformation properties:

$$\left\langle \begin{matrix} \gamma \\ k \end{matrix} \middle| T_i^\alpha \middle| \begin{matrix} \beta \\ j \end{matrix} \right\rangle = \left\langle \begin{matrix} \gamma \\ k \end{matrix} \middle| g^\dagger g T_i^\alpha g^\dagger g \middle| \begin{matrix} \beta \\ j \end{matrix} \right\rangle$$

$$= \sum_{k'} \sum_{i'} \sum_{j'} \mathscr{D}_{k'k}^\gamma(g)^* \mathscr{D}_{i'i}^\alpha(g) \mathscr{D}_{j'j}^\beta(g) \left\langle \begin{matrix} \gamma \\ k' \end{matrix} \middle| T_{i'}^\alpha \middle| \begin{matrix} \beta \\ j' \end{matrix} \right\rangle.$$

One may sum over all group elements without affecting the left side, if we divide by $°G$ and use Eq. (6.2.15):

$$\left\langle \begin{matrix} \gamma \\ k \end{matrix} \middle| T_i^\alpha \middle| \begin{matrix} \beta \\ j \end{matrix} \right\rangle = \sum_{k'} \sum_{i'} \sum_{j'} \left(1/°G \sum_g \mathscr{D}_{k'k}^\gamma(g)^* \mathscr{D}_{i'i}^\alpha(g) \mathscr{D}_{j'j}^\beta(g) \right) \left\langle \begin{matrix} \gamma \\ k' \end{matrix} \middle| T_{i'}^\alpha \middle| \begin{matrix} \beta \\ j' \end{matrix} \right\rangle$$

$$= \sum_\omega \mathscr{C}_{ijk}^{\alpha\beta(\gamma)\omega} \left[(1/l^\alpha) \sum_{i'} \sum_{j'} \sum_{k'} \mathscr{C}_{i'j'k'}^{\alpha\beta(\gamma)\omega} \left\langle \begin{matrix} \gamma \\ k' \end{matrix} \middle| T_{i'}^\alpha \middle| \begin{matrix} \beta \\ j' \end{matrix} \right\rangle \right].$$

The theorem is proved since the bracketed term is not a function of the components.

$$\langle \gamma \| T^\alpha \| \beta \rangle_\omega \equiv \left[(1/l^\alpha) \sum_{i'} \sum_{j'} \sum_{k'} \mathscr{C}_{i'j'k'}^{\alpha\beta(\gamma)\omega} \left\langle \begin{matrix} \gamma \\ k' \end{matrix} \middle| T_{i'}^\alpha \middle| \begin{matrix} \beta \\ j' \end{matrix} \right\rangle \right]. \quad (6.4.15\text{b})$$

The constants $\langle \gamma \| T^\alpha \| \beta \rangle_\omega$ in Eq. (6.4.15a) or (6.4.15b) are usually called REDUCED MATRIX ELEMENTS. They are the "parameters" of the α-tensor operators. For groups with nonsimply reducible products the repetition index (ω) needs to be attached, and the matrix elements are combinations of parameters belonging to different (ω). However, for most of the symmetry groups we will be studying (ω) can be deleted from Eqs. (6.4.15). Then the matrix elements for different polarizations are all simply proportional to one reduced matrix element $\langle \gamma \| T^\alpha \| \beta \rangle$ for each α transition between manifold $\{\ldots,\left|\begin{matrix}\beta\\j\end{matrix}\right\rangle,\ldots\}$ and $\{\ldots,\left|\begin{matrix}\gamma\\k\end{matrix}\right\rangle,\ldots\}$. The proportionality factors are just the coupling coefficients $\mathscr{C}_{ijk}^{\alpha\beta\gamma}$ in each transition $\left|\begin{matrix}\beta\\j\end{matrix}\right\rangle \leftrightarrow \left|\begin{matrix}\gamma\\k\end{matrix}\right\rangle$ induced by (ith) polarization component T_i^α.

The Wigner-Eckart theorem provides a way to factor a matrix element into a geometrical part and a physical part. The coupling coefficient is the geometrical part and for it one only needs to know the group transformation properties of the states and transition operator. The reduced matrix element is the physical part and to evaluate it one needs more detailed knowledge of states and operator. The theorem allows one to still make certain predictions when detailed physical knowledge is lacking.

As an example, let us apply the Wigner-Eckart theorem to evaluate electric-dipole ($V_i^{T_{1u}}$) transition rates between the A_{1g}, T_{1u}, and E_g orbital states of the octahedral tunneling eigenstates in Figure 4.3.2(b). The matrix elements are given by

$$\left\langle \begin{matrix}\gamma\\k\end{matrix}\middle| V_i^{T_{1u}} \middle| \begin{matrix}\beta\\j\end{matrix}\right\rangle = C_{ijk}^{T_{1u}\beta\gamma}\langle\gamma| |V^{T_{1u}}| |\beta\rangle. \tag{6.4.16}$$

The squares $\left|\left\langle \begin{matrix}\gamma\\k\end{matrix}\middle| V_i^{T_{1u}} \middle| \begin{matrix}\beta\\j\end{matrix}\right\rangle\right|^2$ of the matrix elements are proportional to the probability of absorption of light or intensity of emission according to Fermi's Golden Rule. (This will be treated in Chapter 8 Section 6.)

The first thing to notice is that all dipole or (T_{1u}) transitions are FORBIDDEN between the levels A_{1g} and E_g, since $C_{ijk}^{T_{1u}E_g A_{1g}}$ is identically zero. This is an example of a SELECTION RULE. Remember that a product $u \otimes g$ always gives u states and never g. However, this particular matrix element would be zero even if the transition operator was a *magnetic* dipole ($V_i^{T_{1g}}$) with even (g) parity since the product $T_1 \otimes E$ does not contain A_1. The next possible avenue for an ($E_g \leftrightarrow A_{1g}$) transition is by a quadrupole operator ($V_i^{E_g}$). Note that ($E_g \leftrightarrow E_g$) and ($T_{1u} \leftrightarrow T_{1u}$) transitions are electric-dipole forbidden, also.

The ($A_{1g} \leftrightarrow T_{1u}$) and ($T_{1u} \leftrightarrow E_g$) transitions are electric-dipole allowed, however. The Wigner-Eckart theorem gives the matrix elements for all possibilities in terms of two reduced matrix elements $\langle T_1| |T_1| |E\rangle$ and $\langle T_1| |T_1| |A_1\rangle$. All the combinations are given in the following. The O_h coupling coefficients in the Appendix F.3.1d are used.

$$\left\langle \begin{matrix}T_1\\1\end{matrix}\middle| V_1^{T_1} \middle| \begin{matrix}E\\1\end{matrix}\right\rangle = -\frac{1}{2}\langle T_1| |T_1| |E\rangle, \qquad \left\langle \begin{matrix}T_1\\2\end{matrix}\middle| V_2^{T_1} \middle| \begin{matrix}E\\1\end{matrix}\right\rangle = -\frac{1}{2}\langle T_1| |T_1| |E\rangle,$$

$$\left\langle \begin{matrix}T_1\\3\end{matrix}\middle| V_3^{T_1} \middle| \begin{matrix}E\\1\end{matrix}\right\rangle = \langle T_1| |T_1| |E\rangle,$$

$$\left\langle \begin{matrix}T_1\\1\end{matrix}\middle| V_1^{T_1} \middle| \begin{matrix}E\\2\end{matrix}\right\rangle = \frac{\sqrt{3}}{2}\langle T_1| |T_1| |E\rangle, \qquad \left\langle \begin{matrix}T_1\\2\end{matrix}\middle| V_2^{T_1} \middle| \begin{matrix}E\\2\end{matrix}\right\rangle = -\frac{\sqrt{3}}{2}\langle T_1| |T_1| |E\rangle,$$

$$\left\langle \begin{matrix}T_1\\3\end{matrix}\middle| V_3^{T_1} \middle| \begin{matrix}E\\2\end{matrix}\right\rangle = 0,$$

$$\left\langle \begin{matrix}T_1\\1\end{matrix}\middle| V_1^{T_1} \middle| A_1\right\rangle = \langle T_1| |T_1| |A_1\rangle, \qquad \left\langle \begin{matrix}T_1\\2\end{matrix}\middle| V_2^{T_1} \middle| A_1\right\rangle = \langle T_1| |T_1| |A_1\rangle,$$

$$\left\langle \begin{matrix}T_1\\3\end{matrix}\middle| V_3^{T_1} \middle| A_1\right\rangle = \langle T_1| |T_1| |A_1\rangle. \tag{6.4.17}$$

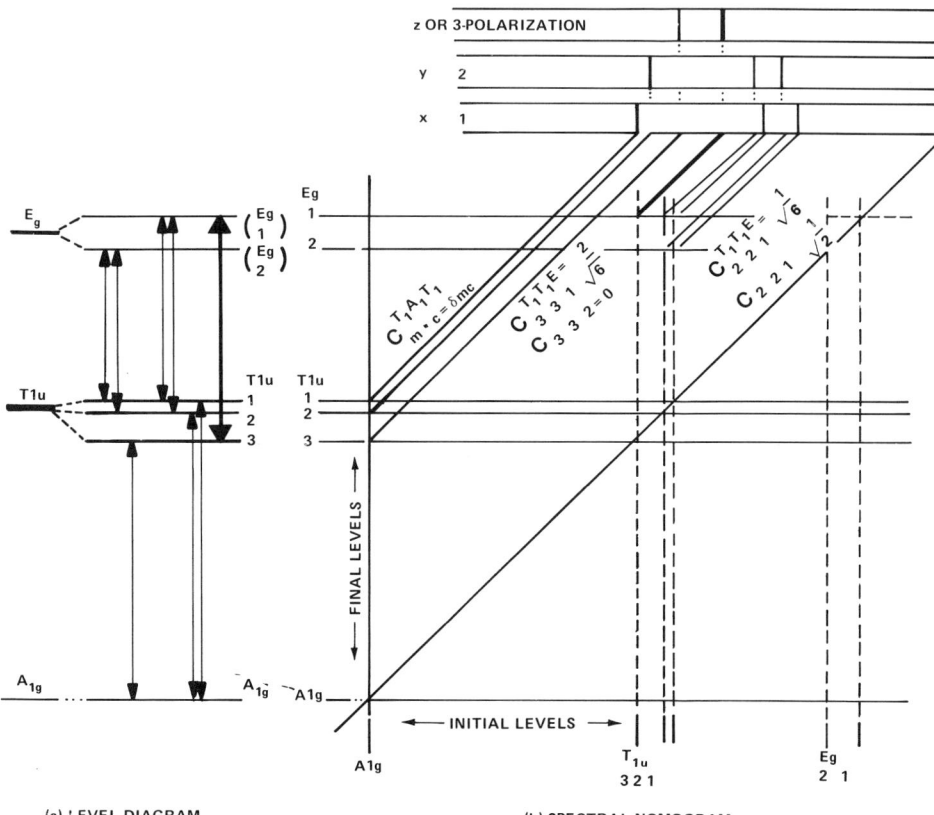

Figure 6.4.1 Intracluster transitions predicted by Wigner-Eckart theorem. (a) Level diagram. Transitions are indicated by arrows drawn between levels. (b) Spectral nomogram. Spectra are indicated by lines drawn at 45° from each intersection that represents an allowed transition.

In the $(T_{1u} \leftrightarrow E_g)$ transition one predicts an intensity ratio of $|1/2|^2 : |\sqrt{3}/2|^2 = 1:3$ between the $\begin{pmatrix} E_{1g} \\ 1 \end{pmatrix}$ and $\begin{pmatrix} E_{1g} \\ 2 \end{pmatrix}$ levels, respectively, using polarization 1 or 2 and a ratio of $1:0$ using polarization 3. In order for a spectroscopist to observe such intensity ratios it would be necessary to partially split the E_g and T_{1u} degeneracies as indicated on the left-hand side of Figure 6.4.1. This could be done using some external perturbation such as a weak Q field as discussed in Section 4.3.D. (Recall the splitting indicated near the center of Figure 4.3.5.) It is very important to polarize or otherwise prepare the octahedral system by some external means. It will not make any difference which polarization component $V_1^{T_1}$, $V_2^{T_1}$, or $V_3^{T_1}$ causes the $(T_{1u} \leftrightarrow E_g)$ transition if one cannot distinguish the three substates $\begin{vmatrix} T_{1u} \\ 1 \end{vmatrix}$, $\begin{vmatrix} T_{1u} \\ 2 \end{vmatrix}$, or $\begin{vmatrix} T_{1u} \\ 3 \end{vmatrix}$

within the T_{1u} manifold or the two substates $\begin{vmatrix} E_g \\ 1 \end{vmatrix}$ and $\begin{vmatrix} E_g \\ 2 \end{vmatrix}$ within the E_g manifold. The detailed Wigner-Eckart intensity predictions are useless unless the system is oriented somehow with respect to the polarization of the dipole field causing the transition.

In case the system is polarized one would obtain a very different spectrum from 3-polarized light than from 1- or 2-polarized light. This is indicated by the 1, 2, and 3 bands of spectral lines shown in the upper portion of Figure 6.4.1. The *lines* are obtained from the *levels* according to geometrical nomogram drawn below in the central portion of the figure. Here the final levels are plotted along the x axis and are extended by lines parallel to the y axis. The final levels are plotted along the y axis and extended by lines parallel to the x axis. (This trick is useful even if the manifold of final levels is completely different from the initial ones.) Whenever an initial level intersects a final level of higher (lower) energy that represents a possible absorption (emission) transition. If the transition is allowed the spectral line is located by a 45° line drawn upward and to the right from that point. The parallel displacement of the 45° lines is proportional to the *difference* between levels and hence models the spectrum.

In this way the ($A_{1g} \to T_{1u}$) and ($T_{1u} \to E_g$) absorption spectra are drawn in Figure 6.4.1. The leftmost lines belong to the ($A_{1g} \to T_{1u}$) absorption. The others belong to various allowed ($T_{1u} \to E_g$) transitions with darker lines indicating larger $\mathscr{C}_{ijk}^{T_{1u}T_{1u}E_g}$ coefficients and greater intensity I where

$$I_i\begin{pmatrix} T_{1u} & & E_g \\ j & \to & k \end{pmatrix} \sim \left(\mathscr{C}_{ijk}^{T_{1u}T_{1u}E_g}\langle E| |T_1| |T_1\rangle\right)^2. \qquad (6.4.18)$$

Note that if the $\begin{vmatrix} T_{1u} \\ 1 \end{vmatrix}$ and $\begin{vmatrix} T_{1u} \\ 2 \end{vmatrix}$ levels become degenerate then the 1- and 2-polarization spectral patterns become indistinguishable. Finally, if the $\begin{vmatrix} T_{1u} \\ 3 \end{vmatrix}$ level joins the other T_{1u} pair and the splitting of E_g is zero then all three polarizations look the same. Each spectrum will have two lines with intensity ratios given by

$$I(A_{1g} \to T_{1u})/I(T_{1u} \to E_g) = (\langle T_1| |T_1| |A_1\rangle/\langle T_1| |T_1| |E\rangle)^2, \qquad (6.4.19)$$

and the Wigner-Eckart theorem predicts nothing of the detail. In the next chapter there will be situations in which symmetry analysis can be used to evaluate reduced matrix elements. However, the reduced matrix elements in Eq. (6.4.19) are simply undetermined physical parameters in this example.

Note that individual intensities such as given by Eq. (6.4.18) may not agree with the spectral experiment if too much splitting is present. As shown in

Section 4.3.D a large Q splitting mixes the $\left|{}^{E_g}_{1}\right\rangle$ and $\left|{}^{A_{1g}}_{\cdot}\right\rangle$ states. When this occurs one says that one line "borrows" intensity from another. The effect of changing Q on the 3-spectrum would be a change of relative intensity as well as position. As the $\left|{}^{E_g}_{1}\right\rangle$ picked up more or less $|A_{1g}\rangle$ the intensity of the transition $\left[\left|{}^{T_{1u}}_{3}\right\rangle \to \left|{}^{E_g}_{1}\text{ (modified)}\right\rangle\right]$ would depend on the square of the matrix element

$$\left\langle {}^{T_{1u}}_{3}\left|V_3^{T_{1u}}\right|{}^{E_g}_{1}\text{ (modified)}\right\rangle = \varepsilon\left\langle {}^{T_{1u}}_{3}\left|V_3^{T_{1u}}\right|{}^{E_g}_{1}\right\rangle + \alpha\left\langle {}^{T_{1u}}_{3}\left|V_3^{T_{1u}}\right|A_{1g}\right\rangle$$

$$= \varepsilon\langle T_1|\,|T_1|\,|E\rangle + \alpha\langle T_1|\,|T_1|\,|A\rangle,$$

where α is the mixing amplitude of the $|A_{1g}\rangle$ and $|\varepsilon|^2 = 1 - |\alpha|^2$. In this way the two large induced dipole moments can interfere.

One should try to get a qualitative physical picture of the significance of a matrix element whenever possible. For example, the fact that the pair of states $\left|{}^{E_g}_{1}\right\rangle$ and $\left|{}^{T_{1u}}_{3}\right\rangle$ would have a large matrix element of the dipole operator $V_3^{T_{1u}}$ implies that their mixture $\alpha\left|{}^{E_g}_{1}\right\rangle + \beta\left|{}^{T_{1u}}_{3}\right\rangle$ corresponds to large charge displacement or dipole moment in the 3-direction as shown in Figure 6.4.2. Indeed, if the two states have different phases in time, say $e^{i\omega_{E_g}t}$ and $e^{i\omega_{T_{1u}}t}$, this dipole moment will oscillate with angular frequency $(\omega_{E_g} - \omega_{T_{1u}})$, and radiate or absorb accordingly. In the following section we discuss the significance of oscillating dipoles in some detail.

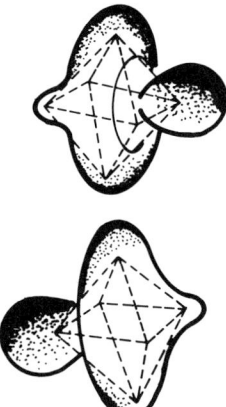

Figure 6.4.2 Sketch of time behavior of wave function for mixture of $\left|{}^{E_g}_{1}\right\rangle$ and $\left|{}^{T_{1u}}_{3}\right\rangle$.

6.5 CLASSICAL APPROACH TO OPTICAL RESONANCE AND SELECTION RULES

We consider now the classical theory of resonance and its application to optical spectroscopy. Resonance may very well be the single most important phenomenon involved in the observation and study of our immediate physical world. Without it we might all be blind, deaf, and dumb, since seeing, hearing, and speaking all involve resonant transfer of energy containing high-quality (i.e., coherent) signals and information. Most physical stimuli in nature are very weak and need to be amplified in order to cause observable responses. Resonance provides a way for a weak stimulus to cause a relatively large response.

Spectroscopic lines or peaks all involve some form of resonance. To understand many types of spectroscopy it is sufficient to know just the classical theory of spectroscopic resonance as developed by Lorentz and others in the late 1800s. In any case classical resonance and radiation theory is a very important prerequisite to the understanding of semiclassical and quantum theory of resonance, which will be taken up in Chapter 8, Section 8.6B. In view of the importance of classical resonance theory, it is surprising how little coverage it receives in many physics or chemical physics texts, particularily the beginning ones. Here we shall try to make up for this.

This section will include an introduction to concepts of polarization, polarizability, susceptibility, absorption, and indices of refraction. A review of some graphical aids such as phasors (first discussed in Sections 2.3.B, 2.6, and 2.7), Smith charts, and Feynman's radiation lever will be introduced. Two well-known forms of spectroscopy, (1) infrared or optical absorption, and (2) Raman spectroscopy, will be discussed in this section. Physical interpretations of infrared and Raman selection rules will be given along with the group-theoretical methods of derivation.

A. Introduction to the Effect of Light on Matter

The Lorentz theory was based on a simple model for an atom or molecule. In this model electrons are imagined to be bound to more massive nuclei in such a way that they behave like harmonic oscillators with certain natural frequencies $\{\omega_0 \omega_0' \cdots \}$. The small mass of the electron ($\mu_e = 9.1 \times 10^{-31}$ kg) compared to that of the nuclei (a single proton has a mass of $M_p \cong 1836 \mu_e \cong 6\pi^5 \mu_e$) means that electrons would usually respond more easily to an externally applied electric field than the nuclei. The electronic charge of $q = e = -1.6 \times 10^{-19}$ coulomb is equal but opposite to that of the proton, so that the electric forces $\mathbf{F} = q\mathbf{E}$ would be of the same magnitude.

One should keep in mind how large is the so-called "tiny" charge of an electron or proton. A mole of an element with atomic number $Z(= 1 - 103)$ has $N_0 Z$ electrons where $N_0 = 6.02 \times 10^{23}$ is Avogadro's number. This amounts to nearly $-10^5 Z$ coulombs of electronic charge often confined to within a few cubic centimeters. According to Coulomb's law the force ($F = qq'/(4\pi\varepsilon_0 r^2)$) or energy ($V = qq'/(4\pi\varepsilon_0 r)$) between a pair of charges of just 1 coulomb each separated by one meter is about nine billion newtons or joules, respectively! ($1/4\pi\varepsilon_0 = 9.0 \times 10^9$ Nm^2/C^2.) This explains why so much energy can arise if atomic ions of the same charge are formed with separations of a few angstroms ($r \simeq 1$ Å $= 10^{-10}$ m) or (much worse!) if nuclei split into parts that are separated by a few Fermis or femtometers ($r \simeq 1$ F $= 10^{-15}$ m). The atomic bomb is really an electric bomb!

Optical resonance provides a way to excite the negative or electronic part of the enormous charge that resides in atoms and molecules. Many of the hues we see in nature involve resonantly excited electrons in molecules that reside in the objects as well as in our retinas. On the other hand, infrared or microwave resonance allows one to "see" nuclear vibrational or rotational excitations using other kinds of detectors.

The Lorentz model for resonance assumes that an electron's spatial coordinate x obeys a forced damped harmonic oscillator equation of the form

$$\ddot{x} + 2\Gamma\dot{x} + \omega_0^2 x = (q/\mu)E(\mathbf{r}, t), \qquad (6.5.1)$$

where ω_0 represents a natural oscillation frequency, Γ is a phenomenological decay constant, and the radiation stimulus is represented by the time-dependent acceleration $(q/\mu)E(\mathbf{r}, t)$ of an electric field $\mathbf{E}$ on a charge q and mass μ. It will be shown in Section 8.4c that the Lorentz oscillator model agrees with the quantum theory for small oscillations of x. It worked very well until the laser was invented and oscillations with much greater amplitudes could be excited.

In discussing the solutions of (6.5.1) it is helpful to consider a monochromatic stimulus, i.e., an E field oscillating at one frequency ω_s. It is also convenient to represent oscillating quantities by complex numbers or phasors. The stimulating field will be written

$$E(t) = E_0 e^{-i\omega_s t + i\phi}. \qquad (6.5.2)$$

One first considers the solutions to the zero-field or homogeneous equation (6.5.1) by substituting the trial expression

$$x_T = ae^{-i(\omega t - \alpha)} \qquad (6.5.3)$$

into (6.5.1) with ($E = 0$). Solving the resulting equation,

$$\left(-\omega^2 - 2i\Gamma\omega + \omega_0^2\right)\alpha e^{-i(\omega t - \alpha)} = 0,$$

yields

$$\omega = -i\Gamma + \left(\omega_0^2 - \Gamma^2\right)^{1/2} \equiv -i\Gamma + \omega_\Gamma$$

and

$$x_T = ae^{-\Gamma t}e^{-i(\omega_T t - \alpha)}. \quad (6.5.4)$$

This is called the transient part of the solution since the damping factor $e^{-\Gamma t}$ will eventually kill it off if $\Gamma > 0$.

The other part of the solution is called the inhomogeneous or steady-state response. Substituting

$$x_S = Re^{-i(\omega_s t - \rho)} \quad (6.5.5a)$$

into (6.5.1) with (6.5.2) yields

$$Re^{i\rho}(-\omega_s^2 - i2\Gamma\omega_s + \omega_0^2)e^{-i\omega_s t} = (q/\mu)E_0 e^{-i\omega_s t + i\phi}. \quad (6.5.5b)$$

Solving this gives the complex response function

$$Re^{i\rho} = \frac{(q/\mu)E_0 e^{i\phi}}{\omega_0^2 - \omega_s^2 - i2\Gamma\omega_s}. \quad (6.5.5c)$$

The response function can be written in Cartesian $(x + iy)$ form

$$Re^{i(\rho - \phi)} = \frac{(q/\mu)E_0(\omega_0^2 - \omega_s^2)}{(\omega_0^2 - \omega_s^2)^2 + (2\Gamma\omega_s)^2} + i\frac{(q/\mu)E_0(2\Gamma\omega_s)}{(\omega_0^2 - \omega_s^2)^2 + (2\Gamma\omega_s)^2}, \quad (6.5.5d)$$

or in the original polar form, where

$$R = \frac{(q/\mu)E_0}{\left[(\omega_0^2 - \omega_s^2)^2 + (2\Gamma\omega_s)^2\right]^{1/2}} \quad (6.5.5e)$$

and

$$\rho = \tan^{-1}\frac{2\Gamma\omega_s}{\omega_0^2 - \omega_s^2} + \phi \quad (6.5.5f)$$

are the magnitudes and phase lags, respectively, of the response function $Re^{i\rho}$. The general complex solution takes the form

$$x(t) = x_S(t) + x_T(t)$$
$$= ae^{-\Gamma t}e^{-i(\omega_T t - \alpha)} + Re^{-i(\omega_s t - \rho)}, \quad (6.5.6)$$

where the transient amplitude a and phase lag α depend upon the initial conditions $x = \text{Re}(x(0))$ and $v_0 = \text{Re}(\dot{x}(0))$. It should be remembered [recall Section 2.3(B)] that only the real part $(\text{Re}(x))$ of the complex coordinate represents the real physical position. The same applies to complex velocity and force.

Setting ($t = 0$) in the real part of (6.5.6) and its derivative gives

$$x_0 = a \cos \alpha + R \cos \rho,$$
$$v_0 = a(\omega_\Gamma \sin \alpha - \Gamma \cos \alpha) + R\omega_s \sin \rho.$$

Solving for a and α gives the transient amplitude

$$a = (x_0 - R \cos \rho)/\cos \alpha$$
$$= -R \cos \rho / \cos \alpha \quad \text{(for } x_0 = 0 = v_0\text{)} \quad (6.5.7)$$

and phase lag

$$\alpha = \tan^{-1}\left[\frac{(v_0 - R\omega_s \sin \rho)}{(x_0 - R \cos \rho)\omega_\Gamma} + \frac{\Gamma}{\omega_\Gamma}\right]$$
$$= \tan^{-1}\left[\frac{\omega_s \sin \rho}{\omega_\Gamma \cos \rho} + \frac{\Gamma}{\omega_\Gamma}\right] \quad \text{(for } x_0 = 0 = v_0\text{)}. \quad (6.5.8)$$

Figure 6.5.1 shows plots of the response (6.5.6) of an initially cold oscillator ($x_0 = 0 = v_0$) to a ($\cos \omega_s t$) stimulus turned on at $t = 0$. The oscillator parameters are $\Gamma = 0.5$ s^{-1}, $\omega_0 = 10\pi$, and ω_s is set at (a) 8 π, (b) 9π, (c) 9.8π, (d) 10π, (e) 10.5π, and (f) 12 π. The first six or so seconds of each plot shows one or more "beats" (recall also Figure 2.3.2), which occur with a frequency equal to the difference $(\omega_0 - \omega_s)/2\pi$ between the interfering natural and stimulating frequencies. This interference is called transient behavior and it is analogous to "quantum beats" which occur in quantum excitations, as will be discussed in Chapter 8.

Lorentz theory is generally concerned with steady-state oscillator behavior after the transient amplitude $ae^{-\Gamma t}$ has died off. In Figure 6.5.1 $e^{-\Gamma t}$ decays to less than 5% ($e^{-3} = 0.05$) after $t = 6$ s. Then the steady-state response x_s, which has the same frequency (ω_s) as the stimulus, is all that remains after the natural frequency (ω_0) part dies away. The magnitude R and relative phase ρ of the responding oscillator coordinate x_S is of prime importance for Lorentz theory.

In particular we know that the stimulating force transfers the maximum power to the oscillator when the phase ρ of x_S is 90° behind the stimulus. This was explained in Section 2.3.A. According to (6.5.5f) this occurs when $\omega_s = \omega_0$, i.e., when the stimulus is exactly in resonance with the oscillator. This is the point where the real part of the response function (6.5.5d) changes sign as shown in Figure 6.5.2(a) and the imaginary part (Figure 6.5.2(b)) is near its maximum value. For a stimulus frequency below resonance ($\omega_s < \omega_0$) the response is almost in phase or only lagging slightly

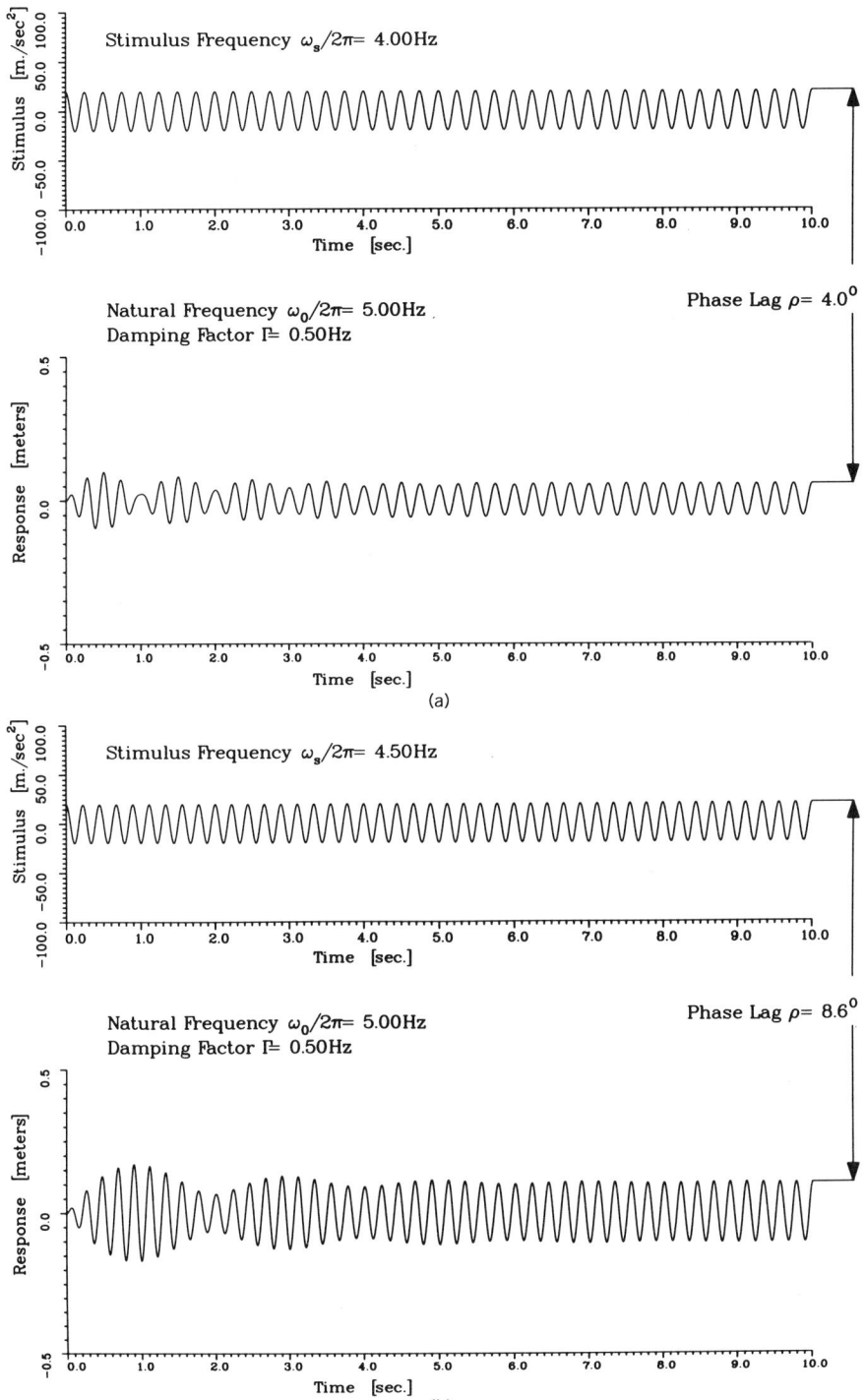

Figure 6.5.1 Time plots of stimulus and response of oscillator for various values of the stimulus frequency. (a)–(c) Below resonance; (d) resonance; (e) and (f) above resonance.

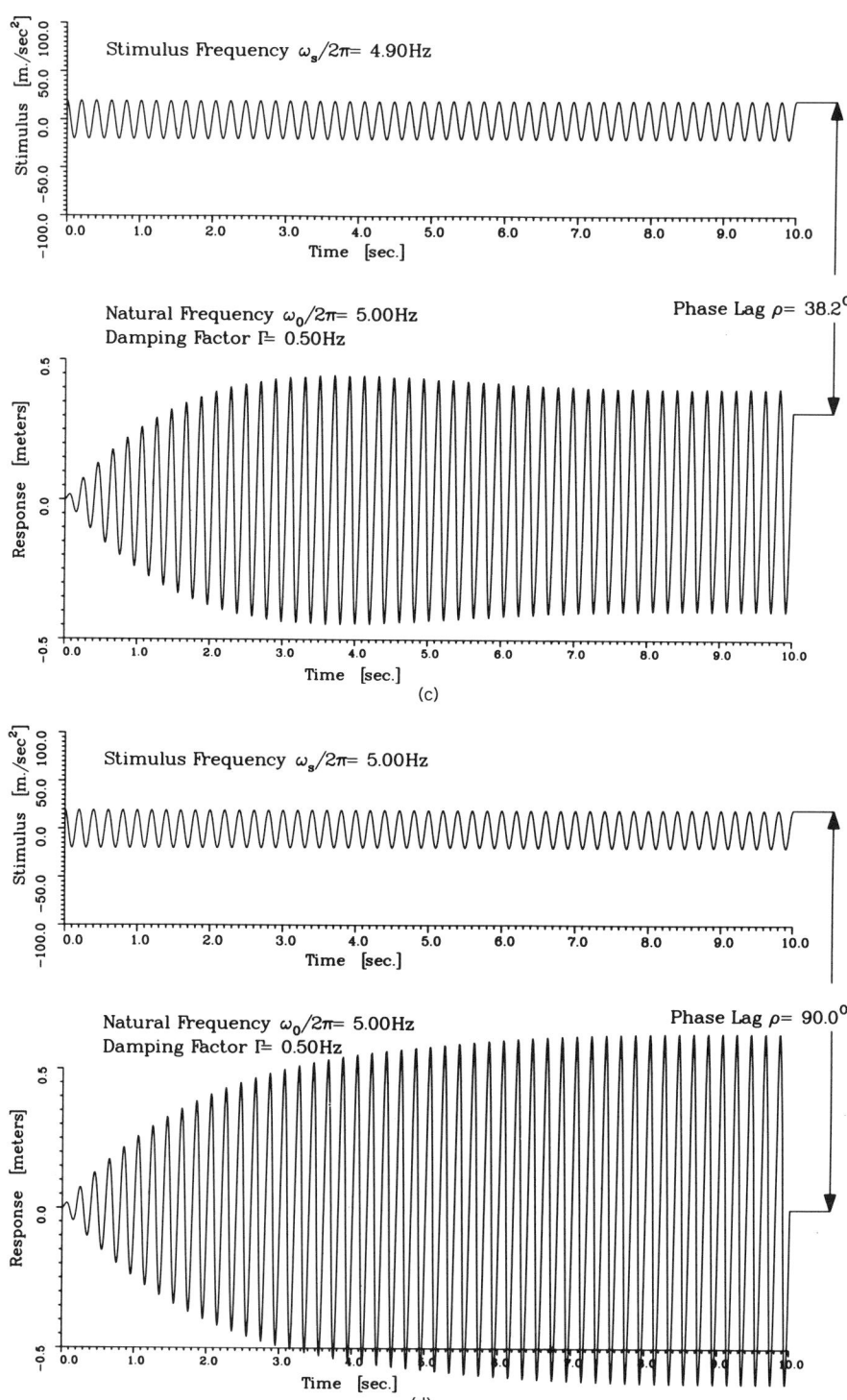

Figure 6.5.1 Time plots of stimulus and response of oscillator for various values of the stimulus frequency *(Continued)*.

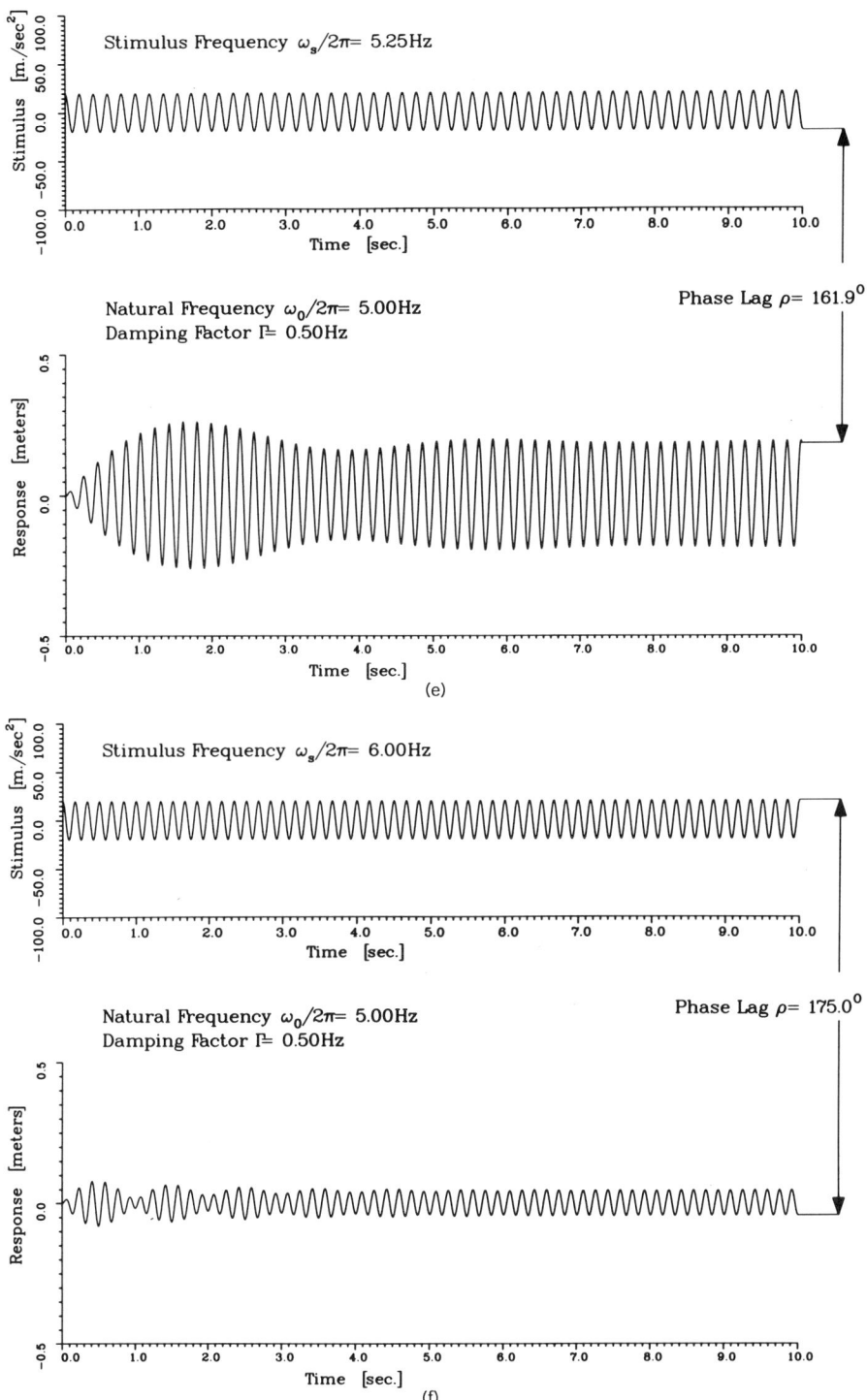

Figure 6.5.1 *(Continued)*.

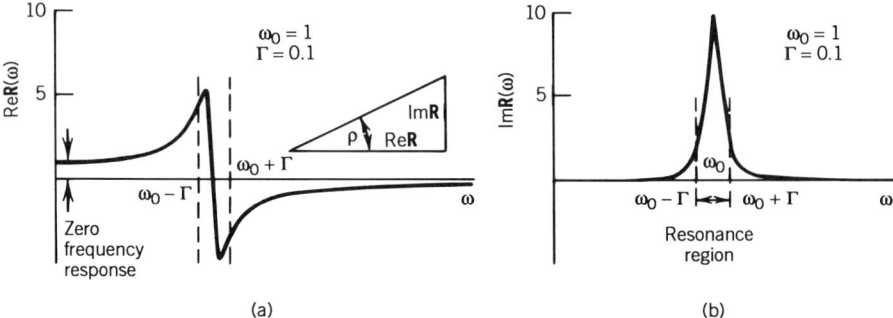

Figure 6.5.2 Classical oscillator response function $R(\omega)$. (a) Real part has a zero value and maximum slope at resonance. (b) Imaginary part has maximum value at resonance.

behind the stimulus ($\rho = 0^+$). Above resonance ($\omega_s > \omega_0$) the response falls almost completely out of phase ($\rho = 180^-$). Only near resonance ($\omega_s = \omega_0$) does the stimulus lead by the magic 90° and feed maximum power into the oscillator, which promptly wastes it through some damping mechanism characterized by the coefficient Γ. (More about this shortly.)

For reasonably low values of Γ (say $\Gamma < \omega_0/10$) it may be helpful to display the response quantities using an approximation and a monogram known as the Smith chart. The near-resonance approximation allows us to write

$$\omega_0^2 - \omega_s^2 = (\omega_0 - \omega_s)(\omega_0 + \omega_s)$$
$$\simeq (\omega_0 - \omega_s) 2\omega_0,$$

so that the response function (6.5.5c) may be approximated by

$$Re^{i\rho} \simeq \frac{(q/\mu)E_0}{(\omega_0 - \omega_s - i\Gamma)2\omega_0} = \frac{S}{\Delta - i\Gamma}, \quad (6.5.9a)$$

where we define

$$S = (q/2\mu\omega_0)E_0 \quad (6.5.9b)$$

as the stimulus strength and

$$\Delta = \omega_0 - \omega_s \quad (6.5.9c)$$

as the resonance detuning. Now the complex response can be written

$$Re^{i\rho} = R\cos\rho + iR\sin\rho = \frac{S\Delta}{\Delta^2 + \Gamma^2} + i\frac{S\Gamma}{\Delta^2 + \Gamma^2}, \quad (6.5.10a)$$

and its absolute square is a Lorentzian function of Δ:

$$R^2 = S^2/(\Delta^2 + \Gamma^2). \qquad (6.5.10b)$$

Equating real and imaginary parts leads to the relation

$$R = (S/\Delta)\cos\rho = (B/\Gamma)\sin\rho. \qquad (6.5.11)$$

From Figure 6.5.3 it is easy to see that the R phasor lies at the intersection of two circles with diameters (S/Δ) and (S/Γ) placed along the real and imaginary axes, respectively, and intersecting at origin. Note that for constant Γ and S the $Re^{i\rho}$ phasor would follow the Γ circle to the top as $\Delta \to 0$ and $\rho \to \pi/2$. For negative detuning ($\omega_s > \omega_0$ or $\Delta < 0$) it would continue into the second quadrant as $|R|$ passed its maximum value, and finally the phase lag ρ would approach 180°. Graphs whose coordinates are (Δ, Γ) circles are called SMITH CHARTS and are used for many different purposes. The amplitude (6.5.9a) and Lorentzian function (6.5.10) will turn out to be quantum resonance functions in Chapter 8, and the Smith constructions may be used to help understand line shapes.

One should note that the Smith chart approximation (6.5.9) is completely wrong for the classical response at low frequency. The ($\omega_s \sim 0$) or DC response given by (6.5.5c),

$$R_{\rm DC} \simeq (q/\mu\omega_0^2)E_0, \qquad (6.5.12)$$

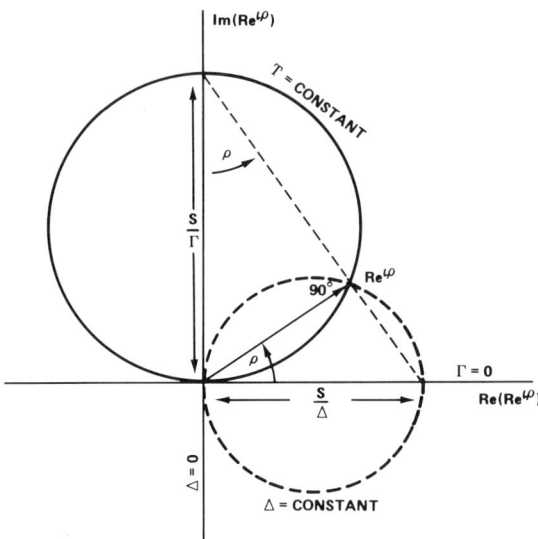

Figure 6.5.3 Geometry of a Smith chart. A nomogram for displaying response $|R|e^{i\rho}$ near resonance is based upon intersecting circles whose diameters are inversely proportional to damping (Γ) and detuning (Δ).

represents the displacement in equilibrium caused by a static or slowly moving field. [See the left-hand side of Figure 6.5.2(a).] It is also wrong for very high frequencies, since the correct R (6.5.5) vanishes as $(1/\omega_s^2)$ not as $(1/\omega_s)$.

B. Introduction to the Effect of Matter on Light

The displacement and separation or oscillation of electronic charge from the nuclei gives rise to dipole moments and currents which affect Maxwell's equations and light propagation. In the Lorentz model the dipole moments are taken to be directly proportional to the oscillator coordinate x, and the volume polarization is

$$\mathbf{P} = N\mathbf{p} = Nq\mathbf{x} \tag{6.5.13}$$

for N dipoles per unit volume. This enters as an extra current term $\mathbf{i} = Nq(\partial \mathbf{x}/\partial t)$ in Ampere's equation,

$$\nabla \times \mathbf{B} = \mu_0 \left(\varepsilon_0 \frac{\partial \mathbf{E}}{\partial t} + \mathbf{i} \right) = \mu_0 \left(\varepsilon_0 \frac{\partial \mathbf{E}}{\partial t} + \frac{\partial \mathbf{P}}{\partial t} \right).$$

Using Faraday's equation $(\nabla \times \mathbf{E} = -(\partial \mathbf{B}/\partial t))$ and the vector identity $\nabla(\nabla \cdot \mathbf{E}) - \nabla^2 \mathbf{E} = \nabla \times (\nabla \times \mathbf{E})$ one obtains the electric wave equation

$$\nabla(\nabla \cdot \mathbf{E}) - \nabla^2 \mathbf{E} = -\frac{\partial}{\partial t}[\nabla \times \mathbf{B}]$$

$$= -\mu_0 \left(\varepsilon_0 \frac{\partial^2 E}{\partial t^2} + \frac{\partial^2 P}{\partial t^2} \right). \tag{6.5.14}$$

This can be simplified by relating volume dipole response to the $\mathbf{E}$ field using (6.5.5) and (6.5.13) and writing

$$\mathbf{P} = N\alpha\varepsilon_0 \mathbf{E} = \chi\varepsilon_0 \mathbf{E} \tag{6.5.15a}$$

where the POLARIZABILITY (α) and VOLUME SUSCEPTIBILITY (χ) are defined by

$$\chi = N\alpha = (Nq^2/\mu\varepsilon_0)/(\omega_0^2 - \omega_s^2 - i2\Gamma\omega_s). \tag{6.5.15b}$$

Dropping the $\nabla \cdot \mathbf{E}$ term (assuming equal positive and negative charge density) in (6.5.14), one obtains the wave equation:

$$\nabla^2 \mathbf{E} = \mu_0 \varepsilon_0 (1 + \chi) \frac{\partial^2 \mathbf{E}}{\partial t^2}. \tag{6.5.16}$$

CLASSICAL APPROACH TO OPTICAL RESONANCE AND SELECTION RULES

The susceptibility χ has the steady-state form if the stimulating electric field has a steady oscillation such as the plane wave

$$\mathbf{E} = E_0 \hat{y} e^{i(k_s z - \omega_s t)}. \qquad (6.5.17)$$

Substituting this into the wave equation yields the following complex dispersion relation [henceforward we shall drop the subscript (s) from ω or k]:

$$k^2 = \mu_0 \varepsilon_0 (1 + \chi) \omega^2 = (1 + \chi) \omega^2 / c^2. \qquad (6.5.18)$$

The resulting wave phase velocity is

$$\omega/k = c/(1 + \chi)^{1/2} = c/\varepsilon^{1/2}, \qquad (6.5.19)$$

where the complex DIELECTRIC CONSTANT (ε) and INDEX OF REFRACTION (n) are defined by

$$n = \varepsilon^{1/2} = (1 + \chi)^{1/2}. \qquad (6.5.20)$$

From (6.5.15b) one has the following expression for the dielectric constant:

$$\varepsilon(\omega) = \left[1 + \frac{(Nq^2/\mu\varepsilon_0)(\omega_0^2 - \omega^2)}{(\omega_0^2 - \omega^2)^2 + 4\Gamma^2\omega^2}\right] + i\left[\frac{\Gamma(Nq^2/\mu\varepsilon_0)\omega}{(\omega_0^2 - \omega^2)^2 + 4\Gamma^2\omega^2}\right]. \qquad (6.5.21)$$

For zero damping $(\Gamma = 0)$ ε and $\varepsilon^{1/2}$ are positive and real, respectively, everywhere except between the resonance frequency $\omega = \omega_0$ and the frequency ω_c. The ω_c is called the CUTOFF FREQUENCY and is defined by

$$\omega_0^2 - \omega_c^2 + Nq^2/m\varepsilon_0 = 0$$

or

$$\omega_c^2 = \omega_0^2 + \omega_p^2, \qquad (6.5.22a)$$

where

$$\omega_p = (Nq^2/\mu\varepsilon_0)^{1/2} \qquad (6.5.22b)$$

is called the PLASMA FREQUENCY. Inside the region $\omega_0 < \omega < \omega_c$ the value of ε is negative. Therefore, $\varepsilon^{1/2}$ and k [recall Eq. (6.5.19)] are pure imaginary. This means that the wave will not propagate along z since it is damped according to $e^{-|k|z}$. For a free plasma where there is no mechanical restoring force; i.e., ω_0 vanishes. However, there is still a Coulomb electrostatic restoring force which gives the plasma a natural frequency of ω_p, and it will therefore respond 180° out of phase to a field oscillating at $\omega < \omega_p$ and tend to cancel the field.

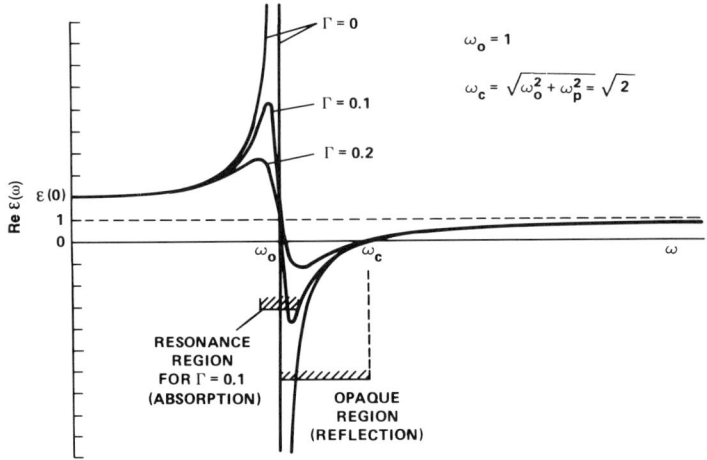

Figure 6.5.4 Real part of dielectric function for various values of damping Γ.

For charges bound to atoms, the wave can propagate unattenuated (in the limit $\Gamma \to 0$) with phase velocity c/n on either side of the cutoff region between ω_0 and ω_c. The phase velocity is less than the vacuum speed of light ($c = 3 \times 10^8$ m/s) on one side ($\omega < \omega_0$), and greater than c on the other side ($\omega > \omega_c$). The group velocity $d\omega/dk$ is less than c on either side.

If damping Γ is nonzero, then ε and $\varepsilon^{1/2}$ are complex. Whenever the wave propagates, it does so with an attenuation factor $e^{-|\text{Im } k|z}$. However, we have seen that the attenuation or absorption due to the imaginary part of ε is appreciable only for frequency ω in the neighborhood ($|\omega - \omega_0| < \Gamma$) of the resonance. Figure 6.5.4 shows the form of Re $\varepsilon(\omega)$ for different values of the damping Γ. The different curves overlap except near the resonance frequency ω_0.

(a) Polarization Mechanisms The polarizability of materials can be due to three basic mechanisms which change the dipole moment. The three categories are sketched in Figure 6.5.5.

The first mechanism involves the rotation of a molecule having a permanent dipole moment. Detailed discussion of this phenomenon really requires an extension of the quantum-mechanical treatment of Chapter 5, which will be pursued again in Chapters 7 and 8. The resonant frequencies associated with rotation are small, usually about 5–50 cm^{-1}.

The second method for changing a dipole moment involves motion of the ions, i.e., the normal modes of vibration in a molecule or solid as indicated in the figure. There will be a resonant frequency for each normal mode generally between 100 and 1000 cm^{-1}—although strong covalent molecules like NH_3 (recall Section 3.7) will resonate above 3000 cm^{-1}.

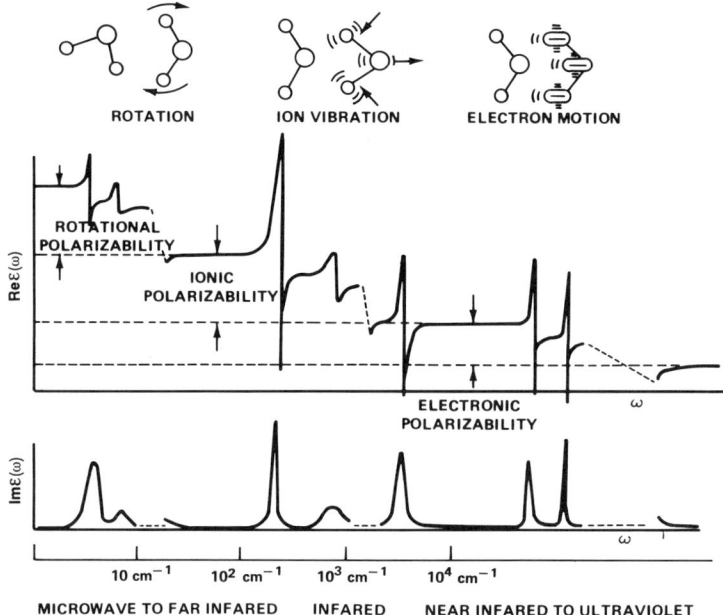

Figure 6.5.5 Classical sketch of three categories of response and dielectric behavior for molecular media.

Finally, the third mechanism of dipole change involves the motion of the electrons with respect to the nuclei. The resonant frequencies for this type of motion will usually be around 10,000 cm^{-1} or higher. Again, the details concerning this type of resonance really require a quantum-mechanical discussion, as seen later.

Nevertheless, in Figure 6.5.5 it is imagined that the entire spectrum of some material could be understood by simply summing dielectric response functions of the form in Eq. (6.5.21) for each "oscillator" or each normal mode, electronic transition, etc., which the material system has. Although the details of the frequency dependence of $\varepsilon(\omega)$, particularly near resonances, require a quantum analysis, it turns out that a classical analysis is qualitatively a good one. In addition, it can give a semiquantitative estimate for polarizability of a given molecule at a frequency that is not near resonance.

For example, if we want to estimate the electronic polarizability of atomic hydrogen at low frequencies ($\omega \to 0$) we may substitute the ultraviolet ionization energy (13.6 eV) in units of frequency ($\omega_0 = 2 \times 10^{16}$ rad/s) into Eq. (6.5.15):

$$\alpha_e \sim \frac{e^2/m\varepsilon_0}{\omega_0^2 - \omega^2} = 7.95 \times 10^{-30} \text{ m}^3.$$

Electronic mass ($m = 9.11 \times 10^{-31}$ kg) charge ($e = 1.60 \times 10^{-19}$ coulomb), and electrostatic constant ($\varepsilon_0 = 8.84 \times 10^{-12}$) are used.

To estimate the zero-frequency polarizability of a system of ions, suppose we take an ionic mass of 30 amu ($m = 5.0 \times 10^{-26}$ kg), the same charge as above, and an infrared resonance frequency of $\omega_0 = 9.4 \times 10^{13}$ rad/s (500 cm^{-1}) with ($\omega \to 0$):

$$\alpha_v \sim \frac{q^2/m\varepsilon_0}{\omega_0^2 - \omega^2} = 6.6 \times 10^{-30} \text{ m}^3. \tag{6.5.23}$$

On the other hand, at optical frequencies ($\omega = 3.7 \times 10^{15}$ rad/s $\sim$ 5000 Å), the polarizability of the same ionic system is negligible by comparison; i.e., $\alpha_v \sim -4 \times 10^{-33}$ m^3, while the electronic polarizability will be growing, particularly if ω is approaching an electronic resonance at ω_0.

Note that a good unit for polarizability is the cubic angstrom $(10^{-10} \text{ m})^3$. Indeed, the polarizability of something is proportional to its volume, as a general rule.

(b) Effect of Matter on Matter For dense materials such as solids, liquids, and even high-pressure gases, the electric field $\mathbf{E}_{local}$ affecting a given molecule or atom may differ appreciably from the "applied" $\mathbf{E}$ field. In texts on electromagnetism one finds the following expression for $\mathbf{E}_{local}$ in isotropic materials and cubic solids:

$$\mathbf{E}_{local} = \mathbf{E} + \frac{1}{3\varepsilon_0}\mathbf{P}.$$

If the polarizable atom or molecule responds to $\mathbf{E}_{local}$, we have

$$\mathbf{P} = \chi\varepsilon_0 \mathbf{E}_{local} = \chi\varepsilon_0\left(\mathbf{E} + \frac{1}{3\varepsilon_0}\mathbf{P}\right)$$

or, solving for $\mathbf{P}$,

$$\mathbf{P} = \frac{\chi\varepsilon_0}{1 - \chi/3}\mathbf{E},$$

where χ is the same susceptibility as before [Eq. (6.5.9)]. This changes the dielectric constant ε defined by

$$\mathbf{D} = \varepsilon\mathbf{E} = \mathbf{E} + \mathbf{P}/\varepsilon_0$$

to the following:

$$\varepsilon = (1 + \chi/(1 - \chi/3)). \tag{6.5.24}$$

Note that if $\chi \ll 1$, we may ignore these corrections.

(c) Visualizing Radiation Coupling Using Feynman's Lever

The detailed solutions of Newton's and Maxwell's equations for coupled particles and em fields are complicated. However, for small numbers of particles there is a graphical construction given in the Feynman Lectures (Section II-21) which is very instructive. It provides a way to tell exactly what the fields will be around an arbitrarily moving charge.

Imagine that you are holding a charge and moving it back and forth. Let the charge be attached to a ring which can slide on a long lever arm as shown in Figure 6.5.6(a). Let the lever have a unit vector $(-\hat{\mathbf{e}}_r)$ or pointer pointing in the opposite direction of the lever $\mathbf{r}$ on the other side of its swivel point (0) at origin. Feynman has shown that the $\mathbf{E}$ field at origin at time t depends on the position of the pointer $\hat{\mathbf{e}}'_r$ and lever $\mathbf{r}'$ at a slightly earlier time ($t' = t - r/c$). The time delay is just the time it would take a signal traveling at c to propagate from r at t' to origin at t. The $\mathbf{E}$ field is given by

$$\mathbf{E}(0,t) = \frac{q}{4\pi\varepsilon_0} \left\{ \frac{-\hat{\mathbf{e}}'_r}{(r')^2} + \frac{r'}{c}\frac{d}{dt}\left[\frac{-\hat{\mathbf{e}}'_r}{(r')^2}\right] + \frac{1}{c^2}\frac{d^2}{dt^2}[-\mathbf{e}'_r] \right\}$$

$$= \underset{\text{term}}{\text{Coulomb}} + \underset{\text{term}}{\text{induction}} + \underset{\text{term}}{\text{radiation}} . \quad (6.5.25a)$$

The first term is just the usual Coulomb field. The second term gives rise to a magnetic induction field,

$$\mathbf{B}(0,t) = (\hat{\mathbf{e}}'_r \times \mathbf{E})/c, \quad (6.5.25b)$$

at origin if the charge has velocity transverse to $\mathbf{r}$. Finally, the third radiation term contributes to $\mathbf{E}(0,t)$ and $\mathbf{B}(0,t)$ in (6.5.25) if the charge has acceleration transverse to $\mathbf{r}$. It is interesting to note that in some ways this term is the reverse of Newton's law. For Newton's law one is given a field $\mathbf{E}$ or

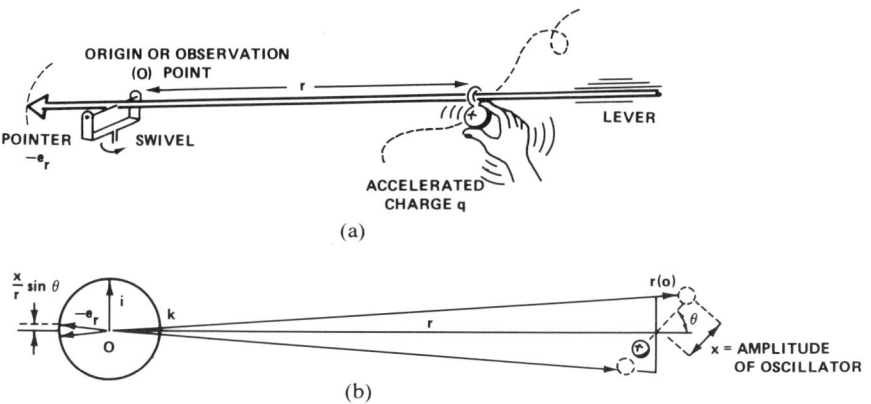

Figure 6.5.6 Feynman's lever. This construction provides a convenient way to visualize the field due to an accelerated or moving charge.

force $\mathbf{F} = q\mathbf{E}$ and computes the double integral $\mathbf{r} = \iint (dt^2(q/\mu)\mathbf{E}$ of $\mathbf{E}$ to obtain the particle's position $\mathbf{r}$ in space. For the third term of (6.5.15a) one is given the position $\mathbf{r}$ and computes $\mathbf{E}$ from the double *derivative* of $\hat{\mathbf{e}}'_r$. Note that the radiation term is the only significant term at large distances since it has no $(1/r')^2$ factor to kill it off.

Feynman's lever can tell you a number of things immediately. For example, one can easily see that a charge rotating around in a circle gives rise to circularly polarized radiation for observers along the axis of the circle. One can also see that a uniform *ring* of charge gives *no* radiation no matter how fast it rotates since all the charges in the ring yield canceling levers. We use the lever now to derive the most common radiation problem: a single oscillating charge giving electric dipole radiation.

As seen in Figure 6.5.6(b) a charge oscillating at an angle θ to the radius vector causes the pointer lever to oscillate with an amplitude $(x/r)\sin\theta$. The unit vector is given by

$$-\hat{\mathbf{e}}_r(t) = -\hat{\mathbf{k}} - \hat{\mathbf{i}}\frac{x}{r}\sin\theta e^{-i\omega t}$$

for $r \gg x$. The second derivative required for (6.5.25a) is

$$\frac{d^2}{dt^2}(-\hat{\mathbf{e}}_r(t)) = \hat{\mathbf{i}}\omega^2\frac{x}{r}\sin\theta e^{-i\omega t},$$

and the resulting E field is

$$\mathbf{E}(t) = \hat{\mathbf{i}}\frac{q\omega^2 x \sin\theta}{4\pi\varepsilon_0 c^2}\left(\frac{e^{i(kr-\omega t)}}{r}\right), \qquad (6.5.26)$$

where the retardation $(\omega r/c = kr)$ has been put in the exponent.

From this one can find the total energy lost by a oscillating charge to Poynting flux,

$$\mathbf{S} = \mathbf{E} \times \mathbf{B}/\mu_0 = \hat{\mathbf{e}}_r |E|^2/\mu_0 c = \hat{\mathbf{e}}_r \frac{q^2\omega^4 x^2 \sin^2\theta \cos^2(kr-\omega t)}{16\pi^2 c^3 \varepsilon_0 \; r^2}, \qquad (6.5.27)$$

and derive the natural decay constant Γ_0 of a free oscillator. The surface integral,

$$P(t) = \iint d\mathbf{A} \cdot \mathbf{S} = \int_0^{2\pi} d\phi \int_0^{\pi} d\theta \, r^2 \sin\theta |S|$$

$$= \frac{q^2\omega^4 x^2}{16\pi^2 c^3 \varepsilon_0}\cos^2(kr-\omega t) 2\pi \int_0^{\pi} d\theta \sin^3\theta$$

$$= \frac{q^2\omega^4 x^2}{6\pi c^3 \varepsilon_0}\cos^2(kr-\omega t) \qquad (6.5.28)$$

is the total power being pumped into radiation at each instant. The time average of the squared cosine factor is $\frac{1}{2}$, and this gives the average power loss

$$\langle P \rangle = \frac{q^2\omega^4 x^2}{12\pi c^3 \varepsilon_0}. \tag{6.5.29}$$

We notice that $\langle P \rangle$ is proportional to the fourth power of frequency and to the square of the amplitude x. Since the energy of a damped harmonic oscillator is $(m\omega^2|x|^2/2)$ its power loss by (6.5.4) is also proportional to x^2; i.e.,

$$\langle P \rangle = -\frac{d}{dt}\langle m\omega^2 a^2 e^{-2\Gamma t}/2\rangle = 2\Gamma\langle m\omega^2 x^2/2\rangle.$$

Comparing this to (6.5.29) one derives

$$\Gamma = \frac{q^2\omega^2}{12\pi c^3 \varepsilon_0 m} \tag{6.5.30}$$

for the decay constant just due to radiation.

For an electron in an atom that oscillates at the 600 THz frequency of green light for which $\omega = 3.77 \times 10^{15}$ rad · Hz one has $\Gamma = 4.4 \times 10^7$ s^{-1}. The atom's energy will decay by 95% in time $t = 3/(2\Gamma) = 3.4 \times 10^{-8}$ s. This is a typical atomic lifetime. It sounds short but that is enough time for about 20 million oscillations! For a 6 THz infrared transition ($\omega = 3.77 \times 10^{13}$ rad · Hz) the lifetime is 3.4×10^{-4} s; still pretty short but enough for two billion oscillations. In terms of heartbeats that is about as long as the average human life.

Because more rapidly oscillating charges radiate more quickly, it follows that higher-frequency stimulating radiation will be scattered more quickly by oscillators they stimulate. This is why higher-frequency light is scattered more in air. The scattering process which makes our sky blue is called Rayleigh scattering, and we consider it in more detail now. The Rayleigh scenario goes as follows. First an atomic oscillator responds according to (6.5.5a) and (6.5.5c) to incoming E field $E_0 e^{ikz-\omega t}$ with an amplitude

$$x(z,t) = \frac{qE_0/\mu}{\omega_0^2 - \omega^2 - 2i\Gamma\omega} e^{i(kz-\omega t+\rho)} \tag{6.5.31a}$$

$$\simeq (qE_0/\mu\omega_0^2)e^{i(kz-\omega t)} \quad \text{(for } \omega \ll \omega_0\text{).} \tag{6.5.31b}$$

The DC approximation (6.5.12) is given in the second line. Then the oscillator radiates or Rayleigh-scatters this energy at the rate (6.5.29)

$$\langle P \rangle = \frac{q^2\omega^4|x|^2}{12\pi c^3\varepsilon_0} = \frac{q^2\omega^4}{12\pi c^3\varepsilon_0}\left|\frac{qE_0/\mu}{\omega_0^2 - \omega^2 - 2i\Gamma\omega}\right|^2. \tag{6.5.32}$$

Scattering theorists like to factor this result in a funny way so that three types of contributions are exposed:

$$\langle P \rangle = \left[\frac{1}{2}\varepsilon_0 cE_0^2\right] \cdot \left[\pi\left(\frac{q^2}{4\pi\varepsilon_0 mc^2}\right)^2\right]\left[\frac{8\omega^4}{3|\omega_0^2 - \omega^2 - 2i\Gamma\omega|^2}\right]$$

$$= [S] \cdot [\pi r_{\text{classical}}^2]\left[\frac{8\omega^4}{3(\omega_0^2 - \omega^2)^2 + 12\Gamma^2\omega^2}\right]. \quad (6.5.33)$$

The first factor is Poynting energy flux per unit area. The second factor is the so-called classical electronic cross-section $\pi r_{\text{classical}}^2 = 2.5 \times 10^{-29}$ m^2, which involves the classical electron radius $q^2/(4\pi\varepsilon_0 mc^2)$. (This is the radius at which dipolar electronic coulomb energy equals the rest mass mc^2 of the electron.) The third factor is a dimensionless frequency enhancement factor for the Rayleigh cross-section, which is approximated by $8\omega^4/3\omega_0^4$ when $\omega \ll \omega_0$. It indicates how an electron can appear to "swell up" and scatter more radiation when its oscillation amplitude is amplified.

(d) Four Points of the Response Phasor Using Feynman's methods one can visualize some of the *mechanisms* behind the Lorentz susceptibility formula (6.5.15) and Figure 6.5.4. It is possible to see (1) how light gets slowed down when its frequency ω is below resonance ($\omega < \omega_0$), (2) how it gets absorbed near resonance ($\omega \simeq \omega_0$), and (3) how it speeds up above resonance ($\omega > \omega_0$). It is also possible to understand a fourth case involving *l*ight *a*mplification by *s*timulated *e*mission or LASER processes. These four cases correspond to four values 0°, 90°, 180°, and 270°, respectively, for the response phasor lag ρ. These four points of the phasor are as important to modern spectroscopy as the four points of the compass are to navigation.

Figure 6.5.7 shows an incremental slab containing N dipoles per cubic meter responding according to (6.5.31a) to the incoming radiation. We need to find the electric field $(z_0 - z)$ meters downstream at an observation point

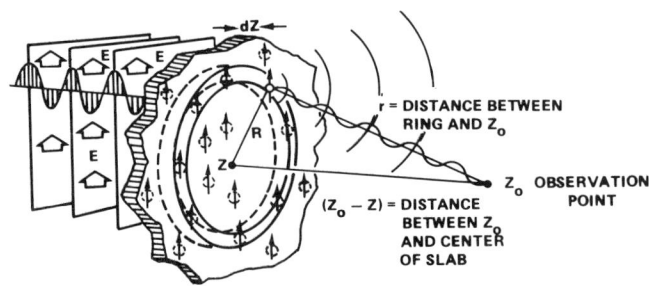

Figure 6.5.7 Incremental slab responding to plane-wave radiation.

CLASSICAL APPROACH TO OPTICAL RESONANCE AND SELECTION RULES   521

z_0. This would be the sum of the original stimulating field $E_0 e^{i[k(z-z_0)-\omega t]}$ plus all the extra little fields broadcast by the dipoles in the slab. Let us label the extra response field by E^R. The contribution dE^R to E^R due just to the dipoles in the ring of volume $2\pi R\, dR\, dz$ in Figure 6.5.7 is

$$dE^R(z_0) = \frac{q\omega^2 x(z, t - r/c)}{4\pi\varepsilon_0 c^2 r} N 2\pi R\, dR\, dz$$

$$= \frac{q\omega^2 |x| e^{i(kz-\omega t)+\rho+\omega r/c)}}{4\pi\varepsilon_0 c^2 r} N 2\pi R\, dR\, dz,$$

according to (6.5.26). Here we take $x \sin\theta \simeq x$ and include the retardation $\omega r/c$ in the exponent.

The figure shows that $r^2 = R^2 + (z_0 - z)^2$. If z is fixed at slab center then $2r\, dr = 2R\, dR$. This simplifies the integral over the ring radius R to give the response generated at z_0.

$$E^R(z_0, t) = \frac{2\pi N\, dz\, q\omega^2 |x|}{4\pi\varepsilon_0 c^2} e^{i(kz-\omega t+\rho)} \int_{R=0}^{R=\infty} \frac{R\, dR\, e^{i\omega r/c}}{r}$$

$$= \frac{N\, dz\, q\omega^2 |x|}{2\varepsilon_0 c^2} e^{i(kz-\omega t+\rho)} \int_{r=z_0-z}^{\infty} dr\, e^{ikr}$$

$$= \frac{N\, dz\, q\omega^2 |x|}{2\varepsilon_0 c^2} e^{i(kz-\omega t+\rho)} \frac{e^{ikr}}{ik}\bigg|_{z_0-z}^{\infty}. \qquad (6.5.34)$$

Now we pull a well-known swindle and set $e^{ik\infty} = 0$. This can be justified by more careful analysis which shows how the distant oscillators cancel. The result is

$$E^R(z_0, t) = |E^R| e^{i(kz-\omega t+\rho)} e^{\pi i/2} e^{ik(z_0-z)}$$
$$= |E^R| e^{i(kz_0-\omega t+\rho+\pi/2)}, \qquad (6.5.35a)$$

where

$$|E^R| = \frac{N\, dz\, q\omega^2 |x|}{2\varepsilon_0 c^2 k}. \qquad (6.5.35b)$$

The main result in (6.5.35a) is the extra phase lag of $\pi/2$ beyond that of the response lag ρ. (The response phase lag is given by (6.5.5f).)

This result is used to draw the four cases in Figure 6.5.8. Each case shows phasors for the stimulating E wave, the polarization wave of the responding slab oscillators with phase lag ρ, the resulting downstream response generated wave E^R with phase lag $\rho + \pi/2$, and, finally, the sum wave $E + E^R$.

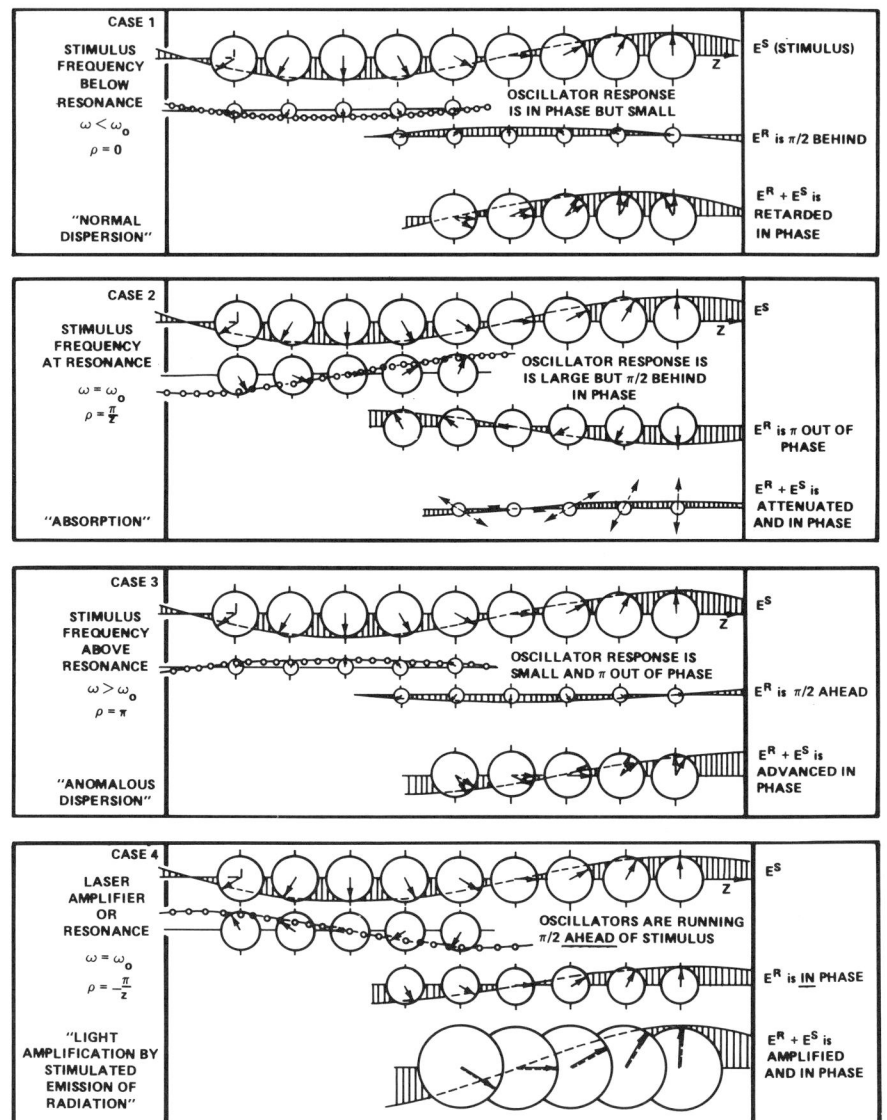

Figure 6.5.8 The Four points for the response compass.

In normal dispersion (case 1) the oscillators are responding weakly but in phase ($\rho \simeq 0$). They broadcast a weak E^R signal which adds a small component at 90° to the original E-field phasors and so the sum is a slightly retarded wave of the same amplitude. This retardation accumulates arithmetically or linearly from slab to slab and so it amounts to a reduction in phase velocity.

In case 2 the oscillators are resonating strongly and 90° behind the stimulus ($\rho \simeq 90°$). This means that the E^R tends to attenuate the E field since the extra 90° lag swings it around to be 180° out of phase. This attenuation accumulates geometrically or exponentially from slab to slab since the response in each slab is proportional to the total field coming from the preceding one. This exponential decay of amplitude is sometimes stated as Beer's law. Note that the original E field is not simply eaten up. It keeps going forever, but it stimulates traveling companions which are out of phase with it. Destruction is accomplished through creation.

Just above resonance ($\omega > \omega_0$) corresponds to anomalous dispersion as shown in case 3 of Figure 6.5.8. Here the phase velocity exceeds that of light in a vacuum. It should be noted that the plasma effects mentioned after Eq. (6.5.22) are not accounted for by the thin slab model which ignores the surface charges produced by dipoles on the edges. Such effects are important in condensed matter but usually ignorable in the gas phase.

In case 4 of Figure 6.5.8 the oscillating dipoles are feeding energy into the E field since they have negative phase lag $\rho - 90° = 270°$; i.e., they *lead* the stimulating field by 90°. Consequently, the E field grows exponentially as in a laser amplifier. It is interesting to note that classical Lorentz theory can account for many laser effects. It is only necessary to replace the oscillator number N by the difference $N = N_l - N_u$ between the number N_l of atoms in the lower laser level (l) and the number N_u of atoms in the upper level (u). When N_u exceeds N_l this is called a population inversion and N in (6.5.15) becomes negative as does the phase lag ρ. The resonance frequency ω_0 is the quantum energy difference $(E_u - E_l)/h$ in all these classical analyses. In Chapter 8 we will see how the Lorentz model and phasor picture is generalized when we introduce the quantum phasor.

C. Two Types of Spectroscopy for Vibration Analysis

(a) Infrared Absorption Spectroscopy An experimental apparatus which gives infrared spectra is sketched in Figure 6.5.9. The output of the

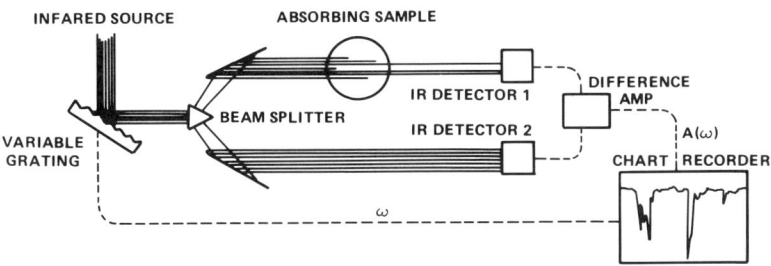

Figure 6.5.9 Infrared spectrometer.

machine is the difference in intensity between the absorbed beam and a reference beam as a function of the input frequency ω.

As explained in the preceding sections, light is absorbed when its frequency is near that of a resonance of some motion that gives an oscillating dipole moment. For a given frequency, the absorption coefficient Im k increases with the magnitude of the induced dipole moment $\mathbf{p} = \alpha \varepsilon_0 \mathbf{E}$, but vanishes if $\mathbf{p}$ is zero. For infrared frequencies, it is important to see what dipole moments can arise from the motion of ions for a few simple examples and to derive some general rules about them.

Suppose the atoms of UF_6 (recall Section 4.4) are rigid ions, and the F ions have charge q_F while the U has charge q_U. Then the dipole moment vector is the vector sum of the charge positions $\mathbf{r}_i$ with respect to the center of charge at that position:

$$\mathbf{p} = \sum_i q_i \mathbf{r}_i. \tag{6.5.36}$$

Now suppose the UF_6 octahedron is distorted according to the base state $|c_3^{T_{1u}}\rangle$ as pictured in Figure 6.5.10. (Recall Figure 4.4.36, which shows $|c_1^{T_{1u}}\rangle$ with amplitude A.) Then the dipole components are, using Eq. (6.5.36),

$$p_x = 0, \qquad p_y = 0, \qquad p_z = 6q_F(MA) - q_U(6mA).$$

Note that p_z is proportional to the amplitude A of this distortion, and that all components of the dipole vector are zero when UF_6 is the octahedral equilibrium configuration. Furthermore, similar calculations show that the dipole moments due to $|A_{1g}\rangle$, $\left|\begin{array}{c} E_g \\ m \end{array}\right\rangle$, $\left|\begin{array}{c} T_{2g} \\ m \end{array}\right\rangle$, and $\left|\begin{array}{c} T_{2u} \\ m \end{array}\right\rangle$ are all identically zero.

However, it is usually easier to tell by symmetry analysis which motions are capable of giving a particular dipole moment component. This is done by

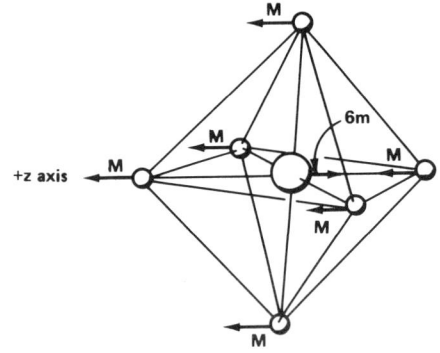

Figure 6.5.10 First-order infrared activity. This is an example of a vibrational motion giving rise to a dipole moment. The moment varies linearly with the normal coordinate, but also depends upon the charges on the central ion and the six octahedrally located ions.

studying the expansion of symmetry-defined dipole components p_m^γ in terms of the normal coordinates x_i^α.

$$p_m^\gamma(\cdots x^\alpha \cdots) = [p_m^\gamma(\cdots 0 \cdots)] + \sum_{\alpha, i} x_i^\alpha \left[\frac{\partial p_m^\gamma}{\partial x_i^\alpha}(\cdots 0 \cdots)\right]$$
$$+ \frac{1}{2} \sum_{\alpha, i} \sum_{\beta, j} x_i^\alpha x_j^\beta \left[\frac{\partial^2 p_m^\gamma}{\partial x_i^\alpha \partial x_j^\beta}(\cdots 0 \cdots)\right] \cdots. \quad (6.5.37)$$

Symmetry definition of dipole components is the same as that of ordinary vector components according to the definition given in Eq. (6.5.36). For O_h we have $p_z \equiv p_3^{T_{1u}}$, $p_y \equiv p_2^{T_{1u}}$, and $p_x \equiv p_1^{T_{1u}}$. For C_{3v} we have $p_z = p^{(1)}$, $p_x = p_1^{(3)}$, and $p_y = p_2^{(3)}$.

All the constant factors in the brackets in Eq. (6.5.37) are derived from the equilibrium configuration ($x_i^\alpha \equiv 0$) and therefore are invariant to symmetry operations. One needs to find which of these constants or which combinations of these constants are independent or nonzero. It is precisely the same problem we solved when writing tensor relations in Section 6.3, only now we have one such problem for each order $N = 0, 1, 2, \ldots$ in the expansion.

The zeroth-order problem is easy. Clearly, $[p_m^\gamma(\cdots 0 \cdots)]$ must vanish unless there is a dipole component p^γ that is a scalar (p^{A_1}). The O_h vector irrep T_{1u} accounts for all three dipole components; none are scalar. Therefore, nothing with O_h symmetry has a permanent electric moment. On the other hand, for C_{3v} we have $p_z = p^{(1)} = p^{A_1}$; that is, the z component of the dipole belongs to the scalar irrep. Therefore, NH_3 could have a dipole moment along its z axis, and this is observed.

The first-order problem is also easy. In order to be invariant, the quantity $[\{\partial p_m^\gamma / \partial x_i^\alpha\}(\cdots 0 \cdots)]$ must have $\alpha = \gamma$. This tells us whether it is possible for a mode x^α to have an associated dipole moment. For UF_6, which has O_h symmetry, this is possible only for $\alpha = T_{1u}$. This is called an INFRARED SELECTION RULE. Only the T_{1u} modes of UF_6 can be seen when the spectrograph is tuned to their eigenfrequencies. All other modes are "first-order infrared forbidden." For C_{3v} symmetry (viz., NH_3) and (1) and (3) type modes, which account for all NH_3 modes (recall Figures 3.7.2 and 3.7.3) are infrared active. Only (2) type modes would be forbidden.

The second- and higher-order terms are a little more difficult to understand. These will be needed to account for the variation of ionic charge with interatomic separation. Generally, the dipole moment will not increase linearly with x_i^α forever. The only higher order terms that can be nonzero are those which can make the vector irrep (γ) through a Clebsch-Gordon coupling. The quadratic terms must be of the form

$$p_m^\gamma = \cdots + k^\gamma \left(\sum_{i,j} \mathscr{C}_{ijm}^{\alpha\beta\gamma} x_i^\alpha x_j^\beta\right) + \cdots, \quad (6.5.38)$$

526 THEORY AND APPLICATIONS OF SYMMETRY REPRESENTATION PRODUCTS

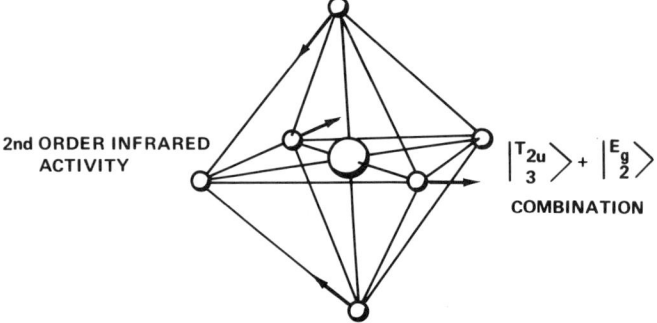

Figure 6.5.11 Second-order infrared activity. This is an example of combination of normal coordinates giving rise to dipole moment. It depends on the product of the normal coordinates. It is also necessary that the charge difference between an octahedral ion and the central ion should depend on the distance between them.

where

$$k^\gamma = \frac{1}{2}\sum_{i,j} \mathscr{C}_{ijm}^{\alpha\beta\gamma}\left[\frac{\partial^2 p_m^\gamma}{\partial x_i^\alpha \partial x_j^\beta}(\cdots 0 \cdots)\right]$$

is independent of m.

In this way a particular combination of normal coordinates may give rise to a dipole moment even though neither one of them could do it separately. Consider, for example, the infrared inactive modes $x_3^{T_{2u}}$ and $x_2^{E_g}$ of UF_6. Since $\mathscr{C}_{323}^{T_{2u}E_g T_{1u}} = -1$ it may be possible for the combination of these two to make a $p_z(p_3^{T_{1u}})$ dipole component. Indeed, from Figure 6.5.11 we may visualize this possibility. It should be clear that this second-order dipole moment depends on how much the charge on the F ions varies with their distance from the U ion.

The frequencies associated with second- or higher-order products $x_i^\alpha x_j^\beta$ $\cdots$ may be the sums or difference $|\pm\omega^\alpha \pm \omega^\beta \pm \cdots|$ of the respective eigenfrequencies and are called OVERTONES or COMBINATION TONES. Understanding their detailed behavior requires quantum mechanics.

(b) Rayleigh and Raman Scattering We mentioned that the polarizability of electrons by optical frequency radiation is much greater than that of the vibrational modes. This is particularly so for larger molecules. Now we will give a more detailed discussion of the polarizability of the electron clouds around various arrangements of ions or atoms.

For objects more complicated than the simple charge-ball model it is necessary to discuss other components of polarizability. The polarizability of nonspherical charge clouds may vary with the direction of **E**, and the induced dipole moment **p** may not point in the same direction as **E**.

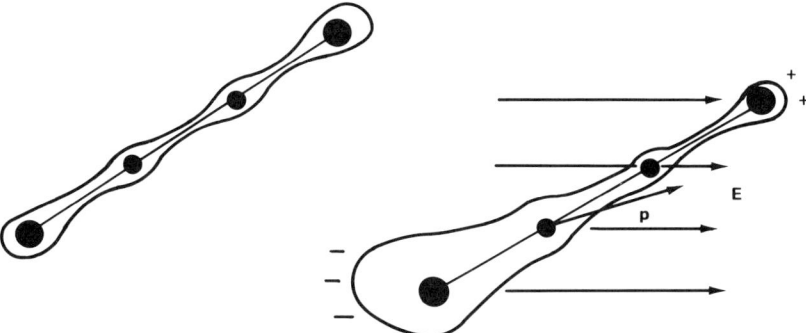

Figure 6.5.12 Electronic polarizability may be anisotropic.

For example, consider a long thin molecule with some electrons that can move more easily along its z axis than in the x or y directions. An electric field **E** applied at a random angle to the axis will produce a dipole vector nearly parallel to the axis as shown in Figure 6.5.12.

In order to describe the general linear polarization, the simple relation $p = \alpha \varepsilon_0$ is replaced by the following tensor relation:

$$\mathbf{p} = \varepsilon_0 \overset{\leftrightarrow}{\boldsymbol{\alpha}} \cdot \mathbf{E},$$

$$\begin{pmatrix} p_x \\ p_y \\ p_z \end{pmatrix} = \varepsilon_0 \begin{pmatrix} \alpha_{xx} & \alpha_{xy} & \alpha_{xz} \\ \alpha_{yx} & \alpha_{yy} & \alpha_{yz} \\ \alpha_{zx} & \alpha_{zy} & \alpha_{zz} \end{pmatrix} \begin{pmatrix} E_x \\ E_y \\ E_z \end{pmatrix}. \qquad (6.5.39)$$

The radiation arising from the oscillating dipole **p** induced by an oscillating **E** through a changing $\overset{\leftrightarrow}{\boldsymbol{\alpha}}$ is the RAYLEIGH AND RAMAN-SCATTERED LIGHT. A basic experimental setup which may detect this is sketched in Figure 6.5.13.

The vector of the **E** field coming to the collecting optics from a single dipole $\mathbf{p}(t)$ can be given by rewriting Feynman's lever formulas (6.5.25) and (6.5.26) as follows:

$$\mathbf{E} = \frac{1}{4\pi\varepsilon_0 c^2} \frac{(\ddot{\mathbf{p}}(t - r/c) \times \hat{\mathbf{r}}) \times \hat{\mathbf{r}}}{r} e^{ikr}, \qquad (6.5.40)$$

or if we assume $\mathbf{p}(t - rc) = \mathbf{p}(+)e^{-i\omega t}$, by

$$\mathbf{E} = -\frac{\omega^2}{4\pi\varepsilon_0 c^2} \frac{(\mathbf{p}(+) \times \hat{\mathbf{r}}) \times \hat{\mathbf{r}}}{r} e^{ikr}. \qquad (6.5.41)$$

The energy flux S was given in (6.5.27).

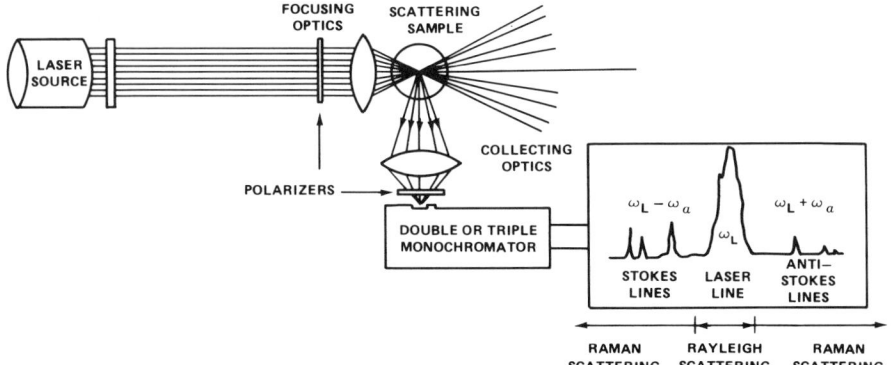

Figure 6.5.13 Raman spectrometer.

The frequency spectrum, polarization, and intensity of scattered light depends upon the incoming **E** field and the polarizability tensor $\overleftrightarrow{\alpha}$ according to the foregoing and Eq. (6.5.39). The most important effects come from the changing of the $\overleftrightarrow{\alpha}$ components as the scattering molecules expand and contract while vibrating. One may express this dependence on atomic motion by the following expansion of $\overleftrightarrow{\alpha}$ in terms of normal coordinates:

$$\alpha_{uv}(\cdots x_i^\alpha \cdots) = [\alpha_{uv}(\cdots 0 \cdots)] + \sum_{\alpha,i} x_i^\alpha \left[\frac{\partial \alpha_{uv}}{\partial x_i^\alpha}(\cdots 0 \cdots) \right]$$
$$+ \frac{1}{2} \sum_{\alpha,i} \sum_{\beta,j} x_i^\alpha x_j^\beta \left[\frac{\partial^2 \alpha_{uv}}{\partial x_i^\alpha \partial x_j^\beta}(\cdots 0 \cdots) \right]. \quad (6.5.42)$$

For symmetry analysis it is more convenient to use the symmetry-defined tensor components α_m^γ instead of the Cartesian components α_{uv},

$$\overleftrightarrow{\alpha} = \sum_{u,v} \alpha_{uv} \hat{x}_u \hat{x}_v = \sum_{\gamma,m} \alpha_m^\gamma \mathbf{T}_m^\gamma, \quad (6.5.43)$$

just as was done for the stress or strain tensors. The expansion of each of these,

$$\alpha_m^\gamma(\cdots x_i^\alpha \cdots) = [\alpha_m^\gamma(\cdots 0 \cdots)] + \sum_{\alpha,i} x_i^\alpha \left[\frac{\partial \alpha_m^\gamma}{\partial x_i^\alpha}(\cdots 0 \cdots) \right]$$
$$+ \frac{1}{2} \sum_{\alpha,i} \sum_{\beta,j} x_i^\alpha x_j^\beta \left[\frac{\partial^2 \alpha_m^\gamma}{\partial x_i^\alpha \partial x_j^\beta}(\cdots 0 \cdots) \right], \quad (6.5.44)$$

can be treated term by term using the same ideas that worked for the dipole expansion in the preceding section.

The first term, like all the coefficients in brackets evaluated at $x = 0$, is invariant to symmetry operations. It can be nonzero only for the components α_m^γ for which $(\gamma) =$ (scalar). This is called the RAYLEIGH-SCATTERING term. According to Eqs. (6.5.39) and (6.5.41), it gives a dipole moment and outgoing light with the same frequency as the incoming (laser) source. For O_h symmetry we have that $[\alpha^{A_{1g}}(\cdots 0 \cdots)]$ is the only nonzero component, and so the Rayleigh polarizability tensor is a unit tensor $\vec{1}$. In this case the Rayleigh polarization must be the same as that of the incoming light.

The second terms give what is called the FIRST-ORDER or FUNDAMENTAL RAMAN SCATTERING. Each oscillating normal coordinate

$$x_i^\alpha = x_i^\alpha(+)e^{i\omega_\alpha t} + x_i^\alpha(-)e^{-i\omega_\alpha t} \tag{6.5.45}$$

gives the following contribution to the oscillating dipole:

$$\mathbf{p} = \sum_{\gamma, m} \alpha_m^\gamma \mathbf{T}_m^\gamma \cdot \mathbf{E} e^{i\omega t}$$

$$= \sum_{\gamma, m} \cdots + \left[\frac{\partial \alpha_m^\gamma}{\partial x_i^\alpha}(\cdots 0 \cdots)\right]$$

$$\times \left(x_i^\alpha(+)e^{i(\omega+\omega_\alpha)t} + x_i^\alpha(-)e^{i(\omega-\omega_\alpha)t}\right)\mathbf{T}_m^\gamma \cdot \mathbf{E} + \cdots. \tag{6.5.46}$$

This depends on the constants in the brackets being nonzero for some irrep (γ). If for some (γ), $[(\partial \alpha_m^\gamma/\partial x_i^\alpha)(\cdots 0 \cdots)]$ is nonzero, then the spectrograph in Figure 6.5.13 may record a bump at frequency $\omega - \omega^\alpha$ called the α-STOKES LINE, and possibly another bump at $\omega + \omega^\alpha$ called the α-ANTI-STOKES LINE. However, if no tensor component α_m^γ exists such that $\gamma = \alpha$, then the bracket constants must be zero for that mode $\binom{\alpha}{i}$. In this case we say that the spectral lines $\omega \pm \omega^\alpha$ are RAMAN INACTIVE or FORBIDDEN.

For example, there are three triply degenerate frequencies of UF_6 which cannot show up in the ordinary Raman spectrum, namely, $\omega_\alpha^{T_{2u}}$, and the two $\omega^{T_{1u}}$ vibrations. To be Raman active, an O_h vibration mode x_i must belong to one of the $\binom{\gamma}{m}$ irreps that are used to define a second-rank symmetric tensor $\vec{\alpha}$, namely A_{1g}, E_g, or T_{2g}.

Consider now the detailed form of some Raman-active modes. In Figure 6.5.14 we show how the expansion of the polarizability tensor gives various contributions. We can visualize various effects in the drawings below.

The main idea is that if the molecule expands in some direction then the polarizability for that direction increases. The symmetry analysis keeps track of the precise ratios of the polarizability components for each case, and tells which components must vanish. We have shown only one component of each

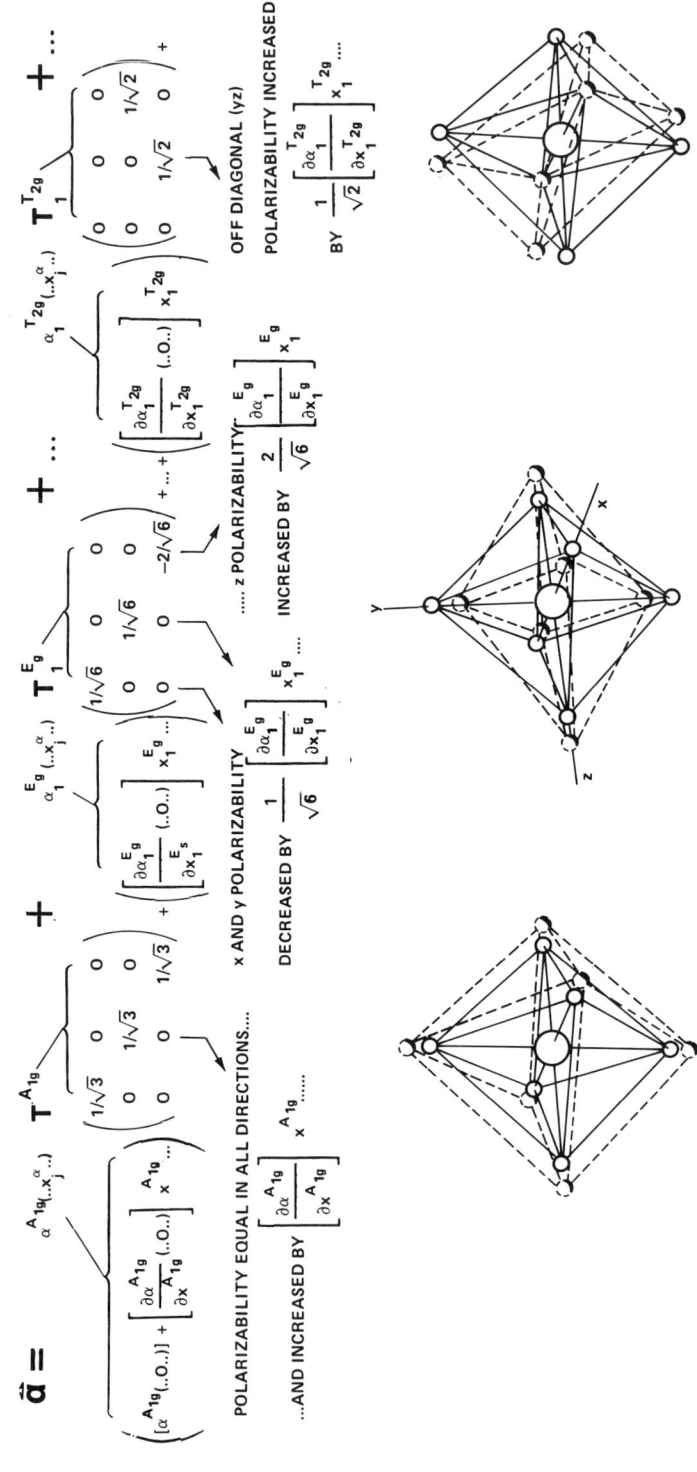

Figure 6.5.14 Linear (first order) effects of modes A_{1g}, $E_g(1)$, and $T_{2g}(1)$ on polarizability.

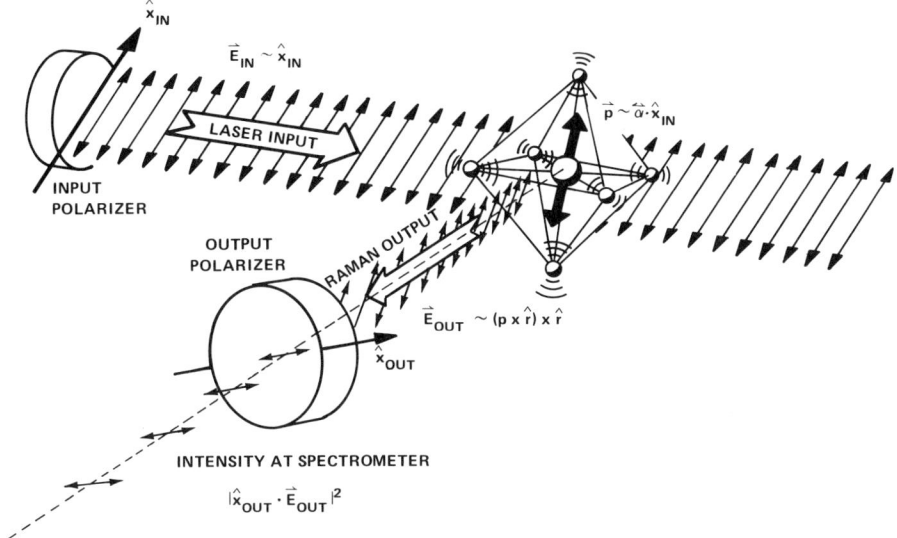

Figure 6.5.15 Geometry of raman polarization experiment.

Raman-active mode in the figure. You might find it instructive to look at some of the others, particularly $\binom{E_g}{2}$. Note that the bracket constants such as $[\partial \alpha_2^{E_g}/\partial x_2^{E_g}(\cdots 0 \cdots)]$ for the other components must equal those of the first component. (Recall the tensor theorem.)

The second-order terms in Eq. (6.5.44) or any higher-order terms may be treated in exactly the same way for the analogous expansion of the dipole moment in the preceding section. The main difference here is that any mode x_i^α can contribute to the polarizability of a second-order term of the form $(\sum_i C_{ii}^{\alpha\alpha^{A_1}} x_i^\alpha x_i^\alpha T^{A_1})$. However, during one vibration period for the rise and fall of x_i^α, this contribution rises twice. Hence, these terms lead to spectral lines of overtone Raman scattering with frequency $\omega_{\text{laser}} \pm 2\omega^\alpha$. In this way inactive modes can be observed, although their contributions are usually quite weak unless they are encouraged by stimulation or resonance techniques.

Each of the different Raman processes have more or less distinctive forms for their contribution to the polarizability tensor, as shown by the examples in Figure 6.5.14. This means it is possible to distinguish one type of mode from another by adjusting polarizations at the laser and at the slit in Figures 6.5.13 and 6.5.15. This is certainly true for a solid for which the crystal axis remains fixed with respect to the experiment. However, it is also true for gas and liquid molecules even though they are rotating.

Figure 6.5.15 shows how the output polarization will depend upon the polarization of the input beam and the polarizability tensor $\vec{\alpha}$. This is derived

532 THEORY AND APPLICATIONS OF SYMMETRY REPRESENTATION PRODUCTS

in the following, where the proportionality constants in Eq. (6.5.41) are deleted:

$$\text{output amplitude} \sim \hat{\mathbf{x}}_{\text{out}} \cdot \mathbf{E}_{\text{out}} \sim \hat{\mathbf{x}}_{\text{out}} \cdot [(\hat{\mathbf{p}} \times \hat{\mathbf{r}}) \times \hat{\mathbf{r}}]$$
$$= (\hat{\mathbf{x}}_{\text{out}} \cdot \hat{\mathbf{r}})(\mathbf{p} \cdot \hat{\mathbf{r}}) - (\hat{\mathbf{x}}_{\text{out}} \cdot \mathbf{p})(\hat{\mathbf{r}} \cdot \hat{\mathbf{r}})$$
$$= -(\hat{\mathbf{x}}_{\text{out}} \cdot \mathbf{p})$$
$$\sim -(\hat{\mathbf{x}}_{\text{out}} \cdot \overleftrightarrow{\alpha} \cdot \hat{\mathbf{x}}_{\text{in}}). \quad (6.5.47)$$

If you can ignore all but one type of symmetry-defined tensor $\mathbf{T}^\gamma$ in the expansion of α, then you will observe the polarization dependence that is peculiar to the γ-type modes. For example, if only the T_{2g} modes are excited, and you only observe the light scattered with the $\omega^{T_{2g}}$ frequency shift, then the output intensity is the square of the following output amplitude:

$$\langle \hat{\mathbf{x}}_{\text{out}} \cdot \overleftrightarrow{\alpha} \cdot \hat{\mathbf{x}}_{\text{in}} \rangle^{T_{2g}} = \overbrace{x_{\text{out}} y_{\text{out}} z_{\text{out}}} \begin{pmatrix} 0 & \alpha & \alpha \\ \alpha & 0 & \alpha \\ \alpha & \alpha & 0 \end{pmatrix} \begin{pmatrix} x_{\text{in}} \\ y_{\text{in}} \\ z_{\text{in}} \end{pmatrix}. \quad (6.5.48)$$

By adjusting the polarization and beam angles between $\hat{\mathbf{x}}_{\text{out}}$ and $\hat{\mathbf{x}}_{\text{in}}$ it is possible to distinguish T_{2g} resonances from others. The A_{1g} resonances, for example, have an isotropic polarization response since their α tensor is proportional to the unit matrix.

6.6 MULTIPLY EXCITED QUANTUM VIBRATION STATES

The quantum description of molecular vibrations was introduced in Section 4.4. That discussion contained a derivation of the eigenstate

$$\frac{\cdots (a_i^{\dagger \alpha})^{n_i^\alpha} \cdots (a_j^{\dagger \beta})^{n_j^\beta} \cdots | \cdots 0 \cdots 0 \cdots \rangle}{\sqrt{\cdots n_i^\alpha! \cdots n_j^\beta! \cdots}} \equiv | \cdots n_i^\alpha \cdots n_j^\beta \cdots \rangle \quad (6.6.1)$$

of the harmonic Hamiltonian: $H = \sum_{i,j} \hbar\omega_\alpha (a_i^{\dagger\alpha} a_i^\alpha + \frac{1}{2}\mathbf{1})$ where the quantum numbers $n_i^\alpha = 1, 2, \ldots$ are the number of excitations of the mode of type $\binom{\alpha}{i}$. [In solid-state physics one speaks of n_i^α as the number of $\binom{\alpha}{i}$ PHONONS.]

An introduction to the excited or multiphonon vibration spectrum was given in Section 4.4.D. The first, second, third, ... excited states of a mode belonging to a one-dimensional irrep (for example, $\alpha = A_{1g}$ in XF_6) are all singly degenerate with energies $E_0 + n^\alpha \hbar \omega_\alpha$. This is just like the ordinary one-dimensional oscillator spectrum. The excited states of a mode belonging to a two-dimensional irrep (for example, $\alpha = E_g$ in XF_6) have degeneracies $(2, 3, 4, \ldots, n+1)$ for levels $E_0 + (\hbar\omega_\alpha, 2\hbar\omega_\alpha, 3\hbar\omega_\alpha, \ldots, n\hbar\omega_\alpha)$. A mode belonging to a triply degenerate irrep (for example, $\alpha = T_{1u}, T_{2u}, \ldots$, etc. in XF_6) has excited-state energies of $E_0 + \hbar\omega_\alpha$, $E_0 + 2\hbar\omega_\alpha$, $E_0 + 3\hbar\omega_\alpha, \ldots,$ $E_0 + n\hbar\omega_\alpha$, with degeneracies $3, 6, 10, \ldots, (n + 1)(n + 2)/2$, respectively. This is the three-dimensional oscillator spectrum.

This can be generalized for any irrep dimension l^α. The resulting degeneracies are the combination coefficients in Pascal's triangle as given by Eq. (4.4.64). Now Clebsch-Gordon coupling analysis can be used to discuss the form of the wave functions and spectral splitting of multiphonon configurations.

Anharmonicity Splitting

The large degeneracies of excited degenerate vibration modes come from a harmonic Hamiltonian. However, for any molecular or solid complex, the harmonic part is only the first of many terms in an approximation that must have higher-order potentials in order to describe what happens when the system is very excited. The $\sum_{\alpha,i}(x_i^\alpha)^2$ terms by themselves are the harmonic oscillator potential, while any higher terms $(x_i^\alpha)^3, (x_i^\alpha)^4, \ldots$ are said to make the oscillator ANHARMONIC.

Now, only those anharmonicities are permitted which have the same (G) symmetry of the physical system. (The harmonic terms obviously have G symmetry.) By systematically constructing all independent scalar polynomials,

$$\sum_{j,k,\ldots} \mathscr{C}_{i_1 i_2 j}^{\alpha_1 \alpha_2 \gamma} \mathscr{C}_{jj_3 k}^{\gamma \alpha_3 \beta} \cdots \mathscr{C}_{l i_n}^{\delta \alpha_n 0} \cdot x_{i_1}^{\alpha_1} x_{i_2}^{\alpha_2} x_{i_3}^{\alpha_3} \cdots x_{i_n}^{\alpha_n}, \quad (6.6.2)$$

using G coupling coefficients, one will find all the independent anharmonic terms of a given order. The final coupling ($\mathscr{C}_{l i_n}^{\delta \alpha_n 0}$) must give a scalar.

One effect of anharmonic terms will be the splitting of the degeneracy of the multiply excited (overtone or combination tone) vibration levels. Consider, for example, the six states of harmonic eigenvalue $E_0 + 2\hbar\omega_{T_{2u}}$,

$$|0 \cdots (2)_{T_{2u}} \cdots 0\rangle = a_i^{\dagger T_{2u}} a_j^{\dagger T_{2u}}|0 \cdots 0 \cdots 0\rangle/\sqrt{2} \quad (6.6.3)$$

in an O_h system such as XF_6. These belong to the $2v_6$ level in Figure 4.4.8. From these we can make states of definite O_h symmetry using coupling coefficients:

$$\left|(2)_{T_{2u}}{}^\gamma_m\right\rangle = \sum_{i,j} \mathscr{C}_{ijm}^{T_{2u} T_{2u} \gamma} a_i^{\dagger T_{2u}} a_j^{\dagger T_{2u}}|0 \cdots 0 \cdots 0\rangle/\sqrt{2}. \quad (6.6.4)$$

To first order, these will be eigenstates in the presence of anharmonic potentials. Note that we obtain just six symmetrized-outer-product bases A_{1g}, E_g (1 and 2), and T_{2g} (1, 2, and 3) because of the commutation symmetry $a_i^\dagger a_j^\dagger = a_j^\dagger a_i^\dagger$. By adjoining one more $a_k^{\dagger T_{2u}}$ factor, we see that 10 third-order states appear. The resulting anharmonic splittings are indicated in Figure 6.6.1(a).

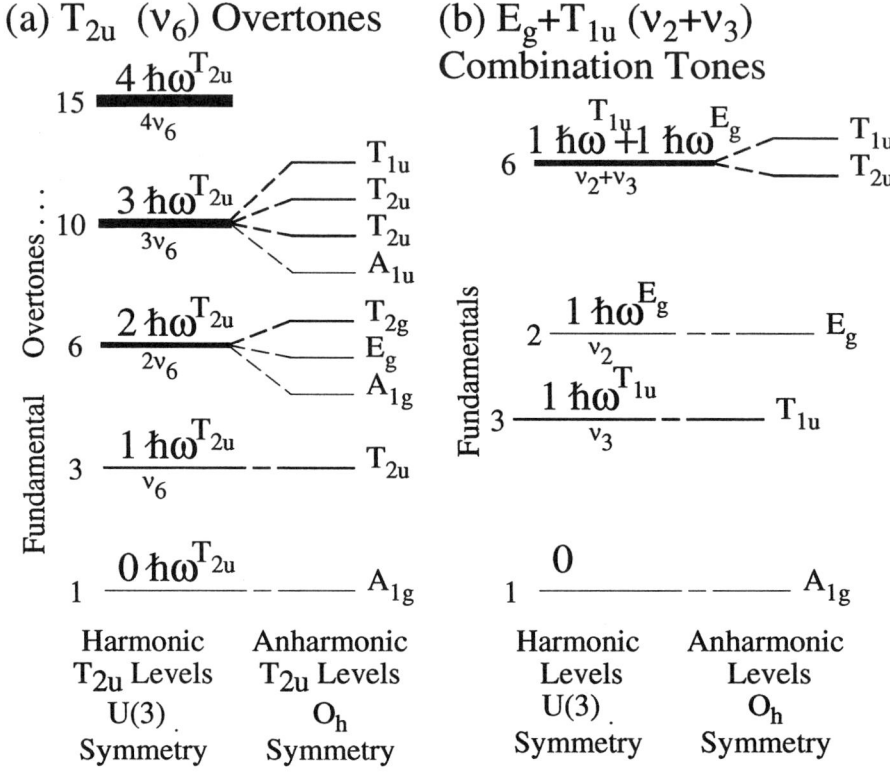

Figure 6.6.1 Examples of multiple vibration levels. Harmonic and anharmonic eigenvalues are indicated. (a) Overtone levels. (b) Combination tone levels.

The combination-tone levels arise from the coupled products in the same way, as seen in Figure 6.6.1(b). Note that the commutation symmetry $a_i^{\dagger\alpha}a_j^{\dagger\beta} = a_j^{\dagger\beta}a_i^{\dagger\alpha}$ does not eliminate any terms when $\alpha \neq \beta$.

To gain some appreciation of the complexity of overtone and combination levels one should examine the level diagrams for SF_6, UF_6, and SiF_4 in Figures 6.6.2(a–c) by R. S. McDowell and B. J. Krohn at Los Alamos. The spectroscopy upon which they are based is described in some of the references given at the end of Chapter 4 and in Chapter 7. The spectroscopic notation for the fundamental levels discussed in Chapter 4 (recall Figures 4.4.5 and 4.4.6) is $\nu_1 = A_{1g}$, $\nu_2 = E_g$, $\nu_3 = T_{1u}(\text{high})$, $\nu_4 = T_{1u}(\text{low})$, $\nu_5 = T_{2g}$, and $\nu_6 = T_{2u}$. The center of the figure is devoted to the ν_3 overtone ladder which provides a pathway for infrared laser dissociation of the molecule. The need to investigate dissociation and apply it to isotope separation was the driving force behind much of the initial high-resolution spectroscopy of UF_6, SF_6, and related molecules. The overtone levels for UF_6 are shown for comparison in Figure 6.6.2(b), and SiF_4 levels are displayed in Figure 6.6.2(c).

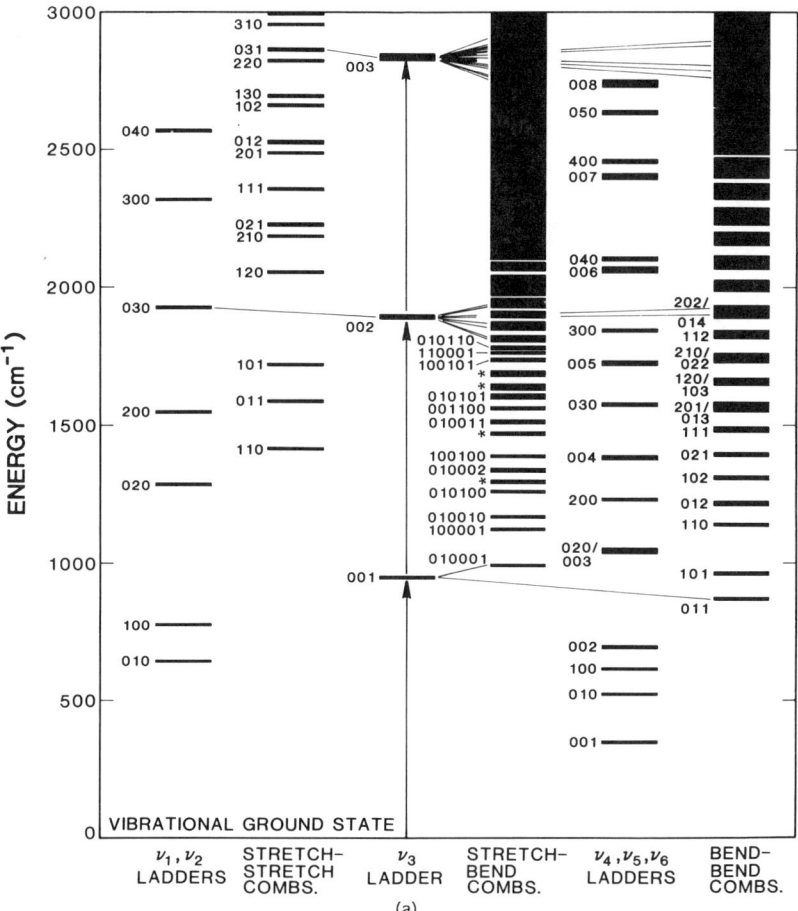

Figure 6.6.2a Vibrational levels in SF_6 below 3000 cm^{-1}, illustrating intramolecular energy transfer in multiple photon excitation of the lower ν_3 ladder. The level energies are based on the ν_i and x_{ii} constants of McDowell and Krohn. In the fourth column the levels are labeled with the vibrational quantum numbers $v_1v_2v_3v_4v_5v_6$; in the first three columns with $v_1v_2v_3$ (v_4, v_5, and v_6 understood to be zero); and in the last two columns with $v_4v_5v_6$ ($v_1 = v_2 = v_3 = 0$). The four levels marked with asterisks are (in order of decreasing energy) 010020/010003, 100011/001002/020001, 001010/100002, and 100010/001001; also, in column six, 031 falls between 202 and 014. Anharmonic splittings of higher vibrational manifolds were arbitrarily assumed to be 1 cm^{-1} between sublevels, except for $2\nu_3$ and $3\nu_3$, for which the sublevel positions of Patterson, Krohn, and Pine were used. Each sublevel was given a width of 12 cm^{-1} to indicate rotational broadening. There are approximately 327 vibrational levels between 0 and 3000 cm^{-1}, comprising about 5540 sublevels. Also shown are typical near-resonant collisional pathways out of the ν_3 ladder for which the centers of the vibrational manifolds are detuned by less than 80 cm^{-1}. (After R. S. McDowell and B. J. Krohn, unpublished.)

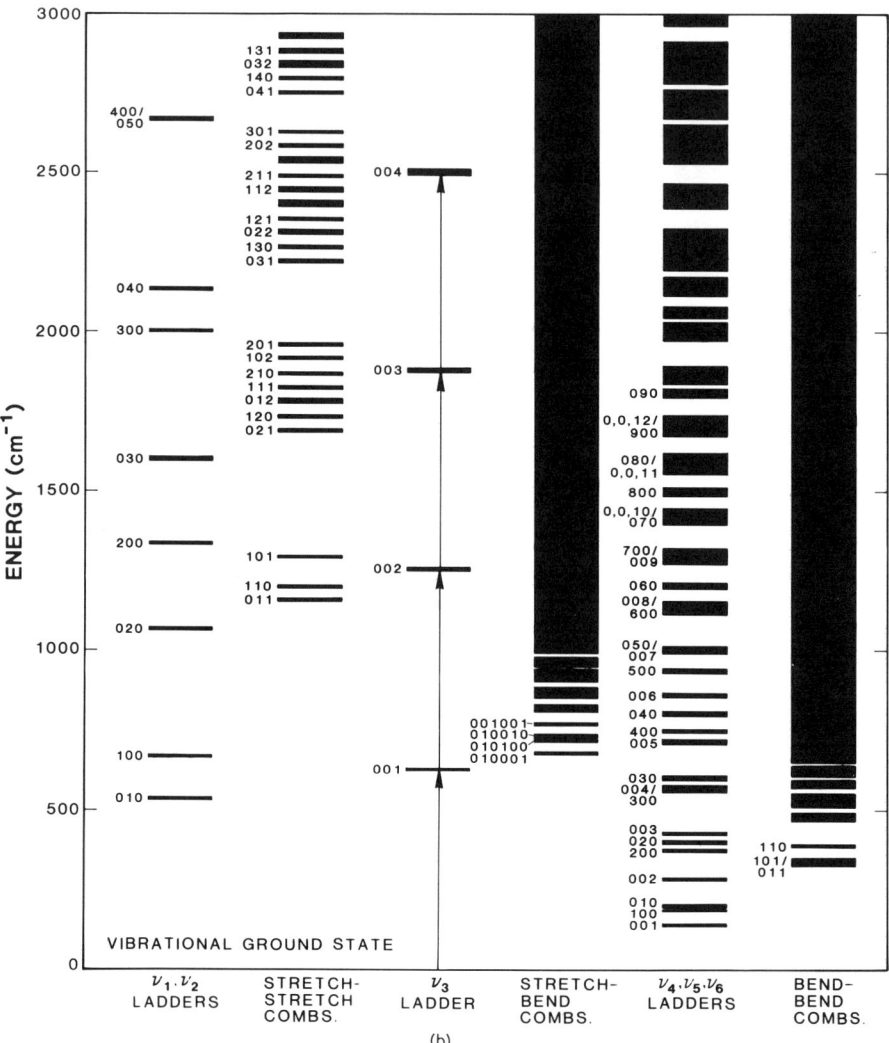

Figure 6.6.2b Vibrational energy levels of UF_6 below 3000 cm^{-1}. The level energies are based on the ν_i and x_{ij} constants of Adlridge et al., except that $x_{33} = -0.7$ cm^{-1}. The levels are labeled as in Figure 6.6.1. Anharmonic splittings were assumed to be 1 cm^{-1} between sublevels, except for the ν_3 ladder, for which the sublevel positions of Krohn et al., were used. Each sublevel was given a width of 12 cm^{-1} to indicate rotational broadening. There are approximately 1015 vibrational levels between 0 and 2000 cm^{-1}, comprising about 120,700 sublevels. For a given amount of ν_3 excitation, higher bending levels are more accessible for UF_6 than for SF_6. (After R. S. McDowell and B. J. Krohn, unpublished.)

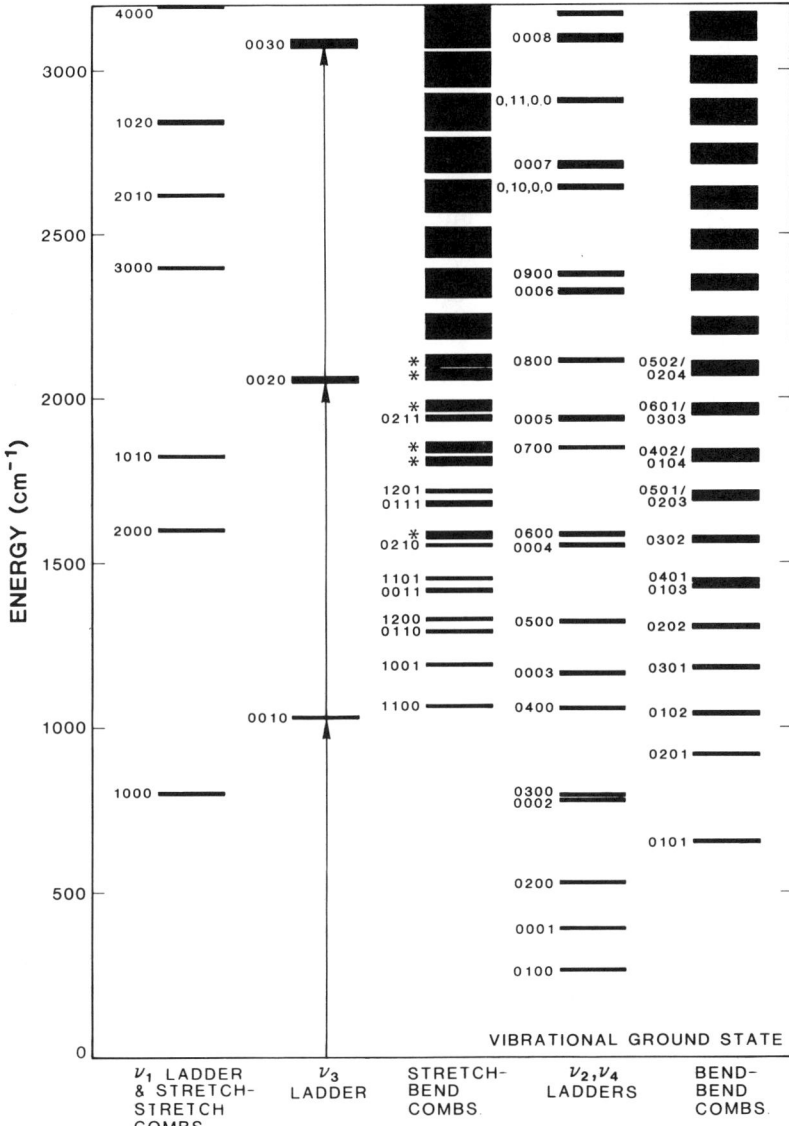

Figure 6.6.2c Vibrational energy levels of SiF$_4$ below 3200 cm^{-1}. The level energies are based on the ν_i and x_{ij} constants of McDowell et al., and are labeled with the vibrational quantum numbers $v_1v_2v_3v_4$. The six levels marked with asterisks in column three are (in order of decreasing energy) 2200/1500/1202, 1110/0410/0112, 2001/1301/1003, 2100/1400/1102, 0310/0012, and 1300/1002. Anharmonic splittings were assumed to be 1 cm^{-1} between sublevels, except for $2\nu_3$ and $3\nu_3$, for which the sublevel positions of Patterson and Pine were used. Each sublevel was given a width of 12 cm^{-1} to indicate rotational broadening. There are approximately 185 vibrational levels between 0 and 3200 cm^{-1}, comprising about 2185 sublevels. (After R. S. McDowell and B. J. Krohn, unpublished.)

A desire to understand the general problem of intramolecular vibrational energy redistribution (IVER) has stimulated research on vibrational overtone and combination states of many molecules. In order to reduce the complexity of the problem, many researchers have chosen to examine simpler molecules and anharmonic vibrational models. Since diatomic molecules seem to be too simple to show IVER effects one must consider triatomic molecules—such as H_2O, H_2S, or O_3, or weakly bound triatomic clusters such as HFAr.

One of the most famous triatomic anharmonic vibrational potential models is the third-order C_{3v} symmetric potential constructed from the two-vector normal coordinates $\{x \equiv q_1^E, y \equiv q_2^E\}$. Using coupling coefficients for the group C_{3v} we have

$$V = \lambda(x^3 - 3xy^2).$$

This is known as the Henon-Heiles potential function. Breaking the symmetry to C_2 yields the Barbannis potential:

$$U = \lambda x^3 - 3uxy^2.$$

This recovers the C_{3v} symmetry when $u = \lambda$. Some effects of nonlinear and anharmonic potentials are discussed in Chapter 7 and references.

6.7 INTRODUCTION TO THEORY OF SYMMETRY STABILITY

So far it has been assumed that the Hamiltonian or equations of motion of various systems had a certain symmetry. By making various models subject to the symmetry constraint one could relate and predict various properties of the systems which the models represent. However, such an approach cannot give you much of a clue about how the symmetry came about in the first place, i.e., why the electronic and electromagnetic forces hold some configuration in a symmetric form. Now by relaxing the assumptions about these Hamiltonians somewhat it is possible to gain a little insight into what forms may or may not be stable under various conditions.

This involves an interaction between nuclear (or vibrational) and electronic motion which is called VIBRONIC motion. This section contains an introduction to vibronic Hamiltonians and their symmetry properties. This leads to the Jahn-Teller theory of symmetry stability.

A. Symmetry Analysis of Vibronic Hamiltonians

In the models of molecular electronic orbitals given in Chapters 2–5 and later in Chapter 7 Section 7.6 some approximate electronic eigensolutions are derived for electrons tunneling or orbiting around several fixed potential wells. The potential wells represent the nuclei of a molecule, and so the electronic levels were derived for one fixed position of nuclei. On the other hand, the models of nuclear motion in molecular vibrations were based on the assumption of internuclear potential functions or "spring constants." The internuclear potential represents the combined effect of nuclear Coulomb repulsion and electronic bonding energy. The bonding energy depends on the eigenvalue of the electronic ground-state eigenfunction.

In Section 4.3.D various changes of energy or positions of octahedral potentials were seen to shift or split the energy eigenvalues. (Recall Figures 4.3.5 and 4.3.7.) One can imagine that each electronic energy eigenvalue ε^γ of a molecule is plotted as a function $\varepsilon^\gamma(\cdots x_i^\alpha \cdots)$ of the nuclear positions or normal coordinates x_i^α. If the internuclear repulsion potential $Z(\cdots x_i^\alpha \cdots)$ is added, there results an effective potential

$$V^\gamma(\cdots x_i^\alpha \cdots) = \varepsilon^\gamma(\cdots x_i^\alpha \cdots) + Z(\cdots x_i^\alpha \cdots)$$

for each electronic state. The BORN-OPPENHEIMER APPROXIMATION amounts to picking just one of these effective potentials for the nuclear vibration Hamiltonian. For molecules in their electronic ground states one picks the ground state V^γ. This is a good approximation as long as the splitting between different $V^\gamma, V^{\gamma'}, \ldots$ is much greater than the fundamental vibration energies resulting from each effective potential. In other words, a single electronic level provides an effective nuclear potential as long as the nuclei move slowly compared to the electrons.

Now it is clear that the Born-Oppenheimer approximation may break down for any ε^γ belonging to degenerate electronic levels or multidimensional ($l^\gamma > 1$) irreps. The possibility of electronic degeneracy signals the need for a vibration-electronic or vibronic Hamiltonian which strongly effects the stability of the molecule. The question of stability is: "What values of x_i^α are stable local minima for $V^\gamma(\cdots x_i^\alpha \cdots)$?" or more to the point: "Is a stable minimum of $V^\gamma(\cdots x_i^\alpha \cdots)$ found at the symmetry configuration where $x_i^\alpha = 0$?"

To answer this, imagine that the total Hamiltonian $H(\cdots \mathbf{r}_e \cdots x_i^\alpha \cdots)$ operator for the nuclei and electrons can be expanded in powers of the nuclear coordinates x_i^α.

$$H(\cdots \hat{r}_e \cdots x_i^\alpha \cdots) = H^0(r) + \sum_{\alpha, i} H_i^\alpha(r) x_i^\alpha$$
$$+ \sum_{\gamma, m} H[\alpha\beta]_m^\gamma(r)[x^\alpha x^\beta]_m^\gamma + \cdots. \quad (6.7.1a)$$

Here each $H(r)$ operator factor is a function only of the electronic operators, and we use a bracket notation to abbreviate the quadratic Clebsch-Gordon combinations of nuclear coordinates:

$$[x^\alpha x^\beta]_m^\gamma = \sum_{i,j} \mathscr{C}_{ijk}^{\alpha\beta\gamma} x_i^\alpha x_j^\beta. \quad (6.7.1b)$$

Higher-order (anharmonic) terms in this expansion will be important in general, but, for now, we will only consider terms of second order or less. The Hamiltonian $H(\cdots \mathbf{r} \cdots \mathbf{x} \cdots)$ being considered contains all the kinetic energy terms ($\cdots P_e^2/2m_e \cdots P_N^2/2M_N \cdots$) for n_e electrons and n_N nuclei, all the Coulomb potentials of interaction between then $1/|\mathbf{r}_e - \mathbf{r}_{e'}| \cdots 1/|\mathbf{x}_N - \mathbf{x}_N'| \cdots 1/|\mathbf{r}_e - \mathbf{x}_N|$), and nuclear and electronic spin operators, too, if they are important.

Assuming the external environment is isotropic, such a Hamiltonian is completely invariant to all rigid spatial rotations from the group O_3 discussed in Chapter 5. Also, it is invariant to all $n_e!n_{N_1}!n_{N_2}! \cdots$ permutations of identical particles. Together, this amounts to an enormous symmetry, in general, which is larger than the ordinary molecular point symmetry G which the nuclei will have if $(x_i^\alpha = 0)$ is a stable minimum.

Indeed, there is more physical symmetry in fundamental descriptions of nature than is readily apparent from most of the objects which we observe. Hidden symmetry is a fascinating subject because it usually involves more fundamental and simpler principles which explain complex spectral effects. Elementary examples of hidden symmetry were introduced in connection with spontaneous symmetry breaking in Section 4.3.C. The "global" symmetry discussed there corresponds to hidden symmetry, while the "local" symmetry corresponds here to the molecular point symmetry G of a stable molecule. Most molecules become "frozen" into stable G-symmetric forms and never get to realize the full freedom of their hidden symmetry.

It is important to establish some criteria for determining whether a particular molecular symmetry G or a G-symmetric arrangement is stable. To do this we imagine, again, putting the nuclei in a G-symmetric position long enough for all the electrons to settle down into some ground-state orbital that transforms according to irrep $\mathscr{D}^\gamma$ of G. Then, using this assumption of a (γ) orbital, the fact that $H(\cdots \mathbf{r} \cdots, \cdots x_i^\alpha \cdots)$ has G symmetry (at least!), and Eq. (6.7.1a), we may derive an energy submatrix for the electronic substates $|_m^\gamma\rangle$:

$$\begin{pmatrix} \left\langle \begin{matrix} \gamma \\ 1 \end{matrix} \middle| H \middle| \begin{matrix} \gamma \\ 1 \end{matrix} \right\rangle & \left\langle \begin{matrix} \gamma \\ 1 \end{matrix} \middle| H \middle| \begin{matrix} \gamma \\ 2 \end{matrix} \right\rangle & \cdots \\ \left\langle \begin{matrix} \gamma \\ 2 \end{matrix} \middle| H \middle| \begin{matrix} \gamma \\ 1 \end{matrix} \right\rangle & \left\langle \begin{matrix} \gamma \\ 2 \end{matrix} \middle| H \middle| \begin{matrix} \gamma \\ 2 \end{matrix} \right\rangle & \cdots \\ \vdots & \vdots & \end{pmatrix}. \qquad (6.7.2)$$

Since $H(\cdots \mathbf{r} \cdots, \cdots x_i^\alpha \cdots)$ has G symmetry, we must have that the purely electronic operator factors $H_i^\alpha(r), H[\alpha\beta]_n^\delta(r), \ldots$ transform as irreducible tensorial operators with irrep labels $\binom{\alpha}{i}, \binom{\delta}{n}, \ldots$, respectively. That is, each term in Eq. (6.7.1a) is a G scalar. This allows one to expand and evaluate each energy submatrix component using the Wigner-Eckart theorem:

$$\left\langle \begin{matrix}\gamma\\m\end{matrix}\Big|H\Big|\begin{matrix}\gamma\\j\end{matrix}\right\rangle = \left\langle \begin{matrix}\gamma\\m\end{matrix}\Big|H^A\Big|\begin{matrix}\gamma\\j\end{matrix}\right\rangle + \sum_{\alpha,i}\left\langle \begin{matrix}\gamma\\m\end{matrix}\Big|H_i^\alpha\Big|\begin{matrix}\gamma\\j\end{matrix}\right\rangle x_i^\alpha + \sum_{\delta,n}\left\langle \begin{matrix}\gamma\\m\end{matrix}\Big|H[\alpha\beta]_n^\delta\Big|\begin{matrix}\gamma\\j\end{matrix}\right\rangle [x^\alpha x^\beta]_n^\delta \cdots$$

$$= \mathscr{C}_{\cdot jm}^{A\gamma\gamma}\langle\gamma\|A\|\gamma\rangle + \sum_{\alpha,i}\mathscr{C}_{ijm}^{\alpha\gamma\gamma}\langle\gamma\|\alpha\|\gamma\rangle x_i^\alpha$$

$$+ \sum_{\delta,n}\mathscr{C}_{njm}^{\delta\gamma\gamma}\langle\gamma\|[\alpha\beta]\delta\|\gamma\rangle[x^\alpha x^\beta]_n^\delta + \cdots . \qquad (6.7.3)$$

B. The Jahn-Teller Theorem

Let us pick once again the example of the XF_6 molecule which we assumed in Section 4.4 had O_h symmetry. The normal cordinates were $x^{A_{1g}}$, x^{E_g}, $x^{T_{1u}}$ (two kinds), $x^{T_{2g}}$, and $x^{T_{2u}}$, as shown in Figure 4.4.4.

Some possibilities for XF_6 electronic orbitals were briefly discussed in Section 4.3. However, the approximate molecular orbital states $|A_{1g}\rangle$, $|_m^{E_g}\rangle$, and $\left|\begin{matrix}T_{1u}\\m\end{matrix}\right\rangle$ displayed in Figure 4.3.2 were made just for one electron shared between the F atoms. In a more detailed analysis in later chapters involving more electrons and more orbitals, it will be clear that all irreps can show up as orbitals in general. So one must consider each possible (γ) in Eq. (6.7.3).

Case 1. $\gamma = A_{1g}$, A_{1u}, A_{2g}, or A_{2u}

Substituting $\gamma = A_x$ into Eq. (6.7.3), we find that most of the coupling coefficients are zero. All linear terms except $\alpha = A_{1g}$ are ruled out:

$$\langle A_x|H|A_x\rangle = \langle A_x\|A\|A_x\rangle + \langle A_x\|A_{1g}\|A_x\rangle x^{A_{1g}}$$

$$+ \langle A_x\|[\alpha\alpha]A_{1g}\|A_x\rangle[x^\alpha x^\alpha]^{A_{1g}} \cdots . \qquad (6.7.4)$$

Of the quadratic terms only $\delta = A_{1g}$ or the harmonic terms

$$[x^\alpha x^\alpha]^{A_{1g}} = (1/\sqrt{l^\alpha})\sum_{i=1}^{l^\alpha}(x_i^\alpha)^2$$

remain. The reduced matrix coefficients $\langle A_x\|[\alpha\alpha]A_{1g}\|A_x\rangle$ must now all be positive in order for the molecule to be stable in a symmetric shape. Then the molecule will alter its $x^{A_{1g}}$ coordinate until a minimum is reached for the energy

$$E = \langle A_x\|A_{1g}\|A_x\rangle x^{A_{1g}} + \langle A_x\|[A_{1g}A_{1g}]A_{1g}\|A_x\rangle(x^{A_{1g}})^2$$

at equilibrium position

$$x^{A_{1g}}(0) = -\langle A_x \| A_{1g} \| A_x \rangle / 2 \langle A_x \| [A_{1g} A_{1g}] A_{1g} \| A_x \rangle.$$

However, this shift involves no reduction of symmetry, just a change of size for the molecule.

Case 2. $\gamma = E_g$ or E_u

In the 2×2 energy matrix for an E electronic doublet, we find that two motions x_i^α give possibly nonzero linear terms. These are the motions $x^{A_{1g}}$ and $x_j^{E_g}$:

$$\langle H \rangle = \langle E \| A \| E \rangle \begin{pmatrix} 1 & 0 \\ 0 & 1 \end{pmatrix} + \langle E \| A_{1g} \| E \rangle \begin{pmatrix} x^{A_{1g}} & 0 \\ 0 & x^{A_{1g}} \end{pmatrix}$$

$$+ \frac{\langle E \| E_g \| E \rangle}{\sqrt{2}} \begin{pmatrix} x_1^{E_g} & -x_2^{E_g} \\ -x_2^{E_g} & -x_1^{E_g} \end{pmatrix}.$$

$$+ \langle E \| [EE] A_{1g} \| E \rangle \cdot \begin{pmatrix} [x^E x^E]^{A_{1g}} & 0 \\ 0 & [x^E x^E]^{A_{1g}} \end{pmatrix}$$

$$+ \frac{\langle E \| [EE] E \| E \rangle}{\sqrt{2}} \begin{pmatrix} [x^E x^E]_1^E & -[x^E x^E]_2^E \\ -[x^E x^E]_2^E & -[x^E x^E]_1^E \end{pmatrix}. \quad (6.7.5)$$

Extra quadratic terms show up, too; however, just the linear terms in $x_j^{E_g}$ are sufficient to spoil the O_h symmetry in the ordinary sense. As we will see in the following, the equilibrium position is not at $x_j^{E_g} = 0$ if the reduced matrix element $\langle E \| E_g \| E \rangle$ is finite. Of course, this reduced matrix element might turn out to be zero anyway, even though symmetry does not require it; however, this is improbable.

Case 3. $\gamma = T_{1g}, T_{1u}, T_{2g},$ or T_{2u}

In the 3×3 energy matrix for any T electronic orbital we find that the motions $x^{A_{1g}}$, $x_j^{E_g}$, and $x_i^{T_{2g}}$ all are capable of giving nonzero linear terms, in addition to some nonscalar quadratic terms. Again, we find that the equilibrium position is not necessarily $x_j^{E_g} = 0 = x_i^{T_{2g}}$. Of the Cases 1, 2, and 3 we see that of all the electronic states, only the nondegenerate ones (A) are likely to yield an O_h symmetric XY_6 molecule. This is the content of the following theorem.

Jahn-Teller Theorem There will be at least one possible linear term involving a symmetry-breaking motion in the energy expansion for a degenerate electronic orbital, except for (a) the electronic spin degeneracy of 2 and (b) the axial angular-momentum degeneracy of 2 for linear molecules.

The theorem was first proved by exhaustive study of each symmetry structure possible with all possible molecular configurations: XY_2, XY_3, $XY_4, \ldots, X_2Y_2$, and so on. Since then more elegant proofs have been devised.

C. Dynamic Jahn-Teller and Renner Effects

We now investigate the $(\gamma = E)$ case or Case 2 in more detail. Before beginning, one should note that the E-case mathematics serves as a solution for at least two distinct problems. It pertains to the stability of an O_h-symmetric XF_6 molecule with an E_g or E_u orbital, but it also applies to the triangular C_{3v} molecule displayed in Figure 3.3.4. It may be a model for ozone (O_3). This coincidence happens because the coupling coefficients involving the O_h irreps E_g, A_{2g}, and A_{1g} are precisely the same as those for C_{3v} irreps $(3) \equiv E$, $(2) \equiv A_2$, and $(1) \equiv A_1$. This is fortunate because it is easier to visualize the ozone molecule when trying to get a physical feeling for the results.

Let us rewrite the matrix in Eq. (6.7.5) as follows:

$$\langle H \rangle = \begin{pmatrix} jx_1 + k(x_1^2 + x_2^2) + r(x_1^2 - x_2^2) & -jx_2 + 2rx_1x_2 \\ -jx_2 + 2rx_1x_2 & -jx_1 + k(x_1^2 + x_2^2) - r(x_1^2 - x_2^2) \end{pmatrix}$$

(6.7.6a)

by dropping the constant term and the irrep superscripts ($x_j^E \equiv x_j$), and by letting

$$j = \langle E \| E \| E \rangle / \sqrt{2}$$ (6.7.6b)

be the coefficients of what are the JAHN-TELLER terms, by letting

$$r = \langle E \| [EE] E \| E \rangle / 2$$ (6.7.6c)

be the coefficients of what are called the RENNER terms, and by letting

$$k = \langle E \| [EE] A_1 \| E \rangle / \sqrt{2}$$ (6.7.6d)

be the coefficient of the HARMONIC terms. One can leave the harmonic terms out of the matrix while diagonalizing since they just are proportional to the unit matrix. In that case the secular equation is simply

$$\lambda^2 - \left[\left(jx_1 + r(x_1^2 - x_2^2) \right)^2 + (jx_2 - 2rx_1x_2)^2 \right] = 0,$$

and has the following solution:

$$\lambda = \pm \left(j^2(x_1^2 + x_2^2) + 2jr(x_1^3 - 3x_1 x_2^2) + r^2(x_1^2 + x_2^2)^2 \right)^{1/2}.$$

It is convenient to switch to polar coordinates in the normal-mode space, where

$$x_1^E = \rho \cos \phi, \qquad x_2^E = \rho \sin \phi. \tag{6.7.7}$$

The total energy eigenvalues (we add the harmonic term back now) have the following form:

$$\varepsilon_\pm = k\rho^2 \pm (j^2\rho^2 + 2jr\rho^3 \cos 3\phi + r^2\rho^4)^{1/2} = k\rho^2 \pm j\rho \qquad \text{(for } r = 0\text{)}. \tag{6.7.8}$$

On the right we have eliminated the Renner terms.

We therefore find some rather curious potential-energy functions of the normal coordinates x_1^E, x_2^E or ρ, ϕ, which are sketched in Figures 6.7.1a and 6.7.1b) for $r = 0$ and $r > 0$, respectively.

The electronic eigenstates can be expressed very nicely in terms of the normal coordinate polar angle for the case $r = 0$

$$|\varepsilon_+\rangle = \cos(\phi/2) \left| \begin{matrix} E \\ 1 \end{matrix} \right\rangle - \sin(\phi/2) \left| \begin{matrix} E \\ 2 \end{matrix} \right\rangle,$$

$$|\varepsilon_-\rangle = \sin(\phi/2) \left| \begin{matrix} E \\ 1 \end{matrix} \right\rangle + \cos(\phi/2) \left| \begin{matrix} E \\ 2 \end{matrix} \right\rangle. \tag{6.7.9}$$

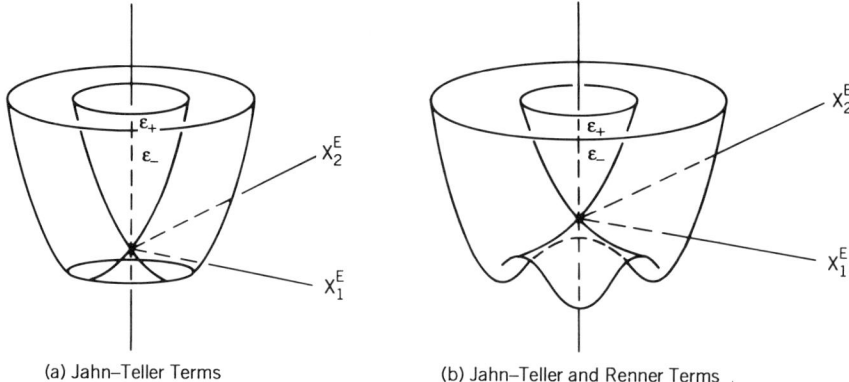

(a) Jahn–Teller Terms   (b) Jahn–Teller and Renner Terms

Figure 6.7.1 Potential energy functions of the E-type coordinates. (a) Jahn-Teller terms ($r = 0$). (b) Jahn-Teller and Renner terms ($r > 0$).

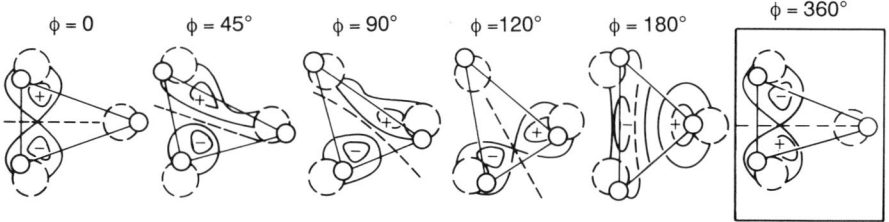

Figure 6.7.2 Electronic state $|\varepsilon_-\rangle$ for various nuclear positions allowed by varying E coordinates.

These are eigenvectors of the electronic matrix [Eq. (6.7.6a)] with $r = 0$:

$$\langle H \rangle = \begin{pmatrix} j\rho \cos \phi & -j\rho \sin \phi \\ -j\rho \sin \phi & -j\rho \cos \phi \end{pmatrix} + k\rho^2 \mathbf{1}.$$

In order to visualize this kind of eigenstate, one may imagine the ozone-molecule motions of type (3) or E which were drawn in Figure 3.4.5. Figure 6.7.2 contains sketches of the form of the electronic eigenwave $\langle \cdots \mathbf{r} \cdots |\varepsilon_-\rangle$ for several values of angle ϕ. (We do not know the exact shapes without solving a Schrödinger equation, but symmetry theory gives us a rough idea of form, as explained in Section 2.12.) As we rotate the nuclear distortion $(x_1^E = \rho,\ x_2^E = 0)$ into $(x_1^E = \rho \cos \phi,\ x_2^E = \rho \sin \phi)$ we see what happens to the electronic wave as shown in Figure 6.7.2. Its progress can be traced by the position of the nodal plane indicated by the dotted line.

One thing to notice is that the electronic wave rotates half as much as the nuclear distortion. A complete rotation $\phi = 360°$ leaves the electronic wave with a minus sign. In Section 5.7.A it was shown that such transformation behavior corresponds to a RAY representation of the symmetry. Half-integral spin states or the states of any two-state system will exhibit this sort of double-valued behavior.

A similar sketch could be made for the electronic state $|\varepsilon_+\rangle$. At $\phi = 0°$ it would be the same as $|\varepsilon_-\rangle$ for $\phi = 180°$. This gives another way to see the multivalued structure. A cross-section of Figure 6.7.1(a) is shown in Figure 6.7.3(a). Note that by going from $\phi = 0°$ to $\phi = 180°$ via the valley, one effectively jumps from the right parabola belonging to state $\left|\begin{smallmatrix}E\\2\end{smallmatrix}\right\rangle$ to the left one belonging to orthogonal electronic state $\left|\begin{smallmatrix}E\\1\end{smallmatrix}\right\rangle$. We have been describing this function as consisting of a top sheet (ε_+) and a bottom sheet (ε_-). This description is probably more suitable, particularly if some perturbation like spin-orbit separates the two sheets as indicated by dotted lines in Figure 6.7.3(b).

By looking again at the various nuclear distortions in Figure 6.7.2 we get some idea what the meaning is of the minima in the (ε_-) energy sheet at

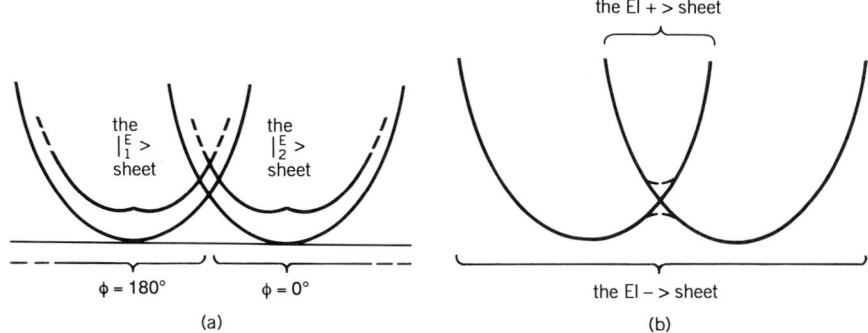

Figure 6.7.3 Different ways to label the cross-sections of the potential surfaces in Figure (6.7.1a). (a) Parabolic potentials due to electronic states $\begin{pmatrix} E \\ 1 \end{pmatrix}$ and $\begin{pmatrix} E \\ 2 \end{pmatrix}$. (b) Separate potential sheets due to excited (+) and ground (−) electronic states.

$\phi = 60°$, $180°$, and $300°$ when the Renner coefficient r is positive [see Figure 6.7.1(b)] or at $0°$, $120°$, and $240°$ when r is negative. In particular, the $\phi = 0°$ and $\phi = 120°$ distortions are pictured in Figure 6.7.2. An acute isosceles triangle pointing in one of three directions is stable when $r < 0$. For $\phi = 60°$, $180°$, and $300°$ we find an obtuse isosceles triangle pointing away from these directions.

It might appear that a model Hamiltonian with nonzero j and r terms reduces the molecular symmetry to the Abelian C_{2v} and splits all the electronic and vibrational levels into singlets. This is *not* the case. Let us go back to the beginning by first assuming j and r are zero. With only the scalar harmonic (k) Hamiltonian one has essentially independent electronic and vibration states. One can be rotated without affecting the other.

Suppose the electronic states and classical vibration coordinates being considered transform according to E [or (3)] irreps as before. Then the quantum vibration states transform according to $\beta = A_1$; E; E, A_1;... for excitation number $N = 0; 1; 2; \ldots$, respectively, according to an accounting explained in the preceding Section 6.6. The quantum states for the whole molecule are then the products,

$$\left| \begin{matrix} E & \beta \\ i & j \end{matrix} \right\rangle = \left| \begin{matrix} E \\ i \end{matrix} \right\rangle_{\text{electron}} \left| \begin{matrix} \beta \\ j \end{matrix} \right\rangle_{\text{nuclei}}, \qquad (6.7.10)$$

as indicated on the left of Figure 6.7.4 [see columns (a) and (b)].

In the presence of any electron-vibration interaction represented by j, r, or other terms, it is appropriate to couple the separate electronic and vibrational states to make what are called VIBRONIC states:

$$\left| \begin{matrix} \gamma \\ m \end{matrix} \right\rangle_{\text{vibronic}} = \sum_{i,j} \mathscr{C}_{ijm}^{E\beta\gamma} \left| \begin{matrix} E \\ i \end{matrix} \right\rangle_{\text{electronic}} \left| \begin{matrix} \beta \\ j \end{matrix} \right\rangle_{\text{nuclei}}, \qquad (6.7.11)$$

INTRODUCTION TO THEORY OF SYMMETRY STABILITY

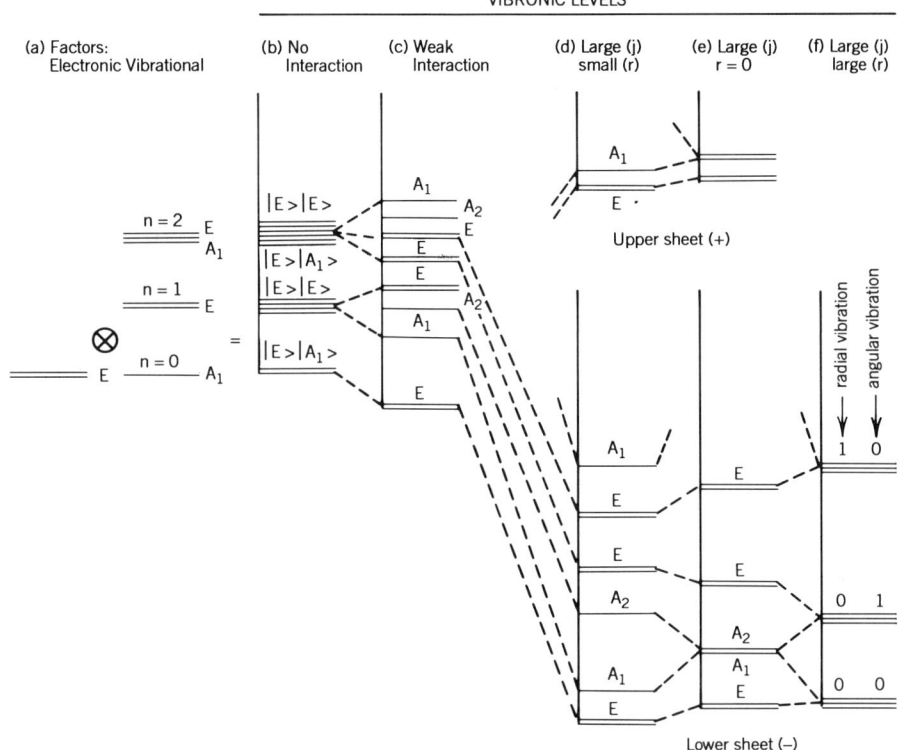

Figure 6.7.4 Vibronic levels arising from the combinations of E-type vibrations with E-type electronic orbitals.

which will, to first order, be eigenstates of the interaction operator. The resulting levels shown in Figure 6.7.4(c) will keep their irrep labels $\binom{\gamma}{m}$ and degeneracy no matter what r or j may be. The E levels will not split, and each vibronic state can be mixed only with others which have the same irrep labels.

Now let us consider the extreme case in which both r and j are large and the nuclear coordinates are essentially "stuck" near the the bottom of one of the three "subvalleys" in Figure 6.7.1(b). The vibration frequency for oscillation in the radial (ρ) direction will be different (presumably higher) than that for the angular (ϕ) direction. Therefore, each state should be singly degenerate. However, this description forces us to consider the corresponding (single) states for each of the configurations in the other two valleys. We then have a total degeneracy of 3 for each level as indicated in Figure 6.7.4(f).

Indeed, we may follow the levels as the constants are changed. For small r, but large j, the angular bumps or saddle points will be reduced; so the three species of vibrational states get mixed with a resulting tunneling and splitting of the triple degeneracy as indicated in Figure 6.7.4(d). This is similar to the hindered rotation mechanics discussed in Section 2.12. [Recall Figure 2.12.4(a).] This is a good example of spontaneous symmetry breaking. With little or no tunneling the dynamic Jahn-Teller levels collapse into $A_1 + E$ or $E + A_2$ clusters which belong to the $O_1 \uparrow C_{3v}$ or $O_2 \uparrow C_{3v}$ induced representations. In the limit of no tunneling the molecule becomes "frozen" into a local C_v symmetry with an isosceles triangular shape. This is the static Jahn-Teller limit, and the clusters are degenerate. Someone unaware of the hidden C_{3v} symmetry will treat the clusters as singlet levels.

For $r = 0$ the effective potential takes the shape shown in Figure 6.7.1(a). Now a type of free rotation is possible with spectra of the form shown in Figure 6.7.4(e). One may derive a differential equation for the vibronic state in this case and discuss the form of the wave function.

Consider a wave function of the form

$$\Psi_{nl}(\cdots \mathbf{r} \cdots \rho, \phi) = \langle \cdots \mathbf{r} \cdots |\varepsilon_-\rangle \phi_{nl}^-(\rho,\phi) + \langle \cdots \mathbf{r} \cdots |\varepsilon_+\rangle \phi_{nl}^+(\rho,\phi), \quad (6.7.12)$$

where the electronic parts

$$\langle \cdots \mathbf{r} \cdots |\varepsilon_-\rangle = \sin(\phi/2)\langle \cdots \mathbf{r} \cdots \left|\begin{matrix}E\\1\end{matrix}\right\rangle + \cos(\phi/2)\langle \cdots \mathbf{r} \cdots \left|\begin{matrix}E\\2\end{matrix}\right\rangle,$$

$$\langle \cdots \mathbf{r} \cdots |\varepsilon_+\rangle = \cos(\phi/2)\langle \cdots \mathbf{r} \cdots \left|\begin{matrix}E\\1\end{matrix}\right\rangle - \sin(\phi/2)\langle \cdots \mathbf{r} \cdots \left|\begin{matrix}E\\2\end{matrix}\right\rangle$$

follow from Eq. (6.7.9), and the nuclear parts

$$\phi_{nl}^\pm(\rho\phi) = R_n^\pm(\rho)e^{il\phi} \quad (6.7.13)$$

are separated into a radial and angular part. The effective nuclear Hamiltonian is $H = T + V$, where the kinetic part is

$$T = -\frac{\hbar^2}{2}\left[\frac{\partial^2}{\partial\rho^2} + \frac{1}{\rho}\frac{\partial}{\partial\rho} + \frac{1}{\rho^2}\frac{\partial^2}{\partial\phi^2}\right], \quad (6.7.14)$$

and the potential part represents the effect of the electronic states:

$$V = \begin{cases} k\rho^2 + j\rho, & \text{acting on } \phi^+, \\ k\rho^2 - j\rho, & \text{acting on } \phi^-. \end{cases}$$

The angular derivatives in the kinetic part lead to the terms which will mix up

INTRODUCTION TO THEORY OF SYMMETRY STABILITY 549

the states from the two energy sheets + and −:

$$\frac{\partial^2}{\partial \phi^2} \langle \cdots \mathbf{r} \cdots | \varepsilon_- \rangle \phi_{nl}^-(\rho, \phi)$$

$$= \frac{\partial^2}{\partial \phi^2} \left(\sin \phi \langle \cdots \mathbf{r} \cdots \left| \begin{matrix} E \\ 1 \end{matrix} \right\rangle + \cos \phi \langle \cdots \mathbf{r} \cdots \left| \begin{matrix} E \\ 2 \end{matrix} \right\rangle \right) R_n^-(\rho) e^{il\phi}$$

$$= \left(-(l^2 + \tfrac{1}{4}) \langle \cdots \mathbf{r} \cdots | \varepsilon_- \rangle + (il) \langle \cdots \mathbf{r} \cdots | \varepsilon_+ \rangle \right) R_n^-(\rho) e^{il\phi},$$

$$\frac{\partial^2}{\partial \phi^2} \langle \cdots \mathbf{r} \cdots | \varepsilon_+ \rangle \phi_{nl}^+(\rho, \phi)$$

$$= \left(-(l^2 + \tfrac{1}{4}) \langle \cdots \mathbf{r} \cdots | \varepsilon_+ \rangle - (il) \langle \cdots \mathbf{r} \cdots | \varepsilon_- \rangle \right) R_n^+(\rho) e^{il\phi}.$$

Using the electronic orthogonality $\langle \varepsilon_+ | \varepsilon_- \rangle = 0$, we derive the following equations for the radial functions:

$$\frac{1}{2} \left[-\frac{\partial^2}{\partial \rho^2} - \frac{1}{\rho} \frac{\partial}{\partial \rho} + \frac{l^2 + \tfrac{1}{4}}{\rho^2} + k\rho^2 + j\rho \right] R_n^+(\rho) + i \frac{l}{2\rho^2} R_n^-(\rho) = E_n R_n^+,$$

$$\frac{1}{2} \left[-\frac{\partial^2}{\partial \rho^2} - \frac{1}{\rho} \frac{\partial}{\partial \rho} + \frac{l^2 + \tfrac{1}{4}}{\rho^2} + k\rho^2 - j\rho \right] R_n^-(\rho) - i \frac{l}{2\rho^2} R_n^+(\rho) = E_n R_n^-.$$

(6.7.15)

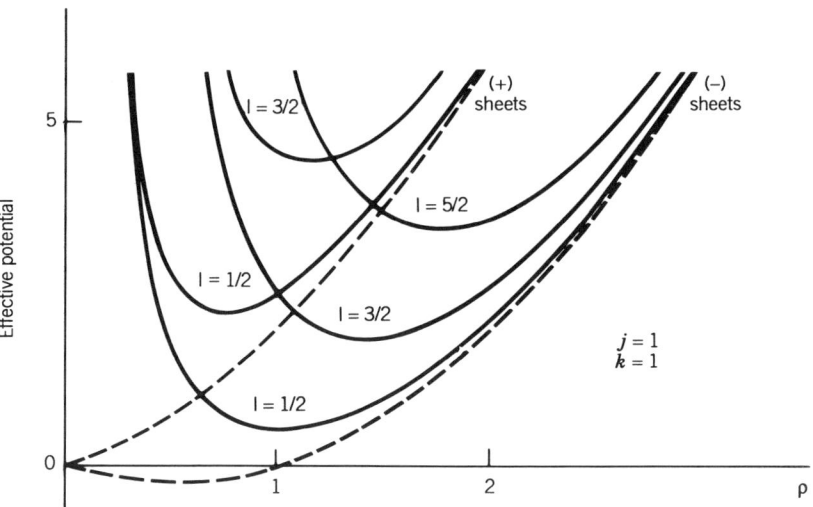

Figure 6.7.5 Effective potential energy surfaces for Jahn-Teller system with various fixed values of vibrational angular momentum ($l = \tfrac{1}{2}, \tfrac{3}{2}, \ldots$).

The angular quantum number is restricted to half-integral values $l = 1/2, 3/2, \ldots$ so that wavefunction (6.7.12) will be single valued. [Recall the behavior of the electronic part shown in Fig. (6.7.2)].

For low excited states, we may ignore the coupling terms in Eq. (6.7.15) and obtain the separate solutions

$$\Psi^- = \langle \cdots \mathbf{r} \cdots |\varepsilon_-\rangle R_n^-(\rho)e^{il\phi}, \qquad \Psi^+ = \langle \cdots \mathbf{r} \cdots |\varepsilon_+\rangle R_n^+(\rho)e^{il\phi}$$

for each energy sheet. (These are Born-Oppenheimer approximate solutions.) Examples of the effective potentials which would give the $R^\pm$ functions are plotted in Figure (6.7.5).

ADDITIONAL READING

A standard text on the tensor properties of crystals is by Nye.

J. F. Nye, *Physical Properties of Crystals* Clarendon, Oxford, 1957).

A short elementary introduction to crystal tensor analysis (without much group theory) is the following:

D. R. Lovett, *Tensor Properties of Crystals* (Adam Hilger, Philadelphia, 1989).

An advanced text on crystalline optical properties is the following:

W. A. Wooster, *Tensors and Group Theory for the Physical Properties of Crystals* (Clarendon, Oxford, 1973).

Modern nonlinear optics problems are treated in books by Yariv.

A. Yariv and P. Yeh, *Optical Waves in Crystals* (Wiley, New York, 1984).

A. Yariv, *Quantum Electronics* (Wiley, New York, 1975).

A standard text on light scattering in solids is

M. Born and K. Huang, *Dynamic Theory of Crystal Lattices* Clarendon, Oxford, 1962).

A well-known work on infrared and Raman scattering complements the Herzberg volumes cited earlier at the end of Chapter 3.

E. B. Wilson, J.C. Decius, and P.C. Cross, *Molecular Vibrations: The Theory of Infrared and Raman Spectra* (McGraw-Hill, New York, 1955).

The following are some modern treatments of Raman scattering.

J. Lascombe and P.V. Huong, *Raman Spectroscopy* (*Linear and Nonlinear*) (Wiley, New York, 1982).

A. Weber (Ed.), *Raman Spectroscopy of Gases and Liquids* (Springer, Berlin, 1978).

G. L. Eesley, *Coherent Raman Spectroscopy* (Pergamon, Oxford, 1981).

The Jahn-Teller effect was first described in the following papers.

H. A. Jahn and E. Teller, "Stability of Polyatomic Molecules in Degenerate Electronic States: I. Orbital Degeneracy," *Proc. R. Soc. London Ser. A*, **161**, 220 (1937). "II. Spin Degeneracy," *Proc. R. Soc. London Ser. A*, **164**, 117 (1938).

These were reprinted in the following volume (which is worth its weight in other precious papers as well).

R. S. Knox and A. Gold, *Symmetry in Solid State* (W. A. Benjamin, New York, 1964).

The Jahn-Teller problem arises in the treatment of crystal fields and orbital bonding which are the subject of the following well-known text. (See also end of Chapter 7).

C. J. Ballhausen, *Introduction to Ligand Field Theory* (McGraw-Hill, New York, 1962).

The Herzberg volume III contains references and discussions of Jahn-Teller and Renner effects.

PROBLEMS

6.1 Use the subgroup chain methods to help compute all the Clebsch-Gordon (CG) coefficients for irreps labeled by subgroup chains:
  (a) $D_3 \supset C_2$ (standing wave).
  (b) $D_3 \supset C_3$ (moving wave).
  (c) $D_4 \supset C_2$.
  (d) $D_4 \supset C_4$.

6.2 (a) Use the $(D_3 \supset C_2)$ Clebsch-Gordon coefficients (see Problem 6.1) to construct all the possible irreducible tensorial sets of rank- 1 (vectors) and rank- 2 (tensors) that can be made from D_3 base vectors $\hat{x}$, $\hat{y}$, and $\hat{z}$. (Let $\hat{x}$ be along the triangular symmetry axis and vertex, while $\hat{z}$ is normal to plane.)
  (b) Construct all the irreducible D_3 polynomials (trigonal harmonics) of degree 1 and 2 using Cartesian coordinates $\hat{x}, \hat{y}, \hat{z}$.
  (c) Construct a third-degree scalar A_1 (invariant) polynomial in x and y. Write it in polar coordinates and sketch its level curves. (This is called the Henon-Heiles potential function.)
  (d) Do the same for a third-degree *pseudo*scalar A_2 polynomial in x and y.

6.3 Recall the four fundamental vibrational levels and six states of the X_3 model in Chapter 3. From the ground state ...
  (a) Which could be excited by x-polarized electric dipole radiation ($E1_x$)?
  (b) Which could be excited by z-polarized magnetic dipole radiation ($M1_z$)?
  (c) Which could be excited by xy-polarized electric quadrupole radiation ($E2_{xy}$)?

6.4 Answer Problem 6.3 for the X_4 model.

6.5 The spectral diagrams for the $(A_{1g}T_{1u}E_g)$ level system in Figures 6.4.1(a) and 6.4.1(b) represent dipole transitions in the presence of a small $O_h \downarrow D_{4h}$ symmetry breaking. Redo these figures for a small *trigonal* $O_h \downarrow D_{3d}$ symmetry breaking.

6.6 Write out the character table of D_{6h} and label group elements and irreps using conventional A, B, E, 1, 2, u, and g labels.

(a) Which irreps transform like polar vectors x, y, and z?

(b) Which irreps transform like axial vectors J_x J_y and J_z?

(c) Which irreps will belong to *genuine* vibrational fundamentals of a hexagonal benzol-like X_6 radical. Account for all motions in 3-space which could have non-zero frequency.

(d) Answer (c) for the benzene molecule, and tell what size matrices would show up in the classical symmetry vibration problem.

(e) Which modes in (d) are infrared-active? Raman-active?

CHAPTER 7

THEORY AND APPLICATION OF SYMMETRY REPRESENTATION PRODUCTS (CONTINUOUS ROTATION GROUPS)

In many ways the basic symmetry analysis with continuous Lie groups R_3, O_3, R_4, U_2, and U_3 is similar to that of finite groups. In either case there exist irreducible representations (irreps) of finite dimension l^α which correspond to l^α-fold degenerate energy levels of a symmetric physical system. It will now be shown that the basic structure and application of irrep outer products of the continuous groups is similar to that of finite groups which was described in Chapter 6. In fact, it will not be necessary to repeat many of the details and motivation which were given in Sections 6.1 and 6.2.

The main difference between the finite and continuous symmetry analyses lies in the methods used to derive and record their structures. For the rotation group R_3 or the closely related SU_2 there exist formulas for the group structure (recall (Eqs. 5.5.10)], irreps [(Eq. 5.4.45)], characters [(Eq. 5.6.4)], and now, as we will see, the coupling coefficients. The finite group theory, on the other hand, does not always provide such convenient formulas, but then there are only a finite number of things to compute.

7.1 INTRODUCTION TO R_3 AND U_2 COUPLING COEFFICIENTS

The R_3 and U_2 coupling coefficients are introduced in this section by reviewing the simplest problems which use angular-momentum coupling. These and other problems which involve coupling will be discussed in detail, later, after a more complete mathematical treatment has been given.

A. Two-Particle Spin States: Hydrogen Hyperfine Structure

The electron and the proton each have spin-$\frac{1}{2}$ and a magnetic moment. The qualitative effects of the electron-proton spin-spin interaction in atomic hydrogen will be discussed as an example of coupling and symmetry. The spin-spin interaction is very weak compared to the Coulomb interaction, so one may assume the electron is fixed in a zero-orbit ($1s$) ground state and consider only the spin states of the electron and the proton nucleus.

If no spin-spin interaction existed the following four outer product states

$$\left|\begin{array}{c}\frac{1}{2}\\ \frac{1}{2}\end{array}\right\rangle^{\text{proton}}\left|\begin{array}{c}\frac{1}{2}\\ \frac{1}{2}\end{array}\right\rangle^{\text{electron}} \rightarrow \begin{pmatrix}1\\0\\0\\0\end{pmatrix}, \quad \left|\begin{array}{c}\frac{1}{2}\\ \frac{1}{2}\end{array}\right\rangle^{\text{proton}}\left|\begin{array}{c}\frac{1}{2}\\ -\frac{1}{2}\end{array}\right\rangle^{\text{electron}} \rightarrow \begin{pmatrix}0\\1\\0\\0\end{pmatrix},$$

$$\left|\begin{array}{c}\frac{1}{2}\\ -\frac{1}{2}\end{array}\right\rangle^{\text{proton}}\left|\begin{array}{c}\frac{1}{2}\\ \frac{1}{2}\end{array}\right\rangle^{\text{electron}} \rightarrow \begin{pmatrix}0\\0\\1\\0\end{pmatrix}, \quad \left|\begin{array}{c}\frac{1}{2}\\ -\frac{1}{2}\end{array}\right\rangle^{\text{proton}}\left|\begin{array}{c}\frac{1}{2}\\ -\frac{1}{2}\end{array}\right\rangle^{\text{electron}} \rightarrow \begin{pmatrix}0\\0\\0\\1\end{pmatrix}$$

would be degenerate eigenstates. As explained in Section 6.1, one could rotate either particle independently without changing the energy if the interaction is zero. All pairs of rotations $(R(\alpha_p\beta_p\gamma_p), R(\alpha_e\beta_e\gamma_e))$ form a symmetry group $G = R_3(\text{proton}) \times R_3(\text{electron})$. The base states $\left|\begin{array}{c}\frac{1}{2}\\m_1\end{array}\right\rangle\left|\begin{array}{c}\frac{1}{2}\\m_2\end{array}\right\rangle \equiv \left|\begin{array}{cc}\frac{1}{2}&\frac{1}{2}\\m_1&m_2\end{array}\right\rangle$ are a basis of the irrep $\mathscr{D}^{\frac{1}{2}}\otimes\mathscr{D}^{\frac{1}{2}} = \mathscr{D}^{\frac{1}{2}\frac{1}{2}}$ of G as shown by the following:

$$(R(\alpha_p\beta_p\gamma_p), R(\alpha_e\beta_e\gamma_e))\left|\begin{array}{cc}\frac{1}{2}&\frac{1}{2}\\m_1&m_2\end{array}\right\rangle$$

$$= R(\alpha_p\beta_p\gamma_p)\left|\begin{array}{c}\frac{1}{2}\\m_1\end{array}\right\rangle R(\alpha_e\beta_e\gamma_e)\left|\begin{array}{c}\frac{1}{2}\\m_2\end{array}\right\rangle$$

$$= \sum_{m'_1}\mathscr{D}^{\frac{1}{2}}_{m'_1 m_1}(\alpha_p\beta_p\gamma_p)\left|\begin{array}{c}\frac{1}{2}\\m'_1\end{array}\right\rangle \sum_{m'_2}\mathscr{D}^{\frac{1}{2}}_{m'_2 m_2}(\alpha_p\beta_p\gamma_p)\left|\begin{array}{c}\frac{1}{2}\\m'_2\end{array}\right\rangle$$

$$= \sum_{m'_1}\sum_{m'_1}\mathscr{D}^{\frac{1}{2}}_{m'_1 m_1}(\alpha_p\beta_p\gamma_p)\mathscr{D}^{\frac{1}{2}}_{m'_2 m_2}(\alpha_e\beta_e\gamma_e)\left|\begin{array}{cc}\frac{1}{2}&\frac{1}{2}\\m'_1&m'_2\end{array}\right\rangle$$

$$= \sum_{m'_1 m'_2}\mathscr{D}^{\frac{1}{2}\frac{1}{2}}_{m'_1 m'_2: m_1 m_2}(\alpha_p\beta_p\gamma_p; \alpha_e\beta_e\gamma_e)\left|\begin{array}{cc}\frac{1}{2}&\frac{1}{2}\\m'_1&m'_2\end{array}\right\rangle.$$

Using the notation of Eq. (6.1.8) the representation $\mathscr{D}^{\frac{1}{2}\frac{1}{2}}$ is written as follows:

$$\mathscr{D}^{\frac{1}{2}\frac{1}{2}}(0\beta_p 0 0 \beta_e 0)$$
$$= \begin{pmatrix} \mathscr{D}^{\frac{1}{2}\frac{1}{2}}_{\frac{1}{2}\frac{1}{2}}(0\beta_p 0)\mathscr{D}^{\frac{1}{2}\frac{1}{2}}_{\frac{1}{2}\frac{1}{2}}(0\beta_e 0) & \mathscr{D}^{\frac{1}{2}\frac{1}{2}}_{\frac{1}{2}\frac{1}{2}}\mathscr{D}^{\frac{1}{2}\frac{1}{2}}_{\frac{1}{2}-\frac{1}{2}} & \mathscr{D}^{\frac{1}{2}\frac{1}{2}}_{\frac{1}{2}-\frac{1}{2}}\mathscr{D}^{\frac{1}{2}\frac{1}{2}}_{\frac{1}{2}\frac{1}{2}} & \mathscr{D}^{\frac{1}{2}\frac{1}{2}}_{\frac{1}{2}-\frac{1}{2}}\mathscr{D}^{\frac{1}{2}\frac{1}{2}}_{\frac{1}{2}-\frac{1}{2}} \\ \mathscr{D}^{\frac{1}{2}\frac{1}{2}}_{\frac{1}{2}\frac{1}{2}}\mathscr{D}^{\frac{1}{2}\frac{1}{2}}_{\frac{1}{2}\frac{1}{2}} & \mathscr{D}^{\frac{1}{2}\frac{1}{2}}_{\frac{1}{2}\frac{1}{2}}\mathscr{D}^{\frac{1}{2}\frac{1}{2}}_{-\frac{1}{2}-\frac{1}{2}} & \mathscr{D}^{\frac{1}{2}\frac{1}{2}}_{\frac{1}{2}-\frac{1}{2}}\mathscr{D}^{\frac{1}{2}\frac{1}{2}}_{-\frac{1}{2}\frac{1}{2}} & \mathscr{D}^{\frac{1}{2}\frac{1}{2}}_{\frac{1}{2}-\frac{1}{2}}\mathscr{D}^{\frac{1}{2}\frac{1}{2}}_{-\frac{1}{2}-\frac{1}{2}} \\ \mathscr{D}^{\frac{1}{2}\frac{1}{2}}_{-\frac{1}{2}\frac{1}{2}}\mathscr{D}^{\frac{1}{2}\frac{1}{2}}_{\frac{1}{2}\frac{1}{2}} & \mathscr{D}^{\frac{1}{2}\frac{1}{2}}_{-\frac{1}{2}\frac{1}{2}}\mathscr{D}^{\frac{1}{2}\frac{1}{2}}_{\frac{1}{2}-\frac{1}{2}} & \mathscr{D}^{\frac{1}{2}\frac{1}{2}}_{-\frac{1}{2}-\frac{1}{2}}\mathscr{D}^{\frac{1}{2}\frac{1}{2}}_{\frac{1}{2}\frac{1}{2}} & \mathscr{D}^{\frac{1}{2}\frac{1}{2}}_{-\frac{1}{2}-\frac{1}{2}}\mathscr{D}^{\frac{1}{2}\frac{1}{2}}_{\frac{1}{2}-\frac{1}{2}} \\ \mathscr{D}^{\frac{1}{2}\frac{1}{2}}_{-\frac{1}{2}\frac{1}{2}}\mathscr{D}^{\frac{1}{2}\frac{1}{2}}_{-\frac{1}{2}\frac{1}{2}} & \mathscr{D}^{\frac{1}{2}\frac{1}{2}}_{-\frac{1}{2}\frac{1}{2}}\mathscr{D}^{\frac{1}{2}\frac{1}{2}}_{-\frac{1}{2}-\frac{1}{2}} & \mathscr{D}^{\frac{1}{2}\frac{1}{2}}_{-\frac{1}{2}-\frac{1}{2}}\mathscr{D}^{\frac{1}{2}\frac{1}{2}}_{-\frac{1}{2}\frac{1}{2}} & \mathscr{D}^{\frac{1}{2}\frac{1}{2}}_{-\frac{1}{2}-\frac{1}{2}}\mathscr{D}^{\frac{1}{2}\frac{1}{2}}_{-\frac{1}{2}-\frac{1}{2}} \end{pmatrix}$$

$$= \begin{pmatrix} \cos\frac{\beta_p}{2}\cos\frac{\beta_e}{2} & -\cos\frac{\beta_p}{2}\sin\frac{\beta_e}{2} & -\sin\frac{\beta_p}{2}\cos\frac{\beta_e}{2} & \sin\frac{\beta_p}{2}\sin\frac{\beta_e}{2} \\ \cos\frac{\beta_p}{2}\sin\frac{\beta_e}{2} & \cos\frac{\beta_p}{2}\cos\frac{\beta_e}{2} & -\sin\frac{\beta_p}{2}\sin\frac{\beta_e}{2} & -\sin\frac{\beta_p}{2}\cos\frac{\beta_e}{2} \\ \sin\frac{\beta_p}{2}\cos\frac{\beta_e}{2} & -\sin\frac{\beta_p}{2}\sin\frac{\beta_e}{2} & \cos\frac{\beta_p}{2}\cos\frac{\beta_e}{2} & -\cos\frac{\beta_p}{2}\sin\frac{\beta_e}{2} \\ \sin\frac{\beta_p}{2}\sin\frac{\beta_e}{2} & \sin\frac{\beta_p}{2}\cos\frac{\beta_e}{2} & \cos\frac{\beta_p}{2}\sin\frac{\beta_e}{2} & \cos\frac{\beta_p}{2}\cos\frac{\beta_e}{2} \end{pmatrix}. \quad (7.1.2)$$

Here all Euler angles are zero except β_p and β_e. (In this section we shall not need to distinguish between representations and ray representations. We call either one an irreducible representation or irrep.)

However, with a nonzero interaction we see that $(R(\alpha_p \beta_p \gamma_p), R(\alpha_e \beta_e \gamma_e))$ is not a symmetry operator unless $\alpha_p = \alpha_e$, $\beta_p = \beta_e$, and $\gamma_p = \gamma_e$. Only the "rigid" rotations which preserve the relative orientation of the two spins are still symmetry operators. Rotation operators such as

$$\left(R(0\beta_p 0), R(0\beta_e 0)\right) = e^{\beta_p J_y^{\text{proton}}/i\hbar} e^{\beta_e J_y^{\text{electron}}/i\hbar} \quad (7.1.3)$$

generated by individual angular-momentum operators J_y^{proton} and J_y^{electron} must have equal angles ($\beta_p = \beta_e \equiv \beta$). When the angles are equal one may write the rigid rotation

$$\left(R(0\beta 0), R(0\beta 0)\right) = e^{(\beta/i\hbar)(J_y^{\text{proton}}+J_y^{\text{electron}})} = e^{(\beta/i\hbar)J_y^{\text{total}}} \quad (7.1.4)$$

in terms of the *total* angular momentum

$$\mathbf{J}^{\text{total}} = \mathbf{J}^{\text{proton}} + \mathbf{J}^{\text{electron}}. \quad (7.1.5)$$

In other words, the individual angular momenta of the proton or the electron may not be conserved when the interaction is on, but the total momentum is constant. By total momentum we mean here total spin angular momentum $\mathbf{J} = \mathbf{S}$. The orbital momentum of an s-state electron is zero. Spin-orbit interactions will be introduced in Section 7.1.C.

The irrep $\mathscr{D}^{\frac{1}{2}} \otimes \mathscr{D}^{\frac{1}{2}}$ of $R_3 \times R_3$ is a reducible representation of the rigid rotation symmetry $R_3^{\text{rigid}} = \{\cdots (R(0\beta 0), R(0\beta 0)) \cdots\}$ generated by $\mathbf{J}^{\text{total}}$ operators. The following transformation (which we derive shortly) of the matrix $\mathscr{D}^{\frac{1}{2}} \otimes \mathscr{D}^{\frac{1}{2}}$ gives a reduced representation:

$$\begin{pmatrix} 1 & 0 & 0 & 0 \\ 0 & \frac{1}{\sqrt{2}} & \frac{1}{\sqrt{2}} & 0 \\ 0 & 0 & 0 & 1 \\ 0 & \frac{1}{\sqrt{2}} & -\frac{1}{\sqrt{2}} & 0 \end{pmatrix} \begin{pmatrix} \cos^2\frac{\beta}{2} & -\sin\frac{\beta}{2}\cos\frac{\beta}{2} & -\sin\frac{\beta}{2}\cos\frac{\beta}{2} & \sin^2\frac{\beta}{2} \\ \sin\frac{\beta}{2}\cos\frac{\beta}{2} & \cos^2\frac{\beta}{2} & -\sin^2\frac{\beta}{2} & -\sin\frac{\beta}{2}\cos\frac{\beta}{2} \\ \sin\frac{\beta}{2}\cos\frac{\beta}{2} & -\sin^2\frac{\beta}{2} & \cos^2\frac{\beta}{2} & -\sin\frac{\beta}{2}\cos\frac{\beta}{2} \\ \sin^2\frac{\beta}{2} & \sin\frac{\beta}{2}\cos\frac{\beta}{2} & \sin\frac{\beta}{2}\cos\frac{\beta}{2} & \cos^2\frac{\beta}{2} \end{pmatrix} \begin{pmatrix} 1 & 0 & 0 & 0 \\ 0 & \frac{1}{\sqrt{2}} & 0 & \frac{1}{\sqrt{2}} \\ 0 & \frac{1}{\sqrt{2}} & 0 & -\frac{1}{\sqrt{2}} \\ 0 & 0 & 1 & 0 \end{pmatrix}$$

$$= \begin{pmatrix} \sin^2\frac{\beta}{2} & \frac{-\sin\beta}{\sqrt{2}} & \sin^2\frac{\beta}{2} & 0 \\ \frac{\sin\beta}{\sqrt{2}} & \cos\beta & \frac{-\sin\beta}{\sqrt{2}} & 0 \\ \sin^2\frac{\beta}{2} & \frac{\sin\beta}{\sqrt{2}} & \cos^2\frac{\beta}{2} & 0 \\ 0 & 0 & 0 & 1 \end{pmatrix}. \quad (7.1.6a)$$

Writing this in matrix notation, we have the following:

$$C^\dagger \mathscr{D}^{\frac{1}{2}}(0\beta 0) \otimes \mathscr{D}^{\frac{1}{2}}(0\beta 0) C = \left(\begin{array}{c|c} \mathscr{D}^1(0\beta 0) & \begin{matrix} 0 \\ 0 \\ 0 \end{matrix} \\ \hline 0 \quad 0 \quad 0 & \mathscr{D}^0 \end{array} \right) = \mathscr{D}^1 \oplus \mathscr{D}^0. \quad (7.1.6b)$$

The standard index notation for this is

$$\sum_{m_1 m_1'} \sum_{m_2 m_2'} C^{\frac{1}{2} \frac{1}{2} J}_{m_1 m_1' M} \mathscr{D}^{\frac{1}{2}}_{m_1 m_2} \mathscr{D}^{\frac{1}{2}}_{m_1' m_2'} C^{\frac{1}{2} \frac{1}{2} J'}_{m_2 m_2' M'} = \delta^{JJ'} \mathscr{D}^J_{mm'}, \quad (7.1.6c)$$

where the transformation coefficients

$$C^{\frac{1}{2}\frac{1}{2}J}_{mm'M} \equiv \left\langle \begin{matrix} \frac{1}{2} & \frac{1}{2} \\ m & m' \end{matrix} \middle| \begin{matrix} J \\ M \end{matrix} \right\rangle \quad (7.1.6d)$$

give states of definite total momentum J

$$\left| \begin{matrix} J \\ M \end{matrix} (\tfrac{1}{2} \times \tfrac{1}{2}) \right\rangle = \sum_{m, m'} \left| \begin{matrix} \frac{1}{2} & \frac{1}{2} \\ m & m' \end{matrix} \right\rangle \left\langle \begin{matrix} \frac{1}{2} & \frac{1}{2} \\ m & m' \end{matrix} \middle| \begin{matrix} J \\ M \end{matrix} \right\rangle$$

$$= \sum_{m, m'} C^{\frac{1}{2}\frac{1}{2}J}_{mm'M} \left| \begin{matrix} \frac{1}{2} & \frac{1}{2} \\ m & m' \end{matrix} \right\rangle. \quad (7.1.6e)$$

At the same time it reduces the product representation $\mathscr{D}^{\frac{1}{2}} \otimes \mathscr{D}^{\frac{1}{2}}$ of the R_3^{Rigid} symmetry as shown in Eqs. (7.1.6a) and (7.1.6b)).

INTRODUCTION TO R_3 AND U_2 COUPLING COEFFICIENTS

The $C^{\frac{1}{2}\frac{1}{2}J}_{mm'M}$ are examples of CLEBSCH-GORDON or COUPLING CO-EFFICIENTS of the rotation group R_3. Coupling coefficients are usually tabulated as they appear in the transformation matrix [see Eq. (7.1.6a)] by arrays of the following form:

$$\left\langle C^{\frac{1}{2}\;\frac{1}{2}\;\cdot}_{\cdot\;\cdot\;\cdot}\right\rangle = \begin{array}{|cc|cccc|} \hline & & 1 & 1 & 1 & 0 \\ \frac{1}{2} \otimes \frac{1}{2} & & 1 & 0 & -1 & 0 \\ \hline \frac{1}{2} & \frac{1}{2} & 1 & \cdot & \cdot & \cdot \\ \frac{1}{2} & -\frac{1}{2} & \cdot & \frac{1}{\sqrt{2}} & \cdot & \frac{1}{\sqrt{2}} \\ -\frac{1}{2} & \frac{1}{2} & \cdot & \frac{1}{\sqrt{2}} & \cdot & -\frac{1}{\sqrt{2}} \\ -\frac{1}{2} & -\frac{1}{2} & \cdot & \cdot & 1 & \cdot \\ \hline \end{array} \quad (7.1.7)$$

The derivation of the coupling coefficients can be done by appealing to the generators $J_z^{\text{total}} = J_z^{\text{proton}} + J_z^{\text{electron}}$ and $J_{\pm}^{\text{total}} = J_{\pm}^{\text{proton}} + J_{\pm}^{\text{electron}}$. Applying J_z^{total} to the state

$$\left|\begin{array}{c}J\\M\end{array}(j_1 \otimes j_2)\right\rangle = \sum_{m_1,m_2} C^{j_1\,j_2\,J}_{m_1 m_2 M}\left|\begin{array}{cc}j_1 & j_2\\m_1 & m_2\end{array}\right\rangle \quad (7.1.8)$$

yields the following:

$$J_z^{\text{total}}\left|\begin{array}{c}J\\M\end{array}(j_1 \otimes j_2)\right\rangle = M\left|\begin{array}{c}J\\M\end{array}(j_1 \otimes j_2)\right\rangle$$

$$= \sum_{m_1,m_2} C^{j_1\,j_2\,J}_{m_1 m_2 M}\left(J_z^{\text{proton}}\left|\begin{array}{c}j_1\\m_1\end{array}\right\rangle\left|\begin{array}{c}j_2\\m_2\end{array}\right\rangle + \left|\begin{array}{c}j_1\\m_1\end{array}\right\rangle J_z^{\text{electron}}\left|\begin{array}{c}j_2\\m_2\end{array}\right\rangle\right)$$

$$= \sum_{m_1,m_2} C^{j_1\,j_2\,J}_{m_1 m_2 M}(m_1 + m_2)\left|\begin{array}{cc}j_1 & j_2\\m_1 & m_2\end{array}\right\rangle.$$

(Here we assume general values j_1 and j_2 of spin for the proton and electron, respectively.) Using Eq. (7.1.6d) and the usual orthonormality

$$\left\langle\begin{array}{cc}j_1 & j_2\\m'_1 & m'_2\end{array}\bigg|\begin{array}{cc}j_1 & j_2\\m_1 & m_2\end{array}\right\rangle = \delta_{m'_1 m_1}\delta_{m'_2 m_2}$$

we see that this implies that either $M = m_1 + m_2$ or $C^{j_1\,j_2\,J}_{m_1 m_2 M} = 0$. The total z

component must be the sum of the z components of the factor states. Therefore the state with the highest z component involving the two factors j_1 and j_2 must be

$$\left|\begin{matrix} j_1 & j_2 \\ j_1 & j_2 \end{matrix}\right\rangle = \left|\begin{matrix} J = j_1 + j_2 \\ M = j_1 + j_2 \end{matrix} (j_1 \otimes j_2) \right\rangle. \qquad (7.1.9)$$

Then we apply the total lowering operator J_-^{total} to this highest state to give

$$J_-^{\text{total}} \left|\begin{matrix} J \\ J \end{matrix} (j_1 \otimes j_2) \right\rangle = J_-^{\text{proton}} \left|\begin{matrix} j_1 \\ j_1 \end{matrix}\right\rangle \left|\begin{matrix} j_2 \\ j_2 \end{matrix}\right\rangle + \left|\begin{matrix} j_1 \\ j_1 \end{matrix}\right\rangle J_-^{\text{electron}} \left|\begin{matrix} j_2 \\ j_2 \end{matrix}\right\rangle,$$

$$\sqrt{2(j_1 + j_2)} \left|\begin{matrix} J \\ J-1 \end{matrix} (j_1 \otimes j_2) \right\rangle = \sqrt{2j_1} \left|\begin{matrix} j_1 & j_2 \\ j_1 - 1 & j_2 \end{matrix}\right\rangle + \sqrt{2j_2} \left|\begin{matrix} j_1 & j_2 \\ j_1 & j_2 - 1 \end{matrix}\right\rangle,$$

$$\left|\begin{matrix} J \\ J-1 \end{matrix} (j_1 \otimes j_2) \right\rangle = \sqrt{\frac{j_1}{j_1+j_2}} \left|\begin{matrix} j_1 & j_2 \\ j_1 - 1 & j_2 \end{matrix}\right\rangle$$

$$+ \sqrt{\frac{j_2}{j_1+j_2}} \left|\begin{matrix} j_1 & j_2 \\ j_1 & j_2 - 1 \end{matrix}\right\rangle, \qquad (7.1.10)$$

where Eq. (5.4.23b) was used. For the proton-electron problem with $j_1 = \frac{1}{2} = j_2$, we have

$$\left|\begin{matrix} J = 1 \\ M = 0 \end{matrix} (\tfrac{1}{2} \otimes \tfrac{1}{2}) \right\rangle = \sqrt{\frac{1}{2}} \left|\begin{matrix} \tfrac{1}{2} & \tfrac{1}{2} \\ -\tfrac{1}{2} & \tfrac{1}{2} \end{matrix}\right\rangle + \sqrt{\frac{1}{2}} \left|\begin{matrix} \tfrac{1}{2} & \tfrac{1}{2} \\ \tfrac{1}{2} & -\tfrac{1}{2} \end{matrix}\right\rangle, \qquad (7.1.10)_x$$

which is the second column of the table of (7.1.7). Lowering this state again gives the third column,

$$\left|\begin{matrix} J = 1 \\ M = -1 \end{matrix} (\tfrac{1}{2} \otimes \tfrac{1}{2}) \right\rangle = \left|\begin{matrix} \tfrac{1}{2} & \tfrac{1}{2} \\ -\tfrac{1}{2} & -\tfrac{1}{2} \end{matrix}\right\rangle.$$

Finally, the fourth column state is obtained by orthogonalizing with the other $M = 0$ state we just derived. The result is

$$\left|\begin{matrix} J = 0 \\ M = 0 \end{matrix} (\tfrac{1}{2} \otimes \tfrac{1}{2}) \right\rangle = \sqrt{\frac{1}{2}} \left|\begin{matrix} \tfrac{1}{2} & \tfrac{1}{2} \\ \tfrac{1}{2} & -\tfrac{1}{2} \end{matrix}\right\rangle - \sqrt{\frac{1}{2}} \left|\begin{matrix} \tfrac{1}{2} & \tfrac{1}{2} \\ -\tfrac{1}{2} & \tfrac{1}{2} \end{matrix}\right\rangle. \qquad (7.1.11)$$

INTRODUCTION TO R_3 AND U_2 COUPLING COEFFICIENTS

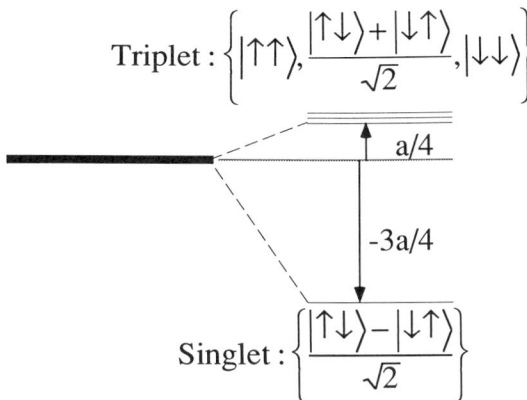

Figure 7.1.1 Singlet-triplet splitting of levels for two interacting spin-$\frac{1}{2}$ particles.

This reduction of a four-by-four representation to a three-by-three and a one-by-one representation implies a splitting of the four spin energy levels of hydrogen into a "triplet" and a "singlet," as shown in Figure 7.1.1.

The observed magnitude of this splitting is small, but very important to radio astronomers. It is one of the more accurately measured quantities: 1,420,405,751.8 ± 0.03 Hz or approximate equivalents 5.88×10^{-5} eV; 0.0474 cm^{-1}, or 1/(21.1 cm). It is the well known 21-cm line which is used to locate atomic hydrogen in intergalactic space.

The spin-spin interaction Hamiltonian which approximately gives this splitting is the Fermi contact interaction.

$$H_{\text{contact}} = a J^{\text{proton}} \cdot J^{\text{electron}}. \qquad (7.1.12a)$$

The name "contact" refers to the dependence of the interaction constant

$$a = (8\pi/3) g_e \beta_e g_p \beta_p |\psi(0)|^2 \qquad (7.1.12b)$$

on the value $\psi(0)$ of the electronic wave function at the proton. For a $(1s)$ wave function one has

$$|\psi_{1s}(0)|^2 = 1/(\pi a_0^3), \qquad (7.1.12c)$$

where $a_0 = \hbar^2/me^2 = 0.5292 \times 10^{-8}$ cm is the Bohr radius. The other constants are the gyromagnetic ratios g_e (= 2.0023) and g_p (= 5.585) and magneton moments β_e (= $e\hbar/2m_e c$ = 0.9273×10^{-20} erg/gauss) and β_p (= $e\hbar/2m_p c$ = 0.50504×10^{-23} erg/gauss) of the electron and proton, respectively. The derivation of the contact interaction follows from the Dirac equation. This may be found in most advanced quantum theory texts.

The eigenvalues of the contact interaction are easy to find if it is rewritten in terms of operators that are diagonal in the bases of triplet and singlet states (7.1.10)$_x$ and (7.1.11):

$$a J^{\text{proton}} \cdot J^{\text{electron}} = \frac{a}{2}\left[(J^{\text{proton}} + J^{\text{electron}})^2 - (J^{\text{proton}})^2 - (J^{\text{electron}})^2\right]$$
$$= \frac{a}{2}\left[(J^{\text{total}})^2 - (J^{\text{proton}})^2 - (J^{\text{electron}})^2\right]. \quad (7.1.13)$$

This trick will be used again many times to evaluate the eigenvalues of interaction operators. The contact eigenvalues are

$$\left\langle \begin{matrix} J \\ M \end{matrix} \middle| H_{\text{contact}} \middle| \begin{matrix} J \\ M \end{matrix} \right\rangle = \frac{a}{2}\left[J(J+1) - \tfrac{1}{2}(\tfrac{1}{2}+1) - \tfrac{1}{2}(\tfrac{1}{2}+1)\right]$$
$$= \begin{cases} a/4 & \text{for the } (J=1) \text{ triplet state,} \\ -3a/4 & \text{for the } (J=0) \text{ singlet state.} \end{cases} \quad (7.1.14)$$

This implies that the magnitude of the singlet-triplet splitting is equal to that of the interaction constant (a). Substituting the magnetic constants given with Eq. (7.1.12) yields the following value:

$$a(\text{calculated}) = 1.500 \times 10^{-18} \text{ erg} = 1422.74 \text{ MHz}. \quad (7.1.15)$$

This agrees with the observed value up to the third decimal place. Further theory of relativistic spin-$\tfrac{1}{2}$ particles is needed to get more accuracy. However, so far no theory has the 10- or 11-place accuracy of the experiment. At the very least this is beyond our present accuracy of knowledge of most fundamental constants.

To continue the coupling analysis for more general values of angular momentum, one needs to finish the lowering job started in Eq. (7.1.10). After N lowerings the result is

$$(J_-)^N \left| \begin{matrix} J = j_1 + j_2 \\ M = j_1 + j_2 \end{matrix} \right\rangle = \sum_{n_1,n_2=0}^{N} \frac{N!}{n_1! n_2!} (J_-^{\text{proton}})^{n_1} \left| \begin{matrix} j_1 \\ j_1 \end{matrix} \right\rangle (J_-^{\text{electron}})^{n_2} \left| \begin{matrix} j_2 \\ j_2 \end{matrix} \right\rangle,$$

where $N = n_1 + n_2$. Using Eq. (5.4.23b) repeatedly one obtains

$$(J_-)^n \left| \begin{matrix} j \\ m \end{matrix} \right\rangle = \sqrt{\frac{(j+m)!(j-m+n)!}{(j-m)!(j+m-n)!}} \left| \begin{matrix} j \\ m-n \end{matrix} \right\rangle. \quad (7.1.16)$$

This gives the desired result

$$\left| \begin{matrix} J = j_1 + j_2 \\ M = m_1 + m_2 \end{matrix} \right\rangle = \sum_{m_1,m_2} C^{j_1 j_2 J}_{m_1 m_2 M} \left| \begin{matrix} j_1 & j_2 \\ m_1 & m_2 \end{matrix} \right\rangle,$$

where $J = j_1 + j_2$ and $M = m_1 + m_2$ are maximal.

$$C^{j_1 j_2 J}_{m_1 m_2 M} = \sqrt{\frac{(J-M)!}{(j_1-m_1)!(j_2-m_2)!} \frac{(J+M)!}{(j_1+m_1)!(j_2+m_2)!} \frac{(2j_1)!(2j_2)!}{(2J)!}}. \quad (7.1.17)$$

These are the coupling coefficients for the cases of highest total momentum $J = j_1 + j_2$. The coefficients for the other possibilities, $J = j_1 + j_2 - 1, j_1 + j_2 - 2, \ldots, |j_1 - j_2|$ are obtained by orthogonalization, or by a generalization of Eq. (7.1.17) which is derived in Section 7.2.D.

B. Two-Electron Atomic Configurations

Consider the elementary electronic structure of the carbon atom which has six electrons in a configuration $(1s)^2(2s)^2(2p)^2$. A very good approximate model can be made by ignoring, at first, the two pairs of electrons in the "closed" $1s$ and $2s$ shells, and treating the atom as though it had only the two $2p$ electrons. The orbital basis of this model is a ninefold product basis $\left|{1 \atop m_1}\right\rangle \left|{1 \atop m_2}\right\rangle$ made from the individual $2p$ orbitals.

If no electrostatic repulsion or interaction of any kind existed between the electrons then these nine states would be degenerate in energy. However, in the presence of electrostatic repulsion one takes the coupled states

$$\left|{L \atop M}(2p)^2\right\rangle = \sum_{m_1, m_2} C^{1\ 1\ L}_{m_1 m_2 M} \left|{1 \atop m_1}\right\rangle \left|{1 \atop m_2}\right\rangle \qquad (7.1.18)$$

to be model eigenstates. They will generally have different energy for different values of total orbit momentum L.

Using Eq. (7.1.17) we find the $L = 2$ coupling coefficients $C^{1\ 1\ 2}_{m_1 m_2 M}$ shown in the left-hand block of the following table:

$1 \otimes 1$		2,2	2,1	2,0	2,-1	2,-2	1,1	1,0	1,-1	0,0
1	1	1	.	.	.	.	.	.	.	.
1	0	.	$\frac{1}{\sqrt{2}}$	.	.	.	$\frac{1}{\sqrt{2}}$	.	.	.
1	-1	.	.	$\frac{1}{\sqrt{6}}$	.	.	.	$\frac{1}{\sqrt{2}}$	.	$\frac{1}{\sqrt{3}}$
0	1	.	$\frac{1}{\sqrt{2}}$	.	.	.	$-\frac{1}{\sqrt{2}}$	.	.	.
0	0	.	.	$\sqrt{\frac{2}{3}}$	.	.	.	.	.	$-\frac{1}{\sqrt{3}}$
0	-1	.	.	.	$\frac{1}{\sqrt{2}}$	.	.	.	$\frac{1}{\sqrt{2}}$	.
-1	1	.	.	$\frac{1}{\sqrt{6}}$	.	.	.	$-\frac{1}{\sqrt{2}}$	.	$\frac{1}{\sqrt{3}}$
-1	0	.	.	.	$\frac{1}{\sqrt{2}}$	.	.	.	$-\frac{1}{\sqrt{2}}$	.
-1	-1	.	.	.	.	1	.	.	.	.

with the left side denoted $\left\langle C^{1\ 1\ \cdot}_{\cdot\ \cdot\ \cdot}\right\rangle =$

(7.1.19)

Using the $L = 2$ results one may orthogonalize and lower to obtain the

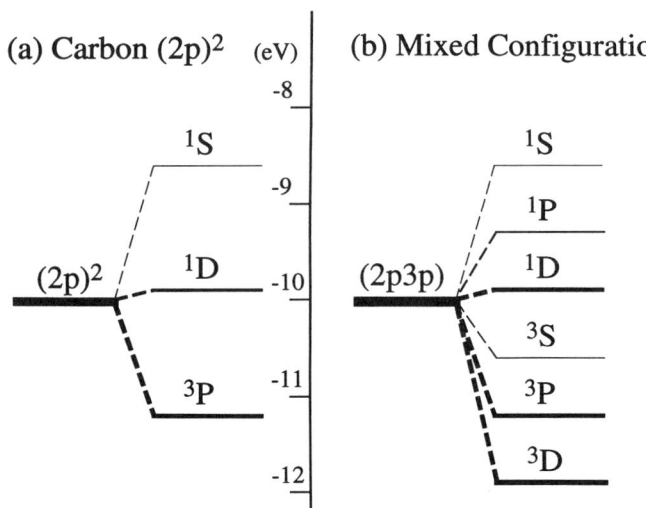

Figure 7.1.2 Atomic ^{2S+1}L multiplet levels for two ($l = 1$) p electrons. (a) Two equivalent electrons. Pauli exclusion principle allows only ^1S, ^1D, and ^3D levels for two p-electrons with the same radial quantum number. (b) Two inequivalent electrons. All combinations of spin and orbit states are allowed.

$L = 1$ and $L = 0$ states. The electrostatic interaction causes a splitting of the nine $\left|{}_{m_1}^{1}\right\rangle\left|{}_{m_2}^{1}\right\rangle$ levels and it results in $L = 2, 1$, and 0 levels. These are labeled by D, P, and S, respectively, in Figure 7.1.2(a), which shows the observed energies for carbon.

Now we should also consider the spins of each $2p$ electron. The correct total spin states have the same form as those derived in Eqs. (7.1.10) and (7.1.11) or Figure 7.1.1. There is a triplet set

$$\left|\begin{matrix}S = 1\\M_S = 1\end{matrix}\right\rangle = \left|\begin{matrix}\tfrac{1}{2} & \tfrac{1}{2}\\ \tfrac{1}{2} & \tfrac{1}{2}\end{matrix}\right\rangle, \quad \left|\begin{matrix}S = 1\\M_S = 0\end{matrix}\right\rangle = \left(\left|\begin{matrix}\tfrac{1}{2} & \tfrac{1}{2}\\ \tfrac{1}{2} & -\tfrac{1}{2}\end{matrix}\right\rangle + \left|\begin{matrix}\tfrac{1}{2} & \tfrac{1}{2}\\ -\tfrac{1}{2} & \tfrac{1}{2}\end{matrix}\right\rangle\right)\Big/\sqrt{2},$$

$$\left|\begin{matrix}S = 1\\M_S = -1\end{matrix}\right\rangle = \left|\begin{matrix}\tfrac{1}{2} & \tfrac{1}{2}\\ -\tfrac{1}{2} & -\tfrac{1}{2}\end{matrix}\right\rangle, \tag{7.1.20a}$$

and a singlet state

$$\left|\begin{matrix}S = 0\\M_S = 0\end{matrix}\right\rangle = \left(\left|\begin{matrix}\tfrac{1}{2} & \tfrac{1}{2}\\ \tfrac{1}{2} & -\tfrac{1}{2}\end{matrix}\right\rangle - \left|\begin{matrix}\tfrac{1}{2} & \tfrac{1}{2}\\ -\tfrac{1}{2} & \tfrac{1}{2}\end{matrix}\right\rangle\right)\Big/\sqrt{2}. \tag{7.1.20b}$$

INTRODUCTION TO R_3 AND U_2 COUPLING COEFFICIENTS 563

The model atomic states are a product of an $\left|{L \atop M}\right\rangle$ orbital state with a triplet spin state to give a triplet atomic term state

$$\left|{}^3LM_LM_S\right\rangle \equiv \left|{L \atop M_L}\right\rangle\left|{S=1 \atop M_S}\right\rangle, \qquad (7.1.21a)$$

and a product of $\left|{L \atop M}\right\rangle$ with a singlet spin state to give a singlet term state

$$\left|{}^1LM_L0\right\rangle \equiv \left|{L \atop M_L}\right\rangle\left|{S=0 \atop 0}\right\rangle. \qquad (7.1.21b)$$

However, it is necessary to consider a rule called the PAULI EXCLUSION PRINCIPLE, which plays an important role. Here it prevents some of these states from existing.

This mysterious principle allows only states which are antisymmetric to permutation of electrons. Note that the orbital states $\left|{2 \atop M}\right\rangle$ and $\left|{0 \atop 0}\right\rangle$ in the table of (7.1.19) are symmetric to the interchange $\left|{1 \atop m_1}{1 \atop m_2}\right\rangle \rightarrow \left|{1 \atop m_2}{1 \atop m_1}\right\rangle$ of the *orbital* states of the electrons. Therefore they can be matched only with the *anti*symmetric $S=0$ singlet *spin* state. Hence, the term states $|^1L=2\rangle$ or (^1D) and $|^1L=0\rangle$ or (^1S) obey the Pauli principle, and occur in Figure 7.1.2(a). Similarly, the antisymmetric orbital states $\left|{L=1 \atop M_L}\right\rangle$ must be matched with the symmetric triplet spin states to give the ^3P term shown in Figure 7.1.2(a). This accounts for all the terms in the ground $(2p)^2$ configuration. In excited configurations like $(2p)(3p)$ other terms can exist, as shown in Figure 7.1.2(b).

Now it is possible to estimate the ordering of ^{2S+1}L terms. Note that the orbital wave function

$$\left\langle \mathbf{x}_1\mathbf{x}_2 \left|{L=1 \atop M_L}\right. (2p)^2\right\rangle = -\left\langle \mathbf{x}_2\mathbf{x}_1 \left|{L=1 \atop M_L}\right. (2p)^2\right\rangle$$

must go to zero as $\mathbf{x}_1 \to \mathbf{x}_2$ because of the antisymmetry of $\left|{1 \atop M_L}\right\rangle$. Therefore, the two electrons in this triplet spin state are never at the same point, and seldom near each other. Therefore, electrostatic repulsion should be less for triplet states than singlets. Indeed ^3P is the ground state of carbon.

Now a classical argument can be made to tell which ^{2S+1}L for a given S should be lowest. One may imagine that to make the greatest L the electrons must orbit in more or less the same direction. Thus they have less chance of colliding and raising the electrostatic energy. Indeed, ^1D is lower than ^1S. These arguments give what are known as HUND'S RULES: The ground state has the highest possible spin and orbital momentum allowed by the Pauli principle.

C. Spin-Orbital Coupling

The outer product is used to describe states corresponding to two properties of a single electron, such as spin and orbit. The states of a single electron in hydrogen can be written

$$\left| \begin{matrix} l \\ m_l \end{matrix} \right\rangle \left| \begin{matrix} \tfrac{1}{2} \\ m_s \end{matrix} \right\rangle \equiv \left| \begin{matrix} l & \tfrac{1}{2} \\ m_l & m_s \end{matrix} \right\rangle, \qquad (7.1.22)$$

and the same coupling formalism can be used to give states

$$\left| \begin{matrix} j \\ m \end{matrix} (l \otimes \tfrac{1}{2}) \right\rangle = \sum_{m_l, m_s} C^{l\ \tfrac{1}{2}\ j}_{m_l m_s m} \left| \begin{matrix} l & \tfrac{1}{2} \\ m_l & m_s \end{matrix} \right\rangle \qquad (7.1.23)$$

of definite total angular momentum j.

Before discussing the spin-orbit interaction, one may predict the form of the splitting of an l level. Several levels of hydrogen are plotted in Figure 7.1.3. They are seen to conform to the reduction

$$\mathscr{C}^\dagger \mathscr{D}^l \otimes \mathscr{D}^{\tfrac{1}{2}} \mathscr{C} = \mathscr{D}^{l+\tfrac{1}{2}} \oplus \mathscr{D}^{l-\tfrac{1}{2}} \qquad (l > 0). \qquad (7.1.24)$$

The splittings are quite small compared to the $(1s)$ energy of -13.6 eV. The shifts and splittings of the excited levels are less than the shift of the

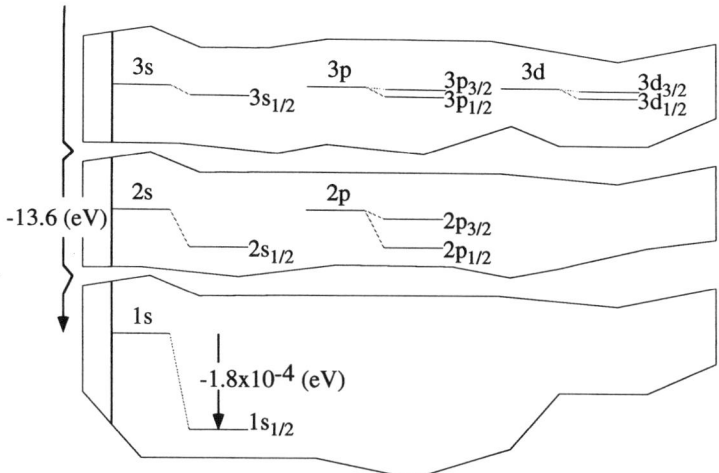

Figure 7.1.3 Fine-structure levels for atomic hydrogen. Hyperfine splittings are not shown.

ground ($1s$) level which is 1.8×10^{-4} eV. The spectral structure which arises from transitions between these levels is called FINE STRUCTURE. Note that splittings are large compared to the hyperfine splittings given by Eq. (7.1.15) in Section 7.1.A. Indeed, each of the fine-electron spin-orbit levels is slightly split into electron spin-nuclear spin or hyperfine levels. Hyperfine splitting is extremely small for p levels, however, since p electrons spend less time near the nucleus.

The spin-orbit Hamiltonian is

$$H_{\text{s.o.}} = a_{\text{s.o.}}(\mathbf{S} \cdot \mathbf{l}), \qquad (7.1.25\text{a})$$

where the interaction constant is the following:

$$a_{\text{s.o.}} = \frac{Z\alpha^2}{2}\langle 1/r^3 \rangle = \frac{Z^4\alpha^2}{2n^3(l+1)(l+\tfrac{1}{2})l}, \qquad (7.1.25\text{b})$$

where $\alpha = e^2/\hbar c \sim \tfrac{1}{137}$ is the fine-structure constant. (Atomic units $\varepsilon = e^2/a_0 = me^4/\hbar^2 \sim 27.21$ eV are used here.) The expectation values for the spin-orbit energies are

$$\left\langle \begin{matrix} j \\ m \end{matrix} (l \otimes \tfrac{1}{2}) \middle| H_{\text{s.o.}} \middle| \begin{matrix} j \\ m \end{matrix} (l \otimes \tfrac{1}{2}) \right\rangle$$
$$= a_{\text{s.o.}}[j(j+1) - l(l+1) - \tfrac{3}{4}] \qquad (j = l \pm \tfrac{1}{2}), \quad (7.1.26)$$

where the total angular momentum $j = l + s$ has been used in the same manner as in the preceding section.

The spin-orbit effects, like most spin interactions, are derived most elegantly using the Dirac equation. This includes all relativistic effects and leads to a very simple expression for energy eigenvalues:

$$\varepsilon_{n,j} = -\frac{Z^2}{2n^2} - \frac{\alpha^2 Z^4}{2n^3}\left(\frac{1}{j+\tfrac{1}{2}} - \frac{3}{4n}\right). \qquad (7.1.27)$$

The Dirac formula predicts that a degeneracy remains between pairs of fine levels such as ($2s_{1/2} - 2p_{1/2}$) and ($3p_{3/2} - 3d_{3/2}$). This degeneracy is lifted by a small quantum electrodynamic perturbation and the splitting is called the Lamb shift. The degeneracy between nl_j and nl'_j states can also be understood in terms of a relativistic generalization of the eccentricity symmetry discussed in Section 5.8.

Fine structure is more easily observed in multielectron configurations. In fact, it dominates the electrostatic energies in some larger atoms since the spin-orbit term in Eq. (7.1.25) varies as Z^4. Model states such as those for

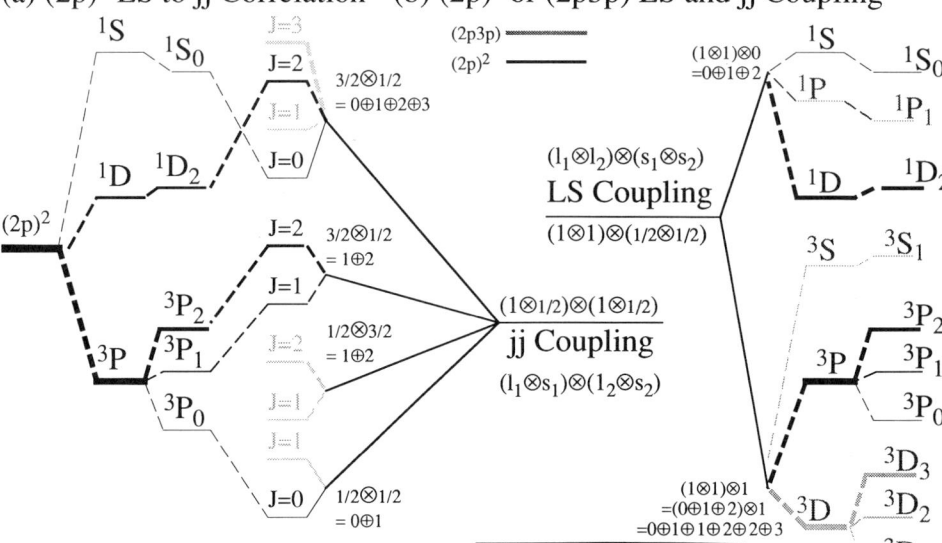

Figure 7.1.4 Fine-structure levels for configurations involving two p electrons. (a) $(2p)^2$ configuration in LS coupling case. (b) Comparison between jj- and LS-coupling cases. In the LS case L and S are good approximate quantum labels and the splitting between levels of different J and the same (L, S) is relative small. In the jj case j and j' are good approximate quantum labels.

carbon can be coupled as follows

$$\left| {}^{2S+1}L_J \ M \right\rangle = \sum_{M_L, M_S} C^{L \ S \ J}_{M_L M_S M} \left| {L \atop M_L} \right\rangle \left| {S \atop M_S} \right\rangle. \quad (7.1.28)$$

This gives, finally, states of definite total electronic angular momentum J. In general, a given ${}^{2S+1}L$ term will split into $2S + 1$ J-states for each $S \leq L$. For example, the triplet levels of carbon split into three, as shown in Figure 7.1.4(a) on the lower right-hand side.

Note that there are *four* angular-momentum factors $(1 \otimes \frac{1}{2} \otimes 1 \otimes \frac{1}{2})$ associated with two p electrons, and one may couple them in different orders. The ordering

$$(1 \otimes 1) \otimes (\tfrac{1}{2} \otimes \tfrac{1}{2}) = (2 \oplus 1 \oplus 0) \otimes (1 \oplus 0)$$
$$= (\cdots \oplus L \cdots) \otimes (\cdots \oplus S \cdots) = (\cdots \oplus J \oplus \cdots)$$

corresponds to what is called LS coupling, and the resulting levels are drawn on the right-hand sides of Figures 7.1.4(a) and 7.1.4(b). [The grayed lines in Figure 7.1.4(b) indicate levels that are present in an $(npn'p)$ configuration but are ruled out of an $(np)^2$ configuration by the Pauli principle.] If spin-orbit splittings are small compared to the electrostatic splittings between ^{2S+1}L terms, then L and S are still useful quantum labels.

On the other hand, a strong spin-orbit perturbation may make the following coupling ordering more appropriate:

$$\left(1 \otimes \tfrac{1}{2}\right) \otimes \left(1 \otimes \tfrac{1}{2}\right) = \left(\tfrac{3}{2} \oplus \tfrac{1}{2}\right) \otimes \left(\tfrac{3}{2} \oplus \tfrac{1}{2}\right)$$
$$= (\cdots \oplus j \cdots) \otimes (\cdots \oplus j' \cdots) = (\cdots \oplus J \oplus \cdots).$$

This corresponds to what is called jj coupling, and the resulting levels are indicated on the left-hand side of Figure 7.1.4(b). For each J term that comes out of the jj combination, there must be a corresponding term with the same J in the LS combination. This is indicated by correlation lines between Figures 7.1.4(a) and 7.1.4(b).

D. Geometrical Interpretation of Angular-Momentum Coupling

In general it is possible to couple two separate angular momenta j_1 and j_2 to make a total momentum equal to $j_3 = j_1 + j_2, j_1 + j_2 - 1, \ldots,$ or $|j_1 - j_2|$.

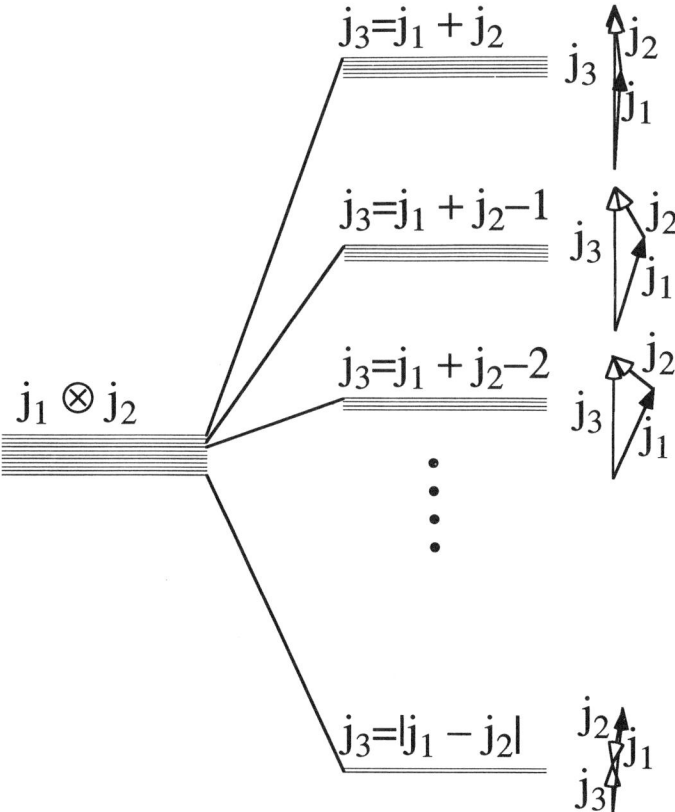

Figure 7.1.5 Vector-addition picture of angular-momentum coupling.

568 THEORY AND APPLICATION OF SYMMETRY REPRESENTATION PRODUCTS

The level diagram corresponding to these coupling possibilities is shown in Figure 7.1.5. Beside each level is sketched a vector triangle composed of sides j_1, j_2, and j_3. The inequality $j_1 + j_2 \geq j_3 \geq |j_1 - j_2|$ is called the TRIANGULAR CONDITION and it must hold in order to have a nonzero coupling coefficient.

It is interesting to look at the vector addition model in terms of the angular-momentum cones which we discussed in Sections 5.4 and 5.5.B. (See Figures 5.4.4 and 5.5.3.) We may imagine that the $C^{j_1\,j_2\,j_3}_{m_1 m_2 m_3}$ is the amplitude for the angular-momentum cone and vector arrangement pictured in Figure 7.1.6.

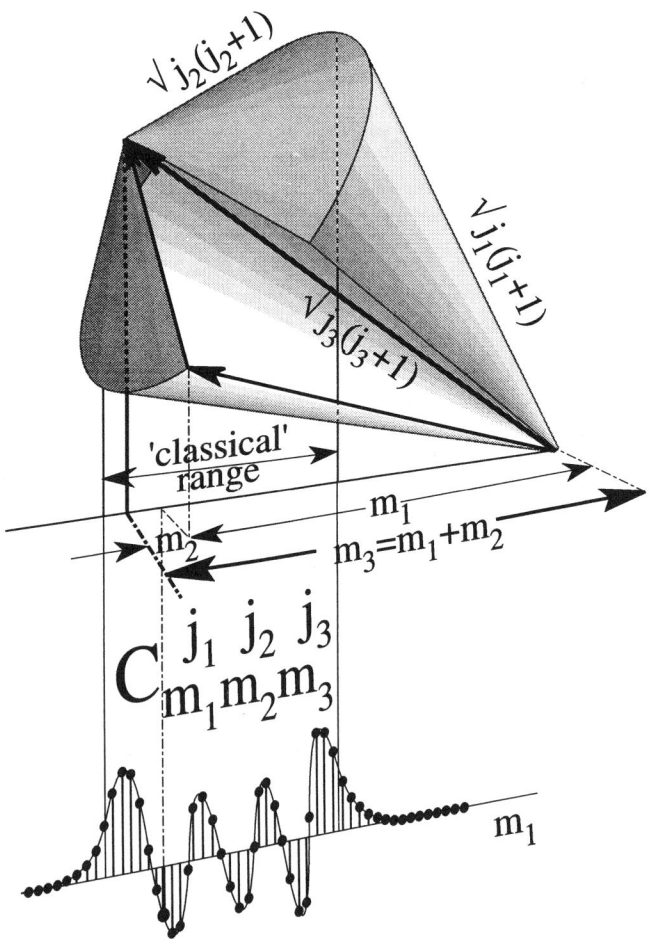

Figure 7.1.6 Angular-momentum cone picture of coupling.

Since z-component conservation requires that $m_1 + m_2 = m_3$ we may plot $C_{m_1 m_2 m_3}^{j_1 \, j_2 \, j_3}$, as a function of a single variable $m_1 = m$ for fixed j_1, j_2, j_3 and m_3. This is done in Figure 7.1.7 for $j_1 = j_2 = 9$ and select values of j_3 and m_3. The plots are quite similar to the ones of $\mathscr{D}_{mm'}^j(0\beta 0)$, which were shown in Figure 5.5.5. Once again we note that the projection of an angular-momentum cone base on the m or z axis defines the "classical" limits, or external inflection points of a "discrete wave." The number of "nodes" of this wave is $j_1 + j_2 - j_3$. We also note that the discrete quantum number m_3 plays the role of the continuous angle β in Figure 5.5.5. Regge and Schulten and Gordon have derived Schrödinger-like differential equations which give solu-

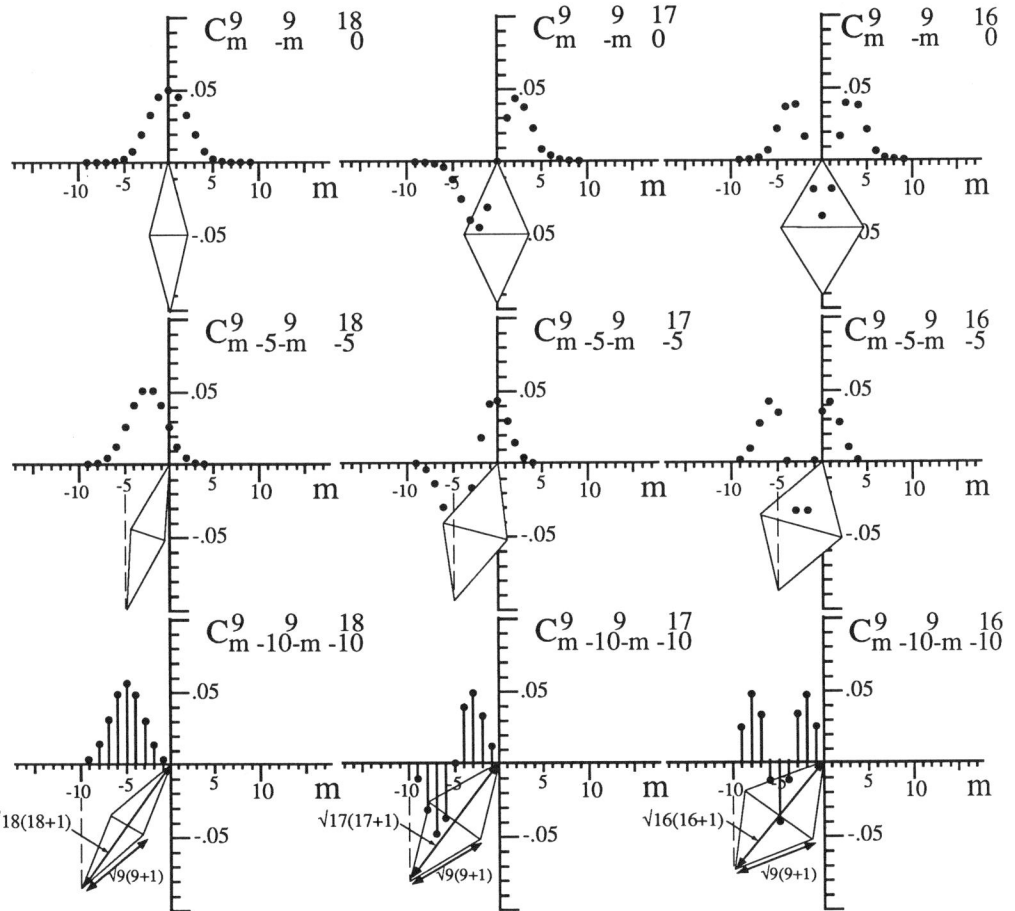

Figure 7.1.7 Plots of coupling coefficients $C_{m \;\; -m-k \;\; -k}^{9 \;\; 9 \;\; J}$ for $J = 18, 17, 16, \ldots$ and $k = 0, -5, -10, \ldots$.

tions that pass near or through those points. (See Additional Reading lists at the end of this chapter and Chapter 5.)

7.2 MATHEMATICAL RELATIONS OF COUPLING AND WIGNER 3j COEFFICIENTS

Before treating more applications of the coupling coefficients we should review the mathematical properties of them. First, the angular-momentum products will be related to common products in vector analysis. Then the fundamental relations between the C coefficients and the D irreps will be derived and used to define symmetry relations. This will lead naturally to the definition of the Wigner 3j coefficients.

A. Scalars, Vectors, and Tensors

Consider two ordinary three-dimensional vectors

$$\mathbf{A} \to \begin{pmatrix} \mathscr{A}_x \\ \mathscr{A}_y \\ \mathscr{A}_z \end{pmatrix} \quad \text{and} \quad \mathbf{B} \to \begin{pmatrix} \mathscr{B}_x \\ \mathscr{B}_y \\ \mathscr{B}_z \end{pmatrix}. \tag{7.2.1}$$

From the theory of vector analysis there are three different types of products between them. There is a SCALAR, DOT, or INNER PRODUCT

$$\mathbf{A} \cdot \mathbf{B} = \mathscr{A}_x \mathscr{B}_x + \mathscr{A}_y \mathscr{B}_y + \mathscr{A}_z \mathscr{B}_z, \tag{7.2.2a}$$

a VECTOR, or CROSS-PRODUCT

$$\mathbf{A} \times \mathbf{B} \to \begin{pmatrix} \mathscr{A}_y \mathscr{B}_z - \mathscr{A}_z \mathscr{B}_y \\ \mathscr{A}_z \mathscr{B}_x - \mathscr{A}_x \mathscr{B}_z \\ \mathscr{A}_x \mathscr{B}_y - \mathscr{A}_y \mathscr{B}_x \end{pmatrix}, \tag{7.2.2b}$$

and a TENSOR, DYADIC, or OUTER PRODUCT

$$\mathbf{AB} \to \begin{pmatrix} \mathscr{A}_x \mathscr{B}_x & \mathscr{A}_x \mathscr{B}_y & \mathscr{A}_x \mathscr{B}_z \\ \mathscr{A}_y \mathscr{B}_x & \mathscr{A}_y \mathscr{B}_y & \mathscr{A}_y \mathscr{B}_z \\ \mathscr{A}_z \mathscr{B}_x & \mathscr{A}_z \mathscr{B}_y & \mathscr{A}_z \mathscr{B}_z \end{pmatrix}. \tag{7.2.2c}$$

It is instructive to note that these three products correspond directly with the three reductions of products between $l = 1(p)$ states which give total $L = 0$, 1, and 2 states, respectively.

The only complication in making this correspondence comes from using two different types of coordinates, namely, CARTESIAN or PLANE-POLARIZATION components ($\mathscr{A}_x, \mathscr{A}_y, \mathscr{A}_z$) on one hand, and R_3 symmetry-defined or circular-polarization components ($\mathscr{A}_1^1, \mathscr{A}_0^1, \mathscr{A}_{-1}^1$) on the other. Let us define the following relations between these sets:

$$\mathscr{A}_1^1 = -(\mathscr{A}_x + i\mathscr{A}_y)/\sqrt{2}, \qquad \mathscr{A}_{-1}^1 = (\mathscr{A}_x - i\mathscr{A}_y)/\sqrt{2},$$

$$\mathscr{A}_x = (\mathscr{A}_{-1}^1 - \mathscr{A}_1^1)/\sqrt{2}, \qquad \mathscr{A}_y = i(\mathscr{A}_{-1}^1 + \mathscr{A}_1^1)/\sqrt{2}, \qquad \mathscr{A}_z = \mathscr{A}_0^1, \quad (7.2.3)$$

so that they conform to those between (x, y, z) and the harmonics or multipole functions $(rY_1^1, rY_0^1, rY_{-1}^1)$ given in Appendix F or Eq. (5.6.19). Then it follows using the table of (7.1.15) that the symmetry-defined product

$$[\mathscr{A}^1 \times \mathscr{B}^1]_0^0 \equiv \sum_{m_1, m_2} C_{m_1 m_2 0}^{1\ 1\ 0} \mathscr{A}_{m_1}^1 \mathscr{B}_{m_2}^1 \quad (7.2.4a)$$

is proportional to the scalar product in Eq. (7.2.2a):

$$[\mathscr{A}^1 \times \mathscr{B}^1]_0^0 = 1/\sqrt{3}\,\mathscr{A}_1^1 \mathscr{B}_{-1}^1 - 1/\sqrt{3}\,\mathscr{A}_0^1 \mathscr{B}_0^1 + 1/\sqrt{3}\,\mathscr{A}_{-1}^1 \mathscr{B}_1^1$$

$$= 1/\sqrt{3}\,(-\mathscr{A}_x - i\mathscr{A}_y)(\mathscr{B}_x - i\mathscr{B}_y)/2 - 1/\sqrt{3}\,\mathscr{A}_z \mathscr{B}_z$$

$$+ 1/\sqrt{3}\,(\mathscr{A}_x - i\mathscr{A}_y)(-\mathscr{B}_x - i\mathscr{B}_y)/2$$

$$= -(\mathscr{A}_x \mathscr{B}_x + \mathscr{A}_y \mathscr{B}_y + \mathscr{A}_z \mathscr{B}_z)/\sqrt{3} = -\mathbf{A} \cdot \mathbf{B}/\sqrt{3}. \quad (7.2.4b)$$

The R_3 symmetry-defined product with $L = 1$ and $M = 0$,

$$[\mathscr{A}^1 \times \mathscr{B}^1]_0^1 \equiv \sum_{m_1, m_2} C_{m_1 m_2 0}^{1\ 1\ 1} \mathscr{A}_{m_1}^1 \mathscr{B}_{m_2}^1, \quad (7.2.5a)$$

is proportional to the $\binom{1}{0}$ or z component of the vector product in Eq. (7.2.2b):

$$[\mathscr{A}^1 \times \mathscr{B}^1]_0^1 = 1/\sqrt{2}\,\mathscr{A}_1^1 \mathscr{B}_{-1}^1 - 1/\sqrt{2}\,\mathscr{A}_{-1}^1 \mathscr{B}_1^1$$

$$= i(-\mathscr{A}_y \mathscr{B}_x + \mathscr{A}_x \mathscr{B}_y)/\sqrt{2} = i(\mathbf{A} \times \mathbf{B})_z/\sqrt{2}. \quad (7.2.5b)$$

Finally, the $L = 2$ products are made of second-rank tensor components:

$$[\mathscr{A}^1 \times \mathscr{B}^1]_m^2 \equiv \sum_{m_1, m_2} C_{m_1 m_2 m}^{1\ 1\ 2} \mathscr{A}_{m_1}^1 \mathscr{B}_{m_2}^1, \quad (7.2.6a)$$

$$[\mathscr{A}^1 \times \mathscr{B}^1]_2^2 = (\mathscr{A}_x \mathscr{B}_x - \mathscr{A}_y \mathscr{B}_y + i\mathscr{A}_x \mathscr{B}_y + i\mathscr{A}_y \mathscr{B}_x)/2, \quad (7.2.6b)$$

$$[\mathscr{A}^1 \times \mathscr{B}^1]_1^2 = -(\mathscr{A}_x \mathscr{B}_z + \mathscr{A}_z \mathscr{B}_x + i\mathscr{A}_y \mathscr{B}_z + i\mathscr{A}_z \mathscr{B}_y)/2,$$

$$[\mathscr{A}^1 \times \mathscr{B}^1]_0^2 = (-\mathscr{A}_x \mathscr{B}_x - \mathscr{A}_y \mathscr{B}_y + 2\mathscr{A}_z \mathscr{B}_z)/\sqrt{6}.$$

572 THEORY AND APPLICATION OF SYMMETRY REPRESENTATION PRODUCTS

All these products are summarized in the following matrix:

$$\mathcal{T}\left\langle xyz \Big| {k \atop q} \right\rangle =$$

$\mathcal{A}_x\mathcal{B}_x$		$[1^2_2]$	$[1^2_1]$	$[1^2_0]$	$[1^2_{-1}]$	$[1^2_{-2}]$	$[1^1_1]$	$[1^1_0]$	$[1^1_{-1}]$	$[1^0_0]$
	$x\ y$	$\dfrac{1}{2}$	.	$-\dfrac{1}{\sqrt{6}}$	.	$\dfrac{1}{2}$	.	.	.	$-\dfrac{1}{\sqrt{3}}$
	$x\ z$	$\dfrac{i}{2}$	.	.	.	$-\dfrac{1}{2}$	.	$\dfrac{i}{\sqrt{2}}$	.	.
	$y\ x$	.	$-\dfrac{1}{2}$	.	$\dfrac{1}{2}$	.	$-\dfrac{1}{2}$	.	$-\dfrac{1}{2}$	.
	$y\ y$	$\dfrac{i}{2}$	.	.	.	$-\dfrac{i}{2}$	.	$\dfrac{i}{\sqrt{2}}$	.	.
	$y\ z$	$-\dfrac{1}{2}$	.	$-\dfrac{1}{\sqrt{6}}$	.	$-\dfrac{1}{2}$	.	.	.	$-\dfrac{1}{\sqrt{3}}$
	$z\ x$	.	$-\dfrac{i}{2}$	.	$-\dfrac{i}{2}$	.	$-\dfrac{i}{2}$	.	$\dfrac{i}{2}$	.
	$z\ y$	.	$-\dfrac{1}{2}$	.	$\dfrac{1}{2}$	.	$\dfrac{1}{2}$	.	$\dfrac{1}{2}$	.
	$z\ z$	.	$-\dfrac{i}{2}$	.	$-\dfrac{i}{2}$	.	$\dfrac{i}{2}$	.	$-\dfrac{i}{2}$	.
		.	.	$\dfrac{2}{\sqrt{6}}$	.	.	.	.	.	$-\dfrac{1}{\sqrt{3}}$

(7.2.7)

This matrix represents a unitary transformation between Cartesian components $\mathcal{T}_{ij} = \mathcal{A}_i \mathcal{B}_j$ and symmetry-defined components $\mathcal{T}^k_q = [\mathcal{A} \times \mathcal{B}]^k_q$ of a second-rank tensor:

$$\vec{T} = \sum_{ij} \mathcal{T}_{ij} \hat{x}_i \hat{x}_j = \sum_{k,g} \mathcal{T}^k_q [xx]^k_q. \qquad (7.2.8)$$

Note that if the two vectors **A** and **B** are the radius vector (i.e., **A** = **B** = **r**) in Eqs. (7.2.4)–(7.2.7), then one obtains the elementary multipole functions such as the following:

$$[r^1 r^1]^2_2 = (x^2 - y^2 + 2ixy)/2, \qquad [r^1 r^1]^2_1 = -(x - iy)z, \ldots \text{etc.}$$
$$= \sqrt{\tfrac{2}{3}}\, X^2_2 \qquad\qquad\qquad = \sqrt{\tfrac{2}{3}}\, X^2_1 \qquad (7.2.9)$$

These polynomial relations were discussed in Section 5.6.C.

Tensor or polynomial algebra can easily be continued up to arbitrary rank or order k, by attaching more vector factors. If we just want the new tensor or polynomial at each stage, i.e., the one which has the highest $L = k$, the necessary coupling formulas are quite simple. For example, to make the kth-order product,

$$\left[\left[[\mathcal{A}^1 \times \mathcal{B}^1]^2 \times C^1 \cdots \right]^{k-1} \times \mathcal{K}^1 \right]^k_q = C^{k-1\,1\,k}_{q-1\,1\,q}\left[\left[\cdots \right]^{k-1}_{q-1}\mathcal{K}^1_1\right.$$
$$+ C^{k-1\,1\,k}_{q0\,q}\left[\left[\cdots \right]^{k-1}_{q}\mathcal{K}^1_0\right.$$
$$+ C^{k-1\,1\,k}_{q+1\,-1\,q}\left[\left[\cdots \right]^{k-1}_{q+1}\mathcal{K}^1_{-1},$$

(7.2.10)

from the $(k-1)$th products one requires the following coupling coefficients:

$$C^{k-1\,1\,k}_{q-1\,1\,q} = \left(\frac{(k+q-1)(k+q)}{(2k-1)(2k)}\right)^{\frac{1}{2}}, \quad C^{k-1\,1\,k}_{q\,\,0\,q} = \left(\frac{k^2-q^2}{k(2k-1)}\right)^{\frac{1}{2}},$$

$$C^{k-1\,1\,k}_{q+1\,1\,q} = \left(\frac{(k-q)(k-q-1)}{2k(2k-1)}\right)^{\frac{1}{2}}. \qquad (7.2.11)$$

These are derived from Eq. (7.1.17).

B. The General R_3 Scalar Coupling

If there is any type of coupling coefficients which should be memorized, it is the ones which make the scalar ($l = 0$) irrep:

$$C^{j_1\,j_2\,0}_{m_1 m_2 0} = \delta_{j_1, j_2} \delta_{m_1, -m_2} (-1)^{j_1-m_1} / \sqrt{2j_j + 1}. \qquad (7.2.12)$$

The product of any two bases belonging to the *same j* can be made combined in the form

$$\left|\begin{matrix}0\\0\end{matrix}\right\rangle = \sum_m \frac{(-1)^{j-m}}{\sqrt{2j+1}} \left|\begin{matrix}j\\m\end{matrix}\right\rangle \left|\begin{matrix}j\\-m\end{matrix}\right\rangle, \qquad (7.2.13)$$

so the result has zero total J and belongs to the scalar irrep $\mathscr{D}^0(R) \equiv 1$. The derivation of this coupling is a little more complicated than it was for real irreps in Section 6.2.A, since the phase is variable.

The derivation is made by using the relation between an irrep $\mathscr{D}^j$, which serves to transform ket vectors:

$$R(\alpha\beta\gamma) \left|\begin{matrix}j\\m\end{matrix}\right\rangle = \sum_{m'} \mathscr{D}^j_{m'm}(\alpha\beta\gamma) \left|\begin{matrix}j\\m'\end{matrix}\right\rangle, \qquad (7.2.14)$$

and its complex conjugate $\mathscr{D}^{j*}$, which does the transformation of bra vectors:

$$\left(R(\alpha\beta\gamma)\left|\begin{matrix}j\\m\end{matrix}\right\rangle\right)^\dagger = \left\langle\begin{matrix}j\\m\end{matrix}\right| R^\dagger(\alpha\beta\gamma) = \sum_{m'} \mathscr{D}^{j*}_{m'm}(\alpha\beta\gamma) \left\langle\begin{matrix}j\\m'\end{matrix}\right|. \qquad (7.2.15)$$

Note that the ket-bra product completeness relation

$$1 = \sum_m \left|\begin{matrix}j\\m\end{matrix}\right\rangle\left\langle\begin{matrix}j\\m\end{matrix}\right| \qquad (7.2.16)$$

is clearly a scalar. One needs to find which combinations of bras $\left\langle\begin{matrix}j\\m'\end{matrix}\right|$ transform like a given ket $\left|\begin{matrix}j\\m\end{matrix}\right\rangle$. This will yield the coefficients that make scalars out of ket-ket products.

To this end let us examine the transpose conjugate,

$$\mathscr{D}^{j\dagger}(\alpha\beta\gamma) = \mathscr{D}^{j^{-1}}(\alpha\beta\gamma) = \mathscr{D}^{j}(-\gamma, -\beta, -\alpha), \qquad (7.2.17)$$

of $\mathscr{D}^j$ and use the fact that it is unitary. By making the substitutions $\alpha \to -\gamma$, $\gamma \to -\alpha$, $m \to -m'$, $m' \to -m$ in Eq. (5.4.45) one finds the following:

$$\mathscr{D}^{j}_{m'm}(\alpha\beta\gamma) = \mathscr{D}^{j}_{-m-m'}(-\gamma, \beta, -\alpha). \qquad (7.2.18)$$

Also, by substituting $\beta \to -\beta$ one obtains

$$\mathscr{D}^{j}_{m'm}(\alpha - \beta\gamma) = (-1)^{-m+m'} \mathscr{D}^{j}_{m'm}(\alpha\beta\gamma). \qquad (7.2.19)$$

Combining the last three equations in turn gives

$$\mathscr{D}^{j\dagger}_{mm'}(\alpha\beta\gamma) = \mathscr{D}^{j}_{mm'}(-\gamma - \beta - \alpha),$$
$$\mathscr{D}^{j*}_{m'm}(\alpha\beta\gamma) = \mathscr{D}^{j}_{-m'-m}(\alpha - \beta\gamma) = \mathscr{D}^{j}_{-m'-m}(\alpha\beta\gamma)(-1)^{-m+m'}. \qquad (7.2.20\text{a})$$

This is substituted into Eq. (7.2.15):

$$\left\langle \begin{matrix} j \\ m \end{matrix} \right| R^{\dagger}(\alpha\beta\gamma) = \sum_{m'} \mathscr{D}^{j}_{-m'-m}(\alpha\beta\gamma)(-1)^{-m+m'} \left\langle \begin{matrix} j \\ m' \end{matrix} \right|,$$

$$(-1)^{-m} \left\langle \begin{matrix} j \\ -m \end{matrix} \right| R^{\dagger}(\alpha\beta\gamma) = \sum_{m'} \mathscr{D}^{j}_{m'm}(\alpha\beta\gamma)(-1)^{-m'} \left\langle \begin{matrix} j \\ -m' \end{matrix} \right|. \qquad (7.2.20\text{b})$$

The resulting equation proves that the bra bases

$$(-1)^{-m} \left\langle \begin{matrix} j \\ -m \end{matrix} \right| \text{ or } (-1)^{j-m} \left\langle \begin{matrix} j \\ -m \end{matrix} \right|$$

transform exactly like the ket bases $\left| \begin{matrix} j \\ m \end{matrix} \right\rangle$. The extra overall phase factor $(-1)^j$ does not affect the transformation. It is conventional to choose $(-1)^{j-m}$ as the phase factor since it is real even when j is half-integral. This completes the proof of Eq. (7.2.12) for scalar coupling.

C. Fundamental Coupling Definitions and Symmetry Relations: The Wigner 3j Coefficient

The R_3 coupling coefficients are defined to be components of orthogonal transformation matrices:

$$\sum_{m_1=-j_1}^{j_1} \sum_{m_2=-j_2}^{j_2} C^{j_1\,j_2\,j_3}_{m_1m_2m_3} C^{j_1\,j_2\,j'_3}_{m_1m_2m'_3} = \delta_{j_3j'_3}\delta_{m_3m'_3}. \qquad (7.2.21)$$

MATHEMATICAL RELATIONS OF COUPLING AND WIGNER 3J COEFFICIENTS

They therefore satisfy completeness relations, too:

$$\sum_{j_3=|j_1-j_2|}^{j_1+j_2} \sum_{m_3=-j_3}^{j_3} C_{m_1 m_2 m_3}^{j_1 j_2 j_3} C_{m_1' m_2' m_3}^{j_1 j_2 j_3} = \delta_{m_1 m_1'} \delta_{m_2 m_2'}. \quad (7.2.22)$$

We shall use these relations with the fundamental irrep orthogonality relation (5.5.28):

$$\int d(\alpha\beta\gamma) \mathscr{D}_{m'n'}^{j'*}(\alpha\beta\gamma) \mathscr{D}_{mn}^{j}(\alpha\beta\gamma) = \delta_{jj'} \delta_{mm'} \delta_{nn'}/2j + 1.$$

Also, we need the reduction equation for $\mathscr{D}^{j_1} \otimes \mathscr{D}^{j_2}$:

$$\sum_{m_1 m_2} \sum_{m_1' m_2'} C_{m_1 m_2 m_3}^{j_1 j_2 j_3} \mathscr{D}_{m_1 m_1'}^{j_1}(\alpha\beta\gamma) \mathscr{D}_{m_2 m_2'}^{j_2}(\alpha\beta\gamma) C_{m_1' m_2' m_3'}^{j_1 j_2 j_3'} = \delta_{j_3 j_3'} \mathscr{D}_{m_3 m_3'}^{j_3}(\alpha\beta\gamma).$$

(7.2.23)

[Recall Eq. (7.1.6c) for an example of this.] The inverse reduction equation,

$$\mathscr{D}_{m_1 m_1'}^{j_1}(\alpha\beta\gamma) \mathscr{D}_{m_2 m_2'}^{j_2}(\alpha\beta\gamma) = \sum_{j_3=|j_1-j_2|}^{j_1+j_2} \sum_{m_3=-j_3}^{j_3} C_{m_1 m_2 m_3}^{j_1 j_2 j_3} \mathscr{D}_{m_3 m_3'}^{j_3}(\alpha\beta\gamma) C_{m_1' m_2' m_3'}^{j_1 j_2 j_3},$$

(7.2.24)

follows from Eq. (7.2.22). Finally, by applying the irrep orthogonality relation one obtains

$$\int d(\alpha\beta\gamma) \mathscr{D}_{m_3 m_3'}^{j_3 *}(\alpha\beta\gamma) \mathscr{D}_{m_1 m_1'}^{j_2}(\alpha\beta\gamma) \mathscr{D}_{m_2 m_2'}^{j_2}(\alpha\beta\gamma) = C_{m_1 m_2 m_3}^{j_1 j_2 j_3} C_{m_1' m_2' m_3'}^{j_1 j_2 j_3}/2j_3 + 1.$$

(7.2.25)

This is an extremely useful result. It is sometimes called the FACTORIZATION LEMMA.

Our first use of this equation will be to suggest a more symmetric form of coupling coefficient. We shall need the conjugation relation,

$$\mathscr{D}_{m_3 m_3'}^{j_3 *} = (-1)^{j_3-m_3-j_3+m_3'} \mathscr{D}_{-m_3 -m_3'}^{j_3} = (-1)^{-j_3+m_3-j_3+m_3'} \mathscr{D}_{-m_3 -m_3'}^{j_3},$$

(7.2.26)

from Eq. (7.2.20a). We also need to become familiar with some of the tricks of phase arithmetic. For example, one can often use the fact that the factor

$$(-1)^{j_3-m_3} = (-1)^{-j_3+m_3} \quad (7.2.27)$$

is positive or negative unity even if j_3 is half-integral. The point is that if j_3 is half-integral then so is m_3, and this means that $j_3 \pm m_3$ is an integer. Also, we will use the fact that for any $\{j_1 j_2 j_3\}$ in $C^{j_1 j_2 j_3}$ one must have

$$1 = (-1)^{2j_1+2j_2+2j_3} = (-1)^{2j_1+2j_2-2j_3} = (-1)^{2j_1-2j_2+2j_3}, \text{ etc.} \quad (7.2.28)$$

This follows since there cannot be an odd number of half-integral j's in $C^{j_1 j_2 j_3}$. Finally, we obtain

$$\int d(\alpha\beta\gamma) \mathscr{D}^{j_3}_{m_3 m_3'}(\alpha\beta\gamma) \mathscr{D}^{j_1}_{m_1 m_1'}(\alpha\beta\gamma) \mathscr{D}^{j_2}_{m_2 m_2'}(\alpha\beta\gamma)$$

$$= \left(\frac{(-1)^{-j_3-m_3} C^{j_1 \ j_2 \ j_3}_{m_1 m_2 -m_3}}{\sqrt{2j_3+1}} \right) \left(\frac{(-1)^{-j_3-m_3'} C^{j_1 \ j_2 \ j_3}_{m_1' m_2' -m_3'}}{\sqrt{2j_3+1}} \right)$$

$$= \left(\frac{(-1)^{j_1-j_2-m_3} C^{j_1 \ j_2 \ j_3}_{m_1 m_2 -m_3}}{\sqrt{2j_3+1}} \right) \left(\frac{(-1)^{j_1-j_2-m_3'} C^{j_1 \ j_2 \ j_3}_{m_1' m_2' -m_3'}}{\sqrt{2j_3+1}} \right). \quad (7.2.29)$$

The last line includes a factor $(-1)^{2j_1-2j_2+2j_3}$ which is always unity. This leads to the conventional WIGNER $3j$ COEFFICIENT:

$$\begin{pmatrix} j_1 & j_2 & j_3 \\ m_1 & m_2 & m_3 \end{pmatrix} \equiv (-1)^{j_1-j_2-m_3} C^{j_1 \ j_2 \ j_3}_{m_1 m_2 -m_3} / \sqrt{2j_3+1}, \quad (7.2.30a)$$

which satisfies

$$\int d(\alpha\beta\gamma) \mathscr{D}^{j_1}_{m_1 m_1'}(\alpha\beta\gamma) \mathscr{D}^{j_2}_{m_2 m_2'}(\alpha\beta\gamma) \mathscr{D}^{j_3}_{m_3 m_3'}(\alpha\beta\gamma)$$

$$= \begin{pmatrix} j_1 & j_2 & j_3 \\ m_1 & m_2 & m_3 \end{pmatrix} \begin{pmatrix} j_1 & j_2 & j_3 \\ m_1' & m_2' & m_3' \end{pmatrix}. \quad (7.2.30b)$$

The idea of this definition is to make a coupling coefficient that has convenient symmetry relations with respect to permutations of the j_1, j_2, and j_3 parts. From Eq. (7.2.30) a number of symmetry relations follow. For example, we have

$$\begin{pmatrix} j_1 & j_2 & j_3 \\ m_1 & m_2 & m_3 \end{pmatrix}^2 = \begin{pmatrix} j_2 & j_1 & j_3 \\ m_2 & m_1 & m_3 \end{pmatrix}^2 = \begin{pmatrix} j_3 & j_2 & j_1 \\ m_3 & m_2 & m_1 \end{pmatrix}^2,$$

$$\begin{pmatrix} j_1 & j_2 & j_3 \\ m_1 & m_2 & m_3 \end{pmatrix} \begin{pmatrix} j_1 & j_2 & j_3 \\ 0 & 0 & 0 \end{pmatrix} = \begin{pmatrix} j_2 & j_1 & j_3 \\ m_1 & m_2 & m_3 \end{pmatrix} \begin{pmatrix} j_2 & j_1 & j_3 \\ 0 & 0 & 0 \end{pmatrix}$$

$$= \begin{pmatrix} j_3 & j_2 & j_1 \\ m_3 & m_2 & m_1 \end{pmatrix} \begin{pmatrix} j_3 & j_2 & j_1 \\ 0 & 0 & 0 \end{pmatrix}.$$

MATHEMATICAL RELATIONS OF COUPLING AND WIGNER 3J COEFFICIENTS   577

However, the phase relations for the individual coefficients are not so obvious, since they depend on the detailed definition of $C^{j_1 \; j_2 \; j_3}_{m_1 m_2 m_3}$ for different products. Wigner's definition is made so the following permutation properties hold:

$$\begin{pmatrix} j_1 & j_2 & j_3 \\ m_1 & m_2 & m_3 \end{pmatrix} = (-1)^{j_1+j_2+j_3} \begin{pmatrix} j_2 & j_1 & j_3 \\ m_2 & m_1 & m_3 \end{pmatrix}$$

$$= (-1)^{j_1+j_2+j_3} \begin{pmatrix} j_3 & j_2 & j_1 \\ m_3 & m_2 & m_1 \end{pmatrix}$$

$$= (-1)^{j_1+j_2+j_3} \begin{pmatrix} j_1 & j_3 & j_2 \\ m_1 & m_3 & m_2 \end{pmatrix}$$

$$= \begin{pmatrix} j_3 & j_1 & j_2 \\ m_3 & m_1 & m_2 \end{pmatrix} = \begin{pmatrix} j_2 & j_3 & j_1 \\ m_2 & m_3 & m_1 \end{pmatrix}. \quad (7.2.31)$$

Also, we have from Eqs. (7.2.30) and (7.2.20b),

$$\begin{pmatrix} j_1 & j_2 & j_3 \\ m_1 & m_2 & m_3 \end{pmatrix} = (-1)^{j_1+j_2+j_3} \begin{pmatrix} j_1 & j_2 & j_3 \\ -m_1 & -m_2 & -m_3 \end{pmatrix}. \quad (7.2.32)$$

The $3-j$ coefficients have easily remembered properties, and one may quickly find various permutation relations for the $C^{j_1 \; j_2 \; j_3}_{m_1 m_2 m_3}$. For example, transposing the first two factors gives the following:

$$C^{j_2 \; j_1 \; j_3}_{m_2 m_1 m_3} = (-1)^{j_2-j_1+m_3}\sqrt{2j_3+1} \begin{pmatrix} j_2 & j_1 & j_3 \\ m_2 & m_1 & -m_3 \end{pmatrix}$$

$$= (-1)^{j_2-j_1+m_3}\sqrt{2j_3+1}(-1)^{j_1+j_2+j_3}\begin{pmatrix} j_1 & j_2 & j_3 \\ m_1 & m_2 & -m_3 \end{pmatrix}$$

$$= (-1)^{j_1+j_2-j_3} C^{j_1 \; j_2 \; j_3}_{m_1 m_2 m_3}. \quad (7.2.33)$$

For $j_1 = j_2 = j$ this gives an important special case of this relation:

$$C^{j \; j \; j_3}_{m_2 m_1 m_3} = (-1)^{2j-j_3} C^{j \; j \; j_3}_{m_1 m_2 m_3}. \quad (7.2.34)$$

This is very useful to have when applying the Pauli principle. Note that when two integral momenta $j_1 = j_2 = n$ are coupled, the even total j_3 states are symmetric, while the odd j_3 states are antisymmetric. For half-integral $j_1 = j_2 = n/2$ the reverse is true.

A permutation of the second two factors gives the following relation:

$$C^{j_1 \; j_3 \; j_2}_{m_1 m_3 m_2} = (-1)^{j_2-j_3+m_1}\sqrt{\frac{2j_2+1}{2j_3+1}} \; C^{j_1 \; j_2 \; j_3}_{-m_1 m_2 m_3}. \quad (7.2.35)$$

D. Clebsch-Gordon and Wigner Coefficient Formulas

In Section 5.4.C the connection between two-dimensional oscillator operators (SU_2) and three-dimensional rotation operators (R_3) was introduced. [Recall Eqs. (5.4.37).] The creation operator algebra of Schwinger and Jordan was used to derive the irrep formula (5.4.45). Now the same methods can be extended to derive coupling coefficient formulas. The methods described here are due to the combined works of many researchers who are developing similar formulas for higher unitary groups $U_3, U_4, \ldots$. [The principal pioneers in this field include Baird, Biedenharn, Bincer, Gelfand, Louck, and Moshinsky. See Additional Reading list at the end of the Chapter.)

Let us begin by representing the product state $\left|\begin{smallmatrix}j_1\\m_1\end{smallmatrix}\right\rangle_A \left|\begin{smallmatrix}j_2\\m_2\end{smallmatrix}\right\rangle_B$ using two pairs of oscillator operators. One pair $\{a_\uparrow^\dagger \equiv a_1^1, a_\downarrow^\dagger \equiv a_2^1\}$ includes the creation operators for momentum j_1 of particle A while the second pair $\{b_\uparrow^\dagger \equiv a_1^2, b_\downarrow^\dagger \equiv a_2^2\}$ includes the creation operators for momentum j_2 of particle B. Note the change in notation in which creation operators a_i^j are indicated without the dagger (†). Destruction operators are denoted by $(a_i^j)^\dagger \equiv \bar{a}_i^j$ in this notation. The product state consists of generalization of Eq. (5.4.38) in which the oscillator analogy is used once for the angular-momentum j_1 state of particle A and then again for the j_2 state of particle B:

$$\left|\begin{matrix}j_1 & j_2\\m_1 & m_2\end{matrix}\right\rangle = \frac{(a_1^1)^{j_1+m_1}(a_2^1)^{j_1-m_1}(a_1^2)^{j_2+m_2}(a_2^2)^{j_2-m_2}|\,\rangle}{[(j_1+m_1)!(j_1-m_1)!(j_2+m_2)!(j_2-m_2)!]^{\frac{1}{2}}} \quad (7.2.36)$$
$$= |j_1+m_1\, j_1-m_1, j_2+m_2\, j_2-m_2\rangle.$$

The empty ket $|\,\rangle = |00,00\rangle$ denotes the vacuum state in which each particle has zero angular momentum.

The problem is to construct the states which are eigenvectors of definite total angular momentum. This may be done by appealing to a generalization of the $\mathscr{D}$-matrix derivation by creation operators. Let us begin with a generalization of Eq. (5.4.41):

$$a_1^{j'} = u_{11}a_1^j + u_{21}a_2^j,$$
$$a_2^{j'} = u_{12}a_1^j + u_{22}a_2^j \qquad (j = 1, 2 \text{ or } A, B). \quad (7.2.37)$$

Here, u_{ij} are components of a general unitary ($u^\dagger = u^{-1}$) unimodular (det $u = 1$) two-by-two matrix. That is, u is an element of SU_2. This transformation of creation operators may be substituted in Eq. (5.4.42) and the entire analysis of the resulting expansion carried out as it was in Section 5.4. The result is an expression for the general irrep $\mathscr{D}^j(u)$ of SU_2:

$$\mathscr{D}^j_{mn}\begin{pmatrix}u_{11} & u_{12}\\u_{21} & u_{22}\end{pmatrix} = \sum_k \frac{[(j+n)!(j-n)!(j+m)!(j-m)!]^{\frac{1}{2}}}{(j-k+m)!(k-m+n)!k!(j-n-k)!}$$
$$\times (u_{11})^{j-k+m}(u_{21})^{k-m+n}(u_{12})^k(u_{22})^{j-n-k}. \quad (7.2.38)$$

MATHEMATICAL RELATIONS OF COUPLING AND WIGNER 3J COEFFICIENTS 579

Since products of representations must be a representation of products we must have $\mathscr{D}^j(u)\mathscr{D}^j(v) = \mathscr{D}^j(uv) \equiv \mathscr{D}^j(w)$, or

$$\sum_n \mathscr{D}^j_{mn}\begin{pmatrix} u_{11} & u_{12} \\ u_{21} & u_{23} \end{pmatrix} \mathscr{D}^j_{nl}\begin{pmatrix} v_{11} & v_{12} \\ v_{21} & v_{22} \end{pmatrix} = \mathscr{D}^j_{ml}\begin{pmatrix} w_{11} & w_{12} \\ w_{21} & w_{22} \end{pmatrix}, \quad (7.2.39a)$$

where

$$w_{ik} = \sum_{j=1}^{2} u_{ij} v_{jk}. \quad (7.2.39b)$$

Now a useful operator results if each u_{ij} component is replaced by a creation operator a_i^j. Let us define a boson polynomial

$$B^j_{mn}(a) = \sum_k \frac{[(j+n)!(j-n)!(j+m)!(j-m)!/(2j)!]^{\frac{1}{2}}}{(j-k+m)!(k-m+n)!k!(j-n-k)!}$$
$$\times (a_1^1)^{j-k+m}(a_2^1)^{k-m+n}(a_1^2)^k(a_2^2)^{j-n-k}, \quad (7.2.40)$$

which is the same as $\mathscr{D}^j_{mn}(a)$ with (a) replacing (u) except for an extra normalization factor $(2j)!$. We will explain the normalization factor shortly. In any case the result is a polynomial that has the correct transformation properties of a total angular-momentum state. To see this consider the same polynomial made of transformed (a') operators given by Eq. (7.2.37). Rewriting this equation as a matrix product, one has

$$a' = \tilde{u} \cdot a, \quad (7.2.41)$$

where

$$\tilde{u}_{ij} = u_{ji}. \quad (7.2.42)$$

Then the representation equation $\mathscr{D}^j(\tilde{u}a) = \mathscr{D}^j(\tilde{u})\mathscr{D}^j(a)$ becomes

$$B^j_{mn}(\tilde{u}a) = \sum_{m'} \mathscr{D}^j_{mm'}(\tilde{u}) B^j_{m'n}(a) \quad (7.2.43a)$$

or

$$B^j_{mn}(a') = \sum_{m'} \mathscr{D}^j_{m'm}(u) B^j_{m'n}(a), \quad (7.2.43b)$$

where the unitarity $(\mathscr{D}^j(u^\dagger) = \mathscr{D}^j(u)^\dagger)$ and conjugation $(\mathscr{D}^j(u^*) = \mathscr{D}^j(u)^*)$ properties of $\mathscr{D}^j$ have been used to write $\mathscr{D}^j_{mn}(\tilde{u})$ as $\mathscr{D}^j_{nm}(u)$. The result shows that the polynomial has correct $\mathscr{D}^j$-transformation properties when the 1 (spin ↑) and 2 (spin ↓) components of the two types of bosons are mixed simultaneously according to Eq. (7.2.37). The transformation corresponds to the rigid rotations of the coupled A and B particles.

The same polynomial also transforms irreducibly under all transformations which mix "A-ness" and "B-ness" of the two different particles. Consider a transformation

$$a_i^{1''} = v_{11} a_i^1 + v_{21} a_i^2,$$
$$a_i^{2''} = v_{12} a_i^1 + v_{22} a_i^2 \quad (i = 1, 2 \text{ or } \uparrow, \downarrow), \quad (7.2.44a)$$

which can be written in matrix notation as follows:

$$a'' = a \cdot v. \quad (7.2.44b)$$

The representation multiplication rules (7.2.39) lead to the following transformation properties:

$$B_{mn}^j(a'') = B_{mn}^j(a \cdot v) = \sum_{n'} \mathscr{D}_{n'n}^j(v) B_{mn'}^j(a). \quad (7.2.45)$$

This is another example which has two commuting groups of transformations for one system. The right- and left-transformation laws (7.2.43) and (7.2.45) are analogous to "laboratory" and "body" transformations (5.5.31) and (5.5.40). Here we have a transformation group $SU_2 = \{ \cdots u \cdots \}$ of states which commutes with a transformation group $SU_2 = \{ \cdots v \cdots \}$ of particles. The combination is often labeled $SU_2 \times SU_2$ or $SU_2 * SU_2$. [The modified cross-product ($*$) notation is used to indicate that the two groups share a common irrep $\mathscr{D}^j$.]

A normalized state results if the boson creation operator (7.2.40) is applied to the vacuum:

$$\left| \begin{matrix} j \\ mn \end{matrix} \right\rangle = B_{mn}^j(a) |00, 00\rangle$$

$$= \left[\frac{(j+m)!(j-m)!}{(2j)!} \right]^{\frac{1}{2}}$$

$$\times \sum_k \left[\frac{(j+n)!}{(n_1^1)!(n_2^1)!} \frac{(j-n)!}{(n_1^2)!(n_2^2)!} \right]^{\frac{1}{2}} |n_1^1 n_2^1, n_1^2 n_2^2\rangle. \quad (7.2.46a)$$

Here the occupation numbers are

$$n_1^1 = j + m - k, \quad n_2^1 = n - m + k, \quad n_1^2 = k, \quad \text{and} \quad n_2^2 = j - n - k. \quad (7.2.46b)$$

The usual oscillator creation rules

$$(a_1^1)^{n_1^1}(a_2^1)^{n_2^1}(a_1^2)^{n_1^2}(a_2^2)^{n_2^2}|00,00\rangle = \sqrt{(n_1^1)!(n_2^1)!(n_1^2)!(n_2^2)!}\,|n_1^1 n_2^1, n_1^2 n_2^2\rangle$$

have been used. [Recall Eqs. (4.4.62).] To check the normalization one may evaluate the scalar product

$$\left\langle \begin{matrix} j \\ mn \end{matrix} \middle| \begin{matrix} j \\ mn \end{matrix} \right\rangle = \frac{(j+m)!(j-m)!}{2j!}$$

$$\times \sum_k \frac{(j+n)!}{(j+m-k)!(n-m+k)!}\frac{(j-n)!}{k!(j-n-k)!},$$

where the orthonormality relations

$$\langle a'b', c'd' | ab, cb \rangle = \delta_{a'a}\delta_{b'b}\delta_{c'c}\delta_{d'd}$$

for oscillator eigenstates are assumed. Then the relation

$$\sum_m \binom{p}{m}\binom{q}{s-m} = \binom{p+q}{s} \qquad (7.2.47)$$

for binomial coefficients $\binom{p}{m} \equiv p!/m!(p-m)!$ may be used. This relation is obtained by equating terms of binomial expansions:

$$(x+y)^{p+q} = \sum_s \binom{p+q}{s} x^s x^{p+q-s} = (x+y)^p(x+y)^q.$$

The desired normalization is then proven.

$$\left\langle \begin{matrix} j \\ mn \end{matrix} \middle| \begin{matrix} j \\ mn \end{matrix} \right\rangle = \frac{(j+m)!(j-m)!}{(2j)!}\sum_k \binom{j+n}{j+m-k}\binom{j-n}{k}$$

$$= \frac{(j+m)!(j-m)!}{(2j)!}\binom{2j}{j+m} = 1.$$

The boson state (7.2.46) is composed of exactly $(2j)$ bosons. This is the minimum number of bosons needed to make a state of total angular momentum j.

Another boson polynomial of interest is the determinantal combination

$$\det(a) \equiv D(a) \equiv a_1^1 a_2^2 - a_2^1 a_1^2. \qquad (7.2.48)$$

This combination is invariant to all $SU_2 \times SU_2$ transformations because of the unimodularity conditions (det $u = 1 =$ det v). The invariance then follows from the elementary properties of determinants:

$$\det(\tilde{u}av) = \det(u)\det(v)\det(a)$$
$$= \det(a). \qquad (7.2.49)$$

A totally scalar ($j = 0$) state of ($N = 2d$) bosons has the form

$$|j = 0 \quad N = 2d\rangle = [(2d)!(2d+1)!]^{-\frac{1}{2}}(a_1^1 a_2^2 - a_2^1 a_1^2)^{2d}|00,00\rangle$$

$$= [(2d)!(2d+1)!]^{-\frac{1}{2}}$$
$$\times \sum_r \binom{2d}{r}(a_1^1 a_2^2)^r(-a_2^1 a_1^2)^{2d-r}|00,00\rangle$$

$$= [(2d)!(2d+1)!]^{-\frac{1}{2}}$$
$$\times \sum_r (2d)!(-1)^r|rr, \quad 2d-r \quad 2d-r\rangle$$

$$= (2d+1)^{-\frac{1}{2}}\sum_r (-1)^r|rr, \quad 2d-r \quad 2d-r\rangle.$$
$$(7.2.50)$$

The sum over r contains exactly ($2d + 1$) terms, so the normalization of the scalar state is seen to be correctly chosen so that $\langle j = 0|j = 0\rangle = 1$.

The Clebsch-Gordon coefficient is the scalar product

$$C^{j_1 j_2 j}_{m_1 m_2 m} = \left\langle \begin{matrix} j_1 & j_2 \\ m_1 & m_2 \end{matrix} \middle| \begin{matrix} j \\ mn \end{matrix} \right\rangle$$

between the uncoupled state (7.2.36) and a coupled state similar to the one in Eq. (7.2.46). However, the scalar product of boson states will vanish unless three criteria are met. First, the eigenvalues of m components of momentum determined by

$$J(m) \equiv J_z^A + J_z^B = \hbar/2\left(\bar{a}_1^1 a_1^1 - \bar{a}_2^1 a_2^1 + \bar{a}_1^2 a_1^2 - \bar{a}_2^2 a_2^2\right) \quad (7.2.51a)$$

must be equal; i.e.,

$$m_1 + m_2 = m. \qquad (7.2.51b)$$

Second, the eigenvalues of

$$J(n) = \hbar/2\left(\bar{a}_1^1 a_1^1 + \bar{a}_2^1 a_2^1 - \bar{a}_1^2 a_1^2 - \bar{a}_2^2 a_2^2\right) \qquad (7.2.52a)$$

for the other commuting SU_2 generator must be equal, i.e.,

$$j_1 - j_2 = n. \tag{7.2.52b}$$

Finally, the total number

$$N = 2(j_1 + j_2) \tag{7.2.53}$$

of bosons must be the same for the states $\left|\begin{matrix} j_1 & j_2 \\ m_1 & m_2 \end{matrix}\right\rangle$ and $\left|\begin{matrix} j \\ mn \end{matrix}\right\rangle$. The last requirement generally means that some number $(N - 2j)$ of bosons in scalar determinantal combinations must be added to $\left|\begin{matrix} j \\ mn \end{matrix}\right\rangle$ to give the following general coupled state:

$$\left|\begin{matrix} j \\ mn \end{matrix} N\right\rangle = \left[\frac{(2j+1)(j+m)!(j-m)!}{(N/2-j)!(N/2+j+1)!}\right]^{\frac{1}{2}}$$
$$\times \sum_k \left[\frac{(j+n)!}{(n_1^1)!(n_2^1)!} \frac{(j-n)!}{(n_1^2)!(n_2^2)!}\right] (a_1^1 a_2^2 - a_2^1 a_1^2)^{N/2-j} \left|n_1^1 n_2^1, n_1^2 n_2^2\right\rangle. \tag{7.2.54}$$

Here the occupation numbers n_j^i from Eq. (7.2.46b) are used. The normalization factors for this state are difficult to prove, and we refer to the work by A. Bincer for their derivation. Expansion of determinantal expression gives

$$(a_1^1 a_2^2 - a_2^1 a_1^2)^{N/2-j} = \sum_r (-1)^r \frac{(N/2-j)!}{r!(N/2-j-r)!} (a_1^1 a_2^2)^{N/2-r} (a_2^1 a_1^2)^r.$$

Insertion of this into the coupled state gives

$$\left|\begin{matrix} j \\ mn \end{matrix} N\right\rangle$$
$$= \left[\frac{(2j+1)(j+m)!(j-m)!}{(N/2-j)!(N/2+j+1)!}\right]^{\frac{1}{2}}$$
$$\times \sum_{k,r} (-1)^r \frac{[(j+n)!(j-n)!(m_1^1)!(m_2^1)!(m_1^2)!(m_2^2)!]^{\frac{1}{2}}}{r!(N/2-j-r)!(n_1^1)!(n_2^1)!(n_1^2)!(n_2^2)!} (N/2-j)!$$
$$\times \left|m_1^1 m_2^1, m_1^2 m_2^2\right\rangle, \tag{7.2.55a}$$

where the new occupation numbers are as follows, according to Eqs.

(7.2.51)–(7.2.53):

$$m_1^1 = n_1^1 + \frac{N}{2} - j - r = j_1 + j_2 + m - k - r \quad (= j_1 + m_1),$$

$$m_2^1 = n_2^1 + r = j_1 - j_2 - m + k + r \quad (= j_1 - m_1),$$

$$m_1^2 = n_1^2 + r = k + r \quad (= j_2 + m_2),$$

$$m_2^2 = n_2^2 + \frac{N}{2} - j - r = 2j_2 - r - k \quad (= j_2 - m_2). \quad (7.2.55b)$$

The equalities written in parentheses on the right must hold when this state is matched with (7.2.36) to derive the coupling coefficient. The r sum is eliminated then, since

$$r = j_2 - k + m - m_1 = j_2 + m_2 - k. \quad (7.2.56)$$

The resulting coupling coefficient formula has $m_3 = m_1 + m_2$, $n = j_1 - j_2$, and $N = 2j_1 + 2j_2$.

$$C_{m_1 m_2 m_3}^{j_1 j_2 j_3} = \left\langle \begin{matrix} j_1 & j_2 \\ m_1 & m_2 \end{matrix} \middle| \begin{matrix} j \\ m_3 n \end{matrix} \quad N \right\rangle$$

$$= (-1)^{j_2 + m_2} \left[\frac{(2j_3 + 1)(j_1 + j_2 - j_3)!(j_3 + j_1 - j_2)!(j_2 + j_3 - j_1)!}{(j_1 + j_2 + j_3 + 1)!} \right]^{\frac{1}{2}}$$

$$\cdot [(j_1 + m_1)!(j_1 - m_1)!(j_2 + m_2)!(j_2 - m_2)!(j_3 + m_3)!(j_3 - m_3)!]^{\frac{1}{2}}$$

$$\cdot \sum_k \frac{(-1)^k}{(j_2 + m_2 - k)!(j_1 - j_3 - m_2 + k)!(j_3 + m_3 - k)!(j_1 - j_2 - m_3 + k)!k!(j_3 - j_1 + j_2 - k)!}.$$

(7.2.57)

The standard formula for the Wigner coefficient

$$\begin{pmatrix} j_1 & j_2 & j_3 \\ m_1 & m_2 & m_3 \end{pmatrix} = (-1)^{j_1 - j_2 - m_3} C_{m_1 m_2 -m_3}^{j_1 j_2 j_3} / (2j_3 + 1)^{\frac{1}{2}}$$

follows and is given in Appendix F.

7.3 ROTATIONAL TENSOR OPERATORS AND THE WIGNER-ECKART THEOREM

The theory and application of R_3 tensor operators will be introduced in this section. The development will be similar to that of Section 6.4, where the tensor operators of finite symmetry were introduced. First, the construction of tensor operators from bra and ket bases will be shown using simple

ROTATIONAL TENSOR OPERATORS AND THE WIGNER-ECKART THEOREM 585

examples. Then the Wigner-Eckart theorem will be discussed and applied to atomic crystal field problems which were first encountered in Section 5.6. Racah coefficients will be introduced in a treatment of a two-electron crystal field splitting.

A. Construction of R_3 Tensor Operators

For each set of irrep bases $\left|{j \atop m}\right\rangle$: $\left\{\left|{j \atop j}\right\rangle, \left|{j \atop j-1}\right\rangle, \ldots, \left|{j \atop -j}\right\rangle\right\}$ of $2j + 1$ ket vectors there is an equal number of bra vectors

$$(-1)^{j-m}\left\langle {j \atop -m}\right|: \quad \left\{\left\langle {j \atop -j}\right|, -\left\langle {j \atop -j+1}\right|, \ldots, (-1)^{2j}\left\langle {j \atop j}\right|\right\},$$

which belong to the same irrep $\mathscr{D}^j$. (Recall Section 7.2.B.) By combining them using coupling coefficients one may construct the IRREDUCIBLE TENSORIAL OPERATORS,

$$T(jj)_q^k = \sum_{m,m'} C_{mm'q}^{j\ j\ k}\left|{j \atop m}\right\rangle\left\langle {j \atop -m'}\right|(-1)^{j-m'}, \tag{7.3.1a}$$

which transform according to irrep $\mathscr{D}^k$ ($2j \geq k \geq 0$) as follows:

$$R(\alpha\beta\gamma)T(jj)_q^k R^\dagger(\alpha\beta\gamma) = \sum_{q'}\mathscr{D}_{q'q}^k(\alpha\beta\gamma)T(jj)_{q'}^k. \tag{7.3.1b}$$

This is similar to the construction which was introduced in Section 6.4.A for finite-group irreps. The only difference is that now we must account for the different transformation behavior of the bra vectors.

If two or more sets $\left\{\left|{j_1 \atop j_1}\right\rangle, \ldots, \left|{j_1 \atop m_1}\right\rangle, \ldots\right\}, \left\{\left|{j_2 \atop j_2}\right\rangle, \ldots, \left|{j_2 \atop m_2}\right\rangle, \ldots\right\}, \ldots$ of angular-momentum states need to be considered, then one may make combinations

$$T(j_1j_2)_q^k = \sum_{m_1,m_2} C_{m_1m_2q}^{j_1\ j_2\ k}\left|{j_1 \atop m_1}\right\rangle\left\langle {j_2 \atop -m_2}\right|(-1)^{j_2-m_2}, \tag{7.3.2}$$

where

$$j_1 + j_2 \geq k \geq |j_1 - j_2|.$$

All combinations are needed to make a complete set of irreducible tensor operators acting on the bases. The first few examples treated in the following will be based on a single set $\{|_m^j\rangle\}$ of angular-momentum states, and so there will only be one combination $T_q^k(jj) = T_q^k$ for each $k = 0, 1, \ldots, 2j$, and q. ($k \geq q \geq -k$).

(a) Tensor Operators for Spin-$\frac{1}{2}$ States From the spin-$\frac{1}{2}$ states $\left\{\left|\frac{1}{2}\right\rangle,\left|-\frac{1}{2}\right\rangle\right\}$ one may construct the tensor operators

$$T_q^k = \sum_{m_1} C_{m_1 m_2 q}^{\frac{1}{2} \frac{1}{2} k} \left|\frac{1}{2} m_1\right\rangle\left\langle\frac{1}{2} -m_2\right|(-1)^{\frac{1}{2}-m_2}, \quad (7.3.3)$$

using $\frac{1}{2} \otimes \frac{1}{2}$ coupling coefficients (7.1.7). The results are given in the following with their representations in the spin-$\frac{1}{2}$ basis:

$$T_{-1}^1 = \begin{pmatrix} 0 & 0 \\ -1 & 0 \end{pmatrix} \qquad T_0^1 = \frac{1}{\sqrt{2}}\begin{pmatrix} -1 & 0 \\ 0 & 1 \end{pmatrix} \qquad T_1^1 = \begin{pmatrix} 0 & 1 \\ 0 & 0 \end{pmatrix}$$

$$= -\left|\frac{1}{2} -\frac{1}{2}\right\rangle\left\langle\frac{1}{2} -\frac{1}{2}\right|, \quad = -\frac{1}{\sqrt{2}}\left[\left|\frac{1}{2} \frac{1}{2}\right\rangle\left\langle\frac{1}{2} \frac{1}{2}\right| - \left|\frac{1}{2} -\frac{1}{2}\right\rangle\left\langle\frac{1}{2} -\frac{1}{2}\right|\right], \quad = \left|\frac{1}{2} \frac{1}{2}\right\rangle\left\langle\frac{1}{2} -\frac{1}{2}\right|,$$

$$T_0^0 = -\frac{1}{\sqrt{2}}\begin{pmatrix} 1 & 0 \\ 0 & 1 \end{pmatrix}$$

$$= -\frac{1}{\sqrt{2}}\left[\left|\frac{1}{2} \frac{1}{2}\right\rangle\left\langle\frac{1}{2} \frac{1}{2}\right| + \left|\frac{1}{2} -\frac{1}{2}\right\rangle\left\langle\frac{1}{2} -\frac{1}{2}\right|\right]. \quad (7.3.3)_x$$

The first three operators form a vector set. Consider the following Cartesian combinations:

$$T_x \equiv -\frac{T_{-1}^1 - T_1^1}{\sqrt{2}} \qquad T_y \equiv -i\frac{T_{-1}^1 + T_1^1}{\sqrt{2}} \qquad T_z \equiv -T_0^1$$

$$= \frac{1}{\sqrt{2}}\begin{pmatrix} 0 & 1 \\ 1 & 0 \end{pmatrix} \qquad = \frac{1}{\sqrt{2}}\begin{pmatrix} 0 & -i \\ i & 0 \end{pmatrix} \qquad = \frac{1}{\sqrt{2}}\begin{pmatrix} 1 & 0 \\ 0 & -1 \end{pmatrix}$$

$$\equiv \frac{1}{\sqrt{2}}\sigma_x \qquad \equiv \frac{1}{\sqrt{2}}\sigma_y \qquad \equiv \frac{1}{\sqrt{2}}\sigma_z$$

$$\equiv \sqrt{2} J_x \qquad \equiv \sqrt{2} J_y \qquad \equiv \sqrt{2} J_z. \quad (7.3.4)$$

Except for an overall minus phase, these relations correspond to Eq. (7.2.3). The resulting Cartesian tensors are proportional to the Pauli spinor operators:

$$\sigma_x \to \begin{pmatrix} 0 & 1 \\ 1 & 0 \end{pmatrix}, \quad \sigma_y \to \begin{pmatrix} 0 & -i \\ i & 0 \end{pmatrix}, \quad \sigma_z \to \begin{pmatrix} 1 & 0 \\ 0 & -1 \end{pmatrix}, \quad (7.3.5)$$

or the spin-$\frac{1}{2}$ angular-momentum operators [recall the original definitions in Eq. (5.5.3)]:

$$J_x = \sigma_x/2, \qquad J_y = \sigma_y/2, \qquad J_z = \sigma_z/2. \quad (7.3.6)$$

An explicit example of the transformation behavior required by Eq. (7.3.1b) is represented in the following:

$$
\begin{array}{cccc}
R(0\beta 0) & T_0^1 & R^\dagger(0\beta 0) & = T_0^{1'} \\
\downarrow & \downarrow & \downarrow & \downarrow
\end{array}
$$

$$
\begin{pmatrix} \cos\frac{\beta}{2} & -\sin\frac{\beta}{2} \\ \sin\frac{\beta}{2} & \cos\frac{\beta}{2} \end{pmatrix} \begin{pmatrix} -1/\sqrt{2} & 0 \\ 0 & 1/\sqrt{2} \end{pmatrix} \begin{pmatrix} \cos\frac{\beta}{2} & \sin\frac{\beta}{2} \\ -\sin\frac{\beta}{2} & \cos\frac{\beta}{2} \end{pmatrix} = -\frac{1}{\sqrt{2}} \begin{pmatrix} \cos\beta & \sin\beta \\ \sin\beta & -\cos\beta \end{pmatrix}
$$

$$
= \mathscr{D}_{10}^1(0\beta 0) T_1^1 \qquad + \mathscr{D}_{00}^1(0\beta 0) T_0^1 \qquad + \mathscr{D}_{-10}^1(0\beta 0) T_{-1}^1
$$

$$
\downarrow \qquad\qquad \downarrow \qquad\qquad \downarrow
$$

$$
= \frac{-\sin\beta}{\sqrt{2}} \begin{pmatrix} 0 & 1 \\ 0 & 0 \end{pmatrix} + \cos\beta \begin{pmatrix} -1/\sqrt{2} & 0 \\ 0 & 1/\sqrt{2} \end{pmatrix} + \frac{\sin\beta}{\sqrt{2}} \begin{pmatrix} 0 & 0 \\ -1 & 0 \end{pmatrix}.
$$

(7.3.7)

The Cartesian form of this equation is simpler. Multiplying the angular-momentum form by $-1/\sqrt{2}$ and using Eq. (7.3.4) yields

$$J_z(\text{rotated}) \equiv R(0\beta 0) J_z R^\dagger(0\beta 0) = \sin\beta J_x + \cos\beta J_z. \qquad (7.3.8)$$

It should be clear now that products of spinor bases form operator quantities that behave like ordinary vectors in 3-space. In this sense spinors are "square roots" of vectors. In order to appreciate the physical meaning of Eq. (7.3.8) one may take its expectation value in an arbitrary state $|\psi\rangle$:

$$\langle\psi|J_z(\text{rotated})|\psi\rangle = \sin\beta\langle\psi|J_x|\psi\rangle + \cos\beta\langle\psi|J_z|\psi\rangle. \qquad (7.3.9)$$

Figure 7.3.1 shows how the average or expectation value of the component of J on the (z-rotated) axis is given by Eq. (7.3.9). However, one should remember that the *actual* values for the component in a Stern-Gerlach spin-$\frac{1}{2}$ analyzer will be $\hbar/2$ for some fraction for events and $-\hbar/2$ for the rest. (Recall discussions at the beginning of Chapter 1 and in Section 5.5.B.) By expanding the expectation matrix, one obtains the result in terms of the fractions or probabilities $\left|\langle\psi|R|{}^{\frac{1}{2}}_m\rangle\right|^2$ (as usual, let $\hbar = 1$):

$$\langle\psi|R(\alpha\beta\gamma)J_z R^\dagger(\alpha\beta\gamma)|\psi\rangle = \sum_m \sum_{m'} \langle\psi|R|{}^{\frac{1}{2}}_m\rangle \langle{}^{\frac{1}{2}}_m|J_z|{}^{\frac{1}{2}}_{m'}\rangle \langle{}^{\frac{1}{2}}_{m'}|R^\dagger|\psi\rangle$$

$$= \sum_m \left|\langle\psi|R|{}^{\frac{1}{2}}_m\rangle\right|^2 m$$

$$= \tfrac{1}{2}\left|\langle\psi|R|{}^{\frac{1}{2}}_{\frac{1}{2}}\rangle\right|^2 - \tfrac{1}{2}\left|\langle\psi|R|{}^{\frac{1}{2}}_{-\frac{1}{2}}\rangle\right|^2. \qquad (7.3.10a)$$

Each quantum amplitude $\langle\psi|R|{}^{\frac{1}{2}}_m\rangle$ depends upon the initial unrotated ampli-

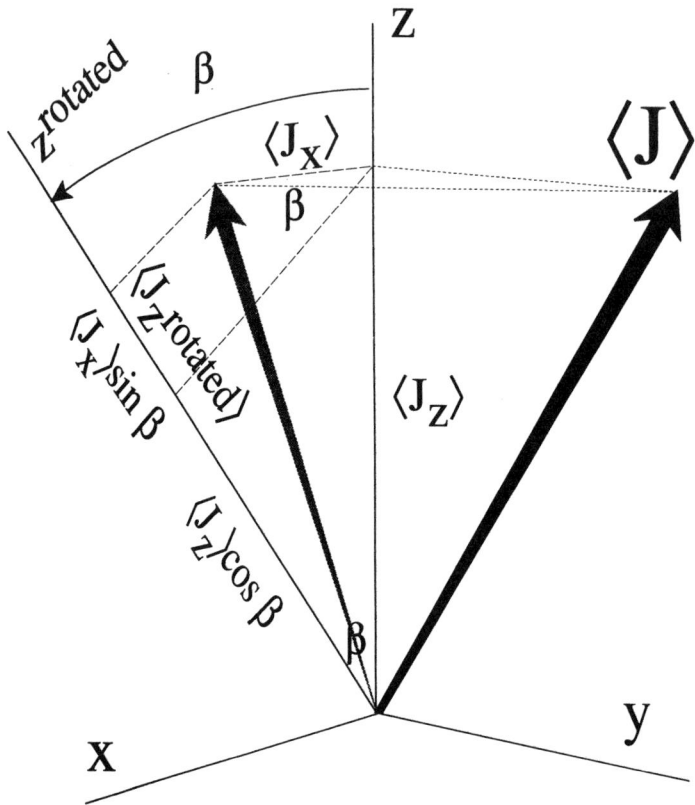

Figure 7.3.1 Geometry of expectation vector $\langle J \rangle$. Component on rotated z axis is determined by ordinary vector geometry. This picture collaborates the tensor equation (7.3.9).

tudes $\psi_{m'} = \left\langle \psi \left| \begin{array}{c} \frac{1}{2} \\ m' \end{array} \right. \right\rangle$ and the rotation matrices:

$$\left\langle \psi | R(\alpha\beta\gamma) \left| \begin{array}{c} \frac{1}{2} \\ m \end{array} \right. \right\rangle = \sum_{m'} \psi_{m'} \mathscr{D}^{\frac{1}{2}}_{m'm}(\alpha\beta\gamma). \tag{7.3.10b}$$

The well-known classical behavior of the angular-momentum vector emerges after many spin-$\frac{1}{2}$ states have passed the analyzer. The "expectation vector" for $|\psi\rangle$ states has components

$$\langle J_x \rangle = \langle \psi | J_x | \psi \rangle, \quad \langle J_y \rangle = \langle \psi | J_y | \psi \rangle, \quad \langle J_z \rangle = \langle \psi | J_z | \psi \rangle, \tag{7.3.11}$$

as indicated in Figure 7.3.1. This vector provides a useful picture of the properties of particles in a given pure spin-$\frac{1}{2}$ state. The vector picture will be studied in detail in Section 7.4.

(b) Tensor Operators for Higher Spin States ($j = 1, \frac{3}{2}, 2, \ldots, 4$) The $1 \otimes 1$ coupling coefficients (7.1.19) may be used to construct a complete set of nine tensor operators for the $j = 1$ basis. The tensor operators and their representations are given in the following, using Eq. (7.3.1a):

$$T^2_{-2} = \left|{1 \atop -1}\right\rangle\left\langle{1 \atop 1}\right|, \quad T^2_{-1} = \frac{\left|{1 \atop 0}\right\rangle\left\langle{1 \atop 1}\right| - \left|{1 \atop -1}\right\rangle\left\langle{1 \atop 0}\right|}{\sqrt{2}}, \quad T^2_0 = \frac{\left|{1 \atop 1}\right\rangle\left\langle{1 \atop 1}\right| - 2\left|{1 \atop 0}\right\rangle\left\langle{1 \atop 0}\right| + \left|{1 \atop -1}\right\rangle\left\langle{1 \atop -1}\right|}{\sqrt{6}}, \quad T^2_1 = \frac{-\left|{1 \atop 1}\right\rangle\left\langle{1 \atop 0}\right| + \left|{1 \atop 0}\right\rangle\left\langle{1 \atop -1}\right|}{\sqrt{2}},$$

$$\rightarrow \begin{pmatrix} 0 & 0 & 0 \\ 0 & 0 & 0 \\ 1 & 0 & 0 \end{pmatrix} \quad \rightarrow \begin{pmatrix} 0 & 0 & 0 \\ 1/\sqrt{2} & 0 & 0 \\ 0 & -1/\sqrt{2} & 0 \end{pmatrix} \quad \rightarrow \begin{pmatrix} 1/\sqrt{6} & 0 & 0 \\ 0 & -2/\sqrt{6} & 0 \\ 0 & 0 & 1/\sqrt{6} \end{pmatrix} \quad \rightarrow \begin{pmatrix} 0 & -1/\sqrt{2} & 0 \\ 0 & 0 & 1/\sqrt{2} \\ 0 & 0 & 0 \end{pmatrix}$$

$$T^2_2 = \left|{1 \atop 1}\right\rangle\left\langle{1 \atop -1}\right|, \quad T^1_{-1} = \frac{\left|{1 \atop 0}\right\rangle\left\langle{1 \atop 1}\right| + \left|{1 \atop -1}\right\rangle\left\langle{1 \atop 0}\right|}{\sqrt{2}}, \quad T^1_0 = \frac{\left|{1 \atop 1}\right\rangle\left\langle{1 \atop 1}\right| - \left|{1 \atop -1}\right\rangle\left\langle{1 \atop -1}\right|}{\sqrt{2}}, \quad T^1_1 = \frac{-\left|{1 \atop 1}\right\rangle\left\langle{1 \atop 0}\right| - \left|{1 \atop 0}\right\rangle\left\langle{1 \atop -1}\right|}{\sqrt{2}},$$

$$\rightarrow \begin{pmatrix} 0 & 0 & 1 \\ 0 & 0 & 0 \\ 0 & 0 & 0 \end{pmatrix} \quad \rightarrow \begin{pmatrix} 0 & 0 & 0 \\ 1/\sqrt{2} & 0 & 0 \\ 0 & 1/\sqrt{2} & 0 \end{pmatrix} \quad \rightarrow \begin{pmatrix} 1/\sqrt{2} & 0 & 0 \\ 0 & 0 & 0 \\ 0 & 0 & -1/\sqrt{2} \end{pmatrix} \quad \rightarrow \begin{pmatrix} 0 & -1/\sqrt{2} & 0 \\ 0 & 0 & -1/\sqrt{2} \\ 0 & 0 & 0 \end{pmatrix}$$

$$T^0_0 = \frac{\left|{1 \atop 1}\right\rangle\left\langle{1 \atop 1}\right| + \left|{1 \atop 0}\right\rangle\left\langle{1 \atop 0}\right| + \left|{1 \atop -1}\right\rangle\left\langle{1 \atop -1}\right|}{\sqrt{3}}$$

$$\rightarrow \begin{pmatrix} 1/\sqrt{3} & 0 & 0 \\ 0 & 1/\sqrt{3} & 0 \\ 0 & 0 & 1/\sqrt{3} \end{pmatrix}. \qquad (7.3.12)$$

The lower four operators are the scalar ($k = 0$) and vector ($k = 1$) operators. T^0_0 is proportional to the identity (1), and T^1_{-1}, T^1_0, and T^1_1 are proportional to the angular-momentum operators J_-, J_z, and J_+, respectively. This follows from a comparison with the original $\mathscr{D}^{j=1}$ representations given by Eqs. (5.4.21)–(5.4.25):

$$T^1_{-1} = J_-/2 \qquad T^1_0 = J_z/\sqrt{2} \qquad T^1_1 = -J_+/2,$$
$$= (J_x - iJ_y)/2 \qquad \qquad = -(J_x + iJ_y)/2. \qquad (7.3.13)$$

The three tensor operators (T^1_{-1}, T^1_0, T^1_1) or (J_x, J_y, J_z) have the same tranformation properties as the corresponding vector operators constructed in the spin-$\frac{1}{2}$ basis. Of course, the scalar operator T^0_0 is invariant.

Beyond the scalar and vector operators in Eq. (7.3.12) there are five more ($k = 2$) operators $\{T^2_{-2}, T^2_{-1}, T^2_0, T^2_1, T^2_2\}$ known as unit quadrupole operators. These are the "tensor" operators which deserve this old-fashioned name. To label a ($j = 1$) state completely one needs a set of quadrupole tensor expectation values $\langle T^2_q \rangle$ as well as the $\langle T^1_q \rangle$ or $\langle J_a \rangle$ vector expectation values. The significance of these values will be discussed when we introduce irreducible density matrices.

590 THEORY AND APPLICATION OF SYMMETRY REPRESENTATION PRODUCTS

Similarly, the quantum mechanics of higher j-states will require in general a complete set of $(2j + 1)^2$ tensor operators. Let us define new tensor operators which differ by an overall phase factor $(-1)^{2j}$:

$$v_q^k = \sum_{m,m'} C_{m\,-m'q}^{j\,\,\,j\,\,\,k}(-1)^{j-m'}\left|\begin{array}{c}j\\m\end{array}\right\rangle\left\langle\begin{array}{c}j\\m'\end{array}\right| = (-1)^{2j}T_q^k. \quad (7.3.14a)$$

The $3-j$ definition (7.2.30a) and symmetry relations (7.2.31) and (7.2.32) yield the following alternative form:

$$v_q^k = \sum_{m,m'} (-1)^{j-m}\sqrt{2k+1}\begin{pmatrix}k & j & j\\q & m' & -m\end{pmatrix}\left|\begin{array}{c}j\\m\end{array}\right\rangle\left\langle\begin{array}{c}j\\m'\end{array}\right|. \quad (7.3.14b)$$

The phase eliminates the annoying minus sign that occurs in T^k for half-integral j. [Recall Eqs. $(7.3.3)_x$.] The v_q^k representations are recorded in Tables 7.1–3 in a condensed form. To understand the condensed form, compare the $(j = 1)$ tensor derived in Eq. (7.3.12) with the $j = l = 1$ tables in Table 7.2(p). Note that each T_q^k has nonzero entries only in particular super- or subdiagonals of the matrix. Each superdiagonal is labeled by a number $q = 1,\ldots,k$ in the tables. The main or center diagonal belongs to $q = 0$ and the subdiagonals belong to $q = -1, -2,\ldots,-k$. At the end of each superdiagonal is a normalization denominator.

Note that each qth superdiagonal in a set of $v_q^q, v_q^{q+1},\ldots$ matrices gives a set of orthonormal vectors. For example, with $j = 2$ the $q = 2$ superdiagonals of Table 7.2(d) are

$$\begin{bmatrix}\dfrac{\sqrt{3}}{\sqrt{14}} & \cdot & \cdot\\ \cdot & -\dfrac{\sqrt{8}}{\sqrt{14}} & \cdot\\ \cdot & \cdot & \dfrac{\sqrt{3}}{\sqrt{14}}\end{bmatrix},\quad \begin{bmatrix}\dfrac{1}{\sqrt{2}} & \cdot & \cdot\\ \cdot & 0 & \cdot\\ \cdot & \cdot & -\dfrac{1}{\sqrt{2}}\end{bmatrix},\quad \begin{bmatrix}\dfrac{\sqrt{2}}{\sqrt{7}} & \cdot & \cdot\\ \cdot & \dfrac{\sqrt{3}}{\sqrt{7}} & \cdot\\ \cdot & \cdot & \dfrac{\sqrt{2}}{\sqrt{7}}\end{bmatrix}$$

(for $k = 4$)   (for $k = 3$)   (for $k = 2$).

Because of orthonormality of coupling coefficients these diagonals are orthonormal vectors. This makes it easy to express any $(2j + 1)$ by $(2j + 1)$ matrix in terms of the v_q^k. For example, using the second numbers from the $q = 2$ superdiagonals we easily find the following elementary matrix or operator:

$$\begin{pmatrix}0 & 0 & 0 & 0 & 0\\0 & 0 & 0 & 1 & 0\\0 & 0 & 0 & 0 & 0\\0 & 0 & 0 & 0 & 0\\0 & 0 & 0 & 0 & 0\end{pmatrix} \rightarrow -\frac{\sqrt{8}}{\sqrt{14}}v_2^4 + 0v_2^3 + \frac{\sqrt{3}}{\sqrt{7}}v_2^2 = E_{24}. \quad (7.3.15)$$

TABLE 7.1 (j) Subshell Tensors

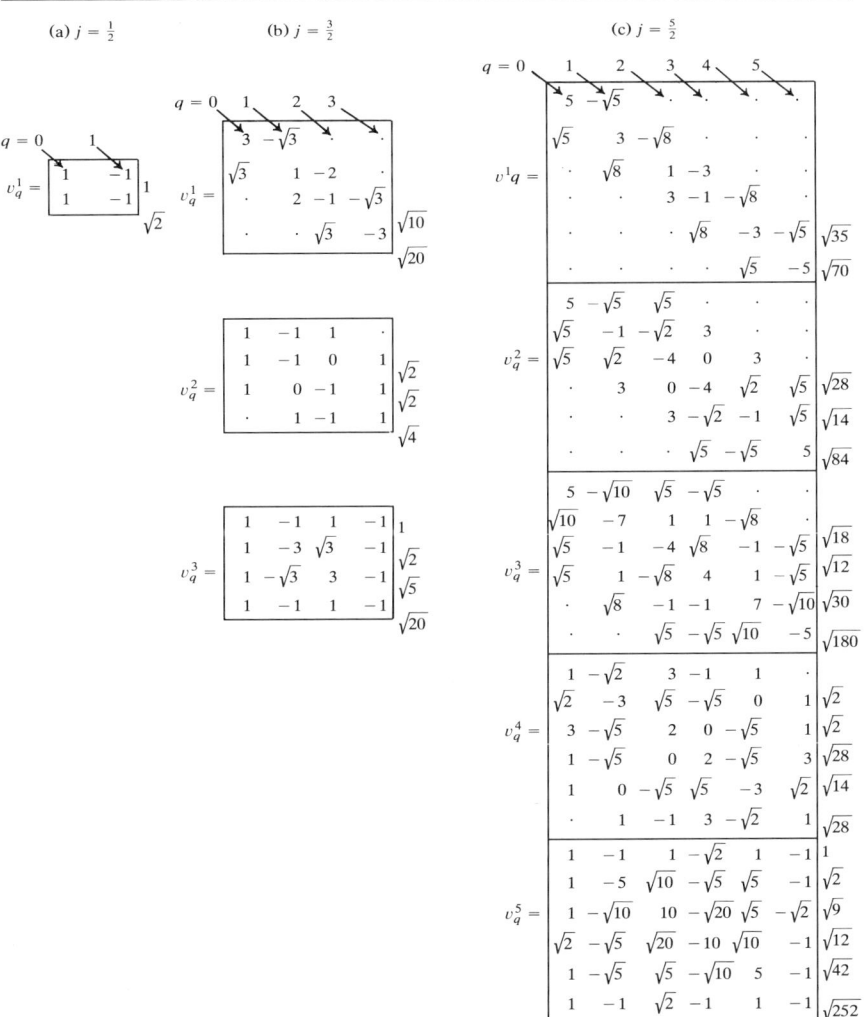

Linear relations between the irreducible tensor operators v_q^k and the elementary unitary operators $E_{m,m+q}$ will be used in later chapters. A simple example of such a relation involves the $q = 0$ operators for $(j = 1)$. From Eq. (7.3.12) [or the diagonals of Table 7.2(p)] one may write

$$v_0^2 = (E_{11} - 2E_{22} + E_{33})/\sqrt{6},$$
$$v_0^1 = (E_{11} - E_{33})/\sqrt{2},$$
$$v_0^0 = (E_{11} + E_{22} + E_{33})/\sqrt{3}. \qquad (7.3.16)$$

592 THEORY AND APPLICATION OF SYMMETRY REPRESENTATION PRODUCTS

TABLE 7.2 (l) **Subshell Tensors**

$$v_q^6 = \begin{pmatrix} 1 & -\sqrt{2} & 1 & -\sqrt{2} & \sqrt{5} & -1 & 1 \\ \sqrt{2} & -6 & \sqrt{30} & -\sqrt{8} & 3 & -\sqrt{12} & 1 \\ 1 & -\sqrt{30} & 15 & -10 & \sqrt{15} & -3 & \sqrt{5} \\ \sqrt{2} & -\sqrt{8} & 10 & -20 & 10 & -\sqrt{8} & \sqrt{2} \\ \sqrt{5} & -3 & \sqrt{15} & -10 & 15 & -\sqrt{30} & 1 \\ 1 & -\sqrt{12} & 3 & -\sqrt{8} & \sqrt{30} & -6 & \sqrt{2} \\ 1 & -1 & \sqrt{5} & -\sqrt{2} & 1 & -\sqrt{2} & 1 \end{pmatrix} \begin{matrix} 1 \\ \sqrt{2} \\ \sqrt{22} \\ \sqrt{22} \\ \sqrt{33} \\ \sqrt{264} \end{matrix}$$

$$\sqrt{924}$$

$$v_q^5 = \begin{pmatrix} 1 & -\sqrt{5} & 1 & -\sqrt{2} & 1 & -1 & \cdot \\ \sqrt{5} & -4 & \sqrt{27} & -\sqrt{2} & 1 & 0 & -1 \\ 1 & -\sqrt{27} & 5 & -\sqrt{10} & 0 & 1 & -1 \\ \sqrt{2} & -\sqrt{2} & \sqrt{10} & 0 & -\sqrt{10} & \sqrt{2} & \sqrt{2} \\ 1 & -1 & 0 & \sqrt{10} & -5 & \sqrt{27} & -1 \\ 1 & 0 & -1 & \sqrt{2} & -\sqrt{27} & 4 & -\sqrt{5} \\ \cdot & 1 & -1 & \sqrt{2} & -1 & \sqrt{5} & -1 \end{pmatrix} \begin{matrix} \sqrt{2} \\ \sqrt{2} \\ \sqrt{6} \\ \sqrt{6} \\ \sqrt{84} \\ \sqrt{84} \end{matrix}$$

$$v_q^4 = \begin{pmatrix} 3 & -\sqrt{30} & \sqrt{54} & -3 & \sqrt{3} & \cdot & \cdot \\ \sqrt{30} & -7 & \sqrt{32} & -\sqrt{3} & -\sqrt{2} & \sqrt{5} & \cdot \\ \sqrt{54} & -\sqrt{32} & 1 & \sqrt{15} & -\sqrt{40} & \sqrt{2} & \sqrt{3} \\ 3 & -\sqrt{3} & -\sqrt{15} & 6 & -\sqrt{15} & -\sqrt{3} & 3 \\ \sqrt{3} & \sqrt{2} & -\sqrt{40} & \sqrt{15} & 1 & -\sqrt{32} & \sqrt{54} \\ \cdot & \sqrt{5} & -\sqrt{2} & -\sqrt{3} & \sqrt{32} & -7 & \sqrt{30} \\ \cdot & \cdot & \sqrt{3} & -3 & \sqrt{54} & -\sqrt{30} & 3 \end{pmatrix} \begin{matrix} \sqrt{11} \\ \sqrt{22} \\ \sqrt{154} \\ \sqrt{154} \\ \sqrt{154} \end{matrix}$$

$$v_q^3 = \begin{pmatrix} 1 & -\sqrt{2} & \sqrt{2} & -1 & \cdot & \cdot \\ \sqrt{2} & -1 & 0 & 1 & -\sqrt{2} & \cdot \\ \sqrt{2} & 0 & -1 & 1 & 0 & -\sqrt{2} \\ 1 & 1 & -1 & 0 & 1 & -1 & -1 \\ \cdot & \sqrt{2} & 0 & -1 & 1 & 0 & -\sqrt{2} \\ \cdot & \cdot & \sqrt{2} & -1 & 0 & 1 & -\sqrt{2} \\ \cdot & \cdot & \cdot & 1 & -\sqrt{2} & \sqrt{2} & -1 \end{pmatrix} \begin{matrix} \sqrt{6} \\ \sqrt{6} \\ \sqrt{6} \\ \sqrt{6} \end{matrix}$$

$$v_q^2 = \begin{pmatrix} 5 & -5 & \sqrt{5} & \cdot & \cdot & \cdot \\ 5 & 0 & -\sqrt{15} & \sqrt{10} & \cdot & \cdot \\ \sqrt{5} & \sqrt{15} & -3 & -\sqrt{2} & \sqrt{12} & \cdot \\ \cdot & \sqrt{10} & \sqrt{2} & -4 & \sqrt{2} & \sqrt{10} \\ \cdot & \cdot & \sqrt{12} & -\sqrt{2} & -3 & \sqrt{15} & \sqrt{5} \\ \cdot & \cdot & \cdot & \sqrt{10} & -\sqrt{15} & 0 & 5 \\ \cdot & \cdot & \cdot & \cdot & \sqrt{5} & -5 & 5 \end{pmatrix} \begin{matrix} \sqrt{42} \\ \sqrt{84} \\ \sqrt{84} \end{matrix}$$

$$v_q^1 = \begin{pmatrix} 3 & -\sqrt{3} & \cdot & \cdot & \cdot & \cdot \\ \sqrt{3} & 2 & -\sqrt{5} & \cdot & \cdot & \cdot \\ \cdot & \sqrt{5} & 1 & -\sqrt{6} & \cdot & \cdot \\ \cdot & \cdot & \sqrt{6} & 0 & -\sqrt{6} & \cdot \\ \cdot & \cdot & \cdot & \sqrt{6} & -1 & -\sqrt{5} \\ \cdot & \cdot & \cdot & \cdot & \sqrt{5} & -2 & -\sqrt{3} \\ \cdot & \cdot & \cdot & \cdot & \cdot & \sqrt{3} & -3 \end{pmatrix} \begin{matrix} \sqrt{28} \\ \sqrt{28} \end{matrix}$$

(f) $l = 3$

$$v_q^4 = \begin{pmatrix} 1 & -1 & \sqrt{3} & -1 & 1 \\ 1 & -4 & \sqrt{6} & -\sqrt{8} & 1 \\ \sqrt{3} & -\sqrt{6} & 6 & -\sqrt{6} & \sqrt{3} \\ 1 & -\sqrt{8} & \sqrt{6} & -4 & 1 \\ 1 & -1 & \sqrt{3} & -1 & 1 \end{pmatrix} \begin{matrix} 1 \\ \sqrt{2} \\ \sqrt{14} \\ \sqrt{14} \\ \sqrt{70} \end{matrix}$$

$$v_q^3 = \begin{pmatrix} 1 & -\sqrt{3} & 1 & -1 & \cdot \\ \sqrt{3} & -2 & \sqrt{2} & 0 & -1 \\ 1 & -\sqrt{2} & 0 & \sqrt{2} & -1 \\ 1 & 0 & -\sqrt{2} & 2 & -\sqrt{3} \\ \cdot & 1 & -1 & \sqrt{3} & -1 \end{pmatrix} \begin{matrix} \sqrt{2} \\ \sqrt{2} \\ \sqrt{10} \\ \sqrt{10} \end{matrix}$$

$$v_q^2 = \begin{pmatrix} 2 & -\sqrt{6} & \sqrt{2} & \cdot & \cdot \\ \sqrt{6} & -1 & -1 & \sqrt{3} & \cdot \\ \sqrt{2} & 1 & -2 & 1 & \sqrt{2} \\ \cdot & \sqrt{3} & -1 & -1 & \sqrt{6} \\ \cdot & \cdot & \sqrt{2} & -\sqrt{6} & 2 \end{pmatrix} \begin{matrix} \sqrt{7} \\ \sqrt{14} \\ \sqrt{14} \end{matrix}$$

$$v_q^1 = \begin{pmatrix} 2 & -\sqrt{2} & \cdot & \cdot & \cdot \\ \sqrt{2} & 1 & -\sqrt{3} & \cdot & \cdot \\ \cdot & \sqrt{3} & 0 & -\sqrt{3} & \cdot \\ \cdot & \cdot & \sqrt{3} & -1 & -\sqrt{2} \\ \cdot & \cdot & \cdot & \sqrt{2} & -2 \end{pmatrix} \begin{matrix} \sqrt{10} \\ \sqrt{10} \end{matrix}$$

(d) $l = 2$

$$v_q^2 = \begin{pmatrix} 1 & -1 & 1 \\ 1 & -2 & 1 \\ 1 & -1 & -1 \end{pmatrix} \begin{matrix} 1 \\ \sqrt{2} \\ \sqrt{6} \end{matrix}$$

$$v_q^1 = \begin{pmatrix} 1 & -1 & \cdot \\ 1 & 0 & -1 \\ \cdot & 1 & -1 \end{pmatrix}$$

(p) $l = 1$

[Here the row-column indices of the elementary operators are simply numbers (1, 2, and 3) rather than angular-momentum quanta ($m = 1, 0$, and -1).] These operators are proportional to the diagonal U_3 operators introduced in Eq. (5.8.41).

The $(2j + 1)^2$ tensor operators $v_q^k(jj)$ are a complete set of generators of the group U_{2j+1}, and so are the elementary operators $E_{m,n}$. Every operator that acts on a $(2j + 1)$-dimensional angular-momentum basis $\{|_j^j\rangle, |_{j-1}^j\rangle, \ldots, |_{-j}^j\rangle\}$ is a linear combination of elementary operators $E_{m,n}$, and hence, also a combination of v_q^k's.

(c) Mixed Angular-Momentum Bases Two or more sets

$$\left\{ \left|\begin{matrix}j_1\\j_1\end{matrix}\right\rangle, \left|\begin{matrix}j_1\\j_1-1\end{matrix}\right\rangle, \ldots \right\}, \left\{ \left|\begin{matrix}j_2\\j_2\end{matrix}\right\rangle, \ldots \right\}, \ldots$$

of angular-momentum bases are connected by generalized tensor operators of the following form:

$$v(j_1 j_2)_q^k = (-1)^{2j_1} \sum_{m_1 m_2} C_{m_1 \, -m_2 q}^{j_1 \, j_2 \, k} \left|\begin{matrix}j_1\\m_1\end{matrix}\right\rangle \left\langle\begin{matrix}j_2\\m_2\end{matrix}\right|(-1)^{j_2+m_2} \qquad (7.3.17a)$$

$$= \sum_{m_1 m_2} (-1)^{j_1-m_1}\sqrt{2k+1} \begin{pmatrix}k & j_2 & j_1 \\ q & m_2 & -m_1\end{pmatrix} \left|\begin{matrix}j_1\\m_1\end{matrix}\right\rangle \left\langle\begin{matrix}j_2\\m_2\end{matrix}\right|. \qquad (7.3.17b)$$

This differs only by a phase $(-1)^{2j_1}$ from the T operator given in Eq. (7.3.2). For $j_1 = j_2$ it reduces to the definition of Eq. (7.3.14). For nonzero "shift" Δ, where

$$\Delta = j_1 - j_2, \qquad (7.3.18)$$

the matrix representations of the operators are rectangular. Some examples are shown in Table 7.4 for integral $j_1, j_2 = 1 - 3$. Two explicit examples are the following:

$$\sqrt{10}\, V_1^1(pd) = \begin{array}{|ccccc|} \hline \cdot & \cdot & 1 & \cdot & \cdot \\ \cdot & \cdot & \cdot & \sqrt{3} & \cdot \\ \cdot & \cdot & \cdot & \cdot & \sqrt{6} \\ \hline \end{array}$$

$$= E_{63} + \sqrt{3}\, E_{74} + \sqrt{6}\, E_{85}$$

$$\sqrt{10}\, V_{-1}^1(dp) = \begin{array}{|ccc|} \hline \cdot & \cdot & \cdot \\ 1 & \cdot & \cdot \\ \cdot & \sqrt{3} & \cdot \\ \cdot & \cdot & \sqrt{6} \\ \hline \end{array}$$

$$= E_{36} + \sqrt{3}\, E_{47} + \sqrt{6}\, E_{58}.$$

TABLE 7.3 (g) $l = 4$

$$V_q^8 = \begin{pmatrix} 1 & -1 & 1 & -1 & \sqrt{5} & -1 & \sqrt{7} & -1 & 1 \\ 1 & -8 & \sqrt{28} & -4 & \sqrt{10} & -\sqrt{32} & 2 & -4 & 1 \\ 1 & -\sqrt{28} & 28 & -14 & \sqrt{70} & -\sqrt{28} & \sqrt{56} & -2 & \sqrt{7} \\ 1 & -4 & 14 & -56 & \sqrt{490} & -\sqrt{112} & \sqrt{28} & -\sqrt{32} & 1 \\ \sqrt{5} & -\sqrt{10} & \sqrt{70} & -\sqrt{490} & 70 & -\sqrt{490} & \sqrt{70} & -\sqrt{10} & \sqrt{5} \\ 1 & -\sqrt{32} & \sqrt{28} & -\sqrt{112} & \sqrt{490} & -56 & 14 & -4 & 1 \\ \sqrt{7} & -2 & \sqrt{56} & -\sqrt{28} & \sqrt{70} & -14 & 28 & -\sqrt{28} & 1 \\ 1 & -4 & 2 & -\sqrt{32} & \sqrt{10} & -4 & \sqrt{28} & -8 & 1 \\ 1 & -1 & \sqrt{7} & -1 & \sqrt{5} & -1 & 1 & -1 & 1 \end{pmatrix} \begin{matrix} 1 \\ \sqrt{2} \\ \sqrt{30} \\ \sqrt{10} \\ \sqrt{130} \\ \sqrt{78} \\ \sqrt{286} \\ \sqrt{1430} \\ \sqrt{12870} \end{matrix}$$

$$V_q^7 = \begin{pmatrix} 1 & -\sqrt{7} & 3 & -5 & \sqrt{5} & -3 & 1 & -1 & \cdot \\ \sqrt{7} & -6 & 10 & -8 & \sqrt{90} & -\sqrt{8} & 2 & 0 & -1 \\ 3 & -10 & 14 & -\sqrt{252} & \sqrt{70} & -\sqrt{28} & 0 & 2 & -1 \\ 5 & -8 & \sqrt{252} & -14 & \sqrt{70} & 0 & -\sqrt{28} & \sqrt{8} & -3 \\ \sqrt{5} & -\sqrt{90} & \sqrt{70} & -\sqrt{70} & 0 & \sqrt{70} & -\sqrt{70} & \sqrt{90} & -\sqrt{5} \\ 3 & -\sqrt{8} & \sqrt{28} & 0 & -\sqrt{70} & 14 & -\sqrt{252} & 8 & -5 \\ 1 & -2 & 0 & \sqrt{28} & -\sqrt{70} & \sqrt{252} & -14 & 10 & -3 \\ 1 & 0 & -2 & \sqrt{8} & -\sqrt{90} & 8 & -10 & 6 & -\sqrt{7} \\ \cdot & 1 & -1 & 3 & -\sqrt{5} & 5 & -3 & \sqrt{7} & -1 \end{pmatrix} \begin{matrix} \sqrt{2} \\ \sqrt{2} \\ \sqrt{26} \\ \sqrt{26} \\ \sqrt{286} \\ \sqrt{286} \\ \sqrt{858} \\ \sqrt{858} \end{matrix}$$

$$V_q^6 = \begin{pmatrix} 4 & -\sqrt{28} & 2 & -4 & \sqrt{40} & -2 & 2 & \cdot & \cdot \\ \sqrt{28} & -17 & 13 & -3 & \sqrt{10} & -1 & -1 & \sqrt{7} & \cdot \\ 2 & -13 & 22 & -\sqrt{63} & 0 & \sqrt{7} & -\sqrt{28} & 1 & 2 \\ 4 & -3 & \sqrt{63} & 1 & -\sqrt{70} & \sqrt{7} & -\sqrt{7} & -1 & 2 \\ \sqrt{40} & -\sqrt{10} & 0 & \sqrt{70} & -20 & \sqrt{70} & 0 & -\sqrt{10} & \sqrt{40} \\ 2 & -1 & -\sqrt{7} & \sqrt{7} & -\sqrt{70} & 1 & \sqrt{63} & -3 & 4 \\ 2 & 1 & -\sqrt{28} & \sqrt{7} & 0 & -\sqrt{63} & 22 & -13 & 2 \\ \cdot & \sqrt{7} & -1 & -1 & \sqrt{10} & -3 & 13 & -17 & \sqrt{28} \\ \cdot & \cdot & 2 & -2 & \sqrt{40} & -4 & 2 & -\sqrt{28} & 4 \end{pmatrix} \begin{matrix} \sqrt{15} \\ \sqrt{10} \\ \sqrt{110} \\ \sqrt{66} \\ \sqrt{33} \\ \sqrt{660} \\ \sqrt{1980} \end{matrix}$$

$$V_q^5 = \begin{pmatrix} 4 & -\sqrt{20} & \sqrt{20} & -\sqrt{80} & \sqrt{8} & -2 & \cdot & \cdot & \cdot \\ \sqrt{20} & -11 & \sqrt{35} & -\sqrt{5} & -\sqrt{2} & \sqrt{5} & -3 & \cdot & \cdot \\ \sqrt{20} & -\sqrt{35} & 4 & \sqrt{5} & -\sqrt{14} & \sqrt{35} & 0 & -3 & \cdot \\ \sqrt{80} & -\sqrt{5} & -\sqrt{5} & 9 & -\sqrt{18} & 0 & \sqrt{35} & -\sqrt{5} & -2 \\ \sqrt{8} & \sqrt{2} & -\sqrt{14} & \sqrt{18} & 0 & -\sqrt{18} & \sqrt{14} & -\sqrt{2} & -\sqrt{8} \\ 2 & \sqrt{5} & -\sqrt{35} & 0 & \sqrt{18} & -9 & \sqrt{5} & \sqrt{5} & -\sqrt{80} \\ \cdot & 3 & 0 & -\sqrt{35} & \sqrt{14} & -\sqrt{5} & -4 & \sqrt{35} & -\sqrt{20} \\ \cdot & \cdot & 3 & -\sqrt{5} & \sqrt{2} & \sqrt{5} & -\sqrt{35} & 11 & -\sqrt{20} \\ \cdot & \cdot & \cdot & 2 & -\sqrt{8} & \sqrt{80} & -\sqrt{20} & \sqrt{20} & -4 \end{pmatrix} \begin{matrix} \sqrt{26} \\ \sqrt{26} \\ \sqrt{234} \\ \sqrt{78} \\ \sqrt{156} \\ \sqrt{468} \end{matrix}$$

with $q = 0, 1, 2, 3, 4, 5, 6, 7, 8$ indicated by arrows above the columns.

$q = 0 \quad 1 \quad 2 \quad 3 \quad 4 \quad 5 \quad 6 \quad 7 \quad 8$

$$V_q^4 = \begin{pmatrix} 14 & -\sqrt{490} & \sqrt{630} & -\sqrt{70} & \sqrt{14} & \cdot & \cdot & \cdot & \cdot \\ \sqrt{490} & -21 & \sqrt{70} & \sqrt{70} & -\sqrt{63} & \sqrt{35} & \cdot & \cdot & \cdot \\ \sqrt{630} & -\sqrt{70} & -11 & \sqrt{360} & -11 & -\sqrt{10} & \sqrt{45} & \cdot & \cdot \\ \sqrt{70} & \sqrt{70} & -\sqrt{360} & 9 & 9 & -\sqrt{360} & \sqrt{10} & \sqrt{35} & \cdot \\ \sqrt{14} & \sqrt{63} & -11 & -9 & 18 & -9 & -11 & \sqrt{63} & \sqrt{14} \\ \cdot & \sqrt{35} & \sqrt{10} & -\sqrt{360} & 9 & 9 & -\sqrt{360} & \sqrt{70} & \sqrt{70} \\ \cdot & \cdot & \sqrt{45} & -\sqrt{10} & -11 & \sqrt{360} & -11 & -\sqrt{70} & \sqrt{630} \\ \cdot & \cdot & \cdot & \sqrt{35} & -\sqrt{63} & \sqrt{70} & \sqrt{70} & -21 & \sqrt{490} \\ \cdot & \cdot & \cdot & \cdot & \sqrt{14} & -\sqrt{70} & \sqrt{630} & -\sqrt{490} & 14 \end{pmatrix} \begin{matrix} \\ \\ \\ \\ \sqrt{143} \\ \sqrt{286} \\ \sqrt{2002} \\ \sqrt{2002} \\ \sqrt{2002} \end{matrix}$$

$$V_q^3 = \begin{pmatrix} 14 & -\sqrt{98} & \sqrt{14} & -\sqrt{14} & \cdot & \cdot & \cdot & \cdot & \cdot \\ \sqrt{98} & -7 & -\sqrt{14} & \sqrt{14} & -\sqrt{35} & \cdot & \cdot & \cdot & \cdot \\ \sqrt{14} & \sqrt{14} & -13 & \sqrt{8} & \sqrt{5} & -\sqrt{50} & \cdot & \cdot & \cdot \\ \sqrt{14} & \sqrt{14} & -\sqrt{8} & -9 & \sqrt{45} & 0 & -\sqrt{50} & \cdot & \cdot \\ \cdot & \sqrt{35} & \sqrt{5} & -\sqrt{45} & 0 & \sqrt{45} & -\sqrt{5} & -\sqrt{35} & \cdot \\ \cdot & \cdot & \sqrt{50} & 0 & -\sqrt{45} & 9 & \sqrt{8} & -\sqrt{14} & -\sqrt{14} \\ \cdot & \cdot & \cdot & \sqrt{50} & -\sqrt{5} & -\sqrt{8} & 13 & -\sqrt{14} & -\sqrt{14} \\ \cdot & \cdot & \cdot & \cdot & \sqrt{35} & -\sqrt{14} & \sqrt{14} & 7 & -\sqrt{98} \\ \cdot & \cdot & \cdot & \cdot & \cdot & \sqrt{14} & -\sqrt{14} & \sqrt{98} & -14 \end{pmatrix} \begin{matrix} \\ \\ \\ \\ \\ \sqrt{198} \\ \sqrt{66} \\ \sqrt{330} \\ \sqrt{990} \end{matrix}$$

$$V_q^2 = \begin{pmatrix} 28 & -14 & \sqrt{28} & \cdot & \cdot & \cdot & \cdot & \cdot & \cdot \\ 14 & 7 & -\sqrt{175} & \sqrt{63} & \cdot & \cdot & \cdot & \cdot & \cdot \\ \sqrt{28} & \sqrt{175} & -8 & -9 & \sqrt{90} & \cdot & \cdot & \cdot & \cdot \\ \cdot & \sqrt{63} & 9 & -17 & -\sqrt{10} & 10 & \cdot & \cdot & \cdot \\ \cdot & \cdot & \sqrt{90} & \sqrt{10} & -20 & \sqrt{10} & \sqrt{90} & \cdot & \cdot \\ \cdot & \cdot & \cdot & 10 & -\sqrt{10} & -17 & 9 & \sqrt{63} & \cdot \\ \cdot & \cdot & \cdot & \cdot & \sqrt{90} & -9 & -8 & \sqrt{175} & \sqrt{28} \\ \cdot & \cdot & \cdot & \cdot & \cdot & \sqrt{63} & -\sqrt{175} & 7 & 14 \\ \cdot & \cdot & \cdot & \cdot & \cdot & \cdot & \sqrt{28} & -14 & 28 \end{pmatrix} \begin{matrix} \\ \\ \\ \\ \\ \\ \sqrt{462} \\ \sqrt{924} \\ \sqrt{2772} \end{matrix}$$

$$V_q^1 = \begin{pmatrix} 4 & -2 & \cdot & \cdot & \cdot & \cdot & \cdot & \cdot & \cdot \\ 2 & 3 & -\sqrt{7} & \cdot & \cdot & \cdot & \cdot & \cdot & \cdot \\ \cdot & \sqrt{7} & 2 & -3 & \cdot & \cdot & \cdot & \cdot & \cdot \\ \cdot & \cdot & 3 & 1 & -\sqrt{10} & \cdot & \cdot & \cdot & \cdot \\ \cdot & \cdot & \cdot & \sqrt{10} & 0 & -\sqrt{10} & \cdot & \cdot & \cdot \\ \cdot & \cdot & \cdot & \cdot & \sqrt{10} & -1 & -3 & \cdot & \cdot \\ \cdot & \cdot & \cdot & \cdot & \cdot & 3 & -2 & -\sqrt{7} & \cdot \\ \cdot & \cdot & \cdot & \cdot & \cdot & \cdot & \sqrt{7} & -3 & -2 \\ \cdot & \cdot & \cdot & \cdot & \cdot & \cdot & \cdot & 2 & -4 \end{pmatrix} \begin{matrix} \\ \\ \\ \\ \\ \\ \\ \sqrt{60} \\ \sqrt{60} \end{matrix}$$

TABLE 7.4 Mixed Subshell Tensors

(f)(d), (f)(p), (d)(p) coupling tables for $v_q^k(l_1 l_2)$ matrices, with V_q^5, V_q^4, V_q^3, V_q^2, V_q^1 blocks.

The numbering for E_{ij} reflects the choice of numbers 1 to 5 for d states ($|1\rangle = |_2^2\rangle$, $|2\rangle = |_1^2\rangle$, ..., $|5\rangle = |_{-2}^2\rangle$) and 6 to 8 for the p states ($|6\rangle = |_1^1\rangle$, $|7\rangle = |_0^1\rangle$, $|8\rangle = |_{-1}^1\rangle$). The tables exhibit the $v_q^k(l_1 l_2)$ matrices for $l_1 - l_2 \equiv \Delta > 0$, and the transpose is found using the symmetry relation

$$v_q^k(l_2 l_1) = (-1)^{l+q} \tilde{v}_{-q}^k(l_1 l_2). \tag{7.3.19}$$

B. Wigner-Eckart Theorem for R_3

The Wigner-Eckart theorem for finite groups was proved in Section 6.4.B, and the same proof works for R_3 representations, too. It is only necessary to replace the sum over group elements $(1/{}^\circ G \Sigma_g)$ by the R_3 integral ($\int d(\alpha\beta\gamma)$) and apply the R_3-factorization lemma (Eq. 7.2.25) instead of Eq. (6.2.15). In fact, the R_3 problem is simpler, since there is no repetition of any irrep $\mathscr{D}^{j_3}$ in the reduction of a product $\mathscr{D}^{j_1} \otimes \mathscr{D}^{j_2}$. We now restate the theorem as it applies to R_3.

R_3 Wigner-Eckart Theorem If an operator T_q^k belongs to a set $\{T_{-q}^k, \ldots, T_q^k\}$ which transforms according to R_3 rotation matrices as follows:

$$R(\alpha\beta\gamma) T_q^k R^\dagger(\alpha\beta\gamma) = \sum_{m_1' = -j_1}^{j_1} T_{q'}^k \mathscr{D}_{q'q}^k(\alpha\beta\gamma),$$

then matrix elements of T_q^k in angular-momentum basis are of the form

$$\left\langle \begin{matrix} j_1 \\ m_1 \end{matrix} \middle| T_q^k \middle| \begin{matrix} j_2 \\ m_2 \end{matrix} \right\rangle = C_{qm_2m_1}^{k\,j_2\,j_3} \langle j_1 || T^k || j_2 \rangle, \qquad (7.3.20a)$$

where the $C_{qm_2m_1}^{k\,j_2\,j_1}$ are coupling coefficients, and the constants

$$\langle j_1 || T^k || j_2 \rangle \equiv \frac{1}{2j_3 + 1} \sum_{q' = -k} \sum_{m_2' = -j_2} \sum_{m_1' = -j_1} C_{q'm_2'm_1'}^{k\,j_2\,j_1} \left\langle \begin{matrix} j_1 \\ m_1' \end{matrix} \middle| T_{q'}^k \middle| \begin{matrix} j_2 \\ m_2' \end{matrix} \right\rangle$$
$$(7.3.20b)$$

are independent of q, m_2, or m_1.

The theorem implies that the representations of different tensor operators which transform according to a given irrep $\mathscr{D}^k$ must be proportional to each other. The proportionality constants $\langle j_1 || T^k || j_2 \rangle$ are called REDUCED MATRIX ELEMENTS.

In particular, a representation of any tensor operator T_q^k must be proportional to the unit tensor v_q^k defined by Eq. (7.3.17):

$$v_q^k = (-1)^{2j_1} \sum C_{m_1 \, -m_2 q}^{j_1 \, j_2 \, k} (-1)^{j_2 + m_2} \left| \begin{matrix} j_1 \\ m_1 \end{matrix} \right\rangle \left\langle \begin{matrix} j_2 \\ m_2 \end{matrix} \right|$$

$$= \sum (-1)^{j_1 - m_1} \sqrt{2k+1} \begin{pmatrix} k & j_2 & j_1 \\ q & m_2 & -m_1 \end{pmatrix} \left| \begin{matrix} j_1 \\ m_1 \end{matrix} \right\rangle \left\langle \begin{matrix} j_2 \\ m_2 \end{matrix} \right|.$$

Indeed, the $3j$ definitions (7.2.30a) and symmetry relations (7.2.31) and

(7.3.32) yield the following:

$$\left\langle \begin{matrix} j_1 \\ m_1 \end{matrix} \middle| V_q^k \middle| \begin{matrix} j_2 \\ m_2 \end{matrix} \right\rangle = (-1)^{2j_1} C^{j_1\ j_2\ k}_{m_1\ -m_2 q}(-1)^{j_2+m_2}$$

$$= C^{k\ j_2\ j_1}_{q m_2 m_1}(-1)^{k+j_1-j_2}\sqrt{(2k+1)/(2j_1+1)}. \quad (7.3.21)$$

This has the Wigner-Eckart form (7.3.20a) with the following reduced matrix element:

$$\langle j_1 | |v^k| |j_2\rangle = (-1)^{k+j_1-j_2}\sqrt{(2k+1)/(2j_1+1)}. \quad (7.3.22)$$

Therefore, we may replace matrix representations of general tensor operators T_q^k with the $\langle |v_q^k|\rangle$ (recall tables 7.1–7.4) multiplied by a constant:

$$\left\langle \begin{matrix} j_1 \\ m_1 \end{matrix} \middle| T_q^k \middle| \begin{matrix} j_2 \\ m_2 \end{matrix} \right\rangle = \left\langle \begin{matrix} j_1 \\ m_1 \end{matrix} \middle| v_q^k \middle| \begin{matrix} j_2 \\ m_2 \end{matrix} \right\rangle \frac{\langle j_1| |T^k| |j_2\rangle}{\langle j_1| |v^k| |j_2\rangle}. \quad (7.3.23)$$

Each representation of T_q^k in angular-momentum bases $|j_1\rangle$ and $|j_2\rangle$ equals a v_q^k matrix multiplied by the following factor:

$$\langle j_1| |T^k| |j_2\rangle / \langle j_1| |v^k| |j_2\rangle = \langle j_1| |T^k| |j_2\rangle (-1)^{k+j_1-j_2}\left(\frac{2j_1+1}{2k+1}\right)^{\frac{1}{2}}. \quad (7.3.24)$$

This factor is proportional to the reduced matrix element of T^k. We now see some applications of the Wigner-Eckart theorem.

C. Evaluation of Crystal Field Splitting

Let us consider an elementary octahedral potential having the form

$$V^{(4)} = D[x^4 + y^4 + z^4 - \tfrac{3}{5}r^4] = D[2(X_4^4 + X_{-4}^4)/\sqrt{70} + \tfrac{2}{5}X_0^4]. \quad (7.3.25)$$

This form was derived first in Eq. (5.6.28) using the multipole expansion. It also follows from the form of the elementary multipole functions (5.6.17) of fourth degree which are tabulated in Appendix F. The $V^{(4)}$ is the fourth-rank octahedral scalar (A_{1g}) function. [See Eq. (5.6.18).]

Let us consider the effect of this potential on a d orbital, i.e., orbitals belonging to total angular momentum $j = 2$. Setting $j_1 = j_2 = 2$ in Eqs. (7.3.23) and (7.3.24) gives

$$\langle V^{(4)}\rangle_{j=2} = D\langle 2(v_4^4 + v_{-4}^4)/\sqrt{70} + \tfrac{2}{5}v_0^4\rangle(\sqrt{5}/3)\langle 2| |X^4| |2\rangle. \quad (7.3.26)$$

ROTATIONAL TENSOR OPERATORS AND THE WIGNER-ECKART THEOREM

From the ($j = 2$) tables [Tables 7.2(d)] the following representation of the potential is derived:

$$\langle V^{(4)} \rangle_{j=2} = (D/\sqrt{70}) \begin{pmatrix} \frac{2}{5} & \cdot & \cdot & \cdot & 2 \\ \cdot & -\frac{8}{5} & \cdot & \cdot & \cdot \\ \cdot & \cdot & \frac{12}{5} & \cdot & \cdot \\ \cdot & \cdot & \cdot & -\frac{8}{5} & \cdot \\ 2 & \cdot & \cdot & \cdot & \frac{2}{5} \end{pmatrix} (\sqrt{5}/3)\langle 2| |X^4| |2\rangle.$$

(7.3.27)

The eigenvectors and eigenvalues of this matrix are easy to find. In fact we derived the eigenvectors $\left\{ \left|\begin{smallmatrix}T_2\\1\end{smallmatrix}\right\rangle, \left|\begin{smallmatrix}T_2\\2\end{smallmatrix}\right\rangle, \left|\begin{smallmatrix}T_2\\3\end{smallmatrix}\right\rangle \right\}$ and $\left\{ \left|\begin{smallmatrix}E\\1\end{smallmatrix}\right\rangle, \left|\begin{smallmatrix}E\\2\end{smallmatrix}\right\rangle \right\}$ in Eqs. (5.6.12) and (5.6.13) by symmetry projection even before introducing the potential. Now the eigenvalues follow by multiplying $\langle V^{(4)} \rangle$ by $\left|\begin{smallmatrix}T_2\\j\end{smallmatrix}\right\rangle$ or $\left|\begin{smallmatrix}E\\j\end{smallmatrix}\right\rangle$. The triply degenerate T_2 eigenvalue is

$$\left\langle \begin{smallmatrix}T_2\\3\end{smallmatrix} \middle| V^{(4)} \middle| \begin{smallmatrix}T_2\\3\end{smallmatrix} \right\rangle = \tfrac{1}{2} \left(\left\langle \begin{smallmatrix}2\\-2\end{smallmatrix}\right| - \left\langle \begin{smallmatrix}2\\-2\end{smallmatrix}\right| \right) V^4 \left(\left|\begin{smallmatrix}2\\2\end{smallmatrix}\right\rangle - \left|\begin{smallmatrix}2\\-2\end{smallmatrix}\right\rangle \right)$$

$$= -8D\langle 2| |X^4| |2\rangle/(15\sqrt{14}), \qquad (7.3.28a)$$

and the doubly degenerate E eigenvalue is

$$\left\langle \begin{smallmatrix}E\\2\end{smallmatrix} \middle| V^{(4)} \middle| \begin{smallmatrix}E\\2\end{smallmatrix} \right\rangle = \tfrac{1}{2} \left(\left\langle \begin{smallmatrix}2\\2\end{smallmatrix}\right| + \left\langle \begin{smallmatrix}2\\-2\end{smallmatrix}\right| \right) V^4 \left(\left|\begin{smallmatrix}2\\2\end{smallmatrix}\right\rangle + \left|\begin{smallmatrix}2\\2\end{smallmatrix}\right\rangle \right)$$

$$= 12D\langle 2| |X^4| |2\rangle/(15\sqrt{14}). \qquad (7.3.28b)$$

Note the $(-2:3)$ ratio of the eigenvalues. This preserves the "center of gravity" of the energy levels, since T_2 has three levels while E has only two. (This splitting was shown in Figure 5.6.3 in the Chapter 5.) In fact, the scalar ($V^{(0)} = v_0^0$) tensor operator is the only one with nonzero trace. No other tensor operator can shift the center of gravity.

Hence, the $j = 2$ example is a little too simple. The Wigner-Eckart results (7.3.28) do not predict anything interesting for the two levels E and T_2 unless one knows the value of the reduced matrix element $\langle 2\|X^4\|2\rangle$. Before we discuss formulas for the reduced matrix elements let us treat examples of crystal field splitting of $j = 3$ levels.

Setting $j_1 = j_2 = 3$ in Eqs. (7.3.23) and (7.3.24) gives the following representation of $V^{(4)}$ after using Table 7.2(f):

$$\langle V^{(4)} \rangle_{j=3} = D \langle 2(v_4^4 + v_{-4}^4)/\sqrt{70} + (2/5)v_0^4 \rangle (\sqrt{7}/3) \langle 3 \| X^4 \| 3 \rangle \quad (7.3.29a)$$

$$= D \begin{pmatrix} 3 & \cdot & \cdot & \cdot & \sqrt{15} & \cdot & \cdot \\ \cdot & -7 & \cdot & \cdot & \cdot & 5 & \cdot \\ \cdot & \cdot & 1 & \cdot & \cdot & \cdot & \sqrt{15} \\ \cdot & \cdot & \cdot & 6 & \cdot & \cdot & \cdot \\ \sqrt{15} & \cdot & \cdot & \cdot & 1 & \cdot & \cdot \\ \cdot & 5 & \cdot & \cdot & \cdot & -7 & \cdot \\ \cdot & \cdot & \sqrt{15} & \cdot & \cdot & \cdot & 3 \end{pmatrix} (2\sqrt{7}/15\sqrt{154}) \langle 3 \| X^4 \| 3 \rangle.$$

(7.3.29b)

The eigenvectors of this $\langle V^{(4)} \rangle$ matrix can be found by symmetry projection as in Section 5.6, by direct solution of the matrix eigenvalue problem, or by inspection of multipole functions. We will use the third method now, since it has not been discussed.

From the $O_3 \supset O_h$ correlation table of (5.6.5b) it is found that $(j = 3)$ splits into $(A_{2u} \oplus T_{1u} \oplus T_{2u})$. From the O_h and O_3 multipole function tables in Appendix F one easily obtains relations between polynomials of O_h and O_3. For A_{2u} one has

$$X^{A_{2u}} = xyz = -i(X_{-2}^3 - X_2^3)/\sqrt{30}. \quad (7.3.30)$$

For the third component of T_{1u} one has

$$X_3^{T_{1u}} = (x^2 - y^2)z = i(X_2^3 + X_{-2}^3)/\sqrt{2}. \quad (7.3.31)$$

Finally, for the third component of T_{2u} one has

$$X_3^{T_{2u}} = (x^2 + y^2)z = -X_0^3/10. \quad (7.3.32)$$

From this we easily deduce three normalized eigenvectors

$$|A_{2u}\rangle = \left(\left| \begin{matrix} 3 \\ 2 \end{matrix} \right\rangle - \left| \begin{matrix} 3 \\ -2 \end{matrix} \right\rangle \right)/\sqrt{2}, \quad \left| \begin{matrix} T_{1u} \\ 3 \end{matrix} \right\rangle = \left(\left| \begin{matrix} 3 \\ 2 \end{matrix} \right\rangle + \left| \begin{matrix} 3 \\ -2 \end{matrix} \right\rangle \right)/\sqrt{2},$$

and $\left| \begin{matrix} T_{2u} \\ 3 \end{matrix} \right\rangle = \left| \begin{matrix} 3 \\ 0 \end{matrix} \right\rangle.$

(7.3.33)

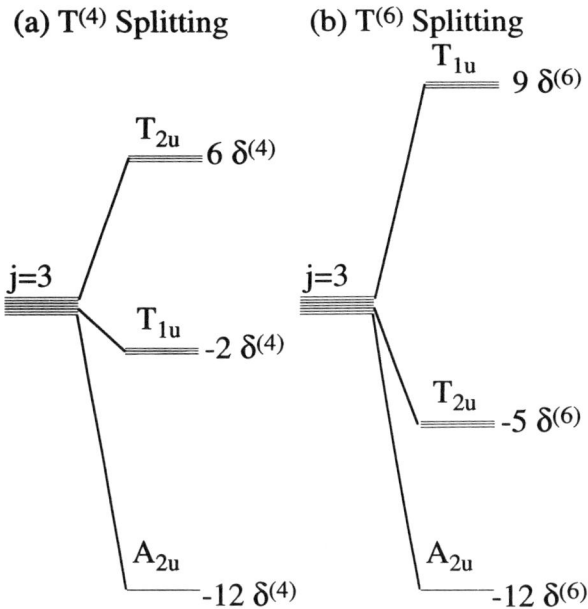

Figure 7.3.2 Octahedral tensor splitting of f-orbital levels. (a) Fourth-rank tensor splitting. (b) Sixth-rank tensor splitting.

Putting these with the $V^{(4)}$ matrix gives the following eigenvalues:

$$\langle A_{2u}|V^{(4)}|A_{2u}\rangle = -12\delta^{(4)}, \quad \left\langle \begin{array}{c} T_{1u} \\ 3 \end{array} \middle| V^{(4)} \middle| T_{1u} \right\rangle = -2\delta^{(4)},$$

and $\left\langle \begin{array}{c} T_{2u} \\ 3 \end{array} \middle| V^{(4)} \middle| T_{2u} \right\rangle = 6\delta^{(4)},$ (7.3.34a)

respectively, where the reduced factor is

$$\delta^{(4)} = D(2/15\sqrt{22})\langle 3\|X^4\|3\rangle. \quad (7.3.34b)$$

From this we predict a $(12 - 2):(6 + 2) = 5:4$ splitting ratio with A_{2u} and T_{2u} levels sandwiching T_{1u}. This is indicated in Figure 7.3.2(a). One should note that this does not imply that all octahedral crystal fields will split all ($j = 3$) levels into the order (A_2, T_1, T_2) with a $5:4$ ratio. Angular-momentum levels with $j = 3$ and higher may be effected by a *sixth*-rank tensor operator:

$$V^{(6)} = E\left[(\sqrt{8}/8)X_0^6 - (2\sqrt{7}/8)(X_4^6 + X_{-4}^6)\right]. \qquad (7.3.35)$$

This is the form of the next term in the octahedral multipole expansion (5.6.28). Its representation for ($j = 3$) follows from Table 7.2(f).

$$\langle V^{(6)} \rangle_{j=3} = E \begin{pmatrix} 1 & \cdot & \cdot & \cdot & -7\sqrt{15} & \cdot & \cdot \\ \cdot & -6 & \cdot & \cdot & \cdot & 42 & \cdot \\ \cdot & \cdot & 15 & \cdot & \cdot & \cdot & -7\sqrt{15} \\ \cdot & \cdot & \cdot & -20 & \cdot & \cdot & \cdot \\ -7\sqrt{15} & \cdot & \cdot & \cdot & 15 & \cdot & \cdot \\ \cdot & 42 & \cdot & \cdot & \cdot & -6 & \cdot \\ \cdot & \cdot & -7\sqrt{15} & \cdot & \cdot & \cdot & 1 \end{pmatrix} \times \langle 3\|X^6\|3\rangle \big/ (4\sqrt{462}).$$

(7.3.36)

This gives a different set of eigenvalues

$$\langle A_{2u} | V^{(6)} | A_{2u} \rangle = -12\delta^{(6)}, \qquad \left\langle \begin{matrix} T_{1u} \\ 3 \end{matrix} \middle| V^{(6)} \middle| \begin{matrix} T_{1u} \\ 3 \end{matrix} \right\rangle = 9\delta^{(6)},$$

and $\left\langle \begin{matrix} T_{2u} \\ 3 \end{matrix} \middle| V^{(6)} \middle| \begin{matrix} T_{2u} \\ 3 \end{matrix} \right\rangle = -5\delta^{(6)}, \qquad (7.3.37a)$

where

$$\delta^{(6)} = E\langle 3\|X^6\|3\rangle / (4\sqrt{462}). \qquad (7.3.37b)$$

A pure $V^{(6)}$ makes the T_{1u} level move into the high position as shown in Figure 7.3.2(b). A 1:2 splitting ratio results for the three levels (A_2, T_2, T_1).

An eighth-rank $V^{(8)}$ octahedral scalar operator exists but cannot have any effect on a ($j = 3$) level. (Its matrix must be zero, since $C_{qmn}^{383} = 0$.) However, another scalar operator is not needed. The fourth- and sixth-rank octahedral operators $V^{(4)}$ and $V^{(6)}$ are sufficient in combination to cause any ordering or splitting of the A_2, T_2, and T_1 sublevels of the $j = 3$ manifold. In fact there are three O_h scalar operators $V^{(6)}, V^{(4)}$, and $V^{(0)}$, but the $V^{(0)}$ only shifts the center of gravity. This is one more example of the parameter theorem in Section 6.4. The number (6.4.11) of independent scalar operators is exactly the number needed to determine all eigenvalues and eigenvectors subject to the constraints of symmetry.

At this point it may be instructive to review the counting of multipole functions and operators which was introduced in Section 5.6.C. Also the treatment of $j = 4, 5$, and 6 octahedral levels should be examined. The analyses for $j = 5$ and 6 are complicated by the fact that octahedral species

T_1 or T_2 are repeated. Then the T_1 or T_2 eigenvectors are not determined totally by symmetry constraints.

For high j the number of repeated species can be large. For $l = j = 50$ one predicts from Eqs. (5.6.5b) and (5.6.6) that 12 T_1 levels and 13 T_2 levels will appear. Since high values of angular quanta are common in molecular spectra it is important to learn how to deal with them. In Section 7.4 we will discuss some efficient methods for analyzing high-j states. The methods are based upon the theory of level clusters and induced representation bases which was introduced in Section 4.3.

Related problems involve very high crystal potentials which cause splittings which are comparable to or greater than the spacing between j levels. A large enough crystal potential could mix states of different angular momentum (j) strongly enough to make j a useless quantum number. This happens in the theory of ions tunneling in solids. The theory of level clusters can be useful then, too.

D. Evaluation of Reduced Matrix Elements

It is possible to use the Wigner-Eckart theorem while treating the reduced matrix elements as undetermined constants. One may derive some information without knowing the values of these constants. Also, the constants can be fitted using experimental results, and then the tensor analysis can be used to predict further results. These approaches require the we know only the basic symmetry properties of the system being studied.

However, in order to compute the numerical value of a reduced matrix element one must define the involved operators and states in more detail. So far we have not assumed much about anything except symmetry properties. It did not matter whether the ($j = 2$) states treated in Eqs. (7.3.28) were d orbitals of one electron in hydrogen or D orbitals made of over 100 electrons in mendelevium. For the Wigner-Eckart theorem it was only important that the states belonged to symmetry irrep $\mathscr{D}^2$ while the operators belonged to irreps $\mathscr{D}^4$ and $\mathscr{D}^{A_{1g}}$ of O_3 and O_h. We now compare the splitting of ($j = 2$) orbitals for one electron with those for two electrons.

(a) Single-Electron Orbitals in Potential Fields Let us assume a single-electron orbital state with a wave function of the form

$$\left\langle r\theta\phi \middle| \begin{matrix} l \\ m \end{matrix} \right\rangle = R_l(r) Y_m^l(\theta\phi),$$

where $R_l(r)$ is the radial wave function and Y_m^l are spherical harmonics. Let us compute the matrix elements for the general potential field multipole expansion:

$$\left\langle \begin{matrix} l' \\ m \end{matrix} \middle| V \middle| \begin{matrix} l \\ m \end{matrix} \right\rangle = \sum_{kq} A_{kq} \left\langle \begin{matrix} l' \\ m \end{matrix} \middle| X_q^k \middle| \begin{matrix} l \\ m \end{matrix} \right\rangle$$

$$= \sum_{kq} A_{kq} \int d\phi \int d\theta \sin\theta$$

$$\times \int r^2 \, dr \, R_{l'}^*(r) Y_{m'}^{l'*}(\theta\phi) r^k \sqrt{\frac{4\pi}{2k+1}} \, Y_q^k(\theta\phi) R_l(r) Y_m^l(\theta\phi).$$

This integral can be written in terms of angular and radial integrals:

$$\left\langle \begin{matrix} l' \\ m \end{matrix} \middle| V \middle| \begin{matrix} l \\ m \end{matrix} \right\rangle = \sum_{kq} A_{kq} \langle r^k \rangle \sqrt{\frac{4\pi}{2k+1}} \int d\phi \int d\theta \sin\theta Y_{m'}^{l'*}(\theta\phi) Y_q^k(\theta\phi) Y_m^l(\theta\phi), \tag{7.3.38a}$$

where the radial integral is denoted by

$$\langle r^k \rangle = \int_0^\infty r^2 \, dr \, |R_l(r)|^2 r^k. \tag{7.3.38b}$$

In the angular integral it is convenient to replace the spherical harmonics Y_m^j by irrep components $\mathscr{D}_{m0}^j$ using Eq. (5.5.65):

$$\sqrt{\frac{4\pi}{2k+1}} \int d\phi \int \sin\theta \, d\theta Y_{m'}^{l'}(\theta\phi) Y_q^k(\theta\phi) Y_m^l(\theta\phi)$$

$$= \frac{\sqrt{(2l'+1)(2l+1)}}{4\pi} \int d\phi \int \sin\theta \, d\theta \mathscr{D}_{m'0}^{l'}(\phi\theta 0) \mathscr{D}_{q0}^{k*}(\phi\theta 0) \mathscr{D}_{m0}^{l*}(\phi\theta 0). \tag{7.3.39}$$

It is useful to get the integral into the form of Eq. (7.2.25) by including the third Euler-angle integral $(1/2\pi) \int d\gamma$. This can be added without any change when all the body quantum numbers are zero. Then the following results for the potential matrix:

$$\left\langle \begin{matrix} l' \\ m' \end{matrix} \middle| V \middle| \begin{matrix} l \\ m \end{matrix} \right\rangle = C_{qmm'}^{k l l'} \left(\sum_{k,q} A_{kq} \sqrt{\frac{2l+1}{2l'+1}} \, C_{000}^{kll'} \langle r^k \rangle \right). \tag{7.3.40a}$$

This in turn gives the reduced matrix elements of the multipole functions.

$$\langle l'\|X^k\|l\rangle = \sqrt{\frac{2l+1}{2l'+1}} C_{000}^{kll'}\langle r^k\rangle = \sqrt{2l+1}(-1)^{k-l}\begin{pmatrix}k & l & l'\\ 0 & 0 & 0\end{pmatrix}\langle r^k\rangle. \quad (7.3.40b)$$

For the crystal splitting example in Eq. (7.3.28) we would need the reduced matrix element

$$\langle 2\|X^4\|2\rangle = \sqrt{\frac{2}{7}}\langle r^4\rangle. \quad (7.3.40c)$$

For hydrogen the radial integral can be shown to be [see K. Bockaster, *Phys. Rev. A* 9, 1087 (1974)]

$$\langle r^4\rangle = (a_0)^4 \frac{n^4}{8}\{63n^4 - n^2[70l(l+1) - 105]$$
$$+ 15(l-1)l(l+1)(l+2) - 20l(l+1) + 12\}$$
$$= (a_0)^4 \frac{63n^4}{8}(n+1)(n-1)(n+2)(n-2) \quad \text{(for } l=2\text{)}.$$

The hydrogen values are often used for approximate theories of other atoms.

(b) Two-Electron Orbitals in Potential Fields We shall now compare the splitting due to a cubic crystal field of a two-electron $L = 2$ level $(d^2 {}^1D)$ with that of the one-electron orbital $(d^1 {}^2D)$. This amounts to a comparison of two separate applications of the Wigner-Eckart theorem and two different reduced matrix elements. For the two-electron case we need energy matrix elements such as

$$\langle [d^2]_{M'}^2 \overset{\text{total}}{|} X_q^4 | [d^2]_M^2\rangle = C_{qMM'}^{4\,2\,2}\langle [d^2]2\| X^4 \|[d^2]2\rangle, \quad (7.3.41a)$$

where the perturbation

$$X_q^4 \overset{\text{total}}{=} X_q^4 \overset{\text{electron 1}}{+} X_q^4 \overset{\text{electron 2}}{} \quad (74.3.41b)$$

is a sum of individual electron operators. For one electron we have

$$\left\langle d^1 \begin{array}{c}2\\M'\end{array}\left|X_q^4\right|d^1\begin{array}{c}2\\M\end{array}\right\rangle = C_{qMM'}^{4\,2\,2}\langle 2 |X^4| 2\rangle, \quad (7.3.42)$$

which is the same as Eq. (7.3.41a) except for the reduced matrix element. So we must compare the reduced matrix elements.

For the sake of generality let us evaluate the matrix element of a general multipole operator X_q^k between general mixed configuration $|[l_1 l_2]L\rangle$ states instead of pure $|l^2 L\rangle$ configuration states.

$$\langle [l_1 l_2]_M^L | \overset{\text{total}}{X_q^k} |[l_1' l_2']_{M'}^{L'}\rangle = \langle [l_1 l_2]_M^L | \overset{\text{electron 1}}{X_q^k} |[l_1' l_2']_{M'}^{L'}\rangle$$
$$+ \langle [l_1 l_2]_M^L | \overset{\text{electron 2}}{X_q^k} |[l_1' l_2']_{M'}^{L'}\rangle. \quad (7.3.43)$$

Treating the first term by the Wigner-Eckart theorem gives

$$\langle [l_1 l_2]_M^L | \overset{\text{electron 1}}{X_q^k} |[l_1' l_2']_{M'}^{L'}\rangle = C_{qM'M}^{kL'L}\langle [l_1 l_2]L|| \overset{\text{electron 1}}{X^k} ||[l_1' l_2']L'\rangle. \quad (7.3.44)$$

However, we can use the fact that each two-electron state is a coupling of the form

$$|[l_1' l_2']_{M'}^{L'}\rangle = \sum_{m_1' m_2'} C_{m_1' m_2' M'}^{l_1' l_2' L'} \left| \begin{matrix} l_1' \\ m_1' \end{matrix} \right\rangle \left| \begin{matrix} l_2' \\ m_2' \end{matrix} \right\rangle \quad (7.3.45)$$

of single-electron states. Inserting these on the left of Eq. (7.3.44) gives

$$\sum_{m_1 m_2} \sum_{m_1' m_2'} C_{m_1 m_2 M}^{l_1 l_2 L} C_{m_1' m_2' M'}^{l_1' l_2' L'} \left\langle \begin{matrix} l_1 \\ m_1 \end{matrix} \right| \left\langle \begin{matrix} l_2 \\ m_2 \end{matrix} \right| \overset{\text{electron 1}}{X_q^k} \left| \begin{matrix} l_1' \\ m_1' \end{matrix} \right\rangle \left| \begin{matrix} l_2' \\ m_2' \end{matrix} \right\rangle$$
$$= C_{qM'M}^{kL'L}\langle [l_1 l_2]L|| \overset{\text{electron 1}}{X^k} ||[l_1' l_2']L'\rangle. \quad (7.3.46)$$

Now $\overset{\text{electron 1}}{X_q^k}$ only acts upon the state vectors of electron (1). Hence one may apply the Wigner-Eckart theorem again just for it:

$$\left\langle \begin{matrix} l_1 \\ m_1 \end{matrix} \right| \left\langle \begin{matrix} l_2 \\ m_2 \end{matrix} \right| \overset{\text{electron 1}}{X_q^k} \left| \begin{matrix} l_1' \\ m_1' \end{matrix} \right\rangle \left| \begin{matrix} l_2' \\ m_2' \end{matrix} \right\rangle = \left\langle \begin{matrix} l_1 \\ m_1 \end{matrix} \right| \overset{\text{electron 1}}{X_q^k} \left| \begin{matrix} l_1' \\ m_1' \end{matrix} \right\rangle \left\langle \begin{matrix} l_2 \\ m_2 \end{matrix} \right| \left. \begin{matrix} l_2' \\ m_2' \end{matrix} \right\rangle$$
$$= C_{qm_1' m_1}^{k l_1' l_1}\langle l_1 || \overset{\text{electron 1}}{X_q^k} ||l_1'\rangle \delta_{l_2 l_2'} \delta_{m_2 m_2'}. \quad (7.3.47)$$

By substituting this last result into Eq. (7.3.46) and using orthonormality [Eq.

(7.2.21)] to bring $C_{qM'M}^{k\,L'\,L}$ to the left-hand side, we obtain

$$\left(\sum_{m_1m_2}\sum_{m'_1m'_2}\sum_{qM'} C_{qm'_1m_1}^{k\,l'_1\,l_1} C_{m_1m_2M}^{l_1\,l_2\,L} C_{qM'M}^{k\,L'\,L} C_{m'_1m'_2M'}^{l'_1\,l'_2\,L'}\right)\langle l_1||\overset{\text{electron 1}}{X^k}||l'_1\rangle$$

$$= \langle[l_1l_2]L||\overset{\text{electron 1}}{X^k}||[l'_1l'_2]L'\rangle. \qquad (7.3.48)$$

The combination of coupling coefficients in the parentheses appears many times in angular-momentum calculations. Up to a factor and a phase it is equal to the RACAH $6j$ coefficient (see Appendix F)

$$\sum_{\substack{m_1,m_2,m_3 \\ m_{12},m_{23}}} C_{m_1m_2m_{12}}^{j_1\,j_2\,j_{12}} C_{m_{12}m_3M}^{j_{12}\,j_3\,J} C_{m_1m_{23}M}^{j_1\,j_{23}\,J} C_{m_2m_3m_{23}}^{j_2\,j_3\,j_{23}}$$

$$= (-1)^{j_1+j_2+j_3+J}$$

$$\times \sqrt{(2j_{12}+1)(2j_{23}+1)} \begin{Bmatrix} j_{12} & j_1 & j_2 \\ j_{23} & j_3 & J \end{Bmatrix}. \qquad (7.3.49)$$

The two-particle calculation requires the numerical values of the $6j$ coefficient. By combining Eqs. (7.3.48) and (7.3.49) one obtains

$$\langle[l_1l_2]L||\overset{\text{electron 1}}{X^k}||[l'_1l'_2]L'\rangle$$

$$= (-1)^{k+l'_1+l_2+L}\sqrt{(2l_1+1)(2L'+1)}\begin{Bmatrix} l_1 & k & l'_1 \\ L' & l_2 & L \end{Bmatrix}\langle l_1||\overset{\text{electron}}{X^k}||l'_1\rangle. \qquad (7.3.50a)$$

By a similar analysis for electron (2) we find

$$\langle[l_1l_2]L||\overset{\text{electron 2}}{X}||[l'_1l'_2]L'\rangle$$

$$= (-1)^{k+l'_2+l_1+L}\sqrt{(2l_2+1)(2L'+1)}\begin{Bmatrix} l_2 & k & l'_2 \\ L' & l_1 & L \end{Bmatrix}\langle l_2||\overset{\text{electron 2}}{X}||l'_2\rangle. \qquad (7.3.50b)$$

Summing these two with the values of momenta for our example gives

$$\langle[22]2||X^4||[22]\rangle = 2\sqrt{5\cdot 5}\begin{Bmatrix} 2 & 4 & 2 \\ 2 & 2 & 2 \end{Bmatrix}\langle 2||X^4||2\rangle$$

$$= \tfrac{4}{7}\langle 2||X^4||2\rangle = 0.57\langle 2||X^4||2\rangle. \qquad (7.3.50)_x$$

This shows that the crystal field effect on a two-d-particle $L=2$ level is 57% that of a single-electron state with the same radial and angular numbers.

7.4 ROTATIONAL LEVEL SPLITTING FOR HIGH J: SEMICLASSICAL ANGULAR MOMENTUM MECHANICS

Rotational or orbital mechanics of atomic electrons in anisotropic potentials is analogous to the quantum mechanics of molecular rotation, and it was first studied in early days of quantum theory. Bethe described the splitting of orbital levels by anisotropic crystalline fields having point symmetries ranging from octahedral (O) to orthorhombic (D_2).

In modern formalism the crystal field orbital eigensolutions are found by first expressing the Hamiltonian H in terms of irreducible (Racah-Wigner) tensors T, and then diagonalizing a representation of H in an orbital basis $\{\cdots |n, L, N\rangle \cdots |n', L', M'\rangle \cdots\}$. The Wigner-Eckart theorem gives the representation of each tensor component as a product of coupling or Clebsch-Gordan coefficients and reduced radial matrix elements:

$$\langle n', L', M'|\mathbf{T}_q^k|n, L, M\rangle = C_{qMM}^{kLL'}\langle n'L'\|T^k\|nL\rangle. \quad (7.4.1)$$

The remainder of the problem (and most of the numerical labor) involves truncating the basis, summing the operators, and matrix diagonalization.

Molecular rotations in a vacuum may be described analogously using anisotropic Hamiltonians. In the simplest cases the rotational Hamiltonians are conveniently expressed as polynomials of angular momentum operators J_x, J_y, and J_z defined with respect to the molecular frame. Pure rotational Hamiltonians conserve J and cannot couple rotational states $|J, K\rangle$ and $|J', K'\rangle$ having different J values. This makes the rotational analysis simpler than the external crystal field problem since numerical diagonalization is limited to treating individual $(2J + 1)$ dimensional block matrices. Even so, heavy molecules tend to have high J. For example, SF_6 spectra with $J = 150$ and higher can be resolved, and so the numerical problem still may be quite formidable.

However, for high-J states it is possible to make approximations. It turns out that for high symmetry the diagonal ($K = K'$) contributions to the tensor matrix elements are dominant, and for high J and K they can be approximated by an asymptotic expression for the Clebsch-Gordan coefficients in terms of a Wigner rotation matrix or a Legendre polynomial. (Here we take the reduced matrix factor to be unity.)

$$\langle J, K|T_0^k|J, K\rangle = C_{0KK}^{kJJ} \cong D_{0,0}^k(0, \Theta_{JK}, 0) = P_k(\cos \Theta_{JK}). \quad (7.4.2a)$$

The polar angle Θ_{JK} is that of the angular-momentum cones introduced in Chapter 5. (Recall Figures 5.4.4, 5.5.3, and 5.5.4.)

$$\cos \Theta_{JK} = K/[J(J+1)]^{1/2}, \quad K = J, J-1, J-2, \ldots . \quad (7.4.2b)$$

The approximation (7.4.2a) is valid in the limit that J and K are both large compared to the tensorial rank ($K \gg k$). The angle Θ_{JK} is the apex half-angle of a cone with a slant height of $\sqrt{[J(J+1)]}$ and altitude of K. The cone is the locus of the quantum angular-momentum vector J subject to the constraints $\langle \mathbf{J} \cdot \mathbf{J} \rangle = [J(J+1)]$ and $\langle J_z \rangle = K$ imposed by the state $|J, K\rangle$. The cone angle Θ_{JK} is a measure of the quantum uncertainty (ΔJ_x) or (ΔJ_y) of transverse components for that state.

The possible motion of a classical angular-momentum J vector can be displayed using a rotational energy (RE) surface. RE surfaces are radial plots of rotational energy as a function of the direction of the J vector in the body frame for a constant magnitude $|\mathbf{J}| = \sqrt{[J(J+1)]}$ of angular-momentum. The classical J vector, while fixed in the laboratory frame, follows a trajectory in the body-frame which conserves both energy E and magnitude $|J|$ of J.

Each classically allowed J trajectory is a topography line on an RE surface, that is the intersection of an RE surface for a given $|J|$ with an energy sphere for a given E. Examples of RE surfaces for D_2 and O symmetric molecules are shown in the following sections. Furthermore, we shall see the quantum eigenvalues can be related to special "quantizing" J trajectories and that these can be approximated by the intersection of the angular momentum cones with the RE surface.

A. Rigid Rotors ($D_\infty \supset D_2$ symmetry)

(a) Rotational Energy Surfaces The Hamiltonian for a rigid rotor or top is

$$H = AJ_x^2 + BJ_y^2 + CJ_z^2. \tag{7.4.3}$$

[Recall Eq. (5.5.53).] The rotational energy (RE) surface of this Hamiltonian follows if we substitute classical body-frame angular-momentum components

$$J_x = -\langle J \rangle \sin \beta \cos \gamma, \quad J_y = \langle J \rangle \sin \beta \sin \gamma, \quad J_z = \langle J \rangle \cos \beta, \tag{7.4.4a}$$

where the J magnitude is constant:

$$\langle J \rangle = \sqrt{J(J+1)} \cong J + \tfrac{1}{2}. \tag{7.4.4b}$$

The resulting energy expression

$$E = J(J+1)\left[A \sin^2 \beta \cos^2 \gamma + B \sin^2 \beta \sin^2 \gamma + C \cos^2 \gamma \right] \tag{7.4.5}$$

is plotted radially in Figure 7.4.1 as a function of body-frame polar coordinates of azimuth ($\phi = -\gamma$) and polar angle ($\theta = \beta$) for the J vector. These

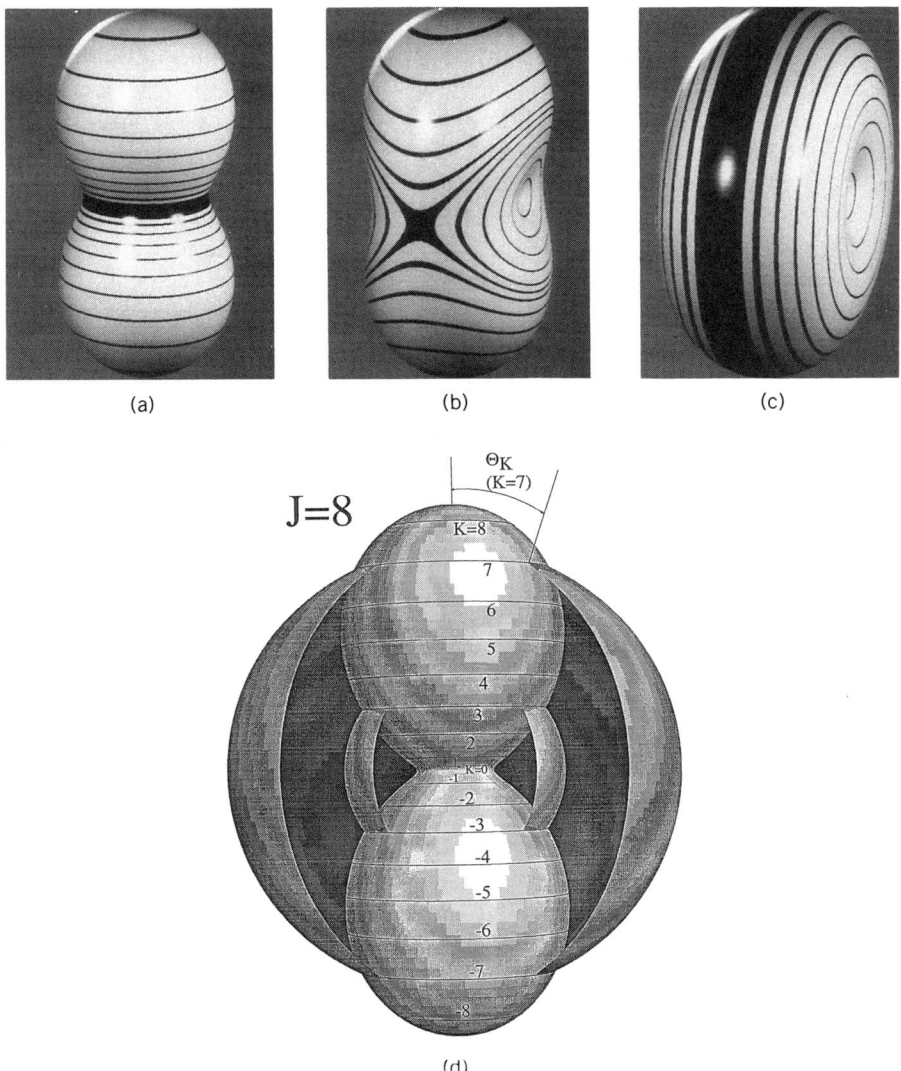

Figure 7.4.1 Rotational energy (RE) surfaces for rigid rotors. (a) Prolate symmetric top ($A = 0.2$, $B = 0.2$, $C = 0.6$). (b) Rigid asymmetric top ($A = 0.2$, $B = 0.4$, $C = 0.6$). (c) Oblate symmetric top ($A = 0.2$, $B = 0.6$, $C = 0.6$). (d) Prolate symmetric top with $J = 10$ quantum energy levels.

angles are two of the three Euler angles (α, β, γ) as explained in Section 5E(b).

Three examples of rigid top RE surfaces are shown in Figure 7.4.1 for the cases of (a) a prolate symmetric top ($A = B < C$), (b) an asymmetric top ($A < B < C$), and (c) an oblate symmetric top ($A < B = C$). Each surface contains 21 contour lines corresponding to the same number of quantum energy levels belonging to $J = 10$ ($2J + 1 = 21$). These are labeled in Figure 7.4.1(d) which is an expanded view of Figure 7.4.1(a). They are quantizing J-phase paths in the sense that the following Bohr quantization condition is satisfied:

$$\int J_z \, d\gamma = \hbar K, \qquad K = J, J - 1, J - 2, \dots . \qquad (7.4.6)$$

Some J paths may be hidden behind a surface. However, time reversal symmetry requires that for each $\mathbf{J}$ on a K path there must be an equivalent path containing $-\mathbf{J}$ or $-K$.

(b) Tensor Operator Mechanics To understand the quantum mechanics better it helps to rewrite the Hamiltonian in terms of tensor operators or as an operator multipole expansion. The multipole functions,

$$T_q^k = D_{0q}^k(0, \beta, \gamma)^* = \sqrt{\frac{4\pi}{2k+1}} \langle J \rangle^k Y_q^k(\gamma, \beta) \qquad (7.4.7)$$

are analogous to the spatial multipole functions defined in Eq. (5.6.17). Examples of quadrupole tensor functions are

$$T_0^0 = \mathbf{J} \cdot \mathbf{J} = \langle J \rangle^2 = \left(J_x^2 + J_y^2 + J_z^2 \right), \qquad (7.4.8a)$$

$$T_0^2 = \tfrac{1}{2} \langle J \rangle^2 (3\cos^2 \beta - 1) = \tfrac{1}{2}\left(2J_z^2 - J_x^2 - J_y^2 \right), \qquad (7.4.8b)$$

$$\left(T_2^2 + T_{-2}^2 \right) = \langle J \rangle^2 \frac{\sqrt{6}}{2} \sin^2 \beta \cos 2\gamma = \frac{\sqrt{6}}{2}\left(J_x^2 - J_y^2 \right). \qquad (7.4.8c)$$

From this one can construct the tensor operator expression for the rigid rotor Hamiltonian:

$$H = \frac{A+B+C}{3} T_0^0 + \frac{2C-A-B}{3} T_0^2 + \frac{A-B}{\sqrt{6}} \left(T_2^2 + T_{-2}^2 \right) \qquad (7.4.9)$$

Only the first term survives for a spherical top ($A = B = C$). The symmetric tops ($A = B \neq C$) have the first two terms.

The asymmetric tops ($A \neq B \neq C$) have all three. The resulting energy expression obtained from Eq. (7.4.8) and (7.4.9) is

$$E = J(J+1)\left[\frac{A+B+C}{3} + \frac{2C-A-B}{6}(3\cos^2\beta - 1) \right.$$
$$\left. + \frac{A-B}{2}\sin^2\beta \cos 2\gamma \right]. \qquad (7.4.10)$$

The preceding multipole expansion is equal to the polynomial expression (7.4.5) but has some quantum mechanical advantages over the first one. Tensor multipole operator expressions like (7.4.9) provide matrix elements immediately via the Wigner-Eckart theorem (7.3.20a) in terms of Clebsch-Gordan coefficients and reduced matrix elements. Only one reduced matrix $\langle J \| T^2 \| J \rangle$ is a problem here, the scalar element of (7.4.8a) is elementary,

$$\langle J \| T^0 \| J \rangle = J(J+1).$$

The tensor element is found by evaluating the easiest component $\langle {}^J_K | T_0^2 | {}^J_K \rangle$ by elementary means which gives

$$\langle {}^J_J | T_0^2 | {}^J_J \rangle = \langle {}^J_J | \tfrac{1}{2}(3J_z^2 - J_x^2 - J_y^2 - J_z^2) | {}^J_J \rangle$$
$$= \tfrac{1}{2}(3J^2 - J(J+1)) = \tfrac{1}{2}(2J^2 - J). \quad (7.4.11)$$

Then the Wigner-Eckart theorem and CG formulas give

$$\langle {}^J_K | T_0^2 | {}^J_J \rangle = C^{2JJ}_{0JJ} \langle J \| T^2 \| J \rangle = \frac{2(2J^2 - J)}{\sqrt{(2J+3)(2J+2)2J(2J-1)}}. \quad (7.4.12)$$

Solving gives the desired reduced matrix element:

$$\langle J \| T^2 \| J \rangle = \sqrt{(2J+3)(2J+2)2J(2J-1)}/4. \quad (7.4.13)$$

Still one might wonder why we deal with tensor operators when J polynomials seem simpler. The reasons for using tensor operators become clearer when comparing the work involved with higher-degree polynomials and corresponding high-rank tensors. Manipulating and computing matrix elements for fourth- or sixth-degree polynomials can be extremely laborious while fourth- or sixth-rank tensors use the same Wigner-Eckart analysis as the T^2 example above.

(c) Symmetric Top Energy Levels ($J - 10$ Example) The trajectories on the RE surface for $A = B = 0.2$ and $C = 0.6$ [see Figures 7.4.1(a) and 7.4.1(d)] are precisely the ones that correspond to exact quantum energy levels for $J = 10$. If the cone-angle cosine formula (7.4.2b) is substituted into the tensor RE surface energy expression (7.4.10) for $A = B$ one obtains

$$E = J(J+1)\left[\frac{2B+C}{3} + \frac{C-B}{3}\left(3\frac{K^2}{J(J+1)} - 1\right)\right], \quad (7.4.14)$$

$$E = BJ(J+1) + (C-B)K^2. \quad (7.4.15)$$

This is the exact symmetric top quantum energy level equation. [Recall Eq.

(5.5.52).] This is an example in which the cone-angle tensor matrix element approximation (7.4.2) gives an exact result. The angular-momentum cones exactly define the quantizing J trajectories shown in Figure 7.4.1a. The same applies to the oblate symmetric top surface shown in Fig. 7.4.1c.

We consider now the asymmetric top for which the J trajectories are more or less distorted from the circular shapes they enjoy in the symmetric case.

(d) Asymmetric Top Energy Levels ($J = 10$ Example) This J-inversion symmetry and the D_2 rotational symmetry of top Hamiltonian (7.4.3) or (7.4.9) combine to give (at least) a D_{2h} symmetry to the RE surface regardless of the symmetry of the rotor which it models. The simplest rigid molecule having the surface shown in Figure 7.4.1(b) would be a bent XY_2 structure like the water molecule.

It is evident that every path on the asymmetric surface in Figure 7.4.1b belongs to a mirror image pair of trajectories with the exception of one path. The exceptional path is the x-shaped separatrix curve which crosses the saddle points on the $\pm y$ axes. The separatrix divides the surface into regions containing two different kinds of trajectory pairs. One kind of trajectory pair encircles the high-energy regions centered on the $\pm z$ axes or C axis. These paths are distorted versions of the paths for the prolate top shown in Figure 7.4.1a. The other pairs encircle the low-energy valley regions around the $\pm x$ axes or A axis, and they are distorted versions of the oblate symmetric top paths in Figure 7.4.1(c).

The separation of regions is manifested in the quantum level spectrum which is shown in the lower center portion of Figure 7.4.2. Here the lower energy quasioblate pairs of trajectories are each identified with quasidegenerate or "clustered" pairs of energy levels below the separatrix level at 44 cm^{-1}. Similarly, the quasiprolate pairs are indicated in the high-energy region on the right-hand side of Figure 7.4.2. The levels belonging to each pair are indicated inside magnifying circles which give D_2 symmetry labels for each level and the magnitude of the splitting between each pair.

The fine structure splitting is the intercluster frequency splitting such as the 150 GHz splitting between the lowest two pairs. This is approximately the frequency of classical precession or the wobbling frequency for the J vector to go once around the lowest energy path. The intracluster splitting such as the 26 kHz splitting of the $A_1 B_1$ pair in the lowest circle is called *superfine structure*. This corresponds to the frequency of a purely quantum mechanical tunneling process between equivalent pairs of semiclassical paths. If the molecule was set initially into a localized nonstationary state with J wobbling around the lowest ($K = 10$) path near the $+x$ axis, then it would gradually evolve into a similar motion around the equivalent ($K = -10$) path near the $-x$ axis after which it would return and (more or less) repeat the whole process at rate of 26 kHz.

The cluster doublets are the angular momentum analogs of inversion doublet levels of a two-well oscillator potential discussed in Chapter 2.

614 THEORY AND APPLICATION OF SYMMETRY REPRESENTATION PRODUCTS

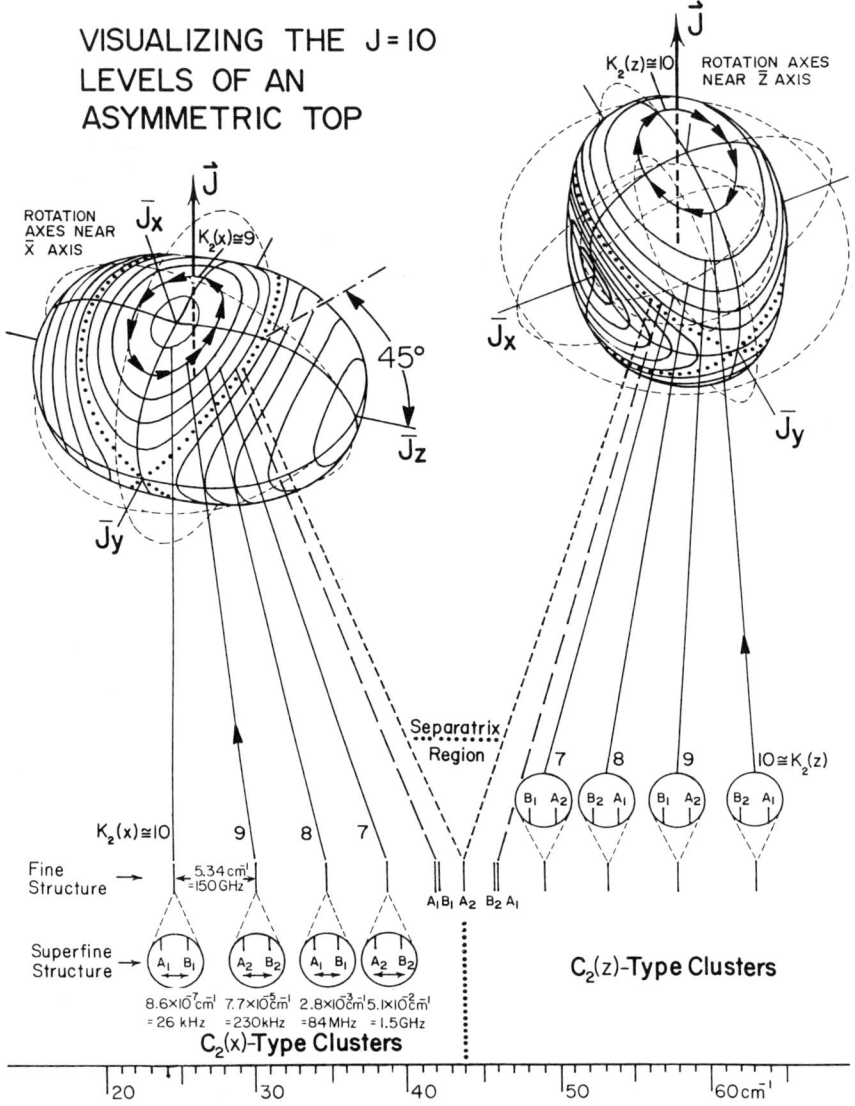

Figure 7.4.2 $J = 10$ asymmetric top energy levels and related RE surface paths ($A = 0.2$, $B = 0.4$, $C = 0.6$). Clustered pairs of levels are indicated in magnifying circles which show superfine splittings.

(Recall Figure 2.12.7.) The stationary A_1 or B_1 eigenstates are, respectively, symmetric or antisymmetric combinations of two separate but equivalent wave functions localized on separate but equivalent paths. The degree of separation or localization is given by the superfine level splitting or tunneling rate. This rate varies exponentially with the magnitude of a path integral between the points of closest approach of the separate semiclassical paths.

For the highest K trajectories which have the greatest separation the precessional motion is more than a million times faster than the tunneling motion. However, near the separatrix the tunneling rate or superfine splitting increases enormously while the classical precession rate or fine structure splitting actually decreases. In this region wave functions cannot remain localized very long, and the distinction between classical and purely quantum motion is blurred.

A classical rotor is always located at just one point on a single phase trajectory at each instant. The quantum rotor, on the other hand, can have nonzero probability spread over many different paths at once. In fact its wave function must be spread out in order to belong to a single irreducible representation such as A_1 or B_1 of the global symmetry group D_2. In order to make a wave function localized on just one trajectory one must add (or subtract) A_1 and B_1 waves. A combination of the two $K = \pm 10$ waves will still have a well defined 0-mod 2 (labeled 0_2) symmetry with respect to the local symmetry subgroup $C_2(x)$ which contains only x-axis rotation. This because $K = 10$ and $K = -10$ are even numbers. The combination states form a space belonging to the induced representation 0_2(of $C_2(x))\uparrow D_2$ of the global D_2 symmetry induced by the even representation 0_2 of the local symmetry $C_2(x)$. The even induced representation is indicated by the first column of the $D_2 \supset C_2(x)$ correlation table in the left-hand side of Figure 7.4.3. The table gives the D_2 species in the even and odd induced representa-

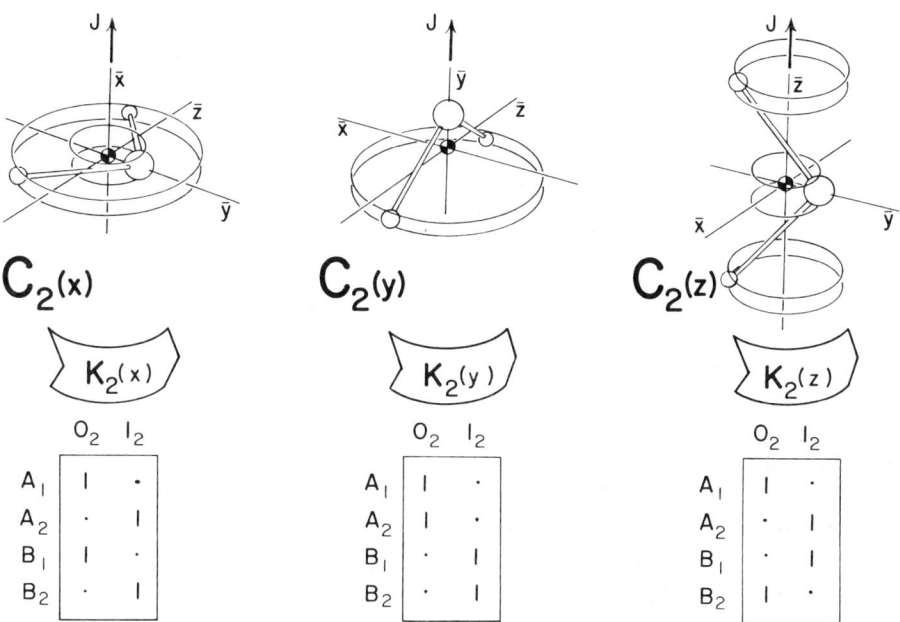

Figure 7.4.3 Correlations between the asymmetric top symmetry D_2 and three dynamical subgroups $C_2(x)$, $C_2(y)$, and $C_2(z)$.

tions according to the Frobenius reciprocity theorem:

$$0_2(\text{of } C_2(x)) \uparrow D_2 = A_1 \oplus B_1, \qquad (7.4.16a)$$

$$1_2(\text{of } C_2(x)) \uparrow D_2 = A_2 \oplus B_2. \qquad (7.4.16b)$$

The even (odd) induced representation labels the even-K (odd-K) clusters which lie below the separatrix in Figure 7.4.2. The clusters above the separatrix have a local symmetry $C_2(z)$, and the clusters corresponding to this region are labeled according to the columns of the $D_2 \supset C_2(z)$ correlation table shown in the right-hand part of Figure 7.4.3.

Sketches of the classical motion correspond to the locally C_2 symmetric trajectories. The $C_2(x)$ motion corresponds to an XY_2 rotating on its side like a boomerang, while $C_2(z)$ motion is like a spinning crankshaft. The $C_2(y)$ motion is around the classically unstable saddle point, and hence no $C_2(y)$ level clusters appear in the spectrum. One should note that the phase portraits describe the precession or "rotation of rotation" rather than rotation itself. Precessionless rotation of a rigid body would occur only if the J vector were precisely localized on one of the principle axes; however, this is not possible for a quantum rotor.

Quantum uncertainty prohibits pure rotation without precession because the transverse components cannot be exactly zero. The transverse components are minimum for the $K = J$ states which have the least cone angle Θ_{JJ}. This corresponds to minimum angular-momentum uncertainty or angular zero-point motion. States with lower z-component quanta $K = J - 1, J - 2, \ldots$ have higher uncertainty angles Θ_{JJ} according to Eq. (7.4.2b). The states $|J, K\rangle$ are eigenstates of the symmetric top ($A = B$) Hamiltonian, and each angular-momentum cone exactly intersects the corresponding semiclassical path on the symmetric top RE surface in Figure 7.4.1(a).

For the asymmetric top in Figure 7.4.1(b) the Θ_{JJ} cones only approximate their corresponding semiclassical trajectories. Asymmetric top trajectories are distorted or "squeezed" so that the projection of J on the local axis of quantization oscillates around the K value which labels each path. The classical precession becomes more and more nonuniform as K decreases and the separatrix is approached. This corresponds to the mixing of more of the states $|J, K \pm 2\rangle$, $|J, K \pm 4\rangle$, and so on into the dominant $|J, K\rangle$ component of the asymmetric top eigenstate. The global and local symmetries for the symmetric top are continuous groups O_2 or $D_{\infty h} \supset R_2$ while the asymmetric top has only a discrete set of symmetries $D_2 \supset C_2$. Hence, one cannot expect the K value to be strictly conserved or the $\pm K$ degeneracy to be perfectly maintained in the latter.

However, the extent of breakdown of R_2 symmetry or K conservation is not necessarily related to the splitting of the cluster doublets. K conservation and cluster splitting are separate phenomena associated with different regions of the RE phase space; the former depends upon the shape of the phase paths, and the latter depends upon the height of the pass or saddle region between the equivalent paths. Furthermore, the symmetry properties of the clusters should be associated with a C_2 induced representation and not an R_2 irreducible representation. This point will be amplified by examples involving the higher octahedral symmetry in the following section.

Another point which arises in the study of higher symmetries concerns the ordering of clusters and the symmetry species inside them. The species ordering in Figure 7.4.2 consists of a repetition of the sequence $A_1 B_1 A_2 B_2$ through the entire spectrum. This remarkably uniform ordering can be related to the number of wave function nodes occurring along and between the semiclassical paths. This sort of ordering was introduced in Chapter 2.

(e) Level Correlation Between $C_2(x)$ and $C_2(z)$ Symmetry

The coefficients A, B, and C determine the symmetry of the rotor Hamiltonian (7.4.3) and its RE surface. The surface represents a rotor that is prolate-symmetric ($A = B < C$) in (a) of Figure 7.4.1, asymmetric ($A < B < C$) in (b), and oblate-symmetric ($A < B = C$) in (c). The two extreme symmetric rotor cases have levels labeled by different $R_3 \supset R_2$ subgroup chains. The prolate case is labeled by $R_2(z)$ and the oblate case by $R_2(x)$. The intermediate asymmetric case is labeled using finite subgroup chains $R_3 \supset D_2 \supset C_2$. Furthermore different subgroups are appropriate for different levels; the levels below the separatrix belong to $C_2(x)$ and those above belong to $C_2(z)$.

In Figure 7.4.4 the $J = 10$ and $J = 20$ levels are plotted as a function of parameter B which ranges between the prolate ($B = 0.2$ cm^{-1}) and oblate ($B = 0.6$ cm^{-1}) cases. Coefficients $A = 0.2$ cm^{-1} and $C = 0.6$ cm^{-1} are fixed. One can see that most $J = 10$ symmetric top doublets tend to stick together for most values of B and even more so for $J = 20$. The $J = 10$ levels in Figure 7.4.2 lie above the point $B = 0.4$ in the $J = 10$ plot of Figure 7.4.4.

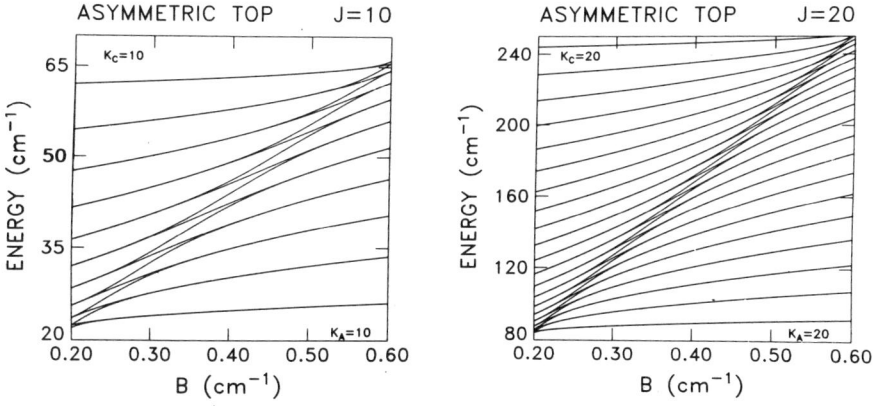

Figure 7.4.4 Rigid rotor energy levels correlations for angular momentum $J = 10$ and $J = 20$.

The separatrix region of the levels in the center of Figure 7.4.2 is the transition region where doublets split and trade levels in Figure 7.4.4. The separatrix or transition region appears to be a small fraction of the $J = 10$ spectrum and even smaller part for $J = 20$.

The doublets in the upper left-hand part of Figure 7.4.4 above the transition region belong to $C_2(z) \uparrow D_2$ induced representations while those in the lower right-hand part belong to $C_2(x) \uparrow D_2$ labeled doublets.

This correlation plot should be compared to the lattice level correlation diagrams in Figure 2.12.5. The latter involves a correlation between bands of doublets of levels belonging to $C_n(z) \uparrow D_n$ induced representations and n-fold degenerate clusters or bands of levels belonging to $C_2(x) \uparrow D_n$ representations. For $n = 2$ the plot in Figure 2.12.5 is more closely analogous to the one in Figure 7.4.4. Then the A_1, A_2, B_1 and B_2 levels (which are the band boundaries plotted in Figure 2.12.5) are the only levels allowed; the E-type levels do not exist for $n = 2$. The transition region occurs at the top of the potential barriers.

By analogy the asymmetric top spectral transition region occurs at the top (or bottom) of the saddles on the RE surface. The saddle points are on the $\pm y$ axes and rise linearly with the coefficient B of J_y^2. Certain of the transition levels are seen to rise rapidly and quasilinearly in Figure 7.4.4 while their doublet partners seen to sail right through the transition region. Wave symmetry determines which of the D_2 species are most sensitive to the y-axis saddle. The correlation table in Figure 7.4.3 for the $C_2(y)$ symmetry shows that only A_1 and A_2 are symmetric (0_2). Therefore only they have wave antinodes and substantial amplitudes on the saddles, and it is therefore A_1 and A_2 levels that "divorce" their partners in the transition region.

Outside the transition region the pairs of levels mostly stick together to form quasidegenerate tunneling doublets. One exception is the $K = \pm 1$ doublet near the lower left-hand side of Figure 7.4.4. It splits immediately, that is, to first order. This is analogous to the first order splitting observed in Figure 2.12.5. Symmetry allows nonzero matrix elements between this pair of states. In this case it is matrix element $\langle K = 1 | T_2^2 | K = -1 \rangle$ and its conjugate that cause the $K = \pm 1$ doublet to split.

B. Semirigid Spherical Tops [Octahedral (O) Symmetry]

We now consider the high-J eigenvalues of octahedrally symmetric tensor Hamiltonians. The fourth-rank tensor Hamiltonian,

$$H = BT_0^0 + 4t_{044}\left[T_0^4 + \sqrt{\frac{5}{14}}\left(T_4^4 + T_{-4}^4\right)\right] \qquad (7.4.17)$$

has the same form as the one introduced in Eq. (7.3.25). Its polynomial form

$$H = B\mathbf{J}^2 + 10t_{044}\left(J_x^4 + J_y^4 + J_z^4 - \tfrac{3}{5}\mathbf{J}\right)^4 \tag{7.4.18}$$

was introduced in Eqs. (5.6.28) and (5.6.30). It is known as the Hecht Hamiltonian after K. T. Hecht who first applied it to the analysis of methane (CH_4) spectra taken by E. Plyler in 1960. The Hamiltonian describes rotation-vibrational distortion of molecules having tetrahedral (T_d) as well as octahedral (O_h) symmetry. Changing the sense of rotation ($\mathbf{J} \rightarrow -\mathbf{J}$) should give a rotor state with the same energy so all rotors must have only pure rotational energy operators of even rank. The third-rank tetrahedral invariant $J_x J_y J_z$ is forbidden to appear alone.

(a) Rotational Energy Surfaces We now express the Hecht Hamiltonian (7.4.17) in terms of body polar angles as was done in the preceding section for the asymmetric rotor. The polynomial

$$E = B\langle J^2\rangle + t_{044}\langle J^4\rangle(35\cos^4\beta - 30\cos^2\beta + 3 + 5\sin^4\beta\cos 4\gamma)/2 \tag{7.4.19}$$

has the form of the harmonic polynomial functions in Eq. (5.6.29). The resulting RE surface is shown in Figure 7.4.5a for positive centrifugal distortion constant t_{044}. This constant is around 5 Hz for SF_6 and is positive for most octahedral XY_6 molecules. It is greatly exaggerated for the figure so that the hill and valleys are clearly visible.

In an octahedral XY_6 molecule rotation about the four-fold XY radial bond axes generally has the highest energy for a given J value since these

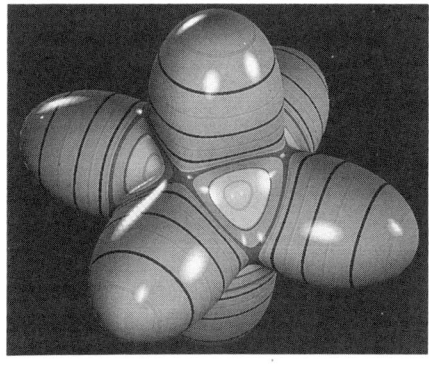
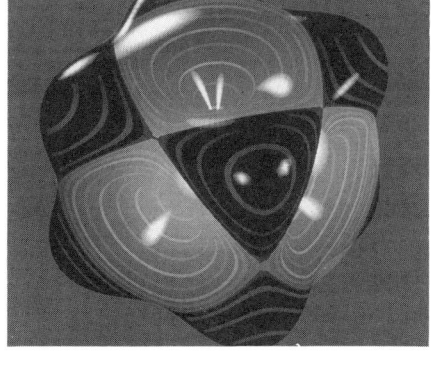

(a)   (b)

Figure 7.4.5 Semirigid rotor RE surfaces with O_h symmetry. (a) $t_{044} > 0$. (b) $t_{044} < 0$.

620   THEORY AND APPLICATION OF SYMMETRY REPRESENTATION PRODUCTS

bonds are stretched relatively little by a longitudinal centrifugal force. However, transverse forces which arise during rotation about the three-fold symmetric axes in between the bonds can bend the molecule relatively easily. Hence, the three-fold symmetry axes lie in RE surface valleys in Figure 7.4.5a while the four-fold (x, y, z) axes are on peaks. For tetrahedral XY_4 or cubic molecules the sign of t_{044} is negative as it is for the surface in Figure 7.4.5(b).

(b) Spherical Top Energy Levels ($J = 30$ Example) The RE topography lines correspond to quantizing J trajectories and to level clusters in the

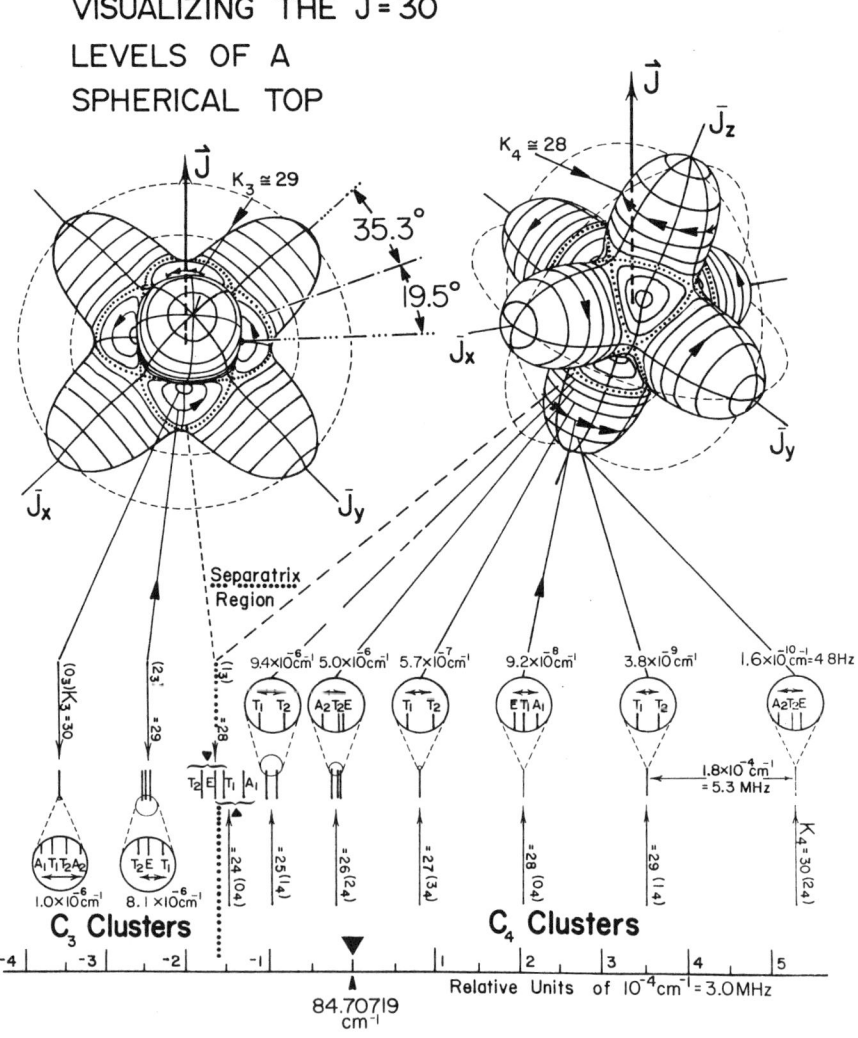

Figure 7.4.6 $J = 30$ octahedral rotor levels and related RE surface paths.

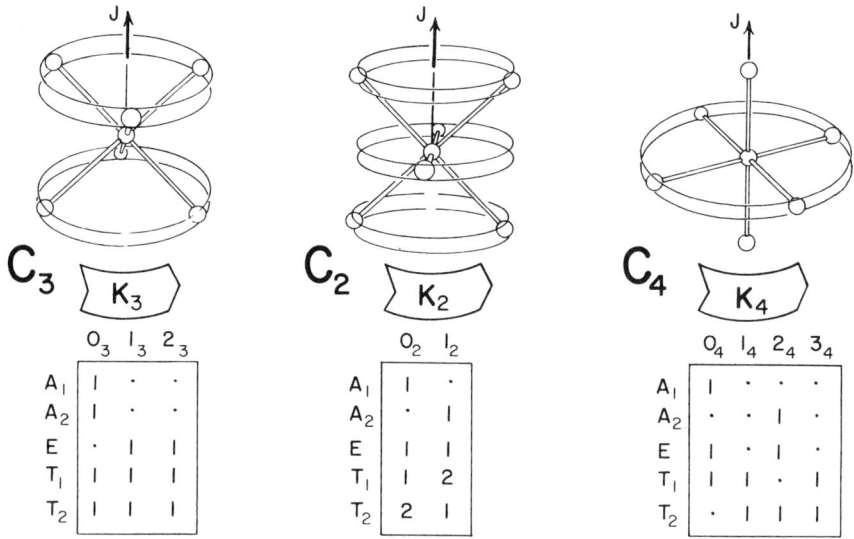

Figure 7.4.7 Different choices of rotation axes for octahedral rotor corresponding to local symmetry C_3, C_2, and C_4. Tables correlate global octahedral symmetry species with the local ones.

energy spectrum as shown by the diagram of the $J = 30$ levels of SF_6 in Figure 7.4.6. This spectrum contains clusters of six and eight rotational levels which are analogous to the rigid rotor doublet clusters in Figure 7.4.2. Above the separatrix region there are repeating sextets $(T_1 T_2)$, $(A_2 T_2 E)$, $(T_1 T_2)$, or $(A_1 T_1 E)$ composed of clustered singlet (A_1 or A_2), doublet (E), or triplet (T_1 or T_2) octahedral symmetry species. Below the separatrix there are two octets $(A_1 T_1 T_2 A_2)$ and $(T_2 E T_1)$. Each set of six or eight clustered levels can be related to the same number of semiclassical J trajectories on the RE surface in Figure 7.4.5 or 7.4.6.

Each set of six rotational levels belongs to one of the C_4 induced representations $0_4 \uparrow O$, $1_4 \uparrow O$, $2_4 \uparrow O$, or $3_4 \uparrow O$ depending upon whether the effective K value is 0, 1, 2, or 3 modulo 4 for the corresponding set of fourfold symmetric semiclassical trajectories. The correlation tables in the lower right-hand part of Figure 7.4.7 tell which O species belong to each K_4 cluster and to each set of trajectories. For example, the minimum uncertainty trajectory has $J = K = 30$ and corresponds to the highest energy $2_4 \uparrow O$ or $(A_2 T_2 E)$ cluster in Figure 7.4.5.

The highest energy semiclassical trajectories are very close to the intersection of the RE surface with the $K = 30$ angular momentum cone which has half-angle $\Theta_{30\,30} = \cos^{-1}(30/\sqrt{(30)(31)}) = 10.3°$. A series of $J = 30$ angular momentum cones are drawn for $K = 30$ down to $K = 24$ in Figure 7.4.8. The next highest $1_4 \uparrow O$ or $(T_1 T_2)$ cluster corresponds to six trajectories which are

622 THEORY AND APPLICATION OF SYMMETRY REPRESENTATION PRODUCTS

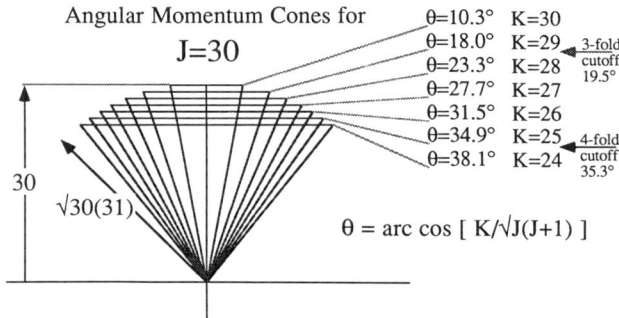

Figure 7.4.8 Quantum angles momentum cone cross sections for $J = 30$ and $K = 30, 29, \ldots, 24$.

localized to within about $\Theta_{30,29} = 18°$ of their respective four-fold symmetry axes. This sequence of clusters ends when $\Theta_{30,K}$ approaches the angle $35.3°$ between the separatrix and the four-fold axes. The $J = 30$ cutoff value is $K_4 = \sqrt{(30)(31)} \cos 35.3° = 24.9$ or about 25 as shown in Figure 7.4.8. This corresponds to a weak $(T_1 T_2)$ cluster just above the separatrix in Figure 7.4.6.

Since the eight three-fold symmetric valley regions of the RE surface are smaller there are fewer clusters associated with the C_3 induced representations. There is only a $19.5°$ angle between the separatrix and the three-fold axes. Hence, the $J = 30$ cut-off value is $K_3 = \sqrt{(30)(31)} \cos 19.5° = 28.7$ or about 29 as indicated in Figure 7.4.8. This allows just two $J = 30$ clusters on the three-fold symmetry side of Figure 7.4.6 corresponding to the induced representations $0_3 \uparrow O = (A_1 T_1 T_2 A_2)$ for $K = 30$ and $2_3 \uparrow O = (T_2 E T_1)$ for $K = 29$.

Two C_3 clusters and five or six C_4 clusters are visible in the infrared spectra of tetrafluorosilane (SiF_4) and cubane ($C_8 H_8$) which is shown in Figure 7.4.9. The spectra are actually due to transitions between level clusters on lower and upper RE surfaces corresponding to ground and vibrationally excited states, respectively. However, the spectra are simply scaled copies of the pure rotational level patterns since the upper and lower RE surfaces have almost the same shape apart from a scale factor. Note that fine structure spectra outside of the separatrix region is relatively insensitive to the J value in that $P(30)$, $P(31)$, and $P(32)$ are quite similar. Note the similarity of $J = 30$ fine structure patterns for quite different molecules having tetrahedral, cubic, and octahedral shapes.

With higher resolution the superfine and even the hyperfine spectral structure can be studied. The hyperfine patterns are very sensitive to the J and K values as well as the detailed structure of the molecule. The number of Pauli-allowed nuclear spin states depends upon the type of arrangement of nuclei, and this affects the relative peak heights for the clusters.

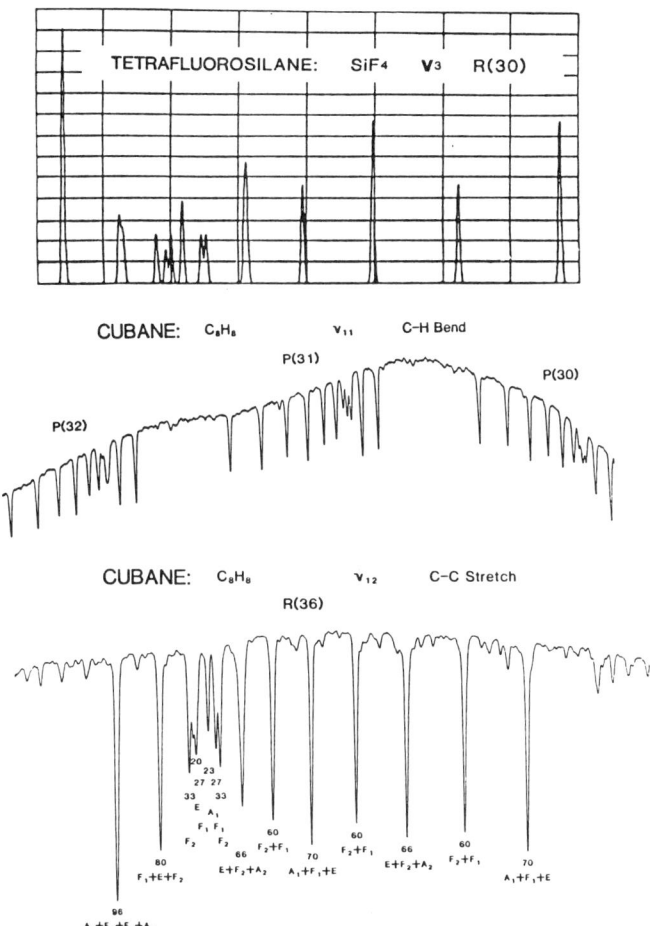

Figure 7.4.9 Infrared spectra showing fine structure clusters. Tetrafluorosilane (SiF$_4$) spectrum from a ν_3 R(30) transition ($J = 30 \to 29$). [After C. W. Patterson, R. S. McDowell, N. G. Nereson, B. J. Krohn, J. S. Wells, and F. R. Peterson, *J. Mol. Spectrosc.* **91**, 416 (1982).] Cubane (C$_8$H$_8$) spectrum from ν_{11} P(30), P(31), and P(32), transitions; cubane (C$_8$H$_8$) spectrum from ν_{12} R(36), transition. [After A. S. Pine, A. G. Maki, A. G. Robiette, B. J. Krohn, J. K. G. Watson, and Th Urbanek, *J. Am. Chem. Soc.*, **106**, 891 (1984).]

However, the superfine structure of intracluster splitting depends only on the shape of the RE surface in the neighborhood of the saddle points. A tunneling factor S can be approximated by an exponential of a phase integral across the saddle region. (Note the rapid decrease of the superfine splitting from about one Megahertz down to just 4.8 Hz as K_4 goes from 25 up to 30 in Figure 7.4.6.) While the magnitude of the splitting may vary by many orders, the splitting patterns have the following invariant form. The first of these was derived before Eq. (4.3.28),

$0_4 \uparrow O$: $\quad\quad\quad\quad$ 1_4 or $3_4 \uparrow O$: $\quad\quad\quad\quad$ $2_4 \uparrow O$:

$\Delta E(A_1) = 4S$, $\quad\quad\quad$ $\Delta E(T_2) = 2S$, $\quad\quad\quad$ $\Delta E(E) = 2S$,

$\Delta E(T_1) = 0$, $\quad\quad\quad\;\,$ $\Delta E(T_2) = 2S$, $\quad\quad\quad$ $\Delta E(T_2) = 0$, $\quad\quad$ (7.4.20)

$\Delta E(E) = -2S$, $\quad\quad\quad\quad\quad\quad\quad\quad\quad\quad\quad\;\,$ $\Delta E(A_2) = -4S$.

The patterns for C_3 clusters can be derived using similar arguments,

$0_3 \uparrow O$: $\quad\quad\quad\quad\quad\quad\quad\quad$ 1_3 or $2_3 \uparrow O$:

$\Delta E(A_2) = 3S$, $\quad\quad\quad\quad\quad\;$ $\Delta E(T_1) = 2S$,

$\Delta E(T_2) = S$, $\quad\quad\quad\quad\quad\quad$ $\Delta E(E) = 0$, $\quad\quad\quad\quad\quad$ (7.4.21)

$\Delta E(T_1) = -S$, $\quad\quad\quad\quad\quad\;\,$ $\Delta E(T_2) = -2S$.

$\Delta E(A_1) = -3S$.

The splitting ratios and ordering hold if tunneling occurs only between nearest neighboring trajectories. The patterns (7.4.20) and (7.4.21) are seen magnified in Figure 7.4.6.

In addition there is an overall ordering that is maintained throughout the fine structure spectrum of Figure 7.4.6 as there was in the case of the asymmetric top. From Figure 7.4.6 one observes the following repeated sequence (recall Figure 5.6.9)

$$0_1 \uparrow O = (A_1 T_1 T_2 A_2 T_2 E T_1 T_2 E T_1). \quad\quad (7.4.22)$$

Taken together this would be the largest possible cubic cluster. It contains just the O regular representation. Giant clusters like (7.4.22) or the C_2 clusters half this size are possible if stable semiclassical orbits are localized around low symmetry points. This occurs for octahedral tensor combination of sixth, eighth, and higher ranks.

(c) Level Correlation between C_3 and C_4 Symmetry So far we have considered only the lowest order rotational tensors which exhibit the symmetries D_2 of the rigid rotor and O_h of the semirigid cubic or octahedral rotor. We consider now the effect of the sixth-rank normalized octahedral tensor operator introduced in Eq. (7.3.35),

$$T^{[6]} = (1/\sqrt{8})\left[T_0^6 - (\sqrt{7}/\sqrt{2})(T_4^6 + T_{-4}^6)\right]. \quad\quad (7.4.23)$$

This will be added in varying amounts to the normalized fourth-rank tensor,

$$T^{[4]} = (\sqrt{7}/\sqrt{12})[T_0^4 - (\sqrt{5}/\sqrt{14})(T_4^4 + T_{-4}^4)], \qquad (7.4.24)$$

introduced in Eqs. (7.3.25) and (7.4.17). A sixth-rank centrifugal distortion may be necessary in the presence of anharmonic and other higher order effects. The magnitude of the $T^{[6]}$ contribution would vary according to a higher power of J than that of $T^{[4]}$ and might be significant at higher J values. Here the magnitudes of their respective contributions are varied artificially through an angle parameter ν in a combination which maintains the overall normalization.

$$T^{4,6}(\nu) = T^{[4]} \cos \nu + T^{[6]} \sin \nu. \qquad (7.4.25)$$

The exact quantum ($J = 30$)-eigenvalues for this mixed [4, 6]-rank tensor operator are plotted as a function of the mixing angle ν in Figure 7.4.10. The plot begins on the left-hand side ($\nu = 0$) with a scaled copy of the $T^{[4]}$ level spectrum in Figure 7.4.6 and ends on the right-hand side ($\nu = \pi$) with the same spectrum inverted. Between these limits the level clusters become completely reorganized.

Certain values of the ν parameter in Figure 7.4.10 are marked (b), (c), (d), and (e). At these values the RE surface of the combination tensor (7.4.25) is

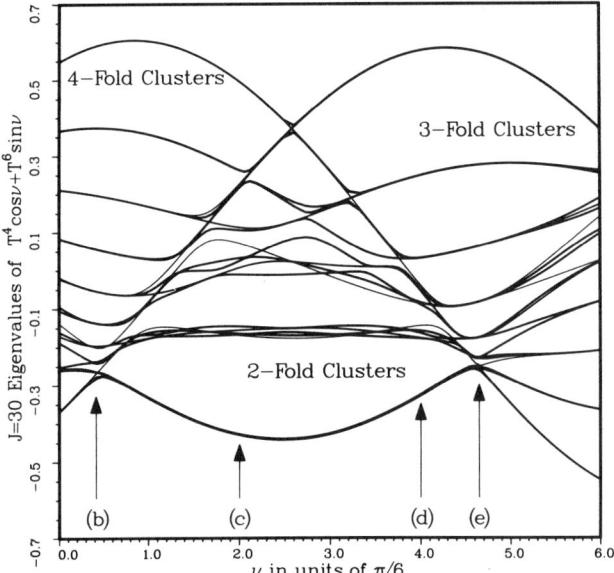

Figure 7.4.10 $J = 30$ eigenvalues of varying mixtures of fourth- and sixth-rank tensors. ($\nu = 0$) corresponds to levels in Figure 7.4.6.

drawn in Figure 7.4.11. The RE surface function used for $T^{[4]}$ is as follows. (Recall (7.4.19)]

$$E^{[4]}(\beta,\gamma) = (7/12)^{1/2}(9/4\pi)^{1/2}(35\cos^4\beta - 30\cos^2\beta + 3 + 5\sin^4\beta\cos 4\gamma)/8 \quad (7.4.26)$$

For $T^{[6]}$ the RE function is as follows,

$$E^{[6]}(\beta,\gamma) = (1/8)^{1/2}(13/4\pi)^{1/2}(231\cos^6\beta - 315\cos^4\beta + 105\cos^2\beta - 5 \\ - 21\sin^4\beta(11\cos^2\beta - 1)\cos 4\gamma)/16. \quad (7.4.27)$$

The tensors $T^{[r]}$ and RE functions $E^{[r]}$ have a spherical harmonic normalization factor $([2r+1]/4\pi)^{1/2}$ that was not included in the previous definition (7.4.19). This factor is used here to slightly enhance the effect of the sixth-rank tensor for this particular example. Also, the $|J|^r$ factors are deleted in (7.4.26) and (7.4.27) so that the higher rank tensor effects are not J dependent.

The eigenlevels marked by (b) in Figure 7.4.10 correspond to the RE surface drawn in Figure 7.4.11(b). The latter shows that the separatrix has taken over the regions that formerly held C_3 symmetric trajectories, and only C_4 trajectories remain. (Note that these are equally spaced contours and are not quantized paths.) The result is the destruction of C_3 clusters in the spectrum which is composed almost entirely of C_4 clusters above the (b) point in Figure 7.4.10.

Beyond this point a remarkable new type of cluster is formed. Just above the points marked (c) and (d) in Figure 7.4.10 lie two clusters which contain twelve levels each. These correspond to trajectories which encircle 12 equivalent valleys which lie on the C_2 symmetry axes in Figures 7.4.11(c) and 7.4.11(d). The symmetry species within each of these clusters are exactly the ones contained in the C_2 correlation table in the center of Figure 7.4.7. The lowest cluster in Figure 7.4.10 would correspond to $K = 30$ and hence to the even local symmetry or 0_2 column of the C_2 table which contains species A_1E, T_1, and $2T_2$. The next cluster has $K = 29$ and contains the five species A_2, E, $2T_1$, and T_2 listed in the odd column 1_2. The superfine splittings between these five levels are actually visible in the scale of Figures 7.4.10. As ν changes the levels as seem to change order within this cluster. This is the result of competition between tunneling mechanisms.

Between the (b) and (d) points in Figure 7.4.10 there is another phenomenon which occurs in the upper energy levels. There are a number of crossings or Fermi-like resonances between accidentally coinciding C_3 and C_4 clusters. This is because there are two kinds of mountains on the RE surfaces in Figures 7.4.11(c) and 7.4.11(d): the C_4 mountains which are

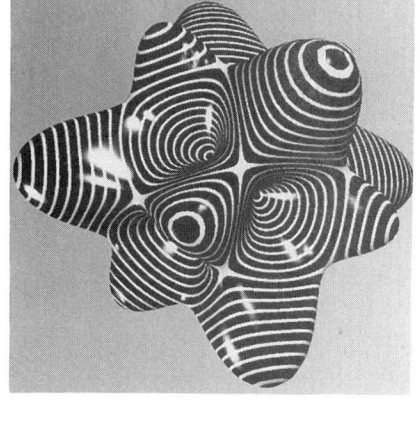
(a)

(b)

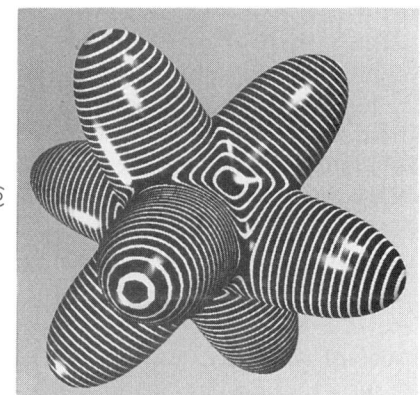
(c)

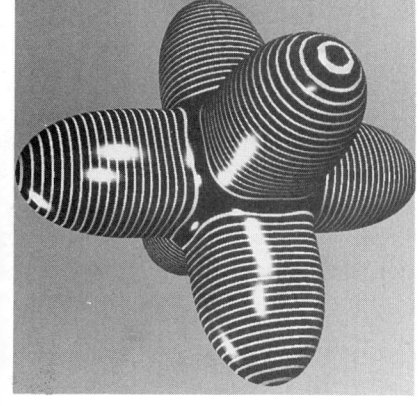
(d)

(e)

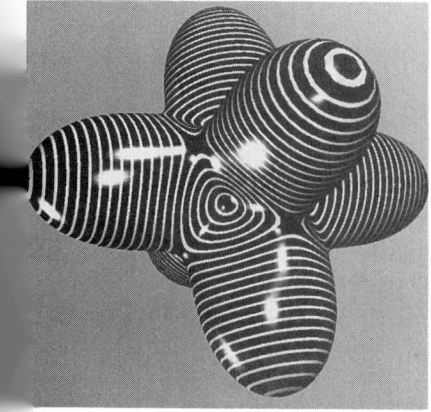

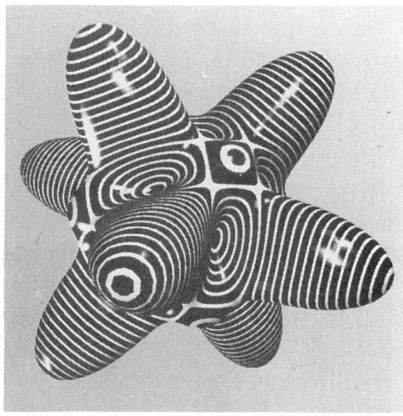
(f)

Figure 7.4.11 *RE* surfaces corresponding to selected ν values in Fig. 7.4.10. (a) $\nu = 0.0$, (b) $\nu = 0.4$ ($\pi/6$), (c) $\nu = 2.0$ ($\pi/6$), (d) $\nu = 4.0$ ($\pi/6$), (e) $\nu = 4.6$ ($\pi/6$), (f) $\nu = 5.0$ ($\pi/6$).

shrinking and C_3 mountains which are growing with ν. For certain values of ν, quantizing paths on one type of mountain are bound to be in resonance with different kinds of paths on the other. The result is an extraordinary kind of tunneling in which eigenfunctions are delocalized over both kinds of paths at once and a peculiar sort of hybrid superfine structure occurs in the eigenlevels.

The spectral region containing the unusual fine structure is bounded on the right-hand side by the (e) point in Figure 7.4.10 which corresponds to the RE surface in Figure 7.4.11(e). At this point the eight C_3 mountains dominate the surface geometry entirely and the eigenlevels are composed entirely of very strong C_3 clusters of eight levels each. The final 7.4.11(f) shows the situation at $\nu = 5.0(\pi/6)$ where the C_4 trajectories begin to return. Now they are occupying the valleys.

7.5 ROTATING SPINOR SYSTEMS AND TWO-DIMENSIONAL OSCILLATOR ANALOGIES

We consider now some physical applications of rotation group theory, spinor algebra, and $U(2)$ operators. Various analogies which use rotational coordinates lead to insight into rotational and vibrational dynamics.

So far we have introduced two physical examples of two-state systems. These were spin-$\frac{1}{2}$ states (Sections 1.1.A and 5.5.A) and the NH_3 inversion states (Section 2.12.B). A third and much older example involves optical polarization or the two-dimensional oscillator, and this will be introduced in this section. The three physical examples are each represented in two different ways by Figures 7.5.1(a) and 7.5.1(b), and the polarization example is shown in the central figure.

We shall discuss two ways to describe two-state systems such as a spin-$\frac{1}{2}$ electron. The first way, as shown in Figure 7.5.1(a1) uses a complex two-dimensional (spin-up and spin-down) space which is a basis of the fundamental representation of the unitary group $U(2)$. A second way, shown in Figure 7.5.1(b1), uses a real three-dimensional (S_x, S_y, S_z) space which is a basis for the vector representation of the rotation group $R(3)$. The spinor space may be less familiar than the vector space, since the latter is more like the one in which we live, and it is easier to visualize a real vector **S**. This spin vector **S** is a nearly complete description of the spin state; however, it turns out to be double valued in the following sense. A 180° three-space rotation of a spin-up vector to spin-down ($R(180°)\mathbf{S} = -\mathbf{S}$) corresponds to only a 90° rotation in the spinor 2-space. A "full" 360° rotation in the 3-space is only a 180° rotation in the 2-space, i.e., |spin-up⟩ goes into *minus* |spin-up⟩.

Spinor spaces therefore provide a more complete, though possibly less intuitive description of the two-state system. As explained in Section 5.5.A, the spinor algebra and geometry lead to simpler and more powerful compu-

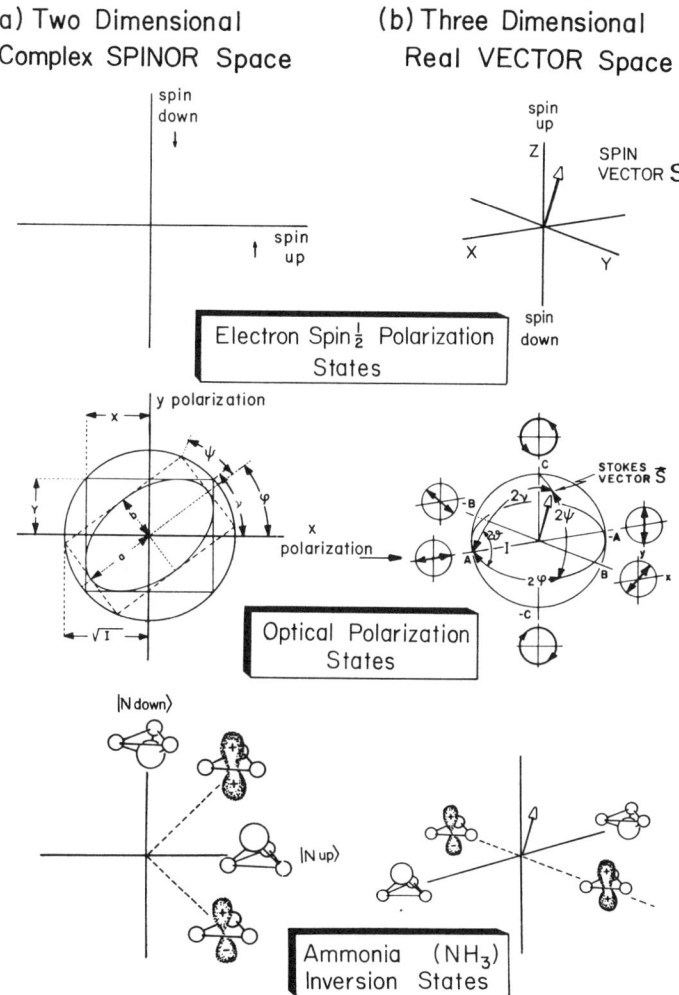

Figure 7.5.1 Two descriptions of three famous examples of two-state systems. (a) The spinor description involves complex vectors in a two-dimensional space. (b) The vector description involves real vectors in a three-dimensional space.

tational aids. To improve the intuitive value of the spinor description it helps to consider the optical polarization formalism developed by Poincaré and Stokes in the last century. In this theory the spinor bases correspond to oscillating x and y components of an electric vector. One can also picture various orbits of a two-dimensional coupled oscillator in the spinor space, and this helps greatly to understand its structure. It also helps to have a corresponding 3-space which is now labeled (ABC) instead of (xyz) in Figure 7.5.1(b2).

A. Euler-Angle Definition of Spinor States

Spinor or spin-$\frac{1}{2}$ states span the basis for the fundamental representation of SU(2), and for every pair of spinor states $|\psi\rangle$ and $|\psi'\rangle$ there exists a unitary operator R which maps one state into the other ($|\psi\rangle = R|\psi'\rangle$). A one-to-one mapping between the states and operators can be established by fixing one of the states, say, $|\psi'\rangle = |1\rangle$, and labeling all states by the mapping operator ($|R\rangle = R|1\rangle$). If the operators are labeled by Euler angles ($R = R(\alpha, \beta, \gamma)$), then so are the states

$$|\psi(\alpha, \beta, \gamma)\rangle = R(\alpha, \beta, \gamma)|1\rangle.$$

The standard Euler coordinate definition of rotation operators involves an ordered product of rotations by α, β, and γ around the x, y, and z axes, respectively, as explained in Sections 5.3 and 5.5.E:

$$R(\alpha, \beta, \gamma) = R(\alpha \cdot \cdot)R(\cdot \beta \cdot)R(\cdot \cdot \gamma). \tag{7.5.1}$$

This leads to the following explicit representation of a general spinor as a mapping of the spin-up state $|1\rangle = \begin{pmatrix} 1 \\ 0 \end{pmatrix}$:

$$\begin{pmatrix} \psi_1 \\ \psi_2 \end{pmatrix} = \begin{pmatrix} e^{-i\alpha/2} & 0 \\ 0 & e^{i\alpha/2} \end{pmatrix} \begin{pmatrix} \cos(\beta/2) & -\sin(\beta/2) \\ \sin(\beta/2) & \cos(\beta/2) \end{pmatrix} \begin{pmatrix} e^{-i\gamma/2} & 0 \\ 0 & e^{i\gamma/2} \end{pmatrix} \begin{pmatrix} 1 \\ 0 \end{pmatrix}$$

$$= \begin{pmatrix} e^{-i\alpha/2} \cos(\beta/2) \\ e^{i\alpha/2} \sin(\beta/2) \end{pmatrix} e^{-i\gamma/2}. \tag{7.5.2}$$

A 3-space mechanical model of the expected spin vector of this state is shown in Figure 7.5.2. The figure displays the expectation values in the state $|\psi(\alpha, \beta, \gamma)\rangle$ of the Pauli spin angular-momentum operators

$$J_x = \sigma_x/2 \qquad J_y = \sigma_y/2 \qquad J_z = \sigma_z/2$$

$$= \begin{pmatrix} 0 & \frac{1}{2} \\ \frac{1}{2} & 0 \end{pmatrix} \qquad = \begin{pmatrix} 0 & -i/2 \\ i/2 & 0 \end{pmatrix} \qquad = \begin{pmatrix} \frac{1}{2} & 0 \\ 0 & -\frac{1}{2} \end{pmatrix}. \tag{7.5.3}$$

From the figure it is seen that the first Euler angle (α) is the relative phase between components ψ_1 and ψ_2, and it is represented by the azimuthal angle of the $\langle J \rangle$ vector. The second Euler angle (β) is the arc-cosine of the relative

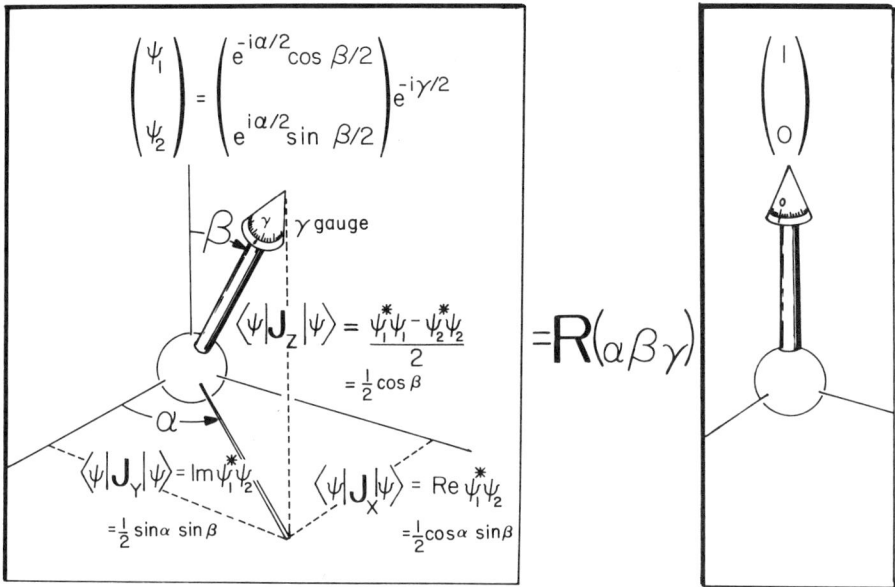

Figure 7.5.2 Detailed relation between spinor state and spin vector. The relation is based upon the fact that all spinor states are rotations of the spin-up state. Euler angles and the mechanical goniometer in Figure 5.3.1 provide a complete description of the pure spinor state.

population $(|\psi_1|^2 - |\psi_2|^2)$, and it is represented by the polar angle of the $\langle J \rangle$ vector. The third Euler angle (γ) is (-2) times the overall phase or gauge of the state $|\psi(\alpha, \beta, \gamma)\rangle$, and it corresponds to the twist of the rigid body whose axis defines the direction of the $\langle J \rangle$ vector in Figure 7.5.2. In the quantum theory of spin this third coordinate is often superfluous. However, for classical applications this angle will help to define the phase of an orbit and should not be ignored. The γ angle is indicated by a dial or gauge in Figures 7.5.2 and 5.3.1.

B. Axis-Angle Definition of Spinor Operators

The general $U(2)$ group operator may be expressed in terms of a single exponential

$$U = U(\omega_0, \omega_x, \omega_y, \omega_z) = e^{-ih},$$

involving a combination of Pauli operators $\sigma = 2J$

$$h = \omega_0 \mathbf{1} + \omega_x J_x + \omega_y J_y + \omega_z J_z = \omega_0 + \frac{\omega_x}{2}\sigma_x + \frac{\omega_y}{2}\sigma_y + \frac{\omega_z}{2}\sigma_z,$$

which has the matrix form

$$\begin{pmatrix} h_{11} & h_{12} \\ h_{21} & h_{22} \end{pmatrix} = \omega_0 \begin{pmatrix} 1 & 0 \\ 0 & 1 \end{pmatrix} + \frac{\omega_x}{2}\begin{pmatrix} 0 & 1 \\ 1 & 0 \end{pmatrix}$$
$$+ \frac{\omega_y}{2}\begin{pmatrix} 0 & -i \\ i & 0 \end{pmatrix} + \frac{\omega_z}{2}\begin{pmatrix} 1 & 0 \\ 0 & -1 \end{pmatrix}. \quad (7.5.4)$$

An important example of this is the time evolution operator

$$U(t) = e^{-iHt} \quad (7.5.5a)$$

for the autonomous (H_{ij} = constant) Schrödinger equation $i|\dot\psi\rangle = H|\psi\rangle$, or

$$i\frac{\partial}{\partial t}\begin{pmatrix}\psi_1 \\ \psi_2\end{pmatrix} = \begin{pmatrix} H_{11} & H_{12} \\ H_{21} & H_{22}\end{pmatrix}\begin{pmatrix}\psi_1 \\ \psi_2\end{pmatrix} \quad (7.5.5b)$$

for a two-level system such as a spin-$\frac{1}{2}$ moment in a magnetic field. The expansion (7.5.4) reduces the solution

$$|\psi(t)\rangle = U(t)|\psi(0)\rangle \quad (7.5.5c)$$

to an exponential expression

$$|\psi(t)\rangle = e^{-i\omega_0 t}e^{-i(\omega_x J_x + \omega_y J_y + \omega_z J_z)t}|\psi(0)\rangle$$
$$= e^{-i\omega_0 t}e^{-i\boldsymbol{\omega}\cdot\mathbf{J}t}|\psi(0)\rangle, \quad (7.5.5d)$$

which involves an overall phase which evolves at the angular rate

$$\omega_0 = (H_{11} + H_{12})/2 \quad (7.5.5e)$$

and an SU(2) rotation with angular velocity $\boldsymbol{\omega}$ whose three Cartesian components are

$$\omega_x = 2\,\mathrm{Re}\,H_{21} \qquad \omega_y = 2\,\mathrm{Im}\,H_{21} \qquad \omega_z = H_{11} - H_{22}$$
$$= \omega \cos\phi \sin\theta, \qquad = \omega \sin\phi \sin\theta, \qquad = \omega \cos\theta. \quad (7.5.5f)$$

In the second line of (7.4.5f) the polar coordinates of the rotational "crank" axis are defined according to Figure 7.5.3(a). The azimuthal angle (ϕ), polar angle (θ), and total rotational angle (ωt) are valid parameters of SU(2) operations $R[\phi, \theta, \omega t]$, and as such they are an alternative to the Euler angle labeling $R(\alpha, \beta, \gamma)$. (We use brackets to denote axis-angle parameters, and parentheses to denote Euler angles.)

ROTATING SPINOR SYSTEMS AND TWO-DIMENSIONAL OSCILLATOR ANALOGIES

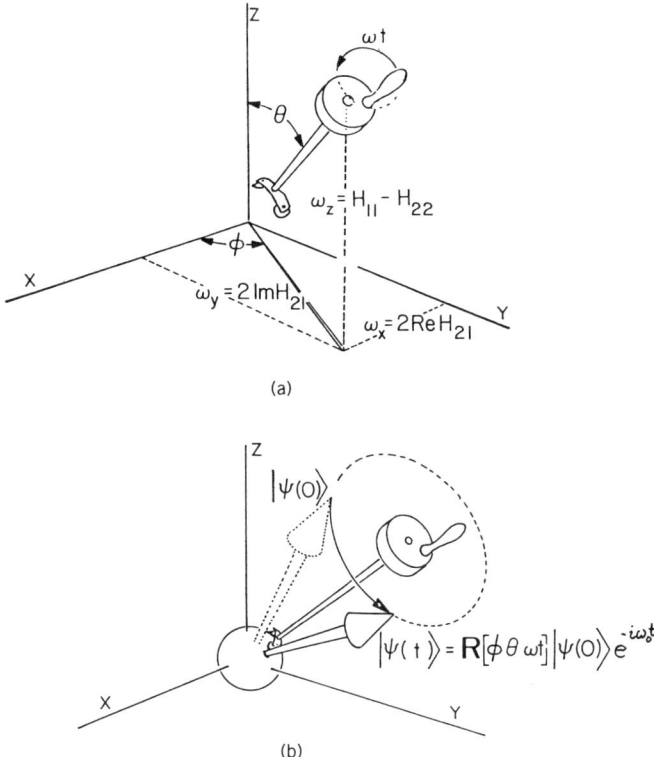

Figure 7.5.3 Mechanical crank analog for Hamiltonian matrix. (a) Components of Hamiltonian determine the direction (ϕ, θ) and rate of turn (ω) of the crank. (b) Effect of Hamiltonian is represented by mechanical analog rotation around crank axis, to within an overall phase.

Generally, it is physically more convenient to label states or rotational position with Euler angles, while Hamiltonians and other operators are more intuitively labeled by axis angles. In any case it is easy to convert one labeling into the other, as explained in Section 5.5.A. Then the time evolution (7.5.5d) is reduced to a product of the Hamiltonian rotation $R[\phi, \theta, \omega t]$, the overall phase factor, and the rotation operator in the initial state definition,

$$|\psi(0)\rangle = R(\alpha(0), \beta(0), \gamma(0))|1\rangle. \qquad (7.5.6)$$

The final-state Euler angles are defined by the first line in the following:

$$\begin{aligned} |\psi(t)\rangle &= R(\alpha(t), \beta(t), \gamma(t))|1\rangle \\ &= R[\phi, \theta, \omega t] R(\alpha(0), \beta(0), \gamma(0))|1\rangle e^{-i\omega_0 t} \\ &= R[\phi, \theta, \omega t] R(\alpha(0), \beta(0), \gamma(0) + 2\omega_0 t)|1\rangle, \end{aligned} \qquad (7.5.7)$$

and they would follow from the group product in the last line.

C. Rotational Angle Parameters for a Two-Dimensional Harmonic Oscillator

The spinor components of an autonomous two-state Schrödinger equation (7.5.5b) of the form

$$i\frac{d}{dt}\begin{pmatrix}\psi_1\\ \psi_2\end{pmatrix} = \begin{pmatrix} A & B-iC \\ B+iC & D \end{pmatrix}\begin{pmatrix}\psi_1\\ \psi_2\end{pmatrix} \quad (7.5.8)$$

may be replaced by separate equations for the real parts ($x_j \equiv \mathrm{Re}\,\psi_j$) and imaginary parts ($p_j = \mathrm{Im}\,\psi_j$) of the spinor components;

$$\psi_1 = x_1 + ip_1 \qquad\qquad \psi_2 = x_2 + ip_2$$
$$= \sqrt{I}\,e^{-i(\alpha+\gamma)/2}\cos\beta/2 \qquad = \sqrt{I}\,e^{-i(\gamma-\alpha)/2}\sin\beta/2. \quad (7.5.9)$$

(An arbitrary but constant normalization factor $\sqrt{I}$ has been included for this discussion.) The resulting equations,

$$\dot{x}_1 = Ap_1 + Bp_2 - Cx_2 = \frac{\partial H_c}{\partial p_1},$$

$$\dot{x}_2 = Bp_1 + Dp_2 + Cx_1 = \frac{\partial H_c}{\partial p_2},$$

$$-\dot{p}_1 = Ax_1 + Bx_2 + Cp_2 = \frac{\partial H_c}{\partial x_1},$$

$$-\dot{p}_2 = Bx_1 + Dx_2 - Cp_1 = \frac{\partial H_c}{\partial x_2}, \quad (7.5.10a)$$

are identical to those of a two-dimensional classical coupled Coriolis harmonic oscillator with the Hamiltonian

$$H_c = \frac{A}{2}(p_1^2 + x_1^2) + \frac{D}{2}(p_2^2 + x_2^2) + B(p_1 p_2 + x_1 x_2) + C(x_1 p_2 - x_2 p_1). \quad (7.5.10b)$$

By exploiting the Euler and axis-angle relations one may label in an intuitive way all possible harmonic Hamiltonians H_c as well as all their possible initial conditions and resulting phase-space trajectories. The phase-space motion can be related though the Euler angles to a precessing rigid top (Figure 7.5.2), and the Hamiltonian can be related through axis angles to a crank [Figure 7.5.3(a)] which rotates the top.

Expansion of (7.5.9) gives the Euler coordinates for each point in phase space,

$$x_1 = \sqrt{I}\cos\left(\frac{\alpha+\gamma}{2}\right)\cos\left(\frac{\beta}{2}\right), \quad p_1 = -\sqrt{I}\sin\left(\frac{\alpha+\gamma}{2}\right)\cos\left(\frac{\beta}{2}\right),$$

$$x_2 = \sqrt{I}\cos\left(\frac{\gamma-\alpha}{2}\right)\sin\left(\frac{\beta}{2}\right), \quad p_2 = -\sqrt{I}\sin\left(\frac{\gamma-\alpha}{2}\right)\sin\left(\frac{\beta}{2}\right).$$

(7.5.11a)

These expressions include the normalization or amplitude factor

$$\sqrt{I} = \left(x_1^2 + p_1^2 + x_2^2 + p_2^2\right)^{1/2} = \left(\psi_1^*\psi_1 + \psi_2^*\psi_2\right)^{1/2}. \quad (7.5.11b)$$

The quantity I is the total intensity or probability in the two-level quantum system and is therefore a constant of motion for the classical system, as well.

A combination of (7.5.5f) and (7.5.8) leads to an axis-angle parametrization of the oscillator Hamiltonian.

$$\begin{aligned}\omega_x &= \omega\cos\phi\sin\theta & \omega_y &= \omega\sin\phi\sin\theta & \omega_z &= \omega\cos\theta \\ &= 2B & &= 2C & &= A - D \\ & & \omega_0 &= (H_{11} + H_{12})/2 & & \\ & & &= (A+D)/2. & & \end{aligned}$$

(7.5.12)

This relates the angular velocity vector $\boldsymbol{\omega}$ with its axis angle (ϕ, θ), angular rate of turn ω, and overall phase rate ω_0 to the four Hamiltonian constants A, B, C, and D. The oscillator Hamiltonian can also be related to a rotor Hamiltonian using angular-momentum variables obtained from the expected J values displayed in Figure 7.5.2:

$$\begin{aligned}\mathcal{J}_x &= \langle\psi|J_x|\psi\rangle = \operatorname{Re}\psi_1^*\psi_2 = x_1 x_2 + p_1 p_2 = (I/2)\cos\alpha\sin\beta,\\ \mathcal{J}_y &= \langle\psi|J_y|\psi\rangle = \operatorname{Im}\psi_1^*\psi_2 = x_1 p_2 - x_2 p_1 = (I/2)\sin\alpha\sin\beta,\\ \mathcal{J}_z &= \langle\psi|J_z|\psi\rangle = (\psi_1^*\psi_1 + \psi_2^*\psi_2)/2 = \left(x_1^2 - x_2^2 + p_1^2 - p_2^2\right)/2 \\ &= (I/2)\cos\beta,\\ 2\mathcal{J}_0 &= \langle\psi|1|\psi\rangle = \psi_1^*\psi_1 + \psi_2^*\psi_2 = x_1^2 + x_2^2 + p_1^2 + p_2^2 = I. \end{aligned}$$

(7.5.13)

The resulting classical variables $\mathcal{J}_a$ are combinations of density matrix components $\psi_i^* \psi_j$ of the two-level quantum problem. The classical Hamiltonian in these variables is the following:

$$H_c = B\mathcal{J}_x + C\mathcal{J}_y + \frac{A-D}{2}\mathcal{J}_z + \frac{A+D}{2}\mathcal{J}_0 \qquad (7.5.14a)$$

$$= \frac{\omega_x}{2}\mathcal{J}_x + \frac{\omega_y}{2}\mathcal{J}_y + \frac{\omega_z}{2}\mathcal{J}_z + \omega_0 J_0 \qquad (7.5.14b)$$

$$= \Omega_x \mathcal{J}_x + \Omega_y \mathcal{J}_y + \Omega_z \mathcal{J}_z + \omega_0 \mathcal{J}_0. \qquad (7.5.14c)$$

These are the action-angle forms ($H = \Sigma \dot\theta J_\theta$) for the classical oscillator and rotor. It should be noted that the angular velocities Ω_α of the classical oscillator are each half of the corresponding components ω_α for the rotor analogy. A factor of 2 is a common feature of spin-vector mappings and will be discussed further in the following.

The rotor Hamiltonian (7.5.14a) describes a rigid spin-moment body with no mass subject to torques applied by a magnetic field. A rigid massive body will have additional quadratic terms $\mathcal{J}_x^2$, etc., and this is analogous to an anharmonic oscillator in rotor-oscillator mapping. This will be discussed later. Nonrigid rotors have still higher-order terms in the angular momentum.

The coordinates conjugate to the momenta $\mathcal{J}_a$ ($a = x$, y, or z) are nonholonomic, i.e., the differentials $d\theta_a = \omega_a \, dt$ are not exact. Since the Euler angles are manifestly holonomic coordinates, it is better for many purposes to use the momenta $\{\mathcal{J}_\alpha, \mathcal{J}_\beta, \mathcal{J}_\gamma\}$ conjugate to these coordinates. Relations between the Euler momenta and the Cartesian quantities are given in Section 5.5.E [Eq. (5.5.83)].

D. Polarization Ellipsometry Coordinates

The classical description of pure states of optical polarization is equivalent to that of a two-dimensional oscillator. The relevant physical quantities in ellipsometry are the x and y electric field strengths defined by

$$E_x = \text{Re}\langle x|\Psi\rangle, \qquad E_y = \text{Re}\langle y|\Psi\rangle, \qquad (7.5.15a)$$

which are the linear polarization amplitudes for a two-component state vector

$$|\Psi\rangle = |x\rangle\langle x|\Psi\rangle + |y\rangle\langle y|\Psi\rangle. \qquad (7.5.15b)$$

The two real quantities E_x and E_y are analogous to the oscillator coordinates x_1 and x_2, respectively, in (7.5.9).

An equivalent description of the same state involves circular polarization amplitudes and bases; i.e.,

$$|\Psi\rangle = |r\rangle\langle r|\Psi\rangle + |l\rangle\langle l|\Psi\rangle, \qquad (7.5.16a)$$

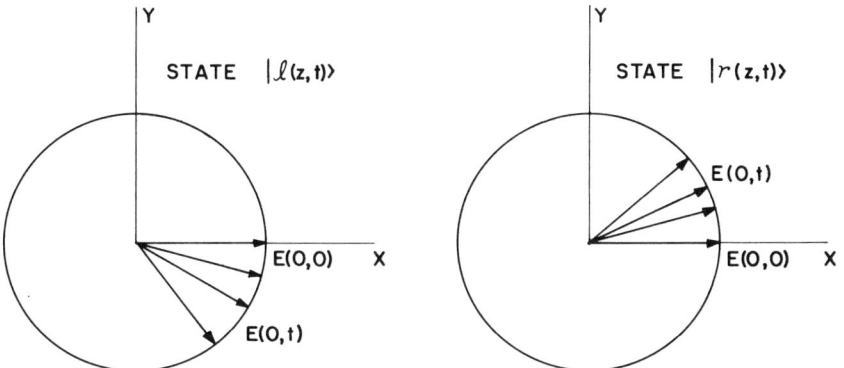

Figure 7.5.4 Circular polarization base states. States $|l\rangle$ and $|r\rangle$ are characterized by right-handed and left-handed time evolution of the E vector.

where

$$|r\rangle = (|x\rangle + i|y\rangle)/\sqrt{2}, \qquad |l\rangle = (|x\rangle - i|y\rangle)/\sqrt{2} \qquad (7.5.16b)$$

represent unit states of right-handed and left-handed circular polarization, respectively. One can visualize these states by noting that the state $|\gamma\rangle = e^{i(kz-\omega t)}|r\rangle$, for example, describes a right-hand or counterclockwise circular time evolution of the electric vector, as shown in the right-hand portion of Figure 7.5.4. Linear and circular polarization bases each give rise to different but equivalent sets of ellipsometry parameters which are conveniently related to Euler angles.

In terms of linear polarization bases the general state may be written as follows:

$$|\Psi\rangle = (Xe^{-i\vartheta}|x\rangle + Ye^{i\vartheta}|y\rangle)e^{-i\theta}. \qquad (7.5.17)$$

According to (7.5.15) the resulting electric vector components are

$$E_x(\vartheta, \nu, \theta) = X\cos(\vartheta + \theta), \qquad (7.5.18a)$$

$$E_y(\vartheta, \nu, \theta) = Y\cos(\vartheta - \theta), \qquad (7.5.18b)$$

where we define

$$X = \sqrt{I}\cos\nu, \qquad (7.5.18c)$$

$$Y = \sqrt{I}\sin\nu. \qquad (7.5.18d)$$

This represents an ellipse trapped in a horizontal box of dimensions $2X$ by $2Y$, as shown in Figure 7.5.5. The angle between the box diagonal and the x

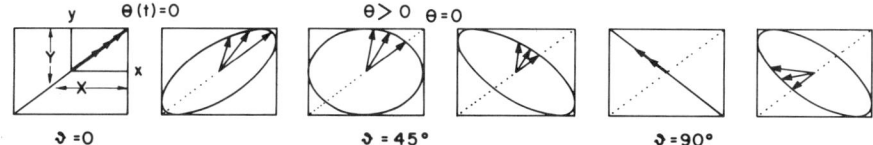

Figure 7.5.5 Examples of **E** (ϑ, ν, θ)-vector paths for various ϑ values. [Angle $\nu = \tan^{-1}(Y/X)$ is fixed.] Overall phase angle θ determines position on the elliptical orbit.

axis is ν [see also Figure 7.5.1(b2)], and the angle ϑ determines the orientation and aspect ratio of the ellipse inside the box. Overall phase θ determines the position of the electric vector on the elliptical orbit.

In terms of circular polarization bases a polarization state may be written as follows:

$$|\Psi\rangle = (Re^{-i\varphi}|r\rangle + Le^{i\varphi}|l\rangle)e^{-i\Phi} \qquad (7.5.19)$$

According to (7.5.15) and (7.5.16b) the electric vector components are

$$E_x(\varphi, \psi, \Phi) = (a\cos\Phi)\cos\varphi - (b\sin\Phi)\sin\varphi, \qquad (7.5.20a)$$

$$E_y(\varphi, \psi, \Phi) = (a\cos\Phi)\sin\varphi + (b\sin\Phi)\cos\varphi, \qquad (7.5.20b)$$

where we define

$$a = (R+L)/\sqrt{2} = I\cos\psi, \qquad (7.5.20c)$$

$$b = (R-L)/\sqrt{2} = I\sin\psi. \qquad (7.5.20d)$$

This represents an ellipse trapped inside a tipped box of dimension $2a$ by $2b$ where a and b are the semimajor and minor axes of the ellipse, as shown in Figure 7.5.6. The angle between the box diagonal and the major axis is ψ [see also Figure 7.5.1(b2)]. The angle φ determines the orientation of the box and its ellipse with respect to the x axis. Overall phase Φ determines the position of the electric vector on the ellipse.

The expectation values of the Pauli momentum operators define a three-dimensional vector which is called the POINCARÉ-STOKES vector. We shall use the letters (A, B, C) to label its Cartesian components since (x, y) are already being used to label the spinor 2-space. Also, the axis of quantization, which is usually labeled by z, will be the C axis for the description based upon circular polarization and the A axis for the description based upon linear polarization. Still other axes will be used in descriptions of "local modes" in later discussions.

By relating right and left circular polarization with spin-up and spin-down, respectively, in the polarization-spin-$\frac{1}{2}$ analogy one is lead to the usual

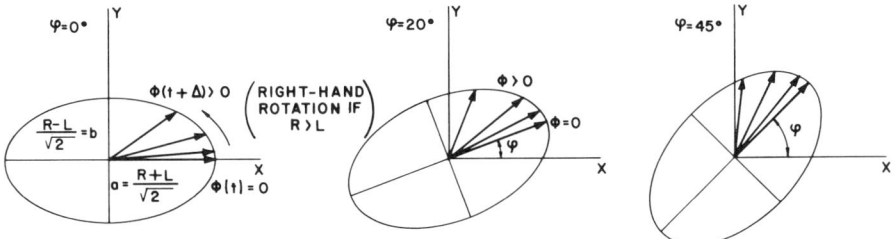

Figure 7.5.6 Examples of E (φ, ψ, Φ)-vector paths for various φ values. [Angle $\psi = \tan^{-1}(b/a)$ is fixed.] Overall phase angle Φ determines position on the elliptical orbit.

ordering $(x \to A, y \to B, z \to C)$ for the three Cartesian components. The three Pauli operators may be represented in the usual way in the $\{|r\rangle, |l\rangle\}$ basis as follows:

$$C(\sigma_A) = \begin{pmatrix} \langle r|\sigma_A|r\rangle & \langle r|\sigma_A|l\rangle \\ \langle l|\sigma_A|r\rangle & \langle l|\sigma_A|l\rangle \end{pmatrix} = \begin{pmatrix} 0 & 1 \\ 1 & 0 \end{pmatrix},$$

$$C(\sigma_B) = \begin{pmatrix} 0 & -i \\ i & 0 \end{pmatrix}, \quad C(\sigma_C) = \begin{pmatrix} 1 & 0 \\ 0 & -1 \end{pmatrix}. \quad (7.5.21)$$

This then implies a different representation for the same operators in the linear $\{|x\rangle, |y\rangle\}$ basis using transformation (7.5.16b):

$$L(\sigma_A) = \begin{pmatrix} \langle x|\sigma_A|x\rangle & \langle x|\sigma_A|y\rangle \\ \langle y|\sigma_A|x\rangle & \langle y|\sigma_A|y\rangle \end{pmatrix} = \tfrac{1}{2}\begin{pmatrix} 1 & 1 \\ i & -i \end{pmatrix} C(\sigma_A) \begin{pmatrix} 1 & -i \\ 1 & i \end{pmatrix}$$

$$= \begin{pmatrix} 1 & 0 \\ 0 & -1 \end{pmatrix},$$

$$L(\sigma_B) = \begin{pmatrix} 0 & 1 \\ 1 & 0 \end{pmatrix}, \quad L(\sigma_C) = \begin{pmatrix} 0 & -i \\ i & 0 \end{pmatrix}. \quad (7.5.22)$$

The diagonal form of σ_A indicates a choice of A-axis quantization with the permuted ordering $(x \to B, y \to C, z \to A)$ in the assignment of Pauli operators.

Points in the (A, B, C) space are identified easily by the expectation values $\langle \psi|\sigma|\psi\rangle$ in the desired representation. This leads to relations between elliptical shape angles (φ, ψ), (ϑ, ν), and the respective Euler angles. In the circular representation one derives

$$\langle\Psi|\sigma_A|\Psi\rangle = \overline{Re^{-i\varphi}Le^{i\varphi}}^* \begin{pmatrix} 0 & 1 \\ 1 & 0 \end{pmatrix} \begin{pmatrix} Re^{-i\varphi} \\ Le^{i\varphi} \end{pmatrix} = 2RL\cos 2\varphi = (a^2 - b^2)\cos 2\varphi,$$

$$\langle\Psi|\sigma_B|\Psi\rangle = \overline{Re^{-i\varphi}Le^{i\varphi}}^* \begin{pmatrix} 0 & -i \\ i & 0 \end{pmatrix} \begin{pmatrix} Re^{-i\varphi} \\ Le^{i\varphi} \end{pmatrix} = 2RL\sin 2\varphi = (a^2 - b^2)\sin^2\varphi,$$

$$\langle\Psi|\sigma_C|\Psi\rangle = \overline{Re^{-i\varphi}Le^{i\varphi}}^* \begin{pmatrix} 1 & 0 \\ 0 & 1 \end{pmatrix} \begin{pmatrix} Re^{-i\varphi} \\ Le^{i\varphi} \end{pmatrix} = R^2 - L^2 = 2ab. \quad (7.5.23)$$

These values are equal to the corresponding linear representation values:

$$\langle\Psi|\sigma_A|\Psi\rangle = \overline{Xe^{-i\vartheta}Ye^{i\vartheta}}^* \begin{pmatrix} 1 & 0 \\ 0 & -1 \end{pmatrix} \begin{pmatrix} Xe^{-i\vartheta} \\ Ye^{i\vartheta} \end{pmatrix} = X^2 - Y^2,$$

$$\langle\Psi|\sigma_B|\Psi\rangle = \overline{Xe^{-i\vartheta}Ye^{i\vartheta}}^* \begin{pmatrix} 0 & 1 \\ 1 & 0 \end{pmatrix} \begin{pmatrix} Xe^{-i\vartheta} \\ Ye^{i\vartheta} \end{pmatrix} = 2XY\cos 2\vartheta,$$

$$\langle\Psi|\sigma_C|\Psi\rangle = \overline{Xe^{-i\vartheta}Ye^{i\vartheta}}^* \begin{pmatrix} 0 & -i \\ i & 0 \end{pmatrix} \begin{pmatrix} Xe^{-i\vartheta} \\ Ye^{i\vartheta} \end{pmatrix} = 2XY\sin^2\vartheta. \quad (7.5.24)$$

Combination of these results with the box-angle definitions (7.5.18) and (7.5.20) yields the following ellipsometry relations:

$$\begin{array}{lll} & \text{(Circular)} & \text{(Linear)} \\ \langle\Psi|\sigma_A|\Psi\rangle = I\cos 2\psi \cos 2\varphi & = I\cos 2\nu, \\ \langle\Psi|\sigma_B|\Psi\rangle = I\cos 2\psi \sin 2\varphi & = I\sin 2\nu \cos 2\vartheta, \\ \langle\Psi|\sigma_C|\Psi\rangle = I\sin 2\psi & = I\sin 2\nu \sin 2\vartheta, \\ \langle\Psi|\Psi\rangle = I = a^2 + b^2 & = X^2 + Y^2. \end{array} \quad (7.5.25)$$

Each of the angles in these relations is indicated on the (ABC) vector diagram in Figure 7.5.1(b) as well as in the spinor diagram. For each angle in the spinor diagram there is a corresponding double angle in the vector diagram. For example, linear polarization states on the AB equator of the vector diagram ($\psi = 0 = \nu$) are characterized by an azimuthal angle 2φ where φ is the inclination angle for the major axis of polarization in a spinor diagram. A complete vector revolution $2\varphi \rightarrow 2\varphi + 2\pi$ corresponds to only a half-revolution of the spinor picture. This is due to the fact that such a revolution maps a polarization state into one that is π out of phase. That spinor states require 4π rotations around any axis in the vector 3-space in order to have no change is seen more easily in the optical polarization analogy. The idea that a vector is a square of spinors (in the outer product sense) is relevant. The amplitude scale for spinors is $\sqrt{I}$, while for vectors it is I in Figure 7.5.1.

By choosing the C axis to be the "z direction" of quantization one picks the following angles:

$$\alpha = 2\varphi, \qquad \beta = \frac{\pi}{2} - 2\psi, \qquad \gamma = 2\Phi, \qquad (7.5.26)$$

to be the Euler azimuthal, polar, and overall phase angles, respectively. This choice favors circular polarization and is convenient if there is a strong Coriolis component or Zeeman field in a vibrational model. If instead one chooses the A axis, then the angles

$$\alpha = 2\vartheta, \qquad \beta = 2\nu, \qquad \gamma = 2\theta \qquad (7.5.27)$$

are the Euler angles. This choice favors linear x and y polarizations and is convenient if there are well-defined normal-mode states $|x\rangle$ and $|y\rangle$ in the vibrational model.

E. Generalized Lissajous Trajectories and Related Dynamics

(a) Examples of Rotation and Oscillation Dynamics The vectorial or three-space description complements the spinorial or two-space picture. While the latter displays more detail the former may be more efficient. Each elliptical trajectory in the 2-space corresponds to a single direction or quasispin vector ($\mathbf{S} = \mathbf{J}$) in the 3-space. This vector only moves if the ellipse changes in size, shape, or orientation.

Consider, for example, the solution to an isotropic oscillator equation (7.5.10) with constants $A = D$ and $B = 0$ for which a Coriolis force ($C \neq 0$) is present. The Coriolis term,

$$C(x_1 p_1 - x_2 p_1) = Cl_z \qquad (7.5.28)$$

in Eq. (7.5.10) describes the effect of adding a magnetic field to a charged oscillating mass or a Foucault pendulum in a rotating frame. The $\mathbf{S}$ vector will precess around the $\boldsymbol{\omega}$ vector $(\omega_x, \omega_y, \omega_z) = (0, 2C, 0)$ according to Eq. (7.5.12). If we use linear polarization bases $|x_1\rangle \equiv |x\rangle$ and $|x_2\rangle \equiv |y\rangle$ for which the z axis of quantization is the A axis then the following coordinate identification is appropriate,

$$(\omega_A, \omega_B, \omega_C) = (\omega_z, \omega_x, \omega_y) = (A - D, 2B, 2C)$$
$$= (0, 0, 2C) \quad \text{for } A = D \text{ and } B = 0. \qquad (7.5.29)$$

An example of the resulting motion is displayed in Figure 7.5.7(a). This involves precession of an ellipse with constant shape angle ψ but variable orientation angle φ. The rotation around the C axis proscribed by (7.5.29) is called *Faraday rotation* of polarization. The states $|r\rangle$ or $|l\rangle$ of pure circular

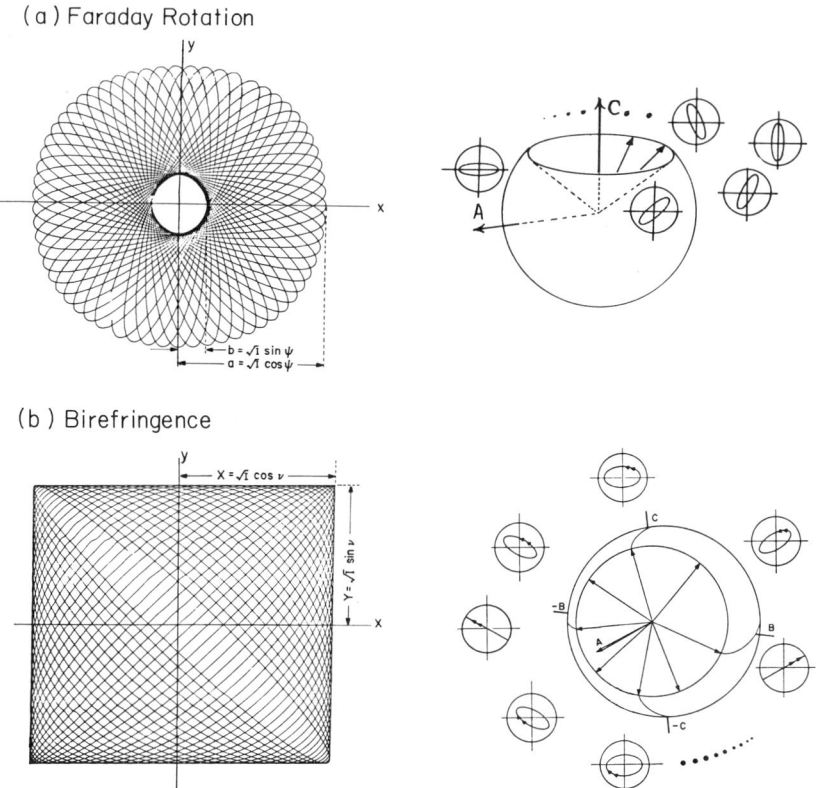

Figure 7.5.7 Analog computer plots of two famous examples of optical activity. (a) Faraday rotation or circular dichroism corresponds to constant $\psi = \tan^{-1}(b/a)$. (b) Birefringence corresponds to constant $\nu = \tan^{-1}(Y/X)$. Note that a small amount of birefringence is present in Figure 7.11(a); i.e., ψ oscillates slightly. Pure Faraday rotation is difficult to achieve on an analog computer.

polarization ($\psi = \pm\pi/4$) correspond to **S** vectors parallel or antiparallel to $\boldsymbol{\omega} = 2C\hat{\mathbf{e}}_C$, and so they represent fixed points in the ABC-vector space for the pure Coriolis Hamiltonian. All other states experience a precession of their principal directions of polarization by angle φ [recall Figure (7.5.6)] while their **S** vectors maintain a constant angle $(\pi/2 - 2\psi)$ with the C axis and rotate by 2φ around it.

A very different type of motion is that of nondegenerate $|x\rangle$ and $|y\rangle$ modes ($A \neq D$) which are uncoupled ($B = 0$) and Coriolis free ($C = 0$). Then the crank vector lies along the A axis:

$$(\omega_A, \omega_B, \omega_C) = (A - D, 0, 0). \qquad (7.5.30)$$

An example of the resulting motion is displayed in Fig. 7.5.7(b). The ellipse deforms and vibrates continuously but remains inside a box of constant diagonal angle ν. (Recall Figure 7.5.5.) The **S** vector maintains a constant angle 2ν with the A axis while rotating by 2ϑ around it. The states $|x\rangle \equiv |x_1\rangle$ and $|y\rangle = |x_2\rangle$ of pure horizontal or vertical polarization represent fixed points in each case since the **S** vector lies along the $+A$ or $-A$ axis ($\nu = 0$ or $\pi/2$). Motion of mixed $|x\rangle$ and $|y\rangle$ states such as is shown in Figure 7.5.8(b) is known as a *birefringence* in polarization optics.

S vectors on the $\pm B$ axis ($2\varphi = \pm\pi/2$) correspond to $\varphi = \pm 45°$ polarization states or equal-mixture states $(|x\rangle \pm |y\rangle)/\sqrt{2}$. If $|x\rangle = |x_1\rangle$ and $|y\rangle = |x_2\rangle$ correspond to two-particle normal mode states of A_1 and A_2 symmetry, respectively, then the ($\varphi = \pm 45°$) mixtures correspond to local mode states in which one particle is oscillating while the other is at rest. The Hamiltonian described by the A vector (7.5.30) will rotate an **S** vector from the $+B$ axis along the BC plane to the $-B$ axis and back to $+B$. During this time the oscillating particle gives all its energy to the one that was stationary and just as quickly takes it all back but suffers a phase shift of π. This is called a "half-beat" in a resonant energy transfer process. (Recall that **S** has to go around *twice* to return the state exactly.) Maximum power transfer occurs each time the **S** vector passes the $\pm C$ axis. Then the phase of the driving particle is $\pi/2$ ahead of the driven particle which has the same amplitude. This corresponds to a circular trajectory in the (x_1, x_2) space if **S** were to remain fixed on C. (Recall Figure 7.5.4.)]

The oscillator half-beat frequency is given by the magnitude of the $\boldsymbol{\omega}$ vector (7.5.30).

$$\omega_{\text{half-beat}} = |\boldsymbol{\omega}| = \left[\omega_A^2 + \omega_B^2 + \omega_C^2\right]^{1/2} = 2\Omega_{\text{beat}}$$

$$= \left[(A - D)^2 + 4B^2 + 4C^2\right]^{1/2}$$

$$= A - D \quad \text{for } B = 0 = C. \tag{7.5.31}$$

This corresponds to the quantum frequency difference between the two-level eigenfrequencies. It is the Rabi frequency in rotating wave version of the two-level problem. The average value of the eigenfrequencies is given by Eq. (7.5.12):

$$\omega_0 = (A + D)/2. \tag{7.5.32}$$

This corresponds to the classical oscillator orbit or carrier frequency, that is, the angular rate at which one elliptical trajectory is orbited if the ellipse is constant. In quantum theory this overall frequency rate is unimportant for an isolated system.

(b) Oscillator Tori and Lissajous Trajectories Trajectories for oscillator Hamiltonians for arbitrary $\{\omega_A, \omega_B, \omega_C\}$ generally appear to form on

644 THEORY AND APPLICATION OF SYMMETRY REPRESENTATION PRODUCTS

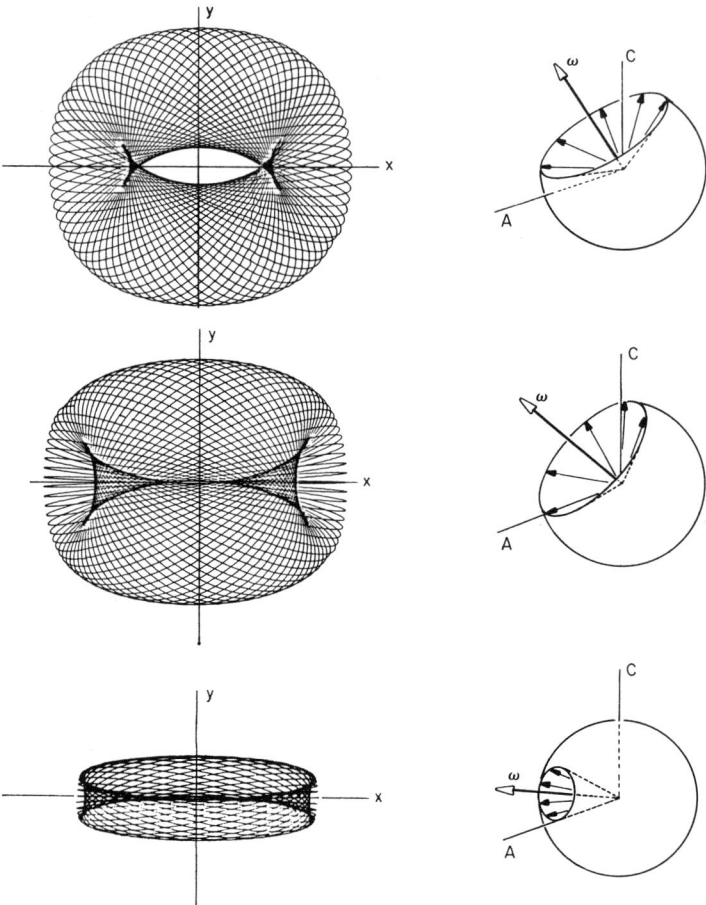

Figure 7.5.8 Evolution of states for various mixtures of A and C components.

toroidal surfaces. Three examples of trajectories with various amounts of birefringent ($\omega_A \neq 0$) and Coriolis ($\omega_C \neq 0$) or Faraday motion are shown in Figure 7.5.8. More detailed views of the oscillator tori are exhibited in Figures 7.5.9 and 7.5.10 using stereo drawings of their phase space, and these will be discussed shortly. We consider first the connection between the oscillator tori and the rotor vectors **S** and **ω**.

Figure 7.5.9 Stereograms of oscillator phase 4-space trajectories for one mixture of ω_A and ω_C components. The 3-space quasispin vector picture of each trajectory is sketched in the upper right-hand corner of each figure. Ratio of high and low eigenfrequencies is $N_H:N_L = 2:1$. (a) Spin vector is about 40° from H end of ω-crank vector. (b) Spin vector is perpendicular to ω vector. (c) Spin vector is about 60° from L end of ω-crank vector. (d) Spin vector is about 40° from L end of ω-crank vector.

ROTATING SPINOR SYSTEMS AND TWO-DIMENSIONAL OSCILLATOR ANALOGIES 645

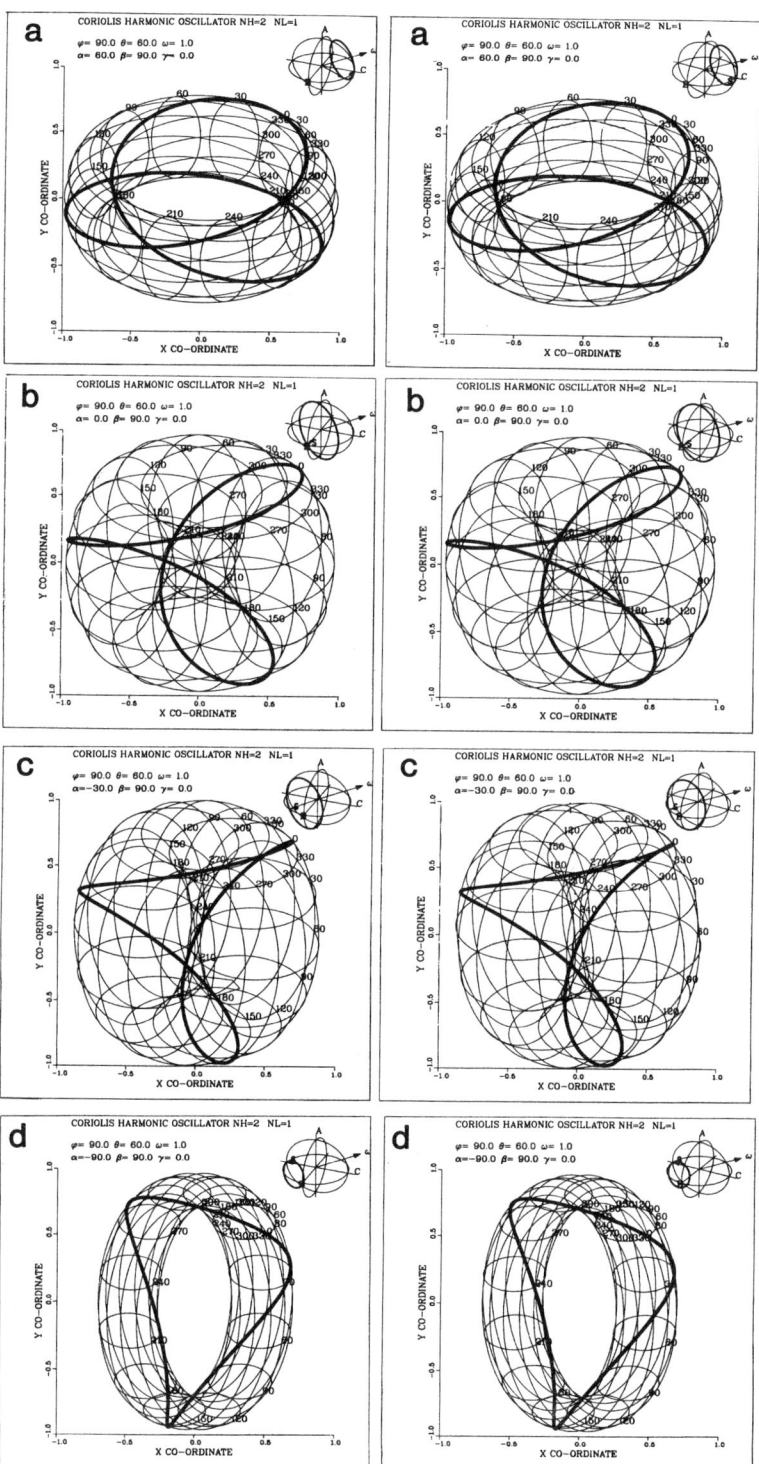

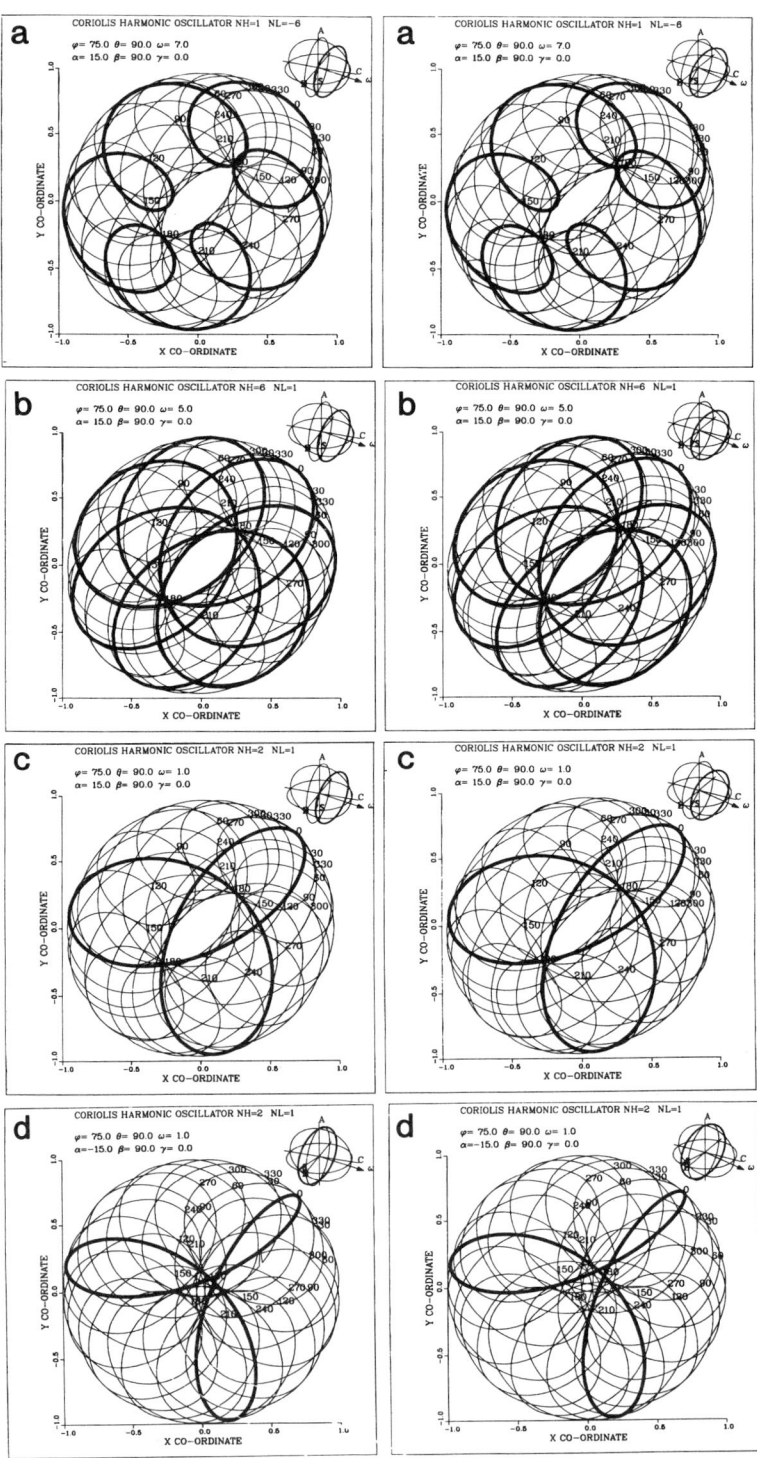

A family of toroidal surfaces exists for each direction $\omega_A:\omega_B:\omega_C$: of the ω-crank vector. Individual tori in each family are distinguished by the direction relative to ω of the initial pseudospin vector $\mathbf{S} = \mathbf{J}$. Its magnitude $|J| = I/2 = J_0$ [recall Eqs. (7.5.11b) and (7.5.13)] determines the amplitude scale and plays no essential role for harmonic motion. The slope of a given trajectory on a particular torus depends upon the ratio of the magnitudes $|\omega| = \omega_{\text{beat}}$ of the crank vector and the overall phase rate ω_0. [Recall Eqs. (7.5.31) and (7.5.32).] Finally, translation of a given trajectory on a torus is achieved by varying the initial value of the third Euler angle γ.

For trajectories in Figures 7.5.7 and 7.5.8 the ratio between elliptical shape evolution rate ($|\omega|$) and orbital rate (ω_0) is equal to a small and essentially irrational number. The trajectories are quasiperiodic and could densely cover their tori. On the other hand, trajectories in Figures 7.5.9 and 7.5.10 have comparable values for ω and ω_0. In addition the ratios ω/ω_0 are chosen to be rational so that each trajectory is absolutely periodic. The resulting closed orbits are generalized or Coriolis-Lissajous figures corresponding to a single closed curve on each torus. Standard Lissajous figures correspond to closed curves in the absence of Coriolis effects ($\omega_c \equiv 0$). Lissajous trajectories only approximate ellipses briefly (i.e., for one orbit period) if the orbital rate is much greater than the shape evolution rate. This is not the case in Figures 7.5.9 and 7.5.10.

The Lissajous figures are plotted in stereo pairs in these figures using a three-dimensional subspace $\{x, y, p_x\}$ of the four-dimensional phase space. They should be viewed with a standard stereopticon or by allowing one's eyes to relax so the left and right eye views the left-hand and right-hand image, respectively. The trajectories are bold line traces which can be seen (in stereo) to reside on surfaces which are similar to the ones which would be traced quasiperiodically by the trajectories in Figure 7.5.8. Each surface is known as an invariant torus, and from the invariance of I in Eq. (7.5.13) it is seen that these examples of tori are four-dimensional spheres. Tori are sketched as doughnut shaped objects in many works, however, the stereo views of them reveal a somewhat different geometry but the same topology.

The coordinate lines on the tori in Figure 7.5.9 are constant angle lines for a choice (θ_H, θ_L) of action-angle coordinates corresponding to the high and low frequency vibrational modes of the classical oscillator. The high and low eigenfrequencies ω_H and ω_L of the two-level quantum Hamiltonian correspond to high and low frequency modes for the classical coupled oscillators.

Figure 7.5.10 Stereograms of oscillator phase space trajectories for various mixtures of ω_A, ω_B, and ω_C and various N_H and N_L. (a) $N_H = 1$, $N_L = 6$. (b) $N_H = 6$, $N_L = 1$. (c) $N_H = 2$, $N_L = 1$. (d) $N_H = 2$, $N_L = 1$.

Suppose that these frequencies are commensurate, that is,

$$\frac{\dot{\theta}_H}{\dot{\theta}_L} = \frac{\omega_H}{\omega_L} = \left(\frac{N_H}{N_L}\right) \qquad (7.5.33)$$

for integers N_H and N_L. Then the quantum transition frequency or classical half-beat frequency will be

$$|\omega| = \omega = \omega_H - \omega_L$$
$$= \omega_L(N_H - N_L)/N_L. \qquad (7.5.34a)$$

The classical orbit frequency will be

$$\omega_0 = \frac{(\omega_H + \omega_L)}{2} = \frac{N_H + N_L}{N_H - N_L}\left(\frac{\omega}{2}\right). \qquad (7.5.34b)$$

For a mode frequency ratio of $N_H/N_L = 2/1$ the ratio orbit and beat frequencies is $\omega_0/\Omega = 3/1$.

The orbits in Figures 7.5.9(a)–7.5.9(d) all have $(N_H, N_L) = (2, 1)$. A perfect recurrence to the initial conditions at $t = 0$ occurs at $t = \Upsilon$ when $(\theta_H, \theta_L) = (\omega_H \Upsilon, \omega_L \Upsilon) = (4\pi, 2\pi)$. During the recurrence period Υ the spin vector precesses through an angle $\omega \Upsilon = 2\pi$ according to (7.5.34a), while the overall phase or orbit angle moves through $\omega_0 \Upsilon = 3\pi$ according to (7.5.34b). That is, an oscillator achieves half of a beat and three halves of an orbit per recurrence with $(N_H, N_L) = (2, 1)$. This recurrence involves two turns along the H direction of the torus and one turn along the L direction. One can clearly see the double winding of the trajectory around the H direction in the upper figure [Figure 7.5.9(a)] and the single winding around the L direction in the lower figure [Figure 7.5.9(d)]. However, the winding topology is the same for all four cases. In Figures 7.5.9(a) or 7.5.9(d) the S vector is closer to the H or L ends, respectively, of the ω-vector. These ends correspond to ellipses with major axes oriented horizontally or vertically, respectively. The extremal ellipses correspond to fixed points or collapsed tori.

It should be noted that the tori for nonzero Coriolis effect ($\omega_C \neq 0$) exhibit curved external caustics as well as internal caustics or holes. Often the holes are clearly evident as in Figures 7.5.9(a) or 7.5.9(d) and Figure 7.5.10(a). Stereo views show that the tori actually have two orthogonal holes which are consistent with their four-dimensional spherical topology. One hole is normal to the H direction and the other is normal to the L direction. Generally, only one hole appears unobstructed in each three-dimensional projection of the four-dimensional spheres. However, Figure 7.5.10(d) shows two osculating holes most clearly.

The trajectory rosette patterns Figures 7.5.9 and 7.5.10 are predictable from the value of ω_0 in (5.8). For $(N_H, N_L) = (1, -6)$ in Figure 7.5.10(a) the orbit angle moves through $\omega_0 \Upsilon = -5\pi$ while for $(N_H, N_L) = (6, 1)$ in Figure 7.5.10(b) it moves through $\omega_0 \Upsilon = 7\pi$ during each recurrence period Υ. There result five curlate and seven prolate rosettes, respectively. For our original example with $(N_H, N_L) = (2, 1)$ one has $\omega_0 \Upsilon / \pi = 3$ and the same number of rosettes as shown in Figures 7.5.10(b) and 7.5.10(c).

F. Rotational Energy (RE) Surface Description of Anharmonic Vibrations

The dynamics and spectral fine structure of quantum rotors was described using the geometry of RE surface trajectories in Section 7.4 (rotational level splitting for high J). An RE surface is a radial plot of a rotor Hamiltonian $H(J_x, J_y, J_z)$ in the space $\{J_x J_y J_z\}$ of its angular momentum for a fixed value of the magnitude,

$$J = |\mathbf{J}| = \left(J_x^2 + J_y^2 + J_z^2\right)^{1/2}.$$

The classically allowed motions for each J then correspond to radial level curves or topography lines formed by the intersection of the RE surface with a sphere whose radius equals the total energy $E = \langle H \rangle$ of the motion.

Here we consider the qualitative features associated with the vibrational analogy of the RE surface dynamics. The analogy is constructed by replacing angular-momentum vector $\mathbf{J}$ of the rotor by the quasispin vector defined by Eq. (7.5.13). The quantum expressions for the quasispin vector result if each phasor $\psi_j = x_j + ip_j$ in Eq. (7.5.9) is replaced by the oscillator boson annihilation operator a_j, and ψ_j^* is replaced with creation operator $a_j^\dagger$ for $j = 1, 2$.

$$\begin{aligned} J_x = J_B &= \left(a_2^\dagger a_1 + a_1^\dagger a_2\right)/2, \\ J_y = J_C &= i\left(a_2^\dagger a_1 - a_1^\dagger a_2\right)/2, \\ J_z = J_A &= \left(a_1^\dagger a_1 - a_2^\dagger a_2\right)/2. \end{aligned} \quad (7.5.35a)$$

Here, the permuted Pauli ordering consistent with Eq. (7.5.29) is used. The momentum conjugate to the overall phase is the total quantum number operator:

$$J_0 = a_1^\dagger a_1 + a_2^\dagger a_2. \quad (7.5.35b)$$

The advantage of the RE surface approach is that qualitative features of the eigenvalue and eigenvector spectra can be visualized relatively easily by plotting $H(\mathbf{J})$ and the classical RE surface trajectories. This will be shown using harmonic and anharmonic examples of vibrational Hamiltonians.

(a) Harmonic RE Surfaces The harmonic oscillator Hamiltonian (7.5.10b) is a linear combination (7.5.14) of the momenta or action variables

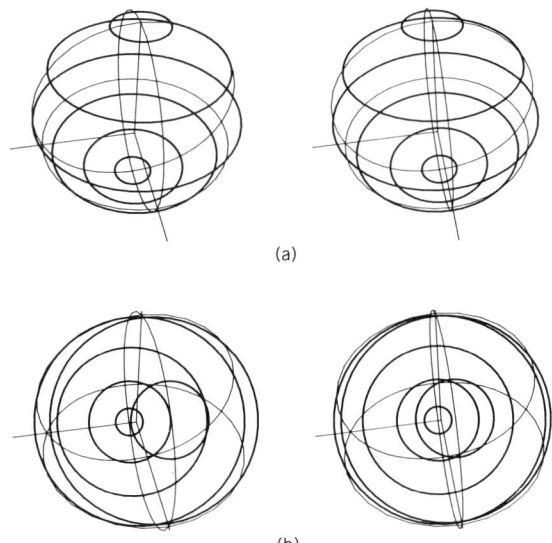

Figure 7.5.11 Quasispin RE surfaces and trajectories for harmonic coupled oscillators. (a) Coriolis free case ($C = 0$). $A = 4.0$, $D = 2.0$, $B = 0$. (b) Coriolis case ($C = 10$). $A = 4.0$, $D = 2.0$, $B = 0$.

J_m. The polar equation for the RE surface is given by substituting the Euler expressions for the momenta (7.5.13) into (7.5.14c):

$$H = (1/2)(\Omega_x \cos\alpha \sin\beta + \Omega_y \sin\alpha \sin\beta + \Omega_z \cos\beta) + J_0\omega_0, \quad (7.5.36a)$$

$$H = (1/2)\left[B \cos\alpha \sin\beta + C \sin\alpha \sin\beta + \left(\frac{A-D}{2}\right)\cos\beta\right] + J_0\omega_0. \quad (7.5.36b)$$

The (ABC) parameters are used in the second expression. The A-axis polar coordinates in (ABC) space are azimuth $\alpha = 2\vartheta$ and polar angle $\beta = 2\nu$ according to (7.5.25):

$$H = (1/4)(\omega_B \cos 2\vartheta \sin 2\nu + \omega_C \sin 2\vartheta \sin 2\nu + \omega_A \cos 2\nu) + J_0\omega_0. \quad (7.5.37)$$

The RE surface topography lines for harmonic examples with $(\omega_A, \omega_B, \omega_C) = (1, 0, 0)$ and $(1, 0, 1)$ are shown in Figures 7.5.11(a) and 7.5.11(b), respectively. The trajectories are families of parallel circles in each case. The circles are perpendicular to the axis of the ω-crank vector. The form of the RE surface is a limacon or cardioid of revolution. For large values of the constant $J_0\omega_0 (\gg \omega I)$ the surface is almost the same as a sphere displaced in the ω direction. This is the simplest type of RE surface, and the topography lines correspond to simple uniform precession around a single direction as shown in the (ABC) parts of Figures 7.5.7 to 7.5.10.

One should note that all the RE surfaces discussed so far represent Hamiltonians which had only even powers of J and were therefore invariant to J inversion $(J \to -J)$ or time reversal of the rotor. This meant that each trajectory with a clockwise direction of precession around a particular axis $(+\hat{a})$ would execute the same precession around the opposite $(-\hat{a})$ axis. In contrast, the trajectories shown in Figure 7.5.11 precess oppositely around $-\omega$ to the way they go around $+\omega$, that is, the vector fields of flow are all in the same direction around ω. There is no separatrix or line of fixed points on these simple RE surfaces to separate one flow field from another.

One should note also that the high-fixed points of the RE surfaces in this work are surrounded by counter-clockwise motion. (In Figure 7.5.11 ω and $-\omega$ are high and low points, respectively.) This is the opposite of the convention used in discussing rotors because the surfaces were drawn in the body frame in that discussion.

(b) Anharmonic RE Surfaces A simple model for vibrational anharmonicity includes the following perturbation operator to the harmonic oscillator:

$$a_x J_x^2 = a_x(x_1 x_2 + p_1 p_2)^2 = a_x(I^2/4)\cos^2 \alpha \sin^2 \beta$$
$$= a_x(a_2^\dagger a_1 + a_1^\dagger a_2)^2/4. \qquad (7.5.38)$$

Anharmonic perturbations can greatly alter the classical dynamics and quantum eigensolutions.

The energy surfaces in Figure 7.5.12 show how the anharmonic perturbation (7.5.38) alters Coriolis-free harmonic normal mode dynamics described by the surface in Figure 7.5.11(a). Figures 7.5.12(a), 7.5.12(b), and 7.5.12(c) show the topography paths for increasing anharmonicity $a_x = 1.0, 2.0,$ and 3.0, respectively, for harmonic values $A = 4.0$, $B = 0 = C$, $D = 2.0$, and $I = 1$. The polar equation for the surfaces in Figure 7.5.12 is

$$H = J_0 \omega_0 + \left(\frac{I}{2}\right)\frac{\omega}{2} \cos \beta + a_2\left(\frac{I^2}{4}\right)\cos^2 \alpha \sin^2 \beta, \qquad (7.5.39a)$$

where

$$\Omega = \frac{\omega}{2} = \frac{A - D}{2} \qquad (7.5.39b)$$

is the true harmonic beat frequency or one-half the classical ω_{beat} frequency in Eq. (7.5.31).

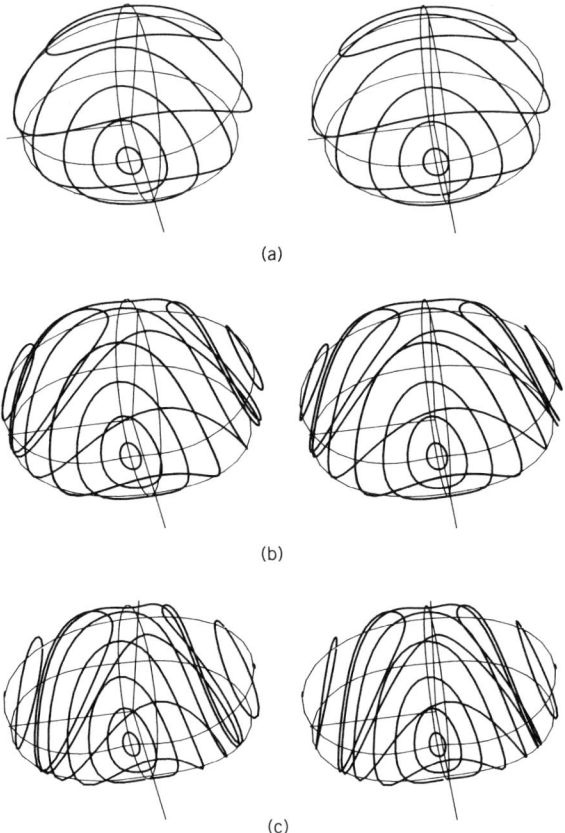

Figure 7.5.12 Quasispin RE surfaces and trajectories for anharmonic Coriolis-free coupled oscillators. (a) $a_x = 1.0$, (b) $a_x = 2.0$, (c) $a_x = 3.0$.

In Figure 7.5.12(a) one can see the first effect of the anharmonicity. The topography paths are no longer circular except in the immediate neighborhood of the normal mode fixed points. In Figures 7.5.12(b) and 7.5.12(c) one sees a more striking effect of the anharmonicity. One of the normal mode fixed points becomes an unstable saddle point at the center of a figure eight separatrix. The separatrix loops surround a pair of equivalent fixed points which move down the AB meridian away from the A point and toward the $\pm B$ axes as a_x increases. These new fixed points correspond to "local modes" and the ellipsometry of the (ABC) space immediately characterizes their trajectories. The elliptical azimuthal Euler angle is $(\alpha = 2\vartheta = 0)$ and the polar angle $(\beta = 2\nu)$ is given by solutions to

$$\frac{dH}{d\beta} = 0 = -\frac{I}{4}\omega \sin \beta + a_x \frac{I^2}{4} 2 \sin \beta \cos \beta. \qquad (7.5.40a)$$

The resulting solutions are

$$\beta = 0, \pi, \cos^{-1}(\Omega/a_x I). \qquad (7.5.40\text{b})$$

For $(\Omega/a_x I) \geq 1$ the only allowed solutions are $\beta = (0, \pi)$ which correspond to normal modes A_1 and A_2. For $a_x I > \Omega$ one mode becomes unstable and two new solutions split off. The quasispin fixed points approach the $\pm B$ axes as $a_x I$ increases. The quasispin orbits near the $\pm B$ axes correspond to oscillations in which the excitation is more or less localized on one or the other of the particles in the two-particle oscillator. This model provides a simple example of an RE surface whose form changes radically when one varies a parameter or a quantum number. The parameter of interest here is the ratio in Eq. (7.5.40) of the half-beat frequency ($\Omega = \omega/2$) to the product of the total quantum number (I) and the anharmonicity strength (a_x).

There are two properties of the RE surface trajectories in Figure 7.5.12 which are different from the ones discussed in connection with rotor motion. First, trajectories in Figure 7.5.12 have lower symmetry. Besides lacking the $(J \rightarrow -J)$ inversion symmetry mentioned previously, the double families of trajectories do not possess rotational symmetry axes at their fixed points, indeed, the fixed points are relocated by changes of the parameters. Second, the two movable stable fixed points appear suddenly, and the $\pm A$ axes are antipodal stable fixed points until $a_x I > \Omega$. Then as $a_x I$ increases, the $+A$ axis becomes an unstable fixed point while the $-A$ axis, which represents the antisymmetric mode (A_2) remains stable for all positive $a_x I$. However, the surrounding domain of stable quasi-antisymmetric trajectories becomes increasingly narrow.

The surface geometry and classical trajectories can be used to determine approximate quantum eigenvalues and wavefunctions as in the analogous rotor RE surfaces. Some qualitative observations can be made immediately. An important new feature is the emergence of equivalent pairs of localized trajectories for $a_x I \gg \Omega$. This signals the onset of a doubling or clustering in the eigenvalue spectrum, and it corresponds to pairs of local mode eigenfunctions. The doublet splitting can be related to tunneling. The quasispin ellipsometry provides a convenient visualization and computational aid for semiclassical analysis.

An advantage of the RE surface description is that it provides a more complete picture of the phase space and each point corresponds to a well defined ellipsometry or trajectory shape. Also, there exist Hamiltonians for which a simple potential curve would be an impossible or misleading description. For example, the fourth degree RE surface which has a six-maxima and eight-minima cannot be described by a simple potential curve since the minima and maxima do not belong to a single curve of section. The precise location of maxima, minima, and saddle points is essential for determining spectral properties such as clustering and tunneling.

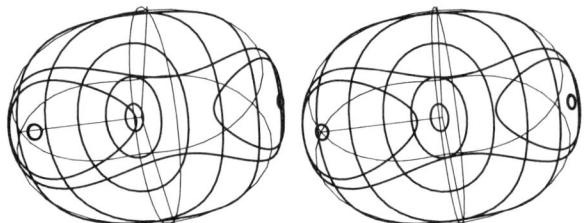

Figure 7.5.13 Quasispin RE surfaces and trajectories for Coriolis anharmonic coupled oscillators.

Finally, the RE surface picture clearly shows the effect of Coriolis terms in the anharmonic Hamiltonian. As seen in Figure 7.5.13 the addition of a nonzero C term tends to bring the local mode fixed points closer to the C axis and to each other. If the harmonic constants are varied so that $(C^2 + A^2)$ remains constant, then the harmonic Coriolis effect is equivalent to a rotation around the B axis. If the local mode points are close to the B axis (strong local mode effect), then the Coriolis effect is small. On the other hand, a large Coriolis effect could coalesce the fixed point pair near the C axis and destroy the local mode effect.

7.6 MOLECULAR ELECTRONIC STRUCTURE

We give now a brief introduction to the electronic eigenvalue problem for molecules. This is a very large and computationally intensive subject, most of which is beyond the scope of this text. Techniques and software for computing electronic structure are developing as rapidly as the computer hardware, and this is likely to continue indefinitely.

In spite of the complexity of this difficult subject there are simply concepts and guiding principles which can be shown using elementary models. We will introduce electronic structure models for the diatomic H_2^+ ion and H_2 molecule, both of which have C_2 internal point symmetry. We will also introduce some aspects of electronic structure of other symmetric molecules such as water (H_2O), ammonia (NH_3), methane (CH_4), benzene (C_6H_6), and sulfur hexafluoride (SF_6), which have C_{2v}, C_{3v}, T_d, D_6, and O_h symmetry, respectively. Symmetry analysis can help greatly to understand and calculate electronic structure.

A. Electronic Models for Diatomic Molecules

The standard electronic structure models begin with an electronic Hamiltonian in which the nuclei are point charges artificially held at fixed locations. Consider a diatomic ion with a single-electron orbiting nucleus A and nucleus B which have charges $Z_A|e|$ and $Z_B|e|$, respectively. This would involve the following electronic Hamiltonian:

$$H_e(R_{AB}) = \frac{p_1^2}{2m} - \frac{k_A}{r_{1a}} - \frac{k_B}{r_{1b}}. \tag{7.6.1}$$

Here the electron-nuclear Coulomb potential constants are

$$k_A = \frac{|Z_A e^2|}{4\pi\varepsilon_0}, \quad k_B = \frac{|Z_B e^2|}{4\pi\varepsilon_0} \tag{7.6.2}$$

The electronic radii r_{1a} and r_{1b} and momentum p_1 are indicated in Figure 7.6.1(a) and m is the electronic mass. For a pair of hydrogen nuclei one has $Z_A = Z_B = 1$.

To account for two electrons orbiting the same nuclei we add the kinetic energy and Coulomb interaction of the second electron with each nucleus and an electron-electron interaction energy with constant $k_{12} = (e^2/4\pi\varepsilon_0)$. (For the H_2 molecule all k's are equal.)

$$H_{e^2}(R_{AB}) = \frac{p_1^2}{2m} + \frac{p_2^2}{2m} - \frac{k_A}{r_{1a}} - \frac{k_B}{r_{1b}} - \frac{k_A}{r_{2a}} - \frac{k_B}{r_{2b}} + \frac{k_{12}}{r_{12}}. \tag{7.6.3}$$

The radii for the two-electron diatomic problem are shown in Figure 7.6.1(b).

Each of these electronic Hamiltonians treat the internuclear separation R_{AB} as a fixed parameter. The nuclear motion is found later using the following approximate nuclear Hamiltonian:

$$H_{\text{nuclear}} = \frac{P_A^2}{2M_A} + \frac{P_B^2}{2M_B} + \frac{Z_A Z_B e^2}{4\pi\varepsilon_0 R_{AB}} + E(R_{AB}). \tag{7.6.4}$$

The effect of the electrons is modeled by an expectation value of electronic energy in some electronic state $|\Psi\rangle$:

$$E(R_{AB}) = \langle\Psi|H_{\text{electron}}(R_{AB})|\Psi\rangle.$$

This is called the Born-Oppenheimer approximation. The electronic energy combined with the internuclear Coulomb potential gives a molecular bonding potential:

$$V_{\text{bond}}(R_{AB}) = E(R_{AB}) + \frac{Z_A Z_B e^2}{4\pi\varepsilon_0 R_{AB}}. \tag{7.6.5}$$

This approximation is valid only if the nuclear R_{ab} motion does not apprecia-

656 THEORY AND APPLICATION OF SYMMETRY REPRESENTATION PRODUCTS

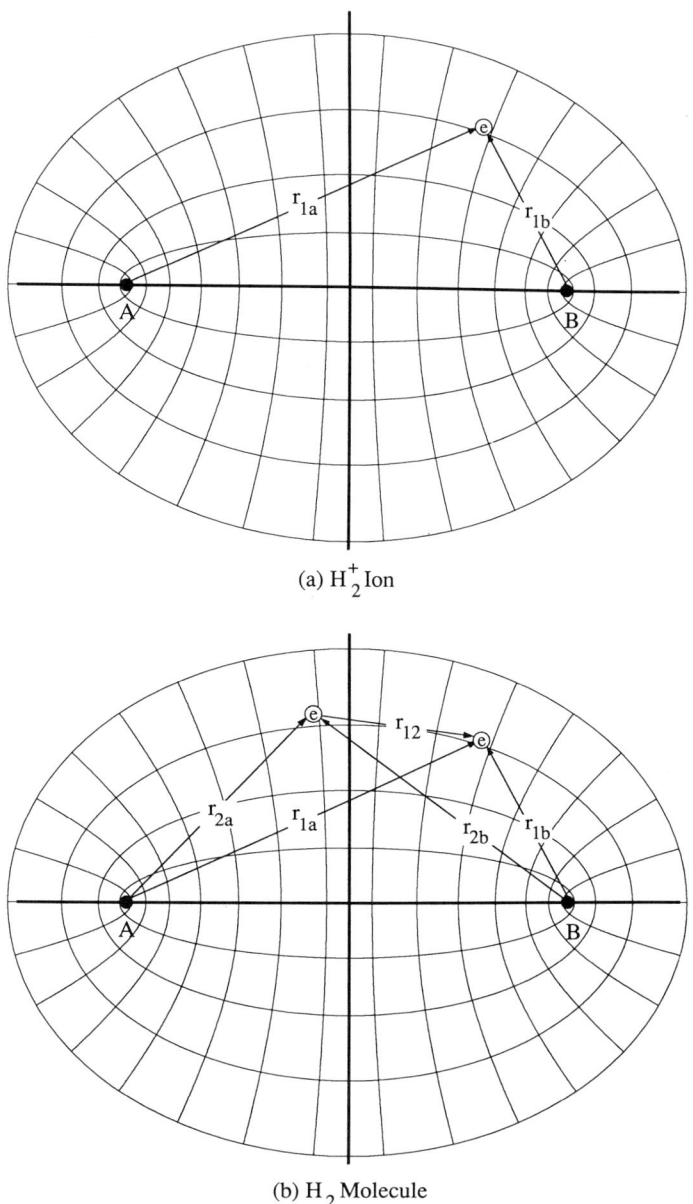

Figure 7.6.1 Coordinates for (a) hydrogen ion and (b) hydrogen molecule.

bly excite the electronic eigenstate $|\Psi\rangle$ into a combination of two or more eigenstates. In the language of Section 2.12.B the state $|\Psi\rangle$ must adiabatically follow the change in R_{ab}.

We consider now several models for electronic energy eigensolutions around a fixed pair of protons or H nuclei. With one electron we will model the bonding of an H_2^+ ion and for two electrons we will model H_2 molecular bonding.

(a) H_2^+: Atomic 1s Orbital Bonding Model The first approximation takes wavefunctions $\alpha(x)$ and $\beta(x)$ to be the lowest (1s) eigenwaves centered around position x_A or else x_B of nucleus A or B, respectively,

$$\alpha(\mathbf{x}) = (\mathbf{x}|a) = \Psi_{1s}(\mathbf{x} - \mathbf{x}_A) \qquad \beta(\mathbf{x}) = (\mathbf{x}|b) = \Psi_{1s}(\mathbf{x} - \mathbf{x}_B)$$

$$= \sqrt{\frac{Z_A^3}{\pi a_0^3}} e^{-Z_A r_a / a_0} \qquad\qquad = \sqrt{\frac{Z_B^3}{\pi a_0^3}} e^{-Z_B r_b / a_0}. \qquad (7.6.6)$$

We shall use parenthetical brackets $|a)$ and $|b)$ to denote these base states to remind us that they are not orthonormal. The overlap matrix $\langle S \rangle$ is defined by

$$S_{ab} = (a|b) = \int d\mathbf{x}\, \alpha^*(\mathbf{x})\beta(\mathbf{x}) = \int d\mathbf{x}(a|\mathbf{x})(\mathbf{x}|b). \qquad (7.6.7)$$

Wave-function overlap plays an important role in bonding theory and usually S_{ab} is *not* δ_{ab}. For the 1s states we still have normality $((a|a) = 1 = (b|b))$ but not orthogonality $((a|b) = (b|a) \neq 0)$.

The H_2^+ bonding model reduces to a C_2 symmetric two-state system. The first state $|a)$ has the electron on nucleus A, while nucleus B is bare and vice versa for state $|b)$. A matrix representation of Hamiltonian H_e given by (7.6.1) is

$$\begin{pmatrix} (a|H_e|a) & (a|H_e|b) \\ (b|H_e|a) & (b|H_e|b) \end{pmatrix} = \begin{pmatrix} h & s \\ s & h \end{pmatrix}, \qquad (7.6.8a)$$

where

$$h = \varepsilon_{1s} - k \int d\mathbf{x}\, \frac{\alpha^*(\mathbf{x})\alpha(\mathbf{x})}{r_{1b}} = \varepsilon_{1s} - \left(a\left|\frac{k}{r_b}\right|a\right), \qquad (7.6.8b)$$

$$s = \varepsilon_{1s} S_{ab} - k \int d\mathbf{x}\, \frac{\alpha^*(\mathbf{x})\beta(\mathbf{x})}{r_{1b}} = \varepsilon_{1s} S_{ab} - \left(a\left|\frac{k}{r_b}\right|b\right), \qquad (7.6.8c)$$

and ε_{1s} is the atomic hydrogen ground state energy value.

However, the representation basis $(|a), |b))$ does not satisfy the axioms of orthonormality and completeness given in Chapter 1. Therefore, the energy eigenstates

$$|E\rangle = \psi_a|a) + \psi_b|b), \qquad (7.6.9a)$$

which satisfy

$$H_e|E\rangle = E|E\rangle \qquad (7.6.9b)$$

are not found by directly diagonalizing (7.6.8a). Instead we must convert (7.6.9) to the following generalized eigenvalue equations:

$$(a|H_e|E\rangle = (a|H_e|a)\psi_a + (a|H_e|b)\psi_b = E[(a|a)\psi_a + (a|b)\psi_b],$$
$$(b|H_e|E\rangle = (b|H_e|a)\psi_a + (b|H_e|b)\psi_b = E[(b|a)\psi_a + (b|b)\psi_b]. \qquad (7.6.10a)$$

In matrix form this becomes

$$\begin{pmatrix} (a|H_e|a) & (a|H_e|b) \\ (b|H_e|a) & (b|H_e|b) \end{pmatrix} \begin{pmatrix} \psi_a \\ \psi_b \end{pmatrix} = E \begin{pmatrix} S_{aa} & S_{ab} \\ S_{ba} & S_{bb} \end{pmatrix} \begin{pmatrix} \psi_a \\ \psi_b \end{pmatrix}. \qquad (7.6.10b)$$

The eigenvalues are roots E of a generalized secular equation

$$\det|\langle \mathbf{H} \rangle - E\langle \mathbf{S} \rangle| = 0. \qquad (7.6.11)$$

For small equations like this one it is easy to invert the overlap matrix and recover a standard eigenvalue problem. The inverse of this overlap matrix is quite simple. (Let $S_{ab} = S = S_{ba}$.)

$$\begin{pmatrix} S_{aa} & S_{ab} \\ S_{ba} & S_{bb} \end{pmatrix}^{-1} = \begin{pmatrix} 1 & S \\ S & 1 \end{pmatrix}^{-1} = \begin{pmatrix} 1 & -S \\ -S & 1 \end{pmatrix} \Big/ (1 - S^2). \qquad (7.6.12)$$

Multiplying it by the H matrix gives a standard eigenvalue equation:

$$\frac{1}{1-S^2}\begin{pmatrix} 1 & S \\ S & 1 \end{pmatrix}\begin{pmatrix} h & s \\ s & h \end{pmatrix}\begin{pmatrix} \psi_a \\ \psi_b \end{pmatrix} = \frac{1}{1-S^2}\begin{pmatrix} h - Ss & s - Sh \\ s - Sh & h - Ss \end{pmatrix}\begin{pmatrix} \psi_a \\ \psi_b \end{pmatrix} = E\begin{pmatrix} \psi_a \\ \psi_b \end{pmatrix}. \qquad (7.6.13)$$

The desired energy eigenvalues are the following:

$$E^+ = \frac{h+s}{1+S}, \qquad E^- = \frac{h-s}{1-S}. \qquad (7.6.14a)$$

These correspond to symmetric and antisymmetric eigenstates:

$$|E^+\rangle = \frac{|a\rangle + |b\rangle}{\sqrt{2(1+S)}}, \qquad |E^-\rangle = \frac{|a\rangle - |b\rangle}{\sqrt{2(1-S)}}. \qquad (7.6.14b)$$

Note that C_2 symmetry projection will give the eigenvalues directly from the generalized eigenvalue equation (7.6.10b). Projection must diagonalize the overlap matrix as well as the H matrix. Note also that the eigenvectors are orthonormal even though the base states $|a\rangle$ and $|b\rangle$ are not.

The Born-Oppenheimer potentials (7.6.5) are a sum of the eigenvalues (7.6.14a) and the nuclear Coulomb potential k/R. Using (7.6.8), we have

$$V^{\pm}(R) = \frac{h \pm s}{1 \pm S} + \frac{k}{R} = \frac{\varepsilon_{1s} - k(a|1/r_b|a) \pm \varepsilon_{1s}S \mp k(a|1/r_b|b)}{1 \pm S} + \frac{k}{R},$$

$$V^{+}(R) = \varepsilon_{1s} + \frac{k}{R} - \frac{k(a|1/r_b|a)}{1 + S} - \frac{k(a|1/r_b|b)}{1 + S}, \qquad (7.6.15a)$$

$$V^{-}(R) = \varepsilon_{1s} + \frac{k}{R} - \frac{k(a|1/r_b|a)}{1 - S} + \frac{k(a|1/r_b|b)}{1 - S}. \qquad (7.6.15b)$$

For small S the $V^{+}(R)$ function must lie below $V^{-}(R)$ because the last term is subtracted rather than added.

To evaluate the energy eigenvalues (7.6.14a) we use confocal elliptic hyperbolic coordinates. These are also known as spheroidal coordinates. These are a generalization of the cylindrical or spherical coordinates used for one nuclear center. They are natural coordinates for a diatomic molecule which has two singularities.

Spheroidal coordinates may be defined a number of ways. The formal Cartesian transformation is

$$x = f \sinh u \sin v \cos \phi,$$
$$y = f \sinh u \sin v \sin \phi,$$
$$z = f \cosh u \cos v, \qquad (7.6.16a)$$

where $R_{AB} = 2f$ is the distance between focal points at the nuclei, and

$$\rho = \sqrt{x^2 + y^2} = f \sinh u \sin v \qquad (7.6.16b)$$

is the radius for cylindrical coordinates (ρ, ϕ, z). Constant coordinates u and v define, respectively, confocal ellipses or hyperbolas of revolution around the z axis (recall Figure 7.6.1):

$$\frac{z^2}{(f \cosh u)^2} + \frac{\rho^2}{(f \sinh u)^2} = 1, \qquad \frac{z^2}{(f \cos v)^2} - \frac{\rho^2}{(f \sin v)^2} = 1. \qquad (7.6.17)$$

The semimajor axes a_e and a_h of the ellipse or hyperbola are, respectively, half the sum and differences of the electronic radii from nuclei A and B:

$$a_e = \frac{r_{1a} + r_{1b}}{2} = f \cosh u \qquad a_h = \frac{r_{1a} - r_{1b}}{2} = f \cos v$$
$$= \frac{R_{AB}}{2} \cosh u, \qquad = \frac{R_{AB}}{2} \cos v. \qquad (7.6.18)$$

Let the ellipsoidal coordinates μ and ν be major axes in units of nuclear separation R_{AB}:

$$\mu \equiv \frac{2a_e}{R_{AB}} = \cosh u, \qquad \nu = \frac{2a_h}{R_{AB}} = \cos v. \qquad (7.6.19)$$

Then the electronic radii are given in terms of μ and ν as follows:

$$r_{1a} = \frac{(\mu + \nu)}{2} R_{AB}, \qquad r_{1b} = \frac{(\mu - \nu)}{2} R_{AB}. \qquad (7.6.20)$$

The Jacobian volume element is found using (7.6.15) and (7.6.19)

$$dx\,dy\,dz = \frac{\partial(xyz)}{\partial(\mu\nu\phi)} du\,dv\,d\phi = \frac{\partial(xyz)}{\partial(uv\phi)} \frac{\partial(uv\phi)}{\partial(\mu\nu\phi)} d\mu\,d\nu\,d\phi$$
$$= \frac{R^3}{8}(\cosh^2 u - \cos^2 v)\,d\mu\,d\nu\,d\phi = \frac{R^3}{8}(\mu^2 - \nu^2)\,d\mu\,d\nu\,d\phi. \qquad (7.6.21a)$$

If the overlap integral is converted to ellipsoidal coordinates it is simplified as follows,

$$S = \int dx\,dy\,dz\,\alpha^*(\mathbf{x})\beta(\mathbf{x}) = \frac{Z^3}{\pi a_0^3}\int_0^{2\pi} d\phi \int_1^\infty d\mu \int_{-1}^1 d\nu\,\frac{R^3}{8}(\mu^2 - \nu^2)e^{-ZR\mu/a_0}$$
$$= \left(\frac{Z^2 R^2}{3a_0^2} + \frac{ZR}{a_0} + 1\right)e^{-ZR/a_0}. \qquad (7.6.21b)$$

Note that the overlap falls off exponentially as atomic number ($Z_A = Z = Z_B$) and internuclear radius $R_{AB} = R$ increase.

The other terms in the eigenvalue and Hamiltonian formulas (7.6.15) are evaluated similarly. Consider Coulomb attraction between electronic charge on nucleus A and nuclear charge B (or vice versa). The potential appears in the third term of (7.6.15):

$$-k\left(a\left|\frac{1}{r_b}\right|a\right) = -k\int d\mathbf{x}\,\frac{\alpha^*(\mathbf{x})\alpha(\mathbf{x})}{r_{1b}}$$
$$= -\frac{kZ^3}{\pi a_0^3}\int_0^{2\pi} d\phi \int_1^\infty d\mu \int_{-1}^1 d\nu\,\frac{\tfrac{1}{8}R(\mu^2 - \nu^2)e^{-(\mu+\nu)ZR/2a_0}}{\tfrac{1}{2}R(\mu - \nu)}$$
$$= -\frac{k}{R} + \left(\frac{k}{R} + \frac{kZ}{a_0}\right)e^{-2ZR/a_0}. \qquad (7.6.22)$$

The main contribution is $-k/R$, which lowers the atomic energy ε_{1s}. (ε_{1s} is due to the interaction of electronic charge on nucleus A and nuclear charge A.)

The crucial bonding energy is the final term in (7.6.15). It appears to be an interaction between nucleus B and the electronic "overlap charge" $\alpha^*(x)\beta(x)$:

$$k\left(a\left|\frac{1}{r_b}\right|b\right) = -k\int d\mathbf{x}\, \frac{\alpha^*(\mathbf{x})\beta(\mathbf{x})}{r_{1b}}$$

$$= -\frac{kZ^3}{\pi a_0^3}\int_0^{2\pi} d\phi \int_1^\infty d\mu \int_{-1}^1 d\nu\, \frac{\frac{1}{8}R^3(\mu^2 - \nu^2)e^{-uZR/a_0}}{\frac{1}{2}R(\mu - \nu)}$$

$$= -\frac{kZ}{a_0}\left[1 + \frac{ZR}{a_0}\right]e^{-ZR/a_0}. \qquad (7.6.23)$$

This term is the main difference between V^+ and V^- in (7.6.15). It represents a fundamentally quantum-mechanical effect, and quite an important one to anyone who wants their molecules to hang together! Like most quantum tunneling or resonance effects it dies off quasiexponentially with separation.

The resulting H_2^+ molecular potentials $V^+(R)$ and $V^-(R)$ reduce to the following functions of R expressed in atomic units ($a_0 = 5.3 \times 10^{-11}$ m):

$$V^\pm(R) = \varepsilon_{1s} + \frac{1}{R}\frac{(1+R)e^{-2R} \pm (1 - \frac{2}{3}R^2)e^{-R}}{1 \pm \left(\frac{1}{3}R^2 + R + 1\right)e^{-R}}. \qquad (7.6.24)$$

The atomic energy unit ($k/a_0 = 4.36 \times 10^{-18}$ J $= 27.21$ eV) is used, as well. A plot of $V^\pm(R)$ is shown in Figure 7.6.2.

The symmetrized (g) orbital state

$$|\sigma_g\rangle = \frac{|a\rangle + |b\rangle}{\sqrt{2(1+S)}} \qquad (7.6.25a)$$

is called a *bonding* orbital since its potential $V^+(R)$ has a stable minimum at about 2.5 a.u. The antisymmetrized state

$$|\sigma_\mu^*\rangle = \frac{|a\rangle - |b\rangle}{\sqrt{2(1-S)}} \qquad (7.6.25b)$$

is called an *antibonding* orbital since its potential $V^-(R)$ is repulsive and would cause the H_2^+ ion to explode. The bonding orbital concentrates electronic charge between the two nuclei, while the antibonding orbital has a

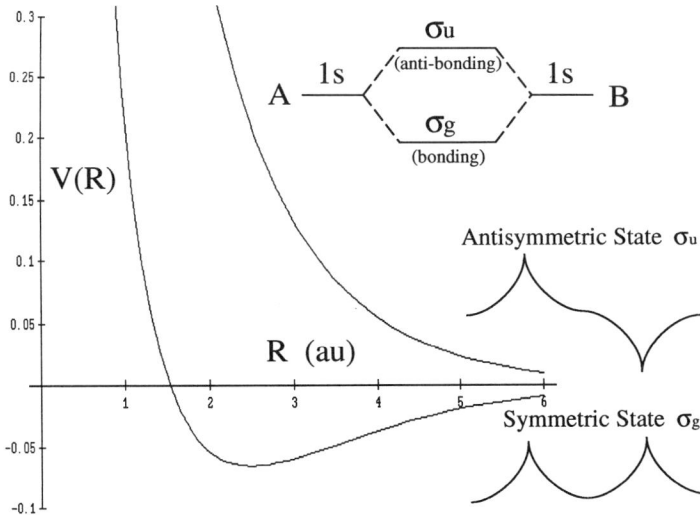

Figure 7.6.2 Antibonding and bonding states of H_2^+ with the corresponding potential curves.

node and tends to exclude charge. Both orbitals are symmetric around the z axis as their σ labels indicate. (Recall $D_{\infty h}$ labels in Section 5.)

This bonding model based upon $1s\sigma$ is qualitatively right but not the whole story. The correct value of the bonding minimum is 25% less, about 2.0 a.u., and the energy is much less, about -0.10 a.u. instead of -0.065 predicated by (7.5.24). What is missing is another quantum effect involving the uncertainty relation ($\Delta x \Delta p \leq \hbar$). An electron that is less confined (greater Δx) has less kinetic energy (less Δp). The H_2^+ electron can "stretch out" over two nuclei and this reduces its energy. As a result it ends up getting closer to the H nuclei and reducing the average potential energy. By using only $1s$ orbitals, this model has prevented this reduction. Exact numerical wave functions of the spheroidal coordinates (7.6.19) show this nuclear electronic concentration.

There is another important piece of physics which is left out of the rigid ($1s$) wave-function model of H_2^+: The wave on each atom cannot be polarized. To make an electronic dipole moment on one of the atoms it needs to be able to mix in some p-wave such as in a $2s2p$ superposition. This is important for modeling H^+ to H scattering potential which has a $1/R^4$ dependence at large R. Polarization of electronic orbitals plays a role in molecular bonding, as we will see later.

(b) H_2: Heitler-London Valence-Bond Model

Suppose two electrons can occupy either of the two atomic $1s$ states $|a\rangle$ or $|b\rangle$, discussed in Section (a). This would yield a basis of four orbital states $|a\rangle|a\rangle$, $|a\rangle|b\rangle$, $|b\rangle|a\rangle$, and

$|b\rangle|b\rangle$). The Heitler-London model chooses to consider only the "covalent" bonding states $|a\rangle|b\rangle$ and $|b\rangle|a\rangle$ in which electrons 1 and 2 are on different nuclei. This model ignores the contribution of "ionic" bonding states $|a\rangle|a\rangle$ or $|b\rangle|b\rangle$ in which both electrons are crowded onto nucleus a or nucleus b, respectively. More complicated models show that the probability for H_2 ionic states is only a few percent.

The H_2 Hamiltonian is (7.6.3) with all k values equal. Its representation in the Heitler-London basis $\{|a\rangle|b\rangle, |b\rangle|a\rangle\}$ has the following diagonal components $h_a = h_b$:

$$h_a = \left(ab\left|\frac{p_1^2}{2m} - \frac{k}{r_{1a}}\right|ab\right) + \left(ab\left|\frac{p_2^2}{2m} - \frac{k}{r_{1b}}\right|ab\right) - \left(ab\left|\frac{k}{r_{2a}}\right|ab\right)$$
$$- \left(ab\left|\frac{k}{r_{1b}}\right|ab\right) + \left(ab\left|\frac{k}{r_{12}}\right|ab\right)$$
$$= \varepsilon_{1s} + \varepsilon_{1s} - \left(b\left|\frac{k}{r_a}\right|b\right) - \left(a\left|\frac{k}{r_b}\right|a\right) + \left(ab\left|\frac{k}{r_{12}}\right|ab\right)$$
$$= 2\varepsilon_{1s} - 2\left(a\left|\frac{k}{r_b}\right|a\right) + D, \qquad (7.6.26a)$$

where the last term D is called the *direct* Coulomb repulsion integral:

$$D = \left(ab\left|\frac{k}{r_{12}}\right|ab\right) = \int d\mathbf{x}_1 \int d\mathbf{x}_2\, \alpha^*(\mathbf{x}_1)\beta^*(\mathbf{x}_2)\left(\frac{1}{r_{12}}\right)\alpha(\mathbf{x}_1)\beta(\mathbf{x}_2).$$
$$(7.6.26b)$$

The other terms were discussed previously. Similarly, the off-diagonal component is as follows

$$h_{ab} = \left(ab\left|\frac{p_1^2}{2m} - \frac{k}{r_{1a}}\right|ba\right) + \left(ab\left|\frac{p_2^2}{2m} - \frac{k}{r_{2b}}\right|ba\right) - \left(ab\left|\frac{k}{r_{2a}}\right|ba\right)$$
$$- \left(ab\left|\frac{k}{r_{1b}}\right|ba\right) + \left(ab\left|\frac{k}{r_{12}}\right|ba\right)$$
$$= \varepsilon_{1s}(a|b)(b|a) + \varepsilon_{1s}(a|b)(b|a) - (a|b)\left(b\left|\frac{k}{r_a}\right|a\right)$$
$$- \left(a\left|\frac{k}{r_{1b}}\right|b\right)(b|a) + \left(ab\left|\frac{k}{r_{12}}\right|ba\right)$$
$$= 2\varepsilon_{1s}S^2 - 2\left(a\left|\frac{k}{r_b}\right|b\right)S + E, \qquad (7.6.27a)$$

where the last term E is called the *exchange* Coulomb repulsion integral

$$E = \left(ab\left|\frac{k}{r_{12}}\right|ba\right) = \int d\mathbf{x}_1 \int d\mathbf{x}_2\, \alpha^*(\mathbf{x}_1)\beta^*(\mathbf{x}_2)\left(\frac{1}{r_{12}}\right)\beta(\mathbf{x}_1)\alpha(\mathbf{x}_2). \tag{7.6.27b}$$

The nonorthogonal Heitler-London representation is

$$\langle H_{e^2}\rangle_{\text{HL}} = \begin{pmatrix} 2\varepsilon_{1s} - 2\left(a\left|\dfrac{k}{r_b}\right|a\right) + D & 2\varepsilon_{1s}S^2 - 2\left(a\left|\dfrac{k}{r_b}\right|b\right)S + E \\ 2\varepsilon_{1s}S^2 - 2\left(a\left|\dfrac{k}{r_b}\right|b\right)S + E & 2\varepsilon_{1s} - 2\left(a\left|\dfrac{k}{r_b}\right|a\right) + D \end{pmatrix}$$

$$= \begin{pmatrix} h_a & h_{ab} \\ h_{ab} & h_a \end{pmatrix}. \tag{7.6.28}$$

To convert this matrix to an orthogonal representation we still need to multiply it by an inverse overlap matrix similar to (7.6.12). Only now each S is replaced by S^2. The orthogonal Heitler-London representation is

$$\langle H_{e^2}\rangle_{\text{OHL}} = \frac{\begin{pmatrix} 1 & -S^2 \\ -S^2 & 1 \end{pmatrix}}{1-S^4}\begin{pmatrix} h_a & h_{ab} \\ h_{ab} & h_a \end{pmatrix}$$

$$= \begin{pmatrix} h_a - S^2 h_{ab} & h_{ab} - S^2 h_a \\ h_{ab} - S^2 h_a & h_a - S^2 h_{ab} \end{pmatrix} \bigg/ (1-S^4). \tag{7.6.29}$$

From C_2 symmetry projection (or direct diagonalization) we obtain a symmetric (g) eigenstate

$$|^1\Sigma_g^+\rangle = \frac{|ab\rangle + |ba\rangle}{\sqrt{2(1+S^2)}}, \tag{7.6.30a}$$

with eigenvalue

$$E(^1\Sigma_g) = \frac{h_a + h_{ab}}{1+S^2} = 2\varepsilon_{1s} + \frac{-2(a|k/r_b|a) + D - 2(a|k/r_b|b)S + E}{1+S^2}, \tag{7.6.30b}$$

and an antisymmetric (u) eigenstate

$$|^3\Sigma_u^+\rangle = \frac{|ab\rangle - |ab\rangle}{\sqrt{2(1-S^2)}}, \tag{7.6.31a}$$

with eigenvalue

$$E({}^3\Sigma_u) = \frac{h_a - h_{ab}}{1-S^2} = 2\varepsilon_{1s} + \frac{-2(a|k/r_b|a) + D + 2(a|k/r_b|b)S - E}{1-S^2}.$$
(7.6.31b)

The notation for singlet ($^1\Sigma$) and triplet ($^3\Sigma$) refers to the Pauli-allowed spin states with spin $S = 0$ and $S = 1$, respectively. (Recall Section 7.1.B.) The uppercase symbols Σ_g^+ and Σ_u^+ label the overall $D_{\infty h}$ symmetry of each electronic orbital.

The energy values (7.6.30b) and (7.6.31b) determine the ordering of singlet ($^1\Sigma_g$) and triplet ($^3\Sigma_g$) according to the relative magnitudes of exchange integral E and overlap factors $2(a|k/r_b|b)S$. The former tends to make the singlet higher than the triplet (as in the $1s2s$ configuration for the He atom), while the latter tends to do the opposite. In H_2 it is the latter which is larger. The Coulomb factor $(a|k/r_b|b)$ plays a decisive role in bonding the molecule H_2 as well as the ion H_2^+. It also guarantees that the ground-state H_2 electronic spin is zero.

The calculation of the two-electron integrals is very complicated even for the simple $(1s)^2$ model. We quote the results tabulated by Atkins:

$$D = \left(ab\left|\frac{k}{r_{12}}\right|ab\right) = \frac{k}{R}\left[1 - \left(1 + \frac{11R}{8} + \frac{3R^2}{4} + \frac{R^3}{3}\right)e^{-2R}\right]$$
(7.6.32a)

$$E = \left(ab\left|\frac{k}{r_{12}}\right|ba\right) = \frac{6k}{5R}\Bigg[(\gamma + \ln R)S^2 - T^2 E_1(4R) + 2STE_1(2R)$$

$$-\left(-\frac{25R}{48} + \frac{23R^2}{24} + \frac{R^3}{2} + \frac{R^4}{18}\right)e^{-2R}\Bigg],$$
(7.6.32b)

where

$$T = \left[\frac{R^2}{3} - R + 1\right]e^R, \qquad E_1(x) = \int_x^\infty du\,\frac{e^{-u}}{u}$$

and $\gamma = 0.577\ldots$ is the Euler constant. The resulting potential energy has a minimum of about $\Delta E_0 = -0.115$ a.u. at $R_0 = 1.6$ a.u. This is above the experimental value of $\Delta E_0^{\text{exp}} = -0.165$ a.u. and $R_0^{\text{exp}} = 1.4$ a.u. The discrepancy is comparable to that of the H_2^+ ion model and perhaps even a little less.

(c) H_2: Improved Valence-Bond Model By including the ionic states $|aa\rangle$ and $|bb\rangle$ one may obtain a more realistic electronic structure model. These states will be particularly important for molecules composed of dissimilar nuclei. The price for improvement is the need to calculate a 4×4 matrix and its eigensolutions in terms of the basis $\{|aa), |bb), |ab), |ba)\}$. This time we will use a biorthogonal bra basis $\{(AA|, (BB|, (AB|, (BA|\}$ which already contains the inverse overlap matrix. The single particle bras are defined using (7.6.12):

$$(A| = (a|\mathbf{S}^{-1} = \frac{(a| - S(b|}{1 - S^2}, \tag{7.6.33a}$$

$$(B| = (b|\mathbf{S}^{-1} = \frac{(b| - S(a|}{1 - S^2}. \tag{7.6.33b}$$

These satisfy biorthonormality: $(A|a) = 1 = (B|b)$, and $(A|b) = 0 = (B|a)$. From this we get a complete set of biorthonormal two-particle bra states which satisfy $(XY|x'y') = \delta_{Xx'}\delta_{Yy'}$.

$$(AA| = ((aa| + S^2(bb| - S(ab| - S(ba|)/(1 - S^2)^2,$$
$$(BB| = (S^2(aa| + (bb| - S(ab| - S(ba|)/(1 - S^2)^2,$$
$$(AB| = (-S(aa| - S(bb| + (ab| + S^2(ba|)/(1 - S^2)^2,$$
$$(BA| = (-S(aa| - S(bb| + S^2(ab| + (ba|)/(1 - S^2)^2. \tag{7.6.34}$$

$D_{\infty h}$ projection gives us two ionic and two covalent basis states of definite parity, permutational symmetry, and corresponding Pauli-allowed and spin multiplicity. The Young tableau notation on the right in the following denotes permutational symmetry by horizontal arrays of boxes and antisymmetry by vertical arrays.

$$\left|{}^1\Sigma_g^+ \text{ ion}\right\rangle = \frac{|aa) + |bb)}{\sqrt{2}} = \frac{\boxed{a|a} + \boxed{b|b}}{\sqrt{2}}, \tag{7.6.35a}$$

$$\left|{}^1\Sigma_g^+ \text{ cov}\right\rangle = \frac{|ab) + |ba)}{\sqrt{2}} = \boxed{a|b}, \tag{7.6.35b}$$

$$\left|{}^1\Sigma_u^+ \text{ ion}\right\rangle = \frac{|aa) - |bb)}{\sqrt{2}} = \frac{\boxed{a|a} - \boxed{b|b}}{\sqrt{2}}, \tag{7.6.35c}$$

$$\left|{}^1\Sigma_u^+ \text{ cov}\right\rangle = \frac{|ab) - |ba)}{\sqrt{2}} = \boxed{\begin{array}{c}a\\b\end{array}}. \tag{7.6.35d}$$

The symmetry-defined kets are written the same way using the uppercase-labeled bra $(AA|, \ldots, (BA|$ in place of kets $|aa), \ldots, |ba)$, respectively.

Use of ionic states means more integrals. Still the C_2 symmetry ($a \leftrightarrow b$) of the nuclei and the S_2 permutational symmetry ($1 \leftrightarrow 2$) of the electronic wave functions reduces the number of different matrix elements to the following five which are derived in the same way as (7.6.26) and (7.6.27):

$$(aa|H|aa) = (bb|H|bb) = 2\varepsilon_{1s} - 2A + C,$$
$$(aa|H|ab) = (aa|H|ba) = (bb|H|ab) = (bb|H|ba) = 2S\varepsilon_{1s} - SA - B + F,$$
$$(aa|H|bb) = (bb|H|aa) = 2S^2\varepsilon_{1s} - 2SB + G,$$
$$(ab|H|ab) = (ba|H|ba) = 2\varepsilon_{1s} - 2A + D,$$
$$(ab|H|ba) = (ba|H|ab) = 2S^2\varepsilon_{1s} - 2SB + E. \qquad (7.6.36)$$

A set of seven $H_2(1s\sigma)^2$ integrals is listed:

$$A = \left(a\left|\frac{k}{r_b}\right|a\right) = \left(b\left|\frac{k}{r_a}\right|b\right), \qquad B = \left(a\left|\frac{k}{r_b}\right|b\right) = \left(a\left|\frac{k}{r_a}\right|b\right),$$

$$D = \left(ab\left|\frac{k}{r_{12}}\right|ab\right) = \left(ba\left|\frac{k}{r_{12}}\right|ba\right), \qquad E = \left(ab\left|\frac{k}{r_{12}}\right|ba\right) = \left(ba\left|\frac{k}{r_{12}}\right|ab\right),$$

$$C = \left(aa\left|\frac{k}{r_{12}}\right|aa\right), \qquad F = \left(aa\left|\frac{k}{r_{12}}\right|ab\right),$$

$$G = \left(aa\left|\frac{k}{r_{12}}\right|bb\right). \qquad (7.6.37)$$

The H_2 Hamiltonian has nonzero matrix elements only between states (7.6.35) which have the same permutational symmetry or spin multiplicity ($2S + 1 = 1$ or 3) and parity (u or g). This leaves a 2×2 matrix involving the first two $^1\Sigma_g$ states (7.6.35a) and (7.6.35b) and diagonal elements for $^1\Sigma_u$ and $^3\Sigma_u$ states. A real symmetric H_2 Hamiltonian has only five independent matrix elements.

Unfortunately, the biorthogonal basis used here does not give a symmetric Hamiltonian. Instead it uses lopsided matrix elements such as the following:

$$\langle ^3\Sigma_g \text{ cov}|H|^3\Sigma_g \text{ cov}\rangle = [(AB|H|ab) - (AB|H|ba) - (BA|H|ab)$$
$$+ (BA|H|ba)]/2$$
$$= (AB|H|ab) - (AB|H|ba)$$
$$= 2\varepsilon_{1s} + \frac{-2A + 2SB + D - E}{1 - S^2}. \qquad (7.6.38)$$

When these matrix elements are expanded using (7.6.34) as in the following:

$$(AB|H|ab) = [-S(aa|H|ab) - S(bb|H|ab) + (ab|H|ab)$$
$$+ S^2(ba|H|ab)]/(1 - S^2)^2, \qquad (7.6.39)$$

a non-Hermitian representation results:

	$\|{}^1\Sigma_g \text{ ion}\rangle$	$\|{}^1\Sigma_g \text{ cov}\rangle$	$\|{}^1\Sigma_u \text{ ion}\rangle$	$\|{}^3\Sigma_u \text{ cov}\rangle$
	$2\varepsilon_{1s} + \dfrac{-2A + 2SB}{(1-S^2)}$ $+ \dfrac{(1+S^2)(C+G) - 4SF}{(1-S^2)^2}$	$\dfrac{-2B + 2SA}{(1-S^2)}$ $+ \dfrac{2(1+S^2)F - 2S(D+E)}{(1-S^2)^2}$	0	0
	$\dfrac{-2B + 2SA}{(1-S^2)}$ $+ \dfrac{2(1+S^2)F - 2S(C+G)}{(1-S^2)^2}$	$2\varepsilon_{1s} + \dfrac{-2A + 2SB}{(1-S^2)}$ $+ \dfrac{(1+S^2)(D+E) - 4SF}{(1-S^2)^2}$	0	0
	0	0	$2\varepsilon_{1s}$ $+ \dfrac{-2A + 2SB + C - G}{1-S^2}$	0
	0	0	0	$2\varepsilon_{1s}$ $+ \dfrac{-2A + 2SB + D - E}{1-S^2}$

(7.6.40)

You might expect the second diagonal matrix element in (7.6.40) to correspond to the Heitler-London eigenvalue (7.6.30b) just as the fourth diagonal element in (7.6.40) equals (7.6.31b). However, the matrix element $\langle ^1\Sigma_g \text{ cov}|H|^1\Sigma_g \text{ cov}\rangle$ in (7.6.40) contains ionic contributions due to the nonorthogonality of our basis. These are not present in (7.6.30b) and so there is no simple correspondence.

(d) H_2: Molecular Orbital Model If the preceding discussion makes you uneasy about the use of nonorthogonal valence-bond orbital bases, you may welcome a discussion of molecular orbital bases which are generally orthonormal. The simplest examples of molecular orbital (MO) states are the approximate H_2^+ (7.6.25), which we relabel below.

$$|\sigma_g\rangle = \frac{|a\rangle + |b\rangle}{\sqrt{2(1+S)}}, \quad |\sigma_u\rangle = \frac{|a\rangle - |b\rangle}{\sqrt{2(1-S)}}. \tag{7.6.41}$$

As long as normalized states $|a\rangle$ and $|b\rangle$ are related by some C_2-like operation **i**,

$$|b\rangle = \mathbf{i}|a\rangle, \quad |a\rangle = \mathbf{i}|b\rangle, \tag{7.6.42}$$

then the MO states will be orthonormal for all values of overlap S with $|S| < 1$. This is guaranteed by the C_2-projection orthogonality ($\mathbf{P}^u \mathbf{P}^g = 0$) as discussed in Chapter 2.

Thus one may choose a variety of trial MO states by redefining $|a\rangle$ or $|b\rangle$. Wave $(\mathbf{x}|a) = \alpha(\mathbf{x})$ may be a polarized mixture of $1s$, $2s$, and $2p$ orbitals, for example, to increase the electronic charge in the overlap region. Or one can simply vary the nuclear charge number Z to fatten $\alpha(x)$ until energy is minimized.

Molecular orbitals are used for heteronuclear diatomic molecules such as LiH or HCl and for polyatomic molecules. XY-diatomic molecular orbitals may have the orthogonal broken-symmetry form

$$|\sigma 1\rangle = \frac{|a\rangle + \lambda|b\rangle}{\sqrt{N_1}}, \quad |\sigma 2\rangle = \frac{\lambda|a\rangle - |b\rangle}{\sqrt{N_2}}, \tag{7.6.43}$$

where amplitude λ depends on the relative attraction of nucleus X versus that of Y.

Given the MO states (7.5.41) we assume the H_2^+ problem is solved and consider H_2. There are four two-electron MO product states:

$$|(\sigma_g)^2\rangle = |\sigma_g\rangle|\sigma_g\rangle, |(\sigma_u)^2\rangle = |\sigma_u\rangle|\sigma_u\rangle, |\sigma_u\sigma_g\rangle = |\sigma_u\rangle|\sigma_g\rangle,$$
$$|\sigma_g\sigma_u\rangle = |\sigma_g\rangle|\sigma_u\rangle. \tag{7.6.44}$$

These are expanded below using (7.6.43) and labeled using standard spectroscopic notation. Permutation symmetrization and antisymmetrization of the latter two are needed to give Pauli-allowed states:

$$|^1\Sigma_g(\sigma_g)^2\rangle = |\sigma_g\rangle|\sigma_g\rangle = [|aa) + |ab) + |ba) + |bb)]/2(1 + S), \quad (7.6.45a)$$

$$|^1\Sigma_g(\sigma_u)^2\rangle = |\sigma_u\rangle|\sigma_u\rangle = [|aa) - |ab) - |ba) + |bb)]/2(1 - S), \quad (7.6.45b)$$

$$|^1\Sigma_g\sigma_u\sigma_g\rangle = [|\sigma_u\rangle|\sigma_g\rangle + |\sigma_g\rangle|\sigma_u\rangle]/\sqrt{2} = [|aa) - |bb)]/\sqrt{2(1 - S^2)}, \quad (7.6.45c)$$

$$|^3\Sigma_u\sigma_u\sigma_g\rangle = [|\sigma_u\rangle|\sigma_g\rangle - |\sigma_g\rangle|\sigma_u\rangle]/\sqrt{2} = [|ab) - |ba)]/\sqrt{2(1 - S^2)}. \quad (7.6.45d)$$

They may also be labeled using Young tableau notation to compare with (7.6.35):

$$|^1\Sigma_g(\sigma_u)^2\rangle = \boxed{g\,g} = \left(\left[\boxed{a\,a} + \boxed{b\,b}\right]/\sqrt{2} + \boxed{a\,b}\right)/(1 + S)\sqrt{2},$$

$$|^1\Sigma_g(\sigma_u)^2\rangle = \boxed{u\,u} = \left(\left[\boxed{a\,a} + \boxed{b\,b}\right]/\sqrt{2} - \boxed{a\,b}\right)/(1 - S)\sqrt{2},$$

$$|^1\Sigma_g\sigma_u\sigma_g\rangle = \boxed{u\,g} = \left(\boxed{a\,a} - \boxed{b\,b}\right)/\sqrt{2(1 - S^2)},$$

$$|^3\Sigma_u\sigma_u\sigma_g\rangle = \boxed{\begin{array}{c}u\\g\end{array}} = \boxed{\begin{array}{c}a\\b\end{array}}/\sqrt{(1 - S^2)}. \quad (7.6.46)$$

A first approximation to MO eigenstates might involve the first $(\sigma_g)^2\ ^1\Sigma_g$ configuration with two electrons in the lowest σ_g orbital. However, an inspection of (7.6.45a) shows that this state is 50-50 mixture of ionic and covalent base states. This is a well-known weakness of the MO bases. They generally fail to account for electronic correlations, i.e., electrons' tendency to avoid each other. [The Heitler-London approximation erred in the opposite direction by completely ignoring the ionic states $|aa)$ and $|bb)$.]

A better MO approximation is obtained by mixing the $(\sigma_g)^2\ ^1\Sigma_g$ configuration with others. The H_2 symmetry allows it to mix with $(\sigma_u)^2\ ^1\Sigma_g$ but not with any of the other configurations listed in (7.6.45). This is a simple example of molecular *configuration interaction* (CI) and involves the eigensolutions of the following Hamiltonian matrix:

	$\lvert {}^1\Sigma_g(\sigma_g)^2\rangle$	$\lvert {}^1\Sigma_g(\sigma_u)^2\rangle$	$\lvert {}^1\Sigma_u\sigma_g\sigma_u\rangle$	$\lvert {}^3\Sigma_u\sigma_g\sigma_u\rangle$
$\lvert {}^1\Sigma_g(\sigma_g)^2\rangle$	$2\varepsilon_{1s} + \dfrac{-2A-2B}{1+S} + \dfrac{\tfrac{1}{2}(C+G)+\tfrac{1}{2}(D+E)+2F}{(1+S)^2}$	$\dfrac{\tfrac{1}{2}(C+G)-\tfrac{1}{2}(D+E)}{(1-S^2)^2}$	0	0
$\lvert {}^1\Sigma_g(\sigma_u)^2\rangle$	$\dfrac{\tfrac{1}{2}(C+G)-\tfrac{1}{2}(D+E)}{(1-S^2)^2}$	$2\varepsilon_{1s} + \dfrac{-2A+2B}{1-S} + \dfrac{\tfrac{1}{2}(C+G)+\tfrac{1}{2}(D+E)-2F}{(1-S)^2}$	0	0
$\lvert {}^1\Sigma_u\sigma_g\sigma_u\rangle$	0	0	$2\varepsilon_{1s} + \dfrac{-2A+2SB+C-G}{1-S^2}$	0
$\lvert {}^3\Sigma_u\sigma_g\sigma_u\rangle$	0	0	0	$2\varepsilon_{1s} + \dfrac{-2A+2SB+D-E}{1-S^2}$

(7.6.47)

To compare the H representation (7.6.47) in the molecular orbital (MO) basis with (7.5.40) in the valence-bond (VB) basis we consider extreme cases. Suppose the ionic Coulomb repulsion integrals C and G of (7.5.37) were so large that all other terms could be neglected. Then we diagonalize the MO submatrix

$$\langle H(^1\Sigma_g) \rangle_{MO} = \frac{1}{2} \begin{pmatrix} \dfrac{C+G}{(1+S)^2} & \dfrac{C+G}{1-S^2} \\ \dfrac{C+G}{1-S^2} & \dfrac{C+G}{(1-S)^2} \end{pmatrix} \quad (C \text{ and } G \text{ large}) \quad (7.6.48a)$$

to get the following eigenvalues and eigenvectors (not normalized):

$$\text{mo}_1 = 0; \quad |\text{mo}_1\rangle = (1+S)|(\sigma_g)^2\rangle - (1-S)|(\sigma_u)^2\rangle = [|ab) + |ba)],$$

$$\text{mo}_2 = \frac{(C+G)(1+S^2)}{(1+S^2)^2}; \quad |\text{mo}_2\rangle = (1-S)|(\sigma_g)^2\rangle + (1+S)|(\sigma_u)^2\rangle$$

$$= [|aa) + |bb)] \qquad (7.6.48b)$$

The VB covalent state $|^1\Sigma_g \text{ cov}\rangle$ and ionic state $|^1\Sigma_g \text{ ion}\rangle$ show up as eigenvectors. It is reassuring to get the same eigenvalues by diagonalizing the VB submatrix from (7.6.40):

$$\langle H(^1\Sigma_g) \rangle_{VB} = \begin{pmatrix} \dfrac{(1+S^2)(C+G)}{(1+S^2)^2} & 0 \\ \dfrac{-2S(C+G)}{(1-S^2)^2} & 0 \end{pmatrix} \quad (C \text{ and } G \text{ large}), \quad (7.6.49a)$$

which gives the following eigensolutions:

$$vb_1 = 0; \quad |vb_1\rangle = |^1\Sigma_g \text{ cov}\rangle = \frac{|ab) + |ab)}{\sqrt{2}},$$

$$vb_2 = \frac{(1+S^2)(C+G)}{(1-S^2)^2}; \quad |vb_2\rangle = (1+S^2)|^1\Sigma_g \text{ ion}\rangle - 2S|^1\Sigma_g \text{ cov}\rangle.$$

$$(7.6.49b)$$

The presence of $|^1\Sigma_g \text{ cov}\rangle$ in the second eigenvector is an artifact of the non-Hermitian VB representation.

The opposite extreme would be to have the single-electron integrals A and B of (7.6.37) large enough to ignore the others. Then the MO submatrix in (7.6.47) is diagonal:

$$\langle H(^1\Sigma_g) \rangle_{MO} = \begin{pmatrix} 2\varepsilon_{1s} - 2\dfrac{A+B}{1+S} & 0 \\ 0 & 2\varepsilon_{1s} - 2\dfrac{A-B}{1-S} \end{pmatrix}. \quad (7.6.50)$$

The eigenvalues are just twice the single-electron potential values (7.6.15a and 7.6.15b) without the nuclear interaction term k/R. The VB submatrix in (7.6.40) gives the same values after diagonalization.

A level correlation diagram between the two extremes is sketched in Figure 7.6.3. The two $^1\Sigma_g$ states which are changed by configuration interaction have levels which are drawn as curves. The other two states $^1\Sigma_u$ and $^3\Sigma_u$ do not change and their levels are indicated by straight lines. The diagram is sketched to show the effects of increasing ionic Coulomb integrals C and G on the left and single-electron integrals A and B on the right. The effect of moderate direct and exchange Coulomb integrals D and E would be to push

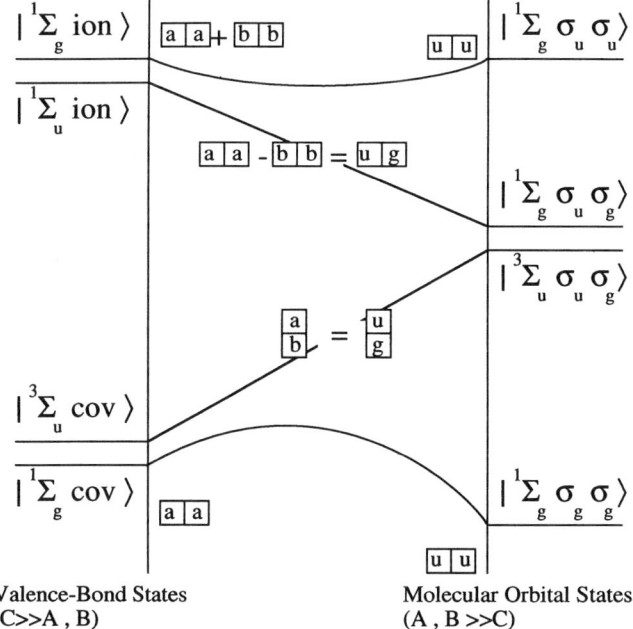

Figure 7.6.3 Sketch of level correlations between valence-bond and molecular orbital bases.

triplet $^3\Sigma_u$ below singlet $^1\Sigma_u$ on the right. Note, however, that Figure 7.6.3 is just a qualitative sketch and a template for the study of whichever parameters one chooses to vary.

B. How to "Point" Electronic Orbitals

So far we have considered scalar s waves and only sigma (σ or Σ) orbitals which have cylindrical symmetry and no component of angular momentum around the bond axis. To build up bonds for polyatomic molecules one needs to combine p, d, and f waves to make π, δ, and ϕ bonds. While σ bonds are the basic glue for most polyatomic molecules, the π bonds and higher-order waves can be important, as well.

We will explain the general method using an example of the octahedral SF_6 molecule discussed in Chapter 4. We will show how to make six equivalent orthogonal orbitals that point along the C_4 symmetric axes of an octahedron. We will apply the subgroup correlation and induced representation theory of Sections 4.2 and 4.3 and the orbital level splitting or crystal field theory of Sections 5.6 and 7.3C.

First we consider how some of the simplest s and p orbitals can make polyatomic molecular bonds or ligands.

(a) Elementary $s^n p^m$ Bonds
The simplest orbitals which "point" are the p orbitals $\{p_x, p_y, p_z\}$. They span the real vector representation R^1 of $R(3)$ [recall (5.6.19)]:

$$\langle \mathbf{r}|p_x\rangle = \frac{-Y_1^1 + Y_{-1}^1}{\sqrt{2}} r\sqrt{\frac{4\pi}{3}}, \quad \langle \mathbf{r}|p_y\rangle = i\frac{Y_1^1 - Y_{-1}^1}{\sqrt{2}} r\sqrt{\frac{4\pi}{3}},$$
$$= x, \qquad\qquad\qquad\qquad = y,$$

$$\langle \mathbf{r}|p_z\rangle = Y_0^1 r\sqrt{\frac{4\pi}{3}},$$
$$= z. \qquad\qquad\qquad\qquad\qquad\qquad (7.6.51)$$

A combination wave function of the form

$$\chi(\mathbf{r}) = a_0\langle \mathbf{r}|s\rangle + a_x\langle \mathbf{r}|p_x\rangle + a_y\langle \mathbf{r}|p_y\rangle + a_z\langle \mathbf{r}|p_z\rangle \qquad (7.6.52)$$

is a wave with a "finger" or *ligand* which points at an atom with coordinates (a_x, a_y, a_z). The relative amount a_0 of the scalar s wave $\langle \mathbf{r}|s\rangle = Y_0^0 f(r)$ determines the degree of polarization or pointing and can be used to adjust normalization and orthogonality with other waves.

A famous example of a set of hybrid molecular orbitals based on (7.6.52) is the set of methane (CH_4) σ bonds which point from carbon to the H atoms located at tetrahedral points $(1, 1, 1)$ $(1, -1, -1)$, $(-1, 1, -1)$, and

$(-1, -1, 1)$, respectively, which form T_d symmetry (recall Figure 4.1.5):

$$|C_1\rangle = (|s\rangle + |p_x\rangle + |p_y\rangle + |p_z\rangle)/2,$$
$$|C_2\rangle = (|s\rangle + |p_x\rangle - |p_y\rangle - |p_z\rangle)/2,$$
$$|C_3\rangle = (|s\rangle - |p_x\rangle + |p_y\rangle - |p_z\rangle)/2,$$
$$|C_4\rangle = (|s\rangle - |p_x\rangle - |p_y\rangle + |p_z\rangle)/2. \quad (7.6.53)$$

This is called the sp^3 hybrid set, since one part $|s\rangle$ and three parts $|p\rangle$ are used to make each state. Carbon's ground configuration is $2s^2 2p^2$. (Recall Figure 7.1.2.) To make states (7.6.53), one $2s$ electron must be moved to a $2p$ orbit. The $2s$-$2p$ energy increase is made up by reduced electronic Coulomb repulsion and localization in order for CH_4 to be stable.

The simplest planar bonding orbitals for H_2O and other XY_2 molecules have the following hybrid form:

$$|X_1\rangle = \lambda|s\rangle + a \cos\phi|p_x\rangle + a \sin\phi|p_y\rangle,$$
$$|X_2\rangle = \lambda|s\rangle + a \cos\phi|p_x\rangle - a \sin\phi|p_y\rangle,$$
$$|X_3\rangle = \lambda'|s\rangle - a'|p_x\rangle. \quad (7.6.54)$$

This gives two bonds at angle $\pm\phi$ with the x axis and a third pointing along the $-x$ axis. For the first two to be orthonormal we must have

$$\lambda^2 + a^2 = 1, \quad \lambda^2 + a^2 \cos^2\phi - a^2 \sin^2\phi = 0.$$

Solving gives

$$a^2 \sin^2\phi = \frac{\sin^2\phi}{1 - \cos 2\phi} = \frac{1}{2}, \quad \lambda^2 = 1 - a^2 = \frac{\cos 2\phi}{\cos 2\phi - 1}. \quad (7.6.55)$$

The angle 2ϕ between the first two bonds must be at least $90°$ in order for the solutions to be useful. At $2\phi = 90°$ there are two orthogonal p waves and no s-wave admixture ($a = 1$, $\lambda = 0$). The third wave must satisfy the following to be orthogonal to the other two:

$$\lambda' = \sqrt{\frac{1 + \cos 2\phi}{1 - \cos 2\phi}}, \quad a' = -\sqrt{2}\lambda = -\sqrt{\frac{-2\cos 2\phi}{1 - \cos 2\phi}}. \quad (7.6.56)$$

For a molecule with bond angle $2\phi = 90°$ we have only two vector ligands and a separate orthogonal scalar wave. We indicate these states in the following using notation $(s^{\lambda^2} p^{a^2})$, where the exponents are the probabilities for each orbital:

$$|X_1(p^1)\rangle = \frac{1}{\sqrt{2}}|p_x\rangle + \frac{1}{\sqrt{2}}|p_y\rangle,$$

$$|X_2(p^1)\rangle = \frac{1}{\sqrt{2}}|p_x\rangle - \frac{1}{\sqrt{2}}|p_y\rangle,$$

$$|X_3(s^1)\rangle = |s\rangle. \qquad (7.6.57)$$

The atomic configuration for these states is $|x_1 p\rangle$, $|x_2 p\rangle$, and $|x_3 s\rangle$. The third wave is 100% s.

For a molecule with bond angle $2\phi = 120°$ we have three equivalently polarized ligands which could form a C_{3v} symmetric bonding. Each of these is a balanced sp^2 atomic configuration:

$$|X_1(s^{1/3}p^{2/3})\rangle = \frac{1}{\sqrt{3}}|s\rangle + \frac{1}{\sqrt{6}}|p_x\rangle + \frac{1}{\sqrt{2}}|p_y\rangle,$$

$$|X_2(s^{1/3}p^{2/3})\rangle = \frac{1}{\sqrt{3}}|s\rangle + \frac{1}{\sqrt{6}}|p_x\rangle - \frac{1}{\sqrt{2}}|p_y\rangle,$$

$$|X_3(s^{1/3}p^{2/3})\rangle = \frac{1}{\sqrt{3}}|s\rangle - \frac{2}{\sqrt{6}}|p_x\rangle. \qquad (7.6.58)$$

Finally, a T-shaped molecule with bond angle $2\phi = 180°$ could use the following sp, sp, and p configurations.

$$|X_1(s^{1/2}p^{1/2})\rangle = \frac{1}{\sqrt{2}}|s\rangle + \frac{1}{\sqrt{2}}|p_y\rangle,$$

$$|X_2(s^{1/2}p^{1/2})\rangle = \frac{1}{\sqrt{2}}|s\rangle - \frac{1}{\sqrt{2}}|p_y\rangle,$$

$$|X_3(p^1)\rangle = -|p_x\rangle. \qquad (7.6.59)$$

Now the third wave is 100% p.

Let us see how an XY_2 molecule like H_2O might be bonded. We will put one electron in each of the X_1 and X_2 states so each may form a σ-pair bond with an electron from one of the Y atoms. We will put two electrons in the X_3 state to make an inert ^1S pair since it will not be used for bonding. This is called a "lone pair." There is another lone pair composed of the third p_z orbital perpendicular to the xy plane which we have ignored. (It is used for π bonding in molecules like ethylene and acetylene.) This soaks up a pair of electrons, too. So the atomic orbital configuration needed is the following:

$$x_1 x_2 x_3^2 p_z^2 = \left(s^{\lambda^2} p^{a^2}\right)\left(s^{\lambda^2} p^{a^2}\right)\left(s^{2\lambda'^2} p^{2a'^2}\right) p^2$$
$$= s^{2\lambda^2 + 2\lambda'^2} p^{2a^2 + 2a'^2 + 2}. \qquad (7.6.60)$$

We use (7.6.55) and (7.6.56) to give the exponents in terms of the bonding angle 2ϕ:

$$x_1 x_2 x_3^2 p_z^2 = s^{2/(1-\cos 2\phi)} p^{(4-6\cos 2\phi)/(1-\cos 2\phi)}$$
$$= \begin{cases} s^2 p^4, & \text{for } 2\phi = 90°, \\ s^1 p^5, & \text{for } 2\phi = 180°. \end{cases} \qquad (7.6.61)$$

Atomic oxygen has a ground configuration of $s^2 p^4$. If it stayed that way for H_2O the water bonding angle would be 90° instead of the observed 104°. The observed value corresponds to a bonding configuration of $s^{1.61} p^{4.39}$. Apparently, H_2O gains more energy by increasing its bond angle than it loses by raising 20% of the $2s$ population into a $2p$ orbital.

(b) π Bonds An example which appears to use the sp^2 model bond states (7.6.58) is the ethylene molecule C_2H_4 sketched in Figure 7.6.4(a). The angle between adjacent H atoms is close to 120°. However, C_2H_4 has additional stability due to the overlap between carbon "lone-pair" p orbitals perpendicular to each CH_2 plane. Their overlap is greatest when the CH_2 triangles are aligned. This gives a torsional stability to the complex which the $\sigma(sp_x)$ bond could never supply. The bond formed by an adjacent parallel p orbitals is called a π bond.

In the linear molecule C_2H_2 (acetylene) sketched in Figure 7.6.4(b) both transverse p orbitals participate in π bonds. These together with the $\sigma(sp)$ orbitals form what is called a triple bond.

The 120° sp^2 bonding combined with π bonds plays a role in quite a number of hydrocarbons. First among these are the *trans* and *cis* structures of butadiene C_4H_6 shown in Figure 7.6.5(a) and 7.6.5(b) and the well-known benzene molecule C_6H_6 shown in Figure 7.6.5(c). The structural pictures are generally drawn with the double bond localized between alternating pairs of carbon atoms. In fact, the π bonds tend to be delocalized and spread out over the carbon chains. This is called *conjugation* of π bonds and provides additional stability to the hydrocarbons.

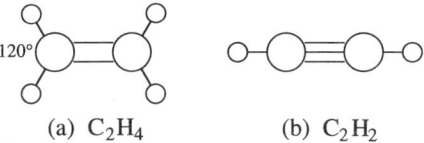

Figure 7.6.4 Examples of sp^2 bonding (a) Ethylene, (b) acetylene.

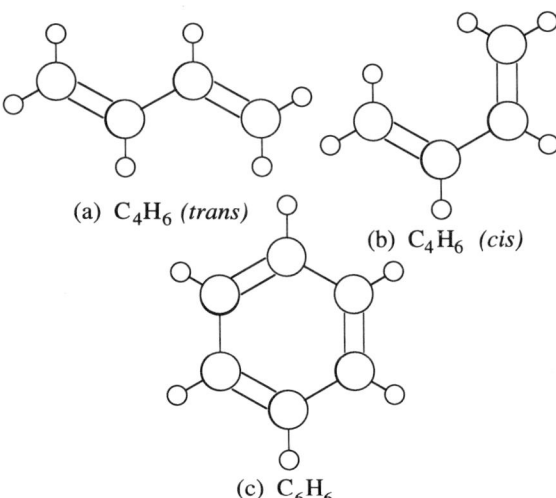

Figure 7.6.5 More examples of sp^2 bonding (a) *trans*-butadiene, (b) *cis*-butadiene, and (c) Benzene.

The increase in stability can be estimated using the Hückel model of conjugated bonds. The Hückel approximation uses a basis of n equivalent π orbitals; one on each of the n C atoms. The Hamiltonian is assumed to have only diagonal matrix elements and off-diagonal matrix elements between nearest-neighbor carbons only. The diagonal matrix elements are all equal to H and off-diagonal matrix elements equal to $-S$. This is precisely the form of the C_n symmetric tunneling problem discussed in Section 2.12 [recall (2.12.23)] and again in Section 3.6 (recall Figure 3.6.5).

For C_6H_6 the symmetry is $C_6 \subset C_{6v} \subset D_{6h}$ and the energy levels are given by (2.12.25), which is now repeated:

$$\varepsilon_k = H - 2S \cos k_n \left(k_n = \frac{2\pi k}{n}; k = 0, \pm 1, \ldots, \leq \frac{n}{2} \right).$$

The π levels labeled A_1, E_1, E_2, B_1 have energies

$$\varepsilon_0^{A_1} = H - 2S,$$
$$\varepsilon_1^{E_1} = H - S,$$
$$\varepsilon_2^{E_2} = H + S,$$
$$\varepsilon_3^{B_1} = H + 2S.$$

Given one electron for each π bond the orbitals will take one spin ($\uparrow \downarrow$) pair in A_1 and two pairs in E_1 for a total energy of $2(H - 2S) + 4(H - S) = H - 8S$. If the bonds had been three localized double bonds (recall Figure 7.6.5) the energy would be $6H - 6S$, so the conjugation gains $2S$ of stability.

The butadiene π orbitals can be modeled in the same way using a trick discussed in Problem 2.7.4. n-equivalent atoms in an open chain can be imagined to occupy one side of a $2n + 2$-member closed ring. [The n standing sine wave solutions will be eigensolutions of the n-atom chain.] The four butadiene π orbitals will use the middle four levels of a C_{10} symmetric ring.

$$\varepsilon^{E_1} = H - 2S\cos(2\pi/10) = H - S(1 + \sqrt{5})/2,$$
$$\varepsilon^{E_2} = H - 2S\cos(4\pi/10) = H - S(\sqrt{5} - 1)/2,$$
$$\varepsilon^{E_3} = H - 2S\cos(6\pi/10) = H + S(\sqrt{5} - 1)/2,$$
$$\varepsilon^{E_4} = H - 2S\cos(8\pi/10) = H + S(1 + \sqrt{5})/2.$$

One ↑↓ pair of electrons goes in each of the lowest two states E_1 and E_2. (Neither level is degenerate now since only sine waves are allowed.) This yields an energy of $4H - 2\sqrt{5}\,S$. Two pairs of localized π bonds would have had energy $4H - 4S$ so the conjugated bonds are $(2\sqrt{5} - 4)S = 0.047S$ better.

(c) Octahedral Bonding The s and p waves can bond no more than four atoms to a central one. Molecules like XY_6 need higher-order waves. We discuss some general symmetry techniques for setting up complex bonding configurations.

The local symmetry of each bond is the first thing to consider. If we desire to place bonds along each tetragonal (C_4) axis of octahedron then C_4 or C_{4v} is the relevant local symmetry group. If we desire σ bonds then their local symmetry belongs to the 0_4 irrep of C_4 or $A' = A_1$ irrep of C_{4v}. Since σ bonds are axially symmetric they belong to the scalar irreps.

If we need to install a pair of $\pi_{\pm 1}$ bonds on each C_4 axis then we would use the $1_4 \oplus 3_4$ irreps of C_4 or the E irrep of C_{4v} to locally label each wave state. The π pair of p_x and p_y transform like x and y bases of the E irrep of C_{4v}.

Next we use the induced representation bases to label the set of equivalent orthonormal bond states that we imagine exist on each C_4 axis. The σ bonds span a six-dimensional $0_4 \uparrow O$ representation induced to octahedral group O or $A' \uparrow O_h$. The π bonds span a 12-dimensional $(1_4 \oplus 3_4) \uparrow O$ or $E \uparrow O_h$ representation.

According to the Frobenius reciprocity theorem described in Section 4.3C the columns of the $C_4 \subset O$ or $C_{4v} \subset O_h$ correlation tables give the octahedral irreps that belong to each induced representation. The first (0_4) column of the table of (4.2.42b) gives

$$0_4 \uparrow O = A_1 \oplus T_1 \oplus E, \tag{7.6.62a}$$

and this is corroborated by the first A' column of (4.2.46c):

$$A' \uparrow O_h = A_{1g} \oplus T_{1u} \oplus E_g. \tag{7.6.62b}$$

680 THEORY AND APPLICATION OF SYMMETRY REPRESENTATION PRODUCTS

The latter includes the inversion-parity labels for the three O_h symmetry species that comprise an XY_6 σ-bonding model.

A π-bonding model of XY_6 would involve the $E \uparrow O_h$ states in the last column of (4.2.46c):

$$E \uparrow O_h = T_{1g} \oplus T_{2g} \oplus T_{1u} \oplus T_{2u}. \quad (7.6.63)$$

The O_h parity labels are particularly important here to distinguish pairs of O irreps.

The final step is the correlation of whichever O_h species we just found with angular-momentum states J^p and the corresponding $O(3)$ species. The columns of (5.6.5b) provide an infinite set of all the atomic orbital states that could possibly contribute to the bonding orbital in question:

$$(A_{1g} \text{ of } O_h) \uparrow O(3) = 0^+ \oplus 4^+ \oplus 6^+ \oplus \cdots = s_g \oplus g_g \oplus i_g \oplus \cdots,$$
$$(T_{1u} \text{ of } O_h) \uparrow O(3) = 1^- \oplus 3^- \oplus 2(5^-) \oplus \cdots = p_u \oplus f_u \oplus 2h_u \oplus \cdots,$$
$$(E_g \text{ of } O_h) \uparrow O(3) = 2^+ \oplus 4^+ \oplus 6^+ \oplus \cdots = d_g \oplus g_g \oplus i_g \oplus \cdots. \quad (7.6.64)$$

As a first approximation we take only the first contributions for each species. In this case we make the singlet A_{1g} using an s wave, the triplet T_{1u} using three p_u waves, and the doublet E_g using two d_g waves. That will be an sp^3d^2 atomic configuration of six octahedrally coordinated ligand orbitals. Later you may find it desirable to "sharpen up" the p ligands with an f_u^3 or h_u^3 configuration, or add some g_g to the s_g and d_g states. If so, the correlation table tells what will work.

The calculation of the states in the octahedral sp^3d^2 configuration follows the steps outlined. The induced representation is labeled and reduced according to Sections 4.3.A and 4.3.B. The resulting A_1 state (4.3.20), E states (4.3.22), and (4.3.23), and T_1 states (4.3.27a)–(4.3.27c) are tabulated backwards so as to give the desired six ligand states $\{|1\rangle, \ldots, |6\rangle\}$

$$|1\rangle = \frac{1}{\sqrt{6}}|A_1\rangle + \frac{1}{\sqrt{3}}|E,1\rangle + \frac{1}{\sqrt{2}}|T_1,z\rangle,$$

$$|2\rangle = \frac{1}{\sqrt{6}}|A_1\rangle + \frac{1}{\sqrt{3}}|E,1\rangle - \frac{1}{\sqrt{2}}|T_1,z\rangle,$$

$$|3\rangle = \frac{1}{\sqrt{6}}|A_1\rangle - \frac{1}{2\sqrt{3}}|E,1\rangle + \frac{1}{2}|E,2\rangle + \frac{1}{\sqrt{2}}|T_1,x\rangle,$$

$$|4\rangle = \frac{1}{\sqrt{6}}|A_1\rangle - \frac{1}{2\sqrt{3}}|E,1\rangle + \frac{1}{2}|E,2\rangle - \frac{1}{\sqrt{2}}|T_1,x\rangle,$$

$$|5\rangle = \frac{1}{\sqrt{6}}|A_1\rangle - \frac{1}{2\sqrt{3}}|E,1\rangle - \frac{1}{2}|E,2\rangle + \frac{1}{\sqrt{2}}|T_1,y\rangle,$$

$$|6\rangle = \frac{1}{\sqrt{6}}|A_1\rangle - \frac{1}{2\sqrt{3}}|E,1\rangle - \frac{1}{2}|E,2\rangle - \frac{1}{\sqrt{2}}|T_1,y\rangle. \quad (7.6.65)$$

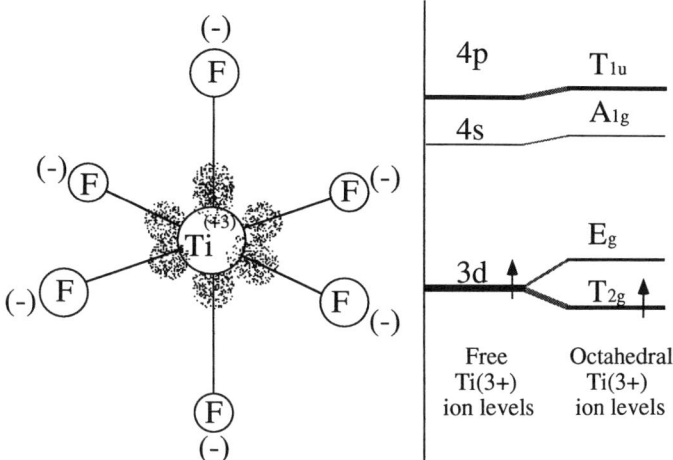

Figure 7.6.6 Octahedral crystal field splitting of 3d level of Ti^{3+} ion.

Finally, one approximates each A_1, E, or T_1 state using combinations of atomic orbitals allowed by correlations such as (7.6.64). For the simplest sp^3d^2 configuration we use combinations as derived in Section 5.6; see (5.6.15d) and (5.6.15e) for E states and the p-orbital relations (7.5.51):

$$|A_1\rangle = |s_0^0\rangle,$$

$$|E,1\rangle = |d_0^2\rangle = (2|d_{z^2}\rangle - |d_{x^2}\rangle - |d_{y^2}\rangle)/\sqrt{6},$$

$$|E,2\rangle = (|d_2^2\rangle + |d_{-2}^2\rangle)/\sqrt{2} = (|d_{x^2}\rangle - |d_{y^2}\rangle)/\sqrt{2},$$

$$|T_1,x\rangle = (-|p_1^1\rangle + |p_{-1}^1\rangle)/\sqrt{2} = |p_x\rangle,$$

$$|T_1,y\rangle = (i|p_1^1\rangle - i|p_{-1}^1\rangle)/\sqrt{2} = |p_y\rangle,$$

$$|T_1,y\rangle = |p_0^1\rangle = |p_z\rangle.$$

For an example of octahedral electronic structure theory we consider a $[Ti^{3+}(Ion^-)_6]$ ionic complex. The Ti atom has a $3d^24s^2$ valence configuration so the ion Ti^{3+} has a single d electron. Suppose it is surrounded by an octahedron of negatively charged ions as in Figure 7.6.6.

If the electrons on the negative ions do not overlap appreciably with the Ti d electron we imagine that latter undergoes a crystal field splitting as described in Section 5.6 (recall Figure 5.6.3). This would be a ground-state spin doublet and orbital triple configuration of $^2t_{2g}$ as shown on the right-hand side of Figure 7.6.6. The magnitude of this splitting was calculated in Section 7.3 [recall (7.3.28)].

If there is electronic overlap, as indeed there must be if the complex is bound, then a more sophisticated molecular orbital picture is needed. This is

682 THEORY AND APPLICATION OF SYMMETRY REPRESENTATION PRODUCTS

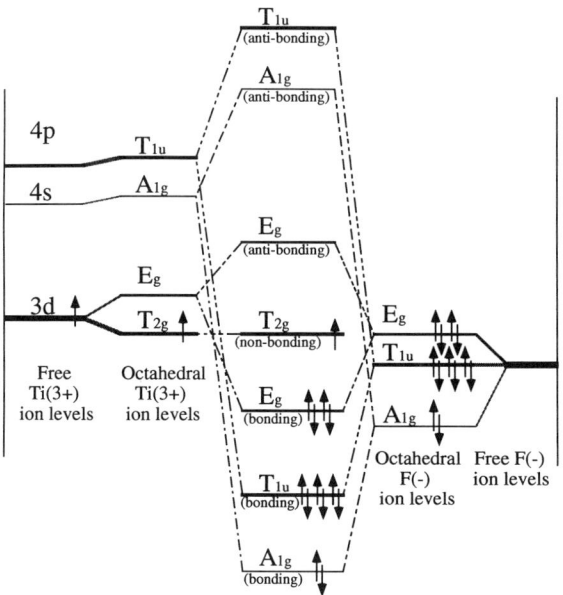

Figure 7.6.7 Bonding and antibonding molecular orbital levels for $[TiF_6]^{3+}$ complex.

sketched in Figure 7.6.7. Here we imagine a coupling between certain combinations of the six σ orbitals and the Ti atomic orbitals of the same symmetry. The six σ orbitals span the induced representation $A' \uparrow O_h = A_{1g} \oplus T_{1u} \oplus E_g$ discussed in the preceding section. Each state from the latter interacts with the nearest Ti level belonging to the same irrep to form a pair of bonding and antibonding orbitals.

Six ($\uparrow\downarrow$) pairs of electrons are needed to give the most stable σ bonding arrangement. If another electron is available it would go in the T_{2g} nonbonding level. From there it could be excited to the antibonding orbitals E_g, A_{1g} of T_{1u} which lie above it. In this sense the picture in Figure 7.6.7 is qualitatively similar to the one in Figure 7.6.6, which neglects the bonding altogether. Inclusion of the electronic bonding structure is necessary to predict the optical and magnetic properties of these complexes.

ADDITIONAL READING

Most of the references given at the end of Chapter 5 treat angular momentum Clebsch-Gordon coefficients and the Wigner-Eckart theorem. In addition there is a well-known text which introduces the idea of the irreducible density operators.

U. Fano and G. Racah, *Irreducible Tensorial Sets* (Academic, New York, 1959).

The derivation of raising and lowering operator matrix elements is the main recurring problem in applied representation theory. The following are a few references that involve $O(n)$ and $U(n)$.

I. M. Gelfand and M. L. Zetlin, *Dokl. Akad. Nauk USSR*, **71**, 1017 (1950).

J. D. Louck and L. C. Biedenharn *J. Math. Phys.*, **11** 2368 (1970).

J. C. Nagel and M. Moshinsky, *J. Math. Phys.*, **6**, 682 (1965).

A. Bincer, *J. Math. Phys.*, **19**, 1173 (1978); **18**, 1870 (1977).

Tables in Clebsch-Gordon coefficients in the Wigner $3j$ form and Racah coefficients in the 6-j form are given in Rotenburg et al. from $J_1 \otimes J_2 = \frac{1}{2} \otimes \frac{1}{2}$ up to $8 \otimes 8$.

M. Rotenburg, R. Bivens, N. Metropolis, and J. K. Wooten, *The 3-j and G-j Symbols* (Technology Press, MIT, 1959).

This book usually is out of print. It is probably better to have coefficients of this sort directly available on your own computer. They can be found fairly easily in numerical or algebraic form in various packages of the *Mathematica* program which is now widely available. (See also W. J. Thompson, Computers in Physics 7, 144 (1993).)

S. Wolfram, *Mathematica: A System for Doing Mathematics by Computer* (Addison-Wesley, Redwood City, CA, 1991).

The asymptotic behavior of Clebsch-Gordon coefficients is explained in the following paper.

K. Schulten and R. G. Gordon, *J. Math. Phys.*, **16**, 1961 (1975).

The first application of Racah tensor analysis to fine spectral structure of methane (CH_4) was by K. T. Hecht and Moret-Bailly.

K. T. Hecht, *J. Mol. Spetrosc.*, **5**, 355 (1960).

J. Moret-Bailly, *Cah. Phys.*, **15**, 237 (1961); *Cah. Phys.*, **178**, 253 (1965).

The first observation of extraordinary tensor eigenvalue degeneracy or clustering involved computer studies of fourth and sixth rank cubic crystal field operators

K. R. Lea, M. J. M. Leask, and W. P. Wolf, *J. Phys. Chem. Solids*, **23**, 1381 (1962).

Computer studies of CH_4 eigenvalues and forbidden transitions to rotational states lead to the next discovery of clustering and the first molecular examples.

A. J. Dorney and J. K. G. Watson, *J. Mol. Spectrosc.*, **42**, 1 (1972).

The most extensive numerical study of tensor eigenvalues and clustering phenomena was published by Fox, Galbraith, Krohn, and Louck.

K. Fox, H. W. Galbraith, B. J. Krohn, and J. D. Louck, *Phys. Rev. A*, **15**, 1363 (1977).

This in turn lead to a semiclassical and quantum mechanical theory for clusters and their superfine structure.

W. G. Harter and C. W. Patterson, *Phys. Rev. Let.*, **38**, 244 (1977).

W. G. Harter and C. W. Patterson, *J. Chem. Phys.*, **66**, 4872 (1977).

C. W. Patterson and W. G. Harter, *J. Chem. Phys.*, **66**, 4866 (1977).

The RE surface picture of the rotor dynamics and spectral structure was developed later. It was first applied to the levels of combined fourth- and sixth-rank tensors of cubic symmetry.

W. G. Harter and C. W. Patterson, *J. Math. Phys.*, **20**, 1453 (1979).

The semiclassical theory of RE surfaces is introduced in the following papers.

W. G. Harter, *Phys. Rev. A*, **24**, 192 (1981).

W. G. Harter and C. W. Patterson, *J. Chem. Phys.*, **80**, 4241 (1984).

A more recent review of semiclassical rotor mechanics and RE surface applications is the following.

W. G. Harter, *Computer Phys. Reps.*, **8**, 319 (1988).

Experimental observations of molecular fine structure, superfine structure, and hyperfine structure were done by developing unique spectroscopic instruments and laser devices.

CH_4:

A. S. Pine, *J. Opt. Soc. Am.*, **66**, 97 (1976).

C_8H_8:

A. S. Pine, A. G. Maki, A. G. Robiette, B. J. Krohn, J. K. G. Watson, and Th. Urbanek, *J. Am. Chem. Soc.*, **106**, 891 (1984).

SF_6:

J. P. Aldridge, H. Filip, H. Flicker, R.f. Holland, R. S. McDowell, N. G. Nereson, and K. Fox, *J. Mol. Spectrosc.*, **58**, 165 (1975).

R. S. McDowell, H. W. Galbraith, C. D. Cantrell, N. G. Nereson, and E. D. Hinkley, *J. Mol. Spectrosc.*, **68**, 288 (1977).

K. C. Kim, W. P. Person, D. Seitz, and B. J. Krohn, *J. Mol. Spectrosc.*, **76**, 322 (1979).

J. Bordé and Ch. J. Bordé, *Chem. Phys.*, **71**, 417 (1982).

The following are recent references which treat the rotor-oscillator analogy and related applications of R(3)-SU(2) coordinates.

W. G. Harter and N. dos Santos, *Am. J. Phys.*, **46**, 251 (1978).

M. E. Kellman, *J. Chem. Phys.*, **76**, 4528 (1982); *J. Chem. Phys.*, **83**, 3843 (1985); *Chem. Phys. Lett.*, **113**, 489 (1985).

K. K. Lehmann, *J. Chem. Phys.*, **79**, 1098 (1983).

W. G. Harter, *J. Chem. Phys.*, **85**, 5560 (1986).

Z. Li, L. Xiao, and M. E. Kellman, *J. Chem. Phys.*, **92**, 2251 (1990).

Two of the original papers on SU(2)-R(3) spin vector models are the following:

I. I. Rabi, N. F. Ramsey, and J. Schwinger, *Rev. Mod. Phys*, **26**, 167 (1954).

R. P. Feynman, F. l. Vernon Jr., and R. W. Hellwarth, *J. Appl. Phys.*, **28**, 49 (1957).

The following are original works on spinors, quaternions, and applications to optics:

W. R. Hamilton, *Lecture on Quaternions* (Dublin, 1853).

G. Stokes, *Proc. Soc. London*, **11**, 547 (1862).

H. Poincaré, *Theorie Mathematique de la Lumiere* (Gauthiers Villars, Paris, 1892).

There is a computer program which demonstrates the U(2)-R(3) analysis of oscillators and polarization states.

W. G. Harter, Color U(2): A Study of Classical and Quantum Resonance Phenomena, Department of Physics, University of Arkansas, Fayettevile, AR 72701.

Development of U(2) vibrational analysis as part of the vibron model is described in the following papers.

O. S. van Roosmalen, F. Iachello, I. Benjamin, R. D. Levine, and A. E. L. Dieperinh, *J. Chem. Phys.*, **79**, 2515 (1983).

O. S. van Roosmalen, I. Benjamin and R. D. Levine, *J. Chem. Phys.*, **81**, 5986 (1984).

The following article reviews the applications of semiclassical geometry and RE surfaces to other problems such as atomic diamagnetism.

T. Uzer, D. Farrelly, J. A. Milligan, P. E. Raines, and J. P. Skelton, *Science*, **253**, 42 (1991).

One of the most readable introductions to molecular electronic orbital theory is by Atkins. More advanced references are contained therein.

P. W. Atkins, *Molecular Quantum Mechanics*, 2nd edition (Oxford University Press, Oxford and New York, 1983).

CHAPTER 8

SYMMETRY ANALYSIS FOR SEMICLASSICAL AND QUANTUM MECHANICS: DYNAMICS WITH HIGH QUANTA

The classical world with all its detailed motion should in principle be described in terms of basic quantum states. In practice, however, the detailed correspondence between classical and quantum descriptions can be fairly subtle and complex. In Chapters 3, 4 and 7 we considered some elementary examples of classical spontaneous symmetry breaking. There, certain combinations of eigenstates were shown to be represented by localized wave packets that corresponded to systems being trapped into quasiclassical configurations with lower symmetry. In this chapter some theory involving wave packets and so-called coherent states will be developed in order to clarify the connection between quantum and classical phenomena. The main application of these theories will be to systems with high quantum numbers. The resulting methodology is loosely referred to as SEMICLASSICAL mechanics.

As we have already noted in Chapters 5–7, states with high quanta (particularly angular quanta with J greater than 5 or 10) can be very complicated, and computations involving them can be extremely laborious. Often this means that the problem is treated numerically and exact eigensolutions are found by computer diagonalization. However, large-scale numerical solutions may not expose interesting physical phenomena or lead one directly toward a better theoretical understanding. One should not be content to just have a computer experiment that parrots some laboratory spectra.

Furthermore, there should be much more to quantum mechanics than the study of individual eigenstates. An eigenfunction is stationary in the sense that only its overall phase ($\psi_m(t) = e^{-iE_m t/\hbar}\psi_m(0)$) is time dependent, while its probability distribution ($\psi_m^*\psi_m$) is forever frozen. By studying individual eigenstates you learn all the ways that a quantum system can play dead! Only by combining two or more eigenstates with different energy phase factors can you get something to actually move. The rate of such motion is determined by the energy differences $(E_m - E_n)$, that is, transition or "beat" frequencies $(\omega_{mn} = (E_m - E_n)/\hbar)$, as explained in Chapter 2, Sections 2.3.A and 2.3.B. In fact, a single eigenstate is really unobservable. Only through combinations of eigenstates or transitions between them can a quantum system exhibit change or dynamics.

Classical mechanics, on the other hand, seems better equipped to describe dynamics or motion. This is because one is better able to visualize how classical objects move even if the equations of motion are difficult to solve analytically. An important part of semiclassical mechanics is to provide ways to visualize and understand quantum dynamics and to compare it to the classical dynamics which approximates it in the limit of high quantum numbers. We shall compare quantum and semiclassical theory for vibrational and rotational dynamics. This will include applications to atomic and molecular spectroscopy. Also, we shall introduce and compare quantum and semiclassical theories of radiation which are fundamental to the theory of spectroscopy in general.

8.1 CONTACT TRANSFORMATIONS, ACTIONS, AND SEMICLASSICAL WAVE FRONTS

The principles involving the action functions are introduced and it is shown how they enter the study of semiclassical dynamics. This treatment includes ways to visualize action transformations geometrically.

A. Contact Transformations

Consider a curve $y(x)$ in a two-dimensional coordinate space (x, y) as shown in Figure 8.1.1(a). This curve may be related to another curve $Y(X)$ in a (generally) different space (X, Y) by what is called a CONTACT TRANSFORMATION. To define a contact transformation one ultimately requires what is called a GENERATOR function $S(x, y : X, Y)$. Then for a fixed value of the generator [say $S(x, y : X, Y) = 10$] one generates a family of curves in the (X, Y) space as shown in Figure 8.1.1. There is one *curve* $S(x_j y_j : X, Y) = 10$ in the (X, Y) space for each *point* (x_j, y_j) on the curve $y(x)$. The envelope(s) or contacting curve(s) of this family comprise the desired contact transformation(s) $Y(X)$ for a particular value of the generator. (Here $S = 10$.) A schematic example of a family and its contact curve are shown in Figure 8.1.1(b).

As we have said, each point $(x, y(x))$ is associated with a curve in (X, Y) space. In addition we shall associate each point $(x_j, y(x_j))$ with the contact point $(X_j, Y(X_j))$ where that curve is tangent to the family envelope $Y(X)$. The points $(X_j, Y(X_j))$ are the ones for which the value of $S(x, y(x): X, Y)$

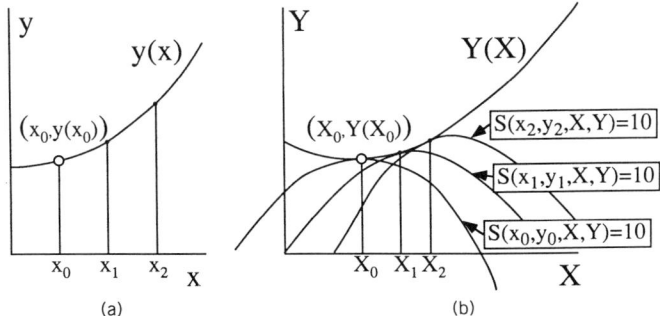

Figure 8.1.1 Geometry of a general contact transformation.

is least sensitive to a small change in x. From Figure 8.1.1(b) you can see that a change of x causes the $S = 10$ curve to slide along the envelope. As the new contact point slides a small distance away from the old contact point $(X, Y(X))$, the latter is still practically on top of the new $S = 10$ curve; that is, S does not change at first near the point $(X, Y(X))$. Therefore we find these points by solving

$$\left.\frac{\partial S(x, y(x) : X, Y)}{\partial x}\right|_{x=x_j} = 0 \tag{8.1.1a}$$

and

$$S(x, y(x) : X, Y) = \text{constant}. \tag{8.1.1b}$$

One should notice that a contact transformation goes either way; each point $(X_j, Y(X_j))$ generates a curve tangent to the $y(x)$ curve at $(x_j, y(x_j))$. Then the following equation would be applicable, too:

$$\left.\frac{\partial S(x, y : X, Y)}{\partial X}\right|_{X=X_j} = 0. \tag{8.1.1c}$$

The LEGENDRE TRANSFORMATION is an example of a contact transformation in which the transformed curve contacts or envelopes a family of straight lines. In its simplest form each point $(x_j, y_j = y(x_j))$ generates a line $Y = x_j X - y_j$ in (X, Y) coordinates, i.e., a line of slope x_j and Y intercept $-y_j$ as shown in Figure 8.1.2. This is equivalent to having the generator relation

$$S(x, y : XY) = y + Y - xX = 0.$$

CONTACT TRANSFORMATIONS, ACTIONS, AND SEMICLASSICAL WAVE FRONTS

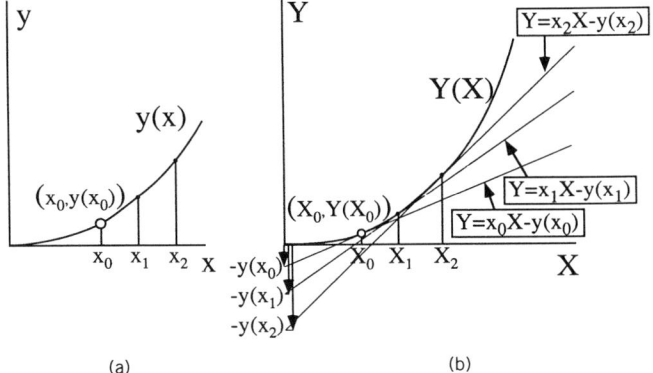

Figure 8.1.2 Geometry of a Legendre contact transformation.

From the derivative equations (8.1.1a) and (8.1.1c) we find

$$\frac{\partial S}{\partial x} = 0, \quad X = \frac{\partial y}{\partial x}, \tag{8.1.2a}$$

$$\frac{\partial S}{\partial X} = 0, \quad x = \frac{\partial Y}{\partial X}. \tag{8.1.2b}$$

This is combined with the generator relation $S = 0$ or

$$Y = xX - y \tag{8.1.2c}$$

to yield the desired Legendre transformation.

One of the best known examples of a Legendre transformation is the transformation between the Lagrangian function $y(x) \equiv L(\dot{q})$ and the Hamiltonian function $Y(X) \equiv H(p)$. Here the independent variables are velocity $(x \equiv \dot{q})$ and momentum $(X \equiv p)$. The transformation equations (8.1.2a)–(8.1.2c) yield the relations

$$p = \frac{\partial L}{\partial \dot{q}}, \tag{8.1.3a}$$

$$\dot{q} = \frac{\partial H}{\partial p}, \tag{8.1.3b}$$

and

$$H(p) = p\dot{q} - L, \tag{8.1.3c}$$

respectively. The slope $\dot{q}$ of the contacting lines which are indicated in Figure

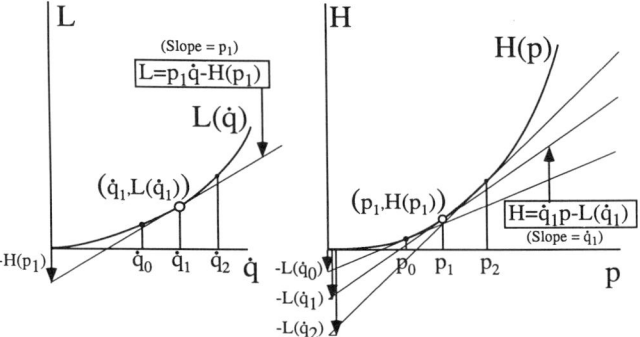

Figure 8.1.3 Geometry of Legendre transformation between Lagrangian and Hamiltonian functions.

8.1.3(b) will be shown to have an important interpretation in semiclassical wave mechanics. Slope p is inversely proportional to a wave phase velocity, and slope $\dot{q}$ corresponds to a wave group velocity. [Given the single-particle de Broglie relations $H \to \hbar\omega$ and $p \to \hbar k$ this geometrical interpretation of phase velocity (ω/k) and group velocity $(d\omega/dk)$ is an elementary consequence of dispersion theory.]

B. Action Functions

An important application of contact transformations involves the transformation of a classical particle or system from one point [say $(x = x_0, y = y_0, \ldots)$] to another point [say $(x = X, y = Y, \ldots)$] which lies along its natural trajectory of motion. The transformations which do this are called "active" (as opposed to "passive") transformations because they represent an actual change of position rather than just a relabeling of coordinates or state variables. The generators of these transformations are called ACTIONS. (This might suggest that the generators of passive transformations should be called "passions.")

The development of the idea of action generators requires one to ask what is so special about a particular classical trajectory or "natural path" followed by a classical system. Part of the answer, as we shall show, is that the action achieves an extreme value (in fact a locally minimum value) for a natural path. Nearby paths must have greater action, but, more importantly, paths which are very close to the natural one must have practically the *same* action. The concept of a stationary action value for a natural path is a key to the understanding of the action generators.

The first type of action function which we shall study is defined by the following time integral of a Lagrangian L which we take to have no explicit

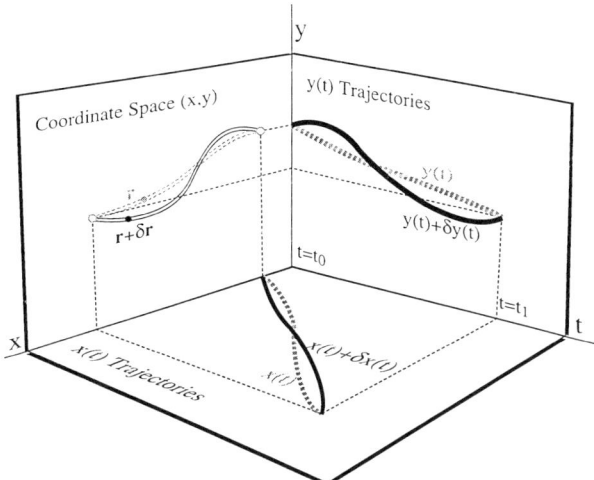

Figure 8.1.4 A comparison of paths appropriate for Hamilton's principle function S_p.

time dependence:

$$S_P(\mathbf{r}_0, t_0 : \mathbf{r}_1, t_1) = \int_{t_0}^{t_1} L(\mathbf{r}(t), \dot{\mathbf{r}}(t))\, dt. \quad (8.1.4)$$

This is called HAMILTON'S PRINCIPLE FUNCTION since it is the subject of Hamilton's principle, which we shall discuss shortly. This discussion involves a comparison of the value of S_P for two nearby paths $\mathbf{r}(t)$ and $\mathbf{r}(t) + \delta \mathbf{r}(t)$. Two such paths are sketched in Figure 8.1.4 for a two-dimensional system with coordinates $\mathbf{r} = (x, y)$. The third dimension of the figure is time.

Here we compare the action for curves whose end points are fixed in space and time, i.e., $\delta \mathbf{r}(t_0) = 0 = \delta \mathbf{r}(t_1)$. A range of positions, velocities, accelerations, jerks, etc., will be considered, but total travel time is fixed. For a small difference function $\delta \mathbf{r}(t)$ one has

$$\begin{aligned}
S_P(\mathbf{r}_0, t_0 : \mathbf{r}_1, t_1)_{\text{PATH}(r+\delta r)} &= \int_{t_0}^{t_1} L(\mathbf{r} + \delta\mathbf{r}, \dot{\mathbf{r}} + \delta\dot{\mathbf{r}})\, dt \\
&= \int_{t_0}^{t_1} L(r, \dot{r})\, dt + \int_{t_0}^{t_1}\left(\frac{\partial L}{\delta \mathbf{r}}\delta\mathbf{r} + \frac{\partial L}{\partial \dot{\mathbf{r}}}\delta\dot{\mathbf{r}}\right) dt + 0(\delta^2) \\
&= S_P(\mathbf{r}_0, t_0 : \mathbf{r}_1, t_1)_{\text{PATH}(r)} + \left.\frac{\partial L}{\partial \dot{\mathbf{r}}}\delta r\right|_{t_0}^{t_1} \\
&\quad + \int_{t_0}^{t_1}\left(\frac{\partial L}{\partial \mathbf{r}} - \frac{d}{dt}\frac{\partial L}{\partial \dot{\mathbf{r}}}\right)\delta\mathbf{r}\, dt, \\
&\quad + 0(\delta^2), \quad (8.1.5)
\end{aligned}$$

where an integration by parts was performed and $O(\delta^2)$ represents presumably negligible terms of second or higher order in $\delta \mathbf{r}$. If (and only if) the path $\mathbf{r}(t)$ is adjusted so that the Lagrange equations $[\partial L/\partial \mathbf{r} - d/dt(\partial L/\partial \dot{\mathbf{r}}) = 0]$ are always satisfied, then Hamilton's principle

$$S_P(\mathbf{r}_0, t_0 : \mathbf{r}_1, t_1)_{\text{PATH}(r+\delta r)} = S_P(\mathbf{r}_0, t_0 : \mathbf{r}_1, t_1)_{\text{PATH}(r)} \qquad (8.1.6)$$

is satisfied to a precision of second order.

Another type of action which we shall consider here is defined by the spatial integral

$$S_H(\mathbf{r}_0 : \mathbf{r}_1) = \int_{\mathbf{r}_0}^{\mathbf{r}_1} \mathbf{p} \cdot d\mathbf{r}. \qquad (8.1.7)$$

This is known as HAMILTON'S CHARACTERISTIC FUNCTION or REDUCED ACTION for reasons that will be mentioned later. Using the Legendre transformation (8.1.3c) one can convert S_H to a time integral and relate it to the action S_P. Using (8.1.3c) we have (with $\mathbf{q} \equiv \mathbf{r}$)

$$S_H(\mathbf{r}_0 : \mathbf{r}_1) = \int_{t_0}^{t_1} \mathbf{p} \cdot \dot{\mathbf{r}} \, dt = \int_{t_0}^{t_1} [H(\mathbf{p}, \mathbf{r}) + L(\mathbf{r}\dot{\mathbf{r}})] \, dt. \qquad (8.1.8)$$

However, it is necessary to carefully define the time limits of integration.

It is convenient to let the time limits be determined by the natural motion for a fixed value of the Hamiltonian: H = constant. (For Lagrangians with no explicit time dependence, the Hamiltonian is a constant of motion.) We shall use the notation $S_H(\mathbf{r}_0 : \mathbf{r}_1)$ to remind us that the travel times between $\mathbf{r}_0$ and $\mathbf{r}_1$ are not fixed but depend upon how much energy $\varepsilon = H$ is given. This means that a comparison of path integrals will have to allow for a variation of travel time $(t_1 - t_0)$. If we fix H and let $t_0 = 0$ then we must expect t_1 to have a different value $t_1 + \Delta t$ for a different path as indicated in Figure 8.1.5. The H-fixed action for a modified path $\mathbf{r}(t) + \Delta \mathbf{r}(t)$ will be

$$S_H(\mathbf{r}_0 : \mathbf{r}_1)_{\text{PATH}(r+\Delta r)}$$

$$= \int_0^{t_1 + \Delta t} [H + L(\mathbf{r} + \Delta \mathbf{r}, \dot{\mathbf{r}} + \Delta \dot{\mathbf{r}})] \, dt$$

$$= Ht_1 + \int_0^{t_1} L(\mathbf{r} + \Delta \mathbf{r}, \dot{\mathbf{r}} + \Delta \dot{\mathbf{r}}) \, dt + H \Delta t +$$

$$\int_{t_1}^{t_1 + \Delta t} L(\mathbf{r} + \Delta \mathbf{r}, \dot{\mathbf{r}} + \Delta \dot{\mathbf{r}}) \, dt$$

$$= S_H(\mathbf{r}_0 : \mathbf{r}_1)_{\text{PATH}(r)} + \int_0^{t_1} \left(\frac{\partial L}{\partial \dot{\mathbf{r}}} \Delta \dot{\mathbf{r}} \right) dt + H \Delta t + L(\mathbf{r}_1, \dot{\mathbf{r}}_1) \Delta t$$

$$= S_H(\mathbf{r}_0 : \mathbf{r}_1)_{\text{PATH}(r)} + \int_0^{t_1} \left(\frac{\partial L}{\partial \mathbf{r}} - \frac{d}{dt}\left(\frac{\partial L}{\partial \dot{\mathbf{r}}}\right) \right) \Delta \mathbf{r} \, dt + \frac{\partial L}{\partial \dot{\mathbf{r}}} \Delta \mathbf{r}(t_1)$$

$$+ H \Delta t + L \Delta t + O(\Delta^2). \qquad (8.1.9)$$

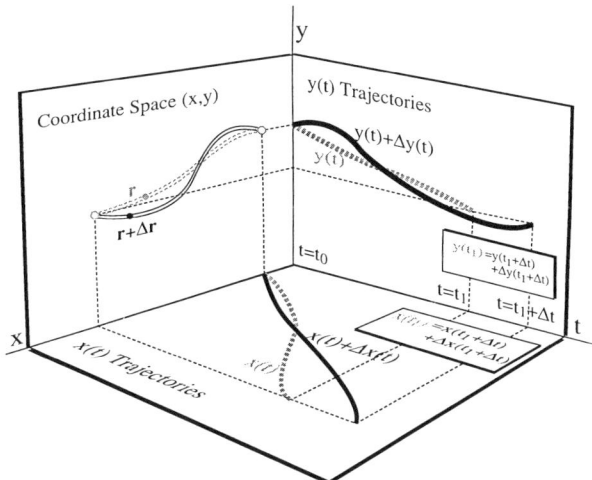

Figure 8.1.5 A comparison of paths appropriate for Hamilton's characteristic function S_H.

Here we have again integrated by parts and let $\Delta \mathbf{r}(t_0) = 0$. The value $\Delta \mathbf{r}(t_1)$ at the upper limit is obtained by solving the equation

$$\mathbf{r}(t_1) = \mathbf{r}(t_1 + \Delta t) + \Delta \mathbf{r}(t_1 + \Delta t)$$

$$= \mathbf{r}(t_1) + \left.\frac{\partial \mathbf{r}}{\partial t}\right|_{t_1} \Delta t + \Delta \mathbf{r}(t_1) + 0(\Delta^2) \qquad (8.1.10)$$

for small Δt. (See Figure 8.1.5, which shows components of this equation.) One then substitutes the result

$$\Delta \mathbf{r} = -\left.\frac{\partial \mathbf{r}}{\partial t}\right|_{t_1} \Delta t = -\dot{\mathbf{r}}(t_1)\,\Delta t \qquad (8.1.11)$$

into (8.1.9). For natural paths $\mathbf{r}(t)$ this yields

$$S_H(\mathbf{r}_0:\mathbf{r}_1)_{\text{PATH}(r+\Delta r)} = S_H(\mathbf{r}_0:\mathbf{r}_1)_{\text{PATH}(r)} + \left(L - \frac{\partial L}{\partial \dot{\mathbf{r}}}\dot{\mathbf{r}} + H\right)\Delta t + 0(\Delta^2)$$

$$= S_H(\mathbf{r}_0:\mathbf{r}_1)_{\text{PATH}(r)} + 0(\Delta^2), \qquad (8.1.12)$$

where (8.1.3c) was used. This shows that the time-independent action S_H reaches an extreme value on natural paths.

Finally, let us consider the change in the action S_P if we vary time *and* position of the destination point by arbitrary amounts dt_1 and $d\mathbf{r}_1$, respectively, but we "aim" the initial momenta so as to follow a natural path $\mathbf{r}(t) + d\mathbf{r}(t)$ to this end point. (As usual let us keep the starting point $\mathbf{r}_0$ and time t_0 the same for both paths.) By following the same procedures that were used in (8.1.9), we have

$$S_P(r_0, t_0 : r_1 + dr_1, t_1 + dt)_{\text{PATH}(r+dr)}$$
$$= \int_{t_0}^{t_1+dt_1} L(r + dr, \dot{r} + d\dot{r})\, dt$$
$$= S_P(r_0, t_0 : r_1, t_1)_{\text{PATH}(r)} + \frac{\partial L}{\partial \dot{r}} dr(t_1) + L(r_1 \dot{r}_1)\, dt + 0(d^2).$$
(8.1.13)

Instead of (8.1.10) we need equations which include the arbitrarily chosen dr_1

$$r(t_1) + dr_1 = r(t_1 + dt_1) + dr(t_1 + dt_1),$$
$$r_1 + dr_1 = r_1 + \dot{r}_1\, dt_1 + dr(t_1),$$
$$dr(t_1) = dr_1 - \dot{r}_1\, dt_1.$$
(8.1.14)

Substituting this result into (8.1.13) yields

$$S_P(r_0, t_0 : r_1 + dr_1, t_1 + dt_1) - S_P(r_0, t_0 : r_1, t_1)$$
$$= p(t_1) \cdot dr_1 + (L(r_1 \dot{r}_1) - p(t_1) \cdot \dot{r}_1)\, dt_1,$$
$$dS_P = p \cdot dr_1 - H\, dt_1,$$
(8.1.15a)

where the Legendre equations

$$p = \frac{\partial L}{\partial \dot{r}} \quad \text{and} \quad H = p \cdot \dot{r} - L$$
(8.1.15b)

from (8.1.3) are used. Note that the changes in S_P due to changes in H are of second order in dr_1 and dt_1 and are therefore neglected.

The differential form (8.1.15a) is called the POINCARÉ-CARTAN IN-VARIANT. From it we derive two very important equations

$$\frac{\partial S_P(r_0, t_0 : r, t)}{\partial r} = p,$$
(8.1.16a)

$$\frac{\partial S_P(r_0, t_0 : r, t)}{\partial t} = -H(p, r).$$
(8.1.16b)

The combination of these two equations leads to the TIME-INDEPENDENT HAMILTON-JACOBI equation for the principal action generator.

$$-\frac{\partial S_P}{\partial t} = H\left(\frac{\partial S_P}{\partial r}, r\right).$$
(8.1.17)

A related set of equations exists for determining the time-independent action generator S_H. From (8.1.8) we have

$$S_H(\mathbf{r}_0:\mathbf{r}_1) = (t_1 - t_0)H + \int_{t_0}^{t_1} L\,dt = (t_1 - t_0)H + S_P(\mathbf{r}_0,t_0:\mathbf{r}_1,t_1), \tag{8.1.18}$$

where we are assuming natural paths in all integrations. Also, the travel time $t_1 - t_0$ depends on the chosen value of energy $\varepsilon = H$ as before. (The initial time t_0 may be chosen to be zero without loss of generality.) Now a variation of the destination point $\mathbf{r}_1$ to $\mathbf{r}_1 + d\mathbf{r}_1$ causes a change,

$$dS_H = dt_1\,H + dS_P = \mathbf{p}\cdot d\mathbf{r}_1, \tag{8.1.19}$$

in which time dependence is removed. (Hence, the name "reduced action" is used for S_H.) This leads to the TIME-DEPENDENT HAMILTON-JACOBI equations

$$\frac{\partial S_H(\mathbf{r}_0:\mathbf{r})}{\partial \mathbf{r}} = \mathbf{p}, \tag{8.1.20a}$$

$$H\left(\frac{\partial S_H}{\partial \mathbf{r}},\mathbf{r}\right) = \varepsilon = \text{constant}. \tag{8.1.20b}$$

C. Generators for Classical Trajectories and Wave Fronts

Let us consider some elementary solutions to the Hamilton-Jacobi equations. Possibly the simplest case is the one treated at the beginning of sophomore mechanics involving a massive body falling in a uniform gravitational field ($f = mg$). Then the Hamiltonian is

$$\varepsilon = H(\mathbf{p},\mathbf{r}) = (1/2m)\mathbf{p}\cdot\mathbf{p} - fy = (1/2m)\left(p_x^2 + p_y^2\right) - fy, \tag{8.1.21}$$

where we neglect the third spatial dimension. The HJ equation (8.1.20) becomes

$$(1/2m)\left[\left(\frac{\partial S_H}{\partial x}\right)^2 + \left(\frac{\partial S_H}{\partial y}\right)^2\right] - fy = \varepsilon. \tag{8.1.22}$$

This example allows a separation of variables $S_H(x,y) = s_x(x) + s_y(y)$ to yield two ordinary differential equations:

$$(1/2m)\left(\frac{ds_x}{dx}\right)^2 = \varepsilon_x, \quad 1/2m\left(\frac{ds_y}{dy}\right)^2 - fy = \varepsilon_y, \tag{8.1.23a}$$

where the separated parts satisfy

$$S_H(x_0, y_0 : x_1, y_1) = s_x(x_1) + s_y(y_1), \quad (8.1.23b)$$

$$\varepsilon = \varepsilon_x + \varepsilon_y. \quad (8.1.23c)$$

The solutions to the separate HJ equations are

$$s_x = [2m\varepsilon_x]^{1/2}(x_1 - x_0)$$
$$= m\dot{x}_0(x_1 - x_0), \quad (8.1.24a)$$

$$s_y = (1/3mf)[2m(\varepsilon_y + fy)]^{3/2}\Big|_{y_0}^{y_1}$$
$$= (m^2/3f)[\dot{y}_1^3 - \dot{y}_0^3], \quad (8.1.24b)$$

where we use the definitions of momentum and velocity which apply to this problem:

$$p_x = m\dot{x} = [2m\varepsilon_x]^{1/2}, \quad (8.1.24c)$$

$$p_y = m\dot{y} = [2m(\varepsilon_y + fy)]^{1/2}. \quad (8.1.24d)$$

The time-dependent action is given by

$$S_P = S_H - \varepsilon T \quad (8.1.25)$$

according to (8.1.18), where the travel time is $T = t_1 - t_0$, and $\varepsilon = H$ is the total energy. We now write the action S_P in various different forms in order to exhibit its physical and mathematical properties using examples. To do this we will use the falling-body time-trajectory solutions,

$$x_1 = x_0 + \dot{x}_0 T, \quad \dot{x}_1 = \dot{x}_0, \quad (8.1.26a)$$

$$y_1 = y_0 + \dot{y}_0 T + (f/2m)T^2, \quad \dot{y}_1 = \dot{y}_0 + (f/m)T, \quad (8.1.26b)$$

obtained by elementary means. (One should not get the impression that action theory is a convenient way to compute specific trajectories. Rather it is a way to analyze *families* of trajectories.)

By combining (8.1.23), (8.1.25) and the second lines of (8.1.26) we have

$$S_P = m\dot{x}_0(x_1 - x_0) + (m^2/3f)\left[(\dot{y}_0 + fT/m)^3 - \dot{y}_0^3\right] - \left(\frac{m}{2}\dot{x}_0^2 + \frac{m}{2}\dot{y}_0^2 - fy_0\right)T,$$

$$S_P = \frac{m\dot{x}_0}{2}(x_1 - x_0) + \frac{mT}{2}\dot{y}_0^2 + fTy_0 + fT^2\dot{y}_0 + \frac{f^2T^3}{3m}. \quad (8.1.27)$$

CONTACT TRANSFORMATIONS, ACTIONS, AND SEMICLASSICAL WAVE FRONTS 697

Solving for $(\dot{x}_0, \dot{y}_0)$ in the first line of (8.1.26) leads to the standard form for S_P:

$$S_P(\mathbf{r}_0, 0: \mathbf{r}_1, T) = \frac{m(x_1 - x_0)^2}{2T} + \frac{m(y_1 - y_0)^2}{2T} + \frac{fT}{2}(y_1 + y_0) - \frac{f^2 T^3}{24m}. \quad (8.1.28)$$

On the other hand, we may write S_P as a function exclusively of time T and initial conditions:

$$S_P(\mathbf{r}_0, 0: \mathbf{r}_1(T), T) = \frac{m\dot{x}_0^2 T}{2} + \frac{m\dot{y}_0^2 T}{2} + fTy_0 + fT^2 \dot{y}_0 + \frac{f^2 T^3}{3m}. \quad (8.1.29)$$

The latter could be obtained most easily by direct integration of the definition (8.1.4) of S_P. However, in so doing, one loses the crucial functional dependence in (8.1.28), which makes S_P a generator. It is instructive to compare the partial and total time derivatives of S_P:

$$\frac{\partial S_P(\mathbf{r}_0, 0: \mathbf{r}_1, T)}{\partial T} = -H, \quad (8.1.30a)$$

$$\frac{dS_P(\mathbf{r}_0, 0: \mathbf{r}_1(T), T)}{dT} = L, \quad (8.1.30b)$$

which are expressions of the HJ equation (8.1.17) and the principle function definitions, (8.1.4), respectively:

$$\frac{\partial S_P}{\partial T} = -\frac{m}{2T^2}\left[(x_1 - x_0)^2 + (y_1 - y_0)^2\right] + \frac{f(y_1 + y_0)}{2} - \frac{f^2 T^2}{8m} = -H, \quad (8.1.30a)_x$$

$$\frac{dS_P}{dT} = \frac{m\dot{x}_0^2}{2} + \frac{m\dot{y}_0^2}{2} + fy_0 + 2fy_0 T + \frac{f^2 T^2}{m} = L. \quad (8.1.30b)_x$$

The consistency of the examples may be verified using solutions (8.1.26). Finally, the following energy derivative formula is important:

$$\frac{\partial S_H}{\partial \varepsilon} = T. \quad (8.1.31)$$

This follows from (8.1.25) and the fact that S_P does not depend explicitly on energy $(\partial S_P/\partial \varepsilon = 0)$.

Examples involving curves with S_H = constant are shown in Figure 8.1.6. The first figure, 8.1.6(a), shows a family of projectile trajectories with $H =$

$\varepsilon = 128$ J(oule) and $F = -4$ N(ewton). The corresponding generator curves $S_H(0,0: x, y) = 50, 100, 150, \ldots, \text{J} \cdot \text{s})$ and so forth are shown superimposed onto the trajectories in the second figure, 8.1.6(b). The curves with $S_H = 50$ to $S_H = 300$, or so, are ovals with increasing radius. They are clearly orthogonal to outgoing projectile trajectories in accordance with the equation (8.1.19):

$$0 = dS_H = \mathbf{p} \cdot d\mathbf{r}.$$

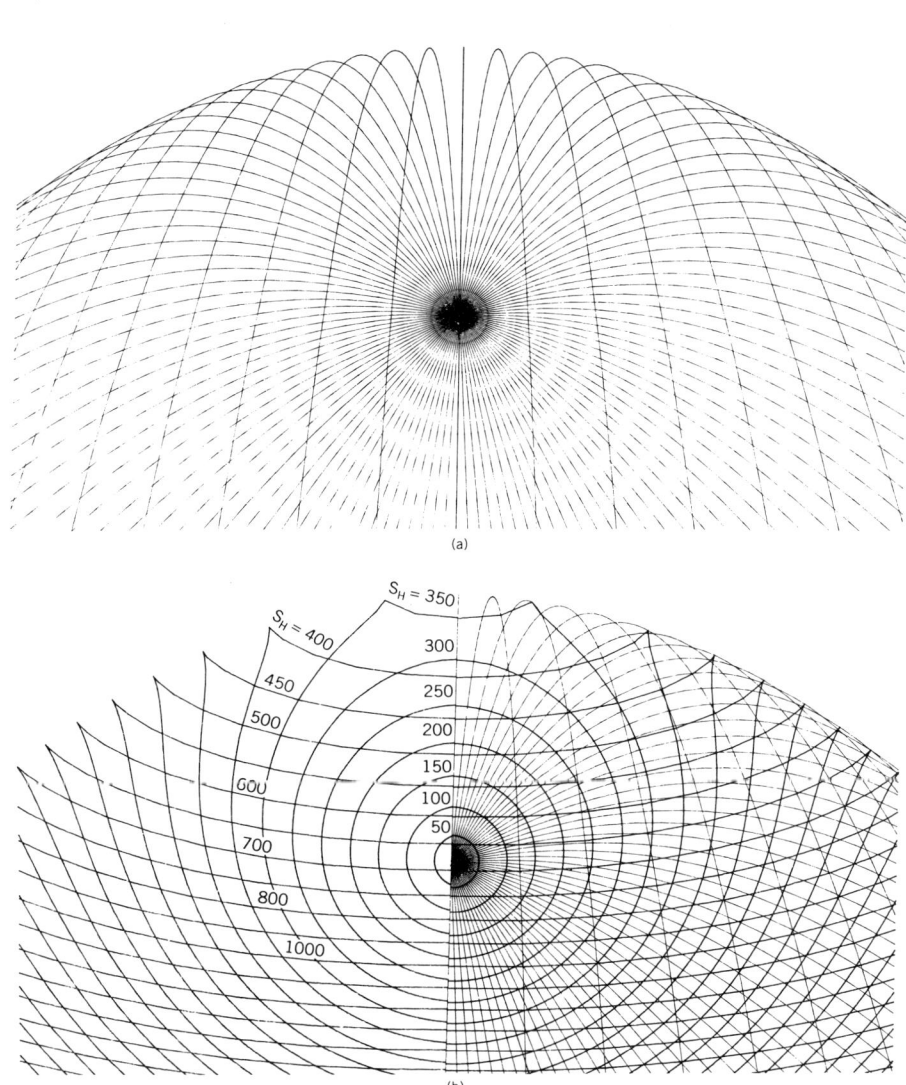

Figure 8.1.6 Classical trajectory families and constant characteristics action curves. (a) Isoenergetic family of trajectories. (b) Isoenergetic family of constant S_H curves.

For $S_H = 350$ or higher the curves vaguely resemble the outline of a bat's head with two ears pointing at the parabolic caustic that marks the classical turning points along the upper boundary of the trajectory region. The ears should be pointed cusps with zero angle at the points. The plot is not precise enough to show this for the first few bat's heads.

The S_H curves can be regarded as wave fronts emerging from the original point and propagating outward until they are reflected by the upper classical turning boundary. The reflected S waves are orthogonal to downward traveling trajectories. The time behavior of these S waves can be deduced by defining them in terms of Hamilton's principle action S_P for constant H:

$$S_P(\mathbf{0}, 0 : \mathbf{r}, T) = S_H(\mathbf{0} : \mathbf{r}) - HT = \text{constant}. \quad (8.1.32)$$

Then a point $\mathbf{r} + d\mathbf{r}$ having the same S_P value at a later time would satisfy the equation

$$S_P(\mathbf{0}, 0 : \mathbf{r} + d\mathbf{r}, T + dT) - S_P(\mathbf{0}, 0 : \mathbf{r}, T) = 0 = dS_P. \quad (8.1.33)$$

For constant $H = \varepsilon$ this becomes

$$dS_P = 0 = dS_H - \varepsilon \, dT,$$
$$0 = p \cdot d\mathbf{r} - \varepsilon \, dT. \quad (8.1.34)$$

For $d\mathbf{r}$ chosen in the direction of the momentum vector $\dot{\mathbf{p}}$ this gives the S-wave phase velocity:

$$\frac{d|\mathbf{r}|}{dT} = \frac{\varepsilon}{|\mathbf{p}|}. \quad (8.1.35)$$

In this simple example the $\mathbf{p}$ vector ($\mathbf{p} = m\mathbf{v} = m\dot{\mathbf{r}}$) points in the direction of the particle velocity $\dot{\mathbf{r}}$. This will not always be so. For a general Hamiltonian, the coordinate velocity

$$\dot{\mathbf{r}} = \frac{\partial H}{\partial \mathbf{p}}$$

has a different direction and magnitude than $\mathbf{p}$. Note that the phase velocity (8.1.35) varies inversely with the momentum and (in this case) particle velocity. One can see in Figure 8.1.6(b) that the spacing between S-wave fronts decreases as the corresponding particles pick up speed. Particle velocity corresponds to group velocity in the wave picture, but one needs more than a single energy ε (or frequency $\nu = \varepsilon/h$) to observe it. The connection between action "wave fronts" and quantum waves will be discussed shortly.

It should be noted that the S-wave fronts do not generally have simple analytic expressions, particularly when one requires constant energy. The "bat's head" wave fronts in Figure 8.1.6(b) may be obtained by solving cubic equations derived from (8.1.24) and (8.1.25) at each point along a trajectory. The S_P equations (8.1.28) are deceptively simple equations for displaced circles; however, that is not for constant energy. Correct $T(\varepsilon)$ dependence leads again to cubic equations. So, even for this simple sophomore mechanics problem the analytic treatment of action is difficult!

Since exact analytic action expressions are going to be difficult or impossible in most problems, one needs to try other approaches. A general numerical approach which turns out to be quite simple involves numerical integration of the first-order Hamilton's equations:

$$\dot{\mathbf{q}} = \frac{\partial H}{\partial \mathbf{p}}, \tag{8.1.36a}$$

$$\dot{\mathbf{p}} = -\frac{\partial H}{\partial \mathbf{q}}, \tag{8.1.36b}$$

$$\dot{S}_P = L = \mathbf{p}\dot{\mathbf{q}} - H. \tag{8.1.36c}$$

In this way an arbitrary Hamiltonian (even a time-dependent one) can be converted into a family of trajectories and action surfaces. The desired constant-S_H surfaces are contours of constant $= (S_P + Ht)$. In fact, this technique for solving the HJ partial differential equation is known as "integration along characteristics." For this reason S_H is also known as a characteristic function or integral.

We will consider how the classical equations of motion may be used to find multidimensional quantum eigenfunctions using coherent wave and wavepacket propagation properties. Now we consider the elementary connection between the S waves of the HJ equation and the ψ waves of the Schrödinger equation.

D. Semiclassical Approximation for Schrödinger Equation

We have seen that the action function S_P behaves very much like a wave phase function. It is instructive to study the substitution

$$\psi(r,t) = \psi_0 e^{iS_P/\hbar} \tag{8.1.37}$$

in Schrödinger's equation:

$$i\hbar \frac{\partial \psi}{\partial t} = H\psi = \left[(1/2m)p^2 + V(\mathbf{r})\right]\psi$$
$$= -(\hbar^2/2m)\nabla^2\psi + V(\mathbf{r})\psi.$$

The result is (we drop the subscript p from the action)

$$-\psi \frac{\partial S}{\partial t} = -\psi(\hbar i/2m)\nabla^2 S + \psi\left[(1/2m)\left(\frac{\partial S}{\partial \mathbf{r}}\right)^2 + V\right]$$

$$(\hbar i/2m)\nabla^2 S = \frac{\partial S}{\partial t} + H\left(\frac{\partial S}{\partial \mathbf{r}}, \mathbf{r}\right) \qquad (8.1.38)$$

In a limit in which the left-hand term vanishes, the Schrödinger equation reduces to the HJ equation (8.1.17). This is the semiclassical limit in which $\hbar \nabla^2 S \ll (\nabla S)^2 = p^2$, or

$$\hbar \frac{d^2 S}{dx^2} = \hbar \frac{dp}{dx} \ll p^2.$$

By using the de Broglie relation $p = \hbar k = h/\lambda$ this becomes

$$\frac{dk}{dx}\bigg/ k \ll k = 2\pi/\lambda. \qquad (8.1.39)$$

This states that the wavelength should be small enough so that the percentage change of momentum over a wavelength is a negligible fraction.

E. Huygen's Principle and Semiclassical Mechanics

The properties of envelope curves generated by contact transformations in semiclassical mechanics are closely connected with Huygen's principle of enveloping wave crests in the study of optics. In both studies there are wave fronts generated by functions $S(\mathbf{r}_0 : \mathbf{r})$ which depend upon extrema or stationary values of path integrals.

Consider a hypothetical action function $S_H(\mathbf{r}_0 : \mathbf{r})$ which would generate the curves (or surfaces) indicated in Figure 8.1.7; i.e., $S_H(\mathbf{r}_0 : \mathbf{r}) = 10$, 20, and 30 for fixed $\mathbf{r}_0$. Consider two points $\mathbf{r}_{10}$ and $\mathbf{r}'_{10}$ on the $S_H(\mathbf{r}_0 : \mathbf{r}) = 10$ wave front. They both required an accumulation of (at least) 10 units of action S_H for a trajectory that started at their common point of origin $\mathbf{r}_0$ with energy H held fixed. From each of these points one might generate a new set of constant action surfaces: $S_H(\mathbf{r}_{10}, \mathbf{r}) = 10$ around $\mathbf{r}_{10}$ and $S_H(\mathbf{r}'_{10}, \mathbf{r}) = 10$ around $\mathbf{r}'_{10}$. Points on these new curves represent a total expenditure of 20 units of action since departure from $\mathbf{r}_0$. But, on each of these intermediate curves there is only one point ($\mathbf{r}_{20}$ and $\mathbf{r}'_{20}$, respectively) for which *at least* 20 units of action is required to arrive there from $\mathbf{r}_0$. These are the contact points of the intermediate curves with the larger $S_H(\mathbf{r}_0 : \mathbf{r}) = 20$ curve. The contact points lie on the natural trajectory through the intermediate points

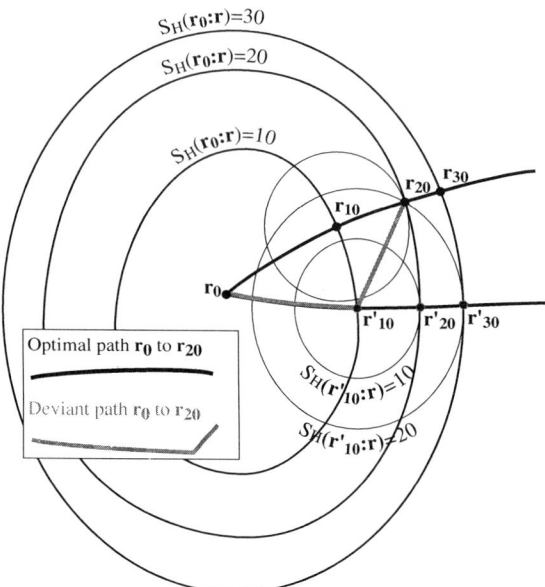

Figure 8.1.7 Comparison of paths and wave fronts for discussion of Huygen's principle. Paths which have locally extreme (i.e., minimal) values of S_H will lead to constructive interference and the formation of wave fronts which are envelopes of the wavelets belonging to nonextreme paths.

r_{10} and r'_{10}, respectively, since they correspond to extreme values of S_H according to the least-action principle.

To arrive at r_{20} by way of r'_{10}, for example, would require more than 20 action units. Figure 8.1.7 indicates the value would be 30; i.e., $S_H(r_0 : r'_{10}) + S_H(r'_{10} : r_{20}) = 30$. Only for trajectories which pass over or very near to $r = r_{10}$ will the action $S_H(r_0 : r_{20})$ achieve this minimum stationary value of 20 units.

In wave optics the same sort of reasoning is applied, only the S_H function is replaced by the optical travel-time function $T(r_0 : r)$

$$T(r_0 : r) = \int_{r_0}^{r} n\, ds,$$
(along optical path)

where $n = n(r)$ is the local index of refraction. The function T is called the INDICATRIX or "slowness" function, since the index n is inversely proportional to optical phase velocity. According to Fermat's principle, travel time $T(r_0 : r)$ is minimum for an optical path. The reason for this is that only the light near a stationary path will undergo constructive interference with other waves following nearby paths. Light waves which take the "wrong paths" interfere destructively; i.e., they "beat" each other to death!

Similarly, there are "wrong paths" in classical mechanics, such as path $(\mathbf{r}_0 \mathbf{r}'_{10} \mathbf{r}_{20})$ in Figure 8.1.7. The right paths, such as path $(\mathbf{r}_0 \mathbf{r}_{10} \mathbf{r}_{20})$, satisfy Newton's, Lagrange's, or Hamilton's equations along their entire lengths and thereby achieve locally minimum (and *stationary*) accumulations of the action S_H. This leads to an accumulation of the wave function due to constructive interference. The resulting waves have crests which parallel the enveloping constant action surfaces. (S_H = constant.)

The wavelike properties of action are one of the ingredients of the theory for Feynman's path integrals. The action functions are the phases of propagators which can be time dependent as in the following:

$$\langle \mathbf{r}_1, t_1 | \mathbf{r}_0, t_0 \rangle \simeq e^{iS_P(\mathbf{r}_0, t_0 : \mathbf{r}_1, t_1)}$$

or time independent with energy fixed as in the following:

$$\langle \mathbf{r}_1 | \mathbf{r}_0 \rangle \simeq e^{iS_H(\mathbf{r}_0 : \mathbf{r}_1)}.$$

Here $\langle A|B \rangle$ is the quantum probability amplitude for going from B to A as introduced in Section 1.1. Note that the completeness relation for amplitudes would require the following:

$$\langle \mathbf{r}_1 | \mathbf{r}_0 \rangle = \sum_{r'} \langle \mathbf{r}_1 | \mathbf{r}' \rangle \langle \mathbf{r}' | \mathbf{r}_0 \rangle \cong \sum_{r'} e^{iS_H(\mathbf{r}_0 : \mathbf{r}') + S_H(\mathbf{r}' : \mathbf{r}_1)}$$

$$\cong e^{iS_H(\mathbf{r}_0 : \mathbf{r}_1)}.$$

This is an algebraic statement of Huygen's principle as it applies to semiclassical paths. It also represents the starting point for the Feynman path-integral approach to quantum mechanics. This approach generally requires sums over *all* paths between end points.

The technology required to perform sums over *all* possible paths has not been developed to the point where it is generally applicable. For this reason the Feynman-Huygen path-integral approach to quantum mechanics is not used as much as it might be. However, for oscillating and rotating molecules there appears to be some simple ways to circumvent some of the difficulties associated with evaluating path sums. These new approaches involve the use of wave-packet or coherent-wave states, and this is discussed in the following sections.

8.2 COHERENT HARMONIC OSCILLATOR STATES

The eigenstates and ladder operators for a harmonic oscillator Hamiltonian of the form

$$H = (1/2\mu)p^2 + (\mu\omega^2/2)x^2 = \hbar(a^\dagger a + \tfrac{1}{2}) \qquad (8.2.1)$$

were derived in Section 4.4 for the case ($\omega = 1$). Excited eigenstates $|n\rangle$ may be defined by

$$|n\rangle = (a^\dagger)^n |0\rangle / \sqrt{n!} \qquad (8.2.2)$$

in terms of action on a ground state $|0\rangle$ by the creation operator

$$a^\dagger = (\mu\omega/2\hbar)^{1/2}(x - ip/\mu\omega). \qquad (8.2.3)$$

The destruction operators $a = (a^\dagger)^\dagger$ and $a^\dagger$ generate an algebra whose representation in the $|n\rangle$ basis is defined by

$$a^\dagger |n\rangle = \sqrt{n+1}\,|n+1\rangle, \qquad (8.2.4a)$$

$$a|n\rangle = \sqrt{n}\,|n-1\rangle, \qquad (8.2.4b)$$

and

$$H|n\rangle = \hbar\omega(n + \tfrac{1}{2})|n\rangle. \qquad (8.2.5)$$

The ground-state wave function

$$\langle x|0\rangle = \psi_0(x) = (\mu\omega/\pi\hbar)^{1/4} e^{-(\mu\omega/2\hbar)x^2}$$
$$= N_0 e^{-y^2/2} \qquad (8.2.6)$$

has the indicated Gaussian form. Using operations (8.2.2) and (8.2.3) one can generate from ψ_0 the other eigenfunctions

$$\psi_n(x) = N_0 (\tfrac{1}{2})^{n/2} H_n(y) e^{-y^2/2} / \sqrt{n!} \qquad (8.2.7)$$

in terms of the well-known Hermite polynomials $H_n(y)$ where the rescaled variable $y = (\mu\omega/\hbar)^{1/2} x$ is used. [Recall Eqs. (4.4.68)–(4.4.70).]

Now we consider other types of states or wave functions which can be generated from $|0\rangle$ or $\psi_0(x)$. Throughout the preceding chapters we have emphasized the use of states of the form $g|0\rangle$ or $P^\mu |0\rangle$ generated by action of group translation or projection operators on a localized state. Here this will lead to the idea of the coherent state and theories of wave-packet propagation.

The idea of translating the ground state is actually a quite natural one. Upon encountering a pendulum, spring-mass system, or other harmonic oscillator sitting quietly somewhere, most people would have the urge to disturb it. Some might simply pull it off its equilibrium position and let it go, while others more prone to violence might deliver an additional hefty impulse of momentum. We now do the same using various operators on the ground state $|0\rangle$.

The operator which moves a wave function by x_0 units in the translation or displacement operator defined by

$$D(x_0) = e^{-x_0(\partial/\partial x)} = e^{-ix_0 p/\hbar}. \tag{8.2.8}$$

It performs the following functional translation:

$$D(x_0)\chi(x) = e^{-x_0(\partial/\partial x)}\chi(x)$$

$$= \chi(x_0) - x_0 \frac{\partial \chi}{\partial x}\bigg|_0 + (x_0^2/2!)\frac{\partial^2 \chi}{\partial x^2}\bigg|_0 - (x_0^3/3!)\frac{\partial^3 \chi}{\partial x^3}\bigg|_0 + \cdots$$

$$= \chi(x - x_0). \tag{8.2.9}$$

Similarly, the operator which boosts momentum by p_0 units is defined by

$$B(p_0) = e^{ip_0 x/\hbar} \quad (=e^{-p_0(\partial/\partial p)}), \tag{8.2.10}$$

where the parenthetical expression is the momentum space representation.

When combining these operations one observes that they do not commute and so the question arises: Should displacement or boost go first? A fair settlement involves defining a symmetric combination as follows:

$$Q(x_0, p_0) = e^{(i/\hbar)(p_0 x - x_0 p)}. \tag{8.2.11}$$

Then we use a special case of the Baker-Campbell-Hausdorf theorem

$$e^{A+B} = e^A e^B e^{-[A,B]/2} = e^B e^A e^{[A,B]/2}, \tag{8.2.12}$$

which holds if

$$[A,[A,B]] = 0 = [B,[A,B]]. \tag{8.2.13}$$

This shows that the Q operator can be factored either way:

$$Q(x_0, p_0) = e^{(i/\hbar)p_0 x} e^{(-i/\hbar)x_0 p} e^{-(x_0 p_0/2\hbar^2)[x,p]}$$

$$= e^{-ix_0 p_0/2\hbar} B(p_0) D(x_0) = e^{ix_0 p_0/2\hbar} D(x_0) B(p_0). \tag{8.2.14}$$

We can also factor this operator when x and p are expressed in terms of $(a, a^\dagger)$ operators by solving (8.2.3) and its conjugate. We have

$$Q(x_0, p_0) = e^{ip_0(a+a^\dagger)/(2\hbar\mu\omega)^{1/2} - x_0(a-a^\dagger)(\mu\omega/2\hbar)^{1/2}}$$

$$= e^{\alpha_0 a^\dagger + \alpha_0^* a} = e^{-\alpha_0^* \alpha_0/2} e^{\alpha_0 a^\dagger} e^{\alpha_0^* a}, \tag{8.2.15a}$$

where we define a phase space position number

$$\alpha_0 \equiv [x_0 + ip_0/(\mu\omega)](\mu\omega/2\hbar)^{1/2}. \tag{8.2.15b}$$

Also, the commutation relation $[a, a^\dagger] = 1$ was used.

Let us define by $|\alpha_0\rangle$ the state obtained by operating with $Q(x_0, p_0)$ on the ground state $|0\rangle$:

$$|\alpha_0\rangle = Q(x_0, p_0)|0\rangle = e^{-|\alpha_0|^2/2} e^{\alpha_0 a^\dagger}|0\rangle$$
$$= e^{-|\alpha_0|^2/2} \sum_n (\alpha_0 a^\dagger)^n |0\rangle/n!. \tag{8.2.16}$$

Here only the creation operator exponential contributes since $(a|0\rangle = 0)$. Using (8.2.2) one finds the expression

$$|\alpha_0\rangle = e^{-|\alpha_0|^2/2} \sum_n (\alpha_0)^n |n\rangle/\sqrt{n!}, \tag{8.2.17}$$

which is the COHERENT or MINIMUM UNCERTAINTY wave-packet state. This type of state was first considered by Schrödinger and later applied and generalized by Schwinger, Glauber, and others. We consider its dynamical properties as time evolves.

The time evolution operator

$$T(t, 0) = e^{-itH/\hbar} \tag{8.2.18}$$

produces solutions $(|\chi(t)\rangle = T(t, 0)|\chi(0)\rangle)$ to the time-dependent Schrödinger equation $(i\hbar|\dot\chi\rangle = H|\chi\rangle)$. The evolution of eigenstates is simple:

$$|n(t)\rangle = T(t)|n(0)\rangle = e^{-i\omega(n+1/2)t}|n(0)\rangle, \tag{8.2.19}$$

that is, no change except for an overall phase factor whose angular frequency is determined by the energy eigenvalue (8.2.5) and Planck's law $(\omega = E/\hbar)$. However, the evolution of a combination of eigenstates such as (8.2.17) may be much more complicated because different frequencies interfere.

Applying the evolution operator to coherent state $|\alpha_0\rangle$ yields

$$T(t)|\alpha_0\rangle = e^{-|\alpha_0|^2/2} \sum_n (\alpha_0)^n e^{-i\omega t(n+1/2)}|n\rangle/\sqrt{n!}$$
$$= e^{-i\omega t/2} e^{-|\alpha_0|^2/2} \sum_n (\alpha_0 e^{-i\omega t})^n |n\rangle/\sqrt{n!}. \tag{8.2.20a}$$

This can be expressed in a simpler form:

$$T(t)|\alpha_0\rangle = e^{-i\omega t/2}|\alpha_t\rangle, \tag{8.2.20b}$$

where the position number simply rotates with frequency ω in the complex plane:

$$\alpha_t = \alpha_0 e^{-i\omega t}. \qquad (8.2.20c)$$

It is instructive to regard the complex quantity α_t as the POSITION PHASOR. A phasor is a common visualization aid which was introduced in Section 2.6 for discussing classical oscillation. Phasor space is the same as phase space except that the momentum or p axis is scaled down by a factor ($\mu\omega$) so that all orbits are circular. It is pedagogically useful to regard the complex α_t-phasor vector as a clock's sweep second hand which, according to (8.2.20c), moves clockwise at the rate of ω rad · Hz.

Therefore, the picture of an evolving coherent state is fairly simple: Imagine a phase space wave packet centered on the tip of the phasor α_t and orbiting around the phase clock as in Figure 8.2.1. The uncertainty $(\Delta p)(\Delta x)$ is the same for this state at all times as it is for the ground eigenstate, i.e., the absolute minimum value allowed by Heigenberg's relation:

$$(\Delta p)(\Delta x) = \hbar/2.$$

The coherent or minimum uncertainty state is just the ground state pulled off center in phase space.

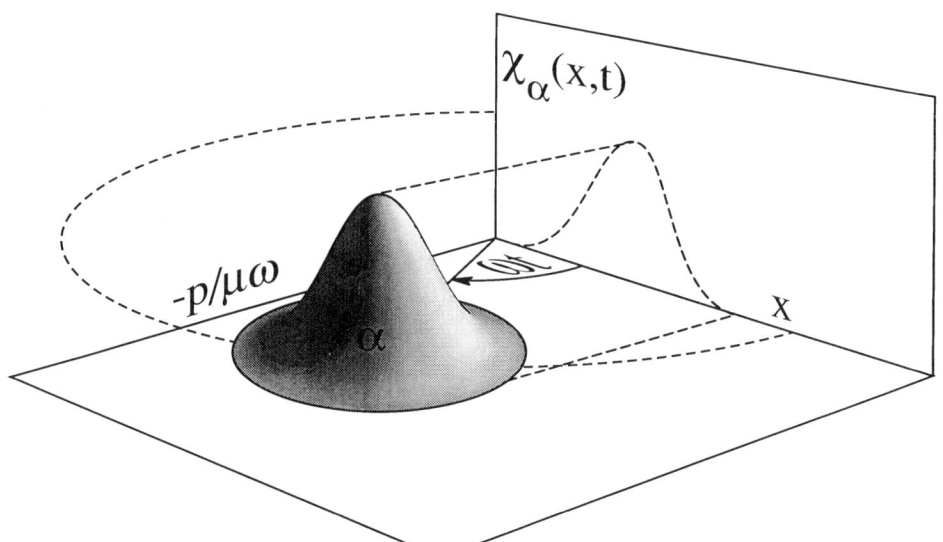

Figure 8.2.1 Sketch of wave packet rotating in phase plane. Corresponding classical phasor is indicated by vector α at the center of the phase plane distribution.

In order to plot the coherent wave packet in coordinate x space one needs the time-dependent translation operator in spatial form. Using the notation $Q(\alpha) = Q(x, p)$ the time transformation of Q is

$$TQ(\alpha_0)T^\dagger = e^{\alpha_0 T^\dagger a T + \alpha_0^* T a T^\dagger} = \exp(\alpha_0 e^{-i\omega t} a^\dagger + \alpha_0^* e^{i\omega t} a)$$
$$= Q(\alpha_t), \qquad (8.2.21)$$

where the creation evolution is $Ta^\dagger T^\dagger = e^{-i\omega t} a^\dagger$. The classical phase components

$$x_t = x_0 \cos \omega t + (p_0/\mu\omega)\sin \omega t, \qquad (8.2.22a)$$
$$p_t = p_0 \cos \omega t - x_0 \mu\omega \sin wt \qquad (8.2.22b)$$

of phasor

$$\alpha_t = [x_t + i(p_t/\mu\omega)](\mu\omega/2\hbar)^{1/2} \qquad (8.2.22c)$$

are determined through (8.2.15b) and (8.2.20c). These components are the $|\alpha_t\rangle$ expectation values of position and momentum. This follows from the action of (a) and $(a^\dagger)$ on $|\alpha_t\rangle$ using (8.2.4):

$$a|\alpha_t\rangle = \alpha_t|\alpha_t\rangle, \qquad (8.2.23a)$$
$$\langle \alpha_t|a^\dagger = \alpha_t^*\langle \alpha_t|. \qquad (8.2.23b)$$

Then one has

$$\langle \alpha_t|x|\alpha_t\rangle = (2\hbar/\mu\omega)^{1/2}(\alpha_t + \alpha_t^*)/2$$
$$= x_t, \qquad (8.2.24a)$$
$$\langle \alpha_t|p|\alpha_t\rangle = (2\mu\omega\hbar)^{1/2}(\alpha_t - \alpha_t^*)/2i$$
$$= p_t. \qquad (8.2.24b)$$

The expectation value for the energy is constant, but the classical and quantum values differ slightly. The classical value is

$$E_c = (p_t^2/2\mu) + \tfrac{1}{2}\mu\omega^2 x_t^2 = |\alpha_t|^2 \hbar\omega = |\alpha_0|^2 \hbar\omega$$
$$= \langle \alpha_t|a^\dagger a|\alpha_t\rangle \hbar\omega = \langle n\rangle \hbar\omega, \qquad (8.2.25)$$

where we note that average n is equal to $|\alpha|^2$. The quantum expectation value

$$E_q = \langle \alpha_t|H|\alpha_t\rangle = \langle \alpha_t|a^\dagger a + \tfrac{1}{2}|\alpha_t\rangle \hbar\omega$$
$$= E_0 + \hbar\omega/2 \qquad (8.2.26)$$

COHERENT HARMONIC OSCILLATOR STATES

is greater than E_c by the zero-point energy. The zero-point energy is the energy of state $|0\rangle$, which is the only state which is both a coherent state ($\alpha = 0$) and an eigenstate ($n = 0$) for the harmonic oscillator.

We can now derive the coherent wave function as a function of time. Combining (8.2.20) and (8.2.21), we have

$$\chi_{\alpha_0}(x,t) \equiv \langle x|T(t,0)|\alpha_0\rangle = \langle x|Q(\alpha_t)|0\rangle e^{-i\omega t/2}.$$

Now rewriting Q using (8.2.14), we obtain

$$\chi_{\alpha_0}(x,t) = e^{-ix_t p_t/2\hbar}\langle x|B(p_t)D(x_t)|0\rangle e^{-i\omega t/2}$$

$$= e^{-ix_t p_t/2\hbar + ip_t x/\hbar} D(x_t)\chi_0(x)e^{-i\omega t/2}, \quad (8.2.27)$$

where (8.2.10) was used. Finally using (8.2.6) and (8.2.9), we get the coherent wave-packet formula:

$$\chi_{\alpha_0}(x,t) = e^{ix_t p_t/2\hbar - i\omega t/2}\left\{N_0 e^{-(\mu\omega/2\hbar)(x-x_t)^2 + ip_t(x-x_t)/\hbar}\right\}. \quad (8.2.28)$$

The real part (Re χ_α) and modulus ($|\chi_\alpha|$) of the coherent wave functions are plotted in Figures 8.2.2(a) and 8.2.2(b). Ten snapshots are shown for equal time intervals following the setting of classical initial conditions ($x = x_0$, $p = 0$). The x_0 value is chosen so that the classical energy $E_c = |\alpha_0|^2 \hbar\omega$ is exactly five quanta ($5\hbar\omega$) in Figure 8.2.2(b) and 20 quanta in Figure 8.2.2(a). Note how the wave packet develops phase wrinkles inside a Gaussian modulus envelope as time increases. The envelope follows a classical oscillator trajectory but does not change its shape. The wrinkles inside are simply a measure of the momentum that packet has at each instant. It is interesting to note that the packet develops a wrinkle long before it has moved even a fraction of its length. This is because momentum and position grow linearly and quadratically, respectively, with time.

Because of this "early wrinkling" the quantum overlap of the $\chi_\alpha(t)$ packet with the initial $\chi_\alpha(0)$ function will vanish long before the packet has moved out of its initial neighborhood. One might view the wave packet as moving uniformly around the phasor clock in Figure 8.2.1, so that initially all its motion is along the imaginary p axis of momentum.

The bracketed factor in the wave function (8.2.28) represents the Gaussian packet with average momentum p_t, and it is centered around position x_t. The phase factor outside the bracket can be shown to be consistent with some fundamental views of quantum theory which are loosely referred to as action integral formalism or Feynman path integrals, which is mentioned in Section 8.3.

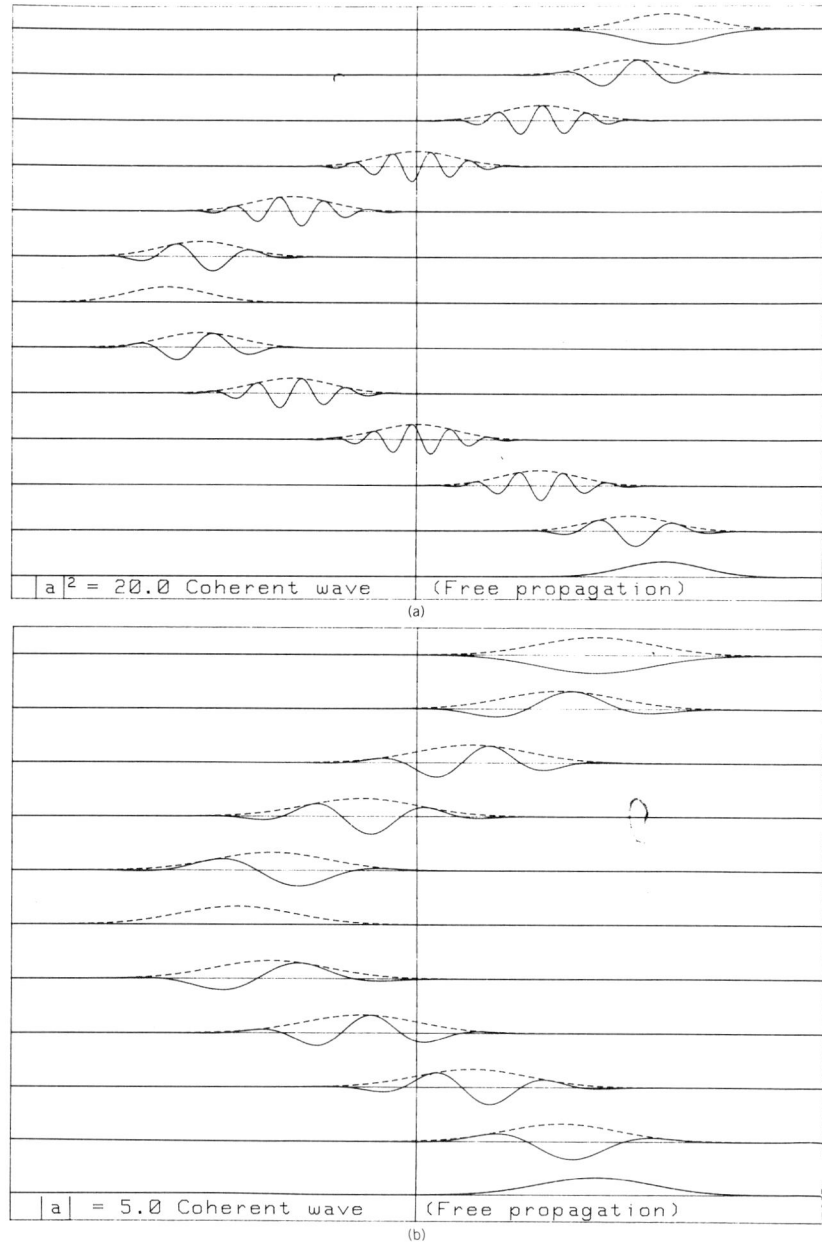

Figure 8.2.2 Coherent state wave packets at various times during oscillation.

According to these views one can associate each classical path with a so-called propagator θ of the form

$$\theta = e^{iS/\hbar}, \tag{8.2.29a}$$

where S_P is the principal action function given by (8.1.4); that is,

$$S_P = \int_0^t L\,dt, \tag{8.2.29b}$$

and L is the Lagrangian function which is related to the Hamiltonian H for the harmonic oscillator as follows:

$$L = p^2/2\mu - (\mu\omega^2/2)x^2 = p\frac{dx}{dt} - H. \tag{8.2.30}$$

If in the integral involving just $p\dot{x}$ [the integral of L without the H term is the "reduced action" S_H given by (8.1.7)] one takes coherent state expectation values of x and p, then one has

$$S_H = \int_0^t p_t\,dx_t = \int \left[2\mu(E_c - (\mu\omega^2/2)x_t^2)\right]^{1/2} dx_t$$

$$= x_t p_t/2\big|_0^t + (E_c/\omega)\sin^{-1}\left(\frac{\mu\omega}{\sqrt{2\mu E_c}}x_t\right)\bigg|_0^t$$

$$= (x_t p_t - x_0 p_0)/2 + E_c t. \tag{8.2.31}$$

Now by including the expected value of the H term (i.e., $E_q = E_c + \hbar\omega/2$) the propagator becomes

$$\theta = e^{iS_P/\hbar} = e^{i(x_t p_t - x_0 p_0)/2\hbar - i\omega t/2}, \tag{8.2.32}$$

which agrees with the phase factor outside of (8.2.28) except for the initial phase $x_0 p_0$, which was set to zero.

8.3 COHERENT WAVE GENERATION OF EIGENFUNCTIONS

The coherent wave $\chi_\alpha(x, t)$ is a combination of oscillator eigenfunctions $\langle x|n\rangle = \chi_n(x)$:

$$\chi_\alpha(x, t) = \sum_{n=0}^{\infty} c_n \langle x|n\rangle e^{-i(n+1/2)\omega t}, \tag{8.3.1a}$$

where the coefficients

$$c_n = e^{-|\alpha|^2/2}(\alpha)^n/\sqrt{n!} \tag{8.3.1b}$$

follow from (8.2.17) and are amplitudes of a Poisson distribution. Examples of this distribution are shown in Figure 8.3.1 for a variety of α values. These are simple cases of what are called Frank-Condon overlap distributions in vibrational spectroscopy. Each distribution peaks for the coefficient c_n for which n is nearest to $|\alpha|^2$. The width of each distribution is roughly determined by

$$\Delta n = (\langle n^2 \rangle - \langle n \rangle^2)^{1/2} = (\langle \alpha | a^\dagger a a^\dagger a | \alpha \rangle - \langle \alpha | a^\dagger a | \alpha \rangle^2)^{1/2} \tag{8.3.2}$$
$$= |\alpha| = \langle n \rangle^{1/2},$$

i.e., the square root of n. For values of n outside of this width the coefficients c_n decrease rapidly. Hence, it is well known that χ_α can be well approximated by a relatively small number ($\sim 2\sqrt{n} = 2|\alpha|$) of terms in its infinite sum representation (8.3.1).

Until recently, it was not known that the "inverse" was true; that is, the same small number of time snapshots of a propagating $\chi_\alpha(x, t)$ wave could be combined to produce eigenfunctions χ_n. Since 1975 Heller, Davis, Tannor, DeLeon, and others have shown how to produce eigenfunctions in a variety of multidimensional potentials using this "inverse" idea. We consider the derivation of these ideas in connection with the one-dimensional harmonic oscillator.

The ideas are based upon an approximate but discrete Fourier inversion of expansion (8.3.1a) in the following form:

$$c_m \langle x | m \rangle \simeq \sum_{t_p=0}^{t_{n-1}} \chi_\alpha(x, t_p) e^{i(m+1/2)\omega t_p} \tag{8.3.3}$$
$$\simeq \sum e^{ix_t p_t(2\hbar)} e^{im\omega t_p} N_0 e^{-(\mu\omega/2\hbar)(x-x_t)^2 - ip_t(x-x_t)/\hbar}.$$

This approximate expression (8.3.3) would be exact if the sum was replaced by an integral $((1/\tau)\int_0^\tau dt)$, but here we are interested in showing that it is still very accurate for just a small number N of order $N \sim 4\sqrt{m}$ if we choose $|\alpha| \simeq \sqrt{m}$. For example, 10 terms ($N = 10$) or "snapshots" of $\chi_\alpha e^{i\phi(t)}$ with $|\alpha|^2 = 5$ and a phase $\phi(t) = (m + 1/2)\omega t_p$ for $p = 0, 1, \ldots, 9$ with $m = 5$ are drawn one above the other in Figure 8.3.2(a). Their sum is shown below in Figure 8.3.2(b) and it is seen to accurately reproduce the fifth excited eigenfunction $\langle x | 5 \rangle$.

This inverse generation process might be described as an accumulation or "painting" of an eigenfunction by an oscillating phased packet. The accumu-

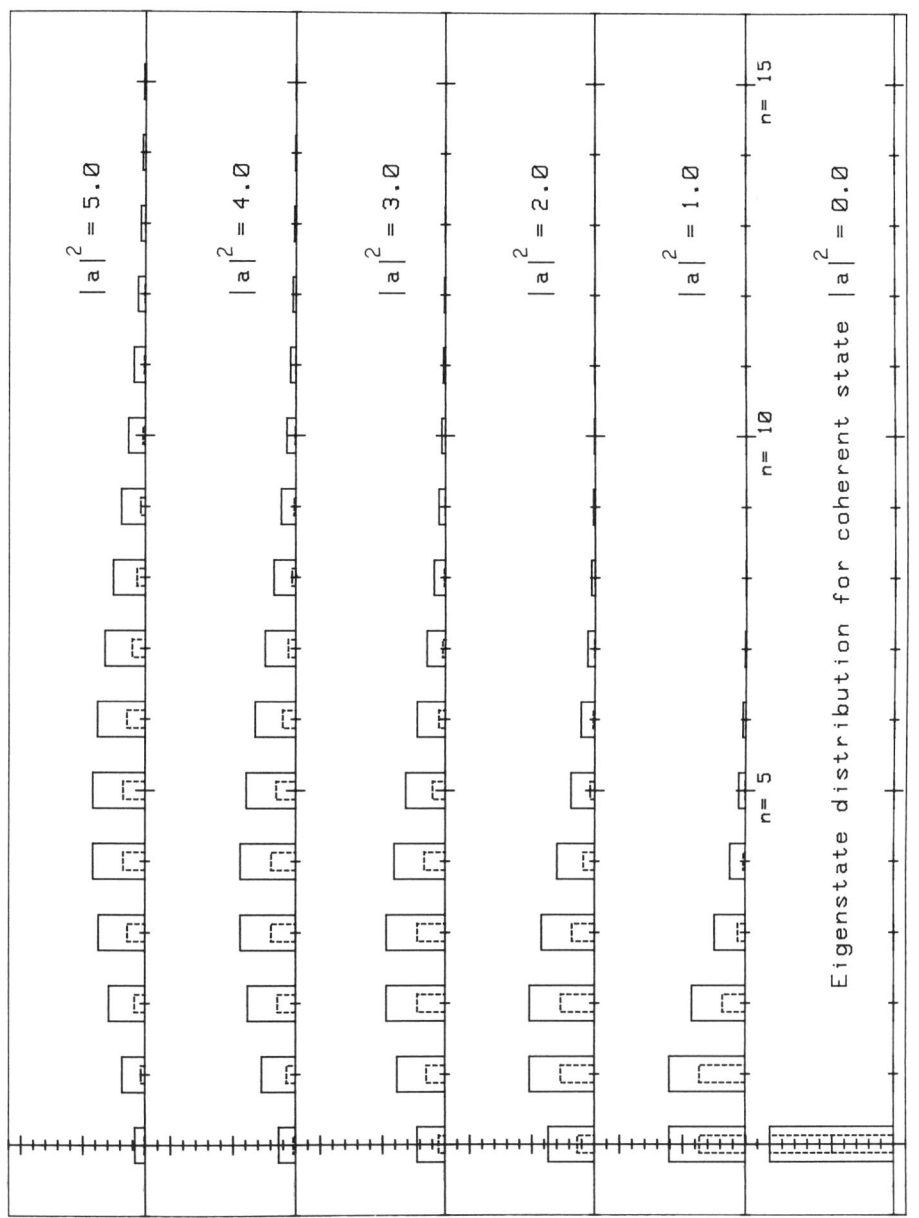

Figure 8.3.1 Poisson eigenstate distribution of amplitudes and probabilities in coherent state.

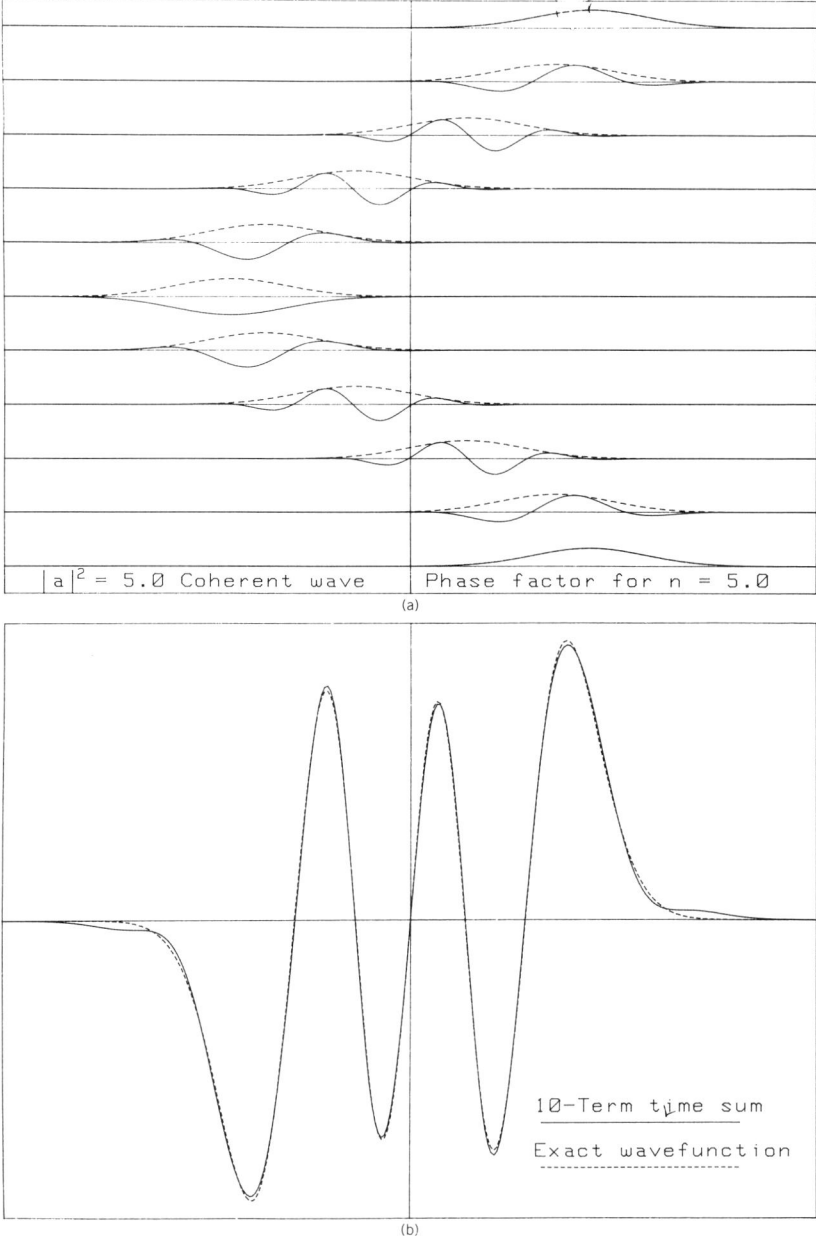

Figure 8.3.2 (a) Phased snapshots of coherent state. (b) Sum of snapshots in part (a) yields approximate eigenfunction.

lation of the same packet (i.e., $|\alpha|^2 = 5$) with a different phase (say one corresponding to $m = 4$) approximates a different eigenfunction (say $\langle x|4\rangle$). In fact, one can try fractional m (say 4.75) in the phase, but such phases will cause the wave to destructively interfere or beat itself to death as time goes on. Only the integer values of m will continue to add constructively and to accumulate more and more of a particular eigenfunction $\langle x|m\rangle$ period after period. For integral m the phased packet is exactly periodic so only one period is needed. Any detuning of m from an integer leads to beating instead of accumulation; that is, the amplitude behaves like an undamped oscillator responding to a detuned force. We will come back to this idea of "tuning" later on.

Let us consider what the error will be in the approximation (8.3.3) for varying $|\alpha|^2$ and m. Consider a sum over N equally spaced time intervals within one oscillator period ($\tau = 2\pi/\omega$) of the form

$$(1/N) \sum_{t_p=0}^{t_{N-1}} \chi_\alpha(x, t_p) e^{i(m+1/2)t_p} = \sum_{n=0}^{\infty} c_n \langle n|x\rangle (1/N) \sum_{t_p=0}^{N-1} e^{i(m-n)2\pi p/N}, \tag{8.3.4a}$$

where

$$t_p = 2\pi p/N \quad (p = 0, 1, 2, \ldots, N-1) \tag{8.3.4b}$$

and (8.3.1a) was used. The sum on the extreme right-hand side of (8.3.4a) is a geometrical series which appears repeatedly in theoretical optics and spectroscopy. Setting $\delta = 2\pi(m-n)/N$ one has for this series the following:

$$\Sigma = \sum_{p=0}^{N-1} e^{ip\delta} = (1 + e^{i\delta} + e^{i2\delta} + \cdots + e^{i(N-1)\delta}), \tag{8.3.5}$$

or

$$e^{i\delta} \Sigma = e^{i\delta} + e^{i2\delta} + \cdots + e^{i(N-1)\delta} + e^{iN\delta}.$$

Subtracting the two foregoing series and solving gives

$$\Sigma = (1 - e^{iN\delta})/(1 - e^{i\delta}) = e^{i(N-1)\delta/2} [\sin(N\delta/2)]/\sin(\delta/2)$$
$$= e^{i(N-1)\pi(m-n)/N} [\sin \pi(m-n)]/\sin(\pi(m-n)/N). \tag{8.3.6}$$

The Σ function is called the spectral function in Section 8.4B(b). Here it determines the relative amounts of various eigenfunctions $\langle n|x\rangle$ which would show up in the wave-packet accumulation (8.3.4). If m is an integer then the following expression results:

$$\Sigma = N\big[\delta_{mn} + e^{i(N-1)\pi}(\delta_{m,n+N} \pm \delta_{m,n-N}) \\ + e^{i2(N-1)\pi}(\delta_{m,n+2N} \pm \delta_{m,n-2N}) + \cdots \big]. \quad (8.3.7)$$

Substituting this into (8.3.4a) yields

$$(1/N) \sum_{t_p=0}^{t_{N-1}} \chi_\alpha(x, t_p) e^{(m+1/2)\omega t_p}$$
$$= c_m \langle m|x\rangle + e^{i(N-1)\pi}(c_{m+N}\langle m+N|x\rangle \pm c_{m-N}\langle m-N|x\rangle) + \cdots. \quad (8.3.8)$$

If N is large enough that $c_{m\pm N}, c_{m\pm 2N}, \ldots$, etc. are small compared to c_m, then (8.3.3) is a good approximation of the mth eigenfunction. Figure 8.3.1 shows the magnitude of the first correction coefficient c_{5+10} on the right-hand side of (8.3.8) for ($N = 10$) and ($m = |\alpha|^2 = 5$). The correction is a fraction

$$c_{15}/c_5 = \big[(5^{15}/15!)/(5^5/5!)\big]^{1/2} = 0.0299 \quad (8.3.9)$$

of the desired result, that is, about 3%. This is about the error that is observed in Figure 8.3.2(b), and it shows up mostly at the turning points. Adding two more terms ($N = 12$) would reduce the error to below 1%, and this shows how rapidly the accumulation procedure converges.

The accumulation procedure may be generalized to generate eigenfunctions for anharmonic potentials. In place of the wave packet in (8.3.3) or (8.2.28), one may accumulate the following wave function:

$$\chi(x, t) = e^{iS_P/\hbar}\{Ne^{-(\mu\omega/2\hbar)(x-x_t)^2 + ip_t(x-x_t)/\hbar}\}, \quad (8.3.10a)$$

where x_t and p_t are determined by *classical* Newton's or Hamilton's equations for the potential. The classical values are used to compute the action function

$$S_P = \int_0^t L\, dt = \int_0^t p_t\, dx_t - \int_0^t \langle \chi|H|\chi\rangle\, dt. \quad (8.3.10b)$$

The Hamiltonian energy $\langle \chi|H|\chi\rangle$ is just the expectation value of H for wave function (8.3.10a) at time t. Eigenfunctions will be obtained in an accumulation of ($\chi e^{i\omega t}$) for those energies $E_q = \hbar\omega$ for which this quantity returns to its original value after one period. This is equivalent to demanding that

$$S = \int p_t\, dx_t - \int \langle H\rangle\, dt = 2\pi n\hbar,$$

which corresponds to the Bohr-Sommerfeld quantization rule.

A. Wave-Packet Propagation and Spectral Quantization

We describe here a semiclassical approximation which yields approximate eigenfunctions as well as eigenvalues. The original spectral quantization method was developed by De Leon and Heller for treating multidimensional anharmonic vibrational potentials. After describing the basic method we shall demonstrate an adaptation of it to rotational wave functions. The wave-packet quantization methods often use a minimum uncertainty state $|\mathrm{MU}(q, p)\rangle$ or wave function $\langle|\mathrm{MU}(q, p)\rangle$ which is localized around a certain point (q, p) in classical phase space. In the original formulation of this problem for vibrational mechanics, this state was chosen to be a (q, p) translation of an oscillator ground state. In this case the corresponding wave function $\langle|\mathrm{MU}(q, p)\rangle$ is a translated complex Gaussian packet such as (8.2.28) but in several dimensions.

The method uses classical equations of motion to derive the time evolution of the phase variables $(p(t), q(t))$. It starts with a judicious choice of initial values $(p(0), q(0))$ which have a classical energy reasonably close to the eigenvalues that are desired. Then a path propagation state

$$|\Psi(t)\rangle = |\mathrm{MU}(q(t), p(t))\rangle e^{(i/\hbar)S_p(t)} \tag{8.3.11}$$

is obtained by numerically integrating the classical equations to obtain the phase variables $(q(t), p(t))$ and the classical action or Hamilton's principle function S_p:

$$S_p(t) = \int_0^t L\,dt = \int_0^t \left(\sum_i p_i \dot{q}_i - E_{cl}\right) dt. \tag{8.3.12}$$

Generally, one defines a classical translation operator **CT** which has the effect of changing the (q, p) parameters of the wave packet to the correct classical values for each time t:

$$|\mathrm{MU}(q(t), p(t))\rangle = \mathbf{CT}(q(t), p(t) : q(0), p(0))|\mathrm{MU}(q(0), p(0))\rangle. \tag{8.3.13}$$

The same **CT** operator can be used to set the initial wave packet, as well. The initial wave-packet expectation value of this operator, including the action phase, is called the time autocorrelation function:

$$\begin{aligned}\mathrm{AC}(t) &\equiv \langle \mathrm{MU}(q(0), p(0))|\mathrm{MU}(q(t), p(t))\rangle e^{(i/\hbar)S_p(t)} \\ &= \langle \Psi(0)|\Psi(t)\rangle.\end{aligned} \tag{8.3.14}$$

This function oscillates or "beats" each time the evolving wave packet returns on a classical path that is close enough to overlap with the initial packet. The energy Fourier transform of the autocorrelation FTAC(E) provides information which can be used to generate eigenfunctions. The values $E = E_{FT}$ which are peaks of the following transform function,

$$\text{FTAC}(E) = \frac{1}{2T} \int_{-T}^{T} dt \, e^{(i/\hbar)Et} \text{AC}(t), \qquad (8.3.15)$$

can be used to obtain an approximate semiclassical eigenfunction by a Fourier integral of the propagation state:

$$|E_{FT}\rangle = \frac{1}{2T} \int_{-T}^{T} dt \, e^{(i/\hbar)E_{FT}t} |\Psi(t)\rangle. \qquad (8.3.16)$$

Finally, the desired semiclassical energy is computed by an expectation value of the quantum Hamiltonian:

$$E_{SC} = \langle E_{FT}|\mathbf{H}|E_{FT}\rangle / \langle E_{FT}|E_{FT}\rangle. \qquad (8.3.17)$$

This last step is necessary since, as we will see, the FTAC peaks $\{E_{FT}, E'_{FT}, \ldots\}$ may differ slightly from the desired quantum energy values $\{E_{SC}, E'_{SC}, \ldots\}$.

One way to understand the wave-packet method is to study the conditions which would cause (8.3.15) to have a peak or to make (8.3.16) keep accumulating a stationary state over longer and longer integration times. Such a peak or accumulation implies that E has been chosen so as to obtain a phase coherence or constructive interference each time the evolving wave packet returns to the neighborhood of the initial packet. This occurs when the action for a closed path satisfies certain quantization conditions. In general, these are the Einstein-Brillouin-Keller (EBK) quantization conditions applied to Hamilton's characteristic function S_H or "reduced action" given according to (8.1.7) by

$$S_H = S_p + Et = \int \mathbf{p} \cdot d\mathbf{q}. \qquad (8.3.18)$$

For a closed path the EBK quantization conditions are

$$\oint \mathbf{p} \cdot d\mathbf{q} = (n + \alpha) 2\pi \hbar, \qquad (8.3.19)$$

where α is called the Maslov constant. In terms of the principle classical action (8.1.8) this gives

$$E_{SC}\tau + S_p = (n)2\pi\hbar, \qquad (8.3.20)$$

where τ is a classical period (or quasiperiod) for a closed (or nearly closed) path and E_{SC} is the desired semiclassical approximation to the energy. This may be rewritten

$$E_{FT}\tau + S_p = (n)2\pi\hbar, \quad (8.3.21)$$

where E_{FT} is a peak value in the FTAC (8.1.15). Let us write this as follows:

$$E_{FT} = E_{SC} + 2\pi\alpha\hbar/\tau. \quad (8.3.22)$$

If the Maslov constant is not zero, then the value of E_{SC} is shifted from that of E_{FT}. From (8.3.14) one sees that condition (8.3.21) makes the phase in (8.3.15) or (8.3.16) come out to a multiple of 2π after classical period τ.

Hence the wave-packet method amounts to a numerical resonance or spectroscopy experiment in which one finds stationary states by "tuning" E to get a peak in (.3.15) and then "painting" the eigenfunction using (8.3.16). The resulting eigenfunction consists of a sum of a series of MU packets set out on a classical trajectory. The last step can be accomplished with a relatively small number of time steps. It is also helpful to "tune" the initial conditions $(q(0), p(0))$ in (8.3.13) to get the strongest peaks.

The rotational adaptation of wave-packet propagation involves a different type of minimum uncertainty state, phase variables, and classical translation operator. The latter is a rotation operator which affects the Euler angles $\{\alpha\beta\gamma\}$, and these angles are the rotational equivalent of phase variables. Recall from Section 7.4 that the angles $-\beta$, and $-\gamma$ are the RE surface coordinates. The wave-packet propagation can be viewed as taking place on the RE surface shown in Figure 7.4.1(b).

A form for the rotational wave packet is suggested by the discussion of angular-momentum cones in Section 7.4. A choice for a rotational minimum uncertainty wave packet is the wave function associated with the narrowest angular-momentum cone; that is,

$$|\text{MU}(0,0,0)\rangle = \left|{J \atop J}\right\rangle. \quad (8.3.23)$$

A good choice for an initial wave packet is one centered on some part of a quantizing RE surface path. Assuming that the point ($\beta = 0$, $\gamma = 0$) is the center of an RE surface hill or valley, then the Kth quantizing path would be near to the point $(\beta_0 = -\Theta_K^J, \gamma_0 = 0)$. A suitable initial wave packet is therefore obtained by the following y rotation of initial state (23):

$$\left|\text{MU}(0, -\Theta_K^J, 0)\right\rangle = \mathbf{R}(0, -\Theta_K^J, 0)|_J^J\rangle. \quad (8.3.24)$$

(Negative Euler angles are used since the rotation is defined with respect to the body fixed frame.) The choice of initial angles is usually not so critical, but a good initial guess can reduce computational time somewhat.

The classically propagating wave-packet state (8.3.11) has the following form for the rotational problem:

$$|\Psi(t)\rangle = \mathbf{R}(\alpha(t),\beta(t),\gamma(t))|^J_J\rangle \exp[(i/h)S_p(t)]. \quad (8.3.25)$$

The Euler angles and the action are obtained by numerical solutions of Hamilton's equations:

$$\dot{\delta} = \frac{\partial H}{\partial J_\delta}, \quad \dot{J}_\delta = -\frac{\partial H}{\partial \delta}, \quad \delta = \alpha, \beta, \gamma, \quad (8.3.26a)$$

and

$$\dot{S}_p = L = J_\alpha \dot{\alpha} + J_\beta \dot{\beta} + J_\gamma \dot{\gamma} - E. \quad (8.3.26b)$$

The initial conditions $(\alpha(0), \beta(0), \gamma(0))$ given by Eq. (8.3.24) may be used along with $S_p(0) = 0$ for the action.

The autocorrelation function is given in terms of the initial and final Euler angles:

$$AC(t) = \langle \Psi(0)|\Psi(t)\rangle = \left\langle \begin{matrix}J\\J\end{matrix}\middle| \mathbf{R}^\dagger(\alpha(0),\beta(0),\gamma(0))\mathbf{R}(\alpha(t),\beta(t),\gamma(t))\middle| \begin{matrix}J\\J\end{matrix}\right\rangle$$
$$\times \exp[(i/h)S_p(t)]. \quad (8.3.27)$$

The rotation product is reduced and represented by a Wigner D function:

$$AC(t) = \left\langle \begin{matrix}J\\J\end{matrix}\middle| \mathbf{R}(\alpha_p, \beta_p, \gamma_p)\middle| \begin{matrix}J\\J\end{matrix}\right\rangle \exp[(i/h)S_p(t)]$$
$$= D^J_{JJ}(\alpha_p, \beta_p, \gamma_p)\exp[(i/h)S_p(t)]. \quad (8.3.28)$$

The group product rule and the D-function formula are given in Chapter 5. An example of a rotor autocorrelation function and its Fourier transform (8.3.15) are plotted in Figures 8.3.3(a) and 8.3.3(b), respectively. The beats due to the returning wave packet are shown in Figure 8.3.3(a). A peak in the transform at energy $E_{FT}(K = 8)$ has been singled out in Figure 8.3.3(b).

Each of the larger peaks $E_{FT}(K)$ indicates a possible eigensolution. According to Eqs. (8.3.16) and (8.3.24) the following state is an approximate eigenstate associated with that peak:

$$|E_{FT}(K)\rangle = \lim_{T\to\infty} \frac{1}{T}\int_{-T}^{T} dt \, \exp[(i/h)(E_{FT}(K) + S_p(t))]$$
$$\times \mathbf{R}(\alpha(t),\beta(t),\gamma(t))\left|\begin{matrix}J\\J\end{matrix}\right\rangle. \quad (8.3.29)$$

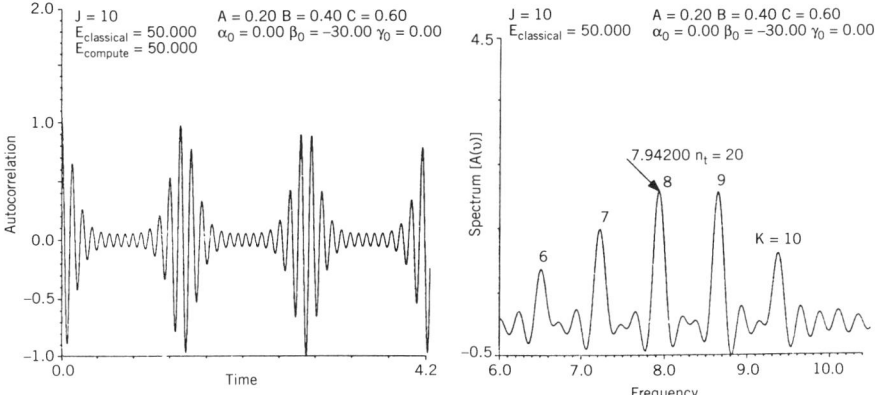

Figure 8.3.3 Wave-packet autocorrelation function and Fourier transform for asymmetric top. (a) Function AC(t) shows beats, (b) function FAC(ω) has peaks which correspond to possible eigenfunctions.

The representation of this state in the angular-momentum basis reduces to a Fourier transform of D functions:

$$\left\langle {J \atop K'} \middle| E_{\mathrm{FT}}(K) \right\rangle = \lim_{T\to\infty} \frac{1}{T} \int_{-T}^{T} dt \, \exp\left[(i/h)\left(E_{\mathrm{FT}}(K) + S_p(t)\right) \right]$$
$$\times D_{K'J}^{J*}(\alpha(t), \beta(t), \gamma(t)). \qquad (8.3.30)$$

It can be shown that the coefficients (8.3.30) are always real. The coefficients for the example chosen in Figure 8.3.3(b) are the following:

$K' =$	10,	8,	6,	4,	2,	0,...
$\langle K' | E_{\mathrm{FT}}(8) \rangle =$	0.118,	0.929,	−0.339,	0.075,	−0.012,	0.001,...	(8.3.31)

These are to be compared with the exact values for a localized state, which are the following:

$$\alpha_{K'} = 0.129, 0.922, -0.347, 0.106, -0.062, 0.018, \ldots. \qquad (8.3.32)$$

Finally, the semiclassical energy value for the wave-packet generated state (8.3.31) is given according to (8.3.17):

$$E_{\mathrm{SC}}(K=8) = \langle E_{\mathrm{FT}}(8) | \mathbf{H} | E_{\mathrm{FT}}(8) \rangle / \langle E_{\mathrm{FT}}(8) | E_{\mathrm{FT}}(8) \rangle = 53.134. \qquad (8.3.33)$$

This is to be compared with the quantum-mechanical values of 53.149 and 53.146 for the ($K = 8$) cluster. This completes an example of the most elementary "bare bones" wave-packet propagation technique. More sophisticated treatments give more accurate wave functions and energy values by taking tunneling into account.

8.4 SEMICLASSICAL RADIATION THEORY FOR SPECTROSCOPY

The quantum theory of radiation and matter is absolutely necessary for an accurate analysis of the most detailed spectroscopic effects. However, such an analysis may be impractical for laser experiments which involve a large number of radiation quanta. In that case semiclassical theory of radiation is used to treat the radiation effects as classical perturbations of the quantum states of matter.

This section contains a description of quantum excitations in atoms or molecules due to classical radiation fields. This semiclassical approach will be compared with the purely classical oscillator resonance approach, which was discussed in Section 6.5. The concept of the oscillator strength of a quantum transition may be clearly stated through such a comparison.

Semiclassical theory of radiation and matter is usually treated using perturbation expansions. When the perturbations are only weakly resonant, one or two terms of such expansions contain most of the relevant information. Perturbation techniques are discussed at the end of this section. When the laser is resonant with a particular transition between two levels the perturbation approach is less convenient. Then one may use the two-level atom semiclassical approach which is discussed in the following Section 8.5. We will set the stage for a comparison of these approaches.

The first part of this section will be devoted to developing the classical formulation of electromagnetic interactions with matter. This includes a rather ticklish and still somewhat controversial question of $\mathbf{A} \cdot \mathbf{p}$ versus $\mathbf{E} \cdot \mathbf{r}$ interaction Hamiltonians.

A. Lagrangians and Hamiltonians for Electromagnetic Interactions

The derivation of the Lagrangian and Hamiltonian equations for the interaction of a particle or mass μ with electric and magnetic fields $\mathbf{E}$ and $\mathbf{B}$ begins with Newton's equation and the electromagnetic force relations:

$$\mu \frac{d\mathbf{v}}{dt} = \mathbf{F} = q[\mathbf{E} + \mathbf{v} \times \mathbf{B}] = q\left[-\nabla \Phi - \frac{\partial \mathbf{A}}{\partial t} + \mathbf{v} \times (\nabla \times \mathbf{A})\right].$$

This expands to the following:

$$\mu \frac{d\mathbf{v}}{dt} = q\left[-\nabla \Phi - \frac{\partial \mathbf{A}}{\partial t} + \nabla(\mathbf{v} \cdot \mathbf{A}) - (\mathbf{v} \cdot \nabla)\mathbf{A}\right]. \quad (8.4.1a)$$

Here the electric and magnetic fields are expressed in terms of scalar and vector potential fields Φ and $\mathbf{A}$, respectively:

$$\mathbf{E} = -\nabla\phi - \frac{\partial A}{\partial t}, \quad \mathbf{B} = \nabla \times \mathbf{A}. \tag{8.4.1b}$$

The total derivative of $\mathbf{A}$ experienced by the moving particle is

$$\frac{d\mathbf{A}}{dt} = \frac{\partial \mathbf{A}}{\partial t} + (\mathbf{v} \cdot \nabla)\mathbf{A}. \tag{8.4.2}$$

Then the Lagrangian form of the force can be written as follows;

$$\mathbf{F} = \mu \frac{d\mathbf{v}}{dt} = q\left[-\frac{d\mathbf{A}}{dt} - \nabla(\Phi - \mathbf{v} \cdot \mathbf{A})\right],$$

$$F = \mu \frac{d\mathbf{v}}{dt} = \frac{d}{dt}\frac{\partial}{\partial \mathbf{v}}(q\Phi - q\mathbf{v} \cdot \mathbf{A}) - \nabla(q\Phi - q\mathbf{v} \cdot \mathbf{A}). \tag{8.4.3}$$

The Lagrangian function defined by

$$L = \mu(\mathbf{v} \cdot \mathbf{v})/2 - (q\Phi - q\mathbf{v} \cdot \mathbf{A}) \tag{8.4.4a}$$

then satisfies the Lagrangian form of Newton's equations; that is,

$$\frac{d}{dt}\frac{\partial L}{\partial \mathbf{v}} - \frac{\partial L}{\partial \mathbf{r}} = 0. \tag{8.4.4b}$$

The canonical form for the Hamiltonian function is found by Legendre transformation of L:

$$H = \Sigma \dot{q}_j p_j - L = \mathbf{v} \cdot \frac{\partial l}{\partial \mathbf{v}} - L \quad \left(H = \frac{\mu}{2}(\mathbf{v} \cdot \mathbf{v}) + q\Phi\right), \tag{8.4.5a}$$

$$H = (1/2\mu)(\mathbf{p} - q\mathbf{A}) \cdot (\mathbf{p} - q\mathbf{A}) + q\Phi, \tag{8.4.5b}$$

$$H = p^2/2\mu - (q/2\mu)(\mathbf{p} \cdot \mathbf{A} + \mathbf{A} \cdot \mathbf{p}) + (q^2/2\mu)\mathbf{A} \cdot \mathbf{A} + q\Phi. \tag{8.4.5c}$$

Here the canonical momentum is defined by

$$\mathbf{p} = \frac{\partial L}{\partial \mathbf{v}} = \mu\mathbf{v} + q\mathbf{A}. \tag{8.4.5d}$$

The Hamiltonian expression in parentheses [Eq. (8.4.5a)] is quantitatively correct but incorrect in form, since Hamiltonians are supposed to be functions of coordinates $\mathbf{q} = \mathbf{r}$ and *momentum* $\mathbf{p}$, while Lagrangians are functions of $\mathbf{q}$ and *velocity* $d\mathbf{q}/dt = \mathbf{v}$.

724 SYMMETRY ANALYSIS FOR SEMICLASSICAL AND QUANTUM MECHANICS

In classical or semiclassical spectroscopic theory the radiation part of the electromagnetic fields is described in terms of plane waves:

$$\mathbf{A} = \hat{\mathbf{e}}(2|a|)\sin(\mathbf{k}\cdot\mathbf{r} - \omega t + \phi), \tag{8.4.6a}$$

$$-\frac{\partial \mathbf{A}}{\partial t} = \mathbf{E}^{\text{rad}} = \hat{\mathbf{e}} E_0 \cos(\mathbf{k}\cdot\mathbf{r} - \omega t + \phi) \quad (\text{where } E_0 = 2|a|\omega), \tag{8.4.6b}$$

$$\nabla \times \mathbf{A} = \mathbf{B}^{\text{rad}} = \hat{\mathbf{k}} \times \hat{\mathbf{e}} B_0 \cos(\mathbf{k}\cdot\mathbf{r} - \omega t + \phi) \quad (\text{where } B_0 = 2|a|k). \tag{8.4.6c}$$

The wave vector $\mathbf{k}$ is transverse to the unit polarization vector $\hat{\mathbf{e}}$. This leads to the well-known transverse gauge conditions:

$$\nabla \cdot \mathbf{E}^{\text{rad}} = 0 = \nabla \cdot \mathbf{A}. \tag{8.4.7}$$

Since the standard quantum representation of the canonical momentum operator is a gradient $\mathbf{p} = (\hbar/i)\nabla$, $\mathbf{p}\cdot\mathbf{A}$ and $\mathbf{A}\cdot\mathbf{p}$ operators then have the same effect:

$$\mathbf{p}\cdot\mathbf{A}\psi(\mathbf{r}) = \mathbf{A}\cdot\mathbf{p}\psi(\mathbf{r}). \tag{8.4.8}$$

The time-dependent Schrödinger equation for a wave function $\psi(\mathbf{r}) = \langle \mathbf{r}|\psi\rangle$ using Hamiltonian (8.45c) is

$$i\hbar\frac{\partial\psi}{\partial t} = H\psi = \left[\frac{((\hbar/i)\nabla - q\mathbf{A})^2}{2\mu} + V(r)\right]\psi. \tag{8.4.9}$$

The potential $V(r)$ includes the electrostatic potential $q\Phi(r)$ and any additional atomic or molecular effective potentials due to forces which are not a direct result of electromagnetic fields. The Schrödinger equation expands to the following form using (8.4.8):

$$i\hbar\frac{\partial\psi}{\partial t} = \left[-\hbar^2\nabla^2/2\mu + V(r) - (q/\mu)\mathbf{A}\cdot(-\hbar i\nabla) + (q^2/2\mu)\mathbf{A}\cdot\mathbf{A}\right]\psi. \tag{8.4.10}$$

It is necessary to distinguish canonical momentum $\mathbf{p}(\alpha)$ [here represented by $(\hbar/i)\nabla$] from particle momentum $\mathbf{p}(\beta)$, where

$$\mathbf{p}(\beta) = \mathbf{p}(\alpha) - q\mathbf{A} \tag{8.4.11}$$

follows from (8.4.5d). Under certain conditions an alternative basis or picture can be obtained in which the particle momentum is also the canonical momentum. One may use a boost transformation B of the form (8.2.10) to subtract $q\mathbf{A}$ from $\mathbf{p}(\alpha)$ as follows:

$$\mathbf{p}(\beta) = B\mathbf{p}(\alpha)B^\dagger = \mathbf{p}(\alpha) - q\mathbf{A}, \qquad (8.4.12a)$$

where

$$B = e^{iq\mathbf{A}\cdot\mathbf{r}/\hbar}, \qquad (8.4.12b)$$

This is possible only if $\mathbf{A}$ satisfies the conditions of the dipole approximation. These conditions require that the radiation wavelength $2\pi/k$ is long compared to atomic dimensions so that one may assume a zero gradient of $\mathbf{A}$ ($\nabla \mathbf{A} \cong 0$). Using the gradient representation of $\mathbf{p}(\alpha)$ it is easy to verify the boost transformation relations (8.4.12a) in the dipole approximation.

(a) Transformation between Pictures The boost transformation B may be used to set up a new set of base states or a new "picture." The old α-picture basis $\{ \cdots |r\rangle_\alpha \cdots \}$ of position states is boosted into the new β-picture basis $\{ \cdots |r\rangle_\beta \cdots \}$, where the following relations hold:

$$|r\rangle_\beta = B|r\rangle_\alpha, \qquad |r\rangle_\alpha = B^\dagger |r\rangle_\beta,$$
$$_\beta\langle r| = {_\alpha\langle r|}B^\dagger, \qquad {_\alpha\langle r|} = {_\beta\langle r|}B. \qquad (8.4.13)$$

Now a given state $|\psi\rangle$ can be represented by an old α-picture wave function $\psi^\alpha(r) = {_\alpha\langle r|\psi\rangle}$ or a new β-picture wave function defined as follows (the conventional script $\mathscr{B}$ denotes a representation of the abstract B operator):

$$\psi^\beta(r) = {_\beta\langle r|\psi\rangle} = {_\alpha\langle r|}B^\dagger|\Psi\rangle$$
$$= \mathscr{B}^\dagger \psi^\alpha(r) = e^{-i\mathbf{A}\cdot\mathbf{r}/\hbar}\psi^\alpha(r), \qquad (8.4.14a)$$
$$\psi^\alpha(r) = \mathscr{B}\psi^\beta(r) = e^{i\mathbf{A}\cdot\mathbf{r}/\hbar}\psi^\beta(r). \qquad (8.4.14b)$$

This transformation is often treated as a *gauge transformation*; however, it may help to picture it physically as a boost. Neither interpretation is emphasized in the original literature. Synder and Richards discuss the general quantum transformation (8.4.14), while Goeppert-Mayer gave an equivalent classical canonical transformation. These papers are mentioned at the end of this chapter.

The first job of the boost operator $\mathscr{B}$ is to reduce the momentum factors $(\mathbf{p}(\alpha) - q\mathbf{A}) = -(hi\nabla + q\mathbf{A})$ in the kinetic term of the α-picture Schrödinger equation (8.4.9) to simple gradients:

$$(\mathbf{p}(\alpha) - q\mathbf{A})\psi^\alpha = \mathscr{B}\mathbf{p}(\alpha)\mathscr{B}^\dagger\mathscr{B}\psi^\beta = \mathscr{B}\mathbf{p}(\alpha)\psi^\beta,$$
$$(\mathbf{p}(\alpha) - q\mathbf{A})^2\psi^\alpha = \mathscr{B}\mathbf{p}(\alpha)^2\psi^\beta.$$

Here (8.4.11) and (8.4.14) were used. The potential term in (8.4.9) is unchanged since $\mathscr{B}$ commutes with all coordinate operators ($\mathscr{B}$ is a function of coordinates only):

$$V(r)\psi^\alpha = V(r)\mathscr{B}\psi^\beta = \mathscr{B}V(r)\psi^\beta.$$

However, the time derivative does not commute with $\mathscr{B}$ if $d\mathbf{A}/dt \neq 0$:

$$i\hbar\frac{\partial \psi^\alpha}{\partial t} = i\hbar\frac{\partial}{\partial t}\left[e^{iq\mathbf{A}\cdot\mathbf{r}/\hbar}\psi^\beta\right]$$

$$= e^{iq\mathbf{A}\cdot\mathbf{r}/\hbar}\left[i\hbar\frac{\partial \psi^\beta}{\partial t} - q\left(\frac{\partial \mathbf{A}}{\partial t}\cdot\mathbf{r}\right)\psi^\beta\right].$$

This gives the following time derivative:

$$i\hbar\frac{\partial \psi^\alpha}{\partial t} = \mathscr{B}\left[i\hbar\frac{\partial \psi^\beta}{\partial t} + q\mathbf{E}^{\text{rad}}\cdot\mathbf{r}\psi^\beta\right].$$

In the last line the vector potential relation (8.4.6b) was used. Assembling the transformed pieces of the α equation yields a new Schrödinger equation in the β picture:

$$i\hbar\frac{\partial \psi^\beta}{\partial t} = \left[-\hbar^2\nabla^2/2\mu + V(r) - q\mathbf{E}^{\text{rad}}\cdot\mathbf{r}\right]\psi^\beta. \qquad (8.4.15)$$

The transformation which gives the new equation is called a change of "picture" rather than just a change of basis. This terminology serves as a reminder that the transformation operator $B = e^{iq\mathbf{A}(t)\cdot\mathbf{r}/\hbar}$ is an explicit function of time through the vector potential. Nevertheless, the usual rules given in Chapter 1 relate representations of operators. For example, the coordinate representation of $\mathbf{p}(\beta)$ in the new β picture is equal to that of $\mathbf{p}(\alpha)$ in the old α picture according to (8.4.12) and (8.4.13):

$$_\beta\langle r'|\mathbf{p}(\beta)|r\rangle_\beta = {}_\alpha\langle r'|B^\dagger\mathbf{p}(\beta)B|r\rangle_\alpha = {}_\alpha\langle r'|\mathbf{p}(\alpha)|r\rangle_\alpha. \qquad (8.4.16)$$

In other words, if $\mathbf{p}(\alpha)$ is represented by $(\hbar/i)\nabla$ in the old picture then the same $(\hbar/i)\nabla$ will represent $\mathbf{p}(\beta)$ in the new picture. Now let us see how a change of picture affects the interaction part of the Hamiltonian.

It is a common practice to write Hamiltonians in the two Schrödinger equations (8.4.10) and (8.4.15) as a sum of a zeroth-order part H_0 and an interaction H_I:

$$H(\mu) = H_0(\mu) + H_I(\mu) \qquad (\mu = \alpha \text{ or } \beta). \qquad (8.4.17)$$

The interaction term for the α picture obviously has a different form than that of the β picture:

$$H_I(\alpha) = m(q/\mu)\mathbf{A} \cdot \mathbf{p}(\alpha) + (q^2/2\mu)\mathbf{A} \cdot \mathbf{A}, \quad (8.4.18a)$$

$$H_I(\beta) = -q\mathbf{E}^{\text{rad}} \cdot \mathbf{r}, \quad (8.4.18b)$$

but the zeroth-order term has the same form in either picture:

$$H_0(\alpha) = p(\alpha)^2/2\mu + V(r), \quad (8.4.19a)$$

$$H_0(\beta) = p(\beta)^2/2\mu + V(r). \quad (8.4.19b)$$

If the (α, β) labeling is deleted (as it almost always is), the two H_0 operators appear to be identical. Indeed, their corresponding representations *are* identical because of (8.4.16). Add to this the fact that the coordinate operators are the same in each picture and you have what may be one of the worst traps in theoretical physics! In extreme cases the unfortunate victims of this trap will fail to distinguish two different pictures and proceed to equate $H_0(\alpha)$ to $H_0(\beta)$ and $H_I(\alpha)$ to $H_I(\beta)$.

The β picture seems to be the most nearly foolproof picture because it allows canonical momentum and particle momentum to be one and the same thing ($\mathbf{p} = \mathbf{p}(\beta) \rightarrow (h/i)\nabla$). Therefore, eigenstates $|n\rangle_\beta$ and eigenfunctions $\phi_n^\beta(r) = {}_\beta\langle r|n\rangle_\beta$ of $H_0(\beta)$ behave "normally" when perturbed by $H_I(\beta)$. (We shall consider examples shortly.) A stationary state such as the ground state $|0\rangle_\beta$ of an oscillator potential has zero expectation value for the *particle* momentum (${}_\beta\langle 0|\mathbf{p}(\beta)|0\rangle_\beta = 0$).

In the α picture, on the other hand, the corresponding oscillator ground eigenstate $|0\rangle_\alpha$ of $H_0(\alpha)$ would have zero *canonical* momentum (${}_\alpha\langle 0|\mathbf{p}(\alpha)|0\rangle_\alpha = 0$). This means that the particle momentum expectation would be ${}_\alpha\langle 0|\mathbf{p}(\beta)|0\rangle_\alpha = -q\mathbf{A}$ according to (8.4.11a). This is consistent with the basis transformation definition in (8.4.13) which takes the following form:

$$|0\rangle_\alpha = B^\dagger|0\rangle_\beta = e^{-iq\mathbf{A}\cdot\mathbf{r}/\hbar}|0\rangle_\beta. \quad (8.4.20)$$

Hence, the state $|0\rangle_\alpha$ is a coherent state with momentum $-q\mathbf{A}$ relative to the $|0\rangle_\beta$ state according to the definitions in Section 8.2.

(b) An Attempt to Visualize and Compare $\mathbf{A} \cdot \mathbf{p}$ and $\mathbf{E} \cdot \mathbf{r}$ Pictures

The β picture has a lot to recommend it. It seems that the interaction Hamiltonian $(-q\mathbf{E} \cdot \mathbf{r})$ is simpler than the expression $[-(q/\mu)\mathbf{A} \cdot \mathbf{p} + (q^2/2\mu)\mathbf{A} \cdot \mathbf{A}]$ needed in the α picture. Also the particle momentum is the canonical $\mathbf{p}$ in the β picture. One wonders if one should not abandon the α picture altogether.

However, it should be remembered that the β picture is only valid under the conditions of the dipole approximation, and that it was derived, after all,

728 SYMMETRY ANALYSIS FOR SEMICLASSICAL AND QUANTUM MECHANICS

from the α picture. If one ever intends to accurately describe quantum optics involving higher multipole transitions, then the α picture will have to be the starting point. Furthermore, one can show that particle momentum can be canonical only in the dipole approximation.

Hence, it may be useful to understand the α and β pictures better, perhaps to even have a picture of the two pictures. Such a picture is attempted in Figures 8.4.1(a) and 8.4.1(b). We emphasize that this graphical portrayal is more of a caricature or mnemonic than a real picture. It should be used with caution. It is based upon a literal interpretation of the B transformation [(8.4.13) and (8.4.14)] as a boost which may be objectionable to some.

The β picture which is shown in Figure 8.4.1(b) is just a simple sketch of a slowly oscillating **E** field perturbing a particle in a fixed oscillator potential $V(r)$. The α picture depicted in Figure 8.4.1(a) shows an atomic potential well $V(r)$ attached to reference frame boosted to velocity $q\mathbf{A}/\mu$ so that the particle fixed in this frame would have momentum $q\mathbf{A}$.

The idea is that $\mathbf{p}(\beta) = \mathbf{p}(\alpha) - q\mathbf{A}$ is always equal to the particle momentum relative to the frame in which the potential $V(r)$ is fixed and $\mathbf{p}(\alpha)$ is the momentum in the α-picture frame literally. The α picture has no manifest electric force; the effect of $\mathbf{E}^{\text{rad}}$ is accomplished by the inertial force associated with an accelerated frame. The β picture is analogous to a martini olive being stirred; in the α picture the same excitation is accomplished by shaking the glass.

Visualizing base states $|r\rangle_\beta = B|r\rangle_\alpha$ or $|n\rangle_\beta = B|n\rangle_\alpha$ in these pictures may also be instructive. States $|r\rangle_\beta$ or $|n\rangle_\beta$ are represented by a delta function or wave packet, respectively, fixed relative to the potential $V(r)$, while $|r\rangle_\alpha = B^\dagger|r\rangle_\beta$ or $|n\rangle_\alpha = B^\dagger|n\rangle_\beta$ are always boosted oppositely to the motion of the $V(r)$ frame so that it remains fixed in the α picture. It appears that the $|n\rangle_\beta$ states will be more convenient for describing perturbations between levels in the potential $V(r)$ since it will not be necessary to "unboost" the initial and final eigenstates.

It is important to note that a boost in momentum does not immediately imply a translation in particle coordinate, since the two variables are independent. In fact the coordinate expectations are the same in either picture, since B and r commute:

$$_\alpha\langle n|\mathbf{r}|n\rangle_\alpha = {}_\beta\langle n|\mathbf{r}|n\rangle_\beta. \tag{8.4.21}$$

This shows that the classical coordinate excursions indicated in Figure 8.4.1(a) are misleading; somehow one needs to picture boost without translation! However, this fact does not invalidate this visual aid for most laser-atom excitations because the actual translations would be so very tiny. The classical translation of the E frame relative to the A frame would be

$$\mathbf{R}(t) = \int (q/\mu)\mathbf{A}\, dt = (q/\mu)(E_0/\omega^2)\cos(\phi - \omega t)\mathbf{e}_A. \tag{8.4.22}$$

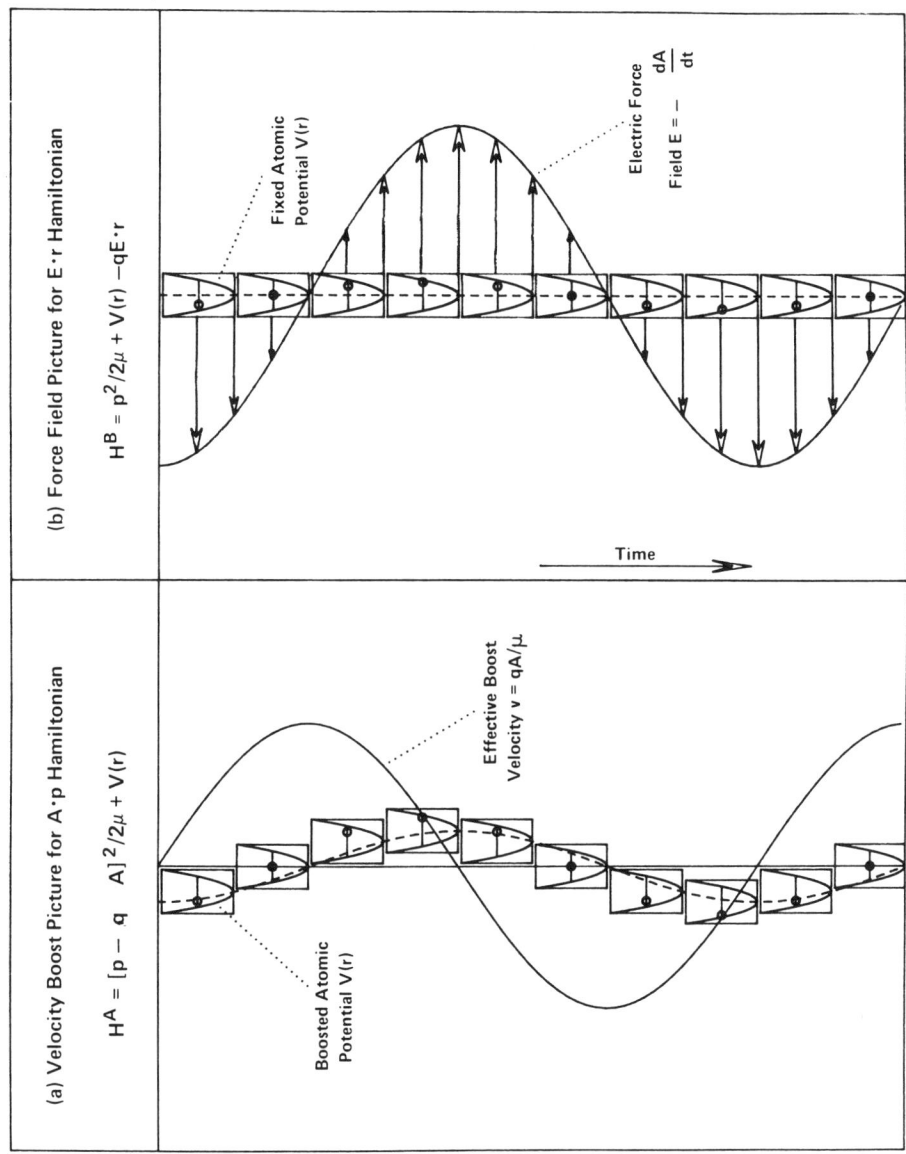

Figure 8.4.1 (a) Velocity boost picture for $A \cdot p$. (b) Force field picture for $E \cdot r$ Hamiltonian.

This is less than one nuclear radius (i.e., about 2×10^{-15} m) for atomic electronic transitions ($q = 1.6 \times 10^{19}$ C, $\mu = 9.1 \times 10^{-31}$ kg, $\omega \cong 10^{16}$) in a laser field of one million volts per meter.

B. Semiclassical Radiation Perturbation Theory

Perturbation theory gives solutions to the Schrödinger equation:

$$i\hbar \frac{\partial}{\partial t}|\psi(t)\rangle = (H_0 + H_I)|\psi(t)\rangle, \qquad (8.4.23)$$

in terms of a basis of eigensolutions $|N\rangle$ of the Hamiltonian H_0 when the coupling perturbation H_I is zero. The desired solution for nonzero coupling is given by

$$|\psi(t)\rangle = \sum_N c_N(t) e^{-i\omega_N t}|N\rangle, \qquad (8.4.24)$$

where the $|N\rangle$ and $\hbar\omega_N$ are (supposedly) known eigenvectors and eigenvalues of uncoupled or zeroth-order Hamiltonian H_0:

$$H_0|N\rangle = \hbar\omega_N|N\rangle. \qquad (8.4.25)$$

The desired coefficients $c_N(t)$ are constants unless the perturbation H_I is nonzero. If H_I is comparatively small, as it must be for standard perturbation theory to work, then the rate of change of $c_N(t)$ will be small compared to frequencies ω_N associated with the excited states. The coefficient of $|N\rangle$ is written $c_N(t)e^{-i\omega_N t}$ so that the unknowns $c_N(t)$ serve as amplitude *modulation* functions rather than whole amplitudes. The $c_N(t)$ modulation will change due to interaction H_I and not due to H_0 alone.

(a) Perturbation Expansion To derive the $c_N(t)$ time dependence we take the time derivative of (8.4.24):

$$i\hbar \frac{\partial}{\partial t}|\psi(t)\rangle = \sum_N \left(\hbar\omega_N c_N e^{-i\omega_N t}|N\rangle + i\hbar \frac{\partial c_N}{\partial t} e^{-i\omega_N t}|N\rangle \right). \qquad (8.4.26)$$

The factor $i\hbar$ was included so it is easy to compare this with the Schrödinger time equation (8.4.23) using the eigenequation (8.4.25):

$$i\hbar \frac{\partial}{\partial t}|\psi(t)\rangle = (H_0 + H_I) \sum_M c_M e^{-i\omega_M t}|M\rangle$$

$$= \sum_M \left(\hbar\omega_M c_M e^{-i\omega_M t}|M\rangle + c_M e^{-i\omega_M t} H_I|M\rangle \right). \qquad (8.4.27)$$

A comparison shows that the right-hand summation terms for (8.4.26) and (8.4.27) are equal:

$$\sum_N i\hbar \frac{\partial c_N}{\partial t} e^{-i\omega_N t} |N\rangle = \sum_M c_M e^{-i\omega_M t} H_I |M\rangle. \quad (8.4.28)$$

Using orthogonality ($\langle N|N'\rangle = \delta_{NN'}$) we see how the time derivative of the amplitude modulation function depends on H_I as follows:

$$i\hbar \frac{\partial c_N}{\partial t} = \sum_M c_M e^{i(\omega_N - \omega_M)t} \langle N|H_I|M\rangle$$

$$= \sum_M c_M V_{NM}. \quad (8.4.29)$$

Here it is convenient to define

$$V_{NM} = e^{i\omega_{NM} t} \langle N|H_I|M\rangle, \quad (8.4.30a)$$

where

$$\omega_{NM} \equiv \omega_N - \omega_M. \quad (8.4.30b)$$

If two or more coefficients c_M and c_N are nonzero the state $|\psi(t)\rangle$ in (8.4.24) will be nonstationary; that is, it will change rapidly even if H_I is zero and the c_N's are constant. The change depends on relative phases, i.e., differences like $\omega_{NM} = \omega_N - \omega_M$ which are the "beat" frequencies of the nonstationary state. In order to most easily see the effect of the perturbation H_I we shall choose the initial state to be a single stationary eigenstate $|\psi(0)\rangle = |B\rangle$. This means that the modulation coefficients at $t = 0$ are given by the following "zeroth approximation":

$$c_N^{(0)} = \delta_{NB}. \quad (8.4.31)$$

To calculate the first approximation this is substituted into the equation (8.4.29) of motion as follows:

$$i\hbar \frac{\partial c_N}{\partial t} = \sum_M \delta_{MB} V_{NM} = V_{NB}.$$

Then the first approximation to the solution to this equation will be

$$c_N^{(1)}(t) = \delta_{NB} + \frac{1}{i\hbar} \int_0^t dt_1 \, V_{NB}(t_1). \quad (8.4.32)$$

732 SYMMETRY ANALYSIS FOR SEMICLASSICAL AND QUANTUM MECHANICS

By repeating the substitution into (8.4.29) we obtain the second approximation, the third, and so forth.

$$c_N^{(2)}(t) = \delta_{NB} + \frac{1}{i\hbar}\int_0^t dt_1 V_{NB}(t_1) + \frac{1}{(i\hbar)^2}\sum_M \int_0^t dt_2 V_{NM}(t_2)\int_0^{t_2} dt_1 V_{MB}(t_1),$$

$$c_N^{(3)}(t) = c_N^{(2)}(t) + \frac{1}{(i\hbar)^3}\sum_M\sum_{M'}\int_0^t dt_3 V_{NM}(t_3)\int_0^{t_3} dt_2 V_{MM'}(t_2)\int_0^{t_2} dt_1 V_{M'B}(t_1).$$

(8.4.33)

The first approximation (8.4.32) becomes

$$c_N^{(1)}(t) = \delta_{NB} + \frac{1}{i\hbar}\int_0^t dt_1\, e^{i\omega_{NB}t_1}\langle N|H_I|B\rangle.$$ (8.4.34)

Here we will use the β picture and the $\mathbf{E}\cdot\mathbf{r}$ interaction (8.4.18b) for a monochromatic plane wave (8.4.6) of angular frequency ω:

$$\langle N|H_I|B\rangle = -q\mathbf{E}\cdot\langle N|\mathbf{r}|B\rangle = -qE_0\cos(\omega t - \phi)r_{NB}$$
$$= -q(i\omega a e^{-i\omega t} - i\omega a^* e^{i\omega t})r_{NB}.$$ (8.4.35a)

The amplitude and phase of the electric field are given by the following definitions

$$E_0 = 2|a|\omega,$$ (8.4.35b)

$$a = -i|a|e^{i\phi}.$$ (8.4.35c)

The dipole expectation value depends on the matrix element of the position operator projected onto the electric polarization vector $\hat{\mathbf{e}}$:

$$r_{NB} = \hat{\mathbf{e}}\cdot\langle N|\mathbf{r}|B\rangle.$$ (8.4.35d)

Inserting these values into (8.4.34) yields the following amplitudes:

$$c_N^{(1)}(t) = \delta_{NB} + q\frac{r_{NB}}{\hbar}\int_0^t dt_1(-\omega a e^{i(\omega_{NB}-\omega)t_1} + \omega a^* e^{i(\omega_{NB}+\omega)t_1}),$$

$$c_N^{(1)}(t) = \delta_{NB} + q\frac{r_{NB}}{\hbar}\left[i\omega a\frac{(e^{i(\omega_{NB}-\omega)t}-1)}{\omega_{NB}-\omega} - i\omega a^*\frac{(e^{i(\omega_{NB}+\omega)t}-1)}{\omega_{NB}+\omega}\right],$$

$$c_N^{(1)}(t) = \delta_{NB} + \frac{qr_{NB}|a|\omega}{\hbar}\left[e^{i\phi}\frac{(e^{i(\omega_{NB}-\omega)t}-1)}{\omega_{NB}-\omega} + e^{-i\phi}\frac{(e^{i(\omega_{NB}+\omega)t}-1)}{\omega_{NB}+\omega}\right].$$

(8.4.36)

(b) Spectral Function   The amplitude (8.4.36) is rewritten as follows

$$c_N^{(1)}(t) = \delta_{NB} + \frac{qr_{NB}E_0}{2\hbar}\left[e^{i\phi}S(\Delta^-,t) + e^{-i\phi}S(\Delta^+,t)\right] \quad (8.4.37)$$

in terms of the *spectral amplitude function*

$$S(\Delta,\tau) = \int_0^\tau dt\, e^{i\Delta t} = (2/\Delta)e^{i\tau\Delta/2}\sin(\tau\Delta/2) \quad (8.4.38a)$$

and angular frequency *detuning parameters*

$$\Delta^- = \omega_{NB} - \omega, \quad (8.4.38b)$$

$$\Delta^+ = \omega_{NB} + \omega. \quad (8.4.38c)$$

It is important to visualize the spectral function since it appears repeatedly in spectroscopy and optics. $S(\Delta,\tau)$ is the DC ($\omega = 0$) component of a running Fourier transform of $e^{i\Delta t}$ from $t = 0$ to $t = \tau$.

The probability or spectral intensity function $|S|^2$ is given by

$$|S(\Delta,\tau)|^2 = (4/\Delta^2)\sin^2(\tau\Delta/2). \quad (8.4.39)$$

This is plotted in Figure 8.4.2 as a function of the detuning (Δ) and time interval (τ). The spectral intensity becomes more and more strongly peaked around the detuning origin ($\Delta = 0$) as time (τ) increases. The frequency integral of the spectral intensity function grows linearly with time:

$$\int_{-\infty}^{\infty} d\Delta\, |S(\Delta,\tau)|^2 = 4\int_{-\infty}^{\infty} d\Delta\, \sin^2(\tau\Delta/2)/\Delta^2 = 2\pi\tau. \quad (8.4.40)$$

This is the basis for the so-called "golden rule" of constant transition rates which is discussed later. Meanwhile, the central peak height grows quadratically with time.

$$|S(0,\tau)|^2 = \tau^2. \quad (8.4.41)$$

The sidebands or fringe peaks on either side grow in a similar fashion but are much smaller. The largest sidebands are located at approximately $\Delta = \pm 3\pi/\tau$ on either side of the central peak. Their height is about $0.045\tau^2$ or less than 5% of the central peak. Concentric hyperbolas in the (Δ,τ) plane are the loci of alternating zeros and sideband peaks. The zeros nearest the central peak fall on the hyperbola

$$\Delta = \pm 2\pi/\tau$$

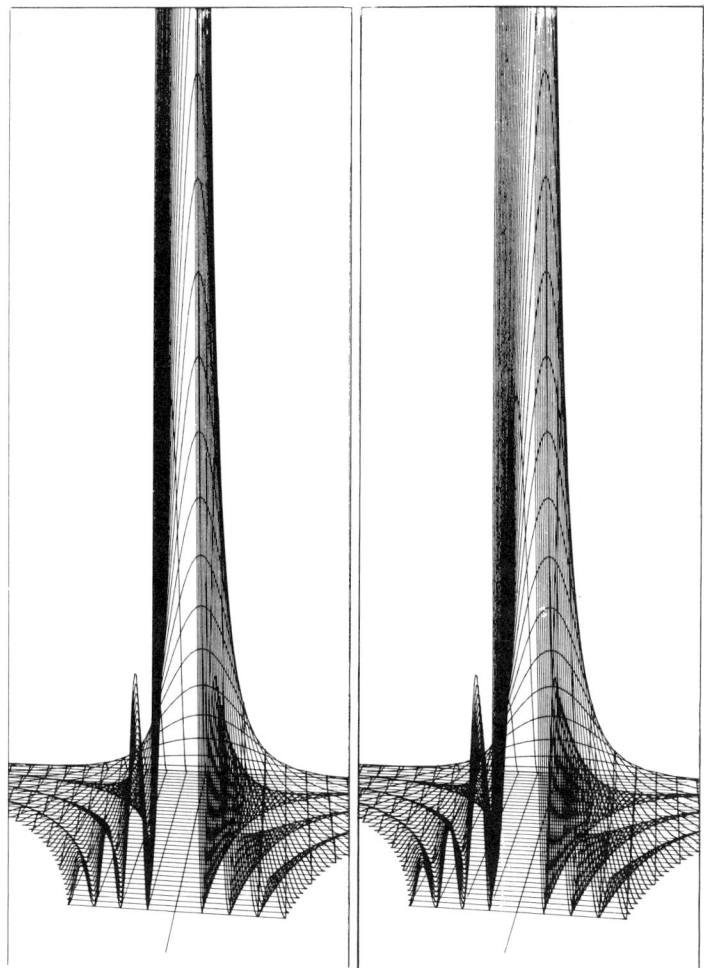

Figure 8.4.2 (Stereoptic pair) Spectral intensity function $|S(\Delta, \tau)|^2$ plotted versus the detuning parameter Δ ($\Delta < 0$ is left and $\Delta > 0$ is right) and elapsed time τ (τ increases coming toward the observer).

or

$$(\hbar\Delta)(\tau) = \pm 2\pi\hbar = \pm h. \qquad (8.4.42)$$

This can be viewed as a restatement of Heisenberg's uncertainty relation between the energy detuning range $(\hbar\Delta)$ and the time interval (τ) allowed for a given transition. One finds most of the probability within the bounds defined by the first hyperbola.

The interpretation of the semiclassical transition amplitude is complicated by the fact that it contains *two* spectral functions. The first function peaks when $\Delta^- = 0$ or

$$\omega = \omega_{NB} = \omega_N - \omega_B. \tag{8.4.43}$$

This peak would correspond to an absorption process in which the electron was stimulated at a precise angular frequency ω to jump from beginning state $|B\rangle$ to a higher state $|N\rangle$. The probability for this jump varies with time τ and frequency ω according to $|c_N^{(1)}(t)|^2$. A good approximation to $c^{(1)}(t)$ for $\omega - \omega_{NB}$ involves just the first spectral function in (8.4.37):

$$\left|c_N^{(1)}(t)\right|^2 \cong \frac{q^2|r_{NB}|^2 E_0^2}{4\hbar^2}|S(\omega_{NB} - \omega, t)|^2_{(\text{Absorption: } \omega \cong \omega_{NB})}. \tag{8.4.44a}$$

On the other hand, if the final state $|N\rangle$ lies below the beginning state $|B\rangle$, then the second term would be the main contribution to an emission probability which would then have the same approximate form:

$$\left|c_N^{(1)}(t)\right|^2 \cong \frac{q^2|r_{NB}|^2 E_0^2}{4\hbar^2}|S(\omega_{NB} + \omega, t)|^2_{(\text{Emission: } \omega \cong -\omega_{NB})}. \tag{8.4.44b}$$

A prominent feature of the probability function in Figure 8.4.2 are the wiggles or "beats" which occur at the difference frequency $\Delta = \omega_{NB} - \omega$ or $(\Delta = \omega + \omega_{BN})$ for fixed nonzero detuning Δ. This may remind you of the response of an undamped classical oscillator of frequency $\omega_0 = |\omega_{NB}|$ driven at frequency $\omega_s = \omega$. The classical equations derived in Section 6.5 [see Eqs. (6.5.5)–(6.6.8)] give the complex position function $x(t)$ for an undamped ($\Gamma = 0$) initially stationary ($x_0 = 0 = v_0$) oscillator of frequency ω_0 which experiences a stimulating force $E_0 q \cos(\omega_s t - \phi)$. For $\phi = 0$, the square of the position is approximately proportional to the square of the spectral function.

$$|x(t)|^2 = \left|(q/\mu)E_0(e^{-i\omega_s t} - e^{-i\omega_0 t})/(\omega_0^2 - \omega_s^2)\right|^2$$

$$= \left|4(q/\mu)E_0\right|^2 \sin^2[t(\omega_0 - \omega_s)/2]/(\omega_0^2 - \omega_s^2)^2$$

$$\cong |qE_0/2\mu\omega_0|^2 |S(\omega_0 - \omega_s, t)|^2.$$

In the last step the near-resonance approximation ($\omega_0 \cong \omega_s$) is used. The exact classical result for all frequencies and for arbitrary E-field phase (ϕ) follows from (6.5.5)–(6.5.8):

$$\text{Re } x(t) = (qE_0/\mu)$$
$$\times \left[\frac{\cos\phi(\cos\omega_s t - \cos\omega_0 t) + \sin\phi(\sin\omega_s t - (\omega_s/\omega_0)\sin\omega_0 t)}{\omega_0^2 - \omega_s^2}\right].$$
$$\tag{8.4.45}$$

(c) Oscillator Strength and Quantum Response An excellent test of semiclassical perturbation theory is a rederivation of the classical response (8.4.45) using a quantum oscillator basis. It can also be used to check $\mathbf{E} \cdot \mathbf{r}$ and $\mathbf{A} \cdot \mathbf{p}$ perturbation schemes. In order to rederive the classical behavior exactly it will be necessary to include both the resonant and nonresonant spectral function terms in the expression (8.4.37) for the amplitudes $c_F(t)$ of the stimulated state. From (8.4.24) and (8.4.31) this state is

$$|\psi(t)\rangle = e^{-i\omega t}\left(|B\rangle + \sum_{F \neq B} e^{-i\omega_{NB}t} c_F(t)|F\rangle\right),$$

and the expectation value of position is

$$\langle \psi|x|\psi \rangle = \sum_{F \neq B} e^{-i\omega_{FB}t} c_F^* x_{FB} + \sum_{F' \neq B} e^{-i\omega_{F'B}t} c_{F'} x_{BF'}.$$

Here we assume zero ground-state position expectation ($x_{BB} \equiv \langle B|x|B\rangle = 0$), and we neglect second-order excited state contributions ($c_F^* c_{F'} \langle F|x|F'\rangle \cong 0$). For the harmonic oscillator states these contributions are exactly zero. From (8.4.36) the expectation value becomes

$$\langle \psi|x|\psi\rangle = 2\,\text{Re}\sum_{F \neq B} e^{-i\omega_{FB}t} c_F x_{BF}$$

$$= (2q|a|\omega/\hbar)\,\text{Re}\sum_{F \neq B} r_{FB} x_{BF}$$

$$\times \frac{\left[e^{i\phi}(e^{-i\omega t} - e^{-i\omega_{FB}t})(\omega_{FB} + \omega) + e^{-i\phi}(e^{i\omega t} - e^{-i\omega_{FB}t})(\omega_{FB} - \omega)\right]}{\omega_{FB}^2 - \omega^2}.$$

(8.4.46)

If the position is measured along the E field so $r_{FB} = x_{BF}^*$ then (8.4.35) gives

$$\langle \psi|x|\psi\rangle = \sum_{F \neq B}\left(qE_0|x_{BF}|^2/\hbar [2\omega_{FB}\cos\phi(\cos\omega t - \cos\omega_{FB}t)\right.$$
$$\left. + \sin\phi(2\omega_{FB}\sin\omega t - 2\omega\sin\omega_{FB}t)\right]/(\omega_{FB}^2 - \omega^2).$$

This may be rearranged to correspond to the form of the classical position response (8.4.45) with $\omega_0 = \omega_{FB}$ and $\omega_s = \omega$:

$$\langle\psi|x|\psi\rangle = \sum_{F \neq B}\left\{\frac{2\omega_{FB}|x_{BF}|^2\mu}{\hbar}\right\}\frac{qF_0}{\mu}$$

$$\times \frac{\cos\phi(\cos\omega t - \cos\omega_{FB}t) + \sin\phi(\sin\omega t - (\omega/\omega_{FB})\sin\omega_{FB}t)}{\omega_{FB}^2 - \omega^2}$$

$$= \sum_{F \neq B} f_{FB} x_{\text{classical}}.$$

(8.4.47a)

The leftover factors are called the transition *oscillator strengths*

$$f_{FB} = 2\omega_{FB}|x_{FB}|^2\mu/\hbar. \quad (8.4.47b)$$

For the harmonic oscillator states $\{|B\rangle = |0\rangle, |F\rangle = |1\rangle, |F'\rangle = |2\rangle,\ldots\}$ only the first oscillator-strength term is nonzero.

$$x_{FB} = \langle F|x|B\rangle = \langle n|a^\dagger + a|0\rangle[\hbar/2\mu\omega_0]^{1/2} = \delta_{n,1}[\hbar/2\mu\omega_{FB}]^{1/2}.$$

For this case the expectation value is precisely the classical one given in (8.4.45) since only the first oscillator strength is unity:

$$f_{n0} = \delta_{n,1}. \quad (8.4.48)$$

It should be noted that a naive replacement of the $(q\mathbf{E}\cdot\mathbf{r})$ perturbation (8.4.18b) with the equivalent $(q\mathbf{A}\cdot\mathbf{p}/\mu)$ perturbation (8.4.18a) would not give the correct oscillator response. The momentum operator is related to position as follows:

$$[H_0, x] = [p^2, x]/2\mu = \hbar p/\mu i, \quad (8.4.49)$$

and the *p*-matrix elements are then derived as follows:

$$p_{NB} = \langle N|p|B\rangle = \mu i \langle F|[H_0, x]|B\rangle/\hbar = \mu i \omega_{NB} x_{NB}. \quad (8.4.50)$$

The matrix element which would replace (8.4.35a) would be

$$\langle N|H_I|B\rangle = -q\mathbf{A}\cdot\langle N|\mathbf{p}|B\rangle/\mu = -q(2|a|\mu)\sin(\mathbf{k}\cdot\mathbf{r} - \omega t - \phi)p_{NB}$$
$$= -q(ae^{-i\omega t} + a^*e^{i\omega t})p_{NB}/\mu$$
$$= -q(i\omega_{NB}ae^{-i\omega t} + i\omega_{NB}a^*e^{i\omega t})x_{NB}. \quad (8.4.51)$$

The relative phase of these terms differs from (8.4.35a) and when $\omega_{NB} \neq \omega$ they differ in amplitude, as well. Naive replacement will not work; the entire picture needs to be transformed, as discussed in Section 8.3. However, the squares of each item in (8.4.51) are equal to corresponding squares for (8.4.35a) at the point $\omega = \omega_{NB}$. This fact alone is sometimes used to "demonstrate" simple equality of $\mathbf{A}\cdot\mathbf{p}$ and $\mathbf{E}\cdot\mathbf{r}$ perturbations. This sets a trap into which many students fall.

An important spectroscopic principle involves the distribution of oscillator strengths among the quantum transitions. The harmonic oscillator transitions are peculiar in that the strength f_{10} of the $(1 \leftarrow 0)$ transition is unity while f_{n0} is zero for $n \neq 1$. In other atomic potentials the strengths f_{FB} could all

738 SYMMETRY ANALYSIS FOR SEMICLASSICAL AND QUANTUM MECHANICS

be nonzero but their sum is unity as shown in the following:

$$\sum_F f_{FB} = \sum_F 2\omega_{FB}\langle B|x|F\rangle\langle F|x|B\rangle \mu\hbar$$
$$= \sum_F 2\langle B|x|F\rangle\langle F|p|B\rangle/\hbar i$$
$$= 2\langle B|xp|B\rangle/\hbar i = -2\langle B|px|B\rangle/\hbar i = 1.$$

The commutation $[x, p] = \hbar i$ and momentum matrix element (8.4.50) were used here. This is the Thomas-Reiche-Kuhn relation for atomic oscillator strengths associated with the motion of a single electron.

The foregoing discussion provides a quantum mechanical foundation for the classical Lorentz model for atomic oscillator response. However, it is based upon first-order perturbation approximations (8.4.36) and (8.4.37). It was noted that the spectral function increases quadratically with time at resonance [recall Eq. (8.4.41)]. Eventually such an approximation yields a probability $|c_F|^2$ greater than one. We now derive an improved theory for an atomic oscillator that is strongly driven by an electric field whose frequency ω is close to the resonant frequency ω_{NB} of the atomic oscillator. This is the two-level atom approximation which is discussed next.

8.5 TWO-LEVEL SYSTEM APPROXIMATION

The Hamiltonian for a two-level model of ammonia (NH_3) inversion was introduced in Section 2.12. A representation (2.12.34) of this Hamiltonian was given using base states $\{|\text{up}\rangle, |\text{dn}\rangle\}$. There $|\text{up}\rangle$ corresponded the N atom being "up" along the direction of a positive z-axial E field, and $|\text{dn}\rangle$ corresponded to the N atom localized against the field. Symmetry-defined eigenvectors of the zero-field Hamiltonian were denoted by $\{|+\rangle, |-\rangle\}$ and given in terms of $|\text{up}\rangle$ and $|\text{dn}\rangle$ by (2.12.32), which is repeated in the following using a modified notation:

$$|1\rangle = |+\rangle = (|\text{up}\rangle + |\text{dn}\rangle)/\sqrt{2}, \qquad (8.5.1a)$$
$$|2\rangle = |-\rangle = (|\text{up}\rangle - |\text{dn}\rangle)/\sqrt{2}. \qquad (8.5.1b)$$

Let us compare the representation of the Hamiltonian in each of these two bases:

$$\begin{pmatrix} \langle +|H|+\rangle & \langle +|H|-\rangle \\ \langle -|H|+\rangle & \langle -|H|-\rangle \end{pmatrix} = \begin{pmatrix} H-S & -pE \\ -pE & H+S \end{pmatrix}, \qquad (8.5.2a)$$

$$\begin{pmatrix} \langle \text{up}|H|\text{up}\rangle & \langle \text{up}|H|\text{dn}\rangle \\ \langle \text{dn}|H|\text{up}\rangle & \langle \text{dn}|H|\text{dn}\rangle \end{pmatrix} = \begin{pmatrix} H-pE & -S \\ -S & H+pE \end{pmatrix}. \qquad (8.5.2b)$$

The change of basis from $\{|up\rangle, |dn\rangle\}$ to $\{|+\rangle, |-\rangle\}$ causes the tunneling parameter (S) or zero-field energy splitting ($2S$) to change places with the electric field potential energy parameter (pE) in the matrix. The states $|+\rangle$ and $|-\rangle$ are approximate eigenvectors when the electric field energy (pE) is small compared to tunneling (S) since the latter is on the diagonal of the $\{|+\rangle, |-\rangle\}$ representation. For large fields the $\{|up\rangle, |dn\rangle\}$ bases are preferred since it is pE which resides on the diagonal in this representation. The effects of strong versus weak constant electric fields are diagrammed in Figure 2.12.8 and these effects are known as the elementary DC-Stark effects.

This section is devoted to so-called AC-Stark effects, which are due to resonant oscillatory electric fields. The methods of analysis were first applied to nuclear magnetic resonance (NMR) experiments by Rabi, Ramsey, and Schwinger. The analogy between spin resonance and resonance of an arbitrary two-level system was pointed out shortly afterwards by Feynman, Vernon, and Hellwarth. These references were mentioned at the end of Chapter 7. The ammonia inversion maser was one of the first examples of coherent stimulated transitions and a forerunner to modern laser technology. The first part of the following describes the analogy and its geometrical interpretation.

A. Two-State Schrödinger Equations

The Schrödinger equation in the $\{|1\rangle, |2\rangle\}$ basis uses the Hamiltonian (8.5.2) as follows:

$$i\frac{\partial}{\partial t}\begin{pmatrix}\psi_1 \\ \psi_2\end{pmatrix} = \begin{pmatrix} H - S & -pE \\ -pE & H + S \end{pmatrix}\begin{pmatrix}\psi_1 \\ \psi_2\end{pmatrix}. \quad (8.5.3a)$$

The long-wavelength (dipole) approximation for an oscillating electric field interaction energy is given by $(-q\mathbf{E} \cdot \mathbf{r})$, where $\mathbf{r}$ is the displacement of effective charge q. If the $\mathbf{E}$ and $\mathbf{r}$ are only in the z direction then the parameter pE is given in terms of the following matrix element:

$$-pE/\hbar = -\langle 2|qE_z z|1\rangle/\hbar = r\cos(\Omega t)$$
$$= (r/2)(e^{-i\Omega t} + e^{i\Omega t}). \quad (8.5.3b)$$

Here the constant r is proportional to the interaction strength

$$r = -pE_z(0)/\hbar, \quad (8.5.3c)$$

where the constant

$$p = q\langle 2|z|1\rangle \quad (8.5.3d)$$

is the transition dipole moment measured along the C_3 symmetry axis (z) of

740 SYMMETRY ANALYSIS FOR SEMICLASSICAL AND QUANTUM MECHANICS

NH_3 which connects the two equilibrium positions for the N atom. The parameters are all expressed in units of $\hbar$ in order to simplify the notation for the Schrödinger equation as much as possible:

$$i\frac{\partial}{\partial t}\begin{pmatrix}\psi_1\\\psi_2\end{pmatrix} = \begin{pmatrix}\frac{\varepsilon}{2} & \frac{r}{2}(e^{-i\Omega t}+e^{i\Omega t})\\\frac{r}{2}(e^{i\Omega t}+e^{-i\Omega t}) & -\frac{\varepsilon}{2}\end{pmatrix}\begin{pmatrix}\psi_1\\\psi_2\end{pmatrix} \quad (8.5.4a)$$

It is also convenient to reset the energy zero ($H = 0$) and let

$$\varepsilon = -2S/\hbar \quad (8.5.4b)$$

be the difference between zero-field atomic or molecular levels 1 and 2 in angular frequency units. (Note that level 1 is below level 2 if $S > 0$ or $\varepsilon < 0$.)

For many applications one may drop either the positive or else the negative frequency component of the interaction $\langle 2|\mathbf{E}\cdot\mathbf{r}|1\rangle$ to obtain the following:

$$i\frac{\partial}{\partial t}\begin{pmatrix}\psi_1\\\psi_2\end{pmatrix} = \begin{pmatrix}\frac{\varepsilon}{2} & \frac{r}{2}e^{-i\Omega t}\\\frac{r}{2}e^{i\Omega t} & -\frac{\varepsilon}{2}\end{pmatrix}\begin{pmatrix}\psi_1\\\psi_2\end{pmatrix}. \quad (8.5.5)$$

This is the equation we will discuss first. It contains a complex interaction term $(r/2)e^{i\Omega t}$ which exactly models a certain type of rotating NMR excitation field as explained in the following. For the analogous NH_3 excitation it is only an approximation, albeit a very good one.

B. Spin and Crank Vector Visualization of Two-State Hamiltonian

Either equation (8.5.4) or (8.5.5) can be visualized more easily by appealing to the three-dimensional ($R(3)$) picture of the two-state ($SU(2)$) system. To do this one may calculate the position of the Hamiltonian crank vector ω indicated in Figure 7.5.3. [See also Eq. (7.5.5).] The result is a crank which flops around at laser angular frequency Ω in addition to its usual cranking at angular frequency ω. The Cartesian crank components are as follows for the Hamiltonian in (8.5.5):

$$\begin{aligned}\omega_x &= 2\,\text{Re}\,H_{21} & \omega_y &= 2\,\text{Im}\,H_{21} & \omega_z &= H_{11}-H_{22}\\ &= r\cos\Omega t & &= r\sin\Omega t & &= \varepsilon.\end{aligned} \quad (8.5.6)$$

If the real Hamiltonian in (8.5.4) is used, then the following crank components result:

$$\omega_x = 2r\cos\Omega t, \qquad \omega_y = 0, \qquad \omega_z = \varepsilon. \qquad (8.5.7)$$

In the latter case only the ω_x component oscillates, but with twice the amplitude. In the former case given by (8.5.6) the ω vector traces out an inverted cone of altitude ε and radius r. One may imagine the tail of the ω vector fixed at origin and its tip rotating at angular velocity Ω counterclockwise around a horizontal circle of radius r. The base circle of the cone is centered at a point on the ω_z axis at ε units above origin.

The well-known interaction between a magnetic moment $\mathbf{m} = g\mathbf{S}$ and field $\mathbf{B}$ is given by the elementary nuclear magnetic resonance Hamiltonian which describes precession of the moment or spin vector. The coefficient g is the gyromagnetic ratio between the spin and moment of a particle:

$$H_{\text{NMR}} = \mathbf{m} \cdot \mathbf{B} = g\mathbf{S} \cdot \mathbf{B}. \qquad (8.5.8)$$

We have seen that any Hamiltonian of the form $\boldsymbol{\omega} \cdot \mathbf{S}$ causes $\mathbf{S}$ to precess at angular velocity $|\boldsymbol{\omega}|$ around vector $\boldsymbol{\omega}$. So for NMR theory the crank or $\boldsymbol{\omega}$ vector is in the direction of the $\mathbf{B}$ field:

$$\boldsymbol{\omega} = g\mathbf{B}. \qquad (8.5.9)$$

In principle, the $\mathbf{B}$ field can be aimed in any direction in space and made to follow an arbitrary curve. For example, it could follow the circle prescribed by (8.5.6). If vectors $\boldsymbol{\omega}$ and $\mathbf{B}$ are stationary or moving very slowly ($\Omega \sim 0$) the spin vector $\mathbf{S}$ precesses around $\boldsymbol{\omega}$ as shown in Figure 7.5.3b. For greater Ω values (particularly for $\Omega \sim \varepsilon$) the spin vector $\mathbf{S}$ may not be able to keep up. Then there is more complicated spin resonance motion which we will describe shortly. For much higher values ($\Omega \gg \varepsilon$) the $\mathbf{S}$ vector will simply process around the average value of $\boldsymbol{\omega}$.

In the NH_3 two-level model the direction of the analogous $\boldsymbol{\omega}$ vector depends on the relative amounts of the real and imaginary parts of the interaction matrix element $H_{21} = \langle 2|q\mathbf{E}\cdot\mathbf{r}|1\rangle$. Normally, the matrix element H_{21} is real and then only x and z components of $\boldsymbol{\omega}$ are nonzero. We need to remember that the three-dimensional quasispin space shown in Figure 7.5.1(b3) is a fiction. Its components should be labeled $\{A, B, C\}$ instead of $\{x, y, z\}$ as they were in Figure 7.5.1(b2) to avoid confusion with ordinary 3-space. The electric-dipole moment or position expectation vector $\langle p_z \rangle = q\langle z \rangle$ is assumed to lie along the electric field direction which is along the molecular z axis, and so this motion is constrained to one spatial dimension. *So, how does one interpret the three dimensions of the spin vectors $\mathbf{S}$?*

The spin vector A or x component $S_x = \langle J_x \rangle$ turns out to be proportional to the position or dipole moment expectation value:

$$\langle z \rangle = \langle \psi | z | \psi \rangle = (\psi_1^* \langle 1 | + \psi_2^* \langle 2 |) z (| 1 \rangle \psi_1 + | 2 \rangle \psi_2)$$
$$= \psi_1^* \psi_2 \langle 1 | z | 2 \rangle + \psi_2^* \psi_1 \langle 2 | z | 1 \rangle$$
$$= \langle 1 | z | 2 \rangle 2 \operatorname{Re} \psi_1^* \psi_2.$$

From Figure 7.5.2 it follows that

$$\langle z \rangle = 2 \langle 1 | z | 2 \rangle S_x.$$

For zero field the C_2 symmetry of the NH_3 model demands that the position matrix element $\langle 1 | z | 2 \rangle$ is real and that diagonal elements are zero $\langle 1 | z | 1 \rangle = 0 = \langle 2 | z | 2 \rangle$.

Furthermore, in the absence of electric field perturbation, the time derivative of position turns out to be proportional to $S_y = \langle J_y \rangle$. Using (8.5.5) we have the following:

$$\frac{d}{dt} \langle z \rangle = \langle 1 | z | 2 \rangle \left(\dot{\psi}_1^* \psi_2 + \dot{\psi}_2^* \psi_1 + \psi_1^* \dot{\psi}_2 + \psi_2^* \dot{\psi}_1 \right)$$
$$= \langle 1 | z | 2 \rangle i \frac{\varepsilon}{2} (\psi_1^* \psi_2 - \psi_2^* \psi_1 + \psi_1^* \psi_2 - \psi_2^* \psi_1)$$
$$= - \langle 1 | z | 2 \rangle \varepsilon 2 \operatorname{Im} \psi_1^* \psi_2.$$

Then from Figure 7.5.2, we deduce

$$\langle \dot{z} \rangle = -2\varepsilon \langle 1 | z | 2 \rangle S_y.$$

The components S_x and S_y of the $\mathbf{S}$ vector form a "shadow" on the horizontal plane, as indicated in Figure 7.4.2. We have just shown that this shadow vector can be visualized as phasor components $(z, dz/dt)$ for the atomic oscillator. (The minus sign on $\langle z \rangle$ indicates the natural phasor direction is *clockwise* in the S_x and S_y plane for $\varepsilon > 0$.)

We recall from Section 8.4B(c) that a classical phasor picture of an atomic oscillator is valid if the perturbing field is weak and Ω is far enough from resonance to keep the responding phasor oscillations small. Then the $\mathbf{S}$ vector oscillates very near to the S_z axis and S_z does not change much. Now we see what happens to this picture as Ω approaches a resonant value. Near resonance there will be a change of the *third* component (S_z) of the spin vector as the first two "phasor" components grow larger with the resonance.

C. Rotating-Wave Solutions

The time-dependent Schrödinger equation (8.5.5) is characterized by a crank vector $\boldsymbol{\omega}$ in (8.5.6) whose direction rotates counterclockwise with angular frequency Ω around the S_z axis (or J_z axis; here we shall use S and J interchangeably to denote spin coordinates). A rotational transformation in the opposite direction is defined as follows:

$$|\psi^R\rangle \equiv R^\dagger(0, 0, -\Omega t)|\psi\rangle = e^{i\Omega t J_z}|\psi\rangle. \tag{8.5.10}$$

It can be applied to the Hamiltonian in (8.5.5) to stop this rotation in its tracks. The result is the following Hamiltonian:

$$H^R = R^\dagger(0, 0, \Omega t) H R(0, 0, \Omega t),$$

whose representation is calculated as follows:

$$\langle H^R \rangle = \begin{pmatrix} e^{i\Omega t/2} & 0 \\ 0 & e^{-i\Omega t/2} \end{pmatrix} \frac{1}{2} \begin{pmatrix} \varepsilon & re^{-i\Omega t} \\ re^{i\Omega t} & -\varepsilon \end{pmatrix} \begin{pmatrix} e^{-i\Omega t/2} & 0 \\ 0 & e^{i\Omega t/2} \end{pmatrix},$$

$$\langle H^R \rangle = \frac{1}{2} \begin{pmatrix} \varepsilon & r \\ r & -\varepsilon \end{pmatrix}. \tag{8.5.11}$$

Note that the resulting H^R is the same except for the off-diagonal components which now are constant. The rotation of the crank has been stopped. If we use the rotating R basis the Hamiltonian will be a constant matrix.

However, there is a small price to pay for using this constant Hamiltonian. The time dependence of the transformation adds a $-\Omega J_z$ term to the new Schrödinger equation as calculated in the following. Recall that a "change of picture" usually comes with an extra term. In this case, at least, the term is constant:

$$i\frac{\partial}{\partial t}|\psi^R\rangle = \left(i\frac{\partial}{\partial t}e^{i\Omega t J_z}\right)|\psi\rangle + e^{i\Omega t J_z}\left(i\frac{\partial}{\partial t}|\psi\rangle\right)$$

$$= -\Omega J_z R^\dagger|\psi\rangle + R^\dagger H|\psi\rangle$$

$$= (-\Omega J_z + R^\dagger H R)|\psi^R\rangle$$

or $\quad i\dfrac{\partial}{\partial t}|\psi^R\rangle = H^{R'}|\psi^R\rangle, \quad$ where $H^{R'} \equiv (-\Omega J_z + R^\dagger H R)$. (8.5.12)

So finally, the resulting equation for amplitudes

$$\psi_j^R = \langle j|R^\dagger|\psi\rangle \quad (j = 1, 2)$$

in the new rotating picture with basis $\{R|1\rangle, R|2\rangle\}$ is the following:

$$i\frac{\partial}{\partial t}\begin{pmatrix}\psi_1^R\\ \psi_2^R\end{pmatrix}=\begin{pmatrix}\dfrac{\varepsilon-\Omega}{2} & \dfrac{r}{2}\\ \dfrac{r}{2} & \dfrac{\Omega-\varepsilon}{2}\end{pmatrix}\begin{pmatrix}\psi_1^R\\ \psi_2^R\end{pmatrix}. \qquad (8.5.13a)$$

In the rotating picture the crank vector $\boldsymbol{\omega}$ is motionless:

$$\begin{array}{lll}\omega_x = 2\,\text{Re}\,H_{21}^{R'} & \omega_y = 2\,\text{Im}\,H_{21}^{R'} & \omega_z = H_{11}^{R'} - H_{22}^{R'}\\ = r & = 0 & = \varepsilon - \Omega = \Delta.\end{array} \qquad (8.5.13b)$$

However, the direction and length of $\boldsymbol{\omega}$ depends on the angular frequency *detuning* parameter which is the difference $\Delta = \varepsilon - \Omega$ between the two-level splitting (ε) and the angular frequency (Ω) of the stimulating laser.

According to (8.5.13b) the $\boldsymbol{\omega}$ vector is nearly aligned to the positive z axis for high positive detuning ($\Delta \gg r$). This case is shown in Figure 8.5.1(a). Then as the driving frequency Ω approaches the resonant frequency ε, the $\boldsymbol{\omega}$ vector becomes shorter and approaches the x axis as shown in Figures 8.5.1(b) and 8.5.1(c). The precessional motion of a spin vector $\mathbf{S}$ which started

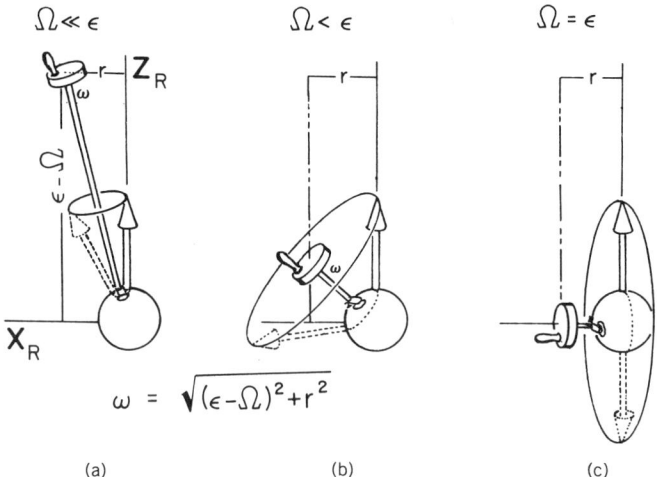

Figure 8.5.1 Motion in rotating frame of quasi-spin vector $\mathbf{S}$ around Hamiltonian $\boldsymbol{\omega}$ vector. (a) Off-resonance. Rotations or "beats" are rapid and of small amplitude. Angular beat frequency $\omega = \sqrt{\Delta^2 + r^2}$ is nearly equal to the detuning ($\omega \approx \Delta = \Omega - \varepsilon$). (b) Approaching resonance. Rotations or beats slow down and increase their amplitude. Angular beat frequency approaches Rabi frequency $\omega \to r$. (c) Resonance. Beats have maximum amplitude and minimum frequency $\omega = r$.

out "up" along the z axis takes place on a greater and greater circle as the detuning decreases. Since ω grows shorter as Δ decreases, the precessional frequency $|\omega|$ decreases. It reaches a minimum value of r (rad/s) at the resonance point ($\Delta = 0$) as shown in Figure 8.5.1(c). [Recall that r is the interaction strength in angular units given by (8.5.3c).] This minimum angular frequency (r) of the resonant motion is known as the RABI frequency, and it is the frequency with which the spin "up" is carried into spin "down" and back again in the resonant case depicted by Figure 8.5.1(c). Note that the Rabi frequency is proportional to the field-dipole interaction energy according to (8.5.3d). It is a remarkable property of the atomic oscillator that its beat frequency on resonance is proportional to the driving field. (Recall that the beat frequency of a classical oscillator simply equals the detuning $\Delta = \varepsilon - \Omega$ which depends on the stimulus frequency Ω and not on its amplitude. Also, the classical beat frequency goes to zero at resonance.)

The precessional frequency $\omega = |\omega|$ due to an off-resonant driving field approaches the detuning value Δ as Δ increases. The magnitude of ω from (8.5.13) is

$$\omega = \sqrt{\Delta^2 + r^2}. \tag{8.5.14}$$

For large detuning ($\Delta \gg r$) this approaches the usual harmonic oscillator beat frequency ($\Delta = \varepsilon - \Omega$) and is independent of the field strength. The spin vector expectation value executes small and rapid circles in the rotating frame as shown in Figure 8.5.1(a). This corresponds to a beating motion in the lab frame of the oscillator variable $\langle x \rangle$. Remember that the rotating frame is revolving around the lab z axis at rate Ω. The beat amplitude is approximately proportional to r (and hence to the field strength) for high Δ, and again this corresponds to the classical driven oscillator.

The precessional motion indicated in Figures 8.5.1(a) and 8.5.1(b) is studied in detail in the series of photos shown in Figure 5.3.5. While the rotational axis vector ω is constant, the Euler phase angles α or γ excite a galloping motion that becomes more pronounced as Δ decreases and the crank axis becomes horizontal. At exactly resonance values ($\Delta = 0$) the Euler phase angles freeze while the population angle β turns at the Rabi frequency. The Euler angles can be derived by applying the axis-angle representation (5.5.1) for an axis vector $\omega = (r, 0, \Delta)$ which has zero azimuthal angle ($\phi = 0$) and polar angle $\theta = \tan^{-1}(r/\Delta)$. The precession of the spin-up state $|1\rangle$ is then given in the rotating picture as follows:

$$\begin{pmatrix} \psi_1^R(t) \\ \psi_2^R(t) \end{pmatrix} = \begin{pmatrix} \cos(\omega t/2) - i\cos\theta \sin(\omega t/2) \\ -i\sin\theta \sin(\omega t/2) \end{pmatrix}$$

$$= \begin{pmatrix} \cos(\omega t/2) - i(\Delta/\omega)\sin(\omega t/2) \\ -i(r/\omega)\sin(\omega t/2) \end{pmatrix}. \tag{8.5.15a}$$

Here the polar angle θ is given in terms of the angular frequencies r and ω:

$$\sin \theta = r/\omega, \qquad (8.5.15b)$$
$$\cos \theta = \Delta/\omega. \qquad (8.5.15c)$$

Now let us undo the rotational transformation (8.5.10) or (8.5.11) which we used to get to the rotating frame. The amplitudes in the original lab picture have different phases but the same magnitudes:

$$\psi_1(t) = e^{-i\Omega t/2} \psi_1^R(t),$$
$$\psi_2(t) = e^{i\Omega t/2} \psi_2^R(t). \qquad (8.5.16)$$

Hence, the probability for a transition from the initial lab state $|1\rangle$ to a final state $|2\rangle$ is given by the square of the 2-component of (8.5.15a):

$$|c_2(t)|^2 = |\psi_2(t)|^2 = [r^2/\omega^2]\sin^2(\omega t/2)$$
$$= [r^2/(\Delta^2 + r^2)]\sin^2(t\sqrt{\Delta^2 + r^2}/2). \qquad (8.5.17)$$

For large detuning ($\Delta \gg r$) this probability agrees with the first-order perturbation result given by (8.4.44):

$$|c_2(t)|^2 \Rightarrow (r^2/\Delta^2)\sin^2(t\Delta/2) = (r^2/4)|S(\Delta, t)|^2 \quad (\text{for } \Delta \gg r). \qquad (8.5.18)$$

Here the Rabi angular frequency is given again by (8.5.3c):

$$|r| = |pE/\hbar| = |qE\langle 1|z|2\rangle|/\hbar.$$

The probability $|c_2|^2$ for the two-level transition is plotted in Figure 8.5.2 using the same format as the plot of the elementary spectral function in Figure 8.4.2.

The main difference between Figures 8.4.2 and 8.5.2 are found near the resonance line ($\Delta = 0$). The elementary spectral probability (8.5.18) is unbounded along $\Delta = 0$ and will eventually exceed the physical limit of unit probability. However, the exact result (8.5.17) is bounded by unity and its value at resonance ($\Delta = 0$) oscillates between unity and zero. It is interesting to note that a perfectly resonant stimulus will actually yield *zero* transition probability at the end of each Rabi period ($\tau_{\text{RABI}} = 2\pi/r$). The spectrum at the end of the first period is indicated by the nearest set of peaks drawn in Figure 8.5.2. The huge peak in the approximate spectral function (Figure 8.4.2) has been replaced by a spectral valley and the neighboring peaks have been beefed up and pulled in toward $\Delta = 0$.

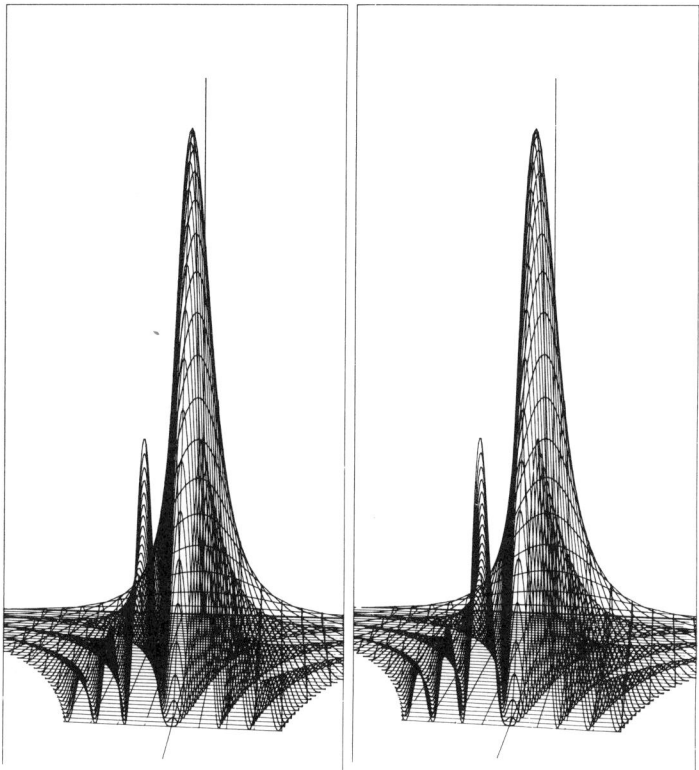

Figure 8.5.2 (Stereoptic pair) Two-level atomic transition probability plotted versus the detuning parameter Δ ($\Delta < 0$ is left and $\Delta > 0$ is right) and elapsed time τ (τ increases coming toward the observer). This is to be compared with approximate probability plotted in Fig. 8.4.2.

In and around the neighborhood between $\Delta = \pm r$ both the approximate quantum theory and the classical phasor picture of the Lorentz model fail. The two-level dipole moment cannot exceed the fixed value of $p = q\langle 2|z|1\rangle$ and therefore cannot grow indefinitely like the classical oscillator without involving more quantum levels. The maximum dipole expectation value is achieved the first time the spin vector is driven to the equator of the spin sphere, that is, when $\beta = \pm \pi/2$ in Eq. (7.4.2) and the states $|1\rangle$ and $|2\rangle$ have equal probabilities. This will happen when the crank polar angle θ in Eq. (8.5.15) exceeds $\pi/4$, and this occurs when $|\Delta|$ is less than the Rabi frequency $|r|$. When the spin vector is horizontal the transition moment is "saturated." Any further excitation or increase in S_z actually causes the $\{S_x S_y\}$ phasor to shrink. Then the system enters the extraordinary regime of "population inversion" in which the excited state $|2\rangle$ has greater probability than the initial state $|1\rangle$. From this discussion one should appreciate how the infinite phasor plane of the Lorentz model has been replaced by a finite sphere. Atomic phasor space is round!

The spectral envelope of the two-level response curve given by (8.5.17) is the following Lorentzian function:

$$P(\Delta) = r^2/(\Delta^2 + r^2).$$

The half-maximum values of $P(\Delta)$ are $\Delta = \pm r$. This means that the width of the envelope equals the Rabi frequency which is proportional to the amplitude of the driving field and to the induced dipole transition moment. This increase in spectral width due to increased pumping is an elementary example of what is known as laser POWER BROADENING. The extra Fourier sidebands are due to the Rabi oscillation which turns the transition on and off at a rate which depends on the field amplitude and detuning.

D. Bloch-Siegert Corrections

An exact solution for the precessing crank Hamiltonian is given by Eqs. (8.5.5) and (8.5.6). However, this is only an approximate solution to the original Hamiltonian in Eq. (8.5.4b) for which the crank vector oscillates in the x direction only. If we apply the same clockwise rotation operator defined by (8.5.10) the resulting Hamiltonian matrix is not constant. Instead of (8.5.11), we get

$$H^R = R^\dagger(0,0,\Omega t) H R(0,0,\Omega t) = \frac{1}{2}\begin{pmatrix} \varepsilon & r + re^{2i\Omega t} \\ re^{-2i\Omega t} + r & -\varepsilon \end{pmatrix}. \quad (8.5.19)$$

Then the time term $-\Omega J_z$ in (8.5.12) is included and the resulting Schrödinger equation is the following:

$$i\frac{\partial}{\partial t}\begin{pmatrix} \psi_1^R \\ \psi_2^R \end{pmatrix} = \begin{pmatrix} \dfrac{\varepsilon - \Omega}{2} & \dfrac{1}{2}(r + re^{2i\Omega t}) \\ \dfrac{1}{2}(re^{-2i\Omega t} + r) & \dfrac{-\varepsilon + \Omega}{2} \end{pmatrix}\begin{pmatrix} \psi_1^R \\ \psi_2^R \end{pmatrix}. \quad (8.5.20)$$

The off-diagonal terms $((r/2)e^{\pm 2i\Omega t})$ which oscillate at twice the driving frequency are neglected in the so-called *rotating-wave approximation*. Then the resulting equation is the same as (8.5.13a). The crank vector for the exact Hamiltonian consists of the usual static vector $(r, 0, \varepsilon - \Omega)$ [recall Eq. (8.5.13b)] added to another vector of length r which is rotating counterclockwise at angular frequency 2Ω in the xy plane. The total crank vector is given by the following:

$$\omega_x = r + r\cos 2\Omega t, \quad \omega_y = r\sin 2\Omega t, \quad \omega_z = \varepsilon - \Omega. \quad (8.5.21)$$

If we had applied the *counter*clockwise rotation operator R instead of $R^\dagger$ we get the same Schrödinger equation as (8.5.20) except everywhere Ω is replaced by $-\Omega$. The resulting Hamiltonian corresponds now to a different static crank vector $(r, 0, \varepsilon + \Omega)$ added to another vector of length r which is rotating *clock*wise in the xy plane at angular frequency 2Ω.

In either case the precessional motion generated by the static part of the crank vector may be perturbed by the rotating r vector. The perturbation will be maximal when the r-vector rotation is synchronized with the precession. The static part of the crank given in (8.5.21) generates a precession angular rate of

$$\omega_{\text{static}} = \pm\sqrt{r^2 + (\varepsilon - \Omega)^2}.$$

Resonance may occur when this is matched by the r-vector rate:

$$-2\Omega = -\left[(\varepsilon - \Omega)^2 + r^2\right]^{1/2}. \tag{8.5.22}$$

In the counterclockwise frame the same reasoning gives a similar relation:

$$2\Omega = +\left[(\varepsilon + \Omega)^2 + r^2\right]^{1/2}. \tag{8.5.23}$$

Either of the two preceding equations yield the same magnitudes for the resonant Ω values; only the sign is different. Rearranging the second equation gives

$$3\Omega^2 - 2\varepsilon\Omega - (r^2 + \varepsilon^2) = 0, \tag{8.5.24}$$

and the possible solutions are

$$\Omega = \left(\varepsilon \pm [4\varepsilon^2 + 3r^2]^{1/2}\right)/3. \tag{8.5.25}$$

For relatively small Rabi frequency $(r \ll \varepsilon)$ the radical may be expanded to give

$$[4\varepsilon^2 + 3r^2]^{1/2} \cong 2\varepsilon + 3r^2/4\varepsilon + \cdots.$$

Then the possible resonances are given approximately as follows:

$$\Omega_1 = \varepsilon\left(1 + (r^2/4\varepsilon^2)\right), \tag{8.5.26a}$$
$$\Omega_2 = -\varepsilon\left(\tfrac{1}{3} + (r^2/4\varepsilon^2)\right). \tag{8.5.26b}$$

The first result is only slightly shifted from the RWA value ($\Omega = \varepsilon$ or $\Delta = 0$). The shifting fraction

$$\delta = r^2/4\varepsilon^2 \tag{8.5.27}$$

is called the Bloch-Siegert shift. For optical laser-induced transitions the Rabi rate is typically $r \equiv 10^{10}$ Hz. This is small compared to the optical frequency which is typically $\varepsilon \equiv 10^{16}$ Hz. This gives a typical shift ratio of $r^2/4\varepsilon^2 \cong 0.25 \times 10^{-12}$, which, until recently, would be considered unobservable. However, if the Rabi rate is significantly increased or if lower-frequency transitions are probed then the RWA predictions must be corrected.

The second solution (8.5.26b) lies at approximately one-third of the primary resonance frequency. This subharmonic resonance is not a strong effect when r is relatively small, but it may lead to interesting effects in other situations.

E. Damping and the Bloch Equations

So far we have considered only the ideal pure states of the pumped two-level system. Now a brief introduction to the real-world effects of decay and noise will be given. This is important for any study of spectroscopy, since rarely, if ever, are these effects totally negligible. In fact, radiative decay and other "damping" mechanisms are so strong that they completely obscured the Rabi two-level optical transition effects until after the laser was invented. Earlier nuclear magnetic spin resonance (NMR) experiments were the first to exhibit two-level oscillation and damping effects, since radiative decay is very small at radio or microwave frequencies. The Bloch theory described below was developed first for NMR phenomenology.

(a) Time Behavior of the Density Matrix Statistical treatment of quantum ensembles rely on the density operator ρ introduced in Section 1.3.B. For a pure state $|\Psi\rangle$ the density operator is simply $\rho = |\psi\rangle\langle\psi|$. For an ensemble of systems in various states, ρ will be a weighted sum of such terms.

A representation of a density operator provides a compact way to record the components of the quasispin or Stokes-vector expectation values. For the general pure state vector represented in Figure 7.5.2 the density matrix is the following:

$$\langle \rho \rangle = \begin{pmatrix} \langle 1|\Psi\rangle\langle\Psi|1\rangle & \langle 1|\Psi\rangle\langle\Psi|2\rangle \\ \langle 2|\Psi\rangle\langle\Psi|1\rangle & \langle 2|\Psi\rangle\langle\Psi|2\rangle \end{pmatrix} = \begin{pmatrix} \Psi_1^*\Psi_1 & \Psi_1^*\Psi_2 \\ \Psi_2^*\Psi_1 & \Psi_2^*\Psi_2 \end{pmatrix}$$

$$= \begin{pmatrix} \cos^2 \dfrac{\beta}{2} & e^{-i\alpha} \sin \dfrac{\beta}{2} \cos \dfrac{\beta}{2} \\ e^{i\alpha} \sin \dfrac{\beta}{2} \cos \dfrac{\beta}{2} & \sin^2 \dfrac{\beta}{2} \end{pmatrix}. \quad (8.5.28)$$

TWO-LEVEL SYSTEM APPROXIMATION

The density matrix may be expanded in terms of the Pauli matrices,

$$\langle \rho \rangle = \frac{1}{2}\begin{pmatrix} 1 & 0 \\ 0 & 1 \end{pmatrix} + \frac{\cos\alpha \sin\beta}{2}\begin{pmatrix} 0 & 1 \\ 1 & 0 \end{pmatrix}$$
$$+ \frac{\sin\alpha \sin\beta}{2}\begin{pmatrix} 0 & -i \\ i & 0 \end{pmatrix} + \frac{\cos\beta}{2}\begin{pmatrix} 1 & 0 \\ 0 & -1 \end{pmatrix}. \quad (8.5.29)$$

The coefficients of this expansion include the expected spin vector components which were given in terms of Euler angles in Figure 7.5.2 (note that overall phase angle γ disappears):

$$\rho = \tfrac{1}{2}\mathbf{1} + S_x\sigma_x + S_y\sigma_y + S_z\sigma_z,$$
$$\rho = S_0\mathbf{1} + \mathbf{S}\cdot\boldsymbol{\sigma}. \quad (8.5.30)$$

In general, the quantities $S_i = \text{Trace}(\rho\sigma_i)$ ($i = x, y, z$) will be the ensemble summed expectation values for the respective components, and S_0 will be the total population of the two levels together throughout the ensemble. Then the summed spin vector magnitude $|\mathbf{S}|$ may be less than S_0. Incoherent or unpolarized ensembles have a zero value for the summed spin vector ($\mathbf{S} = \mathbf{0}$).

The time derivative of a pure state density operator is given using the Schrödinger equation:

$$i\hbar\frac{\partial\rho}{\partial t} = i\hbar\left(\frac{\partial|\psi\rangle}{\partial t}\right)\langle\Psi| + i\hbar|\Psi\rangle\left(\frac{\partial\langle\psi|}{\partial t}\right)$$
$$= H|\Psi\rangle\langle\Psi| - |\Psi\rangle\langle\Psi|H = [H, \rho]. \quad (8.5.31)$$

The Pauli spinor expansion in Eq. (7.5.4) relates the Hamiltonian to its crank vector $\boldsymbol{\omega}$ and overall phase rate ω_0 (recall, also, Figure 7.5.3):

$$H = \omega_0\mathbf{1} + \boldsymbol{\omega}\cdot\boldsymbol{\sigma}. \quad (8.5.32)$$

Substituting the Pauli expansions for H and ρ into (8.5.31) yields

$$i\frac{\partial \mathbf{S}}{\partial t}\cdot\boldsymbol{\sigma} = [(\boldsymbol{\omega}\cdot\boldsymbol{\sigma})(\mathbf{S}\cdot\boldsymbol{\sigma}) - (\mathbf{S}\cdot\boldsymbol{\sigma})(\boldsymbol{\omega}\cdot\boldsymbol{\sigma})]/2$$
$$= i[\boldsymbol{\omega}\times\mathbf{S}]\cdot\boldsymbol{\sigma}, \quad (8.5.33)$$

where the Pauli identity $(\boldsymbol{\sigma}\cdot\mathbf{A})(\boldsymbol{\sigma}\cdot\mathbf{B}) + \mathbf{A}\cdot\mathbf{B} = i(\mathbf{A}\times\mathbf{B})$ has been used. The pure state Bloch equation is then given:

$$\frac{\partial\mathbf{S}}{\partial t} = \boldsymbol{\omega}\times\mathbf{S}. \quad (8.5.34)$$

752 SYMMETRY ANALYSIS FOR SEMICLASSICAL AND QUANTUM MECHANICS

This is consistent with the picture of the crank vector ω causing the spin vector **S** to precess as shown in Figure 7.5.3(b).

(b) Bloch Equations The usual generalization of the Bloch equation to an ensemble involves the addition of phenomenological damping terms as follows. Let us write

$$\frac{\partial \mathbf{S}}{\partial t} = \omega \times \mathbf{S} - \left[\gamma_2(S_x - S_x(0))\mathbf{e}_x + \gamma_2(S_y - S_y(0))\mathbf{e}_y + \gamma_1(S_z - S_z(0))\mathbf{e}_z \right], \tag{8.5.35a}$$

where

$$\gamma_2 = 1/T_2, \qquad \gamma_1 = 1/T_1 \tag{8.5.35b}$$

relate transverse and longitudinal decay rates γ_1 and γ_2 to transverse and longitudinal lifetimes T_1 and T_2, respectively. The quantities $S_i(0)$ ($i = x, y, z$) are the equilibrium values to which the spin vector components tend to decay. Generally, they are ground state values: $S_x(0) = 0 = S_y(0)$, and $S_z(0) = 1/2$.

The Boltzman population factors for an equilibrated two-level system appear on the diagonal of the density matrix,

$$\langle \rho \rangle = \begin{pmatrix} e^{-\varepsilon_1 \beta} & 0 \\ 0 & e^{-\varepsilon_2 \beta} \end{pmatrix} Z^{-1}. \tag{8.5.36}$$

Here the Boltzman factor varies inversely with temperature

$$\beta = 1/kT, \tag{8.5.37}$$

and the partition function

$$Z = e^{-\varepsilon_1 \beta} + e^{-\varepsilon_2 \beta} \tag{8.5.38}$$

depends on energies ε_1 and ε_2 of the two levels. The average value of the transverse spin components for this system are zero ($S_x(0) = 0 = S_y(0)$) while the longitudinal or z component is the following:

$$\langle S_z(0) \rangle = \frac{1}{2} \text{Trace}(\rho \sigma_z) = \frac{1}{2} \frac{e^{-\varepsilon_1 \beta} - e^{-\varepsilon_2 \beta}}{e^{-\varepsilon_1 \beta} + e^{-\varepsilon_2 \beta}} = \frac{1}{2} \frac{1 - e^{-\varepsilon \beta}}{1 + e^{-\varepsilon \beta}}.$$

Here the two-level energy interval $\varepsilon = \varepsilon_2 - \varepsilon_1$ is assumed positive. For high temperatures ($\varepsilon \beta \to 0$) the z-component approaches zero. For low temperatures ($\varepsilon \beta \gg 1$) the z component approaches its maximum value ($S_z(0) = 1/2$).

E. Dressed Eigenstates

In the rotating frame there are two directions in which the spin vector may be placed and will remain fixed while under the influence of the RWA Hamiltonian. These directions are along and against the ω axis of the crank vector defined by (8.5.13b). The states associated with these spin vectors are called the *dressed eigenstates*. They are simply the eigenstates of the RWA matrix $H^{R'}$ which appears in (8.5.13a). [$H^{R'}$ is simply H^R plus the picture changing term $-\Omega J_z$ as given by Eq. (8.5.12).]

$$H^{R'} = \begin{pmatrix} \frac{\Delta}{2} & \frac{r}{2} \\ \frac{r}{2} & -\frac{\Delta}{2} \end{pmatrix}. \quad (8.5.39)$$

The crank axis is tipped by angle $\theta = \tan^{-1}(r/\Delta)$ toward the rotating X_R axis. Hence you can construct dressed eigenstates by simply rotating the spin "up" and "down" states $|1\rangle^R$ and $|2\rangle^R$ by θ around the Y_R axis. The transformation matrix is [from Eqs. (5.4.30) or (7.5.2)]

$$\langle R_D \rangle = \langle e^{-i\theta J_y^R} \rangle = \begin{pmatrix} \cos\frac{\theta}{2} & -\sin\frac{\theta}{2} \\ \sin\frac{\theta}{2} & \cos\frac{\theta}{2} \end{pmatrix} = \begin{pmatrix} {}^R\langle 1|1\rangle^D & {}^R\langle 1|2\rangle^D \\ {}^R\langle 2|1\rangle^D & {}^R\langle 2|2\rangle^D \end{pmatrix} \quad (8.5.40)$$

It is important to remember that the Y_R axis is rotating around the $Z = Z_R$ axis with angular frequency equal to the laser stimulus frequency Ω. The dressed eigenstates are given in terms of the rotating bases as follows:

$$|1\rangle^D = R_D(0, \theta, 0)|1\rangle^R = \cos\frac{\theta}{2}|1\rangle^R + \sin\frac{\theta}{2}|2\rangle^R,$$

$$|2\rangle^D = R_D(0, \theta, 0)|2\rangle^R = -\sin\frac{\theta}{2}|1\rangle^R + \cos\frac{\theta}{2}|2\rangle^R. \quad (8.5.41)$$

We check that this transformation diagonalizes matrix $\langle H_{R'} \rangle$.

$$\langle R_D^\dagger H^{R'} R_D \rangle = \begin{pmatrix} \cos\frac{\theta}{2} & \sin\frac{\theta}{2} \\ -\sin\frac{\theta}{2} & \cos\frac{\theta}{2} \end{pmatrix} \begin{pmatrix} \frac{\Delta}{2} & \frac{r}{2} \\ \frac{r}{2} & -\frac{\Delta}{2} \end{pmatrix} \begin{pmatrix} \cos\frac{\theta}{2} & -\sin\frac{\theta}{2} \\ \sin\frac{\theta}{2} & \cos\frac{\theta}{2} \end{pmatrix}$$

$$= \begin{pmatrix} \frac{\omega}{2} & 0 \\ 0 & -\frac{\omega}{2} \end{pmatrix}. \quad (8.5.42)$$

Here ω is given by (8.5.15) in terms of r, Δ, and θ. This can be rewritten in terms of J_z^D which is the dressed z component.

$$R_D^\dagger H^{R'} R_D = \omega J_z^D. \tag{8.5.43}$$

The dressed-state amplitudes follow from (8.5.41):

$$\psi_1^D = {}^D\langle 1|\psi\rangle = {}^R\langle 1|R_D^\dagger|\psi\rangle = \cos\frac{\theta}{2}\psi_1^R + \sin\frac{\theta}{2}\psi_2^R,$$

$$\psi_2^D = {}^D\langle 2|\psi\rangle = {}^R\langle 2|R_D^\dagger|\psi\rangle = -\sin\frac{\theta}{2}\psi_1^R + \cos\frac{\theta}{2}\psi_2^R. \tag{8.5.44}$$

(a) The AC Stark Shifts The two angular frequencies ε and Ω are fundamental to the theory of two-level resonance. The first frequency ε is the frequency of the unperturbed atomic oscillator which is the energy difference of the two unperturbed levels divided by $\hbar$. [Recall Eq. (8.5.4b).] The second frequency Ω is that of the stimulating laser. It is helpful to regard these as the unperturbed frequency levels for the atom and laser, respectively, and the frequency center of gravity is the average value $(\varepsilon + \Omega)/2$. Now we shall see how the dressed eigenlevels are shifted from the unperturbed values ε or Ω if the laser-atom coupling parameter (or Rabi frequency) r is nonzero. The corresponding shifts are called AC-Stark shifts.

If the average value $(\varepsilon + \Omega)/2$ is added to each of the dressed eigenvalues $(\pm\omega/2)$ then one obtains the following:

$$\frac{\varepsilon + \Omega}{2} + \frac{\omega}{2} = \varepsilon + \frac{\omega - \varepsilon + \Omega}{2} = \varepsilon + \frac{\omega - \Delta}{2} \equiv \varepsilon + \frac{\delta}{2}, \tag{8.5.45a}$$

$$\frac{\varepsilon + \Omega}{2} - \frac{\omega}{2} = \Omega - \frac{\omega - \varepsilon + \Omega}{2} = \Omega - \frac{\omega - \Delta}{2} \equiv \Omega - \frac{\delta}{2}, \tag{8.5.45b}$$

where $\delta/2$ is defined to be the AC-Stark shift with δ given by the first of the following:

$$\delta = \delta(\Delta) = \omega - \Delta = (\Delta^2 + r^2)^{1/2} - \Delta, \tag{8.5.45c}$$

$$\delta' = \delta'(\Delta) = \omega + \Delta = (\Delta^2 + r^2)^{1/2} + \Delta. \tag{8.5.45d}$$

Here, $\Delta = \varepsilon - \Omega$ is the previously defined detuning parameter. When the detuning is negative the "alternate" AC-Stark shift δ' may be more convenient to use. When the stimulus Ω is much greater than resonance frequency ε then $-\Delta \gg r$ and δ' will be a small shift while δ will approach -2Δ.

The shift $\delta/2$ is defined in analogy to the zero-frequency ($\Omega = 0$) shifts or DC-Stark shifts $\delta(\varepsilon)$:

$$\delta(\varepsilon) = \sqrt{\varepsilon^2 + r^2} - \varepsilon. \tag{8.5.46a}$$

A plot of DC-Stark levels versus the field coupling parameter r was given in Figure 2.12.8 for fixed two-level energy ε. The shift $\delta(\varepsilon)/2$ is the difference between the hyperbola and electric field energy asymptotes. The difference is small when the field energy r is large compared to the unperturbed two-level energy ε.

By analogy, a plot of dressed AC-Stark levels (8.5.45a) is shown in Figure 8.5.3, only now it is made versus the detuning parameter Δ for fixed r and ε. Again, the shift $\delta(\Delta)/2$ represents the spacing between hyperbolic eigenlevels and their respective asymptotes. The DC shifts are indicated above the ($\Omega = 0$) point on the left-hand side of the Figure 8.5.3. These shifts grow into

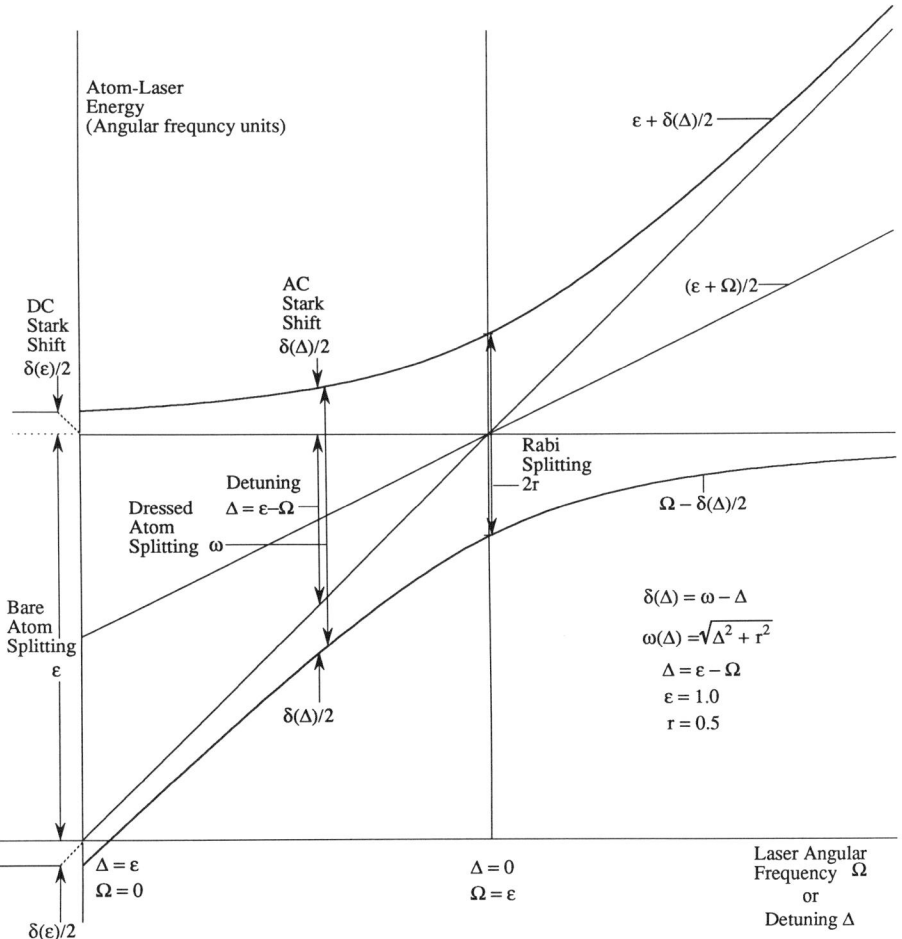

Figure 8.5.3 AC Stark shifts and energy levels for the two-state system. Levels and shifts are plotted as functions of the stimulating laser frequency Ω or detuning parameter $\Delta = \Omega - \varepsilon$ from DC ($\Omega = 0$) to above resonance ($\Omega > \varepsilon$).

larger and larger AC-Stark shifts as the laser frequency increases, and they equal $\delta(0)/2 = r/2$ at resonance. The increasing shift of the excited-state hyperbola above the ε asymptote corresponds to an increasing component of the ground state in the dressed excited state. Similarly, the ground dressed state takes on more of the excited-state character as the lower hyperbola veers away from the Ω asymptote and approaches the horizontal ε asymptote.

To provide a quantitative geometrical interpretation of the mixing let us consider a picture of the quantities Δ, δ, ω, and r in Figure 8.5.4. This is a more detailed representation of the crank diagram in Figure 8.5.1, which shows how to visualize the relevant variables ω and δ as Δ approaches zero. The diagram is based upon the definitions $\delta = \omega - \Delta$, $\omega^2 = \Delta^2 + r^2$, and the relation

$$\delta^2 + r^2 = (\omega - \Delta)^2 + r^2 = 2\Delta^2 + 2r^2 - 2\omega\Delta$$
$$= 2\omega(\omega - \Delta) = 2\omega\delta, \qquad (8.5.46b)$$

and the relations (8.5.15b) and (8.5.15c) involving angle θ. We construct three mutually tangent circles of radii δ, Δ, and δ'. The center of the δ and δ' circles lie at the tip of the ω vector, while the Δ circle is centered at the origin and base of ω. (Recall $\delta + \Delta = \omega = \delta' - \Delta$.) This yields a right triangle with altitude δ, base r, hypotenuse $\sqrt{(2\delta\omega)}$, and angle $\theta/2$, as indicated in Figure 8.5.4(a). According to the dressing or diagonalizing transformation (8.5.40) the components of the dressed eigenstates relative to the rotating frame are proportional to the base and altitude, respectively, of this triangle. The state belonging to the lower hyperbola has the following amplitudes in the $\{|1\rangle^R, |2\rangle^R\}$ basis:

$$|1\rangle^D \rightarrow \begin{pmatrix} \cos\dfrac{\theta}{2} \\ \sin\dfrac{\theta}{2} \end{pmatrix} = \begin{pmatrix} \dfrac{r}{\sqrt{2\delta\omega}} \\ \dfrac{\delta}{\sqrt{2\delta\omega}} \end{pmatrix}, \qquad (8.5.47)$$

while the upper hyperbola belongs to the following orthogonal set of components:

$$|2\rangle^D \rightarrow \begin{pmatrix} -\sin\dfrac{\theta}{2} \\ \cos\dfrac{\theta}{2} \end{pmatrix} = \begin{pmatrix} \dfrac{-\delta}{\sqrt{2\delta\omega}} \\ \dfrac{r}{\sqrt{2\delta\omega}} \end{pmatrix}. \qquad (8.5.48)$$

On the left-hand side of Figure 8.5.2 the upper hyperbola is shifted by a small amount $((\omega - \Delta)/2 = \delta/2)$ above the ε asymptote and a large amount

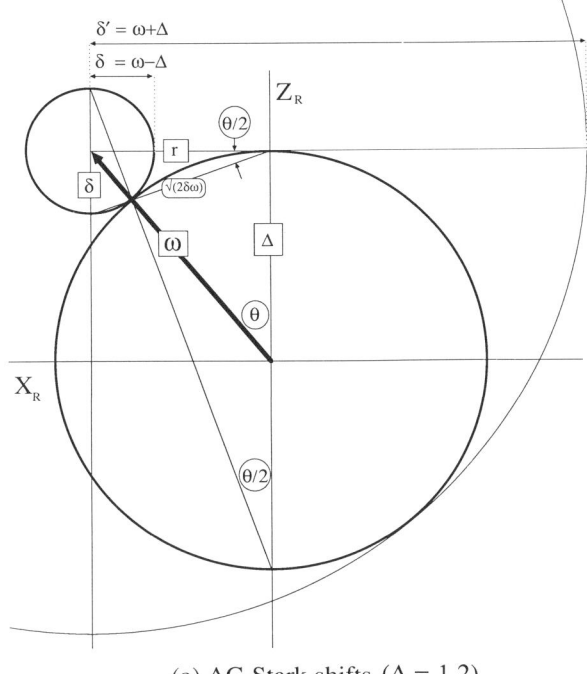

(a) AC-Stark shifts ($\Delta = 1.2$)

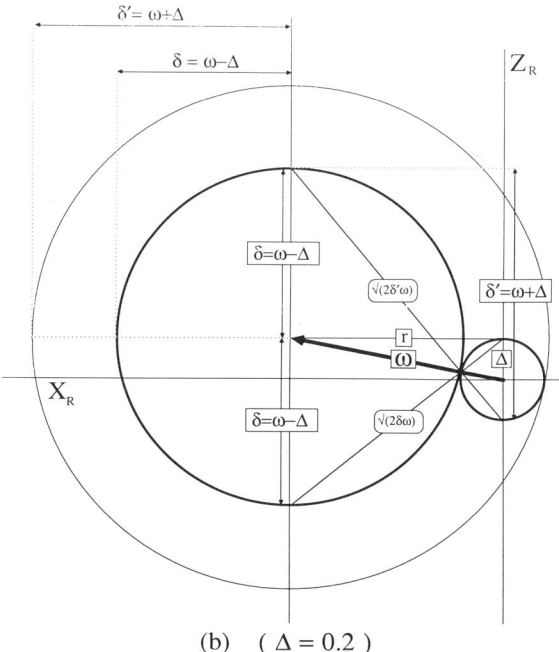

(b) ($\Delta = 0.2$)

Figure 8.5.4 Geometry of ω vector in rotating frame for four values of detuning parameter Δ and fixed Rabi frequency r. This is a detailed version of Figure 8.5.1 which shows the magnitudes of the AC Stark shifts (δ and δ') and the dressed eigenstate amplitudes ($\sqrt{2\delta\omega}$ and $\sqrt{2\delta'\omega}$). (a) Below resonance ($\Delta = 1.2$). (b) Approaching resonance ($\Delta = 0.2$). (c) Just above resonance ($\Delta = 0.2$). (d) Above resonance ($\Delta = -1.2$).

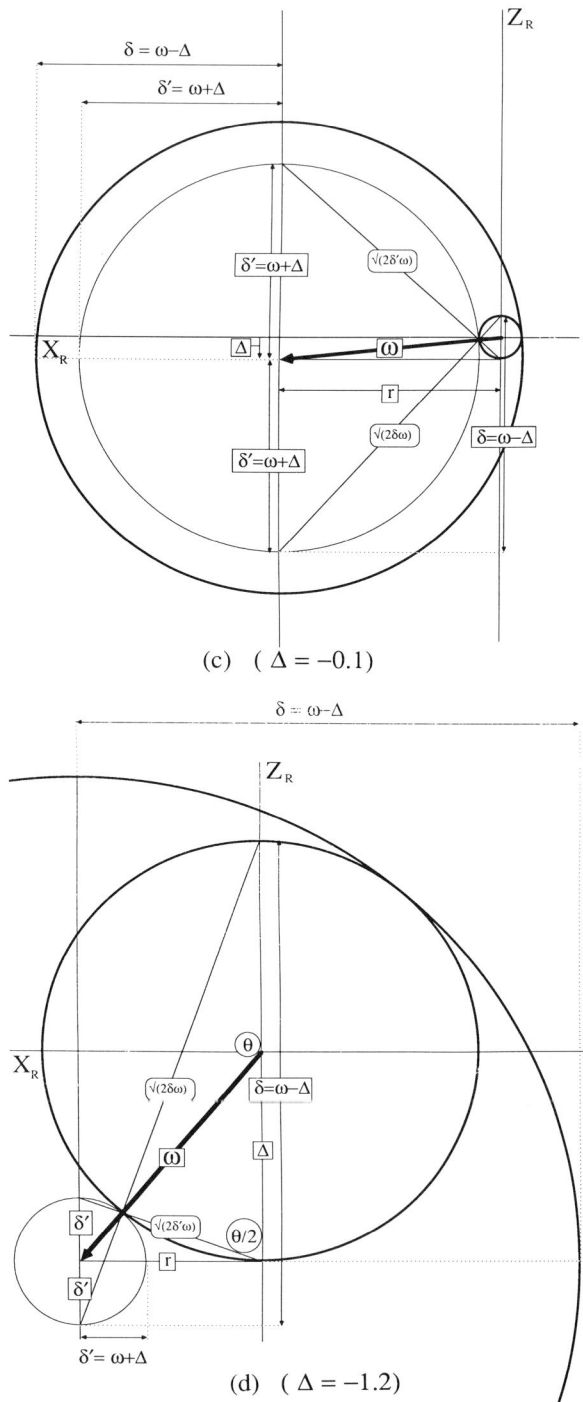

(c) ($\Delta = -0.1$)

(d) ($\Delta = -1.2$)

Figure 8.5.4 *(Continued).*

$((\omega + \Delta)/2 = \delta'/2)$ above the Ω asymptote. The ε asymptote is horizontal while the Ω asymptote has unit slope in the figure. At the resonance point $(\Omega = \varepsilon)$, the detuning parameter Δ changes sign. Then the shift $(\omega - \Delta)/2 = \delta/2$ becomes much larger and approaches Δ while the shift $(\omega + \Delta)/2 = \delta'/2$ becomes smaller as the upper hyperbola moves close to the Ω asymptote.

Meanwhile (for $\Omega > \varepsilon$), the lower hyperbola moves away from the Ω asymptote. The corresponding lower dressed eigenstate represented by (8.5.47) changes from mostly ground state $|1\rangle^R$ ($\delta \ll r$) to mostly excited state $|2\rangle^R$ ($\delta \gg r$) as detuning Δ varies from positive to negative values. If Δ is varied slowly enough, and if damping is negligible, then this process can be used to excite two-level systems to nearly 100% inversion. When this happens the systems are said to adiabatically follow the detuning. The angle between the spin vector and the instantaneous crank vector is nearly invariant to slow changes in r or Δ.

The Schrödinger equation for describing the adiabatic following of a dressed state is derived now. First, the derivative of any dressed state is as follows:

$$i\frac{\partial}{\partial t}R_D^\dagger|\psi\rangle^R = i\frac{\partial R_D^\dagger}{\partial t}|\psi\rangle^R + R_D^\dagger i\frac{\partial|\psi\rangle^R}{\partial t}. \qquad (8.5.49)$$

Suppose we let the rate $d\theta/dt$ be a measure of the change of detuning. Then the time derivative of the dressing transformation (8.5.40) is the following:

$$i\frac{\partial R_D^\dagger}{\partial t} = i\frac{\partial e^{i\theta J_y^R}}{\partial t} = -\dot{\theta}J_y^R R_D^\dagger.$$

The derivative of the rotating-wave ket is repeated from Eq. (8.5.12):

$$i\frac{\partial}{\partial t}|\psi\rangle^R = H^{R'}|\psi\rangle^R.$$

Finally, (8.5.43) and the definition (8.5.44) of dressed kets give the following:

$$i\frac{\partial}{\partial t}R_D^\dagger|\psi\rangle^R = -\dot{\theta}J_y^R R_D^\dagger|\psi\rangle^R + R_D^\dagger H^{R'} R_D R_D^\dagger|\psi\rangle^R.$$

This amounts to a contest between a y^R angular velocity of $-d\theta/dt$ and a z^D angular velocity of ω around the crank:

$$i\frac{\partial}{\partial t}|\psi\rangle^D = \left(-\dot{\theta}J_y^R + \omega J_z^D\right)|\psi\rangle^D. \qquad (8.5.50)$$

When the condition $d\theta/dt \ll \omega$ holds it may be possible for the spin vector to follow the crank vector adiabatically. However, this condition is neither necessary nor sufficient for following.

(b) Dressed Bloch Equations The Bloch equations (8.5.35) in the dressed state basis are obtained by applying a y rotation by angle θ to the density operator in the rotating wave basis. This y rotation is represented in the vector basis by a standard Cartesian 3×3 matrix $\langle x_j^D | x_k^R \rangle = \mathbf{e}_j^D \cdot \mathbf{e}_k^R$, where

$$\mathbf{e}_x^D \cdot \mathbf{e}_x^R = \cos\theta, \qquad \mathbf{e}_x^D \cdot \mathbf{e}_y^R = 0, \qquad \mathbf{e}_x^D \cdot \mathbf{e}_z^R = -\sin\theta, \quad \text{etc.}$$

Applying this matrix and its inverse to Γ^R in the rotating wave basis $\{\mathbf{e}_x^R, \mathbf{e}_y^R, \mathbf{e}_z^R\}$ gives the following transformed damping matrix expressed in the dressed basis $\{\mathbf{e}_x^D, \mathbf{e}_y^D, \mathbf{e}_z^D\}$.

$$\langle \Gamma \rangle^D = \begin{pmatrix} \cos\theta & 0 & -\sin\theta \\ 0 & 1 & 0 \\ \sin\theta & 0 & \cos\theta \end{pmatrix} \begin{pmatrix} \gamma_2 & \cdot & \cdot \\ \cdot & \gamma_2 & \cdot \\ \cdot & \cdot & \gamma_1 \end{pmatrix} \begin{pmatrix} \cos\theta & 0 & \sin\theta \\ 0 & 1 & 0 \\ -\sin\theta & 0 & \cos\theta \end{pmatrix}$$

$$= \begin{pmatrix} \gamma_2 \cos^2\theta + \gamma_1 \sin^2\theta & 0 & -(\gamma_1 - \gamma_2)\cos\theta \sin\theta \\ 0 & \gamma_2 & 0 \\ -(\gamma_1 - \gamma_2)\cos\theta \sin\theta & 0 & \gamma_1 \cos^2\theta + \gamma_2 \sin^2\theta \end{pmatrix}$$

The effective crank vector for constant detuning $\Delta = \varepsilon - \Omega$ has only an $\mathbf{e}_z^D$ component,

$$\boldsymbol{\omega} = \omega \mathbf{e}_z^D,$$

where according to equation (8.5.15c)

$$\omega = \frac{\Delta}{\cos\theta}.$$

If the detuning is changing, the crank vector also has an $\mathbf{e}_y^D$ component according to Eq. (8.5.50),

$$\boldsymbol{\omega} = \omega \mathbf{e}_z^D - \dot{\theta} \mathbf{e}_y^D.$$

The dressed Bloch equations (8.5.35) then become the following:

$$\dot{S}_x^D = -\omega S_y^D - \dot{\theta} S_z^D - (\gamma_2 \cos^2\theta + \gamma_1 \sin^2\theta)(S_x^D - S_x^D(0))$$
$$\quad + (\gamma_1 - \gamma_2)\cos\theta \sin\theta (S_z^D - S_z^D(0)),$$
$$\dot{S}_y^D = \omega S_x^D - \gamma_2(S_y^D - S_y^D(0)),$$
$$\dot{S}_z^D = \dot{\theta} S_x^D + (\gamma_1 - \gamma_2)\cos\theta \sin\theta (S_x^D - S_x^D(0))$$
$$\quad - (\gamma_2 \sin^2\theta + \gamma_1 \cos^2\theta)(S_z^D - S_z^D(0)).$$

These are the equations that replace the Lorentz forced damped oscillator model for atomic spectral transitions. Many texts have been written describing the nature of their solutions as it applies to laser dynamics and quantum electronics.

8.6 QUANTUM ELECTROMAGNETIC FIELDS AND TRANSITIONS

The fully quantum-mechanical treatment of electromagnetic spectral transitions will be given now. It begins by converting the classical em field equations to harmonic oscillator equations for which the quantum states are well known.

A linearly polarized plane wave was described by the classical vector potential (8.4.6) as follows:

$$\mathbf{A} = \mathbf{e}_1 2|a|\sin(\mathbf{k}\cdot\mathbf{r} - \omega t + \phi). \tag{8.6.1a}$$

This gives the following em fields (we neglect the nonradiative or static field $\mathbf{E} = -\nabla\Phi$):

$$\mathbf{E} = -\frac{\partial \mathbf{A}}{\partial t} \qquad\qquad \mathbf{B} = \nabla \times \mathbf{A}$$

$$= \mathbf{e}_1 E_0 \cos(\mathbf{k}\cdot\mathbf{r} - \omega t + \phi) \qquad = (\mathbf{k}\times\mathbf{e}_1)B_0\cos(\mathbf{k}\cdot\mathbf{r} - \omega t + \phi). \tag{8.6.1b}$$

The electric E-polarization vector at zero phase is along unit vector $\mathbf{e}_1$:

$$E_0 \mathbf{e}_1 = 2|a|\omega\mathbf{e}_1. \tag{8.6.1c}$$

At the same time the magnetic B-polarization vector is along a unit vector $\mathbf{b}_1 = \mathbf{e}_2$, which is orthogonal to $\mathbf{e}_1$ and wave vector $\mathbf{k}$:

$$B_0 \mathbf{b}_1 = B_0(\mathbf{k}\times\mathbf{e}_1) = \mathbf{e}_2 2|a|\omega/c \quad \text{(where } k = \omega/c\text{)}. \tag{8.6.1d}$$

In preparation for a quantum-mechanical theory we shall rewrite the vector potential $\mathbf{A}$ as follows:

$$\mathbf{A} = a_{k,1}\mathbf{e}_1 e^{i(\mathbf{k}\cdot\mathbf{r}-\omega t)} + a^*_{k,1}\mathbf{e}_1 e^{-i(\mathbf{k}\cdot\mathbf{r}-\omega t)}, \tag{8.6.2a}$$

where the complex phasor amplitude $a = a_{k,1}$ is given by

$$a_{k,1} = -i|a_{k,1}|e^{i\phi_{k,1}}. \tag{8.6.2b}$$

This will be made to give classical canonical phase-space coordinate of the field and, eventually, the quantum field operators.

It is instructive to calculate the magnitude of the phasor for one quantum of em action. In other words, we need the magnitude of the vector potential for a wave which contains one "photon" in a cavity of volume V. The time averaged em field energy $\langle U \rangle V$ for a plane wave in volume V follows using $\langle \cos^2 \omega t \rangle = \frac{1}{2}$:

$$\langle U \rangle V = \left\langle \frac{\varepsilon_0}{2} \mathbf{E} \cdot \mathbf{E} + \frac{1}{2\mu_0} \mathbf{B} \cdot \mathbf{B} \right\rangle V$$

$$= V \left(\frac{\varepsilon_0}{2} 4|a|^2 \omega^2 + 4 \frac{|a|^2}{2\mu_0} \right) \langle \cos^2(\mathbf{k} \cdot \mathbf{r} - \omega t + \phi) \rangle$$

$$= 2\varepsilon_0 \omega^2 |a|^2 V = 2(k^2/\mu_0)|a|^2 V. \qquad (8.6.3)$$

By equating this to Planck's quantum ($\langle U \rangle V = \hbar \omega$) we derive

$$|a| = \sqrt{\frac{\hbar \omega}{2\varepsilon_0 \omega^2 V}} = \sqrt{\frac{\hbar}{2\varepsilon_0 \omega V}} = A \quad \text{(for one photon)}. \qquad (8.6.4)$$

This is the "photon unit" of quantum field $\mathbf{A}$. Note that it is an inverse root function of frequency which in turn is proportional to the magnitude k of wave vector $\mathbf{k}$ through the vacuum dispersion relation $\omega = ck = k/\sqrt{\mu_0 \varepsilon_0}$.

A. Electromagnetic Fields and Operators

To completely describe an electromagnetic field in a box or "cavity" we need one phasor term like (8.6.2) for every possible value of $\mathbf{k}$ and for each choice $\mathbf{e}_1$ or $\mathbf{e}_2$ of polarization orthogonal to $\mathbf{k}$. The complete expression for the classical $\mathbf{A}$ is a sum over the possible modes:

$$\mathbf{A} = \sum_{\mathbf{k}} \left[(a_{\mathbf{k}1} \mathbf{e}_1 + a_{\mathbf{k}2} \mathbf{e}_2) e^{i(\mathbf{k} \cdot \mathbf{r} - \omega t)} + \text{c.c.} \right]$$

$$= \sum_{\mathbf{k}} \sum_{\alpha=1}^{2} \left[a_{\mathbf{k}\alpha} \mathbf{e}_\alpha e^{i(\mathbf{k} \cdot \mathbf{r} - \omega t)} + a_{\mathbf{k}\alpha}^* \mathbf{e}_\alpha e^{-i(\mathbf{k} \cdot \mathbf{r} - \omega t)} \right]. \qquad (8.6.5a)$$

Here the $\mathbf{k}$ vector satisfies box boundary conditions

$$k_\beta = n_\beta \frac{2\pi}{L} \quad (n_\beta = 1, 2, \ldots, j, \quad \beta = x, y, z), \qquad (8.6.5b)$$

where L is the length of any side of the box. The classical electric field is

$$\mathbf{E} = \sum_{\mathbf{k}} \sum_{\alpha} \left[i a_{\mathbf{k}\alpha} \omega \mathbf{e}_\alpha e^{i(\mathbf{k} \cdot \mathbf{r} - \omega t)} - i a_{\mathbf{k}\alpha}^* \omega \mathbf{e}_\alpha e^{-i(\mathbf{k} \cdot \mathbf{r} - \omega t)} \right]. \qquad (8.6.5c)$$

The magnetic field is

$$\mathbf{B} = \sum_{\mathbf{k}}\sum_{\alpha}\left[ia_{\mathbf{k}\alpha}k\mathbf{b}_\alpha e^{i(\mathbf{k}\cdot\mathbf{r}-\omega t)} - ia^*_{\mathbf{k}\alpha}k\mathbf{b}_\alpha e^{-i(\mathbf{k}\cdot\mathbf{r}-\omega t)}\right], \qquad (8.6.5d)$$

where unit vector $\mathbf{b}_\alpha = \mathbf{k}\times\mathbf{e}_\alpha/k$ is orthogonal to $\mathbf{k}$ and $\mathbf{e}_\alpha$.

(a) Classical Phasor Energy Relations The classical Hamiltonian follows from an integral over volume V of the energy density in (8.6.3). The electric contribution is

$$U_E V = \frac{\varepsilon_0}{2}\int d^3\mathbf{r}\,\mathbf{E}\cdot\mathbf{E}, \qquad (8.6.6a)$$

where

$$\mathbf{E}\cdot\mathbf{E} = \sum_{\mathbf{k}'\alpha'}\sum_{\mathbf{k}\alpha}\left(ia_{\mathbf{k}'\alpha'}\omega'\mathbf{e}_{\alpha'}e^{i(\mathbf{k}'\cdot\mathbf{r}-\omega't)} + \text{c.c.}\right)\cdot\left(ia_{\mathbf{k}\alpha}\omega\mathbf{e}_\alpha e^{i(\mathbf{k}\cdot\mathbf{r}-\omega t)} + \text{c.c.}\right)$$

$$= \sum_{\mathbf{k}'\alpha'}\sum_{\mathbf{k}\alpha}\Big[-a_{\mathbf{k}'\alpha'}a_{\mathbf{k}\alpha}\omega'\omega\mathbf{e}_{\alpha'}\cdot\mathbf{e}_\alpha e^{i(\mathbf{k}'+\mathbf{k})\cdot\mathbf{r}-i(\omega'+\omega)t}$$

$$-a^*_{\mathbf{k}'\alpha'}a^*_{\mathbf{k}\alpha}\omega'\omega\mathbf{e}_{\alpha'}\cdot\mathbf{e}_\alpha e^{-i(\mathbf{k}'+\mathbf{k})\cdot\mathbf{r}+i(\omega'+\omega)t}$$

$$+a^*_{\mathbf{k}'\alpha'}a_{\mathbf{k}\alpha}\omega'\omega\mathbf{e}_{\alpha'}\cdot\mathbf{e}_\alpha e^{i(\mathbf{k}'-\mathbf{k})\cdot\mathbf{r}-i(\omega'-\omega)t}$$

$$+a_{\mathbf{k}'\alpha'}a^*_{\mathbf{k}\alpha}\omega'\omega\mathbf{e}_{\alpha'}\cdot\mathbf{e}_\alpha e^{-i(\mathbf{k}'-\mathbf{k})\cdot\mathbf{r}+i(\omega'-\omega)t}\Big]. \qquad (8.6.6b)$$

This simplifies if we use wave and polarization normalization conditions:

$$\int d^3\mathbf{r}\, e^{i(\mathbf{k}'+\mathbf{k})\cdot\mathbf{r}} = \delta_{\mathbf{k}',-\mathbf{k}}V \quad\text{and}\quad \mathbf{e}_{\alpha'}\cdot\mathbf{e}_\alpha = \delta_{\alpha'\alpha}.$$

The result is

$$U_E V = \sum_{\mathbf{k}\alpha}\frac{\varepsilon_0 V}{2}\left[2|a_{\mathbf{k}\alpha}|^2\omega^2 - a^*_{-\mathbf{k}\alpha}a^*_{\mathbf{k}\alpha}\omega^2 e^{2i\omega t} - a_{-\mathbf{k}\alpha}a_{\mathbf{k}\alpha}\omega^2 e^{-2i\omega t}\right]. \qquad (8.6.7)$$

The magnetic energy $U_B V = \int d^3 r\,\mathbf{B}\cdot\mathbf{B}/2\mu_0$ has the same form as (8.6.6) if we do the following substitutions

$$\mathbf{E}\to\mathbf{B},\quad \frac{\varepsilon_0}{2}\to\frac{1}{2\mu_0},\quad \omega\mathbf{e}_\alpha\to k\mathbf{b}_\alpha \equiv \mathbf{k}\times\mathbf{e}_\alpha,\quad \omega'\mathbf{e}_{\alpha'}\to k'\mathbf{b}_{\alpha'}\equiv\mathbf{k}'\times\mathbf{e}_{\alpha'}.$$

After integration the cross terms have the opposite sign as they did in (8.6.6b). We get $\delta_{\mathbf{k}',-\mathbf{k}}kk' = -k^2$ in the $\mathbf{B}\cdot\mathbf{B}$ integral instead of $\delta_{\mathbf{k}',-\mathbf{k}}\omega\omega' =$

ω^2 which arose in the $\mathbf{E} \cdot \mathbf{E}$ integral. The magnetic energy is

$$U_B V = \sum_{\mathbf{k}\alpha} \frac{V}{2\mu_0} \left[2|a_{\mathbf{k}\alpha}|^2 k^2 + a^*_{-\mathbf{k}\alpha} a^*_{\mathbf{k}\alpha} k^2 e^{2i\omega t} + a_{-\mathbf{k}\alpha} a_{\mathbf{k}\alpha} k^2 e^{-2i\omega t} \right]$$

$$= \sum_{\mathbf{k}\alpha} \frac{\varepsilon_0 V}{2} \left[2|a_{\mathbf{k}\alpha}|^2 \omega^2 + a^*_{-\mathbf{k}\alpha} a^*_{\mathbf{k}\alpha} \omega^2 e^{2i\omega t} + a_{-\mathbf{k}\alpha} a_{\mathbf{k}\alpha} \omega^2 e^{-2i\omega t} \right]. \quad (8.6.8)$$

In the second line of the foregoing we use the dispersion relation for light:

$$\omega^2 = c^2 k^2 = k^2/(\mu_0 \varepsilon_0). \quad (8.6.9)$$

The change of sign makes the electric cross-terms cancel the magnetic ones. The sum of electric and magnetic energies [(8.6.7) and (8.6.8)] is then just a sum of elementary mode energy density values (8.6.3). That simple formula is all we need!

$$UV = (U_E + U_B)V = \sum_{\mathbf{k}\alpha} 2\varepsilon_0 \omega^2 |a_{\mathbf{k}\alpha}|^2 V. \quad (8.6.10)$$

Each mode labeled $\mathbf{k}$ and polarization α is described by a classical complex phasor variable $a_{\mathbf{k}\alpha}$. The real and imaginary parts of this variable can be treated as classical position Q and momentum P of an oscillator as described in Chapter 7. [Recall (7.5.9) and (7.5.10).]

(b) Classical Field Oscillator Variables Let us factor the phasor expression for field energy as follows:

$$UV = \sum_{\mathbf{k}\alpha} 2\varepsilon_0 V \omega^2 a^*_{\mathbf{k}\alpha} a_{\mathbf{k}\alpha} = \sum_{\mathbf{k}\alpha} \tfrac{1}{2} \left[2\omega\sqrt{\varepsilon_0 V} \left(a^{\text{Re}}_{\mathbf{k}\alpha} - i a^{\text{Im}}_{\mathbf{k}\alpha} \right) \right] \left[2\omega\sqrt{\varepsilon_0 V} \left(a^{\text{Re}}_{\mathbf{k}\alpha} + i a^{\text{Im}}_{\mathbf{k}\alpha} \right) \right]$$

$$= \sum_{\mathbf{k}\alpha} \tfrac{1}{2} [\omega Q_{\mathbf{k}\alpha} - i P_{\mathbf{k}\alpha}][\omega Q_{\mathbf{k}\alpha} + i P_{\mathbf{k}\alpha}]$$

$$= \sum_{\mathbf{k}\alpha} \tfrac{1}{2} (P^2_{\mathbf{k}\alpha} + \omega^2 Q^2_{\mathbf{k}\alpha}). \quad (8.6.11)$$

Note that frequency $\omega = \omega(k)$ is a function of k. The canonical phase space variables are

$$Q_{\mathbf{k}\alpha} = 2\sqrt{\varepsilon_0 V} a^{\text{Re}}_{\mathbf{k}\alpha} = \sqrt{\varepsilon_0 V} (a_{\mathbf{k}\alpha} + a^*_{\mathbf{k}\alpha}), \quad (8.6.12a)$$

$$P_{\mathbf{k}\alpha} = 2\omega\sqrt{\varepsilon_0 V} a^{\text{Im}}_{\mathbf{k}\alpha} = \omega\sqrt{\varepsilon_0 V} (a_{\mathbf{k}\alpha} - a^*_{\mathbf{k}\alpha})/i. \quad (8.6.12b)$$

The inverse of the foregoing gives the original phasor variables and their conjugates in terms of P's and Q's:

$$a_{k\alpha} = a_{k\alpha}^{Re} + ia_{k\alpha}^{Im} = \frac{1}{2\sqrt{\varepsilon_0 V}}(Q_{k\alpha} + iP_{k\alpha}/\omega), \qquad (8.6.13a)$$

$$a_{k\alpha}^* = a_{k\alpha}^{Re} - ia_{k\alpha}^{Im} = \frac{1}{2\sqrt{\varepsilon_0 V}}(Q_{k\alpha} - iP_{k\alpha}/\omega). \qquad (8.6.13b)$$

The cavity energy UV in (8.6.11) shall be the classical electromagnetic field Hamiltonian $H = H(Q, P)$. H describes a set of independent harmonic oscillators. To obtain a quantum field theory we make these into quantum oscillators. The situation is very similar to the molecular vibration problem in which a set of classical normal modes were "quantized" in Chapter 4.

(c) Quantum Field Operators Oscillator ladder operations **a** and **a**† were defined in (4.4.50) in terms of coordinate and momentum operators. For each em mode $(\mathbf{k}, \alpha)$ this definition translates to the following:

$$\mathbf{a}_{k\alpha} = \sqrt{\frac{\omega}{2\hbar}}(\mathbf{Q}_{k\alpha} + i\mathbf{P}_{k\alpha}/\omega), \qquad (8.6.14a)$$

$$\mathbf{a}_{k\alpha}^\dagger = \sqrt{\frac{\omega}{2\hbar}}(\mathbf{Q}_{k\alpha} - i\mathbf{P}_{k\alpha}/\omega), \qquad (8.6.14b)$$

where boldface notation $\mathbf{Q}_{k\alpha}$ and $\mathbf{P}_{k\alpha}$ indicates the quantum operators that correspond to the classical phase variables $Q_{k\alpha}$ and $P_{k\alpha}$, respectively.

By comparing (8.6.14) with (8.6.13) we note that the ladder operators are proportional to whatever operator would correspond to the classical phasor amplitude. So with correspondences $Q_{k\alpha} \to \mathbf{Q}_{k\alpha}$ and $P_{k\alpha} \to \mathbf{P}_{k\alpha}$ we have the phasor correspondence relations:

$$a_{k\alpha} = \frac{1}{2\sqrt{\varepsilon_0 V}}(Q_{k\alpha} + iP_{k\alpha}/\omega) \to \frac{1}{2\sqrt{\varepsilon_0 V}}(\mathbf{Q}_{k\alpha} + i\mathbf{P}_{k\alpha}/\omega)$$

$$= \frac{1}{2\sqrt{\varepsilon_0 V}}\sqrt{\frac{2\hbar}{\omega}}\mathbf{a}_{k\alpha} = \sqrt{\frac{\hbar}{2\varepsilon_0 \omega V}}\mathbf{a}_{k\alpha},$$

$$a_{k\alpha}^* = \frac{1}{2\sqrt{\varepsilon_0 V}}(Q_{k\alpha} - iP_{k\alpha}/\omega) \to \frac{1}{2\sqrt{\varepsilon_0 V}}(\mathbf{Q}_{k\alpha} - i\mathbf{P}_{k\alpha}/\omega)$$

$$= \frac{1}{2\sqrt{\varepsilon_0 V}}\sqrt{\frac{2\hbar}{\omega}}\mathbf{a}_{k\alpha}^\dagger = \sqrt{\frac{\hbar}{2\varepsilon_0 \omega V}}\mathbf{a}_{k\alpha}^\dagger. \qquad (8.6.15)$$

The proportionality or scale factor (8.6.15) turns out to be the quantum amplitude derived in (8.6.4). Note that coordinate and momentum operators are observables and are self-conjugate ($\mathbf{Q} = \mathbf{Q}^\dagger$ and $\mathbf{P} = \mathbf{P}^\dagger$). The phasor operator **a** is a complex combination of observables and therefore is not self-conjugate.

The oscillator Hamiltonian operator for the quantum field is the same form as (4.4.52), namely,

$$\mathbf{H} = \sum_{\mathbf{k}\alpha} \hbar\omega \left(\mathbf{a}_{\mathbf{k}\alpha}^\dagger \mathbf{a}_{\mathbf{k}\alpha} + \tfrac{1}{2} \right). \tag{8.6.16}$$

This is the same for the classical energy (8.5.10) or (8.6.11) except for the extra $\hbar\omega/2$ terms which are each mode's quantum zero-point energy. The eigenvalues of the number operator $\mathbf{a}_{\mathbf{k}\alpha}^\dagger \mathbf{a}_{\mathbf{k}\alpha}$ are the number $n_{\mathbf{k}\alpha}$ of photons in mode $(\mathbf{k}, \alpha)$. The creation or destruction operators $\mathbf{a}_{\mathbf{k}\alpha}^\dagger$ and $\mathbf{a}_{\mathbf{k}\alpha}$ raise or lower the photon number:

$$\mathbf{a}_{\mathbf{k}\alpha}^\dagger | \cdots n_{\mathbf{k}\alpha} \cdots n_{\mathbf{k}'\alpha'} \cdots \rangle = \sqrt{n_{\mathbf{k}\alpha} + 1} | \cdots n_{\mathbf{k}\alpha} + 1 \cdots n_{\mathbf{k}'\alpha'} \cdots \rangle,$$

$$\mathbf{a}_{\mathbf{k}\alpha} | \cdots n_{\mathbf{k}\alpha} \cdots n_{\mathbf{k}'\alpha'} \cdots \rangle = \sqrt{n_{\mathbf{k}\alpha}} | \cdots n_{\mathbf{k}\alpha} - 1 \cdots n_{\mathbf{k}'\alpha'} \cdots \rangle,$$

$$\mathbf{a}_{\mathbf{k}'\alpha'}^\dagger | \cdots n_{\mathbf{k}\alpha} \cdots n_{\mathbf{k}'\alpha'} \cdots \rangle = \sqrt{n_{\mathbf{k}'\alpha'} + 1} | \cdots n_{\mathbf{k}\alpha} \cdots n_{\mathbf{k}'\alpha'} + 1 \cdots \rangle,$$

$$\mathbf{a}_{\mathbf{k}'\alpha'} | \cdots n_{\mathbf{k}\alpha} \cdots n_{\mathbf{k}'\alpha'} \cdots \rangle = \sqrt{n_{\mathbf{k}'\alpha'}} | \cdots n_{\mathbf{k}\alpha} \cdots n_{\mathbf{k}'\alpha'} - 1 \cdots \rangle.$$
$$\tag{8.6.17}$$

Again, these relations are the same as before. [See (4.4.62).] Here each additional quanta contributes an increase in **A** amplitude equal to the scale factor $\sqrt{\hbar/2\varepsilon_0 \omega V}$ in the correspondence relation (8.6.15).

The quantum **A**-field operator corresponding to the classical field (8.6.5a) is found by replacing $a_{\mathbf{k}\alpha}$ and $a_{\mathbf{k}\alpha}^*$ according to (8.6.15):

$$\mathbf{A} = \sum_{\mathbf{k}\alpha} \sqrt{\frac{\hbar}{2\varepsilon_0 \omega V}} \left[\mathbf{a}_{\mathbf{k}\alpha} \mathbf{e}_\alpha e^{i(\mathbf{k}\cdot\mathbf{r} - \omega t)} + \mathbf{a}_{\mathbf{k}\alpha}^\dagger \mathbf{e}_\alpha e^{-i(\mathbf{k}\cdot\mathbf{r} - \omega t)} \right]. \tag{8.6.18}$$

The time dependence of the ladder operators is determined by the operator equations: $i\hbar \dot{\mathbf{O}} = [\mathbf{H}, \mathbf{O}]$ [recall (8.5.31)]:

$$i\hbar \dot{\mathbf{a}}_{\mathbf{k}\alpha} = [\mathbf{H}, \mathbf{a}_{\mathbf{k}\alpha}] \qquad i\hbar \dot{\mathbf{a}}_{\mathbf{k}\alpha}^\dagger = [\mathbf{H}, \mathbf{a}_{\mathbf{k}\alpha}^\dagger]$$

$$= -\hbar\omega \mathbf{a}_{\mathbf{k}\alpha} \qquad\qquad = \hbar\omega \mathbf{a}_{\mathbf{k}\alpha}^\dagger.$$

Here we use the $\mathbf{a}^\dagger \mathbf{a}$ form (8.6.16) of the field Hamiltonian and the standard commutation relation (4.4.51) which is repeated in the following [recall also (4.4.53)]:

$$[\mathbf{a}_{\mathbf{k}\alpha}, \mathbf{a}^\dagger_{\mathbf{k}'\alpha'}] = \delta_{\mathbf{k},\mathbf{k}'}\delta_{\alpha,\alpha'}\mathbf{1}. \quad (8.6.19)$$

According to the foregoing à equations the ladder operators have the following time-dependent phases:

$$\mathbf{a}_{\mathbf{k}\alpha} = \mathbf{a}_{\mathbf{k}\alpha}(0)e^{i\omega t}, \qquad \mathbf{a}_{\mathbf{k}\alpha} = \mathbf{a}^\dagger_{\mathbf{k}\alpha}(0)e^{-i\omega t}.$$

The phases cancel time factors in (8.6.18) to give a time-independent field operators

$$\mathbf{A} = \sum_{\mathbf{k}\alpha}\sqrt{\frac{\hbar}{2\varepsilon_0 \omega V}}\left[\mathbf{a}_{\mathbf{k}\alpha}(0)e^{i\mathbf{k}\cdot\mathbf{r}} + \mathbf{a}^\dagger_{\mathbf{k}\alpha}(0)e^{-i\mathbf{k}\cdot\mathbf{r}}\right]\mathbf{e}_\alpha. \quad (8.6.20a)$$

The electric and magnetic quantum field operators follow:

$$\mathbf{E} = \sum_{\mathbf{k}\alpha}\sqrt{\frac{\hbar}{2\varepsilon_0 \omega V}}\left[-i\omega\mathbf{a}_{\mathbf{k}\alpha}(0)e^{i\mathbf{k}\cdot\mathbf{r}} + i\omega\mathbf{a}^\dagger_{\mathbf{k}\alpha}(0)e^{-i\mathbf{k}\cdot\mathbf{r}}\right]\mathbf{e}_\alpha, \quad (8.6.20b)$$

$$\mathbf{B} = \sum_{\mathbf{k}\alpha}\sqrt{\frac{\hbar}{2\varepsilon_0 \omega V}}\left[i k\mathbf{a}_{\mathbf{k}\alpha}(0)e^{i\mathbf{k}\cdot\mathbf{r}} - i k\mathbf{a}^\dagger_{\mathbf{k}\alpha}(0)e^{-i\mathbf{k}\cdot\mathbf{r}}\right]\mathbf{b}_\alpha. \quad (8.6.20c)$$

When atoms are much smaller than the wavelength ($\lambda = 2\pi/k$) of the radiation, the fields can be simplified by the dipole approximation $e^{i\mathbf{k}\cdot\mathbf{r}} \cong 1$.

$$\mathbf{A} \cong \sum_{\mathbf{k}\alpha}\sqrt{\frac{\hbar}{2\varepsilon_0 \omega V}}\left[\mathbf{a}_{\mathbf{k}\alpha} + i\omega\mathbf{a}^\dagger_{\mathbf{k}\alpha}\right]\mathbf{e}_\alpha = \sum_{\mathbf{k}\alpha}\sqrt{\frac{1}{\varepsilon_0 V}}\mathbf{Q}_{\mathbf{k}\alpha}\mathbf{e}_\alpha, \quad (8.6.20d)$$

$$\mathbf{E} \cong \sum_{\mathbf{k}\alpha}\sqrt{\frac{\hbar}{2\varepsilon_0 \omega V}}i\omega\left[\mathbf{a}^\dagger_{\mathbf{k}\alpha} - \mathbf{a}_{\mathbf{k}\alpha}\right]\mathbf{e}_\alpha = \sum_{\mathbf{k}\alpha}\sqrt{\frac{1}{\varepsilon_0 V}}\mathbf{P}_{\mathbf{k}\alpha}\mathbf{e}_\alpha, \quad (8.6.20e)$$

$$\mathbf{B} \cong \sum_{\mathbf{k}\alpha}\sqrt{\frac{\hbar}{2\varepsilon_0 \omega V}}ik\left[\mathbf{a}_{\mathbf{k}\alpha} - \mathbf{a}^\dagger_{\mathbf{k}\alpha}\right]\mathbf{b}_\alpha. \quad (8.6.20f)$$

Note also the simple connection between the approximate **A** and **E** and the canonical field coordinates $\mathbf{Q}_{\mathbf{k}\alpha}$ and momenta $\mathbf{P}_{\mathbf{k}\alpha}$ which follows from (8.6.14).

B. Electromagnetic Quantum States and Transitions

Consider an atom coupled to an electromagnetic cavity. Suppose this system starts in a state in which the atomic state is $|s\rangle$ and all the photon numbers $n^s_{\mathbf{k}\alpha}$ are definitely known. We consider some of the possible final states and their probabilities as a function of time.

The states of the whole system at the start and finish will be labeled $|S\rangle$ and $|F\rangle$, respectively. The starting state $|S\rangle$ is a ket-ket product of atomic $|s\rangle$ and radiation $|\cdots n^s_{\mathbf{k}\alpha} \cdots n^s_{\mathbf{k}'\alpha'} \cdots \rangle$ states:

$$|S\rangle = |s\rangle|\cdots n^s_{\mathbf{k}\alpha} \cdots n^s_{\mathbf{k}'\alpha'} \cdots \rangle.$$

The final state is written in a similar way:

$$|F\rangle = |f\rangle|\cdots n^f_{\mathbf{k}\alpha} \cdots n^f_{\mathbf{k}'\alpha'} \cdots \rangle.$$

One may picture the states by imagining atomic and electromagnetic levels as sketched in Figure 8.6.1. A typical transition which conserves energy (more or less) can be imagined as going from the state on one side of the figure to the other. There we imagine that the atom jumps *up* from level $|s\rangle$ to $|f\rangle$ while simultaneously a mode number jumps *down* one level. This is an atomic *absorption* process. (The atom appears to swallow a photon.) If this is reversed, or if the atom jumps *down* from level $|s\rangle$ to $|f'\rangle$ while a mode number jumps *up* the process is called an *emission*. (The atom appears to spit out a photon.) As we will see these two processes are usually the most likely ones.

The derivation of the probabilities for quantum field atomic transitions of the type shown in Figure 8.6.1 are given now. This derivation uses the first-order perturbation formula, because we only need to create or destroy one photon:

$$c^{(1)}_F(t) = \delta_{FS} + \frac{1}{i\hbar}\int_0^t dt_1\, e^{i\omega_{FS}t_1}\langle F|H_1|S\rangle$$

$$= \delta_{FS} + \frac{1}{i\hbar}S(\omega_{FS}, t)\langle F|H_1|S\rangle. \qquad (8.6.21a)$$

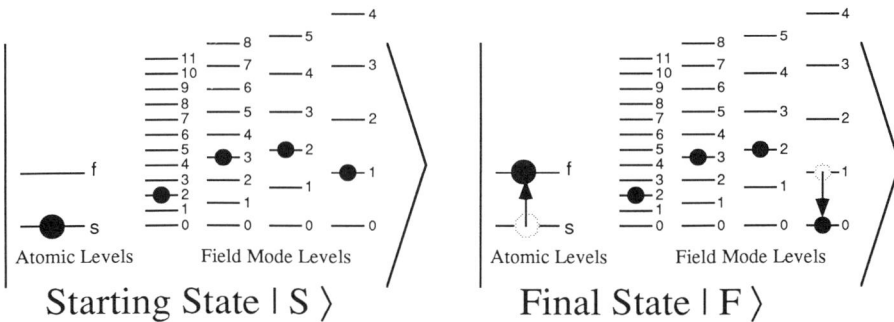

Figure 8.6.1 Atom-field energy levels for initial and final states in an atomic absorption process involving a single photon from a resonant electromagnetic field mode.

Here the spectral function is the following:

$$S(\Delta, t) = 2e^{it\Delta/2} \frac{\sin(t\Delta/2)}{\Delta}. \tag{8.6.21b}$$

[Recall (8.4.34) and (8.4.38) and discussion of Figure 8.4.2.] The detuning parameter Δ must include the difference of both the atomic and radiation energy:

$$\Delta = \omega_{FS} \equiv \omega_F - \omega_S = \left[\omega_f + \sum_{\mathbf{k}\alpha}\left(n_{\mathbf{k}\alpha}^f + \tfrac{1}{2}\right)\omega_{\mathbf{k}}\right] - \left[\omega_s + \sum_{\mathbf{k}\alpha}\left(n_{\mathbf{k}\alpha}^s + \tfrac{1}{2}\right)\omega_{\mathbf{k}}\right]. \tag{8.6.22}$$

For the absorption process depicted in Figure 8.6.1 the mode number has gone down one step for the $(\mathbf{k}\alpha)$ mode $(n_{\mathbf{k}\alpha}^f = n_{\mathbf{k}\alpha}^s - 1)$ while the atom went up. All other quanta stayed the same, so the detuning is

$$\Delta = \omega_f - \omega_s - \omega_{\mathbf{k}} = \omega_{fs} - \omega_{\mathbf{k}}.$$

Here zero detuning corresponds to picking a mode whose frequency $\omega_{\mathbf{k}}$ matches the atomic transition frequency ω_{fs}. This is similar to the semiclassical definition of absorption resonance. Compare (8.4.38b) with the foregoing Δ equation.

Now we see some important differences between quantum field theory calculations and semiclassical ones. For one thing, if you really insist on counting every photon, then the absorption and emission processes become clearly separated.

(a) Single-Mode Atomic Dipole Transitions The first-order S to F transition probability $|c_F|^2$ obtained from (8.6.21) is (assuming $S \neq F$)

$$P_{F \leftarrow S} = |c_F|^2 = |\langle F|H_I|S\rangle|^2 \frac{\sin^2(t\Delta/2)}{\hbar^2(\Delta/2)^2}. \tag{8.6.23}$$

We now evaluate the matrix element of the interaction operator (8.4.18b):

$$H_I = q\mathbf{E} \cdot \mathbf{r} \cong -q\sum_{\mathbf{k}\alpha}\sqrt{\frac{\hbar}{2\varepsilon_0 \omega V}}\left[i\omega a_{\mathbf{k}\alpha}^\dagger - i\omega a_{\mathbf{k}\alpha}\right]\mathbf{e}_\alpha \cdot \mathbf{r}. \tag{8.6.24}$$

The quantized E field (8.6.20b) is used with the dipole approximation $e^{i\mathbf{k}\cdot\mathbf{r}} \cong 1$. The matrix element consists of a field part followed by the usual atomic matrix element:

$$\langle F|H_I|S\rangle = q\sum_{\mathbf{k}\alpha}[\langle \cdots n^f_{\mathbf{k}\alpha} \cdots |i\mathbf{a}^\dagger_{\mathbf{k}\alpha}| \cdots n^s_{\mathbf{k}\alpha}\rangle - \langle \cdots n^f_{\mathbf{k}\alpha} \cdots |i\mathbf{a}_{\mathbf{k}\alpha}| \cdots n^s_{\mathbf{k}\alpha}\rangle]$$

$$\times \sqrt{\frac{\hbar\omega}{2\varepsilon_0 V}}\,\mathbf{e}_\alpha \cdot \langle f|\mathbf{r}|s\rangle.$$

The field part of this matrix element is quite selective. If more than one mode changes its photon number the whole thing is zero. (Recall that $\langle n|n'\rangle = \delta_{nn'}$.) The only possible nonzero elements occur when a single mode goes up or down by exactly one photon. The two possible types of nonzero matrix elements are listed in the following:

$$\langle F|H_I|S\rangle = \left[-i\sqrt{n^s_{\mathbf{k}\alpha}+1}\,+0\right]\sqrt{\frac{\hbar\omega}{2\varepsilon_0 V}}\,q\mathbf{e}_\alpha \cdot \mathbf{r}_{fs},$$

if all $n^f = n^s$ except $n^f_{\mathbf{k}\alpha} = n^s_{\mathbf{k}\alpha} + 1$
(1 PHOTON EMISSION)

$$= \left[0 + i\sqrt{n^s_{\mathbf{k}\alpha}}\right]\sqrt{\frac{\hbar\omega}{2\varepsilon_0 V}}\,q\mathbf{e}_\alpha \cdot \mathbf{r}_{fs},$$

if all $n^f = n^s$ except $n^f_{\mathbf{k}\alpha} = n^s_{\mathbf{k}\alpha} - 1$
(1 PHOTON ABSORPTION)

$$= 0,\text{ otherwise.} \quad (8.6.25)$$

The field matrix elements follow from (8.6.17) and the atomic dipole expectation value is denoted by $\mathbf{r}_{fs} = \langle f|\mathbf{r}|s\rangle$.

If matrix element (8.6.25) allows any transition between $|S\rangle$ and $|F\rangle$ it will be an emission or an absorption but not both. If it allows one then the probability for the other is zero. This is very different from the semiclassical transition amplitude (8.4.37) in which both processes would simultaneously have nonvanishing probability. The semiclassical amplitude is a sum of a resonant and a nonresonant spectral function. The quantum amplitude (8.6.21) has only one spectral function in the c_F expression. Strict photon counting prevents absorption from interfering with emission.

(b) Multimode Atomic Dipole Transitions Suppose that we accept (or are forced to accept) any of a set of possible final states going from $|F\rangle = |f\rangle|\cdots n_{\mathbf{k}\alpha} - 1 \cdots n_{\mathbf{k}'\alpha'} \cdots\rangle$ in which mode $(\mathbf{k}, \alpha)$ lost a photon to $|F'\rangle = |f\rangle|\cdots n_{\mathbf{k}\alpha} \cdots n_{\mathbf{k}'\alpha'} - 1 \cdots\rangle$ in which mode $(\mathbf{k}', \alpha')$ lost a photon. In each case the atom jumps up from state $|s\rangle$ to state $|f\rangle$ but now we let it accept a photon from different cavity modes. It is only necessary that the donor modes have nonzero photon number and a frequency ω that is close enough to the atomic transition frequency $\omega_{fs} = \omega_f - \omega_s$ so the spectral function $|S|^2$ gives a measurable value.

The total probability for the atomic $f \leftarrow s$ transition will be a sum of probabilities $|c_F|^2 + \cdots + |c_{F'}|^2$ as though the contribution of each mode is distinct:

$$P_{f \leftarrow s} = |c_F|^2 + \cdots + |c_{F'}|^2 = \sum_{\mathbf{k}\alpha}^{\mathbf{k}'\alpha'} |c_F|^2 = \sum_{\mathbf{k}\alpha}^{\mathbf{k}'\alpha'} \frac{\sin^2(t\Delta/2)}{\hbar^2(\Delta/2)^2} |\langle H_I \rangle|^2. \quad (8.6.26)$$

Since we are effectively counting the photons from each mode the amplitudes $c_F \ldots c_{F'}$ have random relative phases and interference between them is washed out. Hence the total probability is the sum of their squares $\Sigma |c_F|^2$ instead of the more complicated square of the sum $|\Sigma c_F|^2$. (Recall discussion in Chapter 1, Section 1.1: Axiom 4.)

The sum over mode wave vector $\mathbf{k}$ can be converted to an integral over $k = |\mathbf{k}|$ or over mode frequency $\omega = ck$. According to (8.6.5b) the $\mathbf{k}$ sum is a sum over integer values of photon number $n_\alpha = k_\alpha L/2\pi = 1, 2, \ldots$, or

$$\sum_{\mathbf{k}} = \sum_{n_x} \sum_{n_y} \sum_{n_z} \Delta n_x \Delta n_y \Delta n_z \cong \int dk_x \frac{\Delta n_x}{\Delta k_x} \int dk_y \frac{\Delta n_y}{\Delta k_y} \int dk_z \frac{\Delta n_z}{\Delta k_z}$$

$$= \left(\frac{L}{2\pi}\right)^3 \int dk_x \int dk_y \int dk_z.$$

Here the sum is converted to an integral over Cartesian k components using

$$\Delta n_x = 1 = \frac{L}{2\pi}\Delta k_x, \quad \Delta n_y = 1 = \frac{L}{2\pi}\Delta k_y, \quad \Delta n_z = 1 = \frac{L}{2\pi}\Delta k_z.$$

This sum can then be converted to a polar coordinate integral in k space:

$$\sum_{\mathbf{k}} = \left(\frac{L}{2\pi}\right)^3 \int d^3\mathbf{k} = \frac{V}{(2\pi)^3} \int d\Omega_\mathbf{k} \int k^2 \, dk = \frac{V}{(2\pi)^3} \int d\phi_\mathbf{k} \int d\theta_\mathbf{k} \sin\theta_\mathbf{k} \int k^2 \, dk,$$

(8.6.27)

where $(\phi_\mathbf{k}, \theta_\mathbf{k})$ are azimuth and polar angles of $\mathbf{k}$, $d\Omega_\mathbf{k}$ is the incremental solid angle in k space, and $V = L^3$ is the cavity volume. This reduces the probability sum (8.6.26) to an integral over solid angle and k or $\omega = ck$:

$$P_{f \leftarrow s} = \sum_\alpha \frac{V}{(2\pi)^3} \int d\Omega_\mathbf{k} \int_0^\infty k^2 \, dk |c_F|^2 = \sum_\alpha \frac{V}{(2\pi)^3} \int d\Omega_\mathbf{k} \int_0^\infty \frac{\omega^2}{c^3} d\omega |c_F|^2$$

$$= \sum_\alpha \frac{V}{(2\pi)^3} \int d\Omega_\mathbf{k} \int d\omega \frac{\omega^2}{c^3} \frac{\sin^2(t\Delta/2)}{\hbar^2(\Delta/2)^2} |\langle H_I \rangle|^2. \quad (8.6.28)$$

Now we approximate the integral of the spectral function $|S(\Delta, t)|^2$ by assuming that time t is large enough that $|S|^2$ becomes very narrow. (Recall Figure 8.4.2.) Then most of the probability comes from the neighborhood around zero detuning ($\Delta = 0$). We may set $\omega = \omega_{fs}$ and put all other functions of frequency outside the integral:

$$P_{f \leftarrow s} \cong \sum_{\alpha} \frac{V}{(2\pi)^3} \int d\Omega_{\mathbf{k}} \frac{\omega^2}{\hbar^2 c^3} |\langle H_I \rangle|^2 \int_0^{\infty} d\omega \frac{\sin^2(t\Delta/s)}{\hbar^2(\Delta/2)^2}.$$

The area under the spectral function is simply the elapsed time multiplied by 2π as was noted before equation (8.4.40).

$$P_{f \leftarrow s} \cong \sum_{\alpha} \int d\Omega_{\mathbf{k}} \frac{V\omega}{(2\pi)^3 \hbar^2 c^3} |\langle H_I \rangle|^2 2\pi t \qquad (8.6.29)$$

The peak of the spectral function at $\Delta = 0$ goes up quadratically with time, but the area only increases linearly. This is because the width of the peak decreases linearly with time according to the uncertainty relation (8.4.42). As a result the time derivative or rate $R_{f \leftarrow s} = \dot{P}_{f \leftarrow s}$ of the transition probability is a constant in this approximation. This is known as the *Fermi golden rule* of constant transition rates. We write this as

$$R_{f \leftarrow s} = \sum_{\alpha} \int d\Omega_{\mathbf{k}} \rho(\omega_{fs}) |\langle H_I \rangle|^2 \frac{2\pi}{\hbar^2}, \qquad (8.6.30a)$$

where the spectral density of modes $\rho(\omega)$ is defined here:

$$\rho(\omega) = \frac{V\omega^2}{(2\pi)^3 c^3}. \qquad (8.6.30b)$$

The absorption dipole matrix element (8.6.25) gives the following rate if the photon number in near-resonant modes is $n^s_{\mathbf{k}\alpha} \equiv n$:

$$R_{f \leftarrow s} = n \frac{V\omega^2}{(2\pi)^3 c^3} \frac{2\pi}{\hbar^2} \frac{\hbar\omega}{2\varepsilon_0 V} q^2 \sum_{\alpha} \int d\Omega_{\mathbf{k}} |\mathbf{e}_{\alpha} \cdot \mathbf{r}_{fs}|^2$$

$$= n \left(\frac{\omega^3}{\hbar c^3} \right) \left(\frac{q^2}{4\pi\varepsilon_0} \right) \int d\Omega_{\mathbf{k}} \left(|\mathbf{e}_1 \cdot \mathbf{r}_{fs}|^2 + |\mathbf{e}_2 \cdot \mathbf{r}_{fs}|^2 \right).$$

The polarization sum and integral is simplified by using the vector relation

$$|\mathbf{r}|^2 = |\mathbf{e}_1 \cdot \mathbf{r}|^2 + |\mathbf{e}_2 \cdot \mathbf{r}|^2 + |\hat{\mathbf{k}} \cdot \mathbf{r}|^2.$$

If we let the induced dipole $\mathbf{r} = \mathbf{r}_{fs}$ be along the polar z axis then $\hat{\mathbf{k}} \cdot \mathbf{r}_{fs} = |\mathbf{r}_{fs}| \cos \theta_{\mathbf{k}}$. The sum and integral is then easily evaluated:

$$\sum_\alpha \int d\Omega_{\mathbf{k}} |\mathbf{e}_\alpha \cdot \mathbf{r}_{fs}|^2 = \int d\Omega_{\mathbf{k}} \left(|\mathbf{e}_1 \cdot \mathbf{r}_{fs}|^2 + |\mathbf{e}_2 \cdot \mathbf{r}_{fs}|^2 + |\hat{\mathbf{k}} \cdot \mathbf{r}_{fs}|^2 - |\hat{\mathbf{k}} \cdot \mathbf{r}_{fs}|^2 \right)$$

$$= \int d\Omega_{\mathbf{k}} (1 - \cos^2 \theta_{\mathbf{k}}) |\mathbf{r}_{fs}|^2 = \int d\phi \int d\theta \sin^3 \theta |\mathbf{r}_{fs}|^2 = \frac{8\pi}{3} |\mathbf{r}_{fs}|^2.$$

The resulting absorption rate is

$$R_{f \leftarrow s}(\text{absorption}) \equiv nA = n \frac{4\omega^3}{3\hbar c^3} \frac{q^2}{4\pi\varepsilon_0} |\mathbf{r}_{fs}|^2 = B. \qquad (8.6.31)$$

The corresponding emission rate is the same form except that a factor $(n + 1)$ replaces n in the matrix element (8.6.25):

$$R_{f' \leftarrow s}(\text{emission}) = (n + 1)A = A + B. \qquad (8.6.32a)$$

The first term is the famous Einstein A coefficient, which is the spontaneous decay rate of an excited atom in a vacuum ($n = 0$):

$$A = \frac{4\omega^3}{3\hbar c^3} \frac{q^2}{4\pi\varepsilon_0} |\mathbf{r}_{fs}|^2. \qquad (8.6.32b)$$

The second term is the Einstein B coefficient which is the decay rate induced by the presence of n resonant photons:

$$B = nA = n \frac{4\omega^3}{3\hbar c^3} \frac{q^2}{4\pi\varepsilon_0} |\mathbf{r}_{fs}|^2 \qquad (8.6.32c)$$

B is the only contribution to the absorption rate (8.6.31) since spontaneous excitation is impossible in this approximation.

(c) "Impotence" of Photon Number States We noted that the first-order transition amplitude $c_F^{(1)}$ in (8.6.21) could only have an absorption term or else an emission term but not both. This is because the quantum field transition matrix element (8.6.25) cannot be nonzero for both processes at once. We also noted that the semiclassical transition amplitude (8.4.36) does have terms from both processes. In fact, the derivation of resonant excitation of the oscillator expectation value $\langle x \rangle$ depends upon the precise interference between these two terms to reproduce the classical result. [Recall comparison of (8.4.45) and (8.4.47).]

The calculation of the atomic position expectation value $\langle \Psi | x | \Psi \rangle$ using quantum field number states is quite different and so are the results. The perturbed atom-field state is

$$|\Psi\rangle = |S\rangle + \sum_{F \neq S} e^{-i\omega_{FS}t} c_F |F\rangle,$$

where the first-order approximation to the amplitude is

$$c_F^{(1)} = \delta_{FS} + \frac{1}{i\hbar} \int_0^t dt_1 \, e^{i\omega_{FS}t_1} \langle F|H_I|S\rangle,$$

according to the basic time-dependent perturbation formulas (8.4.34). The matrix element is given by matrix elements (8.6.25) between the initial state $|S\rangle = |s\rangle_{\text{atom}} |n^s_{\mathbf{k}\alpha} \cdots n^s_{\mathbf{k}'\alpha'}\rangle_{\text{field}}$ and final state $|F\rangle = |f\rangle_{\text{atom}} |n^f_{\mathbf{k}\alpha} \cdots n^f_{\mathbf{k}'\alpha'}\rangle_{\text{field}}$. The atom part only requires that the induced dipole moment $q\mathbf{r}_{fs} \equiv \langle f|\mathbf{r}|s\rangle q$ be nonzero. The field part is more restrictive; it requires that exactly one mode gain or lose a single photon. The result is a perturbed state of the form

$$|\Psi\rangle = |S\rangle = |s\rangle |n^s_{\mathbf{k}\alpha} \cdots n^s_{\mathbf{k}'\alpha'}\rangle + d^a_{\mathbf{k}\alpha}(f)|f\rangle |n^s_{\mathbf{k}\alpha} - 1 \cdots n^s_{\mathbf{k}'\alpha'}\rangle + \cdots$$
$$+ d^a_{\mathbf{k}'\alpha'}(f)|f\rangle |n^s_{\mathbf{k}\alpha} \cdots n^s_{\mathbf{k}'\alpha'} - 1\rangle + \cdots$$
$$+ d^e_{\mathbf{k}\alpha}(f')|f'\rangle |n^s_{\mathbf{k}\alpha} + 1 \cdots n^s_{\mathbf{k}'\alpha'}\rangle + \cdots$$
$$+ d^e_{\mathbf{k}'\alpha'}(f')|f'\rangle |n^s_{\mathbf{k}\alpha} \cdots n^s_{\mathbf{k}'\alpha'} + 1\rangle + \cdots,$$

where the transition amplitude to a higher state $|f\rangle$ due to absorption from mode $(\mathbf{k}, \alpha)$ is

$$d^a_{\mathbf{k}\alpha}(f) = \frac{e^{-i\omega_F t}}{\hbar} \int_0^t dt_1 \, e^{it_1 \Delta} w_\mathbf{k} \sqrt{n_{\mathbf{k}\alpha}} \, \mathbf{e}_\alpha \cdot \mathbf{r}_{f's} \qquad (\Delta = \omega_{fs} - \omega_\mathbf{k}),$$

and the transition amplitude to a lower state $|f'\rangle$ due to emission from mode $(\mathbf{k}, \alpha)$ is

$$d^e_{\mathbf{k}\alpha}(f') = \frac{e^{-i\omega_F t}}{\hbar} \int_0^t dt_1 \, e^{it_1 \Delta} w_\mathbf{k} \sqrt{n_{\mathbf{k}\alpha} + 1} \, \mathbf{e}_\alpha \cdot \mathbf{r}_{f's} \qquad (\Delta = \omega_{f's} + \omega_\mathbf{k}),$$

and $w_\mathbf{k}$ is the scale factor (8.6.4) times frequency ω and charge q.

$$w_{\mathbf{k}\alpha} = \omega_\mathbf{k} q \sqrt{\frac{\hbar}{2\varepsilon_0 \omega_\mathbf{k} V}}$$

If $|s\rangle$ is the atomic ground state the emission terms with d^e amplitudes are nonexistent. However, there may be several atomic states $|f\rangle, |f'\rangle, \ldots$ which can be reached by near-resonant absorption. The same goes for state reached by emission if $|s\rangle$ is an excited state.

Now the expectation value $\langle\Psi|x|\Psi\rangle$ can be evaluated. We write the bra and ket on the left and top, respectively, of the box in the following, and we collect the scalar products inside:

$\langle\Psi|x|\Psi\rangle =$

	$|s\rangle|n^s_{\mathbf{k}\alpha}\cdots\rangle + d^a_{\mathbf{k}\alpha}|f\rangle|n^s_{\mathbf{k}\alpha}-1\cdots\rangle + \cdots$
$\langle n^a_{\mathbf{k}\alpha}\cdots|\langle s|x$	$\langle s|x|s\rangle\langle 1\rangle + d^\alpha_{\mathbf{k}\alpha}\langle s|x|f\rangle\langle 0\rangle + \cdots$
$+d^{\alpha*}_{\mathbf{k}\alpha}\langle n^s_{\mathbf{k}\alpha}-1\cdots|\langle f|x$	$+d^{\alpha*}_{\mathbf{k}\alpha}\langle f|x|s\rangle\langle 0\rangle + |d^\alpha_{\mathbf{k}\alpha}|^2\langle f|x|f\rangle\langle 1\rangle + \cdots.$

The sum includes only one field mode and one higher atomic state $|f\rangle$. This is enough to see that all the results must be zero if the atomic states have definite parity ($\langle s|x|s\rangle = 0 = \langle f|x|f\rangle$). The orthogonality ($\langle n_\mathbf{k}|n_\mathbf{k}-1\rangle = 0$) of the photon states kills the possibility of any contribution from the induced moment matrix elements $\langle s|x|f\rangle$ or $\langle f|x|s\rangle$, however large they may be.

So photon number states are "impotent"; they cannot create a coherent excitation of an atom. This is consistent with the idea that classical phase is completely uncertain or random for a field oscillator eigenstate which has probability distributed more or less evenly over its phase space. To have a well-defined phase we need a nonstationary coherent oscillator state $|\alpha_{\mathbf{k}\beta}\rangle$ for a mode $(\mathbf{k},\beta)$ instead of a stationary eigenstate $|n_{\mathbf{k}\beta}\rangle$. This will give a wave packet in $(Q_{\mathbf{k}\beta}, P_{\mathbf{k}\beta})$ phase space or $(\mathbf{A},\mathbf{E})$ space which may have well-defined phase as shown in the next section.

Photon number states may be "impotent" but they are not powerless. They can create large fluctuations in expectation $\langle\Psi|x^2|\Psi\rangle$ even though $\langle\Psi|x|\Psi\rangle$ is identically zero. From the foregoing calculation we get the following:

$$\langle\Psi|x^2|\Psi\rangle = \langle s|x^2|s\rangle + \left|d^a_{\mathbf{k}\alpha}\right|^2\langle f|x^2|f\rangle + \cdots. \qquad (8.6.33)$$

The expectation of x^2 is proportional to the photon intensity $n_{\mathbf{k}\alpha}$, the square $|\mathbf{r}_{fs}|^2$ of the atomic induced moment, and the mean square x for the final state $|f\rangle$.

(d) Coherent Radiation States A much better description of a laser-cavity mode includes the nonstationary coherent or wave packet states $|\alpha_{\mathbf{k}\beta}\rangle$. According to (8.2.17) these may be defined as follows in terms of photon number states $|n_{\mathbf{k}\beta}\rangle$ for a single-cavity mode $(\mathbf{k},\beta)$:

$$|\alpha_{\mathbf{k}\beta}\rangle = e^{-|\alpha_{\mathbf{k}\beta}|^2/2}\sum_{n_{\mathbf{k}\beta}}(\alpha_{\mathbf{k}\beta})^{n_{\mathbf{k}\beta}}|n_{\mathbf{k}\beta}00\cdots\rangle/\sqrt{n_{\mathbf{k}\beta}!}. \qquad (8.6.34)$$

The complex parameter $\alpha_{\mathbf{k}\beta}$ is the field phasor expectation value and the quasieigenvalues of the ladder operators:

$$\mathbf{a}_{\mathbf{k}\beta}|\alpha_{\mathbf{k}\beta}\rangle = \alpha_{\mathbf{k}\beta}|\alpha_{\mathbf{k}\beta}\rangle, \qquad \langle\alpha_{\mathbf{k}\beta}|\mathbf{a}^\dagger_{\mathbf{k}\beta} = \langle\alpha_{\mathbf{k}\beta}|\alpha^*_{\mathbf{k}\beta}. \qquad (8.6.35)$$

Recall (8.2.23) and (8.2.24). We assume here that only the $(\mathbf{k}, \beta)$ mode is excited and all others are in their ground or vacuum states.

The A-field expectation value should equal the classical expression (8.6.2a) which began this section:

$$\langle\alpha_{\mathbf{k}\beta}|\mathbf{A}|\alpha_{\mathbf{k}\beta}\rangle = \sqrt{\frac{\hbar}{2\varepsilon_0\omega V}} \left[\langle\alpha_{\mathbf{k}\beta}|\mathbf{a}_{\mathbf{k}\beta}|\alpha_{\mathbf{k}\beta}\rangle + \langle\alpha_{\mathbf{k}\beta}|\mathbf{a}^\dagger_{\mathbf{k}\beta}|\alpha_{\mathbf{k}\beta}\rangle\right]\mathbf{e}_\beta$$

$$= \sqrt{\frac{\hbar}{2\varepsilon_0\omega V}} [\alpha_{\mathbf{k}\beta} + \alpha^*_{\mathbf{k}\beta}]\mathbf{e}_\beta$$

$$= \left[a_{\mathbf{k}\beta}e^{-i\omega t} + a^*_{\mathbf{k}\beta}e^{i\omega t}\right]\mathbf{e}_\beta.$$

In the last line is the classical value. Note that the dipole approximation $e^{i\mathbf{k}\cdot\mathbf{r}} \cong 1$ is used here. The α has the necessary $e^{-i\omega t}$ time dependence noted in (8.2.20c). This gives the nonstationary phase packet motion described by (8.2.22). The relation between α and the classical amplitude a involves the phasor and the quantum scale (8.6.4):

$$\alpha_{\mathbf{k}\beta} = a_{\mathbf{k}\beta}e^{-i\omega_k t}\sqrt{\frac{2\varepsilon_0\omega V}{\hbar}}.$$

The $\alpha_\mathbf{k}$ expectation value of the $-q\mathbf{E}\cdot\mathbf{r}$ interaction in a coherent state then becomes equal to the classical value quoted in (8.4.35a):

$$\langle\alpha_{\mathbf{k}\beta}|H_I|\alpha_{\mathbf{k}\beta}\rangle = -i\omega_k q\mathbf{e}_\beta\cdot\mathbf{r}\left(a_{\mathbf{k}\beta}e^{-i\omega t} - a^*e^{i\omega t}\right).$$

It has the positive and negative frequency parts needed to coherently excite an atom.

This might lead you to believe that a coherent quantum field can reproduce the effect on an atom of a classical wave. Indeed, for very high $\alpha_\mathbf{k}$ and short enough time the effect of the two are nearly the same. However, eventually the coherent wave produces dephasing and rephasing effects which are quite remarkable. These coherent decays and "revivals" are discussed in works listed at the end of this chapter and are the subject of much ongoing research.

8.7 SPECTRA OF ATOM IN LASER CAVITY

When the atom interacts strongly or resonantly with an electromagnetic field the distinction between the field and the atom is blurred. Observer and the observed become a single entity that is more than just a sum of its parts. It is as though the atom had become part of a molecule in which the excited levels involve excitation of all the constituent parts.

An atom interacting strongly with a single mode of a simple cavity is described by what is called the Jaynes-Cummings model. Here we will give a brief qualitative sketch of states and levels of this model. This is an important model for beginning to understand spectroscopic effects of strong laser fields. It is also a simple solvable example of an atom interacting with something that has a multitude of states. This is the kind of problem one encounters when atomic motions go together to make a molecular rotation or vibration spectrum.

A. Jaynes-Cummings Hamiltonian

In a static electric field the two-level atomic system Hamiltonian has the following representation (8.5.2b) in the basis of atomic eigenstates:

$$H_{\text{atom}} = H\begin{pmatrix} 1 & 0 \\ 0 & 1 \end{pmatrix} + S\begin{pmatrix} 1 & 0 \\ 0 & -1 \end{pmatrix} - pE\begin{pmatrix} 0 & 1 \\ 1 & 0 \end{pmatrix} = H\mathbf{1} + S\sigma_z - pE\sigma_x. \tag{8.7.1a}$$

Generally the electric field-dipole factor pE is folded into the Rabi coefficient

$$r = -pE_z/\hbar. \tag{8.7.1b}$$

In an oscillating electric field the Hamiltonian is transformed into rotating-wave form (8.5.13), which is now rewritten:

$$H_{\text{RW}} = \frac{\Delta}{2}\begin{pmatrix} 1 & 0 \\ 0 & -1 \end{pmatrix} + \frac{r}{2}\begin{pmatrix} 0 & 1 \\ 1 & 0 \end{pmatrix} = \frac{\Delta}{2}\sigma_z + \frac{r}{2}\sigma_x. \tag{8.7.2a}$$

Here the detuning factor

$$\Delta = \varepsilon - \Omega \tag{8.7.2b}$$

is the difference between the atomic transition angular frequency ε and the angular frequency Ω of the stimulating laser.

In preparation for using a quantum field we need to separate the atomic and radiative contributions to the Hamiltonian. In the following the terms which involve the laser frequency Ω have been collected into one term which is labeled H_{field}:

$$H_{\text{RW}} = \Omega \begin{pmatrix} 0 & 0 \\ 0 & 1 \end{pmatrix} - \frac{\varepsilon + \Omega}{2} \begin{pmatrix} 1 & 0 \\ 0 & 1 \end{pmatrix} + \varepsilon \begin{pmatrix} 1 & 0 \\ 0 & 0 \end{pmatrix} + \frac{r}{2} \begin{pmatrix} 0 & 1 \\ 1 & 0 \end{pmatrix}$$

$$= H_{\text{field}} + \frac{\varepsilon}{2}(\sigma_z + 1) + \frac{r}{2}\sigma_x$$

$$= H_{\text{field}} + H_{\text{atom}} + H_{\text{interaction}}.$$

In the Jaynes-Cummings model the classical electric field is replaced by an expression using its quantized form. However, only one mode of the field is considered and only the dipole terms (8.6.20e) are used. We shall drop the unit matrix term. The resulting Hamiltonian is as follows:

$$H_{\text{JC}} = H_{\text{field}} + H_{\text{atom}} + H_{\text{interaction}},$$

$$H_{\text{JC}} = \Omega a^\dagger a + \frac{\varepsilon}{2}(\sigma_z + 1) + i\frac{g}{2}(a^\dagger - a)\sigma_x, \qquad (8.7.3a)$$

where the coefficient

$$g = \sqrt{\frac{\hbar\Omega}{2\varepsilon_0 V}} \frac{q\langle 2|z|1\rangle}{\hbar} \qquad (8.7.3b)$$

is the Rabi factor that would correspond to a one-photon laser field. [Recall (8.6.4).] The field amplitude for an N-photon field is proportional to $\sqrt{N}$, so we have the following relation between the semiclassical and quantum interaction constant:

$$r = g\sqrt{N}. \qquad (8.7.3c)$$

We let $a_{\mathbf{k}\alpha} = a$ since only one laser cavity mode $\mathbf{k}\alpha$ of frequency $\Omega = \omega$ is being considered. By expressing σ_x in terms of spinor raising and lowering operators, $\sigma_x = (\sigma_+ + \sigma_-)$, this becomes

$$H_{\text{JC}} = \Omega a^\dagger a + \frac{\varepsilon}{2}(\sigma_z + 1) + i\frac{g}{2}(a^\dagger - a)(\sigma_+ + \sigma_-). \qquad (8.7.4)$$

A final approximation to the model keeps only the interaction term $a^\dagger \sigma_-$ in which a photon is created while the atom drops from the upper (↑) level 2 to the lower (↓) level 1 and the term $a\sigma_+$ in which the reverse process occurs:

$$H_{\text{JCM}} = \Omega a^\dagger a + \frac{\varepsilon}{2}(\sigma_z + 1) + i\frac{g}{2}(a^\dagger \sigma_- - a\sigma_+). \qquad (8.7.5)$$

Let us apply the Hamiltonian (8.7.5) in turn to the radiation-atom product states with $N = 0, 1, 2, \ldots$ photons which we label $\{|0\rangle|\downarrow\rangle, |0\rangle|\uparrow\rangle, |1\rangle|\downarrow\rangle, |1\rangle|\uparrow\rangle, |2\rangle|\downarrow\rangle, |2\rangle|\uparrow\rangle, \ldots\}$. The following sequence of state transformation equations results:

$$H_{\text{JCM}}|0\rangle|\downarrow\rangle = (0 \cdot \Omega + 0)|0\rangle|\downarrow\rangle + i\frac{g}{2}(0 - 0),$$

$$H_{\text{JCM}}|0\rangle|\uparrow\rangle = (0 \cdot \Omega + \varepsilon)|0\rangle|\uparrow\rangle + i\frac{g}{2}(\sqrt{1}|1\rangle|\downarrow\rangle - 0),$$

$$H_{\text{JCM}}|1\rangle|\downarrow\rangle = (1 \cdot \Omega + 0)|1\rangle|\downarrow\rangle + i\frac{g}{2}(0 - \sqrt{1}|0\rangle|\uparrow\rangle),$$

$$H_{\text{JCM}}|1\rangle|\uparrow\rangle = (1 \cdot \Omega + \varepsilon)|1\rangle|\uparrow\rangle + i\frac{g}{2}(\sqrt{2}|2\rangle|\downarrow\rangle - 0),$$

$$H_{\text{JCM}}|2\rangle|\downarrow\rangle = (2 \cdot \Omega + 0)|2\rangle|\downarrow\rangle + i\frac{g}{2}(0 - \sqrt{2}|1\rangle|\uparrow\rangle),$$

$$H_{\text{JCM}}|2\rangle|\uparrow\rangle = (2 \cdot \Omega + \varepsilon)|2\rangle|\uparrow\rangle + i\frac{g}{2}(\sqrt{3}|3\rangle|\downarrow\rangle - 0), \quad (8.7.6)$$

It is easy to see that an infinite series of two-by-two matrices results in this representation of H_{JCM}:

	$|0\rangle|\downarrow\rangle$	$|0\rangle|\uparrow\rangle$	$|1\rangle|\downarrow\rangle$	$|1\rangle|\uparrow\rangle$	$|2\rangle|\downarrow\rangle \cdots$
$\langle 0|\langle\downarrow|$	$0 \cdot \Omega + 0$	$\cdot$	$\cdot$	$\cdot$	$\cdot$
$\langle 0|\langle\uparrow|$	$\cdot$	$0 \cdot \Omega + \varepsilon$	$\dfrac{-ig\sqrt{1}}{2}$	$\cdot$	$\cdot$
$\langle 1|\langle\downarrow|$	$\cdot$	$\dfrac{ig\sqrt{1}}{2}$	$1 \cdot \Omega + 0$	$\cdot$	$\cdot$
$\langle 1|\langle\uparrow|$	$\cdot$	$\cdot$	$\cdot$	$1 \cdot \Omega + \varepsilon$	$\dfrac{-ig\sqrt{2}}{2} \cdot$
$\langle 2|\langle\downarrow|$	$\cdot$	$\cdot$	$\cdot$	$\dfrac{ig\sqrt{2}}{2}$	$2 \cdot \Omega + 0$

(8.7.7)

The general form of each two-by-two matrix is the following:

$$\langle h_{\text{JCM}}\rangle = \begin{array}{c|cc} & |N-1\rangle|\downarrow\rangle & |N\rangle|\uparrow\rangle \\ \hline \langle N-1|\langle\downarrow| & (N-1)\Omega + \varepsilon & \dfrac{-ig\sqrt{N}}{2} \\ \langle N|\langle\uparrow| & \dfrac{ig\sqrt{N}}{2} & N\Omega + 0 \end{array}. \quad (8.7.8)$$

780 SYMMETRY ANALYSIS FOR SEMICLASSICAL AND QUANTUM MECHANICS

In the classical limit of large N the two-by-two matrix begins to look something like the semiclassical matrix (8.7.2a). We can write the two-by-two part of the Hamiltonian as follows:

$$\begin{aligned}\langle h_{\text{JCM}} \rangle &= (N-1)\Omega \begin{pmatrix} 1 & 0 \\ 0 & 1 \end{pmatrix} + \Omega \begin{pmatrix} 0 & 0 \\ 0 & 1 \end{pmatrix} + \varepsilon \begin{pmatrix} 1 & 0 \\ 0 & 0 \end{pmatrix} + \frac{g\sqrt{N}}{2} \begin{pmatrix} 0 & -i \\ i & 0 \end{pmatrix} \\ &= \left((N-1)\Omega + \frac{\varepsilon + \Omega}{2} \right) \begin{pmatrix} 1 & 0 \\ 0 & 1 \end{pmatrix} + \frac{\varepsilon - \Omega}{2} \begin{pmatrix} 1 & 0 \\ 0 & -1 \end{pmatrix} + \frac{g\sqrt{N}}{2} \begin{pmatrix} 0 & -i \\ i & 0 \end{pmatrix} \\ &= \left((N-1)\Omega + \frac{\varepsilon + \Omega}{2} \right) \begin{pmatrix} 1 & 0 \\ 0 & 1 \end{pmatrix} + \frac{\Delta}{2} \begin{pmatrix} 1 & 0 \\ 0 & -1 \end{pmatrix} + \frac{r}{2} \begin{pmatrix} 0 & -i \\ i & 0 \end{pmatrix}.\end{aligned}$$
(8.7.9)

The Rabi factor $r = g\sqrt{N}$ for N photons is recovered. The factor of (i) is due to our choice of phase for the field operators.

No essential physics is changed if we use the following modified Hamiltonian in which this phase is absorbed:

$$H'_{\text{JCM}} = \Omega a^\dagger a + \frac{\varepsilon}{2}(\sigma_z + 1) + \frac{g}{2}(a^\dagger \sigma_- + a\sigma_+).$$

Then we can apply the semiclassical dressed eigensolutions (8.5.41)–(8.5.46) to the following real two-by-two matrices:

$$\langle h'_{\text{JCM}} \rangle = \begin{array}{c|cc} & |N-1\rangle|\downarrow\rangle & |N\rangle|\uparrow\rangle \\ \hline \langle N-1|\langle\downarrow| & (N-1)\Omega + \varepsilon & \dfrac{g\sqrt{N}}{2} \\ \langle N|\langle\uparrow| & \dfrac{g\sqrt{N}}{2} & N\Omega \end{array} = N\Omega \mathbf{1} + \begin{pmatrix} \Delta & \dfrac{r}{2} \\ \dfrac{r}{2} & 0 \end{pmatrix}$$

$$= \left((N-1)\Omega + \frac{\varepsilon + \Omega}{2} \right)\mathbf{1} + \begin{pmatrix} \dfrac{\Delta}{2} & \dfrac{r}{2} \\ \dfrac{r}{2} & -\dfrac{\Delta}{2} \end{pmatrix}. \quad (8.7.10)$$

These matrices have the same eigenvectors as the semiclassical matrices. The only difference is that the Rabi factor r depends upon photon number N and there is a pair of levels for all N greater than zero. The eigenvalues are the same, too, except for the unit matrix term which yields a ladder of doublet eigenvalues. We examine this ladder of levels now.

B. Jaynes-Cummings Eigensolutions

An attempt to picture the dressed eigenlevels is made in Figure 8.7.1. A column containing stacks of energy levels is shown for each of three cases of detuning: (a) laser tuned below atomic transition ($\Delta > 0$); (b) at resonance ($\Delta = 0$), and (c) laser tuned higher ($\Delta < 0$). Recall that the detuning parameter is $\Delta = \varepsilon - \Omega$.

Two stacks of horizontal lines on each side of the (a), (b), and (c) columns in Figure 8.7.1 indicate what the levels would be without any interaction between atom and field ($r = g\sqrt{N} = 0$). An N-photon level in which the atom is in the first $|1\rangle = |\uparrow\rangle$ or second $|2\rangle = |\downarrow\rangle$ state is labeled $|1, N\rangle$ or $|2, N\rangle$, respectively. Each of these levels for $N > 0$ is connected to a pair of lines in the center of the column that are shifted up and down by $\pm \delta/2$. The quantity δ is the AC-Stark shift discussed in Section 8.5 (recall Eq. (8.5.45c)). The shifted lines are the "dressed" eigenlevels with the interaction turned on ($r = g\sqrt{N} > 0$). The lines that connect the shorter lines to the zero-field levels indicate the relative the greater of the two amplitudes ($\sin \theta/2$ or $\cos \theta/2$) of the zero-field states in each the dressed eigenstates corresponding to that level. Recall Eqs. (8.5.47) and (8.5.48).

Below resonance ($\Delta > 0$) the Hamiltonian rotation vector $\boldsymbol{\omega}$ makes an acute angle ($\theta < \pi/2$) with the z-axis. The lower dressed eigenstate $|1^D N + 1, N\rangle$ indicated at the top left-hand side of Figure 8.7.1 is mostly composed of the atom-field product state $|1, N+1\rangle$ while the higher dressed state $|2^D N + 1, N\rangle$ is mostly composed of $|2, N\rangle$.

As the detuning approaches resonance ($\Delta = 0$) the zero-field levels get lined up, the AC-shifts reach their maximum, and the rotation angle θ approaches $\pi/2$. One may use the diagrams from Section 8.5.E to quantify the variation. However, caution should be used since the Rabi parameter $r = g\sqrt{N}$ is here a function of N. In other words, the Rabi parameter which was a constant in the semiclassical theory is now dependent upon what level you are on. It increases with the laser mode electric field amplitude, which is proportional to the root $\sqrt{N}$ of the photon number.

At resonance ($\Delta = 0$) the rotation angle is $\theta = \pi/2$. Then the Hamiltonian rotation vector $\boldsymbol{\omega}$ makes an angle of $\pi/2$ with the z-axis and has its minimum magnitude of $|\boldsymbol{\omega}| = r$, which is the Rabi frequency. This was shown in Figure 8.5.1c. The resonance values for the dressed eigenstate amplitudes are $\sin \theta/2 = 1/\sqrt{2}$ and $\cos \theta/2 = 1/\sqrt{2}$. This corresponds to 50-50 mixtures of the atom-field product states $|1, N+1\rangle$ and $|2, N\rangle$ in the dressed eigenstates $|1^D N + 1, N\rangle$ and $|2^D N + 1, N\rangle$.

Above resonance ($\Delta < 0$) the Hamiltonian rotation vector $\boldsymbol{\omega}$ makes an obtuse angle ($\theta > \pi/2$) with the z-axis. Now the lower dressed eigenstate $|1^D N + 1, N\rangle$ is mostly composed of $|2, N\rangle$ while the upper dressed state

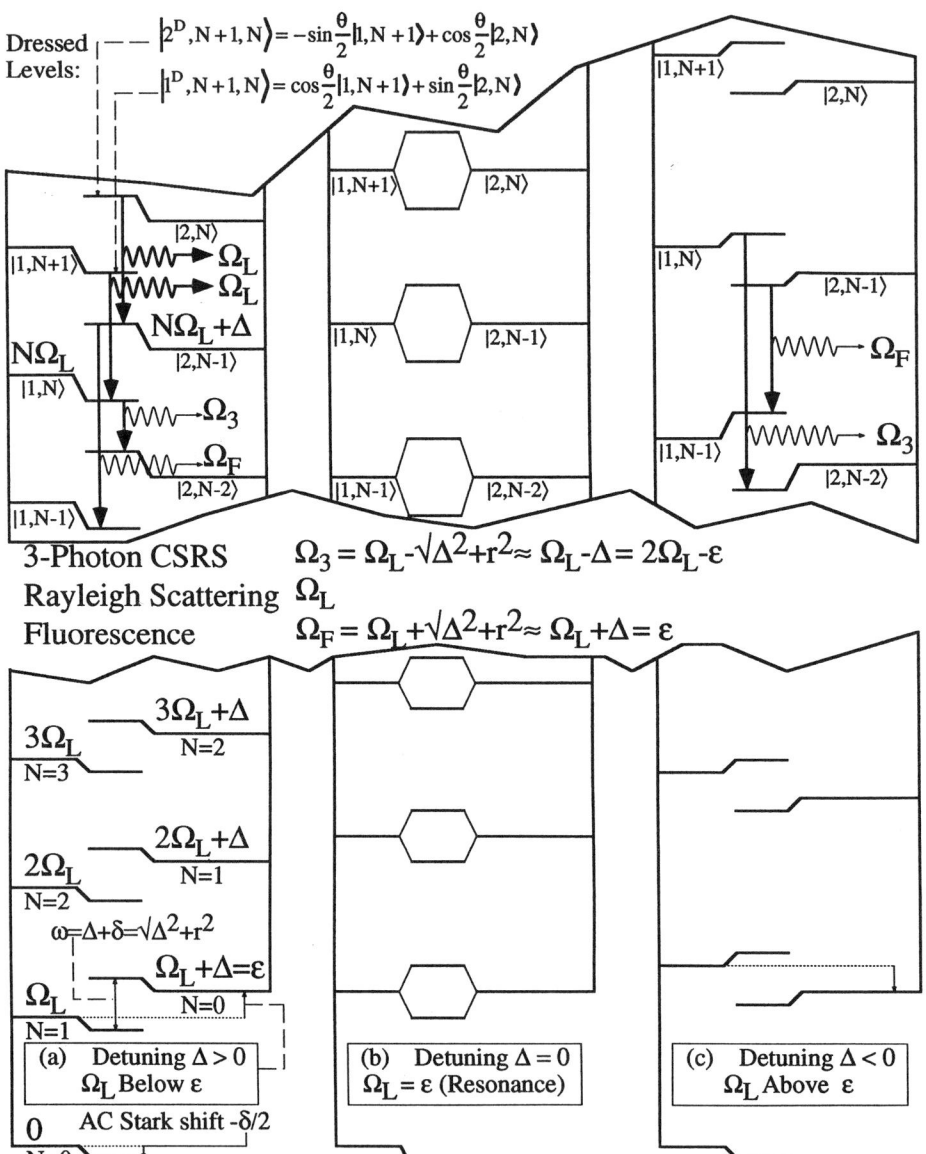

Figure 8.7.1 Level structure of 2-level atom and 1-mode cavity showing elementary transition processes of fluorescence, Rayleigh scattering, and three-photon coherent Stokes-Raman scattering (CSRS). Transitions are between levels belonging to dressed eigenstates.

$|2^D N + 1, N\rangle$ is mostly composed of $|1, N + 1\rangle$, as shown in the upper right-hand side of the figure.

C. Transitions in the Jaynes-Cummings Model

The diagram of levels in Figure 8.7.1 involves just one $\Omega_L = \Omega$ mode interacting with the two-level atom. We have ignored all the other field mode levels such as were sketched in Figure 8.6.1. We have just concentrated on loading photons into one laser cavity mode whose frequency $\Omega_L = \Omega$ is being tuned close to the value $\varepsilon = \omega_f - \omega_s$ of the atomic transition.

However, if these other modes are off-resonance by enough or only have one or two photons, one can treat them using perturbation theory as was discussed in Section 8.6.B. Transition rates between the dressed states of a laser driven atom can be derived using the Fermi golden rule (8.6.30).

Some of the commonly observed transitions are indicated by vertical arrows in the Figure 8.7.1. The strongest transitions involved the so-called RAYLEIGH SCATTERING processes such as $|1, N + 1\rangle \to |1, N\rangle$ for $N = 0, 1, 2, \ldots$ or $|2, N\rangle \to |2, N - 1\rangle$ for $N = 1, 2, \ldots$, where only the photon number changes and the system emits one of its laser-mode photons into an external mode of the same frequency Ω_L. These transitions yield light with the frequency of the laser just like classical Rayleigh scattered light discussed in Section 6.5.B.

The other transitions are more complicated. One called FLUORESCENCE is a transition between dressed states which involve fundamental transitions such as $|2, N\rangle \to |1, N\rangle$ or $|2, N - 1\rangle \to |1, N - 1\rangle$. The latter is the major part of the transition indicated by an Ω_F arrow in the upper left-hand side of the figure since the initial (upper) dressed state $|2^D N, N - 1\rangle$ is mostly composed of $|2, N - 1\rangle$ and $|1^D N, N - 1\rangle$ is mostly composed of $|1, N - 1\rangle$ in the final (lower) dressed state.

The fluorescence transition angular frequency is the difference between the initial and final dressed eigenlevels connected by the Ω_F arrow. The initial and final eigenvalues are

$$\varepsilon^D(2, N, N - 1) = N\Omega_L + \Delta + \delta/2,$$

$$\varepsilon^D(1, N - 1, N - 2) = (N - 1)\Omega_L - \delta/2.$$

The difference is the fluorescence transition frequency,

$$\Omega_F = \Omega_L + \Delta + \delta = \Omega_L + \sqrt{\Delta^2 + r^2}. \tag{8.7.11}$$

For small values of the Rabi factor ($r \ll \Delta$) or large detuning ($\Delta \gg r$) this becomes

$$\Omega_F \to \Omega_L + \Delta = \varepsilon, \qquad r \ll \Delta, \qquad (8.7.12)$$

which is the atomic transition frequency. This transition drops the atom from its upper state $|2\rangle$ to its lower state $|1\rangle$ but takes no photons out of the cavity mode since N stays constant. It emits one photon into an external mode of frequency Ω_F.

Another transition called the three-photon process or COHERENT STOKES RAMAN SCATTERING (CSRS) is a transition between dressed states which mostly involves transitions of the type $|1, N\rangle \to |2, N - 2\rangle$. The latter is the major part of the transition indicated by an Ω_3 arrow in the upper left-hand side of the figure since the initial (upper) dressed state $|1^D N, N - 1\rangle$ is mostly composed of $|1, N\rangle$, and $|2^D N - 1, N - 2\rangle$ is mostly composed of $|2, N - 2\rangle$ in the final (lower) dressed state.

The CSRS transition angular frequency is the difference between the initial and final dressed eigenlevels connected by the Ω_3 arrow. The initial and final eigenvalues are

$$\varepsilon^D(1, N, N - 1) = N\Omega_L - \delta/2,$$
$$\varepsilon^D(2, N - 1, N - 2) = (N - 1)\Omega_L + \Delta - \delta/2.$$

The difference is the CSRS transition frequency.

$$\Omega_3 = \Omega_{\text{CSRS}} = \Omega_L - \Delta - \delta = \Omega_L - \sqrt{\Delta^2 + r^2} \qquad (8.7.13)$$

For small values of the Rabi factor ($r \ll \Delta$) or large detuning ($r \ll \Delta$) this becomes

$$\Omega_3 = \Omega_{\text{CSRS}} \to \Omega_L - \Delta = 2\Omega_L - \varepsilon, \qquad r \ll \Delta, \qquad (8.7.14)$$

which is the difference between two laser photons and the atomic transition frequency. This transition raises the atom from its lower state $|1\rangle$ to its upper state $|2\rangle$. It also takes two photons out of the cavity mode since N decreases by two. It emits one photon into an external mode of frequency Ω_3 which is approximately the difference between 2Ω and the atomic transition frequency ε.

A direct transition of frequency $\omega = \sqrt{\Delta^2 + r^2} \approx \Delta$ between $|1^D N, N - 1\rangle$ and $|2^D N - 1, N - 2\rangle$ is forbidden by C_2 parity. However, in a system that does not have C_2 symmetry it would be possible to have such a transition as is indicated by the small vertical arrow near the bottom of Figure 8.7.1(a).

The three allowed transitions account for three main spectral components that may be observed coming out of the sides of a laser atom cavity. It consists of a strong Rayleigh line at $\Omega_L = \Omega$ and two sidebands Ω_F and Ω_3 as shown in Figure 8.7.2. The triple-pronged spectral line is called the

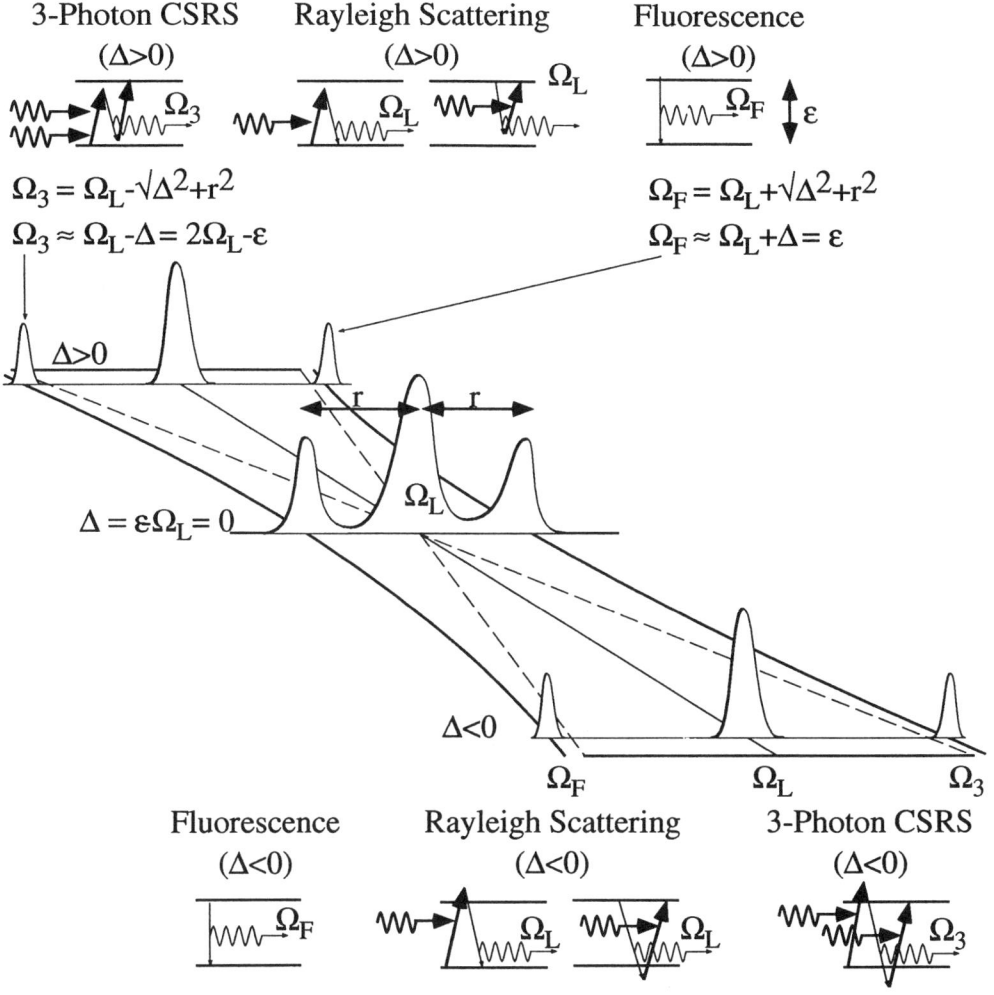

Figure 8.7.2 Structure of the Mollow spectrum and its elementary processes of fluorescence, Rayleigh scattering, and three-photon coherent Stokes-Raman scattering (CSRS).

MOLLOW LINE SHAPE. One sideband is centered at $\Omega_F = \Omega + \omega$ which is approximately $\Omega + \Delta$ far from resonance, and the other is at $\Omega_3 = \Omega - \omega$, which is approximately $\Omega - \Delta$.

Near resonance at $\Delta = 0$ the sidebands will follow AC Stark shift hyperbolic paths given by (8.7.11) and (8.7.13) rather than simply collapsing upon Ω at $\Delta = 0$. The hyperbolic curves in the semiclassical level diagram in Figure 8.5.3 are approximate traces of the spectral sidebands for variable detuning Δ around resonance. At resonance ($\Delta = 0$) there will still be two sidebands but now they will be located at $\Omega \pm r$, where r is the

Rabi parameter. In general, the sidebands are located at $\Omega \pm \omega$, where $\omega = \sqrt{\Delta^2 + r^2} \approx \Delta$ is the frequency of the Rabi precession or crank rotation shown in Section 8.5.

The sidebands correspond roughly to fluorescence and CSRS processes, respectively. With positive detuning ($\varepsilon - \Omega = \Delta > 0$) the upper sideband ($\Omega_F = \Omega + \omega \approx \Omega + \Delta$) is due (mostly) to fluorescence while the lower sideband ($\Omega_3 = \Omega - \omega \approx \Omega - \Delta$) is due (mostly) to the three-photon CSRS process. Above resonance the detuning parameter reverses sign ($\varepsilon - \Omega = \Delta < 0$) and the order is reversed. At resonance ($\Delta = 0$) it is not possible to distinguish the two processes since the initial states $|1, N\rangle$ and $|2, N - 1\rangle$ are mixed 50-50 and so are the final states $|1, N - 1\rangle$ and $|2, N - 2\rangle$.

Below resonance ($\Delta > 0$) the Ω_3 photon from the CSRS process has lower frequency than the Ω_F photon from fluorescence. It also must come earlier in time. The CSRS process pumps the atom from its lower state-1 into its excited state-2. Only then can it emit a fluorescence photon which puts it back into its ground state. Above resonance ($\Delta < 0$) the Ω_3 photon from the CSRS process has higher frequency than the Ω_F photon from fluorescence. Then it is the higher frequency sideband that comes earlier. These time correlations have been observed.

This concludes our introduction to the recent fundamental developments in laser spectroscopy. Many details have been left out of this discussion and many new effects will soon be discovered as this new set of tools becomes more widely used. Perhaps the most important development so far lies in the way we are coming to think about the observed object (atom or molecule) and the observer's tool (radiation). In modern laser spectroscopy the distinction between the observer and the observed has practically disappeared, and the atom-radiation-cavity becomes a single quantum object.

ADDITIONAL READING

Derivation and application of semiclassical quantization and wave-packet propagation techniques are described in the following papers:

E. J. Heller, *J. Chem. Phys.*, **62**, 1544 (1975); *J. Chem. Phys.*, **68**, 3891 (1978).

M. J. Davis and E. J. Heller, *J. Chem. Phys.*, **71**, 3383 (1979).

S. Y. Lep and E. J. Heller, *J. Chem. Phys.*, **71**, 4777 (1979); *J. Chem. Phys.*, **76**, 3035 (1982).

D. J. Tannor and E. J. Heller, *J. Chem. Phys.*, **77**, 202 (1982).

N. DeLeon and E. J. Heller, *J. Chem. Phys.*, **78**, 4005 (1983); *J. Chem. Phys.*, **81**, 5957 (1984).

M. B. Blanco and E. J. Heller, *J. Chem. Phys.*, **83**, 1143 (1985).

J. R. Reimers and E. J. Heller, *J. Chem. Phys.*, **83**, 516 (1985).

N. DeLeon, *J. Chem. Phys.*, **87**, 4722 (1987); *Comp. Phys. Rep.*, **8**, 321 (1988).

An early paper on action quantization which uses color graphics to approximate quantum wave fronts is

M. J. Davis and E. J. Heller, *J. Chem. Phys.*, **75**, 3916 (1981).

The computer program *Color U(2)* mentioned at the end of Chapter 7 uses color quantization and color animation to show the dynamics of quantum wave fronts.
The idea of wave-packet coherent states can be traced back to Schrödinger.

E. Schrödinger, *Naturwissenschaften*, **14**, 664 (1926).

Their introduction in quantum optics is probably due to Glauber.

R. J. Glauber, *Phys. Rev.*, **131**, 2766 (1963).

Other approaches to semiclassical quantization are found in the following papers (this is by no means an exhaustive list of this large and growing field):

I. C. Percival, *Adv. Chem. Phys.*, **36**, 1 (1977).

D. W. Noid, M. L. Kosykowski, and R. A. Marcus, *Ann. Rev. Phys. Chem.*, **32**, 267 (1981).

S. A. Rice, *Adv. Chem. Phys.*, **47I**, 117 (1981).

W. Eastes and R. A. Marcus, *J. Chem. Phys.*, **61**, 4301 (1974).

D. W. Noid and R. A. Marcus, *J. Chem. Phys.*, **62**, 2119 (1975).

I. C. Percival and N. Pomphrey, *Mol. Phys.*, **31**, 97 (1976).

S. Chapman, B. C. Garrett, and W. H. Miller, *J. Chem. Phys.*, **64**, 502 (1976).

K. S. Sorbie and N. C. Handy, *Mol. Phys.*, **33**, 1319 (1977).

C. Jaffe and W. P. Reinhardt, *J. Chem. Phys.*, **71**, 1862 (1979).

R. T. Swim and J. B. Delos, *J. Chem. Phys.*, **71**, 1706 (1979).

R. B. Shirts and W. P. Reinhardt, *J. Chem. Phys.*, **77**, 5204 (1982).

C. C. Martens and G. S. Ezra, *J. Chem. Phys.*, **86**, 279 (19875).

C. W. Eaker and G. C. Shatz, *J. Chem. Phys.*, **81**, 2394 (1984).

W. H. Miller, *J. Chem. Phys.*, **81**, 3573 (1984).

References to the original EBK quantization are as follows:

A. Einstein, *Dent. Ges. Berlin Verh.*, **19**, 9/10 (1917).

M. L. Brillouin, *J. Phys. Paris (Ser. 6)*, **7**, 353 (1926).

J. B. Keller, *Ann. Phys. (N.Y.)*, **4**, 180 (1958).

F. Reiche, *The Quantum Theory*, (Methuen, London, 1922).

A good modern reference to classical, semiclassical, and quantum theory of radiation for spectroscopy is the following:

C. Cohen-Tannoudji, J. Dupont-Roc, and G. Grynberg, *Photons and Atoms* (Wiley Interscience, New York, 1989).

This discusses the $\mathbf{A} \cdot \mathbf{P}$ versus $\mathbf{E} \cdot \mathbf{r}$ perturbations and the Power-Zienau-Wolley transformation. A simplified discussion and other references are in the following paper:

E. A. Power and T. Thirunamachandran, *Ann. J. Phys.*, **46**, 370 (1976).

An early paper which gave a classical transformation between $\mathbf{E} \cdot \mathbf{r}$ and $\mathbf{A} \cdot \mathbf{p}$ Hamiltonians is by Marie Goeppert-Mayer:

M. Goeppert-Mayer, *Ann. Physik (Lpzg.)*, **9**, 273 (1931).

The first paper to give a quantum mechanical transformation of $\mathbf{E} \cdot \mathbf{r}$ to $\mathbf{A} \cdot \mathbf{p}$ is by Richards. H. S. Synder is credited in the paper.

P. I. Richards, *Phys. Rev.*, **73**, 254 (1948).

A restricted version of this transformation for the case of a magnetic field constant in space and time appears in the following paper:

W. E. Lamb, *Phys. Rev.* **85**, 259 (1952).

It was used again in the same restricted context by the following authors:

B. R. Johnson, J. O. Hirschfelder, and K. H. Yang, *Rev. Mod. Phys.*, **55**, 109 (1983).

Another discussion of the problem is in the following paper:

J. R. Ackerhalt and P. W. Milonni, *J. Opt. Soc. Am.*, **B11**, 116 (1984).

For an example of some of the confusion surrounding the $\mathbf{A} \cdot \mathbf{p}$ interaction see the following paper:

D. H. Kobe, *Phys. Rev. Lett.*, **40**, 538 (1978).

Some modern treatments of laser-atom lineshape and two-level atom models are listed below. The first papers are seminal ones by B. R. Mollow:

B. R. Mollow, *Phys. Rev.*, **188**, 1969 (1969); *Phys. Rev. A*, **2**, 76 (1969); *Phys. Rev. A*, **12**, 1919 (1969); *Phys. Rev. A*, **13**, 758 (1969).

A discussion which uses the two-level quasi-spin is by Courtens and Szöke:

E. Courtens and A. Szöke, *Phys. Rev. A*, **15**, 1588 (1977).

The two-level atom is presented as a generalization to classical resonance in the following text:

L. Allen and J. H. Eberly, *Optical Resonance and Two-Level Atoms* (Wiley Interscience, New York, 1975).

Applications of radiation theory to laser dynamics is the subject of the following book, which also relates the classical Lorentz model to modern theory:

P. W. Milonni and J. H. Eberly, *Lasers* (Wiley Interscience, New York, 1988).

Recent developments of the problem of an isolated atom-cavity system are based upon the Jaynes-Cummings model.

E. T. Jaynes and F. W. Cummings, *Proc. IEEE*, **51**, 89 (1963).

The phenomenon of "collapse" and "revival" of Jaynes-Cummings solutions is discussed in the following:

J. H. Eberly, N. B. Narozhny, and J. J. Sanchez-Mondragon, *Phys. Rev. Lett.*, **44**, 1323 (1980).

H. J. Yoo, J. J. Sanchez-Mondragon, and J. H. Eberly, *Phys. Rep.*, **118**, 259 (1985).

Recent discoveries have been made about the behavior of the Bloch vector during collapse and revival.

J. Gea-Banacloche, *Phys. Rev. Lett.*, **65**, 3385 (1990); *Phys. Rev. A*, **44**, 5913 (1991); Optical. Commun. **88**, 531 (1992).

Much of the future work on atoms or molecules in cavities will use so-called driven Jaynes-Cummings models. Some discussions of these have just been published.

P. Alsing and H. J. Carmichael, *Quantum Optics*, **3**, 13 (1991).

P. Alsing, D. S. Guo and H. J. Carmichael, *Phys. Rev. A*, **45**, 5135 (1992).

Time correlations between parts of the reasonance spectrum are described in the following paper.

A. Aspect, G. Roger, S. Reynaud, J. Dalibard, and C. Cohen-Tannoudji, *Phys. Rev. Letters*, **45**, 617 (1980).

APPENDIX F

FORMULAS AND TABLES OF GROUP REPRESENTATIONS AND RELATED QUANTITIES

F.1. THE ORTHOGONAL GROUP O(3) AND UNITARY UNIMODULAR GROUP SU(2)

The multiplication rules for $O(3)$ and $SU(2)$ may be visualized using Hamilton turns. (See Sections 3.1B, 5.3.C, and 5.5.A.) The common choices for parameters are Darboux axis angles $R[\phi\theta\omega]$ and Euler coordinate angles $R(\alpha\beta\gamma)$. (See Sections 5.3.A and 5.3.B.) The following irreducible representations of Wigner D-functions are derived in Section 5.4 using $SU(2)$ boson algebra:

$$D^j_{mn}(\alpha\beta\gamma) = \left\langle \begin{matrix} j \\ m \end{matrix} \middle| R(\alpha\beta\gamma) \middle| \begin{matrix} j \\ n \end{matrix} \right\rangle$$

$$= \sum_{k=0} (-1)^k \frac{\sqrt{(j+m)!(j-m)!(j+n)!(j-n)!}}{(j+m-k)!k!(j-n-k)!(n-m+k)!} e^{-i(m\alpha+n\gamma)}$$

$$\times \left(\cos\frac{\beta}{2}\right)^{2j+m-n-2k} \left(\sin\frac{\beta}{2}\right)^{n-m+2k}. \qquad (F.1.1)$$

Here the rotation operator is expressed in terms of angular momentum generators and Euler angles,

$$R(\alpha\beta\gamma) = e^{-i\alpha J_z/\hbar} e^{-i\beta J_y/\hbar} e^{-i\gamma J_z/\hbar}. \qquad (F.1.2)$$

Representations $D^j(J_\alpha)$ ($x = x, y, z$) are derived in Appendix E. Irreducibility, completeness, and orthogonality properties are derived in Appendix G.

Spherical harmonics Y_m^l and multipole functions X_q^k are the $n = 0$ cases of D-functions. They are functions of polar coordinates ($\alpha = \phi$, $\beta = \theta$) and (ϕ, θ, r) but not the third Euler angle γ:

$$Y_m^l(\phi\theta) = \sqrt{\frac{2l+1}{4\pi}} D_{m0}^{l*}(\phi\theta \cdot), \qquad X_q^k(\phi\theta r) = r^k D_{q0}^{k*}(\phi\theta \cdot).$$

The multipole functions also have a radial k-power dependence and are kth-degree polynomials of $\{x, y, z\}$. In Table F.1.1 these polynomials are denoted by I_q^1, II_q^2, III_q^3, and IV_q^4 for $k = 1, 2, 3$, and 4, respectively. The inverse relations to the kth degree harmonic monomials $x^a y^b z^c$ are also given. The number of harmonic monomials is

$$d(U(3)) = \frac{(k+1)(k+2)}{2},$$

which exceeds the number $(2k + 1)$ of multipole functions in all cases except $k = 0$ and $k = 1$. Therefore the even monomials involve combinations of multipole functions of degree k, $k - 2, \ldots, 2$, and 0 multiplied by r^0, $r^2, \ldots, r^{k-2}$, and r^k, respectively, while the odd monomials combine x^k of degree k, $k - 2, \ldots, 3$, and 1 multiplied by r^0, $r^2, \ldots, r^{k-3}$, and r^{k-1}, respectively. The harmonic monomials can be realized as a basis of a three-dimensional harmonic oscillator and span the symmetric representations of SU(3).

The Clebsch-Gordan and Wigner-$3j$ coupling coefficients are related as described in Section 7.2,

$$\begin{pmatrix} j_1 & j_2 & j_3 \\ m_1 & m_2 & m_2 \end{pmatrix} = \frac{(-1)^{j_1 - j_2 - m_3}}{\sqrt{2j_3 + 1}} C\begin{matrix} j_1 & j_2 & j_3 \\ m_1 & m_2 & -m_3 \end{matrix} \qquad \text{(F.1.3)}$$

The standard CG-Dirac notation is

$$C\begin{matrix} j_1 & j_2 & j_3 \\ m_1 & m_2 & m_3 \end{matrix} = \left\langle \begin{matrix} j_1 & j_2 \\ m_1 & m_2 \end{matrix} \middle| j_1 \otimes j_2 \begin{matrix} j_3 \\ m_3 \end{matrix} \right\rangle. \qquad \text{(F.1.4)}$$

The general formula is similar to the one derived in Section 7.2.D:

$$\begin{pmatrix} j_1 & j_2 & j_3 \\ m_1 & m_2 & m_2 \end{pmatrix} = (-1)^{j_1 - j_2 - m_3} \sqrt{\frac{(j_1 + j_2 - j_3)!(j_1 - j_2 + j_3)!(-j_1 + j_2 + j_3)!}{(j_1 + j_2 + j_3 + 1)!}}$$

$$\times (j_1 + m_1)!(j_1 - m_1)!(j_2 + m_2)!(j_2 - m_2)!(j_3 + m_3)!(j_3 - m_3)!$$

$$\times \sum_k \frac{(-1)^k}{k!(j_1 + j_2 - j_3 - k)!(j_1 - m_1 - k)!(j_2 + m_2 - k)!(j_3 - j_2 - m_1 + k)!(j_3 - j_1 - m_2 + k)!}$$

(F.1.5)

TABLE F.1.1 $R(3)$ Multiple Functions and $SU(3)$ Harmonic Monomials

$I_1^{(1)} = -\dfrac{1}{\sqrt{2}}(x+iy)$

$I_{-1}^{(1)} = \dfrac{1}{\sqrt{2}}(x-iy)$

$I_0^{(1)} = z$

$x = \dfrac{1}{\sqrt{2}}(I_{-1}^{(1)} - I_1^{(1)})$

$iy = -\dfrac{1}{\sqrt{2}}(I_{-1}^{(1)} + I_1^{(1)})$

$z = I_0^{(1)}$

$II_2^{(2)} = \sqrt{\dfrac{3}{8}}(x+iy)^2$

$II_1^{(2)} = -\sqrt{\dfrac{3}{2}}z(x+iy)$

$II_0^{(2)} = \dfrac{1}{2}(3z^2 - r^2)$

$II_{-1}^{(2)} = \sqrt{\dfrac{3}{2}}z(x-iy)$

$II_{-2}^{(2)} = \sqrt{\dfrac{3}{8}}(x-iy)^2$

$x^2 = \dfrac{1}{6}(II_2^{(2)} + II_{-2}^{(2)}) - \dfrac{1}{3}II_0^{(2)} + \dfrac{1}{3}r^2$

$y^2 = -\dfrac{1}{6}(II_2^{(2)} + II_{-2}^{(2)}) - \dfrac{1}{3}II_0^{(2)} + \dfrac{1}{3}r^2$

$z^2 = -\dfrac{2}{3}II_0^{(2)} + \dfrac{1}{3}r^2$

$xy = \dfrac{i}{\sqrt{6}}(II_2^{(2)} - II_{-2}^{(2)})$

$xz = \dfrac{1}{\sqrt{6}}(II_1^{(2)} - II_{-1}^{(2)})$

$yz = \dfrac{i}{\sqrt{6}}(II_1^{(2)} + II_{-1}^{(2)})$

$III_3^{(3)} = -\dfrac{\sqrt{5}}{4}(x+iy)^3$

$III_2^{(3)} = \sqrt{\dfrac{15}{8}}z(x+iy)^2$

$III_1^{(3)} = -\dfrac{\sqrt{3}}{4}(x+iy)(5z^2 - r^2)$

$III_0^{(3)} = \dfrac{1}{2}z(5z^2 - 3r^2)$

$III_{-1}^{(3)} = \dfrac{\sqrt{3}}{4}(x-iy)(5z^2 - r^2)$

$III_{-2}^{(3)} = \sqrt{\dfrac{15}{8}}z(x-iy)^2$

$III_{-3}^{(3)} = \dfrac{\sqrt{5}}{4}(x-iy)^3$

$x^3 = \dfrac{1}{2\sqrt{5}}(III_{-3}^{(3)} - III_3^{(3)}) + \dfrac{\sqrt{3}}{10}(III_1^{(3)} - III_{-1}^{(3)}) + \dfrac{3}{5\sqrt{2}}(I_{-1}^{(1)} - I_1^{(1)})r^2$

$iy^3 = \dfrac{1}{2\sqrt{5}}(III_{-3}^{(3)} + III_3^{(3)}) + \dfrac{\sqrt{3}}{10}(III_1^{(3)} + III_{-1}^{(3)}) - \dfrac{3}{5\sqrt{2}}(I_{-1}^{(1)} + I_1^{(1)})r^2$

$z^3 = \dfrac{2}{5}III_0^{(3)} + \dfrac{3}{5}I_0^{(1)}r^2$

$ix^2y = -\dfrac{1}{2\sqrt{5}}(III_{-3}^{(3)} + III_3^{(3)}) + \dfrac{1}{10\sqrt{3}}(III_1^{(3)} + III_{-1}^{(3)}) - \dfrac{1}{3\sqrt{2}}(I_1^{(1)} + I_1^{(1)})r^2$

$xy^2 = \dfrac{1}{2\sqrt{5}}(III_{-3}^{(3)} - III_3^{(3)}) + \dfrac{1}{10\sqrt{3}}(III_1^{(3)} - III_{-1}^{(3)}) + \dfrac{3}{3\sqrt{2}}(I_{-1}^{(1)} - I_1^{(1)})r^2$

$x^2z = \dfrac{1}{\sqrt{30}}(III_2^{(3)} + III_{-2}^{(3)}) - \dfrac{1}{20}III_0^{(3)} + \dfrac{1}{5}I_0^{(1)}r^2$

$y^2z = -\dfrac{1}{\sqrt{30}}(III_2^{(3)} + III_{-2}^{(3)}) - \dfrac{1}{20}III_0^{(3)} + \dfrac{1}{5}I_0^{(1)}r^2$

$xz^2 = -\dfrac{2}{5\sqrt{3}}(III_1^{(3)} - III_{-1}^{(3)}) + \dfrac{1}{5\sqrt{2}}(I_{-1}^{(1)} - I_1^{(1)})r^2$

$iyz^2 = -\dfrac{2}{5\sqrt{3}}(III_1^{(3)} + III_{-1}^{(3)}) - \dfrac{1}{5\sqrt{2}}(I_{-1}^{(1)} + I_1^{(1)})r^2$

$ixyz = \dfrac{1}{\sqrt{30}}(III_{-2}^{(3)} - III_2^{(3)})$

$$IV_4^{(4)} = \sqrt{\frac{35}{128}}(x+iy)^4$$

$$IV_3^{(4)} = -\frac{\sqrt{35}}{4}z(x+iy)^3$$

$$IV_2^{(4)} = \frac{\sqrt{5}}{4\sqrt{2}}(x+iy)^2(7z^2-r^2)$$

$$IV_1^{(4)} = -\frac{\sqrt{5}}{4}(x+iy)(7z^3-3zr^2)$$

$$IV_0^{(4)} = \frac{1}{8}(35z^4-30z^2r^2+3r^4)$$

$$IV_{-1}^{(4)} = \frac{\sqrt{5}}{4}(x-iy)(7z^3-3zr^2)$$

$$IV_{-2}^{(4)} = \frac{\sqrt{5}}{4\sqrt{2}}(x-iy)^2(7z^2-r^2)$$

$$IV_{-3}^{(4)} = \frac{\sqrt{35}}{4}z(x-iy)^3$$

$$IV_{-4}^{(4)} = \sqrt{\frac{35}{128}}(x-iy)^4$$

$$x^4 = \frac{1}{\sqrt{70}}(IV_4^{(4)}+IV_{-4}^{(4)}) - \frac{2}{7\sqrt{10}}(IV_2^{(4)}+IV_{-2}^{(4)}) + \frac{3}{35}IV_0^{(4)} + \frac{\sqrt{6}}{7}(II_2^{(2)}+II_{-2}^{(2)})r^2 - \frac{2}{7}II_0^{(2)}r^2 + \frac{1}{5}r^4$$

$$y^4 = \frac{1}{\sqrt{70}}(IV_4^{(4)}+IV_{-4}^{(4)}) + \frac{2}{7\sqrt{10}}(IV_2^{(4)}+IV_{-2}^{(4)}) + \frac{3}{35}IV_0^{(4)} - \frac{\sqrt{6}}{7}(II_2^{(2)}+II_{-2}^{(2)})r^2 - \frac{2}{7}II_0^{(2)}r^2 + \frac{1}{5}r^4$$

$$z^4 = \frac{8}{35}IV_0^{(4)} + \frac{4}{7}II_0^{(2)}r^2 + \frac{1}{5}r^4$$

$$x^2y^2 = -\frac{1}{\sqrt{70}}(IV_4^{(4)}+IV_{-4}^{(4)}) + \frac{1}{35}IV_0^{(4)} - \frac{2}{21}II_0^{(2)}r^2 + \frac{1}{15}r^4$$

$$x^2z^2 = \frac{\sqrt{2}}{35}(IV_2^{(4)}+IV_{-2}^{(4)}) + \frac{1}{7\sqrt{6}}(IV_2^{(2)}+IV_{-2}^{(2)})r^2 - \frac{8}{70}IV_0^{(4)} + \frac{1}{21}II_0^{(2)}r^2 + \frac{1}{15}r^4$$

$$y^2z^2 = -\frac{\sqrt{2}}{35}(IV_2^{(4)}+IV_{-2}^{(4)}) - \frac{1}{7\sqrt{6}}(IV_2^{(2)}+IV_{-2}^{(2)})r^2 - \frac{8}{70}IV_0^{(4)} + \frac{1}{21}II_0^{(2)}r^2 + \frac{1}{15}r^4$$

$$ix^2yz = -\frac{1}{2\sqrt{35}}(IV_3^{(4)}+IV_{-3}^{(4)}) + \frac{1}{14\sqrt{5}}(IV_1^{(4)}+IV_{-1}^{(4)}) - \frac{1}{7\sqrt{6}}(II_1^{(2)}+II_{-1}^{(2)})r^2$$

$$xy^2z = +\frac{1}{2\sqrt{35}}(IV_3^{(4)}-IV_{-3}^{(4)}) + \frac{1}{14\sqrt{5}}(IV_1^{(4)}-IV_{-1}^{(4)}) + \frac{1}{7\sqrt{6}}(II_1^{(2)}-II_{-1}^{(2)})r^2$$

$$ixyz^2 = \frac{1}{7}\sqrt{\frac{2}{5}}(IV_2^{(4)}-IV_{-2}^{(4)}) - \frac{1}{7\sqrt{6}}(I_2^{(2)}-II_{-2}^{(2)})r^2$$

$$ix^3y = \frac{1}{\sqrt{70}}(IV_4^{(4)}-IV_{-4}^{(4)}) - \frac{1}{7\sqrt{10}}(IV_2^{(4)}-IV_{-2}^{(4)}) - \frac{3}{7\sqrt{6}}(II_2^{(2)}-II_{-2}^{(2)})r^2$$

$$iy^3x = -\frac{1}{\sqrt{70}}(IV_4^{(4)}-IV_{-4}^{(4)}) - \frac{1}{7\sqrt{10}}(IV_2^{(4)}-IV_{-2}^{(4)}) - \frac{3}{7\sqrt{6}}(II_2^{(2)}-II_{-2}^{(2)})r^2$$

$$x^3z = -\frac{1}{2\sqrt{35}}(IV_3^{(4)}-IV_{-3}^{(4)}) + \frac{4}{14\sqrt{5}}(IV_1^{(4)}-IV_{-1}^{(4)}) + \frac{3}{7\sqrt{6}}(II_1^{(2)}-II_{-1}^{(2)})r^2$$

$$iy^3z = \frac{1}{2\sqrt{35}}(IV_3^{(4)}+IV_{-3}^{(4)}) + \frac{3}{14\sqrt{5}}(IV_1^{(4)}+IV_{-1}^{(4)}) - \frac{3}{7\sqrt{6}}(II_1^{(2)}+II_{-1}^{(2)})r^2$$

$$iz^3y = -\frac{2}{7\sqrt{5}}(IV_1^{(4)}+IV_{-1}^{(4)}) - \frac{3}{7\sqrt{6}}(II_1^{(2)}+II_{-1}^{(2)})r^2$$

$$z^3x = -\frac{2}{7\sqrt{5}}(IV_1^{(4)}-IV_{-1}^{(4)}) + \frac{3}{7\sqrt{6}}(II_1^{(2)}-II_{-1}^{(2)})r^2$$

794 APPENDIX F

Coupling coefficients are matrix elements of tensor operators according to the Wigner-Eckart theorem. (See Section 7.3.B.)

$$\left\langle \begin{matrix} j_3 \\ m_3 \end{matrix} \middle| T \begin{matrix} j_1 \\ m_1 \end{matrix} \middle| \begin{matrix} j_2 \\ m_2 \end{matrix} \right\rangle = C \begin{matrix} j_1 & j_2 & j_3 \\ m_1 & m_2 & m_3 \end{matrix} = \langle j_3 \| j_2 \| j_1 \rangle.$$

Unit tensor operator matrices are given in Tables 7.1 through 7.4, at the end of Chapter 7.

The Racah-6j recoupling coefficient used in Section 7.3.D is given by

$$\begin{Bmatrix} j_1 & j_2 & j_3 \\ l_1 & l_2 & l_3 \end{Bmatrix} = (-1)^{l_1+l_2+j_1+j_2} \Delta(l_1 l_2 j_3) \Delta(j_1 l_2 l_3) \Delta(l_1 j_2 l_3) \Delta(j_1 j_2 j_3)$$

$$\times \sum_k \frac{(-1)^k}{k!(j_1+j_2-j_3-k)!(l_1+l_2-j_3-k)!(j_1+l_2-j_3-k)!}$$

$$\times \frac{(j_1+j_2+l_1+l_2-k+1)!}{(l_1+j_2-l_3-k)!(j_3+l_3-j_1-l_1+k)!(j_3+l_3-j_2-l_2+k)!},$$

where

$$\Delta(jkl) = \sqrt{\frac{(j+k-l)!(j-k+l)!(-j+k+l)!}{(j+k+l+1)!}}. \quad \text{(F.1.6)}$$

F.2. THE OCTAHEDRAL GROUPS O AND $O_h = O \times C_i$

The 24 operations of the octahedral O group are shown by Figure 4.1.2. Its multiplication rules can be determined by the Hamilton turns shown in Figures 4.1.3 and 4.1.4. A group multiplication table (Table F.2.1) is given below. It includes $(-)$ signs which are needed to transform half-integral spin particles or spinor bases. Ignore them for vector transformations. The O character table is given in Eq. (4.1.11) and the spinor characters are given in Eq. (5.7.25).

The full octahedral group O_h is simply related to the outer product $O \times C_2$ of O and the inversion subgroup $C_i \approx C_2$. Its 48 elements are displayed in Figure 4.1.5, which shows other cubic symmetry groups as well. The full O_h vector character table is given by Eq. (4.1.16) and in Table F.4.1 in this appendix.

A conventional set of irreducible representations is given in Table F.2.2, and the corresponding multipole functions or "Kubic harmonics" are given in Table F.2.3. The remaining representations given in Tables F.2.4 to F.2.7 are defined by various subgroup chains which are described in Section 4.2.

TABLE F.2.1 O-Group Table

	1	r_1	r_2	r_3	r_4	r_1^2	r_2^2	r_3^2	r_4^2	R_1^2	R_2^2	R_3^2	R_1	R_2	R_3	R_1^3	R_2^3	R_3^3	i_1	i_2	i_3	i_4	i_5	i_6
r_1	r_1^2	$-r_2^2$	$-r_2^3$	$-r_3^2$	$-r_2^3$	-1	$-R_2^2$	$-R_3^2$	$-R_1^2$	$-r_4$	r_3	$-r_2$	i_3	i_6	i_1	$-R_3$	$-R_1$	$-R_2$	R_1^3	i_5	i_3^*	i_2	$-i_4$	R_3^3
r_2	r_3^2	$-r_4^2$	$-r_4^2$	R_2^2	$-r_1^2$	$-R_2^2$	R_2	R_2^3	R_3^2	$-r_2$	r_1	r_4	i_6	i_3	i_2	R_2	R_2^3	R_2^2	i_5	$-R_1$	R_2	$-i_1$	$-i_4$	i_4
r_3	r_4^2	$-r_1^2$	r_3^2	R_3^2	R_1^2	$-R_3^2$	-1	R_1^3	R_1^2	R_1^3	r_4	r_2	i_5	i_1	i_1^*	i_3^*	i_5	$-R_1$	R_3	$-R_1$	i_1	$-i_2$	R_3^3	R_3
r_4	r_2^2	$-r_3^2$	$-r_1^2$	R_1^2	-1	$-R_1^2$	R_3^2	R_2^2	-1	R_1	r_2	r_3	i_4	i_2	i_6	$-R_1^3$	R_3	i_2	i_6	$-i_4$	$-R_3^3$	$-R_2^3$	R_3^3	i_3
r_1^2	-1	R_1^2	R_2^2	R_3^2	r_2	r_4	r_3	-1	r_2	$-r_3$	R_1	$-i_2$	$-R_3$	$-R_1$	$-i_3$	$-R_1$	$-i_3$	$-R_1$	$-i_4$	R_1	i_5	$-i_6$	$-i_2$	$-R_2$
r_2^2	$-R_1^2$	R_2^2	-1	$-R_3^2$	$-R_2^2$	r_3	$-r_4$	-1	r_2^2	r_1	$-R_2^2$	$-R_3$	$-R_2$	R_3^3	$-R_1$	$-i_3$	R_3^3	$-R_3^3$	$-i_2$	i_1	i_5	$-i_6$	$-i_1$	$-i_1$
r_3^2	$-R_2^2$	$-R_2^3$	-1	$-R_2^2$	$-R_3^2$	$-r_4$	$-r_3$	r_2	$-r_4$	R_3^3	R_3	$-i_4$	i_5	$-i_6$	$-R_2^3$	R_3	$-i_4$	R_3	$-R_1$	i_5	R_1	R_2^3	R_2	$-R_3^3$
r_4^2	$-R_3^2$	R_2^2	$-R_1^2$	-1	r_1	$-r_2$	$-r_3$	r_1	$-r_4$	$-R_3^2$	$-i_4$	R_1	$-i_5$	$-i_5$	i_4	$-R_3^2$	$-i_4$	$-R_3$	$-R_3^3$	R_1	$-i_6$	R_2^3	$-i_1$	$-i_2$
R_1^2	$-r_4$	r_3	$-r_2$	r_1	$-r_4^2$	$-r_1^2$	r_4^2	r_3^2	-1	$-R_3^2$	R_3^2	i_5	i_6	$-R_1$	$-R_3$	i_2	i_2	$-i_3$	$-R_2$	r_2	R_3^2	R_3	$-i_6$	R_3
R_2^2	$-r_2$	r_1	r_4	$-r_3$	$-r_2^2$	r_3^2	-1	$-R_3^2$	$-R_2^2$	R_3^2	-1	R_1^2	R_3^3	$-R_3^3$	R_3	$-R_2$	$-R_2$	$-i_4$	i_1	$-i_2$	R_1	R_1^2	R_1	R_3^3
R_3^2	$-r_3$	$-r_4$	r_1	r_2	r_3^2	r_4^2	$-R_1^2$	-1	$-R_3^2$	R_3^2	$-R_1^2$	-1	i_6	$-i_4$	$-i_5$	$-i_1$	$-i_1$	$-R_3$	R_3	$-R_2$	$-R_1$	R_1^3	R_3^3	$-R_1^3$
R_1	i_1	$-R_3^2$	$-i_2$	$-R_3^3$	R_3^2	R_3^3	R_1^3	R_3	$-R_3^2$	R_2^3	i_6	i_1	-1	r_3	$-r_4^2$	-1	$-r_3$	r_2^2	$-r_4$	r_2	r_3	$-i_6$	$-i_3$	$-R_3$
R_2	i_3	R_3	$-R_3^3$	R_1	i_4	$-i_2$	$-R_3^3$	r_1	$-r_2^2$	$-r_2$	R_3^3	$-R_3^2$	$-r_3$	-1	$-r_4$	$-r_2$	$-r_2$	r_2	$-r_1$	r_2	r_3	$-r_4$	$-r_2$	r_1^2
R_3	i_6	i_5	R_1	R_2^3	$-i_1$	R_3^3	$-R_3$	R_3	$-r_1^2$	r_1	$-r_4$	r_2^3	r_4^2	r_4	-1	r_4^2	r_3^2	$-r_1$	r_2	-1	r_4^2	$-r_3$	$-r_2$	$-R_1$
R_1^3	$-R_2$	$-R_2$	$-i_2$	$-R_2$	$-R_1$	R_1^3	i_2	R_1	$-R_2$	r_2	r_1	$-r_3$	-1	r_3	r_4^2	r_2^2	r_2^2	$-r_2^2$	r_4	$-r_3$	r_1	r_4^2	$-r_4$	R_2^2
R_2^3	$-i_2$	i_1	$-R_3$	i_5	R_2^3	R_3^2	$-R_3^3$	$-R_1^3$	$-r_2^2$	r_3	r_2	$-r_4$	r_3	-1	r_4	$-r_3$	r_3	r_3^2	$-r_2$	r_1^2	$-r_4$	-1	$-r_3$	R_1^3
R_3^3	$-R_3$	R_2	$-i_4$	$-i_6$	$-R_3^3$	R_2^3	R_3	R_3^3	$-R_3^2$	r_4	r_3	r_1	r_4	$-r_1$	r_2	r_3^2	-1	$-r_3$	$-r_3$	$-r_2$	$-r_4$	r_1^2	$-r_2^2$	R_3^3
i_1	$-R_1$	$-R_1^3$	i_3	i_3	$-R_2^3$	R_3	R_2^3	R_3^3	i_5	R_1^3	R_1^3	$-R_2^3$	$-R_3^3$	$-r_1$	$-r_2$	$-r_2$	$-r_1^2$	$-r_2$	-1	$-r_3$	$-r_2^2$	r_3^2	r_2^2	R_2^2
i_2	R_3^3	R_3^3	R_3	$-i_5$	$-R_3^3$	i_4	R_1	$-i_3$	$-R_1^3$	R_3^3	R_1	$-R_1^3$	r_3	r_1	R_2^2	$-r_4$	$-r_4$	r_4	r_3^2	$-R_2^2$	r_4	-1	$-r_1^2$	r_2^2
i_3	R_1^3	R_1	$-i_5$	$-R_1^3$	R_3^3	$-R_3^3$	R_3^2	$-R_3$	i_3	$-R_2^3$	R_3	$-R_3^2$	$-r_2$	R_1	R_3^3	$-r_4$	-1	$-r_4^2$	R_3^2	$-r_4$	r_1	$-r_3$	r_3^2	$-r_4$
i_4	$-i_5$	i_6	$-R_1^3$	R_2^3	$-R_3^2$	$-R_3^3$	$-R_1^3$	$-R_1$	i_4	i_5	i_4	i_5	$-r_4$	$-i_2$	i_4	r_3	$-r_4$	r_3	$-r_3$	r_1^2	r_3^2	R_3^2	r_1	$-r_2$
i_5	i_2	$-R_2$	i_1	$-R_3^2$	$-R_1^3$	R_1^3	R_3^3	R_1	i_3	$-R_3^3$	$-i_1$	i_4	$-i_3$	$-r_2$	i_1	r_2	r_2	r_2^2	r_2^2	-1	R_1^2	R_2	$-r_3$	R_2^2
i_6	R_2^3	i_1	R_2	$-R_1$	$-R_3^2$	R_1^3	R_1	R_1^3	$-R_2$	R_3^2	$-i_1$	$-R_3$	$-R_3^3$	$-r_1$	$-R_2^3$	r_4	$-r_3$	R_2^2	R_3^2	$-r_4$	-1	$-r_3$	R_1^2	-1

TABLE F.2.2 Conventional O Irreducible Representations. (Cartesian Fourfold Axial Bases)

(a) Vector Representation T_1

$$\mathscr{D}^{T_1}(1) = \begin{vmatrix} 1 & \cdot & \cdot \\ \cdot & 1 & \cdot \\ \cdot & \cdot & 1 \end{vmatrix} \quad R_1^2 = \begin{vmatrix} 1 & \cdot & \cdot \\ \cdot & -1 & \cdot \\ \cdot & \cdot & -1 \end{vmatrix} \quad r_1 = \begin{vmatrix} \cdot & \cdot & 1 \\ \cdot & 1 & \cdot \\ 1 & \cdot & \cdot \end{vmatrix} \quad r_2 = \begin{vmatrix} 1 & \cdot & \cdot \\ \cdot & \cdot & 1 \\ \cdot & 1 & \cdot \end{vmatrix} \quad r_1^2 = \begin{vmatrix} \cdot & 1 & \cdot \\ 1 & \cdot & \cdot \\ \cdot & \cdot & 1 \end{vmatrix} \quad r_2^2 = \begin{vmatrix} \cdot & 1 & \cdot \\ \cdot & \cdot & -1 \\ -1 & \cdot & \cdot \end{vmatrix}$$

$$\mathscr{D}^{T_1}(R_3^2) = \begin{vmatrix} -1 & \cdot & \cdot \\ \cdot & 1 & \cdot \\ \cdot & \cdot & -1 \end{vmatrix} \quad R_2^2 = \begin{vmatrix} -1 & \cdot & \cdot \\ \cdot & -1 & \cdot \\ \cdot & \cdot & 1 \end{vmatrix} \quad r_4 = \begin{vmatrix} \cdot & -1 & \cdot \\ -1 & \cdot & \cdot \\ \cdot & \cdot & 1 \end{vmatrix} \quad r_3 = \begin{vmatrix} -1 & \cdot & \cdot \\ \cdot & \cdot & 1 \\ \cdot & 1 & \cdot \end{vmatrix} \quad r_3^2 = \begin{vmatrix} \cdot & -1 & \cdot \\ 1 & \cdot & \cdot \\ \cdot & \cdot & 1 \end{vmatrix} \quad r_4^2 = \begin{vmatrix} \cdot & \cdot & -1 \\ \cdot & 1 & \cdot \\ 1 & \cdot & \cdot \end{vmatrix}$$

$$\mathscr{D}^{T_1}(R_3) = \begin{vmatrix} \cdot & -1 & \cdot \\ 1 & \cdot & \cdot \\ \cdot & \cdot & 1 \end{vmatrix} \quad i_4 = \begin{vmatrix} \cdot & -1 & \cdot \\ -1 & \cdot & \cdot \\ \cdot & \cdot & -1 \end{vmatrix} \quad i_1 = \begin{vmatrix} \cdot & \cdot & 1 \\ \cdot & -1 & \cdot \\ 1 & \cdot & \cdot \end{vmatrix} \quad i_2 = \begin{vmatrix} -1 & \cdot & \cdot \\ \cdot & \cdot & -1 \\ \cdot & -1 & \cdot \end{vmatrix} \quad R_1^3 = \begin{vmatrix} \cdot & \cdot & -1 \\ \cdot & 1 & \cdot \\ -1 & \cdot & \cdot \end{vmatrix} \quad R_1 = \begin{vmatrix} 1 & \cdot & \cdot \\ \cdot & \cdot & -1 \\ \cdot & 1 & \cdot \end{vmatrix}$$

$$\mathscr{D}^{T_1}(R_3^3) = \begin{vmatrix} \cdot & 1 & \cdot \\ -1 & \cdot & \cdot \\ \cdot & \cdot & 1 \end{vmatrix} \quad i_3 = \begin{vmatrix} \cdot & 1 & \cdot \\ 1 & \cdot & \cdot \\ \cdot & \cdot & -1 \end{vmatrix} \quad R_2 = \begin{vmatrix} \cdot & \cdot & 1 \\ \cdot & 1 & \cdot \\ -1 & \cdot & \cdot \end{vmatrix} \quad R_2^3 = \begin{vmatrix} \cdot & \cdot & -1 \\ \cdot & 1 & \cdot \\ 1 & \cdot & \cdot \end{vmatrix} \quad i_6 = \begin{vmatrix} -1 & \cdot & \cdot \\ \cdot & \cdot & 1 \\ \cdot & 1 & \cdot \end{vmatrix} \quad i_5 = \begin{vmatrix} -1 & \cdot & \cdot \\ \cdot & \cdot & -1 \\ \cdot & 1 & \cdot \end{vmatrix}$$

$$\left. \begin{array}{l} O: \ |T_1\rangle \ |T_1\rangle \ |T_1\rangle \\ D_4: \ |E\rangle \ |E\rangle \ |A_2\rangle \\ C_2: \ |B_1\rangle \ |B_2\rangle \ |A_2\rangle \end{array} \right\} \text{basis}$$

(b) Second-Rank Tensor $T_2 = T_1 \otimes A_2$ Representation (T_2 is T_1 with R, R^3, and i Operations Negated.)

$$\mathscr{D}^{T_2}(1) = \begin{array}{|ccc|} 1 & \cdot & \cdot \\ \cdot & 1 & \cdot \\ \cdot & \cdot & 1 \end{array} \quad R_1^2 = \begin{array}{|ccc|} 1 & \cdot & \cdot \\ \cdot & -1 & \cdot \\ \cdot & \cdot & 1 \end{array} \quad r_1 = \begin{array}{|ccc|} \cdot & \cdot & 1 \\ \cdot & 1 & \cdot \\ 1 & \cdot & \cdot \end{array} \quad r_2 = \begin{array}{|ccc|} \cdot & 1 & \cdot \\ 1 & \cdot & \cdot \\ \cdot & \cdot & -1 \end{array} \quad r_1^2 = \begin{array}{|ccc|} \cdot & \cdot & 1 \\ \cdot & -1 & \cdot \\ 1 & \cdot & \cdot \end{array} \quad r_2^2 = \begin{array}{|ccc|} \cdot & 1 & \cdot \\ -1 & \cdot & \cdot \\ \cdot & \cdot & -1 \end{array}$$

$$\mathscr{D}^{T_2}(R_3^2) = \begin{array}{|ccc|} -1 & \cdot & \cdot \\ \cdot & -1 & \cdot \\ \cdot & \cdot & 1 \end{array} \quad R_2^2 = \begin{array}{|ccc|} -1 & \cdot & \cdot \\ \cdot & 1 & \cdot \\ \cdot & \cdot & 1 \end{array} \quad r_4 = \begin{array}{|ccc|} \cdot & \cdot & -1 \\ \cdot & 1 & \cdot \\ -1 & \cdot & \cdot \end{array} \quad r_3 = \begin{array}{|ccc|} \cdot & -1 & \cdot \\ -1 & \cdot & \cdot \\ \cdot & \cdot & -1 \end{array} \quad r_3^2 = \begin{array}{|ccc|} \cdot & \cdot & -1 \\ \cdot & -1 & \cdot \\ -1 & \cdot & \cdot \end{array} \quad r_4^2 = \begin{array}{|ccc|} \cdot & -1 & \cdot \\ 1 & \cdot & \cdot \\ \cdot & \cdot & -1 \end{array}$$

$$\mathscr{D}^{T_2}(R_3) = \begin{array}{|ccc|} \cdot & 1 & \cdot \\ -1 & \cdot & \cdot \\ \cdot & \cdot & 1 \end{array} \quad i_4 = \begin{array}{|ccc|} -1 & \cdot & \cdot \\ \cdot & \cdot & 1 \\ \cdot & 1 & \cdot \end{array} \quad i_1 = \begin{array}{|ccc|} \cdot & \cdot & -1 \\ \cdot & 1 & \cdot \\ 1 & \cdot & \cdot \end{array} \quad i_2 = \begin{array}{|ccc|} \cdot & \cdot & 1 \\ \cdot & 1 & \cdot \\ -1 & \cdot & \cdot \end{array} \quad R_1^3 = \begin{array}{|ccc|} -1 & \cdot & \cdot \\ \cdot & \cdot & -1 \\ \cdot & 1 & \cdot \end{array} \quad R_1 = \begin{array}{|ccc|} -1 & \cdot & \cdot \\ \cdot & \cdot & 1 \\ \cdot & -1 & \cdot \end{array}$$

$$\mathscr{D}^{T_2}(R_3^3) = \begin{array}{|ccc|} \cdot & -1 & \cdot \\ 1 & \cdot & \cdot \\ \cdot & \cdot & 1 \end{array} \quad i_3 = \begin{array}{|ccc|} \cdot & -1 & \cdot \\ -1 & \cdot & \cdot \\ \cdot & \cdot & 1 \end{array} \quad R_2 = \begin{array}{|ccc|} \cdot & \cdot & 1 \\ \cdot & 1 & \cdot \\ 1 & \cdot & \cdot \end{array} \quad R_2^3 = \begin{array}{|ccc|} \cdot & \cdot & -1 \\ \cdot & 1 & \cdot \\ -1 & \cdot & \cdot \end{array} \quad i_6 = \begin{array}{|ccc|} 1 & \cdot & \cdot \\ \cdot & \cdot & -1 \\ \cdot & -1 & \cdot \end{array} \quad i_5 = \begin{array}{|ccc|} 1 & \cdot & \cdot \\ \cdot & \cdot & 1 \\ \cdot & 1 & \cdot \end{array}$$

$$\begin{array}{l} O: \\ D_4: \\ C_2: \end{array} \left. \begin{array}{l} |T_2\rangle \\ |E\rangle \\ |B_1\rangle \end{array} \right| \left. \begin{array}{l} |T_2\rangle \\ |E\rangle \\ |B_2\rangle \end{array} \right| \left. \begin{array}{l} |T_2\rangle \\ |B_2\rangle \\ |A_2\rangle \end{array} \right\} \text{basis}$$

TABLE F.2.2 (Continued)

(c) Second-rank Tensor E Representation

$\mathscr{D}^E(1)$

$R_1^2 = \begin{vmatrix} 1 & 0 \\ 0 & 1 \end{vmatrix}$

$r_1 = \begin{vmatrix} -1/2 & -\sqrt{3}/2 \\ \sqrt{3}/2 & -1/2 \end{vmatrix}$
$r_2 = \begin{vmatrix} -1/2 & -\sqrt{3}/2 \\ \sqrt{3}/2 & -1/2 \end{vmatrix}$
$r_1^2 = \begin{vmatrix} -1/2 & \sqrt{3}/2 \\ -\sqrt{3}/2 & -1/2 \end{vmatrix}$
$r_2^2 = \begin{vmatrix} -1/2 & \sqrt{3}/2 \\ -\sqrt{3}/2 & -1/2 \end{vmatrix}$

$\mathscr{D}^E(R_3^2)$

$R_2^2 = \begin{vmatrix} 1 & 0 \\ 0 & 1 \end{vmatrix}$

$r_4 = \begin{vmatrix} -1/2 & -\sqrt{3}/2 \\ \sqrt{3}/2 & -1/2 \end{vmatrix}$
$r_3 = \begin{vmatrix} -1/2 & -\sqrt{3}/2 \\ \sqrt{3}/2 & -1/2 \end{vmatrix}$
$r_3^2 = \begin{vmatrix} -1/2 & \sqrt{3}/2 \\ -\sqrt{3}/2 & -1/2 \end{vmatrix}$
$r_4^2 = \begin{vmatrix} -1/2 & \sqrt{3}/2 \\ -\sqrt{3}/2 & -1/2 \end{vmatrix}$

$\mathscr{D}^E(R_3)$

$i_4 = \begin{vmatrix} 1 & 0 \\ 0 & -1 \end{vmatrix}$

$i_1 = \begin{vmatrix} -1/2 & \sqrt{3}/2 \\ \sqrt{3}/2 & 1/2 \end{vmatrix}$
$i_2 = \begin{vmatrix} -1/2 & \sqrt{3}/2 \\ \sqrt{3}/2 & 1/2 \end{vmatrix}$
$R_1^3 = \begin{vmatrix} -1/2 & -\sqrt{3}/2 \\ -\sqrt{3}/2 & 1/2 \end{vmatrix}$
$R_1 = \begin{vmatrix} -\sqrt{3}/2 & -1/2 \\ 1/2 & -\sqrt{3}/2 \end{vmatrix}$

$\mathscr{D}^E(R_3^3)$

$i_3 = \begin{vmatrix} 1 & 0 \\ 0 & -1 \end{vmatrix}$

$R_2 = \begin{vmatrix} -1/2 & \sqrt{3}/2 \\ \sqrt{3}/2 & 1/2 \end{vmatrix}$
$R_2^3 = \begin{vmatrix} -1/2 & \sqrt{3}/2 \\ \sqrt{3}/2 & 1/2 \end{vmatrix}$
$i_6 = \begin{vmatrix} -1/2 & -\sqrt{3}/2 \\ -\sqrt{3}/2 & 1/2 \end{vmatrix}$
$i_5 = \begin{vmatrix} -\sqrt{3}/2 & -1/2 \\ 1/2 & -\sqrt{3}/2 \end{vmatrix}$

$O: \begin{vmatrix} T_2 \\ A_1 \end{vmatrix} \begin{vmatrix} T_2 \\ A_1 \end{vmatrix}$ basis
$D_4: \begin{vmatrix} A_2 \\ A_1 \end{vmatrix} \begin{vmatrix} B_1 \\ A_1 \end{vmatrix}$
$C_2: \begin{vmatrix} A_1 \\ A_1 \end{vmatrix} \begin{vmatrix} A_1 \\ A_1 \end{vmatrix}$

For scalar $(A_1) \begin{vmatrix} A_1 \\ A_1 \\ A_1 \end{vmatrix}$ and pseudoscalar $(A_2) \begin{vmatrix} A_2 \\ B_1 \\ A_1 \end{vmatrix}$ representations, see the O character table.

APPENDIX F 799

TABLE F.2.3   Octahedral Multipole Functions and Harmonic Monomials

$I_1^{T_{1u}} = x \quad I_1^{T_{1g}} = J_x$

$I_2^{T_{1u}} = y \quad I_2^{T_{1g}} = J_y$

$I_3^{T_{1u}} = z \quad I_3^{T_{1g}} = J_z$

$II^{A_{1g}} = x^2 + y^2 + z^2$ $\qquad x^2 = \frac{1}{3} II^{A_{1g}} - \frac{1}{6} II_1^{E_g} + \frac{1}{2\sqrt{3}} II_2^{E_g}$

$II_1^{E_g} = -x^2 - y^2 + 2z^2$ $\qquad y^2 = \frac{1}{3} II^{A_{1g}} - \frac{1}{6} II_1^{E_g} - \frac{1}{2\sqrt{3}} II_2^{E_g}$

$II_2^{E_g} = \sqrt{3}(x^2 - y^2)$ $\qquad z^2 = \frac{1}{3} II^{A_{1g}} + \frac{1}{3} II_1^{E_g}$

$II_1^{T_{2g}} = yz$ $\qquad yz = II_1^{T_{2g}}$

$II_2^{T_{2g}} = xz$ $\qquad xz = II_2^{T_{2g}}$

$II_3^{T_{2g}} = xy$ $\qquad xy = II_3^{T_{2g}}$

$III^{A_{2u}} = xyz$ $\qquad x^3 = III_1^{T_{1u}}$

$III_1^{T_{1u}} = x^3$ $\qquad y^3 = III_2^{T_{1u}}$

$III_2^{T_{1u}} = y^3$ $\qquad z^3 = III_3^{T_{1u}}$

$III_3^{T_{1u}} = z^3$ $\qquad xy^2 = \frac{1}{2} III_1'^{T_{1u}} + \frac{1}{2} III_1^{T_{2u}}$

$III_1'^{T_{1u}} = xy^2 + xz^2$ $\qquad xz^2 = \frac{1}{2} III_1'^{T_{1u}} - \frac{1}{2} III_1^{T_{2u}}$

$III_2'^{T_{1u}} = yz^2 + yx^2$ $\qquad yx^2 = \frac{1}{2} III_2'^{T_{1u}} - \frac{1}{2} III_2^{T_{2u}}$

$III_3'^{T_{1u}} = zx^2 + zy^2$ $\qquad yz^2 = \frac{1}{2} III_2'^{T_{1u}} + \frac{1}{2} III_2^{T_{2u}}$

$III_1^{T_{2u}} = xy^2 - xz^2$ $\qquad zx^2 = \frac{1}{2} III_3'^{T_{1u}} + \frac{1}{2} III_3^{T_{2u}}$

$III_2^{T_{2u}} = yz^2 - yx^2$ $\qquad zy^2 = \frac{1}{2} III_3'^{T_{1u}} - \frac{1}{2} III_3^{T_{2u}}$

$III_3^{T_{2u}} = zx^2 - zy^2$ $\qquad xyz = III^{A_{2u}}$

TABLE F.2.3 (*Continued*)

$$\mathrm{IV}_{\cdot}^{A_{1g}} = x^4 + y^4 + z^4$$

$$x^4 = \frac{1}{3}\mathrm{IV}_{\cdot}^{A_{1g}} - \frac{1}{6}\mathrm{IV}_1^{E_g} + \frac{1}{2\sqrt{3}}\mathrm{IV}_2^{E_g}$$

$$\mathrm{IV}_{\cdot}'^{A_{1g}} = x^2y^2 + x^2z^2 + y^2z^2$$

$$y^4 = \frac{1}{3}\mathrm{IV}_{\cdot}^{A_{1g}} - \frac{1}{6}\mathrm{IV}_1^{E_g} - \frac{1}{2\sqrt{3}}\mathrm{IV}_2^{E_g}$$

$$\mathrm{IV}_1^{E_g} = -x^4 - y^4 + 2z^4$$

$$z^4 = \frac{1}{3}\mathrm{IV}_{\cdot}^{A_{1g}} + \frac{1}{3}\mathrm{IV}_1^{E_g}$$

$$\mathrm{IV}_2^{E_g} = \sqrt{3}(x^4 - y^4)$$

$$y^2z^2 = \frac{1}{3}\mathrm{IV}_{\cdot}'^{A_{1g}} - \frac{1}{6}\mathrm{IV}_1'^{E_g} + \frac{1}{2\sqrt{3}}\mathrm{IV}_2'^{E_g}$$

$$\mathrm{IV}_1'^{E_g} = 2x^2y^2 - x^2z^2 - y^2z^2$$

$$x^2z^2 = \frac{1}{3}\mathrm{IV}_{\cdot}'^{A_{1g}} - \frac{1}{6}\mathrm{IV}_1'^{E_g} - \frac{1}{2\sqrt{3}}\mathrm{IV}_2'^{E_g}$$

$$\mathrm{IV}_2'^{E_g} = \sqrt{3}(-x^2z^2 + y^2z^2)$$

$$x^2y^2 = \frac{1}{3}\mathrm{IV}_{\cdot}'^{A_{1g}} + \frac{1}{3}\mathrm{IV}_1'^{E_g}$$

$$\mathrm{IV}_1^{T_{1g}} = y^3z - z^3y$$

$$xy^3 = \frac{1}{2}\mathrm{IV}_3'^{T_{2g}} - \frac{1}{2}\mathrm{IV}_3^{T_{1g}}$$

$$\mathrm{IV}_2^{T_{1g}} = z^3x - x^3z$$

$$yx^3 = \frac{1}{2}\mathrm{IV}_3'^{T_{2g}} + \frac{1}{2}\mathrm{IV}_3^{T_{1g}}$$

$$\mathrm{IV}_3^{T_{1g}} = x^3y - y^3x$$

$$xz^3 = \frac{1}{2}\mathrm{IV}_2'^{T_{2g}} + \frac{1}{2}\mathrm{IV}_2^{T_{1g}}$$

$$\mathrm{IV}_1^{T_{2g}} = x^2yz$$

$$zx^3 = \frac{1}{2}\mathrm{IV}_2'^{T_{2g}} - \frac{1}{2}\mathrm{IV}_2^{T_{1g}}$$

$$\mathrm{IV}_2^{T_{2g}} = xy^2z$$

$$yz^3 = \frac{1}{2}\mathrm{IV}_1'^{T_{2g}} - \frac{1}{2}\mathrm{IV}_1^{T_{1g}}$$

$$\mathrm{IV}_3^{T_{2g}} = xyz^2$$

$$zy^3 = \frac{1}{2}\mathrm{IV}_1'^{T_{2g}} + \frac{1}{2}\mathrm{IV}_1^{T_{1g}}$$

$$\mathrm{IV}_1'^{T_{2g}} = y^3z + z^3y$$

$$x^2yz = \mathrm{IV}_1^{T_{2g}}$$

$$\mathrm{IV}_2'^{T_{2g}} = z^3x + x^3z$$

$$xy^2z = \mathrm{IV}_2^{T_{2g}}$$

$$\mathrm{IV}_3'^{T_{2u}} = x^3y + y^3x$$

$$xyz^2 = \mathrm{IV}_3^{T_{2g}}$$

TABLE F.2.4 $O \supset D_4 \supset D_2$ **Subgroup Chain Labeled Irreducible Representations (Fourfold Standing-Wave Bases)**

(a) Vector Representations

$$\left. \begin{array}{l} O: |T_1\rangle |T_1\rangle |T_1\rangle \\ D_4: E \rangle |E\rangle |A_2\rangle \\ D_2: B_1 \rangle |B_2\rangle |A_2\rangle \end{array} \right\} \text{basis}$$

TABLE F.2.4 *(Continued)*

(b) Second-Rank Tensor Representations

$\mathcal{D}^{T_2}(1) = $ [matrices for R_1^2, r_1, r_2, r_1^2, r_2^2]

$\mathcal{D}^{T_2}(R_3^2) = $ [matrices for R_2^2, r_4, r_3, r_3^2, r_4^2]

$\mathcal{D}^{T_2}(R_3) = $ [matrices for i_4, i_1, i_2, R_1^3, R_1]

$\mathcal{D}^{T_2}(R_3^3) = $ [matrices for i_3, R_2, R_2^3, i_6, i_5]

$\left.\begin{array}{l}O: \left|T_2\right\rangle \left|T_2\right\rangle \left|T_2\right\rangle \\ D_4: \left|E\right\rangle \left|E\right\rangle \left|B_2\right\rangle \\ D_2: \left|B_1\right\rangle \left|B_2\right\rangle \left|A_2\right\rangle\end{array}\right\}$ basis

$\mathscr{D}^E(1) =$ $\begin{vmatrix} 1 & 0 \\ 0 & 1 \end{vmatrix}$

$R_1^2 = \begin{vmatrix} 1 & 0 \\ 0 & 1 \end{vmatrix}$

$r_1 = \begin{vmatrix} -1/2 & -\sqrt{3}/2 \\ \sqrt{3}/2 & -1/2 \end{vmatrix}$

$r_2 = \begin{vmatrix} -1/2 & -\sqrt{3}/2 \\ \sqrt{3}/2 & -1/2 \end{vmatrix}$

$r_1^2 = \begin{vmatrix} -1/2 & \sqrt{3}/2 \\ -\sqrt{3}/2 & -1/2 \end{vmatrix}$

$r_2^2 = \begin{vmatrix} -1/2 & \sqrt{3}/2 \\ -\sqrt{3}/2 & -1/2 \end{vmatrix}$

$\mathscr{D}^E(R_3^2) =$ $\begin{vmatrix} 1 & 0 \\ 0 & 1 \end{vmatrix}$

$R_2^2 = \begin{vmatrix} 1 & 0 \\ 0 & 1 \end{vmatrix}$

$r_4 = \begin{vmatrix} -1/2 & -\sqrt{3}/2 \\ \sqrt{3}/2 & -1/2 \end{vmatrix}$

$r_3 = \begin{vmatrix} -1/2 & -\sqrt{3}/2 \\ \sqrt{3}/2 & -1/2 \end{vmatrix}$

$r_3^2 = \begin{vmatrix} -1/2 & \sqrt{3}/2 \\ -\sqrt{3}/2 & -1/2 \end{vmatrix}$

$r_4^2 = \begin{vmatrix} -1/2 & \sqrt{3}/2 \\ -\sqrt{3}/2 & -1/2 \end{vmatrix}$

$\mathscr{D}^E(R_3) =$ $\begin{vmatrix} 1 & 0 \\ 0 & -1 \end{vmatrix}$

$i_4 = \begin{vmatrix} 1 & 0 \\ 0 & -1 \end{vmatrix}$

$i_1 = \begin{vmatrix} -1/2 & -\sqrt{3}/2 \\ \sqrt{3}/2 & 1/2 \end{vmatrix}$

$i_2 = \begin{vmatrix} -1/2 & \sqrt{3}/2 \\ \sqrt{3}/2 & 1/2 \end{vmatrix}$

$R_1 = \begin{vmatrix} -1/2 & -\sqrt{3}/2 \\ -\sqrt{3}/2 & 1/2 \end{vmatrix}$

$R_1^3 = \begin{vmatrix} -1/2 & -\sqrt{3}/2 \\ -\sqrt{3}/2 & 1/2 \end{vmatrix}$

$i_5 = \begin{vmatrix} -1/2 & -\sqrt{3}/2 \\ -\sqrt{3}/2 & 1/2 \end{vmatrix}$

$\mathscr{D}^E(R_3^3) =$ $\begin{vmatrix} 1 & 0 \\ 0 & -1 \end{vmatrix}$

$i_3 = \begin{vmatrix} 1 & 0 \\ 0 & -1 \end{vmatrix}$

$R_2 = \begin{vmatrix} -1/2 & \sqrt{3}/2 \\ \sqrt{3}/2 & 1/2 \end{vmatrix}$

$R_2^3 = \begin{vmatrix} -1/2 & \sqrt{3}/2 \\ \sqrt{3}/2 & 1/2 \end{vmatrix}$

$i_6 = \begin{vmatrix} -1/2 & -\sqrt{3}/2 \\ \sqrt{3}/2 & 1/2 \end{vmatrix}$

For scalar (A_1) $\begin{vmatrix} A_1 \\ A_1 \\ A_1 \end{vmatrix}$ and pseudoscalar (A_2) $\begin{vmatrix} A_2 \\ B_1 \\ A_1 \end{vmatrix}$ representations, see the O character table.

$O: \begin{vmatrix} E \\ A_1 \\ A_1 \end{vmatrix} \begin{vmatrix} E \\ B_1 \\ A_1 \end{vmatrix}$ basis
$D_4:$
$D_2:$

TABLE F.2.5 $O \supset D_3 \supset C_2$ Labeled Irreducible Representations (Yamanouchi Threefold Standing-Wave Bases)

(a) Vector Representation in Bases $\left. \begin{array}{l} O: \\ D_3: \\ C_2: \end{array} \right| \left. \begin{array}{l} T_1 \\ E \\ B \end{array} \right\rangle \left. \begin{array}{l} T_1 \\ E \\ B \end{array} \right\rangle \left. \begin{array}{l} T_1 \\ A_2 \\ B \end{array} \right\rangle$

$\mathscr{D}^{T_1}(1) = $

$i_4 = [12]$

$R_1^2 = [13][24]$

$R_3 = [1423]$

$$\begin{vmatrix} 1 & \cdot & \cdot \\ \cdot & 1 & \cdot \\ \cdot & \cdot & 1 \end{vmatrix} \qquad \begin{vmatrix} 1 & \cdot & \cdot \\ \cdot & -1 & \cdot \\ \cdot & \cdot & -1 \end{vmatrix} \qquad \begin{vmatrix} \cdot & \frac{\sqrt{3}}{3} & \frac{\sqrt{6}}{3} \\ \frac{\sqrt{3}}{3} & \frac{-2}{3} & \frac{\sqrt{2}}{3} \\ \frac{\sqrt{6}}{3} & \frac{\sqrt{2}}{3} & \frac{-1}{3} \end{vmatrix} \qquad \begin{vmatrix} \cdot & \frac{-\sqrt{3}}{3} & \frac{-\sqrt{6}}{3} \\ \frac{\sqrt{3}}{3} & \frac{2}{3} & \frac{-\sqrt{2}}{3} \\ \frac{\sqrt{6}}{3} & \frac{-\sqrt{2}}{3} & \frac{1}{3} \end{vmatrix}$$

$r_1 = [132]$

$i_5 = [13]$

$r_4 = [234]$

$i_6 = [24]$

$$\begin{vmatrix} \frac{-1}{2} & \frac{-\sqrt{3}}{2} & \cdot \\ \frac{\sqrt{3}}{2} & \frac{-1}{2} & \cdot \\ \cdot & \cdot & 1 \end{vmatrix} \qquad \begin{vmatrix} \frac{-1}{2} & \frac{-\sqrt{3}}{2} & \cdot \\ \frac{-\sqrt{3}}{2} & \frac{1}{2} & \cdot \\ \cdot & \cdot & -1 \end{vmatrix} \qquad \begin{vmatrix} \frac{1}{2} & \frac{-\sqrt{3}}{6} & \frac{\sqrt{6}}{3} \\ \frac{-\sqrt{3}}{2} & \frac{-1}{6} & \frac{\sqrt{2}}{3} \\ \cdot & \frac{-\sqrt{8}}{3} & \frac{-1}{3} \end{vmatrix} \qquad \begin{vmatrix} \frac{-1}{2} & \frac{\sqrt{3}}{6} & \frac{-\sqrt{6}}{3} \\ \frac{\sqrt{3}}{6} & \frac{-5}{6} & \frac{-\sqrt{2}}{3} \\ \frac{-\sqrt{6}}{3} & \frac{-\sqrt{2}}{3} & \frac{1}{3} \end{vmatrix}$$

$r_1^2 = [123]$

$i_2 = [23]$

$r_2^2 = [142]$

$R_2^3 = [1342]$

$$\begin{vmatrix} \frac{-1}{2} & \frac{\sqrt{3}}{2} & \cdot \\ \frac{-\sqrt{3}}{2} & \frac{-1}{2} & \cdot \\ \cdot & \cdot & 1 \end{vmatrix} \qquad \begin{vmatrix} \frac{-1}{2} & \frac{\sqrt{3}}{2} & \cdot \\ \frac{\sqrt{3}}{2} & \frac{1}{2} & \cdot \\ \cdot & \cdot & -1 \end{vmatrix} \qquad \begin{vmatrix} \frac{-1}{2} & \frac{-\sqrt{3}}{6} & \frac{\sqrt{6}}{3} \\ \frac{\sqrt{3}}{6} & \frac{5}{6} & \frac{\sqrt{2}}{3} \\ \frac{-\sqrt{6}}{3} & \frac{\sqrt{2}}{3} & \frac{-1}{3} \end{vmatrix} \qquad \begin{vmatrix} \frac{1}{2} & \frac{\sqrt{3}}{6} & \frac{-\sqrt{6}}{3} \\ \frac{-\sqrt{3}}{2} & \frac{1}{6} & \frac{-\sqrt{2}}{3} \\ \cdot & \frac{\sqrt{8}}{3} & \frac{1}{3} \end{vmatrix}$$

$R_2^2 = [14][23]$

$R_3^3 = [1324]$

$R_3^2 = [12][34]$

$i_3 = [34]$

$$\begin{vmatrix} \cdot & \frac{-\sqrt{3}}{3} & \frac{-\sqrt{6}}{3} \\ \frac{-\sqrt{3}}{3} & \frac{-2}{3} & \frac{\sqrt{2}}{3} \\ \frac{-\sqrt{6}}{3} & \frac{\sqrt{2}}{3} & \frac{-1}{3} \end{vmatrix} \qquad \begin{vmatrix} \cdot & \frac{\sqrt{3}}{3} & \frac{\sqrt{6}}{3} \\ \frac{-\sqrt{3}}{3} & \frac{2}{3} & \frac{-\sqrt{2}}{3} \\ \frac{-\sqrt{6}}{3} & \frac{-\sqrt{2}}{3} & \frac{1}{3} \end{vmatrix} \qquad \begin{vmatrix} -1 & \cdot & \cdot \\ \cdot & \frac{1}{3} & \frac{-\sqrt{8}}{3} \\ \cdot & \frac{-\sqrt{8}}{3} & \frac{-1}{3} \end{vmatrix} \qquad \begin{vmatrix} -1 & \cdot & \cdot \\ \cdot & \frac{-1}{3} & \frac{\sqrt{8}}{3} \\ \cdot & \frac{\sqrt{8}}{3} & \frac{1}{3} \end{vmatrix}$$

$r_2 = [124]$

$R_1 = [1234]$

$r_3 = [143]$

$R_1^3 = [1432]$

$$\begin{vmatrix} \frac{-1}{2} & \frac{\sqrt{3}}{6} & \frac{-\sqrt{6}}{3} \\ \frac{-\sqrt{3}}{6} & \frac{5}{6} & \frac{\sqrt{2}}{3} \\ \frac{\sqrt{6}}{3} & \frac{\sqrt{2}}{3} & \frac{-1}{3} \end{vmatrix} \qquad \begin{vmatrix} \frac{1}{2} & \frac{-\sqrt{3}}{6} & \frac{\sqrt{6}}{3} \\ \frac{\sqrt{3}}{2} & \frac{1}{6} & \frac{-\sqrt{2}}{3} \\ \cdot & \frac{\sqrt{8}}{3} & \frac{1}{3} \end{vmatrix} \qquad \begin{vmatrix} \frac{1}{2} & \frac{\sqrt{3}}{2} & \cdot \\ \frac{\sqrt{3}}{6} & \frac{-1}{6} & \frac{-\sqrt{8}}{3} \\ \frac{-\sqrt{6}}{3} & \frac{\sqrt{2}}{3} & \frac{-1}{3} \end{vmatrix} \qquad \begin{vmatrix} \frac{1}{2} & \frac{\sqrt{3}}{2} & \cdot \\ \frac{-\sqrt{3}}{6} & \frac{1}{6} & \frac{\sqrt{8}}{3} \\ \frac{\sqrt{6}}{3} & \frac{-\sqrt{2}}{3} & \frac{1}{3} \end{vmatrix}$$

$r_3^2 = [134]$

$i_1 = [14]$

$r_4^2 = [243]$

$R_2 = [1243]$

$$\begin{vmatrix} \frac{1}{2} & \frac{\sqrt{3}}{6} & \frac{-\sqrt{6}}{3} \\ \frac{\sqrt{3}}{2} & \frac{-1}{6} & \frac{\sqrt{2}}{3} \\ \cdot & \frac{-\sqrt{8}}{3} & \frac{-1}{3} \end{vmatrix} \qquad \begin{vmatrix} \frac{-1}{2} & \frac{-\sqrt{3}}{6} & \frac{\sqrt{6}}{3} \\ \frac{-\sqrt{3}}{6} & \frac{-5}{6} & \frac{-\sqrt{2}}{3} \\ \frac{\sqrt{6}}{3} & \frac{-\sqrt{2}}{3} & \frac{1}{3} \end{vmatrix} \qquad \begin{vmatrix} \frac{1}{2} & \frac{-\sqrt{3}}{2} & \cdot \\ \frac{-\sqrt{3}}{6} & \frac{-1}{6} & \frac{-\sqrt{8}}{3} \\ \frac{\sqrt{6}}{3} & \frac{\sqrt{2}}{3} & \frac{-1}{3} \end{vmatrix} \qquad \begin{vmatrix} \frac{1}{2} & \frac{-\sqrt{3}}{2} & \cdot \\ \frac{\sqrt{3}}{6} & \frac{1}{6} & \frac{\sqrt{8}}{3} \\ \frac{-\sqrt{6}}{3} & \frac{-\sqrt{2}}{3} & \frac{1}{3} \end{vmatrix}$$

TABLE F.2.5 *(Continued)*

(b) Tensor Representation T_2 in Bases $\begin{array}{c} O: \\ D_3: \\ C_2: \end{array} \left. \begin{vmatrix} T_2 \\ A_1 \\ A \end{vmatrix} \right\rangle \left. \begin{vmatrix} T_2 \\ E \\ A \end{vmatrix} \right\rangle \left. \begin{vmatrix} T_2 \\ E \\ B \end{vmatrix} \right\rangle$ and E in Bases $\begin{array}{c} O: \\ D_3: \\ C_2: \end{array} \left. \begin{vmatrix} E \\ E \\ A \end{vmatrix} \right\rangle \left. \begin{vmatrix} E \\ E \\ B \end{vmatrix} \right\rangle$

$\mathscr{D}^{T_2}(1) =$

$$\begin{vmatrix} 1 & \cdot & \cdot \\ \cdot & 1 & \cdot \\ \cdot & \cdot & 1 \end{vmatrix}$$

$i_4 = [12]$

$$\begin{vmatrix} 1 & \cdot & \cdot \\ \cdot & 1 & \cdot \\ \cdot & \cdot & -1 \end{vmatrix}$$

$R_1^2 = [13][24]$

$$\begin{vmatrix} -\frac{1}{3} & -\frac{\sqrt{2}}{3} & \frac{\sqrt{6}}{3} \\ -\frac{\sqrt{2}}{3} & -\frac{2}{3} & -\frac{\sqrt{3}}{3} \\ \frac{\sqrt{6}}{3} & -\frac{\sqrt{3}}{3} & \cdot \end{vmatrix}$$

$R_3 = [1423]$

$$\begin{vmatrix} -\frac{1}{3} & -\frac{\sqrt{2}}{3} & -\frac{\sqrt{6}}{3} \\ -\frac{\sqrt{2}}{3} & -\frac{2}{3} & \frac{\sqrt{3}}{3} \\ \frac{\sqrt{6}}{3} & -\frac{\sqrt{3}}{3} & \cdot \end{vmatrix}$$

$r_1 = [132]$

$$\begin{vmatrix} 1 & \cdot & \cdot \\ \cdot & -\frac{1}{2} & -\frac{\sqrt{3}}{2} \\ \cdot & \frac{\sqrt{3}}{2} & -\frac{1}{2} \end{vmatrix}$$

$i_5 = [13]$

$$\begin{vmatrix} 1 & \cdot & \cdot \\ \cdot & -\frac{1}{2} & -\frac{\sqrt{3}}{2} \\ \cdot & -\frac{\sqrt{3}}{2} & \frac{1}{2} \end{vmatrix}$$

$r_4 = [234]$

$$\begin{vmatrix} -\frac{1}{3} & \frac{\sqrt{8}}{3} & \cdot \\ -\frac{\sqrt{2}}{3} & -\frac{1}{6} & \frac{\sqrt{3}}{2} \\ \frac{\sqrt{6}}{3} & \frac{\sqrt{3}}{6} & \frac{1}{2} \end{vmatrix}$$

$i_6 = [24]$

$$\begin{vmatrix} -\frac{1}{3} & -\frac{\sqrt{2}}{3} & \frac{\sqrt{6}}{3} \\ -\frac{\sqrt{2}}{3} & \frac{5}{6} & \frac{\sqrt{3}}{6} \\ \frac{\sqrt{6}}{3} & \frac{\sqrt{3}}{6} & \frac{1}{2} \end{vmatrix}$$

$r_1^2 = [123]$

$$\begin{vmatrix} 1 & \cdot & \cdot \\ \cdot & -\frac{1}{2} & \frac{\sqrt{3}}{2} \\ \cdot & -\frac{\sqrt{3}}{2} & -\frac{1}{2} \end{vmatrix}$$

$i_2 = [23]$

$$\begin{vmatrix} 1 & \cdot & \cdot \\ \cdot & -\frac{1}{2} & \frac{\sqrt{3}}{2} \\ \cdot & \frac{\sqrt{3}}{2} & \frac{1}{2} \end{vmatrix}$$

$r_2^2 = [142]$

$$\begin{vmatrix} -\frac{1}{3} & -\frac{\sqrt{2}}{3} & -\frac{\sqrt{6}}{3} \\ -\frac{\sqrt{2}}{3} & \frac{5}{6} & -\frac{\sqrt{3}}{6} \\ \frac{\sqrt{6}}{3} & \frac{\sqrt{3}}{6} & -\frac{1}{2} \end{vmatrix}$$

$R_2^3 = [1342]$

$$\begin{vmatrix} -\frac{1}{3} & \frac{\sqrt{8}}{3} & \cdot \\ -\frac{\sqrt{2}}{3} & -\frac{1}{6} & -\frac{\sqrt{3}}{2} \\ \frac{\sqrt{6}}{3} & \frac{\sqrt{3}}{6} & -\frac{1}{2} \end{vmatrix}$$

$R_2^2 = [14][23]$

$$\begin{vmatrix} -\frac{1}{3} & -\frac{\sqrt{2}}{3} & -\frac{\sqrt{6}}{3} \\ -\frac{\sqrt{2}}{3} & -\frac{2}{3} & \frac{\sqrt{3}}{3} \\ -\frac{\sqrt{6}}{3} & \frac{\sqrt{3}}{3} & \cdot \end{vmatrix}$$

$R_3^3 = [1324]$

$$\begin{vmatrix} -\frac{1}{3} & -\frac{\sqrt{2}}{3} & \frac{\sqrt{6}}{3} \\ -\frac{\sqrt{2}}{3} & -\frac{2}{3} & \frac{\sqrt{3}}{3} \\ -\frac{\sqrt{6}}{3} & \frac{\sqrt{3}}{3} & \cdot \end{vmatrix}$$

$R_3^2 = [12][34]$

$$\begin{vmatrix} -\frac{1}{3} & \frac{\sqrt{8}}{3} & \cdot \\ \frac{\sqrt{8}}{3} & \frac{1}{3} & \cdot \\ \cdot & \cdot & -1 \end{vmatrix}$$

$i_3 = [34]$

$$\begin{vmatrix} -\frac{1}{3} & \frac{\sqrt{8}}{3} & \cdot \\ \frac{\sqrt{8}}{3} & \frac{1}{3} & \cdot \\ \cdot & \cdot & 1 \end{vmatrix}$$

$r_2 = [124]$

$$\begin{vmatrix} -\frac{1}{3} & -\frac{\sqrt{2}}{3} & \frac{\sqrt{6}}{3} \\ -\frac{\sqrt{2}}{3} & \frac{5}{6} & \frac{\sqrt{3}}{6} \\ -\frac{\sqrt{6}}{3} & -\frac{\sqrt{3}}{6} & -\frac{1}{2} \end{vmatrix}$$

$R_1 = [1234]$

$$\begin{vmatrix} -\frac{1}{3} & \frac{\sqrt{8}}{3} & \cdot \\ -\frac{\sqrt{2}}{3} & -\frac{1}{6} & \frac{\sqrt{3}}{2} \\ -\frac{\sqrt{6}}{3} & -\frac{\sqrt{3}}{6} & -\frac{1}{2} \end{vmatrix}$$

$r_3 = [143]$

$$\begin{vmatrix} -\frac{1}{3} & -\frac{\sqrt{2}}{3} & -\frac{\sqrt{6}}{3} \\ \frac{\sqrt{8}}{3} & -\frac{1}{6} & -\frac{\sqrt{3}}{6} \\ \cdot & -\frac{\sqrt{3}}{2} & \frac{1}{2} \end{vmatrix}$$

$R_1^3 = [1432]$

$$\begin{vmatrix} -\frac{1}{3} & -\frac{\sqrt{2}}{3} & -\frac{\sqrt{6}}{3} \\ \frac{\sqrt{8}}{3} & -\frac{1}{6} & -\frac{\sqrt{3}}{6} \\ \cdot & \frac{\sqrt{3}}{2} & -\frac{1}{2} \end{vmatrix}$$

$r_3^2 = [134]$

$$\begin{vmatrix} -\frac{1}{3} & \frac{\sqrt{8}}{3} & \cdot \\ -\frac{\sqrt{2}}{3} & -\frac{1}{6} & -\frac{\sqrt{3}}{2} \\ -\frac{\sqrt{6}}{3} & -\frac{\sqrt{3}}{6} & \frac{1}{2} \end{vmatrix}$$

$i_1 = [14]$

$$\begin{vmatrix} -\frac{1}{3} & -\frac{\sqrt{2}}{3} & -\frac{\sqrt{6}}{3} \\ -\frac{\sqrt{2}}{3} & \frac{5}{6} & -\frac{\sqrt{3}}{6} \\ -\frac{\sqrt{6}}{3} & -\frac{\sqrt{3}}{6} & \frac{1}{2} \end{vmatrix}$$

$r_4^2 = [243]$

$$\begin{vmatrix} -\frac{1}{3} & -\frac{\sqrt{2}}{3} & \frac{\sqrt{6}}{3} \\ \frac{\sqrt{8}}{3} & -\frac{1}{6} & \frac{\sqrt{3}}{6} \\ \cdot & \frac{\sqrt{3}}{2} & \frac{1}{2} \end{vmatrix}$$

$R_2 = [1243]$

$$\begin{vmatrix} -\frac{1}{3} & -\frac{\sqrt{2}}{3} & \frac{\sqrt{6}}{3} \\ \frac{\sqrt{8}}{3} & -\frac{1}{6} & \frac{\sqrt{3}}{6} \\ \cdot & -\frac{\sqrt{3}}{2} & -\frac{1}{2} \end{vmatrix}$$

TABLE F.2.5 *(Continued)*

$\mathscr{D}^E(1) = \begin{vmatrix} 1 & 0 \\ 0 & 1 \end{vmatrix}$

$i_4 = [12]$
$\begin{vmatrix} 1 & 0 \\ 0 & -1 \end{vmatrix}$

$R_1^2 = [13][24]$
$\begin{vmatrix} 1 & 0 \\ 0 & 1 \end{vmatrix}$

$R_3 = [1423]$
$\begin{vmatrix} 1 & 0 \\ 0 & -1 \end{vmatrix}$

$r_1 = [132]$
$\begin{vmatrix} -\frac{1}{2} & -\frac{\sqrt{3}}{2} \\ \frac{\sqrt{3}}{2} & \frac{1}{2} \end{vmatrix}$

$i_5 = [13]$
$\begin{vmatrix} -\frac{1}{2} & -\frac{\sqrt{3}}{2} \\ -\frac{\sqrt{3}}{2} & \frac{1}{2} \end{vmatrix}$

$r_4 = [234]$
$\begin{vmatrix} -\frac{1}{2} & -\frac{\sqrt{3}}{2} \\ \frac{\sqrt{3}}{2} & -\frac{1}{2} \end{vmatrix}$

$i_6 = [24]$
$\begin{vmatrix} -\frac{1}{2} & -\frac{\sqrt{3}}{2} \\ -\frac{\sqrt{3}}{2} & \frac{1}{2} \end{vmatrix}$

$r_1^2 = [123]$
$\begin{vmatrix} -\frac{1}{2} & \frac{\sqrt{3}}{2} \\ -\frac{\sqrt{3}}{2} & -\frac{1}{2} \end{vmatrix}$

$i_2 = [23]$
$\begin{vmatrix} -\frac{1}{2} & \frac{\sqrt{3}}{2} \\ \frac{\sqrt{3}}{2} & \frac{1}{2} \end{vmatrix}$

$r_2^2 = [142]$
$\begin{vmatrix} -\frac{1}{2} & \frac{\sqrt{3}}{2} \\ -\frac{\sqrt{3}}{2} & -\frac{1}{2} \end{vmatrix}$

$R_2^3 = [1342]$
$\begin{vmatrix} -\frac{1}{2} & \frac{\sqrt{3}}{2} \\ \frac{\sqrt{3}}{2} & \frac{1}{2} \end{vmatrix}$

$R_2^2 = [14][23]$
$\begin{vmatrix} 1 & 0 \\ 0 & 1 \end{vmatrix}$

$R_3^3 = [1324]$
$\begin{vmatrix} 1 & 0 \\ 0 & -1 \end{vmatrix}$

$R_3^2 = [12][34]$
$\begin{vmatrix} 1 & 0 \\ 0 & 1 \end{vmatrix}$

$i_3 = [34]$
$\begin{vmatrix} 1 & 0 \\ 0 & -1 \end{vmatrix}$

$r_2 = [124]$
$\begin{vmatrix} -\frac{1}{2} & -\frac{\sqrt{3}}{2} \\ \frac{\sqrt{3}}{2} & -\frac{1}{2} \end{vmatrix}$

$R_1 = [1234]$
$\begin{vmatrix} -\frac{1}{2} & -\frac{\sqrt{3}}{2} \\ -\frac{\sqrt{3}}{2} & \frac{1}{2} \end{vmatrix}$

$r_3 = [143]$
$\begin{vmatrix} -\frac{1}{2} & -\frac{\sqrt{3}}{2} \\ \frac{\sqrt{3}}{2} & -\frac{1}{2} \end{vmatrix}$

$R_1^3 = [1432]$
$\begin{vmatrix} -\frac{1}{2} & -\frac{\sqrt{3}}{2} \\ -\frac{\sqrt{3}}{2} & \frac{1}{2} \end{vmatrix}$

$r_3^2 = [134]$
$\begin{vmatrix} -\frac{1}{2} & \frac{\sqrt{3}}{2} \\ -\frac{\sqrt{3}}{2} & -\frac{1}{2} \end{vmatrix}$

$i_1 = [14]$
$\begin{vmatrix} -\frac{1}{2} & \frac{\sqrt{3}}{2} \\ \frac{\sqrt{3}}{2} & \frac{1}{2} \end{vmatrix}$

$r_4^2 = [243]$
$\begin{vmatrix} -\frac{1}{2} & \frac{\sqrt{3}}{2} \\ -\frac{\sqrt{3}}{2} & -\frac{1}{2} \end{vmatrix}$

$R_2 = [1243]$
$\begin{vmatrix} -\frac{1}{2} & \frac{\sqrt{3}}{2} \\ \frac{\sqrt{3}}{2} & \frac{1}{2} \end{vmatrix}$

For scalar (A_1) $\begin{vmatrix} A_1 \\ A_1 \\ A \end{vmatrix}$ and pseudoscalar (A_2) $\begin{vmatrix} A_2 \\ A_2 \\ B \end{vmatrix}$ representations, see the O character table.

TABLE F.2.6 $O \supset D_4 \supset C_4$ Subgroup Chain Labeled Irreducible Representations (Fourfold Moving-Wave Bases)

(a) Vector T_1 Representation

$$\mathcal{D}^{T_1}(1) = \begin{pmatrix} 1 & \cdot & \cdot & \cdot \\ \cdot & 1 & \cdot & \cdot \\ \cdot & \cdot & 1 & \cdot \\ \cdot & \cdot & \cdot & 1 \end{pmatrix} \quad R_1^2 = \begin{pmatrix} \cdot & \cdot & -1 & \cdot \\ \cdot & -1 & \cdot & \cdot \\ \cdot & \cdot & \cdot & -1 \end{pmatrix} \quad r_1 = \begin{pmatrix} \frac{-i}{2} & \frac{i}{2} & \frac{-1}{\sqrt{2}} & \cdot \\ \frac{-i}{2} & \frac{-i}{2} & \frac{1}{\sqrt{2}} & \cdot \\ \frac{-i}{\sqrt{2}} & \frac{-i}{\sqrt{2}} & \cdot & \cdot \end{pmatrix} \quad r_2 = \begin{pmatrix} \frac{-i}{2} & \frac{-1}{\sqrt{2}} & \frac{i}{2} & \frac{1}{\sqrt{2}} \\ \frac{-i}{2} & \frac{-1}{\sqrt{2}} & \frac{-i}{2} & \frac{-1}{\sqrt{2}} \\ \frac{i}{\sqrt{2}} & \cdot & \frac{i}{\sqrt{2}} & \cdot \end{pmatrix} \quad r_1^2 = \begin{pmatrix} \frac{i}{2} & \frac{-i}{2} & \frac{i}{\sqrt{2}} & \cdot \\ \frac{i}{2} & \frac{-i}{2} & \frac{-i}{\sqrt{2}} & \cdot \\ \frac{-1}{\sqrt{2}} & \frac{-1}{\sqrt{2}} & \cdot & \cdot \end{pmatrix} \quad r_2^2 = \begin{pmatrix} \frac{i}{2} & \frac{-i}{2} & \frac{-i}{\sqrt{2}} & \cdot \\ \frac{i}{2} & \frac{-i}{2} & \frac{-i}{\sqrt{2}} & \cdot \\ \frac{-1}{\sqrt{2}} & \frac{-1}{\sqrt{2}} & \cdot & \cdot \end{pmatrix}$$

$$\mathcal{D}^{T_1}(R_3^2) = \begin{pmatrix} -1 & \cdot & \cdot & \cdot \\ \cdot & -1 & \cdot & \cdot \\ \cdot & \cdot & 1 & \cdot \\ \cdot & \cdot & \cdot & -1 \end{pmatrix} \quad R_2^2 = \begin{pmatrix} \cdot & \cdot & 1 & \cdot \\ \cdot & 1 & \cdot & \cdot \\ \cdot & \cdot & \cdot & -1 \end{pmatrix} \quad r_4 = \begin{pmatrix} \frac{i}{2} & \frac{-i}{2} & \frac{-1}{\sqrt{2}} & \cdot \\ \frac{i}{2} & \frac{-i}{2} & \frac{1}{\sqrt{2}} & \cdot \\ \frac{i}{\sqrt{2}} & \frac{i}{\sqrt{2}} & \cdot & \cdot \end{pmatrix} \quad r_3 = \begin{pmatrix} \frac{-i}{2} & \frac{-i}{2} & \frac{1}{\sqrt{2}} & \cdot \\ \frac{-i}{2} & \frac{-i}{2} & \frac{-1}{\sqrt{2}} & \cdot \\ \frac{i}{\sqrt{2}} & \frac{-i}{\sqrt{2}} & \cdot & \cdot \end{pmatrix} \quad r_3^2 = \begin{pmatrix} \frac{-i}{2} & \frac{i}{2} & \frac{-i}{\sqrt{2}} & \cdot \\ \frac{-i}{2} & \frac{i}{2} & \frac{-i}{\sqrt{2}} & \cdot \\ \frac{1}{\sqrt{2}} & \frac{1}{\sqrt{2}} & \cdot & \cdot \end{pmatrix} \quad r_4^2 = \begin{pmatrix} \frac{-i}{2} & \frac{i}{2} & \frac{i}{\sqrt{2}} & \cdot \\ \frac{-i}{2} & \frac{i}{2} & \frac{-i}{\sqrt{2}} & \cdot \\ \frac{1}{\sqrt{2}} & \frac{-1}{\sqrt{2}} & \cdot & \cdot \end{pmatrix}$$

$$\mathcal{D}^{T_1}(R_3) = \begin{pmatrix} -i & \cdot & \cdot & \cdot \\ \cdot & i & \cdot & \cdot \\ \cdot & \cdot & -i & \cdot \\ \cdot & \cdot & \cdot & -1 \end{pmatrix} \quad i_4 = \begin{pmatrix} \cdot & \cdot & -i & \cdot \\ \cdot & i & \cdot & \cdot \\ \cdot & \cdot & \cdot & -1 \end{pmatrix} \quad i_1 = \begin{pmatrix} \frac{-1}{2} & \frac{-1}{2} & \frac{-1}{\sqrt{2}} & \cdot \\ \frac{-1}{2} & \frac{-1}{2} & \frac{1}{\sqrt{2}} & \cdot \\ \frac{-1}{\sqrt{2}} & \frac{1}{\sqrt{2}} & \cdot & \cdot \end{pmatrix} \quad i_2 = \begin{pmatrix} \frac{-1}{2} & \frac{-1}{2} & \frac{1}{\sqrt{2}} & \cdot \\ \frac{-1}{2} & \frac{-1}{2} & \frac{-1}{\sqrt{2}} & \cdot \\ \frac{i}{\sqrt{2}} & \frac{-i}{\sqrt{2}} & \cdot & \cdot \end{pmatrix} \quad R_1^3 = \begin{pmatrix} \frac{i}{2} & \frac{-1}{2} & \frac{-1}{2} & \frac{i}{\sqrt{2}} \\ \frac{i}{2} & \frac{-1}{2} & \frac{-1}{2} & \frac{-i}{\sqrt{2}} \\ \frac{i}{\sqrt{2}} & \frac{-1}{\sqrt{2}} & \cdot & \cdot \end{pmatrix} \quad R_1 = \begin{pmatrix} \frac{1}{2} & \frac{-1}{2} & \frac{-i}{\sqrt{2}} & \cdot \\ \frac{1}{2} & \frac{-1}{2} & \frac{-i}{\sqrt{2}} & \cdot \\ \frac{-i}{\sqrt{2}} & \frac{-i}{\sqrt{2}} & \cdot & \cdot \end{pmatrix}$$

TABLE F.2.6 (Continued)

$$\mathscr{D}^{T_1}(R_3^3) = \quad i_3 = \quad R_2 = \quad R_2^3 = \quad i_6 = \quad i_5 = $$

$$\begin{array}{c} O: |T_1\rangle \\ D_4: |T_1\rangle |T_1\rangle \\ C_4: |E\rangle |E\rangle |A_2\rangle \\ \text{basis} \quad 1_4 \quad 3_4 \quad 0_4 \end{array}$$

(b) Tensor T_2 Representation

$$\mathscr{D}^{T_2}(1) = \quad R_1^2 = \quad r_1 = \quad r_2 = \quad r_1^2 = \quad r_2^2 = $$

$$\mathscr{D}^{T_2}(R_3^2) = \quad R_2^2 = \quad r_4 = \quad r_3 = \quad r_3^2 = \quad r_4^2 = $$

$$\mathscr{D}^{T_2}(R_3) = i_4 = \begin{vmatrix} -i & \cdot & \cdot & \cdot \\ \cdot & i & \cdot & \cdot \\ \cdot & \cdot & -i & \cdot \\ \cdot & \cdot & \cdot & 1 \end{vmatrix}$$

$$i_1 = \begin{array}{|cc|cc|c|} \hline -\frac{1}{2} & \frac{1}{2} & -\frac{1}{2} & \frac{1}{2} & \frac{1}{\sqrt{2}} \\ -\frac{1}{2} & -\frac{1}{2} & -\frac{1}{2} & -\frac{1}{2} & \frac{1}{\sqrt{2}} \\ \hline \frac{1}{\sqrt{2}} & -\frac{1}{\sqrt{2}} & -\frac{1}{\sqrt{2}} & -\frac{1}{\sqrt{2}} & \cdot \\ \end{array}$$

$$i_2 = \begin{array}{|cc|cc|c|} \hline \frac{1}{2} & -\frac{1}{2} & -\frac{1}{2} & \frac{1}{2} & \frac{1}{\sqrt{2}} \\ \frac{1}{2} & \frac{1}{2} & \frac{1}{2} & -\frac{1}{2} & -\frac{1}{\sqrt{2}} \\ \hline \frac{1}{\sqrt{2}} & -\frac{1}{\sqrt{2}} & \frac{i}{\sqrt{2}} & \frac{i}{\sqrt{2}} & \cdot \\ \end{array}$$

$$R_1^3 = \begin{array}{|cc|cc|c|} \hline -\frac{1}{2} & -\frac{1}{2} & -\frac{1}{2} & \frac{1}{2} & \frac{-i}{\sqrt{2}} \\ -\frac{1}{2} & -\frac{1}{2} & \frac{1}{2} & -\frac{1}{2} & \frac{-i}{\sqrt{2}} \\ \hline \frac{-i}{\sqrt{2}} & \frac{i}{\sqrt{2}} & \frac{i}{\sqrt{2}} & \frac{i}{\sqrt{2}} & \cdot \\ \end{array}$$

$$R_1 = \begin{array}{|cc|cc|c|} \hline \frac{-1}{2} & \frac{1}{2} & \frac{1}{2} & \frac{1}{2} & \frac{-i}{\sqrt{2}} \\ \frac{-1}{2} & \frac{1}{2} & -\frac{1}{2} & -\frac{1}{2} & \frac{-i}{\sqrt{2}} \\ \hline \frac{-i}{\sqrt{2}} & \frac{-i}{\sqrt{2}} & \frac{-i}{\sqrt{2}} & \frac{-i}{\sqrt{2}} & \cdot \\ \end{array}$$

$$\mathscr{D}^{T_2}(R_3^3) = i_3 = \begin{vmatrix} i & \cdot & \cdot & \cdot \\ \cdot & -i & \cdot & \cdot \\ \cdot & \cdot & i & \cdot \\ \cdot & \cdot & \cdot & 1 \end{vmatrix}$$

$$R_2 = \begin{array}{|cc|cc|c|} \hline -\frac{1}{2} & -\frac{1}{2} & -\frac{1}{2} & \frac{1}{2} & \frac{1}{\sqrt{2}} \\ -\frac{1}{2} & -\frac{1}{2} & \frac{1}{2} & -\frac{1}{2} & -\frac{1}{\sqrt{2}} \\ \hline \frac{-1}{\sqrt{2}} & \frac{1}{\sqrt{2}} & -\frac{1}{\sqrt{2}} & -\frac{1}{\sqrt{2}} & \cdot \\ \end{array}$$

$$R_2^3 = \begin{array}{|cc|cc|c|} \hline -\frac{1}{2} & -\frac{1}{2} & -\frac{1}{2} & \frac{1}{2} & \frac{1}{\sqrt{2}} \\ -\frac{1}{2} & -\frac{1}{2} & \frac{1}{2} & -\frac{1}{2} & -\frac{1}{\sqrt{2}} \\ \hline \frac{1}{\sqrt{2}} & -\frac{1}{\sqrt{2}} & \frac{1}{\sqrt{2}} & \frac{1}{\sqrt{2}} & \cdot \\ \end{array}$$

$$i_6 = \begin{array}{|cc|cc|c|} \hline -\frac{1}{2} & -\frac{1}{2} & -\frac{1}{2} & \frac{1}{2} & \frac{i}{\sqrt{2}} \\ -\frac{1}{2} & -\frac{1}{2} & \frac{1}{2} & -\frac{1}{2} & \frac{i}{\sqrt{2}} \\ \hline \frac{-1}{\sqrt{2}} & \frac{1}{\sqrt{2}} & -\frac{i}{\sqrt{2}} & -\frac{i}{\sqrt{2}} & \cdot \\ \end{array}$$

$$i_5 = \begin{array}{|cc|cc|c|} \hline -\frac{1}{2} & -\frac{1}{2} & -\frac{1}{2} & \frac{1}{2} & \frac{-i}{\sqrt{2}} \\ -\frac{1}{2} & -\frac{1}{2} & \frac{1}{2} & -\frac{1}{2} & \frac{-i}{\sqrt{2}} \\ \hline \frac{i}{\sqrt{2}} & \frac{i}{\sqrt{2}} & \frac{i}{\sqrt{2}} & \frac{i}{\sqrt{2}} & \cdot \\ \end{array}$$

$$\text{basis} \begin{array}{l} O: \\ D_4: \\ C_4: \end{array} \left| \begin{array}{c} T_2 \\ E \\ 1_4 \end{array} \right\rangle \left| \begin{array}{c} T_2 \\ E \\ 3_4 \end{array} \right\rangle \left| \begin{array}{c} T_2 \\ B_2 \\ 2_4 \end{array} \right\rangle$$

The tensor E representation is identical to that of $D_4 \supset D_2$ standing-wave basis.

TABLE F.2.7 $O \supset D_3 \supset C_3$ Subgroup Chain Labeled Irreducible Representations (Threefold Moving-Wave Bases)

(a) Vector T_1 Representation

$$\mathscr{D}^{T_1}(1) = \begin{pmatrix} 1 & \cdot & \cdot \\ \cdot & 1 & \cdot \\ \cdot & \cdot & 1 \end{pmatrix} \qquad i_4 = [12] = \begin{pmatrix} \cdot & -1 & \cdot \\ -1 & \cdot & \cdot \\ \cdot & \cdot & -1 \end{pmatrix} \qquad R_1^2 = [13][24] = \begin{pmatrix} -1 & -1+\frac{\sqrt{3}}{3} & -i-\frac{\sqrt{3}}{3} \\ \frac{-1}{3} & \frac{-1}{3} & \frac{i}{3}+\frac{\sqrt{3}}{3} \\ \frac{-i}{3}-\frac{\sqrt{3}}{3} & \frac{\sqrt{3}}{3} & \frac{-1}{3} \end{pmatrix} \qquad R_3^2 = [1423] = \begin{pmatrix} \frac{-1}{3}-i\frac{\sqrt{3}}{3} & \frac{1}{3} & \frac{-i}{3}+\frac{\sqrt{3}}{3} \\ \frac{1}{3} & \frac{-1}{3}+i\frac{\sqrt{3}}{3} & \frac{-i}{3}-\frac{\sqrt{3}}{3} \\ \frac{\sqrt{3}}{3} & \frac{i}{3}+\frac{\sqrt{3}}{3} & \frac{-1}{3} \end{pmatrix}$$

$$r_1 = [132] = \begin{pmatrix} \frac{-1}{2}-i\frac{\sqrt{3}}{2} & \cdot & \cdot \\ \cdot & \frac{-1}{2}+i\frac{\sqrt{3}}{2} & \cdot \\ \cdot & \cdot & 1 \end{pmatrix} \qquad i_5 = [13] = \begin{pmatrix} \cdot & \frac{1}{2}-i\frac{\sqrt{3}}{2} & \cdot \\ \frac{1}{2}+i\frac{\sqrt{3}}{2} & \cdot & \cdot \\ \cdot & \cdot & -1 \end{pmatrix} \qquad r_4 = [234] = \begin{pmatrix} \frac{-2}{3} & \frac{1}{6}-i\frac{\sqrt{3}}{6} & \frac{-1}{6}-i\frac{\sqrt{3}}{6} \\ \frac{1}{6}+i\frac{\sqrt{3}}{6} & \frac{2}{3} & \frac{-1}{6}+i\frac{\sqrt{3}}{6} \\ \frac{-1}{6}+i\frac{\sqrt{3}}{6} & \frac{-1}{6}-i\frac{\sqrt{3}}{6} & \frac{2i}{3} \end{pmatrix} \qquad i_6 = [24] = \begin{pmatrix} \frac{-2}{3} & \frac{-1}{6}-i\frac{\sqrt{3}}{6} & \frac{-i}{3}+\frac{\sqrt{3}}{3} \\ \frac{-1}{6}-i\frac{\sqrt{3}}{6} & \frac{-2}{3} & \frac{i}{3}-\frac{\sqrt{3}}{3} \\ \frac{i}{3}-\frac{\sqrt{3}}{3} & \frac{i}{3}+\frac{\sqrt{3}}{3} & \frac{-1}{3} \end{pmatrix}$$

$$r_1^2 = [123] = \begin{pmatrix} \frac{-1}{2}+i\frac{\sqrt{3}}{2} & \cdot & \cdot \\ \cdot & \frac{-1}{2}-i\frac{\sqrt{3}}{2} & \cdot \\ \cdot & \cdot & 1 \end{pmatrix} \qquad i_2 = [23] = \begin{pmatrix} \cdot & \frac{1}{2}+i\frac{\sqrt{3}}{2} & \cdot \\ \frac{1}{2}-i\frac{\sqrt{3}}{2} & \cdot & \cdot \\ \cdot & \cdot & -1 \end{pmatrix} \qquad r_2^2 = [142] = \begin{pmatrix} \frac{1}{6}-i\frac{\sqrt{3}}{6} & \frac{2}{3} & \frac{-i}{3}-\frac{\sqrt{3}}{3} \\ \frac{2}{3} & \frac{1}{6}+i\frac{\sqrt{3}}{6} & \frac{i}{3}+\frac{\sqrt{3}}{3} \\ \frac{-i}{3}+\frac{\sqrt{3}}{3} & \frac{i}{3}-\frac{\sqrt{3}}{3} & \frac{-1}{3} \end{pmatrix} \qquad R_2^3 = [1342] = \begin{pmatrix} \frac{-1}{3}+i\frac{\sqrt{3}}{3} & \frac{1}{3} & \frac{-i}{3}-\frac{\sqrt{3}}{3} \\ \frac{1}{3} & \frac{-1}{3}-i\frac{\sqrt{3}}{3} & \frac{-i}{3}+\frac{\sqrt{3}}{3} \\ \frac{-i}{3}+\frac{\sqrt{3}}{3} & \frac{i}{3}-\frac{\sqrt{3}}{3} & \frac{-1}{3} \end{pmatrix}$$

$$R_2^2 = [14][23] = \begin{pmatrix} \frac{-1}{3} & \frac{-1}{3}+i\frac{\sqrt{3}}{3} & \frac{i}{3}+\frac{\sqrt{3}}{3} \\ \frac{-1}{3}-i\frac{\sqrt{3}}{3} & \frac{-1}{3} & \frac{-i}{3}-\frac{\sqrt{3}}{3} \\ \frac{-i}{3}+\frac{\sqrt{3}}{3} & \frac{i}{3}-\frac{\sqrt{3}}{3} & \frac{-1}{3} \end{pmatrix} \qquad R_1^3 = [1324] = \begin{pmatrix} \frac{1}{3}+i\frac{\sqrt{3}}{3} & \frac{\sqrt{3}}{3} & \frac{-i}{3}+\frac{\sqrt{3}}{3} \\ \frac{1}{3} & \frac{1}{3}+i\frac{\sqrt{3}}{3} & \frac{i}{3}+\frac{\sqrt{3}}{3} \\ \frac{-i}{3}-\frac{\sqrt{3}}{3} & \frac{-i}{3}+\frac{\sqrt{3}}{3} & \frac{-1}{3} \end{pmatrix} \qquad R_3^2 = [12][34] = \begin{pmatrix} \frac{-1}{3} & \frac{2}{3} & \frac{-2i}{3} \\ \frac{2}{3} & \frac{-1}{3} & \frac{2i}{3} \\ \frac{2i}{3} & \frac{-2i}{3} & \frac{-1}{3} \end{pmatrix} \qquad i_3 = [34] = \begin{pmatrix} \frac{-2}{3} & \frac{1}{3} & \frac{2i}{3} \\ \frac{1}{3} & \frac{-2}{3} & \frac{-2i}{3} \\ \frac{-2i}{3} & \frac{2i}{3} & \frac{-1}{3} \end{pmatrix}$$

$r_2 = [124]$

	$|x_1\rangle$	$|x_2\rangle$	$x_3\rangle$
$\langle 1_3|$	$\frac{1}{6} + i\frac{\sqrt{3}}{6}$	$\frac{2}{3}$	$\frac{i}{3} + \frac{\sqrt{3}}{3}$
$\langle 2_3|$	$\frac{2}{3}$	$\frac{1}{6} - i\frac{\sqrt{3}}{6}$	$\frac{-i}{3} - \frac{\sqrt{3}}{3}$
$\langle 0_3|$	$-\frac{i}{3} - \frac{\sqrt{3}}{3}$	$\frac{i}{3} - \frac{\sqrt{3}}{3}$	$-\frac{1}{3}$

$R_1 = [1234]$

$\frac{1}{3} - i\frac{\sqrt{3}}{3}\quad \frac{-1}{6}+i\frac{\sqrt{3}}{6}\quad \frac{i}{3}+\frac{\sqrt{3}}{3}\quad -\frac{2}{3}$

$\frac{-1}{6}-i\frac{\sqrt{3}}{6}\quad \frac{1}{6}+i\frac{\sqrt{3}}{6}\quad \frac{i}{3}-\frac{\sqrt{3}}{3}\quad \frac{\sqrt{3}}{3}$

$\frac{-2i}{3}\quad \frac{-2i}{3}\quad -\frac{1}{3}\quad \frac{\sqrt{3}}{3}$

$r_3 = [143]$

$\frac{1}{3}-i\frac{\sqrt{3}}{3}\quad \frac{-1}{6}+i\frac{\sqrt{3}}{6}\quad \frac{-1}{6}+i\frac{\sqrt{3}}{6}\quad \frac{2i}{3}$

$\frac{-1}{6}-i\frac{\sqrt{3}}{6}\quad \frac{1}{3}\quad \frac{-1}{3}\quad \frac{\sqrt{3}}{3}\quad \frac{2i}{3}$

$\frac{-i}{3}-\frac{\sqrt{3}}{3}\quad \frac{-i}{3}+\frac{\sqrt{3}}{3}\quad \frac{1}{3}\quad -\frac{1}{3}$

$R_1^3 = [1432]$

$\frac{1}{3}+i\frac{\sqrt{3}}{3}\quad \frac{-1}{6}+i\frac{\sqrt{3}}{6}\quad \frac{-1}{6}+i\frac{\sqrt{3}}{6}\quad \frac{2i}{3}$

$\frac{-1}{6}-i\frac{\sqrt{3}}{6}\quad \frac{1}{3}\quad -\frac{\sqrt{3}}{3}\quad \frac{2i}{3}$

$\frac{i}{3}-\frac{\sqrt{3}}{3}\quad \frac{\sqrt{3}}{3}\quad \frac{1}{3}\quad -\frac{1}{3}$

$r_3^2 = [134]$

$\frac{1}{6}-i\frac{\sqrt{3}}{6}\quad \frac{-1}{3}+i\frac{\sqrt{3}}{3}\quad \frac{i}{3}-\frac{\sqrt{3}}{3}\quad \frac{\sqrt{3}}{3}$

$\frac{-1}{3}\quad \frac{-1}{6}+i\frac{\sqrt{3}}{6}\quad \frac{i}{3}+\frac{\sqrt{3}}{3}\quad -\frac{\sqrt{3}}{3}$

$\frac{2i}{3}\quad \frac{2i}{3}\quad -\frac{1}{3}\quad \frac{\sqrt{3}}{3}$

$i_1 = [14]$

$\frac{-1}{3}\quad \frac{-1}{6}+i\frac{\sqrt{3}}{6}\quad \frac{i}{3}+\frac{\sqrt{3}}{3}\quad -\frac{\sqrt{3}}{3}$

$\frac{-1}{6}-i\frac{\sqrt{3}}{6}\quad \frac{1}{6}+i\frac{\sqrt{3}}{6}\quad \frac{i}{3}-\frac{\sqrt{3}}{3}\quad \frac{\sqrt{3}}{3}$

$\frac{i}{3}-\frac{\sqrt{3}}{3}\quad -\frac{i}{3}+\frac{\sqrt{3}}{3}\quad -\frac{1}{3}\quad -\frac{1}{3}$

$r_4^2 = [243]$

$\frac{1}{6}-i\frac{\sqrt{3}}{6}\quad \frac{-1}{3}-i\frac{\sqrt{3}}{3}\quad \frac{-i}{3}-\frac{\sqrt{3}}{3}\quad -\frac{2i}{3}$

$\frac{-1}{6}+i\frac{\sqrt{3}}{6}\quad \frac{1}{6}-i\frac{\sqrt{3}}{6}\quad \frac{-i}{3}+\frac{\sqrt{3}}{3}\quad -\frac{2i}{3}$

$\frac{i}{3}+\frac{\sqrt{3}}{3}\quad \frac{\sqrt{3}}{3}\quad \frac{1}{3}\quad -\frac{1}{3}$

$R_2 = [1243]$

$\frac{1}{3}+i\frac{\sqrt{3}}{3}\quad \frac{-1}{6}-i\frac{\sqrt{3}}{6}\quad \frac{-i}{3}+\frac{\sqrt{3}}{3}\quad \frac{2i}{3}$

$\frac{-1}{6}-i\frac{\sqrt{3}}{6}\quad \frac{1}{6}-i\frac{\sqrt{3}}{6}\quad \frac{-i}{3}-\frac{\sqrt{3}}{3}\quad \frac{2i}{3}$

$\frac{i}{3}+\frac{\sqrt{3}}{3}\quad \frac{i}{3}-\frac{\sqrt{3}}{3}\quad -\frac{1}{3}\quad -\frac{1}{3}$

$$O: \left.\begin{array}{l}T_1 \\ D_3; E \\ C_3; \end{array}\right\rangle \left|\begin{array}{l}T_1 \\ E \\ 1_3\end{array}\right\rangle \left|\begin{array}{l}A_2 \\ 2_3 \\ 0_3\end{array}\right\rangle$$

$$|x_1\rangle = \left|\begin{array}{l}T_1 \\ E \\ B_1\end{array}\right\rangle \quad |x_1\rangle = \left|\begin{array}{l}T_1 \\ E \\ B_2\end{array}\right\rangle \quad |x_1\rangle = \left|\begin{array}{l}T_1 \\ A_2 \\ A_2\end{array}\right\rangle \left|\begin{array}{l}O \\ D_4 \\ D_2\end{array}\right.$$

$$|v_1\rangle = \left|\begin{array}{l}T_1 \\ E \\ A\end{array}\right\rangle \quad |v_2\rangle = \left|\begin{array}{l}T_1 \\ E \\ B\end{array}\right\rangle \quad |v_3\rangle = \left|\begin{array}{l}T_1 \\ A_2 \\ B\end{array}\right\rangle \left|\begin{array}{l}O \\ D_3 \\ C_2\end{array}\right.$$

$$|1_3\rangle = \left|\begin{array}{l}T_1 \\ E \\ 1_3\end{array}\right\rangle \quad |2_3\rangle = \left|\begin{array}{l}T_1 \\ E \\ 2_3\end{array}\right\rangle \quad |0_3\rangle = \left|\begin{array}{l}T_1 \\ A_2 \\ 0_3\end{array}\right\rangle \left|\begin{array}{l}O \\ D_3 \\ C_3\end{array}\right.$$

The $O \supset D_3 \supset C_3 T_1$ representation is obtained from that of $O \supset D_4 \supset D_2$ by the following transformation matrix:

	$|v_1\rangle$	$|v_2\rangle$	$|v_3\rangle$
$\langle 1_3|$	$\frac{-1}{\sqrt{2}}$	$\frac{i}{\sqrt{2}}$	0
$\langle 2_3|$	$\frac{1}{\sqrt{2}}$	$\frac{i}{\sqrt{2}}$	0
$\langle 0_3|$	0	0	1

$=$

	$|x_1\rangle$	$|x_2\rangle$	$|x_3\rangle$
$\langle v_1|$	$\frac{1}{\sqrt{2}}$	$\frac{-1}{\sqrt{2}}$	0
$\langle v_2|$	$\frac{1}{\sqrt{6}}$	$\frac{1}{\sqrt{6}}$	$\frac{-2}{\sqrt{6}}$
$\langle v_3|$	$\frac{1}{\sqrt{3}}$	$\frac{1}{\sqrt{3}}$	$\frac{1}{\sqrt{3}}$

, where

TABLE F.2.7 *(Continued)*

(b) Tensor Representation T_2

$\mathscr{D}^{T_2}(1) =$

$$\begin{vmatrix} 1 & \cdot & \cdot \\ \cdot & 1 & \cdot \\ \cdot & \cdot & 1 \end{vmatrix}$$

$i_4 = [12]$

$$\begin{vmatrix} 1 & \cdot & \cdot \\ \cdot & \cdot & -1 \\ \cdot & -1 & \cdot \end{vmatrix}$$

$R_1^2 = [13][24]$

$$\begin{vmatrix} \frac{-1}{3} & \frac{1}{3} - i\frac{\sqrt{3}}{3} & \frac{-1}{3} - i\frac{\sqrt{3}}{3} \\ \frac{1}{3} + i\frac{\sqrt{3}}{3} & \frac{-1}{3} & \frac{1}{3} - i\frac{\sqrt{3}}{3} \\ \frac{-1}{3} + i\frac{\sqrt{3}}{3} & \frac{1}{3} + i\frac{\sqrt{3}}{3} & \frac{-1}{3} \end{vmatrix}$$

$R_3 = [1423]$

$$\begin{vmatrix} \frac{-1}{3} & \frac{1}{3} + i\frac{\sqrt{3}}{3} & \frac{-1}{3} + i\frac{\sqrt{3}}{3} \\ \frac{1}{3} + i\frac{\sqrt{3}}{3} & \frac{-1}{3} + i\frac{\sqrt{3}}{3} & \frac{1}{3} \\ \frac{-1}{3} + i\frac{\sqrt{3}}{3} & \frac{1}{3} & \frac{-1}{3} - i\frac{\sqrt{3}}{3} \end{vmatrix}$$

$r_1 = [132]$

$$\begin{vmatrix} 1 & \cdot & \cdot \\ \cdot & \frac{-1}{2} - i\frac{\sqrt{3}}{2} & \cdot \\ \cdot & \cdot & \frac{-1}{2} + i\frac{\sqrt{3}}{2} \end{vmatrix}$$

$i_5 = [13]$

$$\begin{vmatrix} 1 & \cdot & \cdot \\ \cdot & \cdot & \frac{1}{2} - i\frac{\sqrt{3}}{2} \\ \cdot & \frac{1}{2} + i\frac{\sqrt{3}}{2} & \cdot \end{vmatrix}$$

$r_4 = [234]$

$$\begin{vmatrix} \frac{-1}{3} & \frac{-2}{3} & \frac{2}{3} \\ \frac{1}{3} + i\frac{\sqrt{3}}{3} & \frac{1}{6} + i\frac{\sqrt{3}}{6} & \frac{1}{3} + i\frac{\sqrt{3}}{3} \\ \frac{-1}{3} + i\frac{\sqrt{3}}{3} & \frac{1}{3} - i\frac{\sqrt{3}}{3} & \frac{1}{6} - i\frac{\sqrt{3}}{6} \end{vmatrix}$$

$i_6 = [24]$

$$\begin{vmatrix} \frac{-1}{3} & \frac{1}{3} - i\frac{\sqrt{3}}{3} & \frac{-1}{3} - i\frac{\sqrt{3}}{3} \\ \frac{1}{3} + i\frac{\sqrt{3}}{3} & \frac{2}{3} & \frac{-1}{6} + i\frac{\sqrt{3}}{6} \\ \frac{-1}{3} + i\frac{\sqrt{3}}{3} & \frac{-1}{6} - i\frac{\sqrt{3}}{6} & \frac{2}{3} \end{vmatrix}$$

$r_1^2 = [123]$

$$\begin{vmatrix} 1 & \cdot & \cdot \\ \cdot & \frac{-1}{2} + i\frac{\sqrt{3}}{2} & \cdot \\ \cdot & \cdot & \frac{-1}{2} - i\frac{\sqrt{3}}{2} \end{vmatrix}$$

$i_2 = [23]$

$$\begin{vmatrix} 1 & \cdot & \cdot \\ \cdot & \cdot & \frac{1}{2} + i\frac{\sqrt{3}}{2} \\ \cdot & \frac{1}{2} - i\frac{\sqrt{3}}{2} & \cdot \end{vmatrix}$$

$r_2^2 = [142]$

$$\begin{vmatrix} \frac{-1}{3} & \frac{1}{3} + i\frac{\sqrt{3}}{3} & \frac{-1}{3} + i\frac{\sqrt{3}}{3} \\ \frac{1}{3} + i\frac{\sqrt{3}}{3} & \frac{1}{6} - i\frac{\sqrt{3}}{6} & \frac{-2}{3} \\ \frac{-1}{3} + i\frac{\sqrt{3}}{3} & \frac{-2}{3} & \frac{1}{6} + i\frac{\sqrt{3}}{6} \end{vmatrix}$$

$R_2^3 = [1342]$

$$\begin{vmatrix} \frac{-1}{3} & \frac{-2}{3} & \frac{2}{3} \\ \frac{1}{3} + i\frac{\sqrt{3}}{3} & \frac{-1}{3} - i\frac{\sqrt{3}}{3} & \frac{-1}{6} - i\frac{\sqrt{3}}{6} \\ \frac{-1}{3} + i\frac{\sqrt{3}}{3} & \frac{-1}{6} + i\frac{\sqrt{3}}{6} & \frac{-1}{3} + i\frac{\sqrt{3}}{3} \end{vmatrix}$$

$R_2^2 = [14][23]$

$$\begin{vmatrix} \frac{-1}{3} & \frac{1}{3} + i\frac{\sqrt{3}}{3} & \frac{-1}{3} + i\frac{\sqrt{3}}{3} \\ \frac{1}{3} - i\frac{\sqrt{3}}{3} & \frac{-1}{3} & \frac{1}{3} + i\frac{\sqrt{3}}{3} \\ \frac{-1}{3} - i\frac{\sqrt{3}}{3} & \frac{1}{3} - i\frac{\sqrt{3}}{3} & \frac{-1}{3} \end{vmatrix}$$

$R_3^3 = [1324]$

$$\begin{vmatrix} \frac{-1}{3} & \frac{1}{3} - i\frac{\sqrt{3}}{3} & \frac{-1}{3} + i\frac{\sqrt{3}}{3} \\ \frac{1}{3} - i\frac{\sqrt{3}}{3} & \frac{-1}{3} - i\frac{\sqrt{3}}{3} & \frac{1}{3} \\ \frac{-1}{3} - i\frac{\sqrt{3}}{3} & \frac{1}{3} & \frac{-1}{3} + i\frac{\sqrt{3}}{3} \end{vmatrix}$$

$R_3^2 = [12][34]$

$$\begin{vmatrix} \frac{-1}{3} & \frac{-2}{3} & \frac{2}{3} \\ \frac{-2}{3} & \frac{-1}{3} & \frac{-2}{3} \\ \frac{2}{3} & \frac{-2}{3} & \frac{-1}{3} \end{vmatrix}$$

$i_3 = [34]$

$$\begin{vmatrix} \frac{-1}{3} & \frac{-2}{3} & \frac{2}{3} \\ \frac{-2}{3} & \frac{2}{3} & \frac{1}{3} \\ \frac{2}{3} & \frac{1}{3} & \frac{2}{3} \end{vmatrix}$$

$r_2 = [124]$

$$\begin{vmatrix} \frac{-1}{3} & \frac{1}{3} - i\frac{\sqrt{3}}{3} & \frac{-1}{3} - i\frac{\sqrt{3}}{3} \\ \frac{1}{3} - i\frac{\sqrt{3}}{3} & \frac{1}{6} + i\frac{\sqrt{3}}{6} & \frac{-2}{3} \\ \frac{-1}{3} - i\frac{\sqrt{3}}{3} & \frac{-2}{3} & \frac{1}{6} - i\frac{\sqrt{3}}{6} \end{vmatrix}$$

$R_1 = [1234]$

$$\begin{vmatrix} \frac{-1}{3} & \frac{-2}{3} & \frac{2}{3} \\ \frac{1}{3} - i\frac{\sqrt{3}}{3} & \frac{-1}{3} + i\frac{\sqrt{3}}{3} & \frac{-1}{6} + i\frac{\sqrt{3}}{6} \\ \frac{-1}{3} - i\frac{\sqrt{3}}{3} & \frac{-1}{6} - i\frac{\sqrt{3}}{6} & \frac{-1}{3} - i\frac{\sqrt{3}}{3} \end{vmatrix}$$

$r_3 = [143]$

$$\begin{vmatrix} \frac{-1}{3} & \frac{1}{3} + i\frac{\sqrt{3}}{3} & \frac{-1}{3} + i\frac{\sqrt{3}}{3} \\ \frac{-2}{3} & \frac{1}{6} + i\frac{\sqrt{3}}{6} & \frac{1}{3} - i\frac{\sqrt{3}}{3} \\ \frac{2}{3} & \frac{1}{3} + i\frac{\sqrt{3}}{3} & \frac{1}{6} - i\frac{\sqrt{3}}{6} \end{vmatrix}$$

$R_1^3 = [1432]$

$$\begin{vmatrix} \frac{-1}{3} & \frac{1}{3} + i\frac{\sqrt{3}}{3} & \frac{-1}{3} + i\frac{\sqrt{3}}{3} \\ \frac{-2}{3} & \frac{-1}{3} - i\frac{\sqrt{3}}{3} & \frac{-1}{6} + i\frac{\sqrt{3}}{6} \\ \frac{2}{3} & \frac{-1}{6} - i\frac{\sqrt{3}}{6} & \frac{-1}{3} + i\frac{\sqrt{3}}{3} \end{vmatrix}$$

$r_3^2 = [134]$

$$\begin{vmatrix} \frac{-1}{3} & \frac{-2}{3} & \frac{2}{3} \\ \frac{1}{3} - i\frac{\sqrt{3}}{3} & \frac{1}{6} - i\frac{\sqrt{3}}{6} & \frac{1}{3} - i\frac{\sqrt{3}}{3} \\ \frac{-1}{3} - i\frac{\sqrt{3}}{3} & \frac{1}{3} + i\frac{\sqrt{3}}{3} & \frac{1}{6} + i\frac{\sqrt{3}}{6} \end{vmatrix}$$

$i_1 = [14]$

$$\begin{vmatrix} \frac{-1}{3} & \frac{1}{3} + i\frac{\sqrt{3}}{3} & \frac{-1}{3} + i\frac{\sqrt{3}}{3} \\ \frac{1}{3} - i\frac{\sqrt{3}}{3} & \frac{2}{3} & \frac{-1}{6} - i\frac{\sqrt{3}}{6} \\ \frac{-1}{3} - i\frac{\sqrt{3}}{3} & \frac{-1}{6} + i\frac{\sqrt{3}}{6} & \frac{2}{3} \end{vmatrix}$$

$r_4^2 = [243]$

$$\begin{vmatrix} \frac{-1}{3} & \frac{1}{3} - i\frac{\sqrt{3}}{3} & \frac{-1}{3} - i\frac{\sqrt{3}}{3} \\ \frac{-2}{3} & \frac{1}{6} - i\frac{\sqrt{3}}{6} & \frac{1}{3} + i\frac{\sqrt{3}}{3} \\ \frac{2}{3} & \frac{1}{3} - i\frac{\sqrt{3}}{3} & \frac{1}{6} + i\frac{\sqrt{3}}{6} \end{vmatrix}$$

$R_2 = [1243]$

$$\begin{vmatrix} \frac{-1}{3} & \frac{1}{3} - i\frac{\sqrt{3}}{3} & \frac{-1}{3} - i\frac{\sqrt{3}}{3} \\ \frac{-2}{3} & \frac{-1}{3} + i\frac{\sqrt{3}}{3} & \frac{-1}{6} - i\frac{\sqrt{3}}{6} \\ \frac{2}{3} & \frac{-1}{6} + i\frac{\sqrt{3}}{6} & \frac{-1}{3} - i\frac{\sqrt{3}}{3} \end{vmatrix}$$

$$\begin{array}{l} O: \\ D_3: \\ C_3: \end{array} \left|\begin{array}{c} T_2 \\ A_1 \\ 0_3 \end{array}\right\rangle \left|\begin{array}{c} T_2 \\ E \\ 1_3 \end{array}\right\rangle \left|\begin{array}{c} T_2 \\ E \\ 2_3 \end{array}\right\rangle$$

The $O \supset D_3 \supset C_3$ T_2 representation is obtained from that of $O \supset D_4 \supset D_2$ by the following transformation matrix:

$$\begin{array}{c|ccc} & |x_1\rangle & |x_2\rangle & |x_3\rangle \\ \hline \langle 0_3| & \frac{\sqrt{3}}{3} & \frac{-\sqrt{3}}{3} & \frac{\sqrt{3}}{3} \\ \langle 1_3| & \frac{i}{2} + \frac{\sqrt{3}}{6} & \frac{i}{2} - \frac{\sqrt{3}}{6} & \frac{-\sqrt{3}}{3} \\ \langle 2_3| & \frac{i}{2} - \frac{\sqrt{3}}{6} & \frac{i}{2} + \frac{\sqrt{3}}{6} & \frac{\sqrt{3}}{3} \end{array} = \begin{array}{c|ccc} & |v_1\rangle & |v_2\rangle & |v_3\rangle \\ \hline \langle 0_3| & 1 & 0 & 0 \\ \langle 1_3| & 0 & \frac{-1}{\sqrt{2}} & \frac{i}{\sqrt{2}} \\ \langle 2_3| & 0 & \frac{1}{\sqrt{2}} & \frac{i}{\sqrt{2}} \end{array} \cdot \begin{array}{c|ccc} & |x_1\rangle & |x_2\rangle & |x_3\rangle \\ \hline \langle v_1| & \frac{1}{\sqrt{3}} & \frac{-1}{\sqrt{3}} & \frac{1}{\sqrt{3}} \\ \langle v_2| & \frac{-1}{\sqrt{6}} & \frac{1}{\sqrt{6}} & \frac{2}{\sqrt{6}} \\ \langle v_3| & \frac{1}{\sqrt{2}} & \frac{1}{\sqrt{2}} & 0 \end{array},$$

where

$$|x_1\rangle = \left|\begin{array}{c} T_2 \\ E \\ B_1 \end{array}\right\rangle \quad |x_2\rangle = \left|\begin{array}{c} T_2 \\ E \\ B_2 \end{array}\right\rangle \quad |x_3\rangle = \left|\begin{array}{c} T_2 \\ B_2 \\ A_2 \end{array}\right\rangle \begin{array}{l} :O \\ :D_4 \\ :D_2 \end{array}$$

$$|v_1\rangle = \left|\begin{array}{c} T_2 \\ A_1 \\ A \end{array}\right\rangle \quad |v_2\rangle = \left|\begin{array}{c} T_2 \\ E \\ A \end{array}\right\rangle \quad |v_3\rangle = \left|\begin{array}{c} T_2 \\ E \\ B \end{array}\right\rangle \begin{array}{l} :O \\ :D_3 \\ :C_2 \end{array}.$$

$$|0_3\rangle = \left|\begin{array}{c} T_2 \\ A_1 \\ 0_3 \end{array}\right\rangle \quad |1_3\rangle = \left|\begin{array}{c} T_2 \\ E \\ 1_3 \end{array}\right\rangle \quad |2_3\rangle = \left|\begin{array}{c} T_2 \\ E \\ 2_3 \end{array}\right\rangle \begin{array}{l} :O \\ :D_3 \\ :C_3 \end{array}$$

TABLE F.2.7 (Continued)

(c) Tensor Representation E

$\mathscr{D}^{T_2}(1) = \begin{vmatrix} 1 & \cdot \\ \cdot & 1 \end{vmatrix}$

$i_4 = [12]$

$\begin{vmatrix} \cdot & -1 \\ -1 & \cdot \end{vmatrix}$

$R_1^2 = [13][24]$

$\begin{vmatrix} 1 & \cdot \\ \cdot & 1 \end{vmatrix}$

$R_3 = [1423]$

$\begin{vmatrix} \cdot & -1 \\ -1 & \cdot \end{vmatrix}$

$r_1 = [132]$

$\begin{vmatrix} -\frac{1}{2} - i\frac{\sqrt{3}}{2} & \cdot \\ \cdot & -\frac{1}{2} + i\frac{\sqrt{3}}{2} \end{vmatrix}$

$i_5 = [13]$

$\begin{vmatrix} \cdot & -\frac{1}{2} - i\frac{\sqrt{3}}{2} \\ -\frac{1}{2} + i\frac{\sqrt{3}}{2} & \cdot \end{vmatrix}$

$r_4 = [234]$

$\begin{vmatrix} -\frac{1}{2} - i\frac{\sqrt{3}}{2} & \cdot \\ \cdot & -\frac{1}{2} + i\frac{\sqrt{3}}{2} \end{vmatrix}$

$i_6 = [24]$

$\begin{vmatrix} \cdot & -\frac{1}{2} - i\frac{\sqrt{3}}{2} \\ -\frac{1}{2} + i\frac{\sqrt{3}}{2} & \cdot \end{vmatrix}$

$r_1^2 = [123]$

$\begin{vmatrix} -\frac{1}{2} + i\frac{\sqrt{3}}{2} & \cdot \\ \cdot & -\frac{1}{2} - i\frac{\sqrt{3}}{2} \end{vmatrix}$

$i_2 = [23]$

$\begin{vmatrix} \cdot & -\frac{1}{2} + i\frac{\sqrt{3}}{2} \\ -\frac{1}{2} - i\frac{\sqrt{3}}{2} & \cdot \end{vmatrix}$

$r_2^2 = [142]$

$\begin{vmatrix} -\frac{1}{2} + i\frac{\sqrt{3}}{2} & \cdot \\ \cdot & -\frac{1}{2} - i\frac{\sqrt{3}}{2} \end{vmatrix}$

$R_2^3 = [1342]$

$\begin{vmatrix} \cdot & -\frac{1}{2} + i\frac{\sqrt{3}}{2} \\ -\frac{1}{2} - i\frac{\sqrt{3}}{2} & \cdot \end{vmatrix}$

$R_2^2 = [14][23]$

$$\begin{pmatrix} 1 & \cdot \\ \cdot & 1 \end{pmatrix}$$

$r_2 = [124]$

$$\begin{pmatrix} \cdot & \frac{-1-i\sqrt{3}}{2} \\ \frac{-1+i\sqrt{3}}{2} & \cdot \end{pmatrix}$$

$r_3^2 = [134]$

$$\begin{pmatrix} \cdot & \frac{-1+i\sqrt{3}}{2} \\ \frac{-1-i\sqrt{3}}{2} & \cdot \end{pmatrix}$$

$R_3^3 = [1324]$

$$\begin{pmatrix} \cdot & -1 \\ -1 & \cdot \end{pmatrix}$$

$R_1 = [1234]$

$$\begin{pmatrix} \cdot & \frac{-1-i\sqrt{3}}{2} \\ \frac{-1+i\sqrt{3}}{2} & \cdot \end{pmatrix}$$

$i_1 = [14]$

$$\begin{pmatrix} \cdot & \frac{-1+i\sqrt{3}}{2} \\ \frac{-1-i\sqrt{3}}{2} & \cdot \end{pmatrix}$$

$R_3^2 = [12][34]$

$$\begin{pmatrix} 1 & \cdot \\ \cdot & 1 \end{pmatrix}$$

$r_3 = [143]$

$$\begin{pmatrix} \frac{-1-i\sqrt{3}}{2} & \cdot \\ \cdot & \frac{-1+i\sqrt{3}}{2} \end{pmatrix}$$

$r_4^2 = [243]$

$$\begin{pmatrix} \frac{-1+i\sqrt{3}}{2} & \cdot \\ \cdot & \frac{-1-i\sqrt{3}}{2} \end{pmatrix}$$

$i_3 = [34]$

$$\begin{pmatrix} \cdot & -1 \\ -1 & \cdot \end{pmatrix}$$

$R_1^3 = [1432]$

$$\begin{pmatrix} \frac{-1}{2} - i\frac{\sqrt{3}}{2} & \cdot \\ \cdot & \frac{-1}{2} + i\frac{\sqrt{3}}{2} \end{pmatrix}$$

$R_2 = [1243]$

$$\begin{pmatrix} \frac{-1}{2} + i\frac{\sqrt{3}}{2} & \cdot \\ \cdot & \frac{-1}{2} - i\frac{\sqrt{3}}{2} \end{pmatrix}$$

$$O: \left|\begin{array}{c}E\\E\end{array}\right\rangle \quad D_3: \left|\begin{array}{c}E\\E\end{array}\right\rangle \quad C_3: \left|\begin{array}{c}E\\1_3\end{array}\right\rangle 2_3$$

The $O \supset D_3 \supset C_3$ E representation is obtained from that of $O \supset D_4 \supset D_2$ by the following transformation matrix:

	$\|x_1\rangle$	$\|x_2\rangle$
$\langle 1_3\|$	$\frac{-1}{\sqrt{2}}$	$\frac{i}{\sqrt{2}}$
$\langle 2_3\|$	$\frac{1}{\sqrt{2}}$	$\frac{i}{\sqrt{2}}$

$$= \begin{array}{c|cc} & \|v_1\rangle & \|v_2\rangle \\ \hline \langle 1_3\| & \frac{-1}{\sqrt{2}} & \frac{i}{\sqrt{2}} \\ \langle 2_3\| & \frac{1}{\sqrt{2}} & \frac{i}{\sqrt{2}} \end{array} \cdot \begin{array}{c|cc} & \|x_1\rangle & \|x_2\rangle \\ \hline \langle v_1\| & 1 & 0 \\ \langle v_2\| & 0 & 1 \end{array}$$

, where

$$|x_1\rangle = \left|\begin{array}{c}E\\B_1\end{array}\right\rangle \quad |x_2\rangle = \left|\begin{array}{c}E\\B_2\end{array}\right\rangle = \left|\begin{array}{c}E\\E\end{array}\right\rangle \left|\begin{array}{c}:O\\:D_4\\:D_2\end{array}\right.$$

$$|v_1\rangle = \left|\begin{array}{c}E\\A\end{array}\right\rangle \quad |v_2\rangle = \left|\begin{array}{c}E\\B\end{array}\right\rangle \left|\begin{array}{c}:O\\:D_3\\:C_2\end{array}\right.$$

$$|1_3\rangle = \left|\begin{array}{c}E\\1_1\end{array}\right\rangle \quad |2_3\rangle = \left|\begin{array}{c}E\\2_3\end{array}\right\rangle \left|\begin{array}{c}:O\\:D_3\\:C_3\end{array}\right.$$

F.3. CLEBSCH-GORDAN COEFFICIENTS

These coupling coefficients in Table F.3.1 and F.3.2 belong to the representations and bases listed in Tables F.2.1 and F.2.2. To obtain $O \supset D_4 \supset D_2$ labeled coefficients, change the sign of the second component of T_2. [Compare Tables F.2.1(b) and F.2.3(b).]

TABLE F.3.1 Standard Fourfold Axial Octahedral Clebsch-Gordan Coefficients

(a) $T_1 \otimes T_1$

T_1	T_1	A_1	E 1	2	T_1 1	2	3	T_2 1	2	3
1	1	$\frac{1}{\sqrt{3}}$	$\frac{1}{\sqrt{6}}$	$\frac{-1}{\sqrt{2}}$	.	.	.	.	.	.
1	2	.	.	.	.	$\frac{1}{\sqrt{2}}$	.	.	.	$\frac{1}{\sqrt{2}}$
1	3	.	.	.	.	.	$\frac{-1}{\sqrt{2}}$	.	$\frac{1}{\sqrt{2}}$	.
2	1	.	.	.	.	$\frac{-1}{\sqrt{2}}$	.	.	.	$\frac{1}{\sqrt{2}}$
2	2	$\frac{1}{\sqrt{3}}$	$\frac{1}{\sqrt{6}}$	$\frac{1}{\sqrt{12}}$	.	.	.	.	.	.
2	3	.	.	.	$\frac{1}{\sqrt{2}}$	.	.	$\frac{1}{\sqrt{2}}$	.	.
3	1	.	.	.	.	$\frac{1}{\sqrt{2}}$	.	.	$\frac{1}{\sqrt{2}}$	.
3	2	.	.	.	$\frac{-1}{\sqrt{2}}$	.	.	$\frac{1}{\sqrt{2}}$	.	.
3	3	$\frac{1}{\sqrt{3}}$	$\frac{-2}{\sqrt{6}}$	0	.	.	.	.	.	.

(b) $T_1 \otimes T_2$

T_1	T_2	A_2	E 1	2	T_1 1	2	3	T_2 1	2	3
1	1	$\frac{1}{\sqrt{3}}$	$\frac{1}{\sqrt{2}}$	$\frac{1}{\sqrt{6}}$	.	.	.	.	.	.
1	2	.	.	.	.	$\frac{1}{\sqrt{2}}$	.	.	.	$\frac{1}{\sqrt{2}}$
1	3	.	.	.	.	.	$\frac{1}{\sqrt{2}}$	.	$\frac{-1}{\sqrt{2}}$	.
2	1	.	.	.	.	$\frac{1}{\sqrt{2}}$	.	.	.	$\frac{-1}{\sqrt{2}}$
2	2	$\frac{1}{\sqrt{3}}$	$\frac{-1}{\sqrt{2}}$	$\frac{1}{\sqrt{6}}$	.	.	.	.	.	.
2	3	.	.	.	$\frac{1}{\sqrt{2}}$	.	.	$\frac{1}{\sqrt{2}}$	.	.
3	1	.	.	.	.	$\frac{1}{\sqrt{2}}$	.	.	$\frac{1}{\sqrt{2}}$	.
3	2	.	.	.	$\frac{1}{\sqrt{2}}$	.	.	$\frac{-1}{\sqrt{2}}$	.	.
3	3	$\frac{1}{\sqrt{3}}$	0	$\frac{-2}{\sqrt{6}}$	.	.	.	.	.	.

(c) $T_2 \otimes T_2$

T_2	T_2	A_1	E 1	2	T_1 1	2	3	T_2 1	2	3
1	1	$\frac{1}{\sqrt{3}}$	$\frac{1}{\sqrt{6}}$	$\frac{-1}{\sqrt{2}}$	.	.	.	.	.	.
1	2	.	.	.	.	.	$\frac{1}{\sqrt{2}}$	.	.	$\frac{1}{\sqrt{2}}$
1	3	.	.	.	.	$\frac{-1}{\sqrt{2}}$	.	.	$\frac{1}{\sqrt{2}}$	.
2	1	.	.	.	.	.	$\frac{-1}{\sqrt{2}}$	.	.	$\frac{1}{\sqrt{2}}$
2	2	$\frac{1}{\sqrt{3}}$	$\frac{1}{\sqrt{6}}$	$\frac{1}{\sqrt{2}}$	.	.	.	.	.	.
2	3	.	.	.	$\frac{1}{\sqrt{2}}$	.	.	$\frac{1}{\sqrt{2}}$	.	.
3	1	.	.	.	.	$\frac{1}{\sqrt{2}}$	.	.	$\frac{1}{\sqrt{2}}$	.
3	2	.	.	.	$\frac{-1}{\sqrt{2}}$	.	.	$\frac{1}{\sqrt{2}}$	.	.
3	3	$\frac{1}{\sqrt{3}}$	$\frac{-2}{\sqrt{6}}$	0	.	.	.	.	.	.

TABLE F.3.1 (*Continued*)

(d) $T_1 \otimes E$

T_1	E	T_1 1	2	3	T_2 1	2	3
1	1	$\frac{-1}{2}$	.	.	$\frac{\sqrt{3}}{2}$	.	.
1	2	$\frac{\sqrt{3}}{2}$	.	.	$\frac{1}{2}$	.	.
2	1	.	$\frac{-1}{2}$	.	.	$\frac{-\sqrt{3}}{2}$	.
2	2	.	$\frac{-\sqrt{3}}{2}$	.	.	$\frac{1}{2}$	.
3	1	.	.	1	.	.	.
3	2	.	.	.	.	.	-1

(e) $T_2 \otimes E$

T_2	E	T_1 1	2	3	T_2 1	2	3
1	1	$\frac{\sqrt{3}}{2}$	.	.	$\frac{-1}{2}$	.	.
1	2	$\frac{1}{2}$	.	.	$\frac{\sqrt{3}}{2}$	.	.
2	1	.	$\frac{-\sqrt{3}}{2}$	.	.	$\frac{-1}{2}$	.
2	2	.	$\frac{1}{2}$	.	.	$\frac{-\sqrt{3}}{2}$	.
3	1	.	.	.	.	.	1
3	2	.	.	-1	.	.	.

(f) $T_1 \otimes A_2$

T_1	A_2	T_2 1	2	3
1		1	.	.
2		.	1	.
3		.	.	1

(g) $T_2 \otimes A_2$

T_2	A_2	T_1 1	2	3
1		1	.	.
2		.	1	.
3		.	.	1

(h) $E \otimes E$

E	E	A_1	A_2	E 1	2
1	1	$\frac{1}{\sqrt{2}}$	.	$\frac{1}{\sqrt{2}}$	.
1	2	.	$\frac{1}{\sqrt{2}}$	.	$\frac{-1}{\sqrt{2}}$
2	1	.	$\frac{-1}{\sqrt{2}}$	.	$\frac{-1}{\sqrt{2}}$
2	2	$\frac{1}{\sqrt{2}}$	.	$\frac{-1}{\sqrt{2}}$	.

(i) $E \otimes E$ and $A_2 \otimes A_2$

E	A_2	E 1	2
1		.	1
2		1	.

A_2	A_2	A_1
		1

TABLE F.3.2 $O \supset D_3 \supset C_3$ Subgroup Labeled Clebsch-Gordan Coefficients

(a) $T_1 \otimes T_1$

T_1	T_1	A_1	E 1	E 2	T_1 1	T_1 2	T_1 3	T_2 1	T_2 2	T_2 3
1	1	$\frac{1}{\sqrt{3}}$	$\frac{1}{\sqrt{6}}$	·	·	·	·	$\frac{1}{\sqrt{6}}$	$\frac{1}{\sqrt{3}}$	·
1	2	·	·	$\frac{-1}{\sqrt{6}}$	·	·	$\frac{1}{\sqrt{2}}$	·	·	$\frac{-1}{\sqrt{3}}$
1	3	·	·	$\frac{-1}{\sqrt{3}}$	·	$\frac{-1}{\sqrt{2}}$	·	·	·	$\frac{1}{\sqrt{6}}$
2	1	·	·	$\frac{-1}{\sqrt{6}}$	·	$\frac{-1}{\sqrt{2}}$	·	·	·	$\frac{-1}{\sqrt{3}}$
2	2	$\frac{1}{\sqrt{3}}$	$\frac{-1}{\sqrt{6}}$	·	·	·	·	$\frac{1}{\sqrt{6}}$	$\frac{-1}{\sqrt{3}}$	·
2	3	·	·	$\frac{1}{\sqrt{3}}$	·	$\frac{1}{\sqrt{2}}$	·	·	·	$\frac{-1}{\sqrt{6}}$
3	1	·	·	$\frac{-1}{\sqrt{3}}$	·	$\frac{1}{\sqrt{2}}$	·	·	·	$\frac{1}{\sqrt{6}}$
3	2	·	$\frac{1}{\sqrt{3}}$	·	$\frac{-1}{\sqrt{2}}$	·	·	$\frac{-1}{\sqrt{6}}$	·	·
3	3	$\frac{1}{\sqrt{3}}$	·	·	·	·	·	$\frac{-2}{\sqrt{6}}$	·	·

(b) $T_1 \otimes T_2$

T_2	T_1	A_2	E 1	E 2	T_2 1	T_2 2	T_2 3	T_1 1	T_1 2	T_1 3
1	1	·	$\frac{1}{\sqrt{3}}$	·	·	$\frac{-1}{\sqrt{2}}$	·	$\frac{1}{\sqrt{6}}$	·	·
1	2	·	·	$\frac{1}{\sqrt{3}}$	·	·	$\frac{-1}{\sqrt{2}}$	·	$\frac{1}{\sqrt{6}}$	·
1	3	$\frac{1}{\sqrt{3}}$	·	·	·	·	·	·	·	$\frac{-2}{\sqrt{6}}$
2	1	·	$\frac{-1}{\sqrt{6}}$	·	$\frac{1}{\sqrt{2}}$	·	·	$\frac{1}{\sqrt{3}}$	·	·
2	2	$\frac{-1}{\sqrt{3}}$	·	$\frac{1}{\sqrt{6}}$	·	·	·	·	$\frac{-1}{\sqrt{3}}$	$\frac{-1}{\sqrt{6}}$
2	3	·	·	$\frac{-1}{\sqrt{3}}$	·	$\frac{-1}{\sqrt{2}}$	·	·	$\frac{-1}{\sqrt{6}}$	·
3	1	$\frac{1}{\sqrt{3}}$	·	$\frac{1}{\sqrt{6}}$	·	·	·	·	$\frac{-1}{\sqrt{3}}$	$\frac{1}{\sqrt{6}}$
3	2	·	$\frac{1}{\sqrt{6}}$	·	$\frac{1}{\sqrt{2}}$	·	·	$\frac{-1}{\sqrt{3}}$	·	·
3	3	·	$\frac{1}{\sqrt{3}}$	·	·	$\frac{1}{\sqrt{2}}$	·	$\frac{1}{\sqrt{6}}$	·	·

(c) $T_2 \otimes T_2$

T_2	T_2	A_1	E 1	E 2	T_1 1	T_1 2	T_1 3	T_2 1	T_2 2	T_2 3
1	1	$\frac{1}{\sqrt{3}}$	·	·	·	·	·	$\frac{2}{\sqrt{6}}$	·	·
1	2	·	$\frac{1}{\sqrt{3}}$	·	$\frac{1}{\sqrt{2}}$	·	·	$\frac{-1}{\sqrt{6}}$	·	·
1	3	·	·	$\frac{1}{\sqrt{3}}$	·	$\frac{1}{\sqrt{2}}$	·	·	·	$\frac{-1}{\sqrt{6}}$
2	1	·	$\frac{1}{\sqrt{3}}$	·	$\frac{-1}{\sqrt{2}}$	·	·	$\frac{-1}{\sqrt{6}}$	·	·
2	2	$\frac{1}{\sqrt{3}}$	$\frac{1}{\sqrt{6}}$	·	·	·	·	$\frac{-1}{\sqrt{6}}$	$\frac{1}{\sqrt{3}}$	·
2	3	·	·	$\frac{-1}{\sqrt{6}}$	·	·	$\frac{1}{\sqrt{2}}$	·	·	$\frac{-1}{\sqrt{3}}$
3	1	·	·	$\frac{1}{\sqrt{3}}$	·	$\frac{-1}{\sqrt{2}}$	·	·	·	$\frac{-1}{\sqrt{6}}$
3	2	·	·	$\frac{-1}{\sqrt{6}}$	·	·	$\frac{-1}{\sqrt{2}}$	·	·	$\frac{-1}{\sqrt{3}}$
3	3	$\frac{1}{\sqrt{3}}$	$\frac{-1}{\sqrt{6}}$	·	·	·	·	$\frac{-1}{\sqrt{6}}$	$\frac{-1}{\sqrt{3}}$	·

TABLE F.3.2 (Continued)

(d) $T_1 \otimes E$

T_1	E	T_2 1	2	3	T_1 1	2	3
1	1	$\frac{1}{\sqrt{2}}$	$\frac{-1}{2}$	·	$\frac{1}{2}$	·	·
1	2	·	·	$\frac{1}{2}$	·	$\frac{-1}{2}$	$\frac{-1}{\sqrt{2}}$
2	1	·	·	$\frac{1}{2}$	·	$\frac{-1}{2}$	$\frac{1}{\sqrt{2}}$
2	2	$\frac{1}{\sqrt{2}}$	$\frac{1}{2}$	·	$\frac{-1}{2}$	·	·
3	1	·	·	$\frac{1}{\sqrt{2}}$	·	$\frac{1}{\sqrt{2}}$	·
3	2	·	$\frac{-1}{\sqrt{2}}$	·	$\frac{-1}{\sqrt{2}}$	·	·

(e) $T_2 \otimes E$

T_2	E	T_2 1	2	3	T_1 1	2	3
1	1	·	$\frac{1}{\sqrt{2}}$	·	$\frac{1}{\sqrt{2}}$	·	·
1	2	·	·	$\frac{1}{\sqrt{2}}$	·	$\frac{1}{\sqrt{2}}$	·
2	1	$\frac{1}{\sqrt{2}}$	$\frac{1}{2}$	·	$\frac{-1}{2}$	·	·
2	2	·	·	$\frac{-1}{2}$	·	$\frac{1}{2}$	$\frac{-1}{\sqrt{2}}$
3	1	·	$\frac{-1}{2}$	·	$\frac{1}{2}$	$\frac{1}{\sqrt{2}}$
3	2	$\frac{1}{\sqrt{2}}$	$\frac{-1}{2}$	·	$\frac{1}{2}$	·	·

(f) $T_1 \otimes A_2$

T_1	A_2	T_2 1	2	3
1		·	·	1
2		·	-1	·
3		1	·	·

(g) $T_2 \otimes A_2$

T_2	A_2	T_1 1	2	3
1		·	·	1
2		·	-1	·
3		1	·	·

(h) $E \otimes E$

E	E	A_1	A_2	E 1	2
1	1	$\frac{1}{\sqrt{2}}$	·	$\frac{1}{\sqrt{2}}$	·
1	2	·	$\frac{1}{\sqrt{2}}$	·	$\frac{-1}{\sqrt{2}}$
2	1	·	$\frac{-1}{\sqrt{2}}$	·	$\frac{-1}{\sqrt{2}}$
2	2	$\frac{1}{\sqrt{2}}$	·	$\frac{-1}{\sqrt{2}}$	·

(i) $E \otimes E$ and $A_2 \otimes A_2$

E	A_2	E 1	2
1		·	1
2		1	·

A_2	A_2	A_1
		1

F.4. CHARACTER TABLES FOR OCTAHEDRAL SUBGROUPS

Tables F.4.1 through F.4.7 list the characters of the important types of subgroups of octahedral O and O_h symmetry. The characters of O and O_h are given first and then subgroups are listed.

TABLE F.4.1 Octahedral Characters for Groups O and $O_h = O \times C_i$

O_h	1	r, r^2	R^2	R, R^3	i	I	Ir, Ir^2	IR^2	IR, IR^3	Ii
A_{1g}	1	1	1	1	1	1	1	1	1	1
A_{2g}	1	1	1	-1	-1	1	1	1	-1	-1
E_g	2	-1	2	0	0	2	-1	2	0	0
T_{1g}	3	0	-1	1	-1	3	0	-1	1	-1
T_{2g}	3	0	-1	-1	1	3	0	-1	-1	1
A_{1u}	1	1	1	1	1	-1	-1	-1	-1	-1
A_{2u}	1	1	1	-1	-1	-1	-1	-1	1	1
E_u	2	-1	2	0	0	-2	1	-2	0	0
T_{1u}	3	0	-1	1	-1	-3	0	1	-1	1
T_{2u}	3	0	-1	-1	1	-3	0	1	1	-1

TABLE F.4.2 Characters for Dihedral Groups D_4, D_3, and Two Kinds of D_2.

D_4	1	R_3^2	R_3, R_3^3	R_1^2, R_2^2	i_3, i_4
A_1	1	1	1	1	1
B_1	1	1	-1	1	-1
A_2	1	1	1	-1	-1
B_2	1	1	-1	-1	1
E	2	-2	0	0	0

D_3	1	r_1, r_1^2	i_2, i_4, i_5
A_1	1	1	1
A_2	1	1	-1
E	2	-1	0

D_2'	1	R_3^2	R_1^2	R_2^2
D_2	1	R_3^2	i_3	i_4
A_1	1	1	1	1
B_1	1	-1	1	-1
A_2	1	1	-1	-1
B_2	1	-1	-1	1

TABLE F.4.3 Characters for Cyclic Groups $C_4 \cong C_{2s}$, C_3, and Two Kinds of C_2.

C_{2s}	1	IR_3	R_3^2	IR_3^3
C_4	1	R_3	R_3^2	R_3^3
0_4	1	1	1	1
1_4	1	$-i$	-1	i
2_4	1	-1	1	-1
3_4	1	i	-1	$-i$

C_3	1	r_1	r_1^2
0_3	1	1	1
1_3	1	$e^{\frac{-2\pi i}{3}}$	$e^{\frac{2\pi i}{3}}$
2_3	1	$e^{\frac{2\pi i}{3}}$	$e^{\frac{-2\pi i}{3}}$

C_2'	1	R_3^2
C_2	1	i_3
0_2	1	1
1_2	1	-1

TABLE F.4.4 Characters for C_{4v}, C_{3v}, and Two C_{2v} Groups That Are Isomorphic to D_4, D_3, and D_2, Respectively.

C_{4v}	1	R_3^2	R_3, R_3^3	IR_1^2, IR_2^2	Ii_3, Ii_4
A'	1	1	1	1	1
B'	1	1	-1	1	-1
A''	1	1	1	-1	-1
B''	1	1	-1	-1	1
E	2	-2	0	0	0

C_{3v}	1	r_1, r_1^2	Ii_2, Ii_4, Ii_5
A'	1	1	1
A''	1	1	-1
E	2	-1	0

C_{2v}'	1	R_3^2	IR_2^2	IR_1^2
C_{2v}	1	i_4	IR_3^2	Ii_3
A'	1	1	1	1
B'	1	-1	1	-1
A''	1	1	-1	-1
B''	1	-1	-1	1

TABLE F.4.5 Characters for D_{2d} and D_{3d} Groups That Are Isomorphic to D_4 and $D_3 \times C_i$, Respectively.

D_{2d}	1	R_3^2	IR_3, IR_3^3	R_1^2, R_2^2	Ii_3, Ii_4
A_1	1	1	1	1	1
B_1	1	1	−1	1	−1
A_2	1	1	1	−1	−1
B_2	1	1	−1	−1	1
E	2	−2	0	0	0

D_{3d}	1	r_1, r_1^2	i_2, i_4, i_5	I	Ir_1, Ir_1^2	Ii_2, Ii_4, Ii_5
A_{1g}	1	1	1	1	1	1
A_{2g}	1	1	−1	1	1	−1
E_g	2	−1	0	2	−1	0
A_{1u}	1	1	1	−1	−1	−1
A_{2u}	1	1	−1	−1	−1	1
E_u	2	−1	0	−2	1	0

D_{4d} is isomorphic to D_8, is not contained in O_h, and is not an allowed crystal point symmetry.

TABLE F.4.6 Characters for $D_{4h} = D_4 \times C_i$ and $D_{2h} = D_2 \times C_i$ ($D_{3h} = D_3 \times C_h \cong D_6$ is not contained in O_h.)

D_{4h}	1	R_3^2	R_3, R_3^3	R_1^2, R_2^2	i_3, i_4	I	IR_3^2	IR_3, IR_3^3	IR_1^2, IR_2^2	Ii_3, Ii_4
A_{1g}	1	1	1	1	1	1	1	1	1	1
B_{1g}	1	1	−1	1	−1	1	1	−1	1	−1
A_{2g}	1	1	1	−1	−1	1	1	1	−1	−1
B_{2g}	1	1	−1	−1	1	1	1	−1	−1	1
E_g	2	−2	0	0	0	2	−2	0	0	0
A_{1u}	1	1	1	1	1	−1	−1	−1	−1	−1
B_{1u}	1	1	−1	1	−1	−1	−1	1	−1	1
A_{2u}	1	1	1	−1	−1	−1	−1	−1	1	1
B_{2u}	1	1	−1	−1	1	−1	−1	1	1	−1
E_u	2	−2	0	0	0	−2	2	0	0	0

D'_{2h}	1	R_3^2	R_1^2	R_2^2	I	IR_3^2	IR_1^2	IR_2^2
D_{2h}	1	R_3^2	i_3	i_4	I	IR_3^2	Ii_3	Ii_4
A_{1g}	1	1	1	1	1	1	1	1
B_{1g}	1	−1	1	−1	1	−1	1	−1
A_{2g}	1	1	−1	−1	1	1	−1	−1
B_{2g}	1	−1	−1	1	1	−1	−1	1
A_{1u}	1	1	1	1	−1	−1	−1	−1
B_{1u}	1	−1	1	−1	−1	1	−1	1
A_{2u}	1	1	−1	−1	−1	−1	1	1
B_{2u}	1	−1	−1	1	−1	1	1	−1

TABLE F.4.7 Characters for $C_{4h} = C_{4i} = C_4 \times C_i$, $C_{3i} = C_3 \times C_i$, and $C_{2h} = C_{2i} = C_2 \times C_i$.

C_{4h}	1	R_3	R_3^2	R_3^3	I	IR_3	IR_3^2	IR_3^3
0_{4g}	1	1	1	1	1	1	1	1
1_{4g}	1	$-i$	-1	i	1	$-i$	-1	i
2_{4g}	1	-1	1	-1	1	-1	1	-1
3_{4g}	1	i	-1	$-i$	1	i	-1	$-i$
0_{4u}	1	1	1	1	-1	-1	-1	-1
1_{4u}	1	$-i$	-1	i	-1	i	1	$-i$
2_{4u}	1	-1	1	-1	-1	1	-1	1
3_{4u}	1	i	-1	$-i$	-1	$-i$	1	i

C_{3i}	1	r_1	r_1^2	I	Ir_1	Ir_1^2
0_{3g}	1	1	1	1	1	1
1_{3g}	1	ε^*	ε^*	1	ε^*	ε
2_{3g}	1	ε	ε^*	1	ε	ε^*
0_{3u}	1	1	1	-1	-1	-1
1_{3u}	1	ε^*	ε	-1	$-\varepsilon^*$	$-\varepsilon$
2_{3u}	1	ε	ε^*	-1	$-\varepsilon$	$-\varepsilon^*$

where $\varepsilon = e^{2\pi i/3}$

C'_{2h}	1	R_3^2	I	IR_3^2
C_{2h}	1	i_4	I	Ii_4
0_{2g}	1	1	1	1
1_{2g}	1	-1	1	-1
0_{2u}	1	1	-1	-1
1_{2u}	1	-1	-1	1

The symmetry $C_{3h} = C_3 \times C_h$ is distinct from $C_{3i} = C_3 \times C_i$ but isomorphic to it:

C_{3h}	1	r_1	r_1^2	σ	σr_1	σr_1^2
$0'_3$	1	1	1	1	1	1
$1'_3$	1	ε^*	ε	1	ε^*	ε
$2'_3$	1	ε	ε^*	1	ε	ε^*
$0''_3$	1	1	1	-1	-1	-1
$1''_3$	1	ε^*	ε	-1	$-\varepsilon^*$	$-\varepsilon$
$2''_3$	1	ε	ε^*	-1	$-\varepsilon$	$-\varepsilon^*$

σ = reflection through a mirror plane normal to r_1 axis

where $\varepsilon = e^{2\pi i/3}$

TABLE F.4.8 Characters for $C_h = C_v \cong C_i \cong C_2$

C_i	1	I
C_h	1	σ
C_2	1	R_3^2
0_2	1	1
1_2	1	-1

$\sigma = IR_1^2$ = reflection through a mirror-plane normal to R_1 axis
$R_1^2 = 180°$ x-axis rotation.

F.5. CORRELATION TABLES FOR OCTAHEDRAL SUBGROUPS

Tables F.5.1 through F.5.5 list the correlations of representations of select subgroups of octahedral O and O_h symmetry. Rows have subduced representations $\Gamma \downarrow H$. Columns have induced representations $\gamma(H) \uparrow O$. Details and applications of tables are described in Chapter 4.

APPENDIX F   823

TABLE F.5.1 Correlations of O with Dihedral Groups D_4, D_3, $D_2 = \{1, R_3^2, i_3, i_4\}$, and $D_2' = \{1, R_3^2, R_1^2, R_2^2\}$.

$O \supset D_4$	A_1	A_2	B_1	B_2	E
$A_1 \downarrow D_4$	1	·	·	·	·
$A_2 \downarrow D_4$	·	·	1	·	·
$E \downarrow D_4$	1	·	1	·	·
$T_1 \downarrow D_4$	·	1	·	·	1
$T_2 \downarrow D_4$	·	·	·	1	1

$O \supset D_3$	A_1	A_2	E
$A_1 \downarrow D_3$	1	·	·
$A_2 \downarrow D_3$	·	1	·
$E \downarrow D_3$	·	·	1
$T_1 \downarrow D_3$	·	1	1
$T_2 \downarrow D_3$	1	·	1

$O \supset D_2$	A_1	B_1	A_2	B_2
$A_1 \downarrow D_2$	1	·	·	·
$A_2 \downarrow D_2$	·	·	1	·
$E \downarrow D_2$	1	·	1	·
$T_1 \downarrow D_2$	·	1	1	1
$T_2 \downarrow D_2$	1	1	·	1

$O \supset D_2'$	A_1	B_1	A_2	B_2
$A_1 \downarrow D_2'$	1	·	·	·
$A_2 \downarrow D_2'$	1	·	·	·
$E \downarrow D_2'$	2	·	·	·
$T_1 \downarrow D_2'$	·	1	1	1
$T_2 \downarrow D_2'$	·	1	1	1

TABLE F.5.2 Correlations of O with Cyclic Groups C_4, C_3, $C_2 = \{1, i_3\}$, and $C_2' = \{1, R_3^2\}$.

$O \supset C_4$	0_4	1_4	2_4	3_4
$A_1 \downarrow C_4$	1	·	·	·
$A_2 \downarrow C_4$	·	·	1	·
$E \downarrow C_4$	1	·	1	·
$T_1 \downarrow C_4$	1	1	·	1
$T_2 \downarrow C_4$	·	1	1	1

$O \supset C_3$	0_3	1_3	2_3
$A_1 \downarrow C_3$	1	·	·
$A_2 \downarrow C_3$	1	·	·
$E \downarrow C_3$	·	1	1
$T_1 \downarrow C_3$	1	1	1
$T_2 \downarrow C_3$	1	1	1

$O \supset C_2$	0_2	1_2
$A_1 \downarrow C_2$	1	·
$A_2 \downarrow C_2$	·	1
$E \downarrow C_2$	1	1
$T_1 \downarrow C_2$	1	2
$T_2 \downarrow C_2$	2	1

$O \supset C_2'$	0_2	1_2
$A_1 \downarrow C_2'$	1	·
$A_2 \downarrow C_2'$	1	·
$E \downarrow C_2'$	2	·
$T_1 \downarrow C_2'$	1	2
$T_2 \downarrow C_2'$	1	2

TABLE F.5.3 Correlations of D_4 with Subgroups C_4, $D_2 = \{1, R_3^2, i_3, i_4\}$, and $D_2' = \{1, R_3^2, R_1^2, R_2^2\}$

$D_4 \supset C_4$	0_4	1_4	2_4	3_4
$A_1 \downarrow C_4$	1	·	·	·
$B_1 \downarrow C_4$	·	·	1	·
$A_2 \downarrow C_4$	1	·	·	·
$B_2 \downarrow C_4$	·	·	1	·
$E \downarrow C_4$	·	1	·	1

$D_4 \supset D_2$	A_1	B_1	A_2	B_2
$A_1 \downarrow D_2$	1	·	·	·
$B_1 \downarrow D_2$	·	·	1	·
$A_2 \downarrow D_2$	·	·	1	·
$B_2 \downarrow D_2$	1	·	·	·
$E \downarrow D_2$	·	1	·	1

$D_4 \supset D_2'$	A_1	B_1	A_2	B_2
$A_1 \downarrow D_2'$	1	·	·	·
$B_1 \downarrow D_2'$	1	·	·	·
$A_2 \downarrow D_2'$	·	·	1	·
$B_2 \downarrow D_2'$	·	·	1	·
$E \downarrow D_2'$	·	1	·	1

TABLE F.5.4 Correlations of D_3 with Subgroups C_3 and C_2.

$D_3 \supset C_3$	0_3	1_3	2_3
$A_1 \downarrow C_3$	1	·	·
$A_2 \downarrow C_3$	1	·	·
$E \downarrow C_3$	·	1	1

$D_3 \supset C_2$	0_2	1_2
$A_1 \downarrow C_2$	1	·
$A_2 \downarrow C_2$	·	1
$E \downarrow C_2$	1	1

TABLE F.5.5 Correlations of O_h with Groups C_{4v}, C_{3v}, $C_{2v} = \{1, i_4, \mathrm{IR}_3^2, \mathrm{Ii}_3\}$, and $C'_{2v} = \{1, \mathrm{R}_3^2, \mathrm{IR}_1^2, \mathrm{IR}_2^2\}$.

$O_h \supset C_{4v}$	A'	B'	A''	B''	E
$A_{1g} \downarrow C_{4v}$	1	·	·	·	·
$A_{2g} \downarrow C_{4v}$	·	1	·	·	·
$E_g \downarrow C_{4v}$	1	1	·	·	·
$T_{1g} \downarrow C_{4v}$	·	·	1	·	1
$T_{2g} \downarrow C_{4v}$	·	·	·	1	1
$A_{1u} \downarrow C_{4v}$	·	·	1	·	·
$A_{2u} \downarrow C_{4v}$	·	·	·	1	·
$E_u \downarrow C_{4v}$	·	·	1	1	·
$T_{1u} \downarrow C_{4v}$	1	·	·	·	1
$T_{2u} \downarrow C_{4v}$	·	1	·	·	1

$O_h \supset C_{3v}$	A'	A''	E
$A_{1g} \downarrow C_{3v}$	1	·	·
$A_{2g} \downarrow C_{3v}$	·	1	·
$E_g \downarrow C_{3v}$	·	·	1
$T_{1g} \downarrow C_{3v}$	·	1	1
$T_{2g} \downarrow C_{3v}$	1	·	1
$A_{1u} \downarrow C_{3v}$	·	1	·
$A_{2u} \downarrow C_{3v}$	1	·	·
$E_u \downarrow C_{3v}$	·	·	1
$T_{1u} \downarrow C_{3v}$	1	·	1
$T_{2u} \downarrow C_{3v}$	·	1	1

$O_h \supset C_{2v}$	A'	B'	A''	B''
$A_{1g} \downarrow C_{2v}$	1	·	·	·
$A_{2g} \downarrow C_{2v}$	·	1	·	·
$E_g \downarrow C_{2v}$	1	1	·	·
$T_{1g} \downarrow C_{2v}$	·	1	1	1
$T_{2g} \downarrow C_{2v}$	1	·	1	1
$A_{1u} \downarrow C_{2v}$	·	·	1	·
$A_{2u} \downarrow C_{2v}$	·	·	·	1
$E_u \downarrow C_{2v}$	·	·	1	1
$T_{1u} \downarrow C_{2v}$	1	1	·	1
$T_{2u} \downarrow C_{2v}$	1	1	1	·

$O_h \supset C'_{2v}$	A'	B'	A''	B''
$A_{1g} \downarrow C'_{2v}$	1	·	·	·
$A_{2g} \downarrow C'_{2v}$	1	·	·	·
$E_g \downarrow C'_{2v}$	2	·	·	·
$T_{1g} \downarrow C'_{2v}$	·	1	1	1
$T_{2g} \downarrow C'_{2v}$	·	1	1	1
$A_{1u} \downarrow C'_{2v}$	·	·	1	·
$A_{2u} \downarrow C'_{2v}$	·	·	1	·
$E_u \downarrow C'_{2v}$	·	·	2	·
$T_{1u} \downarrow C'_{2v}$	1	1	·	1
$T_{2u} \downarrow C'_{2v}$	1	1	·	1

F.6. HEXAGONAL SYMMETRIES

All hexagonal and trigonal symmetry groups are subgroups of D_{6h} (Table (F.6.1). They are all isomorphic to outer products involving only C_2, C_3, and D_3. D_{6h} itself is isomorphic to $D_3 \times C_2 \times C_2$. See Chapter 3.

TABLE F.6.1 Characters of D_{6h}.

D_{6h}	1	h, h^5	ρ_1, ρ_2, ρ_3	h^3	h^2, h^4	$\rho'_1, \rho'_2, \rho'_3$	I	Ih, Ih^5	$\sigma_1, \sigma_2, \sigma_3$	σ	Ih^2, Ih^4	$\sigma'_1, \sigma'_2, \sigma'_3$
A_{1g}	1	1	1	1	1	1	1	1	1	1	1	1
A_{2g}	1	1	-1	1	1	-1	1	1	-1	1	1	-1
E_{2g}	2	-1	0	2	-1	0	2	-1	0	2	-1	0
B_{1g}	1	1	1	-1	-1	-1	1	1	1	-1	-1	-1
B_{2g}	1	1	-1	-1	-1	1	1	1	-1	-1	-1	1
E_{1g}	2	-1	0	-2	1	0	2	-1	0	-2	1	0
A_{1u}	1	1	1	1	1	1	-1	-1	-1	-1	-1	-1
A_{2u}	1	1	-1	1	1	-1	-1	-1	1	-1	-1	1
E_{2u}	2	-1	0	2	-1	0	-2	1	0	-2	1	0
B_{1u}	1	1	1	-1	-1	-1	-1	-1	-1	1	1	1
B_{2u}	1	1	-1	-1	-1	1	-1	-1	1	1	1	-1
E_{1u}	2	-1	0	-2	1	0	-2	1	0	2	-1	0

F.7. ICOSAHEDRAL AND PENTAGONAL SYMMETRIES

The rotational symmetry Y of the icosahedron was mentioned after Figure 4.1.6. It has 60 elements in 5 classes of rotations: one of $0°$, 12 each of $72°$ and $144°$, 20 of $120°$, and 15 of $180°$. These are listed at the top of its character table (Table F.7.1). The largest subgroup is the pentagonal dihedral group D_5. The D_5 characters are given in Table F.7.2.

Representations and correlation tables involving these symmetries as well as applications to the C_{60} rotation and vibration problem can be found in the sequel to this text and in the references [1]–[5] listed below. Reference [4] contains the entire Y-group table.

TABLE F.7.1 Icosahedral (Y) Group Characters

Y classes	$0°$	$72°$	$144°$	$120°$	$180°$
$°c_g$	1	12	12	20	15
A	1	1	1	1	1
T_1	3	$\dfrac{1+\sqrt{5}}{2}$	$\dfrac{1-\sqrt{5}}{2}$	0	-1
T_3	3	$\dfrac{1-\sqrt{5}}{2}$	$\dfrac{1+\sqrt{5}}{2}$	0	-1
G	4	-1	-1	1	0
H	5	0	0	-1	1

TABLE F.7.2 Pentagonal Dihedral (D_5) Group Characters

D_5 classes	0°	72°	144°	180°
$°c_g$	1	2	2	5
A_1	1	1	1	1
A_2	1	1	1	-1
E_1	2	$\dfrac{-1+\sqrt{5}}{2}$	$\dfrac{-1-\sqrt{5}}{2}$	0
E_2	2	$\dfrac{-1-\sqrt{5}}{2}$	$\dfrac{-1+\sqrt{5}}{2}$	0

REFERENCES

1. W. G. Harter and D. E. Weeks, *Chem. Phys. Lett.* **132**, 387 (1986).
2. D. E. Weeks and W. G. Harter, *Chem. Phys. Lett.* **144**, 366 (1988).
3. W. G. Harter and D. E. Weeks, *J. Chem. Phys.* **90**, 4727 (1989).
4. D. E. Weeks and W. G. Harter, *J. Chem. Phys.* **90**, 4744 (1989).
5. W. G. Harter and T. C. Reimer, *Chem. Phys. Lett.* **194**, 230 (1992).

APPENDIX G

SCHUR'S LEMMA AND IRREDUCIBLE REPRESENTATION ORTHOGONALITY

G.1. COMMUTATION AND SCHUR'S LEMMA

Suppose an all-commuting operator $\mathbb{C}$ commutes with all rotations $R = R(\alpha\beta\gamma)$, that is, $R\mathbb{C} = \mathbb{C}R$ for all R. Then it must also commute with all linear combinations of R including the generators or angular momentum operators J_x, J_y, and J_z which generate infinitesimal rotations. [For example, $R(\varepsilon, 0, 0) = 1 + i\varepsilon J_z$.] This means the magnitude operator $J^2 = J_x^2 + J_y^2 + J_z^2$ must commute with $\mathbb{C}$, as well as the raising and lowering operators $J_\pm = J_x + iJ_y$.

Commutation of $\mathbb{C}$ with J^2 and J_z means that their eigenstates $\left|\begin{matrix}j\\m\end{matrix}\right\rangle$ must also be eigenstates of $\mathbb{C}$. Using $J_z\mathbb{C} = \mathbb{C}J_z$ and $J^2\mathbb{C} = \mathbb{C}J^2$, we have

$$\left\langle\begin{matrix}j'\\m'\end{matrix}\right|J_z\mathbb{C}\left|\begin{matrix}j\\m\end{matrix}\right\rangle - \left\langle\begin{matrix}j'\\m'\end{matrix}\right|\mathbb{C}J_z\left|\begin{matrix}j\\m\end{matrix}\right\rangle = 0 = [m' - m]\left\langle\begin{matrix}j'\\m'\end{matrix}\right|\mathbb{C}\left|\begin{matrix}j\\m\end{matrix}\right\rangle$$

$$\left\langle\begin{matrix}j'\\m'\end{matrix}\right|J^2\mathbb{C}\left|\begin{matrix}j\\m\end{matrix}\right\rangle - \left\langle\begin{matrix}j'\\m'\end{matrix}\right|\mathbb{C}J^2\left|\begin{matrix}j\\m\end{matrix}\right\rangle = 0 = [j'(j' + 1) - j(j + 1)]\left\langle\begin{matrix}j'\\m'\end{matrix}\right|\mathbb{C}\left|\begin{matrix}j\\m\end{matrix}\right\rangle$$

for J-eigenstates satisfying standard angular momentum relations derived in Appendix E,

$$J_z\left|\begin{matrix}j\\m\end{matrix}\right\rangle = m\left|\begin{matrix}j\\m\end{matrix}\right\rangle, \qquad J^2\left|\begin{matrix}j\\m\end{matrix}\right\rangle = \sqrt{j(j+1)}\left|\begin{matrix}j\\m\end{matrix}\right\rangle.$$

Therefore the matrix $\left\langle\begin{matrix}j'\\m'\end{matrix}\right|\mathbb{C}\left|\begin{matrix}j\\m\end{matrix}\right\rangle$ is zero unless $j' = j$ and $m' = m$. Hence, $\left|\begin{matrix}j\\m\end{matrix}\right\rangle$ are *eigenvectors* of $\mathbb{C}$,

$$\left\langle\begin{matrix}j'\\m'\end{matrix}\right|\mathbb{C}\left|\begin{matrix}j\\m\end{matrix}\right\rangle = c_m^j \delta^{jj'}\delta_{mm'} \quad \text{or} \quad \mathbb{C}\left|\begin{matrix}j\\m\end{matrix}\right\rangle = c_m^j\left|\begin{matrix}j\\m\end{matrix}\right\rangle. \tag{G.1}$$

Commutation ($J_\pm \mathbb{C} = \mathbb{C}J_\pm$) of $\mathbb{C}$ with raising and lowering operators and their irreducible representations $J_\pm \left|{j \atop m}\right\rangle = D^j_{m\pm1,m}\left|{j \atop m\pm1}\right\rangle$ gives

$$J_\pm \mathbb{C}\left|{j \atop m}\right\rangle - \mathbb{C}J_\pm\left|{j \atop m}\right\rangle = 0 = (c^j_m - c^j_{m\pm1})\left|{j \atop m\pm1}\right\rangle D^j_{m\pm1,m}(J_\pm).$$

Since $D^j_{m\pm1,m}(J_\pm)$ are nonzero for allowed values of m ($-j \le m \le j$) [recall (5.4.23)], we have $c^j_m = c^j_{m\pm1} = c^j_{m\pm2} = \cdots \equiv c^j$. Therefore $\mathbb{C}$ reduces to a multiple c^j of the unit matrix:

$$\left\langle {j' \atop m'}\right|\mathbb{C}\left|{j \atop m}\right\rangle = c^j \delta^{jj'}\delta_{mm'}. \tag{G.2}$$

Schur's Lemma: Only a multiple of the unit matrix commutes with *all* irreducible matrices or irreps $D^j_{m,m'}(R) = \left\langle {j \atop m}\right|R\left|{j \atop m'}\right\rangle$ of a group.

The same holds for finite group representations $\left\langle {\alpha \atop a}\right|\mathbb{C}\left|{\beta \atop b}\right\rangle$ of an all-commuting operator $\mathbb{C}$ since $\mathbb{C}$ must commute with elementary operators P^α_{ab} in analogy to ladder operators $J_\pm$,

$$P^\alpha_{ab}\left|{\gamma \atop c}\right\rangle = \delta^{\alpha\gamma}\delta_{bc}\left|{\alpha \atop a}\right\rangle.$$

Therefore $\mathbb{C}$ is represented by a multiple of the unit matrix in that case, as well:

$$\left\langle {\alpha \atop a}\right|\mathbb{C}\left|{\beta \atop b}\right\rangle = c^\alpha \delta^{\alpha\beta}\delta_{ab}. \tag{G.3}$$

G.2. REPRESENTATION ORTHOGONALITY

All-commuting operators can be constructed using sums or integrals over all group operators. The following operator integral is an example:

$$\mathbb{C} = \int_{\text{all } R} RER^{-1} = \int_{\text{all } T} TET^{-1}. \tag{G.4}$$

Here E is arbitrary and the integral is over the appropriate range of Euler angles for $R(3)$ or $SU(2)$ or it is a sum for finite groups:

$$\int_{\text{all } R} \equiv \begin{cases} \int_0^{2\pi} d\alpha \int_0^{\pi} \sin\beta \, d\beta \int_0^{2\pi} d\gamma = 8\pi^2, & \text{for } R(3), \\ \int_0^{2\pi} d\alpha \int_{-\pi}^{\pi} \sin\beta \, d\beta \int_0^{2\pi} d\gamma = 16\pi^2, & \text{for SU(2)} \\ \sum_{\text{all } R} = {}^\circ G, & \text{for finite group } G \text{ of order } {}^\circ G. \end{cases}$$
(G.5)

Any $R(3)$ rotation or $SU(2)$ transformation S commutes with $\mathbb{C}$ because the sum over group operators is supposed to be invariant to a change of summation operator from R to $T = SR$,

$$S\mathbb{C} = \int_{\text{all } R} SRER^{-1} = \int_{\text{all } T} TET^{-1}S = \mathbb{C}S, \tag{G.6}$$

where $R^{-1} = T^{-1}S$ was used. Now consider a representation of $\mathbb{C}$ is a basis $\left| \begin{matrix} j \\ m \end{matrix} \right\rangle$ for which matrix

$$\left\langle \begin{matrix} j \\ m \end{matrix} \middle| R \middle| \begin{matrix} j' \\ m' \end{matrix} \right\rangle = \delta^{jj'} D^j_{mm'}(R) \tag{G.7}$$

is irreducible and unitary,

$$D^j_{mn}(R^{-1}) = D^{j*}_{nm}(R). \tag{G.8}$$

The representation of operator products is the matrix product of their representations:

$$\left\langle \begin{matrix} j \\ m \end{matrix} \middle| RS \middle| \begin{matrix} j \\ n \end{matrix} \right\rangle = \sum_{k=-j}^{j} \left\langle \begin{matrix} j \\ m \end{matrix} \middle| R \middle| \begin{matrix} j \\ k \end{matrix} \right\rangle \left\langle \begin{matrix} j \\ k \end{matrix} \middle| S \middle| \begin{matrix} j \\ n \end{matrix} \right\rangle,$$

$$D^j_{mn}(RS) = \sum_{k=-j}^{j} D^j_{mk}(R) D^j_{kn}(S). \tag{G.9}$$

This gives

$$\left\langle \begin{matrix} j \\ m \end{matrix} \middle| \mathbb{C} \middle| \begin{matrix} j' \\ m' \end{matrix} \right\rangle = \int_{\text{all } R} \left\langle \begin{matrix} j \\ m \end{matrix} \middle| RER^{-1} \middle| \begin{matrix} j' \\ m' \end{matrix} \right\rangle$$

$$= \int_{\text{all } R} \sum_{k, k'=-j}^{j} \left\langle \begin{matrix} j \\ m \end{matrix} \middle| R \middle| \begin{matrix} j \\ k \end{matrix} \right\rangle \left\langle \begin{matrix} j \\ k \end{matrix} \middle| E \middle| \begin{matrix} j' \\ k' \end{matrix} \right\rangle \left\langle \begin{matrix} j' \\ k' \end{matrix} \middle| R^{-1} \middle| \begin{matrix} j' \\ m' \end{matrix} \right\rangle. \tag{G.10}$$

By Schur's lemma (G.2), this reduces to

$$c^j \delta^{jj'} \delta_{mm'} = \int_{\text{all } R} \sum_{kk'=-j}^{j} D^j_{mk}(R) E^{jj'}_{kk'} D^{j'}_{k'm'}(R^{-1}). \qquad (G.11)$$

We now make an elementary choice, $E = e_{nn'}$ for the arbitrator operator E:

$$E^{jj'}_{kk'} = \left\langle \begin{matrix} j \\ k \end{matrix} \middle| E \middle| \begin{matrix} j' \\ k' \end{matrix} \right\rangle = \delta_{kn} \delta_{k'n'}. \qquad (G.12)$$

This gives the orthogonality relation

$$c^j \delta^{jj'} \delta_{mm'} = \int_{\text{all } R} D^j_{mn}(R) D^{j'}_{n'm'}(R^{-1}). \qquad (G.13)$$

The constant c^j is found by summing over all $l^j = 2j + 1$ states for $j = j'$ and $m = m' = -j$ to j. Setting $S = R^{-1}$ in Eq. (G.9) gives the identity representation, which must be a unit matrix δ_{mn}:

$$\delta_{mn} = D^j_{mn}(1) = \sum_k D^j_{mk}(R) D^j_{kn}(R^{-1}).$$

This is used to reduce the following:

$$l^j c^j = \sum_{m=-j}^{j} c^j = \int_{\text{all } R} \sum_{m=-j}^{j} D^j_{n'm}(R^{-1}) D^j_{mn}(R) = \int_{\text{all } R} \delta_{n'n} = \delta_{n'n} \, G \text{ volume},$$

$$(G.14)$$

where the group volume is given by (G.5),

$$G \text{ volume} = \begin{cases} 16\pi^2 & \text{for SU(2)}, \\ 8\pi^2, & \text{for } R(3), \\ {}^\circ G, & \text{for finite group of order } {}^\circ G, \end{cases} \qquad (G.15)$$

and l^j is the dimension of the irreducible representation D^j. For $R(3)$ and SU(2) it is

$$l^j = 2j + 1 \qquad (j = 0, \tfrac{1}{2}, 1, \tfrac{3}{2}, 2, \ldots). \qquad (G.16)$$

Finding the constant

$$c^j = \delta_{n'n} \frac{G \text{ volume}}{l^j} \qquad (G.17)$$

completes the derivation of the following *grand orthogonality relation*:

$$\int_{\text{all } R} D^j_{mn}(R) D^{j'}_{n'm'}(R^{-1}) = \frac{G \text{ volume}}{l^j} \delta^{jj'} \delta_{mm'} \delta_{nn'}. \tag{G.18}$$

For unitary D-matrices we have

$$\int_{\text{all } R} D^j_{mn}(R) D^{j'*}_{m'n'}(R) = \frac{G \text{ volume}}{l^j} \delta^{jj'} \delta_{mm'} \delta_{nn'}. \tag{G.19}$$

For the $R(3)$ group this becomes

$$\int_0^{2\pi} d\alpha \int_0^{\pi} \sin\beta \, d\beta \int_0^{2\pi} d\gamma \, D^j_{mn}(R) D^{j'*}_{m'n'}(R) = \frac{8\pi^2}{2j+1} \delta^{jj'} \delta_{mm'} \delta_{nn'}. \tag{G.20a}$$

For SU(2), the $8\pi^2$ is doubled to $16\pi^2$ and the limits of the β-integral are doubled to $-\pi \le \beta \le \pi$. For a finite group G of order $^\circ G$ with irreducible representations D^α of dimensions l^α, we have

$$\sum_{R \in G} D^\alpha_{aa'}(R) D^{\beta*}_{bb'}(R) = \frac{^\circ G}{l^\alpha} \delta^{\alpha\beta} \delta_{ab} \delta_{a'b'}. \tag{G.20b}$$

G.3. IRREDUCIBLE PROJECTION OPERATORS

The group projection operators are defined as follows:

$$P^j_{mn} = \frac{l^j}{G \text{ volume}} \int_R D^{j*}_{mn}(R) R. \tag{G.21}$$

The product of project operators will be evaluated now:

$$P^j_{mn} P^{j'}_{m'n'} = \frac{l^j l^{j'}}{(G \text{ volume})^2} \int_R \int_S D^{j*}_{mn}(R) D^{j'*}_{m'n'}(S) RS. \tag{G.22}$$

Let $RS = T$ so $S = R^{-1}T$ and use matrix multiplication (G.9),

$$P^j_{mn} P^{j'}_{m'n'} = \frac{l^j l^{j'}}{(G \text{ volume})^2} \int_R \int_S D^{j*}_{mn}(R) D^{j'*}_{m'n'}(R^{-1}T) T$$

$$= \frac{l^j l^{j'}}{(G \text{ volume})^2} \sum_k \int_R D^{j*}_{mn}(R) D^{j'*}_{m'k}(R^{-1}) \int_T D^{j'*}_{kn'}(T) T. \tag{G.23}$$

Now orthogonality (G.18) simplifies the product:

$$P^j_{mn} P^{j'}_{m'n'} = \frac{l^j}{G \text{ volume}} \sum_k \delta^{jj'} \delta_{nm'} \delta_{mk} \int_T D^{j*}_{kn'}(T) T$$

$$= \delta^{jj'} \delta_{nm'} \frac{l^j}{G \text{ volume}} \int_T D^{j*}_{mn'}(T) T. \tag{G.24}$$

An elementary matrix algebra product rule results from (G.21),

$$P^j_{mn} P^{j'}_{m'n'} = \delta^{jj'} \delta_{nm'} P^j_{mn'}. \tag{G.25}$$

So diagonal operators like P_{11} are idempotent ($P_{11} P_{11} = P_{11}$) while others like P_{12} are nilpotent ($P_{12} P_{12} = 0$). P-products involving different j vanish ($P^j_{mn} P^{j'}_{m'n'} \equiv 0$ if $j' \neq j$).

The same algebra may be used to derive the effect of multiplying group operators and P-operators:

$$RP^j_{mn} = \frac{l^j}{G \text{ volume}} \int_S D^{j*}_{mn}(S) R, \tag{G.26}$$

$$RP^j_{mn} = \frac{l^j}{G \text{ volume}} \int_T D^{j*}_{mn}(R^{-1}T) T$$

$$= \sum_{m'} D^{j*}_{mm'}(R^{-1}) \frac{l^j}{G \text{ volume}} \int_T D^{j*}_{m'n}(T) T. \tag{G.27}$$

Finally, unitarity $[D^{j*}_{mm'}(R^{-1}) = D^j_{m'm}(R)]$ gives the left multiplication rule,

$$RP^j_{mn} = \sum_{m'} D^j_{m'm}(R) P^j_{m'n}. \tag{G.28}$$

The right multiplication is derived in a similar way:

$$P^j_{mn} R = \sum_{n'} D^j_{nn'}(R) P^j_{mn'}. \tag{G.29}$$

G.4. COMPLETENESS AND SPECTRAL DECOMPOSITION

Orthogonality relation (G.18) makes it possible to invert the definition (G.21) of the projection operator to the following:

$$R = \sum_j \sum_m \sum_n D^j_{mn}(R) P^j_{mn}. \tag{G.30}$$

This is called the group spectral decomposition and applies to all group operators R and their linear combinations including the P-operators, themselves:

$$P_{m'n'}^{j'} = \sum_j \sum_m \sum_n D_{mn}^j(P_{m'n'}^{j'}) P_{mn}^j. \tag{G.31}$$

Because of the orthogonality (G.25) of P-operators this implies that each representation of P_{mn}^j must be comprised of all zeros except for a single 1 in the (m, n) position of the $D^j(P_{mn}^j)$ matrix,

$$D_{mn}^j(P_{m'n'}^{j'}) = \delta^{jj'}\delta_{mm'}\delta_{nn'}. \tag{G.32}$$

The spectral decomposition of (G.30) of the unit operator **1** is called the *group completeness* relation,

$$\mathbf{1} = \sum_j \sum_m \sum_n D_{mn}^j(1) P_{mn}^j \tag{G.33}$$

Since each representation of the unit operator is a unit matrix $[D_{mn}(1) = \delta_{mn}]$, the completeness relation reduces to a sum of all the idempotents, that is, all diagonal P-operators:

$$\mathbf{1} = \sum_j \sum_m \sum_n \delta_{mn} P_{mn}^j = \sum_{j,m} P_{mm}^j. \tag{G.34}$$

This can be written as follows

$$\mathbf{1} = \sum_j \mathbb{P}^j, \tag{G.35}$$

where

$$\mathbb{P}^j = \sum_m P_{mm}^j \tag{G.36}$$

is a central or all-commuting idempotent. According to (G.32), irreducible representation $D^j(\mathbb{P}^j)$ is a unit matrix, as indeed it must be to satisfy Schur's lemma. However, the matrix $D^j(\mathbb{P}^k)$ is a zero matrix if $j \neq k$.

G.5. CHARACTER RELATIONS AND ALL-COMMUTING PROJECTORS

For a finite group G the all-commuting projectors are defined by

$$\mathbb{P}^\alpha = \sum_m P_{mm}^\alpha = \frac{l^\alpha}{°G} \sum_R \sum_m D_{mm}^{\alpha*}(R) R \tag{G.37}$$

or

$$\mathbb{P}^\alpha = \frac{l^\alpha}{{}^\circ G} \sum_R \chi^{\alpha*}(R) R, \qquad (G.38)$$

where

$$\chi^\alpha(R) = \text{Trace } D^\alpha(R) \qquad (G.39)$$

is the αth *irreducible character* of group G. For infinite groups such as $R(3)$ and SU(2), a similar expression is used with the integral $\int_R$ replacing the sum Σ_R.

However, the application to finite groups is more common since the sum is easily reduced to be just over classes $C_R = \{R, SRS^{-1}, TRT^{-1}, \ldots\}$ of equivalent elements related to R by operator transformation URU^{-1}, where U is in G. Let the order of class C_R be ${}^\circ C_R$. Then

$$\mathbb{P}^\alpha = \frac{l^\alpha}{{}^\circ G} \sum_{\substack{\text{classes}\\R}} \chi_R^{\alpha*} \mathbb{C}_R, \qquad (G.40a)$$

where class sum (recall discussion surrounding Equation (3.2.7) on pages 161–162.)

$$\mathbb{C}_R = R + S^{-1}RS + T^{-1}RT + \cdots = \frac{{}^\circ C_R}{{}^\circ G} \sum_{U \in G} URU^{-1} \qquad (G.40b)$$

is the sum of all elements equivalent to element R and

$$\chi_R^\alpha = \chi^\alpha(R) = \chi^\alpha(TRT^{-1}) = \cdots \qquad (G.40c)$$

is the αth character of any element in class R. Characters of R and TRT^{-1} are equal because the trace of equivalent operators are equal. Class sums $\mathbb{C}_R$ are invariant and all-commuting because (G.40b) has the form (G.4):

$$\mathbb{C}_R = T\mathbb{C}_R T^{-1} \quad \text{or} \quad \mathbb{C}_R T = T\mathbb{C}_R \quad (\text{for all } T \text{ in } G).$$

Therefore their irreducible representations must be multiples of the unit matrix according to Schur's lemma (G.3),

$$D_{ab}^\alpha(\mathbb{C}_R) = c_R^\alpha \delta_{ab}. \qquad (G.41)$$

This implies that their spectral decomposition (G.30) involves just the all-commuting idempotents

$$\mathbb{C}_R = \sum_\alpha \sum_a \sum_b D^\alpha_{ab}(\mathbb{C}_R) P^\alpha_{ab}$$
$$= \sum_\alpha \sum_a c^\alpha_R P^\alpha_{aa}$$
$$= \sum_\alpha c^\alpha_R \mathbb{P}^\alpha. \tag{G.42}$$

This means the constants c^α_R are the *eigenvalues* of $\mathbb{C}_R$. These are related to characters χ^α_R by taking the trace of (G.42),

$$\text{Trace } D^\alpha(\mathbb{C}_R) = \sum_a D^\alpha_{aa}(\mathbb{C}_R) = c^\alpha_R \sum_a \delta_{aa} = c^\alpha_R l^\alpha. \tag{G.43}$$

However, this trace is also a sum over a number ${}^\circ C_R$ of equal characters,

$$\text{Trace } D^\alpha(\mathbb{C}_R) = \chi^\alpha(R) + \chi^\alpha(TRT^{-1}) + \cdots$$
$$= {}^\circ C_R \chi^\alpha_R. \tag{G.44}$$

Equating the last two gives the class eigenvalues

$$c^\alpha_R = \frac{{}^\circ C_R \chi^\alpha_R}{l^\alpha} \tag{G.45a}$$

and the class-sum spectral decomposition

$$\mathbb{C}_R = \sum_\alpha {}^\circ C_R \frac{\chi^\alpha_R}{l^\alpha} \mathbb{P}^\alpha, \tag{G.45b}$$

which is given in terms of the order (${}^\circ C_R$) of class R, irrep dimension (l^α), and αth character χ^α_R of any element R in class sum $\mathbb{C}_R$. Expression (G.45b) is the inverse of (G.40a). Note that ${}^\circ G$, ${}^\circ C_R$, and l^α are integers. Some of the characters are integers, too (in particular, $\chi^\alpha_1 \equiv l^\alpha$). In general χ^α_R is irrational and complex.

Substituting (G.45b) into (G.40a) gives the character orthogonality relation,

$$\frac{1}{{}^\circ G} \sum_R {}^\circ C_R \chi^{\alpha'}_R \chi^\beta_R = \delta^{\alpha\beta}. \tag{G.46a}$$

The corresponding completeness relation is

$$\frac{1}{{}^\circ G} \sum_\alpha {}^\circ C_R \chi^{\alpha*}_R \chi^\alpha_S = \delta_{RS}. \tag{G.46b}$$

INDEX

Abelian groups:
    definition, 69
    products of C_n, 110
Abelian point symmetries, 112
Absorption:
    classical, 522
    quantum, 768, 770
    semiclassical, 735
Acceleration matrix:
    definition, 32
    for C_2 systems, 64, 83
    for C_3 systems, 84
    for C_n systems, 92, 97
    for D_2 systems, 104
    for NH_3 systems, 213
    for SF_6, 293, 296
Ackerhalt, J. R., 306, 788
Action, 140, 690
    Hamilton's characteristic function, 692
    Hamilton's principle function, 691
    reduced action, 692
AC-Stark shifts, 754–760
Addition theorem for R_3, 377
Adiabatic following, 131
    in resonance detuning, 759
ADJ or adjunct matrix, 18, 27, 49. *See also* Appendix B
Adjoint operator, *see* Transpose conjugate
Adjoint representation, *see* Regular representation
Algebra, 71

All-commuting idempotents, *see* Class idempotents
All-commuting operators, *see* Class operators
Ammonia (NH_3):
    inversion doublet, 131
    in resonance, 741
    as two-state system, 629
    vibrational modes, 205–216
Amplitude:
    classical, 31
    probability, 2
Angular momentum:
    in body frame, 366–368
    commutation, 344, 368, 382
    cones, 347, 359–361, 568–569, 608, 622
    conservation, 344
    in lab and body, 380–382
    matrices, 345–346, 434–436
    operators, 345, 434, 436
Angular velocity, in lab and body, 380–382
Anharmonicity, 36
    bifurcation in 2D model, 651–654
    splitting, 573
Anti-bonding orbital, 661, 662
Anti-Hermitian matrix, 53
Asymmetric rotor or top:
    high-J states, 609–618
    $J = 10$ levels, 614
    matrix elements, 371
Avoided crossing, 132, 274, 280

837

Axioms:
　for quantum theory, 4–5
　for symmetry group operators, 65–67
Axis angles, *see* Darboux angles; Omega vector

Baker-Campbell-Hausdorf theorem, 705
Banach spaces, 37
Band gap, 99, 122
Band spectra:
　for electrons, 115–129
　for vibrations, 91–100
Band splitting, 123–128. *See also* Clusters and tunneling
　approximate tunneling amplitude, 141
Beats, 74, 91, 687
　in quantum transition, 735
　in two-level resonance, 744–747
　of 2D-model, 642–643
　of forced classical oscillator, 506
Bethe, Hans, 383
Biedenharn, L. C., 437, 578, 683
Bincer, A., 578, 683
Birefringence, 76, 638, 642
Bloch equations, 750–752
　dressed, 760
Bloch-Siegert corrections, 748–750
Block waves, 123, 125
　for D_4, 194
　for D_6, 199
"Bodyguards," 168
Body reference frame, 157, 329, 365–367
Bohr orbitals, 115
　for D_4, 194
　for D_6, 199
Bohr quantization:
　EBK-quantization, 718
　for rotor, 611, 716
Bonding orbitals, 661–662
Boost operator, 705, 725
Born-Oppenheimer approximation, 539, 655
Boson operators, 303–305
　for oscillator, 704
　in derivation of coupling coefficients, 578–584
　in derivation of rotation matrices, 351
Bra and ket vectors, *see* Dirac notation
Brillouin zone, 96, 116, 126
　band boundary states, 120, 194, 199
　symmetry properties, 270
Buckminsterfullerene (C_{60}), 237
Bulk modulus, 482

C_2 group, 69, 658
C_2H_2, C_2H_4, 677
C_3 group, 84
C_{3v}, 151
C_4H_6, C_6H_6, 678
C_{4v}, 158
C_8H_8 (spectra by Pine), 623
CH_4 (spectra by Pine), 264
C_n groups, 91
C-matrix, *see* Crossing matrix
Canonical coordinates, 36, 301
Central operators, *see* Class operators
CG coefficients, *see* Clebsch-Gordan
Change of basis, 14
Character formulas, 184–189
　for eigenvalues, 189
　for irrep frequency, 187
Characters, *see also* Appendix F.4 and G.5
　of C_1 to C_6, 94
　of C_2, 80, 114
　of C_3, 86
　of C_{3v}, 186
　of $C_{4v} \sim D_4$, 191
　of C_n, 91–94
　of D_2, 107, 111
　of $O \sim T_d$, 233
　of O_h, 235
　of R_3, 384
　of spin-1/2 D_2, 411
　of spin-1/2 D_6, 409
　of spin-1/2 O, 414
Circular polarization, *see also* Moving waves
　coordinates R, L, 638
　related to plane, 183, 250, 349, 393, 571, 637
Class algebra:
　of $C_{3v} \sim D_3$, 159
　of $C_{4v} \sim D_4$, 190
　of $O \sim T_d$, 232
Classes, 159
　of $C_{3v} \sim D_3$, 153, 159
　of $C_{4v} \sim D_4$, 158, 190
　of $O \sim T_d$, 229
Class idempotents, 162, 163
　relation to characters, 185, 829
Class operators, 159, 160
Clebsch-Gordan coefficients (finite group):
　chain labeled for $O \supset D_3 \supset C_2$, 461–463, 818–819
　chain labeled for $O \supset D_4 \supset D_2$, 454–461, 816–817
　defined for finite groups, 452
　for $T_{1u} \otimes T_{1u}$, 454
　involving scalar, 464
　orthonormality, 466
　simply reducible, 486
　symmetry, 465
Clebsch-Gordan coefficients (R_3 or SU_2):
　1 ⊗ 1, 561
　½ ⊗ ½, 557
　formula, 584, 816
　introduction, 553

INDEX   839

Clusters of levels, *see also* Bands
  for C_{12}, 116, 120
  for C_3 and C_4, 125
  for $D_4 \sim C_{4v}$, 194
  for $D_6 \sim C_{6v}$, 199
  for O, 262–264
  for rigid rotor, 614–615, 617
  for semi-rigid rotor, 620–627
  splitting amplitude, 141
Coherent states:
  of oscillator, 706
  of radiation, 775
  wavepacket, 709
Coherent Stokes Raman scattering, 784–786
Completeness, *see also* Orthonormality
  conditions for amplitudes, 8, 25
  of projectors, 23–25
  of wavefunctions, 28, 29
Configuration interaction, 670
Conjugate subgroups, 255
Conjugation:
  complex, 3
  of differential operators, 29
  transpose conjugate (†), 8
Conservation of angular momentum, 344
  of total, 555
Contact transformation, 687–690
Coriolis coupling model, 634, 641
Coriolis splitting, *see* Vibrational angular momentum
Correlation between irreps, 196, 201
Correlation tables, *see also* Appendix F.5
  for $D_2 \downarrow C_2$, 615
  for $D_6 \downarrow C_6$ or C_2, 201
  for $O \downarrow D_2$, 254
  for $O \downarrow D_3$ or C_3, 253, 621
  for $O \downarrow D_4$ or C_4, 252, 621
  for $O_h \downarrow C_{2v}$, C_{3v} or C_{4v}, 245
  for $R_3 \downarrow D_6$, 401, 402
  for $R_3 \downarrow O_h$, 385, 403
Coset factorization, 251–263
Cosets, 161, 258
Coulomb eigenstates, 425, 433
Coulomb symmetry, 415
Coupled oscillators, *see* Oscillator
Coupling coefficients, *see* Clebsch-Gordon
Creation operators, *see* Boson operators
Crossing matrix, 137
Cross product (×) of groups, *see* Outer product of groups
Crystal field splitting, 382–415
  compared to MO, 682
  details for $j = 2, 3$, 598–602
  for half-integer j, 404–415
  for high j, 402–403, 415
CSRS, *see* Coherent stokes raman scattering

D_2, 62, 105
D_3, 151
D_4, 158
D_5, 825
D_6, 196–202
D_n, 202–205
D_{2d} to D_{4d}, 202–204
D_{2h} to D_{4h}, 202–204
D-functions, *see* Wigner D-Functions
Darboux angles, 330–333
  as 2D oscillation parameters, 635
  as two-state parameters, 632, 740
  conversion to Euler, 336–337, 354
Davis, M. J., 712, 786, 787
DC-Stark shifts, *see* Stark splitting
DeLeon, N., 712, 786
Delta functions, *see* Kronecker or Dirac
Density matrix, 30
Determinants, 47–51, 318
Diagnolizable matrices, 52, 53
Diagonalization, 15–28
Dielectric constant, 513
Differential operators, 40
Dipole approximation, 497
Dipole operators, *see also* Vector operators
  electric, 276, 349, 497
Dirac delta functions $\delta(x,y)$, 38
Dirac notation:
  bras and kets, 9–11
  for classical problems, 32–36
  wavefunctions, 37
Direct product ($\otimes$), 446–450. *See also* Kronecker product
Direct sum ($\oplus$), 82, 443
Dispersion relations, 95, 122
Dorney, A. J., 683
Double groups, 407. *See also* Spin-½ algebras
Dressed eigenstates, 753–756
Dual space, 21

Eccentricity vector, 416
Eigenchannel, 134
Eigenvalues, 15
  average derived by characters, 188
  by wavepacket propagation, 711, 717
  from secular equation, 17
  generalized equation for, 34
Eigenvectors, 15
  from adjunct matrix, 27
  from projection operator, 19–23
Einstein A and B, 773
Electromagnetic field:
  classical theory, 503–523
  quantum theory, 761–776
  semiclassical theory, 722–738
Electrostriction, 491
Elementary matrix operators, 13

Elementary matrix operators (*Continued*)
  in U_3, 428
  of group algebra, 167
Energy bands, *see* Band spectra
Euler angles, 324–330
  as 2D oscillator coordinates, 635
  as optical parameters, 641
  as two-state coordinates, 630–631
  conversion to Darboux by formula, 354
  conversion to Darboux by slide rule, 336–337
  mechanical definition, 325–327, 378–379
Exchange integrals, 664
External symmetry breaking, *see also* Stark; Zeeman
  $O_h \supset C_4$, 281
  $O_h \supset C_{4v}$, 276
  $O_h \supset D_{4h}$, 271

Factored P-operators, 264–269
  in SF_6 analysis, 288–294
Factorization lemma, 575
Fano, U., 682
Faraday rotation, 642
Fermi's golden rule, 733, 772
Feynman, R. P., 54, 517, 633, 634
Feynman's lever, 517
Feynman-Vernon-Hellwarth picture, 633–634
Fine structure:
  for atoms, 565–566
  for rotors, 614, 620
Floquet's theorem, 123
Fluorescence, 783
Force matrix, 31. *See also* Acceleration matrix
Foucault precession, 641
Fourier analysis compared, 104, 118
Fourier coefficients of potential, 118, 126
Fourier transformation matrix, 39
Franck-Condon distribution, 712
Frequency of irreps, 84, 184
  applied to two-particles, 451
  character formula for, 187
Frobenius reciprocity theorem, 264–268

Gaussian wavepacket, 709–710
  generation of eigenstates, 711–716
Gelfand, I. M., 578, 683
Generalized eigenvalue equation, 34
  for molecular orbitals, 658, 664, 669
Generators:
  of contact transformation, 687
  of finite groups, 192, 308
  of R_3, 338–344
Genuine modes, 182, 208. *See also* Normal coordinates
Golden ratio, 237

Golden rule of Fermi, 733, 772
Gordon, R. G., 360, 569, 683
Group, *see also* Examples C_2; C_3; ... D_2; D_3; ...O; T_d etc.
  axioms, 65–67
  multiplication, 62
  number of finite groups, 68
  simplest non-Abelian, 151
Group multiplication table, *see also* Hamilton turns
  for C_2, 69
  for C_3, 85
  for $C_{3v} \sim D_3$, 67, 172
  for $C_{4v} \sim D_4$, 190
  for D_2, 62
  for O, 231, 794
  formula for $R_3 \sim U_2$, 355
Group velocity, 122, 690

H_2, 554, 654–673
H_2^+, 654–662
H_2O, 676–677
Hamilton's characteristic function, 692, 711
Hamilton's equations, 36, 302, 700
  related to two-state Schröedinger, 634
Hamilton's principle function, 691, 711, 716, 717
Hamilton-Cayley:
  equation (HC Eq), 18, 21–23
  theorem, 18
Hamilton-Jacobi equations, 694–695
Hamiltonian:
  function, 12, 13
  matrix, 12
  operator, 35
Hamilton's turns, 154–157
  nomogram for D_2, 411
  nomogram for D_3, 157
  nomogram for D_4, 191
  nomogram for D_6, 405, 406
  nomogram for O, 230, 231
  $R_3 \sim SU_2$ rotational slide rule, 333–338
  related to spinors, 353–357
  showing class equivalence, 161
Harmonic oscillator symmetry, 426
Harmonic wavefunctions, *see also* Spherical harmonics
  for cubic symmetry, 386–394, 398–400, 470
  for D_6, 400–402
Harmonics, *see* Beats and vibrational overtones
Hecht, K. T., 619, 683
Hecht Hamiltonian, 618–619
Heitler-London model, 662–665
Heller, E. J., 712, 786, 787
Hermitian matrix, 15

eigensolution properties, 38
Hexagonal symmetry, see D_6
Hückel approximation, 678
Huygen's principle, 6, 8
  semiclassical view, 701–703
Hyperfine states, H_2, 554–560

I, 237, 825
I_h, 237
Icosahedron, see I
Idempotent operators, see also Projection operators
  commuting, 26, 109, 833–835
  definition, 20
Idempotent splitting, see Splitting of idempotents
Identity matrix, 4, 48
Index of refraction, 513
Induced representations, 255–270, 310–311
  $0_4 \uparrow O$ example, 259
  $1_4 \uparrow O$ example, 265
  for molecular bonding, 679
Induction $\uparrow$, 255–270. See also Correlation
Inertia matrix, for vibrations, 32
Infinitesimal rotations, 338
Infrared spectroscopy, 523
  selection rules, 525
Inner product, see Scalar product
Internal symmetry breaking, see Spontaneous symmetry breaking
Invariant, see Scalar
Inverse:
  formula for (Appendix B), 50–51
  of matrix, 5, 9, 48
Inversion doublet, 130. See also Two-level system
  related to O cluster, 270
  U_2 theory, 628–629
Inversion operation, 113, 317
Irreducible idempotents $P_i^\alpha \equiv P_{ii}^\alpha$:
  for C_{3v}, 166
Irreducible projectors P_{ij}^α:
  for C_{3v}, 167–170
  for O, 237–245
  for R_3, 362–365
  general formula, 180, 831–832
  general properties, 217–221, 831–833
Irreducible representations, see also Appendix F.2
  complex conjugate of, 464
  derivation for $C_{3v} \sim D_3$, 175–177
  derivation for $C_{4v} \sim D_4$, 191–193
  derivation for $C_{6v} \sim D_6$, 197–200
  derivation for $O \sim T_d$, 237–255
  derivation for R_3, 338–353
  derivation for R_4, 424

derivation for U_3, 431
introduction for C_2, 79–82
introduction for C_3, 86
introduction for $C_{3v} \sim D_3$, 167
orthogonality, 363, 828–831
Irreducible tensorial sets, see Tensor operators
Irreps, see Irreducible representations
Isomorphic, 114
Isomorphism:
  between C_2 groups, 114
  between C_{3v} and D_3, 154
  between C_{4v} and D_4, 158
  between O and T_d, 236
Isotope shift (SF_6), 299
Isotropic solids, 485, 488
Isotropic space, see R_3 symmetry

j-cones, see Angular momentum cones
j-polarization, 357
Jahn-Teller theorem, 541–542
  dynamic effects, 543
  potential, 542
Jaynes-Cummings Hamiltonian, 777
  eigensolutions, 781
  levels, 782
  modified, 780
  transitions, 783
Judd, B. R., 437

Ket and bra vectors, see Dirac notation
Kramer's rule for equations, 51
Krohn, B. J., 535, 536, 537, 623, 683
Kronecker delta δ_{ab}, 38
Kronecker product $\otimes$, see also Tensor product
  of irreps, 446–450
  of vectors, 13, 21
Kronig-Penney potential, 128

Lab and body reference, 167, 328–330, 378
  for C_{3v} projectors, 173, 174
  for quantum rotors, 367
Lab and body symmetry, 372
Lagrange's theorem for classes, 161
Lagrangian function, 689, 700
  of em field, 723
Lagrangian interpolation, 24, 25
Laser, classical response, 520
Legendre functions, 376
  related to CG coefficient, 608
Legendre transformation, 688
  of Hamiltonian, 689
Lenz-Runge vector, 416
Level splitting, see Splitting of levels

Lie algebras, 315
 of R_3, 341
 of R_4, 418–420
 of U_3, 427
Lie groups, 315
Ligand orbitals, 674–677
Linear independence, 46
Linear polarization, *see* Plane polarization
Lissajous figures, 76, 644–647
Local symmetry, 174, 258
Local symmetry conditions, 258, 265, 287–289
 in R_3, 373
Lorentz model of spectra, 503
 compared to quantum models, 738
Louck, J. D., 437, 578, 683
Lowering operators, *see* Boson or raising operators

Magnetostriction, 491
Malette, Vincent, 325, 331, 332
Maslov constant, 140, 718
Mass matrix, *see* Inertia matrix
Mathieu's equation, 127
Matrices:
 inner multiplication, 8
 introduction to, 1
 transformation, 2
Matrix diagonalization, *see* Diagonalization
Matrix representation, *see* Representation
McDowell, R. S., 306, 535, 536, 537, 623
Minimal equation (MEq), 21–22
Minimum uncertainty states, *see also* Coherent states
 of oscillator, 706
Minors, 49
Mnemonic wheels for crystal j-splitting:
 D_6 integral j, 402
 D_6 integral j, 410
 O integral j, 403
 O half-integral j, 415
Modes, *see* Normal coordinates
Molecular orbital (MO) model, 669
 compared to VB, 673
 sp-orbitals, 674
 spd-octahedral orbitals, 679
Mollow lineshape, 785
Moshinsky, M., 578, 683
Moving waves, 89, 90, 103. *See also* Circular polarization
 in C_{3v} symmetry, 184
 in O_h symmetry, 300
Multipole expansions, 394
Multipole functions, *see also* Spherical harmonics
 of O_3, 382, 611
 of O_h, 383

NH_3, 131, 205–216
Nilpotent, irreducible, 167
Nilpotent operator, 52, 164
Nonsingular matrix, 51
Normalizer, 161
Normal modes or coordinates, *see also* Resonant modes
 C_2 defined, 72–74
 C_3 defined, 90
 C_{3v} defined, 182, 183
 C_n defined, 94, 95
 D_2 defined, 108
 introduction, 32–33
 of NH_3, 208
 of SF_6, 290, 300
Normal subgroups, 255

O, 227–235
O_h, 234–235
$O(2)$ or O_2, 315, 321–323
$O(3)$ or O_3, 315–321
$O(4)$ or O_4 and R_4, 321, 418–427
Octahedral symmetry (O), 227–235
Omega vector, *see also* Darboux angles
 for 2-state Hamiltonian, 631–633
 for 2-state resonance, 740–745
 for $2D$ oscillator, 635–636, 641–646
 geometry during resonance, 744, 757–758
 in Bloch equations, 751
 infinitesimal rotations, 338
 rotation axis angles, 330–333
 rotor angular velocity, 380–382
Optical polarization, 629
 circular polarization coordinates, 636–637
 linear polarization coordinates, 636
Orbits of symmetry operations, *see* Symmetry orbits
Orthogonal:
 groups, 315. See also O_2; O_3; etc.
 matrix, 9
Orthonormality:
 introduction to, 5
 of irreducible characters, 833–835
 of irreducible representations, 363, 828–831
 of projectors, 23–25, 831–832
 of wavefunctions, 38, 39
Oscillator, *see also* Resonance, normal modes
 coherent states, 706
 coupled, 73–79
 forced damped, 504–516
 harmonic two-dimensional, 634–636, 641–649
 in em field, 764–767
 Lorentz, 503
 quantum, 704
 strengths, 736–738

INDEX **843**

Outer product (×) of groups:
  applied to $D_2 \sim C_2 \times C_2$, 111
  applied to $D_{2h} \sim D_2 \times C_i$, 204
  applied to $D_{3d} \sim D_3 \times C_i$, 204
  applied to $D_6 \sim C_3 \times C_2$, 197
  applied to $O_3 \sim R_3 \times C_i$, 319
  applied to $O_h \sim O \times C_i$, 235
  defined, 110
  of two-particle symmetries, 446, 450
Outer product ⊗, *see* Tensor product; Kronecker product
Overtones, *see* Vibrational overtones

Parameter of Lie group O_n, 321
  of R_3, *see* Euler angles
Parameter theorem, 497
Parity (O_3), 352, 353
Patterson, C. W., 378, 379, 623, 683, 684
Pauli exclusion principle, 563, 666
Pauli sponors, *see* Spinor operators; Quaternions
Permutations, 228
Perturbations, *see* Symmetry breaking
  time dependent theory, 730–732
Peterson, F. R., 623
Phase velocity, 122, 690, 699
Phasors:
  for em field, 763
  for forced oscillator, 504
  for galloping waves, 103
  in 2-state resonance, 742
  in 2D oscillator model, 634
  in C_3 oscillator, 90
  in C_n oscillators, 94
  in radiation response, 522
  introduction in C_2 oscillator, 78, 79
  related to coherent state, 706–711
Phonon bands, 96, 99
Photon "impotence", 773
Photon unit, 762
Pi-bonds, 677
Pine, A. S., 264, 673, 683
Plane polarization, *see also* Standing waves
  coordinates x, y, 637
  related to circular, 183, 250, 349, 393, 571, 637
Plasma frequency, 513
Plyler, E., 619
Poincaré-Stokes vector, 629, 638
  components, 640
Point symmetry, 111
  Abelian crystal, 112
  non-Abelian crystal, 152
Poisson's ratio, 484
Polarizability, 512, 514–516
Polarizability, tensor, 527

  for Raman, 529–532
Polarization coordinates, 636–640
Potential energy, *see also* Acceleration matrix; Force matrix
  introduction, 35–37
  of C_2 oscillator, 75
Power broadening, 748
Projection operators, *see also* Idempotents
  commuting, 26, 109, 833–835
  eigenvector projectors, 20–27
  elementary definition, 13
  general formula for group projectors, 180, 360
  general properties, 217–220, 831–832
  of C_2, 69–71, 79–81
  of C_3, 86
  of $C_{3v} \sim D_3$, 175
  of $C_{4v} \sim D_4$, 192
  of C_n, 93
  of D_2, 106
  of $O \sim T_d$, 237–255
  of R_3, 362–365
Projective representations, *see* Ray representations

Quadrupole:
  operator, 350, 497, 589
  perturbation, 271
Quantum rotor, *see* Rotor
Quaternions, *see also* Spinors
Quenching of orbitals, 122

$R(3)$ or R_3, 315–321
$R(4)$ or R_4, 415–425
Rabi frequency (r), 739, 743, 744, 746, 755
Rabi-Ramsey-Schwinger spin picture, 633–634
Rabi rotation, 643, 746
Racah coefficients, 607, 794
Racah, G., 682
Radiation, *see also* Electromagnetic field
  classical dipole, 518–519
  mechanism (Feynman's lever), 517
Raising operators, 304, 435. *See also* Boson operators
  in body frame, 367
  in lab frame, 365, 436
Raman scattering, 527–532
  CSRS, 784–786
  in JC model, 783
  selection rules, 929
Ramsauer-Townsend effect, 137
Rank of group algebra, 166
Rank of Lie algebra, 422
Ray algebras, *see* Spin-½ algebras

844 INDEX

Rayleigh scattering, 514–520, 527
   in JC model, 783
Ray representations, 349, 404, 408
RE or RES, *see* Rotational energy surfaces
Reciprocity relations, 481
Reduced action, *see* Hamilton's characteristic function
Reduced matrix element:
   for finite symmetry, 498
   for R_3, 597
   for rotor, 612
   for single electron, 603
   for two electrons, 605
Reducible representations, 8, 178
Reflections, 113, 317
   and Hamilton's turns, 154
   in C_{3v}, 153
Regular representation, 170, 171
   of $D_3 \sim C_{3v}$, 170, 171
   of O, 270, 624
   of R_4, 420
Renner effects, 543–544
Representation, *see also* Irreducible representations
   of bra-ket vectors, 10–11
   of elementary $\mathbf{e}_{ij}$, 13
   of operators, 12
   reduction, 81, 82, 88, 178–180, 386, 451
Resolvants, 41
Resonance:
   for 2-state system, 743–748
   for classical harmonic oscillator, 504–512
   in C_2 scattering, 135
   in C_2 system, 74–79
   in quantum transition, 735
Resonant modes, *see also* Normal coordinates
   introduction, 32, 72–74
Response function, 505, 510, 520–523
   for quantum transition, 736–738
Root vectors, 423
   of R_4, 423
   of U_3, 430
Rotating wave solutions, 743
Rotational energy surfaces, 609–610
   of anharmonic 2D oscillator, 651–654
   of harmonic 2D oscillator, 649–650
   of rigid rotor, 610, 614
   of semi-rigid XY_6 rotors, 620, 627
Rotational product formula, 355
Rotation groups, *see* R_2 or R_3
Rotation-inversion, 114
Rotation matrices:
   irreducible (Wigner D-functions), 348–349
   xyz-basis, 379
Rotor:
   level splitting of hexagonal rotor, 399–403
   level splitting of octahedral rotor, 397–398
   quantum levels, 365–372
   quantum wavefunctions, 362–365
   rigid rotor energy, 609–618
   semi-rigid octahedral, 618–627
   wavefunction propagation, 717–721

SF_6, 286, 289, 300, 306, 535
SiF_4, 314, 537, 623
S-matrix, *see* Scattering matrix
Scalar operators, for O_h symmetry, 494
Scalar product:
   as coupling, 570
   elementary review, 45
   of bra-ket vectors, 11
Scalar representation, 348
   in products, 464, 573
Scattering matrix, 29
   in C_2 system, 133, 138
Schröedinger equations, 15, 115, 126, 128, 130
   in radiation field, 724, 726, 739
   related to 2D oscillator, 634
   semiclassical approximation, 700–701
   time dependent, 15, 115, 129, 260
Schulten, K., 360, 569, 683
Schur's lemma, 827–831
Schwinger, J., 350, 437, 578
Secular equation, 17
   problem 1.2.3, 56, 57
Selection rules, 279, 499–502, 503
   for infrared, 525
   for Raman, 529
Semiclassical quantization, 686, 700–722
Shear modulus, 483
Singular matrices, 51
Smith chart, 511
Spectral decomposition:
   of a matrix, 25–26
   of commuting matrices, 26
   with resolvants, 42
Spectral function, 715
   for 2-state resonance, 746–747
   from perturbation theory, 733–734
   in quantum field, 769, 771
Spectral nomogram, 500
Spherical harmonics Y_m^ℓ, 373–377. *See also* Multipole functions
   related to Legendre functions, 376
   related to Wigner-D functions, 374
Spherical tops, *see* Rotors; SF_6, SiF_4; UF_6
Spin vector, 588, 631
   for 2D oscillator, 635
   as Stokes vector, 629
Spin-½ algebras:
   for D_2, 411–412
   for D_6, 404–410
   for O, 413–415

Spinor operators of Pauli, 354
  axis angle parameters, 631–633
  for polarization, 639–640
  tensor derivation, 586–588
Spin-orbit coupling, 564–567
Spinor representation $D^{1/2}$, 348, 630
Spinor states, 2–7, 350, 357, 554, 629
  Euler angle parameters, 630–631
Spin polarization, see Stern-Gerlach experiments
Splitting of idempotents:
  for Abelian factor groups, 109
  for class idempotents in C_{3v}, 165
  for commuting operators, 27
  related to level splitting, 251–252
  using O subgroup chains, 237–255
Splitting of levels, 252, 271. See also Stark; Zeeman
  O_h crystal field, 383–386, 391, 601
  for high J, 608
Spontaneous symmetry breaking, 255, 269–270
  in Jahn-Teller effect, 547–548
Standing wave ratio (SWR), 100
Standing waves, see also Plane polarization
  $D_3 \sim C_{3v}$, 89
  in D_4 symmetry, 192
  in D_6 symmetry, 198
  in O_2 symmetry, 322
  in scattering, 134
  in symmetry, 182
  related to moving waves, 192, 321, 624
Stark splitting, 132
  AC-Stark shifts, 754–760
  DC-Stark shifts, 132
  dipole, 276, 313
  in molecular model, 372
  quadrupole, 271, 313
State vectors, 9–11
Stern-Gerlach experiments:
  ($j = \frac{1}{2}$) introduction, 2–7, 357
  ($j = 2$) example, 358
  ($j = 6$) example, 359
  ($j = 20$) example, 361
Stokes lines, 529
Stokes-Poincaré vector, 629, 638. See also Spin vector
  components, 640
Strain tensor, 474
Stress tensor, 471
Structure constants, of class algebra, 162
Subduced representations, 195, 255–270
Subduction (↓), 195, 386. See also Correlation
Subgroup chain labeling, 240
  octahedral, 237–255
Subgroups, 113
  for Abelian groups, 112
  for non-Abelian groups, 152
  for octahedral groups, 234
Superellipse, 392
Superfine structure:
  for rigid rotor, 613–617
  for semi-rigid rotor, 620–627
Superposition principle, 10
Susceptibility, see Polarizability
Symmetric rotor or top:
  eigenvalues, 370, 612
  levels, 369
  RE-surface, 610
Symmetry breaking:
  applied vs. spontaneous, 255, 269–285
  external vs. internal, 255, 269–285
  in C_2 Stark splitting, 132, 271, 276
  in O_h crystal field, 383–386
  in Zeeman splitting, 196, 281
  introduction in C_{12}, 97
Symmetry-defined:
  elasticity tetradic, 479–480
  em fields, 489
  operators, 493
  polynomials, 470
  strain tensor, 474–476
  stress tensor, 472–474
  tensors, 469–470, 476, 493
  vectors, 467–469
Symmetry group, 66, 67
Symmetry operator, 65, 66
Symmetry orbits:
  in NH_3, 206
  in SF_6, 287

T, 234, 236
T_d, 234, 236
T_h, 234, 236
Tannor, D. J., 712, 786
Tensor operators, see also Appendix F
  4th rank (O_h), 390, 393, 395, 397, 598, 618, 625
  6th rank (O_h), 602, 624
  for O_h symmetry, 493–495
  for R_3 symmetry, 585–590
  tables, 591–596
Tensor or quadrupole representation, 349, 350
Tensor product ($\otimes$):
  in coupling analysis, 570
  of ket and bra vectors, 13, 443
  of ket and ket vectors, 444
  of matrices, 446–450
Tensor theorem, 478
Tetragonal axes, 244
Tetragonal moving-wave irreps, 284–250, 794
Tetragonal standing-wave irreps, 239–243, 794

Tetragonal symmetry, *see* $C_{4v} \sim D_4$
Tetrahedral bonding, 675
Tetrahedral symmetry, *see* T_d
Thomas-Reiche-Kuhn sum principle, 738
Transformation matrix, 2
Transient oscillations, 506
Translation operator, 705
Transpose (or transpose conjugate):
 of differential operators, 40
 of matrix, 9
Triangular condition, 568
Trigonal axes, 245
Trigonal bonding, 676
Trigonal moving-wave irreps, 251, 794
Trigonal standing-wave irreps, 244–248, 794
Trigonal symmetry, *see* $C_{3v} \sim D_3$
Tunneling amplitude:
 in C_2 system, 130
 in C_n system, 124
 in O_h system, 257, 624
 semiclassical, 141
Two-level system:
 C_2-like system, 129
 parametrized by Euler angles, 630–631
 related to O system, 270
 relation to $U_2 \sim R_3$, 629
 subject to resonant field, 738–747
Two particle states, 444
 atomic $(2p)^2$, 561
 spin ½, 554

U_2 or $U(2)$, 315, 355
U_3 or $U(3)$, 315, 427
UF_6, 297, 305, 536
Unitarity of S-Matrix, 29
Unitarity of transformation matrix, 9
Unitary:
 eigensolutions of, 38, 39
 groups, 315. *See also* U_2 and U_3
Unit matrix, *see* Identity matrix
Unit tensor operators, *see also* Elementary operators
 matrix, 9
 properties, 15

Valence-bond (VB) model, 662–669
 Heitler-London, 662
 related to MO, 673
Vector or dipole operators, 586, 589

Vector or dipole representation:
 of C_{3v} or D_3, 177, 178
 of O or O_h (T_{1u}), 241, 244
 of R_3, 349
 tensorial properties, 467–469
Vector product, 570
Vector spaces:
 elementary review, 44–47
Vibrational angular momentum:
 in 2D-model, 634
 in SF_6, 301
 in X_3 molecule, 184
Vibrational modes, *see* Normal modes
Vibrational overtones, 306, 526
 in SF_6, 535
 in SiF_4, 537
 in UF_6, 536
 quantum basis, 532
Vibronic motion, 538, 546

Watson, J. K. G., 623, 683
Wave packets, *see* Coherent states
Waves:
 galloping, 100–104
 moving, 89, 90
 standing, 87
Weight vectors:
 of R_4, 425
 of U_3, 431
Wigner 3-j coefficients, *see also* Clebsch-Gordan
 definition, 576, 584
 symmetry, 577
Wigner D-functions, *see also* Rotor wavefunctions; Spherical harmonics
 as transformations, 364–366
 formula, 352, 790
 orthogonality, 363, 828–831
Wigner-Eckart theorem:
 for finite symmetry, 497–498
 for R_3, 597, 608
WKBJ approximation, 138

Y or Y_h, see I or I_h
Young's modulus, 484
Young tableau (for C_2), 666, 670

Zeeman splitting, 195, 196
 in molecular model, 371
Zero-point energy, 709

OHIO UNIVERSITY LIBRARY

Please return this book as soon as you have finished with it. In order to avoid a fine it must be returned by the latest date stamped below. All books are subject to recall after two weeks or immediately if needed for reserve.

APR 0054 2001

JAN 2 4 2001

APR 0 1 2010

JUN 1 1 2010

RECEIVED

NOV 0 4 1991